T0190072

Designing Audio Circuits and Systems

Designing Audio Circuits and Systems is a comprehensive guide to audio circuits and systems design. Beginning with analog audio circuit design basics that a novice can understand, this book offers insight all the way through to in-depth design techniques for many different audiophile and professional audio circuits and functions.

Develop and hone your audio design skills with in-depth coverage of these and other topics:

- Low-noise amplifier design
- Understanding and applying negative feedback
- Filter and equalizer design
- Preamplifiers for moving magnet and moving coil phono cartridges
- Preamplifiers for dynamic, condenser, electret and ribbon microphones
- Fader and panning circuit design
- Balanced line driver and line receiver circuits
- DC servo design and application
- Design of headphone amplifiers and crossfeed circuits
- Self-powered loudspeaker design and active crossovers
- Digital-to-analog converters (DACs)

Bridging the analog and digital worlds, *Designing Audio Circuits and Systems* is essential reading for those in the professional audio engineering community, as well as students and enthusiasts who wish to design audio circuits and functions for pro audio or audiophile applications, and live sound or studio mixing consoles.

Bob Cordell is an electrical engineer who has been deeply involved in audio since his adventures with vacuum tube designs in his teen years. He is an equal opportunity designer to this day, having built amplifiers and preamplifiers with vacuum tubes, bipolar transistors, JFETs and MOSFETs, and is a prolific designer of audio test equipment, including a high-performance THD analyzer and many purpose-built pieces of audio test gear. He has published numerous articles and papers on power amplifier design and distortion measurement in the popular press and in the *Journal of the Audio Engineering Society*. In 1983 he published a power amplifier design combining vertical power MOSFETs with error correction, achieving unprecedented distortion levels of less than 0.001% at 20 kHz. He previously published the book *Designing Audio Power Amplifiers*.

Bob is an avid DIY loudspeaker builder and has combined this endeavor with his electronic interests in the design of powered audiophile loudspeaker systems. He also consults in the audio and semiconductor industries and has presented audiophile listening and measurement workshops with his colleagues at the Rocky Mountain Audio Fest and the Home Entertainment Show. As an electrical engineer, Bob has worked at Bell Laboratories and other related telecommunications companies, where his work has included design of integrated circuits and fiber optic communications systems. Bob holds 17 patents and has presented numerous papers at the International Solid State Circuits Conference (ISSCC), the Convention of the Audio Engineering Society and other related professional meetings. Bob maintains an audiophile website at www.cordellaudio.com where diverse material on audio electronics, loudspeakers and instrumentation can be found.

Designing Audio Circuits and Systems

Bob Cordell

LONDON AND NEW YORK

Designed cover image: e-crow/Shutterstock.com

First published 2024
by Routledge
605 Third Avenue, New York, NY 10158

and by Routledge
4 Park Square, Milton Park, Abingdon, Oxon, OX14 4RN

Routledge is an imprint of the Taylor & Francis Group, an informa business

Library of Congress Cataloging-in-Publication Data
Names: Cordell, Bob, author.
Title: Designing audio circuits and systems / Bob Cordell.
Description: New York, NY: Routledge, 2024. |
Includes bibliographical references and index.
Identifiers: LCCN 2023047746 | ISBN 9781032010908 (hardback) |
ISBN 9781032010892 (paperback)
Subjects: LCSH: High-fidelity sound systems—Design and construction. |
Electric circuits. | Audio amplifiers. | Acoustical engineering.
Classification: LCC TK7881.7 .C67 2024 | DDC 621.389/332—dc23/eng/20240329
LC record available at https://lccn.loc.gov/2023047746

ISBN: 978-1-032-01090-8 (hbk)
ISBN: 978-1-032-01089-2 (pbk)

Typeset in Times New Roman
by codeMantra

This book is dedicated to my son Jon Cordell and his family, of whom I am so proud.

Contents

Preface

Designing Audio Circuits and Systems is a sequel to *Designing Audio Power Amplifiers* that covers a very broad range of audio design topics – all but power amplifiers. As such, this book provides broad, in-depth coverage of circuit and system design for audio. While there is some coverage of digital audio presented in the chapter on digital-to-analog convertors (DACs), the primary goal of this book is to cover virtually all analog circuit design pertaining to audio.

Designing Audio Circuits and Systems is written to address many advanced topics and important design subtleties. At the same time, however, it has enough introductory and tutorial coverage to allow designers relatively new to the field to absorb the material of the book without being overwhelmed. It is targeted to professionals in the audio field as well as enthusiasts and students learning electronics. To this end, the book starts off at a relaxing pace that helps the reader develop an intuitive feel and understanding for analog audio design. Although it covers advanced subjects, highly involved mathematics is kept to a minimum – much of that is left to the academics. Design choices and decisions are explained and analyzed. Practical circuits for numerous different audio processing functions in many different ways are described in depth.

This is not just a cookbook; it is intended to teach the reader how to think about audio circuit design and the implementation of the most important functions used in pro audio and consumer audio systems. The book helps readers understand the many concepts and nuances, then analyze and synthesize the many possible variations of the design of audio circuits that underlie most audio functions, like preamplifiers, equalizers, digital-to-analog converters, compressors and mixing consoles, just to name a few. Multiple topologies and circuits for each of the covered audio functions are described.

The design of modern high-performance audio circuits touches on most aspects of electronic design, including solid-state devices, feedback theory, low-noise design, switching power supplies and laboratory measurement, to name a few. As such, skills acquired in audio circuit design can provide a sound educational basis for the study of a wide spectrum of other areas in electronics. Analog circuit design is covered broadly and in depth.

The early chapters in the book introduce audio circuit design including the basics. This part is designed to be readable and friendly to those with less technical experience while still providing a very sound footing for the more detailed design discussions that follow. Even experienced designers may gain valuable insights here. Chapter 2 covers the design and construction of a complete audio preamplifier as an example to provide some circuit design context for the discussions that follow. A large chapter is devoted to transistor characteristics and circuit design, including many key circuit building blocks. The early chapters of the book cover circuit building blocks, negative feedback principles, low-noise circuit design and filter design.

A group of four chapters covers low-noise preamplifiers for moving magnet phono cartridges, moving coil phono cartridges, tape-head playback and microphones. Dynamic, condenser, electret and ribbon microphones are explained and preamplifiers for each are covered in depth. The design of mixing console preamplifiers is also covered.

Playing a very important role in modern audio systems, the design of DACs and use of DAC chips is discussed in detail, from the S/PDIF input to the low-distortion analog output. Coverage includes R2R DACs and ΣΔ DACs, and the signal processing that makes them work, including oversampling, interpolation, decimation, aliasing and reconstruction.

Numerous approaches to distortion measurement are also explained. Some techniques for achieving the high sensitivity required to measure the low-distortion designs discussed in the book are described. Less well-known distortion measurements, such as TIM, PIM and IIM are also described. In the quest for meaningful correspondence between listening and measurement results, other sources of sound quality degradation are also explained. These include nonlinearity in passive components, EMI ingress, ground loops and power supply noise.

In summary, many of the following topics covered in *Designing Audio Circuits and Systems* should prove especially interesting to readers familiar with earlier texts:

- Ultra-low-distortion circuit topologies
- Low-noise amplifier design
- High-performance feedback circuits
- Balanced interconnects, including line drivers and line receivers
- Switch-mode power supplies (SMPS)
- Integrated circuit functions
- Headphone amplifiers and crossfeed circuits
- Active filter and equalizer design
- Digital-to-analog converters (DACs)
- Self-powered loudspeakers and active crossovers

It is my hope that an experienced designer, a student or an enthusiast who seeks to learn more about high-performance circuit design will find this book most helpful. I also hope that this text will provide a sound basis for those wishing to learn analog circuit design techniques.

Acknowledgments

My Lord and Savior Jesus Christ has given me the peace and guidance that has allowed me to complete this undertaking. My mother inspired me by writing her autobiography. My father supported me in my audio and electronics activities beginning before my teen years, helping me to purchase electronic kits and showing me how to put them together.

Those many authors of other texts on audio that have gone before me have truly been an inspiration and have shown how good engineering can be applied to audio while making their writings understandable to those without a formal engineering background.

I owe a special debt of gratitude to Rick Savas, who read every chapter and provided valuable feedback. Many others in the audio field also provided insight and encouragement, including Jan Didden, Gary Galo and Chuck Hansen. Gene Pitts, past Editor of *Audio* magazine, was pivotal in enabling me to get my start in writing about audio. There are too many other friends and colleagues to list, but I wish especially to thank the members of the Audio Engineering Society and the DIYaudio Forum (www.diyaudio.com) for all that I have learned from them. The audiophile fraternity is alive and well.

Finally, I wish to thank Hannah Rowe and Emily Tagg and the rest of the professional group at Taylor & Francis for turning my dream into reality.

Chapter 1

Introduction

Preamplifiers, mixers and signal processors are fundamental to the consumer and professional audio businesses. In this book I cover the fundamentals to designing this equipment and all of its modules. After a brief introduction to preamplifiers and mixers, a simple example preamplifier that can be built is described in detail. Following that, several chapters cover the underlying basics that include transistor operation, circuit building blocks, noise, negative feedback, filter design and numerous related subjects necessary to understanding and designing virtually all consumer and professional audio circuits. After the in-depth chapters covering the design basics, most other key functions in audio electronics are covered in detail. These include microphone and other preamplifiers, equalizers and tone controls, headphone amplifiers, DACs and analog dynamic signal processors like compressors. In this book I use the term mixer to include the numerous functions in a mixing console for recording or live-sound applications.

Here in the Introduction, a few of these functions will be touched upon briefly to provide a sense of what follows later in the book.

1.1 The Basics

The first order of business is coverage of the electronics theory and practice that underlies audio electronics design. This material is quite universal and covers most of the other fields of analog circuit design.

Circuit Building Blocks

Basic building-block circuits are used to build discrete preamplifier, signal processing and mixing console circuits. Many of these building blocks can also be found in integrated circuits, like operational amplifiers. Understanding of the design of these building blocks will provide a good foundation for the detailed analysis of the circuits that follow in later chapters. A basic knowledge of passive components and transistors is assumed, but there is some introductory material on bipolar transistors (BJT) and junction field effect transistors (JFET) to lay a foundation for understanding the workings of the building blocks. These include single-ended amplifiers, differential pairs, current mirrors, current sources and voltage references, among numerous other functions.

Passive Components

The major passive components making up preamplifiers and mixers are resistors, capacitors and inductors. Many of their characteristics that can influence circuit performance, reliability and sound quality are discussed, including distortion-producing nonlinearities. Of particular interest are the many different dielectrics used for capacitors. Other passive components are also described, including potentiometers, transformers, switches and relays.

Surface-Mount Technology

The great majority of electronics for all applications are now implemented with surface-mount printed circuit board technology (SMT). This approach to assembly provides great economy of space and manufacture as compared to traditional through-hole technology. Indeed, a growing number of integrated circuits and other parts are no longer available in through-hole packages. SMT provides great benefits and some challenges. The latter include temperature control and dealing with lead-free soldering ROHS restrictions (Restriction of Hazardous Substances).

Networks, Poles and Zeros

Passive components like resistors, capacitors and inductors are combined to form attenuators, simple filters and resonant circuits. The concept of poles and zeros will be introduced in simple terms along with complex numbers and the complex plane. Bode plots are used to display frequency and phase responses as they are influenced by poles and zeros.

Semiconductors

Semiconductors are described in some detail. These include bipolar transistors, JFETs, MOSFETs (metal oxide field effect transistor), diodes and other solid-state devices. This provides a good foundation for the detailed analysis of the basic audio circuit functions that follow.

Operational Amplifiers

Operational amplifiers (op amps) are high-gain circuits whose larger function is defined by negative feedback. In essence, they constitute a universal gain block that can be used in a great many different ways. Virtually all op amps are implemented conveniently in integrated circuit form. Medium-performance ones are remarkably inexpensive and come in many different varieties to suit different applications, but the operating and application principles are generally the same for all of them. Operational amplifiers are fundamental to most modern audio applications. The chapter on op amps provides some background and describes numerous common applications of these devices to support their ubiquitous use in the chapters that follow.

Negative Feedback

Negative feedback is employed in virtually every kind of amplifier. It acts to reduce distortion and make gain more predictable and stable. We describe and explain the proper use and application of negative feedback with adequate depth to make its use in audio applications optimal. This includes the important issues of frequency compensation and stability. Negative feedback is a powerful tool, but it is not a cure-all and its improper use can lead to unstable or otherwise poor-performing circuits.

Noise

All passive and active circuits create noise, and noise is undesirable. Noise limits the dynamic range of a system. Low-noise amplifier design is fundamental to the design of many audio circuit functions. Topics covered here include thermal noise (Johnson noise), shot noise, $1/f$ noise, generation-recombination noise and impact-ionization noise. Voltage- and current-noise density and noise spectrum, noise bandwidth, input-referred noise and noise figure are explained. Noise in bipolar transistors, JFETs and operational amplifiers is covered. In order to design circuits for the lowest noise, the noise-generating mechanisms must be understood. Noise contributors and the amount of noise in different circuit blocks are discussed. The concept of noise gain is explained. SPICE circuit simulation to predict noise in circuits is also discussed. Other topics include excess resistor noise, noise in Zener diodes and noise in LEDs (light-emitting diodes).

Passive and Active Filters

Passive and active filters are covered in a more general sense than in their ordinary role in playback equalization, tone controls and equalizers. Here filters will be discussed in terms of their basic type, such as low-pass, high-pass, bandpass or band reject. They will also be described with regards to filter shape and Q, such as Butterworth, Chebychev and Bessel, and their order. Different circuit implementations will also be discussed, such as Sallen-Key and state variable filters. Coupled higher-order state variable filters are also described. Component values for second- and third-order filters of common shapes centered at 1 kHz will be shown as examples that can be scaled to different frequencies. A good example is a filter design for a third-order Butterworth high-pass infrasonic filter for a phono preamp. A different example is a passive Twin-T notch filter used to attenuate the fundamental frequency in distortion analysis. Bainter notch filters are also explained. All-pass filters and implementation of time delay are also covered.

Distortion

There are a great many potential sources of distortion in audio amplifiers, some fundamental and others subtle. Distortions caused by Early effect and junction capacitance nonlinearity in bipolar transistors are explained. The origins of distortion in JFET circuits are also discussed. Passive components can sometimes create distortion as well. Distortion from electrolytic capacitors is a special concern, and means of reducing it are discussed. Resistors can also be a source of distortion as a result of both voltage and power dissipation. In general, whenever a device parameter changes as a function of signal, distortion can be created. Types of distortion are often named by the ways in which they are measured, such as harmonic and intermodulation distortion. However, it is very important to recognize that a given nonlinearity can create virtually all of these named distortion types in different amounts. The ways in which distortion are measured are also explained.

Electronic Switches

While very basic line-level preamplifiers require little more than source selection and a volume control, full-function preamplifiers may require more switching, and mixers will require still more switching functions. In many cases an output muting circuit may be needed as well, so that pops and thumps at turn-on and turn-off do not get through. The various switching means described here will also apply to volume controls based on switched attenuators. Design considerations for various kinds of switching options in audio equipment including solid-state approaches that employ JFETs, MOSFETs, CMOS transmission gates and opto-isolators are covered.

Power Supplies and Grounding

Power supply quality and the grounding architecture can play a large role in the sonic performance of a preamplifier or mixing console. The importance of power and ground layout is discussed, especially with regard to the avoidance of ground loops. The importance of equivalent series resistance (ESR) and equivalent series inductance (ESL) of decoupling and reservoir capacitors is explained, and means for reducing their impact are illustrated. Rectifier speed and rectifier noise are also discussed, noting that high peak currents can flow in the rectifiers and that the speed and "softness" with which the rectifiers can turn off (recovery time) can have a large influence on noise that is created. Rectifier snubbing for noise mitigation is also described. Low-noise voltage regulators can be especially important to the performance of the circuits they power. Prevention of EMI ingress is also discussed, recognizing that there are antennas and conduction paths everywhere that can contribute to conducted and radiated emissions. Finally, the matter of inboard vs outboard power supplies and switching supplies for equipment is considered. Powering smaller equipment from USB (universal serial bus) supplies is also discussed, including the use of inexpensive wall-plug phone chargers.

1.2 Audio Signal Processing and Control

Below are the major topics concerning audio signal processing and control that are covered in this book.

Equalizers and Tone Controls

Equalizers and tone controls are used to achieve a desired tonal balance from source material played through a given system, including its loudspeakers. Tone controls and equalizers can be used to remove anomalies or to compensate for non-flat frequency regions. They are usually adjusted to taste. They can also be used to compensate for frequency response effects due to the room. Multi-band or parametric equalizers are often used to tame feedback in sound reinforcement systems. We cover the design of several different tone controls and equalizers, from the Baxandall bass/treble tone control to multi-band, graphic and parametric equalizers. Constant-Q (quality or inverse of bandwidth) and proportional-Q equalizers are both described.

Headphone Amplifiers

High-performance headphones are capable of reproducing exceptional sound quality. For this reason, the headphone amplifier often included in preamplifiers, mixers and other equipment should be more than just an afterthought. Important design goals and several different design approaches for headphone amplifiers are covered. A good headphone amplifier shares many things in common with a quality power amplifier. Although it does not need to produce as much power, it must operate with a wide range of reactive headphone loads whose impedance may vary over frequency and headphone models. The nominal impedance of most models falls between 30 and 300 Ω. The nominal power sensitivity and voltage sensitivity also varies over a fairly wide range among different headphones.

The normal headphone listening experience includes the sound-in-head effect and other effects that result from each ear hearing only the sound of its stereo channel, rather than each ear hearing some of the opposite channel as in the real world with loudspeakers. Some cross-feed circuits that may mitigate this effect and improve the listening experience are also described. Finally, there is some discussion of hearing loss protection circuits.

Volume Controls, Faders, Balance Controls and Panning Circuits

Volume controls are central to the operation of a preamplifier or mixer (where they are often called faders), and in fact some preamplifiers comprise little more than a volume control augmented by input switches and buffers. The traditional audio taper potentiometer volume control is still a workhorse, but its disadvantages include imperfect matching among channels, expense in multi-channel systems and inability to be automated unless it is motorized. Potentiometers can also create wiper noise and wiper distortion. These many issues and active circuit implementation alternatives to the traditional audio taper potentiometer are explained. Similarly, balance controls in preamplifiers and panning controls in mixing consoles are covered.

Digital-to-Analog Converters (DACs)

With more and more program sources being digital, it only makes sense that the digital-to-analog converter (DAC) takes its place among preamplifier input interfaces like the phono preamp. The ability to download or stream high-resolution audio further cements the DAC as a mainstay of audiophile reproduction. For quite some time, some CD/SACD players have incorporated DACS that can accept an external digital input. Many preamplifiers and headphone amplifiers now incorporate built-in DACs. The role of stand-alone DACs may be migrating to the high-end audiophile space, but they may be replacing preamps in all-digital setups. Much of the magic in DACs lies inside the DAC integrated circuit. Most of what can be done by an audio circuit designer that affects sound quality is in how the circuitry external to the DAC is executed. This is especially true for the balanced-to-single-ended conversion circuit that marries the digital source to the analog world. Nevertheless, how audio DAC chips (ICs) work is discussed, including the oversampling and sigma-delta techniques that are employed in most modern DAC chips.

Active Crossovers and Loudspeaker Equalization

The great majority of consumer loudspeaker systems employ passive crossovers to direct signals in different frequency bands to the appropriate drivers, like woofers and tweeters and often mid-range drivers as well. Passive crossovers have limitations, including large passive components like capacitors and inductors, all potential contributors to distortion and frequency anomalies. Active crossovers operating at line level and feeding multiple amplifiers offer an appealing alternative where a system-level approach can be used to achieve the highest sound quality. Active circuits can also provide driver equalization, low-frequency extension and baffle step compensation. While such active crossovers do not frequently appear in consumer electronics, major exceptions include subwoofers and computer speakers. The use of active crossovers in self-powered professional loudspeaker systems is widespread, both in sound reinforcement and studio monitor applications. Often, quality class-D power amplifiers are used here, making heat removal from the cabinet less of a problem.

Voltage-Controlled Amplifiers

The highly versatile voltage-controlled amplifier (VCA) is widely used in professional audio equipment, especially as automated faders in mixing consoles. The term VCA is usually associated with integrated circuits whose variable-gain functionality is based on what are called translinear techniques. These designs are most often implemented with bipolar transistors in integrated circuits. Such functions are also referred to as log-antilog circuits. These techniques are also used to implement analog multipliers. The gain of a VCA is usually dependent on a control voltage in a fashion that is linear in dB, so the control function for gain is described in dB/volt. These circuits can control gain over a very wide range, often 80 dB or more, so they are well-suited to volume control and fader applications. The logarithmic control characteristic is an almost ideal type of audio taper. Most VCAs have a significant tradeoff between distortion and signal-to-noise ratio. It is largely due to imperfect log conformance of the transistors used in the gain cell. This design challenge is discussed in some detail.

Compressors and Other Dynamic Processors

Compressors can be used to prevent overload or to increase the average perceived loudness level of the program material. They reduce dynamic range. Limiters provide a "hard" form of compression that has a steep onset and will prevent the output signal from exceeding a predetermined level no matter how high the input. Limiters generally do not distort the signal waveform, as clippers do. Compressors will sometimes be used to reduce the dynamic range of a given band of frequencies. Compressors are also valuable tools in managing gain levels and audibility among multiple inputs in recording studios. Sometimes compressors are linked together in some way, or the control circuitry of a compressor will respond to the audio level in a different channel while compressing its channel. Compression and expansion are used in most noise reduction systems. We will cover different kinds of compressors and limiters and describe their circuitry.

Level Displays and Metering

Level displays are crucial to the recording, sound reinforcement and broadcasting industries. Level displays help prevent clipping and help ensure reasonable and consistent listening levels. They are especially important in setting recording levels so as not to allow overload, as with analog tape recorders and DACs. The most common display is the VU (volume unit) meter, often using a traditional mechanical meter movement or a bar graph display. VU meters do not respond instantly to signal level, so they are of limited use in preventing clipping, especially with sources having high dynamic range. For this reason, various forms of peak, average and RMS displays are often used as well. Some displays are designed to indicate perceived loudness so as to enforce certain broadcasting standards.

Microcontrollers and Microcomputers

Notwithstanding this being a largely analog text, there is a chapter that touches on the use of microcontrollers (μC) and microcomputers (MCU) in audio preamplifiers and mixing consoles. Microcontrollers are in everything, from kid's toys to home theater receivers. Most consumer items that you purchase today have some kind of microcontroller in them. If nothing else, microcontrollers implement outstanding human interfaces at much less cost than traditional approaches. A decent microcontroller can cost less than \$2–\$8 – sometimes far less when purchased in quantity. Non-volatile FLASH memory built into many modern microcontrollers makes them especially useful and flexible. In fact, many audio integrated circuits, from volume control chips to DACs, depend on digital interfaces to microcontrollers for their settings. Microcomputers are used for more complex tasks. While microcontrollers are often limited to running one task or thread at a time, microcomputers usually have an operating system that enables them to execute multiple threads simultaneously.

Mixing Consoles

At the heart of the recording studio or the live-sound reinforcement system is the mixing console, be it anywhere from a simple mixer to a complex console with a large number of channels and functions. The analog mixer or mixing console is the subject of Chapter 29. The discussion is arranged so that the material progresses from the circuits and architecture of a simple generic 16-channel mixer through that of far more capable mixers that may be used in professional live-sound applications, broadcasting and recording. Technical challenges are discussed that are unique to analog mixing of a large number of channels, such as growth of mixing amplifier noise gain with channel count. With today's powerful DSP (digital signal processing) and computing power, digital mixing consoles can provide all of the traditional functions and more. As a result, digital consoles have replaced analog consoles in many applications.

DI Boxes and Microphone Splitters

The need to drive the input of a stage or monitor mixer brings up the issues of signal splitting, impedance transformation, buffering and ground isolation. The last chapter explains the functionality and circuitry used to implement these two important devices. A microphone splitter is often required to drive both a front-of-house (FOH) mixer and a monitor mixer. A guitar or other instrument may require a type of splitter to enable it to drive both its own amplifier and the FOH mixer. Although not always directly applicable to splitting off a signal for a monitor mixer, understanding the well-known DI (direct inject) box for this latter kind of splitting is a good starting point.

1.3 Line-Level Preamplifiers

Figure 1.1 shows a simplified block diagram of a consumer audio preamplifier. It is important to recognize that not all preamplifiers include all of these functions. Some preamplifiers include as little as input switching, input buffering, a volume control and an output buffer. For purposes here, line-level signals are those that are typical of what would come out of a phono preamp, a CD player or a DAC, and those that would be fed to an audio power amplifier. Typical full-scale levels are on the order of 0.5–2 V_{RMS}. Full-scale line-level signals in mixers are typically higher. The volume control is the central function to a preamplifier and will be discussed in a devoted chapter later. A key consideration in volume controls is tracking among channels.

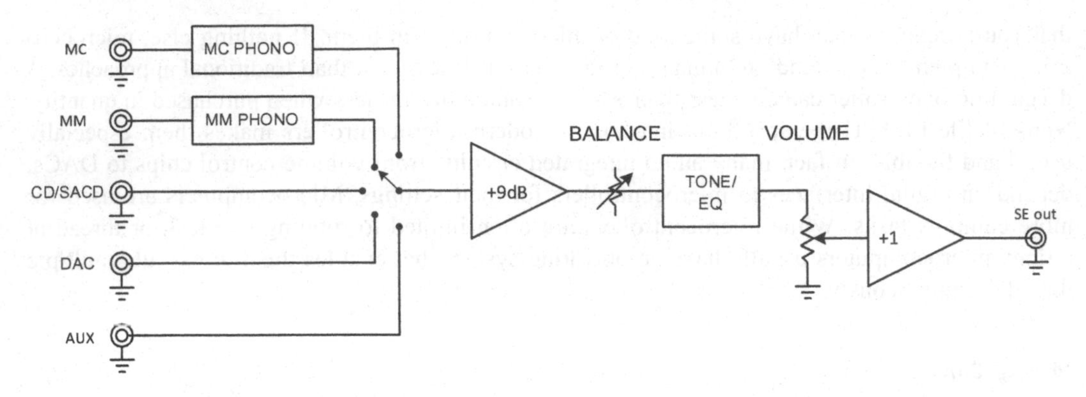

Figure 1.1 Block Diagram of an Audio Preamplifier.

Signal Switching

Signal switching in a preamp is often little more than an input selector switch, usually implemented with a rotary switch, relays or transmission gates. Some preamps include a "tape loop" or monitor function that allows the insertion of another external device into the signal path, such as a graphic equalizer, reel-to-reel tape recorder or a PC-based digital recorder.

Volume and Balance Controls

Every preamplifier must have a volume control. In mixing consoles these are called faders. Most traditional preamplifiers employ a multi-gang audio taper pot, where matching of attenuation among gangs is important. Many contemporary designs use passive or active switched attenuators, where the tracking among channels is usually quite good, but granularity can be important (minimum step size in dB). In active switched attenuators, like those implemented in integrated circuits with functions like transmission gates, distortion and dynamic range are important distinguishing performance features.

Balance controls were once routinely included in preamplifiers, but are often absent in modern preamplifiers, presumably based on the assumption that the devices in the audio chain are sufficiently well-matched among channels. This is not always a good assumption, especially for transducers like phono cartridges and loudspeakers. When included, the balance control may have as small a range as ±3 dB and will often have a center detent defining the 0–dB balance point. Balance controls can be implemented in numerous different ways at different places in the signal path.

Tone Controls and Equalizers

Like balance controls, tone controls have largely disappeared from audio preamplifiers, on the presumption that a good system should not require tone controls that will detract from neutrality and invite abuse. The classic Baxandall tone control is by far the most widely used circuit when tone controls are provided. Tone controls usually provide ±10–15 dB of shelving response at the low and high ends of the spectrum. More sophisticated equalizers, like graphic equalizers, can be found in consumer audio systems, but are a critical element in mixing consoles.

Muting

Depending on the details of the circuit design, some preamplifiers can produce a click or a thump at turn-on or turn-off. In these cases a muting circuit can be included at the output to give the preamp circuits time to settle and stabilize before the signal is allowed to pass to the power amplifier. These circuits will often be implemented with a circuit that short-circuits or open-circuits the output, such as a relay or a JFET.

1.4 Phono and Tape Preamplifiers

Vinyl has made a significant recovery among audiophiles, and phono preamplifiers are required to provide the necessary gain and RIAA equalization between the cartridge and the line-level inputs of the preamp. While at one time phono preamps were included in preamplifiers, integrated amplifiers and receivers as a matter of course, they are not always included in contemporary equipment. Often, they are incorporated as a separate unit in audiophile systems, just as a DAC is often included as an added external box that connects to the line input of a preamplifier.

Moving Magnet (MM) Phono Preamps

Conventional phono preamps are designed to accept the input from a moving magnet (MM) cartridge, whose output may be on the order of a few mV. Noise and accuracy of RIAA equalization are important performance characteristics. Input overload and gain are also important.

Moving Coil (MC) Preamps

Moving coil (MC) cartridges are very popular among audiophiles and are usually believed to produce the best sound, often with extended high-frequency response and fewer resonance effects in the upper end of the audio band. MC cartridges produce smaller nominal output levels, on the order of hundreds of microvolts, on average about 20 dB less than a typical MM cartridge. A very low noise boost amplifier is thus usually needed in front of a conventional MM phono preamp, with about 20 dB of gain. Step-up transformers have been employed in the past for this task, but are no longer widely used. Low noise is the single most important performance characteristic for MC preamps. Input-referred noise levels on the order of 0.5–2.0 nV/√Hz are usually required.

Most outboard phono preamps can accommodate both MC and MM inputs. Some have separate inputs for each and employ a switch, while others have a common input for both and may even pass the signal through a low-noise phono preamp with switchable gain.

Infrasonic Filters

Often incorrectly referred to as subsonic filters, infrasonic filters are sometimes incorporated into a phono preamp to sharply cut off frequencies below 20 Hz. These are often realized as third-order Butterworth filters that are down 3 dB at 10–20 Hz. They attenuate very low frequencies so that signals resulting from record warp and the like are suppressed so that they do not waste amplifier power and cause large, unnecessary woofer cone excursions. This can be especially important when vented loudspeakers are used, since the low-frequency drivers are unloaded at very low frequencies.

1.5 Microphone Preamps

Microphone preamplifiers are critical to the professional audio business, including numerous types such as dynamic, condenser, ribbon and electret, each requiring a different preamp capability for optimum performance, with preamplifiers sometimes included within the microphone body, such as for condenser and electret microphones. In recent years, tiny MEMS (micro-electromechanical systems) microphones, implemented with integrated circuit processing technology, are playing an increasingly important role in many audio applications. A large chapter is devoted to microphone preamplifier design.

Virtually every mixing console incorporates several general-purpose microphone preamplifiers, usually as part of a channel strip. Some mixers may not incorporate a microphone preamp into every channel. For example, the Mackie CR-1604 16-channel mixer incorporates a microphone preamp into only six of its 16 channels [1]. By "general purpose" we do not mean of less than the highest quality, but rather that it can serve the needs of most microphones. For example, it will serve the needs of virtually all dynamic microphones without any other electronics in the path. Condenser, ribbon and electret microphones should also be supported. However, some microphone preamps built into mixers may not have as much gain or low enough noise as would be desirable for some dynamic and ribbon microphones that have very low sensitivity.

Dynamic Microphone Preamps

The standard reference level for microphones is the Pascal (Pa), which corresponds to −94 dB SPL (sound pressure level). Microphone sensitivity is specified in mV at 1 Pa (mV/Pa) or dB below 1 V at a sound level of 1 Pa (dB/Pa). The sensitivity of most microphones lies in the range of 0.5–30 mV/Pa. Maximum acoustic levels that some microphones can handle is on the order of 135 dB/SPL. Microphone self-noise is usually specified as dB-A/SPL, with the A designating A weighting (which accounts for the human ear's differing ability to hear noise as a function of frequency). Self-noise of most microphones falls in the range of 10–30 dB-A. Dynamic range of a microphone is simply the difference between maximum SPL and self-noise SPL. A phenomenal microphone with maximum SPL of 135 dB and with self-noise of only 5 dB-A would have an enormous dynamic range of 130 dB.

A key aspect of the microphone preamps in mixing consoles is the range over which its gain can be varied. This is because microphone sensitivity spans a very wide range, often from as little as 0.5 mV/Pa to as much as 40 mV/Pa.

In addition to standard microphone preamplifiers, microphone preamplifiers that are incorporated into the microphone itself will be covered. Many types of microphones other than dynamic ones must incorporate active circuitry within the microphone to retrieve the signal before sending it out through the microphone cable. The most common occurrence of this is with condenser microphones. Some ribbon microphones also incorporate active preamplifiers into the microphone itself to locally amplify the tiny signal produced by the metalized ribbon suspended in a strong magnetic field.

Phantom Powering

Microphones with active electronic circuitry in them, like condenser microphones, need a source of power for this circuitry. Mixing console microphone preamplifiers provide this power over the balanced microphone cable via what is called phantom powering. A +48 V supply is usually connected to the hot and cold microphone lines through a pair of 6.81 kΩ 0.1% resistors. These lines are then AC-coupled to the input circuit of the mic preamp. The DC on the hot and cold balanced wires of the microphone cable with respect to ground (shield) can then be merged with a pair of resistors or a transformer in the microphone to power the microphone electronics. The source resistance of this supply is about 3.4 kΩ, so if the microphone draws 10 mA, a drop of 34 V will occur, leaving only 14 V at the microphone for the electronics. For this reason, microphone electronics rarely require more than about 7 mA from the phantom supply.

Condenser Microphone Preamps

A condenser microphone consists of a diaphragm in close proximity to a back plate, creating a capacitance typically between 10 and 100 pF. The capacitor thus formed is charged to about 60 V by a polarizing voltage. As sound pressure moves the diaphragm closer to or further from the back plate, the capacitance changes. The charge on the capacitor must be conserved, so as the capacitance decreases, the voltage across it increases. The resulting signal is then amplified by a preamplifier with a JFET first stage with extremely high input impedance. Polarizing and gate return resistors are often on the order of GΩ. This is necessary to achieve satisfactory performance at low frequencies because of the time constant formed by these resistances and the capacitance of the diaphragm. Extremely high impedance is also required to minimize the contribution of thermal noise from the bias resistors.

Electret Microphone Preamps

An electret microphone is much like a condenser microphone, but the backplate includes an electret membrane, which is a permanently polarized piece of dielectric material like a metalized Teflon® foil. A thin spacer then separates the membrane from a pickup plate. The resulting capacitor naturally contains a permanent charge. As with a condenser microphone, when the diaphragm moves with sound pressure, the capacitance changes and a signal voltage is created. A JFET is usually included in the electret microphone capsule. It operates in the common source mode, with the electret transducer connected between its source and gate. The pickup plate is connected to the gate. The source and drain connections are brought out, and an external load resistor connected between the drain and a low-voltage power supply completes the amplifier. The microphone described above is called a back-electret microphone. There also exist front-electret microphones, where the moving pickup plate is polarized.

Ribbon Microphone Preamps

A ribbon microphone consists of a very thin metal ribbon suspended within a strong magnetic field. Sound pressure moves the ribbon and creates a tiny voltage across the ribbon. The microphone sensitivity as measured at the ribbon is typically on the order of −86 dB/Pa, which means that at 94 dB SPL the signal level may be on the order of 50 μV. A step-up transformer with a typical turns ratio on the order of 37:1 is used to boost the signal level by about 31 dB. This results in a microphone sensitivity of about −55 dB. With a ribbon resistance of 0.5 Ω, the impedance at the transformer secondary will be about 500 Ω. A larger turns ratio like 100:1 to achieve 10 dB better sensitivity will result in output impedance of perhaps 5000 Ω, and that is too high. In some cases an active ribbon microphone uses such a transformer and buffers the output with JFET source followers or some similar active circuit arrangement. Active ribbon microphones are then powered by the phantom supply.

MEMS Microphones

Micro-electro-mechanical systems (MEMS) are made with integrated circuit processing and usually include a moving piece of silicon material. This technology is thus suited to making extremely small condenser microphones. Such microphones have dimensions on the order of millimeters, diaphragm spacing on the order of microns and polarizing voltages of a few volts. The capacitance is very small and a CMOS ASIC (application-specific integrated circuit) is located in the same package as the MEMS chip. The ASIC provides the necessary amplification and generates the required polarizing voltage. The concept of scaling is key to the practicality of these microphones. Some ASICs include an ADC (analog-to-digital converter) in order to produce a digital output. MEMS microphones are so small and inexpensive that it is economical to fabricate them in one- or two-dimensional arrays to achieve high-performance directionality in arrangements similar to phased array radar.

1.6 Mixing Consoles

Multi-channel mixing consoles of all levels of complexity and number of channels are fundamental to recoding and live-sound reinforcement, especially in professional audio systems. Figure 1.2 shows a block diagram of a very simple 16-channel mixer. There are three main sections to most mixers: the input-channel strips, the buses where channel signals are mixed and the output module where the mixing amplifiers, master output controls, output amplifiers and metering are located.

Channel Strips

Medium- and large-format mixers are usually organized into channel strips, often one separate plug-in unit for each channel that implements several functions for that channel, including the fader (level control), some equalization that may be as modest as a three-band tone control and some switching for routing of signals associated with that channel. The channel strip will usually include a pan pot that can place the signal from that channel somewhere in the mix between the left and right channels in a stereo mixer. The channel strip may also include a VU meter or some other device to indicate signal levels. At minimum, a clip indicator is usually included.

Some or all of the channel strips will also include a microphone preamplifier. A channel strip that does not include a microphone preamplifier may just accept line-level signals. External microphone preamplifiers will sometimes be used to feed mixer line-level inputs when additional

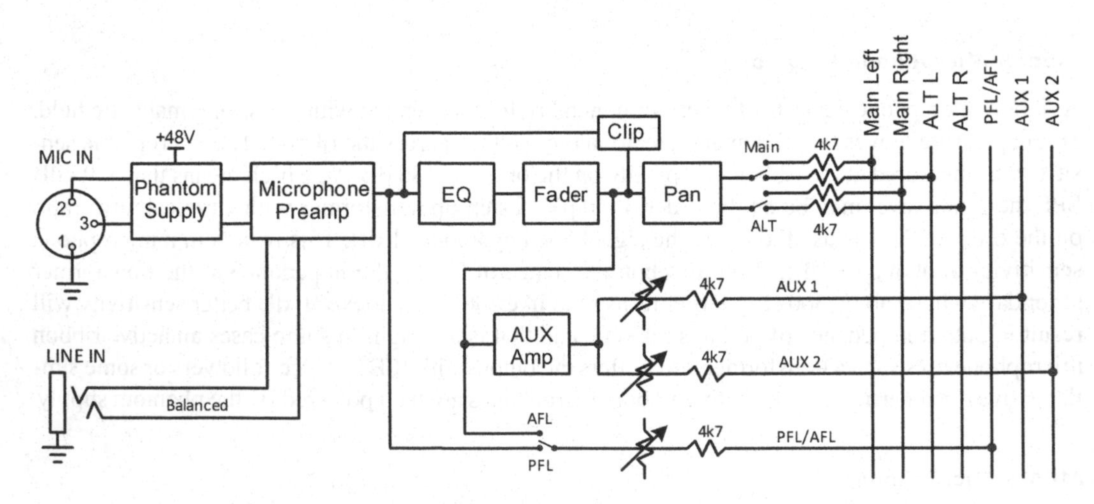

Figure 1.2 Block Diagram of a Mixer Channel Strip.

microphones are employed or a more specialized or high-performance microphone preamplifier is needed.

Buses

The buses are usually passive lines that are accessible to all of the channel strips. Signal currents applied to the buses are added on the bus and applied to the virtual ground of a mixing amplifier in the output module. Although in principle a stereo mixer needs only two buses, most mixers have a multiplicity of buses to support all of their features, such as numerous auxiliary mixer outputs. It is not unusual for a mixer to have 12–24 or more buses.

Output Module

A separate portion of the mixing console creates and controls the outputs of the mixer. This section is where the outputs of the channel strips are mixed together to form the outputs. The mixing amplifiers are inverting op amps that present a virtual ground to the bus feeding them, so that the signals on the bus are mixed on a current basis. The output section will include a master fader and perhaps a stereo balance control, and master tone control or equalization. A headphone amplifier and VU meter are provided as well. The output module will usually include numerous line-level inputs and outputs (Figure 1.3).

A key performance consideration in multi-channel mixers is noise accumulation. When many channels are mixed together in a simple op amp-mixing circuit, the noise gain of that stage increases in proportion to the number of channels being mixed together. Low-noise circuit design is important here, and the mixing architecture and arrangement of mixing buses can also influence noise performance. Unused channel strips will sometimes be disconnected from the bus so as not to increase mixer noise gain.

Faders

Faders control the signal level for each channel and for the main outputs. They must cover a wide attenuation range, from *off* to full output, with a user-friendly and convenient attenuation taper. Faders are usually implemented with slider potentiometers, or VCAs. Most medium and large analog mixing consoles employ VCAs that are controlled by a DC voltage from a linear pot (which may be motorized) or from a DAC or some combination. VCA-based fader arrangements are especially attractive in consoles that support automation. The circuit-level workings of VCAs are covered in Chapter 25.

Pan Pots and Cross-Faders

Pan pots and cross-faders are often incorporated into a channel strip to control placement of that channel's signal within the stereo image. These can also be implemented with VCAs.

Equalizers

While some degree of equalization is usually available in each channel strip, an insert capability is often available to insert an external device, like a multi-band or parametric equalizer, into the signal path of that channel.

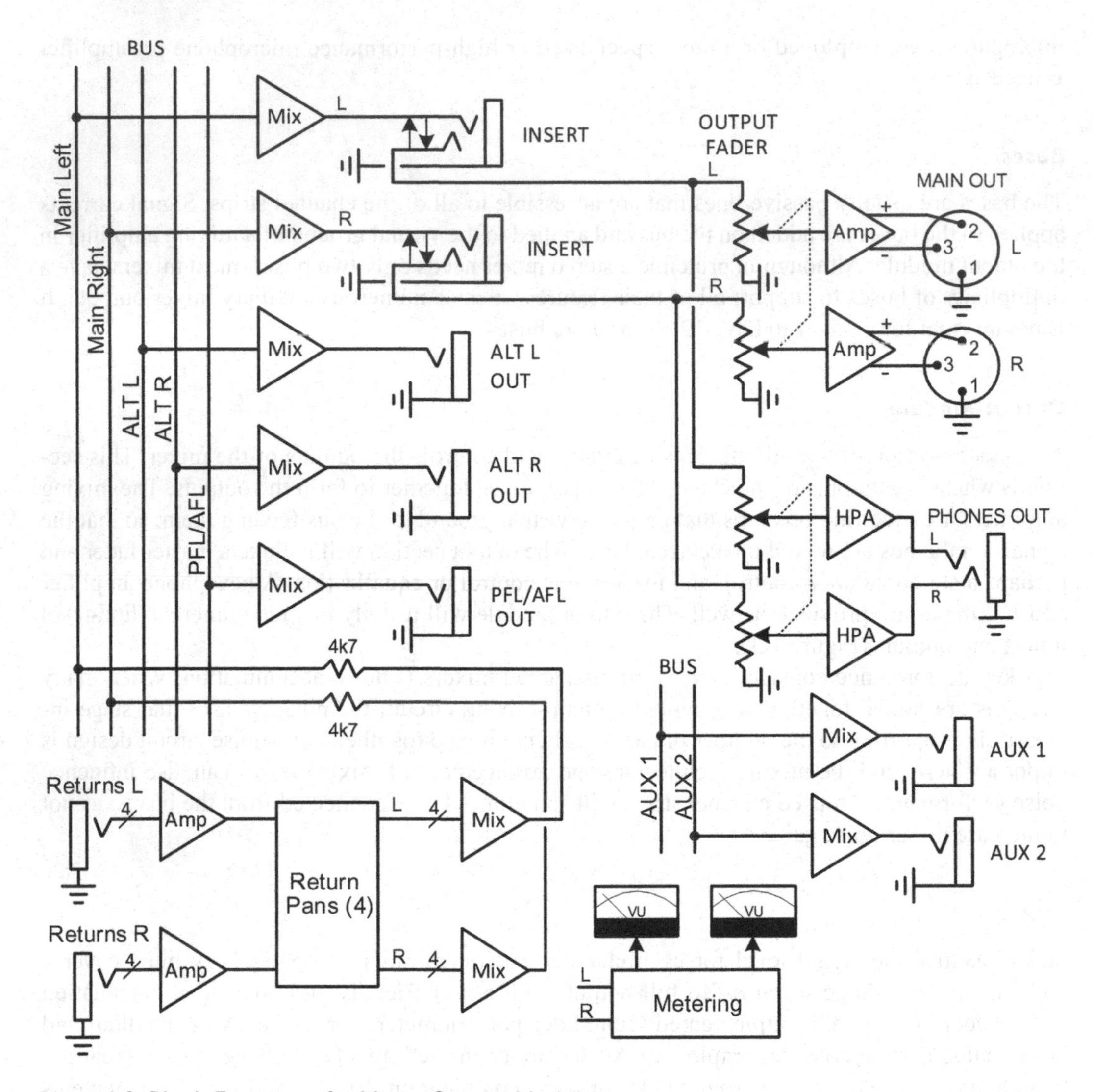

Figure 1.3 Block Diagram of a Mixer Output Module.

Compressors and Limiters

Like equalizers, compressors and limiters of varying capability and complexity can be built into mixing consoles or can be inserted into channel or output paths as external units. These types of dynamic signal processing devices are covered in Chapter 26.

Reference

1. Mackie, CR-1604 Owner's Manual, 1995, http://www.mackie.com.

Chapter 2

A Simple Preamplifier Design

Here we describe a simple full-function audio preamplifier, called the BC-2, that can be built as an example. This chapter is intended to familiarize the reader with audio circuit design before the more detailed chapters that follow. It includes the line-level buffering and gain, volume control, balance control and tone controls. It also includes a moving magnet (MM) phono preamplifier and a moving coil (MC) head amplifier. Any of these functions can be deleted for a more basic preamplifier. The design is a straightforward one that is based on IC operational amplifiers, with the exception of the moving coil head amplifier. The description presumes some basic knowledge of operational amplifier circuits (see Chapter 8). The preamplifier delivers quite respectable performance and employs quality components.

Functional Features

- Simple, but good performance and feature-rich
- ALPS brand volume control
- Balance control
- Baxandall tone controls
- MM phono preamp
- Optional infrasonic filter
- MC head amp
- Auxiliary monitor loop

Design Features

- Audio-quality JFET op amps
- High input impedance >100 kΩ
- Low-noise MM and MC phono preamps
- Multiple DC servos to minimize use of coupling capacitors
- No electrolytic capacitors in the signal path
- Relay-controlled input and mode select (optional)
- Compatible with microcomputer control (optional)
- Balanced XLR output (optional)

2.1 Preamplifier Block Diagram

The block diagram of one channel of the preamplifier is shown in Figure 2.1. One of five inputs can be selected by S1, which can be implemented with relays. All inputs have impedance greater than 100 kΩ.

- CD/SACD
- DAC
- Auxiliary
- Moving magnet phono
- Moving coil phono

When the MC input is selected, the 20-dB MC preamp is switched in ahead of the MM preamp. The MM and MC inputs have separate input jacks. The Aux input also serves the function that used to be called the Tape Monitor input, here called the auxiliary monitor loop. An auxiliary output provides a copy of the selected input. This enables the insertion into the signal path of an analog tape recorder, a digital recorder or other equipment like an external equalizer.

Following the input select switching, the line-level signal path begins with a non-inverting 8-dB input buffer whose gain is controlled over a ±3-dB range by the stereo balance control. This is followed by an optional Baxandall tone control. The subsequent volume control pot feeds the output buffer. A muting relay interrupts the output signal path for 3 seconds after turn-on and quickly after turn-off.

Figure 2.1 shows a block diagram of the preamplifier. It includes a 5-position selector switch S1 that can select MC or MM inputs and three different line-level inputs. When the MC input is chosen, the 20-dB (10×) MC head amp is inserted by S2 in front of the MM preamp. The MC preamp has 10×, 30× and 50× gain taps, as shown in Figure 2.7.

If the auxiliary loop mode is chosen by S3, the selected input is routed to an auxiliary output, while the auxiliary input is routed to the line-level section of the preamp instead of the output from the input selector switch. This enables the insertion of an external auxiliary device in the signal path, such as an equalizer.

The line-level signal path begins with a non-inverting input buffer whose nominal gain is 8 dB, but whose gain on either channel can be changed by the associated balance control pot from 6 to 12 dB. This results in a modest ±3-dB balance control range. The input buffer is followed by a non-inverting Baxandall tone control that can be bypassed by S4. For those who eschew tone controls, this circuit is easily deleted.

The tone control is followed by a quality ALPS volume control that feeds the unity-gain output buffer. An optional balanced output is provided by an inverter and an XLR connector. A muting relay disconnects the signal from the output connector and grounds the output.

The input buffer and output stage are implemented with audio-quality OPA2134 JFET-input dual operational amplifiers. Use of the JFET op amps allows for a high 100-kΩ input impedance. This preserves low-end frequency response by only lightly loading the AC-coupled outputs of input sources. In a preferred implementation, all of the switches in the signal path are realized with subminiature relays that are controlled with DC signals from front-panel switches or a microcontroller. In light of the modular nature of this preamplifier, each schematic figure may have its own component designators.

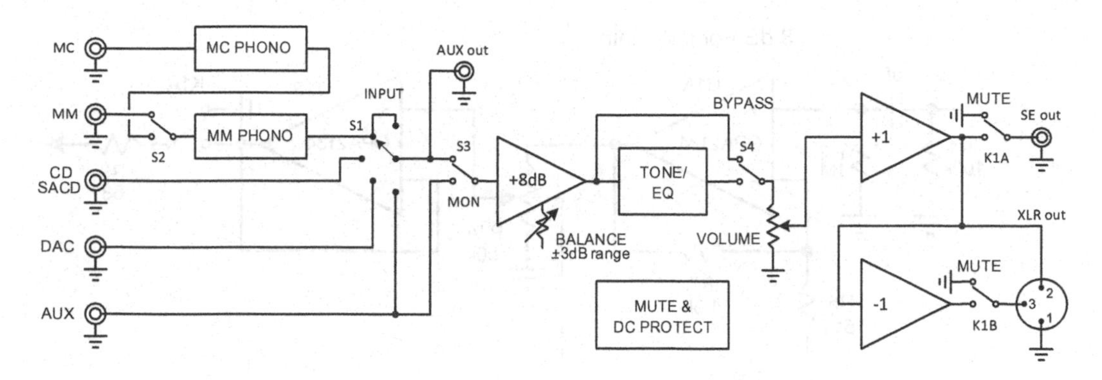

Figure 2.1 Preamplifier Block Diagram.

2.2 Line-Level Signal Path

Figure 2.2 shows the line-level signal path schematic absent the optional tone control circuit that, if used, is inserted between the output of U1A/U2A and the volume control. The line-level circuit is quite simple, employing just a dual JFET OPA2134 op amp for each channel.

Input Buffer and Balance Control

The input buffers are implemented with U1A and U2A for the left and right channels. They are configured as non-inverting gain stages whose gain can be altered by the balance control pot P1. Balance controls are often not found in preamplifiers, but they can be useful in adjusting the image to compensate for slight differences in gain in the left and right audio paths. Accurate balance of left and right levels is important to good imaging. Many cartridges specify level balance to only within 1 dB or so (sometimes worse). Loudspeaker sensitivities can have channel-channel differences as well. Needed balance adjustments are usually small, so the balance control in this simple design has a range of only ±3 dB. With linear taper pot P1 centered, gain of the input stage is +8 dB. At either CW or CCW extreme, one channel has gain of +5 dB while the other channel has gain of +11 dB.

All pots have some wiper resistance, which will allow a tiny bit of crosstalk between channels. The resulting simplicity makes this tradeoff worthwhile. Effective wiper resistance can differ from one pot type or manufacturer to another, and is usually not specified. Because the resistance track is three-dimensional, a pot whose wiper covers the entire surface of the track can still have some effective wiper resistance due to the small leakage path sneaking beneath the surface of the track under the wiper.

Volume Control and Output Buffer

The volume control is implemented with a quality ALPS RK27 dual-gang 50-kΩ audio-taper pot [1, 2]. The unity-gain output stages, comprising U1B and U2B, buffer the pot wipers. Output stage impedance is 68 *Ω*. The 68-*Ω* resistors isolate the op amp outputs from capacitive interconnect loads and prevent oscillation. They also provide a compromise source termination for coaxial interconnect that usually has a characteristic impedance of 50 or 75 *Ω*. A non-coaxial

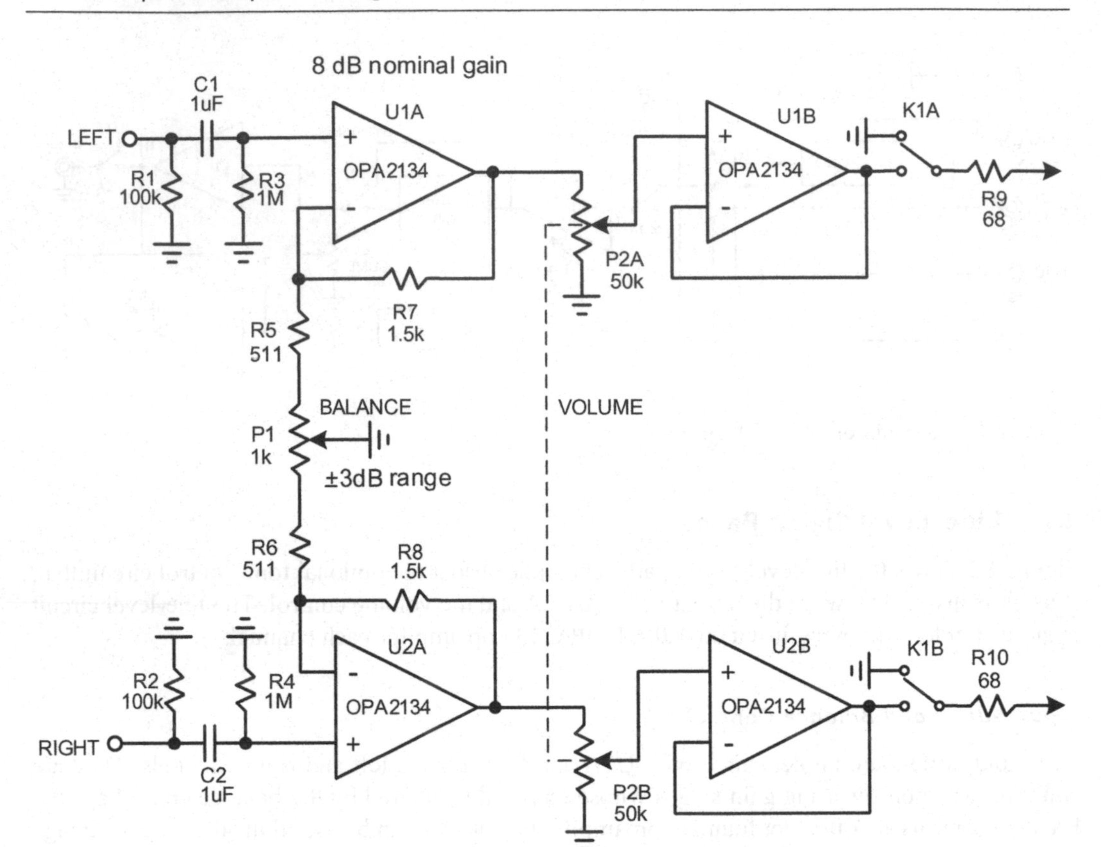

Figure 2.2 Line-Level Signal Path.

interconnect usually has a characteristic impedance of 100–120 Ω, and is still not badly terminated. Approximate matching of the line impedance out to fairly high frequencies can reduce EMI susceptibility in some situations. The muting relay opens the signal path from the output buffer and shunts the preamplifier output to ground through the 68-Ω resistor when the relay is not activated.

Balanced Outputs

While a balanced output is shown in the block diagram, it is an optional feature of this design. If implemented, an op amp inverter is connected to the output of U1B (and U2B) to create the inverted version of the signal. The inverter can be implemented with half of an OPA2134 with 10-kΩ input and feedback resistors. The positive output of U1B is connected to pin 2 of the XLR connector through the muting relay and output terminating resistor. The output of the inverter is similarly connected to pin 3 of the XLR connector. Note that in this case a double-pole, double-throw (DPDT) relay is required for the mute relay for each channel. Balanced outputs can be helpful when the user's power amplifier can accommodate them if the power amplifiers are monoblocks and are distant from the preamplifier.

2.3 Tone Controls

Most modern preamplifiers do not include tone controls. In fact, tone controls are often frowned upon by serious audiophiles, as if putting ketchup on steak. However, there are situations where tone controls can be helpful, especially with regard to making the sound more pleasing in light of frequency response characteristics of the loudspeakers and room acoustics. The tone control circuit sits directly in front of the volume control. It is non-inverting and includes a bypass switch.

By far the most common type of tone control is the Baxandall feedback tone control that provides Bass and Treble control as described in Chapter 20 [3]. The Baxandall tone control circuit used in this preamplifier is shown in Figure 2.3. It is inserted in the preamplifier signal path just ahead of the volume control pot. The Baxandall tone control inverts the signal, so Figure 2.3 includes an inverter at the input of the tone control to make the tone control circuit non-inverting. This makes it easy to incorporate a bypass of the tone control (S4).

In less modular designs where a tone control is always employed, the necessary inversion can be implemented in another stage that may serve a different function. For example, the Baxandall volume control described in Chapter 22 is also inverting [4]. In preamplifiers incorporating both a Baxandall tone control and a Baxandall volume control the inversions cancel.

The input to the tone control is first inverted by U1A to make the conventional inverting Baxandall tone control non-inverting. Central to the Baxandall tone control is U1B, half of an OPA2134 op amp operating in the inverting mode with its positive input connected to ground. Its negative input is thus a virtual ground.

Separate signals for bass and treble frequencies are summed at the virtual ground of U1B. The treble control pot P1 is connected to the input from U1A and to the output of U1B through C1 and C2 that only pass the higher frequencies. Its wiper is connected to the virtual ground of U1B through R5. When the pot is at its center position, equal amounts of high frequencies are fed forward and back to the virtual ground, resulting in a flat response with unity gain. For treble boost, the wiper is moved toward the input end, so that there is more feedforward of high frequencies than feedback, resulting in increased HF response. For treble cut, the wiper is moved toward the right

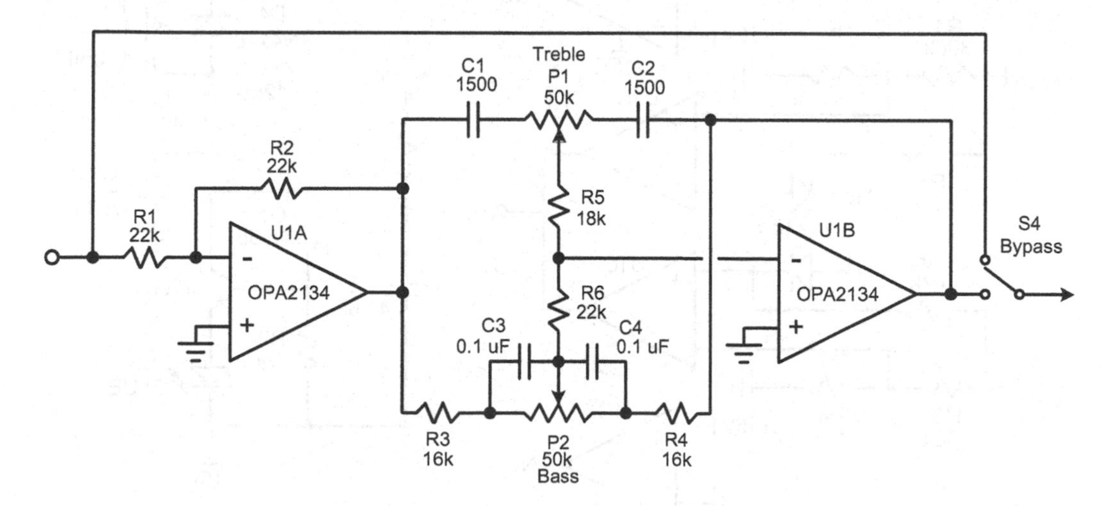

Figure 2.3 Non-Inverting Baxandall Tone Control.

end of the pot, receiving more negative feedback from the output of the op amp. This results in less feedforward and greater feedback for high frequencies and an overall cut in high frequencies.

The feedforward/feedback action is also used for the bass control. In this case, resistors in series with each end of the bass control pot allow capacitors C3 and C4 across P2 and its wiper to attenuate the presence of mid and high frequencies across the ends of the pot, so that the pot position has no effect at mid and high frequencies. The pot wiper goes through a resistor, R6, to the virtual ground to provide feedback at low and mid-frequencies. When the pot is at its center, this part of the circuit also causes gain to be flat and unity at low frequencies. The pots should be dual-gang linear taper with optional center detents. A switch, S4, is included in the design to allow the tone controls to be bypassed. More detail on the Baxandall and other tone controls can be found in Chapter 20.

While it is customary for Baxandall tone controls to produce up to ±15 dB of boost or cut at the high-frequency extremes, this design is more conservative and less intrusive, producing a maximum boost or cut of ±12 dB at the frequency extremes. The component values have also been chosen to produce less boost or cut in a wider relatively flat midrange. This reduces boom when some bass boost is used and reduces harshness when some treble boost is used.

2.4 Muting Circuit

When power is applied or removed, circuits may often create a spurious noise or thump as circuit voltages settle or become uncontrolled. For this reason, a muting circuit is included. The circuit mutes the output for 3 seconds after power is turned on and mutes the output quickly when the power is turned off and before the power supply collapses. The muting device is a relay that opens the signal path and shorts the output when it is not energized. The muting relay contact arrangement is shown in Figure 2.1.

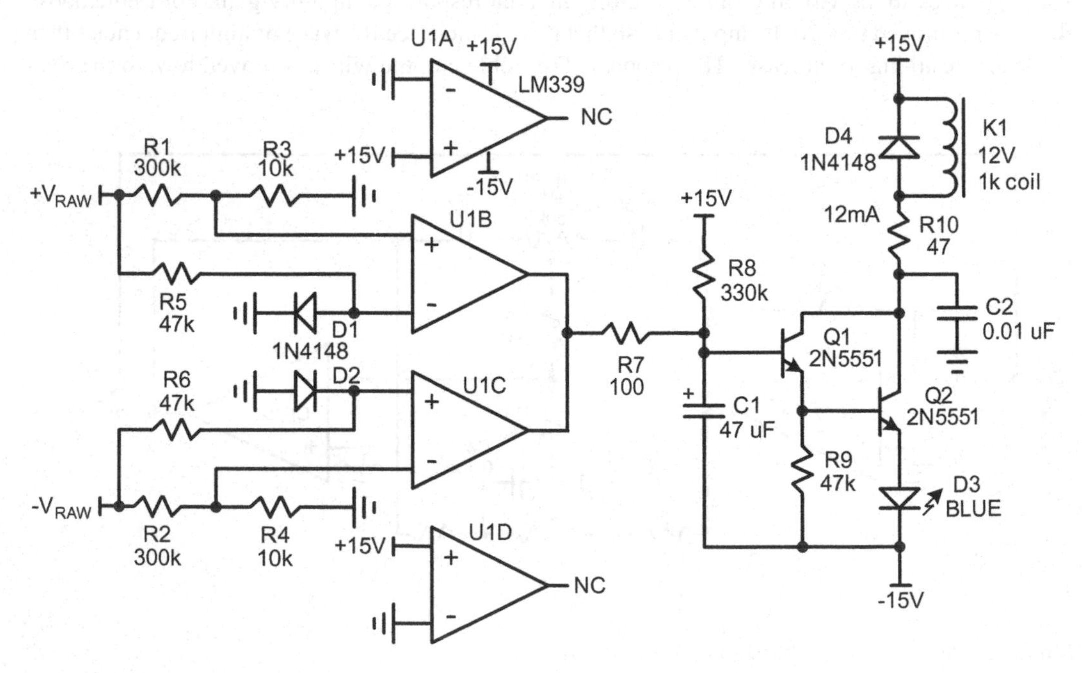

Figure 2.4 Muting Control Circuit.

The muting control circuit is shown in Figure 2.4. It is implemented with an LM339 quad comparator and a relay driver. The comparator monitors the unregulated power supply voltage $+V_{RAW}$ and $-V_{RAW}$ fed to the voltage regulator. When the $+V_{RAW}$ and $-V_{RAW}$ exceed the regulated voltage by 2 V (about 17 V), the comparator threshold is reached and the open collector outputs of the comparator allow R8 to charge C1. A time delay of 3 seconds then begins. After the delay, the voltage on C1 becomes high enough to turn on the Darlington relay driver and close the muting relay contacts.

When the pre-regulator power supply voltage falls below the 17-V threshold in either polarity, the comparator collectors discharge C1 and the relay immediately opens. The same arrangement is used for the negative supply. The output signal will be muted whenever either regulator has less than 2 V of headroom. The voltage drops across D1 and D2 (with the diodes conducting about 330 μA) set the comparator threshold to about 0.55 V. When the input to the regulators is 17 V, the voltage dividers of 300 kΩ and 10 kΩ apply 0.55 V to the comparator to establish the switching point.

2.5 Moving Magnet Phono Preamplifier

The MM phono preamp circuit is shown in Figure 2.5. It is a very simple op amp-based design. It provides needed gain and RIAA equalization for the MM phono cartridge input. Typical MM phono cartridges produce a nominal output on the order of only about 5 mV at 1 kHz. About 40 dB of gain is thus needed at mid-frequencies to bring the signal up to line-level amplitude. Two op amp stages are used in this design, provided by an OPA2134 dual op amp. The U1A stage provides a flat gain of 20 dB and determines the achievable signal-to-noise ratio. The first stage is followed by a passive low-pass network. It comprises R6 in parallel with R7 and C5 and provides the high-frequency 75-μs roll-off part of the RIAA equalization. Op amp U1B provides the remaining 15 dB of mid-band gain and the increased gain at low frequencies for the low-frequency portion of the RIAA equalization. Total MM preamp gain is 35 dB at 1 kHz. This arrangement requires less maximum gain from each op amp than does the more traditional single op amp solution discussed in Chapter 15.

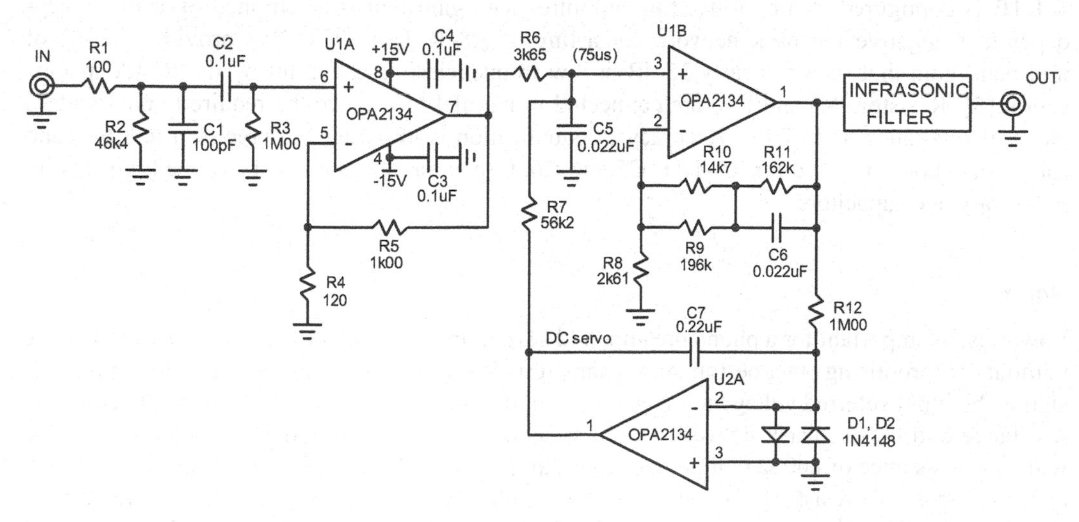

Figure 2.5 Moving Magnet Phono Preamplifier.

DC Servo

Instead of using an electrolytic capacitor in series with R8 in the shunt arm of U1B's feedback network to control output DC offset by reducing gain to unity at DC, a DC servo is used in this design. U2A, in combination with R12 and C7, creates the necessary integrator for the DC servo. The output of the DC servo is applied to the input of U1B via R7. For example, if the average DC at the output of U1A is positive, the integrator will produce a negative-going DC level to be applied to U1B, correcting the offset and driving it down to the input offset voltage of the op amp. Note that use of the DC servo permits DC coupling between the two amplifier stages. The increased negative feedback at low frequencies created by the DC servo causes attenuation of frequencies below 1.5 Hz.

RIAA Equalization

In very simple terms, relative to 1 kHz, the RIAA playback equalization provides about 20 dB of shelving boost at low frequencies and about 20 dB of continuing loss at 20 kHz. More specifically, RIAA equalization is defined by several time constants in the equalization network, as described in more detail in Chapter 15. These time constants are 3180, 318 and 75 μs.

The cartridge input is loaded with a common default load of 46.4 kΩ and 100 pF by R2 and C1. Series resistor R1 against C1 provides some attenuation at very high frequencies to reduce EMI sensitivity. JFET op amps also tend to be more immune to RFI than bipolar types. The input to the gate in the former is a reverse-biased junction, while input to the base in the latter is a forward-biased junction, which is more prone to rectification. The signal is then AC-coupled to the input of U1A by C2. Due to the use of a JFET op amp, no DC current will flow through the cartridge if DC coupling is used at the input. C2 is thus optional. However, potentially damaging DC current will flow in the cartridge under certain fault conditions in U1A. U1A is configured as a simple non-inverting stage with gain of 10. Fairly low-value resistances R4 and R5 are used in the feedback network to minimize thermal (Johnson) noise.

The RIAA 75-μs low-pass time constant is realized passively by C5 and the parallel combination R6 and R7 (3.65 kΩ in parallel with 56.2 kΩ for 3.427 kΩ) [5, 6].

U1B is configured as a non-inverting amplifier with gain that is determined by a frequency-dependent negative feedback network consisting of R8–R11 and C6. This provides 15 dB of mid-band gain that rises to nearly 35 dB at low frequencies, as called for by the RIAA specification [5]. Resistors R9 and R10 are connected in parallel to achieve the required non-standard value of resistance of 13.7 kΩ. Accuracy of equalization is important, so precision resistors and capacitors should be used for R6–R11, C5 and C6. Capacitors C5 and C6 should be high-quality polypropylene capacitors.

Noise

Low noise is important for a phono preamp, and several tradeoffs are needed to achieve low noise without compromising other performance characteristics. The major contributor to noise in this design is the input-referred voltage noise of U1A, but thermal noise contributions from the cartridge resistance and the feedback network resistors cannot be completely ignored. An MM cartridge with coil resistance of 800 Ω will produce thermal noise of 3.8 nV/√Hz. Noise from the feedback network whose net resistance is 100 Ω will be 1.3 nV/√Hz, negligible when RMS summed with the larger cartridge thermal noise. Input noise for the OPA2134 is specified as 8 nV/√Hz. When summed on an RMS basis with the cartridge noise, we have 8.9 nV/√Hz. JFET op amps with lower input noise, like the OPA1642, can be had for a bit more money. Importantly, the 20-dB of gain provided by U1A ahead of the U1B circuit makes the latter's contribution to noise insignificant. Because of this, higher impedances can be used in the frequency-dependent feedback network without any significant impact on overall noise. Those higher impedances allow the use of smaller values for the precision RIAA equalization capacitors.

If the phono preamp frequency response were flat (which it is not) the A-weighted input-referred voltage noise of the op amp could be compared to the 5-mV cartridge input reference level if the op amp voltage noise was the dominant contributor. Noise analysis will be treated in depth in Chapter 10. For now, the A-weighted input-referred noise voltage of an 8 nV/√Hz op amp is 8 nV/√Hz times the square root of 13 kHz, the latter of which is the effective noise bandwidth of the A-weighting curve. This comes to 1.04 μV. This corresponds to a signal-to-noise ratio (SNR) of about 74 dB for a 5-mV cartridge. This would be the SNR if the phono preamp frequency response was flat.

The RIAA equalization improves this number because it acts like a low-pass filter and reduces the overall effective noise bandwidth of the preamplifier taken as a whole. Put another way, it reaps the SNR benefit of the pre-emphasis built into the RIAA playback equalization standard.

Signal Levels and Overload

Nominal signal levels from an MM cartridge are on the order of 5 mV at 5 cm/s stylus velocity at 1 kHz, but MM cartridge outputs can exceed that significantly under some conditions, especially when there are pops, ticks and clicks on the LP. For this reason, input and output overload of the preamp cannot be ignored. In this circuit topology, the main overload concern is the maximum phono preamp output level, and this overload limit would also apply to subsequent line-level circuits. The flat gain of 20 dB before equalization also contributes an overload limit because it sees the output of the cartridge directly, without any high-frequency roll-off. The input amplifier allows for 850 mV_{RMS} maximum cartridge signals without pre-equalization overload in this stage. This corresponds to 8.5 V_{RMS} at the output of U1B.

Infrasonic Filter

An optional infrasonic filter can be added at the output of the MM preamp in accordance with the discussion in Chapter 15. The optional third-order Butterworth infrasonic filter circuit is shown in Figure 2.6. It is a Sallen-Key design that is down 1.0 dB at 20 Hz and down 16.5 dB at 8 Hz. The infrasonic filter does introduce some phase distortion at very low frequencies, however, so a switch can be included that allows it to be bypassed.

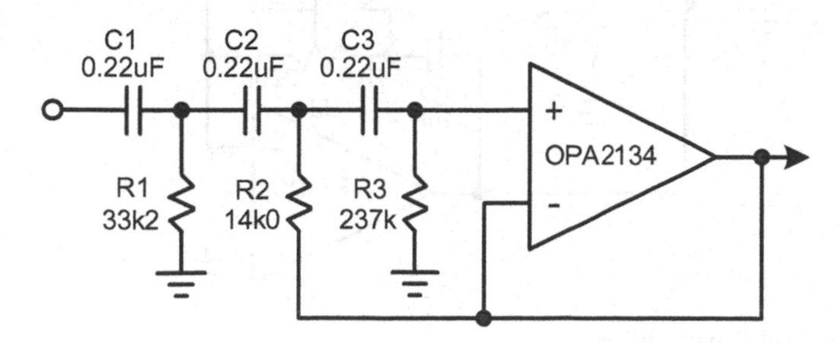

Figure 2.6 Infrasonic Filter.

2.6 Moving Coil Phono Preamplifier

The MC inputs have their own connectors, and are always wired to the MC preamp. Referring back to Figure 2.1, when the MC input is selected, S2 switches the inputs of the MM preamp from the MM cartridge inputs to the output of the MC head amp. The main portion of the input selector switch routes the output of the MM preamp to the line stage for both the MC and MM selections.

The MC preamplifier is shown in Figure 2.7. The typical output of an MC cartridge is about 20 dB lower than that from an MM cartridge, on the order of 200–600 μV for most MC cartridges at 5 cm/s at 1 kHz. For this reason, low noise is very important. Here a low-noise amplifier with a gain of 10 (20 dB) is used in front of the MM phono preamp to accommodate MC cartridges. Figure 2.7 shows that there are actually 10×, 30× and 50× gain taps.

The MC cartridge input is terminated with 100-Ω resistor R1 and AC-coupled to the JFET gates. The use of the high-impedance JFET input allows the use of a relatively small film coupling capacitor C1 at the input and a large-value gate bias resistor R2.

The moving coil preamplifier employs an LSK389C discrete dual monolithic JFET for its input stage, with both JFETs connected in parallel. Each section of the LSK389C operates at a bias current of 3 mA. This provides a very low-noise device with input-referred noise on the order of 0.7 nV/√Hz when running each JFET at 3 mA. I_{DSS} for the LSK389C is 10–20 mA. If an LSK389B with a lower I_{DSS} grade is used, slightly higher noise will result. The value of R4 determines the total 6 mA operating current of the JFET stage. This common-source stage feeds an LM4562 op amp that provides the output and feedback to establish the gain.

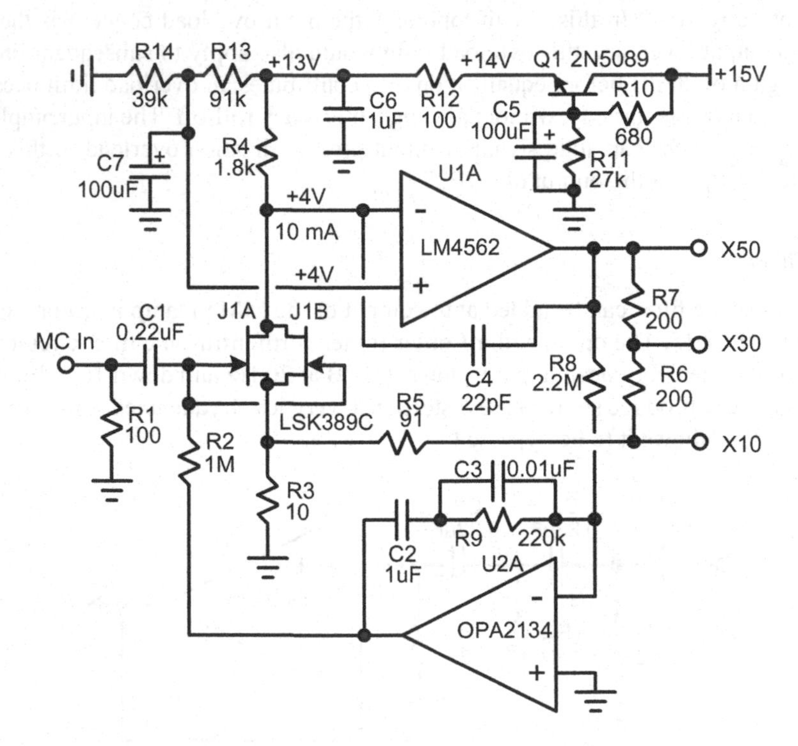

Figure 2.7 Moving Coil Preamplifier.

The total transconductance of the two JFETs is about 32 mS at the total 6 mA drain current. In combination with the 10-Ω resistance of R3, input stage *gm* is about 24 mS. Open-loop gain of the first stage with the 1.5k-Ω load is thus about 36. This amount of gain in front of U1A means that the input-noise voltage of U1A is of no consequence.

The closed-loop gain of the circuit is actually set to 50 by the string of R5, R6 and R7. This is done so that the feedback network load on U1A will not be less than 500 Ω. Three taps on the string of feedback resistors (R5, 6, 7) allows one to choose gains of 10, 30 and 50.

A DC servo controls gate bias through R2 to achieve near-zero output offset. The DC servo feedback loop includes the integrator and a pole at 0.7 Hz created by R2 and C1. The significant amount of lagging phase shift around the loop can lead to instability. For this reason, resistor R9 is inserted into U2A's feedback path to create additional phase margin for the loop. As a result of the zero, at audio frequencies the integrator has a gain of 0.1 due to the ratio of R9 to R8.

The +4-V reference voltage to U1A forces the JFET drain voltage to be +4 V. The 1.5-kΩ load resistor R4 sees 9 V across it, setting the total JFET drain current to 6 mA. The LSK389C is used, each of whose two JFETs has I_{DSS} that is between 10 and 20 mA.

A capacitance multiplier provides a +13-V supply with very low noise to power the drains of the JFETs, as that is a sensitive node. The capacitance multiplier was chosen over a low-noise low-dropout (LDO) regulator because the former provides very low noise while providing isolation from the +15-V supply to very high frequencies. The +4-V reference voltage is also obtained from that supply through a divider consisting of R13 and R14. Capacitors C5, C6, and C7 filter noise from this supply. The ±15-V supplies for the op amps should also have low noise.

2.7 Relay Switching

The preamplifier is illustrated with mechanical switches, labeled as S1–S4, but significant advantages can be had if some or all of the switches are replaced with relays. Putting a relay on the board close to the signal path it is controlling can greatly reduce point-to-point shielded interconnect and potentially reduce hum and noise.

All of the switches including the five-position input selector switch S1 can be replaced with miniature relays. The relays can be activated by simple mechanical switches on the front panel or by a microcontroller. Replacing input selector S1 requires five relays controlled by a five-position rotary switch on the front panel or a 1/5 decoder from a microcontroller. Simple diode logic associated with S1 can be used to automatically control the MM/MC selector switch S2.

Miniature telecom relays have a footprint of about 0.8 by 0.4 inches. Coil power for various such relays typically runs from 150 to 500 mW, corresponding to 12 to 42 mA at 12 V. Subminiature relays like the Panasonic AGQ200S12Z have a footprint of 0.42 by 0.28 inches and a nominal 12-mA coil. These are available for about $2 in small quantities. They save space and current.

With S1–S4 and K1, as many as five relays can be energized simultaneously, corresponding to 60–210 mA. The increase in the preamplifier power budget by up to 3 W is the price paid for full relay control, unless the more sensitive relays like the Panasonic AGQ series are employed. The relays should be run off of their own regulated +15-V supply that can be implemented with an inexpensive 7815 regulator. Its maximum dissipation dropping 20 V to 15 V at 60 mA for the AGQ series will be 0.3 W. Each relay needs an appropriate-value series resistor to absorb the 3-V drop from 15 to 12 V. This reduces the maximum dissipation in the 7815 regulator. These resistors will each dissipate a maximum of about 36 mW. With a 12-mA relay coil, each of these resistors would be 250 Ω. The coil should be bypassed with a 0.1 μF capacitor.

2.8 Power Supply

As described later in the power supply chapter, there are numerous ways to power a preamp. These include use of a traditional transformer-based linear supply, use of an external dongle to produce both polarities from an AC wall transformer, and use of a USB-powered dual-polarity switching supply followed by LDO voltage regulators. Many of these alternatives are discussed in Chapter 14. Powering an on-board dual-polarity switching supply from a 5-V phone charger is tempting, especially because 12- or 20-W chargers are readily available. However, here a conservative path will be followed by using a 24-V_{AC} wall transformer feeding a dongle that creates about ±20 V to be fed to on-board regulators. This eliminates line voltage hum and other worries related to safety and line filtering.

Power Budget

The five miniature DPDT small-signal relays rated at 12 V at 12 mA will draw about 60 mA when all are activated. Each relay serves both stereo channels. The relays are all powered from the positive rail.

The dual op amps consume about 12 mA, or about 6 mA per op amp. Each channel employs two op amps for the MC preamp, four op amps for the MM preamp (including infrasonic filter), one op amp for the input buffer, two op amps for the tone control and one op amp for the output amplifier. The stereo preamp will thus have 20 op amps (10 duals). Total op amp current consumption will then be 120 mA from the ±15-V rails. The discrete JFETs in the MC head amp will consume about 12 mA from the positive supply. With the maximum 210 mA positive rail relay load, the total load on the positive rail will rise to a maximum of about 342 mA. Total preamp dissipation from the 20-V input rails will be 2.4 W from the negative rail and 6.8 W from the positive rail, for total power consumption of 9.2 W. It is advantageous to employ relays that require as little as 12 mA each.

On-Board Linear Power Supply

The choice of an on-board linear supply fed by a dongle supply and a wall transformer was conservative. This approach is probably the lowest-noise approach, since only regulated DC enters the preamplifier and there is no rectifier diode switching in the preamplifier. Figure 2.8 shows the fairly simple on-board ±15-V power supply regulator. It receives ±20 V from the dongle supply via a 4-pin mini-DIN connector (often used for S-video in the past). After preliminary R-C low-pass filtering with a corner at 48 Hz, the rails are regulated by LM317/LM337 regulators. Regulators with lower noise than the LM317/LM337 are available. If relay switching is used, the relays should be powered from a second 7815 regulator (not shown).

Dongle Supply

Figure 2.9 shows the dongle power supply, which can be housed in a small project box (like the power supply for a laptop). It employs a 20-V_{AC} wall transformer that feeds the dongle power supply. That supply employs half-wave rectifiers to produce approximately 30 V_{DC} that is regulated down to ±20 V_{DC}. Those voltages are supplied to the preamp via a four-conductor mini-DIN connector and S-video cable. Within the preamp, that voltage is again regulated down to ±15 V_{DC}. Alternatives to this approach are described in Chapter 14.

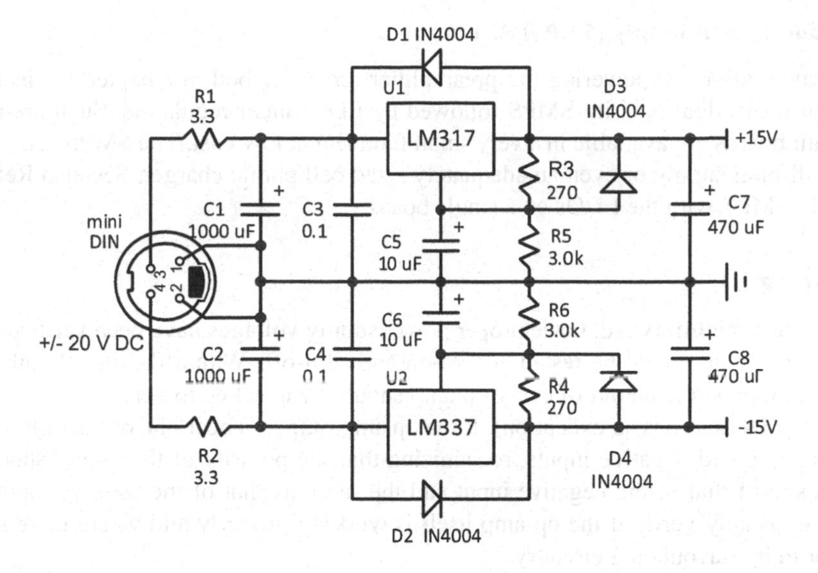

Figure 2.8 Power Supply.

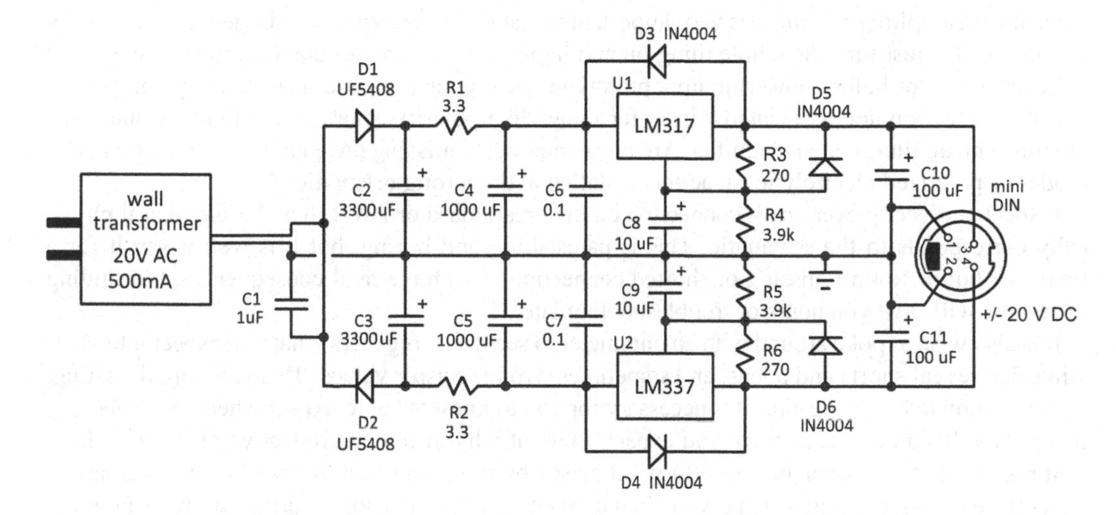

Figure 2.9 ±20 V_{DC} Dongle Supply.

Ground Isolation

It is notable that preamps that do not have line voltage entering them do not need to be grounded through a safety ground. In other words, a ground-lift switch is safe in these instances. In some cases, allowing the preamp to float and be grounded by other equipment to which it is connected can mitigate ground-loop situations. Here the 20-V_{AC} wall transformer provides ground isolation. Other such sources could be a 2.1-A 5-V_{DC} USB charger (10 W), a common 9-V_{DC} wall transformer or even a 19-V_{DC} laptop computer power brick.

Switch-Mode Power Supply (SMPS) Alternatives

Numerous alternatives to powering the preamplifier are described in Chapter 14, including the use of an on-board dual-polarity SMPS followed by LDO linear regulators. Such pre-assembled SMPS circuit boards are available in a very small footprint at low cost. The SMPS can be powered by a DC wall-plug supply or even an adequately sized cell phone charger. See also Ref. [7] for a dual-polarity SMPS with the LDOs on a single board.

2.9 Testing

Testing is quite straightforward. Once proper power supply voltages have been verified, operation of the muting circuit should be tested at turn-on and turn-off. Within the signal path, and with shorted signal inputs, the output of every op amp should be checked to verify that it is at or close to zero volts (a DC servo is an exception). If the op amp output is near one of the rails, one should probe its positive and negative inputs, recognizing that the polarity of the output should always be the opposite of that of the negative input and the same as that of the positive input. Such an approach can usually verify if the op amp itself is working properly and where there might be a wiring error in its surrounding circuitry.

Initial Inspection and Passive Tests

Once the preamplifier is built, it is very important to test it with patience and diligence in an orderly fashion. Don't just turn the whole thing on and hope there will be no magic smoke. Inspect and poke around a lot before powering up a prototype. Take your time and inspect every component and its connection and associated wiring for things like polarity, shorts, open joints, values, etc. Do this with an illuminated magnifier. Are any components missing? Are any transistors, op amps, diodes or polarized electrolytic capacitors installed in the wrong orientation?

Inspect and verify every path connection on the breadboard or PWB to make sure that it physically corresponds to the schematic. This is painstaking and boring, but it is well worth it for a first-time build. Not all missing or shorted connections will have fatal consequences, but finding them now will save you hours of troubleshooting later.

It is also wise to poke around with an ohmmeter to see that it registers what you expect it to. This will often reveal shorts and opens, and sometimes wrong resistor values. This is in-circuit testing, so some knowledge of the circuit is necessary for you to know what to expect when you probe any two points. If you don't get what you expect, find out why. It may be reflective of a circuit fault or it may be that a junction has been forward-biased by the ohmmeter current. In any case, satisfy yourself that you are getting what you should, even if it requires some further study of how the circuit should work. Sometimes an old-fashioned battery-operated ohmmeter works better for this than a DVM; see how the meter you are using reacts to reading a diode by itself in both directions.

Power Supply Test

The dongle power supply should be tested first. It should put out about ±20 V_{DC}. It is best to test it with a moderate load on both output polarities. Use 100-Ω 5-W load resistors to draw 200 mA from each rail. Measure the input voltages to the regulators and make sure they both have headroom no less than 2 V.

Energize the preamplifier with the dongle supply and verify that the on-board supply puts out ±15 V_{DC}. At this point the supply should be loaded with most of the current consumed by the active circuitry. If relay switching is used with a dedicated +15-V regulator, verify that voltage as well. At this point, there might be few if any relays activated.

DC Voltage Tests

Now probe numerous nodes in the preamplifier to see that the voltages are as expected, based on voltages that one should expect from the schematics. Check the voltage at the output of every op amp. In most cases it should be near zero. If it is close to the rail, check the positive and negative inputs to the op amp to see if the output polarity is consistent with the net input voltage polarity. This helps verify whether the op amp or the surrounding circuitry is at fault. Check the power supply and node voltages within the MC preamp. Both inputs to U1A should be at about +4 V and its output should be close to 0 V. The output of the DC servo op amp, U2A, should be negative by 0.1–1.0 V, reflecting the reverse bias applied to the JFET gates.

Mute Circuit and Relay Test

Apply power. The relay should activate after about 3 seconds. Remove power and verify that the mute relay opens as soon as either 20-V rail falls to 17 V or less. If relays are being used for signal switching (S1–S4), verify that they open and close as expected with different input and mode selections.

Signal Path Testing

Tone, volume and balance control circuits should be checked by running a gain and frequency response test from a line input to the output with the tone controls bypassed if they are implemented. Verify that balance control and volume control behavior is as expected. Check gain with the balance control centered and the volume control at maximum. If the tone controls are implemented, engage them and check gain and frequency response as the treble and bass controls are rotated. Check the clipping levels with a 2.2 kΩ load by increasing the input-signal amplitude. Peak output levels at the onset of clipping should be no less than ±13 V.

Test the MM preamp with the tone controls bypassed and a 5-mV, 1 kHz signal applied to the MM input. Gain at 1 kHz should be verified to be about 35 dB. Gain at 20 Hz should be about 60 dB, and gain at 20 kHz should be about 20 dB. The oscillator signal should then be reduced by 20 dB and applied to the MC input with the MC switch engaged with the gain jumper in the 10× position. Output level and frequency response should then be the same as measured for the MM preamp.

Trimming for Volume Pot Tracking

Ganged audio-taper pots don't track well. Before installing the volume control pot, check its tracking with a 1-kHz signal applied across both pot's terminals at either end, with the signal applied at the clockwise terminals. Measure the output signal levels at the wipers at various volume control positions, including 9:00, 12:00 and 3:00. If trimming is needed this procedure will help reduce tracking error by trimming the tracking error to zero when the pot is centered at 12:00. This is done by connecting a high-value shunt resistor to ground from the wiper of the gang that has less attenuation.

1 Center the pot, and connect both ends of each gang together and measure the resistance to the wipers. This is the source resistance of the pot R_s when the pot is centered.
2 Connect both gangs of the pot to a 1-kHz, 1-V_{RMS} source and measure the voltages at both wipers. This voltage will be on the order of 157 mV (−16 dB) with the audio-taper pot centered.
3 Divide the smaller wiper voltage by the larger wiper voltage to obtain ratio K. A 1-dB tracking error will yield a fraction of about 0.9. Define this fraction as K, the gain of the pot with the lower signal compared to the gain of the other pot.
4 Calculate the needed shunt resistance R_{shunt} with:

$$R_{shunt} = R_s\left(K/\left(1-K\right)\right)$$

where R_s is the source resistance of the gang with the higher wiper voltage.

5 Take this resistance R_{shunt} and use it to shunt the wiper of the pot with the larger wiper signal to ground. Verify that the tracking error is near 0 dB when the pot is centered.

2.10 Troubleshooting

If the preamplifier does not work at any point in the testing process, troubleshooting is necessary. A great deal of the testing outlined above is also applicable to troubleshooting. This is particularly true for voltage measurements. If the preamplifier is built and passes the initial testing in the order described above, it is very likely that it will work as designed. Of course, if problems occur at one of those steps, troubleshooting is called for at that point. The sequencing of the tests above helps to localize the problem.

Always start troubleshooting by carefully inspecting the preamplifier, often with an illuminated magnifying glass. Look for parts that appear to be damaged in any way. Look for shorts or opens. Verify proper part values and part orientation.

References

1. ALPS RK271 Potentiometer Datasheet, https://tech.alpsalpine.com.
2. Parts Express, ALPS 50-kΩ Dual Audio Taper Pot, *50k ohm Audio Taper Stereo Potentiometer* # 50kAX2.
3. Peter Baxandall, *Negative Feedback Tone Control – Independent Variation of Bass and Treble without Switches*, Wireless World, vol. 58, October 1952, p. 402.
4. Peter Baxandall, *Audio Gain Controls*, Wireless World, November 1980, pp. 79–81.
5. Stanley Lipshitz, *On RIAA Equalization Networks*, JAES, June 1979.
6. Gary A. Galo, Disc Recording Equalization Demystified, *ARSC Journal*, vol. 27, no. 2, Fall 1996, pp. 188–211. Reprinted in *The LP Is Back*, 2nd edition, Audio Amateur Press, Peterborough, NH, 2000, pp. 44–54. (The reprint is available online at www.smartdevicesinc.com/chpt14.pdf [PDF version for download] or http://www.smartdevicesinc.com/riaa.html [text version].) A copy is also available by contacting the author at galoga@potsdam.edu.
7. Jan Didden, *The Silent Switcher*, www.linearaudio.nl/silentswitcher.

Chapter 3

Circuit Building Blocks

3.1 Introduction

In this chapter we'll look at the basic building block circuits that are used to implement discrete preamplifier and mixer circuits. Many of these building blocks can also be found in integrated circuits, like operational amplifiers. This will provide a good foundation for the detailed analysis of the circuits that follow in later chapters. This chapter presumes a basic knowledge of passive components and transistors. Details of these devices and how they operate are provided in the chapters that follow. This chapter concludes with a section on circuit simulation using LTspice.

3.2 Transistors

This first section of the chapter serves as a very brief introduction to the workings and characteristics of transistors. It will serve as a foundation for the circuit block discussions. Much more information on semiconductors is covered in Chapter 7.

The Bipolar Junction Transistor (BJT) is the primary building block of most audio circuits. If a small current I_b is sourced into the base of an NPN transistor, a much larger current I_c flows in the collector. The ratio of the collector and base currents is the current gain, commonly called beta (β) or h_{fe}. Similarly, if one sinks a small current from the base of a PNP transistor a much larger current flows in its collector. The current gain for a typical small-signal transistor often lies between 50 and 200, but can go as high as 500 or more. Beta can vary quite a bit from transistor to transistor and is also a mild function of the transistor current and collector voltage.

Transistor Characteristics

Figure 3.1 illustrates the collector current of a transistor as a function of collector-emitter voltage (V_{ce}) for several different values of base current. The upward slope of each curve with increasing V_{ce} reveals the mild dependence of β on collector-emitter voltage. The spacing of the curves for different values of base current reveals the current gain. Notice that this spacing tends to increase as V_{ce} increases, once again revealing the dependence of current gain on V_{ce}, called the *Early effect.* The spacing of the curves may be larger or smaller between different pairs of curves. This illustrates the dependence of current gain on collector current. The transistor shown has a β of about 50.

Because transistor β can vary quite a bit, circuits are usually designed so that their operation does not depend heavily on the particular value of β for its transistors. Rather, the circuit is designed so that it operates well for a minimum value of β and for very high β. Because β can sometimes be very high, it is usually bad practice to design a circuit that would misbehave if β became very high.

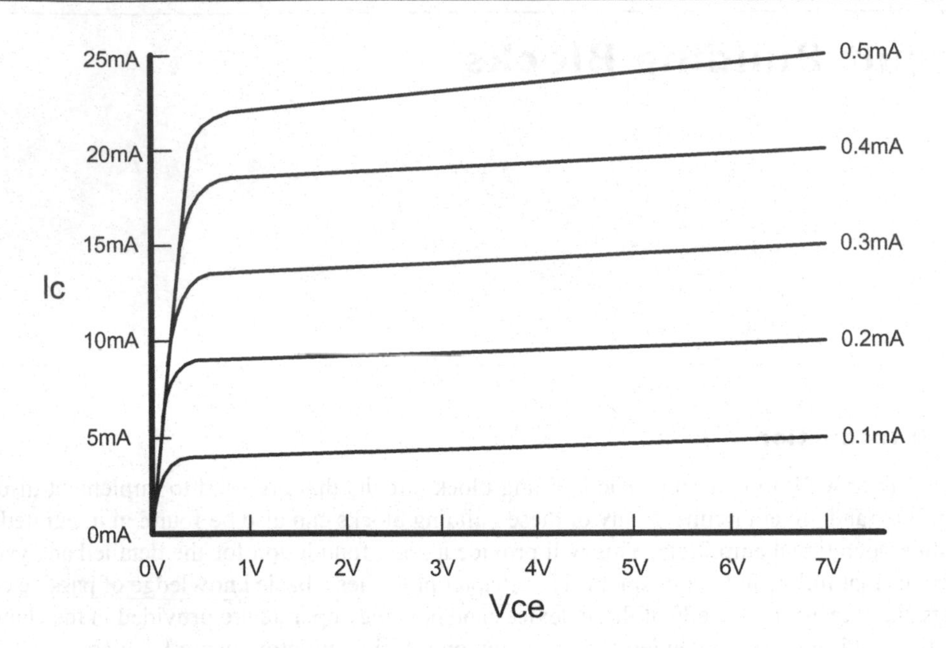

Figure 3.1 Transistor Collector Current Characteristic.

The bipolar junction transistor requires a certain forward bias voltage at its base-emitter junction to begin to conduct collector current. This turn-on voltage is usually referred to as V_{be}. For silicon transistors, V_{be} is usually about 0.6 V. The actual value of V_{be} depends on the transistor device design and the amount of collector current (I_c).

The base-emitter voltage increases by about 60 mV for each decade of increase in collector current. This reflects the logarithmic relationship of V_{be} to collector current. For the popular 2N5551, for example, V_{be} = 600 mV at 100 μA and rises to 720 mV at 10 mA. This corresponds to a 120-mV increase for a two-decade (100:1) increase in collector current.

Collector Current vs Base-Emitter Voltage

Small amounts of collector current actually begin to flow at quite low values of forward bias (V_{be}). Indeed, the collector current increases exponentially with V_{be}. That is why it looks like there is a fairly well defined turn-on voltage when collector current is plotted against V_{be} on linear coordinates. It becomes a remarkably straight line over many decades of collector current when the log of collector current is plotted against V_{be}. Some circuits, like multipliers, make great use of this logarithmic dependence of V_{be} on collector current. Transistors with good log conformance increase the precision of such circuits. Collector current increases exponentially with base-emitter voltage, and we have the familiar relation:

$$I_c = I_s\left[e^{(Vbe/VT)} - 1\right] = I_s\left[\exp\left(V_{be} / V_T\right) - 1\right] \tag{3.1}$$

V_T is called the *thermal voltage* and is equal to kT/q [1]. Here V_T is about 26 mV at room temperature, and is proportional to absolute temperature (PTAT). I_s is called the saturation current and is a characteristic of the transistor. It also has a fairly strong positive temperature coefficient.

k is Boltzmann's constant, with $k = 1.38e^{-23}$ Joule/K
T is absolute temperature in Kelvin
q is the electron charge, with $q = 1.6e^{-19}$ Coulomb

In practice, if $V_{be} = 600$ mV and $V_T = 26$ mV, $V_{be}/V_T = 23.1$, and exp(23.1) equals approximately10 billion, significantly larger than 1, so the −1 term in (3.1) will be dropped for simplicity. We then have the approximation

$$I_c = I_s e^{(Vbe/VT)} \tag{3.2}$$

If I_s is 1e–[14] and V_{be}/V_T, is about 1e[10], then I_c is 1e–[4], or 0.1 mA. V_T plays a role in the temperature dependence of V_{be}. However, the major cause of the temperature dependence of V_{be} is the strong increase with temperature of the saturation current I_s. It approximately doubles for every 5°C of increased temperature [1]. This ultimately results in a negative temperature coefficient of V_{be} of about −2 mV/°C.

V_{be} as a Function of Collector Current

With some rearrangement and taking the natural log (ln) of both sides, we have:

$$V_{be} = V_T\left[\ln\left(I_c / I_s\right)\right] \tag{3.3}$$

where ln(x) is the natural logarithm. At $I_c = 1$ mA, a transistor with $I_s = 1e{-14}$ will have ln(1e[11]) = 25.3. Multiplied by $V_T = 26$ mV, we have $V_{be} = 658$ mV. As a sanity check, one can calculate V_{be} at collector current of 0.1 mA. That turns out to be $V_T\ln(e^{10}) = 599$ mV, which is 59 mV less than the previous calculation for 1 mA, as expected. In practice V_{be} increases by about 60 mV/decade of increase in collector current, reliably so often over a range of many decades of collector current change. The fact that V_{be} changes as the log of collector current is important in many applications.

It is helpful to better understand how V_{be} is calculated and how it behaves. V_{be} is arguably the most important parameter in current and voltage references. One of the most important relationships for BJTs is the equation below that defines the collector current as a function of base-emitter voltage V_{be}:

$$I_c = I_s\left[\exp\left(V_{be} / V_T\right) - 1\right] \tag{3.4}$$

V_{be} Temperature Coefficient (TC)

At a given collector current, V_{be} falls with temperature, with a TC of about −2 mV/°C. With $V_{be} = V_T[\ln(I_c/I_s)]$, we see that there are two competing temperature dependencies that determine the TC of V_{be}. The first is V_T with a positive TC, while the second is I_c/I_s, which has a negative TC as a result of the fact that I_s increases dramatically with temperature. While V_T is almost perfectly proportional to absolute temperature (PTAT), the term $\ln(I_c/I_s)$ is not perfectly complementary to absolute temperature (CTAT).

Delta V_{be}

If two identical transistors Q1 and Q2 are operated at different collector currents, their current densities will be different and therefore their V_{be} will be different. V_{be} of Q1 will be higher than that of Q2 if Q1 is operated at a higher current and higher current density. If instead, the transistors are operated at the same collector current, but Q2 has greater emitter area than Q1, V_{be} of Q1 will once again be higher than V_{be} of Q2. All else remaining equal, the relative current densities will determine the difference of the V_{be}. This difference is commonly referred to as delta V_{be} (ΔV_{be}). For example, we already know the rule of thumb that increasing the current of a transistor by a factor of 10 will increase its V_{be} by about 60 mV. Here we discuss the magnitude and temperature coefficient of ΔV_{be}.

It is also true that ΔV_{be} for a pair of identical transistors operating at different current densities is a function of temperature, increasing with temperature. The increase in ΔV_{be} is PTAT. For reference,

$$\Delta V_{be} = V_T\left[\ln\left(I_{Q1} / I_{Q2}\right)\right] \tag{3.5}$$

for identical transistors operating at different current, or

$$\Delta V_{be} = V_T\left[\ln\left(A_{Q2} / A_{Q1}\right)\right] \tag{3.6}$$

For transistors operating at the same current with different emitter areas A_{Q1} and A_{Q2}. V_T is about 26 mV at room temperature, and is PTAT. At 25°C, the temperature in Kelvin is 297 K. This means that a 1°C temperature increase will increase V_T by 298/297 = 1.0034, corresponding to 3400 ppm/°C. It also means that an increase of 1°C will increase V_T by 0.088 mV.

If the emitter area of Q2 is ten times that of Q1 in Eq. 3.5, ln(10) = 2.303. If kT/q equals 26 mV, then ΔV_{be} of Eq. 3.5 becomes 59.9 mV. Not surprisingly, this is about 60 mV. In this case, a 1°C increase in temperature will increase ΔV_{be} by 0.204 mV. The ΔV_{be} of 59.9 mV resulting from a 10:1 current density ratio has a TC of +0.204 mV/°C.

Transconductance

The *transconductance* (*gm*) of the transistor describes how much the collector current changes when the base voltage is changed a small amount, when the transistor is operated at a given DC bias current. Thus $gm = \Delta I_c/\Delta V_{be}$. Transconductance is actually the more predictable and important design parameter for a transistor (as long as β is high enough not to matter much). Transconductance has the units of *Siemens*, S (amps per volt). If the base-emitter voltage of a transistor is increased by 1 mV, and the collector current increases by 39 μA, the *gm* of the transistor is 39 milli-Siemens (mS). Siemens is actually the inverse of ohms, and long ago the term “mhos” was used to describe its value, with mhos being ohms spelled backward.

While transistor current gain is an important parameter and largely the source of its amplifying ability, the transconductance of the transistor is perhaps the most important characteristic used by engineers when doing actual design. It is sometimes easier to visualize circuit operation and design by taking account of finite *gm* by instead using what may be called the intrinsic emitter resistance $re' = 1/gm$ [2]. We can then think of an “internal” emitter that is moving completely with the base voltage, that emitter then being in series with re'.

Hybrid π Transistor Model

Those more familiar with transistors will recognize that much of what has been discussed above contributes to the *hybrid* π small-signal model of the transistor, shown in Figure 3.2 [1]. The

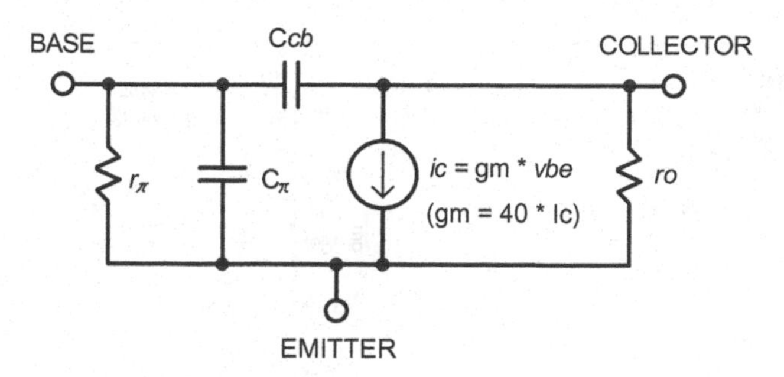

Figure 3.2 Hybrid Pi Small-Signal Transistor Model.

fundamental active element of the transistor is a voltage-controlled current source; namely a transconductance (*gm*). Everything else in the model is essentially a passive *parasitic* component. Small-signal current gain is taken into account by the base-emitter resistance r_π. Early effect is taken into account by r_o. Collector-base capacitance is shown as C_{cb}. Current gain roll-off with frequency (f_T) is modeled by C_π. The values of these elements are as described above. Because this is a small-signal model, element values will change with the operating point of the transistor.

JFET Characteristics

JFETs (junction field effect transistors) operate on a different principle than BJTs. Picture a bar of n-type doped silicon connected from source to drain terminals. This bar will act like a resistor. Now add a p-n junction in the middle of the length of this bar by adding a region with p-type doping. This p-doped region becomes the gate of the JFET. Where the p-doped region meets the p-doped channel, a semiconductor junction diode is formed. As the p-type gate is reverse biased, a *depletion region* will be formed. It will increase in size as the reverse bias of the junction increases. This will begin to pinch off the region of conductivity in the n-type bar. This reduces current flow. This is called a *depletion-mode* device. In this example, the device is an N-channel JFET. The JFET is nominally *on* and its degree of conductance will decrease as reverse bias on its gate is increased until the channel is completely pinched off.

The reverse gate voltage where pinch-off occurs is referred to as the *threshold voltage* V_t. The threshold voltage is often on the order of −0.5 to −4 V for most small-signal N-channel JFETs. Note that control of a JFET is opposite to the way a BJT is controlled. The BJT is normally *off* and the JFET is normally *on*. The BJT is turned on by application of a forward bias to the base-emitter junction, while the JFET is turned off by application of a reverse bias to its gate-source junction.

The reverse voltage V_{dg} that exists between the drain and the gate can also act to pinch off the channel of a JFET. At V_{dg} greater than V_t, the channel will be pinched in such a way that the drain current becomes self-limiting. In this region the JFET no longer acts like a resistor, but rather like a voltage-controlled current source. These two operating regions are referred to as the *linear region* and the *saturation region*, respectively. JFET amplifier stages usually operate in the saturation region.

Figure 3.3(a) shows how drain current changes as a function of gate voltage in the saturation region, while Figure 3.3(b) illustrates how transconductance changes as a function of drain current in the same region. The device shown here is one-half of a Linear Systems LSK489 dual monolithic JFET. Threshold voltage for this device is nominally about −1.8 V. The single version of the LSK489 is the LSK189.

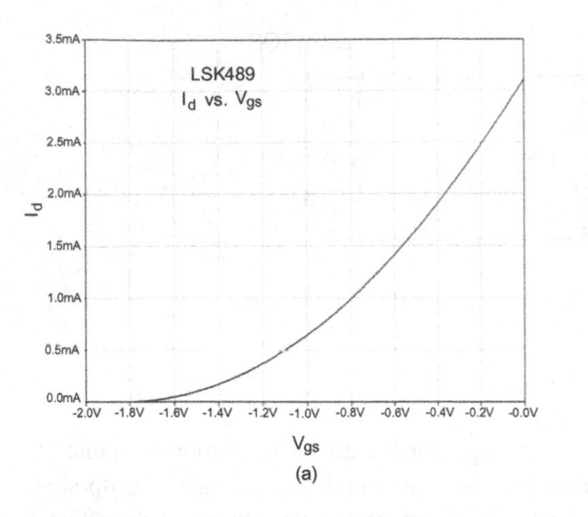

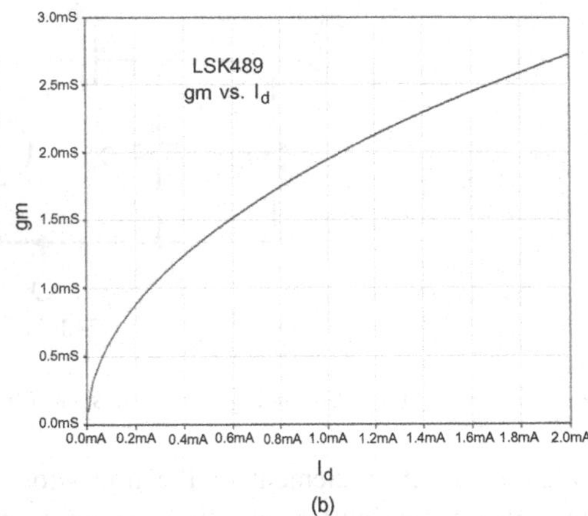

Figure 3.3 JFET Drain Current and Transconductance Curves.

The JFET I-V characteristic (I_d vs V_{gs}) obeys a square law, rather than the exponential law applicable to BJTs [1]. The simple relationship below is valid for $V_{ds} > V_t$ and does not take into account the influence of V_{ds} that is responsible for output resistance of the device.

$$I_d = \beta\left(V_{gs} - V_t\right)^2 \tag{3.7}$$

The equation is valid only for positive values of ($V_{gs} - V_t$). The factor β governs the transconductance of the device. It should not be confused with BJT current gain. β is a JFET SPICE parameter, and for this device β is about 0.01 mS/V. When $V_{gs} = V_t$, the V_{gs}-V_t term is zero and no current flows. When V_{gs} = 0 V, the term is equal to V_t^2 and maximum current flows. The value of this current is called I_{DSS}.

Notice that *gm* for this JFET is about 2 mS at 1 mA. This contrasts with the approximate 39 mS of a BJT at 1 mA. The *gm* of a JFET is often 1/10 or less than that of a BJT operating under similar conditions.

3.3 Basic Amplifier Stages

An audio amplifier is composed of just a few important circuit building blocks put together in many different combinations. Once each of those building blocks can be understood and analyzed, it is not difficult to do an approximate analysis by inspection of a complete amplifier. Knowledge of how these building blocks perform and bring performance value to the table permits the designer to analyze and synthesize circuits.

Common-Emitter (CE) Stage

The common-emitter (CE) amplifier stage is possibly the most important circuit building block, as it provides basic voltage gain. Assume that the transistor's emitter is at ground and that a bias current has been established in the transistor. If a small voltage signal is applied to the base of the transistor, the collector current will vary in accordance with the base voltage. If a load resistance R_L is provided in the collector circuit, that resistance will convert the varying collector current to a voltage. A voltage-in, voltage-out amplifier is the result, and it likely has quite a bit of voltage gain. A simple CE amplifier is shown in Figure 3.4(a).

The voltage gain will be approximately equal to the collector load resistance times the transconductance *gm*. Recall that the intrinsic emitter resistance $re' = 1/gm$. Thus, more conveniently, assuming the ideal transistor with intrinsic emitter resistance *re*′, the gain is simply R_L/re'.

Consider a transistor biased at 1 mA with a load resistance of 5000 Ω and a supply voltage of 10 V, as shown in Figure 3.4(a). The intrinsic emitter resistance *re*′ will be about 26 Ω. The gain will be approximately 5000/26 = 192. This is quite a large amount of gain. However, any loading by other circuits that are driven by the output has been ignored. Such loading will reduce the gain. It will also be reduced slightly by the Early effect output resistance r_o of the transistor.

The Early effect has also been ignored. It acts to place another resistance r_o in parallel with the 5-kΩ load resistance. This is illustrated with the dashed resistor drawn in the figure. As mentioned earlier, r_o for a 2N5551 operating at 1 mA and relatively low collector-emitter voltages will be on the order of 100 kΩ, so the error introduced by ignoring the Early effect here will be about 5%.

Because *re*′ is a function of collector current, the gain will vary with signal swing and the gain stage will suffer from some distortion. The gain will be smaller as the current swings low and the output voltage swings high. The gain will be larger as the current swings high and the output voltage swings low. This results in second harmonic distortion.

If the input signal swings positive so that the collector current increases to 1.5 mA and the collector voltage falls to 2.5 V, *re*′ will be about 17.3 Ω and the incremental gain will be 5000/17.3 = 289. If the input signal swings negative so that the collector current falls to 0.5 mA and the collector voltage rises to 7.5 V, then *re*′ will rise to about 52 Ω and incremental gain will fall to 5000/52 = 96. The incremental gain of this stage has thus changed by over a factor of 3 when the output signal has swung 5 V peak-to-peak. This represents a high level of distortion.

If external emitter resistance is added as shown in Figure 3.4(b), then the gain will simply be the ratio of R_L to the total emitter circuit resistance consisting of *re*′ and the external emitter resistance R_e. Since the external emitter resistance does not change with the signal, the overall gain is stabilized and is more linear. This is called *emitter degeneration*. It is a form of local negative feedback.

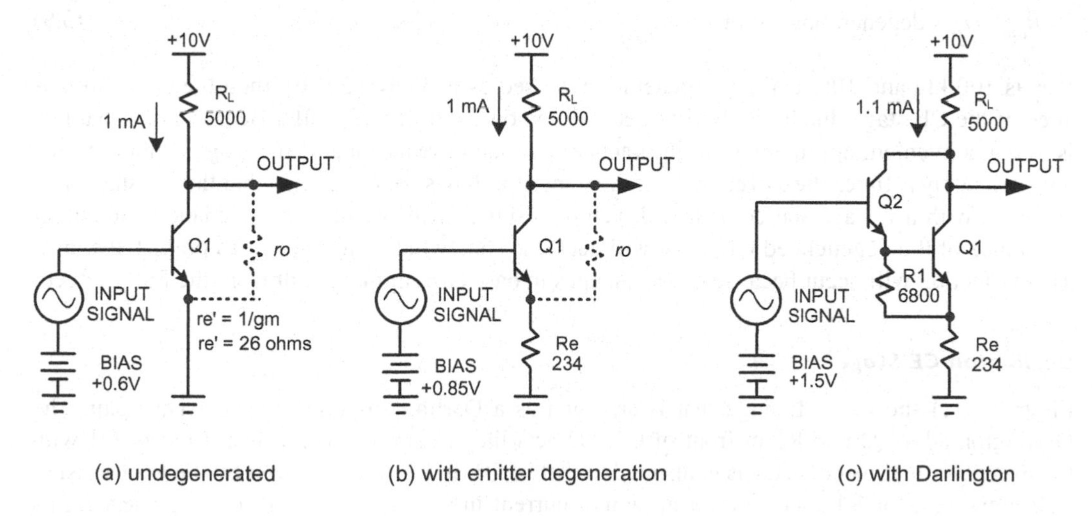

Figure 3.4 Common-Emitter Amplifiers.

The CE stage in Figure 3.4(b) is essentially the same as that in (a) but with a 234 Ω emitter resistor R_e added. This corresponds to 10:1 emitter degeneration because the total effective resistance in the emitter circuit has been increased by a factor of 10 from 26 to 260 Ω. The nominal gain has also been reduced by a factor of 10 to a value of approximately 5000/260 = 19.2.

Consider once again what happens to the gain when the input signal swings positive and negative to cause a 5 $V_{p\text{–}p}$ output swing. If the input signal swings positive so that the collector current increases to 1.5 mA and the collector voltage falls to 2.5 V, total emitter circuit resistance will become 17 + 234 = 251 Ω, and the incremental gain will rise to 5000/251 = 19.9.

If the input signal swings negative so that the collector current falls to 0.5 mA and the collector voltage rises to 7.5 V, then total small-signal effective resistance will rise to about 234 + 52 = 286 Ω and incremental gain will fall to 5000/286 = 17.4. The incremental gain of this stage has now swung over a factor of 1.14:1, or only 14%, when the output signal has swung 5 $V_{p\text{–}p}$. This is indeed a much lower level of distortion than occurred in the un-degenerated circuit of Figure 3.4(a). This illustrates the effect of local negative feedback without resorting to any negative feedback theory.

We thus have, for the CE stage, the approximation

$$Gain = R_L / \left(re' + R_e\right) \tag{3.8}$$

where R_L is the net collector load resistance and R_e is the external emitter resistance. The emitter degeneration factor is defined as $(re' + R_e)/re'$. In this case that factor is 10.

Emitter degeneration also mitigates nonlinearity caused by the Early effect in the CE stage. As shown by the dotted resistance r_o in Figure 3.4(b), most of the signal current flowing in r_o is returned to the collector by way of being injected into the emitter. If 100% of the signal current in r_o were returned to the collector, the presence of r_o would have no effect on the output resistance of the stage. In reality, some of the signal current in r_o is lost by flowing in the external emitter resistor R_e instead of through emitter resistance re' (some is also lost due to the finite current gain of the transistor). The fraction of current lost depends on the ratio of re' to R_e, which in turn is a reflection of the amount of the emitter degeneration. As a rough approximation, the output resistance due to the Early effect for a degenerated CE stage is

$$R_{out} \sim r_o \times \text{degeneration factor} \tag{3.9}$$

If r_o is 100 kΩ and 10:1 emitter degeneration is used as in Figure 3.4(b), then the output resistance of the CE stage due to the Early effect will be on the order of 1 MΩ. Bear in mind that this is just a convenient approximation. In practice, the output resistance of the stage cannot exceed approximately r_o times the current gain of the transistor. It has been assumed that the CE stage here is driven with a voltage source. If it is driven by a source with significant impedance, the output resistance of the degenerated CE stage will decrease somewhat from the values predicted above. That reduction will occur because of the changes in base current that result from the Early effect.

Darlington CE Stage

Figure 3.4(c) shows a CE stage that is arranged as a Darlington-connected transistor pair. The Darlington adds Q2 and R1 in front of Q1. Q2 acts like an emitter follower in front of Q1 with the exception that its collector is connected to that of Q1 and is thus powered by the load resistor. Pull-down resistor R1 establishes a minimum current in Q2, here about 100 µA, preserving its speed. The Darlington essentially forms a super-transistor whose β is on the order of the product of the βs of Q1 and Q2, often in the range of 10,000. If the impedance seen looking into the base of Q1 in Figure 3.4(b) is 23.4 kΩ with β of 100 for Q1, then the impedance seen looking into the base of Q2 in (c) may be over 2 MΩ. Note that the *on* voltage for the Darlington connection is 2 V_{be}, or about 1.3 V.

Bandwidth of the CE Stage and Miller Effect

The high-frequency response of a CE stage will be limited if it must drive any load capacitance. This is no different than when a source resistance drives a shunt capacitance, forming a first-order low-pass filter. A pole is formed at the frequency where the source resistance and reactance of the shunt capacitance are the same; this causes the frequency response to be down 3 dB at that frequency. The reactance of a capacitor is equal to $1/(2\pi fC) = 0.159/(fC)$. The −3 dB frequency f_3 will then be $0.159/(RC)$.

In Figure 3.4(a) the output impedance of the CE stage is approximately that of the 5-kΩ collector load resistance. Suppose the stage is driving a load capacitance of 100 pF. The bandwidth will be dictated by the low-pass filter created by the output impedance of the stage and the load capacitance. A pole will be formed at

$$f_3 = 1/(2\pi R_L C_L) = 0.159/\left(5\text{k}\Omega \times 100\text{pF}\right) = 318\text{kHz} \tag{3.10}$$

As an approximation, the collector-base capacitance should also be considered part of C_L. The bandwidth of a CE stage is often further limited by the collector-base capacitance of the transistor when the CE stage is fed from a source with significant impedance. The source must supply current to charge and discharge the collector-base capacitance through the large voltage excursion that exists between the collector and the base. This phenomenon is called the *Miller effect*.

Suppose the collector-base capacitance is 5 pF and assume that the CE stage is being fed from a 5-kΩ source impedance R_S. Recall that the voltage gain G of the circuit in Figure 3.4(a) was approximately 192. This means that the voltage across C_{cb} is 193 times as large as the input signal (bearing in mind that the input signal is out of phase with the output signal, adding to the difference). This means that the current flowing through C_{cb} is 193 times as large as the current that would flow through it if it were connected from the base to ground instead of base to collector. The input circuit thus sees an effective input capacitance C_{in} that is $1 + G$ times that of the collector-base capacitance. This phenomenon is referred to as *Miller multiplication* of the capacitance. In this case the effective value of C_{in} would be 965 pF.

The base-collector capacitance effectively forms a shunt feedback circuit that ultimately controls the gain of the stage at higher frequencies where the reactance of the capacitor becomes small. As frequency increases, a higher proportion of the input signal current must flow to the collector-base capacitance as opposed to the small fixed amount of signal current required to flow into the base of the transistor. If essentially all input signal current flowed through the collector-base capacitance, the gain of the stage would simply be the ratio of the capacitive reactance of C_{cb} to the source resistance

$$G = X_{Ccb}/R_S = 1/(2\pi fC_{cb})R_S = 0.159/\left(fC_{cb}R_S\right) \tag{3.11}$$

This represents a value of gain that declines at 6 dB per octave as frequency increases. This decline will begin at a frequency where the gain calculated here is equal to the low-frequency gain of the stage. Collector-base capacitance C_{cb} is nonlinear; it decreases as V_{ce} increases. This can cause distortion, since it causes the incremental gain of the stage to vary with signal, especially at high frequencies. The Miller effect is often used to advantage in providing the high-frequency roll-off needed to stabilize a negative feedback loop. This is referred to as *Miller compensation.* Miller effect compensation is usually implemented with a Miller capacitor C_M that is significantly larger than C_{cb}.

Differential Amplifier

The differential amplifier is illustrated in Figure 3.5. It is much like a pair of common-emitter amplifiers tied together at the emitters and biased with a common current. This current is called the *tail current.* The arrangement is often referred to as a *long-tailed pair (LTP).*

The differential amplifier routes its tail current to the two collectors of Q1 and Q2 in accordance with the voltage differential across the bases of Q1 and Q2. If the base voltages are equal, then equal currents will flow in the collectors of Q1 and Q2. If the base of Q1 is more positive than that of Q2, more of the tail current will flow in the collector of Q1 and less will flow in the collector of Q2. This will result in a larger voltage drop across the collector load resistor R_{L1} and a smaller voltage drop across load resistor R_{L2}. Output *A* is thus inverted with respect to Input *A*, while Output *B* is non-inverted with respect to Input *A*.

Visualize the intrinsic emitter resistance *re′* present in each emitter leg of Q1 and Q2. Recall that the value of *re′* is approximately 26 Ω divided by the transistor operating current in mA. With 1 mA flowing nominally through each of Q1 and Q2, each can be seen as having an emitter resistance *re′* of 26 Ω. Note that since $gm = 1/re'$ is dependent on the instantaneous transistor current, the values of *gm* and *re′* are somewhat signal-dependent, and indeed this represents a nonlinearity that gives rise to distortion.

Having visualized the ideal transistor with emitter resistance *re′*, one can now assume that the idealized internal emitter of each device moves exactly with the base of the transistor, but with a

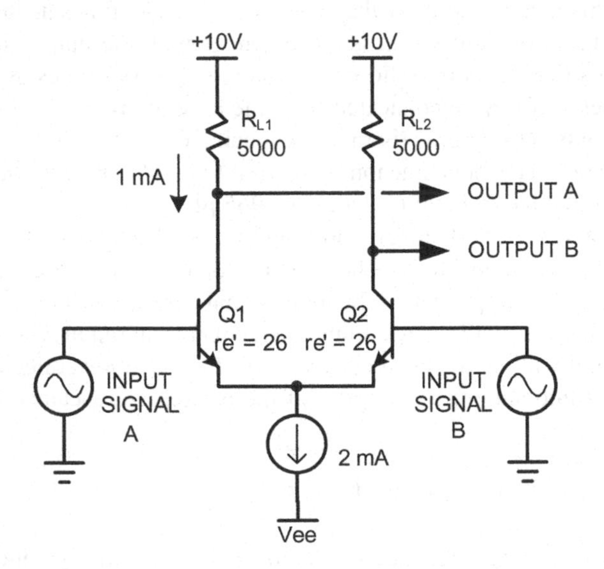

Figure 3.5 Differential Amplifier.

fixed DC voltage offset equal to V_{be}. Now look what happens if the base of Q1 is 5.2 mV more positive than the base of Q2. The total emitter resistance separating these two voltage points is about 52 Ω, so a current of 5.2 mV/52 Ω = 0.1 mA will flow from the emitter of Q1 to the emitter of Q2. This means that the collector current of Q1 will be 100 µA more than nominal, and the collector current of Q2 will be 100 µA less than nominal. The collector currents of Q1 and Q2 are thus 1.1 and 0.9 mA, respectively, since they must sum to the tail current of 2.0 mA (assuming very high β for the transistors).

This 100-µA increase in the collector current of Q1 will result in a change of 500 mV at Output *A*, due to the collector load resistance of 5000 Ω. A 5.2-mV input change at the base of Q1 has thus caused a 500-mV change at the collector of Q1, so the stage gain to Output *A* in this case is approximately 500/5.2 = 96.2. More significantly, the stage gain defined this way is just equal numerically to the load resistance of 5000 Ω divided by the total emitter resistance re' = 52 Ω across the emitters. Taking into account the +500-mV change at output B, the differential gain is 192.4.

Had additional external emitter degeneration resistors been included in series with each emitter, their value would have been added into this calculation. For example, if 48-Ω emitter degeneration resistors were employed, the gain would then become 5000/(26 + 48 + 48 + 26) = 5000/148 = 34. This approach to estimating stage gain is a very important back-of-the-envelope concept in amplifier design. In a typical amplifier design, one will often start with these approximations and then knowingly account for some of the deviations from the ideal. This will be evident in the numerous design analyses to follow.

It was pointed out earlier that the change in transconductance of the transistor as a function of signal current causes distortion. Consider the situation where a negative input signal at the base of Q1 causes Q1 to conduct 0.5 mA and Q2 to conduct 1.5 mA without emitter degeneration. The emitter resistance re' of Q1 is now 26/0.5 = 52 Ω. The emitter resistance re' of Q2 is now 26/1.5 = 17.3 Ω. The total emitter resistance from emitter to emitter has now risen from 52 Ω in the case above to 69.3 Ω. This results in a reduced small-signal gain to output A of 5000/69.3 = 72.15. This represents a reduction in gain by a factor of 0.75, or about 25%. This is an important source of distortion in the LTP. The presumed signal swing that caused the imbalance of collector currents between Q1 and Q2 resulted in a substantial decrease in the incremental gain of the stage. More often than not, distortion is indeed the result of a change in incremental gain as a function of instantaneous signal amplitude.

The gain of an LTP is typically highest in its balanced state and decreases as the signal goes positive or negative away from the balance point. This symmetrical behavior is in contrast to the asymmetrical behavior of the common-emitter stage, where the gain increases with signal swing in one direction and decreases with signal swing in the other direction. To first order, the symmetrical distortion here is third harmonic distortion, while that of the CE stage is predominantly second harmonic distortion.

Notice that the differential input voltage needed to cause the above imbalance in the LTP is only on the order of 25 mV. This means that it is fairly easy to overload an LTP that does not incorporate emitter degeneration. This is of great importance in the design of most amplifiers that employ an LTP input stage.

Suppose the LTP is pushed to 90% of its output capability. In this case Q1 would be conducting 0.1 mA and Q2 would be conducting 1.9 mA. The two values of re' will be 260 Ω and 14 Ω, for a total of 274 Ω. The gain of the stage is now reduced to 5000/274 = 18.25. The nominal gain of this un-degenerated LTP was about 96.2. The incremental gain under these large signal conditions is down by about 80%, implying gross distortion.

As in the case of the CE stage, adding emitter degeneration to the LTP will substantially reduce its distortion while also reducing its gain. In summary we have the approximation

$$\text{Gain} = R_{L1} / 2 \times (re' + R_e) \tag{3.12}$$

where R_{L1} is a single-ended collector load resistance and R_e is the value of external emitter degeneration resistance in each emitter of the differential pair. This gain is for the case where only a single-ended output is taken from the collector of Q1. If a differential output is taken from across the collectors of Q1 and Q2, the gain will be doubled. For convenience, the total emitter-to-emitter resistance in an LTP, including the intrinsic *re'* resistances, will be called R_{LTP}. In the example above,

$$R_{LTP} = 2(re' + R_e) \tag{3.13}$$

Single-ended gain to R_{L1} is then R_{L1}/R_{LTP} and differential gain to R_{L1} and R_{L2} is then $(R_{L1}+R_{L2})/R_{LTP}$.

Emitter Follower (EF)

The emitter follower (EF) is essentially a unity voltage gain amplifier that provides current gain. It is most often used as a buffer stage, permitting the high impedance output of a CE or LTP stage to drive a heavier load without loss of gain.

The emitter follower is illustrated in Figure 3.6(a). It is also called a common collector (CC) stage because the collector is connected to an AC ground. This AC ground is provided by the low (AC) impedance of the power supply to circuit ground. The output pull-down resistor R1 establishes a fairly constant operating collector current in Q1. For illustration, a load resistor R2 is being driven through an AC coupling capacitor C1. For AC signals, the net load resistance R_L at the emitter of Q1 is the parallel combination of R1 and R2. If *re'* of Q1 is small compared to R_L, virtually all of the signal voltage applied to the base of Q1 will appear at the emitter, and the voltage gain of the emitter follower will be nearly unity.

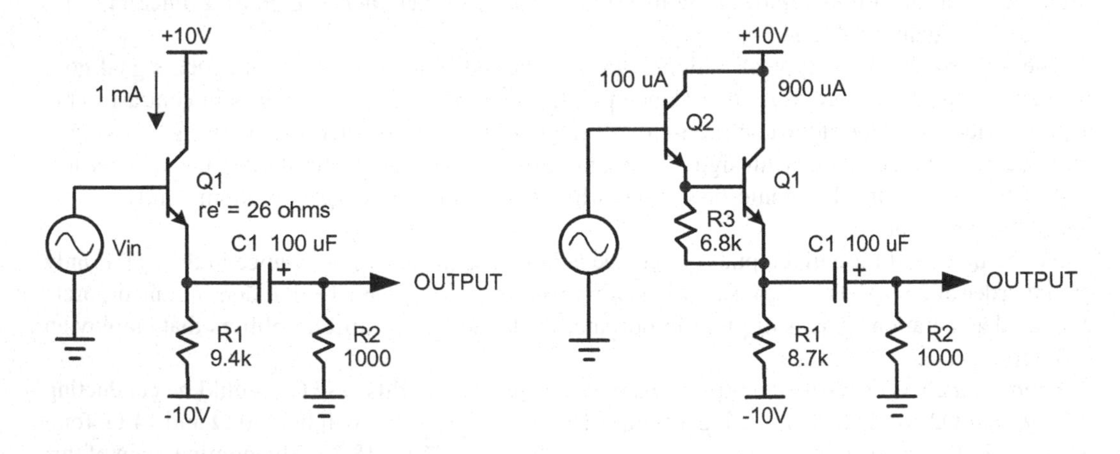

Figure 3.6 Emitter Followers.

The signal current in the emitter will be equal to V_{out}/R_L, while the signal current in the base of Q1 will be this amount divided by the β of the transistor. It is immediately apparent that the input impedance seen looking into the base of Q1 is equal to the impedance of the load multiplied by the current gain of Q1. This is the most important function of the emitter follower.

As mentioned above, the voltage gain of the emitter follower is nearly unity. Suppose R1 is 9.4 kΩ and the transistor bias current is 1 mA. The intrinsic emitter resistance re' will then be about 26 Ω. Suppose R2 is 1 kΩ, making net R_L equal to 904 Ω. The voltage gain of the emitter follower is then approximately

$$G = R_L / \left(R_L + re'\right) = 0.97 \tag{3.14}$$

At larger voltage swings the instantaneous collector current of Q1 will change with signal, causing a change in re'. This will result in a change in incremental gain that corresponds to distortion. Suppose the signal current in the emitter is 0.9 mA peak in each direction, resulting in an output voltage of about 814 mV peak. At the negative peak swing, emitter current is only 0.1 mA and re' has risen to 260 Ω. Incremental gain is down to about 0.78. At the positive peak swing the emitter current is 1.9 mA and re' has fallen to 13.7 Ω; this results in a voltage gain of 0.985.

Voltage gain has thus changed by about 21% over the voltage swing excursion; this causes considerable second harmonic distortion. One solution to this is to reduce R1 so that a greater amount of bias current flows, making re' a smaller part of the gain equation. This of course also reduces net R_L somewhat. A better solution is to replace R1 with a 1-mA constant current source.

The transformation of low-value load impedance to much higher input impedance by the emitter follower is a function of the current gain of the transistor. β is a function of frequency, as dictated by the f_T of the transistor. This means, for example, that a resistive load will be transformed to impedance at the input of the emitter follower that eventually begins to decrease with frequency as small-signal current gain β_{AC} decreases with frequency. A transistor with a nominal β of 100 and f_T of 100 MHz will have an f_β of 1 MHz. The AC β of the transistor will begin to drop at 1 MHz. The decreasing input impedance of the emitter follower thus looks capacitive in nature, and the phase of the input current will lead the phase of the voltage by an amount approaching 90° at high frequencies.

The impedance transformation works both ways. Suppose we have an emitter follower that is driven by a source impedance of 1 kΩ. The low-frequency output impedance of the *EF* will then be approximately 1 kΩ divided by β, or about 10 Ω. However, the output impedance will begin to rise above 1 MHz where β begins to fall. Impedance that increases with frequency is inductive. Thus, Z_{out} of an emitter follower tends to be inductive at high frequencies.

Now consider an emitter follower that is loaded by a capacitance. This can lead to instability, as we will see. The load impedance presented by the capacitance falls with increasing frequency. The amount by which this load impedance is multiplied by β_{AC} also falls with frequencies above 1 MHz. This means that the input impedance of the emitter follower is ultimately falling with the square of frequency. It also means that the current in the load, already leading the voltage by 90°, will be further transformed by another 90° by the falling transistor current gain with frequency. This means that the input current of the emitter follower will lead the voltage by an amount approaching 180°. When current is 180° out of phase with voltage, this corresponds to a *negative resistance*. This can lead to instability, since the input impedance of this emitter follower is a frequency-dependent negative resistance under these conditions. This explains why placing a resistance in series with the base of an emitter follower will sometimes stabilize it; the positive resistance adds to the negative resistance by an amount that is sufficient to make the net resistance positive.

Figure 3.6(b) shows an emitter follower employing a Darlington transistor pair. The input impedance of that emitter follower pair, with the load shown, could be as high as 10 MΩ or more. Notice that R3 is bootstrapped by the output signal. The simplicity of the emitter follower, combined with its great ability to buffer a load, accounts for it being the most common type of circuit used for the output stage of amplifiers. An emitter follower will often be used to drive a second emitter follower in a Darlington arrangement to achieve even larger amounts of current gain and buffering.

Source Follower

The JFET can also be used as common-drain source follower with gain near unity, as shown in Figure 3.7(a). Because the JFET is a depletion-mode device, the output at the source of an N-channel source follower will be positive with respect to the input by the amount of the JFET's V_{gs}. Its source can be pulled down by a resistor or a current source. Since the JFET's *gm* is significantly less than that of a BJT operating at the same current, gain of the source follower when driving a load will typically have gain less than unity by more than that of a BJT emitter follower. A source follower with a JFET *gm* of 1 mS will experience a gain loss of about 1 dB when driving a load of 10 kΩ.

A special case of the JFET source follower is shown in Figure 3.7(b), where the DC voltage offset from input to output can be largely canceled. In this case, an identical JFET J2 is used as a current source to pull down the source follower implemented by J1. A resistor in series with the source of J2 sets the operating current, with V_{gs} appearing across that resistor. If a resistor of the same value is placed in series with the source of J1, it will drop the output voltage by V_{gs} of J1, bringing the output voltage down to the same value as the DC input.

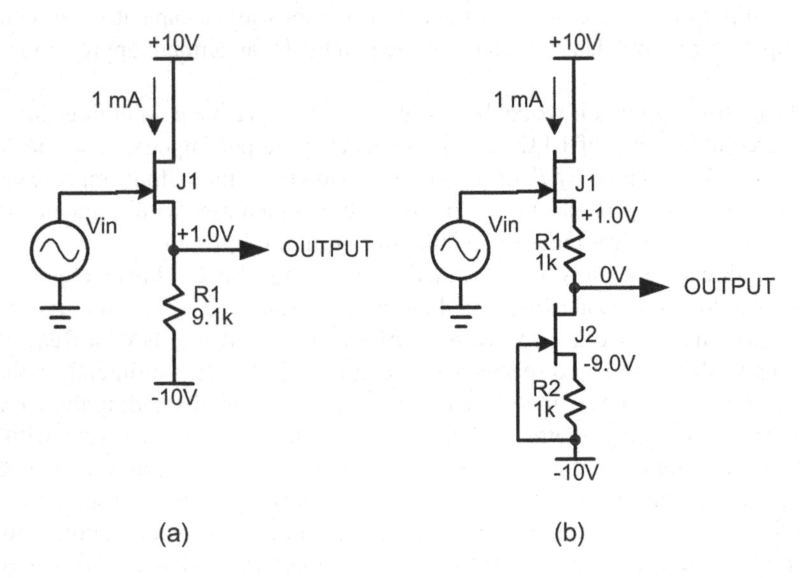

Figure 3.7 Source Followers.

Cascode

A cascode stage is implemented by Q2 in Figure 3.8(a). The cascode stage is also called a common base stage because the base of its transistor is connected to AC ground. Here the cascode is being driven at its emitter by a CE stage comprising Q1. The most important function of a cascode stage is to provide isolation. It provides near-unity current gain, but can provide very substantial voltage gain. In some ways it is like the dual of an emitter follower.

A key benefit of the cascode stage is that it largely keeps the collector of the driving CE stage at a constant potential. It thus isolates the collector of the CE stage from the large swing of the output signal. This eliminates most of the effect of the collector-base capacitance of Q1, resulting in wider bandwidth due to suppression of the Miller effect. Similarly, it mitigates distortion caused by the nonlinear collector-base junction capacitance of the CE stage, since very little voltage swing now appears across the collector-base junction to modulate its capacitance.

The cascode connection also avoids most of the Early effect in the CE stage by nearly eliminating signal swing at its collector. A small amount of Early effect remains, however, because the signal swing at the base of the CE stage modulates the collector-base voltage slightly.

If the current gain of the cascode transistor is 100, then 99% of the signal current entering the emitter will flow in the collector. The input–output current gain is thus 0.99. This current transfer factor from emitter to collector is sometimes referred to as the *alpha* (α) of the transistor.

The Early effect resistance r_o still exists in the cascode transistor. It is represented as a resistance r_o connected from collector to emitter. Suppose r_o is only 10 kΩ. Is the output impedance of the collector of the cascode 10 kΩ? No, it is not.

Recall that 99% of the signal current entering the emitter of the cascode re-appears at the collector. This means that 99% of the current flowing in r_o also returns to the collector. Only the lost 1% of the current in r_o results in a change in the net collector current at the collector terminal. This

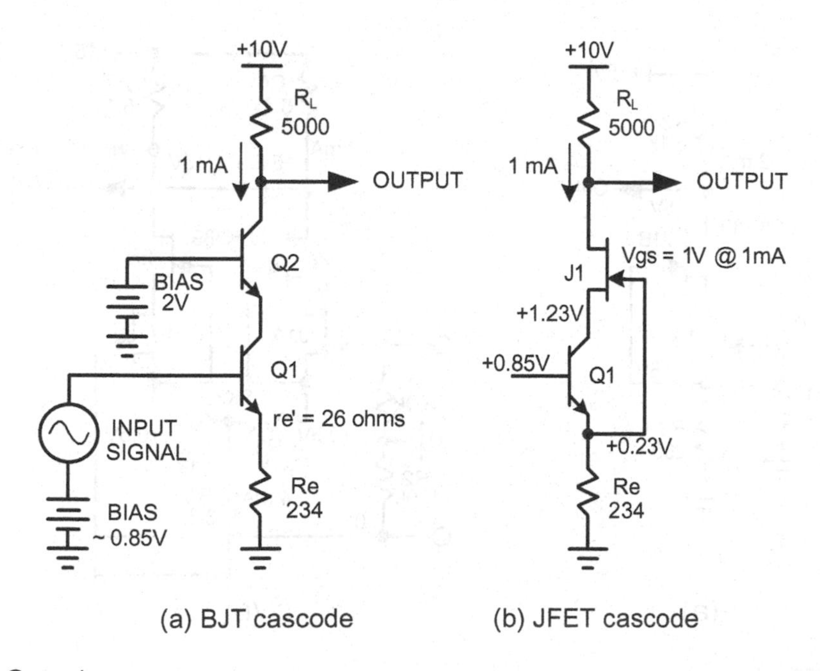

Figure 3.8 Cascodes.

means that the net effect of r_o on the collector output impedance in the cascode is roughly like that of a 1 MΩ resistor to ground. This is why the output impedance of cascode stages is so high even though Early effect is still present in the cascode transistor.

In the equations below, *VA* is the Early voltage of the transistor, and it characterizes the output resistance r_o of the transistor. The higher *VA*, the higher the output resistance. *VA* can be significantly different for different types of transistors, but a typical number is 100 V.

$$R_{out} = \beta \times r_o \tag{3.15}$$

$$r_o = \left(VA + V_{ce}\right) / I_c \tag{3.16}$$

$$r_o \approx VA / I_c \text{ at low } V_{ce} \tag{3.17}$$

$$R_{out} = \beta \times VA / I_c \tag{3.18}$$

The product of β and *VA* is sometimes called the Early effect figure of merit (FOM). The output resistance of a cascode is thus the FOM divided by the collector current.

$$R_{out} = \text{FOM} / I_c \tag{3.19}$$

JFET Cascodes

Figure 3.8(b) illustrates a self-biased cascode using a JFET as the cascode device. Because the N-channel JFET J1 will have its source positive with respect to its gate, V_{ce} for Q1 will equal V_{gs} for J1, here on the order of −1 V. Indeed, V_{ce} of Q1 is nearly constant if the signal current does

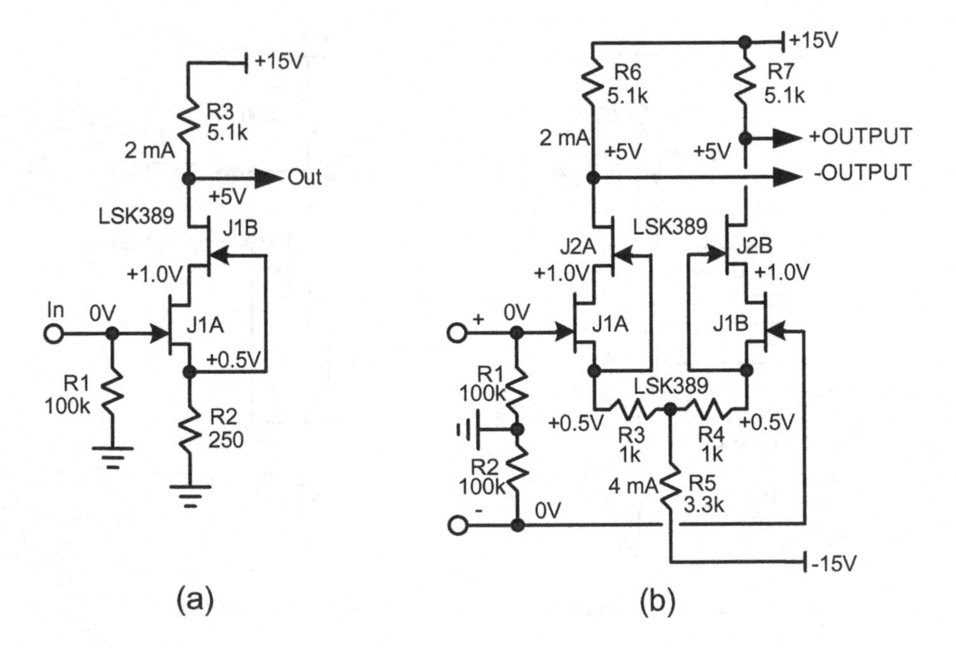

Figure 3.9 JFET Cascodes.

not change V_{gs} very much. With J1 operating in pinch-off, the output impedance will be extremely high. This is especially advantageous in a high-gain stage with a current-source load. Unfortunately, the JFET cascode requires careful JFET selection and suffers from the wide variability of threshold voltage among JFETs of the same part number. A JFET with high threshold voltage will eat into negative swing headroom, while a JFET with low threshold voltage will deprive Q1 of adequate collector voltage and cause it to approach quasi-saturation.

Figure 3.9(a) illustrates a self-biased cascode where both transistors are JFETs. Assume that the input voltage at the gate of J1 is 0 V_{dc}. Assume that the operating point of both identical JFETs is 1 mA with $V_{gs} = -1.0$ V. The source of J1 and the gate of J2 will be at +1.0 V. The source of J2 will then be at +2.0 V. V_{ds} for Q1 will thus be 2.0 V, which will usually be adequate to keep the device in the saturated region. If operation at greater V_{ds} is desired, J2 can be chosen as one with a greater V_{gs} than J1 at the operating current.

Figure 3.9(b) shows a differential pair that comprises two of the cascodes of Figure 3.9(a) and includes four JFETs. It provides balanced inputs and balanced outputs and provides common-mode rejection.

3.4 Current Mirrors

Figure 3.10(a) depicts a very useful circuit called a current mirror [3]. If a given amount of current is sourced into Q1, that same amount of current will be sunk by Q2, assuming that the emitter degeneration resistors R1 and R2 are equal, that the transistor V_{be} drops are the same, and that losses through base currents can be ignored. The values of R1 and R2 will often be selected to drop about 100 mV to ensure decent matching in the face of unmatched transistor V_{be} drops, but this is not critical. In some applications where matching of Q1 and Q2 is naturally very good, as in a matched pair or an integrated circuit (IC), R1 and R2 may be made zero.

If R1 and R2 are made different, a larger or smaller multiple of the input current can be made to flow in the collector of Q2. In practice, the base currents of Q1 and Q2 cause a small error in the output current with respect to the input current. In the example above, if transistor β is 100, the base current I_b of each transistor will be 50 μA, causing a total error of 100 μA, or 2% in the output current.

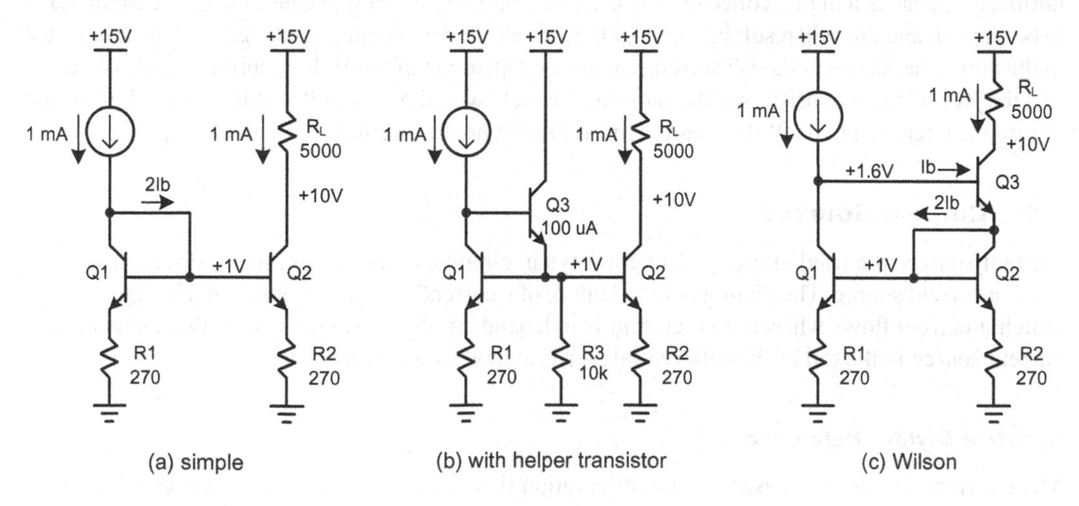

Figure 3.10 Current Mirrors.

Figure 3.10(b) shows a variation of the current mirror that minimizes errors due to the finite current gain of the transistors. Here emitter follower Q3, often called a *helper* transistor, provides current gain to minimize that error. Optional resistor R3 assures that a small minimum amount of current flows in Q3 even if the current gains of Q1 and Q2 are very high. Sometimes a small capacitor, on the order of 470 pF, is connected from base to emitter of helper transistor Q3 to reduce high-frequency peaking by reducing the effective f_T of Q3. Note that the input node of the current mirror now sits one V_{be} higher above the supply rail than in Figure 3.10(a). This also gives Q1 higher V_{ce}, increasing the likelihood of it matching Q2, since it is then operating further from quasi-saturation.

Many other variations of current mirrors exist, such as the *Wilson* current mirror shown in Figure 3.10(c) [4]. The Wilson current mirror includes transistors Q1, Q2 and Q3. Input current is applied to the base of Q3 and is largely balanced by current flowing in the collector of Q1. Input current that flows into the base of output transistor Q3 will turn Q3 on, with its emitter current flowing through Q2 and R2. Q1 and Q2 form a conventional current mirror. The emitter current of Q3 is mirrored and pulled from the source of input current by Q1.

Any difference between the current of Q1 and the input current is available to drive the base of Q3. If the input current exceeds the mirrored emitter current of Q3, the base voltage of Q3 will increase, causing the emitter current of Q3 to increase and self-correct the situation with feedback action. The equilibrium condition can be seen to be when the input current and the output current are the same, providing an overall 1:1 current mirror function.

Notice that in normal operation all three of the transistors operate at essentially the same current, namely the supplied input current. Ignoring the Early effect, all of the base currents will be the same if the betas are matched. Assume that each base current is I_b and that the collector current in Q1 is equal to I. It can be quickly seen that the input current must then be $I + I_b$ and that the emitter current of Q3 must be $I + 2I_b$. It is then evident that the collector current of Q3, which is the output current, will be $I + I_b$, which is the same as the input current. This illustrates the precision of the input–output relationship when the transistors are matched, as on an IC or a monolithic transistor array (e.g., THAT 300 Series). *I/O* precision will not be as good in discrete circuits where the V_{be} and h_{fe} among the transistors are not sufficiently matched.

Transistor Q3 acts much like a cascode, and this helps the Wilson current mirror achieve high output impedance. Transistors Q1 and Q2 operate at a low collector voltage, while output transistor Q3 will normally operate at a higher collector voltage. Thus, the Early effect will cause the base current of Q3 to be smaller, and this will result in a slightly higher voltage-dependent output current. This is reflected in the output resistance of the Wilson current mirror. Operation of the Wilson current mirror presumes that the β of all three transistors is the same, as it largely would be on an IC. This may not be the case in discrete circuits, and the Wilson current mirror might not be as effective in discrete applications.

3.5 Current Sources

Current sources are used in many different ways in circuits, and there are many different ways to make a current source. The distinguishing feature of a current source is that it is an element through which a current flows wherein that current is independent of the voltage across that element. The current source in the tail of the differential pair is a good example of its use.

Resistive Divider Reference

Most current sources are based on the observation that if a known voltage is impressed across a resistor, a known current will flow. A simple current source is shown in Figure 3.11(a). The voltage divider composed of R2 and R3 places 2.7 V at the base of Q1. After a V_{be} drop of 0.7 V, about 2

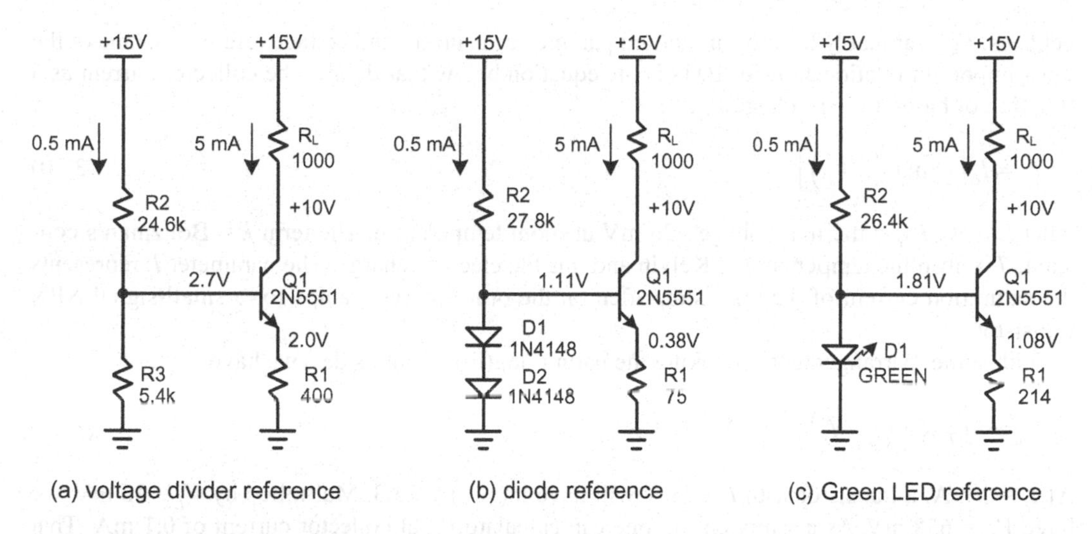

Figure 3.11 Current Sources.

V is impressed across emitter resistor R1. If R1 is a 400-Ω resistor, 5 mA will flow in R1 and very nearly 5 mA will flow in the collector of Q1. The collector current of Q1 will be largely independent of the voltage at the collector of Q1, so the circuit behaves as a decent current source. The load resistance R_L is just shown for purposes of illustration. The output current I_{Q1} will be strongly dependent on the supply voltage.

The output impedance of the current source itself (not including the shunting effect of R_L) will be determined largely by the Early effect in the same way as for the CE stage. Early effect in Q1 causes changes in Q1's base current with Q1's collector voltage, which in turn changes the voltage at the base because of the impedance of the R2–R3 voltage divider. Output impedance will be increased if smaller values for R2 and R3 are used. The output impedance for this current source is found by SPICE simulation to be about 290 kΩ.

Silicon Diode Reference

In Figure 3.11(b), R3 is replaced with a pair of silicon diodes. Here one diode drop is impressed across R1 to generate the desired current (if the diode drop equals the V_{be}, which it often does not). The circuit employs 1N4148 diodes biased with the same 0.5 mA used in the voltage divider in the first example. Together they drop only about 1.1 V, and about 0.38 V is impressed across the 75-Ω resistor R1. The output impedance of this current source is approximately 300 kΩ, about the same as the one above. The diodes can each be replaced by diode-connected versions of Q1. This is often done in IC implementations of this current source.

As temperature increases, the forward voltage of a silicon junction decreases by about 2 mV/°C. While the temperature-dependent voltage drop of one of the diodes is largely canceled by the falling V_{be} of the transistor, the falling junction voltage of the other diode is largely impressed across the external emitter resistor R1. This results in a negative temperature coefficient for the output current.

Transistor V_{be}

If the diodes are replaced with diode-connected transistors of the same type, the voltage drops become transistor V_{be} values. It is helpful to better understand how V_{be} is calculated and how it behaves. V_{be} is arguably the most important parameter in current and voltage references. One of the most important relationships for BJTs is the equation below that defines the collector current as a function of base-emitter voltage V_{be}:

$$I_c = I_o\left[\exp\left(V_{be} / V_T\right)\right] \tag{3.20}$$

where $V_T = kT/q$ = thermal voltage ~26 mV at room temperature. The term k = Boltzman's constant, T = absolute temperature in Kelvin and q is the electron charge. The parameter I_s represents the saturation current of the transistor, often on the order of 1e–^{14} amps for a small-signal NPN transistor.

With some rearrangement and taking the natural log(ln) of both sides, we have:

$$V_{be} = V_T \times \ln\left(I_c / I_s\right) \tag{3.21}$$

At I_c = 1 mA, a transistor with I_s = 1e^{-14} will have ln(1e^{11}) = 25.3. Multiplied by V_T = 26 mV, we have V_{be} = 658 mV. As a sanity check, one can calculate V_{be} at collector current of 0.1 mA. That turns out to be $V_T \times$ ln(e^{10}) = 599 mV, which is 59 mV less than the previous calculation for 1 mA, as expected.

Delta V_{be}

If the diode-connected transistors are identical to Q1 in Figure 3.11(b), notice that the current density in Q1 is ten times higher than the 0.5 mA flowing in the diode-connected transistors. This means the V_{be} of each of the latter transistors will be about 60 mV smaller than that of Q1, an amount that is referred to as delta V_{be} (ΔV_{be}), causing the voltage across R1 to be 120 mV smaller than would be expected if all of the transistors were matched and operating at the same current density. It is also true that ΔV_{be} for a pair of identical transistors operating at different current densities is a function of temperature, its magnitude increasing with temperature. The increase is proportional to absolute temperature (PTAT). For reference,

$$\Delta V_{be} = V_T \ln\left(I_{Q1} / I_{Q2}\right) \tag{3.22}$$

for identical transistors operating at different current, or

$$\Delta V_{be} = V_T \ln\left(A_{Q2} / A_{Q1}\right) \tag{3.23}$$

for transistors operating at the same current with different emitter areas A. V_T is about 26 mV at room temperature, and is PTAT. At room temperature of 25°C, the temperature in Kelvin is 297 K. This means that a 1°C temperature increase will increase V_T by 298/297 = 1.0034, corresponding to 3400 ppm/°C. It also means that an increase of 1°C will increase V_T by 0.088 mV.

If the emitter area of Q2 is ten times that of Q1 in Eq. 3.5, ln(10) = 2.303. If kT/q equals 26 mV, then ΔV_{be} of Eq. 3.5 becomes 59.9 mV. Not surprisingly, this is about 60 mV. In this case, a 1°C increase in temperature will increase ΔV_{be} by 0.204 mV. The ΔV_{be} of 59.9 mV resulting from a 10:1 current density ratio has a TC of +0.204 mV/°C.

Green LED Reference

Turning to Figure 3.11(c), R3 is replaced instead with a Green LED, providing a convenient voltage reference of about 1.8 V, putting about 1.1 V across R1 [5, 6]. Once again, 0.5 mA is used to bias the LED. The output impedance of this current source is about 750 kΩ. It is higher than in the design of 3.11(b) because there is effectively more emitter degeneration for Q1 with the larger value of R1. The temperature coefficient of the forward voltage drop of the LED is about −2.0 mV/°C. This almost completely negates the negative TC of the V_{be} of Q1, making the current source quite temperature-stable [4]. The LED can act as a photo-sensor, so in some circuits and situations hum can be introduced if the LED is exposed to room lighting.

Zener Diode Reference

R3 is replaced with a 6.2-V Zener diode in Figure 3.12(a). This puts about 5.5 V across R1 after the V_{be} drop. The output impedance of this current source is about 2 MΩ, quite a bit higher than the earlier arrangements due to the larger emitter degeneration for Q1. The price paid here is that the base of the transistor is fully 6.2 V above ground, reducing headroom in some applications. The positive temperature coefficient of the 6.2-V Zener diode and the negative *TC* of Q1's V_{be} make the current source have an overall positive *TC*. Zener diodes are noisy, so the Zener should be bypassed with a capacitor or filtered with an *R-C* network before application to the base of Q1. The latter approach is much more effective because the impedance of the Zener is quite low, and would therefore require a larger bypass capacitor.

It should be noted that Zener diodes that operate above about 5.6 V are not true Zeners, but rather are operating as avalanche diodes. This explains their positive TC and their noisy attributes. The positive TC will tend to cancel the TC of a BJT that is in series with it. The predominant breakdown mechanism below about 5.6 V is the Zener breakdown, and it has a negative TC [7]. The Zener-based current source is conveniently cascoded by adding two transistors, as shown in Figure 3.12(b). This results in extremely high output impedance.

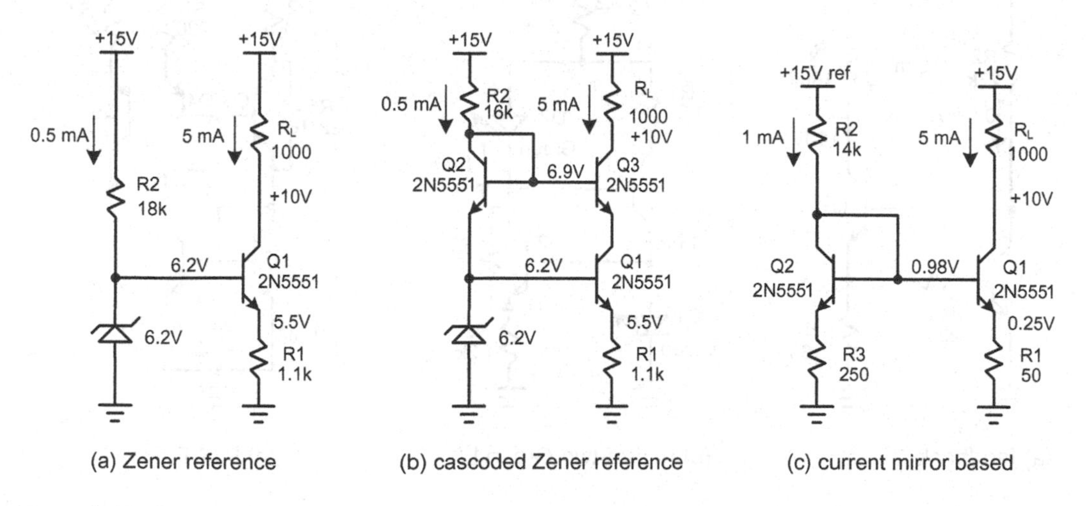

Figure 3.12 Current Sources.

Current Mirror Current Source

In Figure 3.12(c), a current mirror fed from a known supply voltage is used to implement a current source. Here a 1:5 current mirror is used and 1 mA is supplied from the known +15-V power rail. The output impedance of this current source is about 230 kΩ. Only 0.25 V is dropped across R1 (corresponding to about 10:1 emitter degeneration), and the base is only 1 V above ground. Output impedance will be increased if larger values are used for R1 and R3, but this will come at the expense of larger voltage drop in the mirror emitter circuits and less voltage headroom at the output of the mirror. The current created is almost proportional to the power supply voltage.

Feedback Current Source

Figure 3.13(a) illustrates a clever two-transistor feedback circuit that is used to force 1 V_{be} of voltage drop across R1. It does so by using transistor Q2 to effectively regulate the current of Q1. If the current of Q1 is too large, Q2 will be turned on harder and pull down on the base of Q1, adjusting its current downward appropriately. We call this a feedback current source.

A current of 0.5 mA is supplied to bias the current source. This current flows through Q2. The output impedance of this current source is an impressive 3 MΩ, similar to that which a cascode would yield. This circuit achieves higher output impedance than the Zener-based version and yet only requires the base of Q1 to be 1.4 V above ground. The "reference" for this current source is the V_{be} of Q2, which has a temperature coefficient of about −2 mV/°C, so the current source will also have such a negative temperature coefficient. For this reason, circuits with this property are sometimes referred to as a complementary to absolute temperature (CTAT) current source. The V_{be} of Q2 will have a mild dependence on the supply voltage, since the V_{be} of Q2 will change as the log of the current through R2. Ignoring the base current of Q1, if the current through Q2 decreases by a factor of 10, the V_{be} of Q2 will decrease by about 60 mV, resulting in a slightly reduced output current in Q1.

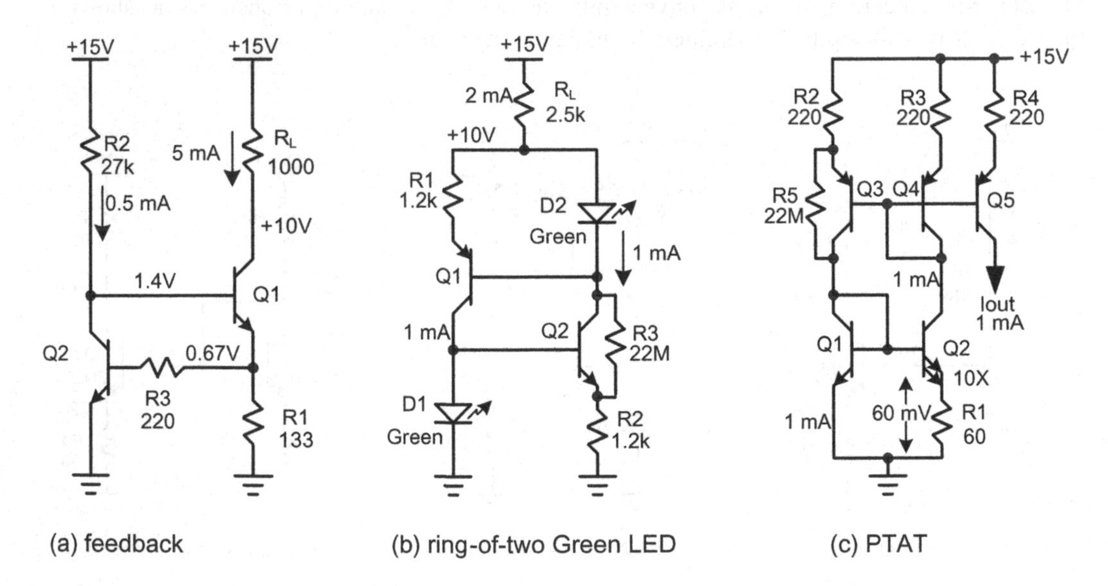

Figure 3.13 Current Sources.

This circuit will work satisfactorily even if less than 0.5 mA (one-tenth the output current) is supplied as bias for Q2, but then the output impedance will fall to a lower value and the "quality" of the current source will suffer somewhat. This happens because at lower collector current, Q2 has less transconductance and its feedback control of the current variations in Q1 as a result of the Early effect is less strong. If the bias current is reduced to 0.1 mA, for example, the output impedance falls to about 1 MΩ.

Because it is a feedback circuit, albeit with a tight loop of only two transistors, stability must be considered. Sometimes a small resistor, R3, is included in series with the base of Q2 to increase the phase margin of the loop.

Ring-of-Two Current Source

Figure 3.13(b) illustrates what is called a ring-of-two current source [8]. It uses two Green LEDs in a positive feedback arrangement while implementing a 2-terminal current source that can float. The current that biases D1 is itself regulated and made quite temperature-independent by D2 and Q1. Notice that the circuit requires a startup resistor, R3. Interestingly, the current flowing in R3 flows back into the emitter of Q2, greatly mitigating the effect of R3 on current source impedance.

PTAT Current Source

A PTAT current source is shown in Figure 3.13(c). It relies on the ΔV_{be} relationship described in Eq. 3.5, where $\Delta V_{be} = V_T \ln(A_{Q2}/A_{Q1})$ if two transistors are operated at the same current but have different effective emitter areas and thus different current densities. The larger-area transistor Q2 is often made with a multiple-emitter transistor. Q1 is connected as a diode and is operated at I_{Q1}. Its larger V_{be} is impressed on the base of Q2. A resistor R1 is connected in series with the emitter of Q2 to ground. The voltage across R1 must obviously equal ΔV_{be} between Q1 and Q2, causing Q2 to conduct I_{Q2}.

If I_{Q2} is fed to a 1:1 current mirror comprising Q3 and Q4, whose output I_{Q3} feeds diode-connected Q1, both transistors will be forced to operate at the same collector current but will have different V_{be}, with the resulting ΔV_{be} impressed across R1. The resulting current will thus be equal to ΔV_{be}/R1. If the emitter area of Q2 is ten times that of Q1, about 60 mV will be impressed across R1 (this corresponds to the fact that V_{be} increases by about 60 mV/decade). If R1 is 60 Ω, the resulting current will be 1 mA PTAT. As a simple example, if a third output Q5 is attached to the 1:1:1 current mirror, the output current will be the PTAT current. However, it is notable that the total supply current of the simple version is also PTAT. Note that the difference in current density between Q1 and Q2 can also be obtained if they are identical but the current mirror has a ratio larger than unity, such as 10:1.

As in the ring-of-two current source in Figure 3.13(b), the ring formed in the PTAT current source requires a startup current. It can be provided by connecting a large-value resistor R5 across Q3.

Combining both the CTAT of V_{be} with the PTAT of ΔV_{be} in the right proportions is central to the operation of a bandgap voltage reference, where the TCs of the former and latter balance each other out.

JFET Current Sources

A JFET current source is shown in Figure 3.14(a). At a given current within its operating range, an N-channel JFET will have its source positive with respect to its gate. If $V_{gs} = -1$ V at $I_d = 1$ mA, 1 V will appear across R1, here 1 kΩ. Since J1 will be operating in its pinch-off region, output impedance will be very high. Unfortunately, the wide variation in threshold voltage and I_{DSS} among JFETs of the same type results in wide variation of the current. Importantly, notice that the JFET current source can float, even though it is not shown to be floating in the figure.

A modification of the JFET current source is shown in Figure 3.14(b). Here an arrangement similar to the feedback current source is used. JFET current source J1 operates as before, with current-setting resistor R2 and D1 in series with its source. BJT Q1 monitors the drop across R2 and thus the current. The gate of J1 is connected to the source through R1. Any voltage dropped across R1 by collector current in Q1 will pull the gate more negative and reduce the current in J1. If the current drops one V_{be} across R2, Q1 turns on and pulls J1's gate more negative, reducing the current by means of feedback. I_d is thus relatively independent of V_{gs}. The current produced will be controlled by V_{be} of Q1, which has a TC of −2 mV/°C, thus the TC of the output current will be about −3300 ppm/°C.

If V_{gs} for J1 at 1 mA is −1.0 V, Q1 will turn on enough to pull the gate more negative by 1 V than the voltage at the source. If the V_{be} of Q1 is 0.6 V at the low 10 μA current needed to accomplish this (implying that R2 is 600 Ω), and the drop across D1 is 1.4 V at 1 mA, the source of J1 will be at +2.0 V relative to Q1's emitter. At 1 mA, the gate of J1 will be at +1.0 V relative to Q1's emitter. Q1 will pull 10 μA through 100 kΩ resistor R1 to drop the needed 1.0 V on the gate to put V_{gs} at −1.0 V. With these numbers, V_{ce} for Q1 is 1.0 V. A JFET requiring V_{gs} of −1.9 V would leave only V_{ce} of 0.1 V for Q1. This is about as far as the correction can go for a current source of 1 mA with a JFET requiring almost −2 V for drain current of 1 mA. However, notice that in the other direction, Q1 can be almost completely off, resulting in the maximum current of I_{DSS} for J1. This would be the case if I_{DSS} of J1 were only 1 mA. This circuit can thus work for JFETs that require V_{gs} of zero to −1.9 V to operate at 1.0 mA.

Notice that here the reference voltage for setting the current is V_{be} of Q1 at a low current of 10 μA, and that the output current will have a negative temperature coefficient. The circuit can be easily modified to use a larger, better-controlled reference voltage. If desired, the circuit can be arranged for Q1 to operate at a higher nominal current, like 100 μA, by reducing the value of R1. The

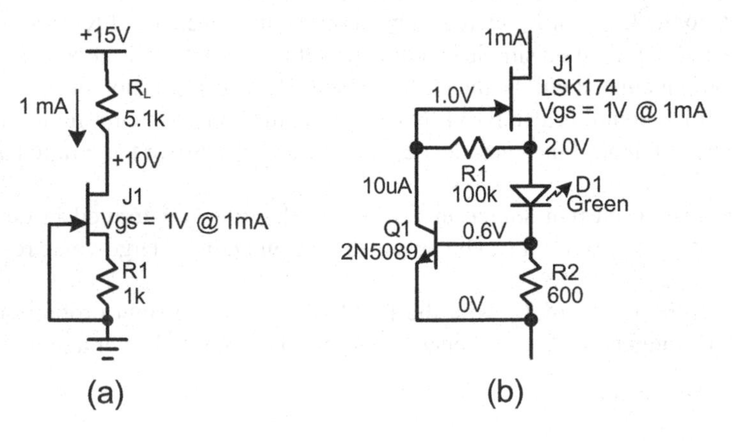

Figure 3.14 JFET Current Sources.

main advantage of this circuit is that it is a 2-terminal current source that can float while providing a predictable current, mainly because of the depletion-mode operation of the JFET. Moreover, because the current is created by the drain current of a JFET, output impedance is high, even up to high frequencies. A small capacitor across R1 can further improve high-frequencyperformance by attenuating the feedback of C_{dg} of the internal JFET to the high impedance node at the gate.

Current-Regulating Diodes

Obviously, a JFET can be connected with its gate to its source to form a 2-terminal current source whose value is I_{DSS} of the JFET. Such "current-regulating diodes" are available in different current values. The J500 series of such diodes from Linear Systems is a good example [9]. The J500 is specified to have a current between 192 and 288 μA with a typical of 240 μA. At the high end of the range, the J511 is specified to have a current between 3.8 and 5.6 mA, with a typical of 4.7 mA. The tolerances are on the order of ±20%, but many useful applications do not require closer tolerances. The J505, rated at 1.0 mA, has typical impedance of 2 MΩ at 25 V. This falls to 400 kΩ with only 6 V across the diode. Capacitance of the current-regulating diode is about 2.2 pF at 25 V across the diode. The advantage of these diodes is their extreme simplicity while providing a known current source that can float.

IC Current Sources

Numerous 3-terminal and 2-terminal IC current sources are available as well, but most seem to have some shortcomings. The LM134 is a 3-terminal programmable reference that applies V_{set} = 64 mV across a current-setting resistor to establish the current value [10]. Voltage V_{set} is PTAT, so temperature stability is not great. It can operate with as little as 1 V across it. Because of the way it operates, the LM134 is fairly noisy, producing 100 pA/√Hz at a current setting of 1 mA. Allowed voltage slew rate across the LM134 for linear operation is only 1 V/μs at 1 mA, and decreases proportionately at lower current settings.

The LT3092 is a temperature-stable programmable current source that can operate from 0.5 mA to 200 mA [11]. As shown in Figure 3.15, the 3-terminal device requires two resistors to set its current and can float as a 2-terminal current source. It can operate from 1.2 to 40 V. A reference current of 10 μA is passed through R_{set}, here 22 kΩ. The resulting voltage drop is impressed across an output resistor R_{out}, here 220 Ω, in series with the load. The current source current is then 10 μA times R_{set}/R_{out}.

Output impedance of the current source decreases with frequency. At 10 mA, output impedance is about 2 MΩ, falling to 300 kΩ at 1 kHz and falling further to 1 kΩ at 100 kHz. The impedance

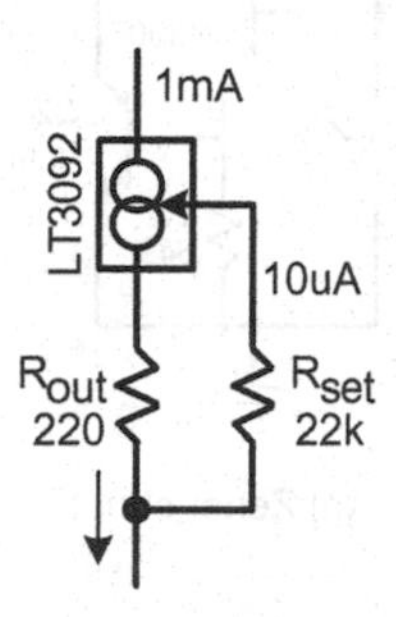

Figure 3.15 LT3092 Programmable Current Source.

in the audio band is about three times higher at a current of 1 mA as opposed to 10 mA. Dropout voltage is typically less than 1.5 V. Higher output impedance can be achieved by cascoding the LT3092 with a JFET whose V_{gs} is greater than 1.5 V at the chosen operating current.

The reference current input noise density is typically 2.7 pA/√Hz. At a setting of 1 mA, current source noise will be 100 times this, or about 270 pA/√Hz. Thermal noise of R_{set}, here 20 nV/√Hz, will also contribute to the noise. Output noise can be reduced by placing a capacitor across R_{set}, but this then requires a compensation network across the current source, which will degrade output impedance at higher frequencies. Testing for stability *in situ* in the application is required for the LT3092.

3.6 Voltage References

Voltage references of various degrees of precision and complexity are important building blocks for many circuit applications. For example, an analog-to-digital converter (ADC) requires an accurate and stable reference voltage. Other applications require less accuracy and temperature stability.

Green LED Shunt Voltage Reference

Figure 3.16(a) shows a 2.5-V active shunt voltage reference based on a Green LED [6]. The Green LED operates at about 0.7 mA because R1 has 1 V_{be} across it, and the LED drops about 1.85 V. Q1 also operates at about 0.7 mA and its V_{be} is about 0.65 V. They are in series, so the shunt drop is about 2.5 V. R3 is an optional trimming resistor. Both the LED and V_{be} have similar negative temperature coefficients, so the TC of the reference is about −4 mV/°C [12, 13]. The action of Q1 and Q2 greatly reduces the impedance of the shunt reference. A small increase in applied voltage will cause the collector current of Q1 to increase and that of Q2 to increase greatly, implying that the output impedance is very low.

Put another way, a small increase in applied current will flow mainly through Q1 to its collector, where it will flow into the base of Q2 and be amplified. Q2 will increase its current as much as necessary to keep the voltage drop across the shunt reference from increasing by all but the smallest amount.

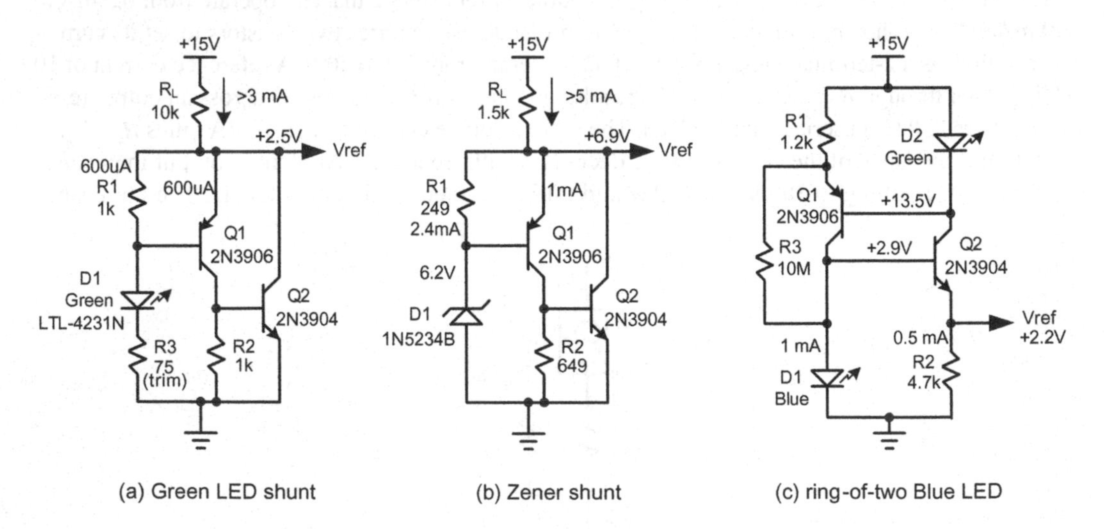

Figure 3.16 Voltage References.

Zener Shunt Voltage Reference

Figure 3.16(b) shows a Zener-based 6.9-V shunt regulator that works on the same principles as (a) in order to provide very low impedance. Here the 6.2-V Zener diode is in series with the 0.7-V V_{be} of Q1 to establish the 6.9-V drop. D1 operates in avalanche mode, and it has a positive temperature coefficient of about +2.0 mV/°C, which largely cancels the V_{be} TC of Q1, resulting in a temperature-stable voltage reference.

Ring-of-Two Voltage Reference

A so-called *ring-of-two* voltage reference is shown in Figure 3.16(c) [6, 10]. Its operation is similar to that of the ring-of-two current source depicted in Figure 3.13(b). Here a Blue LED (D1) with voltage drop of about 2.9 V at 1 mA is used. When the V_{be} of Q2 is subtracted, the resulting reference voltage across R2 is about 2.2 V. The Blue LED has a negative TC of about −4 mV/°C, so the reference TC is about −1.8 mV/°C. If an additional V_{be} is placed in series with the emitter of Q2, the reference voltage across R2 will drop to about 1.5 V and the TC will go positive to about +0.4 mV/°C. Large-value resistor R3 provides startup current for the ring. Note that D2 is Green and with Q1 just provides a feedback current reference.

IC Voltage References

Several IC voltage references are available. The LM329 is a fixed 6.9-V precision shunt reference that acts like a precision Zener diode with very low dynamic resistance of only 1 Ω up to 8 kHz [14]. It can operate from 0.6 to 15 mA. The LM329 has a moderate noise level of 75 nV/√Hz.

The TL431 is a programmable shunt reference for voltage values ranging from 2.5 to 36 V [15]. As shown in Figure 3.17, a simple resistive voltage divider that connects to its V_{ref} terminal sets its voltage. The device operates by shunting current as necessary to keep its V_{ref} terminal at its reference voltage of 2.5 V. For a 2.5-V reference, the V_{ref} terminal is simply connected to its positive terminal. The device can operate from 1 to 100 mA. Its dynamic impedance is typically less than 0.4 Ω up to 50 kHz. The temperature coefficient of its reference voltage is about −0.25 mV/°C. The TL431 is fairly noisy, with input-referred noise at its V_{ref} terminal as high as 125 nV/√Hz.

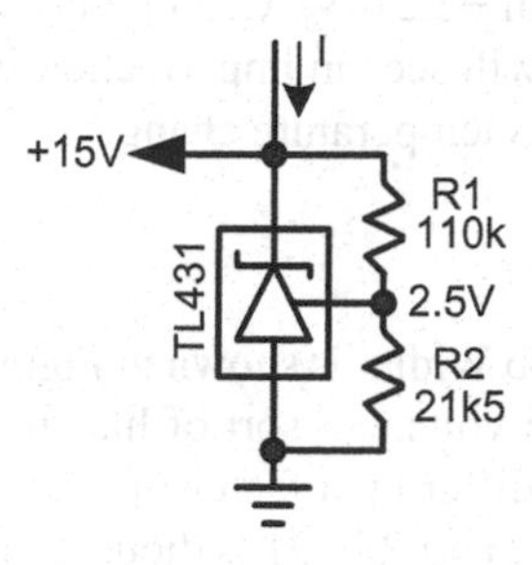

Figure 3.17 The TL431 Programmable Zener.

PTAT Voltage References

Sometimes an application benefits from a voltage reference whose voltage is PTAT. This can be used in applications for temperature compensation or for direct readout of Kelvin temperature. The LM335 is such a device. It acts as a 2-terminal shunt regulator (i.e., an active Zener diode). Its output changes by 10 mV/°C, and its voltage is the Kelvin temperature times 10 mV. At 25°C, its nominal voltage is 2.98 V when operated at 1 mA. Its dynamic impedance at 1 mA is only about 0.5 Ω. The LM335 includes a third terminal that can be used for optional trimming. The pin is connected to the wiper of a 10k trim pot that straddles the other two pins of the device. At 25°C, its "Zener" voltage can be trimmed to 2.982 V.

The LM35 precision centigrade temperature sensor produces an output voltage instead of acting like a Zener shunt. Its power and ground pins are fed a supply voltage between 4 and 20 V. Its output pin provides a voltage of +10 mV/°C over a range of +2 to +150°C. An obvious application is its use on a heat sink to monitor temperature as part of a protection circuit in a power amplifier [2].

3.7 Bandgap Voltage References

Many applications require a fixed, known reference voltage that is independent of temperature. In other words, its temperature coefficient is essentially zero. The common V_{be} voltage reference has a strong CTAT TC of about −2.2 mV/°C, or −3300 ppm/°C. If a V_{be} reference voltage is added to a PTAT reference voltage that changes by +2 mV/°C, then a temperature-independent reference will result. The PTAT voltage created by a ΔV_{be} can be used to accomplish this.

Recall that the ΔV_{be} of 59.9 mV resulting from a 10:1 current density ratio has a TC of +0.204 mV/°C. To achieve a TC of +2 mV/°C, the ΔV_{be} must be multiplied by a factor of about 10, resulting in 599 mV. When a 600-mV V_{be} and the multiplied ΔV_{be} are added to achieve a zero TC, the resulting voltage is 1.205 V, which is approximately the bandgap voltage of silicon.

ΔV_{be} can be obtained by operating the two transistors at different currents or by using transistors of differing emitter area. In the equation below, bear in mind that the transistor with larger emitter area will operate at lower current density if both transistors are operating at the same collector current. A_{Q2} is generally set to be some factor larger than A_{Q1}, making current density and V_{be} of Q2 smaller. That factor is often chosen to be 8:1 or 10:1.

$$\Delta V_{be} = V_T\left[\ln\left(A_{Q2} / A_{Q1}\right)\right] \tag{3.24}$$

A key design task in bandgap references is to amplify the ΔV_{be} by the right amount and add it to a V_{be}. The amplified ΔV_{be} voltage will usually end up being on the order of V_{be} suggesting a gain of about 10. In essence, if V_{be} has a TC of −2.2 mV/°C, ΔV_{be} must be multiplied by a sufficient amount to make its TC +2.2 mV/°C. As we will see, an imperfection in the technique is that the TC of V_{be} deviates slightly from −2.2 mV/°C as temperature changes.

Widlar Bandgap Reference

The original circuit developed by Bob Widlar is shown in Figure 3.18(a) [16]. It operates as a shunt element whose value is the bandgap voltage – sort of like an active Zener. Widlar used identical transistors operated at currents that differ by a factor of 10:1. A constant current is applied to Q1 and Q2 through collector resistors R2 and R3. Q1 is diode-connected to ground and feeds the base of identical transistor Q2 whose emitter goes to ground through R1. Q2 is operated at ten times less current than Q1, so a ΔV_{be} of 59.9 mV is created across its emitter resistor R1. Its 6000-Ω collector

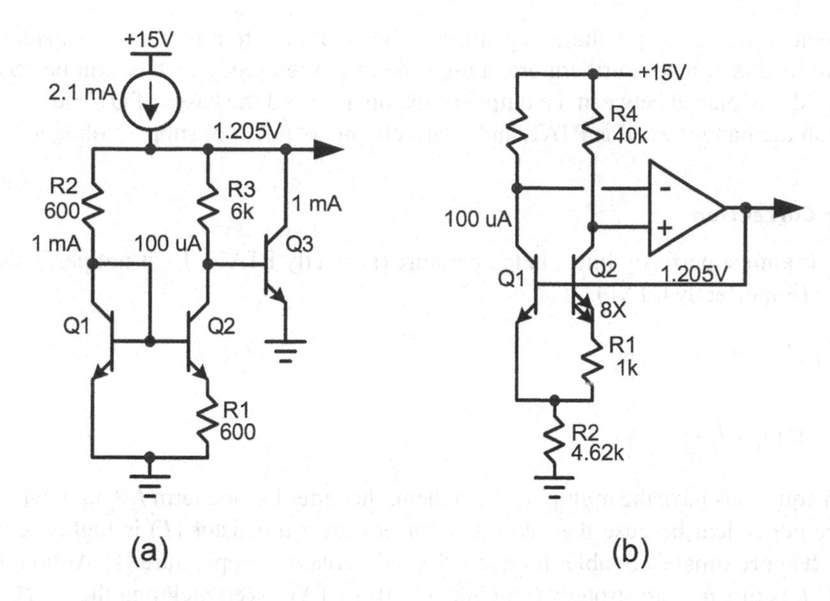

Figure 3.18 Bandgap Voltage References.

resistor R3 is ten times as large as Q1's 600-Ω collector resistor, corresponding to the 10:1 current ratio between Q1 and Q2. This ratio also amplifies ΔV_{be} as seen at the collector of Q2. Transistor Q3 pulls enough current from the input current source to force the collector of Q2 to be at 1 V_{be} above ground, just like the collector of Q1. With 1 V_{be} at the base of Q3 and the proper ΔV_{be} across R3, the collector of Q3 must reside at the sum of V_{be} and ΔV_{be}, which is the desired bandgap reference voltage of approximately 1.205 V.

A drawback of this circuit is that there is some dependence on the input current, since that will affect the V_{be} of Q3. Notice that ΔV_{be} will appear across R2, since the sum of V_{be} and ΔV_{be} is at one end and V_{be} of Q1 is at the other end of R2. If V_{be} of Q1 is 600 mV and R2 is 600 Ω, 1 mA will flow in Q1 and 0.1 mA will flow in Q2, making the total current draw of Q1 and Q2 1.1 mA. Any input current above 1.1 mA will be taken by Q3. For best results, Q3 should operate at the same collector current as Q1, so that their V_{be} will be the same. This suggests that the optimal input current is 2.1 mA. This simplified analysis does not take into account transistor base currents.

Brokaw Bandgap Reference

An improved bandgap reference was subsequently developed by Paul Brokaw [17]. His circuit in Figure 3.18(b) operates the two transistors at the same current but with different emitter areas by a factor of 8:1, with an op amp forcing the collector currents to be the same. A resistor ratio (R2/R1) multiplies the ΔV_{be} voltage across R1 by the proper amount. The value of ln(8) is 2.08, so here ΔV_{be} is 54.1 mV if V_T is 26 mV. It is impressed across R1 in the bandgap cell (Q1 and Q2), so that a PTAT current flows in R1. Since both transistors are operating at the same collector current, the same PTAT current also flows in Q1. As a result, twice the PTAT current flows in R2. The resistance value of R2 multiplies the PTAT voltage by across R1 by the factor 2R2/R1 to achieve the approximate 600 mV PTAT voltage at the emitter of Q2. Q2 also provides the necessary addition of 1 V_{be} to arrive at the output voltage of the op amp that satisfies the relationship that $I_{Q1} = I_{Q2}$.

Conveniently, the bandgap voltage appearing at the op amp output is at a low impedance due to the negative feedback loop. Furthermore, a multiple of the reference voltage can be obtained if a resistive divider is placed between the output of the op amp and the bases of Q1 and Q2. Note that the current in the bandgap cell is PTAT, and relatively independent of supply voltage.

Curvature Correction

While ΔV_{be} is almost perfectly linear in temperature (perfectly PTAT], V_{be} is not perfectly linear in temperature (imperfectly CTAT).

$$\Delta V_{be} = V_T\left[\ln\left(I_{Q1} / I_{Q2}\right)\right] \quad (3.25)$$

$$V_{be} = V_T\left[\ln\left(I_c / I_s\right)\right] \quad (3.26)$$

While both equations have the multiplier V_T in them, the latter has the term I_c/I_s in it, which is itself temperature-dependent because the saturation current for a transistor (I_s) is highly temperature-dependent. It approximately doubles for every 5°C of increased temperature [1]. Although the term V_T is PTAT, I_s is much more strongly (and imperfectly) PTAT, overwhelming the V_T TC contribution, resulting in V_{be} having a net negative TC of −2.2 mV/°C that changes a bit with temperature. This means that the bandgap reference voltage can have a perfectly zero temperature coefficient at a chosen temperature, often room temperature.

At other temperatures it will deviate slightly, causing some bowed curvature in the resulting error. The curve will be such that the voltage from the reference will go down slightly as the temperature deviates on either side of the temperature at which the TC of the reference is zero. For example, if the reference is set to zero TC at 25°C, at −25°C and at +75°C the reference voltage will be down about 2 mV from its value at 25°C. This represents a 0.17% error over 50°C, or an average of 0.0034%/°C or 34 ppm/°C on either side of 25°C [18].

Several curvature correction techniques have been used, but one approach involves making the current density ratio of Q1 and Q2 a bit temperature-dependent by making their current ratio temperature-dependent, introducing another nonlinearity. For example, if the current ratio between Q1 and Q2 is set by a degenerated 1:1 current mirror, while the nominal current density ratio is set by different emitter areas in the bandgap cell, a deliberate temperature-dependent error can be introduced into the current mirror.

Implementation

Although it is possible to implement a bandgap voltage reference discretely and get reasonable performance if a matched pair of transistors is used for the bandgap cell, the bandgap reference is almost always implemented in an integrated circuit because of the matching available and the additional complexity economically available to make the bandgap performance better, such as with curvature correction.

MOSFET Bandgap Voltage References

Bandgap voltage references are also implemented in MOSFET circuits, as in the CMOS process. That is because kT/q plays a role in their I_d vs V_{gs} behavior. MOSFETs conduct a small amount of current in the region where V_{gs} is less than the threshold voltage, called the subthreshold region. In fact, in that region they behave much like BJTs, with their current being an exponential function of V_{gs} and their V_{gs} at a given drain current being a log function of I_d.

The slope of log drain current as a function of gate voltage is constant over several decades of drain current and its slope is approximately q/kT or the inverse of V_T. The presence of V_T in the I_d vs V_{gs} behavior in the subthreshold region makes it possible to implement a bandgap voltage reference.

3.8 Complementary Feedback Pair (CFP)

Another important two-transistor connection that achieves current gain on the order of the product of the β of two transistors is shown in Figure 3.19(a). This circuit is called a Complementary Feedback Pair (CFP), also known as the Sziklai pair after George Sziklai of RCA. It acts like a single transistor and can usually be used as such. Unlike the Darlington, its turn-on voltage is only 1 V_{be}. Like the Darlington, its saturation voltage is on the order of 0.8 V. It is easy to see that collector current in Q1 is multiplied by as much as β of Q2. The CFP is a feedback circuit; Q2 will conduct as much current as necessary to keep Q1's current equal to approximately V_{be} of Q2 divided by R1 (here about 100 μA) plus the collector current of Q2 divided by β of Q2.

If you think of the CFP as an enhanced transistor in an un-degenerated CE configuration, recognize that the gain of the first stage is *gm* of Q1 times load resistor R1 if current gain of Q2 is high. Assuming that Q2 is not degenerated, R1 will always have 1 V_{be} across it. If Q1 is run at higher current, its *gm* will increase in proportion, but the value of R1 will also have to decrease in proportion in order to keep 1 V_{be} across R1. This means that first-stage gain is fairly predictable and largely independent of the current chosen for Q1. At 1 mA for Q1, *re′* of Q1 is 26 Ω and R1 must be about 680 Ω, meaning that first-stage gain is about 26.

The feedback process greatly increases the effective transconductance of Q1 by a factor on the order of β of Q2. The CFP implements an emitter follower in Figure 3.19(b). The very high *gm* of the CFP greatly reduces the distortion of the emitter follower that can result from *gm* variations as a function of signal. Because the CFP is a feedback circuit, oscillation is possible under some

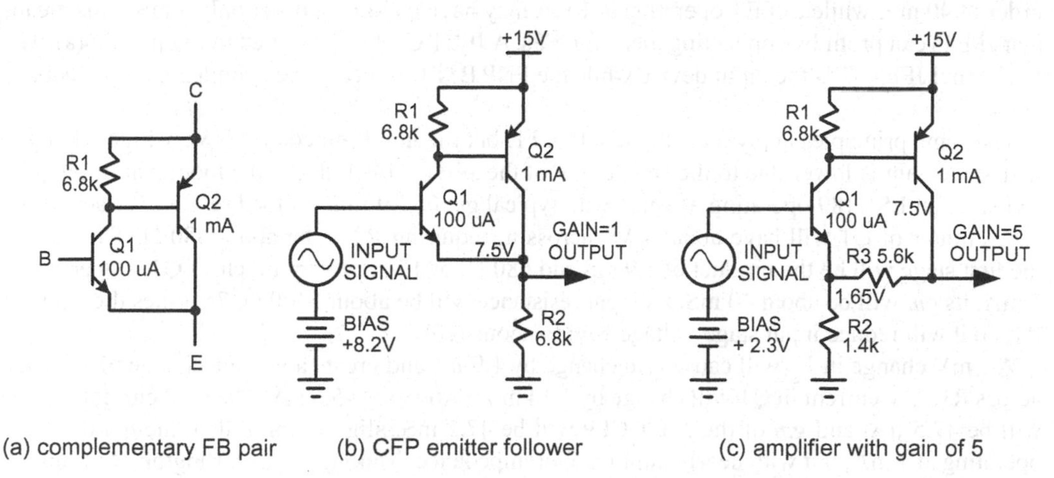

Figure 3.19 Complementary Feedback Pair.

conditions. As with a simple Darlington, Q1 can operate at very small collector current, on the order of 10 μA if Q2 is operating at 1 mA with β of 100. This can cause Q1 to be slow and can make turn-off of Q2 slow. Resistor R1 serves to set a minimum current of 100 μA for Q1, just as in the Darlington.

The CFP can be connected to provide voltage gain, as shown in Figure 3.19(c), where the circuit is set for a gain of 5. The combination of R3 and R2 implements a simple 5:1 voltage divider in the feedback path to the emitter of Q1. Because Q1 forces the input signal to appear at its emitter, if the output is not five times the input, an error current will flow in Q1 that will turn on Q2 harder and correct the output to the proper 5× replica of the input signal. Because R3 creates an error *current*, this arrangement implements a very simplified version of a so-called *current feedback amplifier* (CFA).

Limitations of the CFP and Improvements

The CFP is a feedback circuit, where in Chapter 9 the gain around the feedback loop is referred to as loop gain. Loop gain in (a) and (b) is limited by the small value of R1 because R1 has only 1 V_{be} across it. Loop gain can be doubled by making Q2 a Darlington, allowing 2 V_{be} across R1 and thus a larger value for a given amount of current in Q1. This also reduces the error introduced by base current in Q2. Placing a diode or resistor divider in the Darlington pull-down resistor can further increase the voltage available to R1. Going a step further, R1 can be replaced with a current source, such as a feedback current source. A divider in the Darlington pull-down can also increase the headroom available to the feedback current source. Being a circuit involving negative feedback at high frequencies, circuits like this should always be simulated for stability. In some cases, revisions for stability may have to be included, such as the strategic placement of a capacitor to provide some feedback compensation.

JFET Complementary Feedback Pairs

JFETs have considerably less transconductance than BJTs when operating at the same current. It can be as much as ten times smaller. A BJT operating at 1 mA may have transconductance on the order of 40 mS, while a JFET operating at 1 mA may have *gm* on the order only 4 mS. This means that JFETs can profit by connecting them as CFPs. A JFET CFP is illustrated in Figure 3.20(a). The N-channel JFET J1 is the input device while the PNP BJT Q1 acts as the complementary feedback device.

The same principles apply as with the BJT CFP, but the input impedance is very high while the first-stage gain is lower due to the smaller *gm* of the JFET. This reduces the loop gain of the pair. Consider an LSK189 operating at 1 mA with typical *gm* of 1.9 mS. Its load resistor R1 across the base-emitter of Q1 will have about 1 V_{be} across it, requiring R1 to be about 680 Ω. The gain of the first stage will be the product of 1.9 mS and 680 Ω, or 1.3, not very much. If Q1 is operated at 1 mA, its *gm* will be about 40 mS. Its input resistance will be about 5200 Ω, 7.6 times the value of R1, so it will reduce input stage voltage gain to about 1.14.

A 1 mV change in V_{gs} will cause I_d to change by 1.9 μA and create a voltage change of 1.14 mV across R1. The current in Q1 will change by 1.14 mV × 40 mS = 45.6 μA. The total current change will be 47.5 μA, and *gm* of the JFET CFP will be 47.5 mS, slightly more than the *gm* of a BJT operating at 1 mA, but with nearly infinite input impedance. Making Q1 a Darlington can improve loop gain.

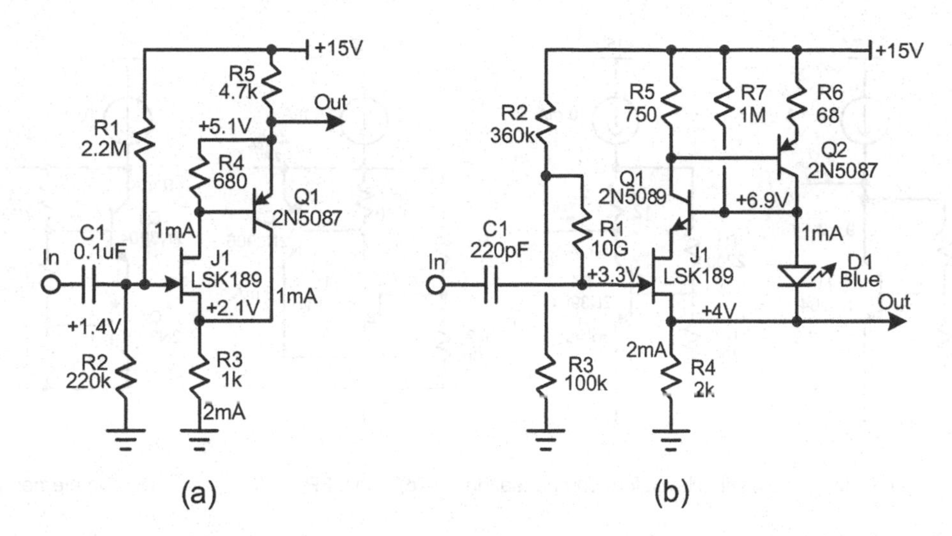

Figure 3.20 JFET Complementary Feedback Pairs.

Cascoded JFET CFP

A JFET CFP can be conveniently cascoded to greatly increase its output impedance. This circuit is shown in Figure 3.20(b). It is configured as a unity-gain source follower with extremely high input impedance, here shown with a small input coupling capacitor and a bias network for the gate of J1. Bias is applied to the gate through a 10-GΩ resistor. The base of the cascode transistor Q1 is biased by an LED or Zener diode D1 connected to the source of J1. Diode D1 passes the current of the complementary feedback transistor Q2, here 1 mA. Because the base of Q1 moves with the signal, it bootstraps the drain of J1. All three terminals of J1 move with the signal, virtually eliminating the effects of C_{gs}, C_{gd} and output resistance of J1. This provides extremely high input impedance and good PSRR for the source follower. If desired, one could also bootstrap R1 with signal, making the input impedance even higher than 10 GΩ. The cascoded JFET CFP can also be used to implement a common-source amplifier.

Bootstrapped circuits usually involve some form of positive feedback that is normally safe from a stability standpoint. However, when bootstrapping can extend to high frequencies, as with the bootstrapping loop involving J1, Q1 and D1, stability should be checked with simulation.

3.9 V_{be} Multipliers

Figure 3.21(a) shows what is called a V_{be} multiplier. This circuit is used when a voltage drop equal to some multiple of V_{be} drops is needed. This circuit is most often used as the bias spreader for complementary push-pull output stages, partly because its voltage is conveniently adjustable.

In the circuit shown, the V_{be} of Q1 is multiplied by a factor of approximately 4 because the voltage divider formed by R1 and R2 places about one-fourth of the collector voltage at the base of Q1. When the voltage at the collector is at four V_{be}, one V_{be} will be at the base, just enough to turn on Q1 by the amount necessary to carry the current supplied. Here about 1 mA flows through the resistive divider while about 9 mA flows through Q1. When the V_{be} multiplier is used as a bias spreader for a power amplifier, R2 will be made adjustable with a trim pot. As R2 is made smaller, the amount

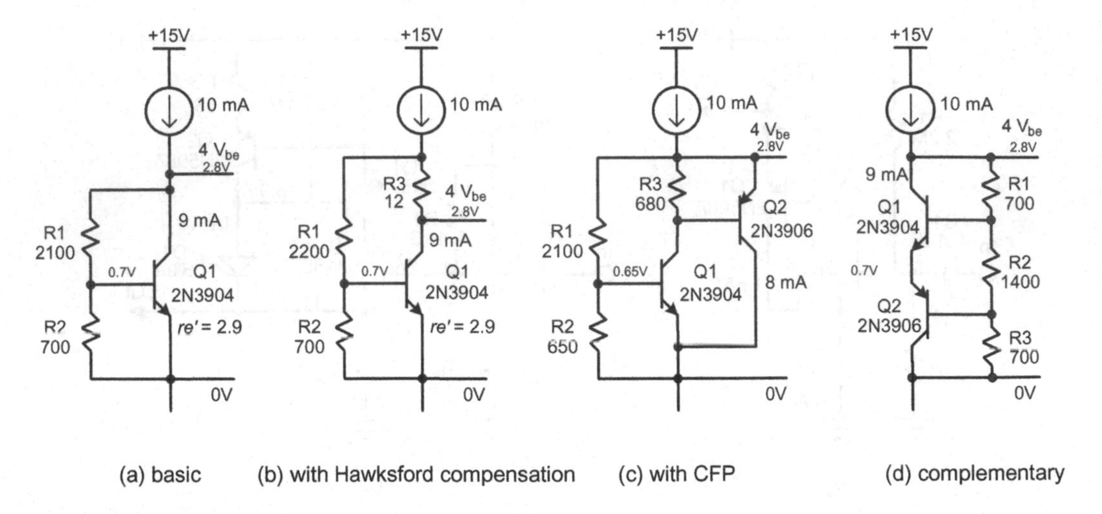

Figure 3.21 V_{be} **Multipliers.**

of bias voltage is increased. Notice that if for some reason R2 fails open, the voltage across the V_{be} multiplier falls to about one V_{be}, failing in the safe direction. In practice the V_{be} multiplication ratio will be a bit greater than 4 due to the base current required by Q1.

The impedance of the V_{be} multiplier depends on the transconductance of Q1. It is about 4 *re′* for Q1. At 9 mA, *re′* is 2.9 Ω, so ideally the impedance of the multiplier would be about 11.6 Ω. In practice, SPICE simulation shows it to be about 25 Ω. This larger value is mainly a result of the finite current gain of Q1.

The impedance of the V_{be} multiplier rises at high frequencies. This is a result of the fact that the impedance depends on a negative feedback process. The amount of feedback decreases at high frequencies and the impedance-reducing effect is lessened. The impedance of the V_{be} multiplier in Figure 3.21(a) is up by 3 dB at about 2.3 MHz and doubles for every octave increase in frequency thereafter. The V_{be} multiplier is usually shunted by a capacitor of 0.1–10 μF. A shunt capacitance of as little as 0.1 μF eliminates the increase in impedance at high frequencies.

The modified V_{be} multiplier shown in (b) includes R3 to reduce the effective impedance of the circuit [19]. As the current applied to the V_{be} multiplier increases, the total drop across the V_{be} multiplier increases, but so does the drop across R3. The voltage drop across R3 acts to reduce the "output" voltage of the circuit. If the value of R3 is the same as the DC impedance of the V_{be} multiplier, the change in output voltage with applied current will be nearly canceled. A CFP V_{be} multiplier is shown in Figure 3.21(c). The very high effective transconductance of the CFP compound transistor greatly reduces the impedance of the V_{be} multiplier.

A complementary V_{be} multiplier is shown in Figure 3.21(d). Here an NPN and a PNP transistor are connected in series with a single three-resistor voltage-dividing network. In some power amplifier designs, one of the transistors will be mounted on the heatsink to provide temperature compensation of a fraction of the total voltage drop, while the other transistor responds only to the ambient temperature.

In all of these V_{be} multipliers, the V_{be} of a transistor is the reference. However, the effective reference voltage can be increased by placing a silicon diode, LED or Zener diode in series with the emitter of Q1. In some cases this may influence the temperature coefficient of the multiplier voltage in a beneficial way.

3.10 Diamond Buffer

The diamond buffer is a unity-gain push-pull and symmetric buffer. It can run in class *A* or class *AB*. It comprises two complementary BJT emitter followers in tandem for each polarity of swing. It is thus a unity-gain buffer with about the same current gain as a Darlington.

Its traditional competition is a full complementary Darlington. But the Darlington needs 4 V_{be} of bias spread. The complementary emitter follower signal path in the diamond buffer essentially means no bias spreader is needed. The minimal amount of bias spread needed to establish idle current in the output stage can be had with the resistors in the emitters of the first-stage transistors. Alternatively, those resistors can be replaced with diodes, so that the voltage drop across the output emitter resistors is about 1 diode drop. Even better, those "diodes" can be made from diode-connected transistors of the same type as used elsewhere in the diamond buffer, resulting in a drop of one V_{be} across each emitter resistor.

The diamond buffer usually provides more headroom in an amplifier circuit, since the first transistor increases the driving voltage by one V_{be}, while in a Darlington the first transistor drops a V_{be}. The current sources for the first transistors can swing nearly to the rail without contributing distortion, since they are looking into the low impedance of an emitter follower. In some cases the current sources can be replaced by resistors. For best performance, a feedback current source is recommended. All of this makes life a lot easier for the Voltage Amplifier Stage (VAS) of an amplifier, giving it more headroom and needing to go not as close to the rail for a given output swing.

Figure 3.22 shows two diamond buffer implementations. In (a), Q1 acts as a PNP emitter follower for the top half of the circuit. Its emitter is pulled up by current source I1. Resistor R1 allows the first-stage output voltage to be a bit greater than +1 V_{be} with respect to the input. Q1's collector is connected to the negative rail. The first-stage output voltage drives the base of output transistor

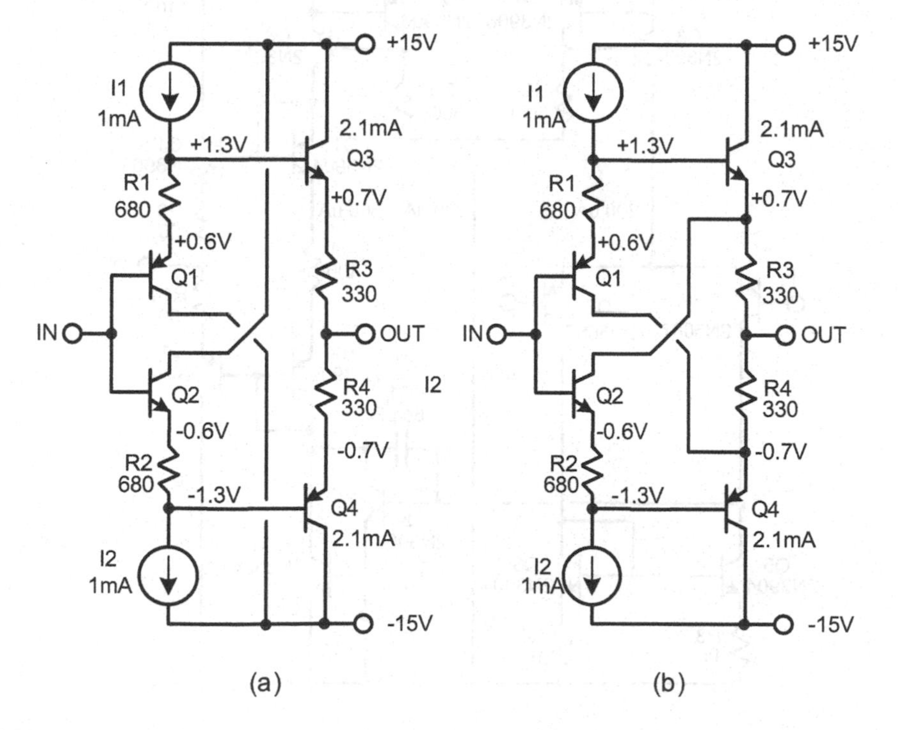

Figure 3.22 Diamond Buffers.

Q3. In the circuit shown, with 1 mA flowing in Q1 and R1 at 680 Ω, the base of Q3 will be at about +1.3 V. If the output node is at 0 V and V_{be} of Q3 is also 600 mV, about 700 mV will appear across R2. With R2 at 330 Ω, Q3's operating current will be about 2.1 mA. This is a simplified explanation, as Q3's V_{be} will likely be different that that of Q1.

Q2, I2, R2 and Q4 perform the same function for the bottom half of the diamond buffer. The bases of Q1 and Q2 are usually connected together at the input node. For analysis, it is assumed that the input node is at 0 V. To the extent that the base bias currents of Q1 and Q2 are similar, some input bias current cancellation will occur.

In (b), the collectors of Q1 and Q2 are connected to the emitters of Q4 and Q3, respectively. This reduces V_{ce} of Q1 and Q2 and reduces dissipation in those transistors. They operate at V_{ce} of about 1.3 V. This connection also bootstraps the collectors of Q1 and Q2 with signal, increasing input impedance and reducing the effects of C_{cb} in Q1 and Q2. Each input transistor will have V_{cb} equal to only 700 mV, or approximately the voltage drop across R1 and R2. Since the collectors of Q1 and Q2 are being fed with the output voltage, a positive feedback path through C_{cb} is formed, so caution should be exercised with regard to stability in this circuit.

3.11 Operational Amplifiers

The very simple discrete operational amplifier of Figure 3.23 can be implemented by a simple combination of some of the circuit building blocks that have been discussed thus far. Q1 and Q2

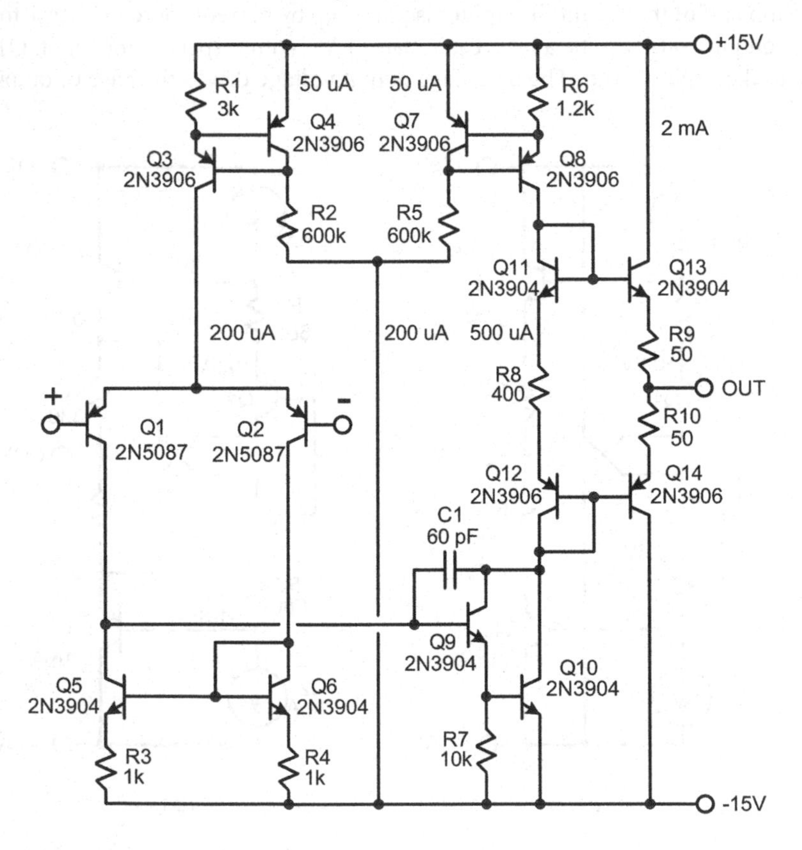

Figure 3.23 Simple Operational Amplifier.

form a differential input stage (IPS) loaded by a current mirror built with Q5 and Q6. The 200-μA tail current for the IPS is provided by Q3, which is part of a feedback current source formed with Q4. The VAS is formed by a Darlington common-emitter transistor loaded by a 500-μA feedback current source consisting of Q7 and Q8.

The output stage (OPS) is a complementary emitter follower biased at 2 mA by the voltage drops of diode-connected transistors Q11 and Q12 and R8. Compensation capacitor C1 (60 pF) sets the gain-bandwidth to about 2.5 MHz and allows slew rate of about 1.8 V/μs to be achieved. Open-loop gain is simulated to be over 100 dB at DC.

Latchup

Long ago some simple op amps like the one above suffered from a phenomenon called latchup. This usually could occur in designs where the closed-loop gain was unity at DC, allowing the full amount of the maximum output voltage to be sent back to the input stage, pushing it out of its common-mode input operating range. This could cause the input stage to malfunction and reverse its gain polarity, resulting in positive feedback. Although virtually no modern op amps suffer from this vulnerability, pointing this out sheds some light on some of the unforeseen effects in circuits that need to be considered.

3.12 Circuit Simulation

The SPICE circuit simulator has been around for over 40 years. Developed at U.C. Berkeley, the acronym stands for *Simulation Program with Integrated Circuit Emphasis* [20]. Because ICs are almost impossible to breadboard and probe, there was a great need to simulate designs before committing to the expensive process of laying out and fabricating an IC. Although that was its original mission, it is equally useful for discrete circuits at the circuit board level. The use of SPICE simulation can save hours in reaching the point where you can build a working circuit. Intuition is not always right when it comes to circuit design, and SPICE helps here.

LTspice

Although there are many SPICE packages available, one that stands out for audio amplifier design is LTspice, originally developed at Linear Technology Corporation (Analog Devices) [21, 22]. The program is free, well supported, widely accepted, and easy to use. It is also one of the best-performing SPICE simulators. It is easily downloaded from the Anaolog Devices site at analog.com. It runs on a PC. The use of SPICE simulation can save hours in reaching the point where you can build a working amplifier.

All of the simulations done for this book were carried out with LTspice. Like most other software tools, LTspice has many options and capabilities (and a very large user manual). The 90-10 rule applies: 90% of what you need to do can be accomplished with 10% of the features. LTspice also comes with a good set of device libraries and a very helpful educational directory where many example designs illustrate how LTspice is used. Two chapters are devoted to LTspice device modeling and simulation in "Designing Audio Power Amplifiers" [2]. A tutorial, a useful device library and some circuit examples can be downloaded at cordellaudio.com [23].

The ADI LTspice web page has an instructional video, tips and articles. There is now an ADI Engineer Zone LTspice support page as well.

References

1. A.S. Sedra and K.C. Smith, *Microelectronic Circuits*, Oxford University Press, Oxford, 2010.
2. Bob Cordell, *Designing Audio Power Amplifiers*, 2nd edition, Routledge, New York, 2019.
3. R.J. Widlar, Some Circuit Design Techniques for Linear Integrated Circuits, *IEEE Transactions on Circuit Theory*, vol. 12, no. 4, December 1965.
4. G.R. Wilson, A Monolithic Junction FET-n-p-n Operational Amplifier, *IEEE Journal of Solid State Circuits*, vol. SC-3, 1968, pp. 341–348.
5. Walt Jung, *Sources 101: Audio Current Regulator Tests for High Performance*, Part 1: Basics of Operation, AudioXpress, audioXpress.com, 2007.
6. Walt Jung, *GLED431: An Ultra Low Noise LED Reference Cell*, December 2015, www.waltjung.org.
7. Ken Walters and Mel Clark, *Zener Voltage Regulation with Temperature*, MicroNotes Series No. 203, Microsemi Corp, www.microsemi.com.
8. P. Williams, Ring-of-Two Reference, *Wireless World*, July 1967, pp. 318–322.
9. Datasheet, *J500 Series Current Regulating Diodes*, Linear Integrated Systems, linearsystems.com.
10. Datasheet, *LM134 3-Terminal Adjustable Current Source*, Texas Instruments/National Semiconductor.
11. Datasheet, *LT3092 200mA 2-Terminal Programmable Current Source*, Linear Technology.
12. E.F. Schubert, *Light Emitting Diodes*, 2nd edition, Cambridge University Press, New York, 2006.
13. Peter A. Lefferts, LED Used as Voltage Reference Provides Self-compensating Temp Coefficient, *Electronic Design*, February 15, 1975, p. 92.
14. Datasheet, *LM329 Precision Reference*, Texas Instruments/National Semiconductor.
15. Datasheet, *TL431 Precision Programmable Reference*, ON Semiconductor, www.onsemi.com.
16. R.J. Widlar, New Developments in IC Voltage Regulators, *IEEE Journal of Solid State Circuits*, vol. SC-6, February 1971, pp. 2–7.
17. A.P. Brokaw, A Simple 3-Terminal IC Bandgap Reference, *IEEE Journal of Solid State Circuits*, vol. 9, no. 6, December 1974, pp. 388–393.
18. Paul R. Gray, et. al., *Analysis and Design of Analog Integrated Circuits*, 5th edition, John Wiley and Sons, Inc., January 2009, ISBN: 978-0-470-24599-6.
19. M. J. Hawksford, Optimization of the Amplified Diode Bias Circuit for Audio Amplifiers, *JAES*, vol. 32, no. ½, January/Febuary 1984, pp. 31–33.
20. L.W. Nagel and D.O. Pederson, *Simulation Program with Integrated Circuit Emphasis*, Proc. Sixteenth Midwest Symposium on Circuit Theory, Waterloo, Canada, April 12, 1973; available as Memorandum No. ERL-M382, Electronics Research Laboratory, University of California, Berkeley.
21. LTspice User's Manual, www.analog.com.
22. Gilles Brocard, *The LTSPICE IV Simulator*, 1st edition, Wurth Electronik, ISBN 978-3-89929-258-9, 2013.
23. Cordell Audio website, cordellaudio.com.

Chapter 4

Passive Components

The major passive components making up audio circuits are resistors, capacitors and inductors. The various types are described here and some of their characteristics that can influence circuit performance, reliability and sound quality are touched upon. Other passive components are also described, including potentiometers, switches, relays and fuses. Surface-mount technology (SMT) is widely used, and this technology will be explained in Chapter 5.

4.1 Resistors

When a constant voltage is impressed across an ideal resistor, a constant current flows immediately. This relationship is described by Ohm's Law. Ideal resistors are simple; real resistors are not, both in performance and application.

Types

There are a great many types of resistors, largely defined by the material used to create the resistive element. Among the earliest was carbon, and it was used to make carbon-composition resistors; little more than a chunk of carbon with wires on each end. Carbon is also used in carbon film resistors wherein a film of carbon is deposited on a typical ceramic (Al_2O_3) body or substrate of the resistor. Sometimes the film is deposited as a spiral to increase and/or control the resistance.

Metal film resistors use a film of metal deposited on the body, usually in a spiral that can be laser trimmed to meet the target resistance value. They are among the most precise and highest quality of small-signal resistors available. They are also available with small temperature coefficients of resistance, on the order of 100 ppm/°C or significantly better. There are two types of metal film, thick and thin. The thin film types are for precision applications that require lower noise and better temperature stability.

Metal oxide film (MOF) resistors tend to be less precise but are available in higher wattages. They are very robust against brief high currents and have very low inductance. Wirewound resistors tend to be larger and are usually used where significant power dissipation is required. They are spiral-wound and exhibit inductance, as expected. Non-inductive wirewound resistors are available wherein the winding topology is arranged to cancel inductance.

Metal foil (MF) resistors outperform most other resistor technologies for applications that require high precision and high stability. The signal-to-noise ratio with these resistors is the best available. Foil resistors comprise a pattern etched in metal. The geometry of the pattern produces cancellation of inductance and very low lumped internal capacitance.

The MELF (metal-electrode leadless film) resistor is among the highest performing resistors [1]. MELF resistors are cylindrical surface-mount (SMD) resistors in which a homogeneous film of a metal alloy is deposited onto a cylindrical Alumina ceramic body. Connection is made by Nickel-plated steel end caps pressed onto the cylindrical body. A laser is used to cut a helical pattern in the deposited metal film, removing just the right amount of material to achieve the target resistance. The resistor elements are covered by a protective coating designed for electrical, mechanical and climatic protection. They are very stable and reliable and can withstand high temperatures. They can be made with very fine tolerances as low as 0.02% and very low temperature coefficients as low as 5 ppm/°C.

Their end-cap construction provides a solderable connection free of interface issues between the end cap and the metal alloy resistive element. Their pulse load capability, the amount of overload current they can be subjected to before significant change in resistance occurs, is about 2.5 times that of a thin film chip resistor. They are more expensive than chip resistors, but their performance is far superior. The standard MMB0207 MELF size is 5.8 mm long by 2.2 mm in diameter and is rated at 1 W and 500 V. The MiniMELF MMA0204 is 3.6 mm × 1.4 mm and is rated at 0.25 W and 200 V. The MicroMELF MMU0102 is 2.2 mm × 1.1 mm and is rated at 0.2 W and 100 V.

Standard Values

Standard IEC resistor values are defined in several E series, where the number indicates the number of different values per decade. For example, the E24 series has 24 different values per decade. Table 4.1 shows the values for one decade of the E12, E24 and E48 series. The E12 series is associated with 10%-tolerance resistors. The E24 series is associated with 5% resistors. The E48 series is associated with 1% or 2% resistors. The E96 series is associated with 1% resistors.

This is important. In the series list below, only 48 values are listed for the E96 series in order to save space. The values shown in the E96 series column are simply the next higher value from the E48 value to the left in the same row. Thus, to the right of the E48 10.0 value is the E96 10.2 value which lies between the E48 10.0 and 10.5 values. The value in between two members of a series, in the next higher series, usually has a value that is approximately the geometric mean of the two values. There is also an E192 series that includes values between the members of the E96 series. The E192 series is for tolerances of 0.5%, 0.25% and 0.1%.

Table 4.1 Standard Resistor Values

E12	*E24*	*E48*	*E96*
10	10	10.0	10.2
		10.5	10.7
	11	11.0	11.3
		11.5	11.8
12	12	12.1	12.4
		12.7	13.0
	13	13.3	13.7
		14.0	14.3
15	15	14.7	15.0
		15.4	15.8
	16	16.2	16.5
		16.9	17.4
18	18	17.8	18.2

(*Continued*)

Table 4.1 (Continued)

E12	*E24*	*E48*	*E96*
		18.7	19.1
	20	19.6	20.0
		20.5	21.0
22	22	21.5	22.1
		22.6	23.2
	24	23.7	24.3
		24.9	25.5
27	27	26.1	26.7
		27.4	28.0
	30	28.7	29.4
		30.1	30.9
33	33	31.6	32.4
		33.2	34.0
	36	34.8	35.7
		36.5	37.4
39	39	38.3	39.2
		40.2	41.2
	43	42.2	43.2
		44.2	45.3
47	47	46.4	47.5
		48.7	49.9
	51	51.1	52.3
		53.6	54.9
56	56	56.2	57.6
		59.0	60.4
	62	61.9	63.4
		64.9	66.5
68	68	68.1	69.8
		71.5	73.2
	75	75.0	76.8
		78.7	80.6
82	82	82.5	83.5
		86.6	88.7
	91	90.9	93.1
		95.3	97.6

It is also notable that the same standard value series is used for other components, including capacitors and inductors. Naming of a value of 19,600 Ω, for example, is usually designated as 19.6 kΩ or simply 19k6. The value 19M6 would represent a 19.6 MΩ resistor. Resistors with 1% tolerance are available in both the E24 and E96 series.

Wattage Ratings and Temperature

Resistors are characterized by a maximum operating temperature, and their rated wattage is usually the amount of power required to heat them to some fraction of this temperature. This applies for many resistors in free-air with a 25°C ambient temperature. De-rating is thus necessary and can have a great benefit in reliability. Getting the heat out of resistors in a high-dissipation situation influences their dissipation ability. For example, a through-hole resistor dissipating anywhere near its rated power should be spaced above the printed circuit board (PCB) when it is mounted. Heat will also flow through the resistor leads to the copper traces on the PCB, helping a bit to remove heat. Some power resistors achieve their higher power ratings by having a metal body that can be attached to a heat sink, just like power transistors. Some SMD power resistors are packaged like a power transistor with a metal tab that is soldered to a copper pad on the PCB.

Resistors also have a heating time constant. This allows some resistors to handle peak dissipation significantly in excess of their rated power under transient non-continuous situations. The time constant also plays a role in their resistance temperature dependence under dynamic conditions. For example, the change in resistance over a cycle with a large audio signal across the resistor will be small at higher frequencies, but may be enough to create some distortion at low frequencies like 20 Hz.

Fortunately, operating power dissipation for most resistors in small-signal applications is very small. Consider a 1-kΩ feedback resistor in a preamp stage with continuous voltage of 10 V_{RMS} across it. Current will be 10 mA and dissipation will be 100 mW. Now consider a 10-kΩ resistor with 1 V_{RMS} across it. Current will be 100 μA and dissipation will be 0.1 mW. Dissipation rises with the square of voltage across a resistor.

Resistors in sensitive parts of a circuit, like feedback networks, should have a dissipation rating that is at least ten times the dissipation occurring at the maximum signal level. A 200-W amplifier with a feedback network of 220-Ω shunt and 4700-Ω feedback resistors for a gain of about 27 dB will have 39 V_{RMS} across it. This will result in current of 8.3 mA, corresponding to dissipation of 0.32 W. The feedback resistor should thus be rated at more than 3 W.

Temperature Coefficients

Most metals and other resistive materials have a small positive temperature coefficient (TC) of around 0.004/°C, such as for copper and aluminum. For Nichrome the number is about 0.0004/°C, explaining why it is often used for wirewound resistors. Depending on the particular materials used in a resistor, TC can often be positive or negative. An ordinary carbon film resistor has a TC that tends to be negative, on the order of −8e-4/°C to +2.5e−^{4}/°C. The TC of metal film resistors is usually between ±50 and ±100 ppm/°C, but thin film resistors can be had with ±25 ppm/°C TC. Thick film resistors tend to have a higher TC, usually between ±50 and ±200 ppm/°C. Metal foil resistors have TC usually between ±0.2 and ±4 ppm/°C.

PTC and NTC Resistors

Some applications require resistors that have a significant and reliable positive or negative temperature characteristic (PTC or NTC). A good example is an inrush control resistor. This device will have a higher resistance when cold and a lower resistance when warm. The inrush current is thus reduced during the turn-on interval when the resistor is cold, while resistance for normal operation under load is small because the resistor is warm. Such resistors may have a TC of −6%/°C. A little bit of power is lost in keeping the resistor warm while in service, so sometimes a relay is used to short the resistor once voltages stabilize.

Thermistors

PTC and NTC resistors are actually in the family of thermistors. Resistors that are specifically referred to as thermistors are more often a bit more precise and are more used for temperature sensing and temperature control. They are more often characterized by a negative temperature coefficient. Thermistors used in these types of applications will usually have nominal 25°C resistance of 5Ω–100 kΩ.

Varistors

Varistors are often used as shunt elements to protect circuitry from high-voltage transients. Their effective resistance is voltage-dependant and quickly decreases as applied voltage exceeds a certain value. The most common varistors are made with metal oxide, and such a device is often called meta-oxide varistor, or MOV.

Photoresistors

Photoresistors are light-dependent resistors, often called LDRs, whose resistance decreases in the presence of light. The most common photoresistor is the cadmium-sulfide (CdS) device. It is popular partly because of its good sensitivity in the visible spectrum. It is what is often referred to as a photocell. Because it employs cadmium, it is not ROHS (restriction of hazardous substances) compliant. Other photoresistors such as the lead-sulfide type (PbS) are most sensitive in the infrared spectrum. In some applications an LDR is mated with a light-emitting diode (LED) to provide a current-controlled resistance, sometimes used in audio compressors. In many applications, where linearity of the device is less important, photodiodes or phototransistors are used instead of photoresistors. Optocouplers are a good example.

Inductance

Ordinary wirewound resistors can have a significant amount of undesired inductance, often between 10 and 50 μH. For circuits that are sensitive to such parasitic inductance, the wirewound resistor can be wound in such a way that the magnetic fields of the coils cancel, reducing inductance to as low as 1 μH or less, but they cost more and are not as widely available. For applications that do not require a large amount of continuous power dissipation, metal oxide film (MOF) resistors can be a good alternative. These will often have inductance between 3 and 200 nH. They will also tolerate higher current for short intervals than their wattage rating would suggest. For example, a 3-W MOF resistor might be used as the emitter resistor in a power amplifier instead of a large non-inductive 5-W wirewound "sand" resistor, even though under some circumstances a 0.22-Ω resistor in such an application might see peak current as much as 7 A, corresponding to 11 W of instantaneous dissipation.

Sources of Distortion and Voltage Dependence

As mentioned above, signal-dependent power dissipation at low frequencies can cause thermal distortion by changing the resistance as a function of instantaneous temperature [2–5]. This can be especially true in feedback network resistors. Figure 4.1 shows measured resistor distortion as a function of frequency for a 1/4-W metal film resistor dissipating 1/4 W as a function of frequency [5]. Total Harmonic Distortion (THD) is about −125 dB at 300 Hz, but rises to −105 dB at 10 Hz.

Resistors also have voltage dependence, denoted as VCR and measured in ppm/V. Metal film resistors usually have a negative voltage coefficient, and it can range from −0.1 ppm/V to as high as −10 ppm/V. Obviously, voltage dependence of resistance represents nonlinearity that can cause distortion, albeit usually quite small.

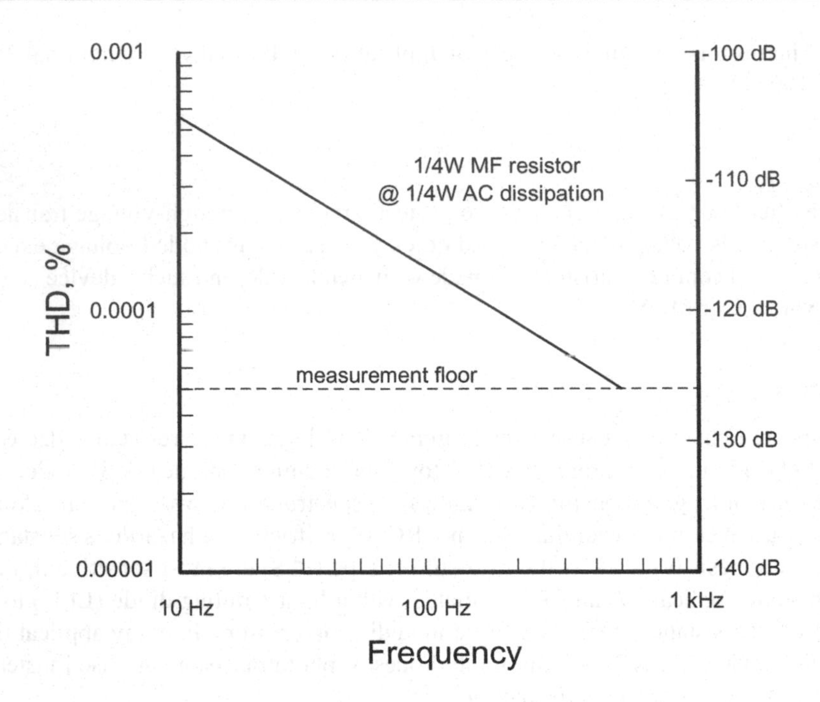

Figure 4.1 Thermal Resistor Distortion.

4.2 Potentiometers

A potentiometer (pot) is essentially a type of variable resistor. Technically, they are a 3-terminal resistor device with a tap (wiper) that moves along a conductive surface (track). As a potentiometer, they conventionally have an input and an output (the wiper), and the potential at the output is a fraction of the input. A common use is as a variable attenuator, like a volume control. The track is usually deposited carbon or a conductive plastic. The electrical position for a linear-taper pot can be defined as the percent of total resistance that lies between the wiper and the counter-clockwise (CCW) terminal. A linear pot at an electrical position of 10% will attenuate the signal by 20 dB if the wiper is not loaded. How the electrical position varies with the rotational position is referred to as the taper. The most common tapers are linear and audio or logarithmic (log).

Potentiometer Tapers

It is easy to envision a potentiometer with a so-called linear taper. This is simply achieved with a conductive track has constant resistivity as a function of wiper rotation. At its center of movement, the resistance from the wiper to either end will be the same. Sometimes, however, it is desirable that the divider ratio so formed will be a nonlinear function of the pot position. The best-known example is the audio taper. Since the ear perceives loudness on a logarithmic scale, it is desirable to have the attenuation from the pot be an approximation of a logarithmic function.

While a true logarithmic taper would yield attenuation linear in dB with rotation, these can only be approximated in the real world and can be expensive and hard to find. Pots with an audio taper usually approximate the log function with two track segments of different resistivity, resulting in two straight-line segments of differing slope when resistance from the wiper to the CCW terminal is plotted linearly as a function of linear pot rotation. At the 12:00 position, the attenuation is usually about 16 dB (0.154) for the audio taper.

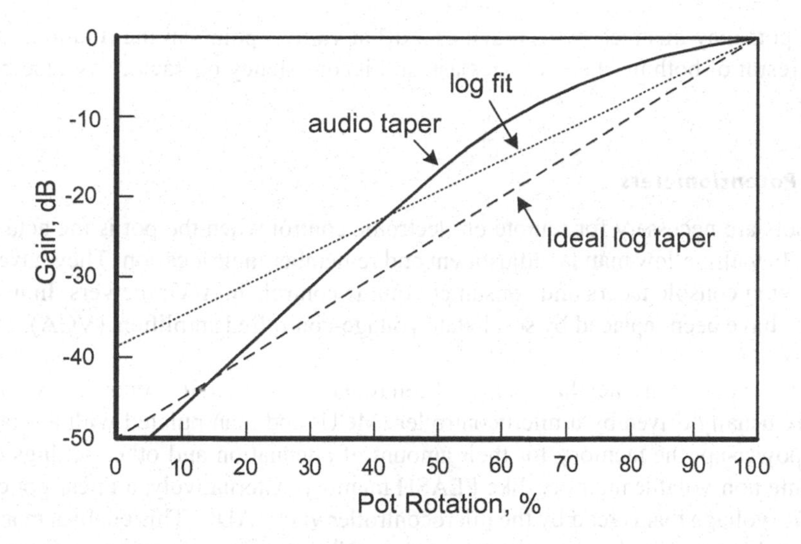

Figure 4.2 Attenuation vs Rotation for Audio and Log Tapers.

Figure 4.2 shows dB attenuation as a function of electrical pot rotation for audio and ideal log tapers. Note that the audio taper is only down 50 dB at 5% rotation, reflecting that the dB axis cannot represent zero gain. A real-world log taper can only be an approximation to an ideal log taper, since the output of the pot must go to zero when the pot reaches its maximum CCW travel. The plot also shows a "log fit" to the audio taper for comparison to show how the audio taper departs from logarithmic behavior. In fitting, the attenuation of the log fit is made equal to that of the audio taper at about 40% of mechanical pot rotation.

Rotation and Travel

The maximum rotational mechanical travel is usually about 300°. Close to either full CCW or full CW rotation there are small regions where the resistance does not change due to the size of the end contacts. The region over which the resistance changes is called the electrical travel, and is usually about 270–290°. At 75% of available mechanical rotation, the pot resistance from full CCW to the wiper with a real-world two-segment audio taper might be about 55% of full resistance. For a log taper it might be about 43%. At 25% rotation, the real-world pot resistance may be at about 5%, while the true log taper might be at about 1.5%.

These errors and differences do not create a serious problem for the listening experience, but can matter in some professional applications as in mixing consoles. Bear in mind that when the knob is at 9:00, this does not correspond to 25% rotation, since 100% mechanical rotation is only about 300°. The same considerations apply to slider potentiometers.

Multi-Gang Potentiometers and Tracking

In a stereo or multi-channel system with a volume control that has multiple sections (gangs), each serving one channel, it is important that the divider ratio provided by each gang is the same at all amounts of rotation so that balance among the channels is not changed as the volume is changed. High-quality pots will maintain matching of 1 dB or better in the vicinity of mid rotation;

inexpensive pots may mistrack by as much as 3 dB at various points in the rotation. Tracking errors are the result of both mechanical backlash and inconsistency of track resistance between the two sections.

Motorized Potentiometers

Motorized pots are necessary for remote or electronic control when the pot is the actual attenuating device. They also allow manual adjustment and remember their location. They have been used widely in mixing console faders and consumer volume controls in A/V receivers. In many mixing consoles they have been replaced by solid-state voltage-controlled amplifiers (VCA), avoiding the use of motors.

In consumer applications they have largely been replaced by digitally controlled volume control chips that are usually driven by a microcontroller (MCU) and manipulated with a rotary encoder control. At power-up, the memory for their amount of attenuation and other settings usually depends on some non-volatile memory, like FLASH memory. Alternatively, a linear pot can be used to create a DC voltage that is read by the microcontroller via its ADC. This enables manual control and a positional indication of amount of attenuation. Often a means of visually displaying volume level with a display is provided.

Wiper Loading Effects

Potentiometer-based attenuators have varying output impedance as seen at the wiper as a function of rotation. At either end, the output impedance is near zero if the input to the attenuator has near zero impedance. When the pot is at the point where resistance is the same from the wiper to the CCW and CW ends of travel, corresponding to 6 dB of attenuation, its output impedance is at its maximum, and is 1/4 the total resistance of the pot. The maximum output impedance of a 50-kΩ pot will be 12.5 kΩ and its thermal noise contribution will be about 15 nV/√Hz.

If the wiper of a pot is loaded, its taper will be altered, be it a linear pot or one with an audio taper. Attenuation will no longer have a ratiometric relationship with the pot resistance below and above the wiper. The effect will be greatest at the point where the pot output resistance as seen at the wiper is greatest. At this point, a load resistance that is the same as the pot resistance will cause a drop to about 80% of its unloaded value, or additional attenuation of about 2 dB.

While this is usually undesirable, the effect can be used to advantage in some situations, with fixed resistors connected from the wiper to the CCW and/or CW terminals. In this way the taper can be altered. Importantly, the accuracy of the result will be affected because of the typical poor ±20% tolerance of the pot resistance.

Wiper Resistance, Noise and Distortion

The wiper makes an imperfect electrical/mechanical contact with the resistance track. This leads to wiper resistance. Obviously, if there is enough dirt to open the contact, wiper noise will result. This can be especially noticeable if there is DC across the pot. There can also be ordinary local wiper resistance, since the contact is with a resistive element, even if the wiper makes contact with the full width of the track. Bear in mind that even though the track is thin, it is still a three-dimensional piece of resistance underneath the wiper. This means that a little bit of signal can sneak around a grounded wiper.

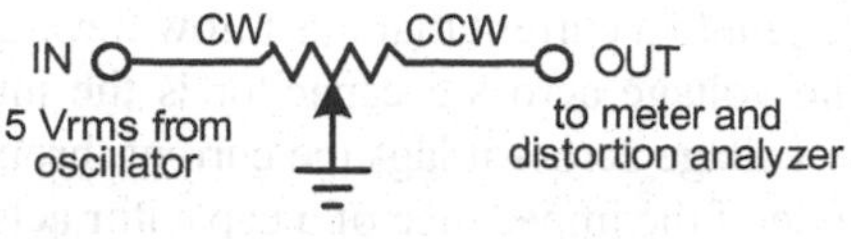

IN CW CCW OUT
5 Vrms from oscillator
to meter and distortion analyzer

Figure 4.3 **Wiper Feedthrough Test.**

This effect can be seen if the wiper is grounded and a test signal is applied to the CW end with the wiper at a mid position for a linear-taper pot, as illustrated in Figure 4.3. With an ideal pot, there will be no signal seen at the CCW end. With a real pot with effective wiper resistance, some attenuated signal will be seen at the CCW end. This test can be used to infer effective wiper resistance. If the pot is used as a single-pot balance control that puts different amounts of shunt loading to ground on the left and right channels, as in the simple preamplifier described in Chapter 2, this effect can lead to a tiny bit of crosstalk.

Finally, the wiper contact resistance might be slightly nonlinear. In this case, the voltage seen at the CCW end may be distorted. In a volume control application where there is loading on the wiper, this effect can result in some distortion.

Several linear-taper pots were subjected to the test in Figure 4.3, using a 5-V, 1-kHz signal and measuring the signal at the CCW end with an AC voltmeter. With a 50-kΩ linear-taper pot that was centered, the signal into the THD analyzer was about 0.1 V, suggesting a feedthrough attenuation of 0.02, or about 36 dB. With the 50-kΩ linear-taper pot centered, the series resistance from the voltage source to the wiper should be 25 kΩ. With a reading of 100 mV at the unloaded end, the effective wiper resistance should be on the order of 25 kΩ × 0.02, or about 500 Ω. This is about 1% of the rated pot resistance.

The amount of wiper feedthrough voltage for several different pots of different values ranged from 100 to 190 mV, with one major exception producing only 2.7 mV. The test described is a pretty tough and rigorous test for a pot, and may be an easy way to distinguish pot quality.

A test was performed on a high-quality 50 kΩ audio-taper potentiometer configured as a volume control. The pot was set to its mid position. A 5-V_{RMS} 1-kHz signal was applied to the input. The signal level measured at the output of the pot was 0.76 V. Distortion measured less than 0.001%. The wiper was then loaded by 10 kΩ to ground. The output then measured 0.45 V. Distortion then measured less than 0.001%. This is indicative of low wiper distortion. This experiment was carried out on two other 50-kΩ audio-taper pots of unknown origin, again with distortion readings with no wiper load and a 10-kΩ wiper load of 0.001% or less. This test of a limited number of sample pots indicates that wiper loading does not create a significant amount of distortion for a pot configured as a volume control.

4.3 Capacitors

When a current is passed through a capacitor, the capacitor resists a change in voltage. If a constant current is passed through a capacitor, the voltage across it will increase linearly, providing an integration function with a constant rate of change, with $V = IT/C$.

A constant current of 1 mA passing through a 1 μF capacitor will cause the voltage across the capacitor to change at a rate of I/C = 1000 V/s. This arrangement is a simple integrator. A voltage-in, voltage-out integrator can be implemented by a resistor in series with the virtual ground of an op amp that has a feedback capacitor from the output to the inverting virtual ground input. The resistor merely converts the input voltage into a current that must flow through the capacitor if the op amp input current is negligible. The voltage across a capacitor is the integral of the current passing through it. This means that the voltage across it lags the current through it by 90°.

For AC signals, the magnitude of the impedance of a capacitor acts somewhat like a resistance, but its impedance is an inverse function of frequency. In fact, its impedance is referred to as its capacitive reactance, X_C. It follows the familiar equation $X_C = 1/(2\pi fC)$. The term $2\pi f$ is the radian frequency ω in radians per second. π radians = 180°, while 1 rad = 57.3°. The reactance is then merely $X_C = 1/\omega C$. In the case of the integrator mentioned above, the inverting gain is simply the ratio of the feedback impedance to the input impedance, which in this case is simply $R_{in}/\omega C$. The gain is inversely proportional to frequency and falls at 6 dB/octave. The output voltage will always lag the input voltage by 90°.

Capacitive Reactance and Imaginary Numbers

When a sine wave voltage is impressed across a capacitor, a current flows, just as with a resistor, but the phase of that current leads the voltage by 90°. Such current flow is referred to as reactive, as opposed to resistive. Just as the inverse of resistance (*R*) is conductance (*G*), the inverse of impedance (*Z*) is admittance (*Y*).

Consider the circuit of Figure 4.4 where two non-inverting integrators are connected in tandem. Each integrator is made non-inverting for clarity of explanation by including an inverting stage at its input. For now, don't worry about DC stability; DC stability would be achieved if an inverter connected the output to the input, forming a negative feedback loop at DC. A voltage V_{in} is applied to the first integrator and V2 appears at its output with a 90° phase lag. V2 is applied to the second integrator and V_{out} appears at its output with an additional 90° phase lag, for a total phase lag of 180° with respect to the input. This is simply a phase inversion for AC signals wherein the output amplitude falls at 12 dB/octave.

If each integrator has a gain of +1 at 1 kHz with a 90° phase lag, the gain from input to output is simply −1 at 1 kHz. Because the integrators are in tandem, the overall gain is the square of the individual integrator gains. This means that the gain of each integrator is the square root of the total gain of −1. But what is the square root of −1? There is no such real number when multiplied by itself that yields −1. So how can the gain of the integrator be described? Such a number would represent unity gain with a phase lag of 90°.

Therefore, a different kind of number is used to represent the 90° phase shift. It is called an "imaginary" number because it is not a "real" number. That is a number whose square is a negative number. In mathematics, it is designated with a prefix of *i*. However, in electronics *i* is used to designate a current. Thus, in electronics the letter *j* is used instead as a prefix to designate the imaginary number. For example, $j1 \times j1 = -1$. Correspondingly, $\sqrt{-1} = j1$ and $\sqrt{-2} = j1.414$ or $j\sqrt{2}$.

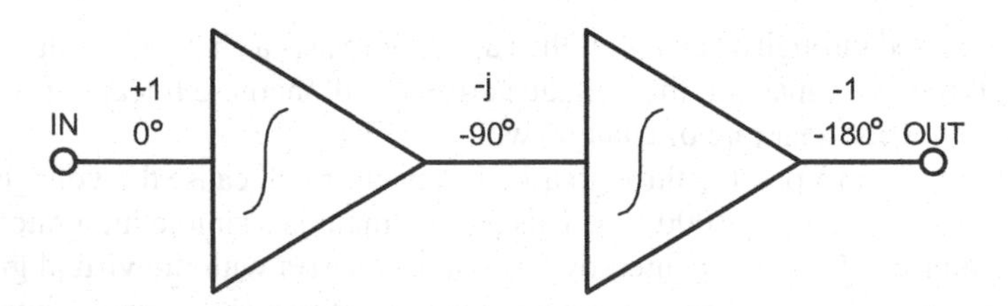

Figure 4.4 Two Non-inverting Integrators in Tandem.

The representation for a unity-gain process with 90° of phase lag is thus $j1$. Correspondingly, the representation for a unity-gain process with 90° of phase lead is $-j1$, since the difference in phase between the two processes is 180°. The impedance of the capacitor, including the representation of its 90° phase shift is thus $Z = 1/j\omega C$. The unit ω is radians per second (rad/s), also called radian frequency or angular frequency, where $\omega = 2\pi f$. It is also convenient to represent jω with a single variable deemed "s", as in $s = j\omega$. The impedance of the capacitor is thus $1/sC$. These terms and relationships will be valuable in Chapter 6 dealing with Bode plots and networks.

Finally, if we connect V_{out} to V_{in} through a unity-gain inverter, we now have a 360° phase shift around the loop, corresponding to +1, and hence positive feedback with unity gain at 1 kHz. We have what is called a state variable oscillator whose frequency is 1 kHz. Obviously, we can delete the two inverters associated with the non-inverting integrators, as their inversions cancel. The oscillator now requires only three op amps. We will see later that some amplitude stabilization means needs to be added to define its oscillation amplitude.

Capacitor Dielectrics

The dielectric of a capacitor is the insulating layer between the plates. The most common dielectric is air. The dielectric constant, denoted as *k*, multiplies the value of the capacitance that would result if a vacuum were the dielectric; the dielectric constant of dry air is very close to that of a vacuum. Air thus has a dielectric constant *k* of unity. The capacitance of a parallel-plate capacitor in a vacuum is:

$$C = \varepsilon_0 \times A / d$$

where A is the area of the plate in m^2 and d is the separation of the plates in meters. The constant ε_0 is the so-called permittivity of free space. With a dielectric other than a vacuum or air, where the dielectric constant is k, the capacitance becomes:

$$C = \varepsilon_0 \times kA / d$$

Note that the term A/d has the unit of meters. For convenience, the value of ε_0 is 8.84 pF per meter. A capacitor with 10-mm by 10-mm plates separated by 1 mm has an A/d value of 100 mm, or 0.1 m. With air as the dielectric, the capacitance will be 0.884 pF. If the plates are separated with

Table 4.2 Dielectric Constants

Dielectric Material	k
Mica	6.8
Polypropylene	2.2
Polycarbonate	2.9
Polyester	3.3
Polystyrene	2.5
Glass	6.8
Ceramic	5.6
Paper	3.8
Tantalum Oxide	28
Polyethylene	2.3
Teflon	2.1

mica with *k* of 7, the capacitance will be 6.2 pF. Typical numbers for several common capacitor dielectrics are listed in Table 4.2 (some have a fairly wide range, so some of the numbers shown are rough).

Dielectric Absorption

Dielectric absorption (DA), sometimes called soakage, is the tendency for a capacitor to retain some charge after it has been shorted for a brief period of time. In other words, if a capacitor is charged to 1 V, the voltage across the capacitor returns to the original voltage by some percentage after the short has been removed. The orientation of the dipoles caused by the electric field when the capacitor is charged relaxes over time when the capacitor is shorted. The time constant represents an exponential decay over time that may not be fully complete when the short is removed.

Dielectric absorption is measured by charging the capacitor to its rated voltage for 1 hour. The capacitor is then discharged with a short for 10 seconds. The voltage to which the capacitor recovers in 15 minutes is measured and compared as a percentage to the original charge voltage. A polypropylene film capacitor may recover to 0.1%. A polyester film capacitor may recover to 0.5%. An X7R ceramic capacitor may recover to 2.5%. An aluminum electrolytic capacitor may recover to as much as 15% [6]. DA is normally treated as a linear process and is modeled as a capacitor with one or more series R-C elements in parallel with it, as depicted in Figure 4.5.

When the dipoles in a dielectric are subjected to an electric field, they will tend to become aligned, and this corresponds to the charge on the capacitor. However, there are different mechanisms at work in aligning the dipoles, and these different mechanisms have different time constants, some very short and some quite long. When the electric field is removed, as by a sudden discharge, some of these mechanisms take much longer to "relax," and that gives rise to the subsequent charge recovery. The simple linear model is just a rough approximation to this complex behavior that occurs at the electron, atom and molecular levels.

Among film capacitors, mylar or polyester capacitors suffer more DA than polypropylene and polystyrene capacitors. Polypropylene capacitors normally represent the best cost/performance tradeoff, and are the most widely used film capacitors in the signal path of quality audio components. Mica capacitors are also characterized by very low DA, but they are quite expensive and large when values exceed 1000 pF. NPO and COG ceramic capacitors have low DA and are a good choice where large values are not required.

DA in signal-path capacitors has been associated with degraded sound quality [6–9]. However, the linear model for DA does not suggest nonlinear distortion; only a very small frequency response error at rather low frequencies. This does not mean that there is no distortion attributable to DA; the model just does not account for it. Obviously, there are many reasons why an electrolytic

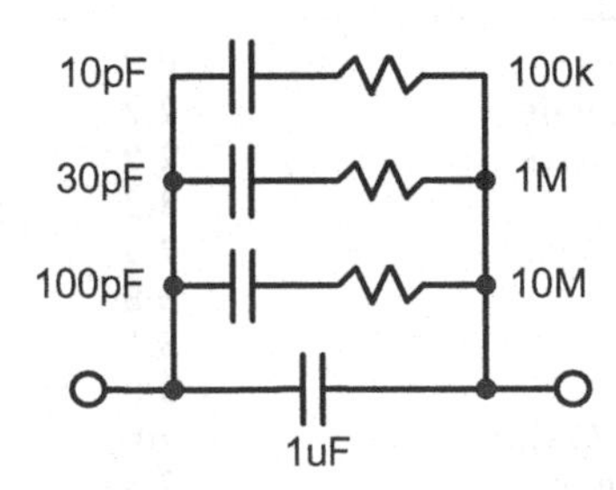

Figure 4.5 Linear Dielectric Absorption Model.

capacitor in the signal path will degrade the sound, and it (perhaps coincidentally) has high dielectric absorption. The same goes for X7R ceramic capacitors. Notably, NPO (EIA letter code COG) ceramic capacitors exhibit low dielectric absorption and are considered OK for audiophile circuits. There is a rough correlation between DA and dielectric constant, but there are significant exceptions to the correlation. Capacitors with high-k dielectrics tend to have higher DA, with electrolytic and X7R capacitors being good examples.

Capacitor Distortion and Dielectric Nonlinearity

When signal current is passed through a capacitor, the voltage across it may become slightly distorted due to capacitor nonlinearity. Similarly, if a voltage is impressed across such a capacitor, the resulting current may be a bit distorted. The tendency for a capacitor to act in a nonlinear fashion depends largely on what dielectric it employs.

The dielectric constant k can be a function of the electric field in the capacitor. This means that the capacitance can be modulated by the voltage across the capacitor, which in turn leads to distortion in circuits where the value of capacitance matters. This voltage dependence of capacitance may be a contributor to the distortion, but may not be the main contributor. It is also the case that capacitors with high DA tend also to have higher voltage dependence of capacitance. Dielectric nonlinearity, where the effective dielectric constant is a function of voltage, does exist, but it is unclear that this is the main contributor to capacitor distortion.

It is quite possible that DA itself is nonlinear with voltage. This effectively means that the resistances and capacitances in the linear model approximation may change with the strength of the electric field. Since DA represents a lossy behavior, this is different than a mere change in capacitance with electric field strength. If DA changes with voltage, then it is plausible that a capacitor with high DA would have higher distortion, apart from any change in capacitance as a function of voltage.

Capacitors for Audio Applications

Capacitor distortion and the importance of capacitor quality in audio circuits has been studied by many [7–11]. Some capacitors change their capacitance as a function of voltage. This is especially true for electrolytic capacitors and is also a problem with certain types of ceramic capacitors. Voltage-dependent capacitance variation will cause distortion at frequencies where the value of capacitance plays a role in setting the gain. For example, variation in the value of a coupling capacitor will change the 3 dB low-frequency cutoff frequency of the circuit, resulting in signal-dependent gain and phase at very low frequencies. Similarly, a small-value Miller compensation capacitor will alter the high-frequency response of an amplifier if its value changes with signal. The Miller compensation capacitor usually has the full signal swing across it. More importantly, a capacitor used in a filter can be especially important in its behavior.

Quality film capacitors have virtually no voltage dependence of capacitance. Ceramic capacitors with an NPO (COG) dielectric have very little voltage dependence, but ceramic capacitors with other dielectrics may have significant voltage dependence and should be avoided for signal passing, and are only recommended for use in supply decoupling applications. Some of these less desirable ceramic capacitors are also known to be slightly microphonic.

The biggest offender in most preamplifiers is the electrolytic capacitor commonly found in the shunt leg of the negative feedback network, often intended to reduce gain to unity at DC. Electrolytic capacitors have numerous forms of distortion. If electrolytic capacitors must be used in this location, large-value non-polarized or bi-polar (NP/BP) types should be employed. The use of a large value will reduce the amount of signal voltage that appears across the capacitor with program material, reducing somewhat the resulting distortion.

Electrolytic capacitors employ a dielectric that enables very high capacitance to be achieved. This approach, however, generally requires that the capacitor be polarized; in other words, voltage of only one polarity can be safely applied to the capacitor. They are especially suitable for use as power supply reservoir capacitors. There are applications where large capacitance is required but the applied voltage must be AC. A good example is when a non-polar electrolytic capacitor is used in a loudspeaker crossover network.

Electrolytic capacitors in the signal path should be bypassed with a film capacitor of at least 1 μF. The low ESR of the film capacitor will effectively short out the electrolytic capacitor at very high frequencies, limiting the damage done by the electrolytic capacitor to lower frequencies. Expectations for improvements in the audio band must be tempered by the realization that the impedance of a 1 μF capacitor at 20 kHz is still 8 Ω.

Curve (a) in Figure 4.6 shows the THD of the voltage across a 100 μF, 16 V polarized electrolytic capacitor as a function of frequency [4]. The capacitor is driven by a 5 kΩ resistor with a 20 V_{RMS} input. Signal current in the capacitor is 4 mA_{RMS}. THD reaches almost 0.01% at 20 Hz. Bear in mind that this is the distortion of the voltage across the capacitor; the resulting distortion of the amplifier stage will be smaller. The signal voltage across the capacitor at 20 Hz is about 300 mV. The frequency response of an amplifier stage with this arrangement would be down by 3 dB at about 6 Hz.

Curve (b) in Figure 4.6 shows the result when a 100 μF, 100-V audio-grade NP electrolytic capacitor is substituted. The capacitor is designed for use in loudspeaker crossovers and is not expensive. Its distortion is about 25 dB less than that of the low-voltage polarized capacitor. If you must use an electrolytic capacitor in the feedback network, use an NP capacitor with a substantial voltage rating.

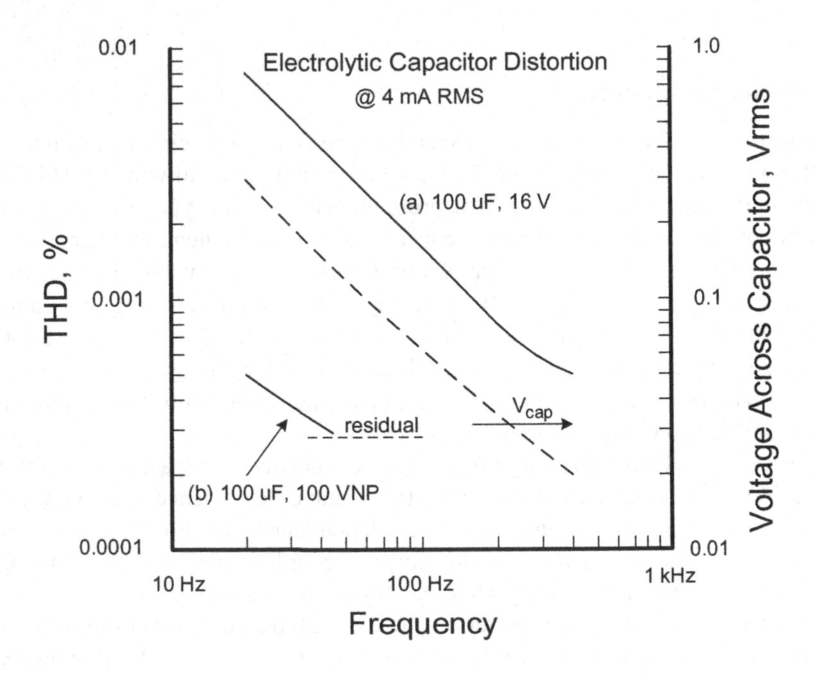

Figure 4.6 THD for 100-μF Electrolytic Capacitors.

Voltage Ratings, Temperature Ratings and Reliability

As long as capacitors are operated well within their voltage rating, they are quite reliable. The exception to this rule is electrolytic capacitors. They can age, dry out and lose capacitance. Their reliability is especially reduced by operation at high temperatures. Electrolytic capacitors are often rated at either 85°C or 105°C. The latter are more reliable for obvious reasons.

Tolerances

Typical "ordinary" film capacitors are usually available in tolerances of ±20%, ±10% and ±5%, and capacitors with much closer tolerances are available at an increased cost. Many applications do not require close tolerance on the capacitance, but passive and active filter applications usually do. Tolerances on electrolytic capacitors are notoriously wide, sometimes −20%/+80%. They are usually employed in applications where more capacitance can't hurt. However, if electrolytic capacitors are used in a filter application, as in a loudspeaker crossover, this can be a problem. The best loudspeaker crossovers employ film capacitors, even for the larger ones. They can be physically large and expensive. Sometimes polypropylene motor-start capacitors are used.

ESR and ESL

Equivalent series resistance (ESR) and equivalent series inductance (ESL) are important parasitic elements in the model of a capacitor. They both limit the effectiveness of capacitors in power supply filters, especially at higher frequencies where the impedance of the capacitor would otherwise be very low. Electrolytic capacitors may have relatively high ESR and ESL as compared to film capacitors of the same value, while ceramic capacitors often have quite low ESR and ESL compared to film capacitors.

Many capacitors are wound, leading to inductance depending on how the parts of the windings are connected to the terminals. Ceramic capacitors are often stacked and achieve very low ESR and ESL, especially important in switching power supplies.

Figure 4.7(a) shows a simplified equivalent circuit for a high-quality 50,000 μF reservoir capacitor. It is important to recognize that the reservoir capacitors have some ESR and that the extremely high-peak currents of the rectifiers will create some voltage drop across the ESR independent of

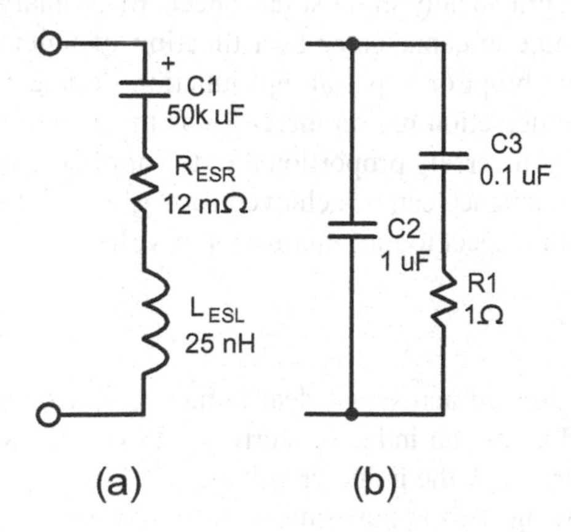

Figure 4.7 Electrolytic Capacitor Model and Zobel Network.

how large the capacitance is. This resistance varies considerably with the type and construction of the capacitor, but a value of 12 mΩ is not unusual. This means that a 7-A rectifier pulse may create a drop of 84 mV across the capacitor. Some evidence of this effect can be seen at the positive peaks of a ripple waveform.

Most capacitors also have a modest amount of ESL. This is often the result of the way they are constructed with windings. This means that at high frequencies the impedance of the capacitors actually starts to rise instead of continuing to decrease. The 25 nH ESL value for the capacitor of Figure 4.7(a) is quite low.

Reservoir capacitors are sometimes also characterized by a resonant frequency, where their impedance dips almost to the value of the ESR when the capacitance and inductance create a series resonance. This is often in the range of kHz to hundreds of kHz. The model in Figure 4.7(a) resonates at about 4.5 kHz. ESR and ESL can both be reduced significantly by employing several smaller capacitors in parallel instead of a single large capacitor. Figure 4.7(b) shows a parallel network that can be used to suppress the resonance and the rise of impedance at high frequencies.

Variable Capacitors and Trimmers

Years ago variable air gap capacitors were used to tune radios. The plates moved in such a way that the rotation caused more or less of the area of each plate to be in close proximity to the other plate. One or both plates were often shaped to create a desired capacitance vs rotation characteristic. They are rarely seen now. Small trimmer capacitors were also used, some with a rotating action by a screwdriver as with tuning capacitors and others with a compression action by a screwdriver on a mica dielectric. Some tuning applications that require high Q trimmers still use variants of these.

Varactors

The most widely used variable tuning element now is the varactor, sometimes referred to as a Varicap. The Varicap was patented by Pacific Semiconductor in 1961 and the name was subsequently trademarked in 1967 [12]. These are silicon junction diodes made in such a way as to emphasize their capacitance dependence on the applied reverse voltage. As in ordinary diodes, when a larger reverse voltage is applied, the depletion region between the n-type and p-type regions increases. This is the equivalent of the spacing between the "plates" increasing, reducing capacitance.

This behavior can be seen graphically in most datasheets of ordinary diodes. Varactors are designed to have a greater change in capacitance as a function of voltage than ordinary diodes or transistor junctions, using an abrupt or hyper-abrupt junction doping profile. For the former, the effective plate spacing of the depletion region increases as the square root of the applied reverse voltage, and the capacitance is inversely proportional to the applied voltage. For the hyper-abrupt junction, a wider range of capacitance can be achieved for a given voltage change, but the capacitance change is nonlinear with respect to the square root of voltage.

4.4 Inductors

When a constant voltage is applied across an ideal inductor, it will resist changes in current. If a constant voltage is applied across an inductor, current will rise linearly with a constant rate of change. If that current is interrupted, the inductor voltage will rise very high very quickly in an attempt to keep the current flowing. This is sometimes referred to as the flyback voltage. This behavior is often used in switch-mode power supplies (SMPS), where the current through the inductor is repeatedly turned on and off. This behavior can also be a problem in relays and switches, damaging a contact with an arc when the circuit path is opened. The flyback from a relay coil can also damage the relay driver transistor by exceeding the transistor's maximum V_{ce} if the flyback is not suppressed.

While the AC current in a capacitor leads the voltage by 90°, the AC current in an inductor lags the voltage by 90°. The impedance of the inductor, including the representation of its 90° phase shift is $Z = X_L = j\omega L$. It is also convenient to represent $j\omega$ with a single variable deemed "s", as in $s = j\omega$. The impedance of the inductor is thus sL. Every inductor has effective series resistance due to the winding, so the simple model for an inductor is an ideal inductor in series with R_L, the resistance of the inductor. The Q of an inductor is simply X_L/R_L, and is a function of frequency.

The permeability of the core around which a coil is wound is somewhat analogous to the dielectric constant of a capacitor; the permeability of air is unity. Air core inductors are very linear, but are large for the amount of inductance they provide. They are used in many RF tuned circuits and sometimes in crossovers in more expensive loudspeakers. Cores made of ferrous materials like ferrite or steel have much higher permeability, so the inductance per turn of the coil can be much higher. A core with higher permeability will increase the level of magnetic flux produced by a given current, corresponding to increased inductance.

The formula for inductance of a single-layer coil is:

$$L = \mu \times N^2 \times A / l$$

where:

L = inductance in Henries
μ = permeability of the core ($\mu_0 = 4\pi e{-7}$ if air core)
$\mu_r = \mu/\mu_0$ (relative permeability)
N = number of turns
A = area of the core in m^2
l = length of the winding in meters

The relative permeability μ_r of ferromagnetic core materials is usually greater than 100. The relative permeability of permalloy is about 8000 or more, while that of μ-metal is in excess of 20,000. Permeability for ferrite cores used in linear applications is on the order of 1000–3000.

Core Materials

The most common core materials are air, various types of ferrites and various types of iron or steel. Air core inductors provide the least inductance for size and amount of copper, but are more linear than other types. Ferrite cores provide a good cost/inductance ratio and good size/inductance ratio. They are very widely used in applications other than traditional linear power supplies. They provide adequate linearity for most applications and are used frequently in passive LRC filters. Their linearity in certain high-quality audio applications is sometimes deemed marginal. The inductance value of some ferrite inductors can be trimmed with a screwdriver. Laminated iron core inductors are often used in power supplies and other applications where higher currents are involved or greater inductance is needed. A good example is their use in the low-frequency portions of loudspeaker crossovers. Inductors with ferrite cores are widely used in switch-mode power supplies (SMPS). In modest SMPS, the inductance is often in the range of 10–100 μH.

Core Saturation

Ferromagnetic materials can support only a finite amount of magnetic flux density, meaning that they can saturate under conditions of high current. Beyond a certain magnetic field force, the magnetic flux does not increase in proportion. This is the equivalent of the permeability of the core decreasing at higher flux densities. The result is that the inductance decreases when higher current is passed through the inductor. This makes the inductance nonlinear and obviously leads to distortion or other undesirable behavior that results from decreased inductance.

If the impedance of an inductor changes as a function of signal current or voltage, distortion may result. The most likely cause is the onset of core saturation. This can be a soft mechanism that begins at amplitudes well below what would be considered outright core saturation. In portions of certain kinds of passive L-C filters, resonance can create larger voltage or current swings than one might expect, pushing the inductor more into the region where nonlinear behavior may occur. Nonlinear effects can also be created when an inductor is in close proximity to parasitic ferrous material that is not deliberately part of the core. An example would be a nearby steel panel of an enclosure.

Ferrite Beads

Ferrite beads comprise a cylinder of ferrite material through which a wire is passed. They create a small inductance of fairly low Q that actually becomes resistive, often on the order of 100 Ω, at very high frequencies [13]. They are widely used to provide noise suppression at high frequencies and/or deliberately introduce damping losses at high frequencies, increasing the stability of some circuits. Ferrite beads are deliberately designed to dissipate *AC* energy at high frequencies. Put another way, the ferrite bead on the wire creates a one-half-turn inductor with very low Q.

The resistance of a ferrite bead is very low at DC, being that of the wire passing through the bead. Depending on the particular bead parameters, a bead may increase its impedance to perhaps 100 Ω at perhaps 10 MHz. These numbers are just illustrative to lend perspective. In reality, ferrite beads are available with a great variety of high-frequency resistance and transition frequency (where they cross over from behaving inductively to behaving resistively). Ferrite beads are also characterized by a small amount of parasitic capacitance, often on the order of 1 pF. Their noise suppression behavior can thus deteriorate at very high frequencies in the range of GHz. Importantly, ferrite beads are subject to core saturation effects at high current, just like inductors. Above a certain current, their magical properties begin to disappear. Many designers do not give enough thought to core saturation in ferrite beads in applications where significant current is involved.

Of course, more than one wire can pass through a larger ferrite bead, the most common example being the large EMI suppression bead on the power cord of many digital devices. These arrangements provide significant common-mode impedance. Ferrite beads are also available in SMD versions. They are useful as radio frequency interference (RFI) filters in many applications.

4.5 Transformers

A transformer magnetically couples a signal or power from typically one primary winding to one or more secondary windings. This is an over-generalization, but it describes the function of a transformer in most applications. Its action is generally bi-directional. The most common application of traditional transformers is in power supplies, although that is being diminished by the increasing use of switch-mode power supplies; of course, some SMPS use smaller transformers for galvanic isolation.

Although important, power transformers are of less interest here as long as they don't radiate hum and noise into nearby small-signal circuitry. There is greater coverage of power transformers in Chapter 14 on power supplies and Refs. [14–19] comprise a six-part series on transformers that covers every aspect of transformer design.

Here the greater interest is in small-signal transformers that are used with moving coil cartridges, microphones and for galvanic isolation in audio applications. Of course, transformers in the signal path are usually avoided as much as possible, since they tend to be large and expensive. There are some applications where they are very difficult to avoid, as will be apparent in Chapter 18; they are usually a must in ribbon microphones due to the extremely small signal produced by the ribbon of those microphones [20]. While there are more practical alternatives to transformers in dynamic and condenser microphones, numerous microphones of these types do employ transformers.

Turns Ratio and Impedance Transformation

The turns ratio 1:N between primary and secondary in a transformer dictates the primary-to-secondary voltage ratio. A 1:30 ribbon microphone transformer will increase the voltage from the ribbon by a factor of about 30. The impedance will be transformed by a factor of N^2, or 900. A ribbon and primary loop impedance of 0.5 Ω will be transformed to a secondary impedance of 450 Ω.

Leakage Inductance and Leakage Reactance

Two key parameters of a transformer are its leakage inductance and its leakage reactance, the latter of which takes into account copper losses (e.g., winding resistances). When the magnetic flux from the primary of a transformer is not 100% transferred to create energy in the secondary, the flux lost creates an effective inductance in series with the primary (or that of the winding with the most turns). This is called the leakage inductance. If the secondary of an ideal transformer is shorted, the primary impedance should go to zero. In a real transformer, the primary impedance will approximate the impedance of the effective winding resistance in series with the leakage inductance. This is referred to as the leakage reactance.

In principle, the effective winding resistance for in-band AC signals with the secondary shorted is the primary winding resistance plus the secondary winding resistance multiplied by the square of the turns ratio. This is the real part of the leakage reactance. However, some additional losses from primary to secondary will make the short-circuit effective winding resistance a bit larger.

Leakage reactance influences the efficiency, regulation and frequency response of the transformer. The input to a real transformer loaded with a pure resistive load will still look slightly inductive. The type of construction of the transformer (e.g., E-I core vs toroid) has a significant influence on its leakage reactance).

Toroid vs Conventional Transformers

Toroidal power transformers tend to be more efficient than transformers of conventional laminated-core construction. They also better confine the noisy magnetic fields that result from large rectifier spikes. Leakage flux is significantly smaller than laminated-core transformers because there is no gap in the core. There is sometimes a small cost premium for toroidal transformers. Some claim that transformers with C-core or EI construction sound better, but this is unsubstantiated. The advantages of toroidal transformers over conventional laminated transformers include [21, 22]:

- High efficiency
- Small size
- Low mechanical hum and buzz
- Low stray magnetic field
- Low no-load losses
- Light weight
- More available turns/layer
- Less copper length due to longer effective bobbin length
- Magnetic field running in grain-oriented direction of the steel
- Ease of adding a winding on the exterior

It is very important that the bolt through the center of the toroid not be allowed to create a shorted turn by having both ends of it conductively tied to the chassis. One end of the bolt should float. It is also undesirable to mount one toroid on top of the other.

Encapsulated toroidal transformers having low VA ratings are available in various voltages and power ratings for PCB mounting. Many versions have dual primaries for 120/240-V operation. Most also have dual isolated secondary windings instead of a single center-tapped secondary.

Split-Bobbin Transformers

Split-bobbin transformers comprise a primary and secondary that are wound next to each other on a common bobbin rather than on top of one another. They offer small size and low cost for applications requiring less than about 60 VA. Their construction provides superior isolation and low capacitive coupling. A deeper discussion of different transformer constructions can be found in Chapter 14 where power supplies are covered.

4.6 Mechanical Switches

Switches are usually mechanical devices wherein electrical contacts are physically moved to come into contact with one another to close an electrical path with a low resistance. A pushbutton will be considered as a switch for the purposes here. Contacts are usually made of silver or silver-cadmium alloy, with good and lasting conductivity. Gold contacts are also used, but are less capable of switching higher currents.

Contact Configurations

The nature of a switch includes the number of throws and poles. The number of throws is the number of different contacts a switch can connect with different positions. The number of poles is basically how many independent switch elements are changed when the mechanical position is moved. A common double-pole double-throw (DPDT) switch will include two independent sets of switch contacts in which the common contact ("swinger") can make contact with either one of two contacts. A rotary stereo selector switch with 5 positions might be described as being double-pole, five-throw.

Most types of switch contacts have a letter designation. A single-throw switch with Form A contacts is normally open. Form B contacts are normally closed. Form C contacts correspond to a double-throw switch. Some double-throw toggle switches have three positions, where the center position corresponds to neither of the two contacts being closed. These are designated as Form K contacts. Sometimes the terms "ON-OFF-ON" or "center-off" are used to describe these.

Contact Resistance

All switches (and relays) have some contact resistance, but it is usually quite small, often in the milliohms. Contact resistance varies depending on the switch model and manufacturer, the switch contact material and sometimes the age of the switch.

Bounce

For many switches and pushbuttons there is some mechanical bounce when the position is moved so that contacts are closed. In other words, contact will be made briefly followed by a brief open interval, followed by closure. The duration of the bounce interval can be very small, but some logic circuits can be confused by bounce because the circuit reacts to the opening and closing much faster than the mechanical changes. For this reason, most logic circuits that are controlled by mechanical contacts must have de-bounce circuitry. Relays also have bounce. If the velocity of the armature during a transition is high, bounce will occur upon contact closure.

Transition Interval

When the setting of a conventional double-throw switch is changed, there will be a brief interval when the switch is open, and this must be considered in some applications. In many cases, its effect can be no different than bounce.

Make Before Break or Break Before Make

Most common switches break their connection with one contact before making connection to the other contact. This prevents the two contacts from ever being shorted together during the positional transition; it also allows a brief transition interval when the switch is completely open. In some applications, even a brief open-circuit condition must be avoided. In this case a make-before-break switch arrangement is used. Rotary switches implement the make-before-break operation by using a rotating contact that is wider than the distance between two adjacent selectable contacts.

Shaft Encoders

A type of rotating switch called a shaft encoder is often used to change selections in a device with a microcontroller. These can be rotated limitlessly, with each change of the shaft by a certain small number of degrees changing the connections to a pair of switches inside the device. Because each set of contacts changes connections at different times, the microcontroller can determine in which direction the shaft is being turned. Such arrangements are often used in conjunction with a menu that indicates what has been changed by the rotation, such as the level of a volume control or the selection of an input. While a mechanical switch retains memory of where it has been set, shaft encoders need to be associated with non-volatile memory in the microcomputer if their setting during prior use is to be retained. For added functionality, some rotary encoders include momentary contact pushbutton switches. Some include illuminated shafts using one or more LEDs, such as the Bourns PEL12 series.

4.7 Electromechanical Relays

Relays comprise a voltage- or current-activated switch with one or more poles. Like mechanical switches, they have contact formats. Current flowing in the coil creates a magnetic field that pulls the armature on which the swinger contacts are mounted, so that they come into contact with the stationary contacts. When not activated, the "normally closed" (NC) contacts are connected. When activated, the "normally open" (NO) contacts are connected. The discussion here will be limited to small-signal relays, like those used to replace mechanical selection switches or for muting circuits. These are sometimes referred to as "telecom" relays.

Not all relays are of the double-throw type. Some are single-throw, where the contact is either NO or NC. Form A contacts are NO, while Form B contacts are NC. Form C contacts are of the double-throw type.

Miniature telecom relays have a footprint of about 0.8 by 0.4 inches. Coil power for various such relays typically runs from 150 to 500 mW, corresponding to 12 to 42 mA at 12 V. Subminiature relays like the Panasonic AGQ200S12Z have a footprint of 0.42 by 0.28 inches and a nominal 11.7-mA, 1028-Ω coil. Typical contact resistance is 50 mΩ. These are available for about $2 in small quantities. They save space and current.

Latching Relays

The Panasonic AGQ series includes latching relays that have a set-reset functionality; the contact closures are changed by changing the polarity of the voltage applied to the coil; they stay that way until a pulse of opposite polarity is applied to the coil. These single-coil latching relays need only be driven for more than 4 ms to change state. Most of the time they consume no power and also "remember" their state. There are also two-coil latching relays with three or four terminals. Each coil can be driven to change the state of the relay from one state to the other. The two-coil relays are useful when the driving circuitry can only supply one polarity of drive.

Relay Drivers

It is very common to employ a simple NPN transistor driven into saturation connected to one end of the relay coil while the other end is connected to the positive supply. In some cases, two transistors in a Darlington connection will be used so that less current is needed from the controlling circuitry. This might be the case when the controlling circuit is the open collector of an LM339 comparator, where the current into the relay driver is only supplied by a pull-up resistor.

The TI ULN2803 is a popular IC Darlington octal relay driver that includes output clamping diodes and can drive (pull down) relays and other electromechanical devices with up to 500 mA and sustain peak voltages of 50 V [23, 24]. It can also function as an LED display driver. The device includes a 2.7 kΩ input series resistor for each Darlington pair, and can be directly driven by 5-V TTL or 3.3-V CMOS sources.

N-channel vertical MOSFETs can also be used. These can result in simpler driver circuitry because they do not require input current to the gate, and the gate is insulated so that it can be allowed to be driven positive by 15 V or more. The internal drain-source diode of these devices is usually rated for the same voltage as the MOSFET. Additionally, it is usually rated to withstand brief reverse current that is not insignificant. Without a suppressor diode across the relay coil, the coil flyback voltage upon de-activation can be allowed to go to a higher voltage, dissipating the energy stored in the coil inductance more quickly.

Minimum Activate and Hold Voltages

For operating margin, relays have a minimum voltage required for activation that is lower than the rated coil voltage. Once activated, there is also a minimum voltage required to hold the relay in the activated state. This minimum hold voltage is less than the minimum activate voltage. For the Panasonic AGQ200S12Z 12-V relay mentioned above, the minimum activate voltage is 75% of the rated coil voltage, while the hold voltage is only 10% of the rated coil voltage, the latter being unusually low. Maximum applied coil voltage is 150% of the rated voltage.

Contact Bounce

Just as with a mechanical switch, the contacts bounce when they come into contact at a relatively high velocity. This means that the contacts may close and open for a few cycles in a very brief interval. If the relay is controlling the input to a digital circuit, that circuit may have to include a de-bounce function so that multiple closures of the contacts are not mistaken for activation or de-activation of the relay.

If one connects one contact of a double-throw relay to a positive voltage and the other to a negative voltage, and connects the swinger to ground through a resistor, one can see the bounce behavior for activation or de-activation on a digital storage oscilloscope (DSO) by looking at the waveform on the swinger node.

Actuate, Release and Transition Times

Whenever a relay is activated or de-activated (operate or release), it takes time to complete the move of the swinger from one contact to the other. This time, often in the millisecond range, is controlled by the mass of the swinger and the strength of the magnetic field when the relay is activated. In contrast, the release time is controlled by the mass of the swinger and the strength of the spring when the relay is de-activated.

On top of this, the inductance of the coil can play a role in transition time. Upon activation, the coil inductance can influence how quickly the coil current ramps up to create the magnetic field. On release, the inductance in the coil can cause the coil current to persist while it flows through a snubber diode across the coil. Maximum operate and release times for the AGQ200S12Z are both 4 ms. Typical operation time is 1.5 ms. Typical release time is 1 ms without a snubber diode and 1.5 ms with a snubber diode.

It takes time for the swinger to complete its journey from one contact to the other. During this time, the relay will be open for both contacts. All of these times can be observed on a DSO connected as described above. It is best if the scope is triggered by the voltage applied to the coil. A pulse generator that creates an activation pulse of about 20–50 ms can make the transition behavior in both directions easy to observe.

Activate time can be reduced by applying a transient over-voltage to the coil that briefly exceeds the rated coil voltage by a significant amount. In some driver arrangements, this can be accomplished by placing a capacitor across a resistor in series with the coil.

The de-activation of the coil can result in a potentially destructive flyback voltage from the coil due to its inductance. This is usually prevented by placing a silicon diode across the coil, preventing the flyback voltage from exceeding about 0.6 V. The coil current will continue to flow through this diode until the energy in the coil is dissipated. This brief current flow after the activation current is removed can lengthen the release time of the relay.

If shorter release time is required, the flyback voltage can be allowed to rise to a higher, but still safe, voltage. This will speed the dissipation of the energy stored in the coil inductance. One approach is to put a Zener diode in series with the snubber diode. A second approach is to use a MOSFET relay driver, relying on the robust reverse breakdown of the MOSFET's internal drain-source diode. A small capacitor can be placed across the coil to reduce the edge rate of the flyback pulse.

References

1. Vishay, *MELF Resistors – The World's Most Reliable and Predictable, High-Performing Film Resistors*, Application Note 28802, vishay.com/docs/28802/melfre.pdf.
2. Bruce Hofer, Designing for Ultra-Low THD+N, Parts 1 and 2, *AudioXpress*, November 2013, pp. 20–23 and December 2013, pp. 18–23, audioxpress.com.
3. Bruce Hofer, *Designing for Ultra-Low THD+N in Analog Circuits*, 139th Convention of the Audio Engineering Society, Session PD3, October 29, 2015.
4. Bruce Hofer, *The Ins and Outs of Audio*, www.aes-media.org/sections/uk/Conf2011/Presentation_PDFs.
5. Bob Cordell, *Designing Audio Power Amplifiers*, 2nd edition, Routledge, New York and London, 2019.
6. Wikipedia, "Dielectrc Absorption", wikipedia.com.
7. Walt Jung and Richard Marsh, Picking Capacitors, *Audio*, February 1980, pp. 52 62 and March, pp. 50–63.
8. Cyril Bateman, Understanding Capacitors, *Electronics World*, December 1997, pp. 998–1003.
9. Cyril Bateman, Capacitor Sound?, *Electronics World*, Parts 1–6, July 2002–March 2003.
10. Cyril Bateman, *Capacitor Sounds II – Distortion v Time v Bias*, Linear Audio.
11. Passive Components Blog, *Capacitance, Dipoles and Dielectric Absorption*, passive-components.eu/capacitors-capacitance-dipoles-and-dielectric-absorption/.
12. Varicap Patent, US 2989671, Pacific Semiconductor.
13. Fair-Rite Products Corp., *How to Choose Ferrite Components for EMI Suppression*, www.fair-rite.com.
14. Chuck Hansen, Power Transformer Parameters, Selection and Testing, Part 1, Transformer Core Materials, *audioXpress*, vol. 53, no. 11, November 2022, pp. 58–61.
15. Chuck Hansen, Power Transformer Parameters, Selection and Testing, Part 2, *audioXpress*, vol. 53, no. 12, December 2022, pp. 55–58.
16. Chuck Hansen, Power Transformer Parameters, Selection and Testing, Part 3, Insulation Materials, Winding Bobbins, and Testing Methods, *audioXpress*, vol. 54, no. 1, January 2023, pp. 58–66.
17. Chuck Hansen, Power Transformer Parameters, Selection and Testing, Part 4, Parameters and Their Losses, *audioXpress*, vol. 54, no. 2, February 2023, pp. 48–52.
18. Chuck Hansen, Power Transformer Parameters, Selection and Testing, Part 5, Losses, *audioXpress*, vol. 54, no. 3, March 2023, pp. 56–61.
19. Chuck Hansen, Power Transformer Parameters, Selection and Testing, Part 6, Quality, Reliability and Design, *audioXpress*, vol. 54, no. 4, April 2023, pp. 56–61.
20. Bill Whitlock, *Design of High Performance Audio Interfaces*, Jensen Transformers, Inc., Chatsworth, CA, 2010.
21. Toroid Corporation of Maryland, Technical Bulletins, No. 1, 3, 4, www.toroid.com.
22. K.L. Smith, *D.C. Supplies from A.C. Sources – 3*, Electronic and Wireless World, February 1985, pp. 24–27.
23. Microcontrollerslab.com, *Introduction to ULN2803 – Relay Driver IC*, microcontrollerslab.com/introduction-uln2803-features/.
24. Texas Instruments, *ULN2803C Darlington Transistor Array*, datasheet, ti.com.

Chapter 5

Surface-Mount Technology

Surface-mount components or devices (SMD) are used in the manufacturing of electronic assemblies to allow for miniaturization and automated assembly. Surface-mount technology (SMT) is a term that encompasses both components (SMD) and assembly.

These SMDs can be mounted on ceramic and FR4 (epoxy/glass) or similar substrates. SMD packages can be both leaded and leadless, using plastic encapsulant (as with DIPs), glass or ceramic (hermetically sealed). It is possible to take a standard package like a dual inline package (DIP) or a miniature relay and lead-form it to sit flat on the surface-mount pads and call it an SMD.

The great majority of electronic devices for all applications are now implemented with SMT, as opposed to through-hole technology (THT). Many integrated circuits (ICs) and other devices are no longer available in through-hole (TH) packages. In this chapter we will briefly cover this important technology, recognizing that many details of the technology are beyond the scope of this book. Fortunately, there are a great many references with much more detail on this technology [1–3].

5.1 Advantages

The most obvious advantage of this technology is space and component density. Resistors, capacitors, discrete semiconductors and ICs are remarkably smaller than their through-hole (TH) peers. Moreover, ICs with many tens or hundreds of leads on all four sides are commonplace. One SMD IC packaging technology, the ball grid array (BGA), can support ICs with upwards of 500–1500 interconnections and in some cases more. SMD components can be placed on both the top and bottom of a PCB. In contrast, TH components take up space on both the top and bottom of a PCB. With TH devices, virtually every component needs vias. With SMT, there are almost no via obstructions on the bottom side of the PCB and there is far less need for drilling. This also leads to better ground planes on the bottom if they are used.

In most cases, the SMD passive components are quite a bit less expensive than their TH counterparts. The smaller-sized assemblies naturally tend to promote higher speed and shorter trace lengths, leading to smaller parasitic capacitance and inductance. Layout approaches can be different as well, with far more wiring traces on the component side of the board and often substantially fewer vias and drill holes. Due to the higher packing density, multi-layer PCBs are sometimes required in dense designs.

SMT permits easier and faster automation of assembly and less cost. One interesting aspect of assembly is that the surface tension of the melted solder tends to accurately pull the device into alignment with the pads to which it is connected.

5.2 Passive Component Packages and Sizes

There are generally accepted standards for the different sizes of passive SMD components. The list below shows some of the more common sizes for resistors, capacitors, diodes, even smaller inductors and ferrite beads. Sometimes these are called leadless or "chip" components.

Compare these sizes and areas to the area often occupied by a ¼-W metal film resistor whose total footprint is on the order of 0.4 × 0.1 in with an area of 0.04 in². A common 0602 resistor requires 20 times less area.

Size (mm)	*Dimensions, in (mm)*	*Area, in² (mm²)*
0201 (0603)	0.024 × 0.012 (0.6×0.3)	0.0003 (0.18)
0402 (1005)	0.039 × 0.020 (1.0×0.5)	0.0008 (0.5)
0603 (1608)	0.063 × 0.031 (1.6×0.8)	0.0020 (1.28)
0805 (2012)	0.079 × 0.049 (2.0×1.2)	0.0039 (2.4)
1206 (3216)	0.126 × 0.063 (3.2×1.6)	0.0079 (5.12)

The 0805 and 0603 sizes are probably the most popular size for resistors and capacitors in manufacture where size is not as critical and power dissipation is not an issue. However, the smaller 0402 size is popular as well. The 0201 size can be found in smartphones. Bear in mind that the common 0603 resistor can dissipate only 100 mW. Further adding to wiring opportunities, there are 0-Ω SMD resistors (jumpers) that can sometimes avoid the need to go down to another layer.

SMD resistors are fabricated using many different processes. The common types are thick and thin film, and wire-wound. Thin film types are for precision, thick film types are general purpose and wire-wound types are for higher power applications. Many are constructed by depositing a resistive material on an alumina substrate. A laser is used to trim off material to meet the target value. Values are printed on the surface using 3 or 4 digits representing value and multiplier.

SMD resistors also come in hermetically sealed packages. Metal electrode leadless face (MELF) packages are cylindrical leadless packages, usually of glass construction. The EN 140401-803 and JEDEC (Joint Electron Device Engineering Council) DO-213 standards describe multiple MELF components. This package is usually used for diodes and precision resistors. The common types are:

- MELF (MMB) SOD-106 (DO-213AB), $L = 5.8$ mm, $D = 2.2$ mm, 1.0 W, 500 V
- MiniMELF (MMA) SOD-80 (DO-213AA), $L = 3.6$ mm, $D = 1.4$ mm, 0.25 W, 200 V
- MicroMELF (MMU), $L = 2.2$ mm, $D = 1.1$ mm, 0.2 W, 100 V

SMD capacitors come in many varieties. Multi-layer ceramic capacitors (MLCC) are the most common. Small 0.1 μF (100 nF) X7R & Z5U type II dielectric are generally used for bypassing power supplies. NPO/COG type I dielectrics are used for precision and decoupling. Lower voltage 100 nF types can easily fit in the 0603 package. In fact, low-voltage 1.0 μF MLCC devices are available in the 0402 size. These are small enough to fit among the balls of a BGA (ball grid array) package to provide extremely close bypassing. Large ICs in BGA packages can have 1000 or more balls. FPGAs (field-programmable logic arrays) with a great many configurable logic elements (LE) and large amounts of memory are but one example of BGAs with a great number of balls. Large system-on-chip (SoC) ICs with 1–10 billion transistors and die sizes approaching one inch on a side usually require BGA packaging.

SMD film types are physically larger than MLCC, requiring larger leadless type packages that look like radial TH capacitors. Packages range from 0603 to 6560 (1608–16,516). Stacked film construction is popular, and Panasonic offers the ECH-U Polyphenylene Sulfide (PPS) types.

SMD aluminum electrolytic capacitors have special packages or case code standards. Sizes are based on the device diameter. They are regular aluminum cans with the leads formed, folded and trimmed over a plastic carrier. Examples are Case code 1010, 10 mm diameter.

SMD tantalum electrolytic capacitors come in EIA (Electrical Industry of America) standard case sizes. These have leads and are constructed like the SMA package where the lead is tight to the body and wraps around the bottom to form the solder pad. Common types use the following Case codes:

- "R" 2012 (0805)
- "A" 3216 (1206)
- "B" 3528–21 (3.5 mm [L] × 2.8 mm [W] × 2.1 mm [H])
- "C" 6032–28
- "D" 7343–31
- "X" 7343–43

5.3 Discrete Semiconductor and IC Packages

There are also generally accepted standardized packages for active components such as discrete transistors (BJT, FET), diodes and IC packages. Each package type is usually available in different numbers of pins or terminals. There is a vast number of different SMD device packages. The majority of packages are plastic encapsulated. There are, however, hermitically sealed packages using glass and ceramic materials. Packages come in leaded and leadless versions.

The SOD (small outline diode) type packages are used for 2-terminal devices such as small signal diodes. The SOD-123 is a popular type that is the current smallest version, being a SOD-1123. The SMA (JEDEC DO-214AC), SMB (DO-214AA), SMC (DO-214AB) packages are used for higher current diodes, rectifiers and other semiconductors.

In April 1966 Piet van de Water started sketching what would become the 23rd Small Outline Transistor or SOT-23 package for Philips. Following pilot line development in Nijmegen during 1968, the first SOT-23 plastic encapsulated device would roll off the Hamburg production line in 1969. The American JEDEC package equivalent to the SOT-23 is the TO-236. The SOT-23 also scales down in package size and lead pitch. The Japanese manufacturers use JEITA registered packages like the SC-59, which is equivalent to a SOT-23 and TO-236.

ICs such as op amps and many other functions use the dual inline SOIC 8-pin package. SOIC stands for small outline IC. Its footprint, including leads, is just 0.24 × 0.20 in (4.9 × 3.9 mm), occupying an area of just 0.048 in^2. SOIC packages generally use a 50-mil (0.050″/1.27 mm) lead spacing or pitch. They come in narrow and wide widths. By comparison, an 8-pin DIP measures about 0.5″ × 0.9″ corresponding to an area of 0.45 in^2. Compare this to an 8-pin SOIC occupying 0.048 in^2.

Going even smaller, the dual inline-leaded SOT-23 packages are available with 3–8 pins and are very common for diodes, transistors and small ICs like dual op amps. The 6-pin variant is commonly used for dual monolithic BJTs and JFETs. The body of the SOT-23 packages is 1.3 × 2.9 mm. Across the leads, it is 2.4 mm. The standard pin pitch is 0.95 mm (74.4 mil) for all of the SOT-23 packages except the 8-pin part, whose pin pitch is 0.65 mm. Yet smaller packages such as

Figure 5.1 SOT-223, SOT-89, SOT-23, DFN1006D.

the shrink small outline (SSOP) use a 25 mil (0.635 mm) lead pitch. The finest leaded pitch packages are usually 0.5 mm.

One of the benefits of using a SOT-23 package is that the pin function assignments are identical between manufacturers of similar devices. Be aware that the power dissipation of a SOT-23 is less than that of a comparable device in a TO-92. Another benefit of SMT is that common parts such as a 1N4148 diode come in a multitude of SMD package types, to fit all sorts of integration and application. Each manufacturer has their own package types and coding systems.

Packages for medium-power devices include the SOT-89 (SC-62) and the SOT-223 (TO-261), as shown in Figure 5.1. The SOT-223 package can dissipate up to 1 W. These packages have a flange area that is used to conduct heat away from the package die into the copper area to which the power pad is soldered. The power dissipation specifications are for stated copper area and thickness on a test PCB. Yet higher power devices, BJTs, diodes or rectifiers use these packages:

- DO-214 (SMA/SMB/SMC) two terminal packages
- DPAK 3 (TO-252-3)
- D2PAK 3 (TO-263-2) is a variation of the TO-220 package

There is a limit on the power dissipation of these SMD packages, and extra heatsinking is sometimes required. The copper to which the device is soldered can be extended to help dissipate heat. Another common approach is to insert a field of *thermal vias* beneath the part that connect to a copper geometry on the wiring side of the board. This also helps the board material itself to draw some heat away.

Higher pin count devices are usually packaged in quad packages with leads on all four sides or in BGAs, (ball grid arrays) where the connections are made on the under-side of the package. Quad packages come in leaded and leadless versions, with plastic and hermetically sealed ceramic packages. Examples of leaded packages include the PLCC and QFP. Leadless types include the DFN and QFN.

5.4 SMT Assembly

There are a few ways SMDs are assembled onto a substrate or PCB. SMDs can be hand-soldered, but the small size of most of the components takes getting used to. Boards made commercially employ automated assembly.

Hand or Manual Assembly

SMDs can be installed and re-worked using traditional hand-soldering techniques. Tools required consist of tweezers, microscope or magnifier, bright light, fine tip soldering iron and extra-fine solder 10–15 mil (0.010–0.015″, ~0.25–0.4 mm).

This manual method can be used for all leaded and leadless parts except BGA. To start, pre-solder one corner pad with a very small quantity of solder. Position and align the package over the solder pads using the tweezers. Re-heat that one pre-tinned pad to fix the component in place. Place the soldering iron tip on the pad not the lead. Proceed to solder the rest of the leads.

There is also the option to dispense solder paste onto the pads and place the components in the paste. Solder paste is basically solder particles in micro ball sizes suspended in flux to form a mud consistency. Some have used a toothpick to dispense and place the solder paste or run a strip along each side to be soldered. There are special dispensing machines available that accurately dispense the correct solder volume. Solder paste comes in a syringe for small batches. It has a finite shelf life and it is usually refrigerated to extend its life. It is messy and expensive to use in small quantities.

Another method is to have a small stencil made up for the PCB or the device you need to solder. A stencil is derived from the solder pad geometries. This stencil is usually made out of a thin sheet of stainless steel. The thickness of the screen and the pad apertures (openings) control the solder volume. The component pad apertures are used to deposit the solder paste. The stencil is accurately registered and held in place against the substrate surface to align it with the component pads in order to deposit the solder paste. Many prototype PCB houses supply stencils.

Place a sufficient amount of solder paste to fill the stencil apertures. Using a hard plastic device (credit card) or a squeegee, pass it over the stencil, drawing the solder paste into the aperture openings. Carefully remove the stencil. Accurately and gently place the components in the paste. This can be done by a steady hand with a tweezer or by using a suction pickup device that holds the device for placement. The solder paste needs to be heated (re-flowed) in order to create the solder joint. This can be done using a hot-air machine (with an appropriately sized nozzle), a hot plate or a toaster oven.

Automated Assembly

There are a few methods to solder SMD components using automatic machinery. One assembly type uses mixed technology, where both SMD and THT parts are employed. This assembly gets soldered using a traditional wave solder machine. SMD components get glued on the back or solder side before soldering. This process is used mainly for passive devices and larger active device packages such as SOT-23, SOIC-8. A dab of glue (thermal cure epoxy) is deposited at the component centroid location. The SMD component is placed in the glue. The glue is cured, then the top-side components are placed, automatically or manually, and finally the PCB assembly is run through a wave solder machine. In this process, the SMDs must be on the bottom side of the board. The pad geometries for wave soldering are larger than those for the next process, which employs infra-red (IR) heating.

Most medium to large assemblers use an IR oven to reflow solder paste that makes contact with the component's leads or pads. A stencil and a squeegee type device are used to screen the solder paste onto the substrate surface. The stencil is carefully removed, not to disturb the paste. The SMDs are placed on the prepared substrate usually using a robotic pick and place machine with a vacuum pickup head or nozzle. Once the components are installed, the assembly is placed on a feeder track or belt to be fed through a hot IR reflow oven. The solder paste is heated up to a precise temperature using a thermal profile developed for that assembly. This melts or reflows the solder paste to form the solder joints. The reflow action has the ability to correct or re-align the SMD in some cases, due to the surface tension of the molten solder.

RoHS

The EU (European Union) was instrumental in creating the RoHS (Restriction of Hazardous Substances) directive. This directive was created to eliminate listed hazardous elements and compounds in electronic assembly. The directive's purpose was to keep these hazardous materials out of landfill sites, which can leach these contaminates into the groundwater. It foreshadowed the end to using lead (Pb) in solder and electronics in general. Lead is used in solder to allow for easy assembly because of its low melting point and ease of soldering.

Lead-free solder consists of alloys that contain Tin (Sn) and other elements such as Copper (Cu), Bismuth (Bi) and Silver (Ag) that have a higher melting point than lead-based solders. Leaded solder (Sn63/Pb37) has a eutectic melting point of 183°C (361°F), while eutectic lead-free solders (96.5Sn/3.5Ag) have a eutectic melting point of 220°C, 37°C higher. Some lead-free solders are so-called off-eutectic solders (e.g., 95Sn/5Ag), and may have an even higher melting point.

The peak reflow temperature for RoHS assemblies must be higher, at least 240°C. This is closer to the temperature where some components may be damaged. As a result, the acceptable reflow temperature has a narrower range and greater accuracy is required in adhering to the time vs temperature reflow profile. Large BGAs with considerable thermal inertia can cause local temperature deviations from the profile, making things more difficult.

RoHS-compliant lead-free solder is more difficult to work with, especially by hand. Lead-free solder joints sometimes look similar to cold solder joints, making inspection more difficult. Most devices manufactured these days are RoHS-compliant using tin-plated leads. On the other hand, enthusiasts assembling their own circuits need not use lead-free solder and are recommended to continue using leaded solder.

5.5 Printed Circuit Board Fabrication

Printed circuit boards (PCBs) can be made from a number of different materials. Hybrid circuits are usually made out of alumina. PCBs are made in different thicknesses; the most common is 1/16″ or 62.5 mil (1.59 mm). Most PCBs use a dielectric or insulator made from a sheet material consisting of woven fiberglass sheets impregnated with an epoxy resin binder. It is a basic epoxy/glass construction. The most common example is the so-called FR4. The term FR4 is a NEMA grade designation for glass-reinforced epoxy laminate material that is flame resistant (self-extinguishing).

FR4 has a dielectric constant (Dk) that is usually greater than 4.2. This results in increased capacitances and greater loss at high frequencis as compared to more sophisticated materials like Rogers 4350 and Panasonic Megtron-6, with Dk of about 3.5. Like discrete capacitors, the capacitances in a PCB related to Dk are characterized by some amount of dielectric absorption (DA). These more sophisticated PCB materials exhibit less DA. PCBs exhibit an undesirable phenomena called "hook", which is where the effective PCB capacitance is a function of frequency. Hook is strongly linked to DA, and is often said to be directly caused by DA. It can more strongly affect sensitive high-impedance circuits.

In the manufacture of a PCB, copper is etched from or deposited on the dielectric layers to create the connections or wires and plated holes or vias. PCBs can be single, double or multi-layers of copper and dielectric. Older PCBs being single sided did not use plated holes. Double-sided and multi-layer PCBs use plated and non-plated holes. PCB materials are specified by the glass transition temperature (TG) and other criteria such as dielectric constants and losses. TG is the temperature in which the substrate transitions from a hard and relatively brittle "glassy" state into a viscous or rubbery state. Standard TG used is 130–140°C, but higher-temperature materials can be used for hotter applications.

Multi-layer PCBs are constructed by stacking alternating layers of core and Prepreg material. Prepreg is a dielectric material that is sandwiched between two cores or between a core and a copper foil in a PCB to provide the required insulation. Prepreg is roughly the same as core material but is in an uncured state [4]. Multilayer PCBs can have 20 or more layers in some applications, with layer thicknesses as small as 4 mils. Such boards will often have a thickness on the order of 0.093 in.

Modern PCBs use a solder mask (SM) or solder resist on one or both sides or surfaces. An SM is used to separate pads, holes and to stop bridging of solder during the assembly process. It is also used to protect the bare exposed copper from oxidizing. A standard PCB fabrication process is known as SMOBC (solder mask over bare copper). The most common SM is liquid photo imageable (LPI). LPI SM is composed of an ink compound that can be either silkscreened or spray-coated (curtain-coated) onto the PCB. The LPI SM technique is commonly used with the hot-air solder leveling (HASL) and electro-less nickel, immersion gold (ENIG) surface finishes. Its application process requires a clean environment, free of particles and contaminants.

After an LPI SM is applied, having the PCB completely covered on both sides with the SM, the next stage in the process is curing. Unlike other types of SM, LPI inks are sensitive to UV light, and can be cured after a short "tack cure cycle," making use of UV light exposure. This curing process cements the solder mask in place permanently, and the durable material is extremely difficult to remove from the board afterward, lending to long shelf life.

Plating is one of the final steps in the PCB fabrication process. As mentioned above, HASL is used as one of the cheaper finishes. Both lead and lead-free solders can be specified. HASL is not recommended for fine pitch components since it creates an un-even surface for the SMD leads to sit on. This can interfere with a lead making proper contact with the pad or solder paste, thus creating an open circuit after the solder process is complete. Finishes like ENIG and organic coat (OCC) are used on PCBs that will employ fine pitch components.

One of the last steps in fabrication of the PCB is to add graphics images and text to the PCB surface after the SM is applied. It is usually called the silkscreen or legend layer. It can be placed on either side. The purpose is to identify component locations that match the references on the schematic, assembly drawings and bill of materials (BOM). This is used as an aid in assembly and testing. Graphics and other items are added to the silkscreen to identify specific information and show locations.

5.6 DIP Adapters and Prototyping

Sometimes during circuit development it is easier to work with through-hole components. But many ICs are no longer available in the TH package. In this case there are small adapter PCBs that have pads for the SMD footprint of the IC and fan out the pins to through-hole pins with the common 0.1-in spacing. ICs on these breakout boards usually take up no more space than the DIP version and are sometimes even socket-able depending the type of header pins used (Figure 5.2).

Figure 5.2 SMD to DIP and SIP Adapters.

5.7 Rework

This can be as simple as using two small soldering iron tips to heat up two solder joints at the same time or using a hot-air machine to reflow the solder on the device pads. It is best to use some liquid flux to aid in removal. A soldering iron tip temperature of ~600°F (315°C) is usually used for rework. Once the device is removed it is possible to re-use the solder that is left on the pads or remove the solder with solder-wick, re-paste, place, reflow or manually re-solder again. If you know a device is dead, you can also chop the leads one by one to remove the body and then un-solder the leads. Be careful to avoid damaging the PCB.

Another method to remove a SOT-23 or a SOIC device is to use a piece of small bare solid copper (Cu) wire. Form it to the leads so it touches all of them at once, as illustrated in Figure 5.3. Have the Cu wire touch the leads in the vicinity of where the heal meets the toe. Add some extra solder to each side to pre-tin the Cu wire and have enough solder to heat the leads on one side simultaneously. Let the solder blob. Have a raised spot on the wire where you can place the hot soldering iron to transfer the heat to the leads and have them reflow, for removal of the device.

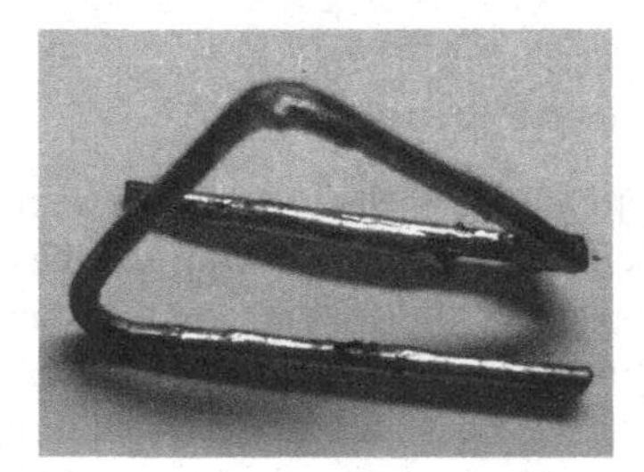

Figure 5.3 SOIC Removal Aid.

References

1. Vincent Himpe, *Mastering Surface Mount Technology*, Elektor International Media, 2012, ISBN 978-1-907920-12-7.
2. *Surface Mount Technology & SMT Devices*, Electronics Notes, https://www.electronics-notes.com/articles/electronic_components/surface-mount-technology-smd-smt/what-is-smt-primer-tutorial.php.
3. *Surface Mount Process*, https://www.surfacemountprocess.com/.
4. Rahul Shashikanth, *What Is a Prepreg in PCB Manufacturing?* Sierra Circuits, 2021, https://www.protoexpress.com/blog/prepreg-the-slice-of-cheese-in-your-pcb/.

Chapter 6

Poles, Zeros, Networks and Bode Plots

In this chapter we'll discuss ways in which passive components like resistors, capacitors and inductors are combined to form attenuators, simple filters and resonant circuits. The concept of poles and zeros will be introduced in simple terms along with complex numbers [1, 2]. Some of these concepts were briefly introduced in Chapter 4 when the concepts of reactance, impedance, radian frequency and imaginary numbers were discussed for passive components. Bode plots will also be explained in the context of frequency response as it is influenced by poles and zeros. Understanding the basics and the role they play in frequency-dependent networks leads to a valuable and intuitive understanding of more complex arrangements. This material will be especially useful in Chapter 11, where passive and active filters will be discussed.

6.1 Resistance and Resistor Networks

Here we'll touch on Ohm's Law and how some combinations of resistances can be formed. This is very simple material to get started. Importantly, we'll see that Ohm's Law is not just applicable to resistance. In fact, with math that includes complex numbers, it can form the basis for treating networks with generalized impedances rather than just resistances. Ultimately, it helps understand the behavior of poles, zeros and the transfer function.

Conductance and Admittance

Conductance is the inverse of resistance and is designated in units of Siemens, S (long ago, it was in units of *mho*, which is *ohm* spelled backward). A conductance of 0.1 S corresponds to 10 Ω. At times, conductance has been designated by Ω^{-1}. The electrical symbol for conductance is G, not to be confused with gain. For this reason we will often use the letter A to denote the magnitude of gain. For a pair of 1-Ω resistors in series, we would write $G = 0.5$ S.

Similarly, admittance is the inverse of impedance, and also has units of Siemens, S. The electrical symbol for admittance is Y. For a series pair of capacitors whose reactance X_C is 1 Ω at a given frequency, we would write $Z = 2\ \Omega$ or $Y = 0.5$ S.

Resistors in Parallel

To determine the resistance of a parallel combination of two resistors, we can just add the corresponding conductance of each resistor and then invert the result. For paralleled devices, it is sometimes easier to deal with conductance or admittance. For R1 and R2 in parallel, we have for conductance:

$$G = \text{G1} + \text{G2} = 1/\text{R1} + 1/\text{R2} = (\text{R1} + \text{R2})/(\text{R1} \times \text{R2})$$
$$R = \text{R1} \times \text{R2}/(\text{R1} + \text{R2})$$

The same manipulations apply to impedances Z and admittances Y. If we desire to find what resistance R_p must be put in parallel with R1 to obtain a target resistance R_T, we have

$$R_P = R_1 \times R_T / (R_1 - R_T)$$

Resistive and Reactive Attenuators

Although resistive attenuators are generally quite simple, understanding them plays an important role in analyzing poles and zeros in a network when reactive elements are included. It is well known that the gain of a resistive attenuator is simply

$$A = \text{R2}/(\text{R1} + \text{R2})$$

where R1 is the series input resistor and R2 is the shunt resistor to ground or common. When R2 is replaced with a capacitor, we have a low-pass filter with a pole and whose transfer function can be described by applying this relationship to the complex quantities of impedance and admittance to be described below. We will see later that the transfer function is referred to as $H(s)$, so we have:

$$H(s) = \text{Z2}/(\text{Z1} + \text{Z2})$$

A capacitance attenuator is a good example of a reactive attenuator. The input attenuator in most oscilloscopes is a resistive-capacitive attenuator where capacitors are placed across each of the resistors. This keeps the attenuation flat with frequency, even with input stage capacitance as a load (it will be taken into account as part of the shunt capacitance in the attenuator).

6.2 Reactance

Reactance is to capacitors and inductors what resistance is to resistors, and also has units of ohms. In other words, inductors and capacitors have reactance, analogous to resistance for resistors. The reactance of a capacitor is:

$$X_C = 1/(2\pi fC) = 1/\omega C$$

where ω is the radian frequency. A 0.1-μF capacitor has reactance of 1592 Ω at 1 kHz. This is also its impedance. The reactance of an inductor is:

$$X_L = 2\pi fL = \omega L$$

A 1-mH inductor has reactance of 6.283 Ω at 1 kHz.

Radian Frequency

One sees the term $2\pi f$ in many equations like those above. This is actually "radian frequency," designated with the character ω, where $\omega = 2\pi f$. Instead of cycles per second or Hz, we have radians per second (rad/s). While one cycle corresponds to 360°, the radian is a different measure of angle in which 2π radians = 360°, or one radian = 57.3°. The reactance of a capacitor then simply becomes $X_L = 1/\omega C$ and the corner frequency of an RC low-pass filter becomes $\omega = 1/RC$. Notice that $R \times C$ is the time constant of the components, measured in seconds.

6.3 Imaginary Numbers and Complex Numbers

In discussing impedance, which is usually a combination of resistive and reactive components, it makes things easier if one understands imaginary numbers and complex numbers. This is not as hard as it sounds. What is an imaginary number and why do we need it?

A sine wave is characterized by its magnitude and phase; two numbers. The gain of an amplifier stage is characterized by its gain and phase shift. Once again, two numbers are required to fully describe it. It is two-dimensional. To represent such quantities in more mathematical ways, we would like to use two numbers that are similar in nature, like coordinates designating an X-Y location. Here is where imaginary numbers come in. The first number, like magnitude or gain, is a "real" number that we are all familiar with. The second number helps us capture phase.

The Square Root of Minus One

Spoiler alert: what is the square root of −1? No real number, which multiplied by itself, can be a negative number. Indeed, any such number is deemed an "imaginary" number. Think of the number 1 as representing 0°. Recognize that the number −1, representing an inversion, corresponds to 180°. How would we represent 90°?

This can be explained with a simple circuit discussion. Consider the circuit of Figure 6.1. It consists of two non-inverting integrators in tandem. A DC voltage applied to the integrator causes the output to rise at a rate of so many volts per second. A sine-wave voltage applied to the integrator creates an output whose magnitude falls inversely with frequency and whose phase lags by 90°, just as the voltage across a capacitor fed with a sine-wave current does. Don't worry about DC offset and the fact that the gain of the arrangement in Figure 6.1 is infinity at DC. Assume that the gain of each integrator is unity at 1 kHz and that it creates a phase lag of 90°. We have already used two numbers to describe the integrator's behavior.

The gain of the two integrators in tandem is unity with a phase angle of −180°; it is simply the product of the two gain magnitudes and the sum of the two phase shifts. The 180° phase shift corresponds to a gain of −1. How can we describe the gain of one integrator stage numerically? We recognize that the gain of two identical stages is the square of the gain of one stage. Here the gain of each integrator is the square root of −1. That is the imaginary number described above, and describing a gain that has 90° of phase shift is one of the reasons we need the imaginary number.

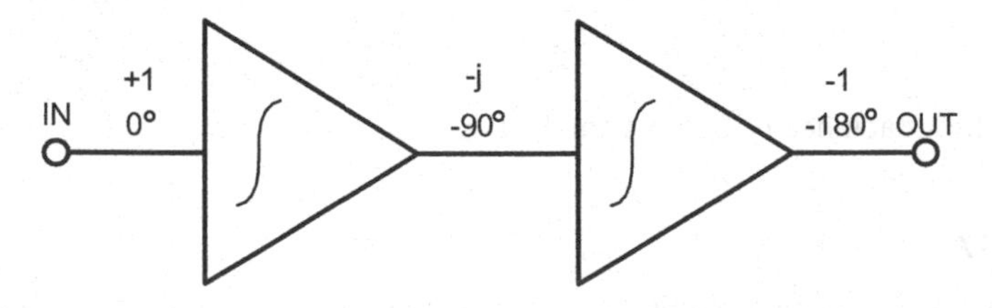

Figure 6.1 Two Integrators in Tandem, Each with Unity Gain at 1 kHz.

Thus, in describing the effect of one of the integrators, we are confronted with a value that is the square root of −1. There is no ordinary (real) number which, when multiplied by itself, results in a negative number. Such a number is what would describe a phase shift of 90°. Therefore, a new kind of number is created to represent the 90° phase shift. It is called an "imaginary" number because it is not a "real" number. It is a number whose square root is a negative number.

In mathematics, this imaginary number is designated with a prefix of *i*. However, in electronics *i* is used to designate a current. Thus, the letter *j* is used instead as a prefix to designate the imaginary number. Here the gain of the integrator is $j1$, and $j1 \times j1 = -1$. Correspondingly, $\sqrt{-1} = j1$ and $\sqrt{-2} = j1.414$ or $j\sqrt{2}$.

If we want to describe the reactance of a capacitor, including the phase relationship of current to voltage, we would describe it as $1/j\omega C$. In electronics, we often want to normalize frequency with shorthand. Thus, we use the term "*s*," as in $s = j\omega$. Actually, $s = j\omega + \sigma$ because it has real and imaginary parts. The term $j\omega$ is the radian frequency in radians per second, while the real term σ is referred to as the Neper frequency in Np per second. We will defer the discussion of σ for now to keep things simple and avoid confusion. Sometimes *s* is referred to as complex frequency, or the *Laplace* variable.

If the reactance of a capacitor is $X_C = 1/(2\pi fC)$, then $X_C = 1/j\omega C = 1/sC$. If an inverting integrator is implemented with a series resistor *R* feeding the virtual ground of an op amp with a feedback capacitor *C*, then the gain of the integrator is $-X_C/R$, so the gain of the integrator can be expressed as $-1/sRC$. If the time constant of *RC* is 1 second, the gain expression becomes $1/s$. The transfer function of a network is referred to as $H(s)$ because the gain is a function of *s*. The transfer function of the inverting integrator is $H(s) = -1/s$.

Complex Numbers

A so-called complex number is a number that includes an imaginary number as one of its components. The number $2 + j2$ is a complex number with a magnitude of 2.828 and a phase angle of 45°. It is a vector. In this case the ratio of the real and imaginary components is unity, so the phase angle is 45°.

Going forward, it is useful to bear in mind that if:

$$j = \sqrt{-1}, \text{ then } j^2 = -1 \text{ and } j = -1/j \text{ or } -j = 1/j$$

6.4 Impedance and the Complex Plane

Impedance is like resistance with a phase angle. If the phase angle is 0°, it is the same as resistance. If the phase angle is 90° or 270°, it is reactance. If the phase angle is somewhere in between, or varies with frequency, impedance is the same as complex impedance. Don't let the term complex make you think this material is complex. Complex impedance is simply impedance that is made up of both resistive and reactive components. The impedance of a loudspeaker is complex impedance. Impedance is just an umbrella term for anything made up of resistors and/or capacitors and inductors or even transmission lines. Impedance has units of ohms.

Impedance can be represented by a complex number, such as $Z = X + jY$. If *Y* is zero, it is the same as resistance. If *X* is zero, it is the same as reactance. If *X* and *Y* are non-zero, the number is a complex number and the impedance is a complex impedance. Impedance that is not complex is just resistance. Note further, the complex number $1 + j$ represents a magnitude of 1.414 with an angle of 45°. This is just a vector representation of the complex number. Impedance can be represented by real and imaginary components *OR* by magnitude and phase. In the general case, impedance has a resistive component and a reactive component.

The Complex Plane

The complex plane is used to represent complex numbers graphically. It is also called the s plane. Consider a plot with Cartesian coordinates. If resistance is plotted on the X axis, called the real axis, and reactance is plotted on the Y axis, which is the imaginary axis, the length of the hypotenuse of the right triangle so formed is the magnitude of the impedance. The angle of the right triangle corresponds to the phase of the impedance. The hypotenuse is just a vector. Figure 6.2 illustrates the impedance vector for a network with impedance of 6 Ω at a phase angle of 60°. The cosine of 60° is 0.5, so the location on the real axis is 3. The sine of 60° is 0.867, so the location on the Y axis is 5.196. The complex number representing the impedance is $(3 + j5.196)$ Ω. The concept of the complex plane hints at simple geometry.

Notice that the expression for impedance here lies in the right half plane, where the real axis is positive. In expressions for poles and zeros, we'll see that most of the action is in the left half plane (LHP).

With $s = j\omega + \sigma$, the term ω controls rotation while the term σ controls magnitude.

$\sigma = 0$ corresponds to a sine wave of constant magnitude; $\sigma > 0$ corresponds to an exponentially growing sine wave; and $\sigma < 0$ corresponds to an exponentially decaying sine wave. This is why we want poles in the LHP, where $\sigma < 0$. For a signal of magnitude X_m, we have:

$$X(t) = X_m \times e^{\sigma t} \cos(\omega t)$$

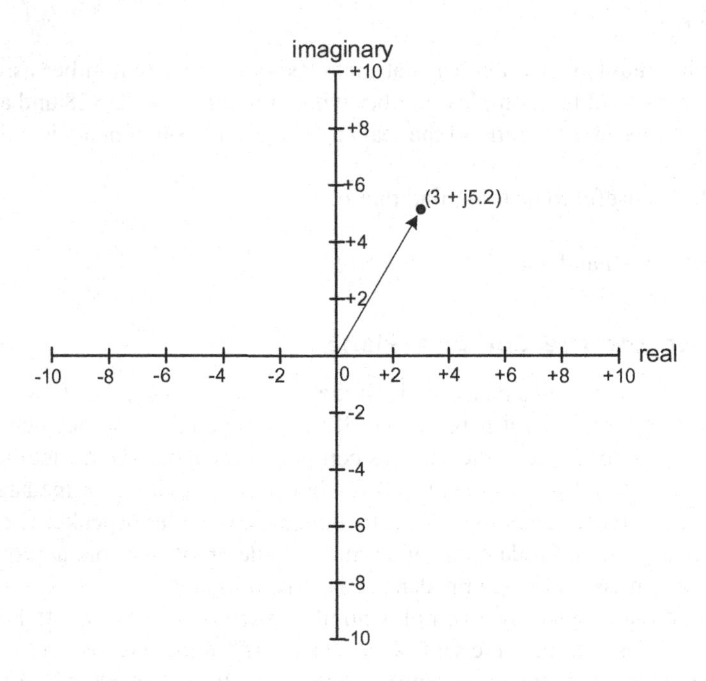

Figure 6.2 Complex Plane Illustrating Impedance of 6 Ω/60°.

When real and imaginary parts are added in the *s* plane, they add like vectors, as illustrated in Figure 6.2. The resulting magnitude, the length of the vector, is simply the square root of the sum of their squares. The term *H*(*s*) represents complex gain that has magnitude and phase. *H*(*s*) is referred to as the transfer function.

$$H(s) = N(s) / D(s)$$

having a numerator *N*(*s*) and a denominator *D*(*s*). The values of *s* where the denominator *D*(*s*) equals zero are the poles of *H*(*s*). The values of s where the numerator *N*(*s*) equal zero are the zeros of *H*(*s*).

So-called real poles lie along the real axis in the LHP. Complex poles come in pairs, with $j\omega$ values mirrored above and below the real axis. *N*(*s*) and *D*(*s*) are often polynomials, like $(s^2 + 3s + 2)$. Note that $(s + 1)$ and $(s + 2)$ are the roots of the factored polynomial $(s^2 + 3s + 2)$. If $D(s) = (s^2 + 3s + 2)$, $(s + 1)$ and $(s + 2)$ designate the locations of the two poles in this second-order system.

A simple second-order system is a parallel-resonant RLC circuit. The center frequency is simply $\omega_0 = 1/\sqrt{(LC)}$. Z_0 is the characteristic impedance of the network, where $Z_0 = \sqrt{(L/C)}$. The *Q* of the circuit is $Q = R/Z_0$.

Conductance and Admittance

Just as conductance is the inverse of resistance, admittance is the inverse of impedance. If two impedances are in parallel, it is sometimes easier to view those impedances as admittances and see their combined effect as the sum of the admittances. This is analogous to how paralleled resistors were dealt with earlier. Remember that conductance is represented by the electrical symbol *G*, not to be confused with gain. Admittance is represented by the electrical symbol *Y*. Conductance and admittance have units of Siemens, *S*. Past representations included mho (ohm spelled backward) and Ω^{-1}. Ohm's Law applies to capacitors, inductors and impedances as long as complex numbers are used to represent the values.

6.5 Poles and Zeros

The concepts of *poles* and *zeros* are fairly simple, and yet central to the understanding of electronic circuits. Ordinary circuits that create high-frequency and low-frequency roll-off contain poles and zeros, respectively (or some combination of them). Poles introduce lagging phase shift (negative), and zeros introduce leading phase shift (positive).

Figure 6.3 illustrates three simple circuits. The first one implements a simple pole, with a 3-dB roll-off at the pole frequency $f_p = 1/2\pi R1C1$, or $\omega_p = 1/R1C1$, here 1 kHz. This is just a simple first-order low-pass filter. Importantly, the −3-dB frequency of an RC network whose time constant $T = RC$, is the radian frequency $\omega = 1/T$. At the 3-dB frequency f_p of a pole, the additional lagging phase shift is 45°. At very high frequencies a pole will contribute 90° of phase shift. The second circuit implements a zero at $f_z = 0$ Hz, causing the gain to rise as frequency increases, resulting in a high-pass filter. The gain can't get any larger than unity, so there is a pole to level off the frequency response at the frequency $f_p = 1/2\pi R1C1$, here 1 kHz (or 6.2832 radian/s).

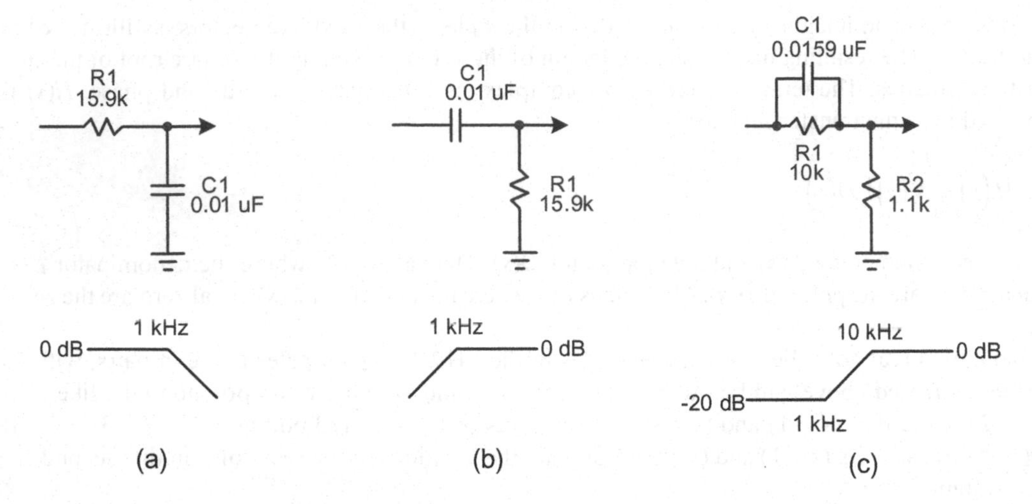

Figure 6.3 RC Circuits Implementing Poles and Zeros.

Recognizing that it is just a complex voltage divider, the low-pass filter in (a) has the transfer function:

$$H(s) = 1/(j\omega CR + 1) = 1/(sCR + 1)$$

If we normalize frequency to 1 rad/s by setting $R = C = 1$, we have

$$H(s) = 1/(s+1)$$

Recall that a pole in the denominator of a transfer function can cause $H(s)$ to be infinite. The frequency of a pole is the frequency at which the denominator goes to zero. That frequency can be a complex number. Here s = −1 when the denominator goes to zero. Negative 1 being a real number in the left half of the complex plane, the pole here is referred to as a real pole in the LHP.

The third circuit implements a *pole-zero pair*, with a resulting frequency response magnitude that begins at a lower value and increases to unity at higher frequencies as C1 begins to act like a short. In this case the zero is not at 0 Hz, but rather at a finite frequency. With C1 = 0.0159 μF and R1 = 10 kΩ, $f_z = 1/2\pi R1C1 = 1$ kHz. Once again, the gain cannot get any larger than unity, so there is a pole at a higher frequency $f_p = 1/2\pi R_p C1$, where R_p is the resistance of the parallel combination of R1 and R2, which is 1 kΩ. This places the pole at 10 kHz. The transfer function is:

$$H(s) = \left(s + 1/(R1C1)\right)/\left(s + 1/(R_pC1)\right)$$

It is easy to see that at very low frequencies where $s = 0$, the gain is f_z/f_p. At very high frequencies, in the limit where the magnitude of s dominates the corner frequencies, the gain is seen to be unity.

6.6 Bode Plots

A *Bode plot* depicts gain or loss in dB as a function of frequency with straight-line approximations that have differing slopes beginning where poles and zeros are located in frequency. It is a log-log plot where the unit dB represents the log value by a linear number. Slopes of the lines are in increments of 6 dB per octave. Each pole adds an additional downward slope of 6 dB/octave. Each zero adds an upward slope of 6 dB/octave. Simple examples of Bode plots are illustrated in Figure 6.3. A *Bode plot* is simply a straight-line approximation to gain magnitude. However, actual magnitude and phase plots are also often referred to as Bode plots.

Figure 6.3 shows *Bode plots* of the three circuits discussed above. The first circuit in (a) forms a low-pass filter. Its frequency response falls at frequencies above the pole frequency f_p at a rate of 6 dB per octave or 20 dB per decade. The phase lag (negative values of phase) increases to −45° at the pole frequency and eventually increases to −90° at high frequencies. The second circuit in (b) forms a high-pass filter. Its frequency response is unity at high frequencies and falls to −3 dB at the frequency of the pole. At this frequency the phase shift is leading at by 45°. As frequency goes lower, its response falls off at 6 dB per octave and the phase lead eventually increases to +90°.

The third circuit in (c) has a rising slope in frequency response beginning at f_z and continuing until f_p. Its gain is equal to R2/(R1 + R2) at low frequencies and rises to unity at high frequencies. This circuit creates a leading phase shift that is at its maximum at the geometric mean of the pole and zero frequencies. At very high and very low frequencies, its phase shift approaches 0°.

Gain and Phase of a Pole vs Frequency

Figure 6.4 shows the actual gain and phase for a pole at normalized frequencies *f/fp* extending far away from the pole. This can be very useful in estimating the impact of multiple additional poles. It illustrates the way that the pole's phase contribution asymptotes to −90° at frequencies far above the pole, while the attenuation from the pole continues to increase. The phase lag for a pole is about 26.6° from its asymptotic value one octave on either side of the pole frequency.

An Example Bode Plot

Figure 6.5 illustrates a Bode plot of the familiar RIAA phono playback equalization curve [3]. The specification includes a pole at 50 Hz, a zero at 500 Hz and a pole at 2122 Hz. It is easily seen that the gain decreases at 20 dB/decade beginning at 50 Hz and ending at 500 Hz. At 2122 Hz, at the frequency of the second pole, the curve resumes its 20 dB/decade decline with frequency.

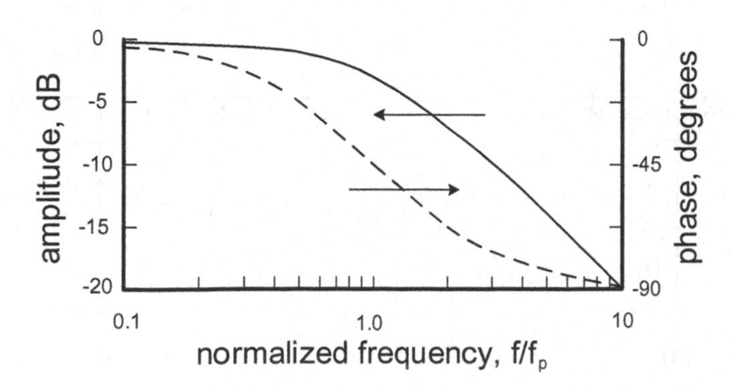

Figure 6.4 Gain and Phase for a Pole at Normalized Frequency.

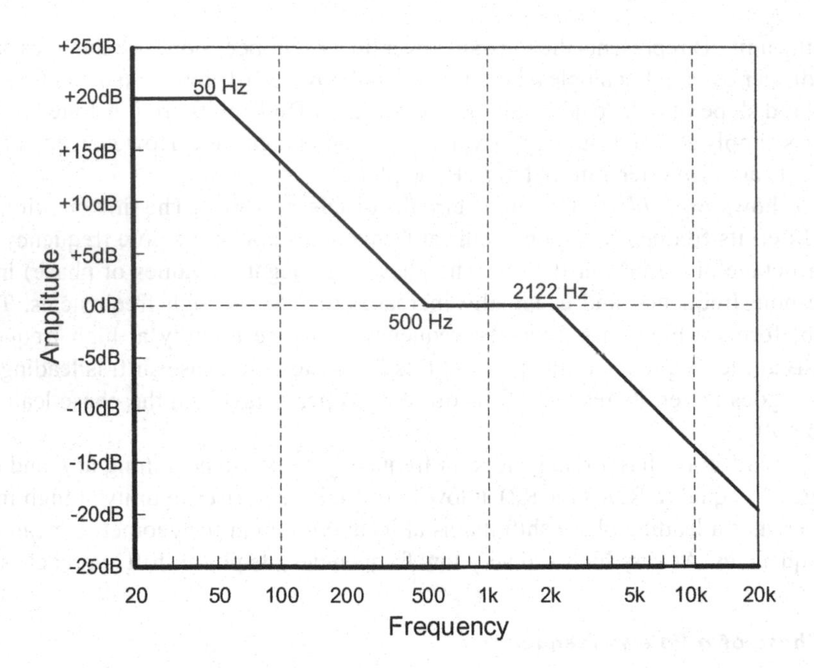

Figure 6.5 Bode Plot for RIAA Equalization.

6.7 The *s* Plane

As discussed earlier, the *s* plane is just a complex plane where poles and zeros are illustrated with Cartesian coordinates. The *x* axis is the real axis and is designated as σ. The *y* axis is imaginary and is designated as $j\omega$. Poles are shown with *x*'s and zeros are shown with *o*'s. Figure 6.6(a) depicts the real pole and zero of Figure 6.3(c) on the real axis in the left half plane. Figure 6.6(b) shows a complex pair of poles corresponding to a resonant circuit. These poles are at $-1 + j1$ and $-1 - j1$.

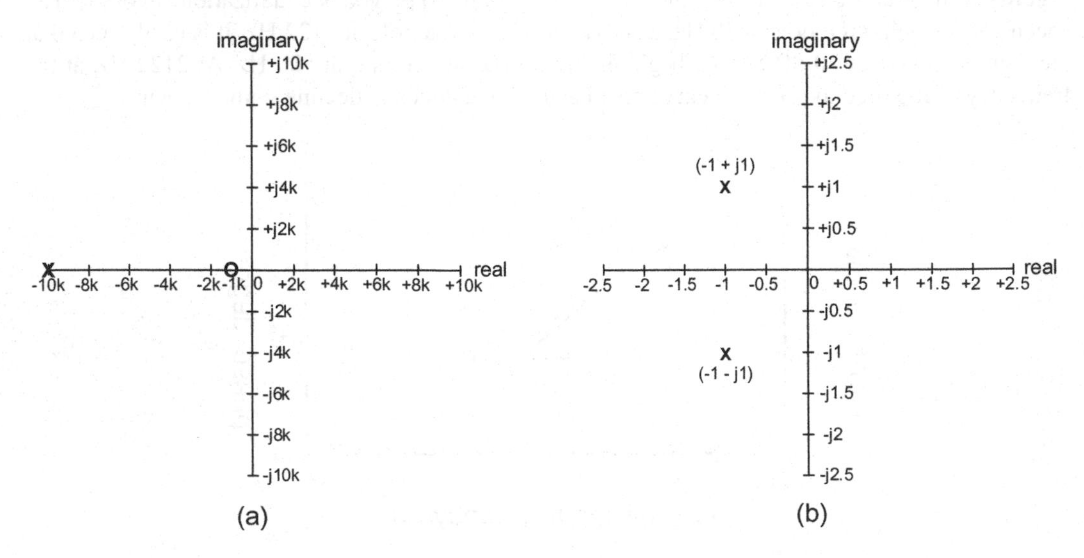

Figure 6.6 Poles and Zeros in the *s* Plane.

The bottom pole is the complex conjugate of the top pole because it differs only in the sign of the complex value. Complex poles must always exist as complex conjugate pairs.

6.8 Transfer Functions

The input–output relationship of a circuit like an amplifier is called its transfer function. This is a more complete description of the circuit's gain. It includes information about phase shift and the dependency of gain on frequency. You don't generally see an expression like G(*f*) to describe the frequency dependence of gain. However, the mathematical description of the gain that includes these effects is called the transfer function, and is referred to as the function *H*(*s*). For the integrator described earlier with an *RC* time constant of 1 second, the transfer function is simply

$$H(s) = 1/s$$

Here the pole is at zero Hz. For the low-pass filter described in Figure 6.3(a) and for R1 = 1 MΩ and C1 = 1 μF the time constant is 1 second and the transfer function is

$$H(s) = 1/(s+1)$$

and the pole is at the radian frequency ω = 1, corresponding to 0.159 Hz.

As described earlier, in the general case, a transfer function *H*(*s*) has a numerator polynomial *N*(*s*) and a denominator polynomial *D*(*s*). The roots of a polynomial in *s* are those values of s for which the polynomial is zero. The values of *s* where the denominator is zero are locations where there are poles. The values of *s* for which the numerator polynomial is zero are locations where there are zeros. In a second-order system, there are two poles in the denominator. In an nth-order system, there are n poles in the denominator. For a system to be stable, all of the poles must be in the left-half of the *s* plane.

Any even-order polynomial can be factored into one or more second-order polynomials of the form ($s^2 + as + b$). These are the roots of the polynomial. Each second-order polynomial contributes two poles. Any odd-order polynomial of $n = 3$ or greater can be factored into one first-order polynomial and one or more second-order polynomials of the form ($s^2 + as + b$). These factorizations into first- and second-order sections play a key role in filter synthesis.

The transfer function below represents a second-order low-pass filter with $\omega_0 = 1$. It is easy to see that at 1 rad/s, the s^2 term equals −1 and cancels the unity term in the denominator, leaving the 1.414*s* term. At ω = 1, the gain is thus 0.707. This is also the *Q* of the filter. The smaller the middle term, the higher the *Q*.

$$H(s) = 1/(s^2 + 1.414s + 1)$$

6.9 Resonant Circuits

Think about a parallel network of a capacitor C and an inductor *L* (a second-order system) and calculate its impedance, knowing that the reactance sign is opposite for the two. It is a parallel combination of reactances X_C and X_L. We use the same formula as for paralleled resistors.

The magnitude of impedance can be represented as |*Z*|. We have:

$$|Z| = X_C X_L / (X_C + X_L) = 1/j\omega C \times j\omega L(1/(j\omega C + j\omega L)$$

Bear in mind that $-j = 1/j$. When $X_C = X_L$, the denominator goes to zero and the impedance of the ideal parallel-resonant circuit goes to infinity, as expected.

Alternatively, we can add admittances to get $j\omega C + 1/j\omega L$. We recognize that $1/j = -j$, so $1/j\omega L = -j/\omega L$. We thus have $j\omega C + 1/j\omega L = j\omega C - j/\omega L$. If C and L are equal we have zero admittance at $\omega = 1$. As expected, the impedance of the parallel-resonant circuit goes to infinity at the resonant frequency, here 1 rad/s. It is very important to recognize that $1/j = -j$. The center frequency is referred to as ω_0.

$$\omega_0 = 1/\sqrt{(LC)}$$

Similarly, think about a series resonant network comprising a capacitor and inductor, and calculate its impedance. Here the reactances of the elements are added together. We find, however, that the reactances are of opposite sign. One decreases with frequency and the other increases with frequency. At some point they will be equal and cancel to zero. This is a series resonance, and its impedance goes to zero at its resonant frequency. Of course, in the real world, at minimum, the inductor and capacitor will have some resistance. In this case, the impedance will fall to approximately the value of the resistance of the inductor if the capacitor is ideal and the inductor is perfect in all respects other than its winding resistance.

Quality Factor, Q

All resonant circuits have resistance, usually in the inductor and sometimes added on purpose. Q, or *Quality Factor*, will be reduced from ideal in a parallel-resonant circuit by series resistance in the inductor (inhibiting circulating current) or by a parallel resistor.

Q will be reduced from ideal in a series resonant circuit by resistance in series, either in the inductor or if it is added in series to the network. Q can also be reduced in a series resonant circuit by putting resistance in parallel with the inductor or the capacitor. In essence, a resistance reduces Q because it dissipates power. ESR in an associated capacitor can also reduce Q.

The definition of Q from a behavioral point of view is the center frequency divided by the difference in frequencies where the impedance has deviated by 3 dB from the impedance at the frequency of resonance. If Q is reduced by a single dominant resistance, the resulting Q will be the ratio of reactance of the tuning capacitor or inductor at resonance to the resistance. The reactance of C or L at resonance is referred to as the characteristic impedance, Z_0.

$$Z_0 = \sqrt{(L/C)} \text{ and } Q = R/Z_0$$

References

1. Hank Zumbahlen, editor, *Linear Circuit Design Handbook*, Analog Devices, Newnes, 2008, ISBN 978-0-7506-8703-4.
2. Marc Thompson, *Intuitive Analog Circuit Design*, 2nd edition, Newnes, 2014, ISBN 978-0-12-405866-8.
3. Stanley P. Lipshitz, On RIAA Equalization Networks, *Journal of the Audio Engineering Society (JAES)*, vol. 27, no. 6, June 1979, pp. 458–481.

Chapter 7

Semiconductors

Here semiconductors are described in some detail. These include bipolar junction transistors (BJTs), Junction Field Effect Transistors (JFETs), Metal Oxide Semiconductor Transistors (MOSFETs), diodes and several other kinds of devices. In this chapter, BJTs and JFETs will first be discussed in detail, followed by more brief discussions of other semiconductors. This will provide a good foundation for the detailed analysis of the basic circuit functions that follow in later chapters.

7.1 Bipolar Junction Transistors

The *Bipolar Junction Transistor* (*BJT*) is the primary building block of most audio circuits. This section is not meant to be an exhaustive review of transistors, but rather presents enough knowledge for you to understand and analyze transistor amplifier circuits. More importantly, transistor behavior is discussed in the context of audio circuit design, with many relevant tips along the way.

Current Gain

If a small current is sourced into the base of an NPN transistor, a much larger current flows in the collector. The ratio of these two currents is the current gain, commonly called *beta* (β) or h_{fe}. Similarly, if one sinks a small current from the base of a PNP transistor, a much larger current flows in its collector.

The current gain for a typical small-signal transistor often lies between 50 and 200. For a power transistor, β typically lies between 20 and 100. Beta can vary quite a bit from transistor to transistor and is also a mild function of the transistor current and collector voltage.

Because transistor beta can vary quite a bit, circuits are usually designed so that their operation does not depend heavily on the particular value of β for its transistors. Rather, the circuit is designed so that it operates well for a minimum value of β and usually better for very high β. Because β can sometimes be very high, it is usually bad practice to design a circuit that would misbehave if β became very high. The *transconductance* (*gm*) of the transistor is actually the more predictable and important design parameter (as long as β is high enough not to matter much). For those unfamiliar with the term, *transconductance* of a transistor is the change in collector current in response to a given change in base-emitter voltage, in units of Siemens, S (amps per volt).

$$gm = \Delta I_c \,/\, \Delta V_{be} \qquad (7.1)$$

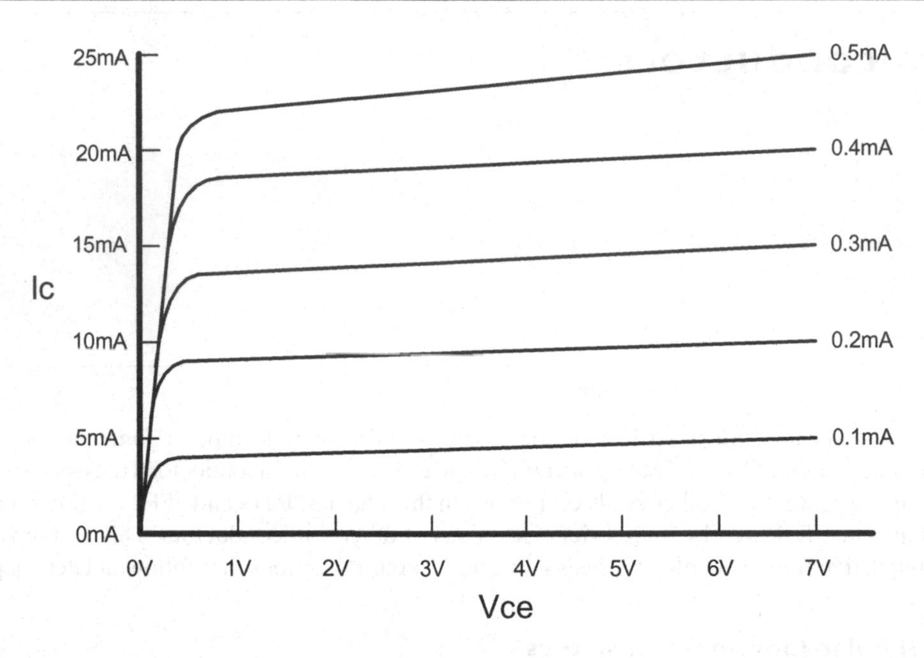

Figure 7.1 Transistor Collector Current Characteristic.

The familiar collector current characteristics shown in Figure 7.1 illustrate the behavior of transistor current gain. This family of curves shows how the collector current increases as collector-emitter voltage (V_{ce}) increases, with base current as a parameter. The upward slope of each curve with increasing V_{ce} reveals the mild dependence of β on collector-emitter voltage. This is called the Early effect. The spacing of the curves for different values of base current reveals the current gain. Notice that this spacing tends to increase as V_{ce} increases, once again revealing the dependence of current gain on V_{ce}. The spacing of the curves may be larger or smaller between different pairs of curves. This illustrates the dependence of current gain on collector current. The transistor shown has β of about 50.

Base-Emitter Voltage

The BJT requires a certain forward-bias voltage at its base-emitter junction to begin to conduct collector current. This turn-on voltage is usually referred to as V_{be}. For silicon transistors, V_{be} is usually between 0.5 and 0.7 V. The actual value of V_{be} depends on the transistor device design and the amount of collector current (I_c). It also depends on the size of the device and the temperature of the device [1–4].

The base-emitter voltage increases by about 60 mV for each decade of increase in collector current. This reflects the logarithmic relationship of V_{be} to collector current. For the popular 2N5551, for example, V_{be} = 600 mV at 100 μA and rises to 720 mV at 10 mA. This corresponds to a 120 mV increase for a two-decade (100:1) increase in collector current.

Tiny amounts of collector current actually begin to flow at quite low values of forward bias (V_{be}). Indeed, the collector current increases exponentially with V_{be}. That is why it looks like there is a fairly well-defined turn-on voltage when collector current is plotted against V_{be} on linear coordinates. It becomes a remarkably straight line when the log of collector current is plotted against V_{be}.

Some circuits, like multipliers, make great use of this logarithmic dependence of V_{be} on collector current.

Put another way, the collector current increases exponentially with base-emitter voltage, and we have the approximation

$$I_c = I_S e^{\left(V_{be}/V_T\right)} \tag{7.2}$$

where the voltage V_T is called the *thermal voltage*. Here V_T is about 26 mV at room temperature and is proportional to absolute temperature. This plays a role in the temperature dependence of V_{be}. However, the major cause of the temperature dependence of V_{be} is the strong increase with temperature of the *saturation current* I_S. This ultimately results in a negative temperature coefficient of V_{be} of about −2.2 mV/°C [1, 4].

Expressing base-emitter voltage as a function of collector current, we have the analogous approximation

$$V_{be} = V_T \ln\left(I_c / I_S\right) \tag{7.3}$$

where ln (I_c/I_S) is the natural logarithm of the ratio I_c/I_S. The value of V_{be} here is the *intrinsic* base-emitter voltage, where any voltage drops across physical base resistance and emitter resistances are not included.

The base-emitter voltage for a given collector current typically decreases by about 2.2 mV for each degree Celsius (°C) increase in temperature. This means that when a transistor is biased with a fixed value of V_{be}, the collector current will increase as temperature increases. As collector current increases, so will the power dissipation and heating of the transistor; this will lead to further temperature increases and sometimes to a vicious cycle called *thermal runaway*. This is essentially positive feedback in a local feedback system.

The V_{be} of power transistors will start out at a smaller voltage at a low collector current of about 100 mA, but may increase substantially to 1 V or more at current in the 1-A to 10-A range. At currents below about 1 A, V_{be} typically follows the logarithmic rule, increasing by about 60 mV per decade of increase in collector current. As an example, V_{be} might increase from 550 mV at 150 mA to 630 mV at 1 A. Even this is more than 60 mV per decade. More discussion about power transistors can be found in Ref. [5].

The Gummel Plot

If the log of collector current is plotted as a function of V_{be}, the resulting diagram is very revealing. As mentioned above, it is ideally a straight line. The diagram becomes even more useful and insightful if base current is plotted on the same axes. This is now called a *Gummel plot* [1, 2]. It sounds fancy, but that is all it is. The magic lies in what it reveals about the transistor. A Gummel plot is shown in Figure 7.2.

In practice, neither the collector current nor the base current plots are straight lines over the full range of V_{be}, and the bending illustrates various non-idealities in the transistor behavior. The vertical distance between the lines corresponds to the β of the transistor, and the change in distance between the lines shows how β changes as a function of V_{be} and, by extension, I_c. The curves in Figure 7.2 illustrate the typical loss in transistor current gain at both low and high current extremes.

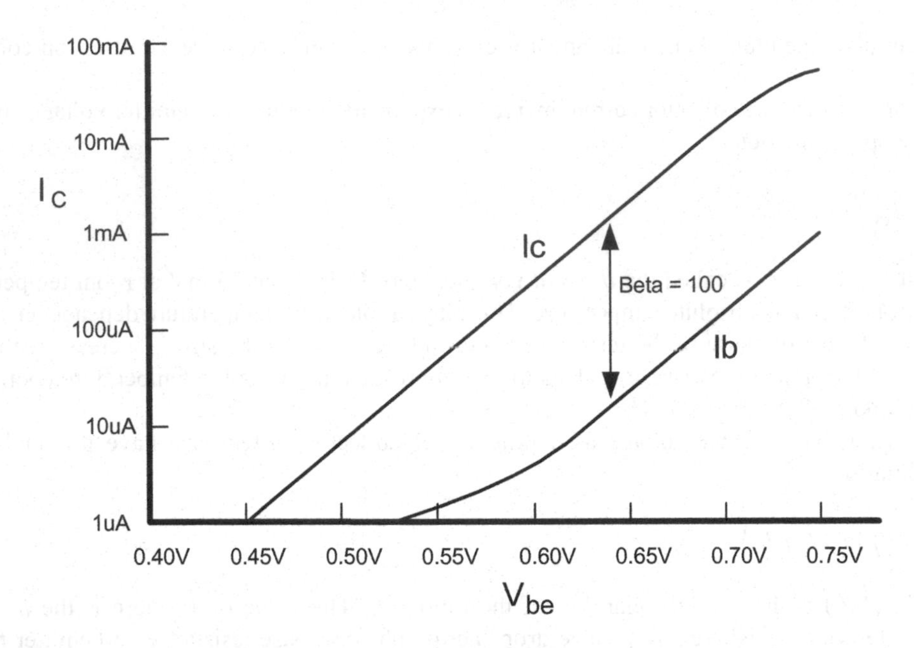

Figure 7.2 Transistor Gummel Plot.

Transconductance

While transistor current gain is an important parameter and largely the source of its amplifying ability, the transconductance of the transistor is perhaps the most important characteristic used by engineers when doing actual design. Transconductance, denoted as *gm*, is the ratio of the change in collector current to the change in base voltage. This represents the voltage-controlled view of transistor behavior, as opposed to the current-controlled view.

The unit of measure for transconductance is the Siemens (S), which corresponds to a current change of 1 A for a voltage change of 1 V. This is the inverse of the measure of resistance, the ohm (it was once called the *mho, ohm* spelled backward). If the base-emitter voltage of a transistor is increased by 1 mV, and as a result the collector current increases by 40 μA, the transconductance of the transistor is 40 milli-Siemens (mS).

The transconductance of a bipolar transistor is governed by its collector current. This is a direct result of the exponential relationship of collector current to base-emitter voltage. The slope of that curve increases as I_c increases; this means that transconductance also increases. Approximate transconductance is given simply as

$$gm = I_c / V_T \tag{7.4}$$

where V_T is the thermal voltage, typically 26 mV at room temperature. At a current of 1 mA, transconductance is 1 mA/26 mV = 0.038 S.

The inverse of *gm* is a resistance. Sometimes it is easier to visualize the behavior of a circuit by treating the transconductance of the transistor as if it were a built-in dynamic emitter resistance *re′*. This resistance is just the inverse of *gm*, so we have

$$re' = V_T / I_c = 0.026 / I_c \left(\text{at room temperature}\right) \tag{7.5}$$

In the above case $re' = 26\ \Omega$ at a collector current of 1 mA.

An important approximation that will be used frequently is that $re' = 26\ \Omega/I_c$ where I_c is expressed in mA. If a transistor is biased at 10 mA, re' will be about 2.6 Ω. The transistor will act as if a small change in its base-emitter voltage is directly impressed across 2.6 Ω; this causes a corresponding change in its emitter current and very nearly the same change in its collector current. This forms the basis of the common-emitter (CE) amplifier.

It is important to recognize that $gm = 1/re'$ is the *intrinsic* transconductance, ignoring the effects of ohmic base and emitter resistance. Actual transconductance will be reduced by emitter resistance R_E and R_B/β being added to re' to arrive at net transconductance. This is especially important in the case of power transistors.

Input Resistance

If a small change is made in the base-emitter voltage, how much change in base current will occur? This defines the effective input resistance of the transistor. The transconductance dictates that if the base-emitter voltage is changed by 1 mV, the collector current will change by about 40 μA if the transistor is biased at 1 mA. If the transistor has a beta of 100, the base current will change by 0.38 μA. Note that the β here is the effective current gain of the transistor for small changes, which is more appropriately referred to as the *AC current gain* or *AC beta* (β_{AC}). It is also known as the small-signal gain, h_{fe}. The effective input resistance in this case is therefore about 1 mV/0.38 μA = 2.6 kΩ. The effective input resistance is just β_{AC} times re'.

Early Effect

The Early effect manifests itself as finite output resistance at the collector of a transistor and is the result of the current gain of the transistor being a function of the collector-base voltage. The collector characteristic curves of Figure 7.1 show that the collector current at a given base current increases with increased collector voltage. This means that the current gain of the transistor is increasing with collector voltage. This also means that there is an equivalent output resistance in the collector circuit of the transistor.

The increase of collector current with increase in collector voltage is called the *Early effect*. If the straight portions of the collector current curves in Figure 7.1 are extrapolated to the left, back to the X axis, they will intersect the X axis at a negative voltage. The value of this voltage is called the *Early voltage, VA*. The slope of these curves represents the output resistance r_o of the device. In practice, these extrapolated lines do not always intersect the x axis at the same point.

Typical values of VA for small-signal transistors lie between 20 and 200 V. A very common value of VA is 100 V, as for the 2N5551. The output resistance due to the Early effect decreases with increases in collector current. A typical value of this resistance for a small-signal transistor operating at 1 mA is on the order of 100 kΩ.

The Early effect is especially important because it acts as a resistance in parallel with the collector and emitter nodes of a transistor. This effectively makes the net load resistance on the collector smaller than the external load resistance in the circuit. As a result, the gain of a common-emitter stage decreases. Because the extra load resistance is a function of collector voltage and current, it is a function of the signal and is therefore nonlinear and so causes distortion.

The Early effect can be modeled as a resistor r_o connected from the collector to the emitter of an otherwise "perfect" transistor [1]. The value of r_o is

$$r_o = \left(VA + V_{ce}\right) / I_c \tag{7.6}$$

It should be pointed out that this equation is a very simplified description of the Early effect and that even many standard versions of SPICE don't support the advanced models needed for accurate modeling of the effect.

For the 2N5551, with a *VA* of 100 and operating at V_{ce} = 10 V and I_c = 10 mA, r_o comes out to be 11 kΩ. The value of r_o is doubled as the collector voltage swings from very small voltages to a voltage equal to the Early voltage.

It is important to understand that this resistance is not, by itself, necessarily the output resistance of a transistor stage, since it is not connected from collector to ground. It is connected from collector to emitter. Any resistance or impedance in the emitter circuit will significantly increase the effective output resistance caused by r_o.

The Early effect is especially important in the voltage amplifier stage (VAS) of an amplifier circuit. In that location the device is subjected to large collector voltage swings and the impedance at the collector node is quite high due to the usual current source loading and good buffering of the load from this node in some circuit topologies.

A 2N5551 VAS transistor biased at 10 mA and having no emitter degeneration will have an output resistance on the order of 14 kΩ at a collector-emitter voltage of 15 V. This would correspond to a signal output voltage of 0 V in an arrangement with ±15 V power supplies. The same transistor with 10:1 emitter degeneration will have an output resistance of about 135 kΩ. The 10:1 emitter degeneration factor means that the presence of the emitter resistor decreases the gain of an ideal common emitter (CE) stage with its emitter connected to AC ground by a factor of 10. A transistor biased at 1 mA and with *re′* = 26 Ω will have 10:1 emitter degeneration if the emitter resistor is nine times *re′*, or 234 Ω.

At a collector-emitter voltage of only 5 V (corresponding to a −10 V output swing) that transistor will have a reduced output resistance of 105 kΩ. At a collector-emitter voltage of 25 V (corresponding to a +10 V output swing), that transistor will have an output resistance of about 165 kΩ. These changes in output resistance as a result of signal voltage imply a change in gain and thus second harmonic distortion.

Because the Early effect manifests itself as a change in the β of the transistor as a function of collector voltage, and because a higher-β transistor will require less base current, it can be argued that a given amount of Early effect has less influence in some circuits if the β of the transistor is high. A transistor whose β varies from 50 to 100 due to the Early effect and collector voltage swing will have more effect on circuit performance in many cases than a transistor whose β varies from 100 to 200 over the same collector voltage swing. The variation in base current will be less in the latter than in the former. For this reason, the product of β and *VA* is an important *figure of merit (FOM)* for transistors. In the case of the 2N5551, with a current gain of 100 and an Early voltage *VA* of 100 V, this FOM is 10,000 V. The FOM for bipolar transistors often lies in the range of 5000–50,000 V.

$$\text{Early effect FOM} = \beta \times VA \tag{7.7}$$

The value of $\beta \times VA$ tends to be constant for a given transistor type; high-β parts have lower *VA*, and vice versa.

Junction Capacitance

All BJTs have base-emitter capacitance (C_{be}) and collector-base capacitance (C_{cb}). This limits the high-frequency response, but also can introduce distortion because these junction capacitances are a function of voltage.

The base, emitter and collector regions of a transistor can be thought of as plates of a capacitor separated by non-conducting regions. The base is separated from the emitter by the base-emitter junction, and it is separated from the collector by the base-collector junction. Each of these junctions has capacitance, whether it is forward biased or reverse biased. Indeed, these junctions store charge, and that is a characteristic of capacitance.

A reverse-biased junction has a so-called *depletion region*. The depletion region can be thought of roughly as the spacing of the plates of the capacitor. With greater reverse bias of the junction, the depletion region becomes larger. The spacing of the capacitor plates is then larger, and the capacitance decreases. The junction capacitance is thus a function of the voltage across the junction, decreasing as the reverse bias increases.

This behavior is mainly of interest for the collector-base capacitance C_{cb}, since in normal operation the collector-base junction is reverse biased while the base-emitter junction is forward biased. It will be shown that the effective capacitance of the forward-biased base-emitter junction is quite high.

The variance of semiconductor junction capacitance with reverse voltage is taken to good use in *varactor diodes*, where circuits are electronically tuned by varying the reverse bias on the varactor diode. Varactors are made with diffusion and doping profiles that emphasize the dependence of junction capacitance on reverse bias voltage. These variable-capacitance diodes were originally called *Varicaps* [6].

In audio circuits, the effect is an undesired one, since capacitance varying with signal voltage represents nonlinearity. It is obviously undesirable for the bandwidth or high-frequency gain of an amplifier stage to be varying as a function of the signal voltage. The collector-base capacitance of the popular 2N5551 small-signal NPN transistor ranges from a typical value of 5 pF at 0 V reverse bias (V_{cb}) down to 1 pF at 100 V. For what it's worth, its base-emitter capacitance ranges from 17 pF at 0.1 V reverse bias to 10 pF at 5 V reverse bias. Remember, however, that this junction is usually forward biased in normal operation.

Speed and f_T

The AC current gain of a transistor falls off at higher frequencies in part due to the need for the input current to charge and discharge the relatively large capacitance of the forward-biased base-emitter junction.

The most important speed characteristic for a BJT is its f_T, or *transition frequency* [1, 2]. This is the frequency where the AC current gain β_{AC} falls to approximately unity. For small-signal transistors used in audio amplifiers, f_T will usually be on the order of 50–300 MHz. A transistor with a low-frequency β_{AC} of 100 and an f_T of 100 MHz will have its β_{AC} begin to fall off (be down 3 dB) at about 1 MHz. This frequency is referred to as f_β.

The effective value of the base-emitter capacitance of a conducting BJT can be shown to be approximately

$$C_{be} = gm / \omega_T \tag{7.8}$$

where ω_T is the radian frequency equal to $2\pi f_T$ and gm is the transconductance.

Because $gm = I_c/V_T$, one can also state that

$$C_{be} = I_c / (V_T \times \omega_T) \tag{7.9}$$

This capacitance is often referred to as C_π for its use in the so-called hybrid pi model. Because transconductance increases with collector current, so does C_{be}. For a transistor with a 100 MHz f_T and operating at 1 mA, the effective base-emitter capacitance will be about 61 pF.

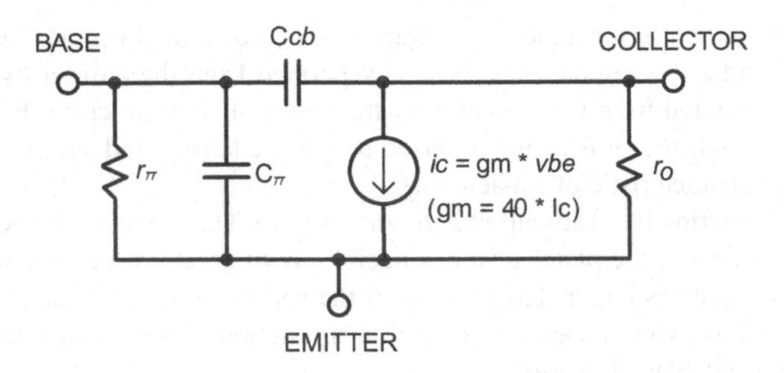

Figure 7.3 Hybrid Pi Model.

The Hybrid Pi Model

Those more familiar with transistors will recognize that much of what has been discussed above is the makeup of the hybrid pi small-signal model of the transistor, shown in Figure 7.3. The fundamental active element of the transistor is a voltage-controlled current source, namely a *transconductance*. Everything else in the model is essentially a passive *parasitic* component. AC current gain is taken into account by the base-emitter resistance r_π. The Early effect is taken into account by r_o. Collector-base capacitance is shown as C_{cb}. Current gain roll-off with frequency (as defined by f_T) is modeled by C_π. The values of these elements are as described above. This is a small-signal model; element values will change with the operating point of the transistor.

The Ideal Transistor

Operational amplifier circuits are often designed by assuming an ideal op amp, at least initially. In the same way a transistor circuit can be designed by assuming an "ideal" transistor. This is like starting with the hybrid pi model stripped of all of its passive parasitic elements. The ideal transistor is just a lump of transconductance. As needed, relevant impairments, such as finite β, can be added to the ideal transistor. This usually depends on what aspect of performance is important at the time.

The ideal transistor has infinite current gain, infinite input impedance and infinite output resistance. It acts as if it applies all of the small-signal base voltage to the emitter through an internal intrinsic emitter resistance *re′*.

V_{be} vs Temperature

The nominal temperature coefficient of V_{be} is −2.2 mV/°C, but this value decreases at higher current density. Delta V_{be} is the difference between V_{be} at nominal current density and V_{be} at a lower or higher current density. When V_{be} is referred to a lower current density, delta V_{be} is a positive voltage that is proportional to absolute temperature (PTAT). The temperature coefficient of V_{be} is sometimes referred to as complementary to absolute temperature (CTAT).

Because V_{be} and ΔV_{be} go in opposite directions with temperature, if a multiplied value of ΔV_{be} is added to V_{be}, the temperature coefficients of the two will cancel, and a temperature-independent voltage will result. This is the basis for the bandgap voltage reference, where the sum of V_{be} and $K \times \Delta V_{be}$ equals the bandgap voltage of silicon [4]. The bandgap of silicon is 1.22 ev at 0°K. In a practical circuit, the voltage created is on the order of 1.25 V. Often, in a bandgap current source, two transistors are run at the same collector current, but at different current densities by creating one transistor with a larger effective emitter area, perhaps four to ten times as much, causing this transistor to operate at a lower current density.

A PTAT current source can also be made using this concept. If a PTAT current source is used as the tail current source of a differential pair (LTP), transconductance will be constant with temperature. The voltage difference between two transistors operated at different current densities, when passed through a resistor, will produce a current that is PTAT.

When used as a V_{be} multiplier, a larger V_{be} multiplier transistor operated at a lower current density will have greater sensitivity to temperature, since ΔV_{be} is inversely proportional to current density. Whereas a transistor operated at nominal current density might have TC V_{be} of −2.2 mV/°C, the transistor operated at low current density may have a TC V_{be} of −2.4 mV/°C. This can make up for some thermal attenuation between the junction of the transistor being monitored and the V_{be} multiplier transistor monitoring it. The use of transistors of different effective emitter area is usually limited to integrated circuits where implementing such differences is much easier.

Safe Operating Area

The *safe operating area* (*SOA*) for a transistor describes the safe combinations of voltage and current for the device. This area will be bounded on the *X* axis by the maximum operating voltage and on the *Y* axis by the maximum operating current. The SOA is also bounded by a line that defines the maximum power dissipation of the device. Such a plot is shown for a power transistor in Figure 7.4, where voltage and current are plotted on log scales and the power dissipation limiting line becomes the outermost straight line.

Unfortunately, power transistors are not just limited in their safe current-handling capability by their power dissipation. At higher voltages they are more seriously limited by a phenomenon called *secondary breakdown*. This is illustrated by the more steeply sloped inner line in Figure 7.4. Safe operating area is a much bigger issue for power amplifier output stages than for preamplifier circuits. A deeper discussion of SOA is presented in Ref. [5].

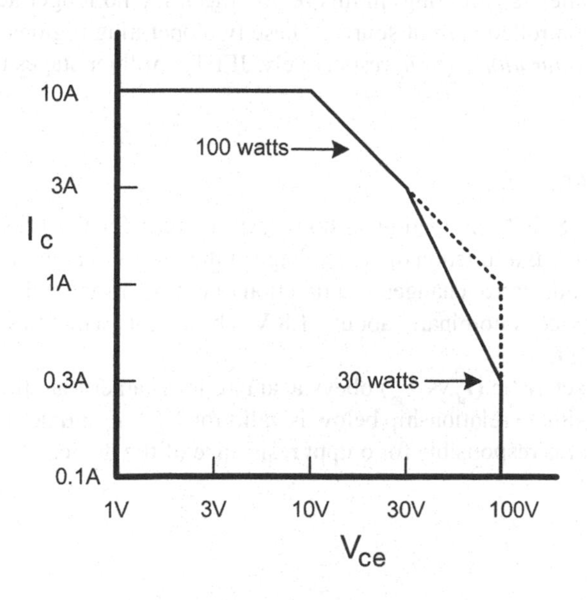

Figure 7.4 Safe Operating Area of a Power Transistor.

Dual Monolithic BJTs

Sometimes a matched pair of transistors with excellent log conformity is needed for translinear and similar circuits. The legendary LM394 Supermatch Pair is no longer available, but the LS312 from Linear Integrated Systems is a close match, with V_{be} matching of 0.2 mV typical and 0.5 mV maximum, minimum current gain of 200 at 10 μA and typical voltage noise of 1.8 nV/√Hz at 100 μA. A similar dual-matched monolithic PNP transistor is also available as the LS352.

7.2 Junction Field Effect Transistors

Junction field effect transistors (JFETs) operate on a different principle than BJTs. Picture a bar of n-type doped silicon connected from source to drain [7–12]. This bar will act like a resistor. Now add a p-n junction in the middle of the length of this bar by adding a region with p-type doping. This becomes the gate. As the gate junction is reverse biased, a *depletion region* will be formed, and this will begin to pinch off the region of conductivity in the n-type bar as the depth of the depletion region grows with increasing reverse bias. This reduces current flow. This is called a *depletion-mode* device. The JFET is nominally *on*, and its degree of conductance will decrease as reverse bias on its gate is increased until the channel is completely pinched off. Ref. [8] provides an excellent overview of the construction, characteristics and applications for a large variety of FET devices, not just JFETs.

The reverse gate voltage where pinch-off occurs is referred to as V_p or as the *threshold voltage* V_t (not to be confused with the thermal voltage V_T). The threshold voltage is often on the order of −0.5 to −4 V for most small-signal N-channel JFETs. Note that control of a JFET is opposite to the way a BJT is controlled. The BJT is normally off and the JFET is normally on. The BJT is turned on by application of a forward bias to the base-emitter junction, while the JFET is turned off by application of a reverse bias to its gate-source junction.

The reverse voltage that exists between the drain and the gate can also act to pinch off the channel. At V_{dg} greater than the threshold voltage, the channel will be pinched in such a way that the drain current becomes self-limiting. In this region the JFET no longer acts like a resistor, but rather like a voltage-controlled current source. These two operating regions are referred to as the *linear region* and the *saturation region*, respectively. JFET amplifier stages usually operate in the saturation region.

JFET I_d vs V_{gs} Behavior

Figure 7.5(a) shows a SPICE simulation of how drain current for the LSK489 dual monolithic N-channel JFET changes as a function of gate voltage in the saturation region [9, 10]. Figure 7.5(b) illustrates how transconductance changes as a function of drain current in the same region. Threshold voltage for this device is nominally about −1.8 V. These plots will be essentially the same for the LSK189 single JFET.

The JFET I-V characteristic (I_d vs V_{gs}) obeys a square law, rather than the exponential law applicable to BJTs. The simple relationship below is valid for $V_{ds} > V_t$ and does not take into account the influence of V_{ds} that is responsible for output resistance of the device.

$$I_d = \beta\left(V_{gs} - V_t\right)^2 \tag{7.10}$$

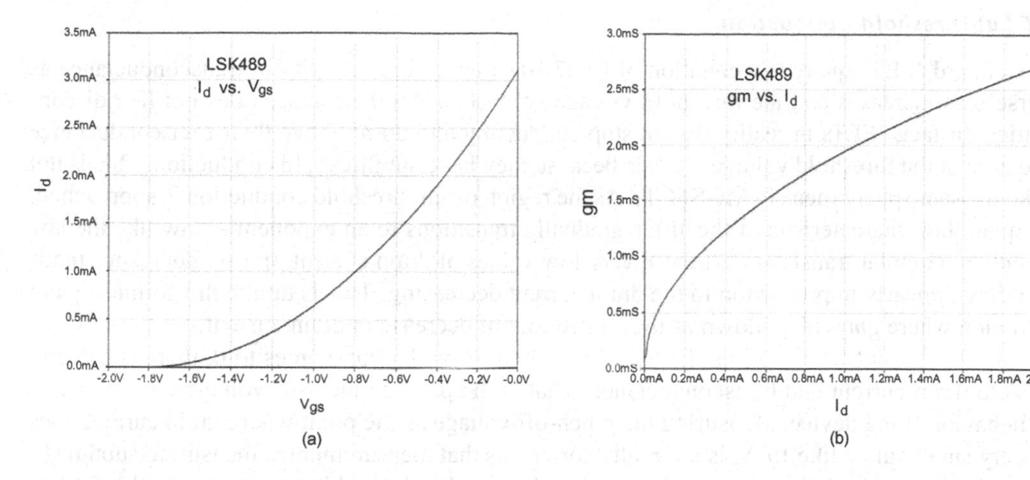

Figure 7.5 JFET Drain Current (a) and Transconductance (b).

The equation is valid only for positive values of $V_{gs} - V_t$. The factor β (not to be confused with BJT current gain) governs the transconductance of the device. When $V_{gs} = V_t$, the $V_{gs} - V_t$ term is zero and no current flows. When $V_{gs} = 0$ V, the term is equal to V_t^2 and maximum current flows.

The maximum current that flows when $V_{gs} = 0$ V and $V_{ds} >> V_t$ is referred to as I_{DSS}, a key JFET parameter usually specified on data sheets. Under these conditions the channel is at the edge of pinch off and the current is largely self-limiting. In this case it is the reverse bias of the gate junction with respect to the drain that is pinching off the channel. The parameter β is the transconductance coefficient and is related to I_{DSS} and V_t. The value of I_{DSS} for the LSK489 is about 3.1 mA, and the value of β is about 0.9 mA/V².

$$\beta = I_{DSS} / V_t^2 \tag{7.11}$$

The parameter β can also be expressed in units of mS/V; this means that if *gm* is plotted as a function of V_{gs}, a straight line will result. With some manipulation of Eq. 7.10, it can be seen that the transconductance of the JFET is proportional to the square root of the drain current.

$$gm = 2\sqrt{\beta I_d} \tag{7.12}$$

This is different from the behavior of a BJT, where *gm* is proportional to collector current. Transconductance for a JFET increases as the square root of drain current. At a given operating current, *gm* of a JFET is smaller than that of a BJT by a factor of 10 or more in many cases. The transconductance for the LSK489 at $I_d = 1$ mA is about 2 mS. The *gm* of a BJT at $I_c = 1$ mA is about 40 mS, greater by a factor of 20. The larger LSK389 has $V_t = -0.54$ V, $I_{DSS} = 8.4$ mA, and *gm* of 11.3 mS at 1 mA. The LSK389 and LSK489 are both low-noise dual monolithic JFETs. The discussion above refers to one of the two matched JFETs in a package [9–12].

The JFET turn-on characteristic is much less abrupt than that of a BJT. Absent of degeneration in a BJT, the input voltage range over which the JFET is reasonably linear is much greater than that of a BJT. The collector current of the BJT increases by a factor of about 2 for every increase of 18 mV in V_{be}. Between 0.75 and 1.5 mA, V_{gs} of an LSK489 changes by about 370 mV.

JFET Subthreshold Conduction

The standard JFET square law equation of Eq. 7.10 shows a discontinuity in transconductance as reverse bias increases and the threshold voltage is reached. Mother nature does not like discontinuities. In fact, JFETs in reality do not stop conducting and do not have their transconductance go to zero at the threshold voltage. This is because they have subthreshold conduction. This is not unlike the same phenomena in MOSFETs. As the region of subthreshold conduction is approached, the square law characteristic of the JFET gradually transitions to an exponential law like the law that governs bipolar transistors. At these very low values of drain current, transconductance tends to go down linearly in proportion to the drain current decreasing. This is unlike the normal square law region where *gm* will go down as the square root of decrease in drain current.

Thus, at the point where extrapolation of the normal I_d vs V_{gs} curve goes to 0, there is actually non-zero drain current and transconductance. That is the proper pinch-off voltage for the square law behavior of the device. Measuring the pinch-off voltage as the point where drain current goes to a very small value, like 1 μA, is not really correct, as that measurement is measuring subthreshold conduction. Indeed, from a behavioral point of view, the channel is never completely pinched off. The presence of the subthreshold region, and the transition into it as drain current decreases, can cause SPICE modeling inaccuracies because the conventional SPICE model equations assume a fully square law device – there is conduction in the real JFET where the SPICE law would say it should be zero. The simulated plots in Figure 7.5 do not show the subthreshold conduction effect because it is not in the SPICE model used for the simulation.

At the other end of the drain current range, where drain current and *gm* are high, there is another source of error. This one is taken into account in the JFET model by the source resistance parameter. It causes a reduction in *gm* as compared to what the square law by itself would predict. Indeed, it causes a gradual transition from the square law region into a linear region, where transconductance will asymptote to a constant value and where drain current will continue to increase. For example, consider a JFET with effective source resistance of 5 Ω at an operating point where drain current and transconductance are high, with transconductance predicted by the square law as being 20 mS, corresponding to rs' = 50 Ω. Actual transconductance at that point will be low by about 10% as compared to what the square law would yield in the absence of source resistance. This can also cause errors in creating SPICE models for the JFET.

Indeed, in some cases the subthreshold and linear regions can encroach on the square law region to the point where there is not a large range of drain current over which the square law behaves with high accuracy. This will especially be the case for JFETs with significant source resistance.

I_{DSS} and Threshold Voltage Variations

One reality of designing with JFETs is that I_{DSS} and threshold voltage can vary over a fairly wide range, even for JFETs of the same type from the same manufacturer. Indeed, there is considerable variation among JFETs from the same wafer. Some JFETs are available in graded I_{DSS} bands [9]. JFETs of the same type with higher I_{DSS} tend to also have higher threshold voltage. Designers must take these variations in I_{DSS} and V_t into account in their designs. In some rare cases, they may have to select and bin the devices themselves.

One useful bit of information is that JFETs of the same type operating at the same current tend to have similar transconductance even if their I_{DSS} and/or V_t are different. This can be helpful if multiple dual monolithic JFET differential pairs are paralleled to achieve ultra low noise [12, 13]. In such a case, each differential pair is provided with its own tail current source so each pair can settle to its own V_{gs} for that value of tail current. This feature is used to advantage in a moving coil phono preamp described in Chapter 16.

Impedance Conversion and Charge Amplifiers

JFETs are especially suited to some applications because of their extremely high input impedance and lack of input bias current. As impedance-converting buffers or amplifiers they are ideal for use with high-impedance sources. Very good examples include condenser and electret microphones, where the source impedance is that of a capacitor ranging from 1 to 100 pF. For the same reasons, JFETs are ideally suited to implementing charge amplifiers and similar functions in instrumentation and sensor applications.

Voltage-Controlled Resistors

JFETs are also used as voltage-controlled resistors (VCRs) in some analog circuits [14–17]. In these applications, there is no drain-source bias voltage applied. Instead, AC signal current of relatively small amplitude flows between the source and the drain. With no reverse bias on the gate, the resistance of the JFET is simply R_{DS_ON}. As reverse bias is applied to the gate, the channel becomes pinched and the resistance increases, eventually going to infinity when the pinch-off voltage is reached.

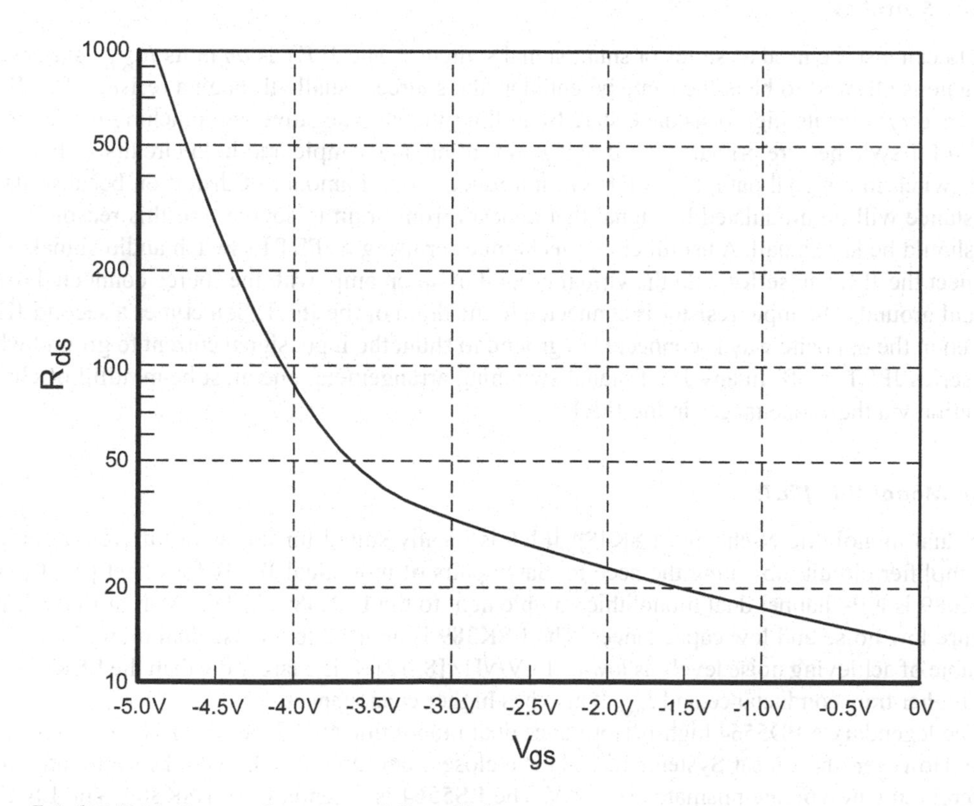

Figure 7.6 2N4391 R_{ds} as a Function of Gate Voltage.

Figure 7.6 shows the small-signal resistance of the 2N4391 JFET as a function of gate voltage. This transistor is often used as the AGC control element in audio oscillators. A control circuit rectifies and filters the output of the oscillator. The resulting DC measure of amplitude is then applied to the gate of the JFET to control its resistance so as to stabilize the amplitude at the desired value. JFET VCRs are also used in some audio compression circuits, where gain is controlled as a function of signal amplitude.

As mentioned earlier, the drain-gate voltage can play a role in pinching off the channel of the JFET. This means that when the JFET is used as a voltage-controlled shunt resistor, the signal swing at the drain, albeit small, can modulate R_{ds}. This modulation of the gain by the signal causes distortion. As the signal at the drain swings positive, the channel becomes a bit more pinched off, increasing R_{ds}. As the signal swing goes negative, the channel becomes less pinched off, decreasing R_{ds}. The behavior is as if half of the drain-source voltage is acting like a negative bias added to the gate control voltage. This causes predominantly second-order distortion. To counter this, many such circuits feed back half the drain-source signal voltage to the gate.

JFETs most suitable for use as VCRs are usually those with high V_t and low R_{DS_ON}. These are less prone to modulation of their resistance by drain voltage. There are some JFETs that are specifically targeted for use as VCRs. These include the LS26VNS N-channel JFET and the LS-26VPS P-channel devices from Linear Systems, Inc. [16]. The dual-matched monolithic N-channel VCR11N VCR is also available [16].

Signal Switches

JFETs can also be used as series or shunt signal switches. The JFET is *on* in its R_{DS_ON} state when the gate is allowed to be at the same potential as the source, usually through a resistor. The JFET is turned *off* into its high-resistance state by pulling the gate negative and pinching off its channel. JFET switches are primarily used in signal routing and sample-and-hold circuits. When used as a switch in a signal path, the JFET will introduce a small amount of distortion because its *on* resistance will be modulated by signal that appears from drain to source. For this reason, signal V_{ds} should be kept small. A useful circuit technique for using a JFET to switch audio signals is to connect the JFET in series with the virtual ground of an op amp, with the source connected to the virtual ground. The input resistor is connected to the drain of the JFET. Sometimes a second JFET driven in the opposite way is connected to ground to shunt the input signal current to ground when the series JFET is off. In any JFET signal switching arrangement, one must be mindful of charge injection via the capacitances in the JFET.

Dual Monolithic JFETs

The dual monolithic N-channel LSK489 JFET is ideally suited for the input differential pair of an amplifier circuit, obviating the need to match pairs of individual JFETs for offset [9, 10]. The LSK689 is a P-channel dual monolithic complement to the LSK489 [9, 11]. Both of these JFETs feature low noise and low capacitances. The LSK389 is an ultra low noise dual monolithic JFET capable of achieving noise levels as low as 1 nV/√Hz [8, 12]. It is a larger die than the LSK489 and has higher transconductance and I_{DSS}. It also has higher capacitances.

The legendary NPD5564 high-performance dual monolithic N-channel JFET is no longer available. However, the Linear Systems LS5564 is a close monolithic match, with the same maximum differential gate voltage mismatch of 5 mV. The LS5564 is essentially an LSK389. The LSK389 has higher maximum V_{gs} offset voltage of 15 mV, but significantly lower noise, especially at low frequencies, where its maximum noise is 4 nV/√Hz at 10 Hz.

Note, there are some dual N-channel JFETs labeled as 2N5564 that are not monolithic but instead are a two-chip design in a single package. The Vishay 2N5564 is an example of a two-chip design.

Constant Current Diodes

Current source diodes are actually JFETs with their gate connected to their source. They conduct a constant current equal to their I_{DSS} as long as there is more than a few volts across the "diodes." These devices are available in numerous current values from about 0.24 mA to about 4.7 mA with tolerance on the order of ±20% [18].

Low Leakage Diodes

Some applications require diodes with ultra-low leakage current, a common example being protection diodes for sensitive circuits. These are often made from JFETs wherein the source and drain are internally connected, forming a 2-terminal device. An example is the Linear Systems DPAD1, a dual-diode with specified leakage current of 1 pA at 25°C [19]. Compare this to the leakage current of the common 1N4148 diode, with reverse current of 25 nA.

7.3 Small-Signal MOSFETs

Metal Oxide Semiconductor Transistors (MOSFETs) are most commonly known for their use in highly complex CMOS-integrated circuits. However, there are important uses for discrete versions of these devices, often as power-switching devices in switch-mode power supplies (SMPS) and some audio power amplifiers. A discussion of lateral and vertical power MOSFETs can be found in Ref. [5].

Here small-signal discrete MOSFETs will be discussed. The key distinguishing feature of MOSFETs is that the gate is insulated from the silicon by a thin silicon dioxide layer, making for extremely high input impedance and very low leakage. The gate controls conduction by creating an electric field that influences the conductivity of the channel. This is in contrast to the JFET where the gate influences conduction in the channel via a junction whose depletion region controls the size of the channel. The term "metal" in the name MOSFET is a misnomer that dates back to the early days of these devices when the gate was made of metal. Since these early days, the gates have been made of conductive polysilicon. They are thus referred to as silicon-gate devices.

MOSFETs are usually enhancement-mode devices, where the gate is forward-biased with respect to the source to turn the device on. The channel under the gate, through which the current flows, is doped in the opposite polarity as the rest of the conductive silicon connecting the source and drain to either end of the channel. When a gate voltage is applied to turn the device on, the resulting electric field inverts the polarity of the channel below and current flows through it, since the path from source to drain is now of one conducting polarity. Small-signal MOSFETs are used in both signal switches and amplifiers with extremely high input impedance.

Lateral MOSFETs

A typical and widely used discrete small-signal lateral MOSFET is the P-channel enhancement-mode 3N163 device [20, 21]. It is a 4-terminal device where the substrate on which the MOSFET is created is brought out as a fourth pin. This pin is often labeled B for body. The substrate is usually connected to the source, but there are applications where it can be used differently.

Sometimes the substrate (body) is reverse-biased with respect to the source. Such a reverse bias can influence the threshold voltage, with a larger negative substrate bias causing the gate threshold voltage to increase.

In N-channel devices, the substrate is p-type and the source and drain are n-type. Body diodes are thus formed between the substrate and the source and drain regions. The substrate must not be allowed to become positive with respect to the source or drain, or these parasitic diodes will conduct. If AC signals of any significant amplitude are being passed through the device, the substrate should be reverse-biased with respect to the source by enough voltage to keep the substrate diodes from conducting.

The 3N163 has a threshold voltage on the order of −2 to −5 V and typical drain current of 10 mA with V_{gs} = −10 V. At *Vgs* of −20 V, drain current can typically be about 40 mA with $V_{ds} = V_{gs}$. The 3N163 has a modest transconductance of about1 mS at 1 mA, rising to 2.7 mS at 10 mA.

These MOSFETs are distinguished by extremely high input impedance and extremely low gate leakage current, typically less than 1 pA (10 pA max.). They also have very low drain-gate capacitance, typically only 0.5 pF. These characteristics make them especially useful for ultra-high input impedance amplifiers, such as for smoke detectors, electrometers and radiation dosimeters [21]. They are also used as fast analog and digital switches, with $r_{ds\text{-}on}$ as low as 300 Ω at V_{gs} = −10 V. Switching applications include audio and video switching. At V_{gs} = −10 V, their typical drain current is 10 mA.

Like JFETs, MOSFETs such as the 3N163 are also useful as voltage-controlled resistors, as shown in Figure 7.7. Its $r_{ds\text{-}on}$ typically has its most useful range of about 2 kΩ at V_{gs} = −3 V to less

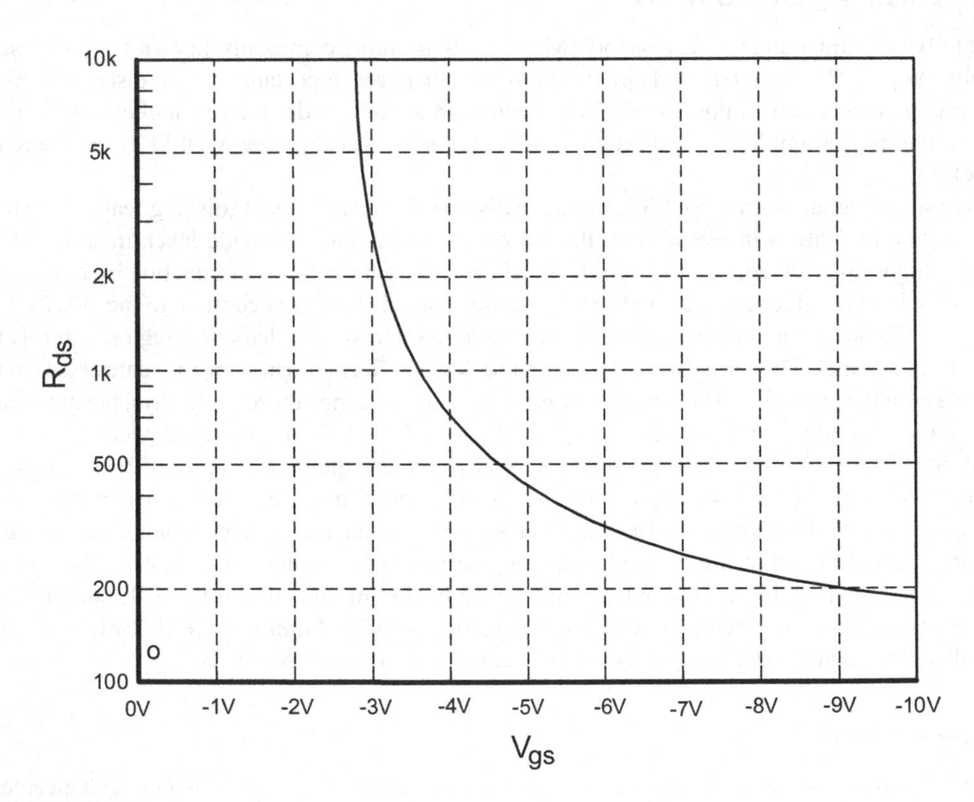

Figure 7.7 **3N163 R_{ds} as a Function of Gate Voltage.**

than 250 Ω at V_{gs} = −10 V. The fact that the gate-source voltage can swing in either direction can simplify control circuits in some applications.

A dual version of the device is available as the 3N165 dual monolithic P-channel MOSFET. A single N-channel MOSFET with similar characteristics is available as the 3N170.

Double-Diffused MOSFETs (DMOS)

The small-signal DMOS transistor is a double-diffused MOSFET, hence the D in DMOS. The device has a lateral current flow, as opposed to vertical DMOS transistors that are often used as power MOSFETs. The channel length of the N-channel lateral DMOS transistor is determined by the depletion region between the two N^+ diffusions that implement the source and drain. The lightly doped p-region between these two n-type regions can be made quite small (short). The resulting small channel length makes it possible to conduct higher current and achieve higher transconductance. The short channel also reduces the area under the gate, reducing C_{gs} and C_{gd} capacitances. An example device is the SD-SST210 enhancement-mode N-channel DMOS transistor [22, 23]. A quad version is available as the SD5000.

DMOS transistors make good switches and ultra-high input impedance amplifiers. An advantage over JFET switches is that there is no junction at the gate, meaning that the gate can have either polarity of control voltage applied to it, allowing significant simplicity of control circuit design and very good isolation between the gate and the signal.

These devices have very low C_{gs} and C_{gd} capacitances, on the order of 1.5 and 0.2 pF, respectively. As a result, there is very low charge injection when the gate voltage is changed to turn the device on or off. This is especially desirable if the device is used in a sample-and-hold or track-and-hold circuit. Because C_{gd} is smaller, the drain should be the one connected to the storage capacitor. The device is also well-suited for use in a DAC de-glitch circuit.

The DMOS transistor has a substrate that is not connected to the source, drain or gate, making it a 4-terminal device. The p-type body in an N-channel device is often connected externally to the source. However, parasitic substrate diodes are formed from the body to the n-type source and drain diffusions, requiring that the body always be negative with respect to the source and drain. Thus, a reverse body bias is needed for negative signal headroom in some applications. A reverse body bias also increases the threshold voltage of the device.

Compared to the 3N163 discussed earlier, the SD210 has a much lower threshold voltage of about +0.8 V and a much lower $r_{ds\text{-}on}$ of less than 45 Ω with V_{gs} at +10 V. With V_{gs} of only +5 V, $r_{ds\text{-}on}$ is still typically only 60 Ω. The SD210 also has much higher transconductance, typically 11 mS at drain current of 20 mA. Maximum gate leakage current is a bit higher, at ±100 pA. Lateral DMOS enhancement-mode FET switches like the SD210 are especially well-suited for fast analog switches and fast sample-and-hold circuits.

MOSFET I_d vs V_{gs}

The threshold voltage V_t is positive for enhancement-mode N-channel devices and negative for P-channel enhancement devices. The parameter *KP* in the following equation governs device transconductance. The drain current obeys a square law relationship, where

$$I_d = \tfrac{1}{2} KP\left(V_{gs} - V_t\right)^2 \tag{7.13}$$

Notice the similarity to the JFET model. The threshold voltage V_t is a positive number, reflecting the enhancement nature of the device. In comparison to the JFET model, the parameter *KP* serves as the same transconductance parameter as β.

Subthreshold Conduction

As with JFETs, MOSFETs exhibit subthreshold conduction. This means that drain current and transconductance do not go to zero at the gate threshold voltage. Instead, the MOSFET transitions from a square law device to an exponential-law device at very low current.

7.4 Diodes and Rectifiers

The forward voltage of a typical small-signal silicon diode like the 1N4148 is on the order of 612 mV at a forward current of 1 mA. The current flowing in a diode is an exponential of the applied voltage, much like that of BJT collector current as a function of base-emitter voltage.

As a result of this relationship, the dynamic impedance of a diode in the forward-biased state decreases as current is increased. The impedance of a 1N4848 diode at a forward current of 1 mA is about 26 Ω. This is not unlike the way in which the transconductance of a BJT increases with increased collector current.

Table 7.1 shows the measured forward voltage drop of several diodes at forward currents of 100 μA and 1 mA. The diodes include germanium, silicon and Schotty diodes as well as a diode-connected transistor. A silicon rectifier diode is included as well.

Notice that the diode-connected 2N3904 has a forward voltage that is increased by 60 mV over one decade of increased current, as expected. The silicon diodes increase by about 108 mV over a decade in current, so the 60-mV/decade rule does not necessarily apply to diodes. Similarly, the 1N4004 rectifier has an increase of 102 mV over a decade. Its overall voltage drop is less than for the smaller diodes, as expected, since its relative current density at 1 mA is lower than that of a 1N4148. The 1N4149 has a slightly higher forward drop than the 1N4148. It has lower capacitance than the 1N4148 and its difference in design for lower capacitance probably explains this.

The germanium diode has the lowest forward drop, and it changes by a relatively large 151 mV for a decade increase in forward current. The Schottky diode has its desirably low forward drop of 250 mV at 100 μA, increasing by a significant 130 mV with a decade increase in forward current.

Table 7.1 Forward Voltage Drop of Several Diodes

Device	*Forward Current*		
	100 μA	*1 mA*	
Germanium	160	311	mV
Schottky BAT41	250	380	mV
Silicon 1N4148	503	612	mV
Silicon 1N4149	513	621	mV
2N3904 diode-connected	603	663	mV
Silicon 1N4004 rectifier	479	581	mV

Junction Capacitance

The maximum rated capacitance of a 1N4148 is 4 pF at reverse voltage (V_r) of 0 V, but typical curves indicate that it can actually be 1 pF at V_r = 0 V, falling to 0.8 pF at V_r = 10 V. The maximum rated capacitance of a 1N4149 is 2 pF at V_r = 0 V, but typical curves indicate 0.86 pF at V_r = 0 V, falling to 0.82 pF at V_r = 10 V.

The p-n junction of a reverse-biased silicon diode acts like the two plates of a capacitor. As the reverse voltage is increased, the depletion region between the junctions increases in size, effectively causing the plates to be further apart, thus reducing the junction capacitance.

The voltage-dependent junction capacitance property of silicon diodes is used to advantage in making varactor diodes. These are voltage-dependent capacitors that are often used in tuning applications. These diodes are designed to have higher capacitance than ordinary small signal diodes, often with a capacitance vs voltage curve that is more suitable to tuning applications. The NXP BB170 varactor diode capacitance is 39 pF at V_r = −1 V, falling to 2.6 pF at V_r = −28 V.

Schottky diodes are a majority carrier device and are significantly faster than silicon diodes, partly because they do not have minority carrier stored charge. However, their junction capacitance is not necessarily smaller than that of a silicon diode.

Rectifier Speed

Rectifiers that are used in linear and switch-mode (SMPS) power supplies are usually required to pass significant current and turn off quickly. During the forward-conduction pulse of rectification, the peak current through the rectifier diode can be very high. As with any p-n junction, there exists a large stored charge in the form of minority carriers during this time. When the driving AC waveform decreases below the DC level stored in the reservoir capacitor on the other side of the diode, the diode must turn off, but it cannot do that instantaneously as a result of the stored charge. This charge must be sucked out of the diode by means of reverse current flow through the diode. The amount of time it takes to pull out this charge is a measure of the diode's speed and is referred to as its *reverse recovery time* (t_{rr}). The pulse of reverse current flow is especially undesirable because it can lead to RF emissions. Rectifier speed and recovery are especially important in switch-mode power supplies.

Soft Recovery and Fast Recovery

Diodes that are designed to have a fast recovery will create less noise because the amount of energy that must be sucked out of them during turn-off is smaller, leading to much shorter duration of the reverse current interval and consequent turn-off spikes. One type of fast recovery diode is the *Fast Recovery Epitaxial Diode* (*FRED*). While a conventional rectifier may have a rated reverse recovery time of 350 ns, a FRED may have a t_{rr} of 35 ns. The trade-off is that FREDs cost more and sometimes have a larger forward voltage drop at a given forward peak current.

At the end of the reverse current interval, the diode finally turns off and its current returns to zero (where it should be in the off state). Once the excess carriers are swept out of the junction, some diodes will have their reverse current go to zero very quickly, almost snapping back. This high rate of change of reverse current can create more noise at high frequencies. Diodes that are designed to have a *soft recovery* have a more gradual return to zero of their reverse current. The smaller rate of change of falling reverse current causes less production of high-frequency noise. Some of the better fast recovery diodes are also characterized by soft recovery. The Vishay HEXFRED® devices are a good example.

Schottky Diode Rectifiers

Schottky diodes are a unipolar majority carrier devices that can operate at very high speeds and with relatively low forward voltage drop. They are sometimes called Schottky-barrier or hot-carrier diodes. One of the largest uses of discrete Schottky diodes is in switch-mode power supplies (SMPS), where speed and low voltage drop play a big role in achieving high efficiency and high switching speeds. Schottky diodes are also well known for their use in TTL (74LS) logic circuits, where their low voltage drop, being less than that of a p-n junction, keeps the logic transistor out of saturation when the Schottky diode is connected from base to collector of the transistor.

A Schottky diode is formed at the junction of a semiconductor and a metal. Unlike semiconductor p-n junctions, which rely on both majority carriers and minority carries, the Schottky junction relies only upon majority carriers, avoiding the charge storage effects of minority carriers and recombination of n-type and p-type carriers that limit the switching speed of other semiconductors. The metal acts as the anode while n-doped silicon usually acts as the cathode. When conducting, electrons travel from the n-type silicon to the metal electrode. Numerous different metals have been used, but the most common contact material is silicide, which is a metal/silicon compound that has good conductivity.

As an example, the surface-mount SS34 Schottky rectifier is widely used in small SMPS in conjunction with SMPS driver chips like the XL6007. It can handle reverse voltage of 40 V and average forward current of 3 A. Its rated maximum forward drop is 0.5 V at 3 A, with typical drop of 0.4 V at 3 A at 25°C. The 1N5822 through-hole Schottky rectifier is also popular in these applications. It can handle reverse voltage of 40 V and average forward current of 3 A. Its maximum rated forward drop is 0.53 V at 3 A, but the typical drop is 0.45 V at 3 A at 25°C.

7.5 Zener Diodes

The Zener diode operates in breakdown mode in the reverse bias direction, and Zeners with various specified breakdown voltages are available. They make a good, simple voltage reference. When a current is passed through them, a reasonably constant voltage appears across them.

Zener Diode Mechanism

There are two different breakdown mechanisms for Zener diodes, and which one is predominant depends on the rated breakdown voltage. For voltages below about 6 V, the Zener effect breakdown mechanism is at work. At higher breakdown voltages, the mechanism at work is actually avalanche breakdown, the same breakdown mechanism at work in a diode subjected to excessive reverse voltage. Interestingly, the base-emitter reverse breakdown mechanism for a BJT is the Zener mechanism. Typical BJT base-emitter breakdown voltages are on the order of 4 V. BJTs can suffer permanent beta degradation if they are subjected to Zener breakdown.

Zener Diode Impedance

Typical small Zener diodes include the 1N52xx half-watt series available in a ±5% tolerance [24, 25]. Zener diodes operating in the avalanche breakdown region have a much sharper V/I curve than those operating in the Zener region. If you plot Zener voltage vs Zener current, the current in the avalanche-mode device will rise much more sharply than that of the Zener-mode device. This means that regulation vs current for low-voltage Zener diodes is not as good. It also means that Zener impedance is higher at a given operating current than for avalanche-mode devices.

One Zener diode in the low-voltage range is the 5.1-V 1N5231B. At its test current of 20 mA (100 mW), its dynamic impedance is 17 Ω. Its impedance rises with reduced current. At only 0.25 mA, impedance is 1600 Ω. As current is increased, the Zener diode's effective impedance decreases, making it a better regulator. Its actual operating voltage is specified at a defined test current, and will be smaller at lower operating current. In contrast, the 15-V 1N5245B at its test current of 8.5 mA (128 mW) has dynamic impedance of 16 Ω, rising to 600 Ω at 0.25 mA. At a lower voltage, the 2.7-V 1N5223B at its test current of 20 mA has dynamic impedance of 30 Ω, rising to 1300 Ω at 0.25 mA.

Zener Diode Temperature Coefficient

Zener diodes with lower breakdown voltages operating in the Zener region have a small negative temperature coefficient (TC) of breakdown voltage [25, 26]. Zener diodes with higher breakdown voltages operating in the avalanche breakdown region have a small positive TC.

The 1N5231B has a rated breakdown of 5.1 V. Its typical temperature coefficient is ±0.03%/°C at 7.5 mA. It lies near the Zener/avalanche TC boundary. The 15-V 1N5245B at its test current of 8.5 mA has a TC of +0.082%/°C. The 7.5-V 1N5236B has an intermediate TC of +0.058%/°C. Conversely, the 1N5223B, with a rated voltage of only 2.7 V, has a TC of −0.080%/°C.

The positive TC of Zeners rated at greater than 5.1 V can be somewhat compensated by connecting them in series with a silicon diode, whose negative TC is about −2.0 mV/°C. For example, a 6.2-V Zener with a +0.045%/°C TC will exhibit a voltage TC of about 2.8 mV/°C, resulting in a net TC of about +0.8 mV/°C. The TC of Zener diodes changes fairly quickly with rated voltage at these low voltages, making temperature compensation more difficult. However, if two silicon diodes are placed in series with the Zener, the desired Zener TC will be +4.0 mV/°C, which is close to the +4.3%/°C TC of the 7.5-V Zener. The negative TC of LEDs can also be used to compensate the positive TC of Zener diodes with ratings above 5.1 V.

Zener Diode Noise

Zener diodes can be noisy, and in fact they are sometimes used as noise sources [27]. Their noise depends on the type of breakdown mechanism in play (i.e., the breakdown voltage). Zener diodes rated at less than about 6.2 V operate in the Zener breakdown region, where the electric field is large because the junction is thin, allowing quantum tunneling. The noise mechanism is shot noise. The Zener noise voltage is the shot noise current times the load resistance. The load resistance is usually dominated by the dynamic impedance of the Zener diode.

Higher-voltage Zener diodes (e.g., 12 V) operate in the avalanche breakdown mode, producing impact ionization noise that dominates over the shot noise. The higher-voltage Zener diodes are quieter as a result of the different breakdown mechanism. Zener diodes should always be thoroughly bypassed if they are used in a noise-critical location.

It is important to know that the effective source impedance of the Zener diode is its dynamic resistance, which can be quite low. The size of the bypass capacitor must be sufficiently large to take account of this reality. For a Zener diode having dynamic impedance of 100 Ω and bypassed with 100 μF, its noise is low-pass filtered beginning at 16 Hz. Alternatively, in low-noise applications, some resistance can be placed between the Zener diode and the filter capacitance, forming a low-pass filter. Of course, this may degrade load regulation.

7.6 LEDs

Light-emitting diodes (LEDs) are widely used as colored indicators, color signage and sources of illumination, but they also have some other uses as voltage references [28, 29]. Just as silicon diodes have a forward voltage drop, so do LEDs, but their forward voltage drop is larger and can be different for different color LEDs.

Use as Voltage References

LEDs are popular as low-voltage voltage references in non-critical applications such as current sources in some circuits. Indeed, low-voltage Zener diodes sacrifice performance and are often not even available below about 3.5 V. Similarly, silicon diodes have too low a voltage drop, on the order of 0.6 V, and have a significant negative voltage coefficient. For reference use between 1.4 and 3.7 V, LEDs can be a good choice. It is best to operate an LED at about 1 mA for use as a reference. An LED can act as photosensor, so it can cause hum if exposed to room light.

Colors and Forward Voltage

The forward voltage drop of the LED is different for different colors because different semiconductor materials are used to create different colors. This is important for use as indicators because they are often powered by a resistor from a fixed-voltage supply. The is particularly the case for Blue and White LEDs used in cases where the voltage supply is low, since their *on* voltage is significantly higher than that of the Red, Yellow and Green LEDs. LED brightness also depends on the amount of current applied and the efficiency of the LED itself. The two most popular LEDs for use as references are the Green and Blue LEDs. As an example, the former measured at 1 mA yielded a forward voltage of 1.8 V. A Blue LED operating at 1 mA measured 2.9 V. Keep in mind, LEDs of the same color may be made with different semiconductor materials, and this can affect the forward voltage drop.

Impedance as a Function of Current and Color

As with ordinary diodes, the impedance of the LED when it is forward biased decreases as current is increased. At 1 mA, the typical impedances of Green and Blue LEDs measured 56 and 190 Ω, respectively.

Temperature Coefficient of Forward Voltage

The temperature coefficient of the forward voltage drop of an LED depends somewhat on the color, but it is typically in the neighborhood of −2.0 mV/°C. The TC of a Green LED with forward voltage of 1.8 V at 0.5 mA is about −1.9 mV/°C. A Blue LED measured −2.8 mV/°C.

LED Noise

LEDs also create voltage noise that largely originates from their shot noise mechanism. LEDs should be bypassed if they are used as a voltage reference in a noise-critical location. Typical forward voltage for a Red LED is about 1.8 V, and its voltage noise is on the order of 2.5 nV/√Hz. A Green LED is similar. Blue LEDs operated at 1 mA have a forward voltage of about 3.3 V and their voltage noise can be as high as 30 nV/√Hz.

References

1. Paul Horowitz and Winfield Hill, *The Art of Electronics*, 3rd edition, Cambridge University Press, New York, 2015.
2. Adel Sedra and Kenneth Smith, *Microelectronic Circuits*, 6th edition, Oxford University Press, New York, 2010.
3. Giuseppe Massobrio and Paolo Antognetti, *Semiconductor Device Modeling with SPICE*, 2nd edition, McGraw-Hill, New York, 1993.
4. Bob Pease, What's All This V_{BE} Stuff Anyhow? in *Analog Circuits: World Class Designs*, Newnes, 2008.
5. Bob Cordell, *Designing Audio Power Amplifiers*, 2nd edition, Routledge, New York, 2019.
6. Pacific Semiconductor Patent, US 298,967, June 1961.
7. Application Note, *Field Effect Transistors in Theory and Practice*, Freescale Semiconductor, AN211A.
8. Ray Marston, *FET Principles and Circuits*, *Nuts & Volts Magazine*, Parts 1–4, May–August 2000, available at www.linearsystems.com.
9. LSK389, LSK489 and LSJ689 JFET Datasheets, Linear Integrated Systems, www.linearsystems.com.
10. Bob Cordell, *LSK489 Ultra Low Noise JFET*, Application Note, Linear Integrated Systems, www.linearsystems.com.
11. Bob Cordell, *LSK689 Ultra Low Noise P-Channel Dual JFET*, Application Note, Linear Integrated Systems, www.linearsystems.com.
12. Bob Cordell, *LSK389 Ultra Low Noise N-Channel Dual JFET*, Application Note, Linear Integrated Systems, www.linearsystems.com.
13. Bob Cordell, *VinylTrak – A Full-Featured MM/MC Phono Preamp*, Linear Audio, Vol. 4, September 2012.
14. Ron Quan, *A Guide to Using FETs for Voltage Controlled Circuits, Part 1*, Application Note, Linear Integrated Systems, www.linearsystems.com.
15. Siliconix/Linear Integrated Systems, *FETs as Voltage-Controlled Resistors*, Application Note AN105, 1997, www.linearsystems.com.
16. LS26VNS, LS26VPS and VCR11N JFET VCR Datasheets, Linear Integrated Systems, www.linearsystems.com.
17. 2N4391 N-Channel JFET Datasheet, Linear Integrated Systems, www.linearsystems.com.
18. J500 Series Current-Regulating Diodes, Linear Integrated Systems, www.linearsystems.com.
19. DPAD1 Picoampere Diode, Linear Integrated Systems, www.linearsystems.com.
20. 3N163 P-Channel Enhancement Mode MOSFET, Datasheet, Linear Integrated Systems, www.linearsystems.com.
21. Mark Stansberry, *Radiation Sensor Design and Applications: The 3N163*, Application Note, Linear Integrated Systems, www.linearsystems.com.
22. SD-SST210/214 N-Channel Lateral DMOS Switch, Datasheet, Linear Integrated Systems, www.linearsystems.com.
23. Linear Integrated Systems, *High-Speed DMOS FET Analog Switches and Switch Arrays*, application note, www.linearsystems.com.
24. NTE 1N5223B through 1N5271B Zener Diode, Datasheet, nteinc.com/specs/original/1N5223B_71B.pdf.
25. ON Semiconductor, 1N5221B through IN5252B Zener Diode Datasheet, onsemi.com.
26. Ken Walters and Mel Clark, *Zener Voltage Regulation with Temperature*, MicroNotes Series No. 203, Microsemi Corp, www.microsemi.com.
27. Maxim Integrated Products (now part of Analog Devices, Inc.), *Build Low Cost White Noise Generator*, Application Note 3469.
28. E.F. Schubert, *Light Emitting Diodes*, 2nd edition, Cambridge University Press, New York, 2006.
29. Peter A. Lefferts, *LED Used as Voltage Reference Provides Self-Compensating Temp Coefficient*, Electronic Design, February 15, 1975, p. 92.

Chapter 8

Operational Amplifiers

Operational amplifiers are high-gain devices whose larger function is defined by negative feedback. In essence, they are universal gain blocks that can be used in a great many different ways [1–4]. They are virtually all implemented conveniently as integrated circuits. Op amps come in many different varieties to suit different applications, but the operating and application principles are generally the same for all of them.

With a few exceptions, operational amplifiers operate in the small-signal domain, where signal voltages are often less than about 10 V_{RMS}. Indeed, many operational amplifiers are operated from ±15 V power supplies. However, some IC op amps can operate from supplies as low as ±5 V or less.

8.1 The Ideal Op Amp

The ideal op amp has differential inputs and infinite gain at DC, with no input DC offset voltage and no input bias current. However, it may be characterized by finite bandwidth. This is the parameter referred to as gain-bandwidth product (GBP). It is the frequency where the voltage gain of the op amp falls to unity, usually at a rate of 6 dB per octave. Most real op amps have very high gain at DC and low frequencies, often on the order of 120 dB or more. A typical modern op amp designed for audio may have a gain-bandwidth of about 10–50 MHz. This means that the op amp open-loop gain (OLG) at 1 kHz will be about 10,000–50,000, or 80–94 dB (GBP divided by 1 kHz, assuming a 6 dB/octave open-loop gain roll-off).

Early op amps had GBP on the order of 1 MHz. This means that OLG at 20 kHz was only about 50. An op amp circuit with closed-loop gain (CLG) of 10 would have a negative feedback (NFB) loop gain of only 5 at 20 kHz, which is very inadequate. In the coming discussions, we will assume an op amp with 100 dB of gain at DC and a GBP of 10 MHz. This assumption makes the numbers easier for analysis. For reference, the popular LM4562 has a typical GBP of 55 MHz and open-loop DC gain of 140 dB.

8.2 Inverting Amplifiers and the Virtual Ground

Consider the simple op amp inverter shown in Figure 8.1(a). Assume its open-loop gain is 100 dB at DC. With GBP of 10 MHz, its OLG at 1 kHz will be 80 dB. When the op amp is delivering +1 V_{DC} at its output, the signal voltage at the inverting input must be only −0.01 mV, since the positive input is connected to ground. This number is so small and close to ground that we refer to the inverting input node of the op amp in such a feedback arrangement as a "*virtual ground.*"

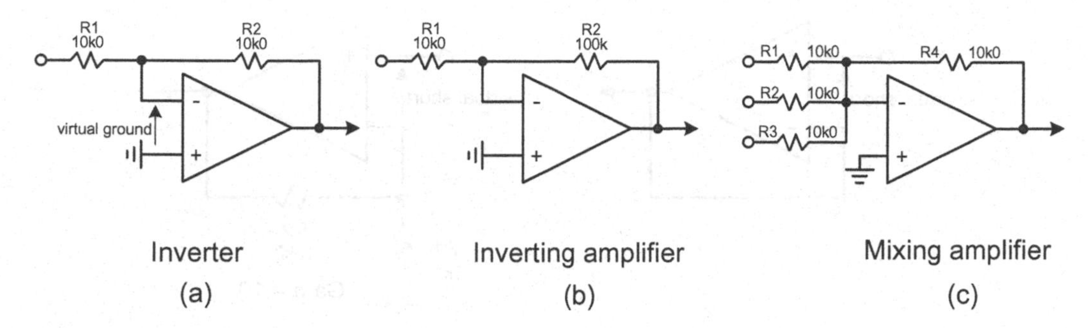

Figure 8.1 Inverting Amplifiers.

The currents in R1 and R2 must be virtually identical. It is then easy to see that a −1 V input to the inverter is required to produce the +1 V output. Virtually all of the current flowing in R1 flows through R2.

In essence, the negative feedback (NFB) forces these relationships to hold. In practice, the voltage at the input will have to be −1.00001 V–0.01 mV that actually must exist at the inverting input of the op amp to produce +1 V at the output. This means that the gain is ever so slightly less than unity, as expected with NFB with less than infinite loop gain. Notice that, for the feedback loop gain, the combination of R1 and R2 forms a 6-dB attenuator. The loop gain in this arrangement is thus 94 dB at DC, corresponding to a ratio of 50,000. The closed-loop bandwidth will be on the order of 5 MHz, since the loop gain is halved by the combination of R1 and R2.

Figure 8.1(b) shows an amplifier with an inverting gain of 10, with R1 = 10 kΩ and R2 = 100 kΩ. Since the currents flowing in R1 and R2 are virtually identical, and since the voltage at the inverting input of the op amp is virtually zero, it can be seen by inspection that the gain is −10. When the output voltage is +1 V, the input voltage will be almost exactly −100 mV. Note here that the combination of R1 and R2 introduces 21 dB of attenuation in the feedback loop: R1/(R1 + R2) = 0.09. This means that the loop gain at 1 kHz is now only 59 dB. The AC voltage error is still about 0.1 mV at the inverting input of the op amp, but it is now larger in comparison to the smaller input signal. The combination of R1 and R2 attenuates the feedback signal by a factor of 11, so the closed-loop bandwidth (CLBW) will be on the order of 10 MHz/11 = 910 kHz. For audio circuits, the loop gain at 20 kHz is one of the most important characteristics.

Figure 8.3 illustrates a simple summing or mixer circuit where three different signal sources can be mixed together in equal amplitudes. In this circuit, each input sees a gain of −1 to the output. If different resistor values are used, different gain can be applied to each signal source. The virtual ground suppresses crosstalk among the sources.

Here the loop gain has been divided by 4 by the action of R4 against the parallel combination of R1, R2 and R3. The CLBW will thus be only about 2.5 MHz. If you visualize all three inputs driven by a single source, you can see that the amplifier looks like a 3× inverter.

Circuits like this are important in pro sound mixers where many inputs are summed together. The reduced loop gain in such situations must be kept in mind, since bandwidth and distortion will be adversely affected with a large number of inputs. Amplifier input noise will also be increased. To amplifier noise, the above arrangement looks like an inverting amplifier with a gain of 4, and the *noise gain* is said to be 4. This makes it all the more important to use low-noise amplifiers in mixing stages with a large number of inputs.

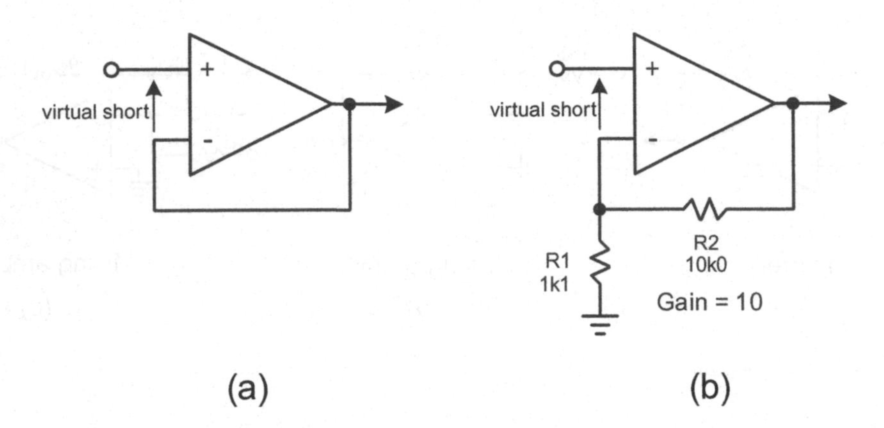

Figure 8.2 Non-Inverting Amplifiers.

8.3 Non-Inverting Amplifiers and the Virtual Short

If both the inverting and non-inverting inputs of the op amp are driven in a feedback arrangement, the very small voltage that will exist across those two inputs allows us to think of there being a "virtual short" between those two inputs, meaning that in a feedback arrangement those two voltages will always be virtually the same. This approximation makes the design of such circuits very easy.

Consider the unity-gain non-inverting buffer in Figure 8.2(a). If there is 1 V at its output at 1 kHz, then with 80 dB of OLG at 1 kHz, there will only be 0.1 mV across the inputs, meaning that the input signal and output signal will differ by only 0.1 mV. In fact, the input will be 1.0 V plus 0.1 mV, or 1.0001 V, reflecting a very small departure in gain from +1. Note that in many of these analyses, we are working backward from the assumed output to the input. This often makes the analysis of feedback circuits easier. This circuit is often called a voltage follower.

Now consider the +10× buffer in Figure 8.2(b). R1 and R2 form a 10× voltage divider in the feedback path. When the output is 1 V, the voltage at the inverting input of the op amp will be 100 mV and the voltage at the non-inverting input will be greater by 0.1 mV due to the finite OLG of 80 dB. The voltage at the main input will thus be 100.1 mV. By inspection, the gain is thus ever so slightly less than 10×. The 10× attenuation in the feedback path means that the loop gain will fall to unity at 10 MHz/10 = 1 MHz, and this will be the approximate CLBW.

8.4 Differential Amplifier

The circuit in Figure 8.3(a) is a popular op amp-based differential amplifier arrangement. It can be analyzed in a number of different ways, but one way is to see it as a combination of an inverting amplifier and a non-inverting amplifier. The concept of the virtual short once again makes analysis easy. With 10-kΩ input resistors (R1 and R2) and a 10-kΩ feedback resistor (R3) and a 10-kΩ shunt resistor (R4) to ground, the amplifier is configured for a differential gain of 1×.

Once again, assume a 1-V output and work your way backward to the input. To make things easy, now ignore the 0.1 mV that must appear across the op amp's input at 1 kHz. If the positive input to the circuit is grounded, it is easy to see that the circuit defaults to an inverter with a gain of −1. If the negative input to the circuit is grounded, it is easy to see that the circuit defaults to a non-inverting amplifier with a gain of +1 (a 2× non-inverting amplifier preceded by a 0.5× attenuator consisting of R2 and R4).

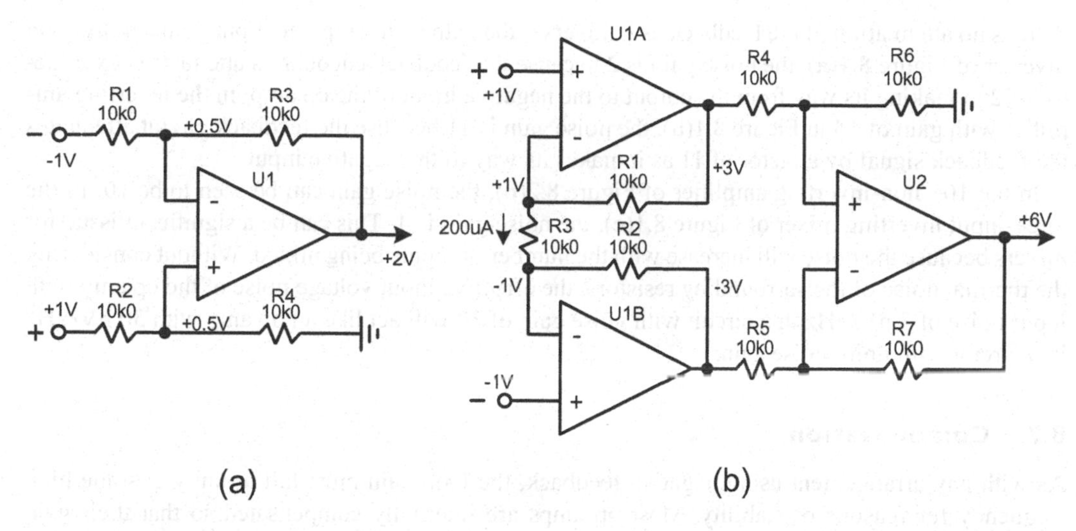

Figure 8.3 Differential Amplifiers.

It is easy to see that if the same signal is applied to both the positive and negative inputs of the circuit, the net gain will be zero. This means that the circuit provides common-mode rejection. Looking at it another way, if the output is zero, both input signals create the same voltage at the inverting and non-inverting inputs of the op amp, so there is no net op amp input signal to create an output signal. R1-R4 must have close tolerances to achieve this.

Notice that the plus and minus input currents to the differential amplifier circuit are not the same if only one or the other input is driven. This means that the input impedances to ground of the plus and minus inputs are not the same. This can be a source of imperfection in the arrangement when it is driven by a source with finite source impedances to ground. This may compromise the common-mode rejection ratio (CMRR) of the circuit.

8.5 Instrumentation Amplifier

The circuit of Figure 8.3(b) is a version of a differential amplifier called an *instrumentation amplifier*. The first two op amps are configured as a non-inverting differential buffer with a gain of 3. Each half can be thought of as a simple non-inverting amplifier with the virtual center-tap of R3 grounded. With a pure differential input this point will be at ground potential anyway. Letting this point float decreases the common-mode gain to unity, thus giving a 3× improvement of differential gain as compared to common-mode gain (e.g., CMRR). The buffering effect of U1 and U2 also means that the input impedance of the circuit at the plus and minus inputs is very high. This preserves the common-mode rejection in the face of finite source impedances. The second half of the instrumentation amplifier is merely a conventional differential amplifier arrangement like that in Figure 8.3(a).

8.6 Noise Gain

Noise gain is the inverse of the amount by which the output of the op amp is attenuated by the feedback network when the input is grounded. It is called noise gain because it is the closed-loop gain of the amplifier as seen by the input voltage noise of the op amp. This is different than closed-loop signal gain. In the non-inverting unity-gain buffer of Figure 8.2(a), the noise gain is unity because there is no attenuation of the feedback as it traverses the path from output to input. In the unity-gain inverter of Figure 8.1(a), the noise gain is 2 because the feedback encounters attenuation by a factor of 2 in making its way from the output to the negative input of the op amp. In the inverting amplifier with gain of 10 in Figure 8.1(b), the noise gain is 11 because the feedback circuit attenuates the feedback signal by a factor of 11 as it makes its way to the negative input.

In the 10× non-inverting amplifier of Figure 8.2(b), the noise gain can be seen to be 10. In the three-input inverting mixer of Figure 8.1(c), the noise gain is 4. This can be a significant issue for mixers because the noise will increase with the number of inputs being mixed. Without considering the thermal noise of the surrounding resistors, the effective input voltage noise of the op amp with input noise of 3 nV/√Hz in a circuit with noise gain of 10 will act like an op amp with 30 nV/√Hz in a circuit with unity noise gain.

8.7 Compensation

As with any arrangement using negative feedback, the loop gain must fall to unity at some high frequency for reasons of stability. Most op amps are internally compensated so that their gain falls to unity at their rated gain-bandwidth frequency. Their forward gain will ideally begin to fall off at their open-loop bandwidth frequency. It will usually fall at 6 dB per octave until the gain-bandwidth frequency is reached. For example, a modern op amp with gain-bandwidth of 10 MHz and DC gain of 120 dB (1 million) will have its gain begin to fall off at 10 Hz. The important objective is that the loop gain will fall to unity well before the lagging phase shift reaches 180°.

Stray capacitance at the inverting input can add a pole that will detract from phase margin. When in doubt, assume 5 pF to ground. If the impedance at that node is 5 kΩ (typical for an inverter using 10 kΩ resistors), a pole at 6.4 MHz will result.

Unity-Gain Compensation

Most modern op amps are designed to be unity-gain stable. This means that if the output is connected directly to the inverting input, as in a non-inverting unity-gain buffer, the arrangement will be stable, with adequate gain and phase margin. This usually means that an op amp with a 10-MHz gain-bandwidth product will have unity gain at 10 MHz with less than perhaps 120° of open-loop phase shift, leaving a 60° phase margin.

Bear in mind that the open-loop gain roll-off creates a near-constant 90° lagging phase shift that will increase somewhat as the OLG approaches unity, so in this example other sources of open-loop phase shift must create less than 30° of phase lag at 10 MHz. The popular NE5532 and LM4562 op amps are both unity-gain stable, as are most op amps. In this context, the term unity gain actually refers to the noise gain. A unity-gain inverter has a noise gain of 2, so it has half the loop gain of a unity-gain non-inverting buffer like that of Figure 8.2(a).

External Compensation

Operational amplifiers that have unity-gain compensation provide sub-optimal performance when they are used in circuits where the gain (specifically *noise gain*) is greater than unity. This is because the closed-loop bandwidth will be less than that which the op amp is capable of based on its achievable phase margin. As a result, high-frequency distortion will be greater than necessary and slew rate will be lower than possible. For example, if a 1-MHz unity-gain-compensated op amp is used in an inverter, where the noise gain is 2, the closed-loop bandwidth will be only 500 kHz and the available feedback loop gain at 20 kHz will be only 28 dB instead of 34 dB, resulting in higher closed-loop distortion.

Some operational amplifiers, like the NE5534, permit external compensation implemented by an external compensating capacitor [5]. In the case of the 5534, the op amp is stable with its internal compensation for noise gain equal to or greater than 3. Adding a 33-pF external compensation capacitor will make it unity-gain stable. A 20-pF compensating capacitor will make it stable for a noise gain of 2, as for an inverter. It is notable that few circuits, other than a non-inverting unity-gain buffer, have unity noise gain. Other operational amplifiers have fixed internal compensation, but are not unity-gain stable. They may be stable for noise gain greater than 2, for example. Alas, op amps with external compensation are rare these days, partly because newer technologies provide op amps with higher gain-bandwidth, so that some loop gain can be sacrificed to achieve lower cost and greater convenience.

8.8 A Simple Op Amp

The very simplified discrete operational amplifier of Figure 8.4 illustrates the basic circuitry of an operational amplifier. It can be implemented by a simple combination of some of the circuit building blocks that were discussed earlier in Chapter 3. Q1 and Q2 form a differential input stage (IPS,

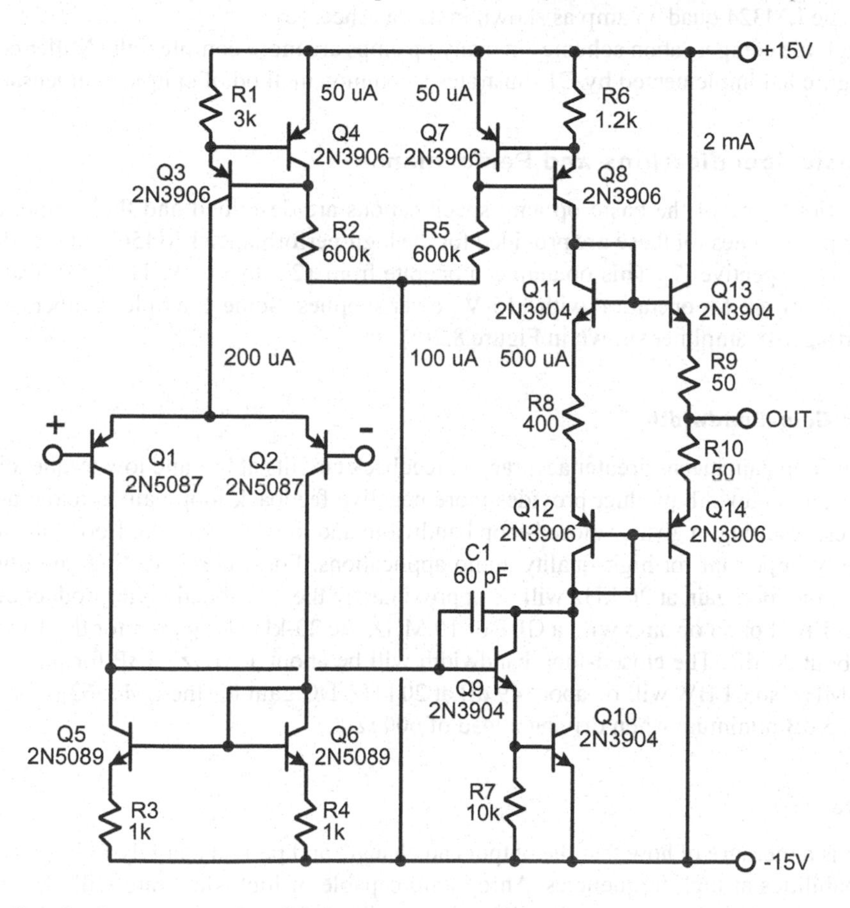

Figure 8.4 Simple Operational Amplifier.

often called a long-tailed pair, LTP) loaded by a current mirror built with Q5 and Q6. The 200-μA tail current for the IPS is provided by a constant current source (CCS) comprising Q3 and Q4. Q1 and Q2 operate at 100 μA and have re' = 260 Ω. The voltage amplifier stage (VAS) is formed by a Darlington common-emitter stage comprising Q9 and Q10. The VAS is loaded by a 200-μA constant current source, comprising Q7 and Q8. The output stage (OPS) is a complementary emitter follower comprising Q13 and Q14. It is biased at 500 μA by the voltage drops of diode-connected transistors Q11 and Q12 and R8.

Compensation capacitor C1 (60 pF) sets the gain-bandwidth product to about 10 MHz and allows slew rate of about 3.3 V/μs to be achieved. Open-loop gain is simulated to be over 100 dB at DC with transistor β averaging 400 and a load of 10 kΩ. Input bias current is about 250 nA.

Figure 8.4 is for illustration purposes only. A real op amp will have quite a few more transistors to increase gain, reduce input bias current, increase output current capability and incorporate short circuit protection, among other things. For example, a pair of emitter followers might precede the IPS to reduce input bias current and increase the input impedances. The output stage might include emitter followers in front of the output transistors to increase the current gain of the output stage to increase output current capability and lighten the load on the VAS to increase voltage gain. Such improvements might result in DC gain of 120 dB with a load of 1 kΩ and input bias current of only 5 nA. These kinds of improvements move this simple circuit in the direction of the simplified circuit of the LM324 quad op amp as shown in its datasheet [6].

Although the compensation schemes in many op amps are more complex, the Miller compensation in Figure 8.4 implemented by C1 illustrates a common method of simple compensation.

8.9 Basic Specifications and Performance

In this section some of the basic op amp specifications are described and their importance discussed. Typical values for these are provided for the high-performance LM4562 bipolar dual audio op amp for perspective [7]. This op amp can operate from ±2.5 to ±17 V. The specification numbers that follow are for operation with ±15-V power supplies. Some example numbers are for the non-inverting 10× amplifier shown in Figure 8.2(b).

Gain and Gain-Bandwidth

High open-loop gain means greater accuracy in feedback circuits at DC and low frequencies, while a higher gain-bandwidth product provides more negative feedback loop gain at audio and higher frequencies, resulting in greater closed-loop bandwidth and lower distortion. Loop gain at 20 kHz is especially important for high-quality audio applications. For a non-inverting amplifier with a gain of 10, the loop gain at 20 kHz will be approximately the gain-bandwidth product divided by 20 kHz and 10. For an op amp with a GBP of 10 MHz, the 20-kHz loop gain for the 10× amplifier will be about 34 dB. The closed-loop bandwidth will be about 1 MHz. GBP for the LM4562 is about 55 MHz, so CLBW will be about 49 dB at 20 kHz. DC gain for the LM4562 is 140 dB typical and 125 dB minimum when driving a load of 600 Ω.

Slew Rate

Slew rate is a measure of how fast the output can change, and pertains largely to large-signal handling capabilities at high frequencies. An op amp capable of high slew rate will also often tend to have less high-frequency distortion. Distortion will often begin to increase quickly when the output voltage rate-of-change approaches the rated slew rate. A conservative design would operate the op amp at less than 1/10th its rated slew rate at 20 kHz when the peak signal amplitude is at its maximum. A useful number is that signal rate of change at 20 kHz is 0.125 V/μs for each peak volt of sinusoidal signal. At 10 V_{RMS}, this comes to about 1.8 V/μs. Ten times this would suggest a desirable op amp slew rate capability of 18 V/μs in a conservative design. The rated typical slew rate for the LM4562 is ±20 V/μs, while the rated minimum is ±15 V/μs.

Input Offset Voltage

Input offset voltage matters most in circuits with high DC gain or in circuits that must make highly accurate DC measurements. When DC offsets are small, many circuits can employ DC coupling instead of requiring a coupling capacitor. Modern bipolar op amps have quite low DC input offset. The LM4562 is specified as having typical input offset voltage of ±0.1 mV, with a maximum value of ±0.7 mV.

Input Bias and Offset Current

The BJT input stages of an op amp draw base current, and this corresponds to input bias current. Input stages are often run at a rather low current with transistors that may have β of several hundred. This leads to fairly low input bias current, but it can still be a concern in high impedance circuits because it will create DC offset at the output of the circuit. The LM4562 is specified to have typical input bias current of 10 nA, and a maximum of 72 nA. These numbers are a bit higher than many other general-purpose op amps because the input stage in the LM4562 is run at a relatively high current to achieve very low noise. In the worst case, 72 nA flowing through an external circuit resistance of 100 kΩ will create a voltage offset of 7.2 mV if there is no compensating offset on the other input. A circuit like this would benefit from the use of a JFET op amp.

Some circuits are designed so that the DC resistance in both the positive and negative input circuits is the same, causing the effect of the input bias current to cancel. This of course depends on the input bias current on the positive and negative inputs of the op amp being the same. They are not always the same, and this corresponds to input offset current current, making such cancellation approaches less effective. Input offset current is usually much less than input bias current, but in the LM4562, input offset current is typically 11 nA, with a maximum of 65 nA. This is a bit unusual. It suggests that the op amp employs input bias current cancellation, where the magnitude of the bias current is reduced because of cancellation circuitry, but the cancellation circuit cannot improve the differential match.

Such op amps utilize an input bias current cancellation scheme where an opposite replica current of the same value is applied to the inputs to cancel input bias current. This is made possible by the good transistor matching that can be achieved on an IC. However, these schemes cannot generally decrease the input offset current. If you see an input offset current specification that is on the order of the stated input bias current, it is likely that op amp is using such a scheme.

In contrast to the LM4562, the NE5532 input bias current is 200 nA typical and 800 nA maximum, while its input offset current is 10 nA typical and 150 nA maximum [8].

Common-Mode Rejection

When the common-mode component of a signal is applied to both the positive and negative inputs of an op amp, it slightly alters the operating conditions of the input stage, so the input offset voltage can change a little. This means that there is a way for the common-mode voltage to create a small output voltage change. The amount by which the op amp is able to ignore this effect is called the common-mode rejection ratio (CMRR). For an op amp with 80 dB of CMRR (10,000:1), a common-mode voltage of 10 V at the input of the op amp would cause an input offset voltage change of 1 mV. If the gain of the circuit were 10, the common-mode input voltage would cause an output change of 10 mV. The LM4562 datasheet specifies typical CMRR of 120 dB, with a minimum of 110 dB. CMRR for the NE5532 is 100 dB typical and 70 dB minimum.

CMRR is also a function of frequency, and tends to decrease as frequency increases. This can be the result of circuit currents flowing through capacitances in the input stage or in those capacitances changing with signal swing. The AC CMRR can also decrease as the impedance of the tail current source decreases with frequency. Typical CMRR for the LM4562 is 120 dB at 1 kHz and 97 dB at 20 kHz. It falls to 90 dB at 100 kHz.

Finally, it is important that the specified maximum input common-mode range not be exceeded. Some op amps will undergo a phase reversal if the common-mode voltage range is exceeded.

Power Supply Rejection

Similarly to CMRR, PSRR refers to how much change in input offset voltage will be caused by a change in the power supply voltage. For an op amp with 80 dB PSRR, a 10 mV_{pk} change in power supply voltage due to ripple, for example, would cause a peak change of input offset voltage of 1 μV. PSRR can also be different depending on which power supply rail is being considered. As with CMRR, PSRR generally decreases with frequency. The LM4562 has DC PSRR of 120 dB typical and 110 dB minimum. At 20 kHz, typical positive-rail PSRR is 102 dB and typical negative-rail PSRR is 70 dB.

Noise

Noise is obviously important in audio applications. Op amp noise is specified as equivalent input voltage noise and equivalent input current noise. If the op amp in a non-inverting amplifier stage with a gain of 10 has input noise density of 5 nV/√Hz, then the output noise will be 50 nV/√Hz if the thermal noise of all of the circuit resistances is ignored. As we will see later, the number of √Hz in a 20-kHz un-weighted audio band is 141 √Hz. This means that the total output noise will be approximately 7.1 μV. If the maximum output signal level is 10 V_{RMS}, then the un-weighted signal-to-noise ratio (SNR) in a 20-kHz bandwidth will be about 123 dB. SNR will only be 83 dB for a 100-mV maximum signal level.

If the input current noise is 1 pA/√Hz and the net resistance at the negative input of the op amp is 5 kΩ (as in a unity-gain inverter with 10 kΩ resistors), and thermal noise of the resistors in the circuit is ignored, then the input noise voltage caused by the input noise current will be about 5 nV/√Hz, similar to the input voltage noise noise of the op amp assumed above. The output noise will again be 50 nV/√Hz (if input noise voltage is ignored). The SNR due to input current noise will then be about 123 dB. If both the input noise voltage and the input noise current were in play in this example, SNR would be degraded by 3 dB down to 120 dB because these two sources of noise are equal and uncorrelated and add on a power basis. Output noise in a 20-kHz bandwidth will be about 10 μV.

To put this into perspective, if the net input resistance at the input of the op amp circuit were 5 kΩ, the resistive thermal noise contribution at the input would be about 9.4 nV/√Hz, producing output noise of about 94 nV/√Hz, corresponding to 13.3 μV in a 20 kHz bandwidth. Taking the

RMS sum of the op amp voltage noise, the op amp current noise and the resistive thermal noise contributions, we have 16.6 μV. The net SNR with a maximum signal level of 10 V_{RMS} is then about 116 dB. This increased noise due to resistive thermal noise is why low-noise circuits are designed with low-impedance circuitry surrounding the op amp.

The LM4562 has rated typical input noise voltage at 1 kHz of 2.7 nV/√Hz, with a maximum of 4.7 nV/√Hz. At 10 Hz, however, typical noise is 6.4 nV/√Hz. Noise is higher at very low frequencies due to so-called 1/*f* or flicker noise, where noise density begins to increase at a rate of 10 dB/decade as frequency goes down, often beginning at around 100 Hz. Rated typical input current noise at 1 kHz is 1.6 pA/√Hz, while at 10 Hz rated typical input current noise is 3.1 pA√Hz. While the typical input voltage noise and input current noises at 1 kHz for the LM4562 are 2.7 nV/√Hz and 1.6 pA/√Hz, respectively, those for the NE5532 are 5 nV/√Hz and 0.7 pA/√Hz. The optimum noiseless source resistance for low noise for an op amp is the ratio of input noise voltage to input noise current. For the LM4562, this number is 1.7 kΩ, while for the NE5532 this number is 7.1 kΩ.

Maximum Output Swing

The maximum peak output voltage swing will determine the available headroom in most applications. With ±15 V power supplies, the LM4562 will typically provide ±13.6 V into a 600-Ω load, corresponding to 9.6 V_{RMS}, with about 0.0001% THD at 1 kHz. Some op amps are designed to be able to swing very close to the power supply voltages. These are referred to as rail-to-rail devices. The OPA 1656 specifies that its output can swing within 0.25 V of either rail for $RL = 2$ kΩ.

Total Harmonic Distortion (THD)

Harmonic distortion is important in audio applications for obvious reasons. At 1 kHz, THD for the LM4562 is extremely low, at 0.00003% typical and a maximum of 0.00009% when driving a 600-Ω load to 3 V_{RMS}. The typical number corresponds to distortion of about −130 dB. In virtually all op amps, THD rises at higher frequencies due to the reduced negative feedback loop gain at higher frequencies. Typical THD+N with 3 V_{RMS} driving 600 Ω is 0.00003% at 1 kHz, rising only modestly to 0.00006% at 20 kHz.

For many op amps, the class-AB nature of the output stage can contribute some crossover distortion when driving a smaller load impedance. This can be mitigated by connecting a pull-down resistor from the output to the negative supply. This will force the output stage to operate in class A for output current swings up to the value of the pull-down current. Of course, the amount of pull-down current will fall with negative output voltage swings. This can be addressed by using a current source for the pull-down function. Fortunately, many modern high-performance op amps have very low distortion driving moderate loads without external pull-down current. Some op amps provide lower crossover distortion if the output is pulled up rather than down.

Input Common-Mode Distortion

The variation of input-referred offset is nonlinear with common-mode voltage. The biggest issue is for non-inverting unity-gain stages where the full output is fed back to the inverting input. Conversely, inverting stages with a virtual ground will not suffer from common-mode distortion. Depending on the specific device, in some cases JFET op amps may be a bit more susceptible to common-mode distortion. Input common-mode distortion is not reduced by negative feedback.

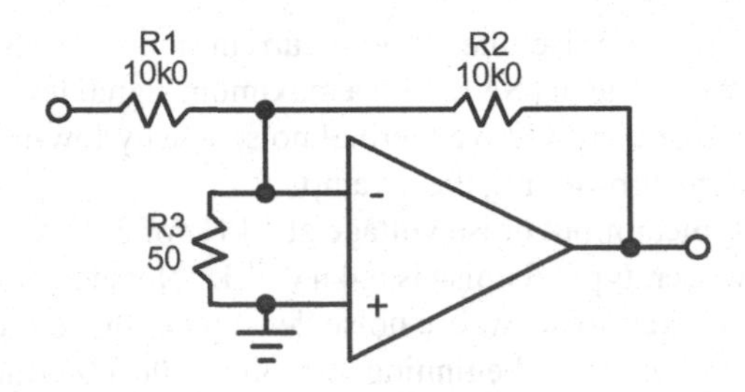

Figure 8.5 Measuring Op Amp Distortion.

Measuring Op Amp Distortion

Measuring the THD of op amps like the LM4562 can be difficult, even with very good distortion analyzers. However, measurement is made easier if the circuit is arranged to magnify the distortion without magnifying the output signal level. This can be done by increasing the noise gain of the circuit. Such an approach using a unity-gain inverter is shown in Figure 8.5. The 50-Ω shunt resistor from the virtual ground to ground works as a 100:1 attenuator against the effective net 5-kΩ resistance of the feedback network, increasing the noise gain to 100 without changing the voltage gain of the inverter. THD sensitivity is thus increased by a factor of 100. The noise is also increased by a factor of 100, so THD measurements should be done with a spectrum analyzer, such as by using a Fast Fourier Transform (FFT).

Technology Advances

For many years, most op amp designs were constrained in performance by the lack of complementary vertical PNP transistors. Instead, they had to employ lateral PNP transistors that had low beta and low f_t. Beginning in the mid-1970s, early full-complementary junction-isolated linear IC processes were developed at Bell Labs and Western Electric for telecommunications applications [9]. These IC processes with vertical PNP transistors were especially valuable in making op amps for use in high-performance active filters.

By the mid-late 1980s, op amps using this full-complementary process were becoming available in some commercial products available on the open market. The full-complementary processes enabled significant improvements in many op amp parameters, not to mention freedom of circuit design approaches. Recognizing that it did not depend on vertical PNPs, the performance achieved by the legendary NE5532/4 was remarkable for its time. A further technological advance was the use of oxide-isolated processes for some full-complementary linear ICs [10].

8.10 FET Op Amps

The other type of popular op amp is the JFET op amp. Typified by the OPA2134 dual op amp, these op amps incorporate JFETs for the input stage, providing extremely high input impedance, virtually no input bias current and virtually no input noise current [11]. A significant circuit design advantage is less worry about offset voltages created by input bias current flowing through the DC resistances of surrounding circuits. Consider a BJT op amp with input bias current of 30 nA in an inverter with a gain of 10, with an input resistor of 100 kΩ and a feedback resistor of 1 MΩ. Most of the input bias current will flow through the feedback resistor, creating an output offset of about

30 mV. Input bias current of the OPA2134 is only 5 pA typical and ±100 pA maximum, resulting in less than 0.1 mV output offset voltage due to input bias current.

Primary disadvantages of JFET op amps compared to BJT op amps include increased input offset voltage and increased input noise voltage density. The classic general-purpose TL072 dual JFET op amp has maximum input offset voltage of ±4 mV [12]. Its typical noise specification is 37 nV/√Hz at 1 kHz. For the OPA2134, the offset numbers have improved to ±2 mV maximum. Input voltage noise of the OPA2134 is 8 nV/√Hz typical. THD+N of the OPA2134 is below 0.0001% at 1 kHz driving 2 kΩ at 10 V_{RMS}. This rises to 0.001% at 20 kHz. This is not quite as good as the BJT LM4562. The OPA1642 is even better, at 5 nV/√Hz typical.

In other respects, the OPA2134 is very competitive with BJT op amps, featuring typical open-loop gain of 120 dB, GBP of 8 MHz, slew rate of ±20 V/μs, maximum output of ±12 V into 600 Ω, 1 kHz THD+N of 0.00015% into 600 Ω at 3 V_{RMS}, DC CMRR of 100 dB and PSRR of 106 dB.

The extremely high input impedance and lack of input bias current make JFET op amps superior to BJT op amps in many applications. They also tend to be less affected by EMI because their input stage p-n junction is reverse-biased as compared to the forward-biased junction of BJT op amps. Subjectively, some audiophiles assert that JFET op amps sound better.

CMOS Op Amps

Until recently CMOS op amps were not considered adequate for audio applications because of their higher noise, high input offset voltages and mediocre distortion performance. That changed dramatically with the introduction of the OPA1656 CMOS Dual FET-input operational amplifier [13, 14].

The OPA1656 can operate over a ±2.25-V to ±18-V supply range. With a ±18-V supply it is able to deliver 3.5 V_{RMS} into 600 *Ω* at 1 kHz with THD+N of 0.000029% (about −130 dB) in a unity-gain non-inverting amplifier. At 20 kHz, this number is still only 0.0001% (−120 dB). Its gain-bandwidth product is a healthy 53 MHz at a gain of 100, while at non-inverting unity gain its closed-loop bandwidth is 20 MHz. Slew rate is a healthy 24 V/μs. Input offset voltage is a maximum of ±1 mV. This is remarkable for a CMOS process. Typical input bias current 10 pA, while its maximum is 20 pA. This is a bit better than the maximum input bias current of the JFET OPA2134. The insulated-gate nature of the CMOS input transistors can be expected to have lower leakage current than the junction-isolated JFET process.

Typical input voltage noise noise for the OPA1656 at 10 kHz is a respectable 2.9 nV/√Hz. However, as is often typical for MOS transistors, there is significant *1/f* noise in the audio band. The *1/f* corner frequency, where voltage noise is up 3 dB is at a fairly high 1 kHz for the OPA1656. From there on down, *1/f* noise will increase at 10 dB/decade. At 1 kHz, typical input noise voltage is 4.3 nV/√Hz. At 100 Hz, noise is 12 nV/√Hz and at 10 Hz noise is 35 nV/√Hz.

8.11 Composite Op Amps

Two op amps can be connected in tandem to achieve what one op amp cannot. Such an arrangement is called a composite op amp (COA). This can allow higher loop gain to be achieved or to use two different op amps to achieve an objective that involves using the best properties of each device. An example of the former would be to achieve higher loop gain than achievable with one op amp, recognizing that most op amps are unity-gain compensated, and that in a higher gain environment loop gain is lost. An example of the latter is an arrangement where the first op amp is a JFET op amp and the second is a high-current BJT buffer.

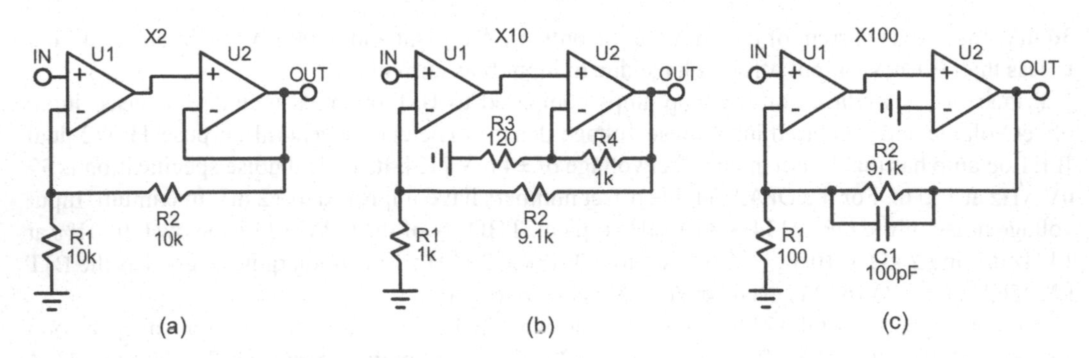

Figure 8.6 Composite Op Amp Arrangements.

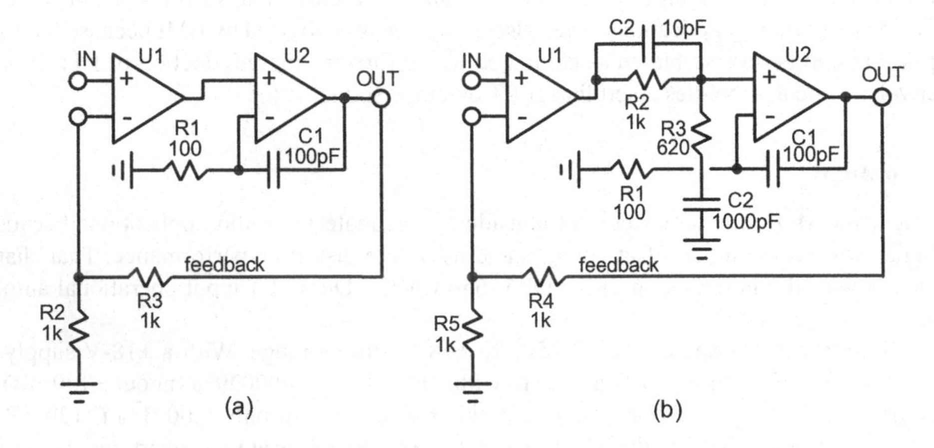

Figure 8.7 COAs with 40 dB/Decade Open-Loop Gain

Figure 8.6 shows three common COA arrangements. In Figure 8.6(a) a COA is configured for a non-inverting gain of 2. A high-performance op amp provides the gain, while a unity-gain op amp provides a higher-current capability by buffering U1 inside the feedback loop. An example might have U1 as a JFET op amp and U2 as a BJT op amp. A key consideration is feedback stability. In Figure 8.6(a), both op amps are assumed to be compensated for unity-gain stability. U2 is assumed to have significantly higher GBW than U1. The fact that the CLG is 2 also helps stability by reducing loop gain and halving the gain crossover frequency.

In Figure 8.6(b), a non-inverting 10× arrangement includes U2 as a non-inverting 10× amplifier to make up for the lost loop gain caused by the overall 10× CLG. Once again, feedback stability must be watched carefully. In Figure 8.6(c), a non-inverting 100× amplifier is implemented with U1 and U2 in tandem to provide very high OLG. Here C1 provides leading phase shift in the feedback loop to stabilize the arrangement.

Sometimes it is desirable to achieve higher loop gain at high audio frequencies for a given gain crossover frequency. This is achieved by causing the open-loop gain to roll off at 40 dB/decade over part of the frequency range rather than just the common 20-dB/decade roll-off found in most op amps. Two such COA arrangements are shown in Figure 8.7.

In Figure 8.7(a), a non-inverting 2× amplifier has very high OLG at audio frequencies, but U2 has its local feedback loop closed at high frequencies so that it becomes a unity-gain amplifier at high frequencies, resulting in an overall 6-dB/octave OLG roll-off at high frequencies to preserve stability. In Figure 8.7(b), the same arrangement includes a lag-lead network between U1 and U2 to reduce loop gain at higher frequencies to further aid stability.

8.12 Op Amp Circuits

Op amps are not just used for amplification. Their applications include oscillators, active filters, integrators and many nonlinear circuits, like precision signal rectifiers, peak detectors and logging circuits. Many such circuits are described in Ref. [2]. A few useful circuits are discussed here.

Full-Wave Rectifiers

Figure 8.8 shows two different precision full-wave rectifier (FWR) circuits, each employing two op amps. By precision, it is meant that ordinary silicon diode voltage drops are not incurred in the overall rectification process. In Figure 8.8(a), D1 and D2 close the loop around U1A so that positive and negative output swings are handled by different feedback paths. If the input swings negative, it appears at the output of U1B via R3 and R5 as a precision unity-gain positive half-wave rectified signal. This signal represents the negative half-wave portion of the input signal. Under these conditions, D1 is reverse-biased and no signal appears at D1's anode.

If the input swings positive, D1 conducts and a negative half-wave rectified signal appears at the anode of D1. That signal has a gain of −2 to the output of U1B via R4 and R5. Were it not for the input signal from R3, a positive half-wave signal representing the positive half-wave portion of the input signal would appear with amplitude twice as large at the output of U1B. However, the positive half of the input signal still passes with unity gain to the output of U1B, canceling half of the signal from the anode of D1. The result is a net unity-gain positive half-wave swing at the output of U2. The end result is a positive full-wave signal at the output of U1B.

A second non-inverting FWR is shown in Figure 8.8(b) [15]. On negative input swings, D1 conducts and closes the feedback loop around U1A, providing a negative half-wave signal at the anode of D1. That signal is inverted by U1B, R1 and R3, producing a positive half-wave signal at the output of U1B.

On positive input swings, the output of U1A goes positive, forward-biasing D2 and causing the output of U1B to go positive to the point where the voltage fed back to U1A's negative input matches the positive input swing of the input signal. The end result is a positive full-wave rectified waveform at the output of U1B with unity gain from the input.

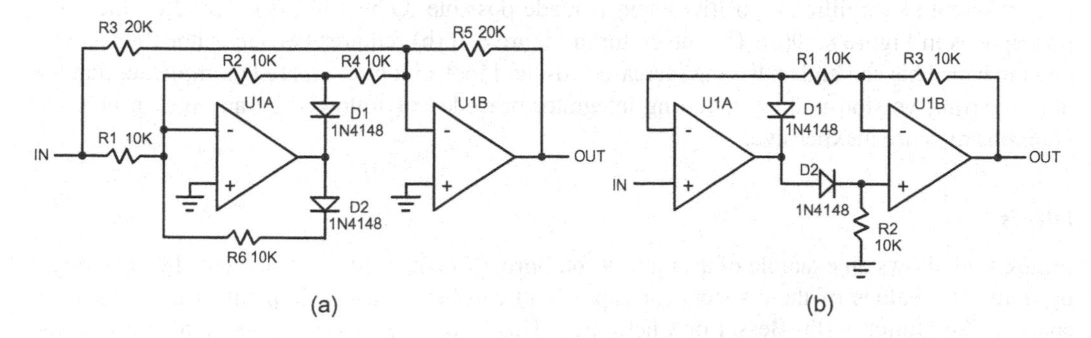

Figure 8.8 Two Full-Wave Rectifier Circuits.

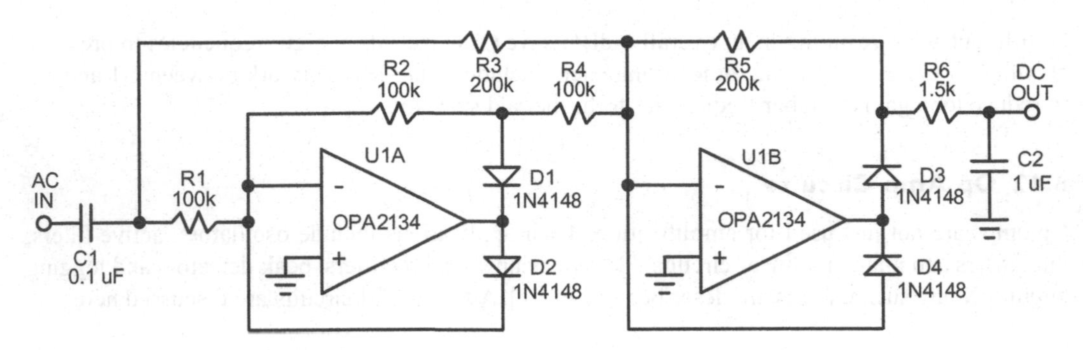

Figure 8.9 Full-Wave Peak Detector.

Full-Wave Peak Detector

Figure 8.9 shows a full-wave peak detector. This is much like the full-wave rectifier in Figure 8.8(a), but with a simple modification to the U1B summer circuit. Diodes D3 and D4 close the loop around U1B with two separate paths.

When the combination of R3 and R4 sinks current from the virtual ground (VG) of U1B, D3 turns on and feeds back enough current to equal the sinking current from R3 and R4 to keep the VG at ground. At the same time, U1B presents a positive full-wave signal to the output filter, but it can only source current to the filter because of D3. Capacitor C2 is then charged to the peak of the full-wave-rectified signal through R6. The rising time constant is relatively fast, and is set by R6 and C2. The voltage on C2 will decay by current flow back through R5 and R6 to the VG. The decay time constant is thus set by R5 + R6 and C2.

Integrators

Figure 8.10 shows two integrators, one inverting and the other non-inverting. In Figure 8.10(a), C1 replaces the feedback resistor in what would normally represent an inverter. Input current flows into the virtual ground through R1 in proportion to the input voltage. That current must flow through C1, causing the output of the op amp to ramp negative to keep the virtual ground at 0 V. With the values of R1 and C1 shown, the output will change at a rate of I/C, which will be −10 V/s per volt of input.

In Figure 8.10(b) a non-inverting integrator is formed by a differential amplifier arrangement where the feedback and shunt resistors are replaced with capacitors. By arranging the integrator as a differential amplifier, a positive input is made possible. Otherwise, it is based on the same principle as in Figure 8.10(a). The integrator in Figure 8.10(b) requires two capacitors that should be matched. This can end up having increased cost and footprint as compared to implementing the non-inverting version with an inverting integrator preceded or followed by an op amp inverter. Dual op amps are inexpensive.

Filters

Figure 8.11 shows an example of a simple second-order low-pass filter implemented with a single op amp. The values of the resistors (or capacitors) can be chosen to implement different filter shapes, like Butterworth, Bessel or Chebyshev. This is the popular Sallen-Key filter topology.

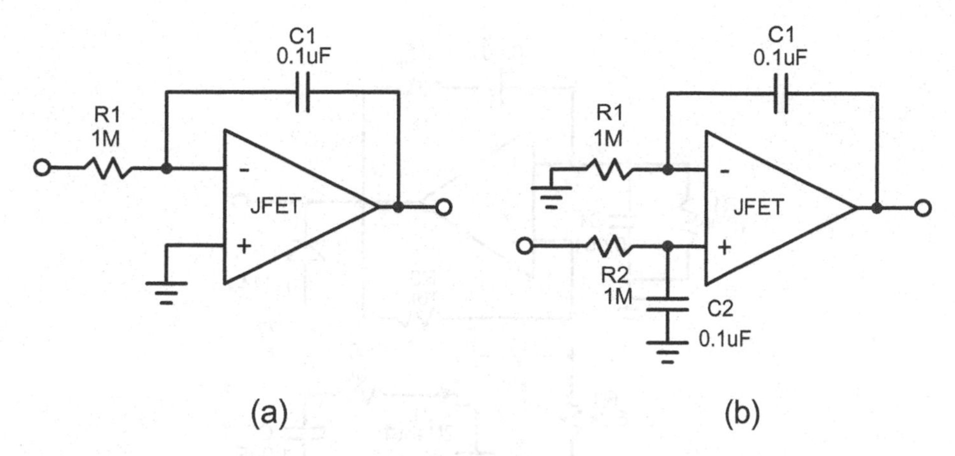

Figure 8.10 Inverting and Non-Inverting Integrators.

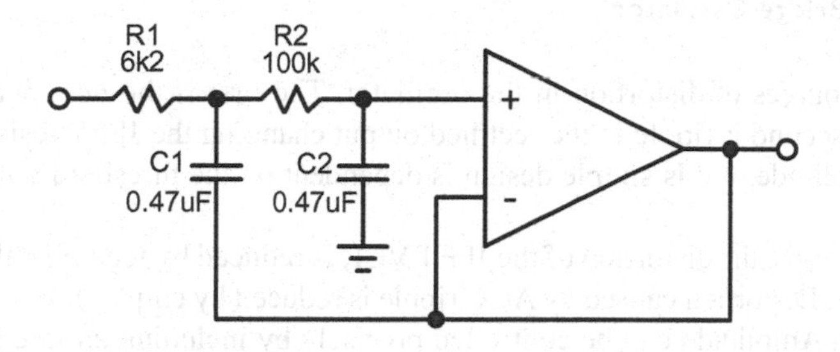

Figure 8.11 Second-Order Active Low-Pass Filter.

These and other filters are explained in Chapter 11. By adding a third *R-C* low-pass section at the front, this circuit can be made to have a third-order characteristic, like the phono preamp infrasonic filter described in Chapter 15.

Sine-Wave Oscillator

The classic Wein bridge oscillator with a JFET amplitude controller, is shown in Figure 8.12. The Wein bridge filter, formed by R1, R2, C1 and C2 has a band pass characteristic with a maximum gain of 1/3 at its center frequency. Positive feedback with unity gain is achieved by feedback through the attenuator formed by R3 and R4 with a nominal gain of 3. The gain of the 3× amplifier must be controlled to make the net positive feedback factor be unity at the center frequency, else the oscillation will eventually fall to zero or grow until the amplifier clips. This is accomplished by J1, which acts as a voltage-controlled resistor (VCR).

The output amplitude is rectified and filtered by D1, R7 and C3. If the amplitude is too large, an increased negative voltage will be fed to the gate of J1, increasing J1's resistance and reducing the non-inverting gain of the amplifier. If the output amplitude is too small, the gain is increased, limited by R5 and the *on* resistance of J1. This is basically an automatic gain control (AGC) loop. The feedback loop formed around the amplifier by the AGC circuit can be unstable if the values of R7 and C3 are not properly chosen.

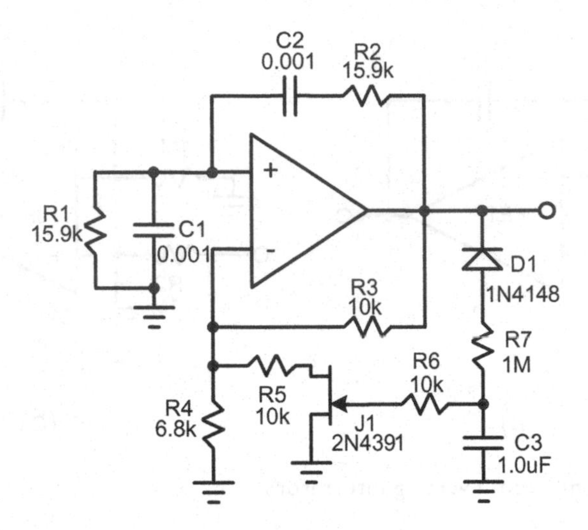

Figure 8.12 Wien Bridge Oscillator.

There are two sources of distortion in the oscillator. The first is the nonlinearity of the JFET resistance and the second is ripple in the rectified output changing the JFET resistance with signal swing. Output amplitude of this simple design is dependent on the threshold voltage of J1, and is not precise.

In improved versions, the distortion of the JFET VCR is reduced by feeding half of the drain signal back to the gate. Distortion caused by AGC ripple is reduced by employing a full-wave rectifier and better filtering. Amplitude can be controlled precisely by including an integrator in the AGC loop and applying a known offset voltage to the integrator to set the peak amplitude of the output signal. A resistor is often put in series with the integrator feedback capacitor to insert a zero into the loop transfer function to achieve stability. A very low distortion oscillator is described in Ref. [16] that employs such an AGC loop (although the oscillator itself is a state variable oscillator).

References

1. Walt Jung, *Audio IC Op-Amp Applications*, Howard W. Sams & Co., Indianapolis, 1987.
2. Walt Jung, *Op Amp Applications Handbook*, Analog Devices Inc., Newness, Burlington, NJ, 2005.
3. Bruce Carter, *Op Amps for Everyone*, 4th edition, Newness, Waltham, MA, 2013.
4. Sergio Franco, *Design with Operational Amplifiers and Analog Integrated Circuits*, 3rd edition, McGraw-Hill, New York, 2001, ISBN: 978-0072320848.
5. NE5534/NE5532 datasheet. www.ti.com.
6. LM324 datasheet. www.ti.com.
7. TI/National LM4562 Datasheet, October 2007.
8. Deane Jensen, *Some Tips on Stabilizing Operational Amplifiers*, Jensen Application Note AN001.
9. P.C. Davis, V.R. Sarri, and S.F. Moyer, High Slew Rate Monolithic Operational Amplifier Using Compatible Complementary PNPs, *IEEE Journal of Solid State Circuits*, vol. SC-9, December 1974, pp. 340–347.
10. Michael Maida, *National Semiconductor Develops New Complementary Bipolar Process for High-Speed High-Performance Analog*, TI Literature Number SNOA843.
11. TI/Burr Brown OPA134. www.ti.com.
12. TL072 op amp datasheet. www.ti.com.

13. OPA1656 op amp datasheet. www.ti.com.
14. Marek Lis, *Trade-Offs between CMOS, JFET and Bopolar Input Stage Technology*, Texas Instruments Application Report SBOA355, May 2019.
15. Ting Ye, Precision Full-Wave Rectifier, Dual Supply, TI Precision Designs TIDU030, 2013.
16. Bob Cordell, Build a High Performance THD Analyzer, *Audio*, vol. 65, July, August, September 1981, available at cordellaudio.com.

Chapter 9

Negative Feedback

Negative feedback is employed in virtually every kind of amplifier. It acts to reduce distortion and make gain more predictable and stable. Invented by Harold Black in 1927, it operates by comparing the output of an amplifier stage to the input, creating an error signal that drives the heart of the amplifier [1].

The basic concepts of negative feedback were discussed briefly in earlier chapters so that the basic amplifier design concepts could be understood. Negative feedback is fundamental to the vast majority of audio circuits and its optimal use requires more understanding. Much of that will be covered in this chapter. There is an enormous amount of literature available covering all aspects of negative feedback [2–4].

9.1 How Negative Feedback Works

Negative feedback was invented to reduce distortions and better control gain and frequency response in telephone amplifiers [1]. Figure 9.1 shows a simplified block diagram of a negative feedback amplifier. The basic amplifier has a forward gain A_{ol}. This is called the *open-loop gain* (OLG or A_{ol}) because it is the gain the overall amplifier would have from input to output if there was no negative feedback.

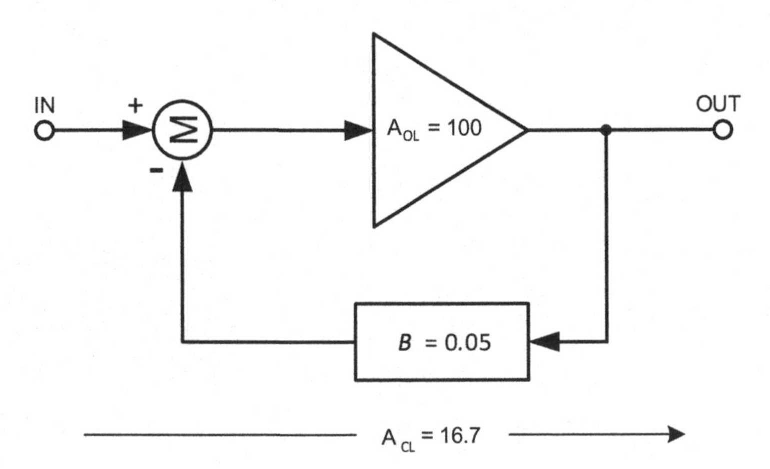

Figure 9.1 Block Diagram of a Negative Feedback Amplifier.

A portion of the output is fed back to the input with a negative polarity. The fraction governing how much of the output is fed back is referred to as *beta* (β), not to be confused with transistor current gain. The negative feedback *loop gain* is the product of A_{ol} and β. The overall gain of the closed-loop amplifier is called the *closed-loop gain* and is designated as CLG or A_{cl}. The action of the negative feedback opposes the input signal and makes the closed-loop gain smaller than the open-loop gain, often by a large factor.

The output of the subtractor at the input of the amplifier is called the *error signal.* It is the difference between the input signal v_{in} and the divided-down replica of the output signal. The error signal, when multiplied by the open-loop gain of the amplifier, becomes the output signal. As the gain A_{ol} becomes large, it can be seen that the error signal will necessarily become small. The output signal will become close to the value v_{in}/β. If A_{ol} is very large and β is 0.05, it is easy to see that the closed-loop gain A_{cl} will approach 20. The important thing to notice here is that the closed-loop gain in this case has been determined by β and not by the open-loop gain A_{ol}. Since β is usually set by passive components like resistors, the closed-loop gain has been stabilized by the use of negative feedback. Because distortion can often be viewed as a signal-dependent variation in instantaneous amplifier gain, it can be seen that the application of negative feedback also reduces distortion.

The closed-loop gain for finite values of open-loop gain is shown in Eq. 9.1. As an example, if A_{ol} is 100 and β is 0.05, then the product $A_{ol}\beta = 5$ and the closed-loop gain (A_{cl}) will be equal to 100/(1 + 5) = 100/6 = 16.7. This is less than the closed-loop gain of 20 that would result for an infinite value of A_{ol}. If A_{ol} were 2000, A_{cl} would be 2000/(1 + 100) = 2000/101 = 19.8. Here we see that if $A_{ol}\beta = 100$, the error in gain from the ideal value is about 1%. The product $A_{ol}\beta$ is called the *loop gain* because it is the gain around the feedback loop. The net gain around the feedback loop is a negative number when the negative sign at the input subtractor is taken into account.

$$A_{cl} = A_{ol} / (1 + A_{ol}\beta) \tag{9.1}$$

Notice that if the sign of $A_{ol}\beta$ is negative for some reason, the closed-loop gain will become greater than the open-loop gain. The closed-loop gain will ultimately go to infinity, and oscillation will result if the product $A_{ol}\beta$ reaches −1. The potential for positive feedback effects is of central importance to feedback compensation and stability.

Negative feedback is said to have a phase of 180°, corresponding to a net inversion as the signal circles the loop. Positive feedback has a phase of 0°, which is the same as 360°.

9.2 Input-Referred Feedback Analysis

A useful way to look at the action of negative feedback is to view the circuit from the input, essentially answering the question, "What input is required to produce a given output?" Viewing negative feedback this way essentially breaks the loop. We will find later that viewing distortion in an *input-referred* way can also be helpful and lend insight.

In the example above with $A_{ol} = 100$ and $\beta = 0.05$, assume that there is 1.0 V at the output of the amplifier. The amount of signal fed back will be 0.05 V. The error signal driving the forward gain path will need to be 0.01 V to drive the forward amplifier. An additional 0.05 V must be supplied by the input signal to overcome the voltage being fed back. The input must therefore be 0.06 V. This corresponds to a gain of 16.7 as calculated above in Eq. 9.1.

Suppose that instead $\beta = -0.009$. This corresponds to positive feedback, which reinforces the input signal in driving the forward amplifier. With an output of 1.0 V, the feedback would be −0.009

V. The drive required for the forward amplifier to produce 1.0 V is only 0.01 V, of which 0.009 V will be supplied by the positive feedback. The required input is then only 0.001 V. The closed-loop gain has been enhanced to 1000 by the presence of the positive feedback. If β were −0.01, the product $A_{ol}\beta$ would be −1 and the denominator in Eq. 9.1 would go to 0, implying infinite gain and oscillation.

9.3 Simple Feedback Circuits

In this section a few simple feedback circuits will be discussed to get a feel for what is involved and how it works in the real world. These discrete circuits are illustrated in Figure 9.2. Some biasing details are omitted.

Emitter Degeneration

The simplest form of negative feedback is emitter degeneration of a BJT common emitter (CE) stage. In general, one gives up gain to obtain the benefits of negative feedback, wherein gain is stabilized and distortion is reduced. Figure 9.2(a) illustrates emitter degeneration. It is a form of series feedback where a feedback signal caused by the output current and appearing across the emitter resistor is effectively in series with the input signal. In very rough terms, the un-loaded gain of the stage is a bit less than R2/R1.

Shunt Feedback

Figure 9.2(b) shows a shunt feedback circuit. Here the feedback opposes the input current by shunting it with current from the output fed back by R2. In the absence of base current and base signal voltage, the current in R1 equals the current in R2 and the gain is slightly less than R2/R1. The ordinary inverting amplifier stage using an op amp is also an example of shunt feedback. The signal current in R4 is ignored because it is quite small.

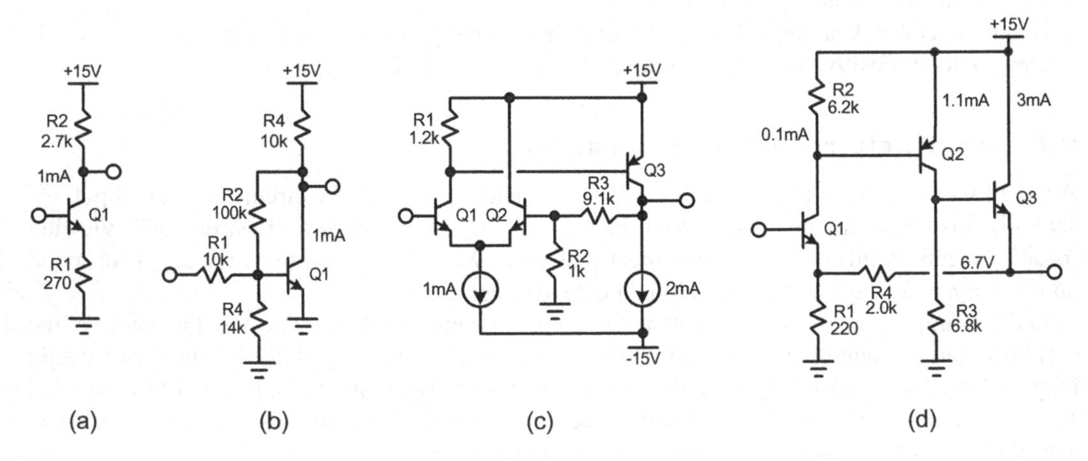

Figure 9.2 Simple Negative Feedback Circuits.

Series Feedback

Figure 9.2(c) shows a discrete series feedback circuit that is non-inverting. A fraction of the output signal is applied to the inverting input of a differential pair, and is effectively placed in series with the input signal, opposing it. This is the same as the series feedback in a non-inverting op amp circuit.

Current Feedback Amplifiers

Figure 9.2(d) shows a current feedback amplifier. The output signal is fed back through R4 to a low-impedance node, in this case the emitter. This is the reason why it is called a current feedback amplifier. Notice that as far as the feedback path goes, Q1 looks like a common-base stage. It is also a form of series feedback circuit. As shown, it is basically a complementary feedback pair (CFP) arranged to have gain. Closed-loop gain is a little less than (R1+ R4)/R1.

9.4 Frequency-Dependent Negative Feedback

If the amount of negative feedback is made frequency-dependent, the closed-loop frequency response will be made to be a function of frequency. This is very useful in implementing tone controls and equalizers. If the amount of negative feedback β increases with frequency, the closed-loop frequency response will fall as frequency increases, for example.

A Phono Preamp

A key point in circuits with frequency-dependent closed-loop gain, as in equalizers, is that β and loop gain are a function of frequency. The simple phono preamp in Figure 9.3 is an example that provides the desired RIAA equalization. Notice that the simple phono preamp is a series-feedback amplifier, with the equalizer *R-C* network connecting the output to the inverting input of the op amp. The use of a JFET op amp allows the circuit to be connected to the cartridge without AC coupling.

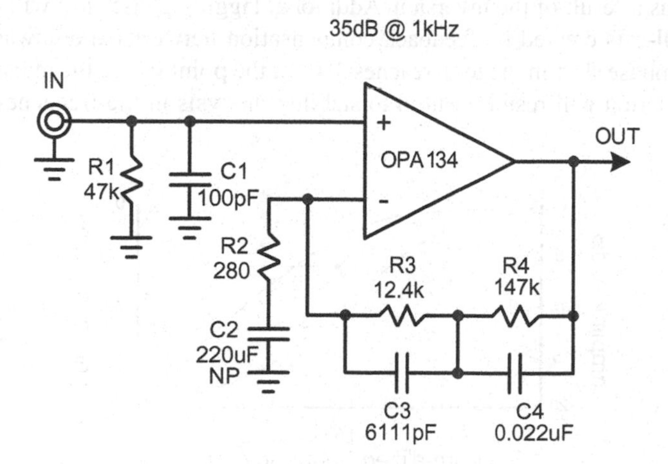

Figure 9.3 Simple Phono Preamp.

It is important to recognize that the feedback loop gain in such a design is frequency-dependant if the open-loop gain of the op amp is constant. However, in the audio band, the open-loop gain of an op amp decreases at 6 dB/octave. It is much higher at low frequencies. The RIAA equalization curve calls for closed-loop gain that falls on average a little less than 6 dB/octave. Therefore, in this example, the amount of distortion-reducing negative feedback is about the same across the audio band. If the amplifier were made discretely with about three transistors, open-loop gain would probably be flat across the audio band, and there would be much less loop gain at low frequencies.

9.5 Feedback Compensation and Stability

Negative feedback must remain negative in order to do its job. Indeed, if for some reason the feedback produced by the loop becomes positive, instability or oscillation may result. As we have seen above, simple negative feedback relies on a 180° phase inversion located somewhere in the loop. High-frequency *roll-off* within the loop can add additional lagging phase shift that will cause the loop phase to be larger than 180°. These excess phase shifts are usually frequency-dependent.

At the 3-dB frequency f_p of a pole, the additional lagging phase shift is 45°, as shown in Figure 9.4. At very high frequencies a pole will contribute 90° of lagging phase shift. If a system has multiple poles this added phase shift may reach 180° at some high frequency. The total phase shift around the loop will then equal 360°, and there will be positive feedback. If the gain around the loop at this frequency is unity, oscillation will occur.

The concept of *poles* and *zeros* is fairly simple and was discussed in Chapter 6. Poles and zeros are central to the understanding of feedback compensation. Ordinary circuits that create high-frequency and low-frequency roll-offs contain poles and zeros, respectively (or some combination of them). Poles introduce lagging phase shift (negative), and zeros introduce leading phase shift (positive).

Phase and Gain Margin

If the accumulation of lagging phase shift causes negative feedback to become positive feedback at some frequencies, instability can result. The nominal phase shift around a simple negative feedback loop is 180° as a result of the inversion. Additional lagging phase shift will be introduced by high-frequency roll-offs created by feedback compensation networks and unwanted poles in the circuit. If the total phase shift in the loop reaches 360° at the point where the magnitude of the loop gain is unity, oscillation will result. Central to stability analysis in the frequency domain are the

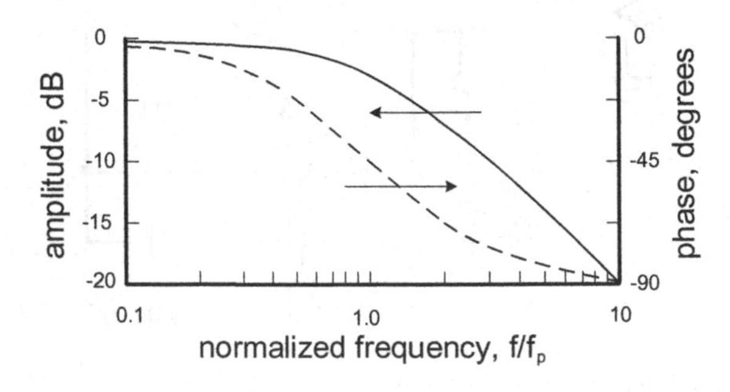

Figure 9.4 Gain and Phase of a Pole.

concepts of *phase margin* and *gain margin.* They describe how much margin a circuit has against instability.

Phase margin enumerates how close the circuit is to having 360° of phase shift at the point in frequency where the loop gain has fallen to unity. This frequency is often referred to as the gain crossover frequency f_c. It is also referred to as the unity loop gain frequency (ULGF). If the phase shift accumulates to a value greater than 360° at a higher frequency than this where there is less than unity gain, oscillation will not result. Since a negative feedback loop has an inversion in the loop corresponding to 180°, if there is 180° of lagging phase shift in the loop, the circuit will oscillate at the frequency where the loop gain has fallen to unity.

Gain margin refers to how much margin exists against the gain becoming unity at a frequency where the loop phase shift has increased to a total of 360°. This is of particular concern because component tolerances can cause a nominal design to have higher gain under some conditions.

Figure 9.5 illustrates the concepts of phase margin and gain margin. The *Y* axis on the left is loop gain in dB. The *Y* axis on the right is lagging phase shift in the loop gain $A_o\beta$. The *X* axis is frequency. As frequency increases, loop gain decreases, falling to unity at 1 MHz. For the frequency range shown, phase lag has already reached −90° at about 100 kHz, and the phase lag is increasing toward −180° as frequency increases. When this lagging phase reaches 180°, oscillation may result.

The hypothetical amplifier for the plot has a gain crossover frequency of 1 MHz. This is where the loop gain has fallen to 0 dB. Loop gain rises by an average of about 6 dB per octave as frequency is reduced, reaching about 30 dB at 50 kHz. Lagging phase shift at low frequencies is 90°. Two poles in the open-loop response are located at 2 MHz. Each contributes 26.6° at 1 MHz, so total lagging phase shift at 1 MHz is 143°. This corresponds to a phase margin of 37°. Loop gain is down to −10 dB at 2 MHz where loop phase shift is 180°. Gain margin is thus 10 dB.

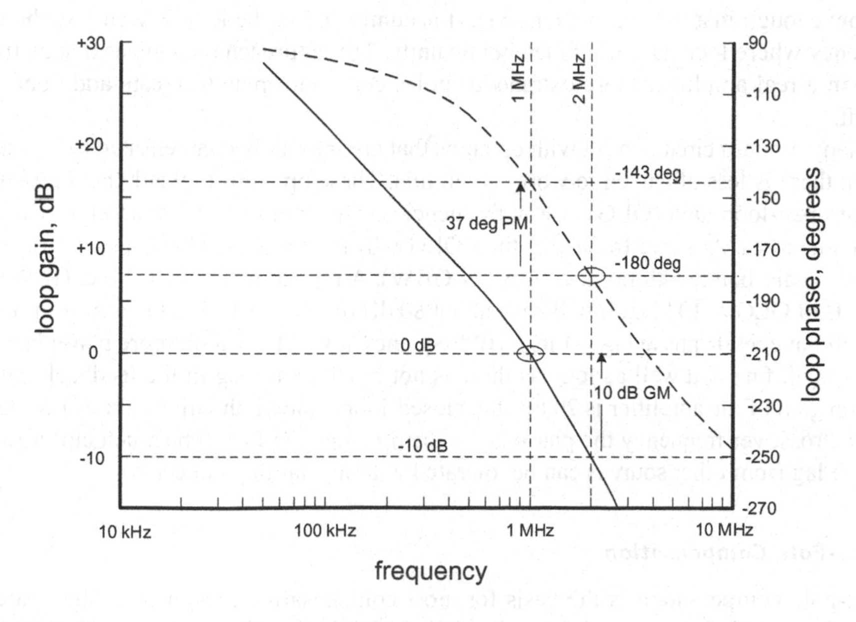

Figure 9.5 Plot Showing Phase Margin and Gain Margin.

Stability against oscillation is not the only reason why adequate phase and gain margin is needed. Amplifiers with inadequate phase or gain margin usually have poor transient response and undesirable peaking in their frequency response. Although some frequency response peaking and overshoot may be acceptable in some applications (no more than 3 dB and 20%), it is good practice to have little or no peaking or overshoot (0.5 dB and 5%). Phase margin of 60° is commonly needed to avoid peaking. About 60° is needed to have no overshoot. A 45° phase margin will typically result in 3 dB of peaking and 50% of overshoot. These rough numbers depend on the origin of the excess lagging phase shift beyond 90° as to whether it is due to one pole or a multiplicity of poles.

Gain and Phase Variation

Component tolerances are not the only contributors to gain and phase variation. Changes in the operating points of active elements like transistors can cause gain and phase changes. One example is the change in collector-base capacitance of a transistor with signal voltage. Another example is change in transistor speed with operating current. Changes in the impedance of the load connected to the amplifier can also affect loop gain and phase.

Capacitive Loads

Capacitive loads on the output of a feedback amplifier can detract from stability margins and may ultimately cause oscillation, either local or due to lagging phase shift introduced into the global feedback loop.

9.6 Feedback Compensation Principles

The simplest way to compensate a negative feedback loop is to roll off the loop gain at a frequency low enough that the lagging phase shift accumulated in the loop is well less than 180° at the frequency where loop gain has fallen below unity. This approach recognizes that as frequency increases in a real amplifier, more extraneous poles come into play to create additional lagging phase shift.

With many modern circuits built with op amps that are internally compensated to be stable with unity gain, there is less of a need to worry about this. These op amps typically begin with 100 dB or more of open-loop gain (OLG) at low frequencies. The gain rolls off at about 6 dB/octave as frequency increases. At some frequency their OLG will fall to unity. This is usually described as the op amp's gain-bandwidth product (GBP or GBW). An op amp with a 10 MHz GBP will have about 60 dB of OLG at 10 kHz. It will have about 80 dB of OLG at 1 kHz. Over most of that range its gain will have a 90° phase lag. At its GBP frequency it will have a bit more phase lag, perhaps 120°. This is all fine and well as long as there is not much phase lag in the feedback path. If the closed-loop gain of the amplifier is 20 dB, the closed-loop bandwidth will be about 1 MHz. At this lower gain crossover frequency the phase lag in the op amp's OLG will be much closer to 90° and more phase lag from other sources can be tolerated without stability concerns.

Dominant-Pole Compensation

Dominant-pole compensation is the basis for most compensation approaches. The strategy is to reduce loop gain with frequency while accumulating as little lagging phase beyond 90° as possible.

The key to achieving this is that a single pole continues to attenuate with frequency while its contribution to lagging phase shift is limited to 90°. For this reason a feedback circuit with a single pole in its loop can never become unstable.

In real circuits there will be many poles, but if the roll-off behavior is strongly dominated by a single pole, the stability criteria will be more straightforward to meet. The effects of all other poles can then be lumped together as so-called *excess phase*. In this case, at the frequency where the gain around the loop has fallen to 0 dB, the phase lag of the dominant pole will be 90°. This means that there is an additional 90° to play with before hitting instability. You might allocate 30° for excess phase from all other sources and keep the remaining 60° in your pocket as phase margin. You then choose the maximum gain crossover frequency f_c as the frequency where the total amount of excess phase does not exceed 30°.

Excess Phase

Excess phase is usually thought of as additional lagging phase shift that would not have been expected based on the amount of high-frequency amplitude roll-off. A pure delay is a very good example of excess phase. If a signal runs through 10 ft of coaxial cable, it will encounter very little loss, but it will experience a time delay due to the speed of light in the cable (flight time or propagation delay). This will be on the order of 1.5 ns per foot, or 15 ns. At 67 MHz, this would correspond to about 360° of phase lag.

More often, excess phase is not truly time delay, but rather the accumulation of phase shift from many high-frequency poles, sometimes called *parasitic poles*. Even together, these poles may not create much attenuation at the unity-gain frequency, but the amount of phase shift they create can accumulate to an amount sufficient to reduce phase margin in a negative feedback circuit. Excess phase in an amplifier can accumulate if there are a significant number of stages within the feedback loop, each of which is creating poles at higher frequencies.

Excess phase can also originate at the feedback input to the differential input stage (LTP). The resistance of the feedback network can form a pole against the input capacitance of the LTP. This input capacitance can be a combination of base-emitter capacitance due to finite f_T and the base-collector capacitance of the transistor. This latter capacitance can be multiplied by the Miller effect if the input stage has voltage gain to its collector at high frequencies. The phase lag created at the input node can be reduced if smaller impedances are employed in the feedback network. Cascoding the input stage can also help by eliminating any Miller effect.

Alternatively, a very small capacitor placed across the feedback resistor can also reduce this effect by forming a capacitance voltage divider with the input capacitance of the input stage. If the input capacitance were 5 pF and the gain of the amplifier were 20, this capacitor would have to be very small, on the order of 0.25 pF, however. Such small amounts of capacitance can be created by running a pair of traces parallel to each other or by running traces above and below each other on opposite sides of the PCB. A slightly larger value of capacitor will provide some phase lead, which may actually help phase margin a bit. Great caution is needed here to avoid using too much capacitance; it can reduce gain margin and provide a sneak path for EMI from the amplifier output back to the input stage.

Lag Compensation

Lag compensation is perhaps the simplest form of frequency compensation, but is usually not the best. It simply involves the introduction of a low-frequency pole by placing a shunt capacitor from the output of the primary voltage amplifier stage (VAS) to ground. Unfortunately, there is also

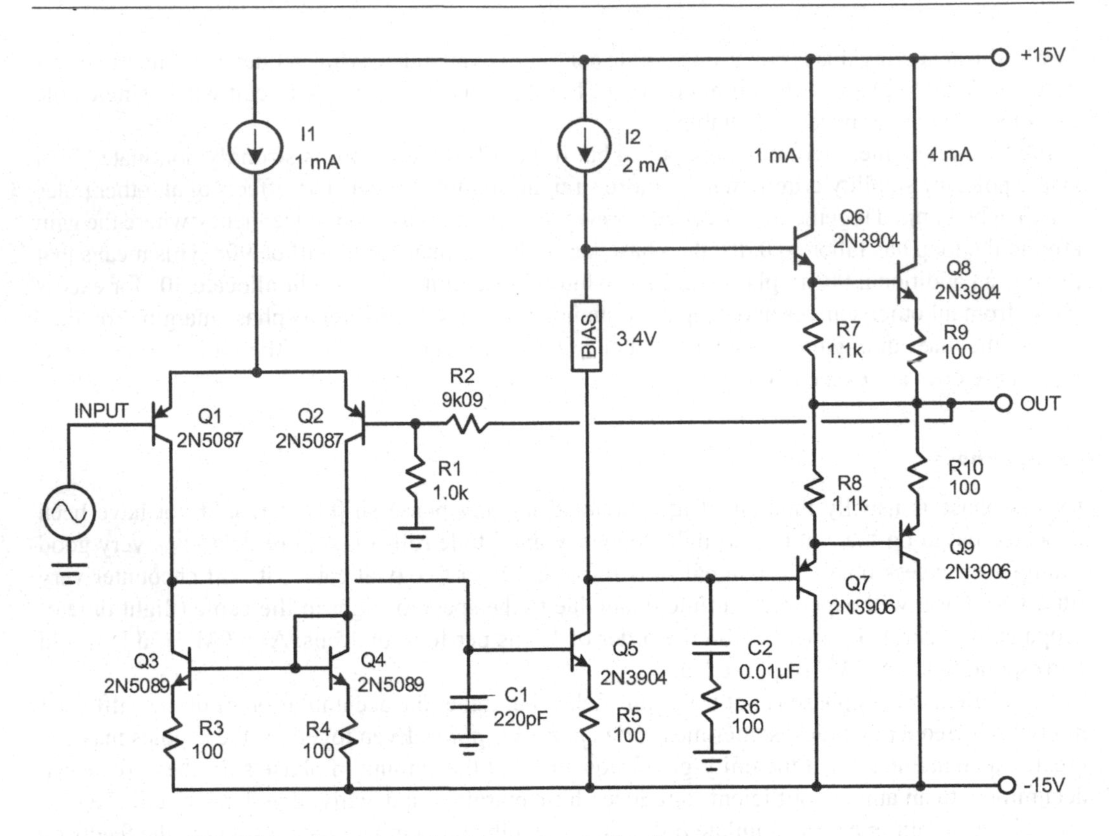

Figure 9.6 An Amplifier where Lag Compensation Is Employed.

a pole in the base circuit of the VAS at a low enough frequency to add a large amount of excess phase. This will often force you to set f_c fairly low. You may have two poles *running together* (close in frequency) over a significant frequency range before the loop gain falls to 0 dB.

Consider the amplifier of Figure 9.6 and ignore compensating components C1, C2 and R6 for the moment. It is a simple amplifier whose closed-loop gain is set to 10 by R1 and R2. It is based on a simple discrete op amp circuit. The uncompensated amplifier has over 96 dB of loop gain at DC, and it does not fall to 0 dB until 16 MHz, at which point the accumulated phase lag is about 140°, which is 50° more than is needed for oscillation. Choose a conservative goal of 1 MHz for the gain crossover frequency f_c. Loop gain of the uncompensated amplifier is about 39 dB at 1 MHz, so the compensation must get rid of 39 dB of loop gain at that frequency.

Using a large 0.02-μF shunt capacitor C2 loading Q5's collector node will provide lag compensation that brings the gain down by the required amount. C2 creates the dominant pole. However, the pole at the base of Q5 is also contributing phase lag, causing the total phase lag at 1 MHz to be 142°, leaving only 38° of margin against oscillation.

The pole at the base of Q5, at about 1 MHz, is heavily dependent on the f_T of Q5. A 220-pF shunt capacitor C1 can be added to the base circuit to bring that pole down to 200 kHz and stabilize it against f_T variations in Q5. That adds attenuation at high frequencies, bringing f_c down to about 513 kHz. C2 can now be reduced to 0.01 µF to bring f_c back up to about 1 MHz. However, phase margin has actually decreased to only 8° because both poles are contributing more phase lag – mainly as a result of moving the pole at the base of Q5 to a lower frequency.

Resistor R6 can now be inserted in series with C2 to create a zero at 200 kHz to cancel the pole at the base of Q5. The result is an approximate single-pole roll-off up to beyond 2 MHz. However, the increased gain resulting from canceling the pole at the base of Q3 has now pushed f_c up to 3.6 MHz. After numerous iterations, C1 is set to 220 pF, C2 is set to 0.01 μF and R6 is set to 100 Ω. A stable amplifier with CLG of 20 dB and bandwidth of 1 MHz results. As to be described later, phase margin is 67° and gain margin is 28 dB.

As described here, this may seem like a tedious and iterative process, but the steps describe the tactics and component interactions for implementing one form of lag compensation. Moreover, SPICE simulation of the circuit makes such iterations go quickly. Lag compensation of this type is somewhat of a brute-force and sub-optimal technique. This is not a very good circuit design, with a slew rate of only about 0.25 V/μs.

Miller Compensation

Miller compensation takes advantage of the *Miller effect*, wherein the effective capacitance seen at the input of an amplifier stage, due to a feedback capacitance from output to input, is approximately equal to that feedback capacitance multiplied by one plus the voltage gain of the stage. The resulting large effective input capacitance pushes the pole frequency at the input node of Q5 down to a much lower frequency. At the same time, the shunt feedback resulting from the feedback capacitance lowers the impedance at the output node of the stage, pushing the pole at the collector

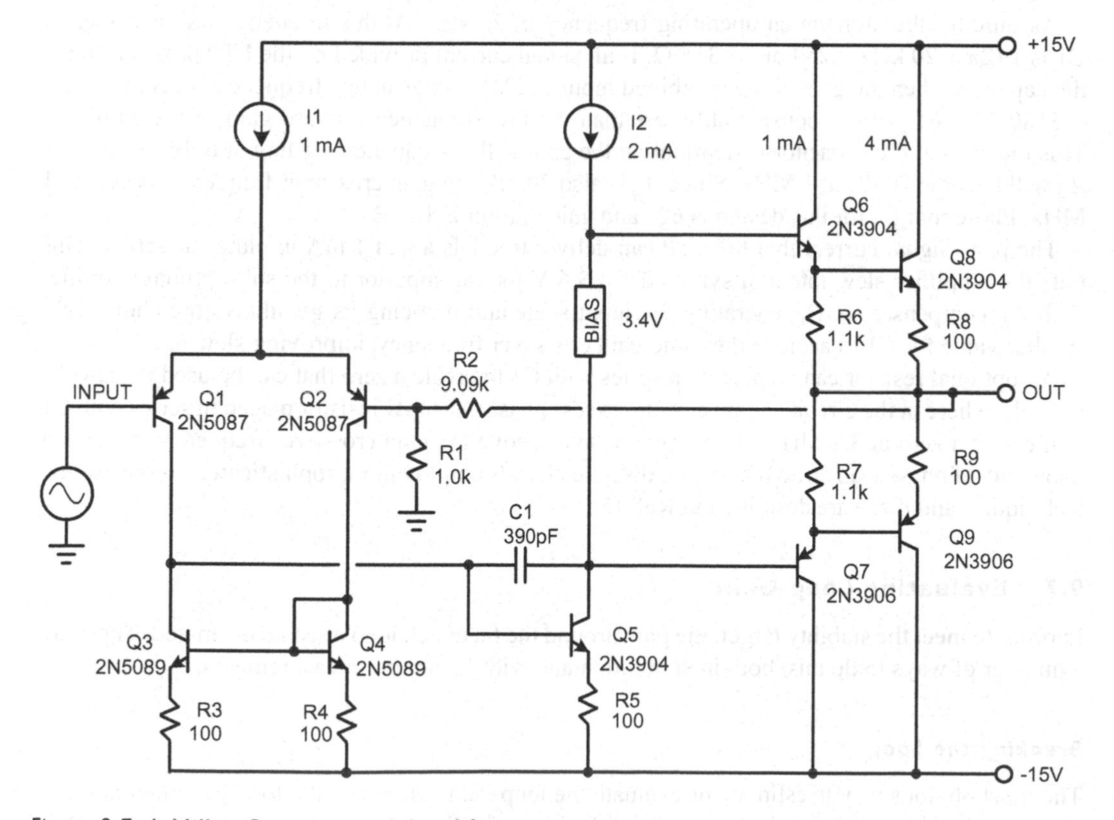

Figure 9.7 A Miller-Compensated Amplifier.

to a higher frequency. The poles at the input and output nodes of the stage are thus pushed apart in frequency. This phenomenon is referred to as *pole splitting*. The first pole is thus in control of the phase response of the stage over a very wide frequency range. Over this frequency range, the phase lag will be a nearly constant 90°.

Miller compensation thus uses local feedback to roll off the high-frequency response of the amplifier. The gain that is "thrown away" acts to linearize the VAS with shunt feedback. The same amplifier as in Figure 9.6 is shown with Miller compensation in Figure 9.7. Capacitor C1 is the so-called Miller compensation capacitor C_M. It stabilizes the global negative feedback loop by rolling off the high-frequency gain of the amplifier so that the gain around the feedback loop falls below unity before enough phase lag builds up to cause instability.

At high frequencies the combined gain of the input stage and the VAS is equal to approximately the product of the transconductance of the input stage (IPS) and the impedance of C1. At high frequencies the gain is thus dominated by C1. Since the impedance of C1 is inversely related to frequency, the gain will decrease at 6 dB per octave as frequency increases. The transconductance *gm* is just the inverse of the total LTP emitter resistance R_{LTP}, here $2r_e$. At $Ic = 0.5$ mA, re' is about 26 Ω. Transconductance *gm* of the input stage is $1/2re'$.

While at low frequencies the gain of the combined input stage and VAS is set by the product of the individual voltage gains of those two stages, at higher frequencies that combined gain is set by the ratio of the impedance of C1 to R_{LTP}. The frequency at which the gain set by R_{LTP} and C1 becomes smaller than the DC gain is where the roll-off of the amplifier's open-loop frequency response begins. R_{LTP} for this design is about 104 Ω. C1 has been chosen to be 270 pF in order to establish f_c at 1 MHz.

Assume for the moment an operating frequency of 20 kHz. At this frequency the reactance of C1 is $1/(2\pi \times 20\text{ kHz} \times 250\text{ pF}) = 318\ \Omega$. If all signal current provided by the LTP passes through the capacitor, then the gain of the combined input and VAS stage at this frequency is about $X_{C1}/2r_{e'}$ $= 3180/52 = 61$. This is considerably less than the low-frequency forward gain of the amplifier. This means that the capacitor is dominating the gain at this frequency. Falling at 6 dB per octave, A_{ol} will become 20 dB at 1 MHz. Since A_{cl} is also 20 dB, the gain crossover frequency occurs at 1 MHz. Phase margin for this design is 62° and gain margin is 12 dB.

The peak signal current that the LTP can deliver to C1 is about 1 mA in either direction. This puts the amplifier slew rate at a symmetrical 3.6 V/μs, far superior to the sub-optimal amplifier with lag compensation. Degenerating the input stage and reducing its *gm* allows the choice of a smaller value for C1 to achieve the same gain crossover frequency, improving slew rate.

An optional resistor can be placed in series with C1 to create a zero that can be used to cancel a pole elsewhere in the circuit or cancel some excess phase. A 160-Ω resistor placed in series with C1 will create a zero at 3.7 MHz, almost two octaves above the gain crossover frequency, providing about 14° of phase lead at 1 MHz. Some discrete circuits employ more sophisticated compensation techniques, and these are described in Ref. [5].

9.7 Evaluating Loop Gain

In order to meet the stability target, the gain around the feedback loop must be estimated. There are a number of ways to do this, both in simulation and with laboratory measurements.

Breaking the Loop

The most obvious way to estimate or evaluate the loop gain is to break the loop in a simulation. A stimulus signal is applied at the input side of the loop. The frequency and phase responses are then

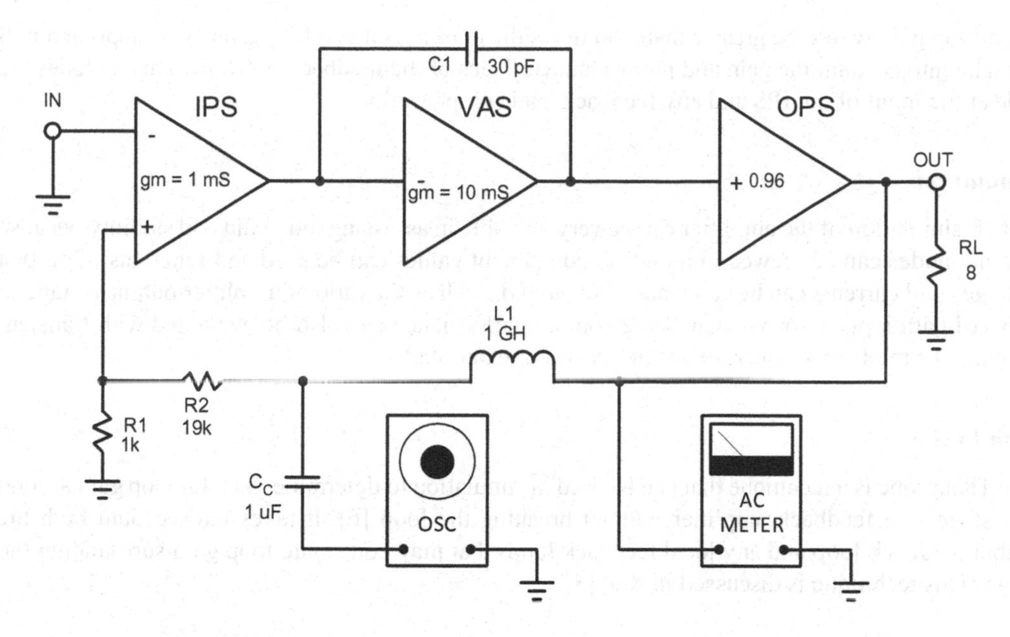

Figure 9.8 **Breaking the Loop with a Large Inductor.**

measured at the output side of the loop. Means must be used to maintain proper biasing and DC levels in the amplifier when this is done. This is not always practical in the real world where very large amounts of gain may be involved.

There is also a caveat: The loading of the output side of the break will often not be identical after the loop is broken, and this will cause some error. If the source on the output side of the break is of very low impedance compared to the load seen looking into the input side of the break, the error will be quite small. This will often be the case in a small-signal amplifier where the loop is broken at the input to the feedback network and the output stage has a relatively low impedance.

In SPICE simulations the loop can be kept closed at DC by connecting an extremely large inductor across the break. It is possible to employ a 1 GH inductor. The source signal is then applied to the feedback network through an AC coupling capacitor. The low-frequency corner of this coupling capacitor against the input resistance of the feedback network will determine how low in frequency the results of this method will be accurate. The loop gain is then inferred by viewing the signal at the output of the amplifier. This is illustrated in Figure 9.8.

In the laboratory, the loop can be kept closed at DC by connecting a non-inverting auxiliary DC servo circuit from the output of the amplifier to the feedback input of the IPS. The low-level test signal from the signal generator can then be applied to the input of the feedback network. This technique is discussed in Ref. [5].

Exposing Open-Loop Gain

Exposing the forward gain A_{ol} by setting closed-loop gain very high is a useful way to estimate the high-frequency open-loop gain roll-off and phase shift. The forward gain is *exposed* by increasing the closed-loop gain by a factor of 100. This usually decreases the loop gain by a sufficient amount over the frequency range of interest that feedback effects do not affect the gain from input to output at the frequencies of interest. The technique will not be accurate at low frequencies where the open-loop gain would be greater than 100 times the nominal closed-loop gain. This approach will not take into account the gain and phase characteristics of the feedback network. This includes the pole at the input of the IPS and any feedback lead compensation.

Simulation

SPICE simulation of the amplifier can be very valuable in assessing loop gain and stability because internal nodes can be viewed, impractical component values can be used and functions of probed voltages and currents can be calculated and plotted, such as the ratio of amplifier output voltage to forward path input-error voltage. Time domain performance can also be evaluated with transient simulations to observe square-wave behavior, for example.

Tian Probe

The Tian probe is a technique that can be used in simulation to determine the total loop gain around any stage in a feedback amplifier without breaking the loop [6]. It takes into account both the global feedback loop and any local feedback loops that may contribute loop gain surrounding the stage. This technique is discussed in Ref. [5].

9.8 Evaluating Stability

One of the most important aspects of feedback amplifier design is assessing its stability. Although it is true that one can design for stability, it still must be assessed in simulation and/or in the actual prototype circuit. Obviously, what can be seen and evaluated is different in simulation than in the real circuit. Each has advantages where the other may be a bit blind. A proper assessment of stability in the real world using the real circuit is a must, but for those who can simulate, it is also desirable and beneficial to assess stability in simulation. Just bear in mind that many real-world parasitics are not taken into account in a typical simulation.

Feedback stability can often be inferred from viewing the closed-loop frequency response and looking for peaking. Peaking of the closed-loop response by more than about 1 dB just prior to final roll-off is a danger sign. However, apparent flatness of the closed-loop frequency response is not always sufficient evidence of adequate stability margin. Transient response must always be checked as well; this is best done with a transient simulation using a square-wave source.

Instability can be either local or in the global feedback loop. Local instability can originate from a local feedback loop or from local circuit instability such as an emitter follower driving a capacitive load. Such local instabilities can often occur in output stages driving a capacitive load. Simply putting a 100-Ω resistor in series with the output of a preamplifier can cure this kind of problem, whether the output amplifier is a discrete design or one implemented with an op amp.

Probing Internal Nodes in Simulation

In assessing stability with *AC* simulations, it is important to look for evidence of peaking (or sometimes sharp dips) at nodes internal to the circuit. Figure 9.9 shows a block diagram of a simplified power amplifier with some suggested probing points. The feedback input P1 of the IPS should look like a unity-gain follower amplifier stage with respect to the amplifier input signal. Probe this point and look for overshoot, ringing, and peaking. This may show behavior that is masked at the output of the amplifier by high-frequency roll-offs. If this is done in a real amplifier, the IPS feedback input should be probed with a high-impedance low-capacitance probe.

Stability of local loops can be assessed with an *AC* simulation in a relatively noninvasive way by injecting an *AC* current at a chosen node and observing the resulting signal voltage. This procedure reveals the impedance of the node as a function of frequency. At frequencies where there is instability, the impedance will rise markedly. In simulation, the current injection is often conveniently carried out with a voltage source in series with a 10-MΩ resistor. The probing locations shown in Figure 9.9 are examples of where this technique can be applied in simulation. If the stability probe is placed at the input node P2 of a Miller-compensated VAS, the results may uncover a local instability. The same can be said for probing at P3. Such local instabilities or oscillations can be in the 20–200-MHz range, and can easily be overlooked.

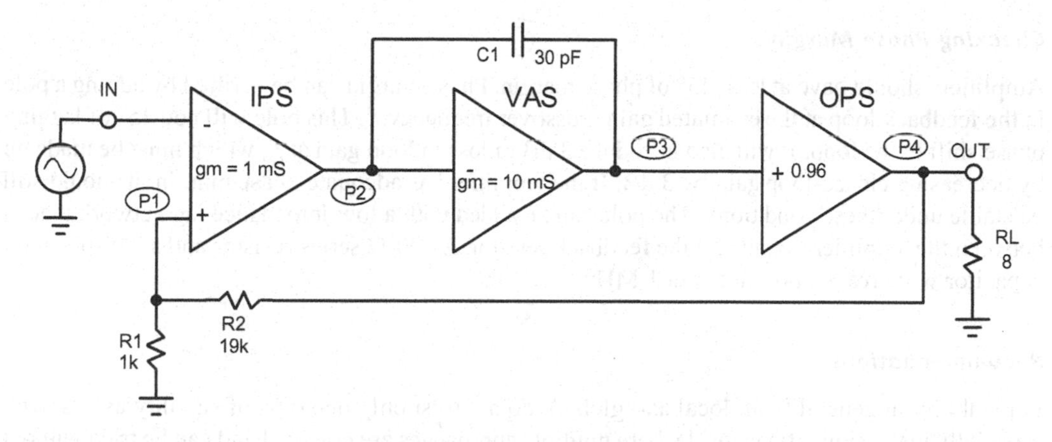

Figure 9.9 **Probing Points for Stability Evaluation.**

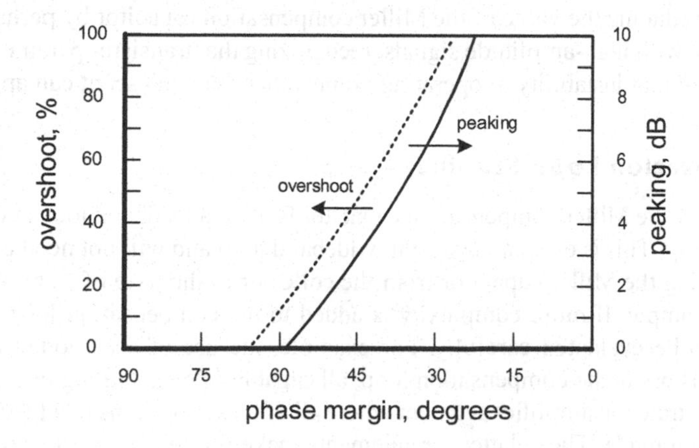

Figure 9.10 **Overshoot and Frequency Response Peaking vs Phase Margin.**

Figure 9.10 is a plot of square-wave overshoot (left) and frequency response peaking (right) as a function of phase margin on the X axis from 90° down to 0°. This data was obtained by simulating an amplifier with a dominant pole and four parasitic poles at a high frequency above the gain crossover frequency. As the frequency of the parasitic poles was reduced, the phase margin, square-wave overshoot and frequency response peaking were recorded and plotted. These numbers will not be accurate for all amplifier designs and roll-off profiles, but they are helpfully representative.

Checking Gain Margin

Amplifiers should have at least 6 dB of gain margin. Gain margin can be verified by reducing the closed-loop gain by 6 dB. In an amplifier with a gain of 20, this would mean reducing the feedback resistor to change the gain to 10. If the amplifier has at least 6 dB of gain margin, it should still be stable under these conditions.

Checking Phase Margin

Amplifiers should have at least 45° of phase margin. Phase margin can be verified by adding a pole in the feedback loop at the estimated gain crossover frequency f_c. This pole will add 45° of lagging phase shift to the loop. It will also introduce 3 dB of loss in loop gain at f_c, which must be made up by decreasing closed-loop gain by 3 dB. If the amplifier has adequate phase margin, it should still be stable under these conditions. The pole can be added with a low-impedance lag network placed between the amplifier output and the feedback resistor. A 470-Ω series resistor and a 330-pF shunt capacitor will create a pole at about 1 MHz.

Recommendations

For stability in general (both local and global), do not trust only one type of stability assessment, especially in the simulation world where multiple approaches are practical and can be tried without great effort. At the breadboard level, don't assume that there is adequate stability just because the circuit does not oscillate. Carefully evaluate frequency response and square-wave response.

Operate the circuit with lighter compensation to assess margin against instability. This can be accomplished by reducing the value of the Miller compensation capacitor by perhaps a factor of 2. Operate the circuit with high-amplitude signals, recognizing that transistor parameters change with operating point and that instability at operating points other than quiescent can appear.

9.9 Compensation Loop Stability

The loop formed by the Miller compensation capacitor is itself a feedback loop and must also obey the rules for stability. This is often a very tight, wideband loop and will not need compensation for stability. Connecting the Miller capacitor from the collector to the base of a simple one-transistor VAS is a good example. If more complexity is added to the compensation loop, stability of this loop may have to be evaluated carefully. For example, the use of a cascoded Darlington VAS places three transistors in the compensation loop, all capable of contributing excess phase.

Figure 9.11 illustrates an amplifier segment where all of these additions to a feedback compensation loop have been made. These latter arrangements make the feedback compensation loop less "local" and in some cases can make it less stable by allowing the introduction of excess phase shift by the added stages.

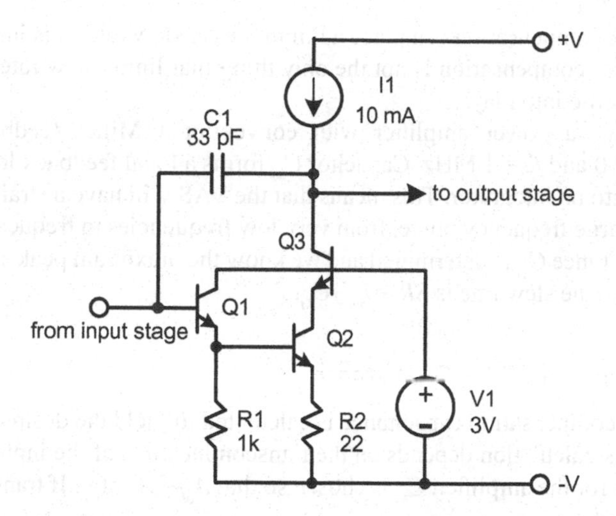

Figure 9.11 Example of a "Longer" Miller Compensation Loop.

9.10 Slew Rate

The maximum rate of voltage change that an amplifier can achieve is usually referred to as the *slew rate*. It is often expressed in volts per microsecond. Virtually all operational amplifier datasheets specify their achievable slew rate. Slew rate in an amplifier is usually limited by the ability of a particular stage to charge a capacitance at a given rate. Indeed, for a circuit that can supply a current I_{max} to a node with capacitance C on it, the slew rate is simply I_{max}/C. If a maximum current of 1 mA is available for charging 100 pF, the slew rate will be limited to 10 V/µs. If an input stage can deliver 2 mA of signal swing to a 50-pF Miller compensation capacitor, achievable slew rate will be about 40 V/μs if there are no other parts of the circuit limiting slew rate.

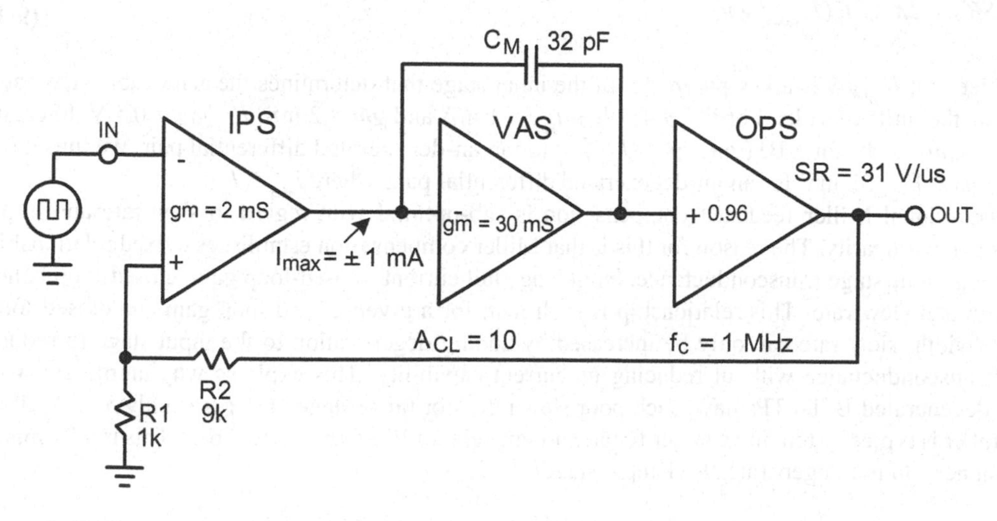

Figure 9.12 Traditional Miller Feedback Compensation.

Negative feedback compensation can place a limitation on slew rate. It is important to point out that negative feedback compensation is not the only thing that limits slew rate, but in many cases it is the first one to come into play.

Figure 9.12 shows a power amplifier with conventional Miller feedback compensation. Closed-loop gain is 10 and $fc = 1$ MHz. Capacitor C_M forms a local feedback loop around the VAS that turns the VAS into an integrator. This means that the VAS will have a straight 6 dB per octave roll-off over a very large frequency range, from very low frequencies to frequencies above the gain crossover frequency. Once C_M is determined and we know the maximum peak signal current output of the input stage I_{max}, the slew rate is $SR = I_{max}/C_M$.

Calculating the Required Miller Capacitance

The required Miller compensation capacitance is calculated to yield the desired unity-gain crossover frequency f_c. This calculation depends on the transconductance of the input stage and the chosen closed-loop gain for the amplifier. C_M is chosen so that $A_{ol} = A_{cl}$ at f_c. If transconductance of the input stage is gm, then

$$A_{ol} = gm / 2\pi C_M f_c \tag{9.2}$$

Setting $A_{ol} = A_{cl}$, we have

$$C_M = gm / (2\pi A_{cl} f_c) \tag{9.3}$$

For $gm = 2$ mS, $A_{cl} = 10$, and $f_c = 1$ MHz, we have $C_M = 32$ pF.

Slew Rate

The slew rate of the amplifier is simply $SR = I_{max}/C_M$. Substituting Eq. 9.3 for C_M and doing some rearrangement, we have

$$SR = 2\pi A_{cl} f_c (I_{\max} / gm) \tag{9.4}$$

Notice that I_{max}/gm is a key parameter of the input stage that determines the achievable slew rate. It has the units of volts. For Eq. 9.4, when $I_{max} = 1$ mA and $gm = 2$ mS, $I_{max}/gm = 0.5$ V. Interestingly, $gm = I_c/V_T$ for a BJT and $gm = I_{tail}/2V_T$ for an un-degenerated differential pair. We thus have $I_{max}/gm = V_T = 52$ mV for an un-degenerated differential pair, where $I_{max} = I_{tail}$.

Traditional Miller feedback compensation is sub-optimal with regard to slew rate and high-frequency linearity. The reason for this is that Miller compensation establishes a fixed relationship among input-stage transconductance, input-stage tail current, closed-loop gain, closed-loop bandwidth and slew rate. This relationship is such that, for a given closed-loop gain and closed-loop bandwidth, slew rate can only be increased by adding degeneration to the input stage to reduce its transconductance without reducing its current capability. This explains why amplifiers with un-degenerated BJT LTPs have such poor slew rate. For these stages I_{max}/gm is only 52 mV. This number is typically ten times larger for an un-degenerated JFET differential pair. This is why many designers do not degenerate JFET input stages.

9.11 Harmonic and Intermodulation Distortion Reduction

How effective is negative feedback at reducing distortion? In general, distortion results in undesired frequencies that were not in the signal in the first place. This is especially easy to comprehend in the case of a continuous sine-wave input. Nonlinearities create distortion products at integer multiples of the fundamental frequency (i.e., harmonics). The reduction in distortion by negative feedback is reflected in a reduction of the amplitude of these various harmonics. What is key to understanding is that the amount of reduction depends on the amount of negative feedback (loop gain) at the frequency of each harmonic, recognizing that the loop gain will often be frequency-dependent, often even in the audio band, as is the case with an op amp.

An op amp circuit with 40-dB of loop gain at a fundamental frequency of 20 kHz will usually only have 34 dB of loop gain at 40 kHz, the second harmonic. In rough terms, this means that the distortion reduction is only 34 dB at the second harmonic, NOT 40 dB. Moreover, the loop gain will only be about 30 dB at the 60-kHz third harmonic, and reduction of this harmonic by feedback will be only about 30 dB. In many cases, the effectiveness of negative feedback reducing distortion decreases with increased frequency of the harmonic being reduced.

Intermodulation Distortion

The way in which negative feedback reduces intermodulation distortion products is a bit different because IM products in the audio band usually appear at similar or lower frequencies than the test frequencies. Bear in mind that IM products appear at sum and difference frequencies of the test tones. They also appear at the sum and difference frequencies of integer multiples of both tones. In general, the tones appear at $f = mf_1 \pm nf_2$. If test tones at 19 and 20 kHz are applied to an amplifier with second- and third-order nonlinearities, IM products will appear at 1, 18 and 21 kHz.

For the common SMPTE IM test, two signals of 60 Hz and 7 kHz are mixed in a 4:1 ratio and fed to the device under test (DUT). A SMPTE IM analyzer connected to the output of the DUT measures the amount of amplitude modulation on the 7 kHz carrier signal and reports it as a percentage. Alternatively, a spectrum analyzer is connected to the output and the spectral components at frequencies other than 60 Hz and 7 kHz are measured and combined on a root-mean-square (RMS) basis and then referred to the RMS value of the 7-kHz carrier.

For the CCIF IM test, two signals of 19 and 20 kHz are mixed in equal amplitudes and fed to the DUT. A spectrum analyzer is connected to the output and the spectral components at frequencies other than 19 and 20 kHz are measured and combined on an RMS basis.

References

1. Black, Harold, U.S. patent 2,102,671, issued 1937.
2. Walt Jung, *Op Amp Applications Handbook*, Analog Devices, Inc., Newnes, Burlington, MA, 2005.
3. "Loop Gain Measurements", Sergio Franco, EDN, September 13, 2014.
4. Marc Thompson, *Intuitive Analog Circuit Design*, 2nd edition, Newnes, Oxford, 2014.
5. Bob Cordell, *Designing Audio Power Amplifiers*, 2nd edition, Routledge, New York, NY, 2019.
6. Michael Tian, V. Visvanathan, Jeffrey Hantgan, and Kenneth Kundert, *Striving for Small-Signal Stability*, Circuits & Devices, January 2001.

Chapter 10

Noise

All passive and active circuits create noise, and noise is undesirable. Noise limits the dynamic range of a system. In order to design circuits for the lowest noise, the noise-generating mechanisms must be understood. There are numerous tradeoffs, and the best low-noise design is usually very dependent on the particular application. More complete discussions on noise as it pertains to audio amplifiers and semiconductors can be found in Refs. [1–13].

The noise characteristics of the circuits in a preamplifier or mixer are increasingly important as the nominal signal level becomes smaller, such as in the case of a microphone preamp. Noise buildup can also be a problem in a mixing circuit that has many inputs, since the noise gain of the circuit increases with the number of inputs being mixed. Here we will explore the ways in which noise is governed by the circuits and discuss ways to minimize it.

10.1 Signal-to-Noise Ratio

Noise is usually specified as being so many dB down from the nominal output of an amplifier stage or function in a stated bandwidth or with a stated noise weighting. If the maximum output is significantly greater than the nominal amount, this difference is often referred to as headroom. The signal-to-noise ratio (SNR) plus the headroom is the available dynamic range.

The noise specification may be un-weighted or weighted. Un-weighted noise for a preamplifier or mixer function will typically be specified over a full 20-kHz bandwidth (or sometimes up to 80 kHz or more). Weighted noise specifications take into account the ear's sensitivity to noise in different parts of the frequency spectrum. The most common one used is *A weighting*, whose frequency response weighting is illustrated in Figure 10.1. The A-weighted noise is measured by placing a filter with this frequency response in front of a true-RMS AC voltmeter. Notice that the weighting curve is up about +1.2 dB at 2 kHz, whereas it is down 3 dB at approximately 500 Hz and 10 kHz.

Consider a microphone preamplifier stage with input-referred noise density of 3 nV/√Hz. Noise density will be discussed in more detail in Section 10.5. Think of a noiseless amplifier driven by a white noise source. That is what it means for the noise to be *input-referred*. All of the noise sources in the preamplifier act together to produce the same noise result as the hypothetical noise generator mentioned above. Assume that the amplifier gain is 100 (40 dB). Output noise will be 300 nV/√Hz. The un-weighted noise in a 20-kHz bandwidth will be 141 times that, since there are 141 √Hz in a 20-kHz noise bandwidth. Output noise will be 42 μV. This is 85 dB below 774 mV (0 dBu), so the un-weighted SNR is 85 dB. Of course, measured SNR will be larger for a higher signal level reference, like 5 V_{RMS}.

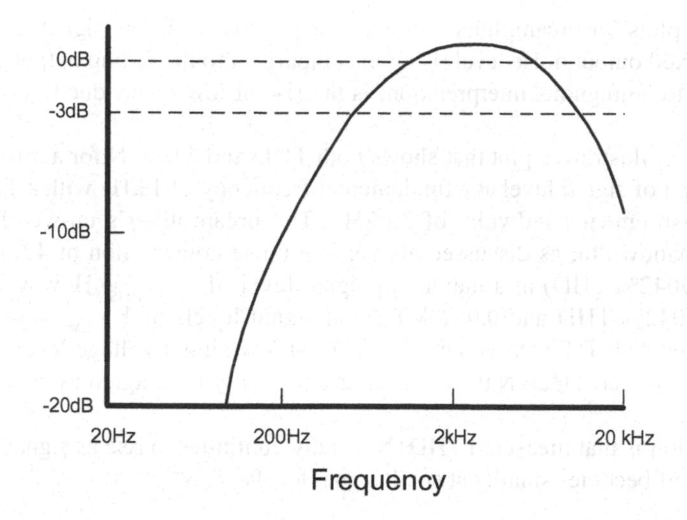

Figure 10.1 A-Weighting Frequency Response.

THD+N Measurements

Noise can introduce confusion into the traditional THD+N measurements made for preamplifier and mixer functions. Analog distortion analyzers report THD+N because they remove the fundamental and display what is left over, which is distortion harmonics and noise. This is also what most specification sheets report. Newer distortion analyzers use the frequency spectrum or FFT analysis and can report THD by itself or THD+N.

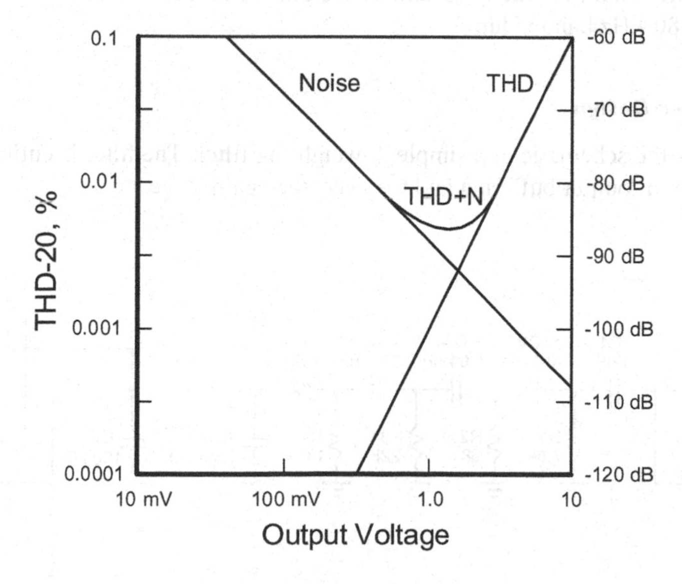

Figure 10.2 20-kHz THD and THD+N vs Power.

Many THD+N plots for preamplifiers show a rise in THD+N at low signal levels. This can be the result of the fixed output noise level rising in comparison to the falling output signal level. The plot is thus open to ambiguous interpretation; is the rise at low levels due to noise or crossover distortion?

Figure 10.2 is an illustrative plot that shows both THD and THD+N for a microphone preamplifier as a function of signal level at a fundamental frequency of 1 kHz with a THD+N analyzer un-weighted measurement bandwidth of 20 kHz. The preamplifier's un-weighted SNR is 85 dB in a 20-kHz bandwidth, as discussed above. The noise contribution of 42 μV is equivalent to −108 dB (0.00042% THD) at a maximum signal level of 10 V_{RMS}. However, note that it is equivalent to 0.0042% THD and 0.042% THD at signal levels of 1 V_{RMS} and 100 mV_{RMS}, respectively. As expected, THD starts out very small at low signal voltage levels, but notice that THD+N starts out higher. THD+N then decreases and finally rises again as the maximum signal level is approached.

A key observation is that measured THD+N usually continues to rise as signal level decreases, while THD by itself becomes smaller at small signal levels.

10.2 A-Weighted Noise Specifications

The frequency response of the A-weighting curve is shown in Figure 10.1. It weights the noise in accordance with the human ear's perception of noise loudness. Note that it is up by 1.2 dB at 2 kHz. The weighting filter includes a second-order high-pass function, causing it to be down by 26 dB at 60 Hz. It also includes a third-order distributed-pole low-pass function that causes it to be down by 10 dB at 20 kHz.

The A-weighted noise specification will usually be quite a bit better than the un-weighted noise number because the A-weighted measurement tends to attenuate noise contributions at higher frequencies and hum contributions at lower frequencies. The A-weighted number will sometimes be as much as 5–15 dB better than the wideband un-weighted number. Sometimes un-weighted noise is measured in an 80-kHz bandwidth.

A-Weighting Filter Design

Figure 10.3 shows the schematic of a simple A-weighting filter. The filter is entirely passive with the exception of input/output buffering and the necessary gain stage.

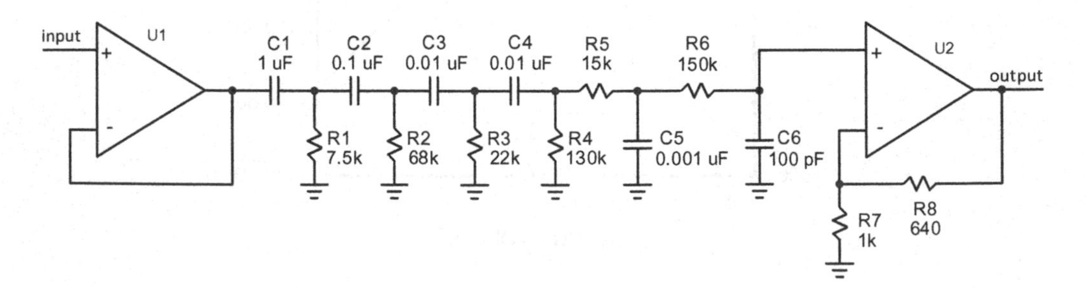

Figure 10.3 A-Weighting Noise Filter.

10.3 Noise Power and Noise Voltage

The noise arising from different sources is usually assumed to be uncorrelated. For this reason, it adds on a power basis. This means that noise voltage adds on an RMS basis as the square root of the sum of the squares of the various sources. Two noise sources each having 10 μV_{RMS} will add to 14.1 μV_{RMS}. Two noise sources, one 10 μV and the other 3 μV, will sum to $\sqrt{(100+9)} = \sqrt{(109)} =$ 10.44 μV. This shows how a larger noise source will tend to dominate over a smaller noise source.

$$V_{noise\_total} = \sqrt{\left(V_1^2 + V_2^2\right)} \tag{10.1}$$

$$V_{noise\_total} = \sqrt{\left(10^2 + 3^2\right)} = \sqrt{109}$$

10.4 Noise Bandwidth

Most noise sources (like thermal noise) have a flat noise spectral density, meaning that there is the same amount of noise power in each Hz of frequency spectrum. This means that total noise power in a measurement is proportional to the bandwidth of the measurement being made. This gives rise to the concept of noise bandwidth.

A perfect brick-wall filter would have a noise bandwidth equal to its signal bandwidth. Because real filters roll off gradually, the noise bandwidth is slightly different than the 3-dB bandwidth of the filter (often slightly more). The term *Equivalent Noise Bandwidth* (*ENBW*) is therefore often used to describe how much white noise will be passed by a real circuit [2]. For example, a 12.7-kHz single-pole low-pass filter has an ENBW of 20 kHz. Conversely, the ENBW of a 20-kHz first-order LPF is 31.4 kHz. The ENBW of a single-pole roll-off is equal to 1.571 times the pole frequency.

Low-Pass Filter Noise Bandwidth

The noise bandwidth of several loss-pass filters of different order is shown in Table 10.1. As the order of the filter increases, the noise bandwidth tends to approach that of a brick-wall filter, as expected. The noise bandwidth of Chebyshev filters is not shown because it depends heavily on the amount of ripple for which the filter is designed. The numbers are multipliers normalized to the 3-dB bandwidth of the filter.

A-Weighting Noise Bandwidth

The ENBW of the *A-weighting* function is 13.5 kHz. This means that a white noise source passing through the A-weighting filter will result in about 4 dB less noise as compared to the same

Table 10.1 Low-Pass Filter Equivalent Noise Bandwidth

Order	*Butterworth*	*Bessel*
1	1.571	1.571
2	1.111	1.560
3	1.047	1.080
4	1.026	1.045
5	1.017	1.039
6	1.012	1.040

source passed through a first-order roll-off at 20 kHz. Compared to passing through a sharp 80 kHz roll-off, it will experience about 1/6 the noise bandwidth, so it will be smaller by a factor of $\sqrt{6}$ = 2.45, or about 8 dB.

Consider the 34-dB microphone preamplifier discussed earlier in Section 10.1. It had 15 nV/$\sqrt{\text{Hz}}$ input noise and an un-weighted SNR of 77 dB in a 20 kHz bandwidth. Its output noise was 750 nV/$\sqrt{\text{Hz}}$. The A-weighted noise in the 13.5-kHz ENBW (116 $\sqrt{\text{Hz}}$) will only be 87 μV, which is 79 dB below 774 mV.

All of these simplified noise discussions have assumed white noise sources and no contributions from hum, line harmonics and electromagnetic interference (EMI). These latter noise sources do not necessarily add on an RMS basis. Only uncorrelated sources add on an RMS basis.

10.5 Noise Voltage Density and Spectrum

White noise has equal noise power in each Hz of bandwidth. If the number of Hz is doubled, the noise power will double, but the noise voltage will increase by only 3 dB or a factor of $\sqrt{2}$. Thus, noise voltage increases as the square root of noise bandwidth, and noise voltage is expressed in nanovolts per root Hertz (nV/$\sqrt{\text{Hz}}$). There are 141 $\sqrt{\text{Hz}}$ in a 20-kHz bandwidth. A white noise source of 100-nV/$\sqrt{\text{Hz}}$ will produce 14.1 μV_{RMS} in a 20-kHz measurement bandwidth.

The most common source of white noise is *Johnson noise*, sometimes also called *thermal noise*. It is typically created in a resistor. A 1-kΩ resistor creates about 4.16 nV/$\sqrt{\text{Hz}}$ of voltage noise at room temperature of 27°C. It is valuable to keep this number in mind. This number goes up as the square root of resistance. The thermal noise of a 50-Ω resistor is 0.93 nV/$\sqrt{\text{Hz}}$. Thermal noise also goes up as the square root of temperature expressed in Kelvin (27°C = 300 K). At 100°C, this number becomes 4.64 nV/$\sqrt{\text{Hz}}$. See Eq. 10.3 in Section 10.8.

Pink Noise

So-called *pink noise* has the same noise power in each octave of bandwidth [9]. Pink noise is usually employed in certain acoustical measurements. Pink noise is created by passing white noise through a low-pass filter that has a roll-off slope of 3 dB per octave. Put another way, the noise power density of pink noise is inversely proportional to frequency. Indeed, so-called *1/f noise* is pink noise.

1/f Noise and Flicker Noise

Noise that increases at low frequencies, and whose power doubles for every decreased octave, is referred to as *1/f noise*. It is usually associated with circuit devices like transistors and diodes. Because it can extend to very low frequencies, it is sometimes called "*flicker noise*" [8–10]. It has the same spectral shape as pink noise. 1/*f* noise does not have a predictable and mathematical basis to it, and is more often the result of imperfections in semiconductor devices. It is notable that many natural processes experience 1/*f* noise.

The Color of Noise

As with white noise and pink noise, other noise having different spectral shapes is sometimes described by other colors [9].

Red noise has a power spectrum that falls off at 6 dB/octave. It has a $1/f^2$ spectrum. One can see that there is some connection between the nature of the noise and the colors in the light spectrum, with the red end of the spectrum corresponding to lower frequencies. Red noise is sometimes called *Brownian noise* or *Brown noise*. That term refers to Robert Brown, who described the noise coming from an erratic process called Brownian motion. Put another way, brown noise (and red noise) have statistics like those of Brownian motion.

Red noise is generated by white noise that is passed through a 6 dB/octave low-pass filter with a very low cutoff frequency. Put more accurately, it is generated by integrating white noise. It is heavier in the lower frequencies.

The white noise present at the input to a phono preamp is converted approximately to red noise by the RIAA equalization that falls at a bit less than 6 dB per octave on average [11]. A condenser microphone capsule and its preamp also create red noise. A typical 30-pF condenser microphone capsule is polarized to about 60 V through a 1-GΩ resistor, resulting in a low-frequency 3-dB point of about 5 Hz. The 4160-nV/√Hz thermal noise of the 1-GΩ resistor is low-pass filtered at 5 Hz. The result is red noise whose amplitude is about 20 nV/√Hz at 1 kHz.

Infrared noise falls off at 9 dB/octave. It can be created when $1/f$ noise is integrated. An example might be the open-loop noise of a DC servo integrator implemented with a JFET op amp that has significant $1/f$ noise. In practice, the infrared noise of a DC servo is quite small.

Grey noise is white noise that is weighted for constant audibility considering the human ear's varying sensitivity to different portions of the audible spectrum. Grey noise is white noise that has been passed through an inverse A-weighting filter. It has a higher spectral density at low frequencies and at high frequencies.

Green noise is in the center of the spectrum and is loosely defined as white noise that is bandpass filtered in the vicinity of 500 Hz with a very low (and undefined) Q. It is sometimes referred to as the sound of nature. It is sometimes used therapeutically to improve sleep. Pink noise is also used in this way as well.

Blue noise is white noise that has been passed through a high-pass filter so that its power density increases at 3 dB/octave. Some forms of dithering in analog-to-digital conversion employ blue noise.

Violet noise is white noise whose power spectral density increases at 6 dB/octave. It is created by differentiation of white noise. A transmission system with a white noise source whose signal is subjected to 6 dB/octave pre-emphasis will create violet noise in the channel at frequencies above the pre-emphasis frequency break-point. The noise put into the channel will be converted back to white noise when the receiving end subjects the signal to de-emphasis.

The Sound of Noise

All of the above noises can be generated electronically and then subjected to listening tests. Which noise sounds more pleasant for a given A-weighted S/N? Hiss at higher frequencies is generally less objectionable. Correlated noise, like that of mains frequency buzz, can be especially annoying. Keeping in mind the A-weighting curve and the ear's frequency-dependent sensitivity, two amplifiers with the same A-weighted S/N can have their respective noises sound different.

10.6 Relating Input Noise Density to Signal-to-Noise Ratio

Most of the noise in an amplifier is usually contributed by the input stage or other early stages. For this reason, the noise of an amplifier is often referred back to the input. Input-referred noise is equal to the measured output noise divided by the gain of the amplifier. A low-noise op amp might have input-referred noise of 2–3 nV/√Hz.

If a preamplifier has 15 nV/√Hz of input-referred noise, what is its *signal-to-noise ratio* (*SNR*)? Assume that the SNR is in an un-weighted 20-kHz bandwidth and that it is referred to an output level of 774-mV_{RMS} (0 dBu). Also assume that the amplifier has a voltage gain of 50. The output noise voltage will be 15 nV/√Hz × 141√Hz × 50 = 106 μV. This is 7300 times smaller than 774 mV_{RMS}, so the un-weighted SNR in a 20-kHz measurement noise bandwidth is 77 dB.

If instead the un-weighted SNR is measured in an 80-kHz noise bandwidth, the SNR will be 6 dB worse, at 71 dB. In all of this, it is important to recognize that contributions of hum and its harmonics are not being considered. Such low-frequency contributions can sometimes increase the un-weighted SNR by a significant amount. A-weighted SNR measurements largely remove the contributions of low-frequency noise.

Now consider a wideband un-weighted noise measurement of the same preamplifier. Assume that the preamplifier has a closed-loop bandwidth of 500 kHz with a single-pole roll-off. The ENBW will be 690 kHz (831 √Hz), and the output noise will be 623 μV_{RMS}. The SNR will be 1242, corresponding to about 62 dB.

10.7 Amplifier Noise Sources

There are many sources of noise in a multi-stage amplifier. Every transistor and every resistor contributes noise, but some contributions are much larger than others [2]. The smaller contributors can often be ignored. Total noise is often referred to the input. The input-referred noise of a given noise source is divided by the gain ahead of it to determine its overall contribution to input-referred noise of the overall amplifier. If the second stage of an amplifier has input-referred noise of 10 nV/√Hz and the first-stage gain is 10, then the net noise contribution of the second stage will be 1 nV/√Hz as seen at the input.

Input Stage

The input stage (IPS) is usually where most of the noise in a preamplifier originates, and a designer first looks to the input stage for improvement of noise performance. A good input stage may have input-referred noise less than 5–10 nV/√Hz. A very low noise input stage may have input noise on the order of 0.5–2 nV/√Hz. As we will see, however, there are numerous other noise sources that can make the amplifier's overall input-referred noise larger.

VAS Noise

The input stage is not the only source of noise in a preamplifier, even though it often dominates. The input stage load and later stages create noise, and their noise can be referred back to the input of the amplifier by the voltage gain that precedes them. For example, a second stage with input noise of 10 nV/√Hz will contribute an amplifier input-referred noise component of 2.5 nV/√Hz if the voltage gain of the input stage is only 4. That will increase the input-referred noise by 3 dB if the input stage noise is 2.5 nV/√Hz. The message here is that second-stage noise cannot be ignored, and may even dominate the noise in some preamplifier designs. This can happen because the second stage is not always designed for low noise and input stage gain is sometimes quite small, sometimes less than 5. This can happen in the case of a heavily degenerated differential input stage with a simple resistive load connected to a one-transistor second stage.

Power Supply Noise and PSRR

The power supply rails in any circuit are often corrupted by numerous sources of noise. These may include random noise and other noises like power supply ripple and EMI. The power supply noise can get into the signal path as a result of the amplifier circuit's finite *power supply rejection ratio* (*PSRR*). If an input stage has 60 dB PSRR, then 1 mV of power supply noise in a 20-kHz ENBW will contribute 1 μV of input-referred noise. This alone corresponds to 7 nV/√Hz in a 20 kHz ENBW.

There are two important ways to control power supply noise. The first approach is to employ circuit topologies that have inherently high PSRR. One example is the use of a current mirror load on the input stage differential pair and a tail current source with high output impedance. Adding a cascode stage to the input pair can also increase PSRR by increasing the output impedance of the input pair. The ability of a circuit to reject power supply noise usually decreases as the frequency of the noise increases. In other words, PSRR degrades at high frequencies. Fortunately, it is often possible to do a more effective job of filtering the power supply rails at higher frequencies. Power supply noise and PSRR are discussed elsewhere in detail.

The second way to reduce noise originating from the power supply is to do a better job filtering the power supply rails. This is especially effective for power supply rails that provide power to the input and other low-level stages. Additional rail decoupling can be used, but it can eat into voltage headroom. A 120-Ω decoupling resistor for an IPS that draws 4 mA will drop 0.5 V. To get 20 dB of additional rejection at 60 Hz will require a 100 μF capacitor to ground.

There is also power supply rectifier noise. This is largely a result of the fact that rectifiers do not turn off instantly due to stored charge that must be removed by reverse current flow. The turn-off process can be a local source of EMI. The time it takes to turn off the diode is usually referred to as the *recovery time*. Sharp current impulses and bursts of ringing with the power transformer inductance result during turn-off. Fast recovery and soft recovery diodes can reduce this, along with rectifier snubbers that damp bursts of ringing during the switching intervals. Good discussions about this can be found in Refs. [12, 13].

Switch-Mode Power Supply Noise

Although power supply noise in conventional linear supplies has always been a concern (ripple, rectifier switching, transformer ringing, etc.), the noise from switch-mode power supplies (SMPS) has usually been thought to be worse. Today, that need not be the case. Modern SMPS operate at much higher frequencies (outside the audio frequency band) and create virtually no audible ripple. High frequencies are often easier to shield and filter out.

Having said that, SMPS are a potential local source of significant EMI. The good news is that they largely eliminate line-rate ripple and potentially large 60-Hz magnetic fields created by the power transformer in a linear supply. Noise that is magnetic in origin is insidious. Careful layout and shielding are imperative in amplifiers employing SMPS. The SMPS is sometimes put in its own shielded sub-enclosure; this is practical because of the small size of the SMPS. In fact, because of the compactness and small size of the switching supply and its inductors and transformers, hum and noise from SMPS power supplies is not much of a problem. Of course, much has been made of EMI as a source of sound quality degradation in conventional preamplifiers with linear power supplies, so one is inclined to suspect that EMI immunity is even more important when there is a local source of it right in the same enclosure. Preamplifier circuits and layouts that are more robust to EMI will probably fare better in such designs. For example, preamplifiers with JFET input stages may be preferred.

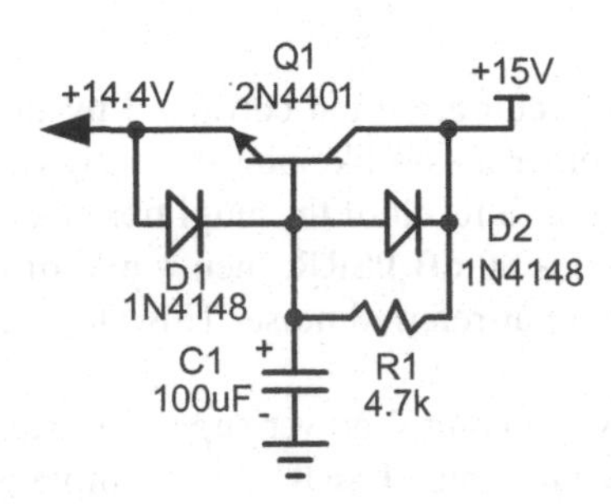

Figure 10.4 Capacitance Multiplier.

Capacitance Multiplier

A more efficient alternative to an R-C filter in the power supply rail is a capacitance multiplier. It is little more than an emitter follower that is fed a version of the power supply rail that has been filtered by an R-C filter with a low cutoff frequency. It may drop as little as 0.6 V even for a moderate current load. A simple capacitance multiplier is shown in Figure 10.4. A small-signal transistor can be used because power dissipation is small. However, the transistor must be rated to withstand the full power supply rail voltage because the base voltage will begin with 0 V when power is applied. The transistor operates at $V_{cb} = 0$ V. For this reason, input ripple should be less than 100 mV. With $I_{load} = 10$ mA and $\beta = 200$, base current will be 50 μA. A 4.7-kΩ decoupling resistor will drop an additional 235 mV for a total drop of almost 1 V. A 100-μF shunt capacitor will now provide 52 dB of attenuation at 120 Hz. The capacitance multiplier approach re-references the rail voltage to the local ground, and greatly decreases the amount of noise current otherwise dumped into the local ground by a conventional large bypass capacitor.

Diodes D1 and D2 protect the transistor in the event that the input rail voltage collapses quickly for some reason. If increased voltage drop is created by the addition of a shunt resistor across C1, a larger amount of power supply ripple can be tolerated on the input rail. Note that a preamplifier having a 1000-μF reservoir capacitor and drawing an average of 250 mA will have about 2 V_{p-p} ripple on its incoming rails from the power supply before the voltage regulator. Care must be taken in the design of such capacitance multipliers to ensure there is enough voltage headroom to allow worst-case rail voltage dips that happen more quickly than C1 can discharge. In other words, in normal operation, D2 should never turn on. The usual use of regulated power supplies in preamplifiers greatly mitigates this issue.

Ground Loops and Noise Pickup

Avoidance of ground loops and other means of noise pickup is important in a preamplifier if best sound quality and SNR are to be achieved. Best practices for grounding and interconnect should be employed in low-level and line-level equipment. Bill Whitlock of Jensen Transformers has published numerous excellent articles on this topic [14–17].

Star Quad Cable

So-called *star quad* cable consists of four wires twisted together inside a shield with diagonally opposite wires connected together at both ends. Manufacturers of star quad cable include Belden,

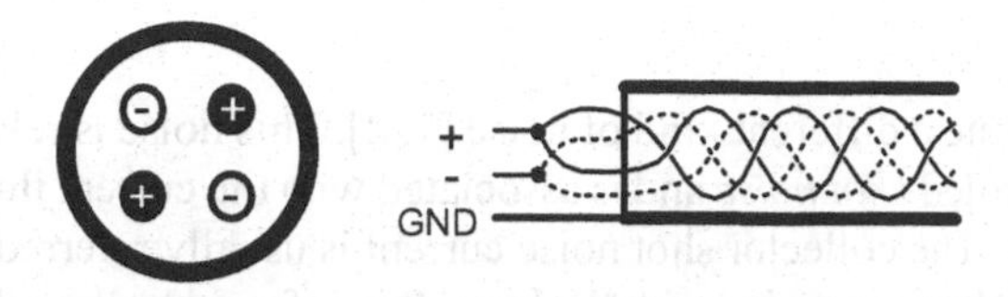

Figure 10.5 **Star Quad Cable Construction.**

Mogami and Canare. This construction results in far superior immunity to externally induced noise from magnetic fields as compared to conventional shielded twisted pair cable [18, 19]. Magnetically induced noise is insidious and difficult to suppress. Cable shielding blocks noise from electric fields but does little for noise originating from magnetic fields. Twisted pairs, whether shielded or not, suppress noise induction from magnetic fields, but the highly symmetrical geometry of star quad cable takes this to a much higher level, sometimes by as much as 20 dB [19]. Of course, the improvement with star quad cable is mainly in effect only with a balanced connection at the receiving end that has a good common-mode rejection ratio (CMRR).

Although most often associated with standard size microphone cable, it is available in miniature sizes as well, down to as small as 30-AWG conductors in a cable of 1/8-in OD that is very flexible. These cables can be especially useful for short runs inside preamplifiers for improved noise immunity from noisy magnetic fields (e.g., from transformers, mains wiring, power supply wiring, etc.). Figure 10.5 illustrates how star quad cable is constructed. In some instances, the star quad geometry has been emulated on printed wiring boards for improved noise immunity. This is accomplished with a particular arrangement of traces on multiple layers.

10.8 Thermal Noise

All resistors generate noise. This is referred to as *Johnson noise* or *thermal noise* [1, 2]. It is the most basic source of noise in electronic circuits. It is most often modeled as a noise voltage source in series with an ideal noiseless resistor. The noise power in a resistor is dependent on temperature.

$$P_n = 4kTB\,\text{W} = 1.66\times10^{-20}\ \text{W/Hz} \tag{10.2}$$

where k = Boltzman's constant = 1.38×10^{-20} J/°K
T = temperature in Kelvin = 300 K @ 27°C
B = equivalent noise bandwidth in Hz
The open-circuit RMS noise voltage across a resistor of value R is simply

$$e_n = \sqrt{4kTRB} \tag{10.3}$$

$$e_n = 0.129\,\text{nV}/\sqrt{\text{Hz}}\ \text{per}\ \sqrt{\Omega}\quad (\text{at}\ 27°\text{C}) \tag{10.4}$$

Noise voltage for a resistor thus increases as the square root of both bandwidth and resistance. A convenient reference is the noise voltage of a 1-kΩ resistor at 27°C:

$$1\ \text{k}\Omega \Rightarrow 4.163\ \text{nV}/\sqrt{\text{Hz}}$$

From this the noise voltage of any resistance in any noise bandwidth can be estimated.

10.9 Shot Noise

Bipolar transistors generate a different kind of noise [1, 2]. This noise is related to the discreteness of current flow. This is called *shot noise* and is associated with the current flows in the collector and the base of the transistor. The collector shot noise current is usually referred back to the base as an equivalent input noise voltage in series with the base. It is referred back to the base as a voltage by dividing it by the transconductance of the transistor. Once again, the resulting input-referred noise is usually measured in nanovolts per root Hz.

The shot noise current is usually stated in picoamperes per root Hz (pA/√Hz) and has the RMS value of

$$I_{shot} = \sqrt{2qI_{dc}B} \quad (10.5)$$

$$I_{shot} = 0.57\,\text{pA}/\sqrt{\text{Hz}}/\sqrt{\mu\text{A}} \quad (10.6)$$

where $q = 1.6e^{-19}$ Coulombs per electron and B = bandwidth in Hz.

It is easily seen that shot noise current increases as the square root of bandwidth and as the square root of current. A 1-mA collector current flow will have a shot noise component of 18 pA/√Hz.

$$1\text{mA} \Rightarrow 18\,\text{pA}/\sqrt{\text{Hz}}$$

10.10 Bipolar Transistor Noise

There are several sources of noise in the bipolar junction transistor (BJT). Collector shot noise current, when referred back to the input of the transistor by its transconductance, represents an effective input noise voltage represented in nV/√Hz. Base shot noise current is simply an input noise current usually represented in pA/√Hz. The third significant noise source in a transistor is base resistance, $r_{bb'}$. This represents a thermal noise voltage source that adds to the effective input noise voltage created by collector shot noise. Base noise current flowing through the base resistance can also create a noise voltage component, but it is usually quite small.

BJT Input Noise Voltage

The transconductance of a BJT operating at 1 mA is 38.5 mS. Dividing the shot noise current by *gm* yields input-referred noise e_n = 0.47 nV/√Hz. From Eq. 10.3 we can see that this is the voltage noise of a 13-Ω resistor.

At the same time, notice that *re'* (1/*gm*) for this transistor is 26 Ω. The noise voltage for a 26-Ω resistor is 0.66 nV/√Hz. The input-referred voltage noise (resulting from collector current shot noise) of a transistor is equal to the Johnson noise of a resistor whose value is half the value of *re'*. This is a very handy relationship.

Base resistance is the second major component of transistor voltage noise. A relatively high base resistance $r_{bb'}$ of 100 Ω will contribute 1.3 nV/√Hz. When added to the collector shot noise component of 0.47 nV/√Hz, we have 1.38 nV/√Hz input noise for a transistor operating at 1 mA. This demonstrates the importance of base resistance in BJT noise. Unfortunately, $r_{bb'}$ is virtually never specified in a datasheet. It can be found in the SPICE model for the transistor as the parameter RB, but this parameter is often an unreliable model. Very careful noise measurements of a transistor can sometimes provide a useful estimate of $r_{bb'}$ if the known shot noise can be accounted for. If a transistor is operating at 10 mA and its $r_{bb'}$ is 10 Ω, the respective shot and $r_{bb'}$ contributors are 0.15 and 0.42 nV/√Hz. If the transistor is operated at 1 mA instead, the respective noise contributors are 0.48 and 0.42 nV/√Hz. Base resistance can be inferred in this way only if transistor shot noise and noise $r_{bb'}$ are only the noise contributors in the test setup.

BJT Input Noise Current

The base current of a transistor also has a shot noise component measured in units of pA/√Hz. Recall that

$$I_{shot} = 0.57\,\text{pA} / \sqrt{\text{Hz}} / \sqrt{\mu\text{A}} \tag{10.7}$$

Consider a BJT biased at 1 mA and with a beta of 100. Base current will be 10 μA. This corresponds to 3.16 √μA. Input noise current will be 1.8 pA/√Hz. Put another way, base shot noise is collector shot noise divided by $\sqrt{\beta}$. It is always important to remember that shot noise current goes down as the square root of the associated current.

If the base circuit includes source resistance, the base shot noise current will develop an equivalent noise voltage across that resistance. If the source impedance driving that transistor's base is 1 kΩ, then the input noise voltage due to input noise current will be 1.8 nV/√Hz. The base resistance of the transistor should be included as part of the source resistance. In comparison, the thermal noise of the 1 kΩ source resistance is much larger, at 4.2 nV/√Hz.

Low-Noise Transistors

There are really two types of bipolar transistors that can be deemed "*low noise.*" The difference relates to the type of application for which they are intended, depending on whether input current noise or input voltage noise is the greater concern. Transistors with very high current gain, like the 2N5089, have very low input current noise due to their high current gain and the low collector current at which they can be operated. The 2N5089 has a typical current gain of over 400 for collector current as low as 10 μA. This makes it possible to achieve very low input noise current. When operated at 10 μA, its base current will be less than about 25 nA and its input current noise will be less than about 0.1 pA/√Hz. Of course, at this low operating current the 2N5089's voltage noise will be about 4.7 nV/√Hz.

For low-impedance applications, transistor voltage noise is much more important. An example application here would be where the source impedance is 50 Ω. Here the source voltage noise is only 0.9 nV/√Hz. A transistor like the 2N4401 performs very well in this application because it has low base resistance on the order of $r_{bb'}$ = 15 Ω, whose thermal noise is only 0.5 nV/√Hz. If the 2N4401 is operated at 5 mA, its transconductance is about 200 mS (re' = 5 Ω) and its input noise voltage due to collector shot noise current will be about 0.2 nV/√Hz. The total of the $r_{bb'}$ noise and the shot noise contributors will be about 0.54 nV/√Hz. Of course, with a minimum specified beta of only 40, base current will be as high as 125 μA and base current noise will be as high as 6.4 pA/√Hz. In an actual circuit with 50-Ω source impedance, this input current noise will add 0.3 nV/√Hz as it flows through the source impedance. The NPN 2N4401 and the PNP 2N4403 have been popular choices for moving coil phono preamps.

For still lower voltage noise, one can consider the NPN ZTX851 and the PNP ZTX951, each with $r_{bb'}$ of only about 2 Ω. These are medium power high current transistors in a TO-92 package. However, they have rather high C_{cb} of 45 and 74 pF, respectively.

Noise Figure

Noise figure (NF) is the amount by which the inherent thermal noise of the source impedance for an amplifier is increased by the amplifier noise [1, 2]. NF is usually associated with the device used for the input stage of the amplifier, since that is where the major noise contribution of the amplifier often originates. As a simple example, consider a source impedance of 1 kΩ, which will have thermal noise of 4.2 nV/√Hz. If the associated amplifier has 4 nV/√Hz input voltage noise and no significant input current noise (like a JFET), then the noise figure for that amplifier will be 3 dB. If instead the voltage noise of the amplifier is 2 nV/√Hz, the total noise will be about 4.5 nV/√Hz and the noise figure will be about 1 dB. The 2N5089 datasheet shows a typical noise figure of 0.5 dB with a source impedance of 15 kΩ when operated at 10 μA.

Now consider the 2N4401 operating at 1 mA with a source impedance of 50 Ω. The source noise is 0.9 nV/√Hz. Voltage noise of the 2N4401 will be about 0.5 nV/√Hz. If the 2N4401 is operating at a typical β of 100, base current will be 10 μA and base current noise will be 1.8 pA/√Hz. Flowing through the 50-Ω source impedance, this will create 0.1 nV/√Hz of additional noise, which is negligible. Total noise with the transistor connected to the 50-Ω source impedance is 1.03 nV/√Hz. This corresponds to NF = 1.1 dB.

Optimum Source Impedance

If the source impedance for a transistor at a given operating point is high, input noise current will dominate the transistor's contribution to noise figure. Conversely, if the source impedance is low, input noise voltage will dominate obtainable NF. This suggests that there is an optimum value of source impedance for best NF, where input current and voltage noise contributions from the transistor are equal. The optimum source impedance for lowest NF for a given device is simply input voltage noise divided by input current noise.

$$Z_{opt} = e_{in} / i_{in} \tag{10.8}$$

For the 2N5089 operating at 10 μA, this works out to about 47 kΩ. For the 2N4401 operating at 5 mA, this works out to about 84 Ω.

10.11 JFET Noise

JFET noise results primarily from *thermal channel noise* [1, 20–23]. That noise is modeled as an equivalent input resistor r_n whose resistance is equal to approximately 0.67/*gm* [24]. If we model the effect of *gm* as *rs'* (analogous to *re'* for a BJT), we have $r_n = 0.67rs'$. This is remarkably similar to the equivalent voltage noise source for a BJT, which is the voltage noise of a resistor whose value is 0.5*re'*. The voltage noise of a BJT goes down as the square root of I_c because *gm* is proportional to I_c, and *re'* goes down linearly as well. However, the *gm* of a JFET increases as the square root of I_d. As a result, JFET input voltage noise goes down as the one-fourth power of I_d. The factor 0.67 is approximate, and SPICE modeling of some JFETs suggests that the number is 0.67.

At $I_d = 0.5$ mA, *gm* for the LSK489 dual monolithic JFET is about 3 mS, corresponding to a resistance *rs'* of 333 Ω. Multiplying by the factor 0.67, we have an equivalent noise resistance r_n of 223 Ω, which has a noise voltage of 2.7 nV/√Hz [24].

At $I_d = 2.0$ mA, *gm* for the lower-noise LSK389 dual monolithic JFET is about 20 mS, corresponding to a resistance *rs'* of 50 Ω. Multiplying by the factor 0.67, we have an equivalent noise resistance r_n of 33.5 Ω, which has a noise voltage of 0.73 nV/√Hz. JFET input voltage noise will also include a thermal noise contribution from ohmic gate and source resistance, but this is often negligible.

JFET Operation

The simplified equation below describes the DC operation of a JFET [1]. The term β is the transconductance coefficient of the JFET (not to be confused with BJT current gain) and V_t is the threshold voltage.

$$I_d = \beta\left(V_{gs} - V_t\right)^2 \tag{10.9}$$

At $V_{gs} = 0$, we have I_{dss}:

$$I_{dss} = \beta\left(V_t\right)^2 \tag{10.10}$$

β can be seen from I_{dss} and V_t to be:

$$\beta = I_{dss} / \left(V_t\right)^2 \tag{10.11}$$

The operating transconductance *gm* is easily seen to be:

$$gm = 2\sqrt{\beta I_d} \tag{10.12}$$

This last relationship is important, because transconductance of the JFET is what is most often of importance to circuit operation [24]. For a given transconductance parameter β, *gm* is largely independent of I_{dss} and V_t, and goes up as the square root of drain current. This is in contrast to a bipolar transistor where *gm* is proportional to collector current. The value of β for JFETs ranges from about $1e^{-3}$ to $100e^{-3}$. A typical low-noise JFET may have $\beta = 50e^{-3}$ and will have $gm = 20$ mS at an operating current of 2 mA. A still lower-noise JFET with $\beta = 81e^{-3}$ and operating at 4 mA will have $gm = 36$ mS.

JFET Noise Sources

In order to understand low-noise JFETs, it is helpful to briefly review the five major sources of noise in JFETs [1, 20, 21, 24].

- Thermal channel noise
- Gate current shot noise
- 1/*f* noise
- Generation-recombination noise
- Impact ionization noise

The first two sources of noise are largely fundamental to the device, while the remaining three sources are largely the result of device imperfections. Examples of such imperfections include lattice damage and charge traps. A major reduction in G-R noise is key to the LSK489's superior noise performance [24].

Thermal Channel Noise

Thermal channel noise, as discussed above, is akin to the Johnson noise of the resistance of the channel. However, it is important to recognize that the channel is not acting like a resistor in the saturation region where JFETs are usually operated. The channel is operating as a doped semiconductor whose conduction region is pinched off by surrounding depletion regions to the point where the current is self-limiting. Conduction is by majority carriers. The constant 0.67 in the equation where $r_n = 0.67/gm$ is largely empirical, and can vary with the individual device geometry. It is often a bit smaller than 0.67.

Gate Shot Noise Current

JFET input current noise results from the shot noise associated with the gate junction leakage current. Shot noise increases as the square root of DC current. A useful relationship is that $I_{shot} = 0.57$ pA/√Hz/√μA [24]. Alternately, $I_{shot} = 0.57$ fA/√Hz/√pA. Gate shot noise is white, and usually has no 1/*f* component [3].

This noise is normally very small, on the order of fA/√Hz. It can usually be neglected. However, in extremely high-impedance circuits and/or at very high temperatures, this noise must be taken into account. Consider a circuit with source impedance of 100 MΩ and a JFET with input noise current of 4 fA/√Hz at 25°C. The resulting voltage noise will be 400 nV/√Hz. Leakage current doubles every 10°C, so at 65°C the leakage current goes up by a factor of 16 and this noise contributor will go up by a factor of 4 to about 1600 nV/√Hz. For comparison, the Johnson noise of the resistive 100-MΩ source is about 1300 nV/√Hz.

1/f Noise

At very low frequencies the input noise power of a JFET amplifier rises as the inverse of frequency. That is why this noise is referred to as 1/*f* noise. When expressed as noise voltage, this means that the noise rises at a rate of 3 dB/octave as frequency decreases. In a good JFET, the 1/*f* spot noise voltage density at 10 Hz may be twice the spot noise at 1 kHz (up 6 dB) when expressed as

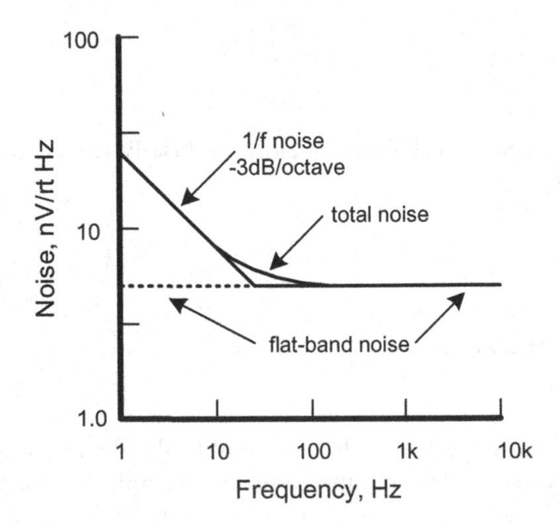

Figure 10.6 1/f Noise Voltage.

nV/√Hz. The noise might typically be up by 3 dB at 40 Hz. 1/*f* noise is associated with imperfections in the fabrication process, such as imperfections in the crystal lattice [21]. Improved processing contributes to reduced 1/*f* noise. In fact, the amount of 1/*f* noise is sometimes considered as an indication of process quality.

By comparison, a JFET IC op amp with input noise of 10 nV/√Hz at 1 kHz may have its noise up by 3 dB at 100 Hz and the spot noise at 10 Hz might be up by 10 dB to 32 nV/√Hz. At 1 Hz that op amp may have spot noise on the order of 100 nV/√Hz.

Figure 10.6 illustrates a plot of voltage noise vs frequency. The 1/*f* corner frequency is defined as the frequency where the 1/*f* noise contribution equals the flat band noise. Put another way, it is where the −3 dB/octave 1/*f* noise line intersects the flat band noise line. At this frequency, the voltage noise will be up by 3 dB.

Generation-Recombination Noise

A less-known source of voltage noise results from carrier generation-recombination in the channel of the JFET. This is referred to as *G-R* noise [21]. This *excess noise* is governed by fluctuation in the number of carriers in the channel and the lifetime of the carriers. *G-R* noise manifests itself as drain current noise. When referred back to the input by the transconductance of the JFET, it is expressed as a voltage noise.

Like 1/*f* noise, *G-R* noise results from process imperfections that have created crystal lattice damage or charge trap sites [21]. In contrast, however, *G-R* noise is not limited to low frequencies. In fact, it is flat up to fairly high frequencies, usually well above the audio band. The *G-R* noise power spectral density function is described in Ref. [21] as:

$$S_{G-R(f)} / N^2 = [\overline{(\Delta N)^2} / N^2] \times [4\tau / (1 + (2\pi f \tau)^2] \tag{10.13}$$

where $\overline{(\Delta N)^2}$ is the variance of the number of carriers N, and τ is the carrier lifetime.

Above a certain frequency, the *G-R* noise power decreases as the square of frequency. When expressed as noise voltage, this means that it decreases at 6 dB/octave. The point where the *G-R* noise is down 3 dB can be referred to as the *G-R* noise corner frequency. That frequency is governed by the carrier lifetime, and in fact is equal to the frequency corresponding to a time constant that is the same as the carrier lifetime [21]. We have,

$$f_{G-R} = 1 / 2\pi\tau \tag{10.14}$$

where τ is the carrier lifetime.

The 6 dB/octave high-frequency roll-off of *G-R* noise is only an approximation because there are normally numerous sites contributing to *G-R* noise and the associated carrier lifetimes may be different. As a result, the *G-R* noise corner frequency is poorly defined and the roll-off exhibits a slope that is usually less than 6 dB/octave over a wider range of frequencies. The inflection in the JFET's noise vs frequency curve may thus be somewhat indistinct. The important takeaway here is that excess *G-R* noise can sometimes exceed the thermal channel noise contribution and thus dominate voltage noise performance of a JFET.

The idealized noise spectral density graph in Figure 10.7 illustrates how the three voltage noise contributors act to create the overall noise vs frequency curve for a JFET [24]. In the somewhat exaggerated case illustrated, *G-R* noise appears to dominate thermal channel noise, but it often just adds a bit to the existing thermal channel noise.

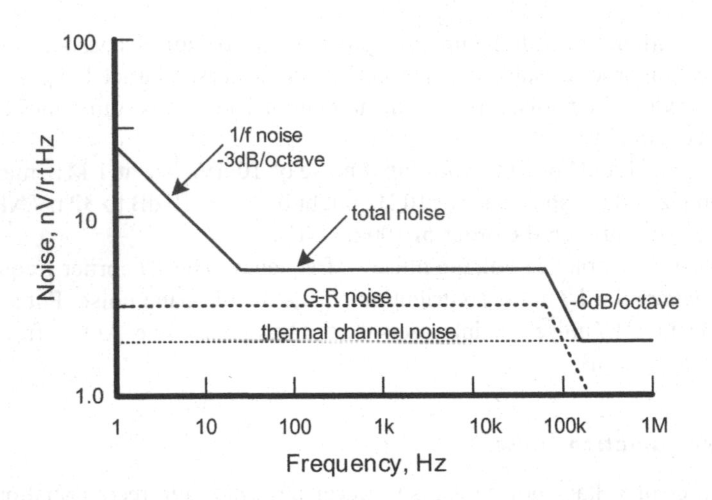

Figure 10.7 JFET Voltage Noise.

Noise improvements derive from process improvements that reduce device imperfections. Those imperfections create *G-R* noise and 1/*f* noise. Such process imperfections include crystal lattice damage and charge trap sites. Put simply, most JFETs are not as quiet as they can be. Process improvements reduce both 1/*f* noise and *G-R* noise.

Impact Ionization Noise

An electron traveling in a strong electric field can be accelerated to the point where it has enough kinetic energy to knock another electron out of its valence band into the conduction band if it impacts an atom in the crystal lattice [21, 23, 25]. For convenience, the electron corresponding to normal current flow can be called a "seed" electron, since it starts the process. This collision creates a new hole-electron pair. This process is called impact ionization. The new hole and electron then act as additional charge carriers and add to the current flow. The new carriers may also be accelerated to the point where they create an impact ionization event, so the process may be multiplied. This is what is called an avalanche effect. Impact ionization often occurs in a p-n junction that is under a high reverse bias voltage that creates a large electric field.

In a JFET, the nominal gate junction saturation current corresponds to the flow of such seed electrons. Impact ionization in the gate-channel junction can create a current flow that results in excess gate-leakage current. This leakage current includes a corresponding amount of shot noise current. The result is impact ionization noise (IIN), sometimes also called avalanche noise.

A second source of IIN originates in the conducting channel of the JFET. Electrons passing along the channel encounter a high electric field due to the drain-source potential difference. The high electric field mainly exists at the drain end of the channel where the channel is pinched off when the JFET is operating in its saturation mode. These carriers are accelerated to a velocity where they can create impact ionization events in the channel and an avalanche multiplication effect.

The "seed" carriers for channel IIN are the carriers associated with the drain current flow, so channel IIN increases with drain current. The minority carriers created by impact ionization are swept across the p-n junction and create excess gate current. Thus, in an N-channel device, the channel impact ionization current flows from the drain to the gate, just as it does for the junction impact ionization current. The majority carriers created by impact ionization merely increase the drain current. The multiplicative effect of the avalanche action means that excess gate current grows exponentially with drain-gate voltage.

P-channel JFETs have much less impact ionization noise in the channel because the majority carriers are holes, which have lower mobility and less kinetic energy. However, gate junction IIN is still present in P-channel JFETs because electron majority carriers exist at the junction.

Impact ionization noise can be minimized in JFET amplifiers by operating at lower drain-gate voltages and lower drain current. In fact, some circuit arrangements have been devised that keep the drain-gate voltage at the smallest value that will keep the JFET in its saturation region [26]. IIN from the channel usually is much larger than that from the drain-gate junction as long as the drain-gate voltage is not close to the breakdown voltage. Because impact ionization noise is primarily a gate noise current, circuits with low source impedance are much less susceptible to IIN. Conversely, circuits with very high source impedance, like condenser microphone preamplifiers, can be affected significantly by IIN. In most circuits, gate noise current due to impact ionization is insignificant, especially if the drain-gate voltage is less than about 10 V.

10.12 Op Amp Noise

Bipolar op amps have both input voltage noise and input current noise, since their input stage employs bipolar transistors [2, 6, 8–10, 22, 27, 28]. JFET op amps have input voltage noise but virtually no input noise current [22]. Below are some examples of op amp noise performance, beginning with the venerable μA741.

Input Noise Voltage

Table 10.2 lists typical input voltage and current noise for several operational amplifiers.

It is notable that several newer JFET op amps have achieved very low voltage noise levels that are competitive with BJT op amps. These include the LT1113 and the OPA1642. The OPA1656 is a CMOS op amp, and has extraordinarily low noise for a CMOS device. Do note, however, that it has a fairly high $1/f$ noise corner frequency of about 1 kHz. Noise at 100 Hz is about 10 nV/√Hz. Although not specified, noise at 10 Hz is about 35 nV/√Hz. For comparison, the noise at 100 Hz for the OPA1642 is only 5.8 nV/√Hz. Distortion for the OPA1656 is remarkably low, even when driving a 600-Ω load.

Table 10.2 Op Amp Noise

Op amp	e_n, *nV/√Hz*	i_n, *pA/√Hz*	*Type*
μA741	22	1.0	BJT
NE5534	3.5	0.4	BJT
LM4562	2.7	1.6	BJT
AD797	0.9	2.0	BJT
TL072	18	0.01	JFET
OPA134	8	0.003	JFET
ADA4627	6.1	0.003	JFET
LT1113	4.5	0.01	JFET
OPA1642	5.1	0.0008	JFET
OPA1656	4.3[a]	0.006	CMOS

All input noise voltages are specified at 1 kHz.
[a]2.9 nV/√Hz at 10 kHz; 10 nV/√Hz at 100 Hz; 35 nV/√Hz at 10 Hz.

Input Noise Current

Operational amplifier input noise current will create noise voltage by flowing through the surrounding resistors in the input and feedback networks. Consider the inverter in Figure 10.8(a) with op amp input noise of 1.6 pA/√Hz, as with the LM4562. The noise current flows through the parallel combination of R1 and R2 (5 kΩ), creating a noise voltage of 8 nV/√Hz, which will be multiplied by the noise gain of 2 to16 nV/√Hz. It will be referred back to the input by unity, resulting in input-referred noise of 16 nV/√Hz. Input noise current dominates input noise voltage in this example, and noise would be lower if R1 and R2 were made smaller. To put matters in perspective, however, bear in mind that R1 and R2 together contribute thermal noise of about 9.4 nV/√Hz (thermal noise of 5 kΩ multiplied by 2).

If there is impedance in the non-inverting input circuit, noise will be contributed there as well. The noise currents of the inverting and non-inverting inputs of the op amp are considered as largely uncorrelated, so this additional noise source will add in an RMS fashion.

Input Noise Impedance

As with transistors, op amps have optimal input noise impedance values. This is simply the input voltage noise divided by the input current noise. For the BJT LM4562, this is 2.3 nV/√Hz divided by 1.6 pA/√Hz, or 1.4 kΩ. This is merely the point where noise contributions from current and voltage are equal. If R1 and R2 were each made 2.8 kΩ, the circuit impedance would be 1.4 kΩ and would be at the optimum value. Because JFET op amps have virtually no input noise current, their input noise impedance is very high. For the OPA134, input noise voltage is 8.0 nV/√Hz and input noise current is 0.003 pA/√Hz, resulting in an input noise impedance of 2.7 MΩ.

Noise Gain

Operational amplifiers are inevitably connected in negative feedback circuits that establish the gain of the circuit. These circuits often do not have the same gain for the signal and the op amp's noise. For this reason the term "*noise gain*" is used to describe the amount by which the

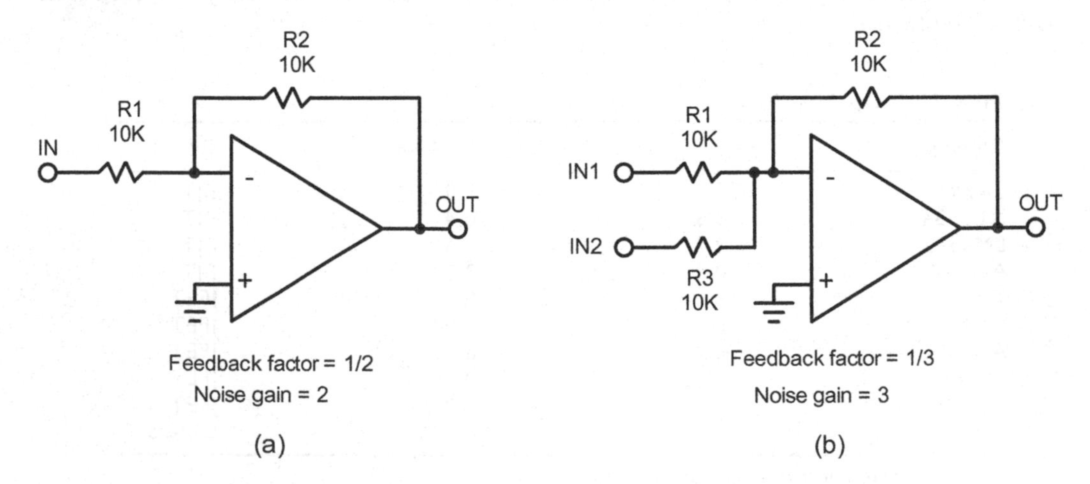

Figure 10.8 Noise Gain of an Inverter.

op amp's input-referred noise is multiplied by the circuit to become output noise [8, 27]. For example, consider a unity-gain inverter with noiseless 10-kΩ input and feedback resistors, as shown in Figure 10.8(a). The gain for the signal is unity. However, with the input shorted, it is easy to see that the input and feedback resistors form a 2:1 attenuator from the output to the input. A noise input to the non-inverting terminal of the op amp will see a non-inverting gain of 2. If the input voltage noise of the op amp is 2.7 nV/√Hz, as with the LM4562, the output noise will be 5.4 nV/√Hz. The input-referred noise of the overall amplifier will be 5.4 nV/√Hz because the output noise has been referred back to the input by dividing it by the gain of the stage, which is unity.

In Figure 10.8(b) an additional 10-kΩ resistor R3 is connected to a second input, as in a mixer application. This does not change the gain of the inverting path for the first input, but does increase the noise gain to 3 because of the increased attenuation in the feedback path. The input-referred noise now becomes 8.1 nV/√Hz.

By contrast, the noise gain of a unity-gain buffer is unity because there is no attenuation of the feedback signal. The main point here is that the noise gain of a circuit is usually not the same as the signal gain, and in fact is usually larger.

10.13 Noise Simulation

With an understanding of the basics of noise and the cause-effect relationships, noise analysis is best handled by SPICE simulations. In a simulation using LTspice®, the noise contribution of any element can be evaluated by clicking on the circuit element [29, 30]. The base current noise in a simulation will show up as a component of the transistor voltage noise contribution. This contribution is the sum of the transistor voltage noise and the transistor base current noise multiplied by the source impedance seen by the base.

Emitter Followers and Source Followers

The emitter follower and source follower circuits shown in Figure 10.9 are easy circuits to simulate for noise in order to obtain a better understanding of how sources of noise contribute to the total noise of a circuit.

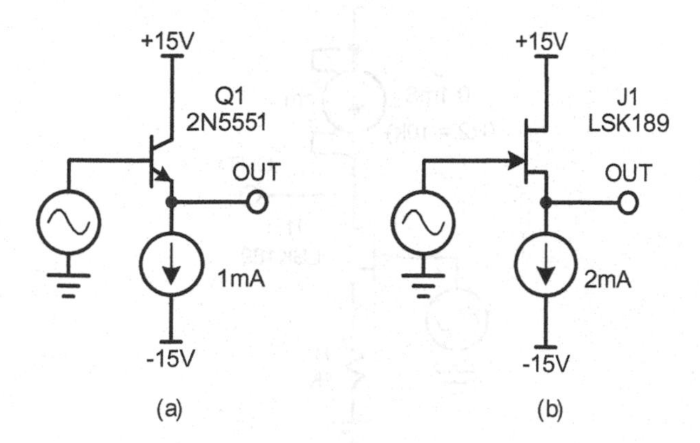

Figure 10.9 Emitter Follower and Source Follower.

The ideal current sources contribute no noise, so as shown the simulations focus only on the voltage noise sources in the transistors. Changing the current can reveal how the noise contributed by the transistor behaves as a function of current. If noiseless resistance is added to the input circuit, the effects of input current noise can be seen as well. This is especially useful in the case of the BJT.

Noiseless Resistors

Sometimes only a certain portion of a circuit needs to be analyzed for noise. Other elements, like load resistors may get in the way because of their thermal noise contributions. In these cases an ideal transconductance element can be used to create a noiseless resistor. The output of the *gm* element is connected back to its input, as shown in Figure 10.10. The arrangement acts as a 2-terminal noiseless resistance whose value is 1/*gm*. Such a noiseless resistor can be used in the input circuits of Figures 10.8 and 10.9 to isolate and evaluate input current noise.

The use of a noiseless load resistor in Figure 8.10 allows the simulation of the common source amplifier circuit noise apart from the effects of any noise contribution from the load.

Noise Sources

Sometimes it is useful to have a reference noise source in a simulation. This is easily achieved with a voltage-controlled voltage source (VCVS) with a resistor from its input to its ground (or reference point, making a floating noise source possible). If a 1-kΩ resistor is used, the noise generated will be 4.2 nV/√Hz if the gain of the VCVS is unity.

A current noise source can be created with a voltage-controlled current source (*gm* element) with a resistor connected across its input terminals. If a 1-kΩ resistor is used, the noise generated will be 4.2 pA/√Hz if the *gm* value is set to 1 mS. If a different value of resistor is used to create different noise amplitude, remember that the noise level changes as the square root of the change in resistance.

Simulating the Noise of a Transistor

Figure 10.11 illustrates a simulation of the noise behavior of a 2N5551 NPN transistor. This is a general-purpose audio transistor popular in audio designs. The circuit operates Q1 in the

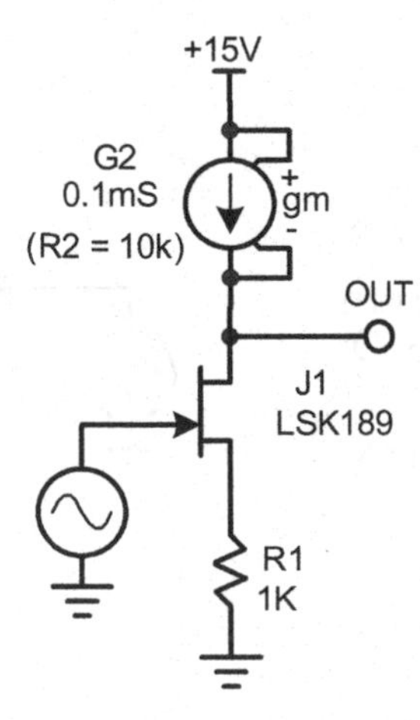

Figure 10.10 Simulation of a JFET with a Noiseless Load Resistor.

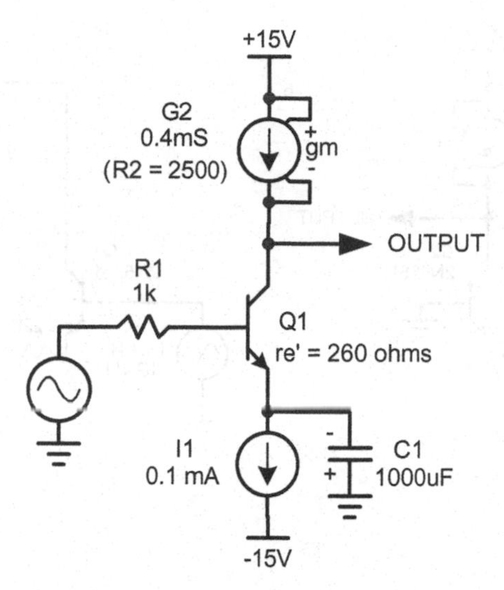

Figure 10.11 2N5551 Noise Simulation.

common-emitter mode with an emitter bias current source that is bypassed with a large capacitance. The base is connected to an AC voltage source through R1, which can be made zero or some chosen value. Load resistor R2 is an ideal noiseless resistor whose value is uncritical. It can be selected to provide a reasonable gain and a nominal voltage drop given the bias current chosen for the transistor.

An *AC* simulation is carried out to measure the gain of the stage. A noise simulation is then done and the noise voltage density (nV/√Hz) reported at the output is referred back to the input by dividing by the voltage gain.

10.14 Preamplifier Circuit Noise

The input stage often dominates the noise in a preamplifier, but there are numerous other sources of noise that can make the overall preamplifier significantly noisier than the input stage by itself would suggest. Here some of the major contributors to the preamplifier system noise will be discussed.

Noise of a Degenerated Differential Pair

Figure 10.12 shows two input stages, each implemented with a BJT differential pair. They are the same with the exception that the one in (b) has about 10:1 emitter degeneration. Each has a 5-kΩ noiseless load resistor so that simulations can be done that focus solely on the noise performance of the differential pair. For BJT input stages that are degenerated, the circuit in (b) is more representative of expected noise performance. A question often asked is to what extent does emitter degeneration degrade the input noise performance of such a circuit, such as a microphone preamplifier that uses emitter degeneration to control the gain.

Each stage is biased with a tail current of 2 mA. The transconductance to the single-ended output has been reduced from about 20 mS in (a) to about 2 mS in (b) by the 10:1 emitter degeneration. Assume that the stage is fed from a voltage source. The voltage noise contributions of each 2N5551 transistor and its degeneration resistor will add to create the input noise because there are

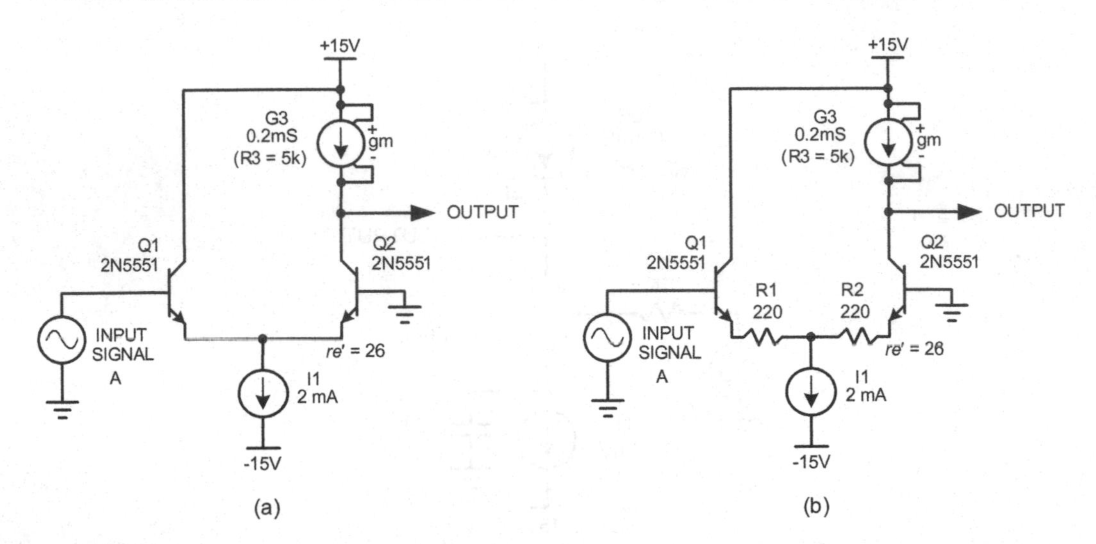

Figure 10.12 Differential Pair Input Stage.

two resistors effectively in series in forming the LTP (long-tailed pair, a differential pair where the emitter current is supplied from a current source). This is responsible for the usual 3-dB noise penalty when a differential pair is used instead of a single-ended stage.

Operating at 1 mA, *re'* for each transistor is 26 Ω and the transistor voltage noise is that of a 13 Ω resistor, so e_n = 0.47 nV/√Hz. Noise from transistor base resistance, about 90 Ω, is about 1.2 nV/√Hz. This results in total transistor noise of 1.3 nV/√Hz for each transistor. Total noise for the differential pair in (a) will be 3 dB higher, at 1.8 nV/√Hz. It is always important to remember that the RMS sum process strongly favors the larger term at the expense of the smaller term. A second noise contributor smaller by half only increases the RMS-summed noise by about 12%.

The thermal noise of each emitter resistor in (b) is 1.9 nV/√Hz. Input-referred voltage noise of each half of the LTP is thus 2.3 nV/√Hz. Input voltage noise for the stage is 3 dB higher, at 3.3 nV/√Hz. At this point, the introduction of the degeneration resistors has increased input stage noise by about 5 dB from 1.8 to 3.3 nV/√Hz.

Now assume that the stage is fed from a 1-kΩ source. This could be from the amplifier feedback network, for example. Source resistance noise is 4.2 nV/√Hz. Assume that transistor β is 100. Base current is 10 μA. Input noise current is 1.8 pA/√Hz. Input noise voltage due to input noise current flowing in the source resistance is 1.8 nV/√Hz. Total input noise voltage across the input impedance is thus 4.7 nV/√Hz, dominated by the resistor noise. Total input noise for the arrangement is the RMS sum of 3.3 and 4.7 nV/√Hz, or 5.7 nV/√Hz.

Current Source Noise

BJT current sources can be made in numerous ways, but the most fundamental is to provide a quiet reference voltage to a common emitter stage, as illustrated in Figure 10.13(a). Transistor Q1 and emitter resistor R1 are both sources of noise in the output current. Output noise current is evaluated by probing the noise voltage across the 1-Ω load resistor (its noise contribution is negligible; a noiseless load resistor of any value could be used instead). Noise current will be numerically equal to the probed noise voltage.

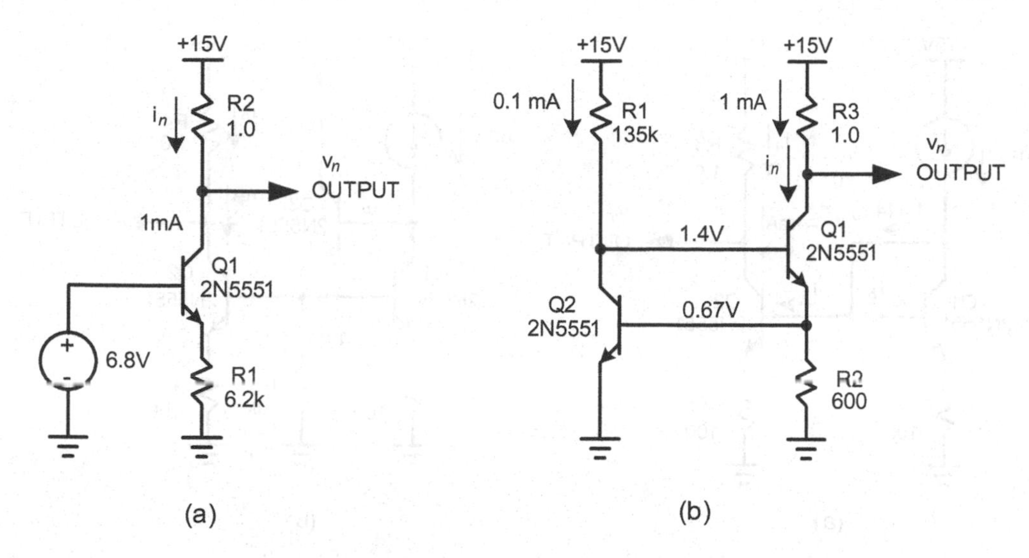

Figure 10.13 Noise in Current Sources.

The noise voltage of R1, approximately 10.2 nV/√Hz, is impressed across the resistance of R1 and *re′* to create noise current of about 1.6 pA/√Hz. The transistor's base current shot noise contributes about 1.7 pA/√Hz. This results in a total noise current of about 2.3 pA/√Hz. If the same 1-mA current source is implemented with a smaller reference voltage and a correspondingly smaller emitter resistor, the noise contribution from the resistor will increase, causing the total noise of the current source to increase. The contribution of the base shot noise will remain about the same, and in this case it will be dominated by the noise contributed by the emitter resistor.

The feedback current source shown in Figure 10.13(b) is a popular choice in many circuits. Its output current noise is influenced by the negative feedback provided by Q2. The feedback holds the base and emitter voltages of Q1 essentially constant, making them low-impedance points. The major noise contributor is the short-circuit current noise of R2, which is just its voltage noise divided by its resistance. This amounts to about 5.2 pA/√Hz. The voltage noise of Q2 operating at 100 μA is about 1.6 nV/√Hz. Impressed across R2, this contributes about 2.7 pA/√Hz. The third major contributor is the base shot noise current of Q1, which is about 2.4 pA/√Hz. These three contributors sum on an RMS basis to about 6.4 pA/√Hz.

The noise contribution of Q2 depends somewhat on the bias current supplied by R1, here 135 kΩ. If R1 is reduced, running Q2 at higher current, Q2's voltage noise will fall, but eventually Q2's base current noise will become a more significant contributor. A broad optimum for the bias current thus results. Here the optimum occurs at R1 = 30 kΩ, corresponding to bias current of 450 μA. This is rather high, and it should be kept in perspective that this noise contribution is not a major contributor to the total. Nevertheless, the bias current should probably never be set less than 1/10th of the current source value. The feedback current source is noisier than the conventional current source in (a) with a 6.8-V noiseless reference, but it requires much less voltage headroom.

Current Mirror Noise

A simple two-transistor current mirror is shown in Figure 10.14(a). The sum of the noise voltages of Q1 and R1 is impressed across R1, creating a noise current in Q1. The same thing happens in Q2, adding another noise current. As the values of R1 and R2 go down, the noise current goes up.

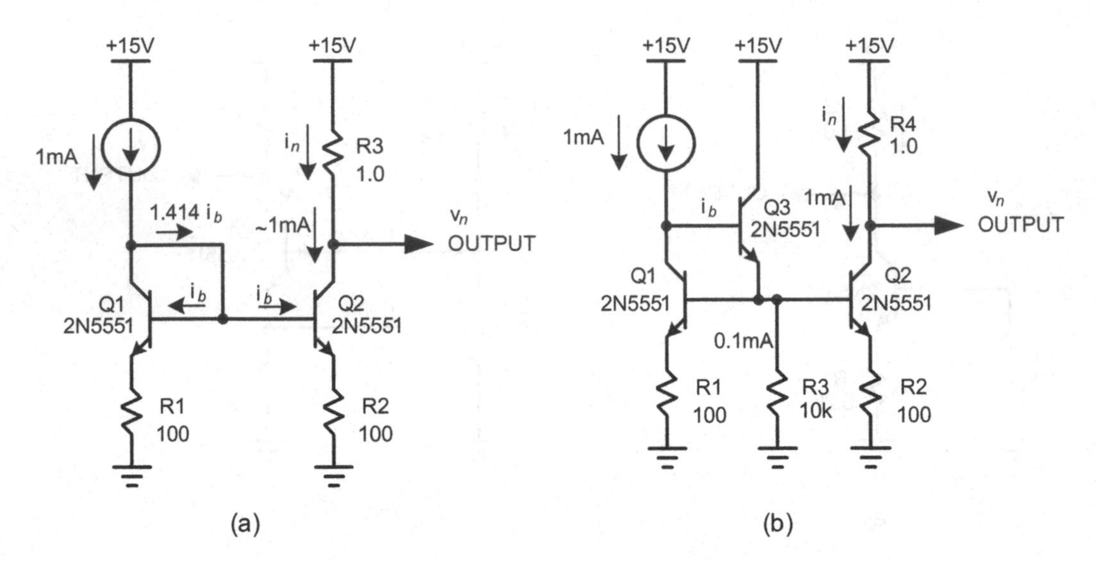

Figure 10.14 Noise in Current Mirrors.

If R1 and R2 are halved, their open-circuit thermal noise voltage goes down by √2, but this reduced thermal noise voltage is impressed across half the resistance, doubling the noise current. Thus, the net thermal noise current from the resistors goes up by √2. At the same time the transistor noise voltage is impressed across half the emitter resistance, doubling its noise current. This means that current mirrors with smaller degeneration resistors will create more noise in their output current. The base current noises of Q1 and Q2 are also injected into the signal path.

A three-transistor current mirror is shown in Figure 10.14(b). The base currents for Q1 and Q2 are supplied by emitter follower Q3. This reduces the contribution of base current noise from Q1 and Q2, while making the current mirror more accurate.

IPS/VAS Noise

Two simple IPS/VAS circuits are shown in Figure 10.15. The negative feedback loop is closed with a buffer with gain of 0.03 so that noise can be evaluated in a circuit with gain of 33. The noise performance of an amplifier is often dominated by the input stage like the degenerated differential pair shown in Figure 10.12. This is because the gain of the input stage is often much greater than 1, meaning that noise created in later stages is divided by that gain when referred back to the input. If the input-referred noise of the VAS is 10 nV/√Hz and the gain of the input stage is 5, then the VAS noise contribution to the amplifier's input-referred noise is only 2 nV/√Hz. Unfortunately, things are usually not this simple.

In the crude IPS/VAS of Figure 10.15(a), the gain of the input stage is only about 2, due to the emitter degeneration by R1 and R2. The 4-nV/√Hz of R3 is referred back to the input as 2 nV/√Hz. The input noise voltage of second-stage transistor Q3 is divided by 2 when it is referred to the input. More significantly, the input current noise of Q3 is referred back to the input as voltage noise by the limited transconductance of the degenerated input stage. This reflects another noise penalty from degeneration of the input stage. Net input noise for this IPS/VAS is 4.3 nV/√Hz.

The IPS/VAS of Figure 10.15(b) includes a current mirror load for the input stage and an emitter follower in front of the VAS transistor. Here the input stage gain is very high, but the current mirror

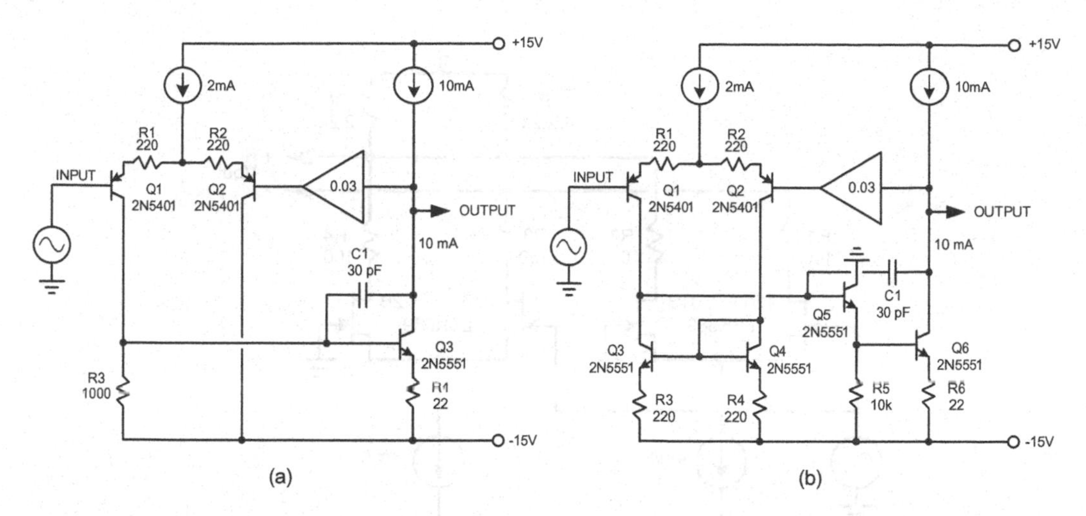

Figure 10.15 IPS/VAS Noise.

creates noise as described earlier. Indeed, at 220 Ω, R3 and R4 create as much input-referred noise as degeneration resistors R1 and R2. Increasing R3 and R4 will reduce this noise contribution, but will reduce voltage headroom available to Q3. At the same time, the current mirror doubles the signal current by enabling the signal currents from both Q1 and Q2 to be used. Put another way, the current mirror load doubles the transconductance of the input stage.

Significantly, voltage and current noise contributions from the VAS are now negligible due to the high input stage gain and the current buffering provided by emitter follower Q5. Net input noise for this IPS/VAS is 4.3 nV/√Hz. This happens to be the same number as in (a), even though the contributing noise sources are different. The differential load presented by the current mirror largely cancels common mode noise from the LTP, such as that contributed by the tail current source.

Paralleled Devices

If two transistors or circuit blocks with current output are connected in parallel, the signal will sum on a voltage basis and the noise will sum on a power basis. This is because the signal contributions are correlated while the noise contributions are uncorrelated. A net improvement of 3 dB in the SNR will thus result. If four such circuit blocks were connected in parallel a net 6-dB improvement would result. An additional 3 dB is gained each time the number of paralleled blocks is doubled.

Two low-noise dual monolithic LSK389 JFETs are connected in parallel to gain a 3-dB noise improvement in Figure 10.16 [11, 31]. Matching is often a problem in connecting JFET transistors in parallel. Tight offset-voltage matching is available in these dual monolithic devices, but this still leaves the issue of differences in I_{dss} or threshold voltages between the two chips. Fortunately, we saw in Eqs. 10.11 and 10.12 that JFETs from the same process have about the same transconductance for a given operating current, independent of their individual I_{dss} or threshold voltage. This means that both pairs will contribute about the same transconductance in the circuit as long as the tail current for each pair is the same. This is the reason why each pair has its own current source in Figure 10.16. The 100-Ω drain resistors isolate the JFETs to prevent *HF* instability. They add no noise. The input-referred noise of this circuit is about 1 nV/√Hz.

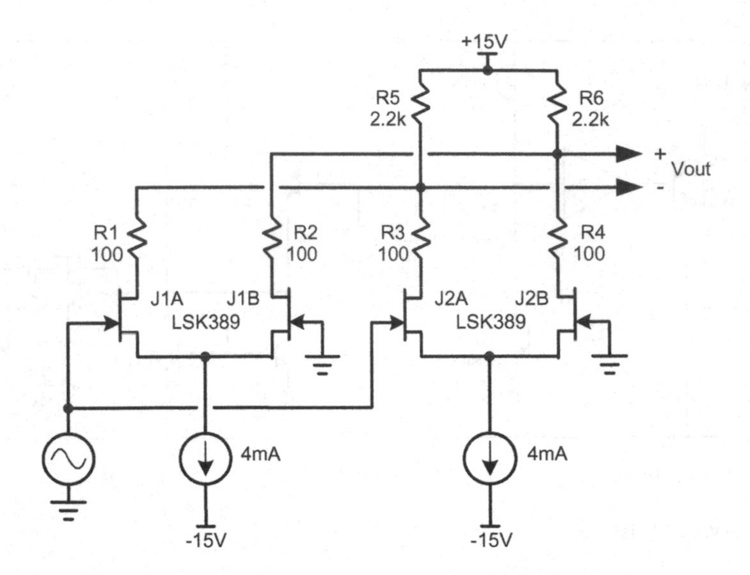

Figure 10.16 JFET Differential Pairs in Parallel.

10.15 Excess Resistor Noise

Resistors can create more noise than just the theoretical thermal noise. This is called excess resistor noise [7]. It usually results from current flow in the resistor, such as when it has a DC voltage across it. It can result from the resistor material itself or from contacts between the resistance material and the leads. Carbon film resistors have the worst excess noise, while metal film resistors have very little excess noise. Thick film resistors fall in between. All else remaining equal, a resistor of larger wattage will create less excess noise. If you use metal film resistors throughout a circuit, excess resistor noise will usually not be a problem.

10.16 Zener Diode and LED Noise

Zener diodes are often used as voltage references, sometimes as part of a current source. Zener diodes can be noisy, and in fact they have sometimes been used to build noise sources. Zener diodes rated at less than about 6.2 V operate in the Zener breakdown mode, while higher-voltage Zener diodes (e.g., 12 V) actually work in the avalanche breakdown mode. The latter are quieter as a result of the different breakdown mechanism. Zener diodes should always be thoroughly bypassed if they are used in a noise-critical location. It is important to know that the effective source impedance of the Zener diode is its dynamic resistance, which can be quite low. The size of the bypass capacitor must be sufficiently large to take account of this reality. For a Zener diode having dynamic resistance of 100 Ω and bypassed with 100 μF, its noise is low-pass filtered beginning at 16 Hz. Alternatively, in low-noise applications, some resistance can be placed between the Zener diode and the filter capacitance, forming a low-pass filter.

Similarly, LEDs have often been used as a convenient low-voltage reference. LEDs also create voltage noise that originates from their shot noise mechanism. LEDs should be bypassed if they are used in a noise-critical location. Typical forward voltage for a red LED is about 1.8 V and its voltage noise is on the order of 2.5 nV/√Hz. A green LED is similar. Blue LEDs operated at 1 mA have a forward voltage of about 3.0 V and their voltage noise can be as high as 30 nV/√Hz.

10.17 Low-Noise Amplifiers for Extremely High Impedance Sources

This section pertains to low-capacitance sources, especially where low-frequency response is important. A condenser microphone or a piezoelectric sensor is an example. A voltage amplifier or a charge amplifier can be used for such transducers. Shunt capacitance can be an issue because it will act to attenuate the signal. Shunt resistance, such as for biasing, can also be an issue, both by introducing thermal noise current and by degrading low-frequency response. The resulting noise can be red because the thermal noise of the bias resistor has been filtered by a first-order low-pass filter formed by the resistance of the bias resistor and the capacitance of the source. A typical example would be a source capacitance of 50 pf and a 1-GΩ bias resistor, with a corresponding low-frequency corner of about 3 Hz. The 4100 nV/√Hz thermal noise of the resistor is low-pass filtered beginning at 3 Hz.

At 1 kHz, the thermal noise of the bias resistor will be attenuated by a factor of 333 by the low-pass filter, resulting in 12.3 nV/√Hz. Noise at 100 Hz will be 123 nV/√Hz. The noise is red noise. This suggests that in some situations the noise of the JFET may not be the largest contributor. It also argues for a bias resistor of higher value than 1 GΩ.

Miller capacitance from the input JFET can act as a shunt capacitance, degrading performance in either a voltage amplifier or a charge amplifier. This can be mitigated by the use of a conventional cascode or a bootstrapped cascode.

Depending on the dominant noise source, the choice of JFET may be one with low voltage noise and high capacitance or one with higher voltage noise but lower capacitance. The advantage of a charge amplifier is that it uses a capacitor in its feedback path, effectively creating an integrator-like stage and creating a virtual ground at the input. This prevents the signal source from moving voltage-wise, and reduces the effect of shunt capacitance as from a connecting cable. Ideally, charge from the signal source is transferred to the feedback capacitor. In some cases, this could lead to the need to use a very small value of feedback capacitor if there is need for gain to the voltage output. However, note that shunt capacitance, be it from a connecting cable or from the input transistor, increases noise gain. This is because the feedback capacitor and the source capacitance form a capacitance voltage divider.

As in an inverter with resistors, if the feedback and source capacitances are equal, a 2:1 divider is formed and noise gain is 2 while the voltage gain is unity. Higher gain and lower noise can be had if the feedback capacitor is smaller as long as this does not result in an unreasonably small feedback capacitor. Still more gain can be had by placing a capacitance voltage divider between the amplifier output and the feedback capacitor, forming a *T* of capacitors.

References

1. Paul Horowitz and Winfield Hill, *The Art of Electronics*, 3rd edition, Cambridge University Press, Cambridge, MA, 2015.
2. W. Marshall Leach, Jr., *Fundamentals of Low-Noise Analog Circuit Design*, Proceedings of the IEEE, vol. 82, no. 10, October 1994, pp. 1515–1538.
3. Burkhard Vogel, *The Sound of Silence*, 2nd edition, Springer, New York, NY, 2011.
4. Burkhard Vogel, *Balanced Phono-Amps*, 2nd edition, Springer, New York, NY, 2016.
5. Gabriel Vasilescu, *Electronic Noise and Interfering Signals*, Springer, New York, NY, 2010.
6. Bruce Carter, *Op Amps for Everyone*, 4th edition, Newness, Burlington, NJ, 2013.
7. Bruce Hofer, Designing for Ultra-Low THD+N, Parts 1 and 2, *AudioXpress*, October-November 2013.
8. *Op Amp Noise Relationships: 1/f Noise, RMS Noise, and Equivalent Noise Bandwidth*, MT-048 Tutorial, Analog Devices, October 2008.

9. Bruce Carter, *Op Amp Noise Theory and Applications*, *Texas Instruments*, Literature Number SLOA082, 2002.
10. Bruce Trump, *1/f Noise – the flickering candle*, EDN, March 4, 2013.
11. Bob Cordell, VinylTrak – A Full-Featured MM/MC Phono Preamp, *Linear Audio*, vol. 4, September 2012, www.linearaudio.net.
12. Morgan Jones, Rectifier Snubbing – Background and Best Practices, *Linear Audio*, vol. 5, April 2013, pp. 7–26.
13. Mark Johnson, Soft Recovery Diodes Lower Transformer Ringing by 10–20×, *Linear Audio*, vol. 10, September 2015.
14. Bill Whitlock and Jamie Fox, *Ground Loops: The Rest of the Story*, presented 5 Nov 2010 at the 129th AES convention in San Francisco. Available from AES as preprint #8234.
15. Bill Whitlock, *An Overview of Audio System Grounding and Signal Interfacing*, Tutorial T5, 135th Convention of the Audio Engineering Society, October, 2013, New York City.
16. Bill Whitlock, *An Overview of Audio System Grounding and Shielding*, Tutorial T2, Convention of the Audio Engineering Society, Star-Quad Page 126.
17. Bill Whitlock, *Design of High Performance Audio Interfaces*, Jensen Transformers, Inc., Chatsworth, CA.
18. Belden, *How Starquad Works*, http://www.belden.com/blog/broadcastav/How-Starquad-Works.cfm.
19. Canare, *The Star Quad Story*, www.canare.com/UploadedDocuments/Cat11_p35.pdf and http://www.canare.com/UploadedDocuments/Star%20Quad%20Cable.pdf.
20. J.W. Haslett, F. N. Trofimenkoff, Thermal Noise in Field-effect Devices, *Proceedings of the IEE*, vol. 116, no. 11, November 1969.
21. Alicja Konczakowska and Bogdan M. Wilamowski, *Noise in Semiconductor Devices*, Chapter 11 in *Industrial Electronics Handbook*, vol. 1, *Fundamentals of Industrial Electronics*, 2nd edition, CRC Press, 2011.
22. *Low-Noise JFETs – Superior Performance to Bipolars*, AN106, Siliconix, March 1997.
23. Ken Yamagouchi and Shojiro Asai, *Excess Gate Current Analysis of Junction Gate FET's by Two Dimensional Computer Simulation*, IEEE Transactions on Electronic Devices, vol. ED-25, no. 3, March 1978.
24. Bob Cordell, *Linear Systems LSK489 Application Note*, www.linearsystems.com, February 2013.
25. J.C.J. Paasschens and R. de Kort, *Modeling the Excess Noise Due to Avalanche Multiplication in (Hetero-Junction) Bipolar Transistors*, Philips Research Laboratories.
26. US Patent 4,538,115, *JFET Differential Amplifier Stage with Method for Controlling Input Current*, James R. Butler, assigned to Precision Monolithics Inc., August 27, 1985.
27. Walter G. Jung, *Op Amp Applications Handbook*, Elsevier/Newnes, Chapter 1, 2005.
28. Glen Brisebois, *Op Amp Selection Guide for Optimum Noise Performance*, Design Note 355, Linear Technology Corporation.
29. Analog Devices, Inc., LTspice® Simulator, analog.com.
30. Gilles Brocard, *The LTSPICE IV Simulator*, 1st edition, Wurth Elektronik, 2011.
31. Bob Cordell, *Linear Systems LSK389 Application Note*, January 2020, www.linearsystems.com.

Chapter 11

Filters

In simple terms, filters are circuits that provide frequency-dependent gain or loss. Most filters fall into the categories of low-pass, high-pass, bandpass, all-pass or notch filters. A simple low-pass filter (LPF) typically passes low frequencies with unity relative gain and attenuates very high frequencies to great loss. They are usually characterized by the frequency at which their loss is 3 dB down from their peak gain. They are also characterized by the steepness of their roll-off, or filter order such as 6 dB per octave (20 dB/decade) being a first-order filter, 12 dB/octave being a second-order filter, etc. Simple high-pass filters (HPFs) have a converse correspondence, with unity relative gain at very high frequencies and great attenuation at very low frequencies. Both low-pass and high-pass filters of greater than first order may have some frequency regions with peaking where their relative gain is more than unity or its nominal value.

Simple bandpass filters (BPFs) have unity relative gain at their center frequencies and have high attenuation at frequencies far above and far below their center frequency. These filters are also denoted by their order, beginning with second-order and higher orders, corresponding to greater and greater steepness in the roll-off of their gain at frequencies above and below their center frequencies. Such filters are often characterized by their so-called "quality factor," or Q. For example, a second-order BPF with high Q has narrow bandwidth and steep slopes above and below its center frequency and rolls off quickly on either side of the center frequency. A notch filter correspondingly rejects frequencies at its center frequency while passing frequencies far from its center frequency with unity relative gain.

This chapter is not intended to make you an expert on filter design. The subject of filters is broad and deep, with a substantial amount of math. There are numerous excellent texts for that [1–7]. In fact, there are numerous online calculators that will calculate component values for filters of different types and orders [8–16]. Of course, this chapter does touch on some of the basic math involved in filter transfer functions and how poles and zeros are placed in the complex plane to achieve a particular filter shape. The primary focus is on many types of filters, how they can be used and how they can be implemented with circuits. Numerous filter designs are presented whose values can be frequency-scaled to provide a desired design.

11.1 Filter Shapes and Characteristics

The most commonly used filters are first-order and second-order low-pass, bandpass and high-pass filters. First- and second-order filters often are the building blocks for higher-order filters. Filters with order two or greater are characterized by different shapes. The most obvious aspect of shape is the Q of a second-order filter, which defines its sharpness in frequency response. Related to Q and some other features like center frequency f_0, there are numerous other shape characteristics that will be discussed. These include the Butterworth, with maximally flat amplitude; the Bessel, with maximally flat delay; and the Chebyshev, with maximum steepness of cutoff for a given allowed amount of passband ripple amplitude. These are the most common types, but there are numerous others, some of which will be discussed in this chapter. Those include notch, elliptic, all-pass and other filters.

Poles and Zeros

One of the most important aspects of filter design and use is the concept of poles and zeros, so a review of Chapter 6 can be helpful here. This includes a good understanding of the role of radian frequency ω, complex frequency s and transfer functions $H(s)$. Recall that $\omega = 2\pi f$ and that in the simple form $s = j\omega$, where j designates the imaginary number that is the square root of -1. In a more complete form, $s = j\omega + \sigma$, where σ is the real part of a pole location in the complex plane. Its units are Nepers. Throughout this chapter, the center frequency (natural frequency) of a second-order filter section will be designated as f_0 in Hz or ω_0 in rad/s. The two terms will be used interchangeably as best suits the context.

The basis for most filter designs is the LPF version, often called an all-pole filter because it consists of numerous poles of various orders in the denominator polynomial (characteristic polynomial) of the transfer function and no zeros in the numerator. The roots of the denominator polynomial are called poles and the roots of a numerator polynomial are called zeros. Their sibling bandpass and high-pass versions are usually derived by transformation from the corresponding "prototype" LPF. The order of a filter is simply the number of poles in the denominator polynomial of the transfer function.

First-order filters, like a simple R-C section, roll off at 6 dB/octave in their stop-band. Second-order low-pass and high-pass filters roll off at 12 dB/octave in their stop bands. Second-order BPFs roll off at 6 dB/octave below and above their center frequency f_0. Depending on Q, all of these filters may roll off at a steeper rate in the vicinity of their 3-dB frequency or center frequency. Some may also have peaking in their passband in the vicinity of their 3-dB frequency. These filter building blocks are simple, and there are many online filter tables that make resort to much of the math unnecessary [8–16].

Second-Order Filters and Q

The sharpness of a filter, characterized by its "quality factor," Q, is often most easily understood in the context of a second-order BPF, where its Q is defined as f_0 divided by the bandwidth (the difference between the -3-dB points below and above f_0). However, all second-order filters are characterized by a quadratic equation (polynomial) in their transfer function's denominator. The Q associated with this quadratic is the Q of the second-order filter, be it a low-pass, bandpass or high-pass filter. The equation below illustrates the transfer function and its denominator for an LPF whose cutoff frequency is normalized to 1 rad/s ($\omega = 1$). It is easy to see that at 1 rad/s, the s^2 term equals -1 and cancels the unity term in the denominator, leaving the $1.414s$ term. At $\omega = 1$, the gain is thus 0.707. This is also the Q of the filter. The smaller the middle term, the higher the Q.

$$H(s) = 1/\left(s^2 + 1.414s + 1\right)$$

In a more general case, we have:

$$H(s) = \left(s^2 + a_1 s + a_0\right)/\left(s^2 + b_1 s + b_0\right)$$

where there is both a numerator and denominator polynomial. If the numerator is 1, we have an LPF. If the numerator is s, we have a BPF. If the numerator is s^2, we have an HPF. If the numerator has more than one non-zero value, the numerator is a quadratic polynomial and the filter transfer function is referred to as biquadratic. In such a filter, the critical (resonant) frequency of the denominator and numerator polynomials can be different. Similarly, the Q of the numerator can be different from the Q of the denominator.

Third-Order and Odd-Order Filters

Third-order filters are usually made up of a second-order filter in tandem with a first-order filter. In fact, all filters that are of odd order (except first) are generally made up of a multiplicity of even-order filters in tandem with one first-order filter. In a high-order filter, the design is characterized by the f_0 and Q of each constituent second-order section. For odd-order filters, the specification of the filter also includes the cutoff frequency of the first-order section. Some third-order filters are implemented with a single circuit, without explicitly separate first- and second-order sections. Sometimes, such a third-order filter can be used as one of the sections of an odd-order filter of higher order. An example would be a fifth-order LPF comprising a third-order filter section in tandem with a second-order filter section.

Butterworth Maximally Flat Filter

The Butterworth filter shape is probably the most popular. Its maximally flat response is well-behaved and a good compromise between filters that give less phase distortion and filters that provide sharper cutoff. It provides the sharpest cutoff without peaking for a given filter order. The filter is 3 dB down at its designated cutoff frequency and its phase lag at that frequency is 45° for each order of the filter. A second-order Butterworth LPF thus has 90° of phase lag at its cutoff frequency.

The Butterworth, Bessel and Chebyshev LPFs are referred to as all-pole filters because they do not have zeros in their transfer functions. While much of the focus on these filters and their shapes is on the low-pass "prototype" version, there are transformations that can convert these filter transfer functions and circuit implementations into bandpass and high-pass filters. The key is that the filters are characterized by the polynomial in the denominator of the transfer function that defines the locations of the poles in the s-plane.

The poles of the prototype Butterworth filter lie on the unit circle in the left-half complex plane, and are thus all at the same frequency. The radius of the circle is the natural frequency, here normalized to $\omega = 1$ rad/s. With poles distributed evenly on the unit circle, an odd-order filter has one pole on the real axis in the left half plane. In fact, all-pole filters have no more than one pole on the real axis.

Transfer Functions

The first-order RC LPF has a very simple normalized transfer function:

$$H(s) = 1/(s+1).$$

The cutoff frequency is normalized to $\omega = 1$ rad/s. The denominator is said to be the characteristic polynomial of the filter function. The first-order filter is of course a Butterworth filter because its roll-off is monotonic and maximally flat.

A second-order Butterworth normalized transfer function is:

$$H(s) = 1/\left(s^2+1.414s+1\right) = 1/\left(s^2+\alpha s+1\right) = 1/\left(s^2+(1/Q)s+1\right)$$

Here the term $\alpha = 1.414$ determines the Q of the second-order section and thus the Butterworth shape. The Q is $1/\alpha$ and is 0.707. The $(s^2 + 1.414s + 1)$ term is the normalized characteristic denominator polynomial $D(s)$ for a second-order Butterworth filter. Because $s = j\omega$, at the normalized frequency of 1 rad/s, it is easy to see that the s^2 term becomes -1 in the polynomial and cancels the $+1$ term. This leaves only the $1.414s$ term in the denominator, resulting in a transfer function ratio of 0.707, corresponding to the 3-dB of attenuation at ω_0. In the general case, the gain at $\omega_0 = Q$.

Q and Peaking

A second-order section will have a peak in its frequency response if Q is greater than 0.707. The amplitude A_{pk} and frequency ω_{pk} of the peak depend on Q. Indeed, Q for a second-order section can be inferred from A_{pk} or ω_{pk}.

$$A_{pk} = Q/\sqrt{\left(1-1/4Q^2\right)}$$

$$\omega_{pk} = \omega_0\sqrt{\left(1-1/2Q^2\right)}$$

As expected, for $Q = 0.707$, A_{pk} is unity (no peaking) and $\omega_{pk} = 0$.
For $Q = 0.8$, $A_{pk} = 1.026$ (+0.22 dB) and $\omega_{pk} = 0.468$.
For $Q = 1$, $A_{pk} = 1.15$ (+1.21 dB) and $\omega_{pk} = 0.707$.
For $Q = 2.0$, $A_{pk} = 2.066$ (+6.3 dB) and $\omega_{pk} = 0.935$.
These equations are not valid for $Q < 0.707$.

Butterworth Filters of Higher Order

Higher-order Butterworth filters are made up of combinations of one or more second-order low-pass denominator polynomials combined with a first-order polynomial for odd-order sections. The factored Butterworth polynomials, available from filter tables, for the second- through sixth-order filters are shown below [3, 4, 7]. The second-order polynomial sections are of the form $(s^2 + \alpha s + 1)$, where ω_0 for the section is normalized to 1 rad/s and $Q = 1/\alpha$. The ω_0 for all of the sections for a Butterworth filter is 1.0, corresponding to the overall filter frequency. However, notice that the Q, as denoted by $1/\alpha$ of each section, is different.

$$\left(s^2+1.414s+1\right)$$

$$\left(s+1\right)\left(s^2+s+1\right)$$

$$\left(s^2+0.765s+1\right)\left(s^2+1.848s+1\right)$$

$$\left(s+1\right)\left(s^2+0.618s+1\right)\left(s^2+1.618s+1\right)$$

$$\left(s^2+0.518s+1\right)\left(s^2+1.414s+1\right)\left(s^2+1.932s+1\right)$$

These denominator polynomials are illustrated in their so-called factored form, where the polynomial consists only of the product of first- and second-order elements. Notice that the Q of the first second-order section is the highest in the filter (smallest α), and that this value of maximum Q increases with the order of the filter. It is easy to see that if one has the polynomials in factored form it is straightforward to create an LPF by cascading individual sections whose component values can be found from online calculators [13]. As we will see later in this chapter, the second-order sections will often be implemented with fairly simple Sallen-Key or multiple-feedback (MFB) filter sections that each employ a single op amp, two capacitors and two or more resistors.

Sometimes the constituent second-order quadratics are described by a frequency scaling factor (FSF) and Q. FSF is just the scale factor for ω_0 of each section with respect to 1 rad/s. For the quadratics of a Butterworth filter, the FSF for all of the complex conjugate poles is unity, since they all lie on the unit circle in the complex plane. However, in other filter functions the poles do not lie on the unit circle, so their FSF for each pole is different. Knowing the FSF and Q for each pole or pair of complex poles and the Q for each quadratic makes it straightforward to design the sections of a filter from filter tables. If a complex pair of poles in a 1-kHz filter section has FSF = 1.2, the f_0 of that second-order section is 1.2 kHz.

Converting Pole Location to Frequency and Q

It is handy to have the frequency and Q for the desired second-order section, but filter functions are often instead provided with their pole locations in the complex plane. Those locations will take the form of $(-\alpha \pm j\beta)$, where α denotes the location along the real axis in the left half plane and $\pm\beta$ denotes the locations of the complex pole pair on the imaginary axis. We have

$$\omega_0=\sqrt{\left(\alpha^2+\beta^2\right)}, \quad Q=\omega_0/2\alpha$$

Bessel Maximally Flat Delay Filter

The Bessel filter has maximally flat group delay as opposed to maximally flat amplitude vs frequency. It is used when a minimal amount of phase distortion is required. It has lower selectivity and no overshoot or ringing in its step response. The price paid is a shallower roll-off. If the Bessel filter's designated cutoff frequency is where it is down 3 dB, the attenuation at high frequencies will be much less and the phase lag at the 3-dB frequency will not be the filter order times 45°. The onset of roll-off will also begin at a much lower frequency than for the Butterworth. The factored Bessel D(s), available from filter tables, and the numerator value of the transfer function, for the second- through sixth-order filters are as follows [3]:

$$\left(s^2+1.733s+1.275\right)$$

$$\left(s+1.237\right)\left(s^2+1.452s+1\right)$$

$$\left(s^2+1.916s+1.419\right)\left(s^2+1.241s+1.591\right)$$

$$\left(s+1.507\right)\left(s^2+1.773s+1\right)\left(s^2+1.091s+1\right)$$

$$\left(s^2+1.961s+1\right)\left(s^2+1.637s+1\right)\left(s^2+0.978s+1\right)$$

While the f_0 of all of the sections of a Butterworth filter are the same, the f_0s of the different sections of a Bessel filter are different. The poles of a Bessel filter lie outside the unit circle. While the real pole of a third-order Butterworth is at −1, the real pole for a corresponding Bessel filter is at −1.33. Gain for the second- through sixth-order Bessel filters above is 1.622, 2.799, 5.100, 11.385 and 26.833, respectively. Gain adjustments to achieve the total gain for the filter can be made by selection of the gain of each section. Often, the gain of each section will be set to unity and the overall filter gain will be in the vicinity of unity.

Chebyshev Equiripple Filter

The Chebyshev filter is used when a steeper roll-off is desired and some ripple in the passband is acceptable. It is called an equiripple filter because the peaks and valleys of all of the ripples are the same dB amount above and below the nominal average filter gain. A Chebyshev filter is characterized by the amount of its ripple, which is the peak-to-peak value of the ripple above and below the average gain. A 1-dB Chebyshev filter has ripple that goes 0.5 dB above and 0.5 dB below the nominal gain (the ripple maxima and minima, respectively). The passband cutoff for a Chebyshev filter is not always defined as the −3-dB point. Rather, it is sometimes defined as the frequency where the response falls below the bottom ripple allowance, here 0.5 dB below the nominal gain. This is called the ripple bandwidth, and is smaller than the −3-dB bandwidth.

For a second-order Chebyshev filter with 1 dB of ripple, the 3-dB bandwidth is 22% wider than the ripple bandwidth. For a fourth-order 1-dB filter, the 3-dB bandwidth is only 5.3% wider than the ripple bandwidth [4]. The 3-dB bandwidth divided by the ripple bandwidth is often tabulated as the ripple bandwidth ratio (RBR) [3]. The smaller the number (closer to unity), the greater is the steepness (selectivity) of the filter. The RBR for a 0.5-dB sixth-order Chebyshev filter is only 1.041.

For purposes of the above, I have defined the nominal gain as the average gain across the pass-band; i.e. that gain above and below which the ripple goes. This seems intuitive. However, at DC, the ripple does not start off at the midpoint of the ripple. The 0-dB gain point in Ref. [3] and other texts is defined as the positive maxima of the ripple. Odd-order filters start off at the ripple maxima, so the DC gain is equal to this definition of gain.

Even-order filters start out at DC with gain at the minima of the ripple. For an even-order filter with 1 dB of ripple, the DC gain is −0.5 dB. As a result, the reference for what is defined as the 3-dB-down bandwidth is the gain at the peaks of the ripple. This can cause confusion. It was brought up in the context of an *LC* implementation. In active implementations, the constituent LPF sections have 0 dB of gain at DC, which would seem to affect the definitions.

The poles of a Chebyshev filter lie inside the unit circle on a vertical ellipse, closer to the $j\omega$ axis. While the real pole of an odd-order Butterworth filter lies at −1, the real pole of a fifth-order Chebyshev filter lies at −0.34.

The Chebyshev 0.5-dB ripple factored denominator polynomials, available from some filter tables, for the second- through sixth-order filters are as follows [3]:

$$\left(s^2+1.157s+1.231\right)$$

$$\left(s+0.627\right)\left(s^2+0.586s+1.069\right)$$

$$\left(s^2+1.418s+0.597\right)\left(s^2+0.340s+1.031\right)$$

$$\left(s+0.362\right)\left(s^2+0.849s+0.691\right)\left(s^2+0.220s+1.018\right)$$

$$\left(s^2+1.462s+0.396\right)\left(s^2+0.552s+0.768\right)\left(s^2+0.154s+1.011\right)$$

The Chebyshev 1.0-dB ripple factored denominator polynomials are as follows:

$$\left(s^2+1.046s+0.862\right)$$

$$\left(s+0.627\right)\left(s^2+0.496s+0.911\right)$$

$$\left(s^2+1.274s+0.502\right)\left(s^2+0.281s+0.943\right)$$

$$\left(s+0.280\right)\left(s^2+0.715s+0.634\right)\left(s^2+0.180s+0.961\right)$$

$$\left(s^2+1.314s+0.346\right)\left(s^2+0.426s+0.727\right)\left(s^2+0.125s+0.973\right)$$

Elliptic Filter (Cauer Filter)

The elliptic filter is used when very sharp cutoff is required, such as in an anti-aliasing filter. It is also called a Cauer filter. It is typically characterized by ripple in both the passband and the stop-band. It is like a Chebyshev filter, but with one or more zeros added in the transfer function numerator $N(s)$ to hasten the roll-off. Zeros produce a notch or band-reject shape. If a zero is added to a Chebyshev filter above the cutoff frequency, the response quickly heads to zero as frequency increases, making for a very sharp cutoff. The elliptic filter is thus not an all-pole filter. Of course, the notch created by the zero "recovers" in its response as frequency increases further, allowing the overall elliptic filter response to come back up to the response that it would have had as a Chebyshev filter without the zero(s). This causes ripple in the stop-band.

Elliptic filters are often designed to have a stated minimum amount of attenuation in the stop-band. Maximum deviation from flatness in the passband is also often specified. The elliptic AES17 measurement filter for analog outputs of digital devices is required to have less than 0.1 dB of ripple up to 20 kHz and then be down by 60 dB at 24 kHz and above. It is similar to an anti-aliasing filter that is put in front of an ADC in the recording process. The actual design of elliptic filters is beyond the scope of this chapter.

11.2 Sallen-Key Filters

The Sallen-Key (S-K) active filter is probably the most popular active filter [17]. It is a positive feedback filter [1]. It is an active second-order section that can be used to build filters of higher order. There are low-pass, bandpass and high-pass versions of this filter, as shown in Figure 11.1. Simplicity is one of its main features. In fact, the active element in most S-K filters is merely a unity-gain non-inverting amplifier, such as a unity-gain op amp buffer or even an emitter follower. The S-K filter is sometimes called a voltage-controlled voltage source *VCVS* filter for this reason. As explained below, sometimes the gain element has greater than unity gain.

The Q of a second-order S-K filter can be chosen as needed for the sections of the higher-order filters with a Butterworth, Bessel or Chebyshev characteristic. There are online calculators and filter tables for designing S-K filter sections of specified f_0 and Q in low-pass, bandpass and high-pass forms [8–15]. Ref. [13] is especially helpful. You can usually plug in the desired f_0 and Q and a starting value for one of the components and the calculator will spit out the remaining component values. The starting value usually establishes the impedance levels in the circuit. Since capacitors are available in far fewer value increments than resistors, a capacitor is often the starting value that is entered. In fact, it is often attractive if equal-value capacitors can be used in a design. As with most filters, if the order of the filter increases, the Q of the second-order section with the highest Q in a filter increases. High-Q filter sections tend to be more sensitive to component values and op amp imperfections [18].

The S-K filter is a positive feedback (PFB) filter. When the amplifier is configured for unity gain, the PFB is just a form of bootstrapping of the *RC* network. The filter depends on positive feedback to achieve its resonance and Q. The natural ("resonant") frequency of this and many other second-order filters is simply:

$$\omega_0 = 1/\sqrt{(\text{R1R2C1C2})}.$$

Sallen-Key Second-Order LPF, HPF and BPF

Figure 11.1 shows example circuit implementations for the S-K low-pass, high-pass and bandpass second-order filters normalized to f_0 of 1 kHz. The filter in (a) is an LPF with a Butterworth characteristic with $Q = 0.707$. The Butterworth HPF can be obtained by swapping the resistors and capacitors in (a). In (b), capacitors of equal value are used with resistor values revised to maintain the Butterworth characteristic.

The BPF in Figure 11.1(c) is designed for $Q = 1.0$ [1]. Notice that the parallel combination of R1 and R2 in (c) equals R3. The appearance of a Wien bridge emerges, and the frequency is set in the same way, with equal values of capacitors and resistors. Component values are easily scaled to other frequencies and impedances. Q is set by amplifier gain k. The BPF requires amplifier gain greater than unity to achieve a Q of 1. The Q achievable for a BPF with a unity-gain amplifier is only about 0.5. It is easier to achieve higher Q for a given amplifier gain with an LPF or HPF than with a BPF. The Sallen-Key BPF will be covered in greater depth further below.

While the S-K design employs positive feedback and is non-inverting, there is a different topology that employs negative feedback available for implementing the LPF and BPF, and it will be covered later. It is called the multiple feedback filter (MFB). It is better in achieving filter sections

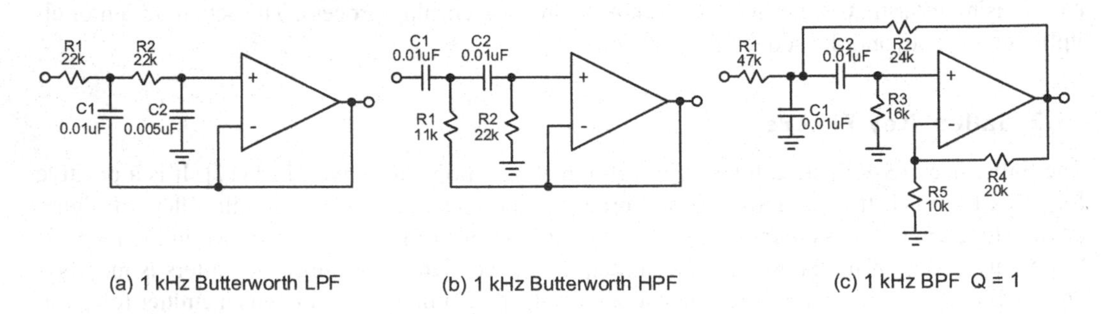

Figure 11.1 Sallen-Key Second-Order Filters.

with higher Q, especially for BPF sections. There is at least one calculator out there that does not even cover the Sallen-Key BPF.

The Low-Pass Prototype

Most filter designs start out as a low-pass version; high-pass and bandpass versions are obtained through a transformation. That is why there is often more attention paid to the low-pass versions of various filter implementations in the literature. The transformation from LPF to HPF is simple: just swap the capacitors and resistors.

Figure 11.2 shows three Sallen-Key LPF versions of particular interest [1]. All can have amplifier gain k to improve performance, manage component sensitivity or increase achievable Q. With $\omega_0 = 1/\sqrt{(R1R2C1C2)}$, it is just a matter of selecting the components to achieve the desired Q.

There are three special cases that make S-K filters easy to design and suitable for most applications. The first is the arrangement of Figure 11.2(a). It operates with a unity-gain amplifier and R1 = R2. Here the ratio of C1/C2 sets Q as:

$$Q = \sqrt{C1/4C2} = 1.58$$

$$C1/C2 = 4Q^2$$

The ratio of C1 to C2 becomes very large if significant Q is required. In (a), C1 = 10C2, and $Q = 1.58$. This filter is very simple and requires only two resistors. It is attractive for LPF and HPF designs requiring Q of less than 1.6.

The second circuit in Figure 11.2(b) is attractive because R1 = R2 = R and C1 = C2 = C. In this case, Q is set by gain k of the amplifier with resistors R3 and R4, as

$$k = 3 - 1/Q$$

Here $Q = 2$ and $k = 2.5$. A nice feature of this design is that $\omega_0 = 1/RC$. Q can be set independently of R1, R2, C1 and C2.

It is often convenient to have second-order filter implementations that employ capacitors all of equal value. This can be especially advantageous in cases requiring more precision or even matched values of capacitance. Some online tables will allow you to specify or arrive at filter designs that can employ equal-value capacitors. In some cases, employing gain elements with greater than unity gain can be helpful in arriving at realizable versions of such designs.

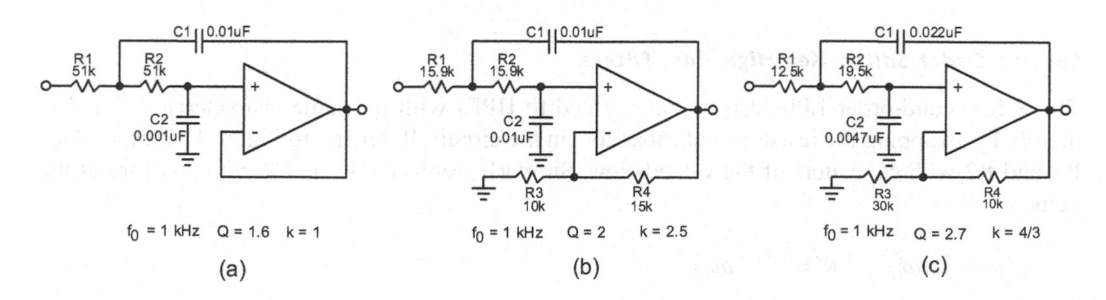

Figure 11.2 Three S-K Low-Pass Implementations.

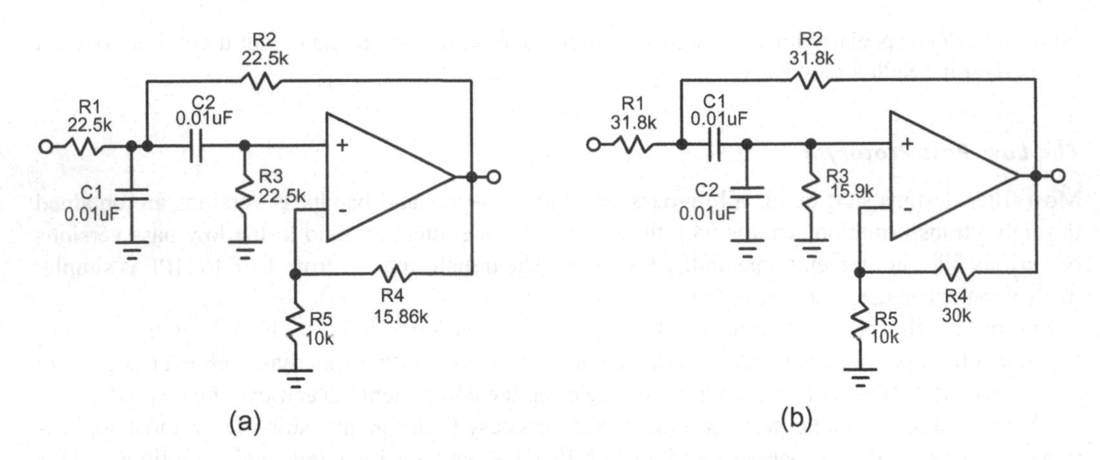

Figure 11.3 Sallen-Key Bandpass Filters with Amplifier Gain.

The third circuit in Figure 11.2(c), by Saraga, provides the lowest component sensitivity for a given Q when $k = 4/3$. Here C1/C2 = $Q\surd 3$ [1]. With the ratio of C1 and C2 set, and with the value of C1 or C2 chosen, it is just a matter of calculating R1 and R2 to set the frequency. Choosing $f_0 =$ 1 kHz, $\omega_0 = 6.283\text{e}3$, $Q = 2.7$, $k = 4/3$ and C2 = 0.0047 µF, we have

$$\text{C2} = 0.0047\,\mu\text{F}$$

$$Q = 2.7$$

$$\text{C1} = \text{C2}Q\surd 3$$

$$\text{R1} = 1/\left(Q\omega_0\right)$$

$$\text{R2} = 1/\left(\omega_0 \surd 3\right)$$

$$\text{R2}/\text{R1} = 1.56$$

$$\text{R1} = 12.5\text{k}\Omega, \quad \text{R2} = 19.5\text{k}\Omega, \quad \text{C1} = 0.022\,\mu\text{F}, \quad \text{C2} = 0.0047\,\mu\text{F}$$

Second-Order Sallen-Key High-Pass Filters

The S-K second-order LPFs can be transformed to HPFs with the same characteristic f_0 and Q merely by swapping the resistors and capacitors in the circuit. Referring to Figure 11.2(b), replace R1 and R2 with capacitors of the value below. Similarly, replace C1 and C2 with resistors of the value

$$C' = 1/\left(R\omega_0\right), \quad R' = 1/\left(C\omega_0\right)$$

Second-Order Sallen-Key Bandpass Filters

So far much of the focus has been on Sallen-Key filters implemented with a non-inverting unity-gain element. These designs are simple and probably the most popular. They result in filters with accurate unity gain in their passband. Sometimes filters are desired to have gain, taking advantage of the op amp that is already there. Low-pass filters like those in Figure 11.2(b) and (c) are good examples of how this can be accomplished with $k > 1$.

A second-order S-K BPF is shown in Figure 11.1(c). Most S-K BPFs with useful Q actually require gain k greater than unity in the gain element. This generalizes the use of the term *VCVS filter* as an alternative to the Sallen-Key name. The expense is two extra resistors.

There are actually two versions of BPF that are referred to as being S-K. They are both non-inverting and nominally require three resistors. The rather subtle difference is in the placement of the shunt capacitor to ground, it being before or after the node where R2 connects, as shown in Figure 11.3. The design equations are different. Interestingly, if one looks carefully at a slight re-arrangement of the circuit in (b), the elements of a Wien bridge network can be seen. The circuit is a bit like a Wien bridge with positive Q-enhancing feedback (as in a Wien bridge oscillator) wherein the input signal is injected by R1 at the junction of C1 and R2.

An S-K BPF with a gain element with gain $k = 3.0$, $Q = 1$ and gain $= 0$ dB is shown in Figure 11.3(a) [1]. The circuit is a very simple extension of the non-inverting S-K filter with the addition of two resistors to add a feedback network to create a gain of 3. Fewer design calculators are available to design S-K BPFs, but the equations below to determine component values for some useful combinations are not difficult.

There is another design that employs the BPF topology of Figure 11.1(c). It is shown in Figure 11.3(a), where C1 = C2 = C and R1 = R2 = R3 = R [1]. One will usually choose a value for C as a start, such as $C = 0.01$ µF for a filter in the 1-kHz range. For $f_0 = 1$ kHz, $Q = 1$ and $C = 0.01$ µF, we have

$$R = \sqrt{2} / \omega_0 C = 22.5\ \text{k}\Omega$$

$$k = 4 - \sqrt{2} / Q = 2.586$$

$$\text{Gain} = k / RC$$

If the resulting filter gain is too high, R1 can be replaced by an attenuator whose output impedance is equal to the value of R1.

A slightly different S-K BPF topology is shown in Figure 11.3(b). It can be informally designed by recognizing its Wien bridge nature, where the resistors and capacitors are of equal values. In this case, the parallel combination of R1 and R2 is made equal to the value of R3. The Q of the filter is set by the gain of the amplifier, as determined by R4 and R5. For the following values of k:

$K = 3.2$, $Q = 0.707$ and gain = 1.16 dB.
$K = 4.0$, $Q = 1.0$ and gain = 6.0 dB.
$K = 5.0$, $Q = 2.0$ and gain = 14 dB.

Yet another version of Figure 11.3(b), with C1 = 2C2, can be designed with the equations below [4] (p. 645). Here $Q = 1$, $f_0 = 1$ kHz, and C1 = 0.01 µF.

$$\text{C2} = \text{C1} / 2$$

$$R1 = 2/(\omega_0 C1)$$
$$R2 = 2/(3\omega_0 C1) = R1/3$$
$$R3 = 4/(\omega_0 C1) = 2R1$$
$$H = (6.5 - 1/Q)/3$$
$$R5 = R4/(H - 1)$$
$$R1 = 31.8\text{k}\Omega, \quad R2 = 10.6\text{k}\Omega, \quad R3 = 63.6\text{k}\Omega, \quad H = 1.833, \quad R5 = 1.2R4$$

Second-Order Sallen-Key Equations

For the simple S-K circuits of Figure 11.1, the equations for ω_0 and Q are summarized below. Note that the filters employing amplifier gain of k will have passband gain of k.

For all of the S-K LPF, HPF and BPF filters,

$$\omega_0 = 1/\sqrt{(R1R2C1C2)}.$$

For the LPF and HPF designs in Figure 11.1(a), with a unity-gain amplifier and R1 = R2,

$$Q = \sqrt{C1/4C2}$$

For the LPF and HPF designs in Figure 11.1(b), with amplifier gain of k and R1 = R2 and C1 = C2,

$$Q = 1/(3 - k), \quad k = 3 - 1/Q$$

For the Saraga BPF design of Figure 11.1(c) with amplifier gain of $k = 4/3$ and $C1/C2 = Q\sqrt{3}$,

$$R1 = 1/(Q \cdot \omega_0), \quad R2 = \omega_0/\sqrt{3}$$

Active Filter Implementations of Order Higher than Two

As suggested earlier, most even-order active filters of order higher than two consist of combinations of second-order filter sections connected in tandem. Odd-order filters will include one first-order section. For example, a fifth-order filter will comprise a first-order *R-C* section in tandem with two second-order sections. Depending on the filter order and shape, each section may have a particular f_0 and Q. Filter tables provide these values for each section of a filter, often for orders up to tenth [19]. The factored polynomials for such filters were shown in Section 11.1. The tandem connection of filter sections should usually begin with the section of lowest Q, since higher-order sections may have considerable gain and be more subject to overload.

11.3 Sallen-Key Third-Order Filters

Figure 11.4(a) shows a 1-kHz third-order Butterworth LPF that requires only a single op amp [20]. As described earlier, a third-order Butterworth LPF can be implemented with a buffered first-order section in tandem with a second-order section, requiring two op amps. Figure 11.4 illustrates how the required first-order pole can be incorporated without buffering. An HPF version of the filter is shown in (b). That circuit is often used as an infrasonic filter in a phono preamp.

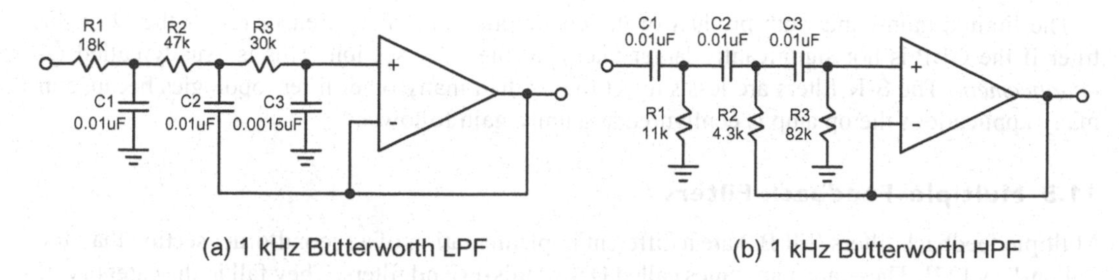

Figure 11.4 Third-Order 1-kHz Sallen-Key Butterworth Filters.

Table 11.1 Third-Order Sallen-Key Filter Components

		R1	*R2*	*R3*	*C1*	*C2*	*C3*
Butterworth		11k0	110k	33k0	0.015	0.0068	0.001
Bessel		9k10	91k0	27k0	0.0068	0.0022	680 pF
Chebyshev	0.2 dB	9k10	130k	39k0	0.022	0.0068	470 pF
Chebyshev	0.5 dB	12k0	130k	36k0	0.022	0.01	470 pF
Chebyshev	1.0 dB	9k10	68k0	36k0	0.033	0.015	680 pF

The values for R1, R2, R3, C1, C2 and C3 for third-order 1-kHz S-K Butterworth, Bessel and Chebyshev LPFs are shown in Table 11.1. Chebyshev filters with ripple values of 0.2, 0.5 and 1 dB are shown. Component values are in kΩ and μF unless noted and were obtained from the online calculator in Ref. [4].

The Butterworth filter is down 3 dB at 1 kHz, as expected, and is down only 1 dB at 800 Hz. The Bessel filter is down only 0.7 dB at 1 kHz, and is not down 3 dB until 1.9 kHz. It has a very shallow roll-off, as expected.

The 0.2, 0.5 and 1.0 dB Chebyshev filters fall below their ripple windows at 971, 995 and 1019 Hz, respectively. They are down by the greatest value of ripple attenuation at 590, 470 and 471 Hz, respectively. At 1 kHz, they are down −0.24, −0.55 and −0.70 dB, respectively. Although they all start out at 0 dB at DC (due to the circuit topolgy used), the ripple in the 1-dB filter is up to +0.3 dB at 861 Hz due to the positive feedback inherent to the filter. The ripple peaks of the 0.2-dB and 0.5-dB filters do not go above 0 dB.

11.4 Filter Sensitivity

Active filters are usually more sensitive to component variations than passive *RLC* filters. Different active filter topologies have different degrees of sensitivity to component value deviations [1, 18, 22]. Active filters also have sensitivity to the characteristics of the op amps used to implement them, particularly the gain-bandwidth product (GBP). The op amp sensitivity also varies with filter topology. Sensitivity is higher for high-*Q* sections, and this means that it is a greater issue for high-order filters. This is easily understood because some form of positive feedback is usually employed to achieve all but the lowest *Q* in second-order sections. In the S-K designs of Figure 11.3, for example, the positive feedback can be seen in the loop where the non-inverting filter output is fed back to the input circuit via R2 and C1.

The limited gain-bandwidth product of the operational amplifier often increases the Q of the filter if the GBP is not significantly larger than f_0 of the filter section. This is usually called *Q-enhancement*. The S-K filters are less subject to this than many other filter topologies because in many applications the op amp is configured as a unity-gain follower.

11.5 Multiple-Feedback Filters

Multiple feedback filters (MFBs) are a different implementation of a second-order section than the Sallen-Key [22]. These are sometimes called Delyiannis-Friend filters. They fall in the category of negative feedback filters [1]. They have somewhat lower sensitivity to component values than the S-K sections. The key difference is that the MFB replaces the flat-gain element in the S-K filter with the open-loop gain of an op amp. This allows the filter architecture to take full advantage of the open-loop gain available from the op amp. The MFB employs the op amp in its inverting mode, with the positive input connected to ground. Thus, the MFB is inverting. Because its functional input is a virtual ground, it may be more immune to common-mode effects.

MFB Low-Pass Filters

Fundamental to the low-pass MFB is that part of it includes an integrator, formed by R1 and C2, as shown in Figure 11.5(a) [21]. The gain of the LPF at ω_0 is to be A_0, not the DC gain. If the LPF has a peak, this is where A0 will be greater than unity. Similarly, a Butterworth LPF with $Q = 0.707$ will have A_0 be −3 dB. Figure 11.5(a) is a Butterworth LPF where A_0 has been chosen to be unity, so its DC gain is actually +3 dB. One tradeoff is that the S-K is quite insensitive to the op amp characteristics, while the MFB is more sensitive due to its use of the integrator and the influence of open-loop output impedance of the op amp at high frequencies. Conversely, the S-K is more sensitive to component tolerances than the MFB [22].

Use the following procedure to design an MFB LPF using the circuit of Figure 11.5(a) [21]. The DC gain of the LPF is to be A_0. Start by choosing the value of C2.

$$K = \omega_0 \mathrm{C2}$$

calculate the gain factor $H = A_0/Q$. $\mathrm{C1} = 4Q^2(H+1)\mathrm{C2}$

$$\mathrm{R1} = 1/\left(2Q(H+1)k\right)$$

$$\mathrm{R2} = 1/\left(2Qk\right)$$

$$\mathrm{R3} = 1/\left(2QHk\right)$$

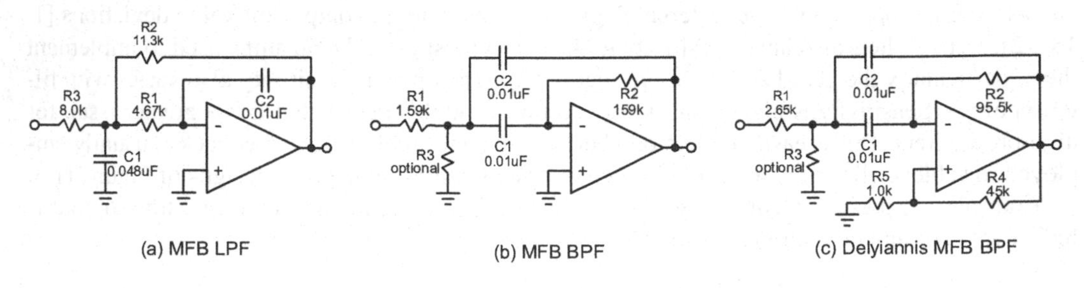

Figure 11.5 Multiple-Feedback Filters.

MFB Bandpass Filters

The Wien bridge and the bridged-T are the most important second-order tuning networks, the former with a peak at f_0 and the latter with a dip at f_0. The Sallen-Key filter places the Wien bridge (or a variant) ahead of a non-inverting amplifier and adds positive feedback to the network, increasing its Q. The MFB uses a bridged-T network at its core in a negative feedback arrangement [1]. The network is placed in the feedback loop of an op amp, and the input signal is applied to the shunt element of the bridged-T, here R1, as shown in Figure 11.5(b). As will be seen in the equation for gain A_{max} below, the filter gain increases as the square of Q, so filters with high Q may have excessive gain. If gain is to be reduced, R1 is replaced with an attenuator with R1′ and R3 and having the same output resistance as the original R1.

Because of its bridged-T heritage, choosing C1 = C2 in an MFB design is usually optimum, and there is little advantage to choosing a different ratio [1]. One seldom sees a bridged-T network where the series capacitors comprising the arms are not equal. This is because C1 = C2 maximizes Q in the passive bridged-T. With C1 = C2 = C having been chosen, we have (from Ref. [1]):

$$\mathrm{R1} = 1/\left(2QC\omega_0\right)$$

$$\mathrm{R2} = 2Q / C\omega_0$$

$$\mathrm{R2} / \mathrm{R1} = 4Q^2$$

$$\omega_0 = 1/(C\sqrt{(1/\mathrm{R1R2}})$$

$$A_{max} = -\mathrm{R2}/\left(2\mathrm{R1}\right) = -2Q^2$$

The R2/R1 ratio and the gain A_{max} at ω_0 both go up as the square of Q. Noting that R2/R1 = $4Q^2$, and confining R2/R1<100, a limit is imposed on maximum Q attainable:

$$Q_{max} = 0.5\sqrt{\mathrm{R2}/\mathrm{R1}} = 5$$

This is a significant limitation on the Q of the MFB for reasonable resistor ratios.

For a 1-kHz section with $Q = 5$ and $C = 0.01$ µF,

$$\mathrm{R1} = 1.59\,\mathrm{k}\Omega$$

$$\mathrm{R2} = 159\,\mathrm{k}\Omega$$

$$A_{max} = -2Q^2 = -50$$

For large A_{max}, the added shunt input resistor R3 can be used to drop the gain at ω_0 down to the desired value by acting as a front-end attenuator to prevent overload at frequencies near ω_0. For comparison, the gain of a state variable filter (SVF) changes only in proportion to Q (if not constant in some arrangements).

A 1-kHz bandpass MFB with Q of 5 is shown in Figure 11.5(b) [1]. As shown, the bandpass MFB has one more resistor (R3) than the S-K bandpass if independent control of the gain is desired. The extra shunt resistor R3 can be omitted in many designs. As shown, without R3 gain at ω_0 is about 34 dB.

Delyiannis MFB Bandpass Filter

Delyiannis further improved the performance of the MFB bandpass filter by adding some positive feedback, as shown in (c), where Q is set to 5 [1, 22]. This allows one to obtain higher Q than Q_{max} above. Gain k of the op amp is set by R4/R5. It provides another degree of freedom and is used to control Q. The new freedom allows one to pick the R2/R1 ratio and set Q by the value of k. If R2/R1 = β, it can be shown that there is an optimum value for β that minimizes sensitivity to passive component tolerances and op amp gain-bandwidth product (GBP) tolerances [1]. The design process is as follows:

Specify C and β, with β usually between 10 and 100. Choosing β will be discussed momentarily. For Figure 11.5(c), we will choose $\beta = 36$ and $Q = 5$. Then,

$$\mathrm{R1} = 1/\left(C\omega_0\sqrt{\beta}\right) = 2.65\,\mathrm{k}\Omega$$

$$\mathrm{R2} = \sqrt{\beta}/C\omega_0 = 95.5\,\mathrm{k}\Omega$$

$$k = \left((\beta+2)Q - \sqrt{\beta}\right)/\left(2Q - \sqrt{\beta}\right) = 46$$

Note that $2Q - \sqrt{\beta}$ must be chosen to be greater than 0.

For $Q = 5$, R1 = 2.65 kΩ

$$\mathrm{R2} = 95.5\,\mathrm{k}\Omega$$

$$K = 46$$

$$\mathrm{R4}/\mathrm{R5} = k - 1 = 45$$

The choice of β controls the sensitivity to passive component variations and op amp gain-bandwidth (GBP) variations. Passive sensitivity decreases as β increases, while active sensitivity increases as β increases. Thus, for given passive and active component tolerances an optimum β_{opt} exists where the two contributors are equal. There is a broad optimum for β centered at approximately $\beta_{opt} = 4(\omega_{GBP}/\omega_0)(\mathrm{tol}_p/\mathrm{tol}_a)$.

In one example in Ref. [1], a change in β by a factor of 2 on either side of β_{opt} increased sensitivity by only 7%. Consider a design having $f_0 = 20$ kHz and op amp GBW = 5 MHz. Assume that passive tolerance is 2% and GBP tolerance is 20%. Then $\beta_{opt} = 100$, allowing a relatively large value of R2/R1. Today's op amps with high GBP and good GBP tolerance of ±20% make op amp sensitivity much less of a problem than it was when GBPs were 1 MHz with a wide tolerance. Given the broad optimum, a good default value is $\beta = 20$.

MFB Advantages and Applications

The MFB has most of its advantage over the S-K when it is used in a higher-Q bandpass section. The MFB is almost never used to implement an HPF. The noise gain of the MFB is about twice that of the S-K. For unity-gain LPF and HPF situations, the S-K is far more accurate in passband gain, since its gain is identically unity for $k = 1$ designs. In the LPF MFB, it can be seen that the integrator portion (R1 and C1) takes care of rolling off the amplitude at high frequencies. The MFB is inverting, while S-K is non-inverting. Both can be designed to have gain, but errors in the gain-setting feedback resistors in the S-K can take away from its high unity-gain accuracy in LPF and HPF designs.

Second-order MFB filters can be put together as building blocks to form higher-order filters, just like S-K sections. Note that a fourth-order MFB with two sections will be non-inverting. S-K and MFB sections can also be mixed in a higher-order application, sometimes using the MFBs for the higher-Q sections. For example, an eighth-order filter might consist of two MFBs for the higher-Q sections and two S-Ks for the lower-Q sections.

In most LPF and all HPF applications, the S-K will be chosen. In many BPF situations, the MFB will be preferred. Sometimes the choice will be made based on whether or not an inversion is desirable.

11.6 State Variable Filters

As shown in Figure 11.6(a), a state variable filter (SVF) is implemented with a ring of three inverting stages, one of which is an adder and two of which are integrators [23–25]. High-pass, bandpass and low-pass filter functions are available at the outputs of U1, U2 and U3, respectively. The center frequency f_0 of the filter, where the bandpass output amplitude is greatest, is at the frequency where the gain around the loop is unity.

The integrators each contribute phase lag of 90°, for a total of 180°. As shown in the 1-kHz example, with 15.92-kΩ resistors and 0.01-μF capacitors implementing the integrators, and with the adder contributing gain of −1 to the loop, all elements in the loop have unity gain at 1 kHz. In fact, with no other passive elements, one has a state variable oscillator (SVO), which just needs some form of amplitude control. With the integrator resistors R1, R2 equal and the integrator capacitors C1, C2 equal, setting the frequency is extremely simple ($\omega_0 = 1/RC$), and the filter lends itself to variable-frequency tuning, often with a dual-gang pot. The loop feedback resistor R6 can be adjusted to trim the frequency.

Four variations of the SVF are shown in Figure 11.6. They have been designated as Types 1–4 for convenience of discussion. In the Type 1 design of Figure 11.6(a), the input signal is applied to the inverting input of the summer U1 through R5. Damping must be added to the filter to set its Q and avoid it being an oscillator. This is accomplished by feeding some of the first integrator's BPF output back through a voltage divider (R6, R7) to the non-inverting input of the summer. Greater attenuation in the damping loop will result in higher Q in the filter. Very high Q can be accurately achieved with the SVF. Bandpass filter gain A_v increases with Q. The Q and f_0 can be adjusted independently.

Notice that the damping loop acts by adding some negative feedback around the first integrator U2 via the summer U1. It is easy to see that such negative feedback damping could be implemented with a simple resistor across the first integrator's capacitor C1. Keep this possibility in mind, because other close relatives of the SVF can be realized this way. For the component values in (a), $f_0 = 1$ kHz, $Q = 0.667$ and gain at $f_0 = -3.52$ dB, non-inverting. All resistors except R1 and R2 are set to 10 kΩ for purposes of illustration. If R6 in (a) were made to be 11.4 kΩ, the filter would have a Butterworth response with $Q = 0.707$ and gain= −2.44 dB. If R6 is set to 20 kΩ, $Q = 1.0$ and gain= 0 dB.

The Type 2 arrangement of the SVF shown in Figure 11.6(b) has the input signal applied through R5 to the non-inverting input of U1. This provides an inverted BPF output. As in (a), changing R5 to increase or decrease gain at f_0 will have an effect on Q, since R5 shunts R7, increasing attenuation in the damping loop and increasing Q. Gain is 0 dB inverting and $Q = 1.5$. Q is higher than in (a) because R5 shunts the feedback in the damping loop, reducing damping.

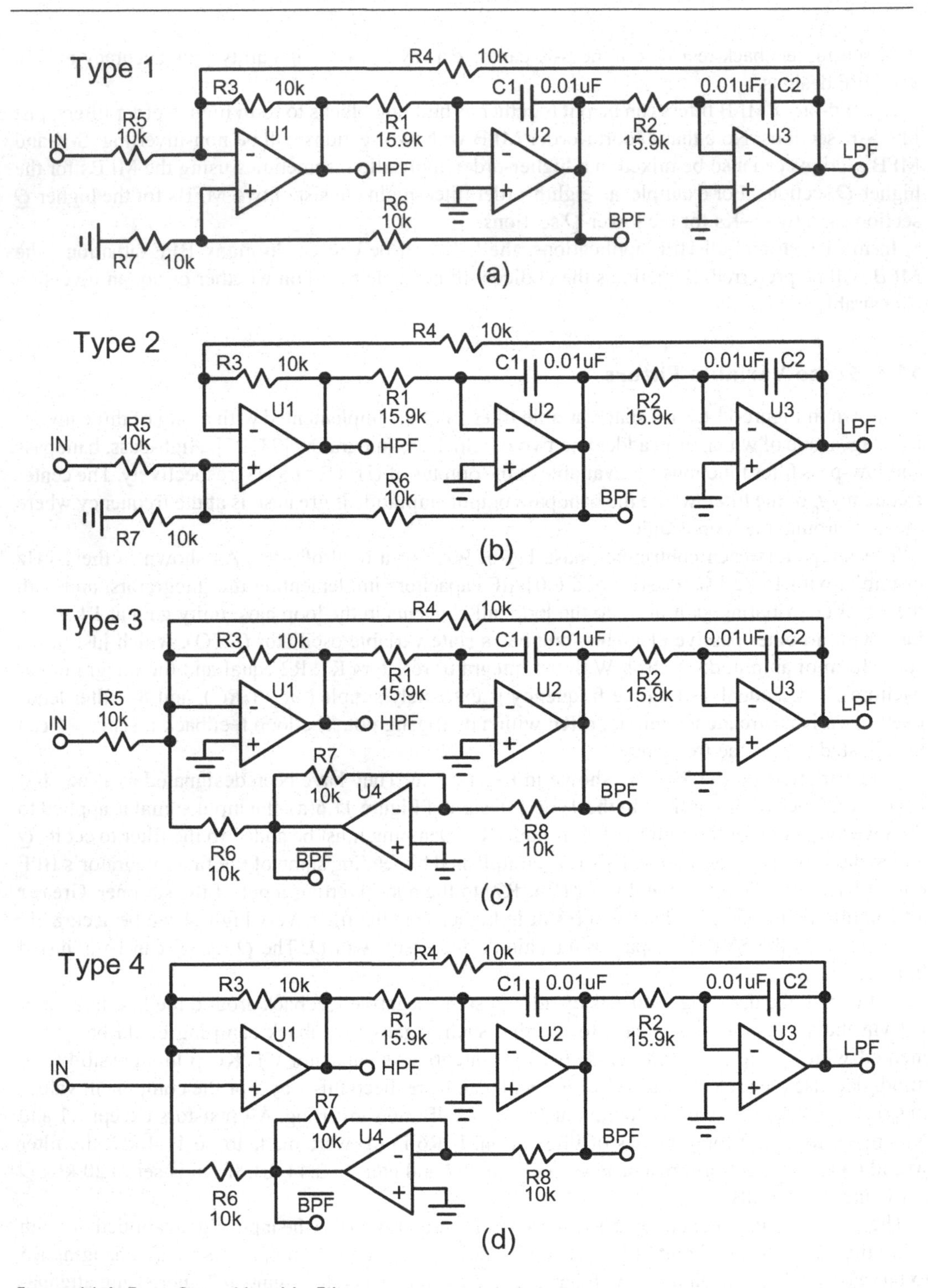

Figure 11.6 Four State Variable Filters.

In the Type 3 SVF of Figure 11.6(c) an inverting unity-gain buffer U4 is placed in the damping loop. This allows the damping signal to be applied to the virtual ground of the summer, eliminating interactions and also eliminating common-mode voltage swings at the input to all of the op amps in the filter. U4 also makes available an inverted version of the BPF output. Gain is 0 dB and $Q = 1.0$.

In many SVF designs, if Q is increased, such as by increasing R6, the gain of the SVF will be increased. In some applications this is undesirable (i.e., gain should be independent of the Q setting). The addition of U4 makes it possible to make gain and Q independent for the inverting BPF output by adjusting Q by changing the gain of U4 while keeping R5 the same. If R8 = 20 kΩ, gain remains at 0 dB but Q increases to 2.0. Gain at the non-inverting BPF output of U2 is no longer independent of Q, and increases as Q is increased. The possibility of signal overload at this node must be watched in high-Q designs.

The Type 4 SVF in Figure 11.6(d) is largely the same as that in (c), but its input is connected directly to the positive input of U1. This inverts the phase of the filter and makes the input a high impedance. This can be very helpful in some applications where the filter is preceded by a pot because it avoids pot loading. Gain is higher, at +9.5 dB, because feedback resistor R3 in the summer sees the loading of R4 and R6, increasing its non-inverting gain. Q remains at 1.0. If R8 is increased to 20 kΩ, Q increases to 2.0 while gain remains at +9.54 dB.

Returning to the Type 1 SVF in Figure 11.6(a), with R1 = R2 = R, and C1 = C2 = C and R3 = R4 = scaling resistance, the behavior of the SVF is dictated by the gains of the various signal paths through the summer U1. R4 and R3 are usually chosen to be the same value so that U1 acts as a unity-gain inverter in the global feedback path. The center frequency is the frequency at which the global loop gain through R4 is unity. The Q of the filter is the inverse of the gain of the damping loop through R6. The gain A_v at the center frequency f_0, is the gain from the input through R5 to the output of U1. These gains are all influenced by R3, the scaling resistance that provides negative feedback around U1. The equations for ω_0, Q and A_v are:

$$\omega_0 = 1/RC$$

$$Q = \left(1 + \mathrm{R6}/R7\right) \times (1/\left(2 + \mathrm{R3}/\mathrm{R5}\right) = \left(1 + \mathrm{R6}/\mathrm{R7}\right)/\left(2 + \mathrm{R3}/\mathrm{R5}\right)$$

$$A_v = \mathrm{R3}/\mathrm{R5}$$

If the tuning frequency is increased by reducing the values of the integrator resistors, the numerical bandwidth in Hz will increase and the Q will remain the same. The bandpass output is non-inverting and its gain increases as Q increases. At high values of Q, gain of the bandpass output at f_0 approaches Q. Gain of the filter for a given Q is controlled by R5. Changing R5 to increase or decrease gain at f_0 will have an effect on Q. Reducing R5 to increase gain reduces Q.

Q of the filter is set by R6 and R7. If the attenuator formed by R6 and R7 is replaced with a pot, Q can be continuously varied without disturbing the center frequency. If R1 and R2 are replaced with ganged variable resistors, the frequency can be continuously changed without disturbing Q. The independence of f_0 and Q in the SVF is a significant feature. This is the basis for a parametric equalizer.

The SVF has low sensitivity to component values and op amp performance. Sensitivity is comparable to that of a passive LC implementation. Of course, the SVF requires two additional op amps and three additional resistors compared to an S-K BPF. In high-order filters with high-Q

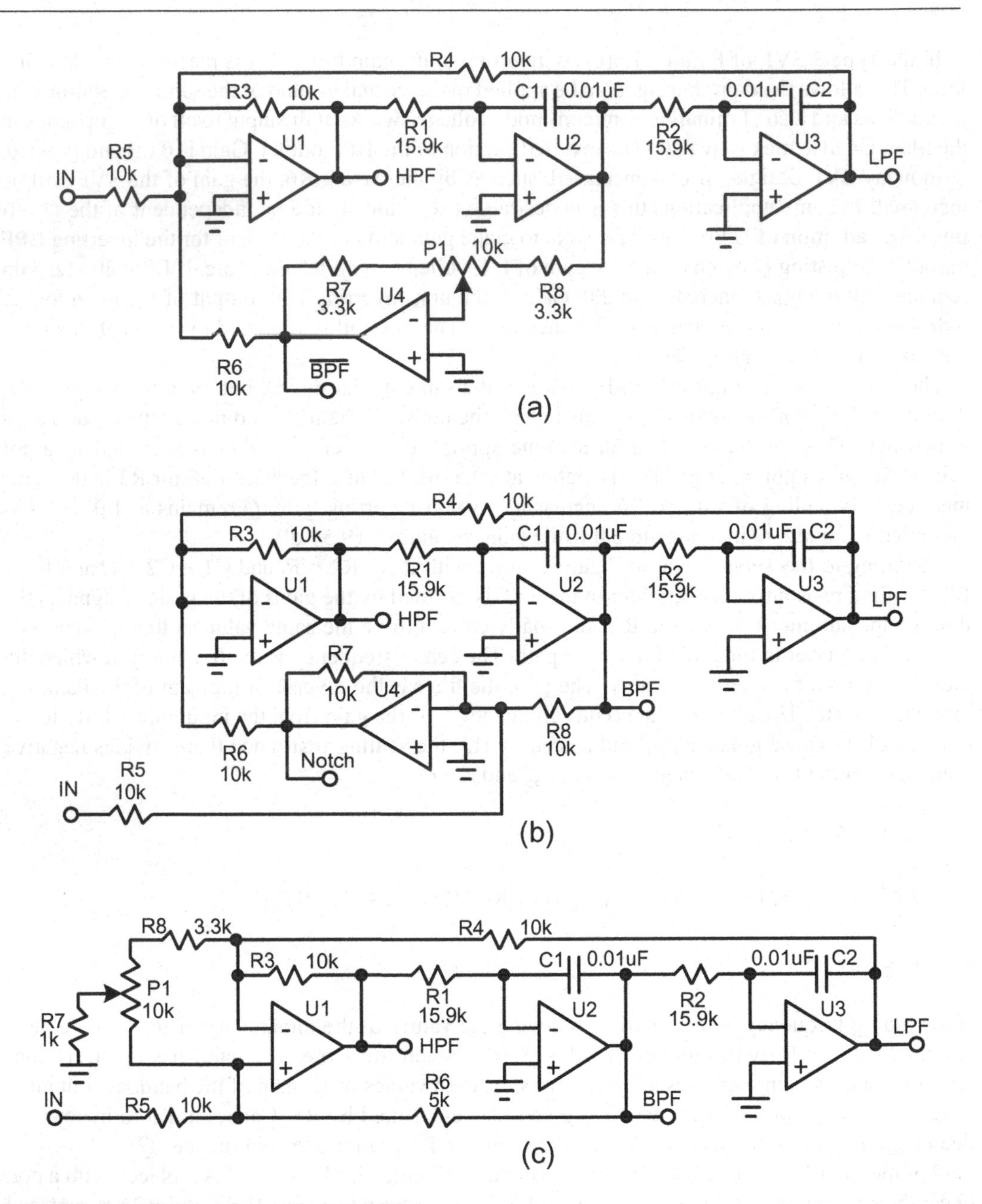

Figure 11.7 Additional State Variable Filters.

sections, an SVF can be used just for those sections. One limitation of this circuit is that there is some interaction between gain-setting via R5 and Q. Decreasing R5 to increase gain reduces Q. This happens because decreasing R5 increases the non-inverting gain of the summer as seen by the damping loop, resulting in reduced Q. Trimming of the center frequency can be accomplished with R4.

Going forward, the integrators in these filters will often be designated by a short-hand integrator symbol with an associated time value that corresponds to the time constant of the integrator's *R-C* product. For example, the time constant for each integrator in a 1-kHz SVF will be 159 μs. Similarly, a unity-gain inverter will be designated by an amplifier symbol with a −1 parameter.

The SVFs in Figure 11.7 bring some further refinements and variations. The first in (a) is a Type 3 SVF wherein inverter U4 has been replaced with a Baxandall level control, which is also inverting. When P1 is centered, the circuit merely acts as an inverter and nominal Q is achieved, in this case 1.0. At full CCW, the gain of U4 is increased to 4, reducing Q to 0.25. At full CW, the gain of U4 is reduced to 0.25 and Q is increased to 4. Change in Q with respect to pot rotation is geometric, a nice feature. Gain at the BPF output is constant at 0 dB over changes in Q. This circuit is attractive for use in a parametric equalizer.

The SVF in (b) is a Type 3 SVF wherein the input signal is applied to the virtual ground of the inverter, U4. The BPF output in this design is inverting, and A_v increases with increased Q. As shown, $Q = 1$ and A_v is 0 dB. A useful feature of this SVF is that it provides a high-quality notch filter response at the output of U4. Nominal gain at the notch output is 0 dB. Notch depth is about 76 dB in simulation at 1 kHz with op amps having 100 dB of gain at this frequency. The notch amplitude is down only 1.6 dB at 2 kHz and 0.6 dB at 3 kHz. If R6 is doubled to 20 kΩ BPF Q becomes 2 and A_v remains at 0 dB. Notch depth falls to 67 dB, with loss of only 0.5 dB at 2 kHz and 0.15 dB at 3 kHz.

This notch filter is superior in some ways to the active twin-T notch. Notch depth appears to be less sensitive to component matching than that of the twin-T. Most importantly, its frequency is easily tuned; the twin-T is difficult to tune, requiring three variable resistors, one of which is half the value of the others. With high-performance op amps, this circuit can be used to increase the dynamic range of a THD analyzer by notching out much of the fundamental prior to the analyzer input.

In such applications it is often desirable to have an accurately controlled notch depth of 40 dB to leave some fundamental for the analyzer to lock onto. This is easily accomplished by adding a 1-Meg resistor from the signal input to the inverting input of U1. Applications using a spectrum analyzer do not need this deliberate feed through. An additional advantage of this design is that all 4 op amps operate in the inverting configuration, eliminating common-mode distortion.

The SVF in (c) is an elegant design in which gain is independent of Q and which only requires 3 op amps. It was used in the UREI parametric equalizer from the 1970s. The design is a Type 2 SVF in which Q control pot P1 straddles the positive and negative inputs of summer U1. Stopper resistors R7 and R8 limit the range of the pot action. As shown, A_v is 0 dB inverting. Q can be adjusted from 0.65 to 6.0, with $Q = 1.2$ at the center of the linear pot. The use of different component values for P1, R7 and R8 can provide different ranges of Q.

State Variable Notch Filters

A high-performance notch filter can be implemented by adding one op amp to sum the high-pass and low-pass outputs of the SVF. With this approach summing errors do not affect notch depth, but do affect the notch frequency.

Alternatively, with a BPF gain of unity, a notch filter can be formed by subtracting the bandpass output from the input. With this approach summing error degrades notch depth. This approach can be used to implement auto-tuning of a notch filter in a THD analyzer [24]. The notch filter provided in Figure 11.7(b) above is also competitive with these more traditional approaches.

State Variable Quadrature Outputs

A useful feature of the SVF is that the LPF output is in phase-quadrature with the bandpass output. This enables the use of in-phase and quadrature synchronous detectors to be used for controlling gain (*Q*) and center frequency of the filter to perfect the null. Gain and frequency can be controlled electronically by using JFETs to trim the effective values of R4 and R6, respectively. This approach can be used in the auto-tune circuitry of a THD analyzer [24].

Ultra-low-distortion state variable oscillators (SVO) can be implemented with similar techniques for stabilization and fine frequency control. The available quadrature outputs in the SVO can be used in many ways. One example is to implement a four-phase amplitude detector with very low ripple for amplitude stabilization of the oscillator. With some additional adders, an eight-phase amplitude detector can be implemented with extremely small ripple. In some applications, a four-phase or eight-phase rectifying detector may be superior to a sample-hold amplitude detector.

11.7 Coupled State Variable Bandpass Filters

Most high-order filters are stagger-tuned, meaning that the multiple first- or second-order sections making up the filter have different center frequencies and/or *Q*. For higher-order filters, the *Q* of the highest-*Q* section may get quite high. For a sixth-order Chebyshev BPF with 1 or 3 dB of ripple, the *Q* of the highest-*Q* section can be approximately 8 or over 12, respectively.

Here we discuss a different way to make some BPFs of moderately high order without resort to stagger tuning. They are based on SVF sections. This can require more operational amplifiers than other approaches. For the coupled filters to be discussed below, the center frequency of every second-order section is the same, namely that of the center of the passband. However, some may have different *Q* values. These filters are useful for the fourth order through at least the sixth order. The *Q* of the individual sections is also relatively low.

Two identical second-order SVF BPFs can be put in tandem to form a fourth-order BPF. Alternatively two SVFs can be stagger-tuned to create a fourth-order BPF with a wider, somewhat flat-topped response. Coupling two second-order SVFs is another way that is better in some applications that require a flat-top bandpass approximation.

The coupled approach is inspired by the way super heterodyne IF filter transformers were made many years ago. Both the primary and secondary of the transformer were parallel-resonant circuits tuned to the same frequency, resulting in a fourth-order BPF. The secret to the flat-top bandpass approximation was the fact that there was magnetic coupling between the two resonant circuits in the

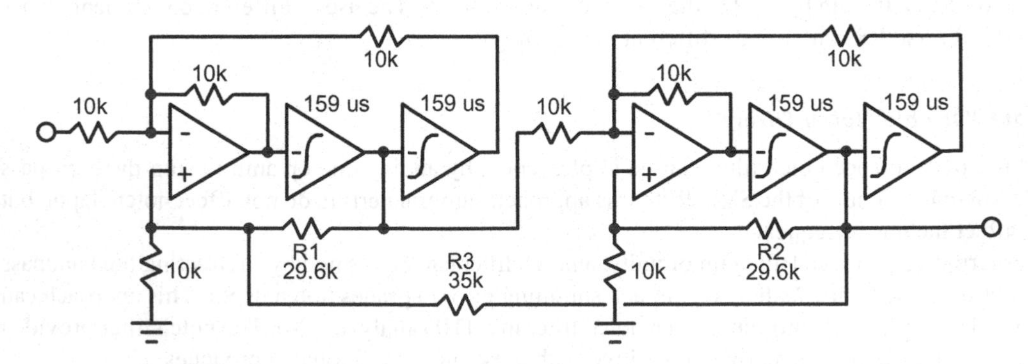

Figure 11.8 Coupled Fourth-Order Butterworth SVF.

transformer. In some cases a small capacitor from the primary to the secondary was also included to control coupling and filter shape. This coupling might be said to create a form of pole-splitting, moving the otherwise-overlapping amplitude peaks apart.

The same approach can be used with a pair of identical SVFs connected in tandem, as shown in Figure 11.8. A single resistor R3 feeds back the signal from the bandpass output of the second SVF to the input of the first SVF. This simple coupling feedback provides a flat-top response. The beauty of this technique is that the two SVFs can be identical in f_0 and Q. Notably, the Q of each of the constituent SVFs alone is only 1.3 for the fourth-order Butterworth filter below. Gain of the complete fourth-order filter is 0 dB.

Coupled Fourth-Order Bandpass Filters

The behavior of coupled filters can be understood by looking at the responses for the uncoupled (a), coupled quasi-Butterworth (b) and coupled quasi-Chebyshev filter (c) approximations plotted in Figure 11.9. The two SVFs in the uncoupled and coupled cases are identical, and have a gain of 2.4 dB and Q of 1.3. Their Q is set by R1 = R2 = 29.6 kΩ. Uncoupled, their tandem combination shows a single sharp peak and high gain of 4.8 dB (normalized to unity in the plot). The uncoupled and coupled Butterworth filters are the same as the filter in Figure 11.8, with the exception that the uncoupled version lacks the coupling resistor R3. Coupling the two sections with R3 = 35 kΩ flattens the combination to a quasi-Butterworth response with unity gain. The passband flatness is created by the negative feedback provided by the coupling, which attenuates the response peak

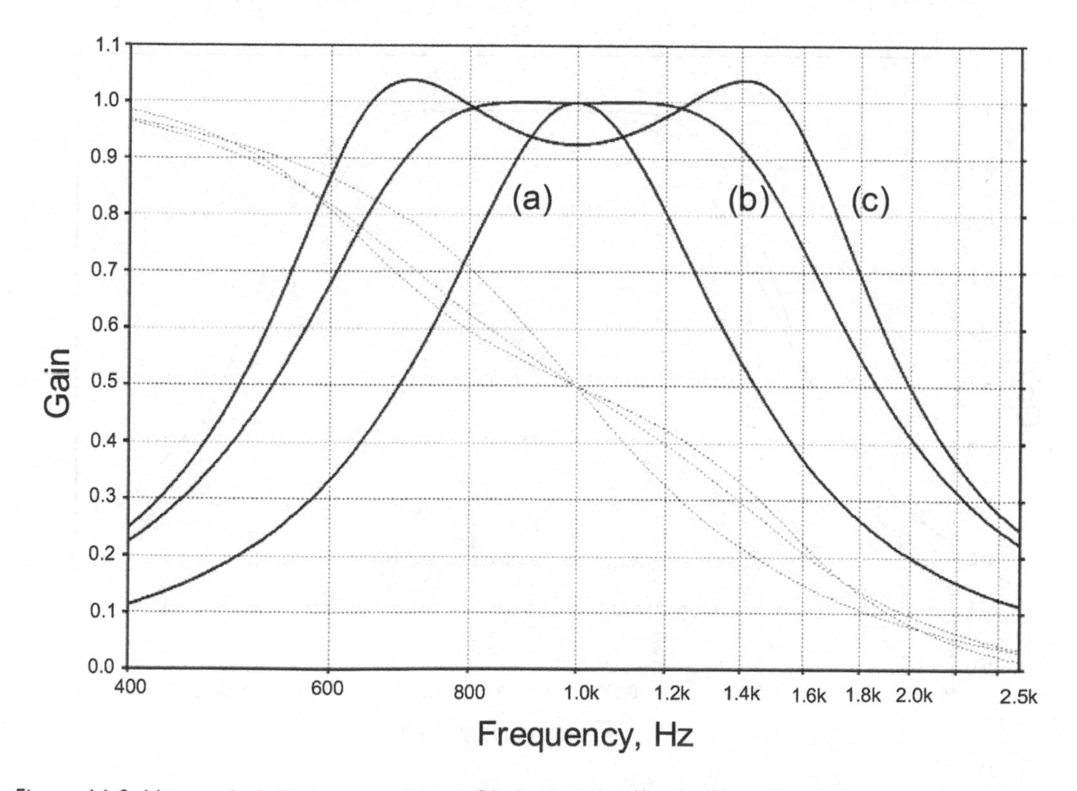

Figure 11.9 Uncoupled, Butterworth and Chebyshev Bandpass Filter Responses.

in a frequency-dependent way – more at the center frequency where the loop gain of the coupling feedback is maximum and has a phase angle of 180°.

The quasi-Chebyshev filter is like Figure 11.8 as well, but with R1 = R2 = 38 kΩ and with R3 = 22 kΩ. The Butterworth and Chebyshev designs thus differ in the amount of coupling negative feedback provided by R3. Their stop-band skirts far from f_0 are identical. Each section of the Chebyshev filter has gain of 4.1 dB and Q of 1.6 before coupling. The design exhibits two bumps and one dip. In a sense, the peak frequencies of the two sections are forced apart by coupling action that is akin to pole-splitting.

The 3-dB bandwidths for the uncoupled, coupled Butterworth and coupled Chebyshev filters are 639, 1233 and 1774 Hz, respectively. The bandwidths at −1 dB are 337, 881 and 1430 Hz, respectively. The ripple bandwidth of the 1-dB Chebyshev is 1360 Hz.

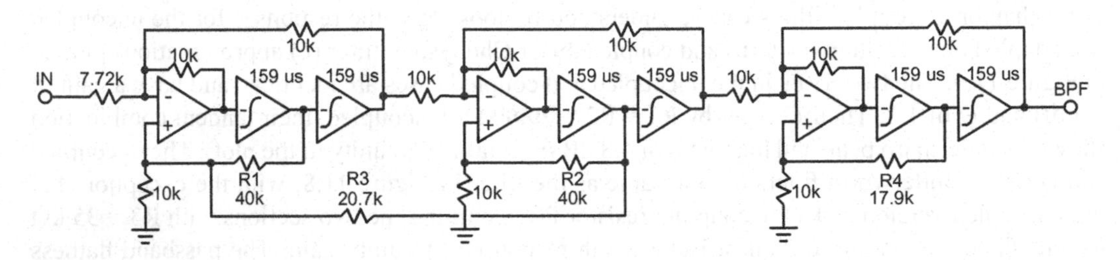

Figure 11.10 Sixth-Order Quasi-Butterworth Coupled SVF.

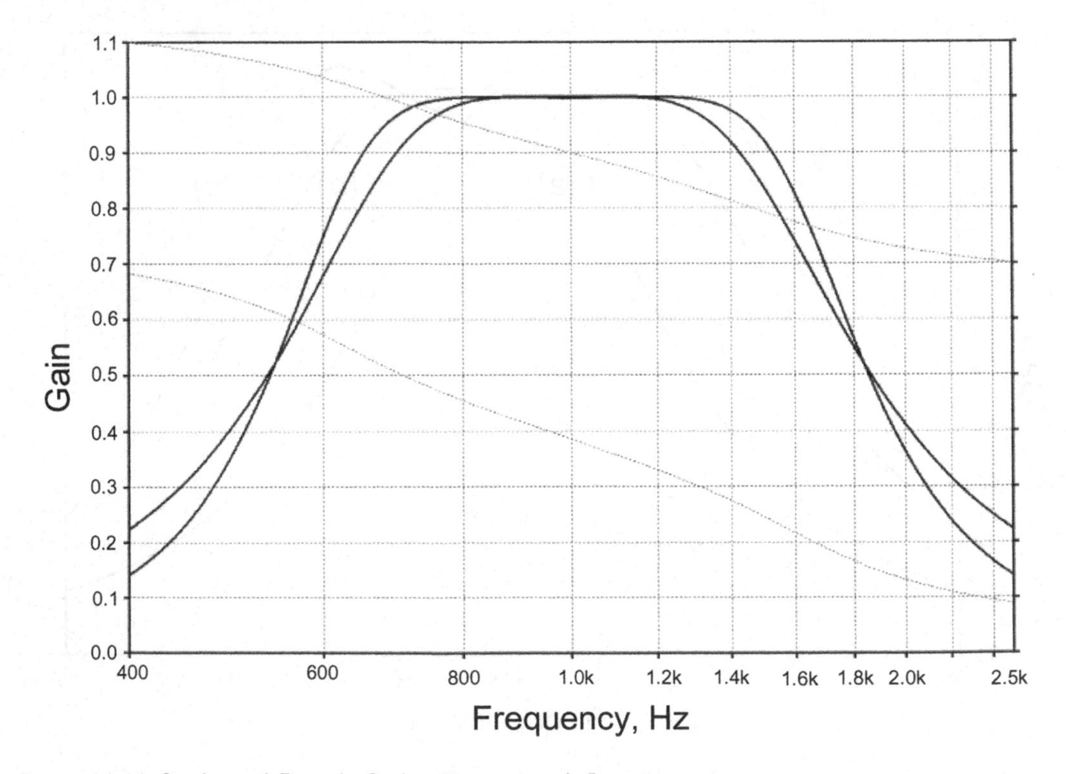

Figure 11.11 Sixth- and Fourth-Order Butterworth Responses.

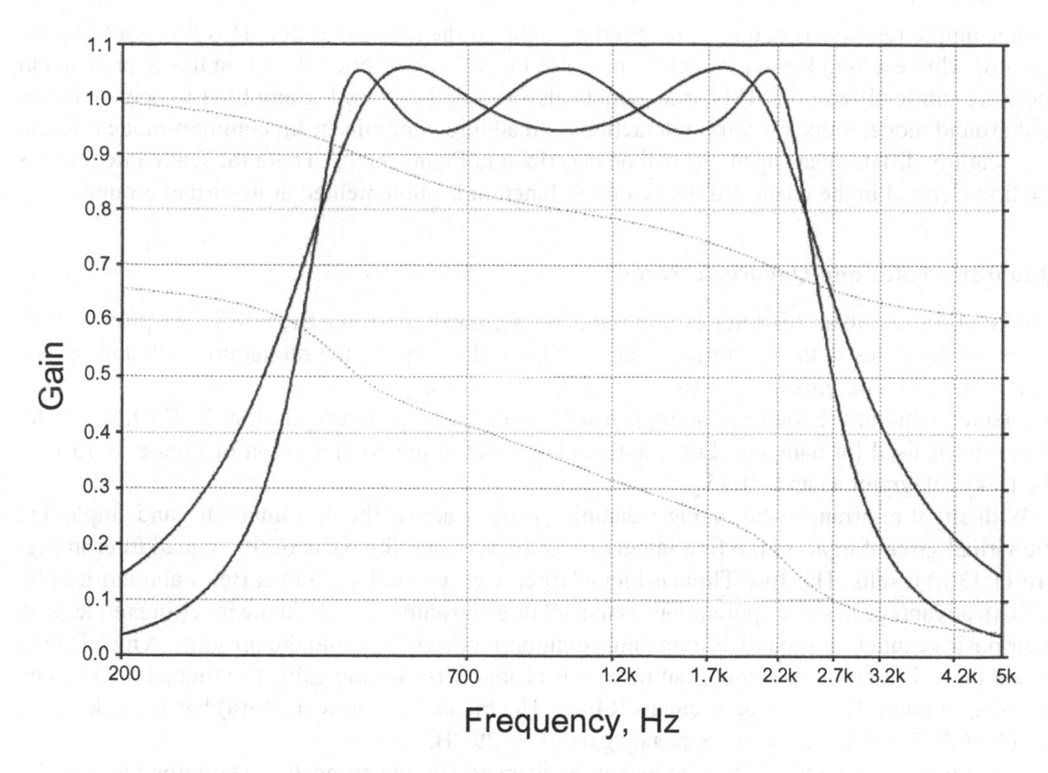

Figure 11.12 Sixth- and Fourth-Order Chebyshev Responses.

Coupled Sixth-Order Bandpass Filter

If an ordinary second-order SVF is added in tandem with a fourth-order coupled SVF, a sixth-order filter is formed having a wider passband and steeper skirts. Doing this is slightly analogous to forming a third-order filter by adding a first-order section to a second-order section to create a third-order filter, as described earlier with a third-order S-K filter. A sixth-order quasi-Butterworth filter implemented in this way is shown in Figure 11.10. The key to this filter is that the dip in the coupled filter at f_0 is largely canceled by the peak in the added uncoupled filter.

The circuit can be made to have a quasi-Chebyshev response by making R1 = R2 = 45 kΩ, R3 = 5.3 kΩ, R4 = 22 kΩ and the input resistor 2.82 kΩ. The sixth-order quasi-Chebyshev filter exhibits three bumps and two dips.

The responses of the fourth- and sixth-order quasi-Butterworth filters are shown in Figure 11.11. They have 3-dB bandwidths of 1021 Hz and 1105 Hz, respectively.

For comparison, fourth- and sixth-order quasi-Chebyshev responses are shown in Figure 11.12 for filters with 1 dB of ripple.

11.8 The Biquadratic Filter

The biquad filter is a close relative of the state variable filter [25–33]. In the SVF of Figure 11.6(a), the Q is controlled by feedback from the output of the first integrator back to the summing amplifier. Seeing this, we can recognize that this is just equivalent to providing negative feedback around the first integrator. In other words, this Q-determining negative feedback can instead be implemented with a simple resistor from output to inverting input of the first integrator. This forms a leaky integrator. The resulting behavior is very similar to that of a conventional SVF in this approach, but there are subtle differences. This arrangement also allows the summing amplifier to operate in virtual ground mode, reducing some interactions and adding immunity from common-mode effects.

A slightly different arrangement will be described in Figure 11.13. There the leaky integrator is the first element in the chain and the summing function is implemented at its virtual ground.

Quadratic Poles and Quadratic Zeros

Unlike all-pole filters, this filter is referred to as a biquadratic filter because it can provide both second-order poles in the denominator and second-order zeros in the numerator with appropriate combining of its outputs using a fourth op amp. Semantically, the filter arrangement, even if implemented without the fourth op amp, is usually called a *biquad* instead of an SVF when a leaky integrator is used for damping. Some authors also refer to the configuration of Figure 11.13 with the leaky integrator as an SVF [3].

With slight re-arrangement, using a damping resistor across the first integrator and employing the virtual ground input of the first integrator as the summer, the basis of the biquad filter in Figure 11.13(a) results. The Tow-Thomas biquad filter was invented in 1969 at Bell Laboratories [26, 27]. An advantage in some applications is that all three op amps operate in the inverting mode, with their positive inputs at ground, eliminating common-mode voltage into the op amps. A nice feature is that the Q is unaffected if the input resistor is changed to alter the gain. The biquad filter has no high-pass output. Bandpass gain equals R4/R3. The biquad of Figure 11.13(a) has been designed for $Q = 0.707$ and has a center frequency gain of +1.29 dB.

In contrast to the SVF, if the frequency is changed (by changing the integrator resistors R2 and R4), the numerical bandwidth in Hz remains constant, but the Q changes. As the tuning frequency is increased, the Q increases, assuming that the damping resistor across the integrator is unchanged. If f_0 is doubled by halving R1 and R2, bandwidth will remain at 1369 Hz and Q will be doubled. The BPF output is inverted from the input signal. Because of the inverter that follows the first integrator, inverting and non-inverting versions of the LPF output are available. If the integrator and inverter are swapped, inverting and non-inverting versions of the BPF output are available.

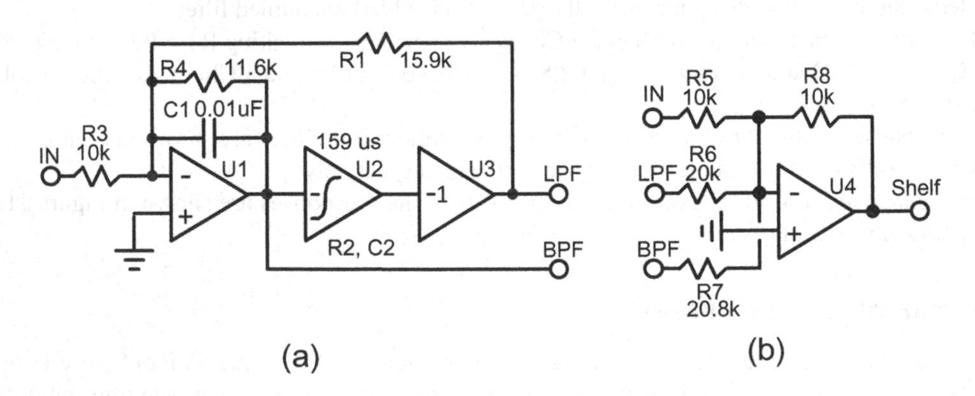

Figure 11.13 Tow-Thomas Biquad Filter.

Forming the Quadratic Numerator

As described thus far, the biquad circuit is an all-pole filter much like the SVF. However, a fourth op amp can be added as an inverting summer to sum the input signal, the bandpass output and the low-pass output in differing amounts to form a biquadratic numerator in the transfer function. This gives the biquad its name. If needed in some applications, the inverted version of the low-pass output is also available. By combining the three outputs in different ways, high-pass, notch, all-pass and shelf equalizer outputs can be realized. The transfer function is of the form [1, 32]:

$$H(s) = \left(s^2 + \omega_z s / Q_{z0} + \omega_z{}^2\right) / \left(s^2 + \omega_p s / Q_{p0} + \omega_p{}^2\right)$$

Recall that the SVF required a fourth op amp to provide a zero in the transfer function to form a notch filter. With a fourth op amp, the biquad can form a more generalized numerator with real and complex zeros in a straightforward way. If desired, a second biquad output can be implemented with a different biquadratic numerator by adding a single op amp. Details of biquad filter design can be found in Refs. [25, 31, 32].

In Figure 11.13(b) an output summing network is shown that creates a shelving frequency response that is down 13.9 dB at low frequencies and 0 dB at high frequencies. Although f_0 of the bandpass is at 1 kHz, the attenuation midpoint for the shelf is at 670 Hz. If the input and BPF signals are combined in equal amounts at the virtual ground of U4, a notch results.

Feedforward Three-Amplifier Biquad

In the biquad of Figure 11.13(a), summing was performed only at the virtual ground input of the leaky integrator. With a slight re-arrangement, a biquad can be formed where summing involving the input signal can be accomplished at the virtual ground input of the inverter and integrator as well. As shown in Figure 11.14, this provides an opportunity to form the numerator coefficients by feedforward means.

The denominator of the biquadratic function is controlled in the usual way by R1, R2, R4, C1 and C2. The numerator is controlled by R3, R7 and R8. Here those latter resistors are just given place keeper values for generality. The calculation approach for those values can be found in Ref. [1].

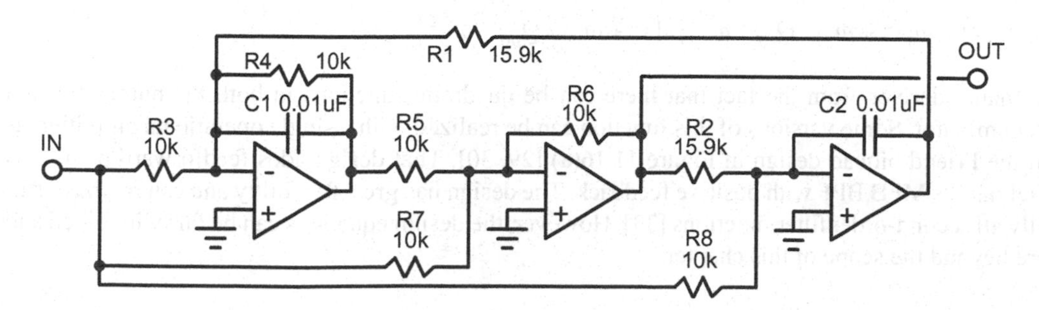

Figure 11.14 Feedforward Biquad Filter.

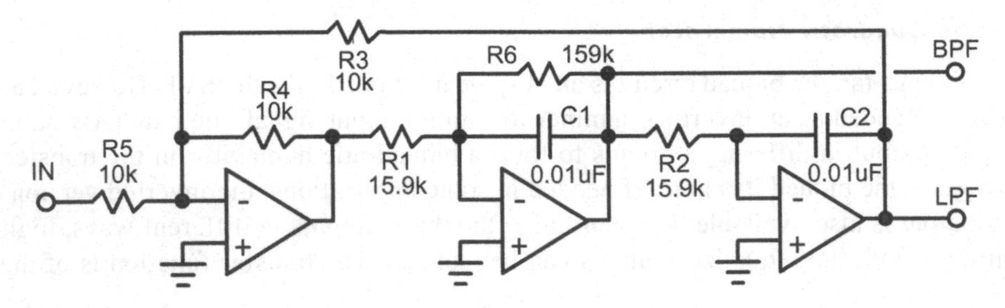

Figure 11.15 A Useful Variant of the Biquad Filter.

A Biquad Variant

A simple change can be made that yields other variants of the biquad arrangement that share some properties with the all-pole portion of the biquad filter. The circuit in Figure 11.15 is like an SVF where the damping loop has been removed and the first integrator is made leaky with a feedback resistor. In contrast to the biquad in Figure 11.13, here the summing is done in a summing amplifier, as in the SVF, instead of at the virtual ground of an integrator. This filter also has all three op amps operating entirely in the inverting mode.

The filter is most useful in the bandpass mode. In such applications, its advantage over the SVF is that Q is adjusted with a single resistor R6, and there is no interaction between gain-setting and Q. For the values shown, $f_0 = 1$ kHz and $Q = 10$. Gain is the ratio of damping resistor R6 to the tuning resistor value R1. If the center frequency is reduced by increased tuning resistor values, bandwidth will remain the same and Q will decrease. This is opposite the behavior of Q with frequency changes for the SVF. Gain and Q are both numerically equal to R6/R1. The BPF output is non-inverting.

It is interesting to note that the denominator Q in many biquads with leaky integrators is the same as the effective Q of the tuning capacitor in the leaky integrator. In other words, the effective Q of C1 is reduced by the shunting action of R6. The Q of the capacitor at the center frequency is equal to the ratio of the resistance of R6 to the reactance of C1 at the tuning frequency. This is also the Q of the filter.

11.9 Single-Amplifier Biquad and Linkwitz Transform

As mentioned earlier, a biquadratic filter like the Tow-Thomas implementation has a transfer function of the form:

$$H(s) = m\left(s^2 + \omega_z s / Q_{z0} + \omega_z^{\,2}\right) / \left(s^2 + \omega_p s / Q_{p0} + \omega_p^{\,2}\right).$$

Its name derives from the fact that there can be quadratic functions in both the numerator and denominator. Some versions of this function can be realized with a single operational amplifier, as in the Friend biquad design in Figure 11.16(a) [29, 30]. That design adds feedforward paths to a Delyiannis MFB BPF with positive feedback. The design has great flexibility and can realize virtually all second-order filter functions [33]. However, the design equations can be fairly involved and are beyond the scope of this chapter.

Linkwitz Transform

The Linkwitz transform is a biquadratic filter with particular relevance to the audio community, as it can be used to extend the bass response of a closed-box woofer [34–39]. Its quadratic

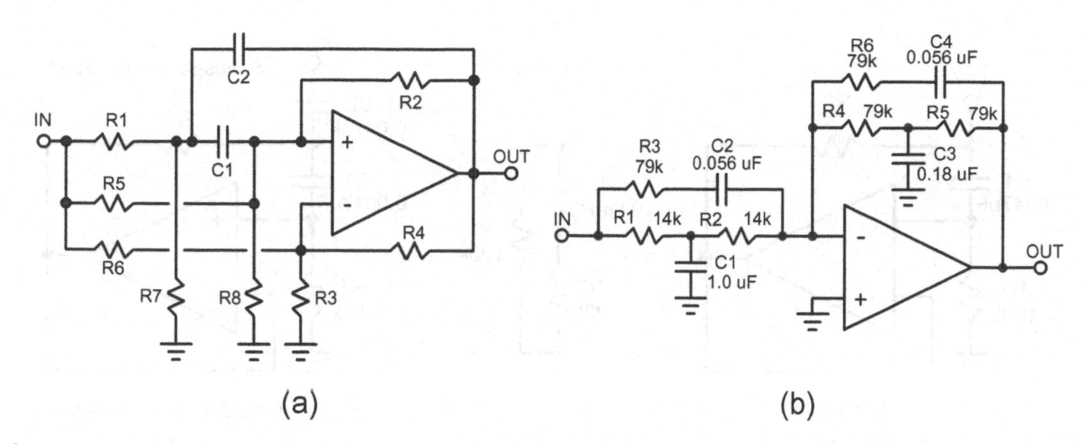

Figure 11.16 Friend Biquad and Linkwitz Transform.

numerator's zeros cancel the quadratic poles of the woofer whose response is being corrected. Its denominator then provides poles that replace those of the woofer, but at a lower frequency. Such a filter provides a second-order, 12 dB/octave shelving transfer function that has higher gain at low frequencies than at high frequencies. A Linkwitz transform single-amplifier biquad is illustrated in Figure 11.16(b). Its detailed use will be described in the active crossover chapter.

Notice that the circuit has a bridged-T network in the feedforward path and a second bridged-T network in the feedback path of an op amp. The gain is thus the ratio of the bridged-T transmission admittances. Each network includes a resistor in series with the bridging capacitor to establish unity gain at high frequencies where capacitors C2 and C4 act like a short. At low frequencies, the gain is established by the ratio of R3 to R1, here 5.6 corresponding to 15 dB.

11.10 Gyrators

Sometimes the functionality of an inductor is desired in order to emulate certain passive filters, most commonly the series-resonant filter to ground. A gyrator is used to create a synthetic inductor to ground using a capacitor, two resistors and an op amp as a non-inverting unity-gain buffer. Most commonly, the gyrator is used in combination with a series capacitor to form a series-resonant circuit to ground [40].

The Gyrator Inductor

The gyrator creates an inductance to ground by using a capacitor. As shown in Figure 11.17(a), the op amp is configured as a unity-gain non-inverting buffer. The buffer is fed by the capacitor in series from the gyrator input node and with a resistor R1 shunt to ground, forming a first-order HPF. Thus, the voltage at the input node of the gyrator is attenuated with decreasing frequency as it goes into the buffer. A second resistor R2 from the output of the buffer connects to the input node of the gyrator. It is responsible for creating the inductive input impedance of the gyrator. The gyrator is a one-port device.

At high frequencies there is little loss in the RC section feeding the op amp, so the output of the op amp and R2 bootstrap the input of the gyrator, resulting in high net input impedance, just like an inductor. At low frequencies there is no input to the buffer as a result of the low-frequency attenuation of the R1C1 section. With no output from the buffer, the input of the gyrator will see the full load of the 220-Ω feedback resistor R2. This results in low net input impedance, just like an

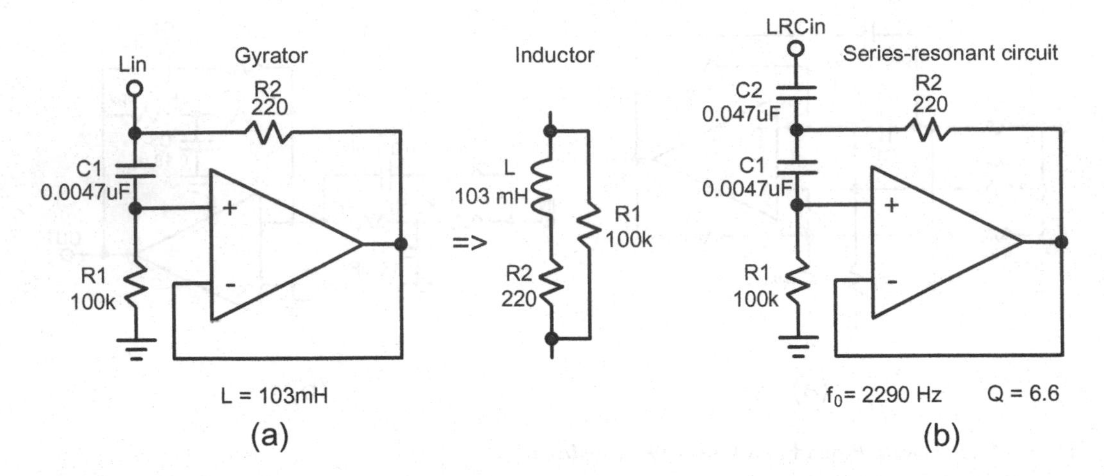

Figure 11.17 The Gyrator.

inductor would behave. The effective coil resistance of the inductor will be 220 Ω. Not a terribly good inductor, but in many applications that amount of resistance is needed to establish the desired Q for the series-resonant filter. The value of the inductor formed by the gyrator is as follows:

$$\mathrm{C1} = L/(\mathrm{R1R2}) \quad \text{or} \quad L = \mathrm{C1R1R2}$$

where R2 is the feedback resistor and R1 is the RC section shunt resistor at the input to the buffer. With R2 = 220 Ω and R1 = 100 kΩ, a 0.0047-μF capacitor will emulate a 103-mH inductor with a series resistance of 220 Ω. As shown in the equivalent circuit of the inductor, R1 acts as a shunt across the inductor at high frequencies, further reducing effective Q.

With a 0.047-μF capacitor (C2) in series with the gyrator in (b), the series-resonant frequency will be about 2.29 kHz. At this frequency the reactance of the inductor and capacitor are both 1477 Ω. The Q of the resulting series-resonant filter is dominated by the series resistance of the inductor:

$$Q = (\omega L)/\mathrm{R2} \quad \text{or} \quad L = \mathrm{R2}Q/\omega$$

At 2.29 kHz, the Q of the resonant circuit will be 1477/220 = 6.6. By using gyrators to accurately emulate inductors, active filters realized with them have low component sensitivity like the passive filters from which they are derived. While it has low sensitivity to component values, the gyrator does have some sensitivity to the gain-bandwidth product of the op amp. A limitation of the conventional gyrator is that it cannot emulate a floating inductor.

Buffered Gyrator

At high frequencies resistor R1 reduces Q, since it acts like a resistor in parallel with the synthesized inductor, limiting how high the impedance of the inductor can ultimately go. This effect can be reduced by making R1 a high resistance and C1 a small capacitance. However, sometimes that can be inconvenient or lead to other concerns, like sensitivity to parasitic elements. Instead, the RC HPF formed by R1 and C1 can be fed from a unity-gain buffer whose input is the inductor's node. The inductor node then only must drive the very high input impedance of the buffer op amp.

In some applications, the small cost of the second op amp justifies the improved performance of the gyrator. However, the finite gain-bandwidth of the added op amp may affect the quality of the inductor at high frequencies.

11.11 Negative Impedance Converters

A negative impedance converter (NIC) is a one-port element that can convert impedance to ground into a negative value of itself [2, 41]. For example, a NIC can convert a 1-kΩ resistor into a 1-kΩ negative resistance. If that negative resistance were connected in parallel with a 1-kΩ load resistor in a circuit, the net load resistance would theoretically go to infinity. A source with resistance of 100 Ω would go from being attenuated by about 1 dB by the load to being attenuated by a miniscule amount. In the general case, a resistor takes energy from a circuit, while a negative resistance delivers energy to a circuit. A 1-kΩ negative resistance in parallel with a lossy parallel-resonant circuit whose resistance at resonance equals 1 kΩ would result in an oscillator.

A NIC is shown in Figure 11.18(a), where the impedance to be inverted is shown connected from the output of the op amp to the input, in a positive feedback manner. The op amp is configured as a closed-loop amplifier with non-inverting gain of 2. It is easy to see that whatever voltage is applied to the input of the op amp will see twice its voltage on the other side of the feedback impedance, meaning that the current will flow in the opposite direction that it would if that impedance were connected from the signal source to ground. If you visualize the impedance as a resistance, you have a negative resistance converter that has created a negative resistance to ground of the same value as the feedback resistor. In the general case, any impedance connected from output to the positive input will be inverted.

The impedance of an inductor is the negative of the impedance of a capacitor, since the current flowing in the inductor lags the voltage by 90° and the current through the capacitor leads the voltage by 90°. In other words, the currents differ by 180°, and thus the impedance of one is the negative of the other. A NIC can therefore convert a capacitor into an inductor – just as the gyrator did. In fact, the circuits look a lot alike, with the exception that the amplifier gain in the NIC is 2 rather than unity. However, in this case, there is no coil resistance, at least in principle.

In a more generalized case, the closed-loop gain of the op amp need not be 2. A different value will just transform the impedance by a different scale factor. For example, if the gain is set to 3, a capacitor of half the value can be used to emulate the same inductance.

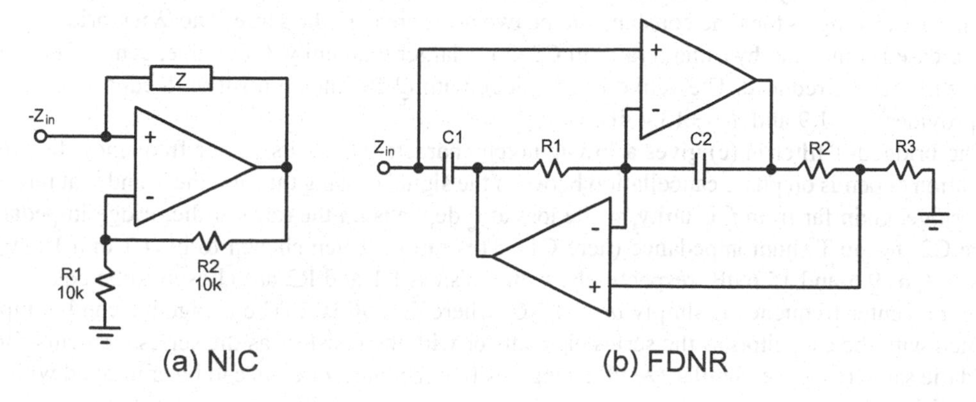

Figure 11.18 NIC and FDNR Circuits.

11.12 Frequency-Dependent Negative Resistors

A frequency-dependent negative resistance (FDNR) is a one-port active circuit element that emulates a resistance to ground (not a reactance) whose value is frequency-dependent [2, 4, 42]. Whereas the phase angle of a resistor is 0°, the phase angle of the FDNR is 180° and its magnitude decreases as the square of frequency. This makes sense when one bears in mind the properties of $j\omega$ and $(j\omega)^2$, the latter equaling −1 at 1 rad/s. Such an element can be used to implement an active filter.

A typical FDNR implementation comprises two capacitors, two op amps and three resistors, as shown in Figure 11.18(b). It makes sense that a circuit responsive to ω^2 would require two capacitors to implement. The details of how this circuit element can be used in active filters can be found in Refs. [2, 4]. As an example, a passive *L-C* ladder filter can be transformed into an active equivalent that does not require inductors.

The FDNR converts resistors into inductors to eliminate inductors in filters. The impedances in an LCR filter can be rotated by the complex variable *s*. An inductor then becomes a resistor, a capacitor then becomes an FDNR and a resistor becomes a capacitor. An LCR ladder filter can thus be emulated.

If C1 = C3 = *C* and $k = R2 \times R4/R5$, the impedance seen looking into the FDNR is:

$$Z = 1/s^2kC$$

The FDNR is an element made from a topology called a general impedance converter (GIC). If all of the capacitors and resistors in Figure 11.17(b) are replaced with impedances Z1–Z5, then in the general case,

$$Z_{in} = Z1Z3Z5/Z2Z4$$

11.13 Wien Bridge and Bridged-T Filters

The 1-kHz Wien bridge and bridged-T filters are shown in Figure 11.19. The Wien bridge filter in (a) gives a low-*Q* bandpass characteristic with gain of 1/3 (−9.54 dB) at the center frequency when the resistors and capacitors are of equal values. The filter shown comprises the two frequency-dependent arms of a full Wien bridge. The center frequency is simply ω = 1/RC. Although the Wien bridge is usually seen with the same RC values in the series and parallel arms, that need not be the case as long as the time constants in the two arms are kept the same. The Wien bridge *Q* can be increased somewhat by using a ratio of C2 to C1 larger than unity. Of course, center frequency gain will then be reduced. The active Wien bridge with *Q*-enhancing positive feedback shown in (b) provides $Q = 1.9$ and $A_v = +13.4$ dB.

The bridged-T filter in (c) gives a low-*Q* notch characteristic at its center frequency. Its notch operation depends on phase cancellation between the signal passing through the T and that through the bridge. Gain far from f_0 is unity, while loss at f_0 depends on the ratio of the bridge impedance (here C2) to the T shunt impedance (here C1). This ratio is often chosen to be 4:1 or 10:1, with dips at f_0 of 9.6 and 15.6 dB, respectively. Arm resistors R1 and R2 are chosen to be equal to *R*. Here the center frequency is simply ω_0 = 1/RC, where $C = \sqrt{C1C2}$. The bridged-T can be implemented with the capacitors as the series elements or with the resistors as the series elements. Both yield the same response. Similar *Q*-enhancing positive feedback circuits can be employed with the bridged-T.

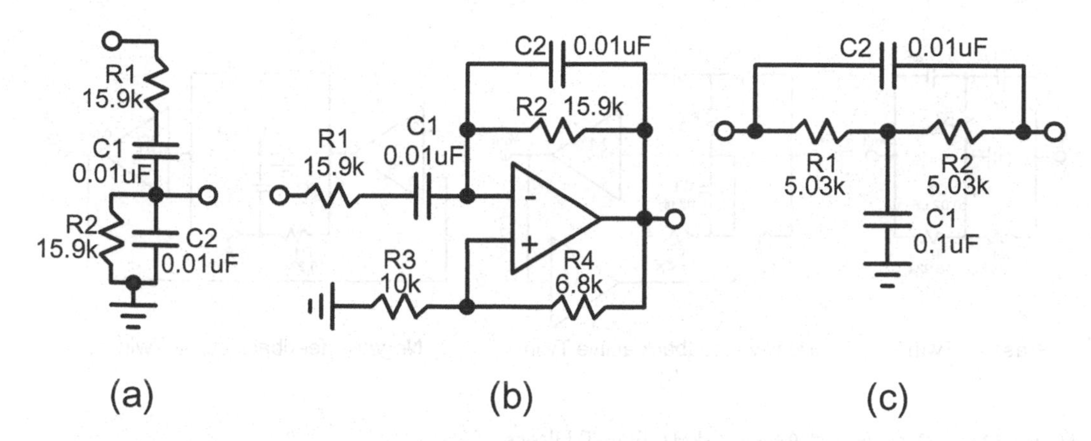

Figure 11.19 Wien Bridge and Bridged-T Filters.

The bridged-T network can feed a low-impedance load, like that of a virtual ground, without losing its band-reject capability or influence on its f_0. This can be accomplished by placing a resistor R3 in series with C2 whose value is the sum of R1 and R2. This allows the admittance of the network to be symmetrical about f_0 and remain finite with frequency. If this network is placed in the feedback loop of an op amp and a large C1/C2 ratio of 100 is employed, a bandpass characteristic is realized whose gain at f_0 depends on the value of an input resistor R4 driving the virtual ground. For a circuit with R1 = R2 = 15.9 kΩ, R3 = 31.8 kΩ, R4 = 415 kΩ, C1 = 0.1 μF and C2 = 0.001 μF, f_0 = 1 kHz, gain at f_0 = 0 dB and Q = 2.5. Maximum attenuation at frequencies far from f_0 is limited to 22 dB.

If a similar network is also placed in the feedforward path of the op amp, in place of R4, a single-amplifier biquadratic filter will result whose numerator and denominator polynomials can be independently specified.

The Wien bridge and bridged-T filters are often used in applications with combinations of negative feedback and positive feedback to increase Q. Both are widely used as the tuning element in oscillators. Wien bridge oscillators are ubiquitous. In an oscillator, the Wien bridge network is used as the positive feedback leg, while the negative feedback leg provides a closed-loop gain of 3 that is controlled by an AGC element to provide amplitude stability. The bridged-T is used in the negative feedback path of an op amp that has a constant-value positive feedback path. The Wien bridge variable equalizer is described in Chapter 20. It is quite popular in mixing consoles, as discussed in Chapter 29.

11.14 Twin-T Notch Filters

Notch filters, or band-reject filters, are often used in instrumentation to sharply reject certain frequencies, like the fundamental of a sine wave to make it easier to see the harmonics. Notch filters can also be used to implement BPFs.

The twin-T passive notch is widely used and is the most popular passive RC notch filter. As shown in Figure 11.20(a), it employs three resistors and three capacitors in a symmetrical configuration. It provides a very deep notch with precision components, but is not easily tuned and its notch depth is sensitive to component mismatch [3, 4, 43, 44]. The resistor and capacitor components to ground with half the impedance are usually made up of two paralleled devices of the same

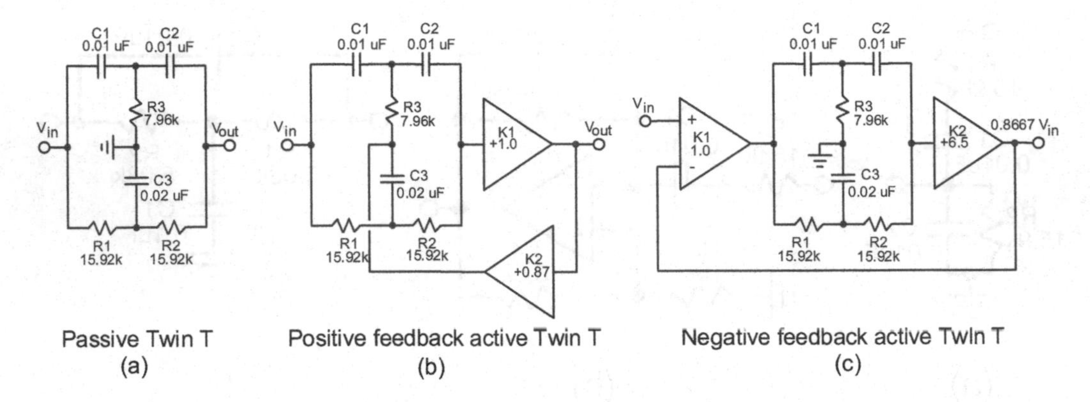

Figure 11.20 Passive and Active 1-kHz Twin-T Filters.

value as used in the arms. These filters are often used in instrumentation, such as for harmonic distortion analysis, where the fundamental frequency is notched out so that a subsequent THD analyzer or spectrum analyzer has greater dynamic range [45].

Such passive notch filters can usually be made so that they introduce very little distortion of their own. However, the twin-T filter does introduce loss at the second and third harmonic frequencies of about 9.4 and 5 dB, respectively [46]. Its center frequency is simply $f_{notch} = 1/(2\pi RC)$. The twin-T filter must be implemented with precision components to maximize the depth of the notch. Moreover, because of this and the fact that there are six tuning elements, it is difficult to implement continuous tuning with the twin-T.

Notch Filter Q

The Q of a notch filter is defined as the difference in −3 dB frequencies divided by the center frequency, just as for a BPF. In spite of the very deep notch, the Q of the passive twin-T filter is fairly low, at 0.25. Passive RC filters usually have low Q. This means that loss at frequencies far from f_0 will not be as low as we would like.

Loss at Harmonic Frequencies

The substantial loss of the passive twin-T notch at the harmonics of the center frequency compromises its use as a THD distortion analyzer. However, knowing these losses allows for corrections to be made in an FFT-based distortion analysis. The losses at the harmonics are as follows:

- Second:9.03 dB
- Third:5.08 dB
- Fourth:3.27 dB
- Fifth:2.27 dB

Notch Depth vs Component Tolerances, Fine Tuning and Trimming

The notch depth of the twin-T filter is quite sensitive to component tolerances. A 1% error in one arm component reduces notch depth in the passive version to −64 dB. If one component is off by only 0.1%, notch depth will be limited to about 84 dB. The twin-T can be fine-tuned over a very narrow frequency range at the expense of notch depth by trimming R3. An increase of 1% in R3 will move the center frequency by about 0.28%, but at the expense of the notch depth being degraded to −58 dB. If instead the notch depth is fine-tuned by adjustment of R3, a design in which R2 was high by 1% with a notch depth limited to 64 dB was brought down to a notch depth of 117 dB by adjusting R3 to 5.0249 kΩ (at least in simulation).

If R1 is high by 0.1%, the notch can be restored to its depth by increasing R3 by 0.05%. Similarly, if C1 is high by 1%, the notch can be restored to its depth by decreasing R3 by 0.5%. Thus, a small-value trimmer in the R3 leg can be quite effective in trimming notch depth to its maximum.

Loading

An interesting question is whether resistive loading of the twin-T filter upsets its notch. It does not degrade the notch at all. Gain remains at unity at high frequencies. A load that is ten times the value of the arm resistors drops the low frequency response by 1.6 dB. The notch thus becomes asymmetrical. The twin-T loss at the harmonic frequencies is increased by about 0.4 dB at the second harmonic frequency and less at higher frequencies.

Active Twin-T Filter with Positive Feedback

An active version of the 1-kHz twin-T, shown in Figure 11.20(b), includes two op amps that introduce positive feedback that greatly reduces the attenuation at the harmonic frequencies at the expense of some reduction in the depth of the notch [46]. The positive feedback acts as a Q multiplier. Amplifier K1 buffers the output of the twin-T network while K2 provides a slightly attenuated and buffered positive feedback signal to the twin-T network with low impedance. The value of K2 sets the Q of the filter. As shown, K2 = 0.87 allows 0.5 dB of attenuation at 2 kHz and 0.16 dB of attenuation at 3 kHz.

The amount of Q multiplication depends on the positive feedback factor P, simply defined as the fraction of output that is fed back to the twin-T filter. Q will be increased by the factor $1/(1 - P)$, so if $P = 0.9$, Q will be multiplied by 10. The passive Q of 0.25 will be multiplied by 10 to 2.5, which reduces loss at the second and third harmonics to 0.3 and 0.01 dB, respectively. At the same time, the Q-multiplying factor of 10 will reduce the non-enhanced notch depth by 20 dB. There is no free lunch.

In principle, a single-amplifier active twin-T filter can be implemented with the op amp connected as a unity-gain buffer, saving an op amp. The arrangement employs near-unity positive feedback to the normally grounded point in the passive twin-T. This causes the attenuation of the twin-T to become almost 0 dB at the second harmonic and above. Unfortunately, this reduces notch depth. It also makes notch depth become more degraded with error in the notch frequency of the twin-T compared with the frequency being notched out.

The fact that the normally grounded point of the active twin-T is connected to the output of an op amp may open the door to some added distortion from that op amp. This distortion is not a result of the very small voltage output signal of the op amp, but rather from the signal current of the twin-T flowing into the non-zero, slightly nonlinear closed-loop output impedance of the op amp. In fact, only 1 Ω of output impedance degrades the notch depth by 8 dB down to 66 dB. A value of 10 Ω degrades the notch by 28 dB down to 26 dB.

Active Twin-T Filter with Negative Feedback

Figure 11.20(c) shows an active twin-T filter that uses negative feedback to reduce attenuation at the harmonic frequencies. K2 provides the loop gain, while K1 implements the feedback summing function. As shown, K2 = 6.5 sets the Q so that 0.5 dB of attenuation is allowed at 2 kHz. The gain of this filter at frequencies far from the notch is slightly less than unity at −1.26 dB because the NFB loop gain is only 6.5. Otherwise, notch depth and relative harmonic attenuation are identical to those of the positive feedback twin-T. This includes notch depth due to a 1% error in C1.

The positive and negative feedback twin-T filters provide equivalent notch depth, Q and tolerance to imperfect component matching. The op amps used for K1 and K2 should have high unity-gain bandwidth to maintain good filter performance at higher frequencies like 20 kHz. The 50-MHz GBP LM4562 op amp is a good choice here. All else remaining equal, the positive feedback version is probably a better choice, since the full signal passes through amplifier K1 in the NFB arrangement and there is a common-mode signal at the input of K1 that may introduce some distortion. However, notch depth degradation of the PFB design due to closed-loop output impedance of K2 as little as 1 Ω must not be overlooked.

Compensated Twin-T Filter

The passive twin-T filter would be best for distortion measurements were it not for its loss errors at the harmonics. It offers a deeper notch and, being passive, has less likelihood of introducing distortion, even with fairly high voltages applied at its input, since it will null the fundamental.

Figure 11.21 shows a compensating filter arrangement where a passive 1-kHz twin-T filter is compensated accurately for its losses at the second and third harmonic frequencies (and above). The compensation is accomplished with a post-filter equalizer that comprises two SVFs. System response is flat to within ±0.1 dB at the second and third harmonics without degrading the depth of the fundamental notch. This approach eliminates any notch depth degradation or distortion introduction that can occur with an active twin-T filter.

The compensated twin-T filter comprises the passive twin-T filter (not shown) followed by two SVFs. The first SVF is configured as an HPF with high enough Q to cause peaking in the vicinity of its high-pass corner frequency of about 1.6 kHz. This peaking is most responsible for correcting the loss of the twin-T filter at the second and third harmonic frequencies. However, the shape of its peaking curve is not enough of a match to correct both the second and third harmonic amplitudes to the desired level of accuracy.

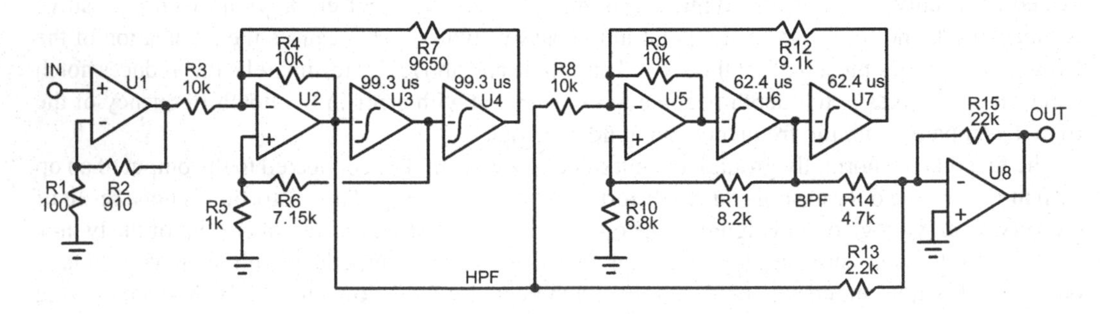

Figure 11.21 Passive Twin-T Filter Compensation Circuit.

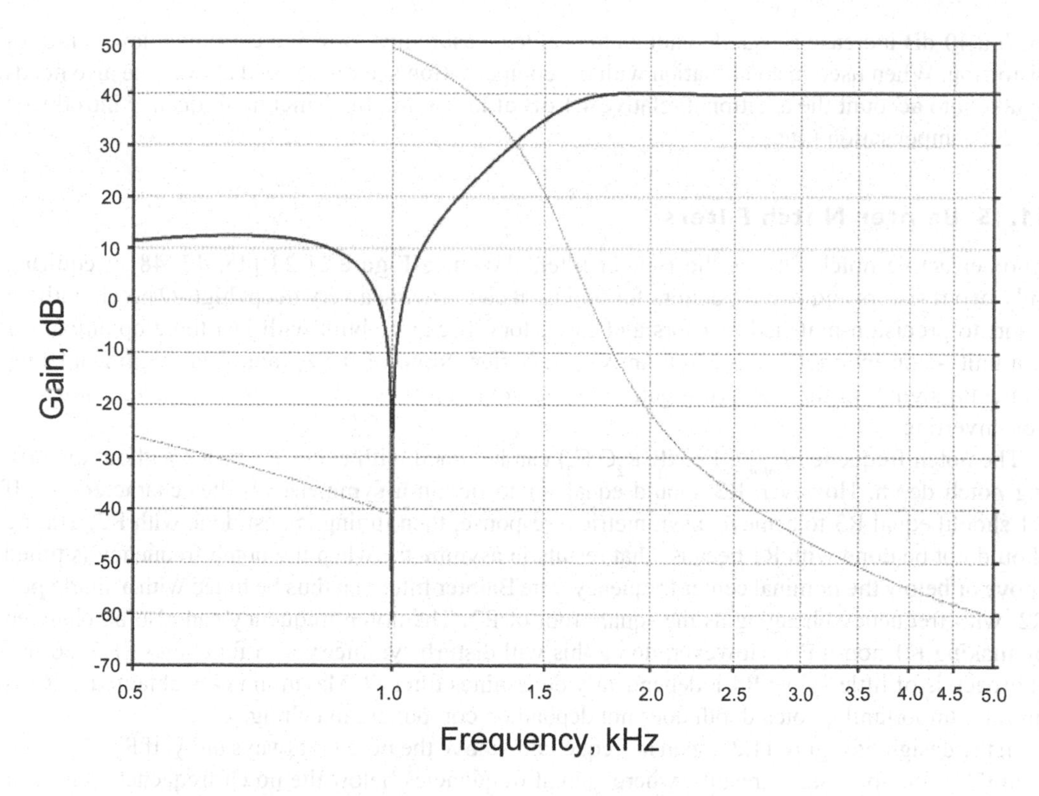

Figure 11.22 Compensated Twin-T Response.

The second SVF is configured as a BPF whose center frequency is in the vicinity of 2.5 kHz. It is arranged so that its output is added to the output of the first HPF SVF. It thus augments the frequency response in the vicinity of the second and third harmonics.

The result is a twin-T filter system that is accurate to better than ±0.1 dB at the second and third harmonic frequencies. The gain of the twin-T post-filter is 40 dB at frequencies well above 1 kHz, yet its gain at 1 kHz is only about 36 dB. Thus, it does not degrade the depth of the passive twin-T notch. In fact, being an HPF, it makes the relative notch about 4 dB deeper. As shown in Figure 11.22, this twin-T system has a net p-p notch depth of 106 dB.

Twin-T Filter with Defined Notch Depth

There are some cases where the deepest notch possible in a twin-T filter is not desired, but rather one with a more accurate, defined notch depth, like 40 dB as an example. In such a case, a known amount of the fundamental is passed through the filter to be used as a reference or as a pilot tone onto which a distortion analyzer can be locked. A reasonably accurate limit on the depth can be implemented by placing a parallel RC network in series between the normally grounded node and ground. The time constant of this RC network is set to be the same as the time constant of R1 and C1 in Figure 11.20(a). For a 40-dB notch the *R-C* network to be added should have 1/100th the impedance of R1 and C1.

Using a twin-T filter with a known notch depth at the center frequency in front of a THD analyzer can increase by 40 dB the dynamic range of the distortion analyzer or spectrum analyzer.

Such a 40-dB increase is usually enough to enable measurement down to extremely low values of distortion. When used in combination with the compensation filter described above, one also needs to take into account the additional relative 4.1 dB of loss at the fundamental frequency introduced by the compensation filter.

11.15 Bainter Notch Filters

Another active notch filter is the Bainter filter, shown in Figure 11.23 [45, 47, 48]. Requiring only two resistors and two capacitors for tuning, it can provide a very deep, high-Q notch without resort to precision-matched resistors and capacitors. It can be built with just three op amps. K1 is a unity-gain inverter and K2 is a unity-gain buffer. None of the op amps has significant fundamental signal on the positive input, reducing common-mode distortion. The Bainter notch is non-inverting.

The notch frequency $f_{notch} = 1/(2\pi R_1 R_2 C_1 C_2)$ can be tuned with either R1 or R2 without disturbing notch depth. However, R3 should equal R1 to obtain a symmetrical filter characteristic. If R1 should equal R3 to achieve a symmetrical response, then tuning is best done with R2. Tuning should not be done with R1 because that results in asymmetry when the notch frequency is tuned above or below the nominal center frequency. The Bainter filter can thus be tuned with a single pot, R2, with frequency changing as the square root of R2. The notch frequency can also be changed by making K1 non-unity. However, doing this will disturb symmetry as a function of K1, so this approach is of little value. R4 independently determines filter Q. Maximum Q is obtained if R4 is infinity. Importantly, notch depth does not depend on component matching.

In the design of Figure 11.23, gain at frequencies above the notch is always unity. If R3 is greater than R1, a low-pass notch results, where gain at frequencies below the notch frequency is greater than unity in proportion to R3/R1. If R3 is less than R1, a high-pass notch results, where gain below the notch is less than unity in proportion to R3/R1.

C1/C2 affects the notch shape and can be any value, reasonably between 0.1 and 10. C1/C2 should be small if high Q and low loss at the second harmonic is desired. In Figure 11.23, with

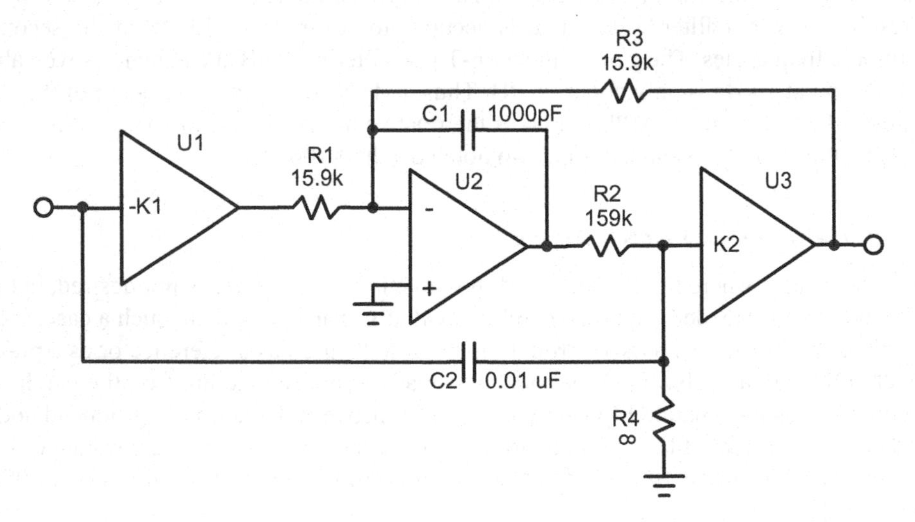

Figure 11.23 1-kHz Bainter Notch Filter.

R4 equal to infinity and C1/C2 = 0.1, the 3-dB points are at 950 and 1052 Hz, notch depth is 82 dB and loss at 2 kHz is 0.02 dB. If R4 is set to 38 kΩ, the 3-dB points are at 775 and 1300 Hz, reflecting lower notch *Q*. Notch depth is then 94 dB and loss at 2 kHz is 0.5 dB. If C1 = C2 and R2 is reduced to 15.9 kΩ, f_0 will remain at 1 kHz, but loss at 2 kHz will increase to 1.6 dB, even with R4 at infinity.

There is a significant caveat to the performance of the Bainter notch in the real world. In the circuit of Figure 11.23, the signal level at the output of U2 is up 20 dB from the input amplitude. This could cause increased distortion in the op amp. On the other hand, if the input signal is reduced by 20 dB to keep distortion down, the SNR will be reduced. If the filter is to be tuned over a wide range, like 10:1, the problem gets worse. If the filter here is tuned to 316 Hz by increasing R2 by a factor of 10, peaking at f_0 increases to 30 dB. If f_0 is tuned to 3160 Hz, peaking decreases to +10 dB.

As mentioned earlier, the Bainter filter is easily tuned with a single pot over a range exceeding a decade by varying R2. However, because frequency changes in proportion to the square root of R2, changing the frequency over a decade requires changing R2 by a factor of 100. This taxes the circuit's ability to perform well. In the circuit here, if R2 is increased by a factor of 10, the maximum notch depth decreases to about 50 dB in simulation, even with ideal components. If R2 is decreased by a factor of 10, notch *Q* becomes compromised and loss at twice the notch frequency increases to 1.6 dB. This performance may still be adequate for many applications. Performance across a full tuning range of only 3:1 is much better. This suggests that in a tunable instrumentation application, frequency bands should be in a 1, 2, 5 or 1:3 sequence.

A 50-MHz GBP op amp (like the LM4562) is recommended for this application, especially for higher notch frequencies like 20 kHz. A small resistor placed in series with C1 can mitigate op amp GBP limitations. Apart from the fact that it is active and can be influenced by op amp distortion, the Bainter notch filter is superior to a precision-tuned twin-T filter for distortion analysis or for increasing the dynamic range for spectral analysis because it is so easily tuned exactly to the fundamental frequency.

Bainter Bandpass Filter

The Bainter notch filter can be used as the basis for a bandpass filter (BPF) with simple tuning and high *Q*, as shown in Figure 11.24. Here the Bainter notch filter is placed in the feedback path of an operational amplifier. Depending on the specific circuit arrangement, the gain at frequencies far away from the notch will not usually go to zero, as it does in a conventional BPF. In the figure, stop-band loss is 0 dB. This is because the gain of the Bainter notch at these frequencies is unity, non-inverting, making the feedback amplifier a voltage follower. Similarly, depending on the application, the gain at the center frequency can be allowed to be very large or as low as unity. The ratio of R6 + R5 to R5 controls the ratio of passband gain to stop-band gain. Changing *Q* does not change f_0 or gain at f_0.

Using the Bainter notch filter of Figure 11.23 and with the values of R5 and R6 in Figure 11.24, filter gain at 1 kHz is 40 dB and *Q* is slightly over 1000 with ideal op amps in simulation. Including R4 = 100 kΩ in the Bainter filter for Figure 11.24 reduces bandpass *Q* to 400. Setting R4 = 10 kΩ reduces *Q* to 60. Setting R4 = 1 kΩ reduces *Q* to 6.4. Component sensitivity for *Q* is quite low.

Headroom on U2 must be watched. The output of U2 in this design is 20 dB above the output of the BPF output at f_0. The Bainter BPF may be an excellent choice for cleaning up an oscillator output, although it is not clear that it is better than a second-order SVF.

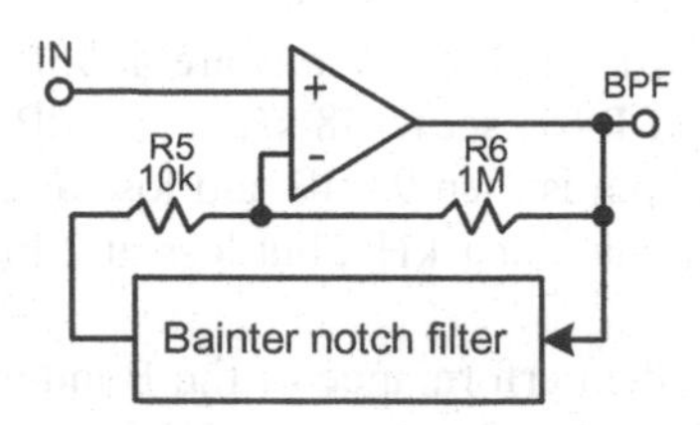

Figure 11.24 Bainter Bandpass Filter.

11.16 Hall Notch Filter

A 1-kHz Hall notch filter is illustrated in Figure 11.25(a) [49–51]. Unlike the twin-T filter, this filter can be frequency-tuned with a single potentiometer. Moreover, the notch depth can be trimmed to account for component tolerances. These are distinct advantages over the twin-T notch filter for some applications. However, since frequency control is achieved by a potentiometer whose total value figures into the notch depth equation, and given that potentiometer resistance tolerances are often no better than ±20%, it is generally required that a notch depth trimming potentiometer be incorporated into R3 and adjusted.

The filter uses phase cancellation of two paths to achieve a deep notch. The frequency-dependent path consists of three capacitors in series, each with a loading resistor, forming a 3-pole HPF that creates leading phase shift. The other path is simply a resistor (R3) that also acts as the load for the third capacitor. When the total phase shift through the series capacitance path is 180° and of the proper amplitude, cancellation of the signal results. This is a bit reminiscent of the three-pole network used in the feedback path of a phase shift oscillator, where the three sections together create a 180° phase shift at the oscillation frequency.

A key aspect of the Hall notch filter is that the required amplitude matching of the two paths for cancellation does not change appreciably with the frequency setting. The single tuning potentiometer alters the phase and amplitude characteristics through the high-pass path in such a way as to create the needed 180° phase shift at a lower or higher frequency.

A simple version of the network has three identical capacitors in series, the first two of which are loaded to ground by R1 and R2, which are the left and right portions of the tuning potentiometer

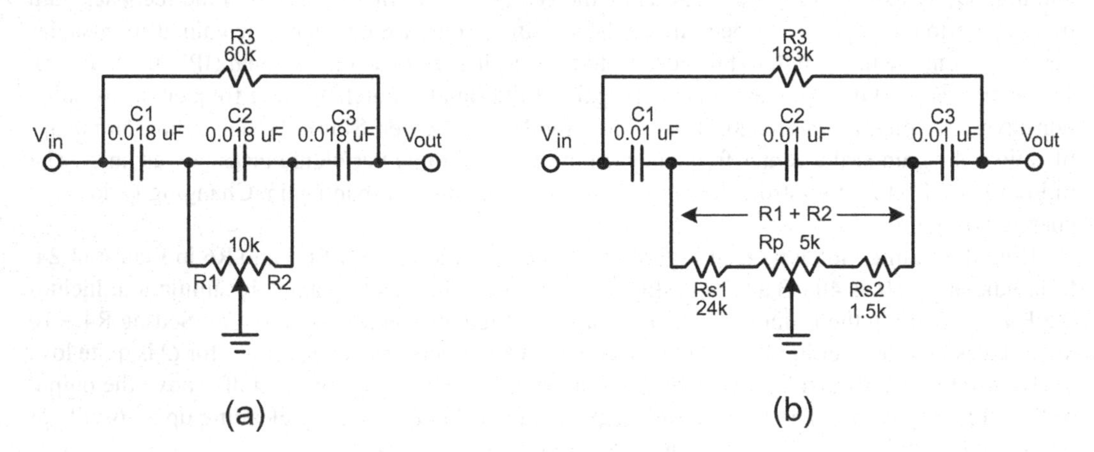

Figure 11.25 Hall Notch Filter.

in (a). A third resistor, R3 straddles the high-pass network and provides a cancellation signal at the output of C3. Its value should always be (see Ref. [50]):

$$R3 = 6(R1 + R2)$$

For this simple version of the network, if C1 = C2= C3 = C and R1 = R2 = R,

$$f_0 = 1/(2\pi RC\sqrt{3})$$

Note that f_0 here is defined as the lowest frequency of the tuning range, not the frequency where the mechanical pot is usually centered in a practical design (see below). If the circuit is tuned by making R1 different than R2 (moving the pot off-center), then

$$f_0 = 1/2\pi C\sqrt{3R1R2}$$

Let the parameter TF be defined as the fraction of pot rotation from CCW to CW, then the normalized notch frequency is:

$$f_{0\_norm} = 0.5/\sqrt{TF(1-TF)}.$$

The rate of change of frequency with pot rotation increases with pot position further from center. Operation of the filter is symmetric about mid-rotation as described here thus far (R1 = R2), and operation is usually with TF greater than 0.5, meaning that R1 > R2. The notch frequency f_0 is lowest when R1 = R2. In practice, only one side of the symmetric behavior is employed (R1 > R2).

Placing the tuning range to one side of the pot range is achieved by adding non-equal end stop resistors Rs1 and Rs2 to the physical pot as shown in Figure 11.25(b). Think of the pot with its end stop resistors being the actual "effective pot." Rs1 is usually made large enough so that R1 > R2 when the pot is full CCW. This means that the lowest tunable frequency will occur when the physical pot is full CCW. The larger the ratio of Rs1 to Rs2, the wider the tuning range. The filter in (b) has its lowest frequency of 736 Hz when the pot wiper is at the left end. The filter notch is at its highest frequency of 1393 Hz when the wiper is at the right end. The notch frequency with the pot centered is 892 Hz.

Q for the notch filter is f_0 divided by the 3-dB bandwidth. It is highest when the effective pot is at 50% rotation (R1 = R2), and is only 0.177, in spite of the deep notch. This Q is smaller than the Q of the twin-T notch (0.25). In general, when R1 > R2,

$$Q = 0.25\sqrt{(1-TF)}$$

and Q decreases as frequency is increased.

Figure 11.25(b) shows a practical Hall notch filter where the pot end stop resistors are included. The larger the ratio of Rs1 to Rs2, the wider the tuning range. The Hall notch can be easily tuned over a fairly wide range, on the order of an octave, if the resulting changes in Q are acceptable. Conversely, the notch filter can be tuned over a small range easily while maintaining high Q and a deep notch. This feature makes it more suitable than the twin-T in situations where the notch may have to be tuned over a range of a few percent to match the incoming fundamental frequency.

In a practical design where a deep notch is required, a trim pot will have to be inserted in series with R3, since potentiometers usually have a tolerance of ±20%, so the value R1 + R2 is not precisely known.

Design Approach

The end resistors are vital to achieving desired tuning range with maximum Q. Bear in mind that pots come only in a 1–2–5 sequence of values. Choose the pot value first. Assume Rs1 will be approximately twice the pot value, since when the pot is full CCW, you want R1 to approximate R2, with R1 being slightly higher than R2. This will be the point of maximum Q. With R2 equal R1, find an available capacitor value to achieve lowest frequency f_0 of the tuning range. For the tuning range ratio you choose, assume that R2 will be smaller than 2R1 by approximately the square of the tuning ratio. Choose R3 = 6(Rs1 + Rp + Rs2). Simulate and see what you get. Tweak from there.

It is still not a bad idea to start with Rs1 equal to the pot value or somewhat greater. Rs2 should be a significantly smaller value than Rs1, since it has a very big influence on the highest frequency reachable at full CW rotation. Consider 1/3 to 1/10 for the value of Rs2/Rs1.

Performance and Applications

The Hall notch filter features great ease of tuning and notch optimization compared to the twin-T. The Hall network is probably the only passive notch filter that can be both tuned and optimized for depth. The twin-T filter requires closely matched components to achieve a deep notch and it cannot be practically tuned. There is very limited ability to optimize the notch depth in the twin-T. However, trimming of the Hall filter is necessary because of pot tolerance issues. As a passive network, the Hall notch can be effective at high frequencies. The ratio of the stopper resistors must be large for there to be a large tuning range. The Hall notch has more loss at the second harmonic than the twin-T. The twin-T notch is down 9 and 5 dB at the second and third harmonics, respectively, while the Hall notch is down 11 and 5 dB at the second and third harmonics, respectively.

The Hall notch is especially well-suited for a design set at its lowest frequency and needing a small tuning range, as with a twin-T application. The Hall notch filter can also be used to create a BPF by including it in the negative feedback loop of an amplifier. Similarly, it can be used to implement a sine wave oscillator if a proper amplitude control circuit is included.

If the Hall notch filter is used in place of a bridged-T notch in an oscillator, frequency control is achieved with a single pot, likely over a 10:1 or 3.16:1 tuning range. The Hall notch frequency changes most slowly with pot rotation at the lowest frequencies, like a log characteristic. This is desirable for oscillator tuning over a decade or half decade. How close the Hall tuning is to logarithmic with a linear pot can be roughly checked by seeing how close the frequency at the pot center position is to the geometric mean of the lowest and highest tuning frequencies. If $f_{0\_norm} = 0.5/\sqrt{\text{TF}(1 - \text{TF})}$, then the maximum and minimum values of TF, as determined by Rs1, Rp and Rs2, can be plugged in and the midpoint value of TF can be determined and plugged in. This allows comparison of the normalized frequencies f_{min}, f_{mid} and f_{max}. If $f_{mid} = \sqrt{f_{min} f_{max}}$, then log conformance should be fairly good.

11.17 All-Pass Filters

As the name implies, all-pass filters (APF) pass all frequencies without any frequency-dependent attenuation [52]. However, they do change the phase as a function of frequency. This feature is useful in many applications, including creation of time delay (delay lines) and correction of nonlinear phase in filters. Group delay is the rate of change of phase with linear frequency. It is thus the derivative of phase as a function of frequency. Since the rate of change of phase at any given frequency corresponds to group delay, APFs modify group delay.

An all-pass filter is like an LPF in the group delay domain; group delay is at its maximum at very low frequencies and decreases as the frequency is increased. A first-order APF has a monotonically falling group delay with frequency. Its phase changes by 180° going from DC to high frequencies. Its phase lag is 90° at its f_0. Its group delay has fallen to half its DC value at f_0.

Second-order APFs have phase shift that changes by 360° going from DC to high frequencies. They have a Q associated with them because they rely on a second-order BPF for their operation, and their group delay frequency shape can be quasi-Butterworth or quasi-Chebyshev, for example. Group delay of the latter will begin at a given value at DC and may decrease or peak and then decrease as f_0 is approached and passed.

First-Order All-Pass Filters

Two 1-kHz first-order APFs are shown in Figure 11.26 [52]. Each consists of an op amp and a simple *RC* section, either a first-order low-pass or a first-order high-pass. Both begin with an op amp connected in the unity-gain inverting mode. If no signal gets to the positive input of the op amp, we just have the inverter. In (a), a low-pass section delivers the input signal to the positive input of an op amp, while the unmodified input is fed though a resistor to the inverting input.

Assume that R3 equals R2. At low frequencies C1 is open and it is easy to see that the input signal sees a non-inverting gain of unity through the filter. At high frequencies, C1 acts like a short, and the input signal is subjected to inverting unity gain. As the frequency goes from very low frequencies to very high frequencies the output phase changes from 0° to −180°. At the 1-kHz 3-dB frequency of R1 and C1, f_0, (ω_0), it can be shown that the gain of the APF is still unity and that the phase shift is lagging by 90°.

In (b), a high-pass section drives the positive input of the op amp instead. At low frequencies C1 is open and no signal gets to the op amp's positive input, so the APF acts like an inverter, causing 180° phase shift in the signal. Conversely, at high frequencies C1 is a short and the full input signal appears at the positive input. It is easy to see that the net gain at high frequencies is unity non-inverting. At the 3-dB frequency of R1 and C1, it can be shown that the gain of the APF is still unity and that the phase shift is leading by 90°.

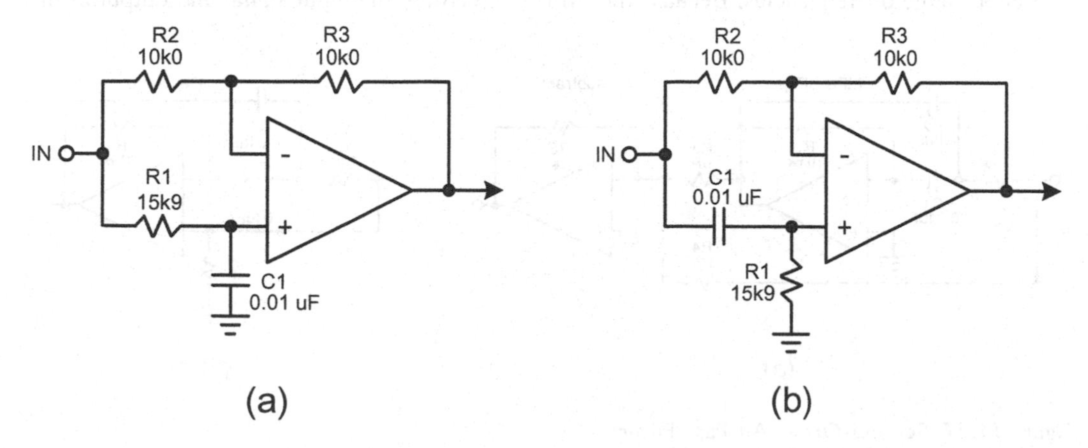

Figure 11.26 First-Order All-Pass Filters.

The phase of the 1-kHz APF in (a) exhibits a lagging phase shift that smoothly transitions from 0° to −180° as frequency increases. The group delay starts out at a relatively constant value and then decreases to a lower value of delay at f_0, where the phase lag is 90°. At higher frequencies, group delay falls to zero. At low frequencies, the group delay is 2R1C1. At f_0, group delay has fallen to R1C1.

Notice that in the above first-order APFs, the filter function was created by subtracting twice the output of a first-order filter from the input signal. For the LPF APF, we have

$$H(s) = 1 - 2/(s+1) = (s-1)/(s+1)$$

which is the transfer function for a first-order APF [49]. Mathematically,

$$\text{phase shift} = -2\tan^{-1}(RC/\omega)$$

$$\text{group delay} = 2RC/\left((\omega RC)^2 + 1\right)$$

Second-Order All-Pass Filters

A second-order APF can be created in the same way, by subtracting twice the output of a second-order BPF from the input signal. In this case, the phase shift increases from zero to 360° of lag at high frequencies, passing through 180° at the center frequency f_0 of the MFB BPF used in Figure 11.27(a). The phase shift is returned to 0° as the output of the BPF falls to zero at high frequencies. Notice that in this case the gain of the BPF is assumed to be unity, so subtracting twice its output from the input at f_0 results in an inverted version of the input at the output, corresponding to 180°, as expected.

The Q of the MFB BPF will determine the shape of the group delay curve as a function of frequency. Peaking of the group delay will occur before f_0 if Q of the BPF is greater than 0.707 (or in some cases, 0.63). In this type of implementation, the curve of group delay vs frequency will look like that of an LPF whose ripple or flatness is controlled by the BPF Q. These designs are useful in implementing delay lines, where group delay is ideally flat with frequency.

If the Q of the MFB is 1, there will be a significant peak in group delay near f_0, giving the group delay curve somewhat the shape of a BPF. This can allow targeting most of the group delay to a particular range of frequencies. Because the MFB is inverting, the input signal and output of the

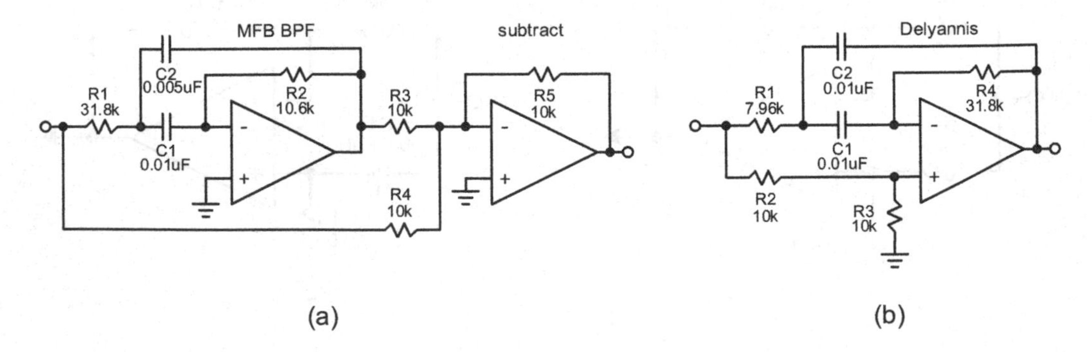

Figure 11.27 Second-Order All-Pass Filters.

MFB can be subtracted by routing them to the virtual ground of the output op amp. If the BPF section has gain different than unity, the values of R3 and R4 can be changed to create the proper subtraction function. Other BPFs, like the S-K, state variable and biquad can also be used as long as the subtraction criteria is satisfied.

Single-Amplifier Second-Order All-Pass Filter

The Delyannis second-order APF requires only a single op amp, as shown in Figure 11.27(b) [22]. At its heart is an MFB BPF. The necessary subtraction is performed by feeding forward some of the input signal to the non-inverting input of the MFB's op amp. The Delyannis APF is non-inverting at DC.

Third-Order All-Pass Filters

Just as with ordinary LPFs, a higher-order APF can be implemented by connecting more than one section in tandem. For example, if a first-order all-pass is combined with a second-order all-pass that has the appropriate Q, the peaking of the group delay of the second-order APF will tend to cancel the droop in the group delay of the first-order APF, resulting in greater group delay that is flatter with frequency. It is possible to implement nearly maximally flat group delay or a group delay characteristic that has some ripple, just as with a Chebyshev LPF. A second-order APF whose Q is unity, preceded by a first-order APF whose f_0 is adjusted to a lower frequency than f_0 of the second-order section, can be used to create a third-order APF whose group delay shape can be modified by adjusting f_0 of the first-order section (Figure 11.28).

Group Delay vs Frequency

The group delay of the first-order 1-kHz APF of Figure 11.26 is shown in Figure 11.29 by the curve labeled (a). The group delay starts out at a relatively constant value and then decreases to a lower value of delay at f_0, where the phase lag is 90°. At higher frequencies, group delay falls to zero. At low frequencies, the group delay is 2R1C1. At f_0, group delay has fallen to R1C1.

The group delay of the 1-kHz second-order APF of Figure 11.27(a) is shown in (b). It is designed for good flatness of group delay. The delay has a slight peak just before the roll-off of the

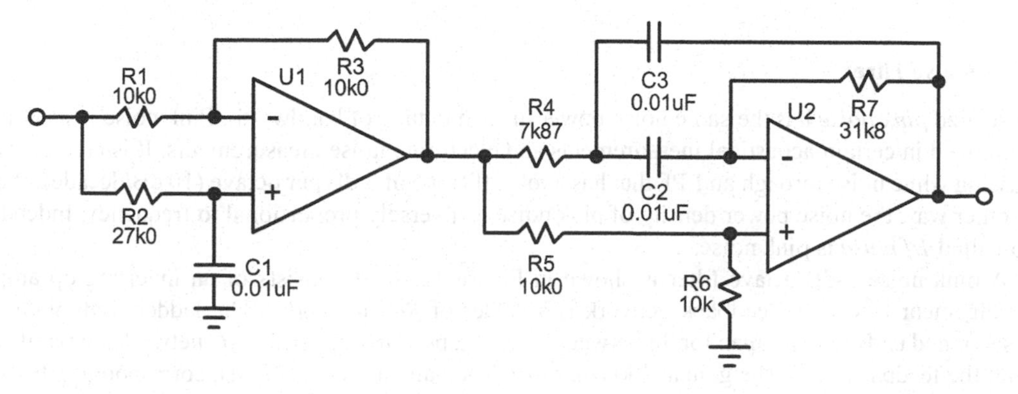

Figure 11.28 Third-Order All-Pass Filter.

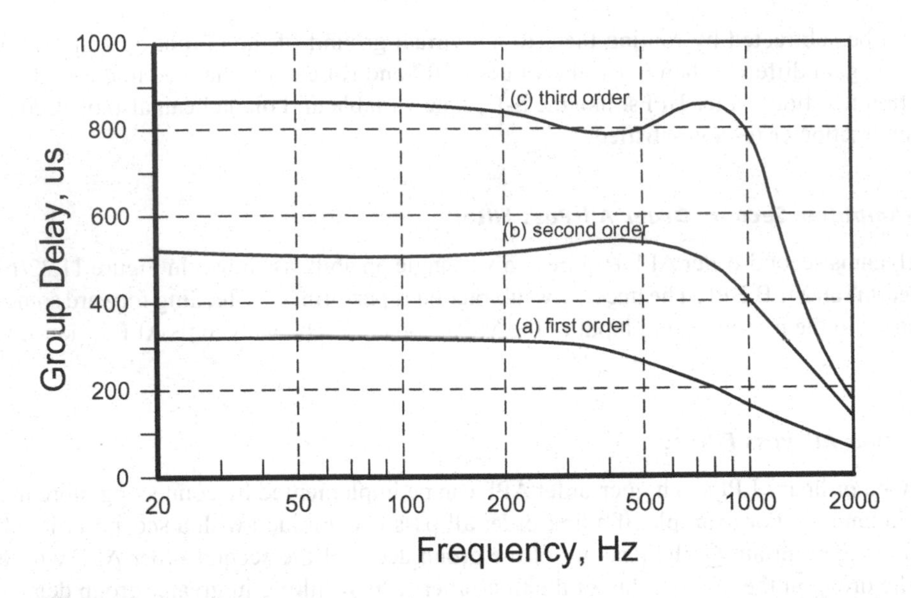

Figure 11.29 Group Delay of the All-Pass Filters.

group delay. Group delay at DC is about 520 μs and peaks at about 550 μs at 450 Hz, falling to half its DC value at 1400 Hz.

The group delay of the third-order APF of Figure 11.28 is shown in (c). It begins with 860 μs at DC and falls to 790 μs at 350 Hz. It then peaks at 860 μs at 750 Hz and falls to half its DC value at 1400 Hz. It has 70 μs of ripple and its group delay is down 3 dB at 1160 Hz.

It should be noted that flat group delay is not always the objective. Flat group delay is desirable if one is making a delay line, but for some applications in delay correction the delay should be maximized at some frequency.

11.18 Application-Specific Filters

To finish the chapter on filters, below are some filters that are designed for specific applications that are important to audio.

Pink Noise Filter

So-called *pink noise* has the same noise power in each octave of bandwidth. Pink noise is usually employed in certain acoustical measurements and electronic noise measurements. It is created by passing white noise through an LPF that has a roll-off slope of 3 dB per octave (10 dB/decade). Put another way, the noise power density of pink noise is inversely proportional to frequency. Indeed, so-called *1/f noise* is pink noise.

A pink noise 3-dB/octave filter is shown in Figure 11.30. It consists of an inverting op amp arrangement where the feedback network is a ladder of *R-C* networks. The ladder starts with a resistor and ends with a capacitor. In between there is a plurality of series *R-C* networks in parallel with the feedback path. The gain at 1 kHz is 0 dB. The gain at 1 Hz is 30 dB, corresponding to 10 dB/decade.

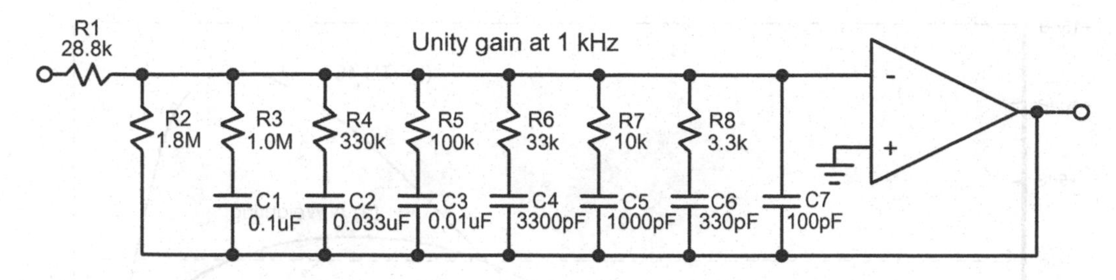

Figure 11.30 Pink Noise Filter.

The white noise source can be a large-value resistor with significant gain after it. It can be calibrated, knowing that thermal noise is 4.16 nV/√Hz for a 1-kΩ resistor. For example, a 100 kΩ resistor at the input of a 100× non-inverting low-noise JFET op amp would create 41.6 nV/√Hz at the input and 4.16 μV/√Hz at the output. The thermal noise of the 100-kΩ resistor will swamp out the input noise of the JFET op amp.

Alternatively, the white noise source can be a Zener diode or a BJT whose base-emitter junction is reverse biased into Zener breakdown. The noise calibration may be questionable.

Noise-Weighting Filters

Weighted noise specifications take into account the ear's perception of noise loudness in different parts of the frequency spectrum. The most common one used is *A-weighting*, whose frequency response weighting is illustrated in Figure 11.31(a). The A-weighted noise is measured by placing a filter with this frequency response in front of a true-RMS AC voltmeter. Notice that the weighting curve is up about +1.2 dB at 2 kHz, whereas it is down 3 dB at approximately 500 Hz and 10 kHz.

Although A-weighting is the most widely used way to specify noise in audio, an alternative was developed in search of a way to better describe the audibility of random noise. Thus was born the ITU-R 468 weighting curve, shown in Figure 11.31(b). The audibility of random noise falls faster at frequencies above 6 kHz than described by the A-weighting curve. At higher frequencies the curve falls at 24 dB/octave, quite a bit faster than A-weighting. The +12-dB peak in the ITU-R 468 curve at 6 kHz is quite notable, as is its steep roll-off at higher frequencies. Both weighting curves are set to have equal weighting at 1 kHz (0 dB). Below 6 kHz the ITU-R 468 curve falls at 6 dB/octave, a shallower slope than A-weighting in the low-frequency region. At 20 Hz, the ITU-R 468 curve is up about 16 dB compared to A-weighting. The ITU-R 468 curve is better known and more commonly used in Europe.

The circuit design of an A-weighting filter is shown in Figure 11.32. The weighting filter includes a low-Q fourth-order high-pass function, causing it to be down by 26 dB at 60 Hz. It also includes a second-order distributed-pole low-pass function that causes it to be down by 10 dB at 20 kHz.

Anti-Alias and Reconstruction Filters

Anti-alias and reconstruction filters are associated with digital audio ADC and DAC functions, respectively. The anti-alias filter for conventional non-over-sampled (NOS) A/D conversion has a very sharp cutoff, with a transition region of only about 4 kHz to −60 dB. This corresponds to

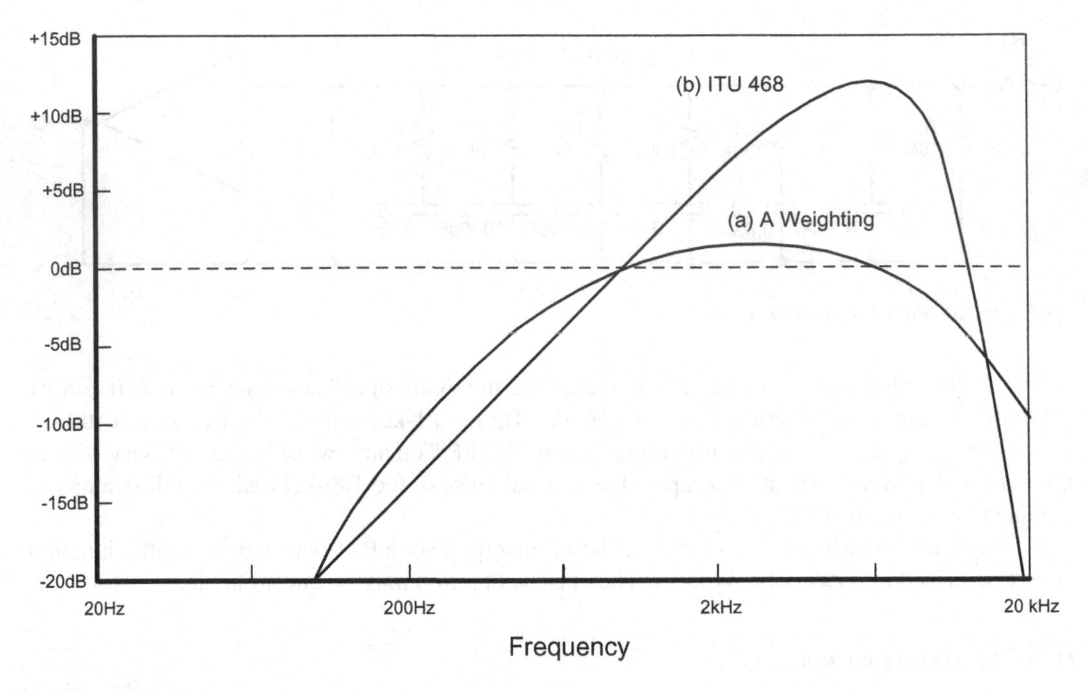

Figure 11.31 A-Weighting and ITU-R 468 Weighting Curves.

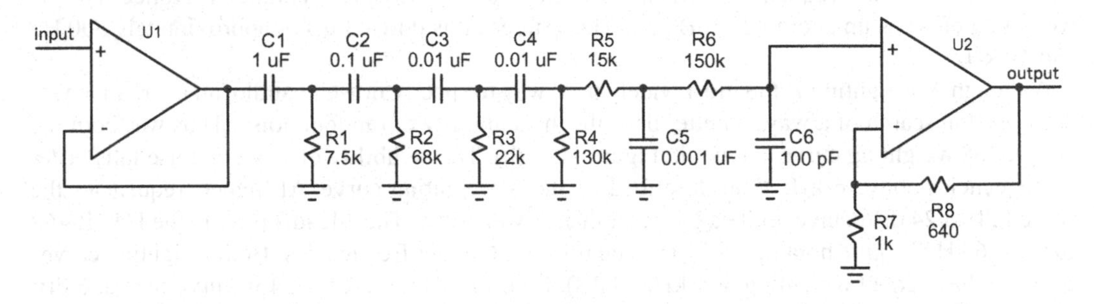

Figure 11.32 A-Weighting Filter.

having high loss at the 24-kHz Nyquist frequency in a system sampled at 48 kHz. These are often implemented with elliptic filters.

The reconstruction filter keeps the high-frequency components created by a DAC out of the subsequent analog path. This is sometimes called an anti-imaging filter because it suppresses images of the audio signal that exist above and below integer multiples of the sampling frequency. The image signals are less harmful, and the reconstruction filter may be as complex as an anti-alias filter or as simple as a first-order LPF (especially in oversampled systems). Reconstruction filters will be covered in greater depth in Chapter 23 on DACs.

AES17 Filters

To deal with the spurious carrier residual and/or quantization noise and other EMI that may be present at the output of digital audio equipment and class D amplifiers, the Audio Engineering Society published a filtering recommendation called AES17 [53, 54]. The 20-kHz brick-wall LPF is placed between the equipment output and measurement instruments like distortion analyzers to prevent overload of their sensitive active input circuits [54]. It may also be placed within a THD analyzer for making distortion measurements.

The filter is very sharp, flat within ±0.1 dB to 20 kHz and then down by 60 dB at 24 kHz. It is much like an anti-alias filter for audio analog-to-digital conversion. Although not required by the standard, a typical AES17 filter will be down at least 75 dB at frequencies above 50 kHz. The 20-kHz brick-wall nature of the filter facilitates reasonably accurate noise measurements in the face of Sigma-Delta converters using noise shaping wherein noise that has been shaped may rise rapidly above 20 kHz. It also establishes a well-defined noise bandwidth of 20 kHz.

The filter usually requires a seventh-order elliptic filter, some or all of which should be implemented with passive components for class D audio amplifier measurements if a pre-filter is not employed. This ensures that neither the filter itself nor sensitive active circuitry in subsequent test instruments is disturbed by the EMI or caused to experience slew rate limiting. In some cases it is sufficient that only the early stages of the filter be passive, but the inductors in at least the early stages of an *L-C* implementation of the filter must be extremely linear (or in the pre-filter, if used there). This can introduce significant expense. Without the filter, measurements at low signal levels will be especially affected in an adverse way.

The AES17 filter's very steep cutoff beginning at 20 kHz is often unnecessarily low in frequency for purposes of measurement where the sensitive input circuits of distortion analyzers and similar equipment must be protected from high-frequency overload. Similarly, the extreme cutoff steepness is also less important in such applications like measurements of class D amplifiers and DACs. Its stringent brick-wall characteristic is often unnecessary for use with much test equipment. The AES17 filter can also be very expensive to implement, especially in passive form where it provides the most protection from high-frequency overload for instrumentation input circuits. Its high order

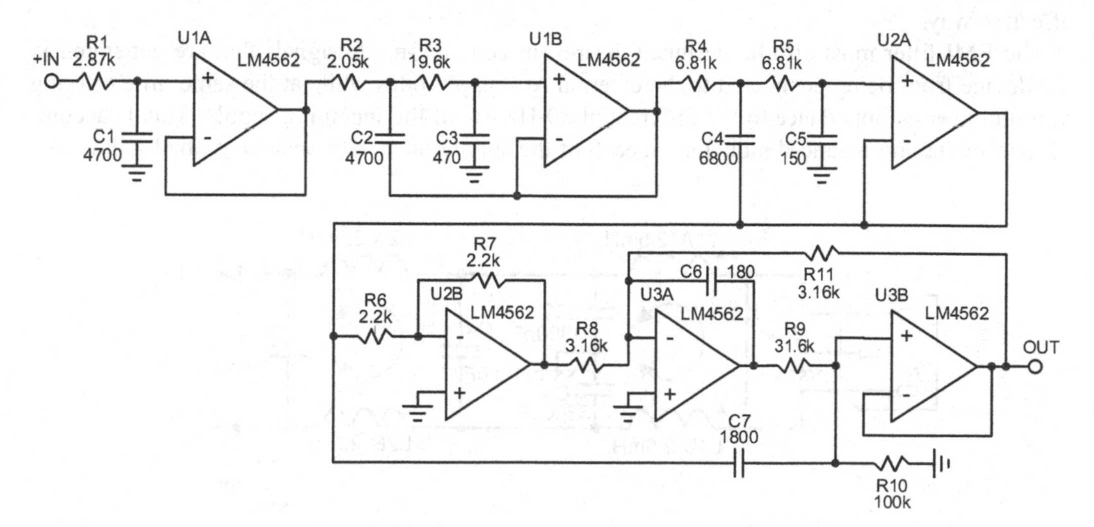

Figure 11.33 Alternative to the AES17 Filter.

requires the use of precision passive components, especially in the active portions of the filter (if present).

There are low-cost and moderate-slope alternatives to the AES17 filter that can be adequate for preservation of the test equipment's performance, even though they do not meet the strict brick-wall cutoff template of the AES17 filter. Active filter approaches where one or two passive real poles are placed at the front-end of an odd-order filter will often suffice. The passive real pole(s) attenuate the high frequencies before the active circuitry. The sharp-cutoff part of the characteristic can be implemented with much less-expensive active filters later in the chain that will perform very well if implemented with high-performance op amps capable of high slew rate and very low distortion, like the LM4562. If a sharper cutoff characteristic is needed, the use of a Bainter active notch filter can be economic and effective in comparison with high-order elliptic filters.

The filter in Figure 11.33 is an approach to approximating the AES17 shape adequately for measurement purposes, but without such an extreme cutoff slope. It is a fifth-order Chebyshev active filter followed by a Bainter filter that places a high-Q notch at 28 kHz. The notch provides increased attenuation and slope in that region, adequate to counteract the sharply increasing noise in that region when Sigma-Delta noise-shaping DACs or class D amplifiers are being tested. The Chebyshev filter has been tweaked slightly to improve flatness given the Bainter filter's roll-off characteristic. The filter is down less than 0.05 dB at 20 kHz and falls to −20 dB at 32 kHz. Attenuation at 200 kHz simulates at over 100 dB.

EMI Filters

Audio equipment often includes an EMI (electromagnetic interference) filter in the AC line of the device. This keeps EMI from the AC line out of the device, while it also keeps EMI from the device escaping into the AC line and being radiated. Many IEC line connectors include a built-in EMI filter.

The EMI filter suppresses differential-mode and common-mode signals on the AC supply lines. A typical EMI filter circuit is shown in Figure 11.34. Differential shunt capacitors C1, C4 and C5, in combination with the inductors, provide differential-mode attenuation in a straightforward and effective way.

The EMI filter must also be designed to prevent common-mode signals that are generated in the device from being conducted to the external AC supply lines while at the same time offering minimum series impedance to the differential 60-Hz AC of the incoming supply. This is accomplished by having balanced inductors in each of the supply lines with positive mutual inductance

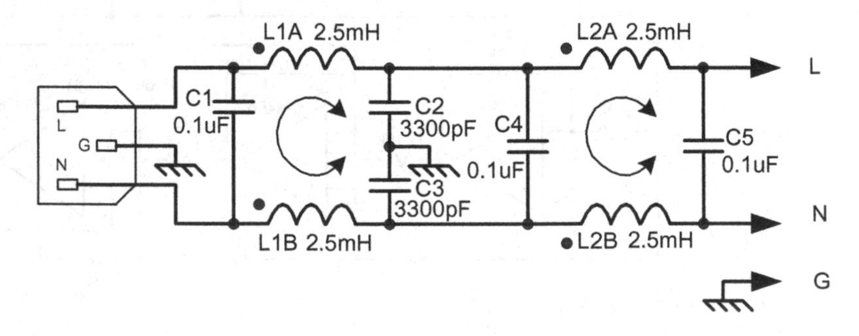

Figure 11.34 EMI Filter.

between the inductors in the line and neutral paths. This arrangement maximizes the inductance for common-mode currents that flow in the same direction in both conductors.

For the oppositely directed differential currents of the 60-Hz mains, the mutual inductance is negative, thus forcing the overall series inductance to a small value. C2 and C3 shunt the common-mode signals to ground. The mutual inductors (called *common-mode chokes*) often have a small amount of remaining series inductance that will provide some differential filtering at high frequencies above 150 kHz.

References

1. Gobind Daryanani, *Principles of Active Network Synthesis and Design*, John Wiley & Sons, New York, NY, 1976.
2. Mac Elwin Van Valkenburg, *Analog Filter Design*, Holt Rinehart and Winston, New York, NY, 1982.
3. Arthur Williams and Fred Taylor, *Electronic Filter Design Handbook*, 4th edition, McGraw-Hill, New York, NY, 2006.
4. Hank Zumbahlen, editor, *Linear Circuit Design Handbook*, Analog Devices, Newnes, Burlington, MA, 2008.
5. Walt Jung, *Op Amp Applications Handbook*, Analog Devices, Newness/Elsevier, Burlington, MA, 2005.
6. Sergio Franco, *Design with Operational Amplifiers and Analog Integrated Circuits*, 4th edition, McGraw-Hill, New York, NY, 2015.
7. Marc Thompson, *Intuitive Analog Circuit Design*, 2nd edition, Newnes, Waltham, MA, 2014.
8. Analog Devices, *Analog Filter Wizard*, tools.analog.com/en/filterwizard/.
9. HiFi Audio Design, *Low-Pass, Band-Pass and High-Pass Filter Calculators*, mh-audio.nl/ChooseAF.html. (based on formulas from *Electronic Filter Design Handbook* by Arthur B. Williams).
10. Martin E. Messerve, *2nd Order Active Filters, Low-Pass.High-Pass*, k7mem.com/Fil_2nd_Active_LP_HP.html.
11. Martin E. Messerve, *2nd Order Active Band-Pass Filter (MFB/VCVS)*, k7mem.com/Fil_2nd_Active_BP.html.
12. Martin E. Messerve, *3rd Order Sallen & Key Low-Pass Active Filter*, k7mem.com/Fil_Third_Order_LP.html.
13. Okawa Electric Design, *Filter Design and Analysis*, sim.okawa-denshi.jp/en/Fkeisan.htm.
14. Texas Instruments, "*Filter Design Tool*," ti.com/design-resources/design-tools-simulation/filter-designer.html.
15. Okawa Electric Design, *3rd Order Sallen-Key Low-Pass Filter Design Tool*, sim.okawa-denshi.jp/en/Sallenkey3Lowkeisan.htm.
16. Okawa Electric Design, *Multiple Feedback Low-Pass Filter Design Tool*, sim.okawa-denshi.jp/en/OPtazyuLowkeisan.htm.
17. R.P. Sallen and E.L. Key, A Practical Method of Designing RC Active Filters, *IRE Transactions on Circuit Theory*, vol. CT-2, May 1955, pp. 74–85.
18. W. Saraga, Sensitivity of 2nd-Order Sallen-Key-Type Active RC Filters, *Electronics Letters*, vol. 3, no. 10, October 1967, pp. 442–444.
19. Jim Karki, *Active Low-Pass Filter Design*, Texas Instruments SLOA049B, September 2002, https://www.ti.com/lit/an/sloa049d/sloa049d.pdf?ts=1712297805628&ref_url=https%253A%252F%252Fwww.google.com%252F.
20. Electronics Tutorials, *Butterworth Filter Design*, https://www.electronics-tutorials.ws/filter/filter_8.html.
21. Hank Zumbahlen, *Multiple Feedback Filters*, Analog Devices Mini Tutorial MT-220, https://www.analog.com/media/en/training-seminars/tutorials/mt-220.pdf.
22. T. Delyannis, High-Q Factor Circuit with Reduced Sensitivity, *Electronic Letters*, vol. 4, no. 26, December 1968, pp. 577–579.
23. J. Tow, Active RC Filters – A State-Space Realization, *Proceedings of the IEEE* (Letter), vol. 56, June 1968, pp. 1137–1139.

24. Robert R. Cordell, Build a High-Performance THD Analyzer, *Audio*, vol. 65, July–September 1981, cordellaudio.com.
25. J. Tow, A Step-by-step Active Filter Design, *IEEE Spectrum*, vol. 6, December 1969, pp. 64–68.
26. L.C. Thomas, The Biquad: Part 1 – Some Practical Design Considerations, *IEEE Transactions on Circuit Theory*, vol. CT-18(3), 1971, pp. 350–357.
27. L.C. Thomas, The Biquad: Part II – A Multipurpose Active Filtering System, *IEEE Transactions on Circuit Theory*, vol. CT-18(3), 1971, pp. 358–361.
28. P.E. Fleischer and J. Tow, Design Formulas for Biquad Active Filters Using 3 Operational Amplifiers, *Proceedings of the IEEE*, vol. 61, no. 5, May 1973, pp. 662–663.
29. J.J. Friend, C.A. Harris and D. Hilberman, STAR: An Active Biquadratic Filter Section, *IEEE Transactions of Circuits and Systems*, CAS-22, no. 2, February 1975, pp. 115–121.
30. P.E. Fleischer, Sensitivity Minimization in a Single Amplifier Biquad Circuit, *IEEE Transactions Circuits and Systems*, vol. CAS-23, no. 1, January 1976, pp. 45–55.
31. W. Emmanuel, S. Yu and F.D. Paluga, *Practical Treatise on the Tow-Thomas Biquad Active Filter*, CE170 Communications Systems Laboratory, August 2001. researchgate.net/publication/2487911_Practical_Treatise_on_the_Tow-Thomas_Biquad_Active_Filter.
32. Hank Zumbahlen, *Biquadratic (Biquad) Filters*, Analog Devices MT-205, analog.com/media/en/training-seminars/tutorials/MT-205.pdf.
33. J. Friend, C. Harris and D. Hilberman, STAR: An Active Biquadratic Filter Section, *IEEE* Transactions *Circuits and Systems*, vol. 22, no. 2, February 1975, pp. 115–121.
34. Sigfried Linkwitz, *12 dB/octave High-Pass Equalization (Linkwitz Transform)*, Linkwitz Lab, linkwitzlab.com/filters.htm.
35. HiFi Audio Design, *Linkwitz Transformation*, mh-audio.nl/Calculators/LinkwitzTransformation.html.
36. *True Audio's Linkwitz Transform Circuit Design Spreadsheet*, trueaudio.com>downloads>linkxfrm.
37. Bob Cordell, *Equalized Quasi-Sealed System (EQSS) – A Method for Achieving Extended Low-Frequency Response in a Loudspeaker System*, 2006, cordellaudio.com.
38. Robert R. Cordell, *System and Method for Achieving Extended Low-Frequency Response in a Loudspeaker System*, US Patent 7,894,614, February 22, 2011.
39. Bob Cordell, EQSS™ Equalized Quasi-Sealed System – Get Extended Low-Frequency Response and Higher SPL with EQSS™, *audioXpress*, September 2014.
40. Rod Elliott, *Active Filters Using Gyrators – Characteristics and Examples*, Elliott Sound Products, 2014, sound-au.com/articles/gyrator-filters.htm.
41. Wikipedia, *Negative Impedance Converter*. wikipedia.org.
42. Wikipedia, *Frequency Dependent Negative Resistor*. wikipedia.org.
43. *Twin T Notch Filter Mini Tutorial*, Analog Devices, MT-225, http://www.analog.com/media/en/training-seminars/tutorials/MT-225.pdf.
44. *The Twin-T notch (band-stop) Filter*, http://fourier.eng.hmc.edu/e84/lectures/ActiveFilters/node4.html.
45. Bob Cordell, *Designing Audio Power Amplifiers*, 2nd edition, Routledge, New York, NY, 2019.
46. Dick Moore, *Active Twin-T Notch Filter – A Path to High-Resolution Distortion Analysis*, December 2014, http://www.tronola.com/moorepage/Twin-T.html.
47. James R. Bainter, Active Filter has Stable Notch, and Response Can be Regulated, *Electronics*, October 2, 1975, pp. 115–117.
48. *Bainter Notch Filters*, MT-203, Analog Devices, www.analog.com.
49. Henry P. Hall, RC Networks with Single-Component Frequency Control, *IRE Transactions*, September 1955, pp. 283–284.
50. Kenneth A. Kuhn, *Design and Applications of the Hall Network*, July 4, 2011, kennethkuhn.com/electronics/design_and_applications_of_the_hall_network.pdf.
51. Kenneth A. Kuhn, *Rediscover the Truly Tunable Hall Network*, Electronic Design, January 24, 2012, electronicdesign.com/technologies/analog/article/rediscover-the-truly-tunable-hall-network.
52. Hank Zumbahlen, *All-Pass Filters*, Analog Devices MT-202, analog.com.
53. AES17, *AES Standard Method for Digital Audio Engineering–Measurement of Digital Audio Equipment*, Audio Engineering Society, 1998.
54. Bruce Hofer, *Measuring Switch-mode Power Amplifiers*, Audio Precision White Paper, October 2003.

Chapter 12

Distortion and Its Measurement

In this chapter we'll look at some distortion theory, some sources of distortion and some approaches to measuring distortion. Throughout the discussion it is important to distinguish between distortion mechanisms and distortion measurement techniques.

12.1 Nonlinearity and Its Consequences

Nonlinearity is the underlying cause of distortion. When a circuit parameter changes as a function of signal, nonlinearity exists. Stimulation of nonlinearity is also necessary for creation of distortion from that nonlinearity. Signal voltage or current is usually the stimulus at the location of the nonlinearity. The resulting distortion will usually be in proportion to the amplitude of the stimulus (or to a power of it). When the collector current of a transistor changes as a function of signal voltage, its transconductance changes and the gain of the stage changes with signal. The resulting distortion has been stimulated by voltage and current swing.

When the value of a capacitance changes as a function of signal voltage, distortion is created that is stimulated by the voltage. The consequences of that capacitance change are what a distortion test measures. If the capacitor is across a very low-impedance source, its signal-dependent capacitance change may not make much difference in the signal and measured distortion will be low. More often, the capacitance will be associated with a resistance, causing a pole in the circuit to move up and down in frequency. One distortion test may measure the time-varying frequency response that results, while another test might measure the time-varying signal phase that results. Yet another test might measure the harmonic frequencies that are created. The same underlying nonlinearity will cause distortion to be seen by many different types of measurement. This is a very important point and is sometimes misunderstood.

This is why it is virtually impossible to have one type of measured distortion without having another type of measured distortion. Having said that, it is important to realize that different distortion measurements can have vastly different sensitivities to the same nonlinearity, depending on how effective they are in stimulating (exercising) that nonlinearity and how sensitive they are in measuring the resulting distortion products. This is why there are numerous different types of distortion tests [1–17]. A given nonlinearity creates THD, SMPTE IM, CCIF IM, TIM, etc.; these are just different ways of stimulating the nonlinearity and measuring its consequences.

High-frequency distortion is a very good example. It is largely a function of the rate of change of the stimulus signal that is exercising the amplifier [7–10]. THD-20 and TIM are measuring the same nonlinearity by stimulating the amplifier with a high rate of change of signal voltage. Indeed, a given type of distortion measurement is really the observation of the symptoms of the nonlinearity. An analogy here would be that in medicine the presence of many diseases is inferred from symptoms or from the presence of antibodies, not necessarily the disease itself.

The Order of a Nonlinearity

As signal level is increased, the increase in distortion percentage is a function of the order of the distortion being considered. For example, with second-order distortion, a 1-dB increase in signal level will cause the magnitude of the second-order distortion product to go up by 2 dB. This means that the distortion expressed as a percentage will go up by 1 dB. For example, if second harmonic distortion is 1% at a fundamental level of 1 V_{RMS}, then the second harmonic will rise to 2% at a fundamental level of 2 V_{RMS}. A simple example of second-order distortion is when the incremental gain of the amplifier is higher for positive signal swings and lower for negative signal swings.

For third-order distortion, a 1-dB increase in fundamental level will result in a 3-dB increase in the product magnitude and a 2-dB increase in the distortion percentage. If the third-order distortion is −80 dB relative to the fundamental at a fundamental level of 0 dBV, it will rise to −78 dB relative to the fundamental at a fundamental level of +1 dBV. Distortion of order n will go up by $(n - 1)$ dB relative to the fundamental when the signal level is increased by 1 dB. A simple example of second-order (second harmonic) distortion is when the incremental gain of the amplifier is higher for positive swing than for negative swing. A simple example of third-order distortion is when the incremental gain of the amplifier is higher for small peak-to-peak signal swings and lower for large peak-to-peak signal swings, as in a compressive effect on a sine wave.

This known behavior can be helpful in inferring whether the distortion component being observed (as on a spectrum analyzer) is from the source or the *device under test* (*DUT*). Increase the level to the DUT by 1 dB. The magnitude of the third harmonic (for example) should go up by 3 dB. If it goes up by only 1 dB, it has probably originated in the source. Expressed in dB relative to the signal amplitude, third harmonic distortion in this case will go up by 2 dB.

12.2 Total Harmonic Distortion

Total harmonic distortion (THD) is one of the most common measures of distortion. It is based on the fact that when a sine wave encounters a nonlinearity, harmonics will be created at integer multiples of the fundamental frequency of the sinusoid. A 1-kHz sine wave fed through a second-order nonlinearity will create harmonics of varying amplitudes at 2 kHz, 4 kHz, 6 kHz, etc. A 1-kHz sine wave fed through a third-order nonlinearity will create harmonics of varying amplitudes at 3 kHz, 5 kHz, 7 kHz, etc.

In measuring THD, the amplifier under test is fed a low-distortion sine wave. The fundamental frequency of the sine wave is removed with a very narrow and deep notch filter. What remains after the notch filter is the residual, consisting of noise and harmonics [1]. This is why this measurement is usually referred to as *THD+N*. The residual is passed through a low-pass filter that limits the measurement bandwidth to typically 80 or 200 kHz. This improves the SNR of the measurement. The amplitude of the residual is displayed on the instrument meter as a distortion percentage and the residual signal is made available at an output jack for viewing with an oscilloscope or a spectrum analyzer [1]. With spectral analysis and computation, THD without noise can be computed.

Interpretation of THD

Single-number THD specifications are of limited use in characterizing the sound quality of an audio circuit because they do not convey the nature of the distortion. For example, they do not distinguish between low- and high-order distortions. Worse, single-number THD will often be quoted at 1 kHz, where it is easy to achieve very small distortion numbers. This is a big part of the reason why it is common in many circles to dismiss THD as having little or no relationship to perceived sound quality. Such a view paints THD with too broad a brush. THD is a much better indicator of amplifier performance when it is measured at high frequencies like 20 kHz (THD-20) and a full spectral analysis is presented of the amplitudes of the individual harmonics.

Advantages of THD Measurements

While not an air-tight guarantee of good sound, a very low THD number for an amplifier leaves little room for most other distortions to be present. By very low THD we mean THD well under 0.01% under all conditions with thorough testing and observation of the amplitudes of the harmonics. However, this is not the only path to good sound. Benign distortions that elevate THD readings may not audibly degrade sound quality. Second harmonic distortion would be an example. Vacuum tube circuits regularly have higher distortion, usually greater than 0.01% or much more, and yet many listeners like the sound.

Very low THD assures exceptional overall circuit linearity under static conditions. Low THD at all frequencies means that most measurements of other distortions will be very small as well. These include CCIF IM, SMPTE IM, TIM and PIM. It is very difficult for these other measured distortions to exist without there also being at least small amounts of THD. This is especially so if the THD residual has been evaluated on a spectrum analyzer and if the THD has been measured in a bandwidth that is at least ten times that of the fundamental.

Very low THD also assures that power supply coupling distortions due to limited PSRR are very small. Low THD also indicates that ground-induced distortions due to imperfect grounding are very small. Very low THD-20 suggests that parasitic oscillations are absent when driving the load used in the test setup. In thorough testing, these THD tests should be done with appropriate capacitive loads (200–500 pF for line-level outputs). The presence of parasitic oscillations usually causes subtle increases in THD. These subtle increases will go unnoticed in amplifiers with higher THD. A parasitic oscillation burst at the peak of a large test sine wave may increase the distortion to 0.01% when the same amplifier without the burst of oscillation might render a distortion measurement of 0.001%.

Very low 20-Hz THD strongly suggests that many low-frequency thermal distortions are absent. These include feedback resistor thermal distortion and transistor junction thermal distortion. It also suggests that some measurable capacitor distortions, such as from electrolytic capacitors at low frequencies, are absent. Very low THD+N at 20 or 50 Hz virtually guarantees that power supply ripple and its harmonics are not entering the signal path. These will show up in the distortion residual even though they are not harmonics of the fundamental signal stimulating the amplifier.

In general, the attention to design detail and implementation necessary to achieve very low THD across the entire audio band will tend to result in a better amplifier or other audio circuit (as long as something stupid is not done to achieve low THD at the expense of something else).

Limitations of THD Measurements

THD-20 is one of the tougher and more revealing distortion measurements that can be made on an audio circuit. However, the higher harmonics lie well above the audio band and many audio spectrum analyzers cannot display spectra above 50–100 kHz. If the THD analyzer has an 80-kHz filter engaged to improve its sensitivity by reducing out-of-band noise, these harmonics will be attenuated. Single-number THD measurements do not tell the whole story because they do not distinguish between the low-order nonlinearities and the more troublesome high-order nonlinearities.

THD+N measurements are of limited value at low signal levels because there is no way of knowing whether the reported THD+N level is primarily noise or crossover distortion. If THD+N is very low, it means that both THD and noise are both low. Spectral analysis is necessary to distinguish THD from noise at low signal levels.

Low THD does not assure the absence of many sonic shortcomings. For example, it does not reveal distortion resulting from thermal bias instability in a power amplifier. It also does not assure that there is no flabby low-end performance due to a sloppy power supply in a power amplifier. Nor does it assure the absence of sonic degradation due to poor transient performance and ringing.

Good THD readings do not assure that the preamplifier is reproducing music faithfully in the presence of EMI ingress. It does not address some linear and nonlinear distortions that are less well understood, such as the influence on sound quality of passive component quality. In fairness to THD, many other distortion tests also do not reveal these sonic shortcomings. Mid-band THD, like THD-1 at 1 kHz, can give a misleading impression of good amplifier performance. This may be the single biggest reason why some eschew THD measurements, claiming that it has little correlation with sound quality. The fact that some amplifiers with relatively high THD sound very good further contributes to this impression.

12.3 SMPTE Intermodulation Distortion

SMPTE intermodulation distortion (SMPTE IM) is another distortion measure that has long been in use. It is based on the observation that nonlinearity can be represented as a change in the incremental gain of a circuit as a function of instantaneous signal amplitude. This measurement is also referred to as amplitude intermodulation distortion (AIM) [18]. The dynamic changes in incremental gain are observed by creating a test signal with a small-amplitude high-frequency *carrier* on top of a large-amplitude lower-frequency signal.

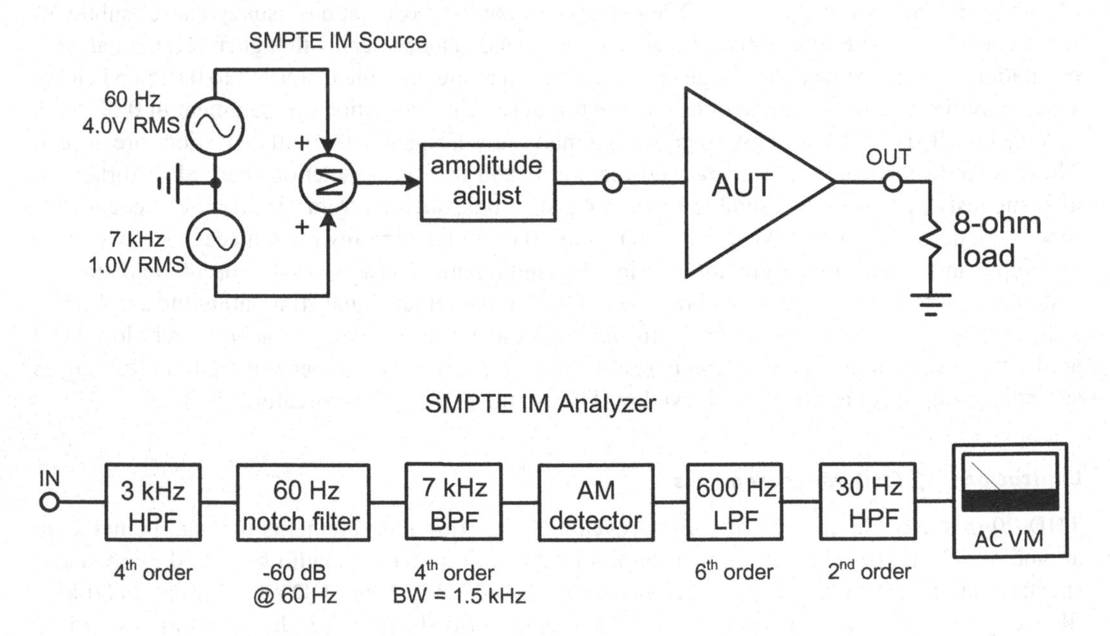

Figure 12.1 Measuring SMPTE Intermodulation Distortion.

After the test signal passes through the device under test (DUT), the low-frequency signal is filtered out and the smaller carrier signal is AM-detected. The SMPTE IM test employs test signals at 60 and 7000 Hz mixed in a 4:1 ratio. A typical SMPTE IM measuring arrangement is shown in Figure 12.1.

The sensitivity of the analyzer is largely determined by the rejection characteristics of the various filters. Many SMPTE IM analyzers use a conventional rectifying AM demodulator, but better ones employ *synchronous detection* where phase-locked loops are required to recover the 7-kHz carrier [18]. Synchronous detection does a far better job of keeping DUT noise out of the measurement.

12.4 CCIF Intermodulation Distortion

The CCIF IM distortion test takes advantage of the fact that if two tones are passed through a nonlinear circuit, spectral components will be created at frequencies $mf_1 \pm nf_2$ where f_1 and f_2 are the frequencies of the two tones and the numbers m and n are integer multiples. This test, often referred to as the twin-tone test, is usually conducted with equal-level sine waves at 19 kHz (f_1) and 20 kHz (f_2).

A second-order nonlinearity will produce a distortion component at 1 kHz, the difference of the two frequencies. Notice here that two frequencies are involved in the calculation of the distortion product frequency and that $m + n = 2$. This characterizes the nonlinearity as second order. The sum of m and n always designates the order of the nonlinearity that the spectral component represents. The second-order nonlinearity will also produce a spectral *line* at 39 kHz, representing the $f_1 + f_2$ component (19 + 20 kHz).

A third-order nonlinearity will produce components at $2f_1 - f_2 = 18$ kHz and $2f_2 - f_1 = 21$ kHz. A fifth-order nonlinearity will produce components at $3f_1 - 2f_2 = 17$ kHz and $3f_2 - 2f_1 = 22$ kHz. It is easy to see the progression as the order of the nonlinearity increases. Figure 12.2 shows a typical CCIF IM spectrum plot that illustrates the result when nonlinearities at second through seventh order are present.

Notice that the fourth-order nonlinearity shows up at $2f_2 - 2f_1 = 2$ kHz and that the sixth-order nonlinearity shows up at $3f_2 - 3f_1 = 3$ kHz. The CCIF IM test reflects even-order nonlinearities down to low frequencies. Early uses of the test simply employed a low-pass filter to attenuate the higher test frequencies so that the lower-frequency products could be measured with an AC

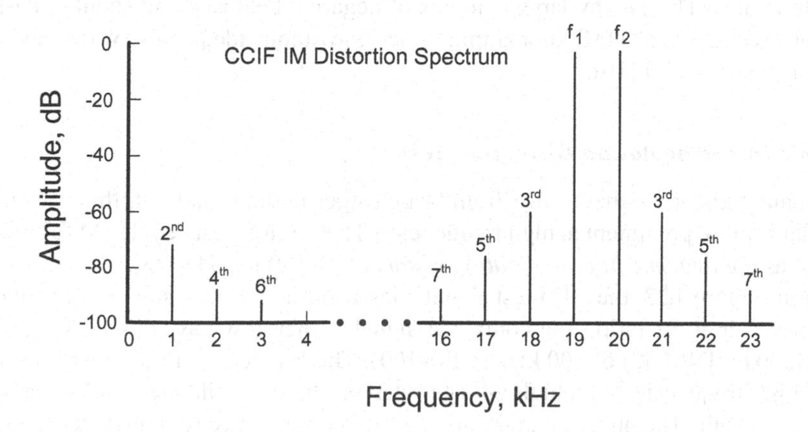

Figure 12.2 A CCIF IM Plot.

voltmeter, with particular emphasis on the second-order product at 1 kHz. Unfortunately this was a very incomplete test. Proper use of the CCIF IM test requires the use of a spectrum analyzer. This enables the odd-order distortion products to be viewed.

The great advantage of the CCIF IM test is that representatives of all of the distortion orders are present in-band, allowing the use of spectrum analyzers of modest high-frequency capability. The individual oscillators do not have to have very low distortion, and this is a major advantage of this test. However, the summing circuit where their outputs are combined to create the test signal must have very low distortion. Sometimes the frequencies 19.5 and 20.5 kHz are used. This frequency offset prevents the odd-order components that are reflected down in frequency from overlapping even-order distortion products that lie on 1-kHz intervals.

12.5 Transient Intermodulation Distortion

Transient intermodulation distortion (TIM) received a great deal of attention in the 1970s and early 1980s [2–10]. It is a distortion mechanism that is often described in time-domain terms. It has wrongly been blamed on the use of large amounts of negative feedback and small open-loop bandwidth. If an input signal to a feedback amplifier changes very quickly – too fast for the output of the "slow" amplifier to respond – the input stage may be overloaded and clip. The stage will be overloaded by the large error signal that arises before the feedback from the output catches up to the input. The overload occurs because the input stage of a feedback amplifier is not usually designed to be able to handle the full amplitude of the input signal, since under normal conditions it need only handle a much smaller error signal.

Slew Rate Limiting and Input Stage Stress

The *slew rate limiting* distortion mechanism was known many years before the term *TIM* was coined. TIM has in fact been described as slewing-induced distortion (SID) [7, 8]. *Hard TIM* occurs when the input stage clips and the amplifier is in slew rate limiting. *Soft TIM* occurs when the stress on the input stage increases as the slew rate limit is approached, resulting in increased input stage nonlinearity. It is important to recognize that TIM usually results from signal stress on the input stage.

Amplifiers with large amounts of negative feedback and small open-loop bandwidth can achieve very high slew rates. This is why large amounts of negative feedback and small open-loop bandwidth are not a root cause of TIM. Poor amplifier design without adequate slew rate and input stage dynamic range causes TIM [10].

The Dynamic Intermodulation Distortion Test

TIM is a dynamic distortion that results from fast changes in the signal. For this reason it is also a distortion that is more prominent at high frequencies. The original test developed for this distortion is referred to as the *dynamic intermodulation distortion* (*DIM*) test [11–13].

As shown in Figure 12.3, the DIM test signal consists of a 3.18-kHz square wave and a 15-kHz sine wave mixed in a 4:1 ratio. The combined signal is then low-pass filtered with a first-order network at 30 kHz (DIM-30) or 100 kHz (DIM-100). The fast edges of the square wave stress the amplifier at high frequencies with high voltage-rates-of-change while the 15-kHz carrier signal is modulated as a result. The output signal must be viewed on a spectrum analyzer, and the amplitudes of all of the relevant spectral lines must be added on an RMS basis and then referred to the

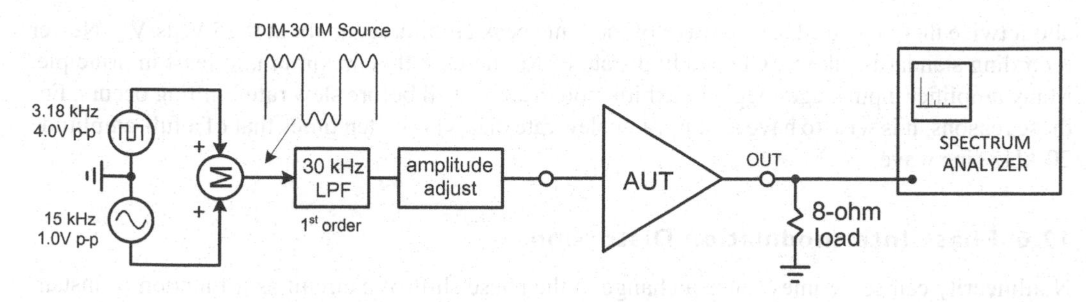

Figure 12.3 **The DIM Test for Transient Intermodulation Distortion (TIM).**

amplitude of the 15-kHz sine wave. Notice that the square-wave portion of the input signal contains components at the fundamental and its odd harmonics. Each of those harmonics can interact with the 15-kHz carrier on an $m \pm n$ basis. This means that nonlinearities of low order will still produce a very rich spectrum.

The dynamic range of the spectrum analyzer limits the measurement floor for the DIM-30 test. The vintage HP 3580A analog audio spectrum analyzer has a dynamic range of 85 dB on a good day. The 15-kHz component of the DIM-30 test signal is down 14 dB from the full p–p amplitude of the composite DIM-30 test signal. This means that the measurement floor is at about −71 dB, or about 0.03%. Many newer spectrum analyzers based on high-performance sound cards can do much better because the spectral components to be measured are in-band and the bandwidth limitations of sound cards that in some cases limit bandwidth to 20 kHz do not come into play.

THD-20 Will Always Accompany TIM

With some caveats, I have never seen an amplifier that had measurable TIM that did not also have measurable THD-20. The essential difference between these tests is that TIM exercises the amplifier with a high peak rate of change that has a small duty cycle, while THD-20 exercises the amplifier with a smaller rate of change but with a much larger duty cycle. The peak slew rate for a 20 kHz sine wave is 0.125 V/μs per volt of peak signal amplitude ($V/\mu s/V_{pk}$). The peak slew rate for the DIM-30 test signal is 0.319 $V/\mu s/V_{pk}$. A 20-kHz sine wave at 10 V_{RMS} has a slew rate of 1.8 V/μs. This is definitely not to suggest that 1.8 V/μs is adequate for a line-level audio circuit.

THD-20 will tend to track DIM fairly well as long as the amplifier is not pushed into slew rate limiting (hard TIM) by the higher peak slew rate of the DIM test signal. THD-20 will often be about 4–7 dB below the DIM-30 number [19]. However, the measurement floor with a good THD analyzer (about 0.001% or −100 dB) lies well below that of the DIM-30 test. Because of the large rate of change applied to the amplifier in order to stimulate the high-frequency nonlinearity, it is possible for a DIM test to cause an amplifier to go into slew rate limiting when a THD-20 test would not. In this case it is possible to have large amounts of DIM with fairly small amounts of THD-20. More often, low THD-20 virtually guarantees the absence of TIM, especially if a line-level amplifier is known to have a healthy slew rate of 20 V/μs or more.

Recommended Slew Rate

The maximum slew rate from a CD source is limited by the very steep anti-alias filtering required by the *Red Book* standard for audio CDs. A square wave recorded on a CD will have a slew rate of about twice that of a 20-kHz sine wave of the same peak amplitude, or about 0.25 V/μs/V_{pk}. Newer recording standards, like SACD and high-rate PCM, increase this maximum, at least in principle. Many amplifier input stages begin to exhibit nonlinearity well before slew rate limiting occurs. For these reasons, it is wise to have an amplifier slew rate that is about ten times that of a full-amplitude 20-kHz sine wave.

12.6 Phase Intermodulation Distortion

Nonlinearity can sometimes cause a change in the phase shift of a circuit as a function of instantaneous signal amplitude. This distortion mechanism is called phase intermodulation distortion (PIM) [15–18]. Some have also called it *FM distortion* (phase modulation is simply the integral of frequency modulation).

The PIM measurement is analogous to SMPTE IM (AIM), but involves phase modulation instead of amplitude modulation. 60-Hz and 7-kHz test signals are mixed in a 4:1 ratio and applied to the DUT. The amplitude of the phase modulation on the 7-kHz carrier is then measured. A simple example of a circuit that creates PIM is a nonlinear capacitance in an R-C low-pass filter arrangement. If the capacitance changes as a function of signal amplitude, the corner frequency of the low-pass filter will change and, correspondingly, the phase shift of the filter will change.

PIM is also created by AIM in a feedback amplifier. If the signal amplitude modulates the gain of the input stage, the changing open-loop gain of the amplifier will result in a changing closed-loop bandwidth, as illustrated in Figure 12.4. Just as in the passive case above without feedback, the movement of the closed-loop −3 dB frequency will cause a change in phase shift, which corresponds to PIM.

This is an example of amplitude-to-phase conversion caused in this case by the action of the negative feedback loop. Only a portion of the AIM is converted to PIM, leaving AIM as well as PIM. For this reason, an amplifier of conventional design cannot exhibit PIM without also exhibiting AIM. The measurement of AIM and PIM is not new to the analog video world. There it is referred to as *differential gain and phase,* referring simply to gain and phase shift that are a function of signal.

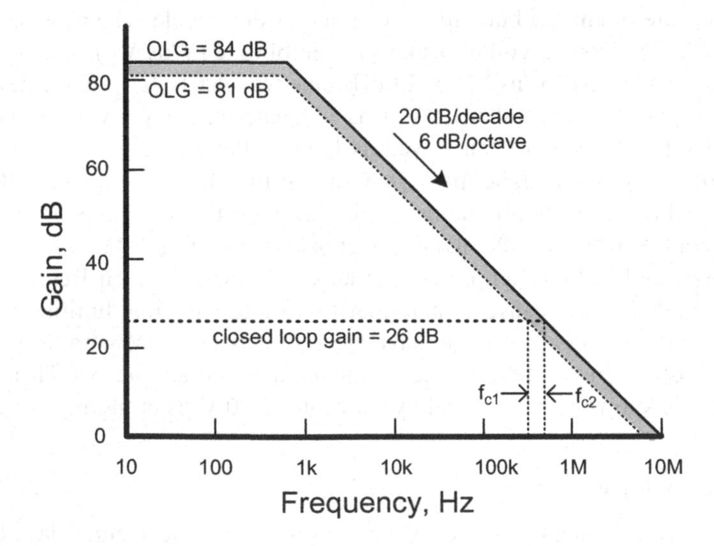

Figure 12.4 Creation of PIM by Modulation of the Open-Loop Gain.

Measuring PIM

PIM is measured in much the same way as AIM, but with a phase detector substituted for the AM detector. The test signal for PIM and AIM is the same [15–17]. A large low-frequency signal, typically 60 Hz, is used to cause the operating points in the amplifier to traverse a large-signal range. A smaller carrier signal, typically four times smaller and at a high frequency like 7 kHz, then has its amplitude and phase modulations measured. Modulation on the smaller 7-kHz signal at 60 Hz or its harmonics is either AIM or PIM depending on whether amplitude or phase detection is used.

This is a more challenging measurement that usually requires a phase-locked-loop and synchronous detection, as described in Ref. [18]. The arrangement is referred to as a coherent IM analyzer [18]. The instrument uses in-phase and quadrature coherent detection to extract both conventional SMPTE IM (AIM) and PIM, respectively. The analyzer that was built had a PIM measurement floor of 1.2 ns_{RMS}.

The 19 + 20 kHz CCIF test is also sensitive to PIM. Phase and amplitude distortions both create spectral lines at frequencies that have an $mf_1 \pm nf_2$ relationship. For this reason, an amplifier with PIM will exhibit spectral lines using the 19 + 20 kHz CCIF test. However, one will not be able to distinguish PIM from AIM with this test.

Negative Feedback and PIM

If the open-loop gain of a feedback amplifier is a function of signal swing (this is AIM), the gain crossover frequency of the global negative feedback around the amplifier will change. If the open-loop gain decreases, the gain crossover frequency and the closed-loop bandwidth of the amplifier will become smaller. When this happens, the amplifier will exhibit slightly higher in-band phase lag. This is PIM. The negative feedback has effectively caused some of the AIM to be converted to PIM. Put simply, movement of the closed-loop pole of a feedback amplifier is the source of feedback-generated PIM. Even though the closed-loop pole is well above the audio band (typically above 100 kHz), it causes a small amount of phase lag in-band, whose change creates PIM. For this reason, PIM has been blamed on negative feedback.

Moreover, PIM has wrongly been blamed on low open-loop bandwidth [15]. It seems intuitive that if movement of the closed-loop pole causes PIM, then the lower the frequency of the open-loop pole, the worse the PIM. This is not the case. Consider a conventional Miller-compensated amplifier. Input stage *gm* and the size of the Miller capacitor set the gain crossover frequency for a given closed-loop gain. If you simply look at the calculation for gain crossover frequency as a function of input stage *gm* and Miller capacitor, you see that the open-loop bandwidth is not in the picture. Open-loop bandwidth does not play a role in determining the gain crossover frequency for a given nominal gain crossover frequency. For this reason, low open-loop bandwidth does not contribute to PIM.

PIM can be expressed in RMS degrees or in RMS nanoseconds. The amount of PIM generated by a given amount of input stage nonlinearity is mathematically derived in Ref. [18].

Input Stage Stress

Like TIM, PIM results from input stage stress that causes changes in the incremental gain of the input stage. In fact, for a given amount of negative feedback at 20 kHz, high feedback and correspondingly low open-loop bandwidth actually reduce input stage stress and therefore AIM and PIM. Keep in mind that the error signal at the input stage of a feedback amplifier is smaller if the open-loop gain is larger.

PIM in Preamplifiers without Negative Feedback

Preamplifiers without negative feedback can also create PIM. One common source of PIM in a preamplifier without negative feedback is nonlinear junction capacitance that changes the bandwidth, and thus the phase shift, of the preamplifier as a function of signal swing. Miller effect in the VAS from nonlinear collector-base junction capacitance is an example. Feedback amplifiers thus have PIM in the open loop even before negative feedback is applied. Interestingly, the application of negative feedback reduces this component of PIM in the same way it reduces distortion from any other open-loop nonlinearity.

The results in Ref. [18] show that PIM is not a problem in most contemporary circuits and that negative feedback reduces total PIM in most cases. The instrument built in Ref. [18] has a PIM measurement floor of 1.2 ns_{RMS}. A power amplifier of ordinary design tested in Ref. [18] had PIM of less than 3 ns. The power amplifier of Ref. [19] was also measured for PIM. It has very high negative feedback and very low open-loop bandwidth and measures less than 0.001% THD-20. It had measured PIM of only 0.04 ns (40 ps). That measurement required the use of a spectrum analyzer connected to the residual output of the coherent IM analyzer [18].

12.7 Multitone Intermodulation Distortion (MIM)

The *multitone intermodulation* (*MIM*) distortion test employs three tones instead of the two tones used in the CCIF test [14]. This allows odd-order distortion products to be reflected to a low frequency in the audio band so that they can be measured without the need for a spectrum analyzer. This is also referred to as a *triple-beat* test because three tones create beat frequencies among them when they interact with nonlinearities. Three equal-level tones at 20.00, 10.05, and 9.00 kHz are fed to the DUT. If these three frequencies are designated f_a, f_b and f_c, then an even-order nonlinearity will produce a beat frequency distortion product at $f_b - f_c = 1050$ Hz and an odd-order nonlinearity will produce a distortion product at $f_a - f_b - f_c = 950$ Hz.

The MIM distortion products can be measured with simple equipment. The output of the DUT is passed through a sixth-order 2-kHz LPF to discard the three high-frequency test tones. The signal is then passed through a fourth-order 1-kHz bandpass filter with a bandwidth of 150 Hz, which is wide enough to pass the distortion product frequencies at 950 and 1050 Hz. This arrangement provides very good sensitivity because the noise bandwidth of the analyzer is quite small. Details of the test setup and how well it compares to THD-20 and DIM-30 tests can be found in Ref. [14]. MIM produces smaller distortion percentage numbers, but it has a correspondingly lower measurement floor. MIM is largely sensitive to distortion products of only second and third order.

12.8 Highly Sensitive Distortion Measurement

Highly sensitive distortion measurements can be made using several pieces of test equipment together. The simplest example is the analysis of the residual output of a THD analyzer with a spectrum analyzer. Figure 12.5 illustrates how a combination of instruments can be used to obtain very good measurements. A distortion magnifier (DM) is placed in front of the THD analyzer to increase its dynamic range and measurement floor by 20 or 40 dB [20]. It does this by subtracting the sine-wave source from the appropriately scaled output of the DUT to produce a deep null at the fundamental frequency. A phase adjustment alters the phase of the source signal to account for phase shift in the DUT.

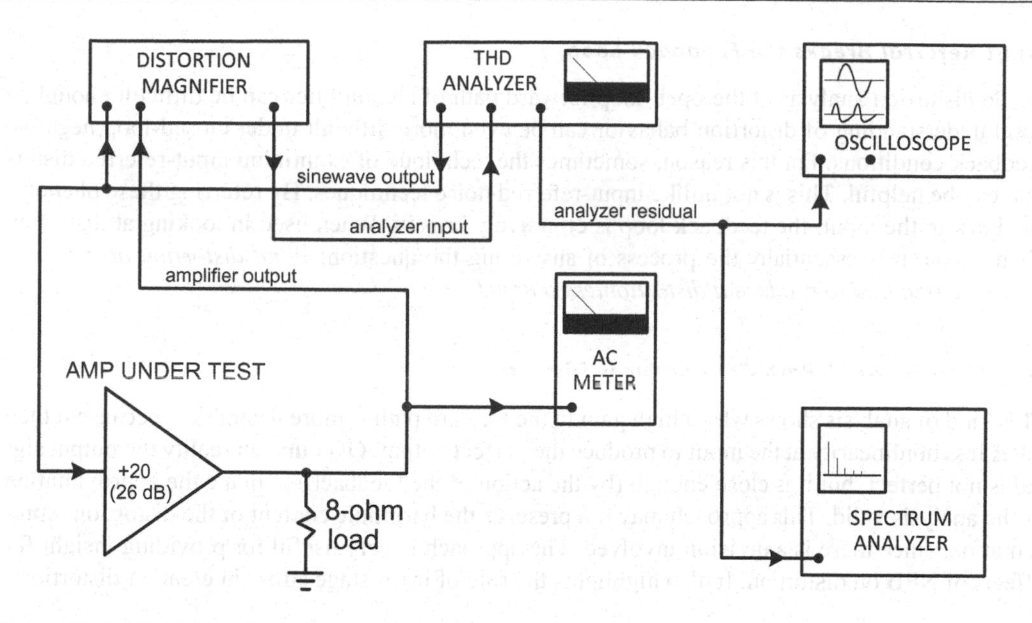

Figure 12.5 Highly Sensitive Distortion Measurement.

A chosen amount of the fundamental is then added to the nulled signal to provide a reference signal for the THD analyzer and to provide an amplitude reference. The amount of the source signal re-introduced is down either 20 or 40 dB, corresponding to distortion magnification of 10× or 100×. However, this does not improve the noise floor as established by noise in the DUT. The spectrum analyzer is connected to the residual output of the THD analyzer. The small noise bandwidth afforded with spectrum analysis eliminates most of the DUT and THD analyzer noise from the measurement. The complete setup can provide a measurement floor well below −120 dB.

The spectrum analyzer can be one based on a PC and soundcard. If the DM is used to test a device with a flat response within and beyond the full audio range, the DM will tend to cancel out harmonics of the test source as well. However, the capabilities of the DM are limited with regard to testing devices with a non-flat frequency response, such as equalizers. Notice in Figure 12.5 that even without the DM in the signal path, the use of a spectrum analyzer on the residual output of the THD analyzer greatly improves dynamic range.

12.9 Input-Referred Distortion Analysis

The effect of negative feedback on distortion is most easily understood by working backward from the output. We assume a perfect output and evaluate the input-referred distortion required to generate that perfect output, just as we do in calculating input-referred noise. Because the feedback signal under these conditions is perfect, the level of the input-referred distortion is the same for either open-loop or closed-loop conditions. Distortion percentage is reduced by feedback simply as a result of the larger pure component of the input signal required under closed-loop conditions. This technique is quite accurate when the referred distortion products are small compared to the total closed-loop input, that is, when closed-loop distortion is small. It is important to remember that the gain involved in referring a distortion product back to the input may be a strong function of frequency.

Input Referral Breaks the Feedback Loop

While distortion analysis of the open-loop forward path of an amplifier can be difficult enough, a good understanding of distortion behavior can be even more difficult under closed-loop negative feedback conditions. For this reason, sometimes the technique of examining input-referred distortion can be helpful. This is not unlike input-referred noise techniques. By referring these phenomena back to the input, the feedback loop is effectively broken. When used in looking at distortion phenomena, it is essentially the process of answering the question: *What distortion at the input would be required to produce a distortionless output?*

Why High Forward-Path Gain Reduces Distortion

This kind of analysis shows why a high gain in the forward path is more desirable – because it then takes less nonlinearity at the input to produce the perfect output. Of course in reality the output signal is not perfect, but it is close enough (by the action of the feedback) to make the approximation of the analysis valid. This approach may not preserve the harmonic content of the distortion representation, since there is a division involved. The approach is very useful for providing insight for effects of NFB on distortion. It also highlights the role of input stage stress in creating distortion.

12.10 Other Sources of Distortion

The discussions on distortion in this and other chapters have so far focused on well-understood distortions that are caused by nonlinearities largely in semiconductor devices. Examples are the nonlinear relationships of current to base or gate voltage, Early effect wherein transistor current gain is a function of collector voltage or junction capacitance that is a function of reverse bias.

These kinds of nonlinearity produce many types of distortions. Harmonic and CCIF IM distortion are just different ways of measuring the symptoms of such nonlinearities. This is also the case with many other distortions, such as TIM [14]. Here we focus on other sources of distortion that must be considered as well. In many cases, these distortions result from the way in which the circuits are implemented, either as a result of other component imperfections or as layout-related matters.

In many cases, minimizing these distortions requires a great amount of attention to detail, such as layout, ground topology and power supply bypassing. It is a shame that in some cases these sources of distortion can ruin an otherwise finely designed audio device. Although some of these sources of distortion have been touched on in other chapters, they can often easily be overlooked. For that reason, each will be discussed here.

Distortion Mechanisms

Nonlinear distortion occurs as a result of a variety of sources, but these sources often share many of the same distortion-creating mechanisms. In the most general terms, distortion occurs when one or more parameters of an active or passive device change as a function of the signal. Examples include changing transistor current gain as a function of signal voltage and changing junction capacitance as a function of signal voltage. If a resistance changes as a result of signal-dependent temperature, this will also create nonlinear distortion.

Another trait of nonlinear distortion is that small-signal characteristics of the circuit change as a function of signal excursion. The most common example is when the incremental gain of the amplifier changes over the signal swing. If the gain is higher on positive swings and lower on negative swings, this corresponds to second-order distortion. If the gain is smaller for large signal deviations from zero, this corresponds to third-order distortion. If instead the frequency response or phase response of the amplifier changes with signal, this is also a form of nonlinear distortion. Anything that changes as a function of signal is a likely source of distortion.

12.11 Early Effect Distortion

The current gain of a transistor is mildly dependent on the collector-base voltage. This is referred to as the Early effect. To the extent that the gain of an amplifier stage is dependent on the current gain of the transistor, the Early effect can cause distortion. The Early effect was discussed in Chapter 7. Early effect distortion is minimized by employing cascodes (which reduce signal-dependent collector-base voltage changes) or by designing stages whose gain is less dependent on transistor current gain. A common-emitter stage preceded by an emitter follower suffers less Early effect distortion because beta changes in the common-emitter (CE) transistor have less influence.

12.12 Junction Capacitance Distortion

The capacitance of p-n junctions in transistors and diodes is a function of voltage. This means that the frequency response of a gain stage may change with signal voltage. This in turn means that the gain at high frequencies can be modulated.

A good example of this distortion is the Miller effect in a gain stage. A larger value of collector-base capacitance will reduce high-frequency gain of the stage. In some amplifiers with simple Miller feedback compensation, the collector-base capacitance of the VAS transistor forms a part of the total compensation capacitance. The gain crossover frequency then becomes a function of signal, resulting in small in-band gain and phase changes that are signal-dependent [18]. Consider the collector-base capacitance of the 2N4401 NPN small-signal transistor. It is 9 pF at 0 V, 4.5 pF at 5 V and 1.5 pF at 50 V.

It is also important to recognize that the gate of a JFET forms a reverse-biased p-n junction with the source-drain channel of the JFET. This junction will also be subject to modulation of its capacitance by the signal voltage.

12.13 Grounding Distortion

It is very important that bypass capacitors not form AC ground loops that can destroy a star-grounding architecture at high frequencies. Always remember that signal currents (including nonlinear signal currents) will tend to take the path of least impedance. Sometimes the strategic placement of a small resistance will effectively break a ground loop and make a great improvement. Sometimes a bypass capacitor can dump nonlinear currents into a sensitive analog ground if grounding topology is not optimized.

12.14 Power Rail Distortion

All of the stages in a preamplifier have some amount of power supply rejection ratio (PSRR). This describes the degree to which the stage can suppress the influence of noise and distortion on the power supply lines on the signal output of the stage. The power supply rail will often contain a mixture of rectifier ripple, noise, hum and distortion components.

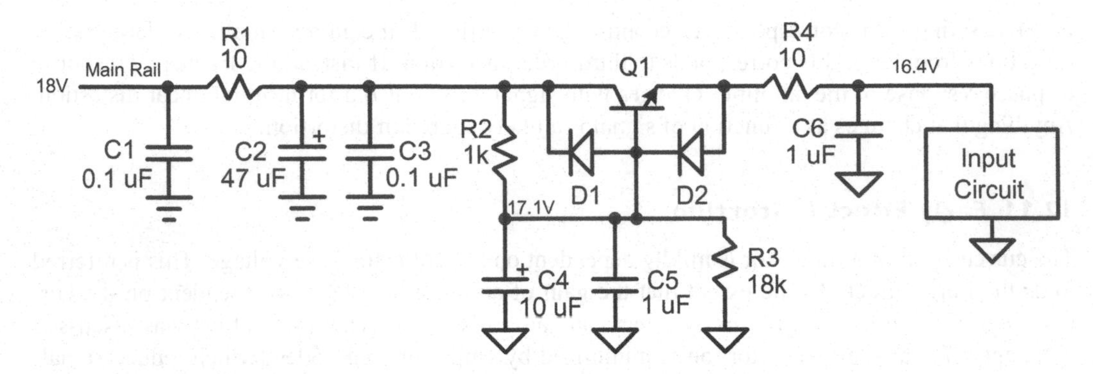

Figure 12.6 **Filtering the Power Supply with a Capacitance Multiplier.**

There are two basic ways to reduce distortion from the power rails. The first is to minimize the garbage appearing on the power rails. This can be done by employing generous R-C filtering in the power supply lines and by regulating the power supply lines. Sometimes a *soft* form of regulation is implemented locally with a pass transistor acting as an emitter follower for the power supply voltage. This will often take the form of a *capacitance multiplier*. Such an arrangement is shown in Figure 12.6. A nice feature of this arrangement is that the filtering capacitors C4 and C5 can have their ground referenced to the quiet ground of the circuit that will be employing the resulting power rail. D1 and D2 protect the transistor in the event that the main rail voltage falls below the regulated rail voltage. R4 isolates emitter follower Q1 from the capacitive load.

The second approach is to design the individual circuit stages to have higher PSRR out to higher frequencies. In many cases the PSRR of a stage will fall with increasing frequency. Differential, push-pull and cascode circuits often exhibit better power supply rejection up to higher frequencies.

12.15 Input Common-Mode Distortion

An often-neglected source of distortion results from the common-mode signal swing on the input differential pair of a non-inverting amplifier stage. The common-mode voltage is simply equal to the input signal swing, which may be on the order of a couple of volts.

The distortion can result from the collector-base voltages of the LTP (long tailed pair differential amplifier stage) changing with signal, causing beta to change via the Early effect. It can also result from changes in the collector-base junction capacitances as a result of their voltage dependence on V_{cb}. Nonlinear output impedance of the tail current source may also create common-mode distortion. Techniques for minimizing input common-mode distortion include using cascodes in the input stage and in some cases bootstrapping those cascodes to move with the common-mode signal. Cascoding the tail current source can also help.

An input stage that has better common-mode rejection will tend to have smaller common-mode distortion. Conversely, steps taken to reduce common-mode distortion will often improve common-mode rejection. Noise on power supply rails is often seen as common-mode noise by the input stage. For this reason, input stages with good common-mode rejection will tend to have better PSRR.

Testing for Common-Mode Distortion

One way to test for the presence of common-mode distortion in an op amp is to drive the amplifier in an inverting configuration, with no signal applied to the conventional non-inverting input. In the inverting mode, there is no common-mode signal swing on the input LTP. If distortion is significantly reduced under these conditions compared to when it is in a non-inverting configuration, it is likely that there is common-mode distortion.

12.16 Resistor Distortion

Some resistors can change their value slightly as a function of the voltage across them or the current through them [21]. These effects are quite small and very difficult to measure in most resistors of reasonable quality. In some cases carbon composition resistors will show this effect. Sometimes the interface between the material of the resistor element and the resistor leads can develop some nonlinear voltage drop.

The feedback network resistors in an amplifier are especially important in any stage because they play a direct role in setting the gain of the stage. There is nearly a one-to-one dependence of gain on resistor value; a 1% change in resistance will usually cause a 1% change in gain. The sensitivity of gain to change in the value of other resistors in the open-loop portions of the amplifier is usually much less (it is usually reduced by the feedback factor).

All resistors have a temperature coefficient of resistance. In many cases this might be on the order of ±100 ppm/°C. Because the body of the resistor has a thermal time constant (its temperature will not change instantly), the effect of signal amplitude on resistance through heating will usually take place at low frequencies. This effect is particularly important in the feedback network resistors because the full signal swing of the output of the stage is across the network. This may cause significant short-term power dissipation and signal-dependent heating of the resistor. This effect is exacerbated when smaller values of feedback network resistance are used in pursuit of lower input stage noise [19]. This effect is found mainly in power amplifiers where the feedback network resistors may be exposed to relatively high voltages or currents.

Suppose a resistor's temperature swings by 10°C peak to peak. If the resistor has a TC of 100 ppm/°C, then the resistance will swing by 0.1% peak to peak, and distortion on the order of 0.05% may result. Suppose we have a feedback resistance of 1 kΩ and the amplifier output swings by 10 V_{pk}. Assume that the gain is high so that voltage drop across the feedback shunt resistor can be ignored. This will result in a peak current of 10 mA and peak power dissipation of 100 mW. Almost all of this power dissipation is in the series feedback resistor. It is easy to see how this could swing the temperature of the resistor by several degrees Celsius at sufficiently low frequencies. In an experiment, a 6.81-kΩ ¼-W metal film resistor subjected to 40 V DC dissipated 230 mW; its resistance changed by 0.06%.

Figure 12.7 shows the measured distortion of an ordinary ¼-W metal film resistor as a function of frequency. The 1-kΩ resistor was subjected to an average AC power dissipation of ¼-W. Distortion was measured by forming two 20 dB attenuators. One employed the resistor under test in the series arm and the other employed a 2-W version of the same resistor in its series arm. The difference of the outputs of the two attenuators was then amplified and analyzed. This technique virtually eliminates the influence of distortion from the driving amplifier and enhances the sensitivity of the spectrum analyzer used for measurement.

Resistor distortion was primarily of third order and decreased with increasing frequency, as expected. The key thing to keep in mind is that the resistor distortion in Figure 12.7 is mainly as a result of the temperature coefficient of resistance in combination with signal-dependent swings in power dissipation. These temperature-dependent effects can be reduced by employing resistors with lower TC and higher power rating in critical portions of a circuit, like the negative feedback path.

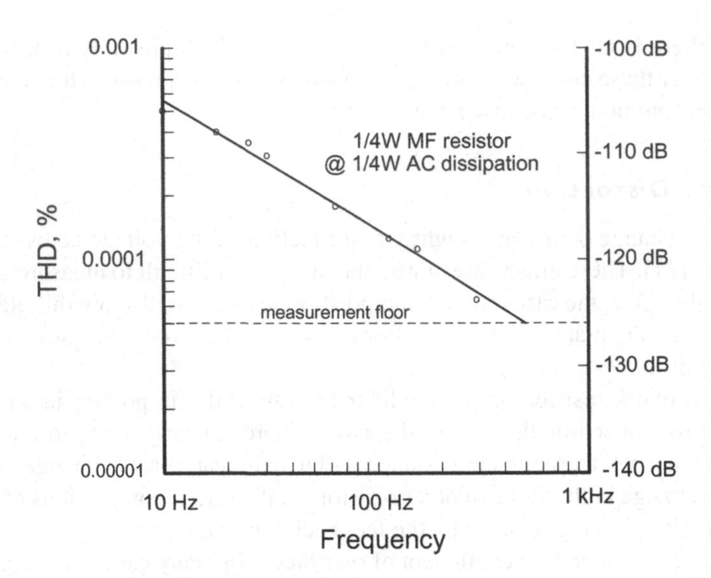

Figure 12.7 THD of ¼ W Resistor vs Frequency at ¼-W Dissipation.

These effects will be much smaller in line-level amplifiers than in power amplifiers. But remember, dissipation with 10 V_{RMS} across a 600-Ω feedback resistor may not be trivial, at approximately 167 mW.

The lesson here is to employ resistors of larger power dissipation in the feedback network, especially in the series feedback resistor. In some cases series-parallel combinations of resistors can be employed to achieve higher total power dissipation. Series combinations are superior because they also reduce the voltage swing on each resistor, mitigating voltage-dependent resistor distortions. Although often not practical, a nearly distortionless N:1 feedback divider can be made with N identical resistors, where $N - 1$ resistors in series form the feedback resistor. This will cancel voltage- and dissipation-related distortions. Such an approach is more likely to be important in a power amplifier design.

12.17 Capacitor Distortion

Capacitor distortion and the importance of capacitor quality in audio circuits has been studied by many [21–25]. Some capacitors change their capacitance as a function of voltage. This is especially so for electrolytic capacitors and is also a problem with certain types of ceramic capacitors. Voltage-dependent capacitance variation will cause distortion at frequencies where the value of capacitance plays a role in setting the gain. For example, variation in the value of a coupling capacitor will change the 3-dB low-frequency cutoff frequency of the circuit, resulting in signal-dependent gain and phase at very low frequencies. Similarly, a small-value Miller compensation capacitor will alter the high-frequency response of an amplifier if its value changes with signal. The Miller compensation capacitor usually has the full signal swing across it.

Quality film capacitors have virtually no voltage dependence of capacitance. Ceramic capacitors with an NPO (COG) dielectric have very little voltage dependence, but ceramic capacitors with other dielectrics, like X7R, may have significant voltage dependence and should be avoided. Capacitors with significant dielectric absorption have also been associated with impaired sound quality [23–25]. Dielectric absorption was discussed in Chapter 4.

The biggest offender in most non-inverting amplifier stages is the electrolytic capacitor commonly found in the shunt leg of the negative feedback network. Electrolytic capacitors have numerous forms of distortion. If electrolytic capacitors must be used in this location, large-value non-polarized (NP) types should be employed. The use of a large value will reduce the amount of signal voltage that appears across the capacitor with program material, reducing somewhat the resulting distortion.

Placing conventional polarized electrolytic capacitors of twice the required value in series instead of in parallel can be better for reducing distortions. This is similar to using an NP electrolytic capacitor, but the polarized capacitors may be more readily available [24]. Of course, one must keep in mind that placing capacitors in series reduces the total value of capacitance. Applying a small bias voltage at the junction of the capacitors through a high-value resistor can sometimes help as well.

These capacitors should be bypassed with a film capacitor of at least 1 μF. The low ESR of the film capacitor will effectively short out the electrolytic capacitor at very high frequencies, limiting the damage done by the electrolytic to lower frequencies. This will also limit EMI developing across the electrolytic due to its ESR/ESL. Expectations for improvements in the audio band must be tempered by the realization that the impedance of a 1-μF capacitor at 20 kHz is still 8 Ω.

Back-to-back silicon diodes will often be placed in parallel across the electrolytic capacitor to protect it and the input stage from a large DC voltage if the circuit output goes to the rail. This is more important in power amplifiers. At very low frequencies some signal voltage will appear across the capacitor and these diodes depending on the cutoff frequency established by the capacitor and feedback network. Some distortion at these frequencies may result from small amounts of conduction by the diodes if peak signal voltages of more than 200 mV or so appear across the diodes. Use of a larger value of capacitance obviously reduces this effect.

Curve (a) in Figure 12.8 shows the THD of the voltage across a 100-μF, 16-V polarized electrolytic capacitor as a function of frequency. The capacitor is driven by a 5-kΩ resistor with a 20-V_{RMS} input, corresponding to a 50-W amplifier with a low-value feedback resistor. Signal current in the

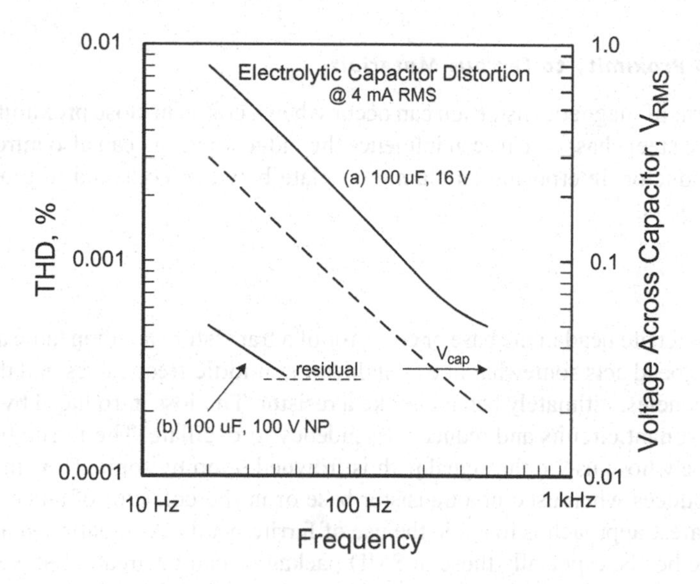

Figure 12.8 THD for a 100-μF Non-Polarized Electrolytic Capacitor.

capacitor is 4 mA_{RMS}. THD reaches almost 0.01% at 20 Hz. Keep in mind that this is the distortion of the voltage across the capacitor; the resulting amplifier distortion will be smaller. The signal voltage across the capacitor at 20 Hz is about 300 mV. The frequency response of an amplifier built with this arrangement would be down by 3 dB at about 6 Hz.

Curve (b) in Figure 12.8 shows the result when a 100-μF, 100-V audio-grade NP electrolytic capacitor is substituted. The capacitor measured is designed for use in loudspeaker crossovers and is not expensive. Its distortion is about 25 dB less than that of the low-voltage polarized capacitor. If you must use an electrolytic capacitor in the feedback network, use an NP capacitor with a substantial voltage rating.

12.18 Inductor and Magnetic Distortions

The use of inductors is frowned upon in audio preamplifiers, so few if any are found in the signal path. An exception is where they are used to implement a passive filter or equalizer. The absence of inductors does not mean that there are no nonlinear magnetic effects, so other magnetic effects will be examined here as well.

Magnetic Core Distortion

Magnetic materials have a magnetization curve that is nonlinear. The degree to which a magnetic material can be magnetized is ultimately limited. In an inductor or a transformer, this is referred to as *core saturation*. If the magnetization of the core approaches the saturation point, the inductance decreases. The simplest example of this is when an inductor is carrying DC current. As the current increases, the inductance will decrease. In general, the inductance of coils with ferrous cores is nonlinear. Great caution should be used if surface-mount (SMD) inductors or ferrite beads are used for filtering or EMI suppression in power supply circuits, as their effectiveness decreases greatly when passing modest amounts of current. Discrete ferrite beads or cores should be made to have zero net DC current flowing through them by passing both polarities of the supply leads through the same bead or core.

Distortion from Proximity to Ferrous Materials

A more subtle form of magnetic distortion can occur when a coil is in close proximity to a magnetic material, such as a steel chassis. This can influence the inductance and can also introduce nonlinear losses into the inductor. Interposing an aluminum plate between coils and ferrous materials can reduce this effect [21].

Ferrite Beads

The insertion of a ferrite bead in the base or collector of a transistor can often tame a parasitic oscillation. The ferrite bead acts somewhat like an inductor at middle frequencies and then experiences loss at high frequencies, ultimately behaving like a resistor. The loss introduced by the ferrite bead tends to damp resonant circuits and reduce the tendency to oscillate. The ferrite bead is a nonlinear passive device whose use in the signal path is frowned upon by some. How much measurable distortion it introduces when used in a transistor base or in the collector of an emitter follower is debatable. The safest approach is to avoid the use of ferrite beads. As mentioned above, the effectiveness of ferrite beads, especially those in SMD packages, can be greatly reduced if DC current is passed through them. Always consult the datasheet when using these devices.

12.19 EMI-Induced Distortion

Amplifier circuits are subject to *radio frequency interference* (*RFI*), more commonly referred to as *electromagnetic interference* (*EMI*). Electric drills, light dimmers, radio stations, WIFI and cell phones are common sources of EMI. Amplifiers are susceptible to EMI through three conductive ports: the input port, the output port and the mains port. If EMI gets into an amplifier, it can show up as noise or buzzing which alone is objectionable. This is not distortion as such. It is additive noise. However, if the EMI disturbs amplifier stage operating points or intermodulates with the audio signal, then nonlinear distortion can result. Any EMI effect that is correlated with the audio signal will be perceived as nonlinear distortion.

12.20 Thermally Induced Distortion

Whenever a temperature change can affect the gain or operating point of a circuit, distortion may result when the program material contains low frequencies or when the short-term program power changes with time [27]. This is sometimes called memory distortion.

This type of distortion was discussed in the case of resistor distortion. If the gain of an amplifier stage changes with temperature, then it can also cause distortion at low frequencies or intermodulation with higher frequencies. Thermal distortion is often much lower at high frequencies because the thermal inertia of the affected elements prevents the temperature from changing too much with faster signal swings. However, it should be kept in mind that the local thermal time constant of a transistor junction can be on the order of milliseconds.

Memory distortion will not show up with conventional steady-state tests like THD-1 and THD-20. Memory distortions can sometimes be unmasked with an intermodulation distortion test where a large, very low-frequency signal is added to a higher frequency signal and the resulting amplitude modulation of the higher frequency is measured. This is how the SMPTE IM test works, where frequencies of 60 and 7000 Hz are mixed in a 4:1 ratio. For thermal distortion, more sensitive results will be obtained if the low frequency is 20 Hz or even much lower. For such studies, I designed and built an IM analyzer that used coherent AM detection and whose low-frequency component could be chosen to be additional frequencies of 1, 5 and 20 Hz [18]. Thermal distortion may also show up as increased THD at low frequencies like 20 Hz. However, these symptoms can be confused with distortion from electrolytic capacitors in feedback networks.

References

1. Robert R. Cordell, Build a High Performance THD Analyzer, *Audio*, vol. 65, July, August, September 1981, www.cordellaudio.com.
2. Matti Otala, Transient Distortion in Transistorized Audio Power Amplifiers, *IEEE Transactions on Audio and Electro-Acoustics*, vol. AU-18, September, 1970, pp. 234–239.
3. Matti Otala and Eero Leinonen, The Theory of Transient Intermodulation Distortion, *IEEE Transactions on Acoustics, Speech and Signal Processing*, vol. ASSP-25, no. 1, February 1977, pp. 2–8.
4. W. Marshall Leach, Transient IM Distortion in Power Amplifiers, *Audio*, February 1975, pp. 34–41.
5. W. Marshall Leach, Suppression of Slew-Rate and Transient Intermodulation Distortions in Audio Power Amplifiers, *Journal of Audio* Engineering Society, vol. 25, no. 7–8, July–August 1977, pp. 466–473.
6. RichardA. Greiner, Amp Design and Overload, *Audio*, November 1977, pp. 50–62.
7. Walter G. Jung, Mark L. Stephens, and Craig C. Todd, *Slewing Induced Distortion and Its Effect on Audio Amplifier Performance – With Correlated Measurement Listening Results*, AES preprint No. 1252 presented at the 57th AES Convention, Los Angeles, CA, May 1977.

8. Walter G. Jung, Mark L. Stephens, and Craig C. Todd, An Overview of SID and TIM, *Audio*, vol. 63, no. 6–8, June–August 1979.
9. Peter Garde, Transient Distortion in Feedback Amplifiers, *Journal of Audio Engineering Society*, vol. 26, no. 5, May 1978, pp. 314–321.
10. Robert R. Cordell, Another View of TIM, *Audio*, February, March 1980, www.cordellaudio.com.
11. Eero Leinonen, Matti Otala, and John Curl, A Method for Measuring Transient Intermodulation Distortion (TIM), *Journal of Audio Engineering Society*, vol. 25, no. 4, April 1977, pp. 170–177.
12. Susumo Takahashi and S. Tanaka, *A Method of Measuring Transient Intermodulation Distortion*, 63rd Convention of the Audio Engineering Society, preprint No, 1478, May 1979.
13. Eero Leinon and Matti Otala, Correlation Audio Distortion Measurements, *Journal of Audio Engineering Society*, vol. 26, no. 1–2, January–February 1978, pp. 12–19.
14. Robert R. Cordell, A Fully In-Band Multitone Test for Transient Intermodulation Distortion, *Journal of the Audio Engineering Society*, vol. 29, September 1981, www.cordellaudio.com.
15. Matti Otala, *Feedback-Generated Phase Modulation in Audio Amplifiers*, 65th Convention of the Audio Engineering Society, London, 1980; preprint 1576.
16. Matti Otala, *Conversion of Amplitude Nonlinearities to Phase Nonlinearities in Feedback Audio Amplifiers*, Proc. of IEEE International Conference on Acoustics, Speech and Signal Processing, Denver, CO, 1980, pp. 498–499.
17. Matti Otala, *Phase Modulation and Intermodulation in Feedback Audio Amplifiers*, 69th Convention of the Audio Engineering Society, Hamburg, 1981; preprint 1751.
18. Robert R. Cordell, Phase Intermodulation Distortion–Instrumentation and Measurements, *Journal of the Audio Engineering Society*, vol. 31, March 1983, www.cordellaudio.com.
19. Robert R. Cordell, A MOSFET Power Amplifier with Error Correction, *Journal of the Audio Engineering Society*, vol. 32, January 1984, www.cordellaudio.com.
20. Bob Cordell, The Distortion Magnifier, *Linear Audio*, September 2010, pp. 142–150.
21. Bruce Hofer, Designing for Ultra-Low THD+N, Parts 1 and 2, *AudioXpress*, November 2013, pp. 20–23 and December 2013, pp. 18–23, audioxpress.com.
22. Bruce Hofer, *The Ins and Outs of Audio*, www.aes-media.org/sections/uk/Conf2011/Presentation_PDFs.
23. Walt Jung and Richard Marsh, Picking Capacitors, *Audio*, February 1980, pp. 52–62 and March, pp. 50–63.
24. Cyril Bateman, Understanding Capacitors, *Electronics World*, December 1997, pp. 998–1003.
25. Cyril Bateman, *Capacitor Sound*, Electronics World, Parts 1–6, July 2002–March 2003.
26. Edward M. Cherry, A New Distortion Mechanism in Class B Amplifiers, *Journal of the Audio Engineering Society*, vol. 29, no. 5, May 1981, pp. 327–328.
27. T. Sato, K. Higashiyama, and H. Jiko, *Amplifier Transient Crossover Distortion Resulting from Temperature Changes in the Output Power Transistors*, presented at the 72d Convention of the Audio Engineering Society, *JAES* (Abstracts), vol. 30, pp. 949–950, December 1982, preprint 1896.

Chapter 13

Switches and Relays

While basic line level preamplifiers require little more than source selection, full-function preamplifiers may require more switching. Mixers will require still more switching functions. In many cases an output muting circuit may be needed as well, so that pops and thumps at turn-on and turn-off do not get through. The various switching means described here will also often apply to volume controls based on switched attenuators.

This chapter will cover design considerations for various kinds of switching options in audio equipment, including mechanical approaches and solid-state approaches that employ JFETs, MOSFETs, transmission gates and opto-isolators. The focus will be mainly on small-signal switching.

13.1 Mechanical Switches and Pushbuttons

Mechanical switches and pushbuttons have been the mainstay of switching, selection and routing for over a century, and are still important and have some characteristics to bear in mind, but advances in technology have diminished their role. Switches take up panel space and modern applications often require that switching be electronically controlled. In some cases, such roles have been replaced by miniature relays, but in a great many applications various solid-state devices have taken over. In many cases small-signal routing has been simplified and shortened by elimination of mechanical switches as well. Pushbuttons continue to play a role in modern user interfaces, but more often are relegated to controlling DC signals to microcontrollers for selection activities.

Conventional mechanical switches are characterized by many different kinds of features and limitations. Toggle switches are usually characterized as number of throws and number of poles, such as double-pole double-throw (DPDT), although some have a center-off position and can be referred to as on-off-on. The small-signal switches relevant to this chapter do not matter much with regard to voltage ratings, current ratings and contact damage due to arcing.

Mechanical switches in general do not have much contact resistance (milliohms) and contact resistance is not a serious concern in small-signal switching. Contact bounce is mainly a concern when controlling digital circuits, such as microcontrollers. However, most of these devices incorporate hardware or software de-bouncing functionality. It is important to realize that a double-throw switch will have a very small interval of being open to both contacts as the switching element is moved.

Rotary switches have traditionally been used to make input and mode selections, frequency selections and for attenuators. They most often have 5 or 11 positions with as many wafers ganged together as needed for the number of poles required. A five-position wafer usually has two poles on the single wafer. These switches come in both break-before-make and make-before-break configurations.

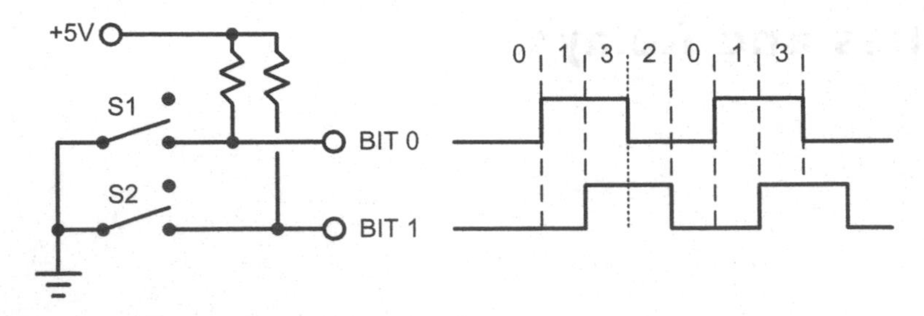

Figure 13.1 A Rotary Encoder Circuit.

Rotary Encoders

Rotary encoders are the exception to the declining use of mechanical switches [1, 2]. These devices communicate incremental rotary action on a knob to a controller that then makes changes electronically to routing, selection, etc. Rotary encoders communicate change and direction of change. They have no inherent memory of position, so the memory function is associated with the microcontroller with which they are usually connected. The controller usually employs FLASH memory or some other means of non-volatile storage to retain the selection. Applications that use rotary encoders also often have a display of the current state as part of the user interface. A common use is as the volume control on a receiver or preamplifier.

A typical rotary encoder can turn without boundaries and has a large number of clicks per turn. It delivers information about how many steps have been taken and in what direction. As shown in Figure 13.1, the encoder includes two switches, S1 and S2, connected to Bit 0 and Bit 1 output terminals. Each output terminal has a pull-up resistor connected to the positive supply of the associated microcontroller. The other side of each switch is connected to a common terminal that is often connected to ground. When the shaft is rotated, one or the other switches will be closed or opened; one or the other or both will connect its terminal to the common terminal. The mechanical position of the switches is staggered so that only one switch transitions at a time. The waveforms produced by rotation are illustrated in Figure 13.1. The waveforms produced by rotation are essentially square waves that are in a quadrature phase relationship. The switch that transitions first indicates direction. These sequences are easily decoded by a microcontroller. Many such encoders also include a push button switch on the shaft. It can provide additional functionality, such as selection or change-in-mode function. Some even have integrated LEDs for display of function or mode purposes. The rotary encoder provides a very efficient user interface.

13.2 Mechanical Relays

The mechanical relay is another mainstay that goes back a great many years, yet to this day is very important in its many forms. Here we will discuss the small-signal relay, but many of the same issues apply to larger relays for AC and DC applications. Speaker relays and their use in amplifier-speaker protection circuits are treated in Ref. [3].

Although mechanical relays are often larger, more expensive and more power-consuming than some modern solid-state alternatives, they have some advantages that are hard to beat. Most prominently among them are galvanic isolation and extremely low *on* resistance. Isolation in the *off* state is also extremely good. They also are available with normally open (NO) and normally closed (NC) functionality without the need to provide inverted control signals. Unlike mechanical switches, they are electrically controlled. Groups of relays are also an obvious replacement for mechanical rotary switches, which are increasingly hard to find in multiple gangs. Relays in such cases are also freely located anywhere on the printed circuit board, thus permitting them to be optimally placed to reduce the distances in the signal paths.

Relay contact configurations are often described as a Form type. A normally-open set of contacts is referred to as Form A, while a normally-closed set of contacts is referred to as Form B. Double-throw contacts that include Form A and Form B functionality are referred to as Form C. A DPDT relay can be referred to as a Form 2C.

Contact Resistance and Self-Cleaning

The contact resistance of small signal relays is very small, measured in the milliohms (mΩ). However, contacts can become dirty or oxidized and cause distortion in some cases, more often when higher currents are involved. Many relays have a wiping action that helps keep the contacts clean [4–7].

The term "dry switching" refers to relays that are switching very low AC current and usually no DC current; these are the types of switching discussed here for small signals. The term "wet switching" refers to relays that are passing AC or DC current. The contact-cleaning mechanisms and characteristics are sometimes different for these two types of situations. Wet switching can play a role in keeping contacts from oxidizing or getting dirty, while dry switching relies on the nature of the contact material (often gold) and possibly some wiping action.

Coil Voltage, Resistance, Current and Dissipation

Relays always have a rated coil-operating voltage and usually coil resistance. This defines the nominal activation current and power dissipation. A minimum voltage for closure and sometimes a maximum voltage for opening may also be specified. As an example, a small conventional miniature 12-V DPDT relay that was tested had coil resistance of 335 Ω, drew 36 mA and dissipated 430 mW. The minimum pull-in coil voltage was 7.0 V_{DC} and release occurred at 5.4 V_{DC}. There is thus hysteresis in their activate and release voltages. Of several tested, both 5- and 12-V types, dissipation ranged from 200 to 500 mW.

A key reality here is that most conventional relays require moderate power to operate. Moreover, if they are operated from unregulated supplies their coil dissipation must be watched. It is also important to realize that the coils of some relays are polarized, in that for proper function the correct polarity of voltage must be applied to the coil. This is not because the relay has a diode in series with the coil but rather different magnetic polarity.

Relays should generally not be powered from the regulated power supply that feeds the analog circuitry. Doing so will add unnecessary power dissipation to the regulators and contribute noise to the power rails. They can be powered from the raw supply that feeds the regulators or by their own regulator. One should also be thoughtful about where the ground current from the relay goes, recognizing that dumping that current into an analog ground may be asking for noise.

Relay Drivers and Flyback Control

Relay coils have inductance and store energy. This means that when the coil is de-energized by opening its driving circuit, it will create a flyback voltage that can be harmful to a transistor driving it. Coil inductance for relays is rarely specified. The conventional 12-V relay discussed above exhibited a 200-V flyback voltage when no flyback suppression was used. This could destroy a driver transistor. As expected, connecting a 1N4148 diode across the coil in the reverse direction solved that problem. This limits the flyback voltage to about 0.6 V. It should be noted that the presence of the diode increased the release time by about 1 ms when it was used, since the diode forms a path for the magnetizing current in the relay coil to continue flowing briefly after driver current is removed.

Vertical MOSFET Relay Drivers

Vertical MOSFETs are readily available with ratings of 100 V or more. In addition, they contain an integral diode whose breakdown is approximately equal to the voltage rating. This diode can withstand considerable current in breakdown without failure. If a small power MOSFET is used as a relay driver, a flyback diode is not needed across the relay coil. Moreover, the flyback voltage is allowed to go as high as the voltage rating of the MOSFET, allowing for much faster release. However, it may be prudent to put a capacitor across the coil to control interference from the resulting voltage spike. In most applications, the slowed release time caused by the conventional diode is not important. The high gate input impedance and absence of a junction permits simpler driver arrangements in some cases.

Activate, Transition and Release Times

All relays have operate and release times that can be important in some applications, and these times are usually governed by the coil voltage applied and the way in which the relays are driven. Relays are most always break-before-make. As a result, relays also have transition times. In a double-throw relay, this is the contact transition time during which the swinger is not connected to either the normally-open (NO) or normally-closed (NC) contacts. Some circuits can be sensitive to this brief open-circuit interval. Transition time is rarely specified. The actuate and release behavior for an SPDT relay is illustrated in Figure 13.2.

As with mechanical switches and pushbuttons, relays can have contact bounce, usually when contact is made. This is usually not an issue for small-signal routing of audio signals, but could be

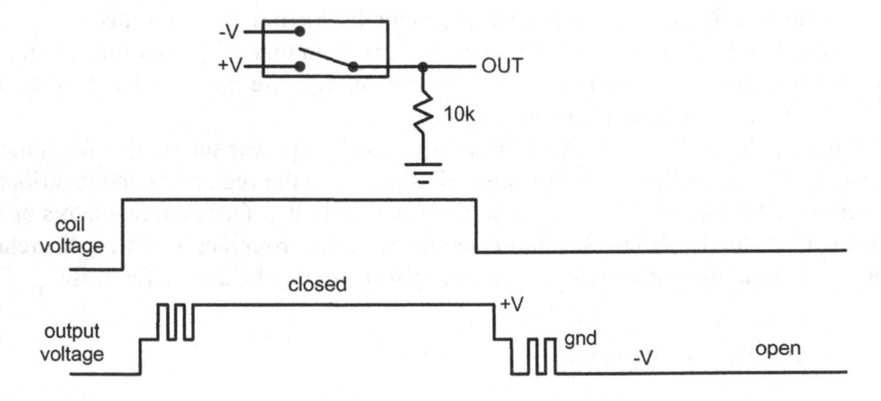

Figure 13.2 Relay Activate, Transition, Release and Bounce Timing.

an issue if the closure of the contacts is being used to control another function, such as to signal a microcontroller that does not have some form of de-bounce functionality.

Relays have an operate time delay before the swinger moves, partly due to inductance in the coil. In the conventional relay mentioned above, this was less than 2.6 ms. Once the swinger moves away from the NC contact, it begins a transition toward the NO contact. During this time, lasting about 0.5 ms in the above relay, the swinger is open-circuited. When contact is made with the NO contact, there is a little bit of bounce when the magnetic force grabs the swinger (about five cycles with a total duration of about 0.2 ms).

Upon release, there was a 2.5 ms delay followed by a 0.5 ms transition period. There was an immediate bounce when the swinger came into contact with the NC contact. The bounce is partly governed by the mechanical resonance of the swinger spring and the swinger mass. The relay tested went though several cycles of bounce over a 0.5 ms period before it settled. The 2.5 ms release delay was largely caused by current continuing to flow through the flyback diode. Without the diode, opening of the NO contact was almost immediate.

Consider as an example the KEMET EC2/EE2 miniature signal relay, costing less than $2. It consumes about 140 mW with a 5-V, 178-Ω coil. It is specified to operate with a coil voltage greater than 3.75 V_{DC} and to release with a coil voltage of less than 0.5 V_{DC}. It is allowed to operate at 150% of its rated coil voltage if the ambient temperature is less than 40°C. Operate time is specified as 2 ms excluding bounce time. Release time is specified as 1 ms if no flyback diode is used across the coil.

Latching Relays

Latching relays are used to retain their state after power is removed [5, 8]. These relays employ permanent magnets or a mechanical arrangement to retain their swinger position. There are two types of latching relays. Single-coil devices are made to change state by applying a coil voltage polarity opposite to that which put it in its current state. Double-coil devices employ one coil to close the relay and another to open the relay. This allows the relay to be opened and closed using only one polarity of applied voltage. There are thus four types of latching relays categorized by whether the latching mechanism is magnetic or mechanical, and whether they have one coil or two coils.

Latching relays have two striking advantages. First, they are non-volatile – meaning that they retain their position without power. The second is that they are power-saving. Their state can be changed with just a brief electrical pulse. They are, however, more expensive than a non-latching relay, often by a factor of at least 2.

Switching and Muting Applications

Although relays are larger, more expensive and consume more power than various solid-state devices, they have many advantages in numerous applications where isolation, low contact resistance or low distortion is required. The low distortion of relays makes them ideal for signal selection, while the low contact resistance makes them highly effective in muting the output of audio devices by shorting the signal to ground with the NC contacts, while being completely non-intrusive in the non-muting mode when they are activated. Comparisons with various solid-state alternatives, and the nuances of driving relays and passing signals through solid-state devices are both very well covered in Refs. [9, 10].

13.3 JFET Switches

JFET switches can be used to select or route signals [11, 12]. They are closed (conductive) when their gate-source voltage is zero. If they are connected in series with the virtual ground of an inverter, with the source connected to the virtual ground, their source potential will always be at ground. This is shown in Figure 13.3(a), where J1, a J112, is placed in series with the input resistor in the inverting arrangement. This makes it easy to turn them on with a control voltage that has a resistance to ground. To turn them off, an open-collector circuit can pull the gate down to a negative potential, often the negative supply rail. An LM339 quad comparator is a convenient means to drive the gate of N-channel JFETs in these applications. It can be provided with a threshold voltage that can be made compatible with just about any logic level. A resistor in series with the gate and a gate capacitor to ground will sometimes be included to slow down the turn-on and turn-off rates from the LM339.

Sometimes it is important to minimize feedthrough via J1's source-drain capacitance or to have the input impedance of the circuit be the same whether the JFET switch is open or closed. This is accomplished in Figure 13.3(b) by adding J2. It shunts the signal at the drain of J1 to ground when J1 is off. The input current through R1 flows to ground either through J2 or through J1 via the virtual ground. J2 is driven with gate control voltage whose polarity is opposite that of J1's control voltage. If multiple JFET input pairs are used to drive a virtual ground, a mixing switch is formed with very little inter-channel crosstalk. In fact, one or more of the inputs can be selected and they will be mixed.

JFETs can also be used to shunt a signal to ground, such as in a muting circuit, as shown in Figure 13.4(a). Here the muting attenuation is limited by the $R_{DS(on)}$ of the JFET divided by the series amplifier output resistance if the latter is significantly larger than the former. It is normally desirable to keep the output series resistance for a preamplifier less than 1 kΩ. The 50-Ω maximum $R_{DS(on)}$ of the J112 in this arrangement will limit mute attenuation to about 26 dB, which may be insufficient.

In Figure 13.4(b), the JFET increases the gain of a non-inverting amplifier stage from unity to 20 dB when in the *on* state. Some distortion will occur when the JFET is in the *on* state for higher gain,

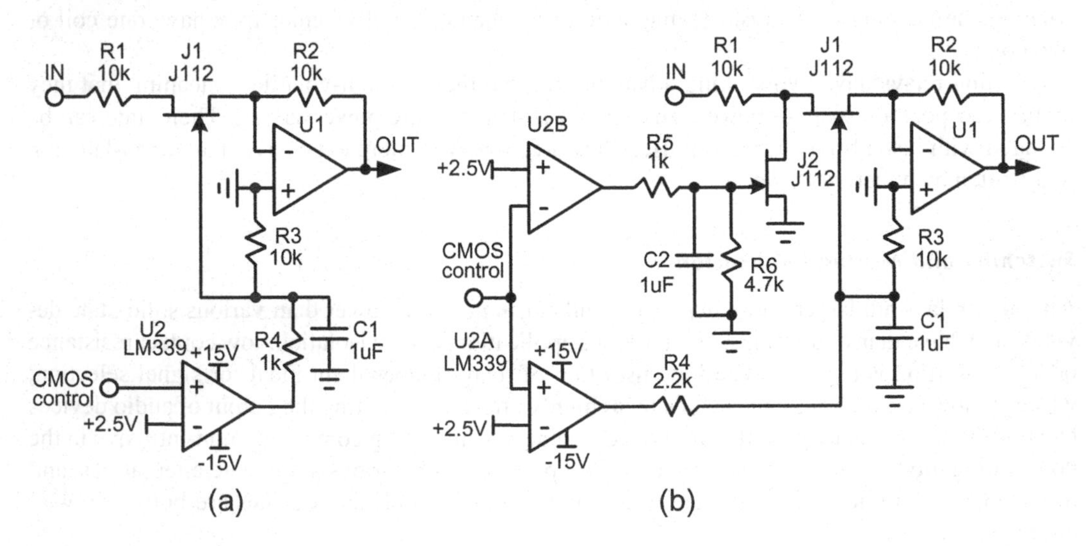

Figure 13.3 JFET Switches.

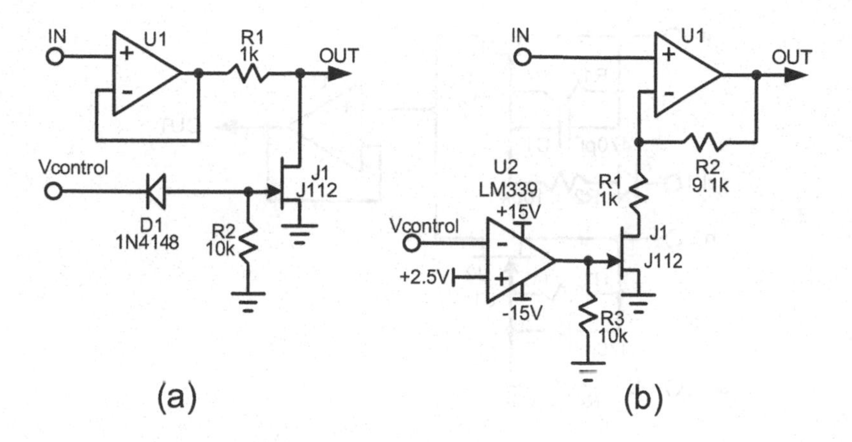

Figure 13.4 JFET Shunt Switches.

since there will then be signal across the JFET. With a 1-V_{RMS} input and output of 10 V_{RMS} in the high-gain state, the 50-Ω maximum $R_{DS(on)}$ of the J112 will permit about 45 mV across it. In some arrangements, half the drain-source signal voltage can be fed back to the gate to reduce distortion.

On Resistance and Capacitance

JFET switching performance is largely governed by $R_{DS(on)}$ and any variation of it with its drain-to-source voltage. In the series arrangement, the drain voltage will be a fraction of the input signal equal to $R_{DS(on)}$ divided by the sum of the input resistance and $R_{DS(on)}$. If the input resistance is not significantly higher than $R_{DS(on)}$, there will be enough signal voltage across the JFET from drain to source to modulate $R_{DS(on)}$ and thus cause distortion. This same effect can occur when the JFET is used to increase the gain of an amplifier by shunting a portion of the feedback resistance to ground. The JFET's $R_{DS(on)}$ should always be much smaller than the resistance in series with it.

If the JFET is used as a muting shunt, its $R_{DS(on)}$ in comparison to the series source resistance will limit the amount of muting attenuation it can provide. If it is desired that the output impedance of a preamplifier be low, on the order of 100–200 Ω, it may be difficult to find a JFET with low enough $R_{DS(on)}$ to achieve adequate muting attenuation. Obviously, a series-shunt arrangement of JFETs driven with opposite control polarity can make this work.

The J112 JFET is often used in this application. On resistance for this device is a maximum of 50 Ω. The related J111 has maximum on resistance of 30 Ω. If the gate is bypassed to ground, feedthrough at high frequencies due to gate-source and gate-drain capacitances (each 5 pF) will be minimized. Maximum required cutoff gate-source voltage can be as high as −5 V_{DC} for the J112 and −10 V_{DC} for the J111. This means that that they should be turned off by pulling the gates to a negative 15-V_{DC} rail or more.

Floating JFET Switches and Biasing Conditions

A JFET can also be used as a series on-off element driving a high-impedance load, such as a non-inverting amplifier or buffer, as illustrated in Figure 13.5. In this case both the source and drain have signal on them, so control of the gate is less straightforward. The input signal can be applied to the source of the JFET and a high-value resistance (R1) from source to gate can bias the JFET

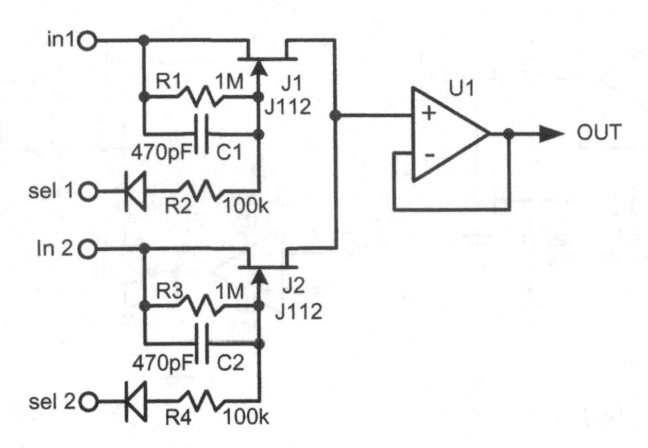

Figure 13.5 Floating JFET Selector Switches.

in a normally *on* state. An open-collector circuit can then pull the gate to a negative value to turn it off. If the JFET is one of several that form a selector, when it is off and another one is on, crosstalk through the first will be attenuated by the shunt resistance of the device that is on.

Sources of Distortion

The main source of distortion when using JFETs in the signal path is that their $R_{DS(on)}$ will change with signal when signal current is passing through the JFET, causing signal voltage to appear from source to drain. Increased reverse bias of the gate-drain junction will further pinch off the channel and increase $R_{DS(on)}$. Signal levels thus should be kept small. In some applications, half of the drain-gate voltage can be fed back to the gate to reduce the resulting second harmonic distortion.

13.4 MOSFET Switches

The normally-off N-channel 3N170 and the P-channel 3N163 enhancement-mode MOSFETs are also used as switches [13–15]. The $R_{DS(on)}$ of the 3N170 is a maximum of 200 Ω with a V_{gs} forward bias of 10 V, while $R_{DS(on)}$ of the 3N163 is a maximum of 250 Ω with a V_{gs} forward bias of −20 V. These devices feature low capacitances and extremely low gate leakage current (<10 pA). The gate is insulated, so that V_{gs} can be of either polarity. This sometimes results in simpler gate control circuitry. These devices also find use in ultra-high input impedance amplifiers, electrometers and smoke detectors.

MOSFETs are implemented on a semiconductor substrate that is the opposite polarity as the source and drain, so substrate diodes are formed to the source and drain that usually have no forward bias. These are also called body diodes, since the substrate is considered to be the body of the device. Some devices like the 3N170 and 3N163 bring out the connection to the body as a fourth pin. In most applications it is just connected to the source. In fact, in ordinary 3-terminal MOSFETs the body is drawn as connected to the source right in the symbol. The body connection is the pointed line that is drawn as pointing into the device for an N-channel MOSFET and pointing away from the device for a P-channel MOSFET. In many applications using a 4-pin discrete MOSFET, the body pin is externally connected to the source. In others, it is not connected at all, allowing the substrate to float. In a few applications it can be connected to a voltage that is called the substrate bias; such a bias will affect the threshold voltage of the MOSFET.

As an aside, dual monolithic JFETs are implemented on a common substrate and therefore have substrate diodes connected to the gates. In the eight-pin packaged versions of these dual JFETs, the substrate is sometimes brought out on the two unused pins [16]. These pins are usually left floating, but if they are connected to a substrate bias, the threshold voltage of the JFETs will be influenced. Dual monolithic JFETs in a six-pin metal package will often have the substrate connected to the metal package [16].

13.5 DMOS Switches

Double-diffused MOS transistors (DMOS) are also widely used as signal switches [17, 18]. Like the JFET, conduction is through a channel, but in this case the channel is normally an open circuit and it must be enhanced (inverted) by an electric field to cause conduction. The electric field is created by an insulated gate in the metal-oxide-semiconductor process. Unlike most JFET transistors, DMOS transistors are asymmetrical; the structures adjacent to the source are different than the structures adjacent to the drain. They have very small gate-drain capacitance (0.5 pF) and small gate-source capacitance (1.5 pF). The insulated gate makes them easy to control with large voltage swings of either polarity allowed between gate and source.

N-channel DMOS switch transistors like the SD210 can be used as switches in the same way as N-channel JFETs, but in some respects they have some advantages [19]. While JFETs are a depletion device, DMOS transistors are enhancement-mode devices, meaning that the channel must be enhanced to cause conduction. With no voltage applied to the gate, they are an open circuit. The insulated gate means that there is no gate semiconductor junction that may conduct if the gate voltage goes positive with respect to the source. The gate-source voltage is fee to go positive or negative by up to 30 V. These transistors are also available in quad versions, like the SD5000 series [20]. Because of the thin polysilicon insulated gate, DMOS transistors are susceptible to electrostatic discharge (ESD) damage. For this reason, they are often available in versions with and without built-in Zener protection diodes from gate to substrate. The Zener diodes can add leakage current and capacitance, so they are left out for sensitive applications.

The N-channel DMOS transistors are fabricated on a p-type substrate with heavily doped *n*+ diffusions for the source and drain. Thus, there are substrate diodes from the substrate to the source and drain regions with the p-type substrate as the anode. The p-type substrate is also referred to as the body, and these diodes are called body diodes. These diodes are usually reverse-biased by applying a voltage negative with respect to the source. Sometimes, however, the body will just be connected to the source. Below the gate there is no conductive channel in the absence of enhancement from a positive voltage applied to the gate. The substrate connection is usually brought out, so a single DMOS transistor is usually a four-pin device.

Below the *n*+ source diffusion lies a lightly doped p-region that extends laterally beyond the *n*+ source region to part-way under the gate. This p-region is what gives rise to the asymmetry of the DMOS device. When the gate is made positive with respect to the source, the entire region under the gate is inverted (enhanced) to become an n-type region to allow conduction. It is worth noting that the amount of reverse bias of the substrate with respect to the source, V_{SB}, has an influence on the threshold voltage $V_{GS(th)}$ of the device, with greater reverse bias increasing the $V_{GS(th)}$; i.e., requiring greater forward voltage on the gate to turn the device on. Threshold voltage at 1 μA for the SD210 is typically +1.0 V_{DC} for V_{SB} = 0 V_{DC}, while for V_{SB} = −15 V_{DC}, $V_{GS(th)}$ = +3.0 V_{DC}.

Similarly, $R_{DS(on)}$ increases with increased V_{SB}. For V_{GS} = 4 V_{DC}, $R_{DS(on)}$ = 50 Ω with V_{SB} = 0 V_{DC} and climbs to 270 Ω with V_{SB} = −15 V_{DC}. However, For V_{GS} = 10 V_{DC}, $R_{DS(on)}$ = 30 Ω with V_{SB} = 0 V and climbs to only 33 Ω with V_{SB} = −15 V. If an analog signal is present at the source, V_{SB} will be changing and $R_{DS(on)}$ will be modulated. This can be a source of distortion if the device is not driving a light load.

The message is clear: in a switch application where low and constant $r_{DS(on)}$ is needed, V_{GS} should be large in the *on* state. It is also important to recognize that V_{SB} reverse bias needs to be greater than the largest peak negative voltage of an analog signal being passed through the device in order not to forward-bias the source body diode. In a typical audio application where the device is in series with the signal, the substrate might be biased at −15 V_{DC} and $V_{GS(on)}$ might be +15 V_{DC}.

Body Diode Restrictions

The source and drain body diodes place a restriction on some applications. For example, it might be tempting to connect the body to the source so that there is no V_{GS} modulation in a series arrangement. However, in the *off* state, where the drain might be at ground, a positive signal swing at the source could forward-bias the body-drain diode. Similarly, in a shunt arrangement where the source and body are connected to ground, a negative signal swing on the drain in the *off* state could forward-bias the body-drain diode. In some applications, however, the body may be allowed to float, mitigating some of these issues.

1/f Noise

Like many MOSFET transistors, $1/f$ noise can be quite a bit higher than for JFETs. This can be important if the DMOS device is used as an amplifier. The $1/f$ noise corner is typically at a significantly higher frequency. For the 3N163, the $1/f$ noise corner is at about 9 kHz. At 100 kHz it is typically 6–12 nV/√Hz, while at 100 Hz it is 100–300 nV/√Hz.

13.6 CMOS Switches

CMOS transmission gates are frequently used as bi-directional analog switches to route audio signals [21, 22]. They are easily controlled by logic signals. The most well-known and early CMOS transmission gate was the CD4016B quad bilateral switch. PMOS and NMOS transistors are connected in parallel so that both polarities of the signal have a low-resistance path, as shown in Figure 13.6. Each channel includes the inverter to drive the gate of the opposite polarity transistor. The NMOS transistor has lower *on* resistance when the input signal goes negative, while the PMOS transistor has lower *on* resistance when the signal being passed goes positive.

Typical on resistance is 280 Ω. The variation in *on* resistance with signal voltage causes some distortion, often odd-order, if the CMOS switch must drive a load. Power supply voltage limitations can place a limitation on maximum signal swings. Depending on the specific version, the CD4016 can have up to 15 or 20 volts from V_{DD} to V_{SS}, meaning that in an analog application its rails would be limited to ±7.5 V_{DC} or ±10 V_{DC}. This limits the peak amplitude of the audio signals. Requiring additional positive and negative rails lower than the usual ±15 V_{DC} is a disadvantage.

The CD4066B quad bilateral switch offers about half the on-state resistance of the CD4016B. Its *on* resistance is closer to constant over the signal swing range, resulting in lower distortion. It is limited to 20 V from V_{SS} to V_{DD}. It uses metal-gate technology. The CD74HC4066 quad bilateral switch uses silicon-gate CMOS technology and achieves significantly reduced *on* resistance, in the range of 15–30 Ω.

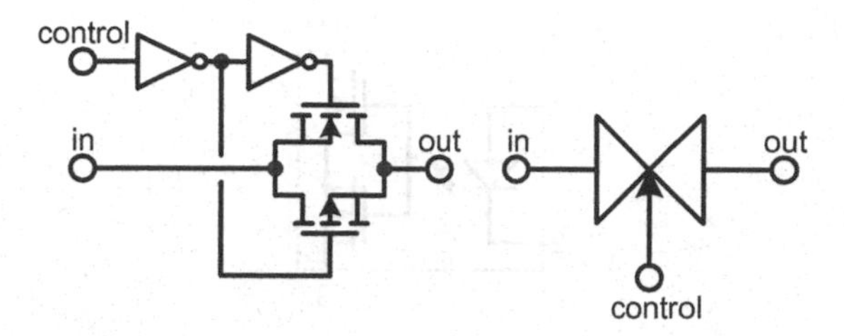

Figure 13.6 CMOS Transmission Gate.

The CD4051B, 52B and 53B are useful variants that are multiplexers that have 1:8, 2 × 1:4 and 3 × 1:2 functionality, respectively. Typical on resistance is about 180 Ω with ±5-V_{DC} rails. When driving a 10-kΩ load with ±5-V_{DC} rails, and a ±5-V signal swing, gain can change by about 0.2% over the signal swing, suggesting moderate distortion. This distortion results from a change in *on* resistance that in turn results from signal swing. If signal swing is kept small, distortion will be kept small. If the voltage swing is made very small by connecting the transmission gate to a virtual ground, distortion will be lower, but not zero because current must still flow through the devices.

13.7 CMOS/DMOS Switches

The Vishay-Siliconix DG611 family of analog switches combines CMOS and DMOS technologies on the same die [23, 24]. Recall that DMOS switches can implement very good analog switches and have very low capacitance, but require significant gate drive voltage for *on/off* control. This means that they cannot be controlled by ordinary logic levels, such as CMOS.

The DG611 is a quad bilateral analog switch that accepts 5-V CMOS logic levels by employing ground and +5-V_{DC} (V_L) powering to its logic input circuit. A level translator and driver then control the gate of an N-channel DMOS analog switch, providing the necessary voltage swings and substrate bias to operate the DMOS device for excellent analog switching performance. This is accomplished with a high-voltage V+ rail, typically +10–15 V_{DC}, and a V-rail, typically negative 5–10 V_{DC}. The maximum voltage from V– to V+ is 21 V_{DC}. An example audio analog switch circuit driven by CMOS logic would have V_L = +5 V_{DC}, V+ = +15 V_{DC} and V– = –5V_{DC}.

In such an arrangement, peak analog voltage swings of ±5 V can be handled. The DG611 achieves typical *on* resistance of 18 Ω with extremely low charge injection of 3 pC, typical. Off-state input and output capacitances are typically 3 and 2 pF, respectively. *On*-state input capacitance is typically 10 pF.

13.8 Photo-Coupled Switches

The Panasonic PhotoMOS® relays are a form of optocoupler in which an LED illuminates a photocell whose output drives the gates of a pair of series-connected MOFET transistors to turn them on. When on, the MOSFETs form a bilateral solid-state switch that can pass AC signals with relatively low *on* resistance [25]. The PhotoMOS® optically isolated solid-state relays are available in standard four-, six- or eight-pin dual inline IC packages, like the SMD-6 (Figure 13.7).

The four-pin Form A SPST AQY280EH has a typical LED operate current I_F of 1.8 mA (3.0 mA max) and has a typical on resistance of 20 Ω at I_F of 5 mA. I/O capacitance is on the order of 1 pF

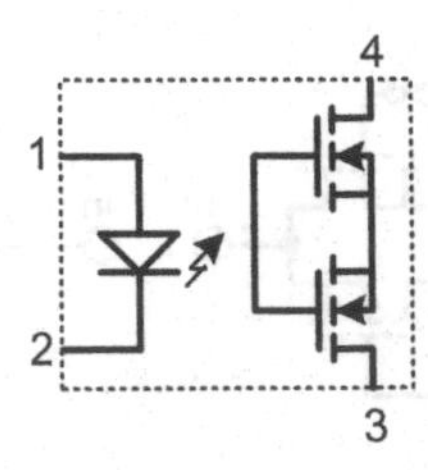

Figure 13.7 AQY280 Photo MOSFET Switch.

and off-state I/O resistance is typically 1 GΩ. They can withstand at least 300 V in the off state and can pass over 100 mA in the on state. The eight-pin AQW280EH provides two independent Form A switches in an eight-pin package.

These devices are not fast, since they employ a photocell. Maximum turn-on time at $I_L = 5$ mA is 3 ms and 0.3 ms typical. Maximum $T_{off} = 2$ mS. These time delays are not a problem for most audio switching applications. They naturally provide galvanic isolation and there is no inter-dependence of gate drive and device resistance on signal voltage, making drive circuitry simple. These devices provide clean, bounce-free switching.

The Vishay VOR1121 device is a similar single Form A optically coupled solid-state switch that can pass small-signal AC [26]. It employs a hybrid combination of a GaAlAs IRED illuminator and a series pair of N-channel DMOS switch transistors. *On* resistance is typically 12 Ω when the LED is driven with $I_F = 5$ mA ($V_F = 0.35$ V). *Off* resistance is claimed to be 5000 GΩ, with input-output capacitance of 0.4 pF. It can handle 250 V in the *off* state and 200 mA in the *on* state. $T_{on} = 0.2$ ms typical, 0.5 ms maximum with LED drive of 5 mA. A similar dual Form A device is available in an eight-pin package as the VOR2142.

Phototransistor

The familiar bipolar phototransistor is shown in Figure 13.8. If the LED illuminates the associated BJT, current will flow from collector to emitter, just as if base current were being applied. The transistor current depends on the degree of illumination and thus the current flowing in the LED. Their coupling performance is described as the DC current transfer ratio (CTR), which is BJT collector current divided by LED current. The 4N35 is rated at a minimum current transfer ratio of 100% at LED current of 10 mA [27]. The CTR is nonlinear, decreasing with reduced LED current. For this device, it may be on the order of 30% at LED current of 1 mA. Rated collector-emitter breakdown voltage for this device is 70 V. The base of the BJT is also brought out on a pin.

The 4N35 can be used as switch or as an analog current-in current-out device that provides galvanic isolation. It is often used in the feedback path of isolated switch-mode power supplies.

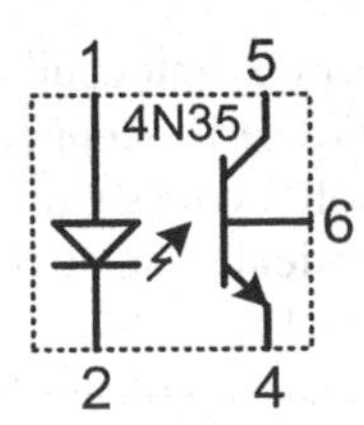

Figure 13.8 4N35 Phototransistor Optocoupler.

References

1. Arduino Tutorials, *How Rotary Encoder Works and How to Use It with Arduino*, howtomechatronics.com/tutorials/arduino/rotary-encoder-works-use-arduino/.
2. T.K. Hareendran, *Rotary Encoders & Arduino*, Electro Schematics, electroschematics.com/rotary-encoder-arduino/.
3. Bob Cordell, *Designing Audio Power Amplifiers*, 2nd edition, Routledge, New York, NY, 2019.
4. Galco, *How Relays Work*, https://www.galco.com/comp/prod/relay.htm.
5. Wikipedia, *Relays*.
6. TE Connectivity, *Relay Contact Life*, Application Note, pandbrelays.com.
7. Kemet Electronics Corp., EC2/EE2 Miniature Signal Relays, Datasheet, kemet.com.
8. Omron, *What Is a Latching Relay?* https://www.omron-ap.com/service_support/FAQ/FAQ02822/index.asp.
9. Rod Elliott, *Preamp Input Switching Using Relays*, Project 163, http://sound.whsites.net/project163.htm.
10. Rod Elliott, *Muting Circuits for Audio*, Elliott Sound Products, 2015, https://sound-au.com/articles/muting.html#s2.
11. Analog Devices, Inc., *Analog Switches and Multiplexors Basics*, MT-088, Tutorial, 2008, analog.com.
12. Ray Marston, *FET Principles and Circuits*, Parts 1–4, Nuts and Volts Magazine, May–August 2000.
13. Linear Integrated Systems, *3N170 N-Channel Enhancement Mode MOSFET*, Datasheet, linearsystems.com.
14. Linear Integrated Systems, *3N163 P-Channel Enhancement Mode MOSFET*, Datasheet, linearsystems.com.
15. Siliconix, *P-Channel Enhancement Mode MOSFET Transistors*, 3N163/3N164 Datasheet.
16. Linear Integrated Systems, *LSK389 Ultra Low Noise Monolithic Dual N-Channel JFET Amplifier*, SOIC 8L and TO-71 6L packages, linearsystems.com.
17. ST Microelectronics, *Understanding LDMOS Device Fundamentals*, Application Note AN1226, July 2000, http://www.st.com.
18. Linear Integrated Systems, *High-Speed DMOS FET Switches and Switch Arrays*, Application Note.
19. Linear Integrated Systems, *SD-SST210/214 N-Channel Lateral DMOS Switch*, Datasheet, linearsystems.com.
20. Linear Integrated Systems, *SD-5000/5001/5400/5401 Quad N-Channel Lateral DMOS Switch Zener Protected*, Datasheet, linearsystems.com.
21. EDN, *Use Analog Switches to Multiplex Your Signals*, https://edn.com/use-analog-switches-to-multiplex-your-signals/.
22. Usama Munir, David Canny, *Selecting the Right CMOS Analog Switch*, Maxim Integrated Application Note 5299, 2013.
23. Vishay Siliconix, *DG611 Family Combines Benefits of CMOS and DMOS Technologies*, AN207, 1999, www.vishay.com.
24. Vishay Siliconix, *High-Speed Low-Glitch D/CMOS Analog Switches*, DG-611/612/613 datasheet, 2007, www.vishay.com.
25. Panasonic AQY280 PhotoMOS® Datasheet, Panasonic Automation Controls Catalog.
26. Vishay VOR1121 and VOR2142 Datasheets, Document numbers 84298 and 84303.
27. Vishay 4N35 Optocoupler with Base Connection, Datasheet, www.vishay.com.

Chapter 14

Power Supplies and Grounding

Power supply design and grounding arrangements are too often the stepchild of the creative process when it comes to audio circuit design. Poor power supply and grounding techniques can ruin an otherwise outstanding design by allowing grunge into the signal path in many different ways.

Power supply quality and the grounding architecture can play a large role in the sonic performance of a preamplifier or mixer, and these two topics are covered here. The importance of power and ground layout is discussed, especially with regard to the avoidance of ground loops. Both traditional linear power supplies (non-switching) and switch-mode power supplies (SMPS) are discussed.

The importance of equivalent series resistance (ESR) and equivalent series inductance (ESL) of reservoir capacitors is explained, and means for reducing their impact are illustrated. Issues related to rectifier speed and rectifier noise are also discussed, noting that high peak currents can flow in the rectifiers and that the speed with which the rectifiers can turn off (recovery time) can have a large influence on noise. Low-noise voltage regulators can be especially important. Prevention of electromagnetic interference (EMI) ingress is also discussed. Finally, the matter of inboard vs outboard power supplies is considered.

Most small-signal circuitry requires relatively low power, and the current drawn by the circuits is relatively constant. This allows the power supplies to be relatively compact, but they must be very clean and they must not radiate hum and noise into the sensitive preamplifier circuits. Some preamplifiers may actually have their power supplies external to the preamplifier box. This may take the form of a separate box or an external dongle powered by a wall transformer.

Switching power supplies (SMPS) that operate at higher frequencies are also popular. Such supplies may receive a single-polarity DC voltage from an external wall transformer or switcher and then convert it to something like ±15 V. Often the external supply in these arrangements will be +5 V at up to 2 A, and may be supplied via a USB connector and phone charger. A power source of 5–10 W is often adequate for modest preamplifier designs. Power supply rails for low-level circuits should be extremely quiet, and power supply rejection ratio (PSRR) should not be relied upon as the sole defense against power supply hum and noise ingress. For this reason, SMPS supplies should be followed by linear pass regulators.

Grounding in preamplifiers and mixers is especially important due to the small signal levels involved. Star grounding and star-on-star grounding architectures should be employed where practical to keep the grounds clean and to avoid ground loops. Printed circuit board (PCB) layout is also very important. Layout topology should reflect intentional control of where current flows, keeping in mind that it flows through the path of lowest impedance. It is very important that power supplies neither allow EMI ingress nor create EMI.

14.1 The Design of the Power Supply

Traditional power supplies for preamplifiers, so-called linear power supplies, consist primarily of a transformer, rectifier, reservoir capacitors and IC voltage regulators. Such linear supplies are gradually being overtaken by switch-mode power supplies (SMPS) as discussed later.

The power supply must convert the AC mains voltage into reliable and stable negative and positive DC rails. It must provide adequately small ripple and sufficient reserve current and regulation. Mains noise must not be communicated to the circuit side of the power supply.

The range of load current for preamplifiers through studio console modules can be very large, typically from ±15 V_{DC} at 100 mA (3 W) to ±18 V_{DC} at 1 A (36 W) for a larger piece of equipment. Studio consoles, of course, consume a lot more power. This chapter will focus mainly on power supplies that can deliver ±15 V_{DC} at about 200 mA for the examples, recognizing that straightforward scaling can be applied for other applications.

Many designs mainly comprise a plurality of op amps as the main current consumers. Dual op amps like the NE5532 and LM4562 usually consume less than 16 mA (8 mA per op amp). A fairly complex preamplifier or equalizer with ten dual op amps would require about ±160 mA for the op amps. Of course, other power consumers like LEDs, displays and relays must also be taken into account. The latter could be substantial for designs that use a significant number of PCB-mounted relays in place of mechanical switches. In many cases, some of these other functions not in the signal path may often be powered from a separate 5-V_{DC} power supply.

Bleeder resistors are often connected across the reservoir capacitors, assuring that that the rail voltages fall to zero in a reasonable time after power is removed. So-called *X* capacitors are often connected across the primary and/or the secondary windings of the power transformer to reduce noise. *X* capacitors used on the mains side must be of the type designed for continuous connection across the mains. They should be mounted on the mains side of the power switch close to the mains wiring entrance and should be shunted by a bleeder resistor to discharge any residual voltage that may be present after power is disconnected. IEC power inlet connectors with integral EMI filters usually include *X* capacitors. Such connectors are highly recommended, since the EMI design and implementation has already been done. IEC connectors that also include fuses and/or power switches are also very attractive.

Y capacitors are required whenever there is a connection from the AC line and/or AC neutral to user-accessible AC ground points, such as a metal chassis. This can occur when the equipment power cord is two-prong rather than 3-prong. The Y capacitors are usually connected from the line and neutral to the safety ground.

14.2 The Power Transformer

The power transformer in audio equipment can play a critical role in noise performance and cost, and there are many types, including the EI laminated, the toroid and the R-Core [1, 2]. A great deal of information on transformers can be found in the six-part series in audioXpress by Chuck Hansen [3–8]. Many details on how they work, how they are made, the different types and how they perform are provided.

Toroidal Transformers

Toroidal power transformers tend to be more efficient than transformers of conventional EI construction. They also better confine the noisy magnetic fields that result from large rectifier spikes. There is little cost premium for toroidal transformers. Some claim that transformers with C-core or

EI construction sound better, but this is unsubstantiated. The advantages of toroidal transformers over conventional laminated transformers include [1, 2, 9]:

- High efficiency
- Small size
- Low mechanical hum and buzz
- Low stray magnetic field
- Low no-load losses
- Less weight
- More available turns/layer
- Less copper length due to longer effective bobbin length
- Magnetic field running in grain-oriented direction of the steel

It is very important that the mounting bolt through the center of the toroid not be allowed to create a shorted turn by having both ends of it conductively tied to the chassis. One end of the bolt should float. It is also undesirable to mount one toroid on top of the other. C-core, O-core and R-core transformers are also popular because of their low magnetic radiation.

The Telema PC-mount toroidal transformers are a good example. The 70064K has dual 115-V_{AC} primaries and dual 18-V_{AC} secondaries, each rated at 12.5 VA for a total of 25 VA. The 72464K is identical except for having 120-V_{AC} primaries.

The O-core transformer is a variation of the toroidal transformer concept. While the toroidal transformer core has a rectangular cross section created by winding a grain-oriented silicon steel strip, the O-core is wound with steel strips whose width changes as the core being created moves from the inside, through the middle to the outside. This results in a circular cross section as opposed to a rectangular cross section. The O-core thus ends up looking like a donut. This more idealized core results in a transformer that usually has a smaller height profile and slightly larger outside diameter (OD). The circumference of a circle is about 12% less than that of a square and 18% less than that of a rectangle, all with the same cross-sectional area. The windings are distributed evenly over the core circumference, resulting in a shorter mean turn length. The rounded core also has advantages in that the inner portions of the winding are not subjected to rectangular edges.

R-Core Transformers

The R-core transformer is growing in popularity for use in audio and similar applications [1, 10]. It shares many of the desirable properties of a toroid, but is often less expensive. Some audiophiles actually prefer the R-core power transformers over toroids but the debate is certainly unsettled.

The R-core transformer was developed in Japan. It employs a magnetic core made from a continuous strip of silicon steel that looks like an elongated rectangular donut. Picture an O-core with a circular cross section where the circumference has been shaped into a rectangle with straight sides and somewhat rounded corners. Like a toroid, the core has no gap. The primary and secondary windings are wound on opposite elongated sides with two separate bobbins. This leads to high isolation between the primary and secondary, resulting in lower noise.

The core has no gap, resulting in low leakage flux. The flow of current in the primary and secondary can be made to buck out most of the magnetic fields from the two windings, resulting in very low escape of magnetic flux. The winding of the coil on a bobbin permits each winding to be very uniform and densely wound. This is in contrast to a toroid transformer, where the wires are close to each other on the inside perimeter and spaced considerably further apart on the outside perimeter. The R-core design thus tends to use less copper per turn. R-core transformers usually cannot be mounted on a PCB.

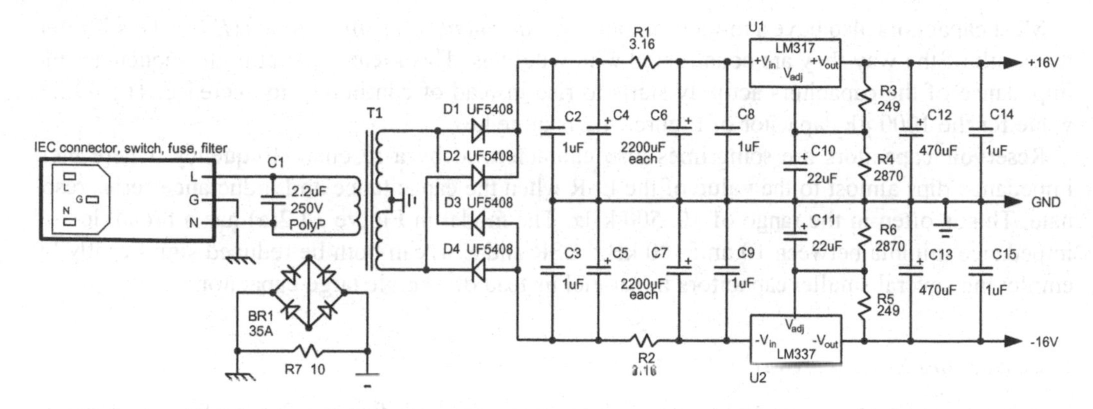

Figure 14.1 A Typical Preamplifier Power Supply.

14.3 The Reservoir Capacitors

The primary job of the reservoir capacitors is to maintain the output voltage during the time between charging pulses from the rectifier; this will reduce ripple. Calculating the ripple for a given load current and reservoir capacitor is fairly straightforward. The peak-to-peak 120-Hz ripple will be a sawtooth waveform at 120 Hz. The peak-to-peak ripple voltage can be approximated by recognizing that the voltage developed across a capacitor is $V = I\,T/C$, where T will be 8.3 ms for a full-wave rectifier operating at 60 Hz. For the circuit of Figure 14.1 operating at a load of 200 mA, with 4700-µF reservoir capacitors, the ripple on each rail is estimated to be about 350 $mV_{p\text{-}p}$.

Equivalent Series Resistance (ESR) and Inductance (ESL)

Figure 14.2(a) shows a highly simplified equivalent circuit for a high-quality 1000-µF, 35-V reservoir capacitor. It is important to recognize that the reservoir capacitors have some *equivalent series resistance* (*ESR*) and that the high peak currents of the rectifiers will create some voltage drop across the ESR independent of how large the capacitance is. This resistance varies considerably with the type and construction of the capacitor, but a value of 100 mΩ is not unusual for a small capacitor. This means that a 3-A rectifier pulse may create a drop of 300 mV across the capacitor.

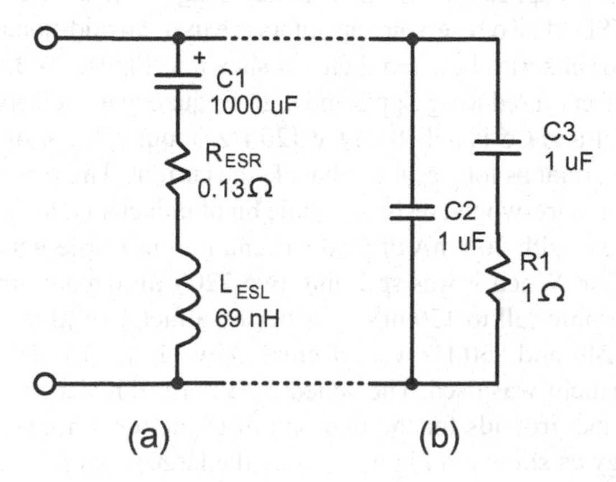

Figure 14.2 (a) Simplified Equivalent Circuit for a 10,000-µF Reservoir Capacitor. (b) A Bypass/Snubber Arrangement.

Most capacitors also have a modest amount of *equivalent series inductance* (*ESL*). This is often the result of the way they are constructed with windings. This means that at high frequencies the impedance of the capacitors actually starts to rise instead of continuing to decrease. The 69-nH value for the 1000 µF capacitor of Figure 14.2 is quite low.

Reservoir capacitors are sometimes also characterized by a resonant frequency, where their impedance dips almost to the value of the ESR when the capacitance and inductance series resonate. This is often in the range of 10–500 kHz. The model in Figure 14.2(a) has a broad, low-Q impedance minima between 10 and 100 kHz. ESR and ESL can both be reduced significantly by employing several smaller capacitors in parallel instead of a single large capacitor.

Bypasses and Zobel Networks for Reservoir Capacitors

Because large reservoir capacitors are far from perfect at high frequencies, performance can be improved by the addition of extra filtering components that have better high-frequency characteristics. One example is shown in Figure 14.2(b). The most common approach is to bypass the reservoir capacitors with a quality film capacitor of 1 µF or more as done with C2 in the figure. This will generally assure that the combined impedance will continue to fall at high frequencies above the resonant frequency of the reservoir capacitor. Even with a bypass capacitor in place there often remains the possibility of impedance peaks due to a resonance between C2 and the ESL of the capacitor. The addition of the 1-µF capacitor in Figure 14.2(b) actually created a resonance at about 600 kHz where the net impedance more than quadrupled.

For this reason a high-frequency Zobel network can be added as well. The Zobel in Figure 14.2(b) consists of R1 and C3. At high frequencies it transitions to a damping circuit with a resistive behavior. Notice that 1 µF and 1 Ω are used here. The transition from capacitive to resistive for this network occurs at about 160 kHz. It is important that the capacitor has low ESR and ESL. This Zobel network will tame the 600 kHz resonance, but not kill it. It will also help at higher frequencies where it looks like a 1-Ω resistance. Finally, connection of a low-ESR/ESL 100-µF electrolytic capacitor across the reservoir capacitor can be very helpful. Its ESR can actually help damp resonances like the one at 600 kHz mentioned here.

Split Reservoir Capacitors

Often two smaller reservoir capacitors will have an advantage over a single reservoir capacitor in terms of both ESR and ESL. If two reservoir capacitors are used an additional advantage can be had by placing a small resistor in series between them, as shown in Figure 14.3. A 2.2-Ω series resistor can have a remarkable effect on reducing ripple and high-frequency noise content. The impedance of the 2200-µF capacitors C1 and C5 is only 0.6 Ω at 120 Hz, about 1/4 that of the 2.2-Ω resistor. The type of resistor is not important as long as it can handle the current. The resistor might typically be a MOF (metal-oxide film) or wire-wound resistor. A little bit of inductance in the resistor can only help.

In a typical arrangement with 200 mA of load current, output ripple was about 350 $mV_{p\text{-}p}$ with a single 4700-µF capacitor. When it was split into two 2200-µF capacitors separated by a 2.2-Ω resistor, 120-Hz output ripple fell to 120 $mV_{p\text{-}p}$, better by a factor of almost 3. The spectral components of the ripple at 240 and 360 Hz were decreased by 11 and 13 dB, respectively, when the split capacitance arrangement was used. The added 2.2-Ω of rail resistance will drop only 0.44 V at 200 mA. Notice that the grounds for the two sets of capacitors should be physically separate in the grounding topology as shown in Figure 14.3 by the larger dots labeled S3 and S2. Node S1 can be the star ground for the system. This will be discussed further in Section 14.9 on grounding.

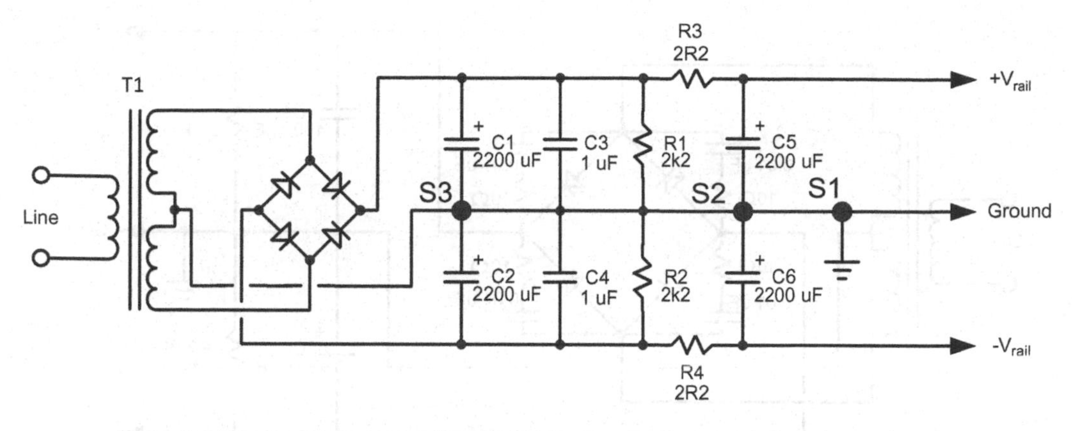

Figure 14.3 Split Reservoir Capacitor Arrangement.

Keep in mind that the power supplies for small-signal circuits will all have voltage regulators. It is helpful to keep ripple down in front of them, since their rejection is not infinite, especially at high frequencies. The split capacitor arrangement helps to keep steep rectification edges away from the regulator, and also helps filter out EMI.

14.4 Rectifier Speed

During the forward conduction pulse the peak current through the rectifier diode can be very high. As with any p-n junction, there exists a large stored charge in the form of minority carriers during this time. When the driving AC waveform decreases below the DC level stored in the reservoir capacitor on the other side of the diode, the diode must turn off, but it cannot do that instantaneously as a result of the stored charge. This charge must be sucked out of the diode by means of reverse current flow through the diode. The amount of time it takes to pull out this charge is a measure of the diode's speed and is referred to as its *reverse recovery time* (t_{rr}). The pulse of reverse current flow is especially undesirable because it can lead to RF emissions [11].

The power supply of Figure 14.1 can use an ordinary 5-A bridge rectifier for the 200 mA load. The reverse recovery current manifests itself as a significant current pulse lasting 30 μs when the rectifier in the power supply turns off. The pulse has very sharp edges with rise times on the order of 1 μs. The load current for this test was 200 mA. Instead, four discrete UF5408 soft recovery ultra fast rectifier diodes are used for faster recovery.

Soft Recovery Rectifiers

Diodes that are designed to have a fast, soft recovery will create less noise because the amount of energy that must be sucked out of them during turn-off is smaller, leading to much shorter duration of the reverse current interval and consequent turn-off spikes [12]. One type of fast recovery diode is the *Fast Recovery Epitaxial Diode* (*FRED*) [13, 14]. While a conventional bridge rectifier may have a rated reverse recovery time of 350 ns, a FRED may have a t_{rr} of 35 ns. The trade-off is that FREDs cost more and sometimes have a larger forward voltage drop at a given forward peak current.

At the end of the reverse current interval, the diode finally turns off and its current returns to zero. Once the excess carriers are swept out of the junction, some diodes will have their reverse

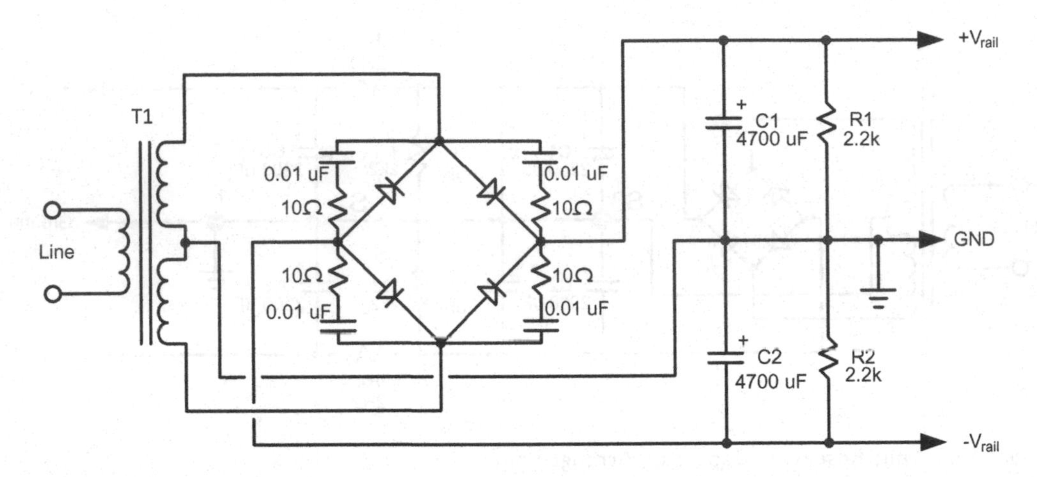

Figure 14.4 A Bridge Rectifier with Snubber Networks.

current go to zero very quickly, almost snapping back. This high rate of change of reverse current can create more noise at high frequencies. Diodes that are designed to have a *soft recovery* have a more gradual return to zero of their reverse current. The smaller rate of change of falling reverse current causes less production of high-frequency noise. Some of the better fast recovery diodes are also characterized by soft recovery.

The Vishay HEXFRED® devices are a good example [14]. The Vishay VS-HFA04TB60-M3 HEXFRED® Ultrafast Soft Recovery 4A, 600-V diode comes in a TO-220 package. Forward drop is 1.5 V at 4 A. Reverse recovery time is 17 ns at $I_F = 1$ A, while peak recovery current is 2.9 A at $I_F = 4$ A. Reverse recovery charge $Q_{RR} = 40$ nC at $I_F = 4$ A.

Rectifier Noise and Snubbers

Rectifier noise due to turn-on and turn-off can be reduced by using snubber networks across each diode in a bridge, as shown in Figure 14.4 [11]. The resistors shown are often not used, but can be very helpful in damping resonance effects. The snubber networks should be mounted right at the rectifier nodes.

14.5 IC Voltage Regulators

All preamplifiers and the like should have voltage regulators to create stable, low-noise positive and negative rails, typically at ±15 or ±18 V, even if they are just the ubiquitous LM317/337. However, low dropout (LDO) regulators with low noise are readily available [15]. The following are just a few examples of available regulators of both polarities.

LM317/337 Adjustable Regulators

The legendary LM317/337 adjustable regulators provide adequate performance for many applications if they are well-bypassed, including bypass of the control pin. Their reputation for circuit noise is actually worse than it should be if they are properly implemented. A key characteristic for voltage regulators is the minimum voltage needed to be dropped across the regulator, often referred to as the dropout voltage. Regulators with low dropout voltage reduce the amount of DC voltage needed from the unregulated power supply, and thus the amount of power dissipated in the regulator. They will also be more tolerant of input ripple voltage briefly causing them to drop out of regulation.

The LM317 and LM337 are not low-dropout (LDO) regulators, as they generally are recommended to have a minimum of 3 V_{DC} across them so that they have adequate regulation headroom. Importantly, if the supplied input voltage has ripple in it, this subtracts from the headroom margin and increases the average drop required across the regulator. For example, a 15-V_{DC} regulator requiring a minimum voltage drop of 3 V_{DC} will actually require a minimum average DC input voltage of at least 3.5 V if the source ripple is 1 $V_{p\text{-}p}$. This will result in a required average input voltage of at least 18.5 V_{DC} under worst-case line voltage, component tolerance and temperature conditions. A full-wave-rectified regulator input in a 60-Hz supply with a 2200 µF reservoir capacitor and supplying 200 mA will have p-p ripple voltage of about 0.75 V.

Noise of the LM317 is specified as 0.003% of its output voltage for a frequency band from 10 Hz to 10 kHz. For a 15-V regulator, this amounts to about 450 μV_{RMS}. With a 22 µF bypass of the V_{adj} terminal, output noise is about 200 nV/√Hz to 100 kHz. PSRR at 120 Hz is 57 dB without a bypass capacitor from the adjust pin to ground (C10 and C11 in Figure 14.1), and improves to 64 dB with a 22 µF capacitor. PSRR falls to 40 dB at 100 kHz and 20 dB at 1 MHz.

It is good practice to install reverse diodes from output to input on the regulator ICs to protect them against reverse current that could flow if there is a short circuit in the power supply ahead of the regulator. Such reverse current could come from the charge remaining on the electrolytic capacitors downstream from the regulator. It is also wise to connect a reverse diode from the regulator's adjust pin to ground. These diodes are not shown in Figure 14.1 for purposes of clarity in the figure.

150-mA Adjustable LDO Regulators

Newer regulators offer low dropout (LDO) performance, meaning that the output voltage can be as little as about 350 mV smaller than the input voltage. Most of the LDOs also feature lower noise than the older regulators. Cost is higher than the legendary LM317/LM337. Two examples are the TPS7A49 positive regulator and the TPS7A30 negative regulator. The TPS7A49 features very low noise of 13 μV_{RMS} in a 20 Hz–20 kHz frequency band, with PSRR of over 52 dB from 10 Hz to 400 kHz.

150-mA Dual-Polarity LDO Regulators

The TPS7A39 is a convenient dual-polarity LDO regulator that can supply ±150 mA. It can save space, but at almost $6 in small quantities, it only costs a little less than separate positive and negative regulators. It features low noise of 21 μV_{RMS} in a 10 Hz–100 kHz frequency band. PSRR is greater than 50 dB up to 2 MHz. It comes only in a WSON (10) leadless surface-mount package with a thermal pad on the bottom. This is not a very user-friendly package.

1-Amp Adjustable LDO Regulators

Two examples of ultra-low-noise 1-Amp LDO regulators are the TPS7A47 positive regulator and the TPS7A3301 negative adjustable regulators. In a bandwidth of 10 Hz to 100 kHz, these regulators feature output noise of just 4 and 16 μV_{RMS}, respectively. It is generally the case that negative regulators are a bit noisier than positive regulators. Maximum dropout voltage at 1 A is 450 mV_{DC}. These regulators are not inexpensive. Some do not come in particularly user-friendly packages.

14.6 Capacitance Multipliers

Sensitive circuits dealing with very small signals sometimes require extra quiet power rails. Capacitance multiplier filters are often used for these circuits. Such a circuit is shown in Figure 14.5. The headroom loss can be designed to be less than 2 V_{DC} if ripple on the main power supply is not excessive. A further advantage is that the filtered output of the capacitance multiplier can be referenced to the quiet ground without dumping a lot of ripple current into that ground. The capacitance multiplier can be located in close proximity to the circuits they feed. These alternatives to a regulator are suitable for powering circuits that do not require very precise voltages. Because they do not depend on negative feedback for regulation, they potentially have better PSRR at high frequencies. Of course, the simple capacitance multiplier provides no short-circuit protection.

Q1 is the pass transistor while R1 and R2 in combination with R3 and R4 drop the input voltage by 1 V so that Q1 operates with 1-V collector to base. An additional 0.7 V is dropped through the base-emitter junction to arrive at the output voltage. R1 and C3 prefilter the input to the capacitance multiplier. C3 is the multiplier capacitor and it forms a pole at 5 Hz for filtering purposes. C3 and C4 will often have smaller film or ceramic capacitors in parallel with them to provide enhanced rejection at high frequencies. Diodes D1 and D2 protect Q1 and Q2 from voltage reversals. Vertical power MOSFETs also make good pass transistors for capacitance multipliers if you are willing to tolerate a bit more voltage drop, which will be on the order of the threshold voltage or a bit more.

One might ask, why not just use another LDO? Voltage regulators have PSRR that decreases with frequency. A capacitance multiplier with a small pass transistor and small C_{cb} can provide isolation to higher frequencies.

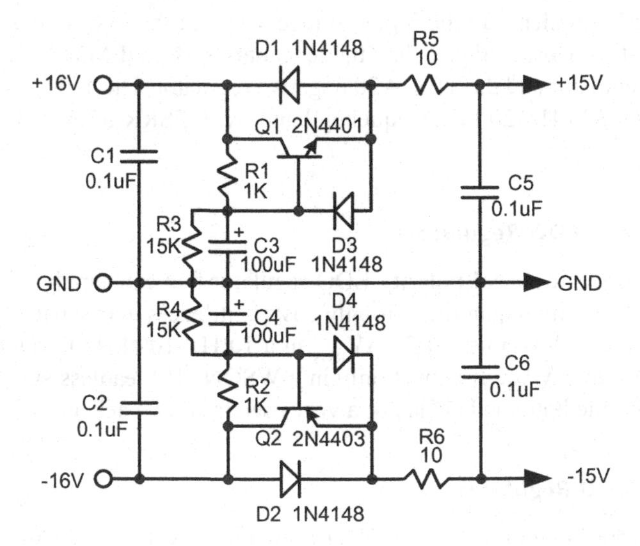

Figure 14.5 A Capacitance Multiplier Filter.

14.7 External Power Supplies

Power supplies that are external to the equipment enclosure can take many forms. Significantly, they avoid having AC mains and power transformers inside the equipment, improving the chances of achieving very low noise in the signals. These supplies also usually provide isolation from the line, eliminating the need for the safety ground, EMI filter and the possibility of ground loops that it introduces. External supplies can be linear or SMPS, and may or may not be tightly regulated.

AC Wall Transformers and External Power Supply Blocks

The simple approach is to use common Class 2 AC wall transformers that produce in the neighborhood of 18–24 V_{AC} to provide isolated low-voltage AC to the device being powered. A better approach is to do the rectification externally in an inline dongle. The AC from the wall transformer can then be fed to half-wave rectifiers and reservoir capacitors of opposite polarities, providing rails of both polarities. This circuitry can be implemented in a small inline dongle. These rails can be fed directly to the device being powered or they can be pre-regulated in the dongle to provide ±18–22 V_{DC} to the device, with final regulation being done in the device.

14.8 Switch-Mode Power Supplies

Switch-mode power supplies (SMPS) are making their way into more preamplifiers and mixers. They offer high efficiency while being compact and lightweight. Their advantages include regulated outputs with almost no penalty in power dissipation.

In its simplest form, an SMPS is a DC–DC converter that outputs a different voltage than the input voltage, either higher or lower. It uses energy transfer with an inductor as the intermediate energy storage device. If one applies a voltage to an inductor for a given amount of time, energy will be delivered to the inductor. The energy will be stored as a magnetic field in the inductor. If one then disconnects the voltage source and connects the inductor to a load, the magnetic field will collapse and deliver its energy to the load.

Simple SMPS are often non-isolated. In a system, the SMPS receiving the mains voltage must be isolated by use of a switching transformer. The design of SMPS is quite specialized but well-documented in several seminal and fairly large textbooks [16–19]. However, there are a number of more easily digestible references that can provide good coverage and understanding [20–22].

AC mains-powered SMPS typically operate by rectifying the mains voltage on the line side and storing the DC on a reservoir capacitor. This DC voltage is then switched at a high frequency (several hundred kHz) to become AC to drive an isolating transformer. The secondary voltage is then rectified and stored on secondary reservoir capacitors [22]. High-frequency transformers are smaller, lighter, and less expensive than transformers that operate at the mains frequency. Many designs incorporate voltage-regulating feedback to the primary side through a phototransistor or similar device that provides galvanic isolation. The design of SMPSs is highly specialized.

There are some caveats when switching supplies are used to power audio circuits. For one, they can be a prolific source of EMI, especially if they are not carefully designed with a good circuit board layout. Secondly, because they operate and rectify at high frequencies, the traditional ripple problem is not as great. This can lead designers to employ smaller reservoir capacitors. This is not always a good idea.

Finally, start-up current can be higher than expected, since the input of an SMPS has a negative resistance characteristic [22]. This is because, apart from efficiency, the input power equals the output power of an SMPS. This means that, for a given load, input current increases as input voltage decreases. This means that as power is applied, the SMPS draws more current initially before the input voltage ramps up to its nominal value as the SMPS works to power its load (including reservoir capacitors). An SMPS driven with a supply of limited current capability may not be able to lift the initial heavier load, and thus never be able to ramp up. This can be so even if the supply is able to support the normal SMPS input current. This can occur more frequently if the SMPS is supplied from a +5-V_{DC} USB smartphone charger, which itself is an SMPS.

This section will begin with a highly simplified discussion of five basic switching converters found in low-power, low-voltage audio equipment power supplies. These are the buck, boost, buck-boost, Ćuk, and SEPIC supplies. Some variants and other configurations are also briefly discussed. A significant issue covered is how to provide the negative supply rail required for virtually all audio circuits while using a non-isolated SMPS.

Basic DC–DC Converters

Figure 14.6 shows four basic and fairly simple non-isolated converters commonly found in electronic equipment. They are the buck, boost, buck-boost and Ćuk circuits [22]. DC–DC converters have one or more switches that are controlled by an IC controller and are usually fed with a pulse-width modulation (PWM) to turn the switches on and off. Most switches are in series with the flow of current or are shunting nodes within the circuit to ground.

Throughout this discussion the switches are depicted as simple switch symbols. The actual electronic switches are usually implemented with N-channel or P-channel vertical MOSFETs or with NPN bipolar transistors, the former being the most common external to the controller chip and the latter being the most common when integrated within the switching controller. All of these converters include one or more inductors to store energy. Advantage is taken of the fact that inductors resist changes in current amplitude or polarity. Energy is exchanged between the inductor(s) and capacitors with little loss in order to convert a DC source voltage to an output voltage of a different value or polarity. Component values are not shown in the drawings because they depend specifically on the converter controller being used.

Consider an inductor connected to a positive voltage source and whose other end is connected to ground through a switch. While the switch is closed, current will flow in the inductor, ramping up with time and charging the inductor with energy. When the switch is opened, the inductor will seek to keep that current flowing, resulting in a positive-going pulse of potentially high amplitude at the inductor's node connected to the switch. This is often referred to as the flyback action, named after the horizontal leftward flyback of the raster in an analog TV that is created by this phenomena in the horizontal deflection circuit. The automotive Kettering ignition system also relies on this behavior with the switch being the "points."

Buck Converter

The buck converter takes a DC input voltage and steps it down to a lower DC voltage. Understanding the buck converter is central to understanding most of the concepts underlying SMPS design. The buck converter uses a transistor switch, a diode, an inductor and a filter capacitor, as shown in Figure 14.6(a). It is basically a PWM arrangement that relies on the fact that an inductor tries to keep current flowing as its magnetic field collapses. It operates with high efficiency because the switching device is only *on* or *off*. The relative *on* time of the switch controls the amount of energy that is delivered to the load and thus the output voltage. In PWM terms, the ratio of switch *on* time to the switching period is referred to as D. In other words,

$$D = T_{on} / T_s \quad (21.1)$$

When S1 closes, current flows through L1 into C1 and the load. The current flowing through L1 charges it with energy as the current in L1 ramps up with time. When S1 opens, the flyback current from L1 flows into C1 and the load, producing a positive output voltage. Schottky diode D1 keeps the V_{sw} node to the left of L1 from going much below ground. Current flows into C1 and the load during both the *on* and *off* phases of the switch. The longer the *on* interval (or duty cycle), the more energy is stored in L1 and a greater amount of current flows into the load. A feedback circuit in the controller adjusts the duty cycle percentage to achieve the desired voltage at the output. A heavier load will require a longer duty cycle to maintain a given voltage at the output.

S1, being in series with the input source and the inductor, is often called the high-side switch. An N-channel MOSFET is most commonly used, with its source connected to the inductor. A boost circuit in the controller may be used to fully turn on the MOSFET by supplying a gate voltage larger than the source voltage to turn on the switch. Alternatively, a P-channel MOSFET with its source connected to the input voltage source may be turned on by pulling the gate voltage low. Finally, an NPN transistor can be connected as an emitter follower from the source to act as the switch, resulting in a 1-V_{be} voltage drop to the switch in the *on* condition. Any voltage drop across the switch creates power dissipation in the switch and reduces conversion efficiency. NPN transistors are generally slower than MOSFETs, but they can be integrated into the controller chip.

A synchronous buck converter (not shown) is implemented by replacing D1 with a shunting switch S2 to ground (low-side switch) that the controller turns on when S1 is off. This reduces losses that would be incurred in D1 and increases efficiency.

The inductor is placed between the source and the load. If $V_{in} - V_{out}$ is ΔV, then the current flowing in L1 when the switch is closed will rise with a slope of ΔV/L1, reaching a maximum at the end of the time that the switch is closed. When the switch opens, the voltage at the input side of the inductor flies negative and flows through the catch diode D1, sometimes called the *freewheeling diode*. Current continues to flow into the load as the magnetic field in the inductor collapses, decreasing with time until the switch turns back on. Most buck converters can be arranged to provide a negative voltage by swapping their ground and output pins [23].

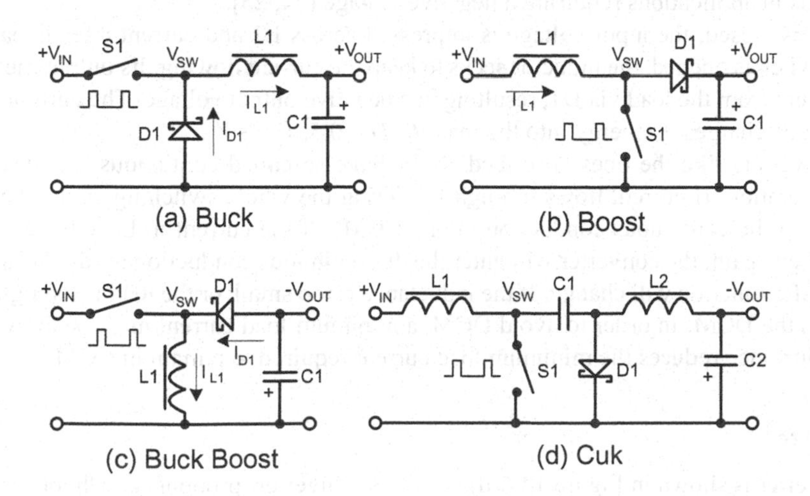

Figure 14.6 Common DC–DC Converters.

The LM2576 is a simple example of an IC buck converter. It includes an internal BJT switch and operates at about 52 kHz. The device requires only four external components, an input capacitor, and output capacitor, a 1N5822 Schottky diode and a 100 μH inductor. The LM2596 is similar, but operates at 150 kHz. A newer device, the LMR33630, is also available.

Boost Converter

A boost converter, as shown in Figure 14.6(b), takes an input voltage and steps it up. This converter includes a series inductor L1 from the voltage source followed by a low-side switch S1. The switch is turned on during the first half-cycle of operation, causing current to flow from the source through L1 to ground. The current rises with time until S1 is turned off. During this time energy is stored in L1. When S1 is opened, L1 seeks to keep the current flowing and feeds current to rectifier D1, producing a positive output. The flyback effect creates the boosted voltage.

The output of the boost converter is the sum of the source voltage and the amount of boost created. This means that the boost converter can never produce a voltage less than the source voltage. When S1 is opened, the current in L1 will decrease with time as its stored energy is transferred to the load. Because there is no inductor between the switch and the load, pulsating current is delivered to the load that can create EMI on the load port.

Buck-Boost Converter

A buck-boost converter is shown in Figure 14.6(c). It is able to produce lower voltages or higher voltages than that of the power source. It produces a negative output. A buck-boost converter can be obtained by a brute-force tandem connection of a buck converter and a boost converter. However, the resulting combination includes a shunt capacitor between two series inductors.

It can be shown that this capacitor creates a third-order low-pass filter (LPF), but that the capacitor is otherwise superfluous [19]. The capacitor can be removed and the inductors merged into one. With some further rearrangement, the output portion can be flipped and the second switch can be replaced with a commutating diode, arriving at the much-simplified arrangement of Figure 14.6(c). As a result of this simplification, the output voltage is inverted and becomes negative. This can be advantageous in applications requiring a negative voltage [24, 25].

When S1 is closed, the input voltage is impressed across L1 and current rises linearly to I_{max}. When the switch is opened, the inductor seeks to keep the current flowing. Its output flies negative, sinking current from the load via D1, resulting in a negative output voltage. The current in L1 falls linearly as it discharges its energy into the load. At $D = 0.5$, $V_{out} = V_{in}$.

Most converters like the ones described so far have so-called continuous and discontinuous modes of operation. If current flows through L1 during the whole switching cycle, the converter is operating in the continuous conduction mode (CCM). If the current in L1 falls to zero before S1 is turned on again, the converter will enter the discontinuous conduction mode (DCM), and the control transfer function will change. If the inductance is too small, or the load is too light, the converter enters the DCM. In order to avoid DCM, a minimum load current must be drawn. A larger value of inductance reduces the minimum load current required to remain in CCM.

Ćuk Converter

A Ćuk converter is shown in Figure 14.6(d). The Ćuk converter (pronounced Chook) is an important member of the boost-buck family, and can be thought of as a reduced form of the boost-buck converter. It was named after Slobodan Ćuk, who presented it in 1976 at the IEEE Power Electronics Specialists Conference, co-authored with Robert Middlebrook [26]. It has sometimes been referred to as the "Optimum Topology Converter," but this assertion has understandably been controversial [27, 28]. Additional understanding of the Ćuk converter can be found in Refs. [22, 29–31].

The Ćuk converter can produce a higher or lower output voltage than the input, but the non-isolated version can produce only a negative output. It requires two inductors and one low-side switch. It is somewhat like a boost converter followed by a buck converter, with the capacitor of the boost converter coupling the energy to the buck converter. Because there are inductors on both sides of the switch, less EMI is transferred to the input and output terminals, making it a quieter converter.

In analyzing operation, first recognize that in the steady state C1 has $V_{in} - V_{out}$ across it, and that C1 is large enough that there is little ripple across it. When S1 is closed, the input voltage is impressed across L1. As current increases, more energy is stored in L1. When S1 is opened, L1 seeks to keep the current flowing, causing V_{sw} to fly positive. This charges C1 through D1. When S1 again closes, V_{sw} is pulled negative, D1 is reverse-biased and current is sunk through L2, storing energy in L2. All of the energy in L1 is thus transferred to L2, via coupling capacitor C1. If the duty cycle for S1 is D, the conversion ratio is $M = -D/(1 - D)$. The converter can be made to create a positive output if an inverting transformer is interposed between two halves of C1 [22].

Inverting Buck Converter

A buck converter as shown in Figure 14.7(a) can be connected to produce a negative output [23]. This is done by swapping the ground and output pins, as shown and re-drawn in Figure 14.7(b). There are three external nodes in a buck converter: A, B and C. Node A is always the positive source input. Node B is normally ground, but now becomes the negative output. Node C is normally the positive output, but now is connected to ground.

When S1 is closed, L1 is charged with positive current flow at its dot end from S1. When S1 opens, L1 tries to maintain the same current flow, and its flyback pulls the cathode of D1 negative, resulting in a negative output at node B. The higher the duty cycle, the more energy is stored in L1. This increases the ability to pull the output negative to the desired voltage. The internal feedback loop of the converter maintains the voltage between B and C at the desired magnitude, just as it does in normal operation.

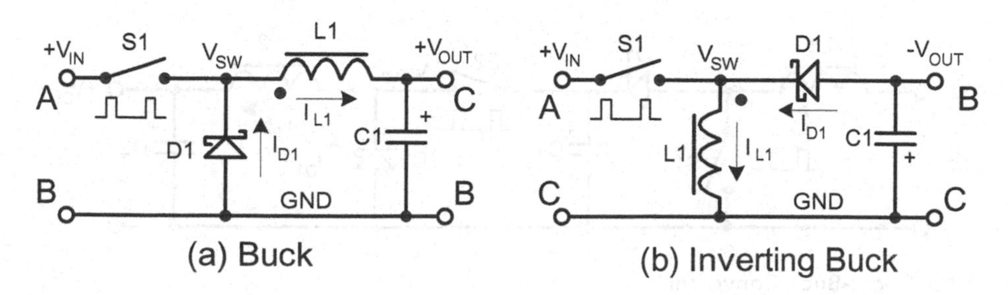

Figure 14.7 Positive and Negative Buck Converters.

The normal maximum rated voltage for the converter is that across A and B. When inverting, the voltage from A to B is the sum of the magnitudes of the input and output voltages. That sum must usually be less than 40 V. With a 20-V_{DC} input and −15-V_{DC} output, that voltage is 35 V_{DC} and there is 5 V_{DC} of margin against an over-voltage stress. With a 24-V_{DC} source, the absolute maximum output is −16 V_{DC}.

Inverting Buck-Boost Converter

There are numerous inexpensive sources of single-polarity DC that can be external to the audio device being powered. These include 24-V_{DC} wall transformers and computer power supply "bricks." The latter is a good example, typically supplying +19 V_{DC} at several Amperes. This power can be brought into the enclosure and regulated to +15 V_{DC} in a straightforward way. But what about the −15-V_{DC} supply that is inevitably needed for op amp circuits? For this, a so-called inverting buck-boost converter can be used. This is a very simple switching converter that is available as an IC and requires little more than an external inductor. It can take the +19-V_{DC} supply and invert it to become a −19-V_{DC} supply that can then be regulated to power the −15-V rail.

The inverting buck-boost converter in Figure 14.6(c) uses a high-side switch to charge an inductor in the *on* state. Then the switch opens the inductor creates a flyback of negative polarity that is then rectified.

Boost-Buck Converter

If a boost converter is followed by a buck converter, the output will be a boosted positive voltage. Such a circuit is shown in Figure 14.8. It requires two switches and two inductors. The boost-buck converter family is topologically somewhat the reverse of the buck-boost family. It literally comprises boost and buck converters wired in tandem. It requires two each of every component. Switches S1 and S2 are opened and closed at the same time. It can provide boost or cut of the input voltage without inversion of the output voltage polarity. A further advantage is that there is no pulsating current present on either the input or output side of the converter as a result of the use of two inductors. This makes the converter relatively quiet. The conversion ratio for the boost-buck converter is $M = D/(1 - D)$.

When S1 closes, L1 is charged with energy by the input source. S2 closes at the same time, delivering the charge energy on C1 to L2. When S1 opens, flyback current from L1 charges C1. Switch S2 also opens at this time, and flyback current from L2 flows into C2 and the load.

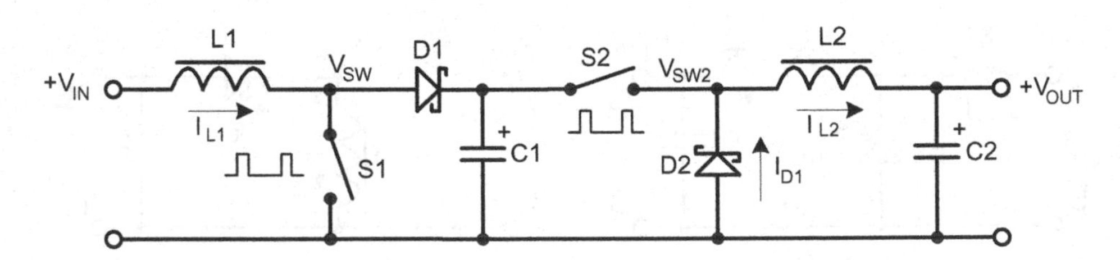

Figure 14.8 Boost-Buck Converter.

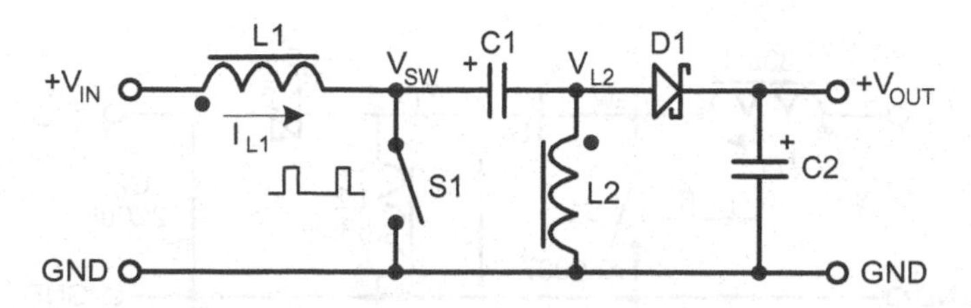

Figure 14.9 SEPIC Converter.

SEPIC Converter

The single-ended primary-inductor converter (SEPIC) can produce a positive output that is less than or greater than the source voltage [32]. In function it is like a buck-boost converter, but one that produces a positive output rather than a negative output voltage. It requires two inductors but only one low-side switch. This can be quite convenient (Figure 14.9).

When S1 closes, L1 is charged positive at its dot. As usual, the longer the duty cycle for S1 being closed, the more energy is stored in L1 as current ramps up. When S1 opens, L1 keeps the current flowing to the right, into C1. The transient current flowing through C1 initially flows through D1 to C2 and the load with a positive output polarity. Because of the positive voltage at D1's anode (approximately the output voltage) current in L2 will begin to ramp up. When the anode voltage of D1 falls below the output voltage, D1 is open and charge will flow back and forth between C1 and L2.

Charge Pumps for Creating a Negative Rail

Charge pumps are often used to create a negative voltage from a positive voltage that alternates between ground and a positive value on opposite half cycles. If you start with a square wave or similar waveform and AC-couple it, and then connect a diode to ground to prevent the voltage from going positive, the resulting voltage will be pulsating from approximately ground to negative on alternate half cycles. It will have the same peak-to-peak amplitude and wave shape as the original waveform, and be net negative. If this voltage is fed through an inductor to a load, it will be filtered and the inductor will seek to have the resulting current into the load continue during the half-cycle when the diode is conducting. The switching node in a positive DC–DC converter can be used to feed the charge pump.

Dual-Polarity Converters

Figure 14.10 shows a SEPIC converter that includes a charge pump so that output voltages of both polarities are generated. The circuit is based on the XLSEMI XL6007, which is depicted simply as a switch in the figure for simplicity. The chip includes an internal MOSFET as the switching device and operates at about 400 kHz. Apart from the ground and switch terminals, the 8-pin chip includes a V_{in} pin and a feedback pin. A voltage divider from the positive output to the feedback pin sets the output voltage against an internal 1.25-V_{DC} reference.

The charge pump for the negative output comprises C3, D2, L3 and C4. The AC waveform at node V_{L2} is AC-coupled to diode D2, causing C3 to be charged on positive swings at V_{L2}. On negative swings, D2 turns off and its anode swings negative by the charge on C3. The resulting half-wave negative voltage is filtered by L3 and C4. Notice that L3 continues to conduct on all or

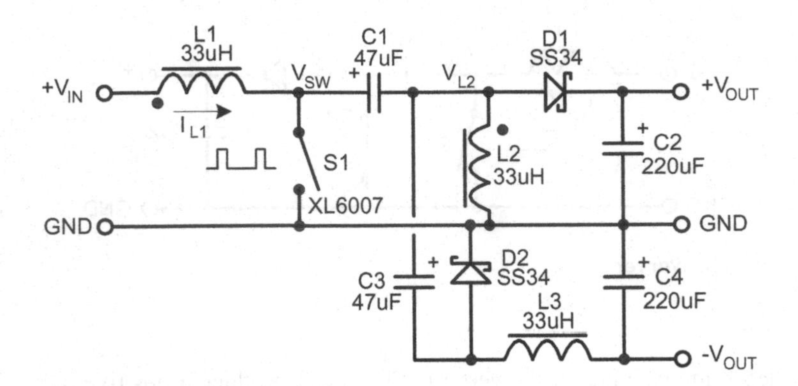

Figure 14.10 Dual-Polarity Converter.

part of the half-cycle when D2 is *on*. The negative output is nominally about 0.5 V less than the positive output and is not tightly regulated.

A small, completely functional board can be had for less than $11 online at places like Amazon. The supply measures only 0.9 by 1.6 in. Using pre-built boards like these frees the designer from the detailed design and layout of SMPS. There is a caveat if these economical SMPS are run from a 5-V_{DC} source to produce higher voltages like ±18 V_{DC}. For a given output voltage and current, SMPS efficiency usually decreases, especially if the SMPS is operating near its maximum output current.

For the SMPS characterized here delivering ±18 V_{DC} at 120 mA, efficiency was 83% when running from 12 V_{DC}. Efficiency dropped to 64% when running from a 5-V source. This can lead to unexpected trouble when the source current available is limited (as running from a USB charger). Never assume that your SMPS is operating at 90% efficiency. In some cases, efficiency may be as low as 50% while the converter is still doing its job, but getting hot doing it.

The XL6007 has a wide input voltage range of 3.6–24 V_{DC}, so it can be powered by a USB source, like that of a smartphone charger. Fed with a USB charger, the supply was able to provide 120 mA at ±18 V_{DC}. Inexpensive implementations of such converters, usually with lossy inductors, can be less efficient than the typical 90% that we often hear about, but that is not a big problem for many applications. Such converters are usually able to supply more output current and higher efficiency when powered from a higher voltage, like 12 V_{DC}. For applications requiring higher current, converters based on the XL6019 are available.

USB Powering

Powering devices from USB has become ubiquitous, with +5 V_{DC} available at computers, cell phone chargers, battery packs and automobiles, even when the serial data connection is not being used. When using a USB wall charger, isolation is provided and no safety ground is needed. The UL (Underwriters Laboratories) safety work has already been done for you. USB wall chargers can easily be found with capability up to 2.1 A or more, corresponding to over 10 W. For even greater isolation, one can power the equipment from a rechargeable cell phone battery charger.

Noise can be a concern with ordinary USB wall chargers. Although they feature galvanic isolation, noise from the line may sneak through. Noise in the common mode on the 5-V_{DC} output can be of special concern. One must ask how well the line-load isolation holds up at high frequencies where EMI lives. How much switching noise escapes the charger in the differential mode? Even if the differential noise on the 5-V_{DC} output can be snuffed out with a large capacitor at the client device input, the switching noise current may cause noise problems for some systems. Obviously, not all wall chargers are the same in these respects.

Wall Transformer Powering

Class-2 wall transformers that provide between 9 and 24 V_{DC} are readily available at low cost. These devices have long been used to power consumer equipment. Some are linear supplies and some are SMPS. One of the first wall transformers appeared in 1959, powering the lighted Princess telephone.

Computer Brick Powering

A handy source of DC voltage that can produce greater power is the ubiquitous computer brick used to power laptop and small desktop computers. These usually provide 19 V_{DC} at between 40 and 140 W. They can be used to power SMPS inside equipment requiring higher power and/or multiple voltages and polarities. Due to their very high-volume production, their cost is quite low.

Post-SMPS Regulation

It is always wise in an audio application to follow an SMPS with low-noise LDO regulators to isolate the SMPS noise. A good off-shelf example is a USB-powered supply, called the *Silent Switcher*, developed by Jan Didden [33]. It includes the SMPS and linear regulators on a single board that is carefully designed and laid out to provide extremely low noise. Using the +5-V_{DC} supplied from an ordinary phone charger (2.1 A) or computer USB port, it supplies ±15 V_{DC} at up to 150 mA to power small electronic devices, which can include preamplifiers of modest complexity. Using a 9-V, 2-A wall-wart will double the available output current. It will also supply +5 V_{DC} at up to 0.5 A, which is enough to power some digital logic, digital displays and even a microcontroller.

14.9 Grounding Architectures

Ground is the reference for all signals in an audio circuit, but not all grounds are the same. Currents flowing through grounds create voltage drops, even across seemingly heavy ground conductors. All ground paths have resistance and inductance, even though the values may be small. Magnetic fields can also induce voltages across ground conductors. The most obvious symptom of poor or inadequate grounding is hum and noise. However, a much more insidious result of poor grounding is distortion. Any ground “noise” that is correlated to the program signal in any way is considered distortion [34–36].

Noisy and Quiet Grounds

In simple terms, there are two kinds of grounds – quiet signal grounds and noisy power grounds. All signal voltages that are single-ended should be referenced to quiet grounds. Quiet grounds are distinguished as grounds that have little or no current flowing through them.

When Ground Is Not Ground

Most ground points in a preamplifier are interconnected with wires, PCB traces or ground planes (never use the chassis). These paths have some impedance due to resistance and *self-inductance.* This is especially a concern for high frequencies or currents with sharp waveform edges (like rectifier spikes). Self-inductance of a straight small-AWG piece of wire is about 25 nH per inch, only a mild function of wire gauge if length is significantly larger than diameter. The impedance of a 75-nH 3″ wire at 20 kHz is about 10 mΩ.

Star Grounding

If all grounds are returned to a single point, then that point can be considered as the ground reference for the entire circuit. This is the idea behind a *star ground* arrangement. The two reservoir capacitors can be connected by a short wide PCB trace. The center of this heavy conductor is defined as the star, and all other ground connections are returned to this point, including the center tap of the power transformer. Separate ground lines can be used for the quiet ground and the power ground.

If no current flows in a quiet ground connected to the center of the star, the potential on the quiet ground will remain unchanged; the same ground reference will exist at all places in that node. This will be the *true* ground for the preamplifier. The exception to this is if there is a voltage that is magnetically induced in the connection of the quiet ground to the star ground point. Noise and ripple currents can pass to the star ground without causing a disturbance to the quiet ground. Note that the "ground" potential at the far end of lines carrying such currents will be moving with respect to the true ground reference at the star point.

However, there are some caveats. With heavy current flowing through it, the star ground node itself can develop some small voltage potentials among the several wires connected to it. In other words, it cannot really be a perfect equipotential point. Never use steel hardware to fasten together the elements of a wired star ground.

Star-on-Star Grounding

A star-on-star grounding architecture like that shown in Figure 14.11 is a better approach [22]. All of the high-current rectifier spikes passing through C1 and C2 are resolved at the auxiliary star-3 ground (S3) independently of the main star ground. Split reservoir capacitor returns from C5 and C6 are merged at star S2. Ground node S2 is physically separate from ground node S3 even though it connects to S3 through a trace. In addition to providing *R–C* filtering, resistors R3 and R4 prevent the formation of an AC ground loop among the reservoir capacitors. This arrangement keeps those dirty high currents away from the main star ground and keeps them from corrupting it.

An important concept is to have dirty currents circulate locally and be resolved before they are passed to a ground node involving other circuitry. If large electrolytic capacitors are used on the circuit board local to noisy loads, their grounds should be tied together before connection to another ground, such as the main star ground. The common ground of these capacitors can be thought of as another auxiliary star ground connected to the main star ground through a spoke.

The star-on-star approach is typical of the general approach to good grounding practice; follow the currents and beware of the voltage drops they can cause. More complex star-on-star grounding arrangements can be imagined when a thorough analysis of current flows is carried out. In many

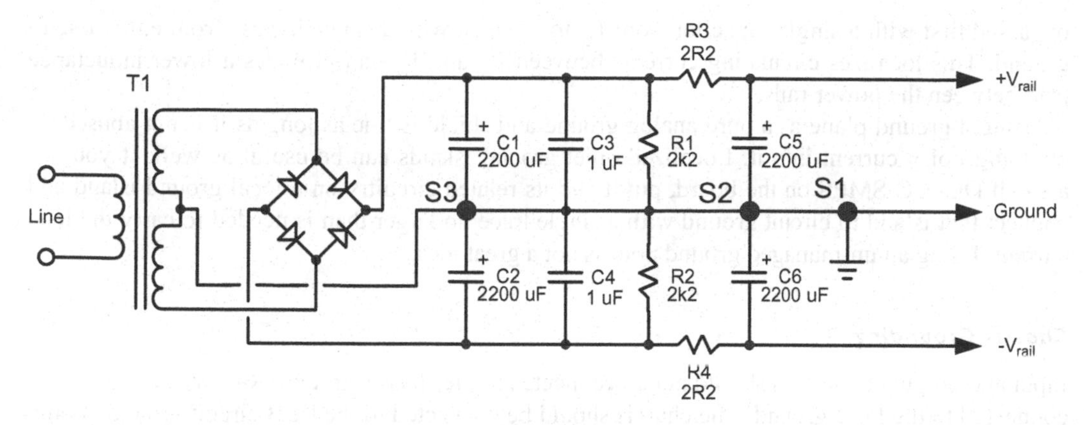

Figure 14.11 Split Reservoir Grounding Arrangement.

cases real-world preamplifiers end up having some sort of a star-on-star grounding architecture without it having been intentionally designed that way.

Some systems contain numerous local switching regulators due to a large number of different power supply voltages required by different sub-circuits, like complex ICs that require two or more power supply voltages, such as 3.3 and 0.9 V, the latter required for portions of the IC that employ deep sub-micron transistors (e.g., a 22-nm process node). In these cases, it is often helpful to put all of the components of the switching regulator on a small ground island that is connected to the main digital ground through a short trace at only one point on the ground island. This forces most of the noisy currents in the regulator circuit to be resolved (circulated) before connection to the digital ground.

Ground Corruption

One source of trouble that is often overlooked is the corruption of a quiet analog ground by currents injected into the ground by power supply bypass capacitors. This can happen when the main rail is routed to sensitive circuits through a small resistor and then bypassed. Bypass capacitors can ruin an otherwise clean star grounding architecture by creating an AC ground loop and providing another path for alternating current flow.

Ground Planes

The indiscriminate use of ground planes and ground pours is not always the best way to go for audio-frequency analog circuits. They interfere with intentional and planned flow of noisy currents. Indeed, just grounding power supply bypass capacitors to a ground plane can inject noise into what is supposed to be the quiet ground. Ideally, no such currents should flow in the ground lines or ground plane. One should be intentional about where any currents flow, recognizing that current flows though the path of least impedance. This can also be a factor in achieving stability.

Of course, local op amp bypass capacitors can and should be grounded to the local signal ground. A value of 0.01 μF is adequate for the intended purpose of HF stability and is less likely to inject power rail noise into the analog ground. Sometimes it is recommended that op amps be bypassed first with a single capacitor from V_{cc} to V_{ee}, then with a single bypass from either one to ground. This localizes circulating currents between V_{cc} and V_{ee} *and* provides a lower-inductance path between the power rails.

Using a ground plane as a pure analog ground and shield is fine as long as it is not abused by dumping noisy current into it. Localized quiet ground islands can be useful as well. If you use a small DC–DC SMPS on the board, put it and its related circuitry on a local ground island and connect that island to circuit ground with a single trace no larger than is needed to carry the load current. Using an un-managed ground pour is not a great idea.

Chassis Grounding

Input and output connectors should not be connected to the chassis ground. Rather, they should be connected to the PCB ground. The chassis should be connected to the PCB circuit ground at only one point. Ground current should never flow in the chassis.

14.10 Radiated Magnetic Fields

The power transformer is usually the single biggest source of radiated magnetic fields. These fields can include both hum and rectifier spikes. However, any wiring that is carrying high current can radiate a magnetic field that will induce voltages in nearby wiring. Toroidal power transformers can sometimes be rotated to minimize the impact of their radiated fields.

The best way to pick up an induced signal from a magnetic field is to build a loop with a fairly large area. Pickup from radiated magnetic fields is reduced if the loop area is small. This means that signal paths and their returns should be close to each other and not form a loop larger than absolutely necessary.

The possibility of magnetic coupling is maximized when wires pass parallel to each other. One wire carrying an aggressor current will create a magnetic field and induce a voltage into the victim wire. This transformer effect can be minimized if wires carrying current are crossed at a 90° angle.

14.11 Safety Circuits

Regardless of how good an audio device sounds, safety must not be ignored. The two main aspects of safety are fire and the hazard of electrocution. The primary defense against fire is the line fuse or circuit breaker. A slow-blow characteristic is desirable for the line fuse. This is required due to the power transformer inrush current at power-on, which can be higher for a toroid transformer.

Safety Ground

Three-wire mains power cords include a safety ground as the third (green) wire. This wire runs right back to the house ground via the electric outlet. This minimizes the danger if the neutral return path becomes open and is energized through the resistance of the load device. The safety ground is connected to the chassis so that if there is a fault to the chassis it will be shunted by the safety ground, likely triggering a GFCI disconnect. A typical fault that this guards against is a primary-secondary short in the power transformer. If the power transformer incorporates an inter-winding screen, it should be connected to the safety ground.

Interestingly, if the AC voltage between the safety ground and neutral is measured, a rough indication of the quality of the AC line under load can be inferred. Ideally, there will be no voltage measured when there is no load on the line. This measurement will give an indication of the voltage drop in the neutral line due to loading, since there should be no voltage drop in the safety ground wire. If the only significant loading on the line is at the location of the audio device, the total AC line drop from the breaker panel to the device can be inferred to be twice this measured voltage. A fairly long 14-AWG mains line in a residential setting feeding a beefy stereo system can show a significant drop.

Breaking Safety Ground Loops

Several pieces of interconnected equipment, each with a safety ground connected to its circuit ground via its chassis ground, may form a ground loop among the various pieces of equipment. This is undesirable. One approach, similar to that used in Ref. [37] is illustrated in Figure 14.12. The safety ground is connected directly to the chassis ground. However, small-valued resistor R1 connects the safety ground to the circuit star ground. This is the only connection in normal operation, and the resistance breaks the ground loop. Diodes D1 and D2 are connected across R1, preventing the voltage from the safety ground to the circuit ground from exceeding one diode drop. The diodes must be properly sized to handle the current resulting from a fault condition.

Alternatively, as seen in Figure 14.1, a 35-A bridge rectifier is connected in parallel with R7. In the event of a fault, the diodes in the bridge rectifier conduct the fault current to the safety ground and prevent the circuit ground from ever deviating from the safety ground by more than two diode drops. The AC terminals of the bridge rectifier are connected across R1 and the positive and negative "output" terminals of the bridge rectifier are shorted together. In this case the maximum allowed voltage drop between safety ground and circuit ground is about 1.4 V.

While not directly related to safety considerations, one should keep in mind that EMI/RFI line inlet filters often have some capacitors connected from the hot and neutral lines to the safety ground. These "*Y*" capacitors are about 5000 pF. These capacitors, although small, can inject some leakage current into the safety ground. If a large number of pieces of equipment are fed from the same line circuit, the sum total of this capacitive leakage current can add up to enough current to trip a GFCI (ground fault circuit interrupter). A GFCI generally trips at about 4 mA of leakage to the safety ground. A capacitance of 5000 pF with 120 V_{AC} across it will conduct about 240 μA.

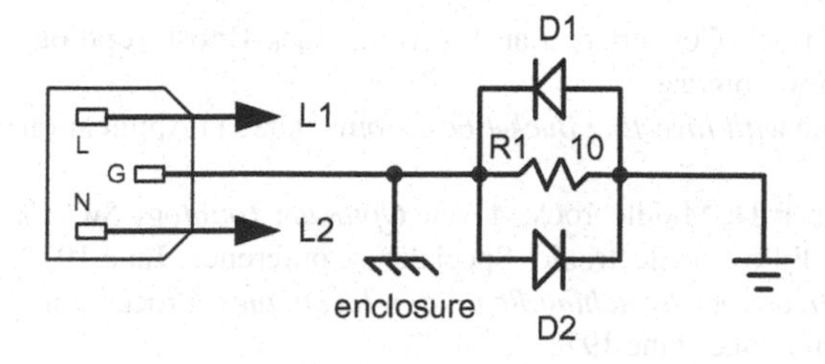

Figure 14.12 Breaking a Safety Ground Loop.

References

1. Pete Millett, Power Transformers for Audio Equipment, *audioXpress*, June 2001, audioxpress.com/article/Power-Transformers-for-Audio-Equipment.
2. D. Baert, Designing Small Transformers, *Electronic and Wireless World*, August 1985, pp. 17–19.
3. Chuck Hansen, Power Transformer Parameters, Selection and Testing, Part 1, *Transformer Core Materials, audioXpress*, vol. 53, no. 11, November 2022, pp. 58–61.
4. Chuck Hansen, Power Transformer Parameters, Selection and Testing, Part 2, *audioXpress*, vol. 53, no. 12, December 2022, pp. 50–56.
5. Chuck Hansen, Power Transformer Parameters, Selection and Testing, Part 3, *Insulation Materials, Winding Bobbins, and Testing Methods, audioXpress*, vol. 54, no. 1, January 2023, pp. 58–66.
6. Chuck Hansen, Power Transformer Parameters, Selection and Testing, Part 4, *Parameters and Their Losses, audioXpress*, vol. 54, no. 2, February 2023, pp. 48–52.
7. Chuck Hansen, Power Transformer Parameters, Selection and Testing, Part 5, *Losses, audioXpress*, vol. 54, no. 3, March 2023, pp. 56–61.
8. Chuck Hansen, Power Transformer Parameters, Selection and Testing, Part 6, *Quality, Reliability and Design, audioXpress*, vol. 54, no. 4, April 2023, pp. 56–61.
9. Toroid Corporation of Maryland, Technical Bulletins, No. 1, 3, 4, www.toroid.com.
10. MPAudio, *Advantages of Toroidal vs EI and R-Core Transformers*, October 25, 2019.
11. Morgan Jones, Rectifier Snubbing – Background and Best Practices, *Linear Audio*, vol. 5, 2013, pp. 7–26.
12. Mark Johnson, Soft Recovery Diodes Lower Transformer Ringing by 10–20×, *Linear Audio*, vol. 10, September 2015, pp. 97–108.
13. IXYS VBE 26-12N07 data sheet for FRED 32A bridge rectifier.
14. Vishay data sheet for VS-HFA04TB60-M3 HEXFRED® Ultrafast Soft Recovery 4A, 600-V rectifier.
15. Steve Hageman, *Simple Circuits Reduce Regulator Noise Floor*, EDN, October 15, 2013, edn.com/simple-circuits-reduce-regulator-noise-floor/.
16. Keith Billings, *Switchmode Power Supply Handbook*, McGraw-Hill, New York, NY, 2010.
17. Pressman, Billings & Morey, *Switching Power Supply Design*, 3rd edition, McGraw-Hill, New York, NY, 2009.
18. Robert Erickson, *Fundamentals of Power Electronics*, Springer, New York, NY, 1997.
19. Rudy Severns and Ed Bloom, *Modern Dc-to-Dc Switch Mode Power Converter Circuits*, Van Nostrand Reinhold, 1985.
20. Linear Technology, *Linear Technology Magazine Circuit Collection Volume II Power Products*, Application Note 66, 1996.
21. Maxim Integrated Products, *DC-DC Converter Tutorial 2031*, https://www.maximintegrated.com/en/app-notes/index.mvp/id/2031, 2001.
22. Bob Cordell, *Designing Audio Power Amplifiers*, 2nd edition, Focal Press, New York, NY, 2019, Chapter 20, "Switching Power Supplies."
23. Len Sherman, *Making a Voltage Inverter From a Buck (Step-Down) DC-DC Converter*, Maxim (ADI) Application Note 3844.
24. John Tucker, Using a Buck Converter in an Inverting Buck-Boost Topology, *TI Analog Applications Journal*, 4Q2007, www.ti.com/aaj.
25. Frank De Stasi, *Working with Inverting Buck-Boost Converters*, TI Application Report SNVA856A, May 2020.
26. Slobodan Cuk and Robert D. Middlebrook, *A New Optimum Topology Switching DC-to-DC Converter*, Proceedings of the IEEE Power Electronics Specialists Conference, June 1977.
27. Rudy Severns, *High Frequency Switching Regulator Techniques*, Proceedings of the IEEE Power Electronics Specialists Conference, June 1978.
28. Rudy Severns and Hal Wittlinger, *High Frequency Power Converters*, Intersil Application Note AN9208, April 1994.

29. Kevin Scott and Jesus Rosales, *Differences between the Cuk Converter and the Inverting Charge Pump Converter*, Analog Devices, Inc., Analog Dialog, analog.com.
30. Electrical Technology, *What Is a Cuk Converter and How It Works*, Blog, September 2021, https://www.electricaltechnology.org/2021/09/cuk-converter.html.
31. Anushree Ramanath, *What Is a Cuk Converter?* EE Power, August 3, 2020, https://eepower.com/technical-articles/into-to-cuk-converters-part-1/#.
32. Wikipedia, "Single-Ended Primary-Inductor Converter." wikipedia.org.
33. Jan Didden, *The Silent Switcher*, www.linearaudio.nl/silentswitcher.
34. Bill Whitlock and Jamie Fox, *Ground Loops: The Rest of the Story*, presented 5 Nov 2010 at the 129th AES convention in San Francisco. Available from AES as preprint #8234.
35. Bill Whitlock, *An Overview of Audio System Grounding and Signal Interfacing*, Tutorial T5, 135th Convention of the Audio Engineering Society, October, 2013, New York City.
36. Bruno Putzeys, The G-Word, or How to Get Your Audio Off the Ground, *Linear Audio*, vol. 5, 2013, pp. 105–126.
37. Bryston 4B-SST amplifier schematic.

Chapter 15

Moving Magnet Phono Preamplifiers

The resurgence of Vinyl phonograph playback in the last 20 years has been remarkable. Vinyl LPs are popular among many audiophiles, and the task of high-performance playback preamplification for them continues to be of great interest. The design of phono preamplifiers includes good examples of technical challenges and diversity in preamplifier circuit architecture. The task of the phono preamplifier is to provide a substantial amount of gain with low noise and accurate RIAA equalization. The phono preamp must also provide cartridge termination.

Phono preamplifiers are required to provide the necessary gain and RIAA equalization between the cartridge and the line-level inputs of the preamp. While at one time phono preamps were included in preamplifiers, integrated amplifiers and receivers as a matter of course, they are not always included in contemporary equipment. Often, they are incorporated as a separate unit in audiophile systems, just as a DAC is often included as an added external box that connects to the line input of a preamplifier.

Conventional phono preamps are designed to accept the input from a moving magnet (MM) cartridge, whose output may be on the order of a few mV. Noise and accuracy of RIAA equalization are important performance characteristics. Input overload and gain are also important.

15.1 Gain and Equalization

Figure 15.1 shows the schematic of a very simple MM phono preamp based on an op amp as the gain element. It is a simple feedback amplifier whose gain and frequency response are established by the feedback network. A JFET op amp is used for simplicity, allowing DC coupling of the cartridge to the amplifier. R1 and C1 provide typical cartridge loading.

C3 and C4, in combination with R3 and R4, comprise the RIAA equalization network. R2 is the shunt part of the RIAA network, and largely determines the 1-kHz gain for a given set of values in the equalization network. C2 is a large-value electrolytic capacitor that allows the gain of the preamp to fall to unity at DC so that amplifier offsets are not multiplied by the high gain at low frequencies. Gain at 1 kHz is 35 dB. RIAA equalization calls for about 20 dB of boost at 20 Hz and 20 dB of cut at 20 kHz. R4 and C4 are primarily responsible for the former, while R3 and C3 are mainly responsible for the latter. R3 and R4 are E96 resistor values. C3 is made up of 5600 pF in parallel with 511 pF.

Lipshitz designated six frequencies that define the poles and zeros that usually exist in the real-world RIAA equalizer [1]. The three most important of these are depicted in Figure 15.2. Not all of the six exist in all RIAA preamps. C2 limits the gain of the stage to unity at DC, preventing large DC offsets that would otherwise result. As the reactance of C2 gets smaller with increasing

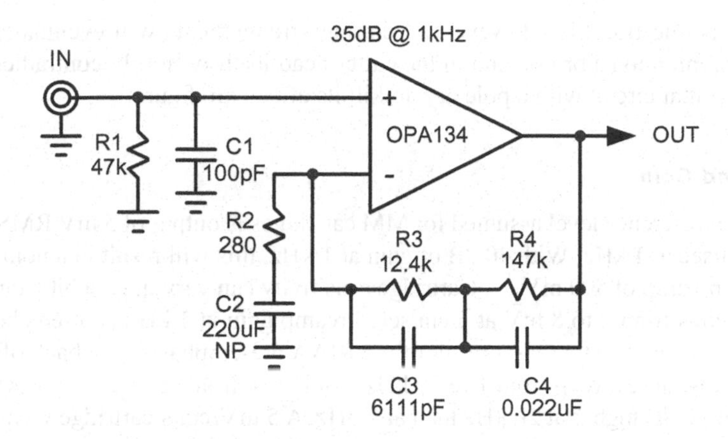

Figure 15.1 A Simple Moving Magnet Preamplifier.

frequency, gain begins to rise above unity, there is a zero located at frequency f_1, here 0.0047 Hz. As the reactance of C2 falls to near the value of R2, there will be a pole at f_2 as the low-frequency gain nears its full value, usually at a frequency between 1 and 20 Hz. Here f_2 is 2.4 Hz.

Frequencies f_1 and f_2 are not part of the current RIAA equalization specification. They exist as a practical matter in many designs. In a now-extinct optional version of the specification, f_2 was specified by the IEC as 20 Hz. This was to limit low-frequency response to reduce the amplitudes of rumble and flutter at infrasonic frequencies. However, the shallow 6 dB/octave high-pass roll-off degraded the in-band low-frequency response and did not provide much attenuation at the dominant rumble and flutter frequencies. The 20-Hz IEC RIAA roll-off has been abandoned and withdrawn. Very few designs place f_2 at 20 Hz, with it often being below 5 Hz. If the IEC 20-Hz roll-off were implemented, it would certainly not be wise to set it with the electrolytic capacitor C2, whose tolerance is inevitably bad, and which would cause distortion.

If you must use an electrolytic, use a large-value one that provides a 3-dB roll-off around 2 Hz. This keeps the in-band audio signal voltage across the capacitor small, leading to less distortion. Use of a non-polarized electrolytic with a higher voltage rating also reduces distortion. In a later section the infrasonic gain issue is optionally dealt with using a proper infrasonic high-pass filter of higher order.

Now we turn to the meat of the RIAA specification. As the impedance of the feedback network begins to fall due to the action of C3, low-frequency gain begins to fall with a pole at $f_3 = 50.05$ Hz. At the higher frequency $f_4 = 500.5$ Hz, the network introduces a zero into the response, bringing it back to flat. At $f_5 = 2122$ Hz, C4 introduces a pole, causing the gain to again decline with frequency at 6 dB/octave. The frequencies f_3, f_4 and f_5 are the key frequencies that define the RIAA equalization. With the exception of the approximate 2-octave flat region, the RIAA curve falls off at about 6 dB/octave. It is important to point out that all of the official break frequencies are specified by time constant. For example, $f_5 = 2122$ Hz corresponds to T5 = 75 µs.

The gain of the circuit of Figure 15.1 cannot go below unity, so there must be a zero at frequency f_6 where the response levels off and becomes flat at unity. Here f_6 is at 119 kHz. There was a time when f_6 was set to about 50 kHz to correct for the so-called Neumann pole in the recording cutter head. The 50-kHz Neumann pole should be ignored, as few if any cutters enforce it; even

Neumann's use is questionable. However, all cutting arrangements will eventually have one or more such poles, intentional or not, and in the cutter head itself or not. In contradiction, there are those who follow that circuit with a pole at f_6 to kill its effect. Go figure.

Input Levels and Gain

The conventional reference level assumed for MM cartridges is output of 5 mV RMS at a recorded velocity of 5 cm/sec at 1 kHz. With 40 dB of gain at 1 kHz, this will result in a nominal line-level output from the preamp of 500 mV_{RMS}. Cartridge sensitivity can vary quite a bit from the assumed nominal, sometimes from 2 to 8 mV at 5 cm/sec. Preamp gain at 1 kHz is often chosen to be between 34 and 40 dB. Notice that, as a result of the RIAA pre-emphasis, playback of a record with a flat frequency response sweep from 1 to 20 kHz would result in a signal before RIAA equalization that is about 20 dB higher at 20 kHz than at 1 kHz. A 5 mV/cm/s cartridge would thus put out about 50 mV at 20 kHz.

RIAA Equalization

Figure 15.2 is a Bode plot showing the prescribed RIAA equalization frequency response [1–5]. Compared to the 1-kHz reference gain of 40 dB, it is up 19.7 dB at 10 Hz and down 19.6 dB at 20 kHz, and continues the 6 dB/octave HF roll-off to higher frequencies. Notice that the gain of the circuit of Figure 15.1 can go no lower than unity, so that circuit does not follow the curve all the way down at higher frequencies. This departure occurs at about 200 kHz for a preamplifier with 40 dB of gain at 1 kHz. This does not cause a very big error in the audio band.

However, some op amps that are not unity gain stable require a resistor in series with the network to allow enough closed-loop gain at high frequencies for the op amp to be stable. This value of gain is usually between 2 and 5. Examples of some such op amps are the NE5534 (without external compensation), the OPA637, the AD745 and the LT1028/1115. Such a resistor will cause a decrease in f_6, leading to greater error. In some preamplifiers with architectures like Figure 15.1, an *R-C* circuit is added after the equalizer to create a pole at f_6, eliminating the error.

Similarly, decoupling capacitor C2 causes the circuit to have a bit less gain than prescribed at very low frequencies. These two sources of error will be discussed further below. If R2 is set to a low value of 100 Ω to keep noise down, and C2 is 470 µF, a zero at about 3 Hz in the response will result. This will result in an error of about −0.3 dB at 20 Hz. Importantly, placing an electrolytic in such a critical portion of the signal path is frowned upon for good reason. For perspective, note that an unwanted zero one decade away from a prescribed frequency will cause an error of about 0.05 dB at that frequency.

The Bode plot shows a response characterized by a pole at 50.05 Hz followed by a zero at 500.5 Hz followed by a pole at 2122 Hz. These frequencies are denoted as f_3, f_4 and f_5, respectively. These frequencies are also defined by time constants at 3180, 318 and 75 µs [1–5]. Gary Galo has created an Excel spreadsheet that allows one to determine the required component values for four different RIAA circuit implementations that will be illustrated in Figure 15.7. Gary has graciously made it available at cordellaudio.com.

Note there are component interactions in the topology of Figure 15.1 that mean direct application of the three RIAA time constants does not yield exact EQ in the end. This means that the time constants of those pieces must be modified from the prescribed values to account for interactions.

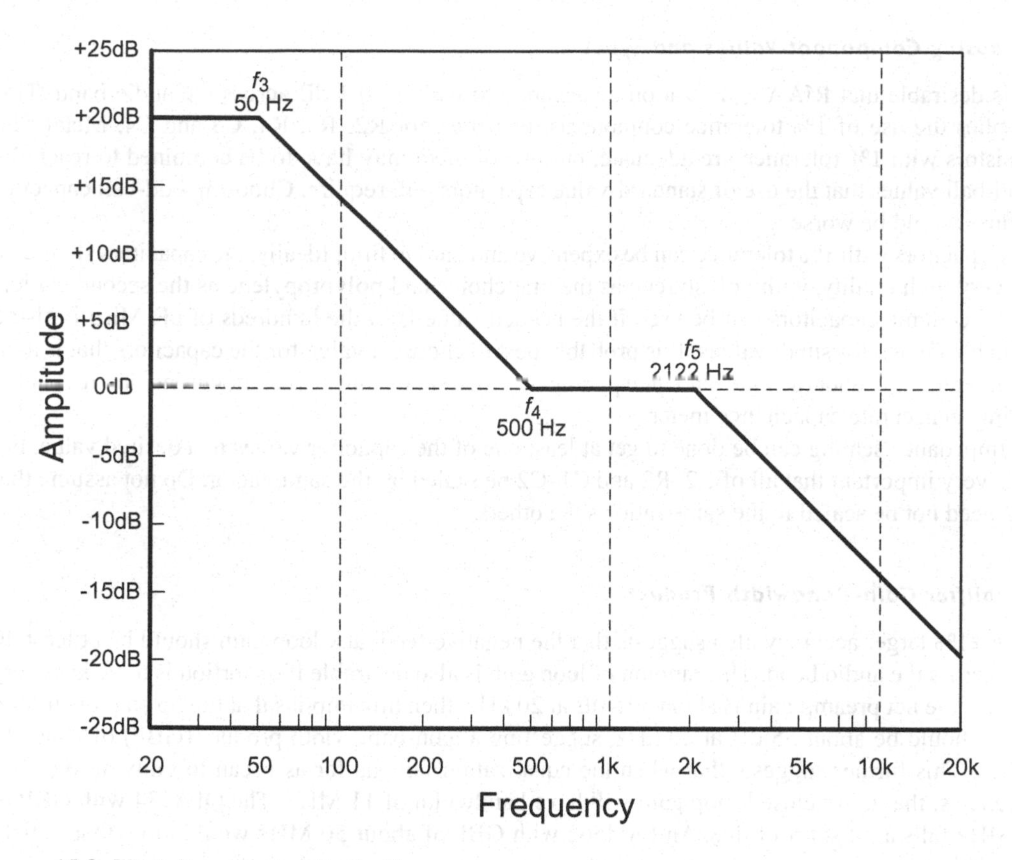

Figure 15.2 RIAA Equalization Bode Plot.

RIAA Equalization Values at Key Frequencies

The RIAA equalization relative gain values are shown for audio frequencies in a 1–2–5 sequence below. Although more detailed tables are available online, knowing the exact values at these frequencies allows adequate assessment of equalization accuracy. These values assume only the 3180, 318 and 75 µs time constants are in play.

Frequency (Hz)	*dB EQ*
20	19.27
50	16.95
100	13.09
200	8.22
500	2.65
1000	0
2000	−2.59
5000	−8.21
10,000	−13.73
20,000	−19.62

Choosing Component Values and Types

It is desirable that RIAA equalization be accurate to within ±0.1 dB across the audio band. This implies the use of 1% tolerance components, or better, for R2, R3, R4, C3 and C4. Metal film resistors with 1% tolerance are adequate, but two of them may have to be combined to reach the odd-ball values that the use of standard-value capacitors will require. Choosing odd-ball capacitor values would be worse.

Capacitors with 1% tolerance can be expensive and hard to find. Ideally, the capacitors should be of very high quality, with polystyrene as the first choice and polypropylene as the second choice. COG ceramic capacitors can be used if the needed value is in the hundreds of pF. Mica is also a suitable choice for small values. It is probably best to choose a value for the capacitors that is a bit smaller than a standard value, and then parallel it with a small trim capacitor, getting the right value using an accurate capacitance meter.

Impedance scaling can be done to get at least one of the capacitor values to a desired value, but it is very important that all of R2–R2 and C1–C2 be scaled by the same factor. Do not assume that R2 need not be scaled in the same ratio as the others.

Amplifier Gain-Bandwidth Product

The ±1% target accuracy also suggests that the negative feedback loop gain should be at least 40 dB across the audio band. This amount of loop gain is also desirable if distortion is to be kept very low. If the net preamp gain is about +15 dB at 20 kHz, then this implies that the op amp open-loop gain should be about 55 dB at 20 kHz, suggesting a gain-bandwidth product (GBP) of about 11 MHz. This further suggests that when the equalization falls as far as it can to unity at high frequencies, that unity closed-loop gain will have bandwidth of 11 MHz. The OPA134 with GBP of 8 MHz falls a bit short of this. An LM4562 with GBP of about 50 MHz would more than satisfy this objective.

Few preamplifiers actually have a gain stage with GBP of 20 MHz. In some cases, designers may tweak the resistor and capacitor values in the equalization network to gain a bit more accuracy in light of less GBP in the amplifier of the phono preamp, be it an op amp or discrete design. It is interesting to note that on average the frequency response of the RIAA equalization falls at a bit less than 6 dB/octave across the audio band. This is largely due to the flat region between 500.5 and 2122 Hz. The open-loop gain of a typically compensated op amp also falls at about 6 dB/octave across the audio band. This means that the loop gain of a phono preamp using an op amp will be roughly constant across the audio band. This tends to mitigate some of these errors. This cannot be usually said for discrete phono preamps, where the discrete amplifier, sometimes consisting of as few as two transistors, may have a relatively flat open-loop gain across the audio band, resulting in loop gain that is smaller at low frequencies.

15.2 Moving Magnet Cartridge Model and Termination

The moving magnet cartridge has been the reliable mainstay of LP reproduction for decades. It is economical and very convenient. Its nominal output of 3–5 mV/cm/s means that it can be used with virtually any conventional phono preamp. However, this amount of sensitivity comes at a significant expense to high-frequency response. It is inevitably associated with more turns in the coils and substantial cartridge inductance, sometimes as high as 500 mH.

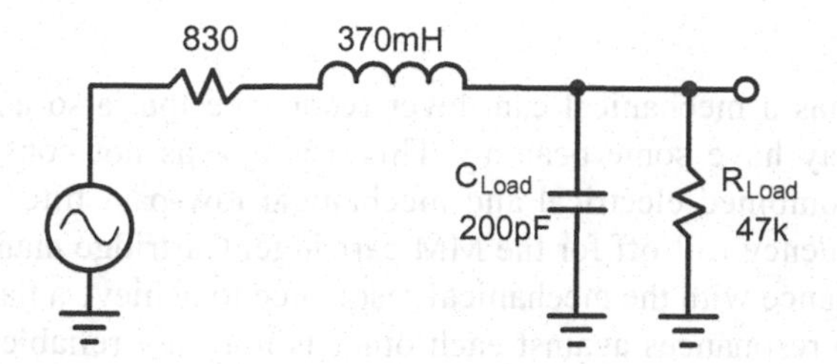

Figure 15.3 V15 Type V Electrical Model Including 47 kΩ and 200 pF Loading.

Cartridge Inductance and Resistance

Figure 15.3 shows a simple electrical model for the discontinued Shure V15 Type V moving magnet cartridge and its "standard" load of 47 kΩ and 200 pF (including turntable wiring) [6, 7]. The V15 is used as a representative example of a typical MM phono cartridge. The primary elements of the cartridge itself are its coil inductance of 370 mH and resistance of 830 Ω. If you calculate the pole frequency of 47 kΩ and 370 mH, you get 20 kHz. Were it not for anything else, the cartridge would be down 3 dB at 20 kHz. A 500-mH cartridge would be down 3 dB at only 15 kHz.

Cartridge Electrical Resonance and Cartridge Loading

Here is where the cartridge-loading capacitance comes into play, and why it is so important. The loading capacitance forms a series resonant circuit with the cartridge inductance. This lifts the frequency response as the resonant frequency is approached and then more sharply cuts the frequency response with a second-order roll-off when the resonant frequency is exceeded. The capacitive loading extends the bandwidth to a higher frequency in exchange for a second-order roll-off. This is why the frequency and Q of the resonance is important, as is the cartridge-loading capacitance. For the typical total loading capacitance of 200 pF, the resonant frequency with 500 mH is only 16 kHz. The real story is a bit more complex, but it is easy to see that there are limitations to the *HF* response of the MM cartridge, and these largely stem from its inductance.

Some phono preamplifiers make provision for selectable cartridge-loading resistance and capacitances. This is simply in recognition that there are differences among cartridge parameters. It would be nice if all cartridge manufacturers designed their cartridges to provide a flat response with the standard 47 kΩ /200 pF loading, but that is not the case. Moreover, different turntables with their different interconnect do not all have the same wiring capacitance.

Cartridge Impedance

Based on Figure 15.3, it can be seen that the V15 Type V impedance at low frequencies is 830 Ω and rises at higher frequencies due to the 377 mH of inductance. The impedance of the coil equals the resistance at 351 Hz, so the cartridge impedance will be up 3 dB to 1174 Ω at that frequency. At 1 kHz, the inductance will dominate and the impedance will be up to about 2.5 kΩ. Without considering the capacitive load and resulting resonance, the impedance will be up to about 47 kΩ at 20 kHz. Parallel resonance effects will increase this further in the frequency region near the resonance.

Cantilever Resonance

The stylus assembly has a mechanical cantilever resonance that also acts as a second-order low-pass filter that may have some peaking. This aspect was not considered in the simple analysis above. The combined electrical and mechanical low-pass filter functions result in a fourth-order high-frequency roll-off for the MM cartridge. Cartridge manufacturers often juggle the electrical resonance with the mechanical resonance to achieve a flatter overall response. However, playing two resonances against each other is not very reliable. One might say that two wrongs don't make a right. Cartridges with lower inductance and a higher cantilever resonance frequency can perform far better because they are less dependent on working the electrical and cantilever resonances against each other. Cartridges with better cantilever structures and materials (such as Beryllium instead of Aluminum) can achieve higher cantilever resonance frequencies. The Shure V15 Type V has a cantilever resonance at just under 30 kHz [6]. With an inductance of 370 μH and loading of 200 pF, the V15 has an electrical resonance at about 18 kHz.

15.3 Noise

Extremely low noise in an MM preamp should not be achieved at the expense of other things that influence sound quality, such as cartridge interaction, distortion, EMI immunity and overload. For example, the use of many JFETs in parallel operating at high currents may compromise distortion and cartridge interaction due to the high input capacitance. Using the same input stage for both MM and MC applications is an example of such a performance compromise. Discussions on noise analysis and tradeoffs can be found in Chapter 10 and in Refs. [8, 9].

Consider the use of an input differential pair instead of a single-ended stage. The long-tailed pair (LTP) has a noise penalty of 3 dB, while using twice as many transistors. However, the LTP has lower distortion, better overload characteristics and greater EMI immunity. These advantages are even greater when JFETs are used instead of BJTs.

Simulations show that a properly loaded Shure V15 Type V cartridge with a completely noiseless preamp will yield an A-weighted *SNR* of only 83 dB. This is mainly due to the thermal noise from the resistance of the cartridge coil, and the thermal noise of the 47 kΩ terminating resistance. With an amplifier input noise of 4 nV/√Hz, this *SNR* decreases to 81 dB.

It is convenient to discuss the noise performance of a phono preamp in terms of input-referred noise in nV/√Hz. In this way, comparisons can be made without reference to the effects of the RIAA equalization and possible A-weighting, as these will be about the same for all phono preamps. Moreover, this approach makes it easier to evaluate the effect on noise when a preamplifier is in the real world and connected to a cartridge and terminating resistance. Such a comparison between shorted-input noise and real-world noise quickly shows the futility of shooting for an MM preamp with 1 nV/√Hz of noise with a shorted input. Measurements of phono preamp noise with the input shorted can be misleading. Measurements with a standardized simulated cartridge impedance like 820 Ω and 350 mH in series would provide much more realistic *SNR* numbers.

Cartridge Thermal Noise

The coil resistance of 830 Ω will create thermal noise of about 3.7 nV/√Hz. This is the best input-referred noise that one can achieve, even with a perfect preamp and no resistive cartridge loading.

Cartridge-Loading Noise

The open-circuit thermal noise of a 47 kΩ loading resistor is about 28 nV/√Hz. This will create short-circuit noise current to flow of about 0.6 pA/√Hz. About the same amount of thermal noise current will flow into a shunting impedance much smaller than 47 kΩ, like that of the cartridge at frequencies below 10 kHz. At 1 kHz, with cartridge impedance of 2.5 kΩ, the current will create thermal noise of 1.5 nV/√Hz. This noise will rise at about 6 dB/octave to 10 kHz and beyond, being 15 nV/√Hz at 10 kHz.

Preamplifier Voltage and Current Noise

The input current noise from a BJT preamp input stage will also create input voltage noise in the same way that the thermal current noise from the 47 kΩ terminating resistor does. For example, the low-noise bipolar LM4562 op amp has typical input current noise of 1.6 pA/√Hz [10]. This exceeds by a factor of 2.7 the current noise created by the 47 kΩ cartridge termination resistor. It will create input-referred voltage noise of 4 nV/√Hz at 1 kHz and about 40 nV/√Hz at 10 kHz. By comparison, note that the typical input voltage noise for the LM4562 is 2.7 nV/√Hz. Interestingly, the NE5534 input noise current is only 0.6 pA/√Hz, creating 1.5 nV/√Hz, the same as the 47 kΩ loading resistor [11].

The virtual absence of input current noise for JFET op amps gives them an advantage in this regard, but many JFET op amps, like the popular OPA2134, have input voltage noise on the order of 8 nV/√Hz, a disadvantage [12]. The OPA1641 series of JFET op amps is an even better choice, with noise of only 5.1 nV/√Hz. An attractive exception is the CMOS FET-input OPA1656, with typical input noise voltage of only 4.3 nV/√Hz at 1 kHz, falling to 2.9 nV/√Hz at 10 kHz [13]. As a word of caution, however, its *1/f* noise climbs to 10 nV/√Hz at 100 Hz and 22 nV/√Hz at 20 Hz. This increased noise below 1 kHz will be further magnified by the LF portion of the RIAA equalization curve that rises at 6 dB/octave as frequency goes down. Owing to the large-area FETs used to achieve low noise, differential input capacitance is 9 pF and common-mode input capacitance is 2 pF.

Low-noise discrete JFET input amplifiers can do better, as in the LSK489-based JFET phono preamp described in Section 15.11 [14, 15]. The LSK389 is also a great part, featuring lower voltage noise and higher *gm*, but larger capacitances [16, 17].

Output Noise Spectrum and the Color of Noise

White noise presented to the input of the preamp, as from cartridge thermal noise and input amplifier voltage noise, will fall at a bit less than 6 dB/octave as a result of the RIAA equalization HF roll-off. This will convert that white noise to red noise, which is much less pleasant than white noise or pink noise for the same amount of noise power in a given bandwidth. This will also likely be the case for record surface noise, if one presumes the surface noise to be white.

Signal-to-Noise Ratio

The resulting phono preamp output SNR assuming shorted-input white noise can be determined by simulation for both 20-kHz flat weighting and A-weighting using ideal noiseless RIAA filters and ideal A-weighting filters. If a large white input noise voltage is used in the simulation, these filters can be simulated using real components, since the former will swamp out their noise contribution.

Noise Weighting

It is notable that the ubiquitous A-weighted noise measurement is not entirely the most appropriate for measuring noise, especially in phono preamps where the noise spectrum is highly non-uniform due to the RIAA equalization curve.

The A-weighting noise measurement specification is very old, and is based on the audibility of single tones at different frequencies [18]. It is related to the Fletcher-Munson equal-loudness contour at a perceived sound level of 40 phon [19]. The A-weighting curve is less representative of the audibility of random noise as a function of frequency [7]. Nevertheless, the A-weighting curve is largely the de-facto standard for specifying SNR ratio in audio equipment and in other areas.

Phons and the Equal-Loudness Contours

The *phon* is a unit of loudness for pure tones as perceived by human hearing [19]. It is based on equal-loudness contours. The earliest equal-loudness contours were the Fletcher-Munson curves developed in the 1930s. The equal-loudness contours were later revised in 1956 by Robinson and Dadson to become the basis for the ISO 226 standard [20]. The differences are largely at frequencies below 300 Hz. The equal-loudness contour was again revised as ISO 226:2003 in 2003. Interestingly, ISO 226:2003 is closer to the original Fletcher-Munson curves than the earlier ISO 2003 created in the 1950s. The phon is in units of dB referenced to perceived loudness at 1 kHz. For example, a 1 kHz tone at 80 dB has a perceived loudness of 50 phons [DIN 45631 and ISO 532]. Just as the threshold of hearing is defined as 0 dB SPL at 1 kHz, the threshold of hearing at any frequency is 0 phons.

In order to better reflect the audibility of random noise as a function of frequency, the ITU-R 468 weighting curve was developed [21]. Compared to A-weighting, this curve reflects more sharply

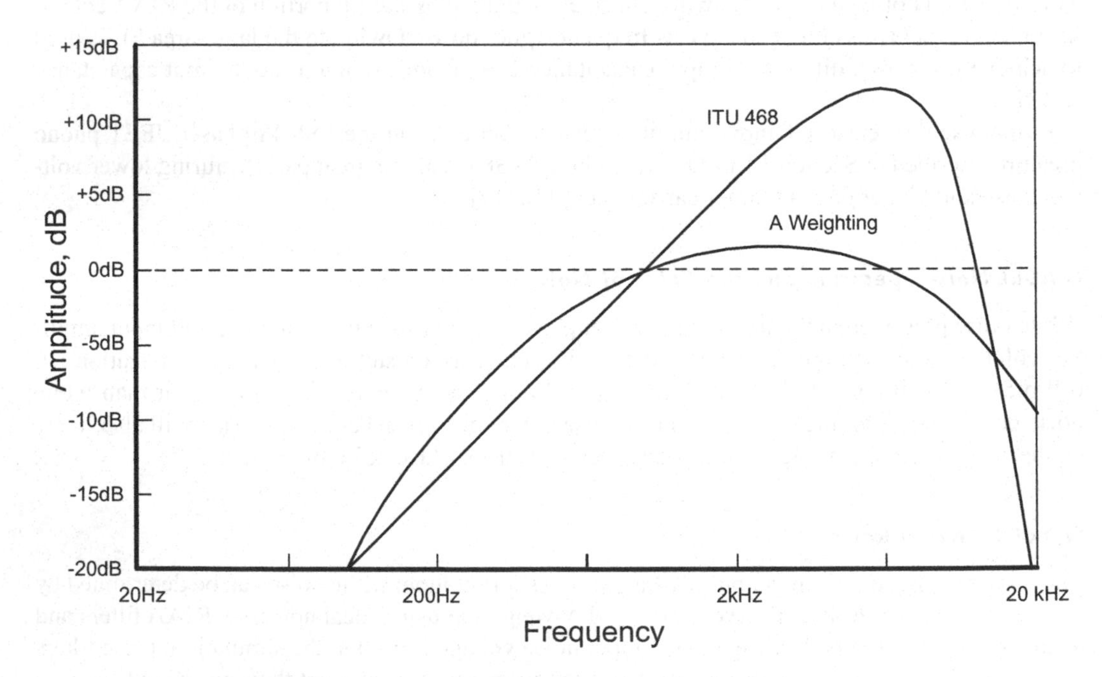

Figure 15.4 A-Weighting and ITU-R 468 Weighting Curves.

falling audibility of random noise at frequencies above 6 kHz. Beyond 6 kHz, audibility falls at more than 24 dB/octave, with peak sensitivity occurring at 6 kHz. Below 6 kHz, audibility falls at 6 dB/octave. The A-weighting curve and the ITU-R 468 curve have equal reference audibility at 1 and 15 kHz. At 6 kHz, the ITU-R 468 audibility is about 12 dB higher than that for A-weighting.

Figure 15.4 shows the A-weighting and ITU-R 468 weighting curves together [7]. The peak in the ITU-R 468 curve at 6 kHz is quite notable, as is its steep roll-off at higher frequencies.

Keeping in mind the A-weighting curve and the ear's frequency-dependent sensitivity, two preamplifiers with the same A-weighted *SNR* can have their respective noises sound different [7].

Real-World Noise with Cartridge Connected

Phono preamps are not operated with their inputs shorted in the real world; they are connected to a cartridge that presents substantial impedance in the case of moving magnet cartridges (830 Ω in series with 370 mH for a Shure V15 Type V) [6]. Because of its inductive component, this impedance can rise to tens of kΩ at high frequencies. As a result, the cartridge-loading resistance, nominally 47 kΩ, can contribute a significant amount of thermal noise at the input. Simulations indicate that even with a perfectly noiseless preamp, the achievable A-weighted SNR with a Shure V15 Type V cartridge is about 83 dB as a result of this effect. With a preamp having 4 nv/√Hz, this *SNR* decreases to 81 dB [6].

EMI Susceptibility

EMI can be a concern because it can intermodulate down into the audio band and possibly even disturb operating points. While the RIAA equalization curve keeps the gain of the preamp low at very high frequencies, most phono preamps have little or no EMI suppression at their inputs other than the cartridge-loading capacitance. Moreover, the load capacitor is not usually preceded by a series resistor. That would increase noise. An exception that increases noise very little is the addition of a small-value resistor, on the order of 68 Ω, in series ahead of the loading capacitor. This approximates a termination of typical interconnect in its characteristic impedance at very high frequencies above about 20 MHz while also creating a low-pass filter at that frequency with the typical 100-pF loading capacitor. JFET inputs, either in IC op amps or discrete designs, will tend to be more tolerant of EMI because the EMI signal will appear at the reverse-biased junction of the JFET rather than at the more rectification-prone forward-biased junction of a bipolar transistor.

15.4 Overload

Significant overload margin is very important in phono preamplifiers. At 1 kHz it is desirable to have at least 20 dB with respect to 5 mV/5 cm/s as a bare minimum. This equates to an overload capability of at least 50 mV RMS at the input. Very good preamps may have an overload of around 200 mV_{RMS}. In many preamplifiers, overload is dictated by the maximum output that can be produced divided by the mid-band gain at 1 kHz. A preamplifier that can produce only 10 V_{RMS} at its output will have a rated overload of 100 mV_{RMS} if its gain at 1 kHz is 40 dB and nothing else in the signal chain overloads before the output stage. Overload is usually declared when THD reaches 1%.

Overload voltage at 20 kHz is also considered important. Ideally, it should be just as high as at 1 kHz, if not higher. One could argue that because the RIAA recording EQ is about +20 dB at 20 kHz, a 50-mV overload capability at 1 kHz should be accompanied by a 500-mV capability at 20 kHz. In practice, the likelihood of an MM cartridge producing 500 mV_{RMS} at any frequency with program material is very remote when one recognizes that the velocity would have to be 500 cm/sec. However, ticks and pops on records can produce fairly high levels on occasion, and their audibility will likely be increased if they drive the preamp into overload. Preamplifiers that implement the 75-μs pole near the output and that have little gain thereafter are more susceptible to overload. The simple and popular series feedback preamp, as in Figure 15.1, has very good overload characteristics, largely because its gain falls to unity at high frequencies.

15.5 Classic Phono Preamp Circuit

The classic single-stage series feedback phono preamp of Figure 15.1 can be implemented in a variety of ways, each of which has advantages and disadvantages.

Discrete BJT Preamp

The discrete BJT implementation of the architecture of Figure 15.1 has been around forever. It can be built with two or three transistors, although much-improved versions may include more transistors [22]. The ones with just two or three transistors suffer from a lack of feedback loop gain and input impedance that is small enough to interact with the cartridge impedance [22]. The simpler designs also tend to have less equalization accuracy for the same reasons. The input noise current of these designs can also add significant noise when the preamp is operated with the high impedance of a cartridge as a source.

Discrete JFET Preamp

The same issues apply to simple discrete JFET input designs except that there will be little or no cartridge interaction. They will also likely be more robust against EMI. They will be sonically more like vacuum tube phono preamplifiers. A lack of adequate feedback loop gain will still put these designs at a disadvantage unless a design with more transistors is employed, as above.

BJT Op Amp Preamp

BJT op amp phono preamps are probably the most popular in modern times, as they are simple and generally have adequate loop gain, even at the very low frequencies where nearly 60 dB of closed-loop gain may be required. They also can yield very low noise, at least when they are measured with a shorted input. A good candidate for such designs is the LM4562, with typical voltage noise of only 2.7 nV/√Hz. As mentioned earlier, its 1.6 pA/√Hz flowing through the cartridge impedance will add noise voltage of about 4 nV/√Hz at 1 kHz. That noise will rise at 6 dB/octave with frequency. Bear in mind that noise voltages add on an RMS basis. The effect of the total input-referred noise in a phono preamp is usually mitigated a bit by the low-pass filtering of the 2122-Hz pole of the RIAA equalization that follows.

JFET Op Amp Preamp

JFET op amps are also widely used for phono preamplifiers, as shown in Figure 15.1. However, their typically higher input voltage noise, on the order of 8 nV/√Hz for an OPA2134, puts them at an SNR disadvantage when compared to BJT op amps when noise is measured with shorted inputs. The SNR difference is less when the measurements are made with a cartridge connected, but few if any manufacturers do that measurement mush less quote it. Their greater EMI immunity is a plus. The OPA1641 is a better choice, with input noise of only 5.1 nV/√Hz.

JFET Hybrid Op Amp Preamp

A hybrid op amp in this context is one that employs a discrete JFET differential pair that drives an op amp [15]. Discrete JFETs generally are capable of much lower noise than those implemented in an IC FET op amp [14, 16]. The op amp used in this arrangement can be either BJT or JFET, and its noise is not important because it is preceded by the gain of the discrete JFET input stage. This best-of-both-worlds approach comes at the expense of greater circuit complexity. Such a hybrid op amp is illustrated in the phono preamp described in Section 15.11 (although that preamp is not of the classic architecture) [15].

15.6 DC Servos

The large electrolytic decoupling capacitor in the feedback network of the classic phono preamp can be a problem for those seeking the highest performance. The solution here is to employ a DC servo to manage the offsets, allowing the preamplifier circuit to operate without a large decoupling capacitor in the feedback network. Figure 15.5 shows one such implementation. The servo can require as few as two resistors, one capacitor and one op amp. It operates on the same concept as DC servos for audio power amplifiers [23].

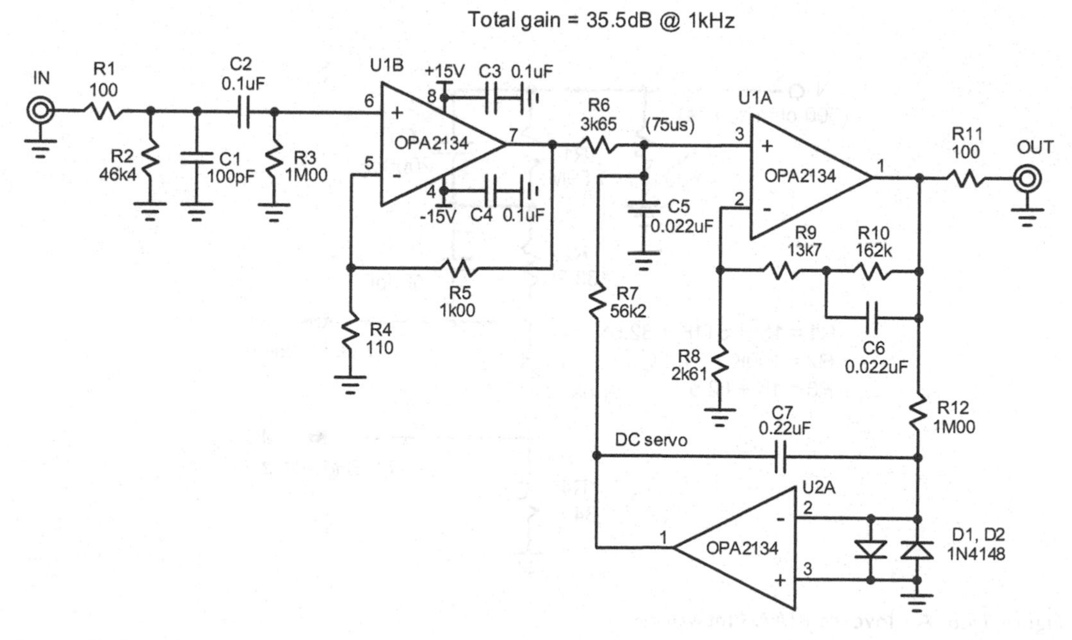

Figure 15.5 A Phono Preamp with a DC Servo.

DC Coupling the Cartridge Input

As in Figure 15.1, the use of a JFET input preamp allows the elimination of the input coupling capacitor without causing input bias current to flow through the cartridge. Moreover, if a coupling capacitor is used, as in Figure 15.5, it can be a relatively small-value, high-quality film capacitor.

15.7 Inverse RIAA Network

The conformance of a phono preamp to the RIAA equalization standard should always be checked. This can be done by inserting a passive inverse RIAA network between a signal generator and the input of the phono preamp under test. The output of the phono preamp should then be flat. Figure 15.6 shows a passive inverse RIAA network that can be used to verify conformance to the required RIAA equalization curve [24].

A passive inverse RIAA like this one by Lipshitz cannot conform perfectly because it is passive and has maximum gain of unity at HF. However, the nominal attenuation it provides between the generator and its output allows it to have considerable relative gain at high frequencies, mitigating this error. That network appears to be down only 0.6 dB at 100 kHz compared to the ideal. The 600-Ω source resistance requirement of that network comes from 600-Ω audio generators that are then terminated in 600 Ω to produce a net source resistance of 300 Ω. Use of a generator with a 50-Ω output impedance requires the insertion of a 250-Ω series resistor.

The network is quite insensitive to the usual 47-kΩ input impedance the phono preamp being tested. Loading of 47-kΩ drops the output by 0.2 dB, but causes only +4 mdB of error at 20 kHz. Further adding a 100-pF load causes a –6 mdB error at 20 kHz. Overall, apart from the flat 0.2-dB drop in gain, the error is negligible. Hagerman developed an inverse RIAA network that compensates for the mythical 3.18 μs (50 kHz) Neumann pole [5].

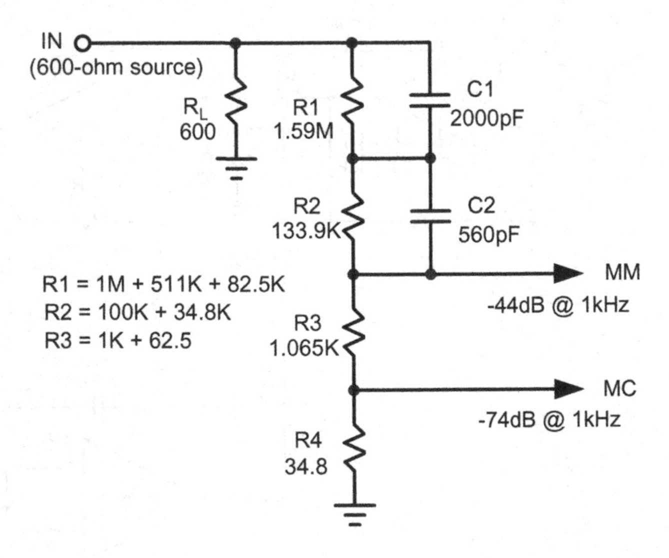

Figure 15.6 An Inverse RIAA Network.

Reference RIAA EQ

A reference RIAA playback EQ can be constructed by using separate R-C time constant circuits, isolated by op amps, so there is no interaction. A buffer feeds a 75 μs LPF that feeds the virtual ground of the second op amp. A parallel *RC* feedback network then provides the 3180 and 318 μs time constants. This can be used to test inverse RIAA networks for accuracy.

15.8 Other Phono Preamplifier Architectures

Good references describing RIAA equalization can be found in Ref. [1–5]. Relative to nominal gain at 1 kHz, RIAA EQ calls for about +20 dB at 20 Hz and −20 dB at 20 kHz, for a total equalization range of about 40 dB. There are numerous circuit arrangements that can implement the required equalization accurately and with adequate gain. Five different arrangements are shown.

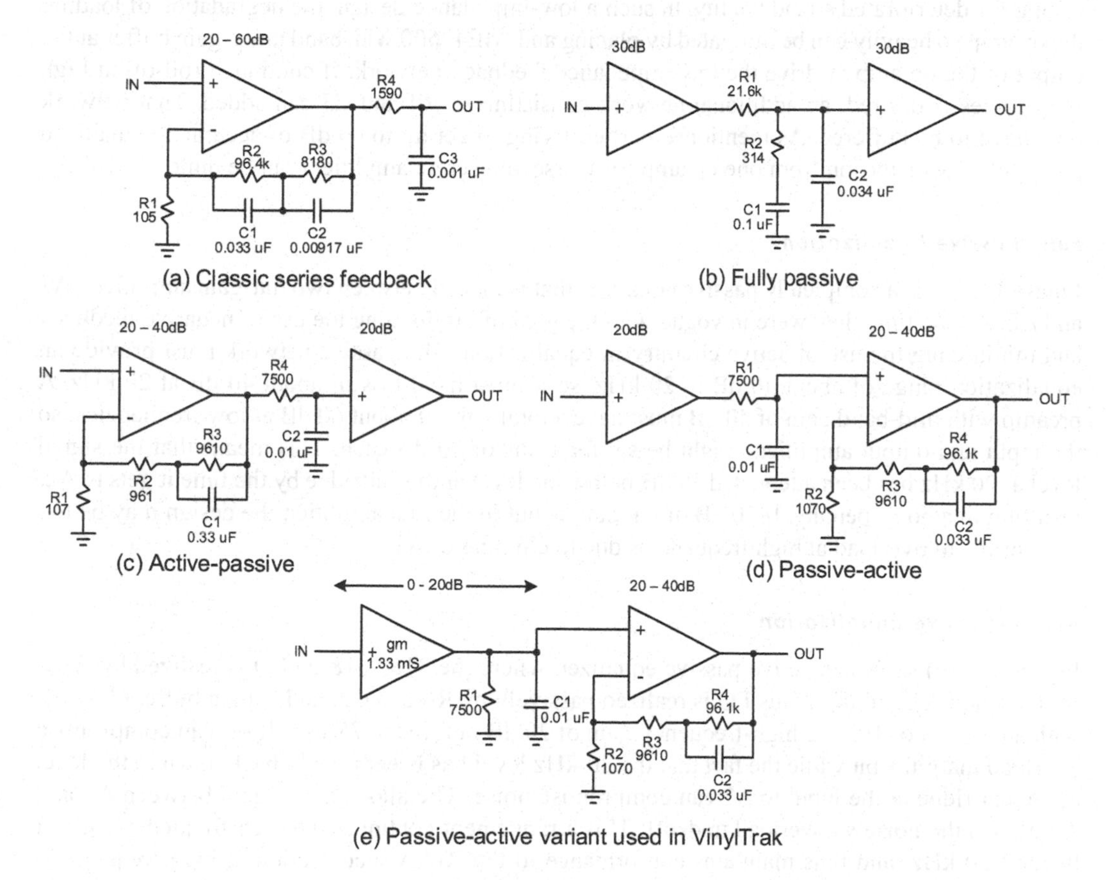

Figure 15.7 Various RIAA Equalizer Arrangements.

Classic Single-Amplifier Series Feedback Design

Figure 15.7 illustrates several RIAA phono preamp arrangements. The classic series feedback design in (a) is by far the most common. It implements all of the gain and equalization with a single op amp. If the nominal 1-kHz gain is 40 dB, over the frequency range it must supply gain from a high of 60 dB at low frequencies to a low of about 20 dB at 20 kHz. This design provides the most bang for the buck while being capable of low noise and high overload margin, even at high frequencies.

There are some disadvantages and limitations to this simple design, many of which were discussed in depth earlier. If R1 is set to 105 Ω, that resistor will contribute 1.3 nV/√Hz. In the quest for low noise, a value of 10 Ω (or less) has been used for R1 in some designs [25]. This requires that all of the capacitances in the network be made quite large, adding to expense and board area if quality capacitors are to be used. The electrolytic decoupling capacitor usually in series with R1 will need to be very large and it will have significant signal current flowing through it, causing distortion due to the nonlinear ESR and dielectric absorption of it. The use of a DC servo can eliminate the decoupling capacitor.

The heavy loading that the feedback network with a small value of R1 presents to the amplifier at high frequencies can force the op amp deeply into class B operation and generally causes distortion to increase substantially. Asking an op amp to deliver such high gain into such a heavy load is asking for deteriorated sound quality. In such a low-impedance design, the degradation of loading the op amp so heavily can be mitigated by placing an LME49600 wideband unity-gain buffer at the output of the op amp to drive the low-impedance feedback network. If continued roll-off at high frequencies is desired, an additional network consisting of R4 and C3 can added. That network may have to be buffered. As mentioned earlier, trying to get up to 60 dB of accurately equalized gain with low distortion from one op amp (or worse, a discrete amplifier) can be quite difficult.

Fully Passive Equalization

Figure 15.7(b) is a completely passive equalizer that is placed between two flat-gain amplifiers, A1 and A2. At one time they were in vogue with the goal of minimizing the use of negative feedback and minimizing the use of active circuitry in equalization. The passive network must provide an equalization range of about 40 dB to 20 kHz, so it must have loss of about 40 dB at 20 kHz. A preamp with mid-band gain of 40 dB must have a total gain of about 60 dB at low frequencies, so the input and output amplifiers might be set for gains of 30 dB each. This means that the signal level at 20 kHz has been attenuated 10 dB below the level at the cartridge by the time it gets to A2, resulting in a noise penalty. If 30 dB of flat gain is put in the first amplifier, the design may be too susceptible to overload at high frequencies due to clipping of A1.

Active-Passive Equalization

Figure 15.7(c) shows an active-passive equalizer, where the 3180/318-μs EQ is realized by feedback around A1 and the 75-μs EQ is realized passively by R4 and C2, and is then buffered by A2 with a flat gain of 10. The high-frequency gain of 20 dB before the 75-μs roll-off can compromise overload margin a bit while the fact that the 20-kHz level has been brought back down to the level at the cartridge at the input to A2 can compromise noise. The allocation of gain between A1 and A2 affects the noise vs overload tradeoff. This arrangement continues the high-frequency roll-off beyond 20 kHz, and thus maintains conformance to the RIAA specification at high frequencies without additional filtering. The use of two amplifiers allows more closed-loop gain in each stage, resulting in lower distortion and more accurate equalization. The fact that the LF and HF portions of the RIAA equalization are independent makes achieving accurate equalization easier.

Calculation of component values is quite straightforward. The feedback amplifier provides gain G_L at low frequencies and G_H at high frequencies, transitioning from the former to the latter with a −6 dB/octave step having a pole at 50 Hz and a zero at 500 Hz. The step is thus 20 dB. Typical values for G_L and G_H would be 100 and 10, respectively. R1 is chosen first. It sets the impedance levels in the network and also influences noise. R2 is chosen as R1(G_H − 1), since C1 is a short at high frequencies. R3 is set so that (R2 + R3) = R1(G_L − 1), since C1 is open at low frequencies. C1 is chosen so that its reactance at 500 Hz is equal to (R1 + R2)||R3. C1 will usually result as an odd-ball value. If a different value, perhaps closer to a standard value, is desired, the network impedances of R1, R2, R3 and C1 can be scaled together. R4 and C2 are chosen so that they have a time constant of 75 μs. Typical values would be 7500 Ω and 0.01 μF.

Passive-Active Equalization

Figure 15.7(d) is a passive-active equalizer that is the reverse of (c). It begins with a flat gain of 20 dB and immediately passes the signal through the passive 75-μs *EQ*. This has the advantage of eliminating the high-amplitude high frequencies early in the signal chain. However, it also delivers an attenuated signal to A2 at high frequencies, requiring that A2 have low input noise in order to achieve a good overall SNR. This is relatively straightforward because A2 is fed from fairly low impedance. Some implementations have significantly smaller gain in A2 and may suffer from high frequency overload due to clipping in A1. The passive-active design in (d) allows the use of significantly higher impedances in the active portion because it is preceded by gain. This allows the use of a smaller capacitor C2.

While the "hybrid" circuits of (c) and (d) may not provide as good a combination of overload and noise as the classic circuit of (a), and require two gain stages, they are capable of lower distortion and more accurate equalization. The independent implementation of the LF and HF portions of the RIAA equalization make achieving accurate equalization more straightforward and eliminates component value interactions. How the gain is distributed between A1 and A2 in these designs controls the tradeoff between overload margin and SNR. When properly designed, the hybrid approaches are capable of providing good overload margin and good SNR.

Passive-Active Variant with Transconductance Input Stage

A variant of the passive-active approach is shown in (e). A low-noise transconductance amplifier is used as the front end and it is loaded by the 75-μs network. This eliminates the output voltage swing overload problem. In simple implementations, the transconductance stage is operating open-loop, with only local degeneration to stabilize its gain and mitigate distortion. This approach was used in the VinylTrak MM preamp to be described later [6, 7].

15.9 Infrasonic (Subsonic) Filters

A common concern is that a rotating record can cause low-frequency flutter below 20 Hz if there is the slightest warp in the record. This can cause large and unnecessary woofer excursions and in some cases even damage. For this reason it is sometimes desirable to employ a low-cut filter to attenuate these signals that are inaudible anyway. The optional IEC RIAA low-frequency roll-off was meant to address this issue. However, its 6 dB/octave roll-off is too shallow to adequately mitigate the problem without causing too much low-frequency attenuation in the audio band. A different approach is to use a so-called "subsonic" filter – which properly should be called an infrasonic filter.

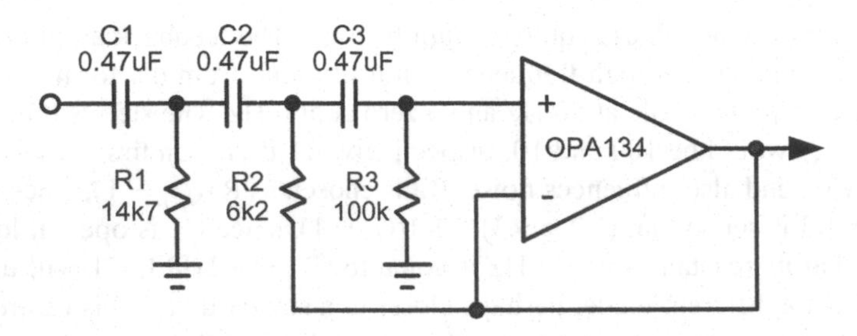

Figure 15.8 Third-Order Butterworth Infrasonic Filter.

Third-Order Butterworth Filter

The more conventional way of reducing warp flutter is to employ an infrasonic filter. This more strongly attenuates the undesired flutter. A good choice is a third-order Butterworth high-pass filter realized as a Sallen-Key filter of the voltage-follower type, as shown in Figure 15.8. The filter cutoff frequency is set at 16 Hz to provide 1 dB of loss at 20 Hz. The filter is down 30 dB at 5 Hz.

15.10 Vertical Flutter Reduction

The *L-R* signal content in a recording represents vertical stylus motion and contains most of the infrasonic warp flutter. Attenuating the *L-R* component at low frequencies can provide a marked improvement in sound quality by reducing subsonic speaker cone motion, especially in vented systems [26]. Those infrasonic disturbances also then do not waste amplifier power. These are especially important considerations when playing music at higher sound levels. The loss of separation at frequencies below 100 Hz is of no listening consequence (it is done with subwoofers all the time).

The *LF* merge is implemented by AC-coupling the L and R signals with 0.47 μF capacitors and then bridging the outputs with a 4.7-kΩ resistor [6]. At high frequencies where the capacitors are a short, full stereo separation is maintained. At low frequencies where the capacitors become an open circuit, the *L* and *R* signals are merged to become monophonic by the resistor. The network feeds a high-impedance load so that LF response is maintained. This circuit is included in Figure 15.15.

15.11 A JFET Phono Preamplifier

The combination of low noise and low input capacitance makes the dual monolithic LSK489 JFET an ideal choice for mitigating some of the disadvantages of the conventional phono preamps described above in an upscale preamp. It brings all of the advantages of a JFET input without the usual compromises in noise and input capacitance. Noise for each of the two LSK489 JFETs is 1.8 nV/√Hz typical at 2 mA. Gate-drain capacitance is only 3 pF maximum.

Note that lower input voltage noise could be achieved by using an LSK389, but this would be at the expense of higher input capacitance and cartridge interaction. LSK389 noise is 1.3 nV/√Hz typical at 2 mA and typical gate-drain capacitance is 5.5 pF maximum. Moreover, such lower voltage noise would be somewhat overkill in the phono application, since the noise created by the cartridge coil resistance and 47-kΩ load resistance set a lower limit on noise voltage anyway.

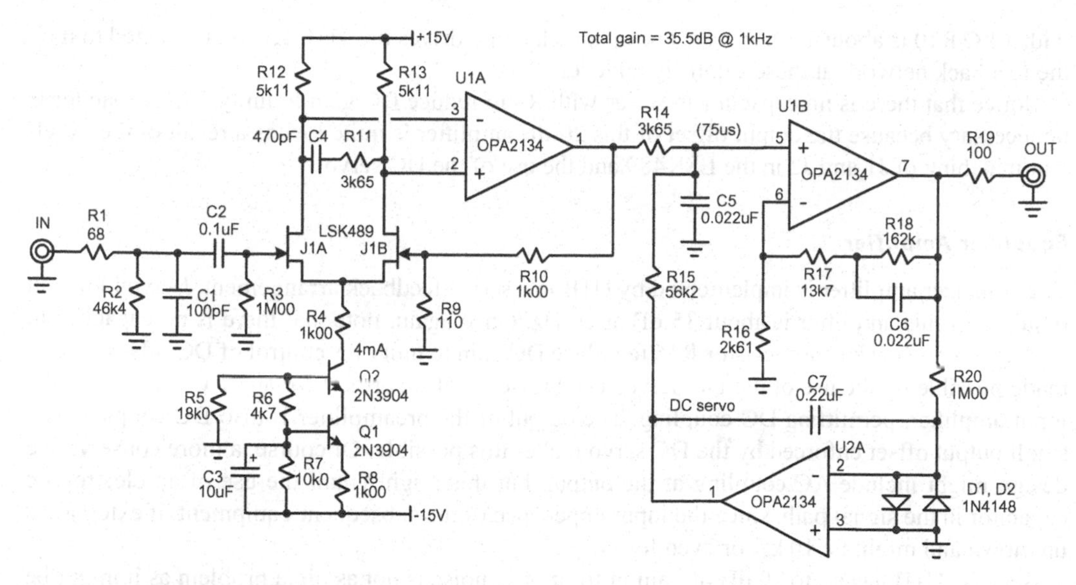

Figure 15.9 Discrete JFET Phono Preamp.

Circuit Description

Figure 15.9 shows the schematic of the preamplifier [15]. The design includes two stages of gain. The first stage provides a flat 20 dB of gain and is implemented with a hybrid op amp using a discrete LSK489 front end. This provides a good interface to the cartridge and boosts the signal well above the noise floor of the second stage.

The second stage provides additional gain and the low-frequency portion of the RIAA curve. The second stage is preceded with a passive single-pole low-pass filter. It provides the 75-μs time constant for the high-frequency portion of the RIAA equalization characteristic. A DC servo circuit implemented by U2 completes the design.

Input Amplifier

The input amplifier provides 20 dB of flat gain, and is implemented with an LSK489 differential pair (LTP) driving one half of an OPA2134 JFET operational amplifier. This arrangement capitalizes on the desirable characteristics of the LSK489 while minimizing the added noise contribution of the LSK489's load resistors and U1A. The tail current for the LTP is provided by the cascoded current source consisting of Q1 and Q2. Each JFET of the LSK489 is operated at 2 mA. The drains sit at about +5 V_{DC}. This is a near-optimum drain voltage for this application for achieving low noise.

The cartridge interface is AC-coupled, although this is probably unnecessary. Note that only 0.1 μF of film capacitance is needed to achieve a low cutoff frequency of 1.6 Hz, since the gate return resistor can be 1 MΩ.

C4 and R11 provide additional frequency compensation so as to accommodate the larger loop gain of the combined amplifier stages of the discrete JFETs and the op amp. The closed-loop gain is established by the feedback network comprising R9 and R10. The impedance of this network is kept low so as to minimize its noise contribution. The noise contribution of 110-Ω R9 in parallel with 1 kΩ R10 is about 1.3 nV/√Hz. The op amp handily drives the 1110-Ω load presented to it by the feedback network at these small signal levels.

Notice that there is no capacitor in series with R9 to reduce DC gain to unity. This capacitor is unnecessary because the output offset of this 20-dB amplifier is quite low as a result of the excellent matching of J1 and J2 in the LSK489 and the use of the DC servo.

Equalizer Amplifier

The equalizer amplifier is implemented by U1B in a series-feedback arrangement. Maximum gain required of this amplifier is about 35 dB at 20 Hz. Once again, note that there is no capacitor in series with feedback shunt resistor R16 to reduce DC gain to unity for control of DC offset. This is made possible by the use of the DC servo. The DC servo also corrects for any DC offset from the input amplifier, permitting DC coupling. The output of the preamplifier is also DC coupled. The small output offset enforced by the DC servo makes this possible. Of course, a more conservative design might include AC coupling at the output, but this might entail the use of an electrolytic capacitor in the signal path, since the input impedance of the subsequent equipment, if external, is unknown and might be 10 kΩ or even lower.

Because U1B has up to 20 dB of gain in front of it, noise is not as big a problem as it might be in other architectures. This allows the use of the JFET OPA2134 op amp, with its 8 nV/√Hz input noise. Moreover, it enables the use of higher impedances in the equalization feedback network, with C6 needing only to be 0.022 μF.

DC Servo

The DC servo is merely an inverting integrator formed with JFET op amp U2A. The use of a JFET op amp allows R20 to be large and integrating capacitor C7 to be small. The DC servo correction signal is injected into the 75-μs low-pass filter. The DC servo forces the output DC offset of the preamplifier to equal the input voltage offset of the DC servo op amp, typically less than 1 mV.

The loop gain and dynamics of the DC servo have a subtle influence on the low-frequency response of the preamplifier, and the DC servo gain and component values have been chosen by simulation to maintain proper conformance to the RIAA equalization curve.

Noise

The LSK489 is specified with a typical voltage noise of 1.8 nV/√Hz at an operating drain current of 2 mA [14]. Since the input stage is a differential pair, the input-referred noise will be 3 dB higher than that of a single LSK489 JFET. This results in an ideally achievable input noise voltage of 2.6 nV/√Hz.

Circuitry before and after the input devices will always degrade the achievable noise performance, and this is where much of the challenge of low-noise circuit design lies. As mentioned in the earlier discussion, for example, the MM coil resistance and the load resistance contribute about 5 nV/√Hz at 1 kHz. It is also easy to overlook the noise contributions of the drain load resistors, and they are an important contributor to the excess noise in this and many other designs.

The input amplifier ultimately achieves input noise performance of 4.8 nV/√Hz with a shorted input and the complete preamplifier has an impressive A-weighted *SNR* of 84 dB with a shorted input.

Power Sequencing and Muting

This phono preamp should only be used with a preamplifier that includes an output muting circuit that enables the output signal only after a few seconds at turn-on and quickly disables the output at turn-off. The reason for this is that the preamplifier can produce a transient at the output as the high-gain circuits stabilize or as the rail voltages collapse at turn-off. This is not uncommon for high-gain audio circuits, and is often dealt with by the inclusion of a simple muting relay circuit.

15.12 Synthetic Cartridge Loading

The thermal noise of the 47-kΩ cartridge-loading resistor can create a noise current that results in additional preamplifier noise. The thermal noise of the resistor is 28 nV/√Hz, which divided by 47 kΩ will produce 0.6 pA/√Hz into a short circuit. This noise source can be reduced by synthesizing the 47-kΩ resistance with negative feedback. If an inverted version of the input signal with a gain of 10 is made available, a feedback resistor of 517 kΩ will create an effective load of 47 kΩ while injecting much less thermal noise current. This is an old idea. One implementation can be found in Ref. [25].

The effectiveness of synthetic loading in the real world is best evaluated with simulation. It is important to recognize that a fundamental source of noise is the winding resistance of the MM cartridge itself. The 830-Ω winding resistance of a Shure V15 Type V creates an input noise level of 3.7 nV/√Hz even in the absence of cartridge load resistance and any amplifier noise.

A 47-kΩ resistor creates thermal noise of 28 nV/√Hz. The impedance of the 370-mH cartridge inductance of the V15 Type V at 2 kHz is about 4.6 kΩ. The peak frequency of the A-weighting curve is about 2 kHz. The attenuation of the 28-nV/√Hz noise of the 47-kΩ load resistor against the 4.6-kΩ cartridge impedance will be about 0.1, resulting in a contribution of about 2.8 nV/√Hz at this most hearing-sensitive frequency. This can be compared to the cartridge resistance thermal noise of 3.7 nV/√Hz. This noise contribution initially climbs at 6 dB/octave towards about 28 nV/√Hz at high frequencies if some other factors are not considered. At about 2640 Hz, the contribution from the terminating resistor becomes 3.7 nV/√Hz, the same as that from the cartridge resistance, thus degrading the SNR by 3 dB to 5.2 nV/√Hz. Keep in mind that this is the best you can do with a V15 Type V loaded with a 47-kΩ resistor, even with perfectly noiseless electronics.

Notice that this +6 dB/octave initial noise increase cancels the −6 dB/octave noise decrease from the 75-μs RIAA roll-off, resulting in approximately flat noise at the preamplifier output. It is a bit unclear that the use of synthetic cartridge loading is worth the extra complexity, especially when A-weighting is considered, given that most of its noise benefits accrue only at high frequencies where the cartridge impedance is much higher. To put things into perspective, if BJT op amps are used for the input amplifier, their input noise current contribution must also be considered, as it will rise with frequency in the same way as the contribution from the cartridge load resistor. The input noise current of the LM4562 is 1.6 pA/√Hz, larger than the approximate 0.6 pA/√Hz of the 47-kΩ resistor into a short.

The two simple circuits shown in Figure 15.10 illustrate how synthetic cartridge loading works. In a nutshell, shunt feedback is used to create smaller impedance with a higher resistance. In (a), the preamplifier employs a flat-gain input stage with a gain of 10 (20 dB). This boosts the signal out of the noise floor for the following circuits and provides a very high impedance input for the MM cartridge. With U2 providing an inversion, the signal voltage across R_L is easily seen to be 11 times that of the input signal. To cause the same signal current to flow as with a real 47-kΩ resistor,

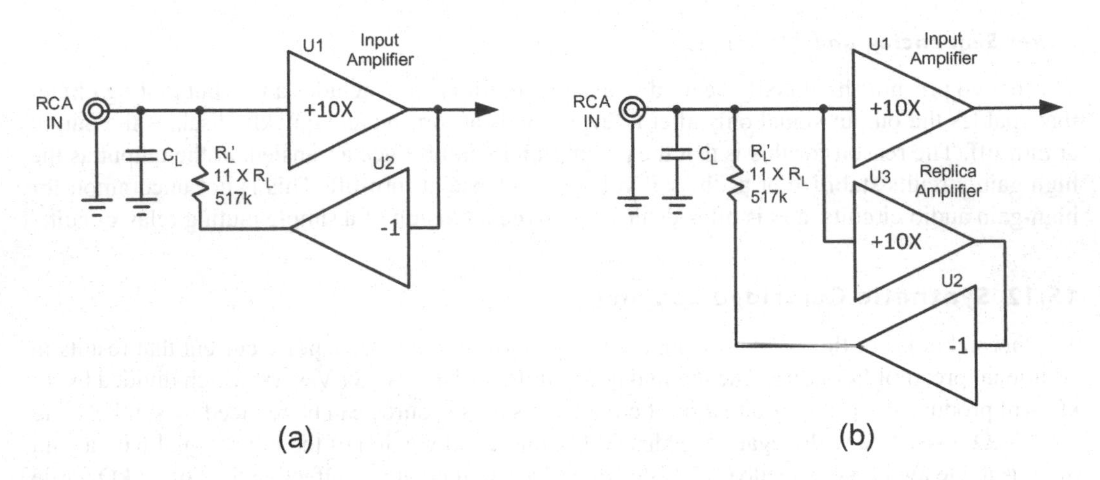

Figure 15.10 Synthetic Cartridge Loading.

R_L must be 11 times the 47-kΩ value, or 517 kΩ. From a noise point of view, however, the much higher resistance of the 517-kΩ resistor injects much less noise current than did the real 47-kΩ resistor. Thus arises the noise advantage. The approach in (a) is what I employed in a phono preamp that I designed circa 1980.

In many preamplifiers the input amplifier includes some or all of the RIAA equalization. This spoils the functioning of the simple circuit in (a). The VinylTrak™ preamplifier is an example of a preamplifier with some EQ in the first stage. In such cases the circuit in (b) can be used. U3 simply creates a replica of the necessary signal that would have been created had U1 been a flat amplifier. Such an approach has been employed in Ref. [25].

The effectiveness of synthetic loading in the real world is best evaluated with simulation. It is important to recognize that a fundamental source of noise is the winding resistance of the MM cartridge itself. The 830-Ω winding resistance of the Shure V15 Type V creates an input noise level of about 3.6 nV/√Hz even in the absence of a cartridge load resistance and any amplifier noise. The promise of reduced noise by load synthesis brings with it some extra complexity and the opportunity to lose the noise advantage in some implementations. This is especially the case in the circuit of (b) where an extra amplifier is connected to the sensitive cartridge input node. If BJT op amps are used for the input amplifier and replica amplifiers in (b), simulations show that their added input current noise may negate most of the noise advantage promised by synthetic loading. JFET input amplifiers are best used in these locations because of their absence of input current noise.

One might also ask if voltage noise in U2 and U3 adds significant noise to the final result, detracting from the noise advantage of the load synthesis circuit. Simulations show that this has very little effect in most designs where the noise of these amplifiers is not large compared to that of the input amplifier. For example, there is negligible degradation if the input amplifier has input noise of 5 nV/√Hz and the two feedback amplifiers in (b) have input noise of 8 nV/√Hz. In fact, even if the noise of U2 and U3 is 12 nV/√Hz, the reduction in noise advantage is only about 0.5 dB. This is good news because it means that JFET op amps can be used for U2 and U3 without compromising the noise advantage.

As an interesting aside, this arrangement allows the cartridge load resistance to be adjusted by changing the gain of the load synthesis feedback amplifier at a lower-impedance, less sensitive point in the circuit. This is feasible as long as the gain needed to achieve the lowest desired load resistance does not cause a potential input overload vulnerability in the load synthesis feedback amplifier. If need be, a smaller load synthesis feedback resistor of 100 or 150 kΩ could be used to mitigate this concern without giving up too much SNR.

15.13 Balanced Inputs

Balanced is beautiful, especially for small signals. Some turntables can be configured to provide a balanced output terminated in two XLR connectors. This helps avoid a host of problems, including ground loops. Correspondingly, some hi-end phono preamps offer XLR balanced inputs.

Implementing a balanced input phono preamp without causing asymmetrical or heavy loading on the cartridge can sacrifice noise performance if it is not done carefully. In fact, in the general case, it will tend to cause a 3-dB noise penalty even if the input amplifier is implemented as a 3-op amp instrumentation amplifier with low resistances. This is because the input-referred noise of each of the two input op amps will add to increase noise by 3 dB over that possible using just a single op amp. As we will see later, there is another way of implementing a balanced input if negative feedback is not used in the input amplifier.

Balanced Input with THAT 1512 Preamp

THAT Corporation makes the 1512 microphone preamplifier IC that has balanced inputs and a single-ended output [27]. It is basically a very good instrumentation amplifier. It can be configured for gain from unity to over 60 dB with a single resistor. The 1512 can be configured for a flat gain of 10. It cannot be configured for a flat gain of more than 20 dB if good phono preamp overload margin is to be retained, since the 75-μs time constant cannot be incorporated into it. Thus, the 75 μs time constant is realized passively after the 1512 and before an equalized output amplifier as in the passive-active arrangement in Figure 15.7(d). Unfortunately, while the 1512 is capable of 1 nV/√Hz input noise at 60 dB of gain and 1.4 nV/√Hz at 40 dB of gain, its input noise increases to 4.6 nV/√Hz at 20 dB of gain. This is normal for amplifiers of this architecture and is directly related to the need for the gain-setting resistor to be a larger value (hence more thermal noise) for the lower gain of 20 dB.

15.14 The VinylTrak Phono Preamp

The VinylTrak preamp was designed and implemented as a high-performance phono preamp for MM and MC cartridges [6, 7]. The design includes an optional implementation of RIAA equalization that mitigates the conventional cartridge loading issue. A block diagram of the VinylTrak phono preamp is shown in Figure 15.11.

The preamp incorporates the following features:

- Independent MM and MC preamplifiers
- JFET MC and MM input stages without NFB
- DC-coupled cartridge interface and signal path
- DC servo control of offset
- Balanced and single-ended outputs
- Optional damped-cartridge RIAA EQ for MM cartridges
- LF mono merge for vertical flutter rejection
- Third-order infrasonic filter
- Left-Right balance control
- Selectable *L-R* output for balance and anti-skate setup

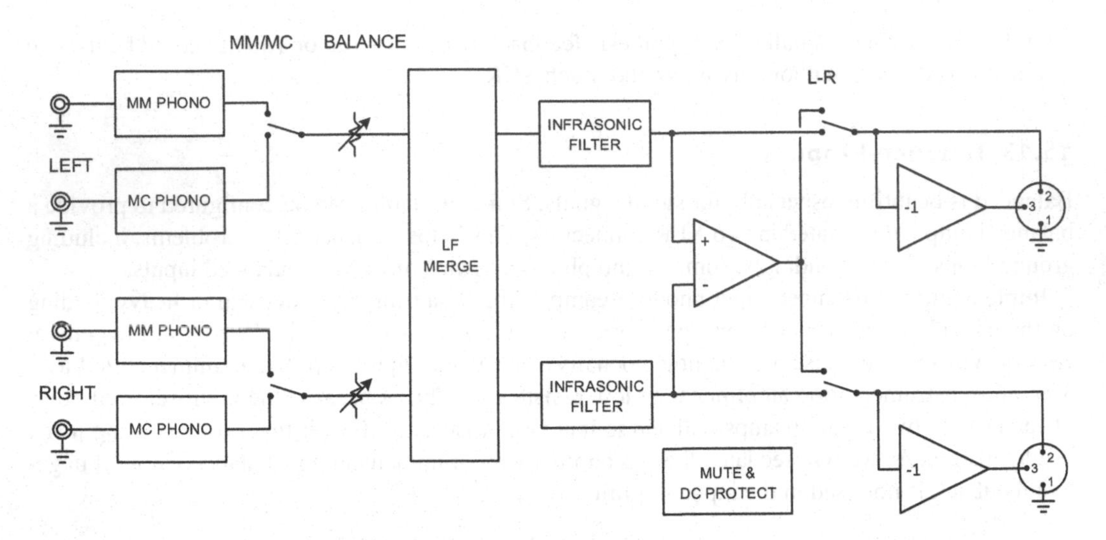

Figure 15.11 VinylTrak Phono Preamp Block Diagram.

The MM and MC RIAA preamps are entirely separate. This allows each function to be optimized for its application without compromise. The vastly different impedances of MM and MC cartridges make approaches with shared circuitry sub-optimal. Each preamp implements the 75-μs roll-off passively within a discrete JFET input amplifier that provides 20 dB of gain at low frequencies. That is followed by an active stage that provides the 3180/318-μs *LF* portion of the RIAA EQ. The active stage is implemented with an OPA2134 op amp. The MC front-end of the VinylTrak preamplifier is described in Chapter 16.

The MM preamp uses a JFET double folded cascode differential input amplifier with no negative feedback. The input amplifier is DC-coupled to the cartridge. The 75-μs RIAA time constant is implemented with a shunt RC load at the high-impedance output node of the folded cascode. A diamond buffer output stage completes the input amplifier. Distortion without feedback is very low due to the small signal levels. High-frequency overload margin is very good because the shunt 75-μs load is implemented inside the first stage. Overload is soft due to the use of JFETs and the absence of negative feedback in the first stage. The JFET inputs provide much higher EMI immunity than do bipolar designs.

The MM input amplifier uses an LSK489 dual JFET and achieves 5 nV/√Hz input noise. The input stage is capable of providing true high-impedance differential inputs. The JFET inputs ensure virtually no cartridge interaction. The MM input amplifier provides a gain of 10 at 1 kHz.

Selectable cartridge loading is provided at the front panel for the MM preamp (68, 47, 33, 22, 15, 10 kΩ; 0, 100, 200 pF). The lower-resistance cartridge-loading values are associated with the over-damped equalization option discussed below.

The MM preamp includes an over-damped approach to RIAA equalization as a switch-selectable option. This approach loads the cartridge with a resistance low enough to form a pole at 8 kHz with the cartridge inductance. An 8-kHz zero is then inserted in the 75-μs roll-off circuit to correct the RIAA roll-off slope above 8 kHz. A selectable cartridge-damping load resistance matches the cartridge inductance. The MM preamp can be set for 35 or 41 dB of gain at 1 kHz with a jumper on the circuit board.

The preamp is DC-coupled and includes a DC servo for control of offset. There are no electrolytic capacitors anywhere in the signal path. Output functionality includes a balance control to compensate for cartridge-channel mismatch, a switched low-frequency left-right signal merge to attenuate vertical flutter, a switched third-order Butterworth infrasonic filter and balanced outputs. An *L-R* signal can be sent to the outputs to aid balance control adjustment using a mono track or a center-channel voice. The *L-R* output mode may also be useful in setting anti-skate.

Gain and Equalization Architecture

The equalizer in Figure 15.7(e) illustrates the approach used in this preamp. It is a variant of the passive-active arrangement. The key difference is that A1 is a transconductance amplifier and the 75-µs EQ is implemented by a shunt RC network comprising R1 and C1. Any large-amplitude high-frequency signals never see the light of day in the voltage domain, so HF clipping of A1 is virtually eliminated. This arrangement also contributes to RFI immunity by attenuating HF signals very early in the signal chain. The transconductance amplifier operates without negative feedback. Amplifier A2 must still have low noise in order to achieve good overall SNR, but it is fed by low impedance.

Damped-Cartridge RIAA Equalization

For over 50 years, RIAA equalization for MM cartridges has been accomplished in basically the same way. The cartridge is assumed to be a perfect constant-velocity transducer, absent of frequency response limitations described above. Textbook RIAA EQ is simply applied to what is assumed to be a voltage source. This makes a joke of RIAA EQ that is accurate to 0.1 dB to 20 kHz (however, matching between the channels is no joke). There may be a better approach that is a significant departure from tradition.

We note with interest that if the cartridge is loaded with a much lower value of resistance than 47 kΩ, the pole formed with the inductance will fall to a rather low frequency and the electrical response will be first-order to very high frequencies. If we load a 500-mH cartridge with 6.7 kΩ, the pole falls to 2120 Hz, which corresponds to the RIAA 75-µs time constant. One can take advantage of the inductance and implement the 75-µs EQ right at the cartridge and simply leave that part of the EQ out of the subsequent RIAA equalization circuit. The reduced impedance at the cartridge output now obviates the issue of the loading capacitance and the absence of a resonance greatly extends the frequency response. The pole of 6.7 kΩ and 100 pF of turntable wiring capacitance is at 240 kHz, where the cartridge electrical response will revert to a second-order roll-off (all ignoring the cantilever resonance). I refer to this as "damped" RIAA equalization because the cartridge electrical resonance is completely damped out. An approach somewhat like this was also described in Ref. [28], where cancellation of the cantilever resonance was an objective.

Alas, part of the RIAA equalization is now dependent on the cartridge inductance, so the required value of loading resistance for a flat response must be different for different cartridges. It must be adjustable and the cartridge inductance must be known. Otherwise, it must be adjusted either by ear or by the use of a test record. There is also a slight SNR penalty. Perhaps a compromise is in order.

In the VinylTrak preamp, the cartridge is loaded to form a pole in the neighborhood of 8 kHz [6]. This decreases frequency response sensitivity to the required adjustment for cartridge inductance. It also reduces the noise penalty and pushes it to higher frequencies, to the point where it does not matter. The 370-µH Shure V15 Type V is loaded by about 18 kΩ and the equalized electrical response becomes flat to about 90 kHz. To account for this added pole at 8 kHz, the 75-µs pole in the preamp EQ is stopped at 8 kHz by inserting a zero, with merely a resistor in series with the EQ

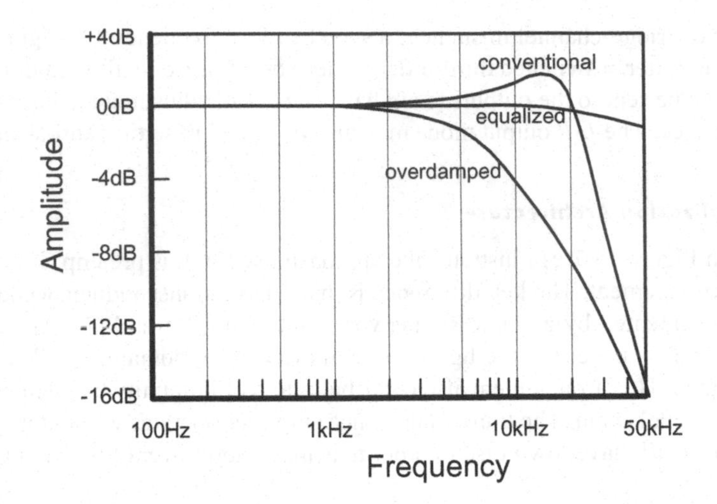

Figure 15.12 Shure V15 Type V Simulated Electrical Response.

capacitor (C1). This leaves a flat electrical response where there once was a peaked second-order roll-off. Any error in the resistive loading of the cartridge vs its inductance will result in a slight shelving equalizer (treble control) characteristic above 8 kHz.

Figure 15.12 is a plot of the simulated electrical response of a Shure V15 Type V cartridge with conventional and damped RIAA equalization, showing the substantial high-frequency extension. Not shown or modeled is the contribution of the cantilever resonance, which is at about 28 kHz for the V15. The damped equalization technique makes the MM cartridge more competitive with the MC cartridge, where both are left only with the cantilever resonance. Notably, the conformance to the RIAA equalization specification cannot be measured in the conventional way with this damped EQ approach. It can be measured if an actual cartridge is placed in series with the input and driven from a low-impedance signal source.

The pole at 8 kHz right at the cartridge provides increased attenuation of high frequencies before they even hit the electronics, improving *HF* overload immunity. A front-panel switch provides for matching cartridge inductances of 200, 250, 300, 400, 500 and 700 mH. If the cartridge inductance is unknown, selection can be made by ear by switching between the conventional and damped modes to obtain the same perceived HF response, or whatever response is preferred. This is no different than the way people adjust R and C cartridge loading on preamps so equipped, except that a two-dimensional adjustment has been reduced to a one-dimensional adjustment here. Get this right and you have clear sailing all the way up to the cantilever resonance.

Damped-Cartridge Equalization and Load Synthesis

The use of lower-value cartridge load resistors in the damped equalization approach increases the thermal noise contribution from the load resistor. Although not used in the VinylTrak™ preamplifier, the load synthesis technique discussed in Section 15.12 can prove valuable when applied to it. Simulations show that it improves the SNR of the damped mode by 2.4 dB to the point where it is virtually the same as that of a conventional EQ approach with a real 47-kΩ load.

Two things appear to be responsible for this. First, the 196-kΩ load synthesis feedback resistor eliminates the noise penalty imposed by the smaller 18-kΩ load resistance in the damped approach.

Secondly, since the 196-kΩ resistor is still substantially larger than the real 47-kΩ resistor in the conventional EQ case, the extra noise benefit tends to offset the remaining noise penalty of the damped approach due to the 8-kHz zero placed in the 75-μs part of the RIAA *EQ*.

As an interesting aside, this arrangement would allow the cartridge load resistance to be adjusted by changing the gain of the load synthesis feedback amplifier at a lower-impedance, less sensitive point in the circuit. This is feasible as long as the gain needed to achieve the lowest desired load resistance does not cause a potential input overload vulnerability in the load synthesis feedback amplifier. If need be, a smaller load synthesis feedback resistor of 100 or 150 kΩ could be used to mitigate this concern without giving up too much SNR. There may also be a tendency for cartridges with a lower inductance, which need a smaller load resistance in the damped approach, to have smaller output amplitude.

Discrete JFET Moving Magnet Input Stage

It is nice to DC-couple to the cartridge, eliminating the coupling capacitor, but the input bias current of a BJT stage should not be allowed to flow through the cartridge. This is where the JFET input shines. It goes without saying that if a coupling capacitor is needed, it should never be an electrolytic. The JFET input does not suffer the input current noise contribution from BJT stages that is conveniently overlooked when measuring preamp SNR with the input shorted. The JFET input also enjoys greatly increased EMI immunity, since the input node sees a reverse-biased semiconductor junction instead of a forward-biased one. Cartridge interaction is essentially absent.

Figure 15.13 shows the MM input amplifier and 75-μs equalization circuit. The amplifier is a discrete differential JFET-input, folded cascode design without negative feedback and without source degeneration. The 4-mA tail current source is cascoded, with the bases of Q3 and Q4 AC-referenced to ground and the negative rail, respectively, for best PSRR. The JFET drains are cascoded prior to their connection to the folded cascode emitters, making this a double folded cascode circuit. This improves power supply rejection by isolating the rail-referenced folded cascode emitters from the ground-referenced JFET drains. Figure 15.13 shows the LSK489 dual monolithic JFET (Q1, Q2), but the LSK489 device is now recommended because it has lower noise. The Q5 through Q14 2N5551/2N5401 devices have since been replaced by the 2N5089/2N5087 transistors because the latter have lower base input noise current, thus improving preamp SNR.

The differential input stage incurs a 3-dB noise penalty, but it is worth it. Excellent SNR is achieved anyway. Input-referred noise is 5 nV/√Hz and A-weighted *SNR* for the complete MM preamp is 83 dB referred to 5 mV.

The output stage is a bootstrapped diamond buffer, which provides exceptional linearity and a very light load to the cascoded VAS node. R20 serves to inject the offset correction voltage from the DC servo. Absent the 75-μs EQ, the amplifier bandwidth is 4 MHz and exhibits a slew rate in excess of 200 V/μs when driven with a full-amplitude square wave.

Amplifier gain is about 20 dB at low frequencies, established by the transconductance of the LSK489 JFET pair and the 3.83-kΩ VAS shunt resistor R18. Importantly, the impedance at the VAS node due to the active circuits is more than three orders of magnitude higher than the value of R18. The output circuit includes a ±1-dB gain trim to account for JFET transconductance differences between the two channels. The absence of NFB permits a very high-impedance true differential input and enables a very soft overload characteristic with great robustness against EMI. At the small input signal levels from an MM cartridge, distortion is very low even without NFB, and with substantial margin for overload. THD is less than 0.005% at 20-mV_{RMS} input, with no

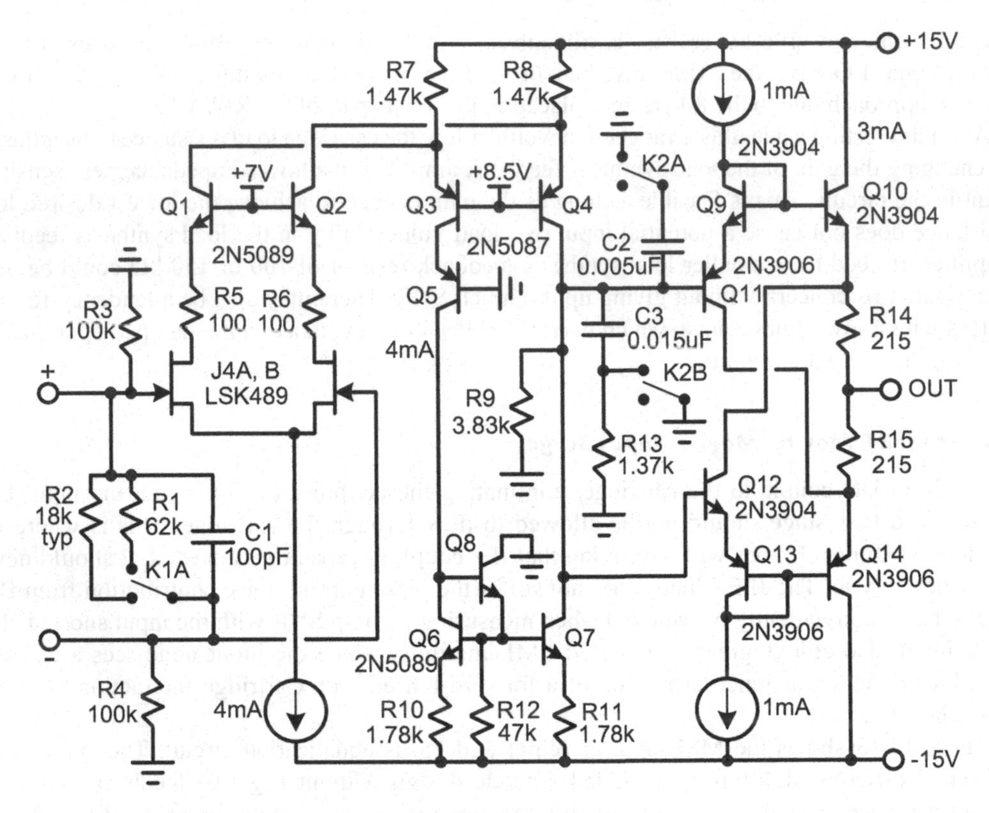

Figure 15.13 MM Input Amplifier and 75-μs Equalization.

visible harmonics above the third. At 200 mV, input distortion is still only 0.5%. Even at 400 mV, distortion is below 3% and the waveform is rounded, not clipped.

The absence of NFB allows the 75-μs RIAA time constant to be implemented with a hand-selected combination of two polystyrene shunt capacitors C5 and C6 totaling 0.02 μF at the high-Z VAS output node. This means that the amplifier need not ever swing the amount of voltage that would otherwise occur at high frequencies if the amplifier had a flat gain of 10. The preamplifier can be easily switched to conventional RIAA equalization by switching the cartridge load resistor to 47 kΩ and shorting out the resistor R19 in series with the 75-μs capacitor. Shorting nodes D1 and D2 are connected to ground through relay contacts for conventional equalization. The contacts are opened for damped *EQ*, inserting a zero at 8 kHz via R19.

JFET Gain Spread

The input stages of the MM and MC preamps operate without negative feedback, so the gain spread between stereo channels due to device-to-device transconductance differences between left- and right-channel JFETs must be corrected. That is why gain trim pots (R27) were put in the signal path of the preamps.

Ten different LSK489 JFETs from two different batches fabricated two years apart showed a gain spread of just ±0.7 dB. This is impressive, especially given that the threshold voltages of the first batch were on the order of 1.6 V and those of the second batch were on the order of 0.6 V. The

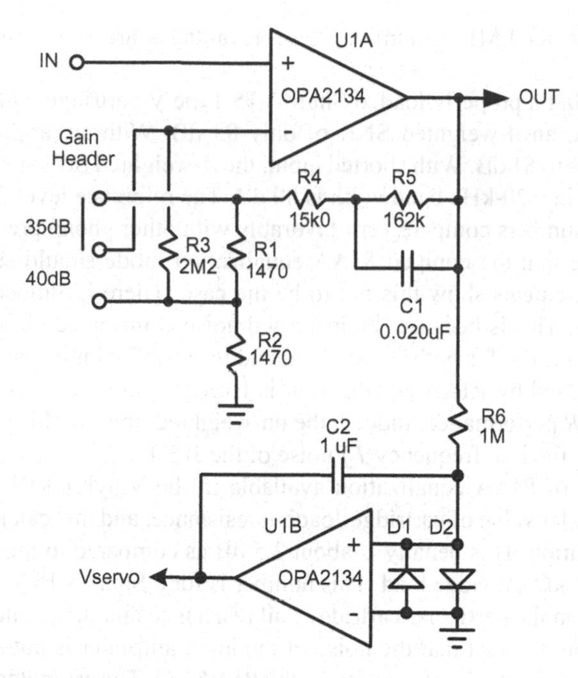

Figure 15.14 RIAA Active 3180/318-μs Equalizer.

key to this small gain spread is that differential pairs of the same type of device tend to have similar gain when they are operated at the same tail current.

RIAA LF Equalization and DC Servo

Figure 15.14 shows the active 3180/318-μs *LF* portion of the RIAA EQ, as implemented with one half of an OPA2134 JFET op amp. It is a simple series-feedback design that employs a 0.02-μF polystyrene capacitor and provides about 20 dB of gain at 1 kHz when the jumper is set for the high-gain mode. The alternate jumper position sets the gain 6 dB lower.

Occupying the other half of the op amp is the DC servo. A simple inverting integrator monitors the output DC level of the preamp and injects an error current via R26 into the VAS high-impedance output node within the input stage. The output offset of the MM preamp will equal the input offset of the DC servo op amp, which in the case of the OPA2134 is less than 2 mV. This means that the signal can be DC-coupled to the MM/MC selector switch without there being a loud pop when the switch is thrown.

Noise

Extremely low noise in an MM preamp should not be achieved at the expense of other things that influence sound quality, such as cartridge interaction, distortion, EMI immunity and overload. The use of many JFETs in parallel operating at high currents may compromise distortion and cartridge interaction due to the higher input capacitance. Using the same input stage for both MM and MC applications is an example of such a performance compromise. Consider the use of an input differential pair instead of a single-ended stage. The LTP has a noise penalty of 3 dB, while using twice as many transistors. However, the LTP has lower distortion, better overload characteristics and greater EMI immunity. These advantages are even greater when JFETs are used instead of BJTs.

Simulations show that a properly loaded Shure V15 Type V cartridge with a completely noiseless preamp will yield an *A*-weighted SNR of only 83 dB. With an amplifier input noise of 4 nV/√Hz, this decreases to 81 dB. With shorted input, the *A*-weighted SNR measures 85.5 dB, while the un-weighted SNR in a 20-kHz bandwidth is 80 dB. The reference level for the MM preamp is a 5-mV input. These numbers compare very favorably with other phono preamps.

While it is intuitive that the damped RIAA equalization mode should suffer a noise penalty, simulation and measurements show this not to be the case. There is almost no noise increase in the damped EQ mode. This is because the increased noise contributed above 8 kHz is swamped by the larger mid-band noise for both *A*-weighted and un-weighted noise in a 20-kHz bandwidth. The noise shape imparted by RIAA equalization is largely responsible for the predominance of mid-band noise in *SNR* performance. Indeed, the un-weighted noise in this preamp, although very good, is dominated by the low-frequency *1/f* noise of the JFETs.

The damped mode of RIAA equalization available in the VinylTrak™ preamp for MM cartridges employs a smaller value of cartridge-loading resistance, and this can lead to a larger overall thermal noise contribution. This penalty is about 2.5 dB as compared to the conventional form of equalization with a 47-kΩ cartridge load. This number is for a Shure V15 Type V cartridge, and is somewhat dependent on the particular cartridge coil resistance and inductance. A small part of the penalty also arises from the fact that the noise of the input amplifier is not attenuated as much at frequencies above 8 kHz by the 75-µs portion of the RIAA EQ. The estimated SNR for the damped mode using a Shure V15 Type V cartridge is 78 dB.

Overload Margin

With its gain set to 35 dB, 1-kHz overload for the VinylTrak preamp is 180 mV. The VinylTrak preamp implements the 75-µs pole very early in the signal chain while still achieving a very good SNR. As a result, input overload is a healthy 450 mV at 20 kHz, with very soft distortion of 1%. Even at 700 mV, distortion is only 3.3% and the waveform is still rather sinusoidal in appearance. Here the HF overload is set by the soft nonlinearity of the no-feedback JFET input stage.

Balance Control and L-R Output

Accurate balance of left and right levels is important to good imaging, yet many cartridges specify level balance tolerance of ±1 dB or worse. This preamp includes a balance control and a means of adjusting it by ear. The balance control provides a ±2-dB range. Figure 15.15 is a simplified schematic of the output circuit, where the balance control pot (R3A, B) can be seen near the input, right after the MM/MC relay contacts (K1).

The VinylTrak™ preamp can send an *L-R* signal to both outputs instead of the normal signals via K4. The *L-R* signal can be used for adjusting level balance by ear if a monophonic test track is played and balance is adjusted for an audible minimum. A center-stage voice will also suffice. The *L-R* output can also be valuable in unmasking tracking distortion and improper anti-skating settings, since these can cause differences between the left and right signals even when they contain the same information. When the anti-skating is wrong, the contact pressure on the two walls of the groove is different, and a distorted channel difference signal is generated.

The *L-R* signal is generated with an extra pair of op amps (U3, U4A) connected as a differential amplifier. A front-panel switch energizes a relay (K4) that is used to route *L-R* to both stereo outputs.

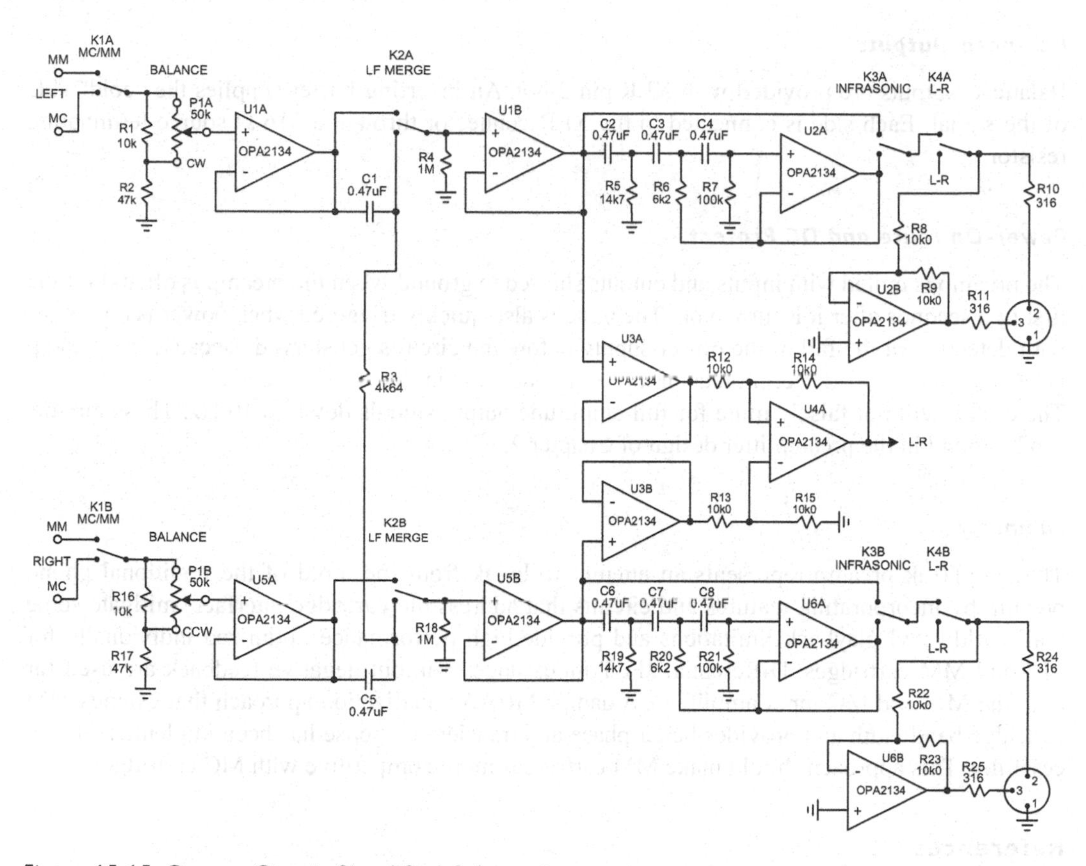

Figure 15.15 Output Circuit Simplified Schematic.

Low-Frequency Monaural Merge

L-R signal content in a recording represents vertical stylus motion and contains most of the subsonic warp flutter. Attenuating the *L-R* component at low frequencies can provide a marked improvement in sound quality by reducing infrasonic speaker cone motion, especially in vented systems [26]. Those infrasonic disturbances also then do not waste amplifier power. These are especially important considerations when playing music at higher levels. The loss of separation at frequencies below 100 Hz is of no listening consequence (it is done with subwoofers all the time). The selectable LF merge (K2) is implemented by AC-coupling the *L* and *R* signals. The outputs are then bridged with a resistor (R4) to mix the signals at low frequencies.

Infrasonic Filter

The more conventional way of reducing warp flutter is to employ an infrasonic filter. This more strongly attenuates the undesired flutter than the LF monaural merge, but it will introduce some phase distortion. The switched infrasonic filter is a third-order Butterworth high-pass realized as a Sallen-Key filter of the voltage-follower type. The filter cutoff frequency is set to provide 1 dB of loss at 20 Hz. When neither the LF merge nor the subsonic filter is used, the preamplifier is fully DC coupled from the cartridge to the output. The obsolete 20-Hz IEC roll-off is not implemented in light of the better alternatives provided.

Balanced Outputs

Balanced outputs are provided with XLR pin 2 hot. An inverting buffer supplies the "cold" side of the signal. Each side is connected to the XLR connector through a 316-Ω source terminating resistor.

Power-On Mute and DC Protect

The preamp is muted with inputs and outputs shorted to ground when the preamp is off and for the first two seconds after it is turned on. The mute is also quickly triggered when power is turned off with detection of the fall of the power supply before the circuits get starved. Because the preamp is DC coupled, a DC detect/protect circuit is included that initiates a mute under a fault condition. The circuit will not falsely mute for full-amplitude output signals down to 10 Hz. These circuits can be found in the preamplifier design of Chapter 2.

Summary

The VinylTrak preamp represents an attempt to break from the mold of the traditional phono preamp by incorporating features and circuits that address the cartridge interface, mitigate some real-world vinyl playback limitations and provide high performance optimized individually for MC and MM cartridges. Differential JFET input stages without negative feedback are used for both the MM and MC input amplifiers. A damped RIAA equalization approach that extends MM cartridge bandwidth and provides better phase and transient response has been implemented successfully. This approach should make MM cartridges more competitive with MC cartridges.

References

1. Stanley Lipshitz, On RIAA Equalization Networks, *JAES*, vol. 27, no. 6, June 1979, pp. 458–481.
2. Gary A. Galo, Disc Recording Equalization Demystified, *ARSC Journal*, vol. 27, no. 2, Fall 1996, pp. 188–211. Reprinted in *The LP is Back*, 2nd edition, Audio Amateur Press, Peterborough, NH, 2000, pp. 44–54 (The reprint is available online at www.smartdevicesinc.com/chpt14.pdf [PDF version for download] or http://www.smartdevicesinc.com/riaa.html [text version]). A copy is also available by contacting the author at galoga@potsdam.edu.
3. Gary A. Galo, The Columbia LP Equalization Curve, *ARSC Journal*, vol. 40, no. 1, Spring 2009, pp. 40–58.
4. Reg Williamson, *Understanding the RIAA Curve*, The Audio Amateur, February 1990.
5. Jim Hagerman, *On Reference RIAA Networks*, hagtech.com/pdf/riaa.pdf.
6. Bob Cordell, VinylTrak – A Full-Featured MM/MC Phono Preamp, *Linear Audio*, vol. 4, September 2012, pp. 130–139.
7. Bob Cordell, *The VinylTrak Phono Preamp*, expanded description, cordellaudio.com.
8. Burkhard Vogel, *The Sound of Silence*, 2nd edition, Springer Heidelberg, 2008.
9. Douglas Self, *Small Signal Audio Design*, Focal Press, Burlington, MA, 2009.
10. Texas Instruments, LM4562 Datasheet, ti.com.
11. Texas Instruments, NE5534 Datasheet, ti.com.
12. Texas Instruments, OPA2134 Datasheet, ti.com.
13. Texas Instruments, OPA1656 Datasheet, ti.com.
14. Linear Integrated Systems, LSK489 Datasheet, linearsystems.com.
15. Bob Cordell, *Linear Integrated Systems LSK489 Application Note*, linearsystems.com, February 2013.
16. Linear Integrated Systems, LSK389 Datasheet, linearsystems.com.

17. Bob Cordell, *LSK389 Application Note – Dual Monolithic JFET for Ultra-Low Noise Applications*, linearsystems.com, January 2020.
18. Rod Elliott, *A-Weighting Filter for Audio Measurements*, Project 17, Elliott Sound Products, sound.au.com.
19. Don Davis, Eugene Patronis, and Pat Brown, *Sound System Engineering*, 4th edition, Chapter 7, Focal Press, New York, NY, 2013.
20. Wikipedia, *Equal Loudness Contour*. wikipedia.org.
21. ITU-R 468, *Measurement of Audio Frequency Noise Voltage Level in Sound Broadcasting*. https://www.itu.int/dms_pubrec/itu-r/rec/bs/R-REC-BS.468-4-198607-I!!PDF-E.pdf
22. Tomlinson Holman, *New Factors in Phonograph Preamplifier Design*, presented October 31, 1975, at the 52nd Convention of the Audio Engineering Society, New York, NY.
23. Bob Cordell, *Designing Audio Power Amplifiers*, 2nd edition, Routledge/Focal Press, New York, NY, 2019.
24. Stanley P. Lipshitz and Walt Jung, *A High Accuracy Inverse RIAA Network*, The Audio Amateur, January 1980.
25. Douglas Self, *Preamplifier 2012*, Elektor, Part 2, May 2012.
26. Jeff Macaulay, Warp Filter, *AudioXpress*, February 2011.
27. THAT Corporation, THAT1512 Microphone Preamplifier datasheet, thatcorp.com.
28. Steven van Raalte, Correcting Transducer Response with an Inverse Resonance Filter, *Linear Audio*, vol. 3, April 2012, pp. 98–104.

Chapter 16

Moving Coil Phono Preamps

The moving coil (MC) cartridge is generally the preferred high-end audio pickup. It is characterized by low output, low impedance and wide bandwidth. It tends to show less mechanical and electrical resonance effects in the audio band than a moving magnet (MM) cartridge. The role of the MC preamplifier (head amp) is quite simple: provide about 20–30 dB of flat gain with extremely low noise. However, this can be a significant technical challenge. A step-up transformer has often been employed to provide this function, and it is the competition for solid-state implementations. However, good MC step-up transformers can be expensive. Shielding is extremely important, and less expensive transformers will likely fall short in this regard.

MC cartridges produce smaller nominal output levels, on the order of hundreds of microvolts, about 20 dB less than a typical MM cartridge. A very low-noise booster amplifier is thus usually needed in front of a conventional MM phono preamp, with about 20 dB of gain. Low noise is the single most important performance characteristic for MC preamps. Input-referred noise levels on the order of 0.5–2.0 nV/√Hz are usually required.

We first explore the details of using transformers instead of active MC preamplifiers. Transformers are the competition in terms of low noise and sound quality. This provides good insight regarding achievable SNRs and how low the noise must be for a head amp to compete with a transformer. One reality is that transformers are not noiseless.

16.1 Signal Levels and MC Cartridge Characteristics

There is a wide variation in MC cartridge resistance and output level at the 5 cm/s reference velocity. In generic descriptions, the resistance is assumed to be 5 Ω and the output is assumed to be 500 μV (20 dB down from the generic number assumed for MM cartridges). Table 16.1 shows some of these numbers for eight different MC cartridges [1]. Sowter Transformers Inc. also has a lengthy list of MC cartridges and their characteristics [2].

The numbers in Table 16.1 are for the load resistance recommended by the manufacturer, or the recommended load for a different cartridge with approximately the same resistance. The cartridges are listed in order of output voltage. All of the output voltages for the cartridges (at 5 cm/sec) are normalized to 5 mV by the calculated required step-up ratio. The "cartridge noise" column represents the thermal noise of the cartridge resistance in nV/√Hz. The "Norm Noise" column represents the normalized noise after step-up for comparison with 5 mV. The SNR Figure of Merit (FM_{sn}) is the output voltage divided by the square root of DC cartridge resistance. This reflects the fact that the SNR is inversely proportional to the square root of cartridge resistance.

It can be seen that the output ranges from 250 to 700 μV, a range of 8.4 dB. The resistance ranges from 4 to 40 Ω, a range of 10:1. This implies a coil resistance self-noise range of 10 dB.

Table 16.1 Moving Coil Cartridges

Cartridge	*Out (μV)*	*Res (Ω)*	*FM_{sn}*	*Load (Ω)*	*Step-Up*	*Cart Noise*	*Norm Noise*
Denon DL-103R	250	14	67	100	20.0	0.45	9.0
Ortofon MC A90	270	4	135	>10	18.5	0.26	4.8
Dynavector XX-2 MKII	280	6	114	30	17.9	0.32	5.7
Koetsu Onyx	300	5	134	30	16.7	0.29	4.8
Denon DL-103	300	40	51	>400	16.7	0.82	13.7
Audio Technica AT-F7	350	12	101	100	14.3	0.45	6.4
Ortofon Vivo Red	500	5	223	>10	10.0	0.29	2.9
Benz Micro LP-S	510	38	83	>400	9.8	0.80	7.8
Lyra Delos	600	8	210	>91	8.3	0.37	3.1
Sumiko Palo Santos	700	12	202	100	7.1	0.45	3.2

Significantly, the coil resistance has little correlation with the output voltage. The self-noise figure of merit FM_{sn} ranges from 83 to 223, with many in the range of 100–200. This is strictly an indication of self-noise and not an indication of sound quality. No "high-output" MC cartridges are considered here.

The numbers after step-up do not take into account any noise contributions of the transformer, so the real-world noise of the system may be greater. One cannot do better than these noise numbers. Note for comparison, the normalized output noise contribution from a step-up amplifier with the required gain and with input-referred noise of 1 nV/√Hz ranges from 7.14 to 18.5 nV/√Hz, showing its dominance.

Required Gain

A typical transformer-based MC signal path is illustrated in Figure 16.1. It shows a 500-μV_{RMS} MC cartridge with its winding resistance driving a 1:10 step-up transformer. The secondary of the transformer then feeds the input of a conventional MM phono preamp that provides gain and RIAA equalization. Not shown is the usual 47-kΩ termination internal to the MM phono preamp.

The required voltage gain from a transformer or head amp is approximately the amount needed to boost the cartridge signal up to or greater than the nominal MM input level of 5 mV at 5 cm/s, but not so much as to overload the MM preamp when a high-output MC cartridge is used. A higher boost can be important in reducing the overall noise degradation caused by the noise of the MM preamp.

Figure 16.1 does not show typical modeling resistors corresponding to a 5-Ω MC cartridge resistance and a 0.3-Ω transformer primary resistance. At the secondary of the transformer, these resistances look like about 500 Ω as a result of the 1:100 impedance transformation from the 1:10 transformer. A typical transformer secondary resistance of 150 Ω is also not shown. Sometimes a loading network is also connected across the secondary to control the frequency response. Such a network might consist of a load resistor R_L in parallel with a Zobel network comprising a series combination of C_Z and R_Z. The MM preamp will normally present an additional load of 47 kΩ.

There are some conflicting recommendations in the literature about where R_L should be put: across the primary or the secondary of the transformer, and these can vary from one transformer manufacturer to another. For purposes here, R_L will be assumed to be loading the secondary in the form of the 47 kΩ input resistance of the MM preamp. It is also sometimes recommended that there be no termination capacitor at the input of the MM preamp that is being fed. Keep in mind

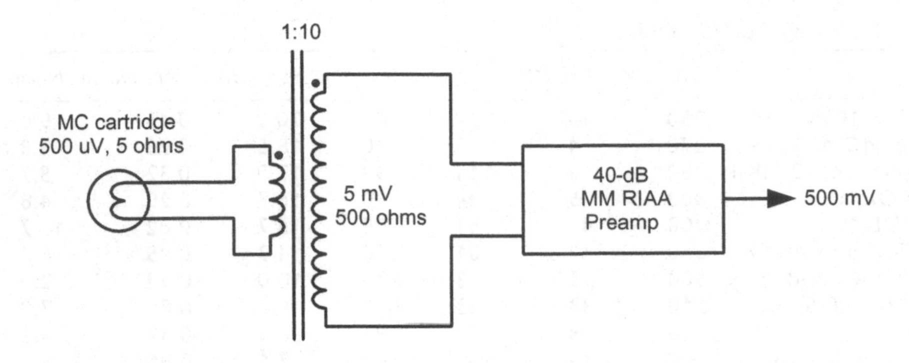

Figure 16.1 Typical Transformer-Based Moving Coil Signal Path.

that a 40-Ω cartridge needing a 1:17 step-up ratio will present a source impedance of 11.6 kΩ at the secondary of even an ideal transformer.

In all of the discussions that follow, remember that the thermal noise density of a 1-kΩ resistor is 4.1 nV/√Hz and that the noise density scales as the square root of the resistance, so that a 10-Ω resistor has noise density of 0.41 nV/√Hz. Total noise in a given bandwidth scales with the square root of the bandwidth. For example, in a 20-kHz audio bandwidth, there are 141 √Hz. The total noise created by a 1-kΩ resistor in that bandwidth is then 141 √Hz times 4.1 nV/√Hz, or 578 nV_{RMS}. Finally, noise adds as the RMS sum (i.e., the square root of the sum of the squares). A noiseless amplifier that has a 1-kΩ resistor at its input will have input-referred noise of 4.1 nV/√Hz. The input-referred noise of a noiseless booster amplifier with gain of 10 driving an RIAA phono preamp with input-referred noise of 4.1 nV/√Hz would have input-referred noise of 0.41 nV/√Hz. An ideal transformer with a turns ratio of 1:10 driving an amplifier with input noise of 5 nV/√Hz would have input-referred noise of 0.5 nV/√Hz as seen at its primary.

Consider a 500 μV_{RMS} cartridge whose thermal noise is only 0.5 nV/√Hz, stepped up by an ideal 1:10 transformer feeding an MM preamp whose input noise is 5 nV/√Hz. Overall system noise will increase by 3 dB, since the MM preamp noise, when referred to the cartridge input, will be 0.5 nV/√Hz. In this case, it is wise to use a 1:20 ratio to make the input-referred noise of the MM preamp only 0.25 nV/√Hz. This makes the overall input noise 0.56 nV/√Hz, only a 0.5-dB loss in SNR due to the MM preamp noise. RMS noise addition suppresses the effect of smaller noise contributors.

To achieve the same result with an MC cartridge with a smaller nominal 200-μV_{RMS} output, the step-up ratio would have to be 2.5 times larger, for a transformer voltage gain of 50. This can be problematic using a transformer because transformer performance tends to degrade with higher turns ratio and a heavier load (19 Ω from the impedance-transformed 47 kΩ MM load) placed on the cartridge. A better solution here is to use a much quieter MM preamp, perhaps with input noise of 2.5 nV/√Hz.

Cartridge Noise

At the most fundamental level, the MC cartridge sets a limit on achievable SNR because of its coil resistance. The nominal output voltage of the cartridge as compared to the thermal noise of the coil resistance sets the achievable SNR even if all subsequent circuitry is noiseless. Consider a cartridge with 200 μV output at 5 cm/s and a coil resistance of 10 Ω. The coil will create 0.41 nV/√Hz of thermal noise. In a flat un-weighted bandwidth of 20 kHz there are 141 √Hz, so the noise voltage will be 58 nV. The un-weighted SNR of 200 μV to 58 nV is 3450, or about 70.5 dB. This number is sobering until RIAA equalization is factored in. The SNR will further improve if A-weighting is employed. In fairness, a more nominal cartridge with a 500 μV_{RMS} output and a 5-Ω coil will have a fundamental un-weighted SNR of about 82 dB.

Cartridge Termination

The MC cartridge termination is a de-facto 100 Ω unless recommended otherwise by a cartridge manufacturer (which may be the case for cartridges with coil resistance significantly larger than 10 Ω). MC cartridges are far less sensitive to cartridge termination than MM cartridges because of their low output impedance. Cartridge termination is straightforward if a head amp is being used to interface the cartridge. However, if a transformer is being used, the effective termination is the MM preamp input resistance (usually 47 kΩ) divided by the square of the turns ratio of the transformer. If a 1:20 transformer is being used, the 47-kΩ MM phono termination is transformed to 118 Ω. If a 1:10 transformer is being used, the effective load becomes 470 Ω. It is OK that this is significantly higher than the 100-Ω norm, but it may not be optimal. Much of this depends on the combination of the specific cartridge and the specific transformer, including the step-up ratio.

If a 1:30 transformer is being used, the transformed MM cartridge load impedance of 47 kΩ will be only 52 Ω, and this is on the low side for many MC cartridges. A cartridge with a 5-Ω coil will see a signal loss of about 1 dB when loaded with 52 Ω, and a corresponding loss in SNR. Any transformer primary resistance will add to the 5-Ω number and further increase the loss. However, none of the cartridges in Table 16.1 require a step-up ratio greater than 20.

Fortunately, most MC transformers have very low primary resistance in the vicinity of 1.5 Ω with a large range from about 0.3 to 6 Ω. The thermal noise contributions of the primary and secondary resistances of five transformers was evaluated at their outputs, assuming a shorted input and using the published data for the transformers. The noise added by the transformer ranged from 2.0 to 5.4 nV/√Hz, with an average of 4 nV/√Hz. The sum of the average MC cartridge and average transformer was 6.2 nV/√Hz. An MM preamp with 5 nV/√Hz would increase input noise to 8.0 nV/√Hz as seen at the output of the transformer.

Matching the Transformer and Cartridge

Further to the issue of cartridge termination is the general issue of the optimum mating of a transformer to different cartridges, and the ease with which that can be accomplished. Some cartridges may not perform well with a given transformer turns ratio. Here is where a major advantage of a head amp lies. A head amp basically does not care about the MC cartridge resistance. It can load the cartridge with the widely accepted and recommended load of 100 Ω (and that can be changed fairly easily).

Finally, there is a myth that must be dispelled regarding cartridge-transformer matching. It is a falsity that the impedance seen by the MC cartridge is optimally the same as the coil resistance of the cartridge. It is not maximum power transfer we are looking for here. Rather, it is the maximum secondary voltage that can be obtained while loading the cartridge in a way that provides the best fidelity and sound. Indeed, the most common load recommended for MC cartridges is 100 Ω, which is 20 times larger than the coil resistance of many MC cartridges. Cartridge load recommendations span a wide range from about 20 to 400 Ω.

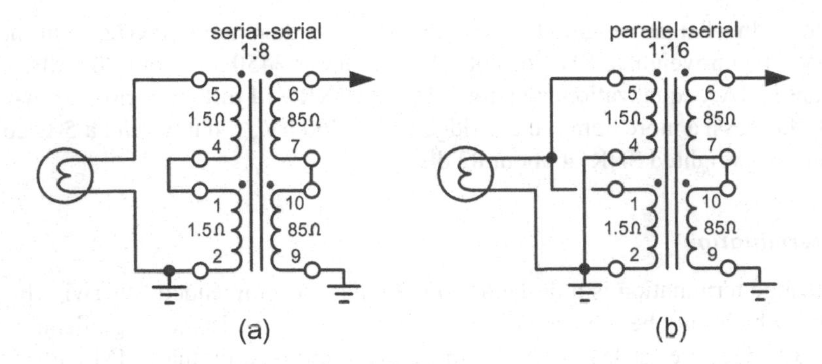

Figure 16.2 A Transformer That Can Be Configured for 1:8 or 1:16 Ratios.

16.2 Moving Coil Transformers

The MC cartridge output level is very small, on the order of 200–600 μV at 5 cm/s. This is about 20 dB lower than that of an MM cartridge. This makes it difficult for vacuum tube and even solid-state amplifiers to amplify these signals with adequate SNR. Tubes were never able to compete on noise, but very carefully designed solid-state MC preamps can come close enough. Fortunately, the source impedance of most MC cartridges is very low, on the order of 3–12 Ω. In the absence of extremely low-noise electronics, transformers have been used to step up the MC signal by a factor of 10 or more without adding a significant amount of noise. They are in principle capable of the best noise performance, but only under the right conditions will they achieve their lowest noise and not compromise the sound [3, 4]. The simplicity of a passive transformer is appealing. However, cost is a deterrent. Good MC transformers cost from $50 to $250 and more.

As alluded to above, MC transformers can be found or configured with a wide variety of step-up ratios, from a low of about 1:4 to a high of about 1:37 [5, 6]. The frequency response of the Jensen *JT-346-AXT* 1:4 transformer is 20 Hz–20 kHz +0, −0.1 dB, while that for the *JT-34K-DX* 1:37 design is down 0.37 dB at 20 Hz and down 0.05 dB at 20 kHz. It should be pointed out that the low-ratio 1:4 transformer is for use in conjunction with a low-noise amplifier ahead of the MM preamp (a hybrid approach).

The Lundahl LL1933 is an example of a transformer that can be configured for two different turns ratios of 1:8 and 1:16 [7]. It is illustrated in Figure 16.2. It is configured by the connection of two coils on the primary and two coils on the secondary. The turns ratios are specified as 1+1: 8+8. A 1:8 ratio is achieved by connecting the primary coil pair and secondary coil pair in series. A 1:16 ratio is achieved by connecting the primary coil pair in parallel and the secondary coil pair in series. Each primary coil has resistance of 1.5 Ω and each secondary coil has resistance of 85 Ω. Its frequency response is 8 Hz to 100 kHz ±1 dB.

Step-up Ratio and Impedance Transformation

As mentioned above, the impedance ratio (transformation) of the primary to the secondary is equal to the square of the turns ratio N. For a 1:20 transformer, a 47-kΩ load on the secondary will be divided by 400 down to the 118-Ω impedance seen looking into the primary, which is about as low as you want to go. With a 350-μV_{RMS} cartridge one gets 7 mV_{RMS} to drive the MM preamp. This is not bad.

Coil Resistance

The resistances of the primary and secondary coils contribute thermal noise, just as does the coil resistance of the cartridge. Together, these resistances place an upper limit on achievable SNR. Primary and secondary resistances R_{pri} and R_{sec} depend on the step-up ratio N and the way the transformer is configured if it is configurable. They vary considerably from manufacturer to manufacturer and among models in a brand. For a 1:20 transformer, the primary resistance may span 0.3–6 Ω and the secondary resistance may span 60–1500 Ω. Input-referred noise contributed by a 5-Ω primary will be 0.29 nV/√Hz, while the input-referred noise from a 500-Ω secondary will be 0.14 nV/√Hz. Higher-quality transformers tend to have lower resistance, providing a better SNR.

If you take $R_{sec} + N^2R_{pri}$, you get the secondary-referred effective noise resistance R_{ref}. The N^2 term arises because any impedance at the primary is multiplied by N^2 at the secondary. For the Sowter 9570, with $N = 10$, $R_{pri} = 3.0$ Ω and $R_{sec} = 1300$ Ω, the resulting R_{ref} is 1600 Ω, implying transformer noise of 5.2 nV/√Hz [8].

Magnetic Shielding and Faraday Screens

Most premium transformers have double magnetic shields (one or both of mu-metal) and Faraday shields. The dual-coil structure of the Lundahl LL1933 transformer is also said to improve immunity to external magnetic fields. These features add to cost, but are very worthwhile in many physical design environments.

Core Material and Core Saturation

Different materials are used for the cores of the transformers. These include amorphous cobalt and nickel laminations. The core material affects the linearity of the magnetization curve and the vulnerability to the onset of early core saturation effects. This is particularly true at 20 Hz. For example, one 1:37 transformer is specified as having total harmonic distortion (THD) of 0.125% at 20 Hz with an input level of −60 dBu, corresponding to about 750 μV_{RMS}. A similar 1:10 transformer exhibits only 0.07% THD at 20 Hz with a higher input level of −50 dBu, corresponding to about 2.4 mV. A premium 1:12 transformer that costs more than twice as much exhibits 0.025% THD at 20 Hz with an input level of −40 dBu, corresponding to about 7.5 mV_{RMS}.

Ringing and Zobel Network Termination

A good number of transformers behave best when they have a carefully chosen Zobel network across the secondary to mitigate ringing caused by the resonance of the leakage inductance with the winding capacitance [1]. Such a network is difficult to design and highly dependent on the details of the transformer. In some cases the transformer manufacturer may suggest such a network and its component values.

A useful measurement approach is to feed the transformer a low-level square wave through a resistive divider that models the cartridge winding resistance for which the transformer should be optimized. One can then experiment with different Zobel R and C to obtain the best square-wave response. Recognizing that the ringing will be at a high frequency, one can start by finding the highest load resistance value that will reasonably damp the ringing. One then places a capacitor in series with the resistance that is as small as possible to retain suppression of the ringing. Iterations of the R and C values can then follow.

Even without a Zobel network, loading the transformer with a resistance less than the usual 47-kΩ value may make the transformer behave better. For example, one might load a 1:10 transformer with 10 kΩ to make the impedance seen by the cartridge be equal to the often-used 100 Ω. If a low-ratio transformer is to be used to drive a dedicated low-noise boost amplifier, it usually must be terminated for proper operation. In the case of the 1:4 transformer mentioned above (Jensen *JT-346-AXT*), the recommended load is 6.8 kΩ.

Target SNR vs Sound Quality

There is limited benefit in achieving noise that is more than about 10 dB below that on a nearly perfect vinyl LP. It is always important that low noise is not pursued exclusively at the expense of other performance that may detract from the sound quality. For example, a higher step-up ratio in a transformer is often accompanied by greater ringing and poorer frequency response. Keep in mind that the annoyance factor for noise in an RIAA-equalized system is frequency-dependent with noise at low frequencies emphasized. White noise at the input of the system will be approximately transformed to Red (Brown) noise at the output by the RIAA equalization.

Balanced Inputs

Sometimes a balanced input is helpful with an MC turntable that can provide a balanced signal with an XLR connector. This has the potential to reduce hum and noise pickup. A balanced input is obviously easy to accomplish with a transformer. It is more difficult to implement with electronics without incurring a noise penalty of about 3 dB, as will be discussed later.

16.3 Noise in the Cartridge-Transformer System

Even with a passive transformer, there are several noise contributors that limit the achievable SNR. These include cartridge coil resistance, transformer primary resistance and transformer secondary resistance. At the system level, the input noise of the subsequent MM preamp also contributes to limiting the achievable SNR. The chosen transformer step-up ratio also influences achievable SNR for a given cartridge.

Achievable Signal-to-Noise Ratio

Many MC cartridges have coil resistance on the order of 5 Ω. The thermal noise of 5 Ω is 0.3 nV/√Hz. If the primary winding resistance of the 1:10 transformer is 2.5 Ω, its thermal noise will be 0.21 nV/√Hz. The secondary winding resistance of the transformer will also create thermal noise. Assume that the secondary winding resistance is 500 Ω. The thermal noise of 500 Ω is 2.9 nV/√Hz. Its noise when referred back to the primary is 0.29 nV/√Hz if the turns ratio is 1:10. The summed input-referred noise at the primary and cartridge output will be 0.46 nV/√Hz.

If the subsequent MM preamp has 5 nV/√Hz of input noise, its noise when referred back to the cartridge input is 0.5 nV/√Hz. In this case, net input-referred noise increases to 0.68 nV/√Hz. This exercise shows that in the real world the input-referred noise of an MC signal chain with a 1:10 transformer will rarely be better than 0.7 nV/√Hz, corresponding to 7 nV/√Hz at the MM preamp input. This also underscores the need for the MC preamp to be followed by an MM preamp with low noise.

These numbers do improve if a 1:20 step-up turns ratio is employed in cases where the MC cartridge works well with the lower impedance and typically lower inductance seen at the transformer primary. A number of MC step-up transformers can be configured for different step-up ratios. The main advantage of a higher step-up ratio will be the reduction of the input-referred noise contribution of the subsequent MM preamplifier.

A significant issue is that the optimal step-up ratios are different for the very wide variety of MC cartridges, whose coil resistance can range from 3 to 40 Ω and whose outputs can range from 180 μV_{RMS} to 1 mV_{RMS} [1]. A user may not want to have to change their $250 transformer when they decide to change to a different MC cartridge. Available fixed-ratio transformers have turns ratios ranging from 1:4 to 1:37. Some low-ratio transformers actually require an additional active gain stage between the transformer and the MM preamplifier. Low-ratio transformers are more likely to provide better performance.

It is notable that a head amp can easily produce more gain so as to make the MM preamp effectively less noisy by being driven with a larger signal. It is much more difficult for a transformer to do this because the transformer with a higher turns ratio may place too heavy a load on the MC cartridge. Active head amps are often incorporated into phono preamps, and outboard head amps are no longer as popular.

16.4 Moving Coil Head Amps

MC step-up transformers can be expensive and require considerable magnetic shielding. They can add frequency response coloration. They are also not noise-free. For thess reasons, MC head amplifiers are an attractive alternative. MC head amps are mostly about achieving very low noise. Fortunately, the MC cartridge is a source with a very low impedance, usually between 2 and 10 Ω. The typical output for an MC cartridge is very roughly in the range of 500 μV_{RMS} at recorded velocity of 5 cm/s. This is about ten times less than that of an MM cartridge for the same velocity. A gain of 10 in the head amp is thus often adequate.

It can be argued that an MC cartridge/preamp system should be competitive in noise performance with a good MM cartridge system. A good MM preamp can achieve input-referred noise of 5 nV/√Hz. With head amp gain of 10, this implies that the head amp should have input noise of 0.5 nV/√Hz. This is a tall order. The thermal noise of a 60-Ω resistor is 1 nV/√Hz. It is common to accept 1 nV/√Hz as adequate for a head amp.

MC cartridges and transformers are not noise-free. It is notable that an MC cartridge that produces 200 μV at 5 cm/s and has a 5-Ω coil has thermal noise of 0.3 nV/√Hz. In a 20-kHz flat bandwidth, the noise voltage will be 42 nV_{RMS}, corresponding to un-weighted SNR of 74 dB.

As mentioned earlier in the previous section, a step-up transformer creates some thermal noise. As calculated earlier, a cartridge with a 5-Ω coil and the example transformer described, the total effective noise at the primary is 0.46 nV/√Hz. This further supports the goal of head amp input noise being about 0.5 nV/√Hz.

Putting Noise in Perspective

An input-referred noise density target of 1.0 nV/√Hz is often cited as a reasonable goal for a head amp. It is quite challenging to achieve, and it is considered adequate. This is best put into perspective by comparing it to the capabilities of an MM phono preamp. A very good MC preamp has input noise density of about 3 nV/√Hz, while a good one has less than 5 nV/√Hz. A fair one is probably about 8 nV/√Hz. Assuming a head amp gain of 10, these would translate to 0.3, 0.5 and 0.8 nV/√Hz. This suggests that a head amp for a 500 μV_{RMS} MC cartridge must have 0.5 nV/√Hz to be on a par with the noise performance of an MM preamp.

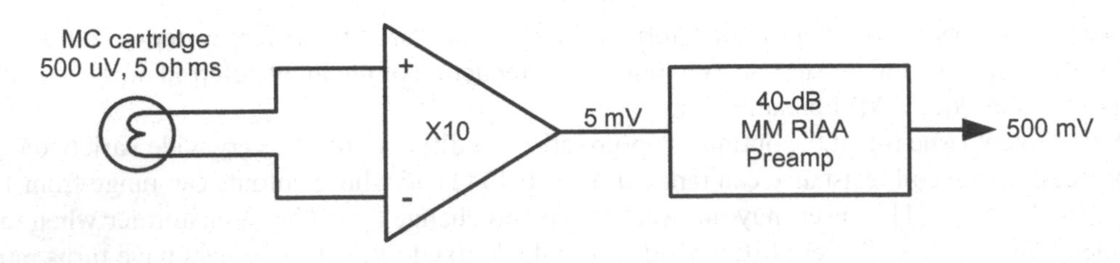

Figure 16.3 MC Playback Arrangement Using Active Head Amp.

SNR of a phono preamp is expressed as the output at 5 mV_{RMS} input divided by the output noise in a 20-kHz noise bandwidth. This is referred to as the un-weighted or wideband SNR. Alternatively, A-weighted SNR is specified by measuring the output noise voltage after passing it through an A-weighting filter. This filter cuts out low-frequency noise and hum, and low-pass filters the higher frequencies. It produces a better SNR number, which manufacturers like to quote.

However, it is important to recognize the effect of the RIAA equalization curve on the noise measurement and its spectrum. For a white noise input, the RIAA curve improves SNR by 14 dB compared to a flat amplifier with the same gain at 1 kHz [9]. The RIAA curve also changes the noise spectrum to be non-white to a noise spectrum that is more red than pink because of its near 6-dB/octave roll-off. This alteration of the spectrum shape also changes how A-weighting affects the SNR measurement.

Figure 16.3 illustrates the typical MC signal flow when an active head amp is used instead of a transformer. The head amp simply replaces the transformer and brings with it advantages and disadvantages. Advantages include flexibility, avoidance of cartridge-matching impedance issues, neutrality, reduced low-frequency distortion and lower cost. The main disadvantage is that it usually cannot quite match the low noise of an optimized transformer arrangement. If the head amp is in a separate box, like a transformer often is, it will need a power source. The head amp may be an outboard device or it may be in the same piece of equipment in which the MM phono preamp resides. It should also be said that there are some cases where the MC and MM preamps are essentially one in the same, meaning that an adequately low-noise MM preamp is used for MC cartridges, where it is just configured for more gain.

MC preamp design seems simple at first – just design a very low noise amplifier with a voltage gain of 10 that has the benefit of a very low-impedance source. However, it is not so simple because of the need for very low feedback network resistances and the resulting difficulty of applying negative feedback and DC stabilization. Achieving input-referred noise much below 1 nV/√Hz can be difficult, often requiring paralleled transistors.

The range of output levels of MC cartridges varies greatly, so it may be desirable to have some gain selection, maybe by means of internal jumpers. An example might be selectable gain of 10 or 30. It is desirable, from a noise perspective, to have the MC cartridge preamp bring the signal up to greater than that normally coming from an MM cartridge, since we want to swamp out the normal noise of the MM preamp. We do not want the MM preamp adding much to the noise from the MC preamp. Every bit counts. Mitigating this is the fact that the MM preamp will tend to operate with less noise because it is being driven from a low-impedance source rather than the higher impedance of an MM cartridge. The thermal noise from the usual 47-kΩ input resistor is also killed. Numerous MC head amp designs will be discussed in this section in order to illustrate the many different choices and the technical challenges that must be addressed.

Two-Transistor CFP Head Amp

Figure 16.4(a) shows a simple two-transistor CFP head amp. The PNP 2N4403 is one of the best BJTs for this application because it has low base spreading resistance $r_{bb'}$ of only about 20 Ω, contributing about 0.6 nV/√Hz. The base spreading resistance can increase if the transistor is operated at a larger V_{cb} (like 12 V_{DC}). A reduced supply voltage of −6 V_{DC} is used here to keep $r_{bb'}$ low. A capacitance multiplier can be used to reduce V_{cb} of Q1 and further quiet the supply.

Recall that the collector shot noise of a BJT, viewed as input voltage noise, is equal to that of a resistor whose value is half that of $r_{e'}$. Operating at 4 mA, $r_{e'}$ is 26/4 = 6.5 Ω, so input noise voltage of the transistor due to shot noise (and driven from a zero-impedance source) is that of a 3.25-Ω resistor, which is 0.24 nV/√Hz. A transistor with $r_{bb'}$ = 20 Ω will have base resistance thermal noise of 0.59 nV/√Hz, dominating the shot noise. This is why $r_{bb'}$ is so important in low-noise transistors. If there is any RF instability at the base of Q1 a ferrite bead should be placed at the base. Do not use the usual 100-Ω series resistor solution, as it will introduce more thermal noise.

A 10-Ω feedback shunt resistor R5 is used in Q1's emitter to keep its thermal noise contribution down. Two feedback resistors, R6 and R7, in series are used to provide output gain taps of 10 or 30. The noise contribution of R5 is about 0.4 nV/√Hz.

Resistors R2 and R3 set Q1's base bias to −2.2 V_{DC}, putting the emitter at about −1.6 V. R4 has 1 V_{be} across it and sets Q1's current to 4 mA. R8 sources 5 mA to Q2 setting its bias. R8 also sources 4 mA to Q1. With the −6 V_{DC} supply, V_{cb} of Q1 is 3.2 V, keeping $r_{bb'}$ small. Loop gain of the circuit is fairly low, at about 16, leading to mediocre performance.

A major issue that is often common to simple head amps is the need for C2, a large electrolytic decoupling capacitor to allow gain to be unity at DC and to allow adequate bias stability without forcing high current through R5. About 4700 μF is required to achieve an LF corner of 3.4 Hz to preserve low-frequency response at 20 Hz. Passing signal through an electrolytic like this is not good practice for the best sound quality. C2 should be of high quality and have low ESR, and should be rated at 50 V or more.

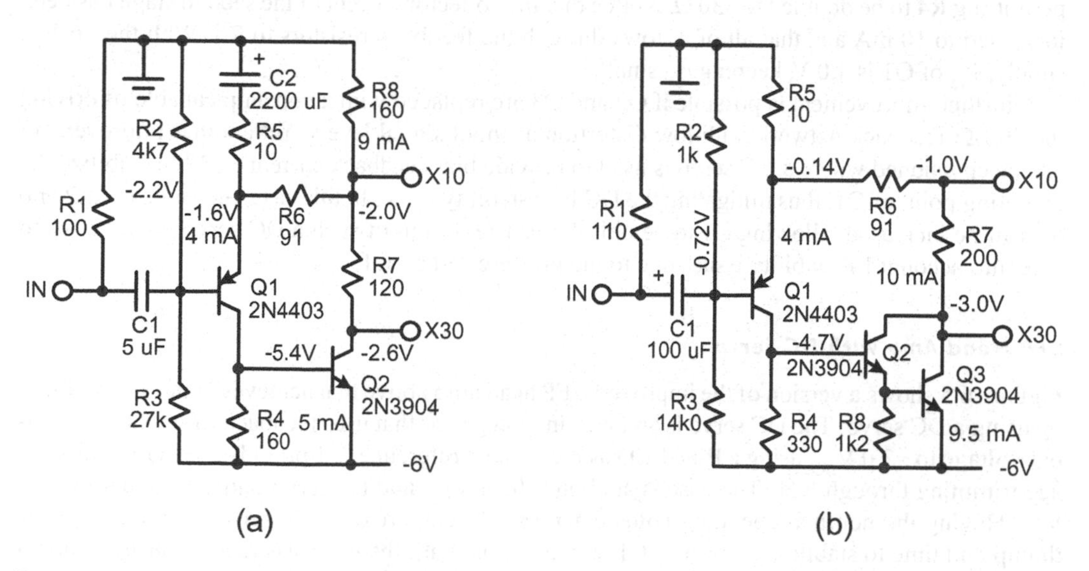

Figure 16.4 Two CFP Head Amplifiers.

C1 is loaded by the 4.2-kΩ parallel combination of R2 and R3. With C1 at 5 μF, it can be a film capacitor. Its transmission will be down by 3 dB at 7.6 Hz and less than 1 dB at 20 Hz. However, LF noise created by Q1's base shot noise current passing through C1's impedance must be considered. A transistor with Beta of 50 and biased at 4 mA will exhibit shot noise base current of 5.1 pA/√Hz. Flowing through a 5-Ω cartridge resistance, this will create very low noise voltage of just 26 pV/√Hz. However, flowing through the impedance of C1 will create noise at low frequencies. Suppose C1 is chosen to be 5 μF. At 100 Hz, the impedance of C1 will be 318 Ω. The 5.1 pA/√Hz of shot noise current flowing through 318 Ω will result in 1.65 nV/√Hz. This is acceptable because of the low frequency of 100 Hz and the fact that this noise contributor goes down at 6 dB/octave.

In the designs of Figure 16.4 and those that follow using a BJT for the input stage, Q1 can comprise two transistors connected in parallel, cutting effective $r_{bb'}$ in half. The V_{be} of the two transistors should be matched to better than 5 mV to achieve acceptable current sharing. The total current of the two transistors can remain the same to realize a significant noise improvement. However, if the total current is doubled some further improvement will be realized. Finally, in circuits like these it is very important that the power supply be very quiet, as PSRR in these designs is not stellar.

Improved CFP Head Amp with Three Transistors

Figure 16.4(b) shows an improved version of the CFP head amp. The large electrolytic capacitor in the emitter circuit of Q1 is eliminated, but this comes at the price of reduced bias stability in the face of beta variations in Q1 and Q2. Bias resistors R2 and R3 must be made smaller than in (a) to reduce the effect of Q1's base current on the operating point. The resulting reduced load resistance of slightly less than 1 kΩ requires that C1 be a 100-μF electrolytic capacitor to achieve satisfactory low-frequency response and minimize low-frequency capacitor distortion. Capacitor C1 should be rated at 50 V or more to further keep its distortion contribution low.

The Darlington pair (Q2, Q3) increases loop gain by reducing the load on Q1's collector and permitting R4 to be doubled to 330 Ω. Notice that the collector current of the second stage has been increased to 10 mA and that all of it flows through the feedback resistors to R5. With the −6 V_{DC} supply, V_{cb} of Q1 is 4.0 V, keeping $r_{bb'}$ small.

A further improvement is possible if Q2 and Q3 are replaced with an op amp capable of driving the 300-Ω feedback network with low distortion at small signal levels. Yet another improvement can be envisioned where a DC servo is used to provide bias feedback current to R3 to stabilize the operating point of Q1, thus mitigating the DC bias stability issue. In this case, the values of R2 and R3 can be increased, allowing C1 to be only 5 μF. The design of such a DC servo would have to take into account LF stability issues due to the pole created by C1.

CFP Head Amp with DC Servo

Figure 16.5 shows a version of the improved CFP head amp above that achieves better DC stability by using a DC servo. The DC servo is an inverting integrator that uses feedback to set the X30 output voltage to −3.0 V_{DC}, using a Blue LED as the voltage reference. U1 provides the base bias voltage trimming through R11. The base is at about −0.74 V_{DC}, and the servo output is at about −3.0 V_{DC}. Having the nominal operating voltage across C2 be approximately zero minimizes turn-on thump and time to stabilize. Capacitor C1 creates a pole with the base bias resistor string at about

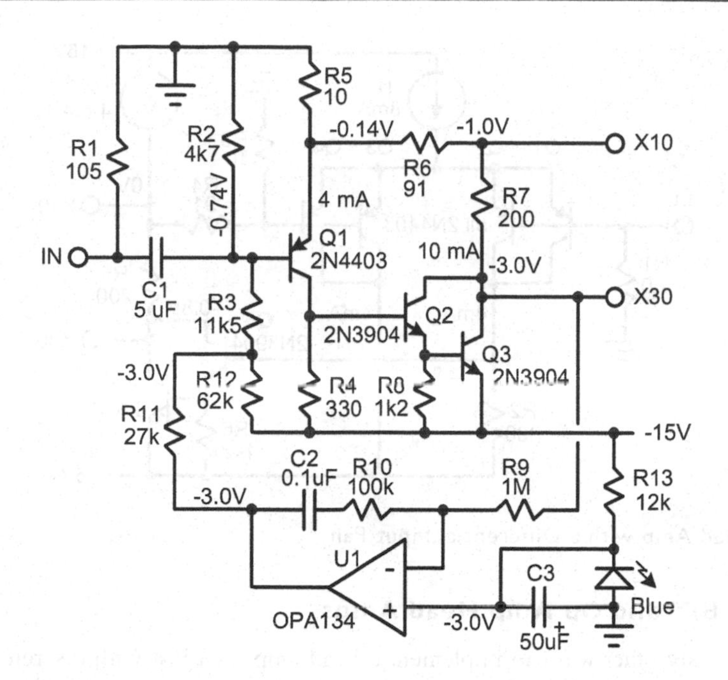

Figure 16.5 Adding a DC Servo for Improved Bias Control.

0.2 Hz. Since the integrator also creates a pole, R10 is inserted in series with C2 to create a zero at 16 Hz to assure feedback stability of the DC servo loop.

Head Amp with Differential Input Stage

Figure 16.6 shows a CFP-like design using a differential pair (LTP) as the input stage. This makes biasing much easier and also eliminates the need for the large emitter-decoupling capacitor of Figure 16.4(a). The input and output can now both reside at 0 V. Notice that the input to the MC cartridge is DC-coupled, allowing base bias current of Q1 to flow in the MC cartridge. This is probably allowable, since it will be nominally less than 100 μA. If not, an input AC-coupling arrangement can be added.

Each side of the LTP comprises two 2N4403 transistors in parallel and is operated at a total of 4 mA to achieve low noise. The input pair thus consumes 8 mA. The 2N4403 transistors must be matched to within 5 mV for adequate current sharing. The ultra low-noise ZTX951, with $r_{bb'}$ of only about 2 Ω can be used without paralleling transistors. In the past, devices with low $r_{bb'}$ included the Rohm 2SB737, Hitachi 2SA1083 and Toshiba 2SA1316.

A current source is used to provide the tail current to minimize power supply noise injection. A capacitance multiplier/regulator (not shown) is used to keep collector voltages to less than 6 V. A current source is also required to pull current from the emitter of Darlington Q5–Q6 to set its bias current at 4 mA, allowing a peak output voltage swing of ±1.2 V. Total current consumption is about 12 mA. The convenience of using the differential pair comes at the expense of 3-dB higher noise contribution from the input transistors compared to a single-ended design. That is the main shortcoming of this design.

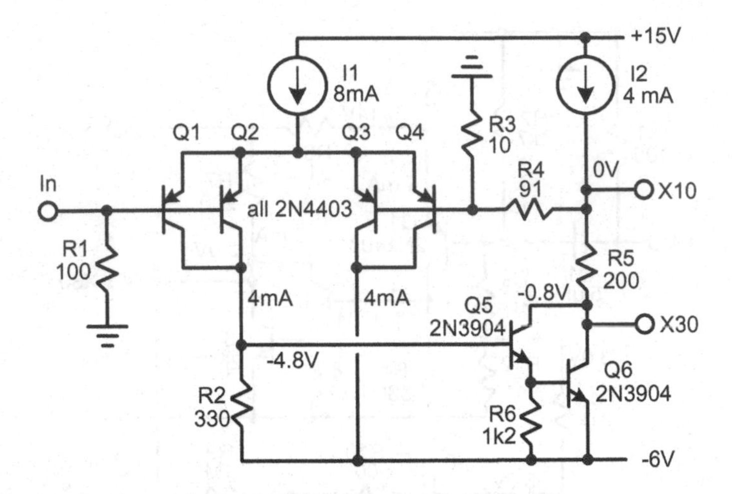

Figure 16.6 Head Amp with a Differential Input Pair.

16.5 Other BJT and Op Amp Head Amps

There are obviously other ways to implement a head amp than just with discrete BJT circuitry. Here we touch on some circuits that involve low-noise discrete front ends combined with op amps and solutions that involve op amps alone. A key point is that op amps are a very simple way to obtain the gain needed for accuracy and low distortion, even if, by themselves, they cannot achieve noise as low as desired.

Hybrid Differential Input Stage with Op Amp Output Stage

Figure 16.7 shows a design with a differential input amplifier like that in Figure 16.6 where an op amp is used for the second stage. This provides much more loop gain and lower distortion. The op

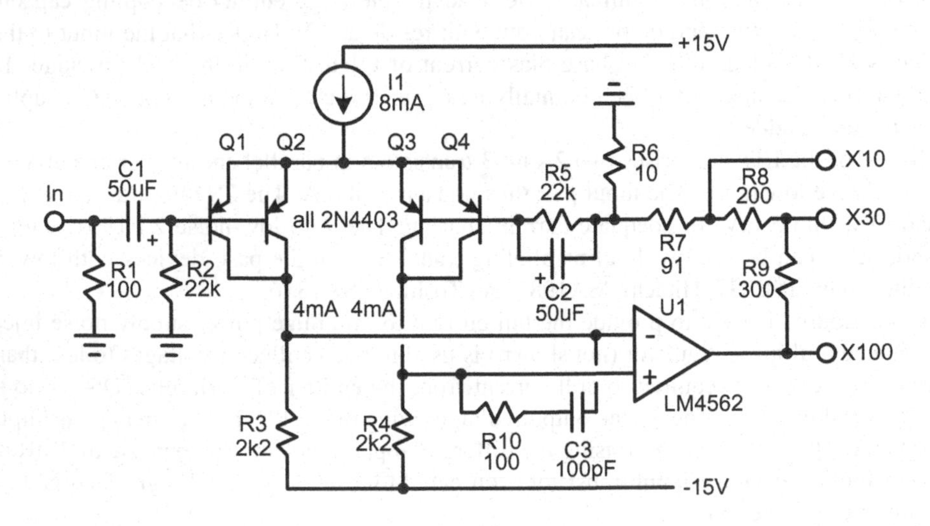

Figure 16.7 Combining a Discrete BJT Input Stage with an Op Amp.

amp is driven differentially from both collectors of the differential pair, improving power supply rejection. A capacitance multiplier/regulator can still be employed to keep the collector voltages of the input transistors low and power supply noise down, but the collector resistors R3 and R4 can instead be chosen to set the collector voltages to the desired $-5\ V_{DC}$.

Frequency compensation is provided by R10 and C3, acting as a lag-lead filter in conjunction with collector resistors R3 and R4. Together they reduce loop gain at high frequencies to assure stability. The higher loop gain provided by the op amp makes it possible to provide an X100 output by adding resistor R9 to the feedback network. More importantly, this resistor also assures that the LM4562 does not see a heavier load than 600 Ω.

Op Amp Head Amps

It is difficult for op amps to compete with the low noise of a discrete single-ended BJT input stage, but op amps are very convenient. A good head amp can be implemented with an extremely low-noise op amp like the AD797 [11].

Figure 16.8 illustrates the use of an AD797 op amp to implement the head amp. This fairly expensive op amp has input noise of only 1 nV/√Hz. Use of an op amp makes biasing easy and also allows direct coupling to the cartridge in some cases, eliminating the input electrolytic capacitor. This head amp has no electrolytic capacitors in the signal path, just like the discrete design above [9–11].

The AD797 is configured as a non-inverting amplifier with a voltage gain of 60 so that a feedback network of 600 Ω can be used with a shunt resistor of 10 Ω to achieve very low noise. Most op amps are not specified for distortion with a load of less than 600 Ω, so the feedback network load resistance is kept this high. In spite of the effective closed-loop gain of 60, THD of the ultra-low-distortion AD797 is still at about −100 dB (0.001%) at 20 kHz with a 3-V_{RMS} output. Taps along the feedback resistor path provide gains of X10 and X30. An X60 tap can also be implemented at the output of U1.

The input is shunted by R1, which in combination with R2 provides an approximate 100-Ω load for the MC cartridge. C1 provides some low-pass filtering to reduce EMI susceptibility. The input is AC-coupled by C2 to prevent the input bias current I_{in} of the AD797 from flowing through the cartridge. This small current is specified as 0.25 μA typical and 1.5 μA maximum. For those who are not concerned about such a small DC current flowing through the cartridge, C2 can be bypassed. In most MC preamps, C2 is a large-value electrolytic capacitor; here a 10-μF polypropylene film

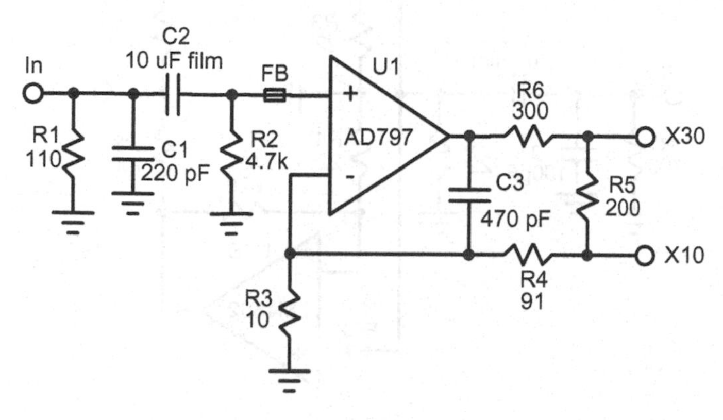

Figure 16.8 A Head Amp Employing an AD797 Op Amp.

capacitor is chosen. In combination with R2, it provides an acceptable low-frequency 3-dB corner of 3 Hz.

The choice of 4.7 kΩ allows the use of a 10-μF film capacitor for C2, providing a 3-Hz LF cutoff. The AD797's maximum 1.5 μA I_{in} flowing through R2 will create a maximum offset voltage of 7 mV, which will be multiplied by 60 to become a maximum offset voltage of 423 mV at the output, which is acceptable assuming that the output is AC-coupled. The worst-case loss of headroom of 423 mV is not of concern in this small-signal circuit. Op amp input noise current and thermal noise current from R2 will flow through C2 and create some noise voltage across C2 at low frequencies that will decrease at 6 dB/octave as frequency increases. At 100 Hz, the 2-pA/√Hz input noise current of the AD797 contributes only 0.3 nV/√Hz across C2. The thermal noise of R2 also contributes only 0.3 nV/√Hz. When these contributors are added on an RMS basis to the 1 nV/√Hz input noise voltage of the AD797, total noise is only increased to 1.06 nV/√Hz at 100 Hz.

The ferrite bead in series with the input of U1 provides further protection against EMI and also reduces the likelihood of instability in the AD797. The small feedback capacitance of C3 defines a closed-loop bandwidth of just over 500 kHz and helps maintain feedback stability.

Figure 16.9 shows a less expensive approach that uses two LM4562 op amp sections in parallel to improve noise by 3 dB. With each LM4562 having typical 2.7 nV/√Hz input noise, this gets the noise down to 1.9 nV/√Hz, adequate for many applications [9]. Gain is fixed at 29 with a pair of 642-Ω feedback networks to keep the load on each op amp above 600 Ω. The 21-Ω (620||22) impedance of each feedback network creates only 0.6 nV/√Hz of thermal noise, small enough in comparison to the typical 2.7 nV/√Hz of each op amp to increase the total to only 2.8 nV/√Hz.

The very low input bias current of the LM4562 (10 nA typical, 72 nA maximum) makes it possible to set R2 to 22 kΩ while keeping output offset voltage to less than 95 mV. This in turn allows the use of a 2-μF film capacitor for C2, allowing a 3.6-Hz low-frequency corner and eliminating an electrolytic capacitor in the signal path.

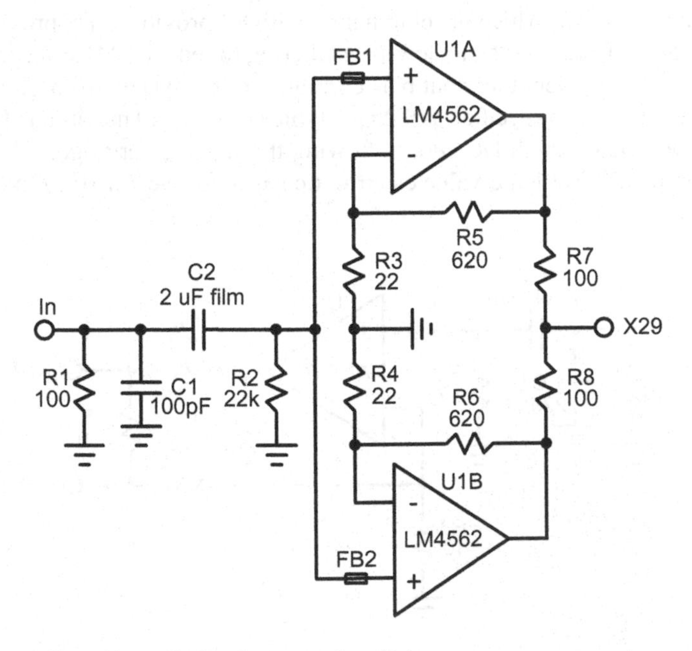

Figure 16.9 A Head Amp Using 2 Op Amps in Parallel.

Dedicated RIAA Head Amp/Preamp

Figure 16.10 illustrates an MC preamplifier with a dedicated signal path all the way to line level. In essence, it comprises an MC head amp with an RIAA MM preamp, but in a combination optimized for an MC cartridge. In an ordinary arrangement, an MM preamp with a nominal midband gain of about 40 dB is preceded by an MC head amp with gain of 20 dB or more.

In an ordinary MM preamp, the feedback circuit must be designed with fairly low impedance to keep the noise down. This can result in uncomfortably large capacitances in the precision feedback path; even worse if an electrolytic capacitor is placed in series with the shunt resistor in the feedback network.

If the MC portion is designed with a gain of perhaps 100, the noise issue in the MM portion is greatly diminished and the precision RIAA equalization components can be of higher impedance. This approach takes advantage of the extra gain usually available at no cost in performance in the MC portion of the circuit. It also takes advantage of the buffering provided by the MC circuit, so the MM circuit does not have to be designed to operate with the high input impedance from an MM cartridge. Such an approach was used in the VinylTrak™ preamp [10].

There can thus be some improvement in performance and gain allocation tradeoffs if the MC preamp is merged with the MM RIAA preamplifier and both are optimized together. This was the approach taken in the VinylTrak preamp discussed below in Section 16.8, where the MC section was operated at a gain of 100 [12, 13]. This was also the approach taken in Ref. [10]. Optimal gain allocation can make a substantial improvement in SNR and distortion performance. A good MM preamp is not necessarily optimal for receiving a signal from a head amp. It is optimized for a high-impedance input.

The MC head amp circuit can be operated at a gain of 100, still with very low distortion if it includes an op amp; that extra 20^{+} dB of gain can lessen the gain needed in the subsequent RIAA section. Moreover, it can be used as a flat-gain element followed by a passive 75-µs filter between it and the rest of the preamp, having the subsequent amplifier implement only the low-frequency part of the RIAA equalization, as in the MM preamp described in Chapter 15. This optimization allows one to completely dominate any noise contributions that would occur with a subsequent MM preamp that must be able to achieve a good SNR. with only 5 mV_{RMS} input.

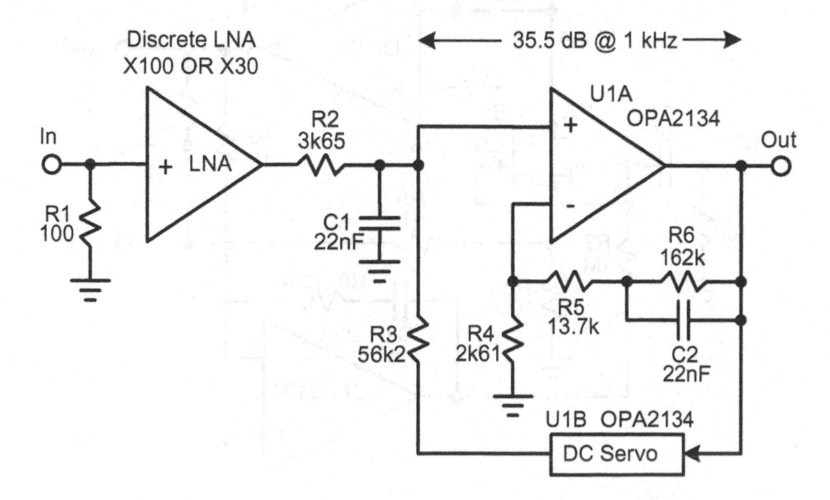

Figure 16.10 A Dedicated MC Signal Path.

This dedicated approach also provides the advantage of a line-level output from the MC signal chain, so MC selection can be done at line level by the ordinary selector switch. This eliminates bypass switching at low signal levels to switch between MC and MM. The MC input will of course be separate from the MM input.

16.6 JFET Head Amp

It is not intuitive that JFETs can compete (or even be better) in the MC application because they often have lower *gm* than BJTs and higher noise. However, they can make biasing much easier, as in the design described here, while still achieving 1-nV/√Hz performance when using ultra-low-noise devices. JFETs definitely have a place in low-noise audio design [14–16]. As mentioned earlier, the JFET is more immune to EMI effects than BJTs.

Operating with Paralleled JFETs

The design in Figure 16.11 employs a dual monolithic low-noise LSK389B JFET with both transistors in parallel, each operating at 9 mA [15, 16]. The use of a dual monolithic device with two matched JFETs makes paralleling them work well. High typical transconductance of 43 mS is achieved and gain of the JFET stage is about 20 when feedback network resistance is included. High transconductance translates to low noise for a JFET. Here the typical input noise is 0.9 nV/√Hz.

The bias-determining feedback is applied to the AC-coupled gate; it is supplied by U2A at the DC servo output. The very high input impedance of the JFET and R2 permits the use of a small 1-μF film coupling capacitor C1. The signal passes through no electrolytic capacitors.

The drains of the LSK389B connect directly to U1A, an LM4562 low-noise op amp that drives the feedback network and produces the output. The gain of the LM4562 is 90 dB at 20 kHz, so

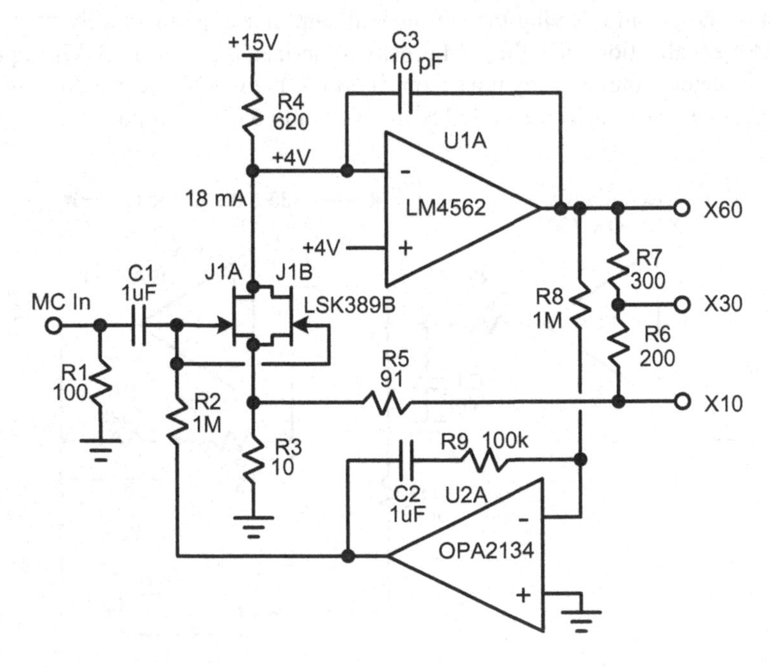

Figure 16.11 JFET Moving Coil Head Amp.

plenty of loop gain is available for distortion reduction. The positive input of the op amp is connected to a +4-V_{DC} reference, setting the drain voltage to 4 V. Operating the JFETs at low V_{ds} minimizes impact ionization noise. A string of feedback resistors totaling 600 Ω in value allows various selections of gain, up to 60. The feedback network is DC-coupled.

The power supply feeds a noise-sensitive node, so it should be rigorously filtered to achieve low noise. The +4-V_{DC} reference (not shown) feeding the positive input of U1A must also be rigorously filtered and free of noise.

16.7 Balanced Input Head Amp

Sometimes a balanced input is helpful with a turntable that can provide a balanced signal with an XLR connector, since this has the potential to reduce hum and noise pickup. A balanced input is obviously easy to accomplish with a transformer. It is more difficult to implement with active circuitry absent a noise penalty. All else remaining equal, a differential amplifier input suffers a 3-dB noise penalty (or worse) as compared to a single-ended input circuit.

Fighting the 3-dB Penalty

One can make a balanced input MC preamp with an un-degenerated JFET differential pair without NFB. A 3-dB penalty is incurred unless a significant number of dual JFETs is used in parallel. This approach was taken with the author's VinylTrak™ preamplifier described below, where four LSK389 dual monolithic JFET differential pairs were connected in parallel to achieve input noise of about 0.8 nV/√Hz [12, 13]. With small signals, distortion is not a big problem without negative feedback, but gain-matching between channels must be dealt with using a gain trimmer, since the amplifier is operating open-loop and differences in operating transconductance in the left and right channel JFETs may result in gain differences.

Hybrid Head Amp with Transformer and Active MC Preamplifier

A more costly design combines the best of transformers and active circuits. The 1:4 low-ratio transformer boosts the cartridge signal by 12 dB, enough to get it to the point where the subsequent very low noise active circuitry does not contribute materially to noise. Transformers can provide exceptional sound quality and fidelity, especially when their turns ratio is small. Figure 16.12 illustrates such a design. The complete MC phono signal path with RIAA equalization is dedicated to the MC function.

In this design, where cost is not a concern, a Jensen JT-346-AXT transformer with a 1:4 ratio (12 dB) is used. This transformer costs about $230.00. It is followed with a low-noise amplifier (U1) with gain of about 27 dB. Net gain prior to the remainder of the circuit is thus about 39 dB. The flat-gain amplifier (U1) is followed by a passive 75-μs RIAA *HF* section, which is then followed by an active series-feedback amplifier to provide the LF RIAA 318-μs and 3180-μs time constants. Its gain at 1 kHz is 16 dB, for a total of 55 dB. Overall gain is easily changed by altering the value of R5. The RIAA LF circuit also includes a DC servo for DC offset management. This resembles the passive-active architecture of the MM preamp of Figure 15.5 in Chapter 15.

The RIAA portion that brings the signal up to line level is thus dedicated to the MC function. Being fed by the low-impedance MC stage it need not contend with high-impedance MM cartridge sources. Obviously, an AD797 would work well here. This hybrid approach to the head amplifier naturally provides a balanced input. The cost is high.

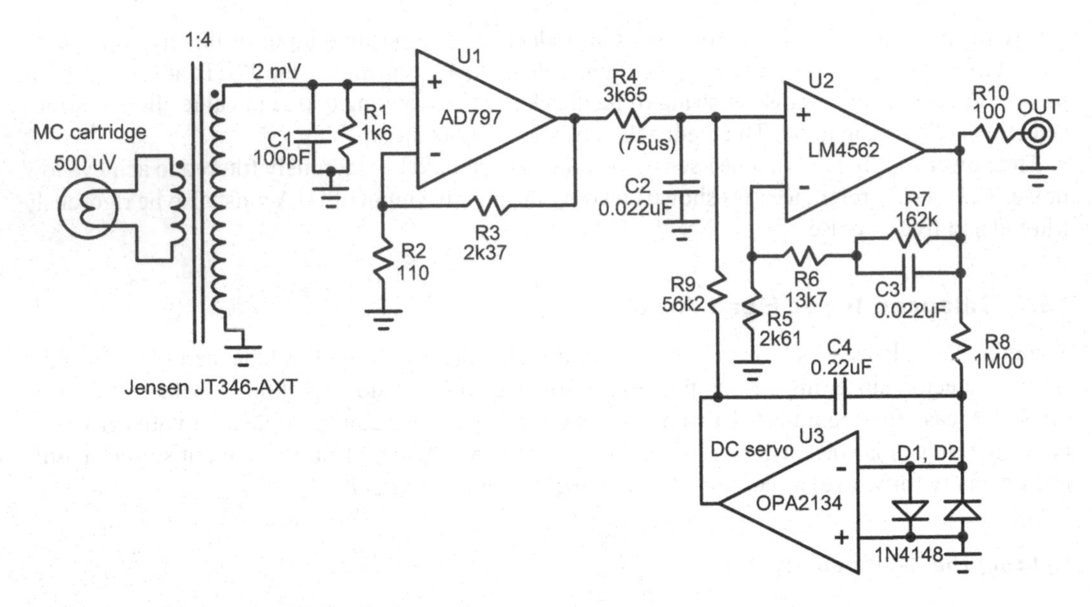

Figure 16.12 Hybrid Head Amp with 1:4 Transformer.

16.8 The VinylTrak™ Dedicated MC Preamp

The VinylTrak™ phono preamp for MM cartridges was described in Chapter 15 [12, 13]. The VinylTrak preamp employs two entirely separate phono sections, one for MC and one for MM. The desired type is selected at line level by a relay at the input of the preamp's line-level circuit. The MM and MC preamps each have their own RIAA EQ networks and DC servo. With the resulting very small DC offset, the switching path can be DC-coupled without introducing pops when the mode is changed. This approach allows total flexibility in optimizing the whole preamp design for its intended purpose (i.e., MM or MC). The active 3180/318-*µs* LF portion of the RIAA EQ is identical for the MM and MC designs, as is the DC servo. See Figure 15.5.

The MC section of the VinylTrak preamp is described here. The MC preamp employs a discrete JFET input stage that includes the 75-*µs* RIAA time constant, just as in the MM input stage. The topology is very similar, except that the first-stage gain is about 100 and the circuit is designed for very low noise by paralleling four LSK389 dual monolithic differential pairs [15, 16].

The Input Amplifier

The MC input amplifier is shown in Figure 16.13. It is a JFET-input, double folded cascode design without negative feedback. Its gain is about 40 dB. Without negative feedback, there is no need for a feedback network with very low impedance (on the order of ohms) to keep the noise down. Such a feedback network can be difficult to drive. The input JFET pair actually consists of four paralleled LSK389 JFET differential pairs, each pair with its own tail current source [16]. This enables input-referred noise of 0.7 nV/√Hz to be achieved. A key advantage is that this approach naturally provides a balanced input.

The NPN cascode (Q1, Q2) sets the drain voltages of the input JFETs to a fairly low +6.2 V_{DC} to reduce impact ionization noise. Q1 and Q2 pull 8 mA from each leg of the PNP differential

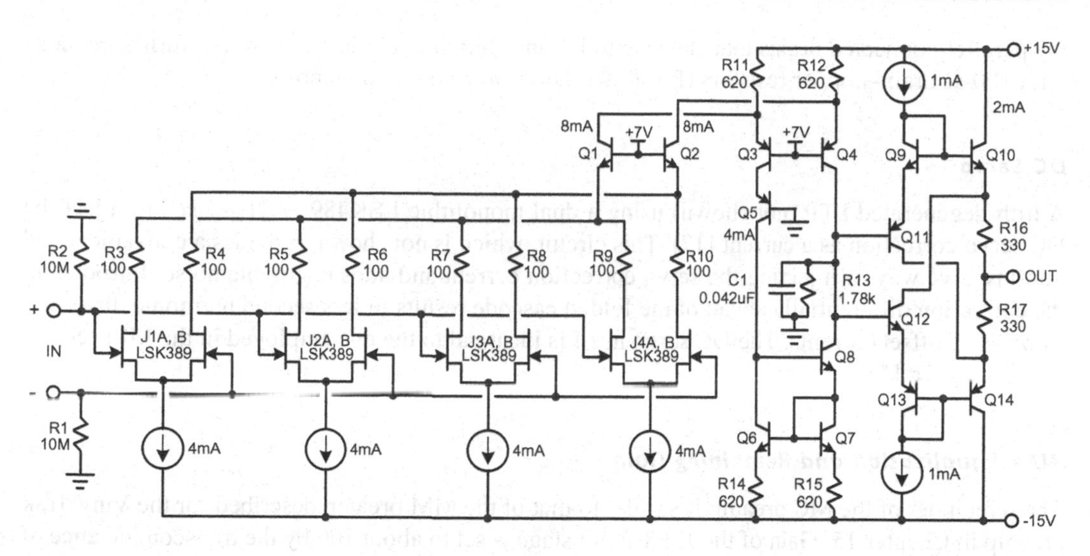

Figure 16.13 VinylTrak Moving Coil Preamplifier.

folded cascode. The emitter resistors of the folded cascode supply 11.2 mA to the emitter nodes, adequately in excess of 8 mA. The differential outputs of the folded cascode are applied to a Wilson current mirror operating from the negative supply rail. The Wilson current mirror was subsequently replaced with a conventional three-transistor helpered current mirror. This change reduced noise slightly.

Gain is set by shunt resistor R13 to ground from the output collectors of the gain stage. The 75-μs RIAA time constant is implemented with a shunt capacitor C1 at this node. The signal on this node is passed to the output through a bootstrapped diamond buffer. At its gain of 100, this amplifier has a bandwidth of about 10 MHz if the 75-μs time constant is disabled.

Paralleling JFET Differential Pairs

Paralleling of transistors is a well-known technique for reducing input voltage noise. Each time the number of transistors is doubled, a 3-dB improvement of SNR results. In the case here, four JFET differential pairs (LTPs) are paralleled for a net improvement of 6 dB over a single pair. Simple paralleling of JFETs can lead to high-frequency instability and sub-optimal biasing due to differences in threshold voltage from pair to pair. If gate stopper resistors are used to mitigate the instability, noise is compromised. The key to paralleling JFET differential pairs is to literally parallel four pairs, each with its own tail current source. This decouples the sources among the pairs and allows each to find its own operating point. The decoupling of the sources also mitigates interaction among the pairs that can lead to instability. This, together with measures taken in the drain circuits, eliminates the need for gate stopper resistors [16].

This approach also takes advantage of the property of JFETs of the same type to have approximately the same transconductance at the same operating current, largely independent of the I_{DSS} and threshold voltages. This means that each of the four pairs will contribute about the same amount of transconductance without close matching of the pairs.

The drains of the differential pairs are parallel-connected to the emitters of a single NPN differential cascode stage. The low input impedance of the cascode helps reduce interactions among the parallel-connected drains that can lead to HF instability. Drain interactions are further reduced with 100-Ω drain-stopper resistors (R3–R10). They have no effect on noise.

DC Servo

A fifth degenerated LTP (not shown) using a dual monolithic LSK489 JFET is used to inject the DC servo correction as a current [17]. This circuit, which is not shown, provides a convenient and non-invasive way of injecting the servo correction current and adds negligible noise. Introducing the correction differentially ahead of the folded cascode results in less second harmonic distortion from JFET offset currents. The DC servo itself is identical to the one employed in the MM section of the VinylTrak™ preamp.

RIAA Equalization and Remaining Gain

The remainder of the MC preamp is similar to that of the MM preamp described for the VinylTrak preamp in Chapter 15. Gain of the JFET input stage is set to about 100 by the transconductance of the input pairs and shunt load resistor R13. The RIAA 75-μs time constant is set by C1 and R13. C1, at 0.042 μF, consists of three hand-selected polystyrene capacitors with stated values of 0.018, 0.018 and 0.005 μF. At an input level of 5 mV_{RMS}, 20 dB above a nominal MC level of 500 μV_{RMS}, THD is only 0.005%, with no harmonics above the third This is very good for a design without negative feedback.

Noise

The dual monolithic LSK389 JFET pair is a very quiet device, with typical 1-kHz noise of about 1 nV/√Hz at an operating current of 2 mA for each JFET [15]. A single differential pair with a tail current of 4 mA will exhibit input-referred noise of about 1.4 nV/√Hz. Placing four pairs in parallel, with total current consumption of 16 mA, will provide a 6-dB reduction in noise, achieving about 0.7 nV/√Hz [16].

Distortion

Distortion is very low simply because the typical MC signal level is very small. The differential circuit largely cancels second harmonic distortion, so that the third harmonic is the only predominant distortion, and its percentage goes down 2 dB for every 1-dB reduction in signal level. The soft square-law characteristic of the JFETs produces very few higher-order harmonics. At an input level of 5 mV_{RMS}, ten times the nominal 500 μV signal from an MC cartridge, THD for the input amplifier is only 0.005%, with no significant harmonics above the third. This measurement was made with the 75-μs RIAA roll-off disabled.

With the 75-μs roll-off engaged, 20-kHz performance of the JFET input stage extends to higher input voltage levels because output stage clipping does not occur until much higher input levels as a result of the lower gain at 20 kHz. At an input level of 10 mV, 26 dB above the 500-μV_{RMS} nominal input reference level, THD-20 is only 0.007%. At an input level of 50 mV_{RMS}, fully 40 dB above 500 μV_{RMS}, THD-20 is only 0.37%. Even at 100 mV_{RMS}, THD-20 is only 3% and the MC stage is into soft clipping.

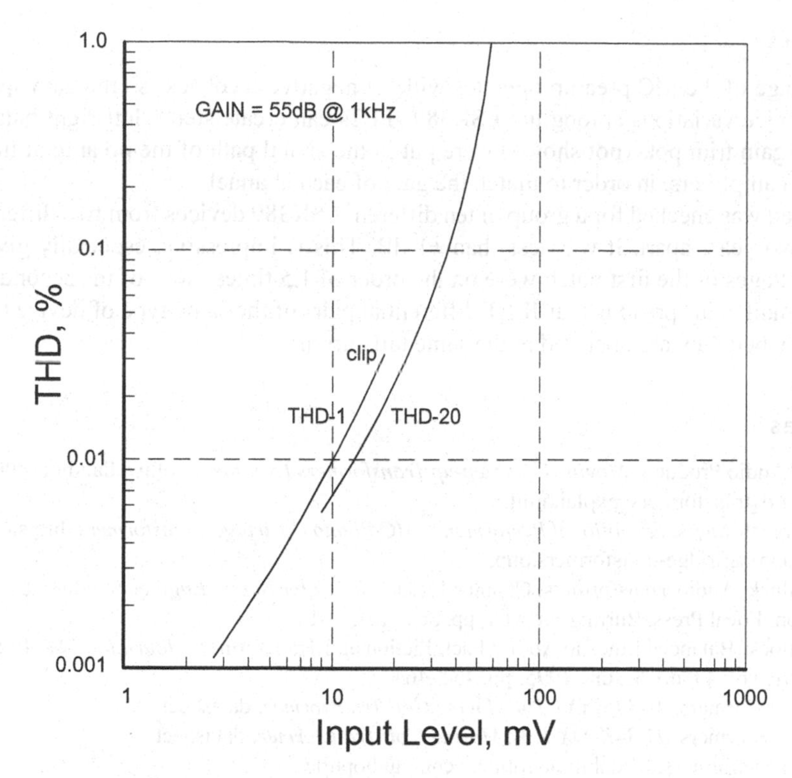

Figure 16.14 Moving Coil Preamp Distortion.

Distortion for the complete MC RIAA preamp is shown in Figure 16.14 [12, 13]. The gain is set to 55 dB at 1 kHz. The maximum input level at 1 kHz is limited by clipping of the output op amp at about 10 V_{RMS}. THD-1 is only 0.001% at an input level of 2.5 mV_{RMS}. THD-1 at a nominal input level of 500 μV_{RMS} is un-measurable. At a 20-dB overload level of 5 mV_{RMS}, THD-1 is still only 0.003%. Just under output clipping of the RIAA stage, at an input level of 17 mV, THD-1 is only 0.03%. THD-20 is lower than THD-1 because the output level at 20 kHz is lower due to the RIAA equalization.

Overload Margin

Overload margin for the MC preamp at 1 and 20 kHz is 17 and 60 mV_{RMS}, respectively. Overload is expressed as the RMS input voltage level when THD reaches 1%. The overload margin at 1 kHz is simply dictated by the maximum output of the output op amp, which is about 10 V_{RMS}. The overload margin for the preamp at 20 kHz is large due to the transconductance-shunt RC implementation of the RIAA 75-µs time constant in the input stage.

The large high-frequency input level tolerance for the MC preamp contributes to especially good EMI immunity as well. This is a direct result of the use of JFETs at the input in combination with the use of the transconductance-shunt *RC* approach for implementing the RIAA 75-µs time constant in the first stage, which keeps the high-frequency voltage swings low right from the beginning of the signal chain. This prevents large-amplitude high-frequency signals from propagating through the signal path.

Gain Balance

The input stage of the MC preamp operates without negative feedback, so the gain spread due to device-to-device variations among the LSK389 JFETs can create stereo left-right balance errors. That is why gain trim pots (not shown) were put in the signal path of the preamp at the output of the front-end amplifiers, in order to match the gain of each channel.

Gain spread was checked for a group of ten different LSK389 devices from two different batches fabricated two years apart. It was less than ±1 dB. This is impressive, especially given that the threshold voltages of the first batch were on the order of 1.5 times those of the second batch. The key to this small gain spread is that JFET differential pairs of the same type of device tend to have similar gain when they are operated at the same tail current.

References

1. Rothwell Audio Products, *Moving Coil Step-up Transformers Explained*, rothwellaudioproducts.uk/html/mc_step_up_transformers_explai.html.
2. Sowter Transformers, *Exceptional Performance MC Phono Cartridge Transformers*, https://www.souter.co.uk/phono-cartridge-transformers.php.
3. Bill Whitlock, Audio Transformers, Chapter 15, *Handbook for Sound Engineers*, edited by Glen Ballou, 5th edition, Focal Press, Burlington, MA, pp. 367–401, 2015.
4. Bill Whitlock, Balanced Lines in Audio: Fact, Fiction and Transformers, *Journal of the Audio Engineering Society*, vol. 43, no. 6, June 1995, pp. 454–464.
5. Jensen Transformers, *JT-346-AXT 1:4 Moving Coil Transformer*, datasheet.
6. Jensen Transformers, *JT-34K-DX 1:37 Moving Coil Transformer*, datasheet.
7. Lundahl Transformers, lundahltransformers.com/audiophile/.
8. Sowter Transformers, Souter Type 9570, Datasheet, https://souter.co.uk/specs/9570.php.
9. Elliot Sound Products, *Moving Coil Phono Head Amplifier*, Project 187, 2019.
10. George Ntanavaras, *The MC100- A High-Quality Moving Coil RIAA Preamplifier*, audioXpress, March, 2014.
11. AD797 Op Amp datasheet, Analog Devices, analog.com.
12. Bob Cordell, *VinylTrak™ – A Full-Featured MM/MC Phono Preamp*, Linear Audio, vol. 4, September 2012, linearaudio.net.
13. Bob Cordell, *The VinylTrak™ Phono Preamp*, cordellaudio.com.
14. *Low-Noise JFETs – Superior Performance to Bipolars*, AN106, Siliconix, March 1997.
15. Linear Systems LSK389 and LSK170 JFETs, linearsystems.com.
16. Bob Cordell, *LSK389 Application Note*, Linear Integrated Systems, January 2020, linearsystems.com.
17. Bob Cordell, *Linear Systems LSK489 Application Note*, linearsystems.com, February 2013, linearsystems.com.

Chapter 17

Tape Preamps and NAB/IEC Equalization

Analog magnetic tape recording was the mainstay of the recording business for many decades until digital recording started to become a viable competitor in the 1980s. In the 1990s analog recording began a decline in usage as advances in technology provided digital recording with higher performance and lower cost. However, just as with vinyl, and vacuum tubes, this vintage technology is still preferred by some. Moreover, it is applicable to playback of existing master tapes to create new digitally re-mastered issues of recordings.

Signal levels from analog tape machine playback heads are similar to those of moving magnet (MM) phono cartridges, and the NAB (National Association of Broadcasters) equalization required is similar to, but simpler than, RIAA phono equalization. Many of the same issues and challenges thus exist with tape head preamps and phono preamps. In fact, many of the same circuit topologies are applicable to phono and tape preamplifiers. For this reason, most of the circuits discussed earlier for MM phono preamps are also applicable for use as tape head preamps when revised to incorporate NAB or IEC (International Electrotechnical Commission) tape equalization.

Typical signal levels from tape playback heads tend to be smaller than those from MM phono cartridges, so noise in the electronics takes on more importance. The key objectives are low noise, accurate equalization, minimal interaction between the tape head and the input characteristics of the tape preamp, and proper loading of the tape head. In many cases of well-respected tape machines, the playback preamplifiers are not as good as they can be. This is especially so of many preamplifiers in otherwise excellent machines where as few as three transistors are used in the playback preamp.

17.1 Signal Levels and Tape Head Characteristics

The playback head is essentially a gapped inductor that creates voltage in response to the rate-of-change of magnetic flux presented to it by the tape moving past the gap. Playback head inductance typically ranges from 50 to 800 mH, with many falling in the 200-mH range. Impedance at 1 kHz ranges from 250 to 5000 Ω. Output levels tend to be about 10–20 dB lower than those from moving magnet (MM) cartridges at nominal 1-kHz reference levels, depending on tape speed and several other factors.

Playback heads with very small gaps are able to reproduce higher frequencies, but the smaller gap tends to result in less sensitivity to the magnetic flux from the tape and lower signal output. This makes for a tradeoff between high-frequency response and signal-to-noise ratio (SNR). Playback heads exhibit self-resonance due to winding capacitance, and it is desirable to keep this resonance high in the audio band for good high-frequency response. Response may be lifted a bit as the resonant frequency is approached, but then it will crash above the resonant frequency. Playback heads with lower inductance will have a higher frequency of self-resonance, but will suffer lower sensitivity. Once again there is a tradeoff between high-frequency response and SNR. For this reason, playback heads with significant inductance are not generally loaded with some capacitance, unlike MM cartridges, which typically require 100–200 pF of capacitive loading in parallel with the recommended 47 kΩ load resistance.

Typical Signal Levels and Required Playback Gain

The typical signal levels from the playback head are smaller than those from a moving magnet (MM) phono cartridge. While 5 mV is often cited as a reference level for an MM cartridge, 1.5 mV is more commonly cited for a playback head with tape speed of 7.5 inches per second (ips). Indeed, with some playback heads producing only 500 μV or less, signals from tape heads can be 10–20 dB smaller than those from an MM cartridge.

While an MM preamp requires 40 dB of gain at 1 kHz to bring a 5-mV cartridge signal up to a 500-mV line level, a tape preamp will require 50 dB of gain to bring a 1.5-mV signal from a tape head up to a 500-mV output. With a tape machine whose playback head produces only 0.5-mV nominal output, playback gain must be 60 dB to reach line level. Signal levels from playback heads can be 500 μV or lower, especially at low speeds with narrow track widths. This poses an obvious SNR challenge for the tape preamp. Tape preamp gain of 54 dB (500) at 1 kHz is a reasonable target. Tape heads generally produce more output if they are operated at higher tape speeds, have higher inductance or have larger gaps.

Tape Head Characteristics and Electrical Model

The Nortronics 1221 playback head is a good example [1]. With inductance of 500 mH and a small gap of 50 micro-inches (μin) it produces 1.3 mV output at 1 kHz at 7.5 ips. Its coil resistance is 315 Ω. It is a relatively high-impedance head, with $Z = 3500\ \Omega$ at 1 kHz and self-resonance that is not much above the audio band. If its distributed winding capacitance plus interconnect and preamplifier input capacitance is 230 pF, it will resonate at 15 kHz. Coil capacitance is on the order of 100 pF. For this head, the in-circuit resonance should probably be above 15 kHz, allowing for external capacitive loading of less than 130 pF. Figure 17.1 shows a simple electrical model for this head that can be used in circuit simulations. The model represents the distributed coil capacitance as a lumped capacitance, so this is somewhat of a simplification.

The head resonance can have significant Q and cause significant peaking in the response, so it is desirable that it be above the audio band and/or have reasonably low Q. The impedance of the 500-mH head inductance at 15 kHz is 47 kΩ and at 20 kHz it is 63 kΩ. With no load resistance and 100 pF of external capacitance, a high-Q resonance will occur at 15.8 kHz. With preamplifier input resistance of 100 kΩ, peaking will be +6 dB at 15 kHz. With only 50 pF of external capacitance and a 56-kΩ preamplifier load, peaking will be a modest 1 dB at 13 kHz and response will be down 1 dB at 19.8 kHz.

Many preamplifiers might load this head with 100 kΩ and may incur only 25 pF of parasitic loading, resulting in a peak of about 4.4 dB at 18 kHz. A preamplifier with 100 pF of deliberate loading capacitance for additional RF suppression and with de-facto loading of 100 kΩ will be up 6 dB at 15 kHz. This may be a bit too much peaking. A load of 56 kΩ will reduce peaking to +2 dB at 13 kHz and have response down 3 dB at 21 kHz. Any RC loading in the tape preamplifier must be chosen with a particular tape head in mind.

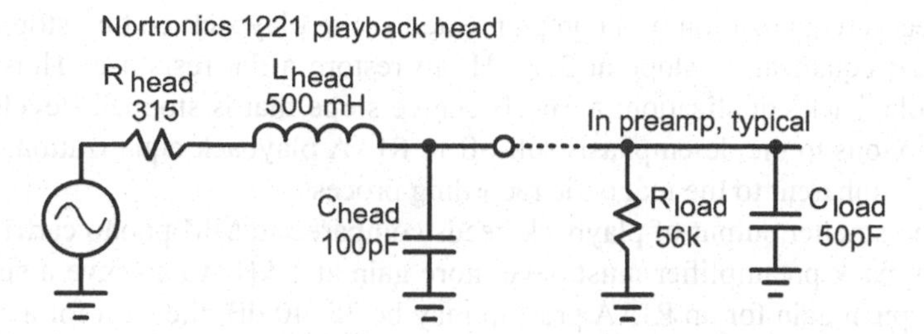

Figure 17.1 Tape Head Electrical Model.

The Nortronics 8607 record/play head has inductance of 200 mH and a 100 µin gap [1]. It produces 1.2 mV, almost as much output as the 1221 playback head with its 50 µin gap. This head and the Nortronics 8608, with 400-mH of inductance, are used in many respected tape machines.

17.2 NAB and IEC Equalization

The basis for NAB (National Association of Broadcasters) tape equalization was devised by Frank Lennert at Ampex in 1948 for the 15-ips tape speed [2]. It was well-suited to the early ferric oxide tape formulations and electronics capabilities of the time.

The concept was straightforward in an idealized context. In recording, the audio signal is applied as a constant current to the record head, creating magnetic flux in proportion to the amplitude of the audio signal. An AC bias signal current of greater amplitude at a frequency between 50 and 100 kHz is also applied to the record head to linearize the transmission process in the face of the nonlinear B-H magnetization curve of the magnetic tape material. It is worth noting that the audio signal voltage across the record head will increase at 6 dB/octave because of the constant-current audio feed into the inductance of the recording head.

At the playback end, the reproduce head responds to the rate-of-change of the magnetic flux presented to its gap by the tape passing by. This is a natural result of the tape head being largely an inductor. This conforms to Faraday's law that induced EMF (electromotive force) increases in proportion to the rate-of-change of flux. It means that the transmission channel from audio recording current to playback head output voltage behaves like a differentiator, with a frequency response that rises at +6 dB/octave with increasing frequency. As a result, the playback equalization should be, to first order, an integrator that creates a low-pass response that falls at −6 dB/octave with increasing frequency. This, ideally, creates a flat end-to-end frequency response.

NAB Equalization

The simple integrator playback equalization is straightforward and is the basis for analog magnetic tape recording. However, the magnetic tape material is subject to earlier saturation at higher frequencies. In other words, there is less magnetic headroom at high frequencies given a magnetic recording flux that is constant with frequency. The ferrous oxide tapes available at the time of the development of the NAB equalization were more subject to high-frequency saturation than modern tape formulations. As a result of the magnetic tape saturation reality, the audio record current at the record end was rolled off at a specified transition frequency, chosen to be 3183 Hz (50 µs) [3]. The equalization transition frequencies are formally specified as time constants in microseconds.

The 3183-Hz record equalization was complemented at the playback end by stopping the −6 dB/octave downward equalization slope at 3183 Hz to restore a flat response. Thus results the essence of NAB playback equalization: a −6 dB/octave slope that is stopped (leveled off) at 3183 Hz. This is analogous to the de-emphasis roll-off in RIAA playback equalization, except that the "pre-emphasis" is inherent to the magnetic recording process.

Because of the smaller output of playback heads compared to MM phono cartridges, by 10–20 dB, the tape playback preamplifier must have more gain at 1 kHz to achieve a similar line-level output. While typical gain for an RIAA preamp may be 35–40 dB, the gain for a NAB tape playback preamp may be on the order of 45–60 dB.

Assume that the tape playback equalization is implemented with negative feedback in the usual way as with RIAA equalization. Given the integrating equalization in both cases, a 40-dB/1-kHz RIAA preamp might be called upon to deliver about 60 dB of closed-loop gain at 20 Hz. Similarly, a 54-dB/1-kHz NAB tape preamp could be called upon to deliver closed-loop gain approaching 74 dB at 20 Hz. For a mere 40 dB of negative feedback loop gain at 20 Hz, open-loop gain of the amplifier would have to approach 114 dB.

This was a tall order for the preamplifier technology at the time when the NAB EQ was developed. To accommodate this reality, the integrating NAB playback curve was stopped from rising further at 50 Hz (3183 μs) by including a pole at 50 Hz. A complementary LF "pre-emphasis" curve at the record side was introduced at 50 Hz to restore a flat response. That LF "pre-emphasis" would be "stopped" at some frequency below 20 Hz. This LF pre-emphasis arrangement also reduced hum and LF noise slightly. Although part of the NAB standard, the 50-Hz breakpoint has not always been taken seriously at either the record or playback end. In fact, it has been argued that the 50-Hz record pre-emphasis was not really beneficial in light of the greater possibility of overload at low frequencies, the lack of musical content below 50 Hz, and the greater ease with which playback electronics could meet the LF gain challenge beginning in the late 1960s [4].

Although optimum tape equalization is usually different for different tape speeds, the 3183 Hz (50 μs) NAB equalization for 15 ips was adopted for the more consumer-oriented 7.5-ips speed as well. However, NAB equalization for 3.75 ips has its high-frequency transition at 1800 Hz (90 μs).

IEC (CCIR) Equalization

Partly enabled by improvements in magnetic tape formulations and playback electronics technology, the IEC equalization curve, used mainly in Europe and in professional applications, was created [5]. Better tape formulations permitted higher levels of magnetic flux at high frequencies before the onset of saturation, permitting the 15-ips 3183-Hz (50 μs) playback transition to be moved up in frequency to 4547 Hz (35 μs). This improved the high-frequency SNR because the −6 dB/octave playback EQ was allowed to continue further up in frequency by a factor of 1.43. At the same time, the IEC equalization dropped the low-frequency pre-emphasis, putting the LF EQ corner at (theoretically) zero Hz. This means that the playback preamplifier should extend its integrating slope down to frequencies well below 50 Hz, often between 10 and 20 Hz.

IEC equalization is also defined for 7.5 ips. The HF IEC equalization transition frequency for 7.5 ips is specified as 2274 Hz (70 μs). This is a reflection of the optimum EQ being a function of tape speed. This contrasts with the NAB equalization corner at 3183 Hz (50 μs).

By modern standards, one might like the playback EQ to be down only 1 dB from the IEC curve at 20 Hz, implying the −3-dB frequency to be at 10 Hz if a first-order roll-off is assumed. Moreover, the preamplifier should then have at least 40 dB of loop gain across its bandwidth down to 10

Table 17.1 Tape Equalization Parameters

Equalization	τ_1 (μs)	f_1 (Hz)	τ_2 (μs)	f_2 (Hz)
NAB 3.75 ips	3183	50	90	1800
NAB 7.5 ips	3183	50	50	3183
NAB 15 ips	3183	50	50	3183
IEC 7.5 ips	10,000	10	70	2274
IEC 15 ips	10,000	10	35	4547
AES1971 30 ips	10,000	10	17.5	9095

Hz. This means that a 60-dB IEC preamplifier should have 100 dB of closed-loop gain at 10 Hz, and open-loop gain of 140 dB at 10 Hz, assuming that the preamplifier is realized with just a single stage having global feedback for equalization. This is still a tall order to this day, best realized with a two-stage preamp. Such a preamp might begin with a very low-noise input stage with 40 dB of closed-loop gain at low frequencies that includes the HF roll-off transition at 2274 or 4547 Hz for 15 and 7.5 ips, respectively. That might be followed by an integrating stage with 20-dB of gain at 1 kHz and 57 dB closed-loop gain at 10 Hz, requiring 100 dB of open-loop gain at 10 Hz.

AES-1971 30 ips Equalization

Although IEC equalization was defined for a tape speed of 30 ips, being the same as that for 15 ips, it became obsolete. The AES proposed a 30-ips equalization standard in 1971. It called for a 9095-Hz transition frequency (17.5 μs) and no LF pre-emphasis [6].

The equalization time constants and turnover frequencies for the various equalizations are summarized in Table 17.1.

In 1971 a standards committee of the AES recommended to continue the then-current practice of Ampex 30-ips test tapes, which incorporate a 9.095-kHz transition frequency (17.5 μs), with no low-frequency pre-emphasis, as an interim standard. It is the "de-facto Ampex standard" that has never been officially recognized. The committee recommended that the de-facto 30-ips standard be called "Proposed AES Recommended Practice Number 1" Indeed, all 30-ips test tapes sold since 1965 had already been in compliance with Proposed Practice Number 1. This de-facto standard is often referred to as AES1971 [6].

17.3 Other Frequency Response Effects

The world is not so simple in analog magnetic tape recording, and quite a few issues beyond basic record/play equalization come into play. Some of these are covered in this section.

Head Loading

Moving magnet phono cartridges provide the smoothest and most extended high-frequency response when they are properly loaded with a parallel R-C combination, often in the neighborhood of 47 kΩ and 100 pF. In combination with the cartridge inductance, a damped resonance at an optimal high frequency is produced that lifts the high-frequency response before it falls off. This is often not the case with playback tape head loading. While the same resonant behavior may occur, it is more often the case that a playback head has the best frequency response with no appreciable amount of deliberate capacitive loading. There is no de-facto standard loading for tape playback heads.

Playback heads have a self-resonant frequency as a result of effective winding capacitance, but reducing this frequency by adding capacitance often does not extend the frequency response. This is especially so for heads with high inductance, like 500 mH. The Nortronics 1221 head has L = 500 mH and R = 315 Ω. Its self-resonant frequency is about 20 kHz.

Referring again to the playback head model of Figure 17.1, the impedance of the 500-mH head inductance at 15 kHz is 47 kΩ. At 20 kHz it is 63 kΩ. With no load resistance and 100 pF of external capacitance, a high-Q resonance will occur at 15.8 kHz. With preamplifier input resistance of 100 kΩ, peaking will be +6 dB at 15 kHz. With only 50 pF of external capacitance and a 56-kΩ preamplifier load, peaking will be a modest 1 dB at 13 kHz and response will be down 1 dB at 19.8 kHz. Many preamplifiers might load this head with 100 kΩ and may have only 25 pF of parasitic loading, resulting in a peak of about 4.4 dB at 18 kHz. A preamplifier with 100 pF of deliberate loading capacitance for additional RF suppression and with de-facto loading of 100 kΩ will be up 6 dB at 15 kHz. This may be a bit too much peaking. A load of 56 kΩ will reduce peaking to +2 dB at 13 kHz and response will be down 3 dB at 21 kHz.

That having been said, a small shunt capacitor on the order of 100 pF is often placed across the input of the playback preamp to trap RFI. A series resistor might be put ahead of the capacitor to enhance RF rejection. However, any such resistance will degrade the noise performance of the preamp. A 470-Ω resistor and a 100-pF capacitor will create a pole at 3.2 MHz if we assume a short-circuit input. That resistor will contribute about 2.9 nV/√Hz of thermal noise. A smaller resistance of 68 Ω might be more appropriate. For purposes of RF rejection, a ferrite bead might be a good alternative.

If a miniature 75-Ω coaxial cable like Belden 8218 is used, capacitance will be 21 pF per foot. A 6-inch run from head to preamp in a tape machine will then only contribute 10.5 pF of stray capacitance. A carefully designed JFET input playback preamplifier may contribute as little as 5 pF or less of stray capacitance.

The situation can be different for some playback heads with lower inductance in some situations. In most cases, the resonant frequency of the head inductance with the sum of the head's distributed capacitance and shunt-circuit capacitance should lie at or above the highest frequency to be passed [7]. As illustrated by the typical top frequencies listed below, playback heads with higher inductance will have a lower top-operating frequency [7].

Head Inductance (mH)	*Top Frequency (kHz)*
800	15
200	25
100	35

External Playback Preamplifiers

In some cases an external playback preamp may be used with a tape deck. While in a tape machine with an integrated playback preamp the wiring can be short between the heads and the preamp, incurring little stray capacitance, wiring from a tape deck to an external preamp will be longer and will incur more stray capacitance. This could be a concern with high-inductance playback heads.

If 93-Ω coaxial cable is used, like Belden 8255 (0.25 in. dia.), capacitance can be kept to 13.5 pF per foot, so a 1-m interconnect would have about 44 pF of capacitance. This can be compared to a miniature 50-Ω RG174 coaxial cable (0.1″ dia.) with capacitance of 30.1 pF/ft or 99 pF/meter. When using an external playback preamplifier there may be a stronger case for use of balanced interconnect. In this case quality microphone cable can be used with an effective hot-cold capacitance of about 26 pF per foot.

Gap Loss at High Frequencies

At the playback end, when the recorded wavelength is the same as the head gap, average flux across the gap will be zero and no output will result – there will be a high-frequency null [8]. As frequency moves up towards this null, there is HF attenuation. In some cases this may be flattened with additional playback EQ (since it is playback-head-dependent) or by shifting upward the NAB equalization zero in frequency. Some NAB preamps have a pot that allows adjustment of the NAB transition frequency to optimize overall high-frequency response when a test tape is played. The pot is usually set to make the 10-kHz response the same as the 1-kHz response when playing a standard test tape.

As a reference, the recorded wavelength of a 20-kHz signal (period = 50 μs) at 7.5 ips is 0.375 mils (9.5 μm). A typical replay head gap is 0.1 mils (2.54 μm). This works out to a ratio of 0.27, implying gap loss of 1.2 dB at 20 kHz. Some Nortronics heads have a 0.05-mil gap (1.27 μm). Gap loss for such heads is not a concern. There is usually no need for gap-loss high-frequency compensation.

Many professional machines that are mainly used at 15 ips have head gaps closer to 200 μin so as to achieve a higher SNR. At 15 ips, that head gap causes insignificant gap loss (less than 0.5 dB at 15 kHz). However, when such a machine is operated at 7.5 ips the gap loss may be about 2 dB at 15 kHz, perhaps bit more than desired. This is often left uncorrected [9]. This can be corrected fairly well by introducing a zero-pole pair that runs for an octave starting at 16 kHz. As Nortronics suggests,

> A rule of thumb is that the playback gap should be between 1/10 and ¼ the wavelength of the highest reproduced frequency. The longer gap of ¼ wavelength will produce gap loss of no more than 1 dB at the shortest wavelength. Typical recommended gap sizes are 200 micro-inches for 15 ips and 100 micro-inches for 7.5 ips.
>
> [8, 10]

Head Bump

Just as there is a high-frequency anomaly due to the ratio of recorded wavelength to head gap, there is also a low-frequency anomaly related to the ratio of recorded wavelength to the length of the pole pieces of the playback head. This phenomenon is referred to as "head bump," sometimes also called contour effect [11]. The primary head bump will occur at the frequency where half of the recorded wavelength equals the effective length of the pole pieces. That will be followed by a dip at twice the frequency. The deepest dip will occur at the frequency where the recorded wavelength equals the effective length of the pole pieces. The low-frequency response will fall very quickly as the frequency goes below the frequency of the head bump. The amplitude of the head bump is usually 1–2 dB. The subsequent dip at a higher frequency is usually smaller than the bump.

Head bump at low frequencies is more of a problem at 15 and 30 ips because the wavelength of low frequencies on the tape is longer with respect to the size (length) of the pole pieces of the head. Head bump exists at 7.5 ips, but is at too low a frequency to matter. While a tape speed of 30 ips provides better high-frequency response than 15 ips, head bump will be worse at 30 ips because the higher tape speed moves the head bump to a higher frequency.

From Ampex Ref. [8]:

> Low frequency response is almost completely a function of the effects generally known as "head bumps". This effect will occur in the reproduce mode at the low frequencies, as the recorded wavelength of the signal on the tape begins to approach the overall dimension of the two pole pieces on either side of the head gap. In effect, the two pole pieces now begin to act as a second gap, because they can pick up magnetic flux on the tape quite efficiently.
>
> As our frequency decreases we may start to notice bumps and dips in the output of the head. The largest bump will occur when one-half wavelength of the recorded signal equals the combined distance across the two pole pieces, but there will be progressively smaller bumps at 1-1/2 wavelengths, 2-1/2 wavelengths, etc. Similarly the largest dip will occur when one complete wavelength of the recorded signal equals the distance across the pole pieces, and again there will be progressively smaller dips at 2 wavelengths, 3 wavelengths, etc. So as our frequency goes lower and lower the bumps and dips will get bigger and bigger. Below the largest bump, at ½ wavelength, the output rapidly falls to zero.
>
> [8]

The wavelength of 50 Hz at 15 ips is 0.30″. If the physical distance of the secondary gap is 0.30″, then this would place the post-bump dip at 50 Hz. However, if the physical distance of the secondary gap is 0.15″ (as would be the case for a center-to-center measurement of the pole pieces), then the upward bump would be at 50 Hz. The average 15 ips bump frequency measured by Endino falls at about 60 Hz on professional machines. At 30 ips, the bump will be one octave higher in frequency [12]. Most playback preamplifiers and tape machines do not equalize head bump.

For reference, the recorded wavelength of a 50-Hz signal (20 ms period) at 7.5 ips is 0.15 inches. At 15 ips, the wavelength is 0.3 inches. At 30 ips, the wavelength is 0.6 inches. The end-to-end length of the pole pieces of the playback head acts somewhat like a magnetic gap, since the pole pieces conduct the magnetic flux well.

A typical head body may be about 0.6″ wide (in this case, long), and the distance between the outer edges of the poles may be about 0.2″. This is close to the recorded wavelength of 50 Hz of 0.15″ at 7.5 ips. As frequency goes down, head bump may go positive, but then at still lower frequencies the response will crash.

If the NAB-specified pole at 50 Hz is moved way down in frequency, head bump may become more predominant, but the low-frequency response may become better. In this case, some head bump equalization may be desirable. It can be implemented with a gentle and shallow band-reject filter or parametric equalizer centered at the frequency of maximum head bump, maybe at 50 Hz. It could probably be implemented with a Wien bridge active filter or a passive-bridged T filter.

To summarize, head bump occurs when the wavelength of the recorded signal gets close to the physical length of the head's pole piece face. The bump is largest when the wavelength is twice the length of the pole pieces, and the depth of the valley is greatest when the wavelength is equal to the length of the pole pieces, but there are smaller effects at higher frequency multiples as well. At higher tape speeds the wavelength on the tape is longer, so higher frequencies will be at the correct length to cause the problem. Once again, the head bump frequency and amplitude will depend on the geometry of the head. The longer wavelength (lower frequency) is where the (first) positive bump is. The shorter wavelength, an octave above that, is where the (first) dip is.

Record Equalization

Although it does not directly pertain to tape playback preamplifiers, a few words about record equalization are useful. It is the responsibility of the record equalization to make the end-to-end record-playback frequency response flat when a NAB-equalized reference playback machine is used. That machine may be one that has been adjusted to be as flat as possible with a reference tape. That machine may in fact be the playback section of the recorder whose record equalization is being designed or adjusted. It is not usual that the record equalization be responsible for correcting playback-specific frequency response deviations, such as head bump and gap loss. Record equalization corrects for the real-world frequency response deviations that result from the recording process at the set level of record bias and for the tape used in the recording.

Record Bias

The AC record bias is usually at a frequency on the order of five to ten times the highest signal frequency, and whose amplitude may be ten times that of the 0 VU reference record level. It acts to linearize the effect of the nonlinear B-H magnetization curve. The bias amplitude affects both distortion and frequency response. Too little bias causes significant distortion and some lifting of high-frequency signals. Too much bias will cause attenuation of high frequencies, and at some point an increase in distortion. Bias is most often set by adjusting it to yield a peak in the 1-kHz playback response. Minimum distortion will often occur at a slightly higher bias level that causes a loss of 1 dB from the peak level. Bias is sometimes also adjusted for equal response at 1 kHz and 10 kHz. With the chosen bias level, any record equalization can be adjusted for a flatter overall response.

17.4 Noise

Noise is one of the greatest challenges in magnetic tape recording. There are usually significant tradeoffs in distortion and high-frequency response with noise. There are two types of noise, the first being that inherent on the tape. That noise can be reduced relative to the signal by using a better tape formulation and/or recording at a higher amplitude. Some tape formulations will allow putting more signal onto the tape to improve SNR. That issue is largely on the record side.

The second type of noise contribution is from the playback electronics, including the thermal noise of the playback head coil resistance. This is not unlike the noise issue for a moving magnet phono preamplifier. Obviously, it is desirable to keep the noise contribution from the playback electronics as small as possible. However, there are tradeoffs in SNR vs bandwidth. A playback head with a very small gap length will better reproduce very high frequencies due to smaller gap loss. However, that head will produce less signal output because of the smaller gap. This in turn may reduce SNR.

Similarly, a playback head with lower inductance will often perform better at high frequencies because of its lower impedance and higher self-resonant frequency. However, the head with lower inductance will typically deliver a smaller output signal, once again possibly at the expense of SNR. In fairness, it is notable that a playback head with lower inductance will tend to have lower winding resistance and thus a smaller thermal noise contribution. It will also be less vulnerable to noise contributions from input current noise of the playback electronics.

For these reasons, it is especially important to employ playback electronics with the lowest possible noise, since that may permit using a playback head that has better high-frequency performance.

The high-frequency roll-off in the NAB and IEC equalization standards improves the SNR in the same way that the de-emphasis does for MM phono playback. This is of greatest importance in the 500–3183-Hz region for NAB equalization. Above the transition frequency, the reproduce curve is essentially flat, so no further noise-reduction benefit occurs. The same applies to the 2274- and 4547-Hz transition frequencies for IEC equalization at 7.5 and 15 ips, respectively. The integrating nature of the equalization turns white noise from playback electronics (e.g., thermal noise and shot noise) into red noise up to the playback equalization transition frequency.

Input Stage Voltage Noise

The input-referred voltage noise of the preamplifier input stage dominates the noise budget in many cases. With mediocre circuitry, this can be up to 10 nV/√Hz or more. With excellent circuitry, this can approach 1 nV/√Hz. The popular BJT LM4562 boasts typical input noise voltage density of 2.7 nV/√Hz. Were this the only source of noise, it would have total input-referred noise of 0.38 μV in a flat 20 kHz bandwidth. Compared to the 1.3 mV nominal output of the Nortronics 1221 playback head, this results in an SNR of 71 dB.

Input Stage Current Noise

Input current noise can be an issue for discrete and op amp designs that employ BJT input transistors. The base current shot noise will create voltage noise across the impedance of the playback head. This impedance can be quite high for a 500-mH head, being about 3500 Ω at 1 kHz, and rising in proportion to frequency. Input noise current of 0.6 pA/√Hz, as for the NE5534, will cause noise voltage across this head of 2.1 nV/√Hz at 1 kHz that rises with frequency. The LM4562 has larger typical input noise current of 1.6 pA/√Hz at 1 kHz that will contribute 5.6 nV/√Hz at 1 kHz. This reality argues for FET-input preamplifiers.

Playback Head Thermal Noise

Even with noiseless electronics, there is still thermal noise from the resistance of the playback head's winding. The thermal noise of the playback head resistance cannot be completely ignored. For a 500-mH playback head with resistance of 315 Ω, such as the Nortronics 1221, the thermal noise is 2.3 nV/√Hz. Nominal reference-level output for this head is 1.3 mV at 1 kHz at 7.5 ips.

A Simple Example

Using an NE5534 whose input noise voltage is 3.5 nV/√Hz, in the above examples of thermal noise and shot noise with a 500-mH playback head, we have at 1 kHz the power sum of 3.5, 2.3 and 2.1 nV/√Hz, which comes to a total of 4.7 nV/√Hz. This is close to the thermal noise of a 1 kΩ resistor. In a flat 20-kHz measurement bandwidth, this noise density will correspond to 0.66 μV. If the signal from the tape head is 1.3 mV, the resulting un-weighted and un-equalized SNR will be 64 dB. Fortunately, the subsequent de-emphasis of the playback equalization will increase this SNR by a significant amount. With respect to 1 kHz, the NAB -6-dB/octave roll-off continues by approximately a half-decade to 3183 Hz before it levels off to be flat. This means that all noise contributions above 3183 Hz will be down by roughly 10 dB, and that those noise spectra between 1 kHz and 3183 Hz will also be attenuated, but by a lesser amount. In considering SNR, it is important to bear in mind that the nominal output of a tape head will be greater by 6 dB at 15 ips as compared to 7.5 ips.

17.5 NAB Playback Preamplifiers

Two playback preamplifiers are described here to illustrate the technical challenges and ways in which improved preamplifiers can be implemented. Many tape machines would profit by having better playback electronics, in terms of noise, distortion and accurate conformance to the equalization standard. Although these preamplifiers are configured for NAB equalization at 7.5 and 15 ips, it is easy to adapt them to IEC equalization and/or to operate at other speeds, like 3.75 and 30 ips. Because their task is much like that of an MM phono preamp, most of the circuit techniques in Chapter 15 are applicable.

An Op Amp-Based Tape Preamp

A NAB playback preamp using an LM4562 op amp is shown in Figure 17.2. It is fairly simple and not very noteworthy, employing a common version of the necessary equalization network. Some manufacturers allow for trimming of R4 for high-frequency alignment. The LM4562 is direct-coupled to the tape head, allowing the very small 72 nA (max) input bias current to flow through the tape head. The low input bias current of the LM4562 is made possible by the use of input bias current cancellation. The input noise current of 1.6 pA/√Hz, which will create about 5.6 nV/√Hz with a tape head whose impedance at 1 kHz is 3500 Ω, is a significant noise contributor. A very low-noise JFET op amp like the OPA1642 would be competitive here as well.

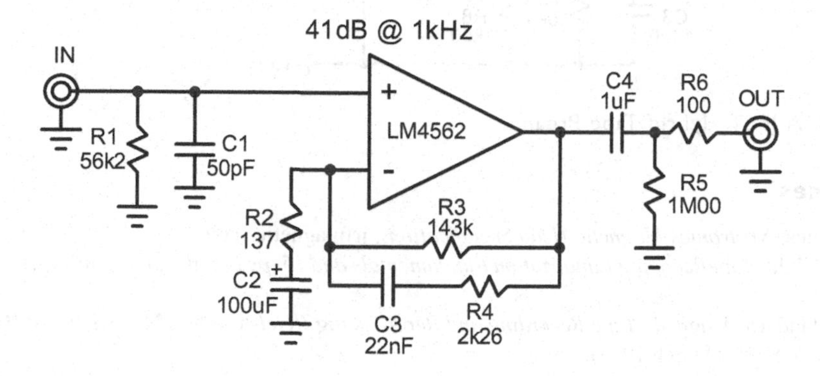

Figure 17.2 Tape Preamp Employing an Op Amp.

A JFET Hybrid Tape Preamp

Low-noise JFETs are a good choice for tape preamps because they have no input bias current or shot noise. The LSK389 is a good choice, with input-referred noise of about 1 nV/√Hz. Figure 17.3 shows such a hybrid tape preamp, where a discrete JFET input stage is followed by an op amp.

The input stage employs a cascoded tail current source for a very high impedance tail. The LSK389 JFETs, operated at 2 mA each, achieve very low input-referred voltage noise. They are operated at fairly low V_{dg} by use of cascodes Q3 and Q4 to minimize impact ionization noise. The cascodes are driven from the JFET-buffered tail so as to minimize effects of the moderate gate-drain capacitance of the LSK389. This minimizes tape head interactions at high frequencies. A Blue LED is used to level-shift the source voltage of buffer J3 to provide adequate V_{ds} for J1 and J2.

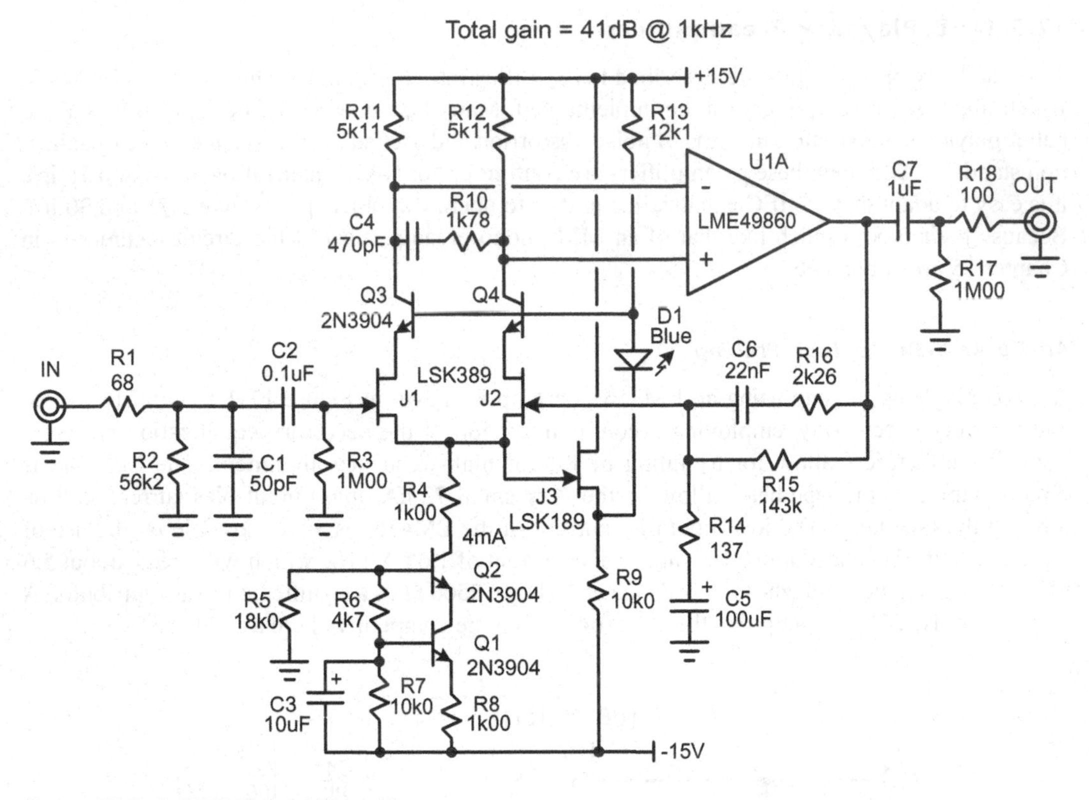

Figure 17.3 A JFET Hybrid Tape Preamp.

References

1. Nortronics, *Nortronics Magnetic Head Specifications*, jrfmagnetics.com.
2. Jay McNight, *TapeRecoding Equalization Fundamentals and 15 ips Equalizations*, mrltapes.com/equaliz.html.
3. NAB Standard: *Magnetic Tape Recording and Reproducing (Reel-to-Reel)*, Nat. Assoc. of Broadcasters, Washington, DC (March 1965).
4. John G. McKnight, The Case against Low-Frequency Pre-emphasis in Magnetic Recording, *JAES*, vol. 10, no. 2, 1962, pp. 106–107.
5. IEC Recommendation 94 Magnetic Tape Recording and Reproducing Systems: Part 1: *Dimensions and Characteristics*, 3rd edition 1968, including Amendment No. 1–1971; and *Part 2: Calibration Tapes*, 1975 International Electrotechnical Commission, Geneva.
6. John T. Mullin and John G. McKnight, Work in Progress: AES Standards Committee, Excerpt of Notes Taken at 1970 Spring Meeting, *JAES*, vol. 19, no. 1, January 1971.
7. Nortronics Design Digest, *Application Factors to Consider for Audio Magnetic Heads*. JRF Magnetic Sciences, pp. 2–9, jrfmagnetics.com/Nortronics_pro/nortronics_silver/Nortronics_silver_catalog.pdf.
8. Ampex, *Basic Concepts of Magnetic Tape Recording*, Bulletin 218, Ampex, Palo Alto, CA, 1960.
9. John G. McKnight, The Frequency Response of Magnetic Recorders for Audio, *JAES*, vol. 8, no. 3, 1960, pp. 146–153.
10. Ampex, *Ampex Tape Recording Basics*, 1960, archive.org/details/JL10157.
11. Klaus Fritzsch, Long-Wavelength Response of Magnetic Reproducing Heads, *IEEE Transactions on Audio Electroacoustics*, vol. AU-14 (also 16.4), December 1968, pp. 486–494.
12. Jack Endino, *Response Curves of Analog Recorders*, endino.com/graphs/.

Chapter 18

Microphone Preamps

Virtually every mixing console incorporates several versatile microphone preamplifiers, usually as part of a channel strip. They must handle all types of microphones, from completely passive dynamic microphones to active microphones like condenser microphones. Ribbon and electret microphones must also be supported. However, some microphone preamps built into mixers may not have as much gain as would be desirable to support optimally some dynamic and ribbon microphones that have very low sensitivity. They must also have low noise working with such small signals.

The preamplifier must provide phantom power for active microphones and be able to handle the high sensitivity and signal levels that some active microphones present. For this reason it must provide an extremely wide gain range while preserving wide dynamic range over that full range of gain settings. A microphone preamp must be capable of extreme dynamic range because the signal at its input has not been compressed in any way. The preamp must have low input noise while having a very high overload tolerance [1].

Because the microphone cable run is probably the most challenging example of interconnect in any sound system, the console microphone preamp must provide very high common-mode rejection over all conditions, especially under high-gain conditions. Regardless of the microphone type or sensitivity rating, the microphone preamp should not compromise the dynamic range of which the microphone is capable, which can sometimes exceed 130 dB.

In addition to standard microphone preamplifiers, preamplifiers that are incorporated into the microphone itself will also be covered. Many types of microphones other than dynamic ones must incorporate active circuitry within the microphone to retrieve the signal before sending it out through the microphone cable. The most common occurrence of this is with condenser microphones. Active ribbon microphones also incorporate preamplifiers into the microphone. Other active microphones include electret and piezoelectric types.

18.1 Acoustic and Electrical Signal Levels and Sensitivity

To best discuss microphone preamps, sensitivity, noise and other challenges, it is important to understand some of the terms used to describe sound intensity, perceived sound as a function of frequency and the units of noise and noise perception. The standard reference level for microphones is the Pascal (Pa), which corresponds to 94 dB SPL (sound pressure level). One Pascal corresponds to the typical sound intensity 1 inch from a person speaking. Sometimes the μbar is also cited; it equals 0.1 Pa. Referring to another term, 1 Pa equals 10 dyn/cm^2. As a reference, 0 dB SPL corresponds to the threshold of hearing.

Sensitivity

Microphone sensitivity is specified in mV at 1 Pa (mV/Pa) or dB below 1 V at a sound level of 1 Pa (dBV/Pa). A typical dynamic microphone with sensitivity of 2 mV/Pa has sensitivity of −54 dBV/Pa. IEC 268-4 measures sensitivity in mV per Pascal at 1 KHz [1]. The sensitivity of most microphones lies in the range of 0.5–30 mV/Pa. Maximum acoustic levels that some microphones can handle is on the order of 135 dB SPL [1–3]. The microphone also has inherent noise, called self-noise. This is expressed as the equivalent A-weighted SPL. A quiet microphone might have self-noise of 14 dBA. If that microphone can handle a maximum SPL of 135 dB, then its dynamic range is 121 dB.

Typical dynamic microphone sensitivity is −60 to −48 dB, corresponding to 1–4 mV/Pa. Condenser microphones are typically more sensitive, in the range of −42 to −30 dB, corresponding to 8–32 mV/Pa. Passive ribbon microphones are typically less sensitive than dynamic microphones, falling in the range of −60 to −54 dB, corresponding to 1–2 mV/Pa. Active ribbon microphones fall in the upper part of the dynamic microphone range of sensitivity.

It is notable that a sensitive condenser microphone (−30 dBV/Pa) exposed to 128 dB SPL would produce 1.6 V if it did not clip. At the other end of the scale, a less sensitive passive ribbon microphone (−60 dB) exposed to 74 dB SPL (a person speaking 12″ from a microphone) would produce only 0.1 mV. This is a whopping 84 dB range in electrical signal levels.

Self-Noise

Microphone self-noise is usually specified as dB-A SPL or just dB-A, with the A designating A-weighting. Microphone self-noise is the electrical noise level produced in the presence of total silence. It is the SPL that corresponds to the A-weighted voltage in dBV that would be produced in total silence. Self-noise is specified in dB-A, and corresponds to the SPL above the hearing threshold at 1 kHz. It includes electrical thermal noise and inherent acoustic noise in the microphone [2].

If a microphone produces 1 mV at 94 dB, then it would produce 80 dB less than that at a self-noise level of 14 dB. The A-weighted noise would thus be 80 dB below 1 mV, which would be 0.1 μV. There is electrical self-noise and acoustic self-noise in microphones. Electrical self-noise is that noise produced electrically, as by the thermal noise of the resistance of the coil of a dynamic microphone. It also includes the noise of the internal amplifier, if one is used. Even in total silence, there is acoustic self-noise produced by random Brownian motion of air molecules in the microphone striking the diaphragm. Uncorrelated noise from multiple sources adds on a power basis (i.e., the square root of the sum of squares of the noise voltage of the individual contributors). The self-noise of most microphones falls in the range of 5–30 dB-A.

The thermal noise density of a 200-Ω coil in a dynamic microphone is about 1.8 nV/√Hz. In a 20-kHz bandwidth there are 141 √Hz, so the RMS noise voltage will be 0.26 μV. However, the effective noise bandwidth of A-weighting is about 13 kHz, or 114 √Hz. The A-weighted noise is thus about 0.21 μV, which is about −134 dBV. If the dynamic microphone sensitivity is 2.1 mV/Pa, the A-weighted self-noise voltage is down by 80 dB from the reference level of 94 dB. The self-noise in this case is then 14 dB-A. In the real world, dynamic microphones do well to achieve self-noise of 18 dB-A if they are mated with a preamp with extremely low noise.

In contrast, a low-sensitivity 1-mV/Pa passive ribbon microphone with 200-Ω impedance will be about 6 dB worse, with self-noise of 20 dB-A. Moreover, it will need to feed an extremely low-noise microphone preamp in order not to make the noise situation worse. Some people wrongly believe that only active microphones with electronic noise have self-noise. All microphones have self-noise, acoustic and thermal. Unfortunately, this wrong belief leads to self-noise only rarely being specified for passive microphones.

Signal-to-Noise Ratio

The SNR of a microphone is the ratio of its signal voltage at 94 dB SPL to the noise voltage corresponding to its self-noise in dB-A. In other words, 1 Pa is the reference level against which the SNR is calculated [3]. Consider a dynamic microphone with sensitivity of −50 dB. It will produce 3.16 mV at an SPL of 94 dB. If the microphone has self-noise of 14 dB-A, its A-weighted noise voltage will be 80 dB below that produced at an SPL of 94 dB. The microphone's SNR is thus 80 dB.

Dynamic Range

The dynamic range for a microphone is the difference between the maximum SPL it can handle without distortion and its self-noise in dB-A SPL. A phenomenal microphone with maximum SPL of 135 dB and with self-noise of only 5 dB-A would have an enormous dynamic range of 130 dB [3].

Typical Sound Levels

The following are some typical sound levels to put things in perspective [4]. The Shure SM58 microphone can handle 150-dB SPL at 100 Hz, where it can produce a remarkable output level of 1 V.

- Quiet recording studio 10 dB
- Quiet average home 30 dB
- Person speaking 12″ from mic 74 dB
- Person speaking 1″ from mic 94 dB
- Person shouting 1″ from mic 135 dB
- Very loud kick drum 140 dB
- Close-mic'd trumpet 145 dB
- Shure SM58 maximum 150 dB

Loudness

Perceived loudness is a function of frequency and SPL, as described by the Fletcher-Munson curves (equal-loudness contours) [5]. The ear's frequency response changes with sound level. A simplified illustration of the equal-loudness contour curves is shown in Figure 18.1. Each curve represents the SPL required for a tone at a given frequency to be perceived to be as loud as a tone at 1 kHz. Thus, the lower the curve, the less SPL is needed to hear it at a given loudness. Low points on the curve indicate increased hearing sensitivity, so it is somewhat like the inverse of a frequency response curve. In particular, notice that the ear is much less sensitive to lower frequencies at lower SPL levels. This is sometimes attributed to the perception that music sounds better when it is played louder.

Fletcher and Munson developed the first curves in 1933. In 1956 Robinson and Dadson developed refined curves that became the ISO 226:2003 equal-loudness contours. There are significant differences in the curves below 200 Hz, especially at 20 Hz and large SPL. The curves indicate the threshold of hearing in dB SPL as a function of frequency for different sound levels in phons (dB SPL at 1 kHz). The official ISO 226:2003 actually only extends down to 30 Hz.

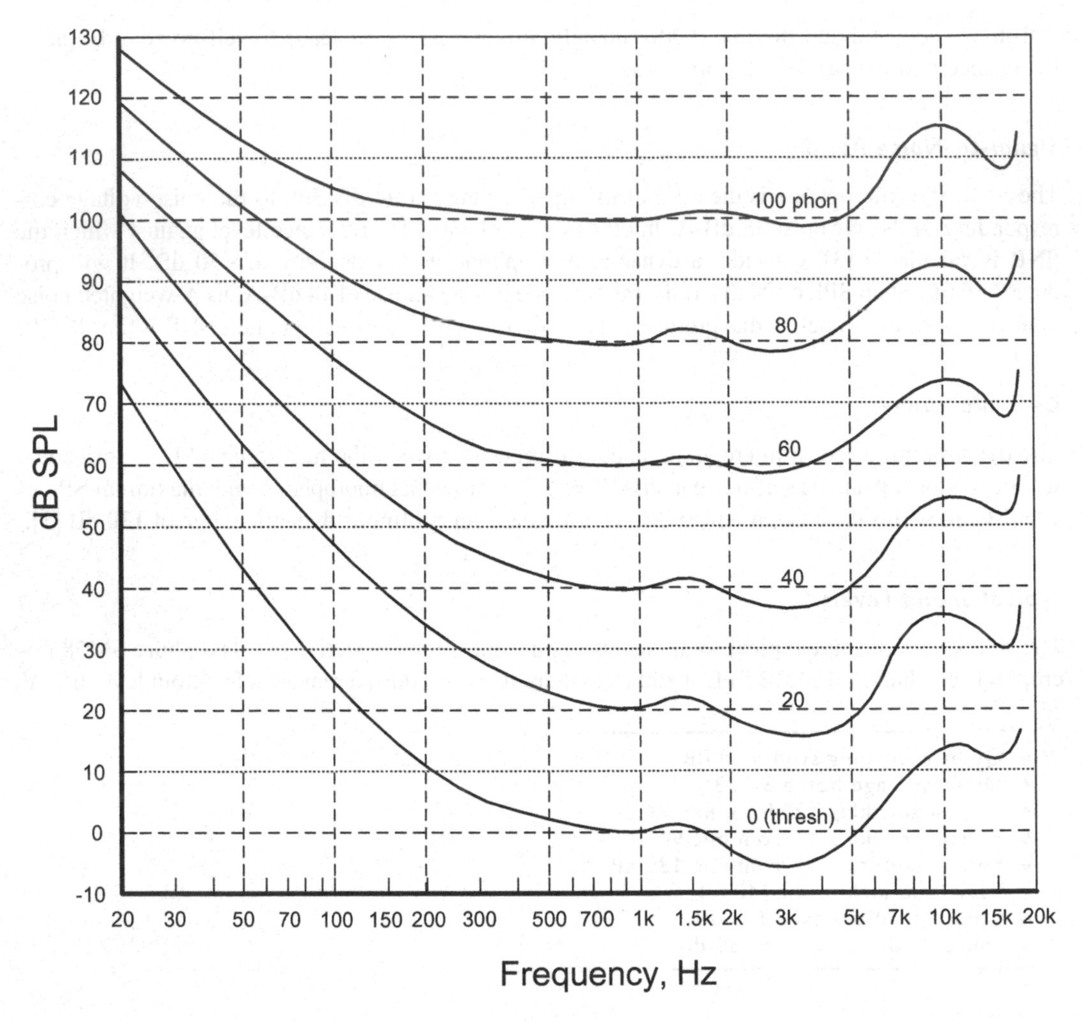

Figure 18.1 Simplified Depiction of the Equal-Loudness Contours.

Phons, Sones and the Equal-Loudness Contours

Sounds with the same SPL of different frequencies are perceived to have different loudness. The *phon* is a unit of loudness for pure tones as perceived by human hearing. It is based on equal-loudness contours. The earliest equal-loudness contours were the Fletcher-Munson curves developed in the 1930s. The currently accepted equal-loudness contours defined in ISO 226:2003 are quite similar.

The phon is defined in units of dB referenced to perceived loudness at 1 kHz. A sound level of 80 phons is the same as the sound level of 80 dB SPL at a frequency of 1 kHz. However, at some lower frequency, 80 dB SPL may have the perceived loudness of only 60 phons because lower-frequency sounds have a perceived loudness less than those at the reference frequency of 1 kHz, in correspondence with the equal-loudness contour. Just as the threshold of hearing is defined as 0 dB SPL at 1 kHz, the threshold of hearing at any frequency is 0 phons.

The *sone* is a linear scale of loudness. One sone is a unit of perceived loudness equal to the loudness of a 1 kHz tone at 40 dB above the threshold of hearing. An increase of 10 phons corresponds to a doubling of the number of sones. Fifty phons would have a loudness of 2 sones, and 60 phons would have a level of 4 sones.

Perceived loudness is also different for noise than it is for single-frequency tones. It is dependent on the spectral shape of the noise, such as white or pink or red. Loudness of noise is described by perceived noise level (PNL) dB. Just as the sone designates the perceived sound level of a tone at any frequency on a linear scale, the *noy* expresses the perceived loudness of noise on a linear scale as dictated by the equal-loudness contours. The PNL and the noy are usually used in the context of measurement of aircraft noise, so they are a bit less applicable to the audio industry.

A-Weighting, ITU-R 468 Weighting and CCIR-ARM Weighting

Although A-weighting is the most widely used way to specify noise in audio, an alternative was developed in search of a way to better describe the audibility of random noise. Thus was born the ITU-R 468 weighting curve, also referred to as CCIR (Consultative Committee on International Radio). The audibility of random noise falls faster at frequencies above 6 kHz than described by the A-weighting curve. Both weighting curves are set to have equal weighting at 1 kHz. The peak of the ITU-R 468 curve lies at 6 kHz, at about +12 dB. At higher frequencies it falls at 24 dB/octave, quite a bit faster than A-weighting. Below 6 kHz the ITU-R 468 curve falls at 6 dB/octave, a shallower slope than A-weighting in the low-frequency region. At 20 Hz, the ITU-R 468 curve is up about 16 dB compared to A-weighting. The ITU-R 468 curve is better known and used more in Europe. Figure 18.2 shows the A-weighting and ITU-R 468 weighting curves together. The peak in

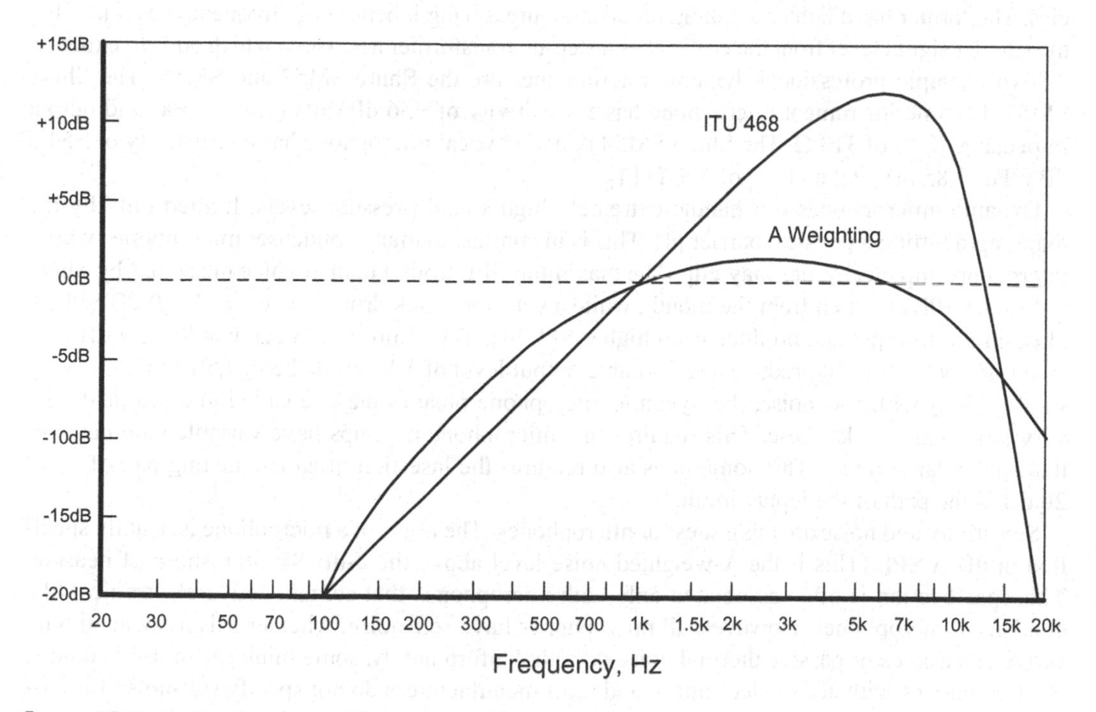

Figure 18.2 A-Weighting and ITU-R 468 Weighting Curves.

the ITU-R 468 curve at 6 kHz is quite notable, as is its steep roll-off at higher frequencies. While the A-weighted measurement is done with a true-RMS meter, the ITU-R 468 measurement is done with a specialized quasi-peak meter, rendering direct comparisons difficult.

The Dolby CCIR-ARM measurement is basically a CCIR-weighted measurement made with an average-responding meter. It employs the CCIR weighting curve, but normalized for weighting of 0 dB at 2 kHz rather than 1 kHz.

18.2 Microphone Types and Characteristics

There are several different types of microphones, each with its own advantages and disadvantages. The most well-known microphone is the dynamic microphone, but other important types include the condenser, electret, ribbon, MEMS and piezoelectric microphones. Some microphones are passive, like the dynamic, and others are active, like the condenser. Active microphones are powered by so-called "phantom power," which is DC power sent down the microphone cable from the console microphone preamp.

Dynamic Microphone

A dynamic microphone operates by the sound causing a coil mounted to its diaphragm to move within a magnetic field, thus creating a voltage. These are the most popular microphones and they are robust and do not require any external power. Many produce a reasonable output level and are not as expensive as some others.

Some dynamic microphones include a step-up transformer while others do not. The latter must have more turns in the coil on the diaphragm, making it heavier and less responsive to high frequencies. The former has a lighter diaphragm/coil structure, giving it better high-frequency response, but the smaller signal level from the coil makes a step-up transformer necessary, which adds to expense.

Two example professional dynamic microphones are the Shure SM57 and SM58. The Shure SM57 dynamic instrument microphone has a sensitivity of −56 dBV/Pa (1.6 mV/Pa) and output impedance (Z_{out}) of 310 Ω. The Shure SM58 dynamic vocal microphone has a sensitivity of −54.5 dBV/Pa (1.85 mV/Pa) and Z_{out} of 300 Ω [4].

Dynamic microphones can handle extremely high sound pressure levels, limited only by the diaphragm hitting a physical barrier [4]. This is in contrast to many condenser microphones whose internal preamp electronics may clip. The maximum SPL from a human voice measured by Shure is 135 dB SPL at 1 inch from the mouth, while a very loud kick drum may exceed 140 dB SPL. A close-mic'd trumpet can produce even higher SPL [4]. The Shure SM58 can handle 150-dB SPL at 100 Hz, where it can produce a remarkable output level of 1 V. While being able to handle very small sounds with low noise, the dynamic microphone preamp may be called upon to deal with very large signals like these. This requires that microphone preamps have variable gain controls that span a large range. This sometimes also requires the insertion of an attenuating pad of 10 or 20 dB in the path of the input circuit.

Sensitivity and noise are big issues for microphones. The noise of a microphone is usually specified in dB-A SPL. This is the A-weighted noise level above the 0-dB SPL threshold of hearing. This specification is often associated only with microphones that contain active electronics, like condenser microphones. However, all microphones have self-noise, whether it is associated with active electronics or passive thermal noise or both. Unfortunately, some think self-noise is unique to microphones with active electronics, and most manufacturers do not specify self-noise for passive microphones. This makes apples-apples comparisons difficult.

The resistance of the passive dynamic microphone ultimately limits how low the self-noise level can be. The Johnson noise of the typical 300-Ω coil resistance of a dynamic microphone is about 2.2 nV/√Hz. The A-weighted noise of this source is 259 nV. This is about 78 dB below a typical 2-mV/Pa sensitivity and the 94-dB reference level. This corresponds to microphone self-noise of about 16 dB-A. The noise can't get lower than this even with perfect electronics. A microphone preamplifier with effective input noise (A-weighted equivalent input noise – EIN) of 259 nV would degrade the overall SNR by 3 dB. Note that EIN for microphone preamplifiers is actually specified with flat weighting in the band 20 Hz to 20 kHz.

The Shure SM57 dynamic microphone includes a capsule and a 1:4.5 step-up transformer. The capsule resistance is 12 Ω and its series inductance is 157 μH. The transformer primary resistance is 1.5 Ω [1]. The transformer transforms the impedance on the primary side by the square of the turns ratio, which in this case is a factor of about 1:20. This brings the total secondary-side resistive component of the impedance up to about 270 Ω plus the 24-Ω resistance of the secondary winding. A simple electrical model of the microphone comprises a small inductance in series with 300 Ω. The mid-band Z_{out} is 310 Ω, but the diaphragm resonance at 150 Hz (not in a simple model) drives the impedance up to 510 Ω at that frequency. The series inductive reactance of 380 Ω at 20 kHz drives the impedance at higher frequencies up to 790 Ω at 20 kHz [1]. Actual measured impedances of a sample SM58 revealed nominal impedance of 270 Ω at 1 kHz and most other frequencies. From 20 to 100 Hz it slowly rose from 270 to 430 Ω, finally peaking at 450 Ω at 120 Hz, then gradually falling to its nominal 270 Ω at 1 kHz. From 5 to 20 kHz it slowly rose to 370 Ω. A low-Q resonance appeared at 175 kHz, where impedance rose to 1700 Ω.

Cable capacitance can influence the frequency response of a dynamic microphone. A cable run of 75 feet of microphone cable with capacitance of 2500 pF forms a resonant circuit with the microphone inductance at about 40 kHz with an SM57 microphone. The input resistance of the microphone preamplifier damps the resonance. Frequency response peaks at about 40 kHz will be 8, 5 and 1 dB with loadings of 10, 3 and 1.5 kΩ, respectively [1]. This results in a lift in response at 20 kHz of 1.6, 1.4 and 0.5 dB, respectively.

Preamps that don't use a transformer have typical input resistances of 3–10 kΩ. Those with transformers may have input resistance more like 1.5 kΩ. Very long cable runs, like 500 feet, can obviously restrict HF response. Belden 8451 microphone cable has differential capacitance of 34 pF/ft. Star-quad cable has about twice the capacitance per foot. A microphone with a resistive source impedance of 300 Ω will be down 3 dB at 21 kHz as a result of cable loading of 2500 pF. In measurements, a 2500 pF load had almost no effect on impedance in the audio band up to 20 kHz. A low-Q resonance caused impedance to rise to a maximum of 920 Ω at 73 kHz.

Condenser Microphone

The condenser microphone operates on the principle that if a charged capacitor changes its capacitance, a voltage change will result, since charge must be conserved. In a condenser microphone, the diaphragm is one plate of a capacitor that is charged (polarized) to about 30–60 V. The other plate of the capacitor is the backplate, which is often grounded. It is denoted by a thicker line in the symbol for the capsule. It may have numerous holes in it to allow sound to enter the rear. This makes the microphone bi-directional, with a figure-8 polar pattern.

The signal voltage from the capsule is proportional to the polarizing voltage on its capacitance. The diaphragm is implemented with a very thin Mylar film with a sputtered thin layer of gold to make it conductive [3]. As incident sound pressure moves the diaphragm, a signal voltage is created. The capsule capacitance is quite small, in the range of 10–100 pF, so the electrical impedance is very high. The voltage produced by the capsule must therefore be buffered or amplified by a JFET amplifier with extremely high input impedance (often GΩ) in the microphone housing. The condenser microphone is thus an active microphone requiring phantom power. Condenser microphones tend to be more costly than dynamic microphones, but usually provide superior sound, and are very popular in studios.

Capsules that are conventionally polarized (positive voltage on the diaphragm) will produce a voltage output that is of negative polarity (positive sound pressure pushing the diaphragm in and increasing its capacitance results in a negative-going change in voltage).

Some condenser microphones with their internal preamps have sensitivity as high as −18 dBV, and the maximum condenser microphone output can sometimes be as high as 0.4–2 V at very high SPL. Many condenser microphones incorporate a switched input attenuator of 10 or 20 dB that is simply implemented by shunting the diaphragm capacitance with a larger capacitor, creating a form of capacitance voltage divider.

A simplified schematic of a condenser microphone circuit is shown in Figure 18.3(a). The circuit merely uses a JFET source follower to buffer the output from the capsule. Notice the very large-value resistors for polarizing the capsule and for the gate return. These high values are required because of the low source capacitance of the capsule, usually on the order of 15–70 pF. For this reason, the front end of the preamplifier for a condenser microphone is often called an impedance converter.

The sensitivity of the capsule can be quite high, and this contributes to the very good noise performance of condenser microphones. A capsule polarized to 60 V might have sensitivity of 5.5 mV/Pa (−45 dB) or more. This is why the unity gain of a JFET source-follower preamp can be adequate. In other cases, gain from capsule to output might be 12 dB. This would result in a microphone sensitivity of −33 dBV, or 22 mV/Pa. The circuit in Figure 18.3(b) acts as a phase splitter to prepare the signal for a balanced output stage. This circuit was introduced by Schoeps with one of the first solid-state condenser microphones.

Although the phantom voltage of +48 V can be used to polarize the capsule, any current drawn by the electronics will reduce that voltage by 3.4 V for every mA that is drawn by the electronics. That is because the effective resistance of the phantom voltage source from the console preamp is 3.4 kΩ (the parallel combination of the two 6.81 kΩ resistors in the console preamplifier). The

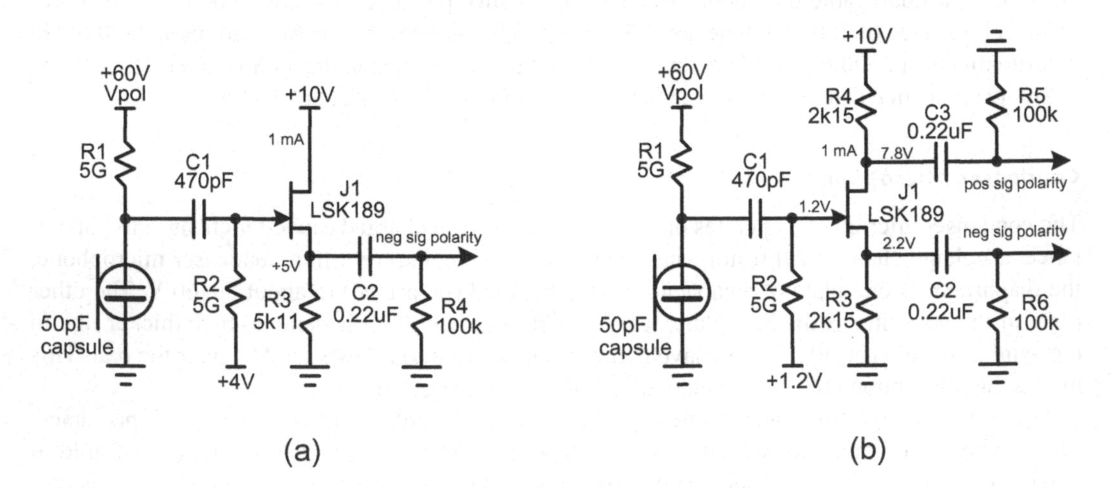

Figure 18.3 Simplified Condenser Microphone Circuits.

result is a smaller polarizing voltage and less output from the capsule. Condenser microphones draw between 1 and 10 mA from the phantom circuit, so the nominal voltage available at the microphone lies between 44.6 and 14 V. With the industry-specified IEC 61938 P48 tolerance of 48 ± 4 V, the available voltage for a microphone drawing 10 mA could be as small as 10 V [6]. Therefore, it is often the case that a small DC-DC converter is used inside the microphone to boost the phantom voltage at the microphone capsule to a higher voltage like 60 V. This results in improved sensitivity and less self-noise.

Condenser microphones come in different diaphragm diameters, often in the range of 12–25 mm. Large-diaphragm condenser microphones are often 1″ (25.4 mm) in diameter (sometimes more). A small-diaphragm condenser microphone is usually about 0.5″ (12.7 mm) in diameter, but may be as small as 0.25″. The large-diaphragm microphones do not respond quite as well to the higher frequencies due to the larger mass of the diaphragm. However, some have a resonance between 10 and 20 kHz that gives them added "air" or "brightness." The AKG C-414 is a classic example.

Small-diaphragm microphones have a frequency response that can extend well above 20 kHz [3]. The higher resonance frequency can provide a smooth, natural sound in the upper octave. Of course, all else remaining equal, a small-diaphragm microphone has significantly lower capacitance, so its electrical output is not as strong; the capsule has much higher source impedance. The impedance of a 50-pF capsule is about 160 MΩ at 20 Hz. A small-diaphragm 12-pF capsule has impedance of about 667 MΩ at 20 Hz.

Large-diaphragm microphones are usually configured as side entry. Some employ dual-diaphragm capsules that allow selection of the polar pattern among figure-8, cardioid and omnidirectional. Small-diaphragm microphones are usually configured as end entry and only provide a cardioid polar pattern, since there is no sound entry to the rear of the diaphragm [2, 7].

The area of the diaphragm, and thus its capacitance, changes as the square of the diameter if the distance between the diaphragm and back plate remains the same. This reduction in capacitance can sometimes be mitigated a bit by allowing a smaller spacing between the plates because that is easier to maintain with a smaller diaphragm. While large-diaphragm microphones may have capacitance from 50 to 100 pF, a small-diaphragm microphone might have capacitance as small as 12 pF or less.

The classic Neumann U87A condenser microphone has sensitivity of 20 mV/Pa with 200-Ω Z_{out} and maximum output voltage of 390 mV. It can handle maximum SPL of 117 dB. A 10-dB attenuation switch allows it to handle 127-dB SPL. It is a very quiet microphone with self-noise of only 15 dB-A, giving it an SNR of 79 dB. This is due in part to its use of a large dual-diaphragm 2 × 50-pF capsule. U87 phantom power is 48 ± 4 V at only 0.8 mA. The original U87 used phantom power directly to polarize the capsule at about 47 V, while the U87A includes a DC-DC converter to polarize the capsule to 60 V [7].

Electret Microphone

The electret microphone is a condenser microphone. Its capacitance is just polarized differently, with a permanent electrostatic charge. In an electret microphone, the diaphragm is one plate of a capacitor that has a permanent electrostatic charge. The electret is a dielectric material with a permanent static charge. The electret is analogous to a permanent magnet, just in the electrostatic domain. A thin electret film, like Teflon, is deposited on one plate of the capacitor. As incident sound pressure moves the diaphragm, a signal voltage is created. As with a condenser microphone, the voltage produced by the capsule must be amplified by a JFET amplifier with extremely high

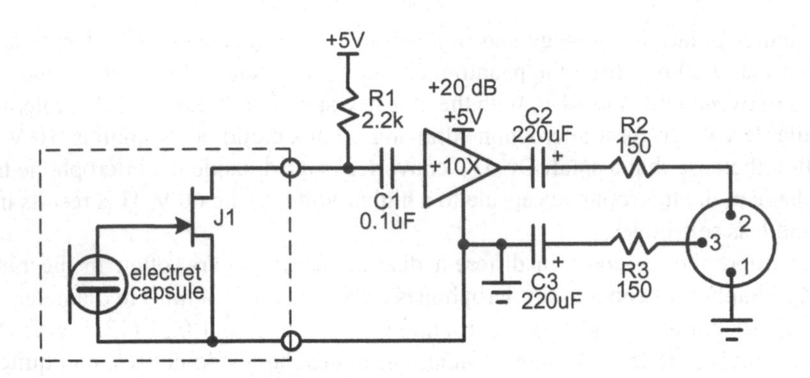

Figure 18.4 Simple Electret Microphone and Preamp.

input impedance in the microphone housing. The electret microphone is thus an active microphone requiring phantom power or similar powering means that can include batteries or low-voltage supplies.

Some electret microphones have a diaphragm with a permanent electrostatic charge, and the diaphragm forms a capacitance with the back plate of the housing. There are also "back-electret" microphones where the charged electret material is put on the back plate instead of the diaphragm. These are more common. This approach makes the diaphragm lighter and improves high-frequency response [2]. The Shure SM-81 cardioid electret microphone is a good example.

Electret microphones are everywhere in consumer electronics, and the capsules are inexpensive – well less than $1. That has sometimes given them a reputation for inferior quality. That is certainly not the case in professional electret microphones, and their manufacturers often just refer to them as condenser microphones. For this reason, externally polarized condenser microphones are sometimes called "true condensers."

The required preamplifier in an electret microphone can be very simple. In a consumer application it can be as simple as a JFET whose gate is connected directly to the diaphragm, as shown in Figure 18.4. The source is grounded and the drain is connected to an external load resistor across which the amplified signal is developed. While some of the preamplifier issues are the same as those in a condenser microphone, there are two major simplifications. First, there is no need for a polarizing voltage and the extremely high resistances and coupling capacitor for biasing the capsule. Secondly, the gate of the JFET floats near ground potential and the JFET operates near its I_{dss} value of current. Operating a JFET at I_{dss} results in the lowest noise possible from the JFET. For this reason, electret microphones can achieve quite low self-noise.

While consumer electret microphones are often powered by 5 V or less, professional electret condenser microphones are usually powered from 48-V phantom power supplied from the microphone preamplifier. They are a bit more complex as well, since they must supply a low-impedance balanced signal to the XLR connector. With the benefit of its amplifier, an electret microphone often has high sensitivity between −46 and −35 dB (5–18 mV/Pa).

Ribbon Microphone

A ribbon microphone operates by the sound pressure causing a very thin corrugated aluminum strip (ribbon) to move within a strong magnetic field, thus creating a very small voltage. It is essentially a dynamic microphone in which the ribbon serves both as the diaphragm and the coil. The ribbon is much lighter than the coil in a dynamic microphone, and therefore can follow the motion of the sound wave more accurately. In exchange for this, the output voltage is quite small. For this reason, ribbon microphones employ a step-up transformer to increase the voltage by a factor of typically 30–37. Even with the transformer, they often produce a smaller output (sensitivity) than dynamic microphones. There are also active ribbon microphones wherein an amplifier further amplifies the output of the transformer. Active ribbon microphones require phantom power.

Ribbon microphones came before condenser microphones, but were largely replaced later by condenser microphones, partly because the ribbon microphones are fragile [1–3]. The ribbons are known for their very smooth sound and lack of peaking in the high-frequency region. They have been enjoying a renaissance in the digital age where systems are less forgiving of high-frequency peaking.

Ribbon microphones are highly valued for their sonic characteristics, but they are bulkier and more fragile than condenser microphones. One should not blow or puff on the ribbon microphone. It is also important to avoid wind damage or air turbulence that may stretch the ribbon diaphragm. The ribbon is fragile, and it is often desirable to use it with a wind screen. Sometimes, when the microphone is not being used, the ribbon can be shorted out on the secondary side of the transformer so as to use back-EMF to stiffen the ribbon's movement a bit.

Ribbon microphones produce a warm and natural sound. Their resonance lies at low frequencies, sometimes below the audio band. AEA Ribbon Mics, inc. tunes their ribbons to 16.5 Hz [8]. Ribbon microphones have a stronger proximity effect (low-frequency boost at closer distance) than other microphones and some performers use this to advantage. Condensers tend to have a resonance in the upper portion of the audio band and can sound brighter. Ribbon microphones preceded the development of condenser microphone, but were largely replaced by condenser microphones because the latter were a bit more robust, lighter, and produced more output. The slightly brighter sound of condenser microphones complemented the inevitable high-frequency roll-offs encountered in the analog magnetic tape era. Digital recording is less forgiving, and the less aggressive sound of ribbons helped them regain popularity.

The classic RCA KU3A ribbon microphone has sensitivity of 2.5 mV/Pa, and impedance of 300 Ω. The motor employs a 1.8-μm, corrugated ribbon 0.082″ wide and 1.25″ long suspended and tensioned in a strong magnetic field [2].

The nominal output voltage at the ribbon motor is tiny – typically 10–20 dB smaller than that from a moving coil (MC) phono cartridge. The microphone sensitivity at the ribbon is often on the order of −80 dB/Pa, corresponding to 100 μV/Pa. Ribbon resistance usually lies in the range of 0.1–0.2 Ω [1]. The signal voltages are generally too small for an active preamplifier to achieve an adequate SNR, so a step-up transformer with a typical ratio of 1:37 must be used. The primary winding of the transformer must also have very low resistance. Higher ribbon and/or transformer resistances will create more thermal noise and result in a higher self-noise number. The transformer

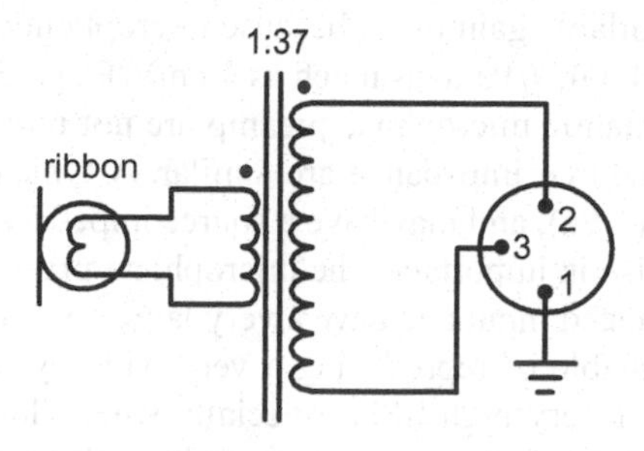

Figure 18.5 Simple Ribbon Microphone Arrangement.

will transform the sum of the ribbon and primary resistance to a larger resistance at the secondary as the square of the step-up ratio, so output impedance can sometimes be higher than that of other microphones. Figure 18.5 illustrates a simple ribbon microphone arrangement.

The sensitivity of ribbon microphones is quite low, often −60 dB (1 mV/Pa), even with the step-up transformer, and they pose a challenge to mixer preamplifiers, both in terms of available maximum gain and achievable SNR. For this reason, phantom-powered active ribbon microphones have been developed. They are able to achieve sensitivities like those in the upper range of dynamic microphones. While condenser microphones often have some resonance in the high-frequency portion of the spectrum, ribbon microphones have a resonance at the low end of the audio spectrum, or even below 20 IIz.

Piezoelectric Microphones and Sensors

A piezoelectric microphone, sometimes called a crystal microphone, operates on the principle of piezoelectricity, wherein some materials, like crystals and certain ceramic materials, create a voltage when subjected to sound pressure. They are often used as so-called "contact microphones" where they are in contact with a vibrating surface, sometimes in an electric guitar. They are also the basis for other types of transducers as well, including inexpensive phono cartridges once used in portable record players. The output impedance is very high, so they also may be accompanied in close proximity with a JFET amplifier with high input impedance [9].

Sound pressure creates a charge on the crystal that results in a changing voltage across the crystal's capacitance. As with condenser-based microphones, the source impedance of the piezoelectric transducer is extremely high, so JFET amplifiers must be used. These can be voltage amplifiers as used in condenser microphones or so-called charge amplifiers, as discussed later. The sound quality of piezoelectric microphones is usually not great, but they can be extremely robust. A great many sensors in a wide variety of applications employ piezoelectric transducers.

18.3 Console Microphone Preamplifiers

The mixer or mixing console microphone preamplifier must be able to handle a great variety of microphone types and sensitivities. They serve dynamic microphones directly, some with sensitivity as low as −63 dB. Similarly, they must also serve low-sensitivity microphones like passive ribbon types. For these cases, they must be able to provide high gain and very low noise to bring the signal up to the line level of 775 mV (0 dBu).

The requirements for a dynamic microphone preamp include low input voltage noise and a wide range of variable gain. A typical dynamic microphone might produce 2 mV at 94 dB SPL (1 Pa). The microphone preamp must accept a balanced input with excellent common-mode rejection. It must have a very large controllable gain range because microphone sensitivity spans a very wide range, often from as little as 0.5 mV/Pa to as much as 40 mV/Pa [1–3].

The requirements for a dynamic microphone preamp are not unlike those of a phono preamp, since the voltage level and source impedance are similar. A typical dynamic microphone may produce 2 mV at 94 dB SPL (1 Pa), and may have a source impedance of about 300 Ω. Low input voltage noise and current noise is important. The microphone preamp does not require equalization, but it must accept a balanced input and have a very large controllable gain range.

Many microphones are capable of reproducing a very wide dynamic range, and many musical instruments are capable of very high SPL, especially when close-mic'd. For this reason the microphone preamp must have wide dynamic range, and the ability to handle large input signals while having low input noise. Of course, in extreme cases the mic preamp should often be able to insert a 20-dB pad. A microphone with 2 mV/Pa sensitivity that is exposed to 134 dB SPL will produce 200 mV. If that microphone has self-noise of 15 dBA, the dynamic range that it can produce under these conditions is 119 dB. Related to dynamic range, the preamplifiers must have large headroom, which is the number of dB from a nominal input level to the clipping point.

The maximum preamp gain can be 60–80 dB, while the minimum gain can be 0–12 dB. A 20-dB input pad may be needed to accommodate some high-output microphones without overload. The activation of the pad should not decrease the input impedance of the preamp, and it is best if it does not change it at all [1]. Some "pad-less" preamplifiers can control their gain continuously over a range that is so wide that a pad or other gain switching means is not needed. Most microphone preamps must also accept a balanced line-level signal from a ¼″ TRS phone jack with an appropriate pad in the signal path.

DC servos can be used in microphone preamplifiers to control DC offsets, allowing the elimination of electrolytic capacitors in the signal path. This approach is used in the Jensen Twin Servo 990 microphone preamplifier [10].

The microphone preamp must also serve other microphone types, like active condenser microphones that typically have higher sensitivity and produce higher voltages. For active microphones and other active sources, the preamp must provide phantom power that is sent down the balanced microphone cable in the common mode, usually supplied by a pair of 6.81-kΩ precision resistors fed from a +48-V supply. The 6.81-kΩ phantom powering resistors limit maximum preamp differential input resistance to be no greater than their sum, 13.6 kΩ. Even if phantom power is turned off, these resistors usually remain in place. Most preamps have additional input shunting resistances that bring the balanced input impedance down to 1.5–3 kΩ. Ribbon microphones tend to perform better with loads of higher impedance if they do not contain an active preamplifier.

The preamplifier dynamic range should be retained even at low-gain settings; this can be difficult because the input-referred noise of the preamplifier can go up when a low-gain setting is used. Most microphone inputs come as balanced signals from XLR connectors, and the microphone preamp must have extremely good common-mode rejection to suppress hum and noise from long microphone cable runs. Most of these microphone preamps must also serve inputs that may originate from electronic instruments, where voltages may be at line level. In these cases, a 20-dB attenuator can be engaged. Such signals can originate from XLR connectors or ¼″ plugs.

Signal Levels, Noise and Overload

Microphone sensitivity can span −60 dBV to as much as −18 dBV, for a range of 42 dB for the same SPL. On the other hand, microphone self-noise can span 5–30 dB-A. Finally, acoustically significant sounds can span 30–145-dB SPL [4]. It is easy to see the large dynamic range that a microphone preamp must deal with, given different environments and equipment. In an ideal world, the microphone preamp does not compromise the SNR from the lowest-sensitivity microphone or overload with the most sensitive microphone under high-SPL conditions. This is a very tall order, and fulfilling it is only approached in practice.

Noise and EIN

Equivalent input noise (EIN) is the input-referred noise voltage in a 20-kHz flat bandwidth. Input noise spectral density is described in nV/√Hz. EIN equals the input noise spectral density of a preamplifier multiplied by the number of √Hz in the 20-kHz bandwidth if the noise is white. There are 141 √Hz in a 20-kHz bandwidth. An op amp with 3 nV/√Hz input noise will have un-weighted EIN of 3 × 141 = 423 nV. With A-weighting, whose equivalent noise bandwidth (ENBW) is 13.3 kHz, this number will fall by about 2 dB to 346 nV. This assumes that the noise source is white and that hum is not present.

In measuring EIN of a microphone preamp, the input should not be shorted, but rather have a 150- or 200-Ω resistor across it, emulating the impedance of a low-impedance dynamic microphone [1]. That takes into account the thermal noise of a mythical microphone whose Z_{out} is 150 or 200 Ω. It also takes into account noise that is created by input current shot noise in the preamplifier flowing through Z_{out}, creating additional noise. A preamplifier with noiseless electronics would have measured EIN corresponding to the noise of a 150- or 200-Ω resistor in a 20-kHz bandwidth. A 200-Ω resistance has thermal noise of 1.88 nV/√Hz and 0.265 μV in a flat 20 kHz bandwidth. Some manufacturers use 150 Ω while others use the slightly stricter value of 200 Ω. EIN is stated in dBu, the number of dB the noise is below 0 dBu (0.774 V_{RMS}, or 0 dBm). With 200 Ω, noiseless electronics would result in an EIN of −129 dB.

A preamplifier that increases the noise level in a 150-Ω measurement by only 3 dB has the same EIN as a 150-Ω resistor, which is 224 nV. That corresponds to input noise of only 1.6 nV/√Hz. A preamp that increases this noise level by only 2 dB is quite good. This is a very stringent way of evaluating preamp EIN, and most dynamic microphones have impedance closer to 300 Ω. Moreover, the actual signal level from the microphone is what matters most. EIN becomes most important with low-output passive ribbon microphones, whose sensitivity may only be −60 dBV.

EIN for most preamplifiers goes up as the gain setting goes down, but it usually does not go up as fast as the input signal being amplified for a given output level. Thus, EIN is usually of greatest importance at maximum gain, and that is where it is usually specified. For a given microphone sensitivity, the effective self-noise of a microphone preamplifier can be calculated in dB-A if its EIN at the corresponding gain setting is A-weighted. This would make it possible to compare the noise contribution of the preamplifier to that of the microphone. That would be nice, but I have not seen it done.

RFI and the Pin 1 Problem

Our environment is loaded with radio-frequency interference (RFI) and other forms of electromagnetic interference (EMI). Cell phones, TV stations and light dimmers are but a few sources. RFI is present over a very broad frequency spectrum, from MHz to GHz. RFI is usually most prominent in the common mode on balanced microphone cables. It can cause intermodulation and rectification in active input circuits and even disturb operating points. It can be difficult to suppress in low-noise circuits because adding series resistors compromises noise performance.

Sometimes capacitors connected from each input conductor to ground can help, but they must be matched in value if they are not to detract from CMRR at high frequencies. Good layout and grounding is important because ground is not always ground at these high frequencies. Ferrite beads can be placed in the lines ahead of the capacitors without compromising noise because they have virtually no resistance at audio frequencies. They have the unique property that their impedance increases at high frequencies and then at much higher frequencies their impedance levels off and becomes resistive. They are somewhat like a low-value inductor with a resistor in parallel. Depending on the particular bead, their high-frequency resistance may be on the order of 100 Ω.

The hum problems created by poor grounding and ground loops are well known, but some grounding arrangements can also allow RFI to create noise and other disturbances. This can be true for balanced XLR connections. The "pin 1 problem" relates to the pin 1 ground of an XLR

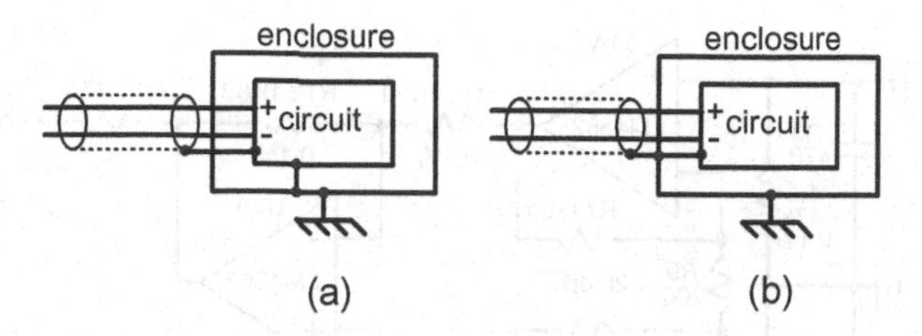

Figure 18.6 Connections with (a) and without (b) the Pin 1 Problem.

connector that connects to the shield of the cable. The problem can occur where RFI current flows in the shield of the cable and passes through common impedance to the equipment enclosure and circuit board [11–14]. Pin 1 should be wired to the metal enclosure directly, not to the circuit board and then to the enclosure. Even short lead lengths have high impedance (ohms or tens of ohms) at frequencies of tens of MHz and above. Sometimes a strategically placed ferrite bead can help because it behaves like a short at audio frequencies and like a resistance at high frequencies. Always remember that current takes the shortest path. Make the interfering current go where you want it to and take the path you want it to. Figure 18.6 illustrates connections with (a) and without (b) the pin 1 problem.

Always assume that noise current is flowing in the shield of microphone cables. It can be hum, noise or RFI at high frequencies. The result is what is called shield current-induced noise (SCIN) [1, 14]. Think of the enclosures of two pieces of interconnected equipment (even one being a microphone) as part of a large shield that shields the system. The shield of the interconnecting cable should be thought of as part of that larger shield, thus enclosing the whole system in one big shield, and treat it as such. This means that the shield should connect directly to the enclosures, and not go anywhere else first. The grounds of the circuits within those enclosures can then be connected to the enclosure with only one wire. Shield current should never be permitted to flow in such a way that it can create a voltage drop between the circuit ground and the enclosure. Many designers instinctively connect pin 1 directly to the circuit ground, and this can be the source of the pin 1 problem.

A Simple IC Dynamic Microphone Preamp

Figure 18.7 illustrates a simple microphone preamp that employs a low-noise BJT op amp. For simplicity, this does not show circuitry to control accumulation of DC offset at high gain. This could include AC coupling or a DC servo. A Baxandall volume control circuit is used for its gain adjust function, resulting in gain that is almost linear in dB with rotation of a linear-taper pot. The Baxandall circuit is discussed in Chapter 22 on volume controls. The input signal passes through a 20-dB low-noise instrumentation amplifier and is then fed to the Baxandall gain control, whose range is from zero transmission to +40 dB of gain.

Resistors R3 and R4 implement a typical phantom powering circuit. These resistors must have 0.1% tolerance to preserve common-mode rejection. Phantom power should be disabled by opening the connection to the +48-V phantom supply, further reducing any CMRR degradation created by mismatch between R3 and R4. Coupling capacitors C1 and C2 should be rated at 100 V or more. The +48-V phantom supply should be well-regulated with extremely low noise. A local capacitance multiplier can be used if a supply voltage of at least +54 V is available for distribution of phantom power to the microphone preamps.

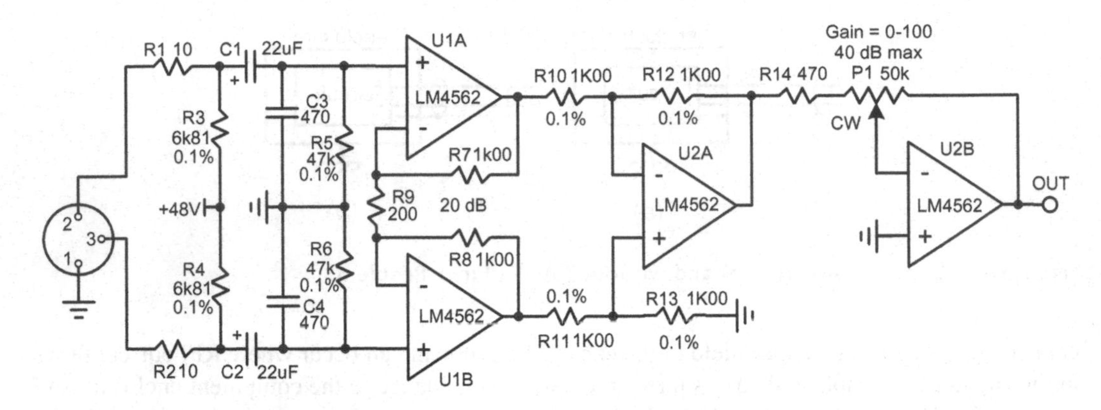

Figure 18.7 Simple Op Amp-Based Microphone Preamp.

Bias resistors R5 and R6 should have a tolerance of 1% or better. Components R1, R2, C3, and C4 implement an RFI filter with a corner frequency of 30 MHz if driven by a voltage source. Differential input impedance is about 12 kΩ. A lower input impedance, typically about 3 kΩ, can be obtained by bridging a shunt resistor between pins 2 and 3 of the input connector. The op amps should be powered by 15–18-V positive and negative rails. The 18-V rails are more commonly used in professional gear as long as all of the op amps are rated at or more than 18 V.

Discrete Microphone Preamp

Figure 18.8 shows a typical discrete BJT-based microphone preamp that exhibits reasonably good performance. It uses a differential complementary feedback pair (DCFP) for the input stage with a gain-adjust pot in the emitter circuit. Its differential output is connected to a quality op amp that provides gain and conversion to single-ended signal format. The very high current gain afforded by the CFPs allows R6 and R7 to be of fairly high value, allowing the use of 1-μF film coupling capacitors at the input, thus avoiding electrolytic capacitors. The larger values of R6 and R7 also contribute to a higher common-mode input impedance of the preamplifier, improving CMRR. The desired differential loading impedance is then achieved by choice of the value of R5.

Gain is controlled by pot P1, and is inversely proportional to the total emitter-emitter resistance. As shown, the gain range is about 18–60 dB. The pot must cover a very large resistance range, so a reverse-log taper is needed to have reasonable gain change as a function of pot rotation. The pot should be of very high quality. Input-referred noise for the preamplifier is about 1.2 nv/√Hz when set for maximum gain. With a 200-Ω input termination (emulating a dynamic microphone), EIN is about 0.18 μV.

In a conventional implementation of this topology, a large electrolytic capacitor of perhaps 1000 μF is placed in series with R10 and P1 in the emitter circuit to control voltage offset. I recommend keeping electrolytic capacitors out of the signal path whenever possible. Here, not shown, a DC servo circuit using an inverting integrator is connected from the output of U1A through a large resistor to the base of Q1. A resistor of the same value is connected from the base of Q2 to ground to keep things balanced.

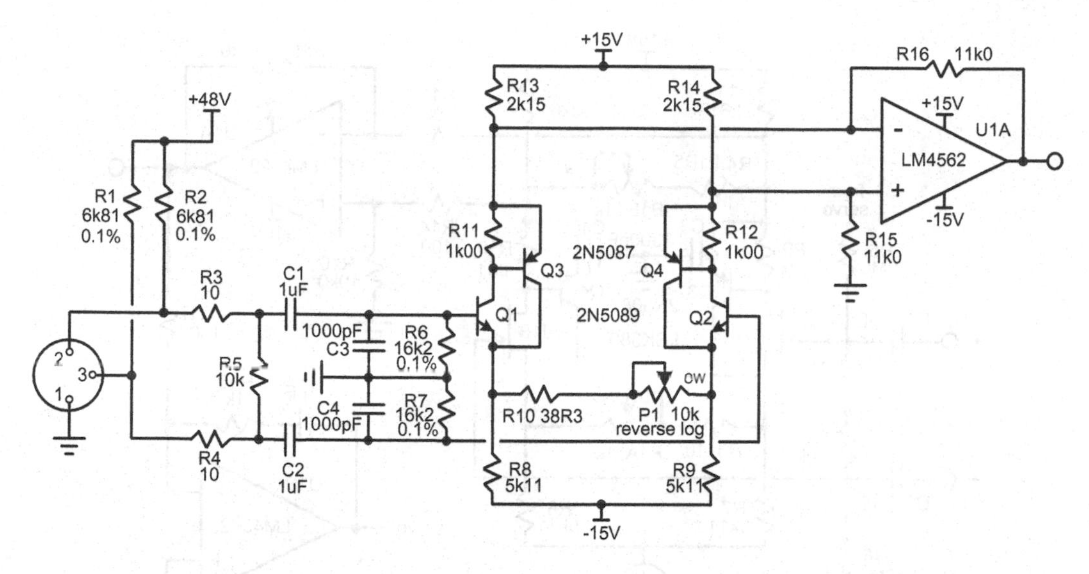

Figure 18.8 Discrete BJT Microphone Preamp.

JFET Input Preamps

Figure 18.9 depicts a microphone preamp with a discrete JFET input stage [15–17]. Many dynamic microphone preamps use front-ends with bipolar transistors because of the stringent low-noise performance needed. However, some listeners prefer the sound of JFETs. Softer overload characteristics and better immunity to electromagnetic interference are also among the advantages of JFETs. The microphone preamp here meets the low-noise requirement while enjoying the advantages of the JFET. Use of the dual monolithic LSK389 low-noise JFET makes this possible. The extremely high input impedance of the JFETs allows the use of large-value gate resistors and the freedom to use quality film input coupling capacitors. Those capacitors should be rated at 100 V because they must handle the 48-V phantom voltage. Input shot noise current is essentially eliminated by the use of JFETs.

Phantom power and other parts of the input circuit are not shown, but are similar to those in Figure 18.8. In particular, a bridging resistor across the differential inputs helps establish the desired balanced input impedance. The very high input impedance of the JFETs and their associated circuits allows for a very high input common-mode impedance, limited mainly by the phantom power resistors. High common-mode input impedance improves common-mode rejection (CMRR). In fact, if phantom power is turned off by disconnecting the junction of the 6.81-kΩ phantom power resistors from the phantom power source, they end up looking like merely a 13.6-kΩ differential bridging resistance. Any mismatch in them cannot then degrade CMRR. Common-mode rejection is most important when using low-output passive microphones.

Each JFET is configured as a complementary feedback pair (CFP) by adding a PNP transistor in the drain circuit. This arrangement increases the effective transconductance of the JFET by a factor of about 50 and provides local distortion-reducing feedback. The JFETs are biased at 1 mA and the BJTs are also biased at 1 mA.

With a single tail current source, its noise is in the common mode, so its influence is mitigated. Nevertheless, care should be taken to minimize the amount of noise from the tail current source, shown here as an ideal current source.

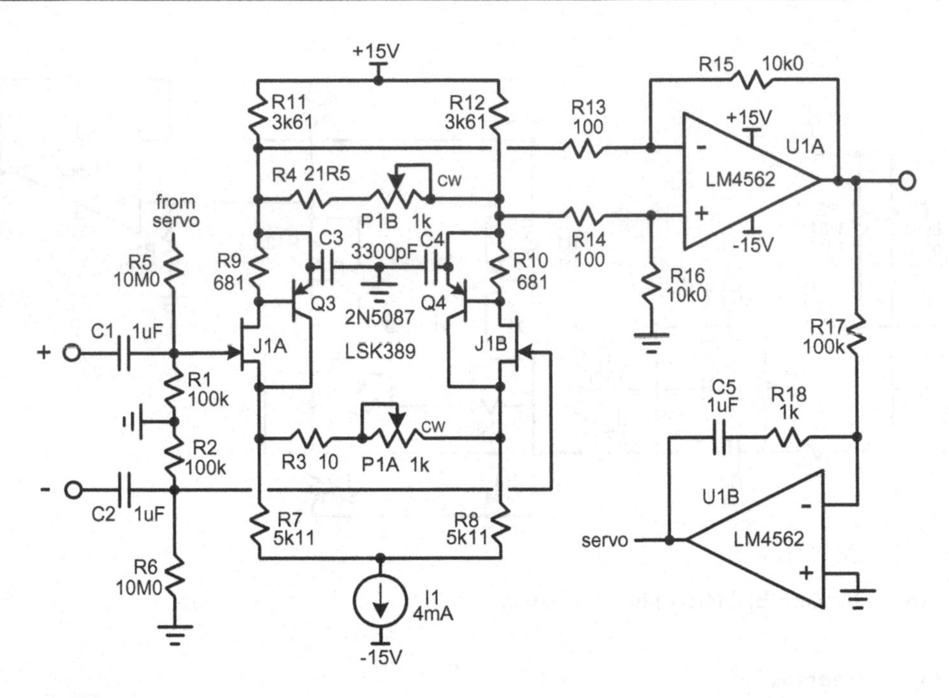

Figure 18.9 JFET Input Microphone Preamp.

As discussed in connection with Figure 18.8, no electrolytic capacitor is placed in series with the gain setting resistor for DC offset control. Instead, a DC servo is used to adjust the bias to the gate inputs to achieve low offset at the output for all gain settings. U1B, R17, R18 and C5 form an inverting integrator that feeds back the required correction voltage to the gate of J1A through a 10-Meg resistor (R5), forming a 100:1 voltage divider. If the output of U1B swings by 13 V, a correction of 130 mV can be applied, far more than will ever be needed. R18 helps servo stability in light of the pole formed by R5 and R1 against C1. R6 makes the circuit symmetrical.

Gain control over a wide range is difficult to achieve with a single variable resistance in the source circuit, so a second pot P1B ganged with P1A is added in a differential shunt arrangement in the drain circuit. A gain adjustment range of 6–60 dB is achieved with a single knob, and is usefully distributed with respect to pot rotation even when using linear pots. Use of log-taper pots can provide an even more uniform distribution of attenuation as a function of rotation.

Input-referred noise is only 1.6 nV/√Hz at the highest gain setting with inputs shorted. With an input test resistor of 200 Ω, an additional 1.9 nV/√Hz will be contributed, for total input noise of 2.5 nV/√Hz, corresponding to EIN of –127 dBu in a flat 20-kHz bandwidth. Harmonic distortion is no more than 0.002% at an output level of 1 V_{RMS} at a gain setting of 20 dB, rising to 0.003% at gain of 40 dB and 0.02% at a gain setting of 60 dB. THD is below 0.05% at 5 V_{RMS} output for gain settings greater than 20 dB. Even though the circuit appears differential and symmetrical, distortion is dominated by the much more benign second harmonic. This is due to the asymmetry created in the drain circuit by the differential op amp configuration. It results in different signal amplitudes at the drains of J1 and J2. Distortion above the third harmonic is virtually absent.

Figure 18.10 shows the gain as a function of percent of electrical pot rotation. Maximum gain is 60 dB. The gain characteristic is quite log-linear from 10% to 90%, corresponding to a gain range of about 18–42 dB. Gain becomes more sensitive to pot rotation for the first and last 18 dB portions of the gain curve.

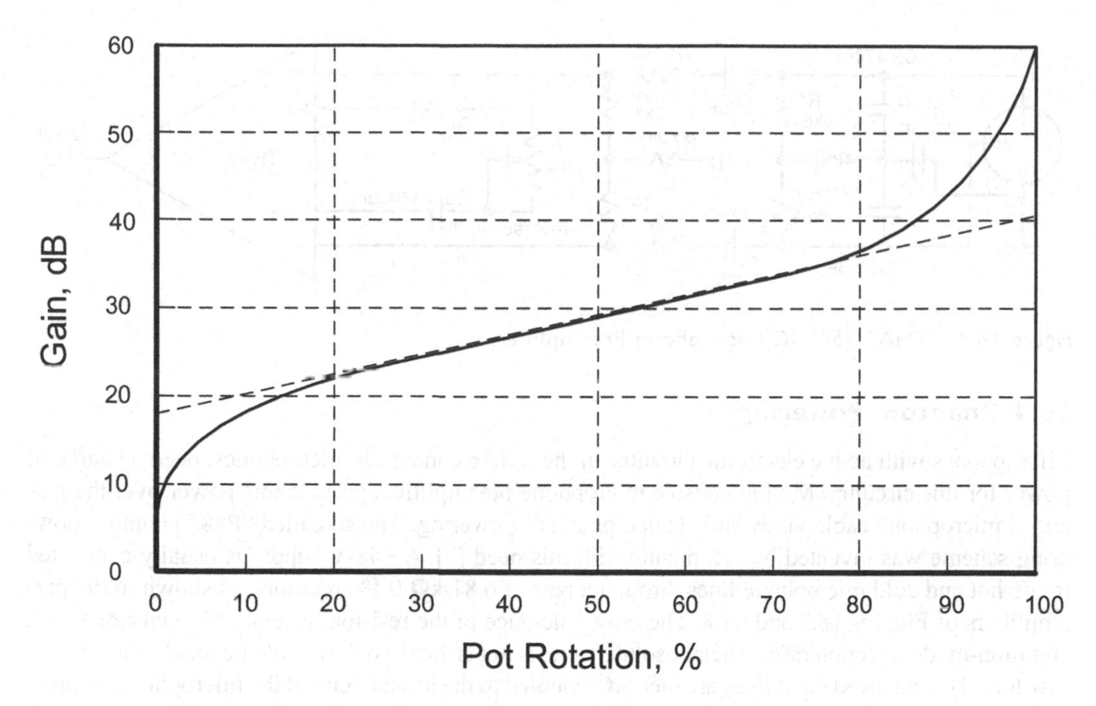

Figure 18.10 Gain vs Pot Rotation.

THAT 1512 Microphone Preamp

Figure 18.11 shows a simplified microphone preamp based on the low-noise THAT 1510 microphone preamplifier IC [18, 19]. This IC functions like an instrumentation amplifier, but achieves remarkably low noise of 1 nV/√Hz at its highest gain setting of 60 dB (input shorted).

The THAT 1510 is essentially an instrumentation amplifier like the Analog Device SSM2019 and the TI INA217, but with better performance. As with other instrumentation amplifiers, gain is set by a single external resistor between the two R_G pins. The smaller the resistor, the higher the gain. In a microphone preamp, a gain of 60 dB can be desirable, while being set by a single potentiometer for the resistor R_G. This resistance can be as little as 10 Ω at the 60-dB setting, and must be AC-coupled. The capacitor, C6 in the figure, must be a very large-value electrolytic. It is shown with arbitrary polarity. Significant signal current can flow through it at high signal levels when gain is set to be high. At the same time this capacitor is in a sensitive location. This can lead to some distortion from the electrolytic capacitor. It would be nice to be able to DC-couple this resistor and use a DC servo to eliminate the resulting amplified offset. The DC servo could inject a very small correction current through a high-value resistor into one of the R_G pins.

Similarly, with a 10-kΩ pot being asked to change its resistance over a 1000:1 range, a reverse-log taper should be used. However, at high-gain settings in the 10-Ω range, significant signal current flows through the wiper, so any wiper distortion may be problematic. This is the price paid for having a single control cover a 60-dB range in this arrangement. An alternative is to use a switch to set the gain, such as a 12-position design that can range from 0 to +60 dB of gain in 5-dB steps, knowing that the fader in the channel strip will allow for more precise level control anyway. A similar approach is employed in the AEA RPQ500 preamplifier.

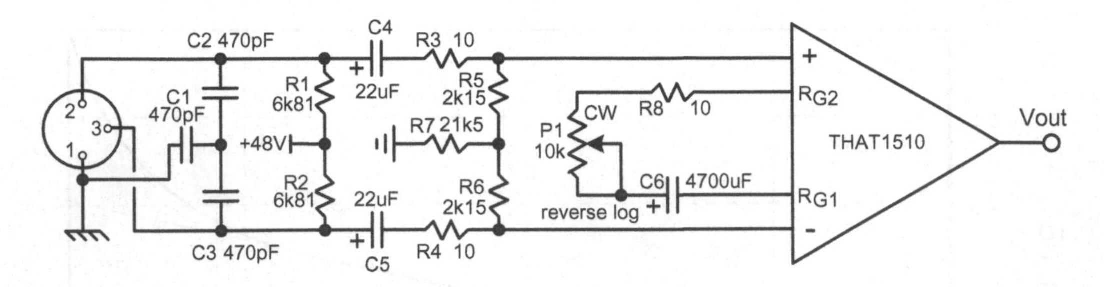

Figure 18.11 THAT 1512 IC Microphone Preamplifier.

18.4 Phantom Powering

Microphones with active electronic circuitry in them, like condenser microphones, need a source of power for this circuitry. Mixing console microphone preamplifiers provide this power over the balanced microphone cable via what is called phantom powering. The so-called "P48" phantom powering scheme was invented by Neumann to fill this need [7]. A +48-V supply is usually connected to the hot and cold microphone lines through a pair of 6.81-kΩ 0.1% resistors, as shown in the preamplifiers of Figures 18.7 and 18.8. The close tolerance of the resistors is required to ensure a high common-mode rejection ratio. Their absolute value is not critical, so they could be hand-matched 1% resistors. The balanced input lines are then AC-coupled to the input circuit of the microphone preamp.

The DC on the balanced signal lines of the microphone cable (with respect to shield) can then be merged with a pair of resistors in the microphone to power the microphone electronics. This approach incurs additional voltage drop. If the microphone has an output transformer, the power can be taken from a center tap on its secondary. Caution must be observed to avoid any DC imbalance in the power pick-off arrangement that might cause core saturation. Other microphone circuits extract power from the phantom supply in different ways, as will be illustrated in some circuits to come.

The source resistance of the phantom supply is about 3.4 kΩ, so if the microphone draws 10 mA, a drop of 34 V will occur, leaving only 14 V at the microphone for the electronics. For this reason, microphone electronics rarely require more than about 7 mA from the phantom supply. With the industry-specified IEC 61938 P48 tolerance of 48 ± 4 V, the available voltage for a microphone drawing 10 mA could be as small as 10 V [6].

Effect on Common-Mode Rejection

Very good common-mode rejection must be maintained at the input of the mixer's preamplifier. Unfortunately, any mismatch in the impedance to ground from the hot and cold balanced lines will degrade common-mode rejection. For this reason, the 6.81-kΩ phantom powering resistors should have ±0.1% tolerance. Even a 0.1% mismatch can sometimes limit CMRR to only 94 dB. A 1% mismatch will limit CMRR to 74 dB [1]. In the worst case, two ±0.1% tolerance resistors can be mismatched by 0.2%. Similar matching of load resistors in the powered devices, like condenser microphones, is also important.

If phantom power is turned off local to the preamplifier by opening the circuit to the 6.81-kΩ resistors, those resistors will no longer have an AC ground reference and will just look like a 13.6-kΩ differential load that cannot degrade CMRR. Microphones that employ phantom power usually have more output, making CMRR less important. Keeping phantom power off for microphones that do not need it, and which likely have lower output, is good practice. This advantage in the off state is usually lost in consoles where phantom power application is not controlled for individual channels. Phantom power should remain off for microphones that do not need it, especially ribbon microphones that can be damaged when "hot-plugged." It is wise to turn on phantom power only after the microphone(s) is plugged in.

Differential Z_{out} of the microphone matters, since if it were zero the differential mode would be shorted and no conversion of common mode to differential mode could occur. Common-mode Z_{out} of the microphone matters as well, since if it were infinity no common-mode current would flow and no common-mode voltage could be developed across the 6.81-kΩ resistors. Common-mode current must flow in those resistors for a common-mode to differential-mode conversion to occur. As a general rule, lower differential Z_{out} and higher common-mode Z_{out} for the source makes for better common-mode rejection in the face of mismatches between the two 6.81-kΩ resistors. In this regard, a microphone with a rather high differential Z_{out} of 600 Ω could have nearly 6 dB less CMRR from this effect.

Consider the case of a 300-Ω condenser microphone with an output transformer whose secondary is center-tapped for phantom power pickoff. Assume that the center tap is connected to ground through a voltage-regulating Zener diode with zero impedance and that the transformer windings have zero resistance and no leakage inductance. The common-mode Z_{out} of such a microphone would be zero Ω. However, the differential Z_{out} would be 300 Ω. Common-mode current would still flow and the results would be about the same as above.

Now consider that same arrangement where a current-regulating diode or JFET is paced in series between the transformer center tap and the Zener diode. The current-regulating diode might be set for 7 mA and have nearly infinite impedance at audio frequencies. Virtually no common-mode AC current can flow and thus CMRR would be very high even if the 6.81-kΩ resistors were mismatched. If at very high frequencies the impedance of the current-regulating diode were to fall, an RF choke could also be placed in series to reduce common-mode current flow at RF.

Similarly, consider a microphone with an active output stage where a differential pair has its collectors or drains connected to the balanced lines to apply its signal. The 6.81-kΩ resistors where phantom power is applied become the load resistors for the output devices. The output stage might have a 300-Ω resistor across those lines to define its differential output impedance. If the differential pair has a current source for its tail, common-mode Z_{out} will be very high if there are no other connections that can pass AC current from the balanced lines to ground. High CMRR would result even in the presence of mismatch between the 6.81-kΩ resistors.

Noise

The 48-V phantom power supply must have extremely low noise and be well-regulated. Otherwise, noise will find its way into the differential signal mode by way of finite common-mode rejection. The phantom power supply should be thoroughly filtered at every preamp in a mixer. A series inductor in the filter can provide high-frequency isolation and filtering without loss of phantom voltage. Although the P48 specification allows a tolerance of ±4 V, it can and should be held much tighter than that, probably like 48–49 V [6].

Enable and Disable

The phantom power should be able to be turned off for some passive microphones that can be damaged by phantom power or on/off transients that it may create. This is sometimes done with a single switch that controls phantom power for all microphone channels, but it is preferable for each channel to have its own phantom power switch.

Although leaving the phantom powering on when using passive microphones that don't need it is generally acceptable, it is wise, when possible, to turn off the phantom power when not needed. It can damage passive dynamic and ribbon microphones under some transient conditions, such as when a microphone is plugged into a jack with phantom power already on and there is an asymmetrical surge or contact to both hot and cold signal lines is not made simultaneously. Also, if there is a fault anywhere in the cable that shorts one signal lead to ground, potentially destructive phantom current will flow.

It is best if every console has an individual phantom power switch for each microphone channel that opens the connection to the phantom supply. This provides safety, but also can improve CMRR and reduce noise when phantom power is turned off for a microphone that does not require it.

18.5 Condenser Microphone Preamplifiers

A condenser microphone consists of a diaphragm in close proximity to a backplate, creating a capacitance typically between 10 and 100 pF, depending on whether the microphone is a large diaphragm or small diaphragm type. The capacitor thus formed is charged to about 60 V by a polarizing voltage through a very large resistance. As sound pressure moves the diaphragm closer to or further from the backplate, the capacitance changes. The charge on the capacitor must be conserved, so as the capacitance decreases, the voltage across it increases. More specifically, $V_{cap} = J_{chg}/C_{cap}$. The resulting signal is applied to a local preamplifier with a JFET first stage. This provides extremely high input impedance [1, 3]. The condenser microphone electronics must be phantom powered. Only a few mA (usually less than 10 mA) are available, so low-power electronics are important. The backplate may be perforated to allow sound from the rear to enter.

The microphone preamp functions mainly as a buffer, since the output voltage of the capsule is often high in comparison to the output voltage of a dynamic microphone [1]. Some condenser microphone preamps drive the balanced microphone cable with a single-ended signal, AC-coupling the "cold" wire to shield ground, sometimes through a terminating resistor (a so-called "impedance-balanced output"). This is adequate but not optimal. A better approach is to drive the hot and cold (pins 2 and 3) lines with the true balanced output from a differential pair of transistors, essentially using the pair of 6.81-kΩ phantom drive resistors at the far end as the collector or drain load. A differential shunt in the microphone can be used to establish the microphone preamp's differential output impedance. Another approach is to configure the input JFET as a phase splitter and drive the output lines through a pair of emitter followers.

The Capsule

The capsule capacitance is quite small, perhaps on the order of 10–100 pF, so the electrical impedance is very high, from 800 to 80 MΩ at 20 Hz. The voltage produced by the capsule must therefore be buffered or amplified by a JFET amplifier with extremely high input impedance. Polarizing and gate return resistors are often on the order of 1–10 GΩ. This is necessary to achieve satisfactory performance at low frequencies because of the time constant formed by these resistances and the capacitance of the diaphragm. A 50-pF diaphragm driving a pair of 330 MΩ resistors will have a corner at about 20 Hz. However, as we will see, these resistances should be larger than what is required to achieve only −3 dB loss at 20 Hz for reasons of thermal noise (to some this may be non-intuitive). Sometimes resistances as high as 10 GΩ are used.

Given the small capacitance of the capsule, the input capacitance of the JFET preamplifier must be small, since a capacitance voltage divider is formed. It is obviously undesirable to lose signal amplitude. This can be more problematic if the input capacitances of the JFET, C_{iss} and C_{rss}, are nonlinear, which they tend to be. This may cause distortion. The problem is more significant with capsules whose capacitance is low, as in the neighborhood of 10 pF. For common-source JFET arrangements where the source is at AC ground, the full amount of C_{iss} at the operating V_{gs} will be in play. For arrangements where there is gain greater than unity and the drain is not cascoded, the effective value of C_{rss} at the operating V_{dg} will be multiplied by the Miller effect. These effects can be significant in the presence of the fairly high capsule signal voltages that can be present with a condenser microphone under high-SPL conditions.

This can lead to a tradeoff in the selection of the JFET to be used. A larger JFET with higher transconductance has lower voltage noise but higher capacitances and gate leakage current. A smaller JFET with lower transconductance will have larger voltage noise but will enjoy lower capacitances. There are other noise sources in the input circuit of a condenser microphone which may be more significant than the JFET voltage noise, so choosing the JFET with the lowest voltage noise at the expense of increased capacitance may not always be a wise choice.

The voltage gain of the preamp will often be less than 10 dB. In fact, for more sensitive capsules and high sound levels, attenuation of the signal may be necessary. A common approach to such attenuation is to employ an input pad that creates a capacitance voltage divider between the capsule source impedance and a shunt capacitance, often with an attenuation of 10–20 dB. A 50-pF capsule with a 450-pF shunt capacitance will see 20 dB of attenuation.

The LSK189 gate leakage current I_{gss} is rated at 25 pA maximum at room temperature and 10 nA at 125°C. Gate leakage current is usually smallest when the device is operated with V_{dg} less than 5 V. The typical value is 2 pA at room temperature [15]. One must beware of the effect of increased junction temperature. Gate leakage current doubles every 10°C. The microphone will never see more than 55°C in hot sunlight, so the temperature will be up by no more than 30°C from 25°C. This causes three doublings in I_{gss}, or a factor of 8, implying typical gate leakage current of about 8 × 2 pA = 16 pA at 55°C. A current of 16 pA flowing through a very large 10-GΩ gate return resistor creates a voltage drop of 0.16 V. This may cause a DC offset. One solution is to employ a DC servo driving the ground end of the gate return resistor if a very large resistor value is being used. Gate leakage current will also produce shot noise current (input current noise).

A Simple Condenser Microphone Preamplifier

Figure 18.12 shows a simple unity-gain source-follower condenser microphone preamplifier circuit that serves as the impedance converter. Not shown are the power supply details, including the boost converter often used to create the polarizing voltage of typically 60 V. Resistor R1 polarizes the diaphragm of the capsule to 60 V with respect to the backplate, which is grounded. C1 AC-couples the diaphragm to the gate of the JFET and R2 provides +3 V bias for the gate from a voltage divider consisting of R3 and R4. If the JFET V_{gs} is 1 V at its operating current, the voltage at the source will be +4 V. R5 then establishes the JFET bias current at 2 mA.

The JFET is connected in a cascode CFP arrangement that increases the effective transconductance of the JFET and provides negative feedback. This brings the source follower to very nearly unity gain and boostraps the source so that C_{gs} of the JFET matters little. The JFET CFP is formed by J1 and Q1, each of which operates at about 1 mA.

At the same time, J1's drain is cascoded by driven cascode Q2, whose base is biased at 3.5 V above the source of J1 via blue LED D1. The LED is biased with the collector current of Q1. As a result, the drain of J1 is also bootstrapped with signal, making C_{dg} of J1 of little consequence. As a result of this very effective source follower, the input capacitance seen by the diaphragm is very

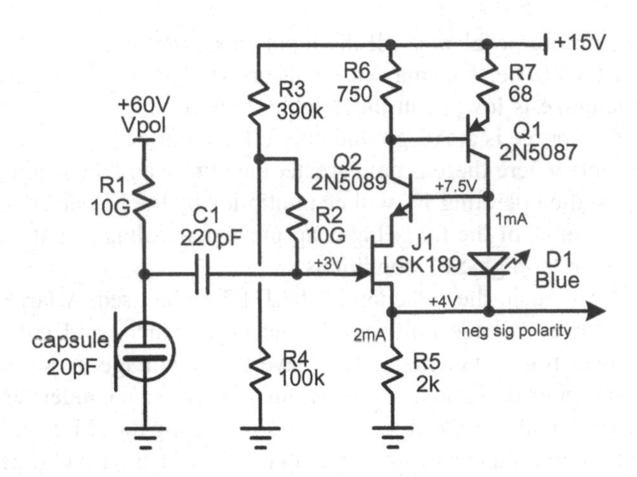

Figure 18.12 Simple Condenser Microphone Preamplifier.

small. As an added advantage, V_{ds} of J1 is fixed at a relatively low voltage equal to the forward voltage of D1 less one V_{be}, or about 2.9 V. This helps minimize impact ionization noise in J1. The JFET cascode CFP forms a kind of super transistor that can also be used in other configurations like a common-source amplifier.

Because of the relatively high output of condenser capsules, sometimes on the order of −35 dBV/Pa for a large-diaphragm device, unity gain is often adequate for a condenser microphone preamplifier. However, this circuit does not complete the preamplifier because it provides only a single-ended output that cannot drive the microphone cable in a balanced fashion. Circuits to do that are still to be discussed.

Noise Sources

Consider a condenser microphone with a 50-pF capsule that is polarized to about 60 V though a 1-GΩ resistor. The capsule diaphragm is AC-coupled to the gate of the JFET input stage. A 1-GΩ resistor returns the gate to AC ground. Assume that the voltage noise of the JFET is 1.7 nV/√Hz and that its typical gate leakage current at 25°C is 2 pA.

There are four significant noise sources in this preamplifier. The first two are the thermal noise contributions from the 1-GΩ polarizing and gate return resistors R1 and R2 (net 500 MΩ). The large-value resistors have high open-circuit thermal noise that is injected into the microphone input node and attenuated by the capacitance of the capsule. This noise contribution thus drops at 6 dB/octave with increasing frequency, and is therefore red noise. The open-circuit thermal noise voltage of a 500-MΩ resistor is 3.0 μV/√Hz. Passed through the 500-MΩ net resistance to the capsule node, the injected current noise is 6.0 fA/√Hz. The impedance of a 50-pF capsule at 100 Hz is 32 MΩ. The result is 192 nV/√Hz. At 1 kHz, this becomes 19.2 nV/√Hz due to the low-pass filtering effect of the capsule capacitance. This is a significantly larger noise contributor than JFET voltage noise at 1 kHz – almost ten times as large as that of an LSK189.

Increasing the effective input resistance from 500 MΩ to 5 GΩ will drop the noise by √10 to 6.1 nV/√Hz at 1 kHz; the open-circuit thermal noise goes up as the square root of resistance, but its ability to create current noise goes down by a factor of 10, resulting in a net decrease of √10. This is why high input resistance is important, even apart from low-frequency response. The frequency response of the 50-pF capsule capacitance loaded by 5 GΩ is down 3 dB 0.64 Hz.

The third noise contributor is the input-referred voltage noise of the JFET. This is white noise plus a *1/f* noise component that rises at 3 dB/octave as frequency goes down, usually beginning around 100 Hz. At frequencies at and above 1 kHz, theoretical JFET voltage noise is approximately 0.67 times the noise of a resistance equal to 1/*gm*. At 2 mA, *gm* for an LSK189 is 3.9 mS, corresponding to resistance of 256 Ω. Multiplied by 0.67, this resistance is 172 Ω. The noise of 172 Ω is 0.41 × 4.2 nV/√Hz = 1.7 nV/√Hz. At 1 kHz, this is well less than the 19.2 nV/√Hz contributed at 1 kHz by the thermal noise of a net input resistance of 500 MΩ. Note that even if R1 and R2 were increased by a decade to 10 GΩ, their noise contribution would only go down by √10 to 6.1 nV/√Hz.

The fourth contributor is the JFET gate leakage current. It creates shot noise current flowing through the impedance of the input circuit.

$$I_{shot\_noise} = 0.57\ \text{pA}/\sqrt{\text{Hz}}/\sqrt{\mu\text{A}}$$

Shot noise current increases as the square root of the DC current. For leakage current of 100 pA, the shot noise current is 5.7 fA/√Hz. The impedance of a 50-pF capsule is 3.2 MΩ at 1 kHz. The 5.7 fA/√Hz flowing into the 3.2 MΩ capsule node impedance will result in 18 nV/√Hz at 1 kHz. That is considerably more than the input voltage noise of the JFET. The shot noise input current of the JFET creates red noise, since that contributor goes down at 6 dB/octave. At 100 Hz, the shot noise contribution will be 180 nV/√Hz. As frequency goes down, it rises twice as fast in dB as the 1/*f* noise of the JFET. Finally, gate leakage current was quoted as only the typical value at 25°C. Leakage current will double for every 10°C rise in temperature, so a condenser microphone in the hot sun may be a bit noisier. Choosing a JFET with low gate leakage current is important.

Knowing which of these noise contributors is dominant is key to some design decisions, such as the choice of JFET. For example, JFET voltage noise or JFET *1/f* noise may not always be the dominant source of noise. If this is the case, one might choose a JFET with lower capacitances, slightly higher voltage noise and less gate leakage current.

Thermal Noise vs Shot Noise

Since the thermal noise and shot noise contributions both increase at 6 dB/octave with decrease in frequency, it can be difficult to determine which one is dominant. Knowing which one is dominant can influence important design choices, such as which JFET should be chosen. Take a JFET amplifier and connect its input gate to ground through a resistor R_g. Choose R_g as a high-value resistor whose thermal noise is at least ten times the JFET's input noise voltage. Thermal noise will increase as the square root of R_g, while shot noise voltage will increase directly with R_g. If these noise sources adequately dominate JFET input voltage noise, they can be differentiated by measuring the noise at three different values of gate resistance differing by a factor of 2. This will indicate the rate of rise with resistance for the noise. A square-root rate of rise will imply thermal noise from R_g. A direct rise will imply shot noise from the JFET's gate leakage current.

If we can ignore JFET voltage noise, quadrupling R_g will double the thermal noise contribution, but will quadruple the shot noise contribution. We also know that thermal noise will increase very slowly with temperature increase, only in proportion to the absolute temperature, while the contribution from shot noise will increase with the square root of the increase of leakage current with temperature. Leakage current will double for every 10°C rise in temperature. This means that the shot noise contribution will double for every 20°C rise in temperature. Heating the preamplifier and viewing the increase in noise can thus help differentiate between the two contributors.

Figure 18.13 shows the JFET and resistor thermal noise contributors for a design using a 50-pF capsule and 2-GΩ polarizing and gate return resistors for net input resistance of 1 GΩ. An LSK189 JFET is assumed. The open-circuit thermal noise of the net 1-GΩ load resistance is 4200 nV/√Hz. This is low-pass filtered by the combination of 1 GΩ and the 50-pF capsule capacitance, with a 3-dB corner frequency of 3 Hz. This low-pass filter will be down 40 dB at 300 Hz, taking the 4200 nV/√Hz down to 42 nV/√Hz at 300 Hz. This is quite a bit of noise when one considers that it corresponds to 420 nV/√Hz at 30 Hz, completely dwarfing any *1/f* JFET noise. Moreover, this corresponds to approximately 13 nV/√Hz at 1 kHz, eclipsing the 2–3 nV/√Hz of flat noise expected of the JFET preamplifier.

Note that this thermal noise will go down as the square root of any increase in the net biasing resistance. In the case at hand, increasing each biasing resistor to 10 GΩ, resulting in a net 5-GΩ resistance, will reduce the noise in the above example by √5, or a factor of 2.2:1. The noise at 1 kHz will then be reduced from 13 to 6.5 nV/√Hz, still well in excess of the JFET voltage noise.

An A-weighted noise measurement will mitigate the contribution of this large low-frequency noise due to its sharp low-frequency weighting roll-off. A-weighting is kind to the condenser microphone when self-noise is measured. CCIR weighting is less kind to condenser microphones because its low-frequency weighting roll-off is more shallow.

Even at the very audible 1 kHz, the noise is 13 nV/√Hz. This is more than one would normally incur from the JFET, since the voltage noise of a low-noise JFET like the LSK189 is typically only 1.7 nV/√Hz at drain current of 2 mA. These considerations illustrate that it is foolhardy to evaluate the noise of a condenser microphone preamp with its input shorted. Using resistors larger than 1 GΩ will reduce their thermal noise contribution, but only by 3 dB for each doubling of value.

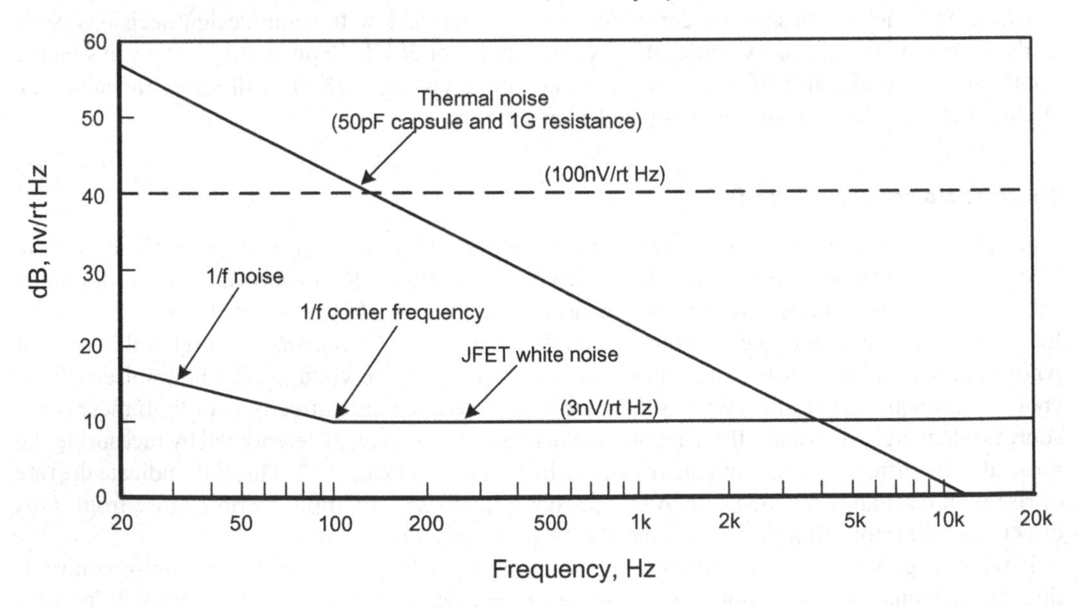

Figure 18.13 Resistor and JFET Noise Contributors.

Nevertheless, we are left with JFET voltage noise as the only significant contributor at high frequencies because the capsule capacitance shorts out the resistor thermal noise and the JFET shot noise contributors at high frequencies, above about 3.5 kHz in the example above. Over much of the audio band, JFET voltage noise may not be the main contributor.

As a sanity check, consider the self-noise in dB-A of a condenser microphone whose only noise contribution is the JFET's input voltage noise. Assume a JFET with white input voltage noise of 3 nV/√Hz and a capsule with sensitivity of 20 mV/Pa. A-weighting has 114 √Hz, so the JFET input noise is 342 nV, or 0.34 μV. The ratio of 20 mV to 0.34 nV is 59e3, corresponding to 95 dB. This corresponds to −1 dBA. Even a 5-dBA microphone is not dominated by 3 nV/√Hz of preamplifier noise if its capsule sensitivity is 20 mV/Pa.

The Sound of Condenser Microphone Self-Noise

If you listen to the sound of a dynamic microphone's self-noise, it will sound somewhat like the hiss of white noise. If you do the same for a condenser microphone, it may sound different, with a lower-frequency sound and not as much hiss. If that is the case, the thermal noise created by the polarizing and gate return resistors may be the dominant noise source, since this noise component has spectral density falling at 6 dB per octave, making it red noise. This will also be true of gate shot noise from the JFET.

The noise of a dynamic microphone and a condenser microphone with the same self-noise dB-A number may not sound the same and may not have the same degree of annoyance. In other words, the noise spectrum of a dynamic microphone will be different from that of a condenser microphone. The noise from a ribbon microphone should also be reasonably white. In all of these discussions the electrical self-noise has been the subject. Acoustic self-noise due to Brownian motion of air molecules hitting the diaphragm is being ignored. That is red noise (sometimes called brown noise).

Frequency Response

The condenser microphone's response at low frequencies can be degraded if the capsule capacitance is small and the preamp load resistance is not extremely high. The load resistance is the parallel combination of the polarizing resistance and the gate return resistance. Consider a 50-pF capsule driving a pair of 2-GΩ polarization and gate return resistors (net 1 GΩ). This filter will be down 3 dB at 3.2 Hz, which is acceptable. However, if the capsule has significantly lower capacitance, the low end will begin to suffer. Bear in mind that the self-noise considerations discussed above may dominate this issue anyway.

Capsule Diameter and Capacitance

For a given capsule diaphragm spacing, capacitance will increase as the area of the diaphragm, or simply as the square of the diaphragm diameter. A capsule with a larger capacitance will tend to have a higher SNR for a given net load resistance. This is largely the case if the thermal noise from the resistance is larger than the input noise of the JFET. At higher frequencies the input noise of the JFET will eventually dominate. A 1/2-inch small-diameter capsule may have only 1/4 or less the capacitance of a large-diameter capsule unless its diaphragm spacing is smaller.

Sensitivity vs Polarizing Voltage

The sensitivity of the condenser microphone capsule depends on the polarizing voltage. If a nominal polarizing voltage of 60 V is reduced to 30 V, output voltage from the capsule will go down by 6 dB and SNR will go down by 6 dB.

There are a number of ways that the polarizing voltage can be derived from the working voltage within the condenser microphone preamplifier. The good news is that the current drawn from the polarizing voltage supply is very close to zero. If the preamplifier circuit draws only 1 mA from the phantom supply, then about 45 V is available to use directly in polarizing the capsule. This will result in capsule sensitivity 2.5 dB lower than if the polarization voltage were 60 V. Microphones whose active circuits draw more than 1 mA will be at a greater disadvantage if the polarizing voltage is not boosted above the available loaded polarization voltage. In either case, microphone sensitivity will probably vary as a result of the ±4-V tolerance on the +48-V phantom supply (absent local regulation of the boosted polarizing voltage) [6]. Details of the polarizing power supply in the microphone will be discussed later.

Input Capacitance of the Preamplifier

For smaller capsules with less capacitance, the input capacitance of the JFET can become a concern, especially if a large-capacitance JFET is being used in the pursuit of lower noise. For example, if an LSK170 JFET is being used in a simple un-degenerated common-source circuit with gain, the input capacitance of the preamplifier will be C_{iss} plus G × C_{rss}, where G is the Miller effect multiplier. For the LSK170, C_{iss} = 20 pF and C_{rss} = 5 pF. It is easy to see how large the input capacitance can be with this transistor in the common-source circuit arrangement. The capacitance seen looking into the JFET gate would significantly dominate over the capacitance of a 10-pF capsule. Condenser microphones employing an output transformer may require significant common-source gain to offset the step-down ratio of the output transformer.

The LSK189 might be a better choice, with C_{iss} = 4 pF and C_{rss} = 2 pF. The 2N3819 has C_{iss} = 2.2 pF and C_{rss} = 0.7 pF, a still better choice. Its transconductance at 1 mA is 2.5 mS, somewhat smaller than the 4 mS of the LSK189. Input noise of the 2N3819 is close to 10 nV/√Hz at 1 mA, while that of the LSK189 is closer to 2 nV/√Hz. For a 10-pF capsule, a different circuit may have to be used in order to reduce preamplifier input capacitance. In the cascoded source-follower circuit of Figure 18.12, both C_{iss} and C_{rss} are bootstrapped by the signal, nearly eliminating their effect.

Capacitance Attenuator (PAD)

If the signal from the condenser microphone is too high, it is easily attenuated by switching in a fixed capacitor in parallel with the condenser capsule, thus forming a capacitance voltage divider. A 450-pF capacitor placed in parallel with a 50-pF capsule will provide 20 dB of attenuation. This will also reduce the thermal noise from the polarizing and gate return resistors by the same amount as the signal.

The ratio of the voltage divider formed by the series source capacitance of the capsule and the shunt attenuator capacitance obviously depends on the two capacitance values. However, in order to produce a signal voltage, the capacitance of the capsule must change. This means that the incremental ratio of this attenuator will change as the capacitance of the capsule changes with signal. This translates to second harmonic distortion. Consider a capsule with 20 mV/Pa sensitivity. If it is exposed to 124 dB SPL it will produce about 600 mV_{RMS}. If the capsule is polarized with 60 V and is producing a large 600-mV signal, its capacitance is changing by about 1% RMS. If that capsule is loaded by an attenuating capacitor of nine times the capacitance of the capsule to create 20 dB of attenuation, it is easy to see that some second-order distortion will result. Its capacitance will be 1% larger under positive sound pressure and 1% smaller under negative sound pressure, changing the attenuation factor of the capacitance voltage divider by about 1% in each direction. However, it is unclear if this effect is small in comparison to the inherent distortion of the capsule itself at high SPL.

To evaluate capacitive attenuator distortion, listen to the sound of the microphone with the same moderately high acoustical input with the attenuator both in and out, and with the console preamp gain control adjusted to offset the amplitude difference. Does the microphone sound different under these two conditions? Just make sure that the high acoustic input level does not overload the microphone when the attenuator is out.

Relative Importance of JFET Input Noise Voltage

As suggested in Figure 18.13, the relative contributions of resistor thermal noise and JFET input voltage noise as a function of frequency must be evaluated, as there is a tradeoff in total noise and its spectrum. The question must always be considered: Is a smaller JFET with slightly higher voltage noise but lower capacitance a better choice? Over-sizing the JFET in pursuit of lower voltage noise may not always be the best choice. Significant JFET input capacitance can cause attenuation of the signal from the capsule and may cause distortion at high signal levels as described above. Using a large-die JFET to obtain low voltage noise might also increase gate leakage current and consequent shot noise.

Dual Monolithic JFETs

A condenser microphone preamp can be made with a single-ended JFET input amplifier or a differential pair input amplifier, with feedback applied to the opposite side of the pair in the latter case. Effective preamplifier input capacitance due to the JFET can be dramatically reduced in the latter case at the expense of a 3-dB increase in total JFET voltage noise and total current. Cascoding the input JFET can eliminate the Miller multiplication of C_{rss}. Example dual monolithic JFETs include the LSK389 and LSK489 [15].

One can also employ a dual monolithic JFET in a different way. If the input gate of the JFET is allowed to float with no input gate return resistor, the gate bias voltage can be controlled by adjusting the gate leakage current from the JFET on the other side of the dual monolithic device. This can be done with a DC servo. In this case, the second device is used as a dummy device, with its source and drain connected together and floating. The bias control voltage is applied to the gate of the dummy device. This approach takes advantage of the fact that there are reverse-biased diodes connected from each gate to the common floating substrate of the dual monolithic device. The input gate voltage will tend to float to the same voltage as the voltage at the gate of the dummy device. In some dual monolithic JFETs the common substrate is brought out, allowing for the possibility of controlling bias by means of that node as well.

DC Servo

The variability of JFET threshold voltage and I_{dss} requires that the source resistor be of a different, possibly hand-selected, value for different JFETs to obtain the desired bias current in many single-ended common-source designs. Alternatively, a bias adjust pot may be needed. Selection of

JFETs may also be required. JFET gate leakage current can also come into play when large gate return resistance values are used. An attractive solution is to use a DC servo to control the biasing. A micropower op amp can be used to implement the DC servo.

An Example JFET Design

The LSK189's combination of low noise and low input capacitance make it a good choice for the input stage of condenser and electret microphones [15–17]. Figure 18.14(a) shows the essence of a basic condenser microphone preamplifier developed by Schoeps in the early days of solid-state condenser microphones. Notably, it is transformerless. Countless condenser and electret microphones use this clever circuit or a variant of it.

J1 performs the impedance conversion and is connected as a phase splitter, thus providing signals of opposite phase to the emitter followers of the output stage. Potentiometer P1 is adjusted to drop about 25% of the supply voltage across the identical source and drain resistors, leaving about 50% of the supply voltage across J1. This provides optimal headroom and accounts for differences among JFETs. The phantom-powered hot and cold microphone lines load the emitter of the output stage and provide their operating current. The current from their collectors flows into a Zener diode to create a regulated voltage to power the input circuit.

The bases of Q1 and Q2 are connected through bias resistors R5 and R6 to the collectors of Q1 and Q2, both of which are connected to the 12-V suppy. If Q1 and Q2 have infinite beta, V_{cb} of Q1 and Q2 will be zero, seriously limiting headroom. In practice, this simple version of the circuit, widely used, depends on the finite beta of Q1 and Q2 to cause base current to flow through R5 and R6, creating voltage drop to obtain V_{cb} greater than zero. This causes the bases of Q1 and Q2 to reside at a more positive voltage than the collectors. Q1 and Q2 should also have about the same beta in order to avoid DC offset being applied to the microphone lines. R7 and R8 help stabilize the emitter followers, since the capacitive load of the microphone cable could otherwise cause instability. They also help define the output impedance. Output impedance at the emitters will be approximately the resistance of R4 or R3 divided by the beta of the output transistors. If beta is 100, that impedance will be about 21 Ω.

The revised design in Figure 18.14(b) addresses the shortcomings of the original circuit. It requires two additional transistors and a micropower op amp. A DC servo implemented with U1, R16 and C5 forces the source of J1 to be at 1/4 the supply voltage by controlling the bias voltage applied to the gate of J1 through R2, eliminating the need for the pot adjustment. The output

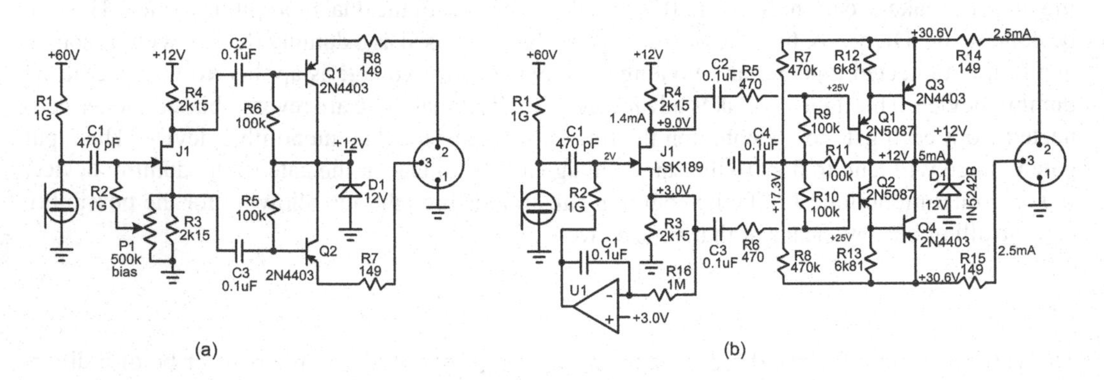

Figure 18.14 Example JFET Condenser Microphone Preamps.

transistors are implemented as Darlingtons by adding two emitter followers, Q3 and Q4 after Q1 and Q2, providing very high effective beta and reducing dependence on beta for biasing. Biasing is now accomplished by creating a positive voltage with respect to the collectors of Q1, Q2, Q3 and Q4 to provide adequate and predictable headroom. This bias is created by the voltage divider consisting of R11, R12 and R13, and filtered by C4, and places the bases about 5 V more positive than the collectors.

Phantom-Derived Microphone Power Supply

Grabbing phantom power from the microphone lines is a fundamental function that must be performed by the circuitry of a phantom-powered microphone. The circuit in Figure 18.15 is a clever way to do this, re-using the power for the output stage to power the other parts of the preamplifier, such as the input stage and polarization supply. Extracting the phantom power can be done in a number of different ways, but it is desirable that the phantom pickoff not degrade common-mode rejection.

If the line is driven with the center-tapped secondary of a transformer, that is a convenient place to pick off the power from the hot and cold microphone wires. However, it is important that little if any net DC current is allowed to flow through the halves of the secondary winding so as to avoid saturation effects. If there is no center-tapped secondary, or if there is no transformer at all, a pair of precision resistors can be bridged across the lines, creating a kind of center tap. In any of these cases, if the tap creates an AC path to ground, there is a chance that CMRR will be reduced – much for the same reason that the pair of 6.81-kΩ phantom powering resistors at the microphone preamplifier should be matched within 0.1%.

Ideally, power would be picked off by two identical current sources of a fixed value. For example, in a mic that requires 7 mA of phantom current, two closely matched 3.5 mA current sources could be used. The resulting summed current could then flow into a Zener regulating diode to provide a fixed operating voltage for the active circuitry. The nice thing about such an idealized approach is that it presents an almost infinite-resistance common-mode path to ground.

Figure 18.15 illustrates a simplified phantom powering arrangement for a transformerless condenser microphone. A differential JFET output stage is used to pick off the phantom power by using the phantom resistors in the main preamp as their load. It provides approximately unity gain with a 300-Ω output impedance. The stage is inverting from the input to pin 2 of the XLR connector. This corrects for the signal inversion that is normally present at the diaphragm of a conventional capsule arrangement. The circuit can be fed from a simple source-follower impedance converter input stage. The tail current is set by U1, a 7-mA constant current source that then feeds active Zener U2 to provide +15 V for the remaining preamplifier signal circuitry and polarization supply. This arrangement allows the current flowing in the output stage to be re-used by the other circuits. The value of 7 mA is usually about the maximum that condenser microphones draw, and is the value at which the maximum power (169 mW) can be extracted from the 3.4-kΩ net phantom supply source impedance.

Note the high-impedance 20-V source created by R7 and R8 merely feeds the gates of the JFET pair, and is bypassed. The stage has about 1 dB of gain, taking into account the finite transconductance of the JFETs. In practice, the JFETs may be connected as CFPs to reduce distortion at higher signal levels. R5 and R6 can be made larger if greater output signal swing or gain is needed. There is nothing sacred about the differential output impedance being only 300 Ω as long as high-frequency response is not significantly impaired by microphone cable capacitance.

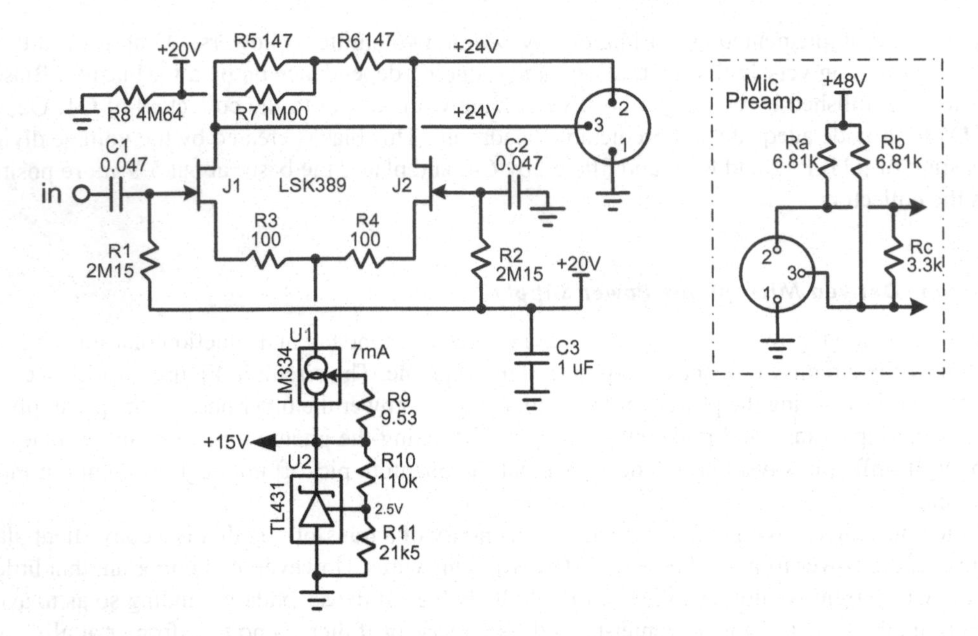

Figure 18.15 Phantom Powering for a Condenser Microphone.

Balanced Interface to the Microphone Cable

Thus far the front end of a condenser microphone has been mainly discussed. It usually produces a single-ended signal that must be converted to a balanced signal for application to the balanced microphone line. A straightforward way to drive the balanced microphone cable is to use a transformer. Some other preamplifiers dispense with the transformer and use an active balanced output.

As shown in Figure 18.15, the transformer can be avoided, and a true-balanced output can be achieved if the line is directly driven from the drains of a JFET (or BJT) differential pair, using the 6.81 kΩ phantom powering resistors at the far end as the load resistors. In the microphone, a resistance (R5 + R6) can bridge the differential pair to establish the desired output impedance for the microphone. If this resistor is center-tapped, DC power can be taken from there. Alternatively, as shown, the tail current of the differential pair can be used to power the other electronics. U1 is a constant current source that establishes the tail current, and also dumps that into electronic Zener U2. As a result, +15 V at up to 7 mA is available to the other circuitry.

Polarizing Power Supply

While some condenser microphones use the phantom voltage directly for the necessary capsule polarizing voltage, the actual voltage available will be well less than 48 V because of the current drawn by the impedance converter circuit. An alternative approach to supplying polarization voltage is to employ a DC-DC converter to boost the available voltage to 60 V or more. This is practical and reasonably efficient because the polarization current is nearly zero [7]. Many condenser microphones use this approach. The converter adds some expense and complexity, but it can enable a significant reduction in self-noise by enabling the capsule to operate at its full sensitivity. A further advantage of the converter is that its output voltage can often be regulated, independent of the phantom voltage. Of course, it is imperative that noise from the converter does not contaminate the audio signal.

A common example of a polarizing DC-DC converter is an L-C Hartley oscillator followed by a voltage doubler, as shown in Figure 18.16(a). In this design, two voltage doublers are employed so that a negative polarizing voltage is also available if needed. Negative polarizing voltages are useful in some multi-pattern microphones to be discussed later. As shown, this oscillator operates at about 1.6 MHz. C1 tunes the pair of inductors while C2 and C3 control the amount of drive voltage applied to the base of Q1. C2 should be chosen so that the oscillation signal at the base does not drive the base more negative than the emitter by about 4 V. The applied power supply voltage determines the amplitude of the oscillation and the resulting DC polarizing voltage. The ratio of L2 to L1 can also be selected to control the value of the polarizing voltage.

There are numerous other options for implementing the polarizing voltage step-up supply, but some may create noise with sharp waveform edges or changing magnetic fields and some do not have great regulation. The use of a stabilized RC sine wave oscillator running from regulated 15 V, driving a 12X voltage multiplier diode string at 80–100 kHz is an attractive option. Although it has more components than an LC supply, this supply can be made very small with surface-mount components. The supply uses no inductors.

Figure 18.16(b) illustrates such an approach. The oscillator is a simple Wien bridge design with an op amp and a JFET AGC element to set the amplitude and keep the signal sinusoidal. It is implemented with a micropower rail-to-rail dual op amp. The frequency is set to about 100 kHz by 100-pF and 18-kΩ components in the bridge. The other half of the dual op amp implements an integrator that drives the JFET gate to control amplitude. A 3-V reference is connected to the integrator and the 64-V polarizing output voltage is divided down to apply 3 V to the AGC integrator. The polarizing voltage is thus well-regulated. In fact, if a pot is placed in the circuit of the integrator, microphone sensitivity can be trimmed. The "ground" reference for the oscillator is the midpoint of the 15-V power supply, created by R2 and R3.

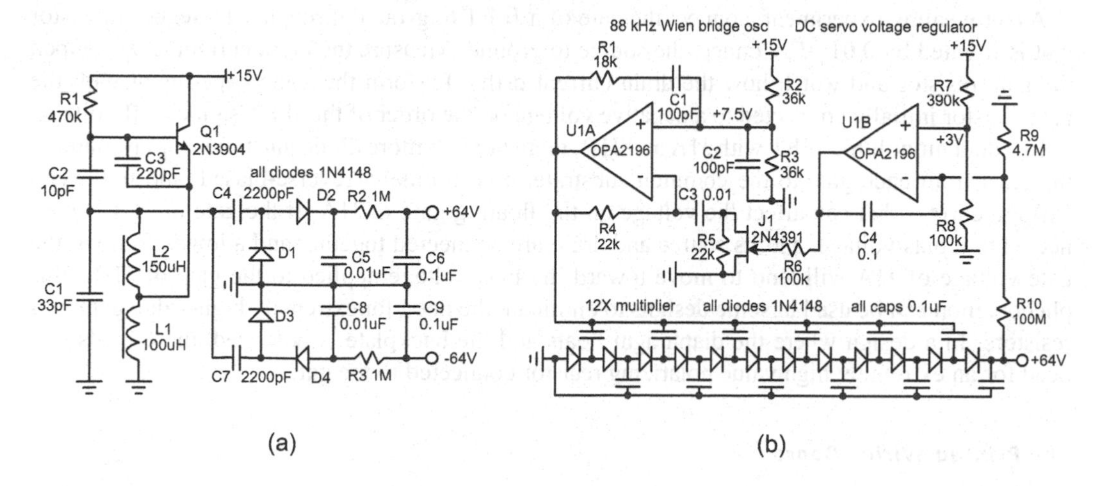

Figure 18.16 L-C and R-C Oscillator Polarizing Supplies.

The price paid for closed-loop voltage regulation is the small current drawn by the voltage divider, implying the need for very high resistance values for voltage divider resistors R9 and R10. A 100 MΩ resistor is used for R10, keeping the current below 1 μA.

The 12X serial voltage multiplier includes a series string of 12 diodes with all of their cathodes pointing in the direction of the positive output. Two groups of six capacitors are connected to the cathodes of the diodes in alternate fashion, one group originating from the AC source and the other from ground, as shown in Figure 18.16(b). The multiplier is essentially six voltage doublers connected in series, with each voltage doubler starting with the positive output of the previous doubler. Alternatively, a Cockroft-Walton multiplier can be used, wherein none of the components needs to be rated at a voltage much greater than the p-p voltage of the input waveform [20].

A 12-$V_{p\text{-}p}$ sine wave from the oscillator produces approximately 64-V output with a 10 MΩ load. With a 15-V supply a rail-to-rail oscillator op amp can provide at least 14-$V_{p\text{-}p}$ (~5 V_{RMS}). The multiplier is quite efficient, requiring just 300 μA_{RMS} from the oscillator to drive a 10 MΩ load to 64 V. Current consumption will be lower when driving the 100 MΩ load of the AGC voltage divider. The multiplier can be made more efficient if Shottky diodes are used. Although the parts count seems high, the tiny surface-mount (SMT) components can be densely packed on both sides of the board. Two such multipliers can be used to create output voltages of both polarities, useful in polar pattern control to be discussed. The OPA2196 micropower op amp draws 0.14 mA per channel. Its maximum input bias current of 20 pA flowing through the approximate 4.7 MΩ source resistance of the divider will cause only a 0.1-mV offset.

JFETs with Gate Floating at DC

Consider a JFET with its gate AC-coupled to the signal source and with no gate return resistor, so the gate floats at DC. At what current will it operate and what will determine that? All JFETs have a very small gate leakage current through the reverse-biased p-n junction between the gate and the channel. This tends to pull the gate voltage toward the source, causing the current to move toward I_{DSS}. However, as the gate is pulled further positive, the junction begins to become forward biased, creating a small current of the opposite polarity. These competing currents cause the gate to float to a tiny positive bias, stabilizing the current at very close to I_{DSS}.

As a revealing experiment, connect the gate of a JFET to ground through a 1-Megohm resistor that is shunted by 0.01 μF. Connect the source to ground. Measure the drain current as I_{DSS}. Open the gate resistor and watch how the drain current drifts. Perform the same experiment with the gate resistor initially connected to a negative voltage on the order of the JFET's pinch-off voltage.

In a dual monolithic JFET with J1A and J1B, there are two more diode junctions in play, namely the ones from each gate to the common substrate, both normally reverse-biased, and both with leakage current that can affect the voltage on the floating gate of J1A. If the gate of J1B is connected to a bias voltage, and its source and drain are connected together and allowed to float, the gate voltage of J1A will tend to move toward the bias voltage applied to the gate of J1B. This phenomenon can be used in some designs to eliminate the need for extremely high-value gate bias resistors. In a design where the diaphragm floats and the backplate is polarized, there is also no need for an extremely high value polarizing resistor connected to the gate.

The Printed Wiring Board

The extremely high input impedances in a condenser microphone dictate that those nodes be carefully guarded and/or shielded in the layout and that high quality printed wiring board material be used to minimize dielectric absorption losses, *hook* effects and parasitic resistance. Teflon, Rogers 4350 or Panasonic Megtron 6 PCB materials are good choices. Meticulous cleaning of the PWB after assembly is also a must to avoid contamination that can introduce leakage resistance. In some cases the use of insulated posts and fly wiring for the diaphragm and JFET gate nodes can help.

Polarizing the Backplate

A common approach is to polarize the diaphragm positive and connect the backplate to ground. This results in an inverted audio signal at the diaphragm (positive sound pressure creates negative-going output). However, the backplate in a capsule does not need to be connected to ground if it is insulated. In fact, many condenser microphones do not connect it to ground [7]. This means that the polarization voltage can be applied to the backplate while the diaphragm can remain at ground potential. This inverts the signal coming from the capsule to become a positive audio signal. Alternatively, the backplate can be polarized with a negative voltage to create the same inverted audio signal as if the diaphragm were polarized positive. Regardless of the approach, because a balanced output must be created anyway, any needed phase reversal of the audio signal is trivial to implement. We will see that these various polarizing options can be used with dual-diaphragm capsules to provide selectable polar patterns.

Polarizing the backplate and allowing the diaphragm to be at ground potential opens up the possibility of direct coupling of the diaphragm to the gate of the JFET, eliminating the large-value polarizing resistor and its thermal noise. In fact, with the backplate polarized, the capsule begins to look like a back-electret microphone, as described below in Section 18.6. In Figure 18.17(a) the diaphragm is direct-coupled to the JFET gate and the gate return resistor R1 sets the gate voltage to +3 V to provide bias for the source follower impedance converter. Measures must be taken to avoid inadvertent damaging transient or DC potentials to reach the JFET gate from the diaphragm.

Following the electret analogy a bit further, many electret microphones do not use a gate return resistor, allowing the JFET gate to float near $V_{gs} = 0$ V, with the JFET operating near its I_{dss} value of current. As mentioned earlier, leakage current will cause the gate to float to near the potential of

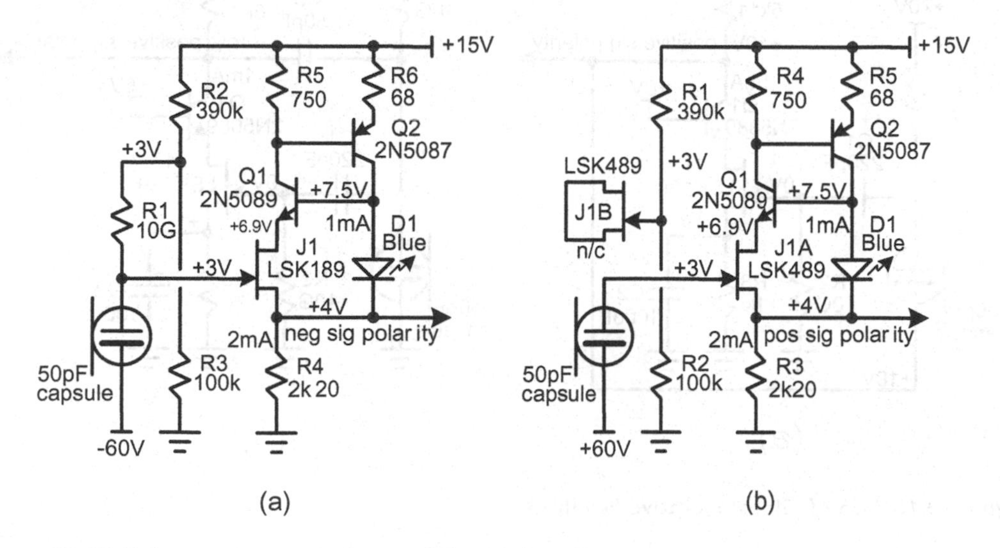

Figure 18.17 Polarizing the Backplate and Establishing Bias.

the source. This opens the possibility of dispensing with the second large-value resistor, the gate return resistor. With both resistors gone, their thermal noise contributions go away.

A simplified version of such a circuit is shown in Figure 18.17(b). In this circuit, the gate of J1A floats and the leakage current of the back-to-back gate-substrate diodes of a dual monolithic JFET is used to pull the gate of J1A toward the same potential as the gate of dummy JFET J1B. With the thermal noise of the two large resistors gone, we are left with the impedance and small leakage current of the JFET junctions, the latter contributing shot noise current.

Driving the Backplate or Diaphragm with Negative Feedback

With the backplate not grounded, and able to be polarized from a series resistance, the backplate can be driven with a signal. This means that negative feedback can be implemented around the JFET amplifier, either for distortion reduction or equalization. In fact, the Neumann U87 employs some frequency-dependent negative feedback to the backplate to equalize the frequency response [7].

Figure 18.18(a) shows such a design wherein negative feedback is applied to the backplate from the drain of the JFET. In this case, the loop gain is sufficient to cause the input of the preamplifier to act somewhat like a virtual ground, where most of the voltage created by the capsule is canceled out by the negative feedback signal applied to the backplate. Obviously variations of this approach can be taken (i.e., different amounts of loop gain and frequency-dependent feedback to implement some equalization).

In the limit with high loop gain, the signal voltage fed to the backplate will cancel the signal voltage generated by the capsule, resulting in little or no signal at the JFET gate. This means that the input capacitance of the JFET becomes much less consequential. This may allow the use of a larger JFET with lower input voltage noise. Figure 18.18(b) shows a circuit in which shunt feedback is applied to the diaphragm.

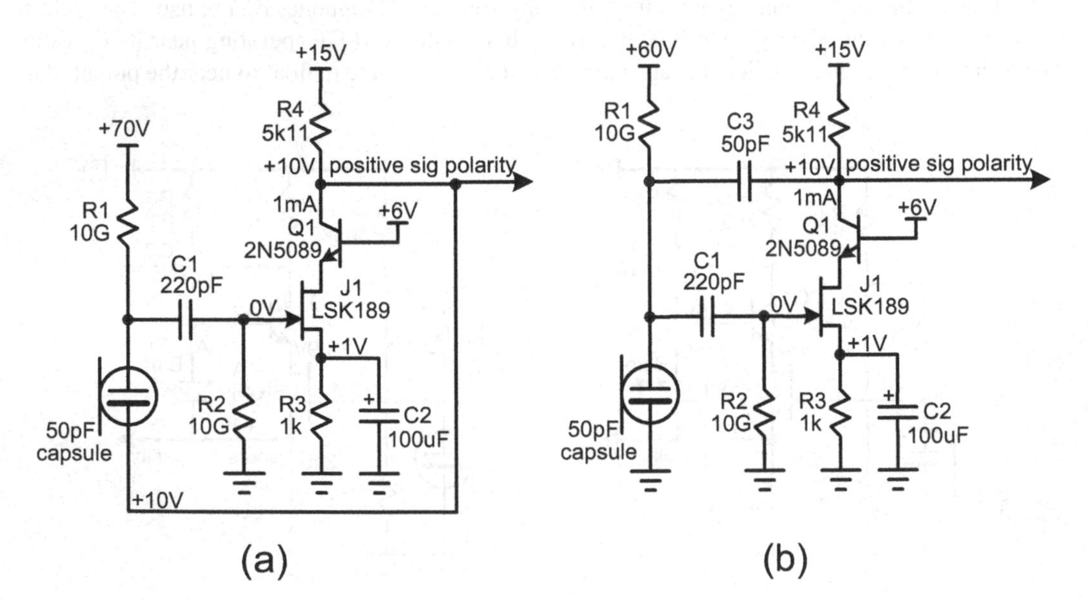

Figure 18.18 Use of Global Negative Feedback.

Polar Pattern Control

The polar pattern of a condenser microphone can be controlled by a switch if the capsule has front and rear diaphragms. The polarization voltage applied to the rear diaphragm controls selection of the pattern [21].

The single-diaphragm condenser microphone has a figure-8 polar pattern due to its bi-directional pickup from front and rear if the backplate is perforated. It will have a cardioid polar pattern if the backplate is solid. Numerous condenser microphones employ capsules that contain dual diaphragms and dual backplates, as if two capsules are put together back-to-back. A non-conductive spacer insulates the two backplates from each other. Such a capsule thus has four wires. Some microphones connect the two backplates together and use a 3-wire arrangement with a common backplate.

The back-to-back construction of the dual-diaphragm, dual-backplate capsule, with its non-conducting insulating disk between the backplates, prevents sound from entering the rear side of either diaphragm. This causes each diaphragm to act with a cardioid polar pattern. If the rear diaphragm is not polarized, the rear diaphragm does not function and the front diaphragm performs with a front-facing cardioid polar pattern.

If the rear diaphragm is polarized in such a way that sound coming from the rear is out of phase with sound coming from the front, then the dual-diaphragm microphone acts like a single-diaphragm microphone with a perforated backplate, where the functionality is bi-directional. This results in the figure-8 polar pattern. If the rear diaphragm is polarized in the opposite fashion, so that sound pressure moves both diaphragms inward together to increase capacitance and create the same polarity of signal, the microphone becomes omni-directional.

This is the basis for a tri-polar (or multi-pattern) condenser microphone, wherein an electrical switch can choose one of the three patterns by applying different polarization voltages to the rear part of the capsule. In the simplest incarnation, the rear backplate is polarized with +60, 0 or −60 V to achieve the three different polar patterns.

In the original Neumann U87, only +47 V was available for polarization. That design used the K87 dual-diaphragm, dual-backplate capsule. The front backplate was always polarized with +47 V and the front diaphragm operated at 0 V. The pattern switch controlled how the rear part of the capsule was polarized in order to select the desired pattern. The switch disabled the positive polarization voltage of the rear backplate and left the rear diaphragm at +47 V for the figure-8 pattern. This caused sound from the rear to be out of phase with sound from the front.

For the omni-directional pattern (pressure-responsive), the switch disabled the positive polarization voltage from the rear diaphragm and left the rear backplate at +47 V. This caused sound from the rear to be in phase with sound from the front.

For cardioid operation, the switch allowed the positive polarization voltage to be applied to both the rear diaphragm and rear backplate, disabling the rear part of the capsule because there was no polarization voltage across the rear capsule. This is why the backplates needed to be insulated from each other and a 4-wire interface was required.

Figure 18.19 shows a simplified diagram of how polarization control works in the Neumann U87A [7]. The Neumann U87A is equipped with a DC-DC converter that produces ±60 V for polarization. The availability of both polarities of polarization voltage simplified pattern selection. Both backplates are connected together and held at ground potential. The front and rear diaphragms are AC-coupled together, but get different DC polarization voltages depending on the pattern selected. The front diaphragm is polarized to +60 V through a 10-GΩ resistor. The rear

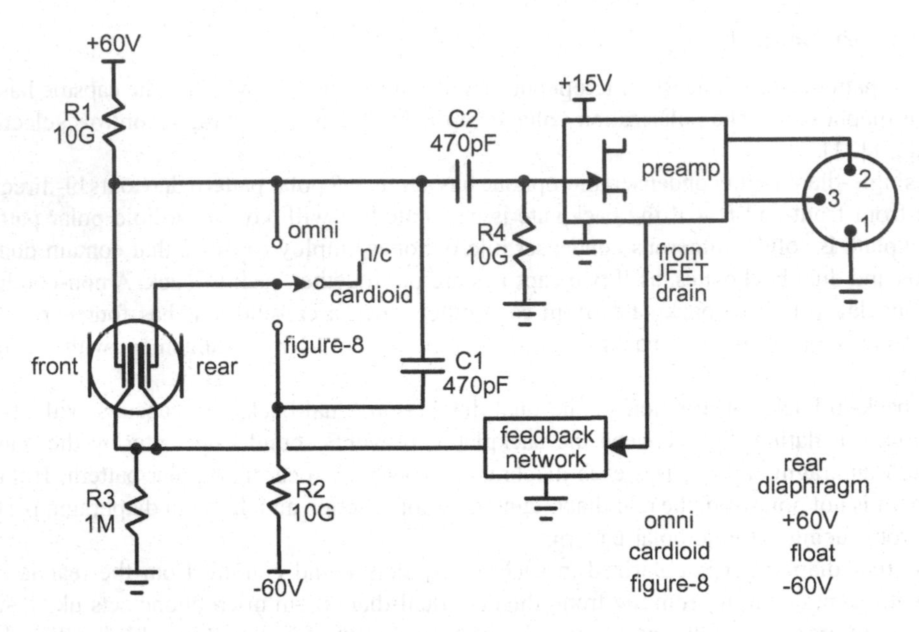

Figure 18.19 Simplified Polarization Control in the Neumann U87A.

diaphragm is also polarized to +60 V by connecting it in parallel with the front diaphragm for the omni-directional pattern. The rear diaphragm is left to float un-polarized for the cardioid pattern. The rear diaphragm is polarized to −60 V through a 10-GΩ resistor for the figure-8 pattern.

Multi-Pattern Feature without Polarizing and Gate Resistors

Here we describe a multi-pattern condenser microphone without large-value polarizing and gate return resistors. A 4-wire K87 dual-diaphragm, dual-backplate capsule is used, as shown in Figure 18.20. The front backplate is polarized positive and the front diaphragm is floated and connected to the JFET gate. It is floated at roughly +3.7 V as determined by the bias on the dummy gate. The rear diaphragm is connected directly to the front diaphragm. The rear backplate is biased in accordance with the selected polar pattern.

For cardioid operation, the rear backplate is held at ground potential. It is non-functional in this state. There is no polarization for the rear diaphragm and there is almost no sensitivity to sound from the rear.

For the figure-8 pattern, the rear backplate is polarized negative. This causes the signal from the rear diaphragm to create the same signal polarity as the rear sound on a single-diaphragm microphone with a perforated backplate. This emulation of the figure-8 polar pattern thus has a null at the sides because the signals from the two diaphragms will cancel in that orientation.

For the omni-directional pattern, the rear backplate is polarized positive (the same as the front backplate) so that pressure increases the capacitance of both the front and rear portions of the capsule and creates in-phase voltages and currents.

As in the original U87, in the cardioid mode, the rear capsule capacitance just acts as a capacitive load on the front diaphragm, since the two diaphragms are connected together.

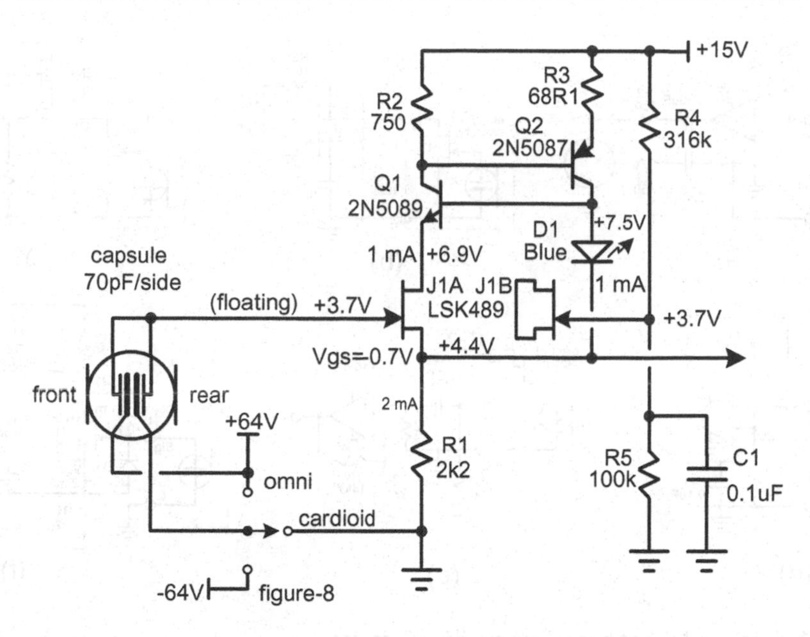

Figure 18.20 Multi-Pattern Microphone without Gate Resistor.

Continuous Electronic Polar Pattern Control

It is easy to see that in Figure 18.20, continuous pattern control can be achieved if the three-position switch is replaced with a potentiometer that can sweep the voltage and polarity applied to the rear backplate from −64 V, through zero, to +64 V. However, this may not always be practical because the potentiometer straddling the positive and negative 60-V supplies may draw more current than those supplies can provide. In many designs, the polarization supply sees very little or no load. In the case here, a 1-MΩ potentiometer would draw 128 μA, which is a lot in this context.

18.6 Electret Microphone Preamplifiers

The electret microphone is much like an externally polarized condenser microphone. It operates on the principle that if a charged capacitor changes its capacitance, a voltage change will result, since charge must be conserved. The previously discussed condenser microphones are sometimes called *true condenser microphones*, while an electret condenser microphone is called an *ECM* (electret condenser microphone).

In an electret microphone, the diaphragm is one plate of a capacitor whose dielectric has a permanent charge. An electret is a dielectric with a permanent static charge. An electret is made by heating a plastic sheet to a high temperature and exposing it to an intense electric field. When the sheet cools down, it retains the electric charge indefinitely [22]. A thin electret film like Teflon is deposited on one plate of the capacitor. As incident sound pressure moves the diaphragm, a signal voltage is created [2]. As with a condenser microphone, the voltage produced by the capsule must be amplified by a JFET amplifier with extremely high input impedance in the microphone. The electret microphone thus requires phantom power or similar powering means.

The electret membrane can be part of the diaphragm or be affixed to the backplate. The *front-electret* microphone employs a thin metalized Teflon foil diaphragm. The Teflon is the charged

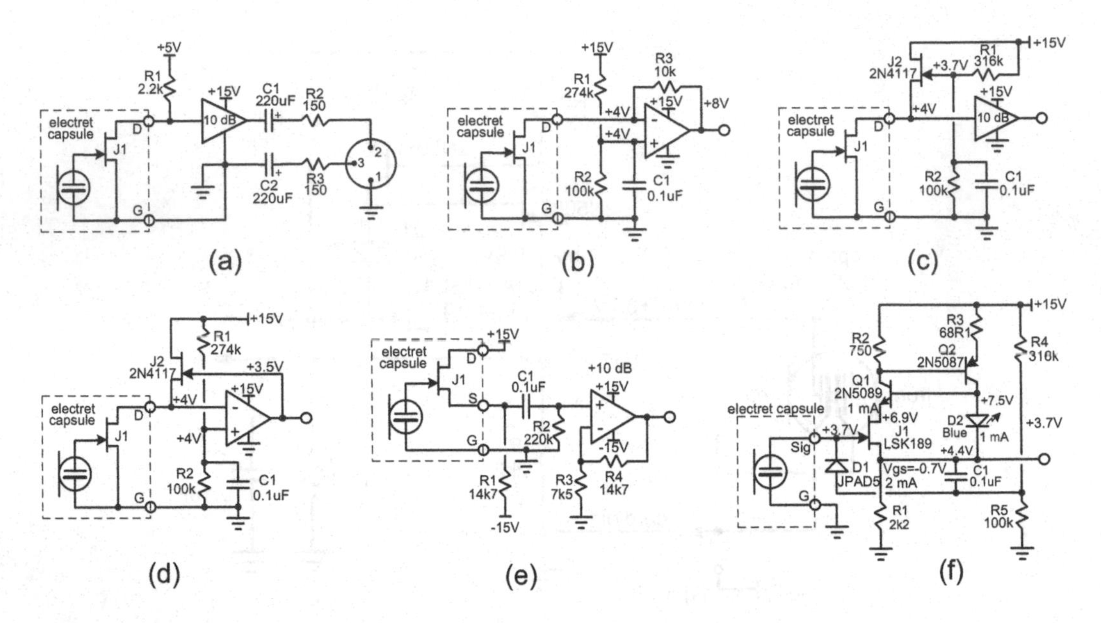

Figure 18.21 Electret Microphone Preamplifier Circuits.

electret element. The *back-electret* microphone employs a foil diaphragm with a fixed electret material on the backplate. The back-electret microphones generally have better performance and are more popular [2, 23]. A thin spacer then separates the membrane from a pickup plate that is connected to the JFET gate. The resulting capacitor naturally contains a charge. As with a condenser microphone, when the diaphragm moves with sound pressure, the capacitance changes and a signal voltage is created.

While no polarizing voltage is required for an electret microphone, power is required for the JFET amplifier associated with the electret transducer. The amplifier, often a single JFET in consumer applications, is located in very close proximity to the diaphragm within the capsule. The ordinary electret microphone usually has a single-ended output and is connected with simple shielded cable. The preamplifier to which it is connected actually includes the load resistor for the capsule's internal JFET, as shown in Figure 18.21(a). Notice that the electret condenser element directly drives the un-degenerated JFET between gate and source. For this reason, most ordinary 2-terminal JFETs distort at high signal levels because of the square-law characteristic of the JFET if the JFET is operating in its saturation region. If the supply voltage is low enough to allow the JFET to operate in its linear (triode) region, as opposed to its saturation region, the square law will not accurately apply and the resulting distortion may be higher or lower.

Notice that there is no gate return resistor for the built-in JFET. At DC its gate is a floating open circuit. Leakage current in the JFET causes the JFET to find a nominal operating point not far from I_{dss}. There is thus no need for a large-value gate return resistor. However, there are some electret capsules that include a built-in very-high-value resistor or reverse-biased diode from gate to ground.

Electret Microphone Types

There are basically three types of electret microphone capsules. The most common is the 2-terminal capsule that includes a JFET as illustrated in Figure 18.21(a–d). Here the electret is connected between ground and the gate of the JFET. The JFET source is also connected to ground. The JFET drain is brought out as the signal line and is externally connected to a load resistor, usually something like 2.2 kΩ to a +5-V supply. These are prolific in consumer applications. Virtually every laptop computer has a jack for this type of microphone. The four different circuits in (a–d) will be discussed in a moment.

The second type in (e) is a 3-terminal electret capsule. It includes the JFET, but separately brings out the source. The three terminals are ground, source and drain. This allows greater circuit flexibility while still retaining the internal JFET. For example, this arrangement allows the JFET to be connected in a source-follower configuration as in (e), or in a phase-splitter configuration as in Figure 18.14. The source-follower arrangement allows control over the bias current of the internal JFET by connecting the source terminal to a negative supply voltage through a source resistor or current source. This arrangement also provides lower distortion.

The third type of electret capsule in (f) has two terminals but does not include an internal JFET. Both terminals of the electret are brought out for connection to an external JFET. This provides the most flexibility for preamplifier design, but brings with it the need to manage stray capacitance carefully. Some electret capsules have large diaphragms and are used just like true condenser studio microphones.

Four different circuits are illustrated for the popular 2-terminal electret microphones of Figure 18.21(a–d). The first in (a) is the very common one discussed earlier. The electret capsule in (b) drives the virtual ground of a transimpedance amplifier, setting the JFET drain at a fixed potential and seeing a low impedance [24]. The design in (c) loads the JFET drain with the source of an external JFET, J2. Loading J1 with a similar JFET with a square-law characteristic results in lower distortion. Although it has an appearance similar to a cascode, the signal is taken from the source of J2 [25]. In (d), the concept is similar with the drain of J1 connected to the source of J2. However, feedback from the op amp to the gate of J2 causes the drain of J1 to see a virtual ground, causing the voltage at the gate of J2 to emulate the internal gate voltage of J1, providing some cancellation of the distortion.

Professional Electret Microphones

Apart from being of higher quality and sometimes having a larger diaphragm with side entry like a true condenser microphone, the professional electret microphone will include a differential output amplifier that drives an XLR connector. For powering the active circuitry, +48-V phantom power is used. Many of the design considerations and approaches will be similar to those preamplifiers for true condenser microphones described earlier, with the exception that high polarization voltages are not needed. Notice that the input circuit in Figure 18.21(f) employs the cascoded JFET CFP circuit discussed earlier for true condenser microphones.

18.7 Ribbon Microphone Preamplifiers

A ribbon microphone operates by incident sound pressure causing a very thin corrugated aluminum ribbon to move within a strong magnetic field, thus creating a very small voltage across the ribbon element [1–3]. The ribbon and magnet structure is sometimes called the ribbon motor. The process is like that of a dynamic microphone, but the implementation is vastly different. The ribbon is much lighter than the coil in a dynamic microphone, and therefore can follow the motion of the sound wave more accurately. The ribbon is often on the order of 1.8 μm thick, making the microphone vulnerable to wind and speech plosives. The extremely small output voltage mandates the use of a step-up transformer.

The traditional passive ribbon microphone does not have a preamplifier, but its low output on the order of −60 dBV/Pa (1 mV @ 94 dB SPL) raises issues for the microphone preamplifier that it feeds. Those issues include the need for very high gain, very low noise and relatively high input impedance. Not all mixing console microphone preamplifiers are up to this challenge. However, there are phantom-powered external booster preamplifiers that mitigate this concern [26, 27]. Such devices, sometimes variously called microphone activators, are usually placed within 10 feet of the ribbon microphone. Active ribbon microphones that have internal preamplifiers have also been developed. They require phantom power.

The Step-Up Transformer

The output voltage of the ribbon is very small and the resistance of the ribbon is very small. The signal level at the ribbon is substantially smaller than that of an MC cartridge. It is generally too small to amplify with an active circuit and still achieve adequate SNR. For this reason, a ribbon microphone employs a step-up transformer to increase the signal voltage, often by a factor of 30–37 [28]. Even with the transformer, ribbon microphones often produce a smaller output than dynamic microphones.

The microphone sensitivity as measured at the ribbon is typically on the order of −86 dBV/Pa, which means that at 94 dB SPL the signal level may be on the order of 50 μV. A step-up transformer with a typical turns ratio on the order of 1:37 is used to boost the signal level by about 31 dB. This results in a microphone sensitivity of about −55 dB with a sensitive motor. With a transformer the impedance transformation factor is the square of the turns ratio, here 1369. With a ribbon resistance of 0.2 Ω, the impedance seen at the transformer secondary will be about 274 Ω. These numbers do not include the winding resistances of the transformer.

The transformer primary resistance must be kept very small for best microphone performance, sometimes as small as 0.05 Ω. Such a low primary resistance is often achieved by employing two or four primary windings in parallel. A larger turns ratio like 100:1 to achieve 8.5 dB better sensitivity will result in output impedance of about 2000 Ω for a 0.2-Ω ribbon, and that is way too high. In some cases an active ribbon microphone uses such a transformer and buffers the output with a JFET source follower or some similar active circuit arrangement. Active ribbon microphones are then powered by the phantom supply.

This all means that the big technical challenge in ribbon microphone preamplifier design is input-referred voltage noise. This is not unlike the challenge faced by MC cartridge preamplifiers, but is worse. While active head amplifiers are practical for MC cartridges, they are largely too noisy for direct use with a ribbon motor.

A quiet condenser microphone may have 14 dB-A self-noise, so this is a reasonable objective for a ribbon microphone. Most ribbon microphones use a step-up transformer to obtain 1–2 mV at 94 dB SPL with a 1:37 transformer [28]. The fundamental self-noise of such a ribbon microphone is thermal noise created by the resistances of the ribbon and the primary and secondary resistances of the transformer. A ribbon microphone with an output impedance of 300 Ω will have thermal noise of about 2.3 nV/√Hz, corresponding to A-weighted noise of 0.26 μV. If the microphone has sensitivity of −60 dBV/Pa, it will produce 1000 μV at 94 dB SPL. The SNR will be 72 dB and the self-noise will be 22 dB-A.

Preamplifiers

Passive ribbon microphones require good mixing console preamplifier performance because of their low sensitivity and sometimes higher-than-usual output impedance. The preamplifier should have high gain, low noise and high input impedance. Preamp gain of 65 dB in an ordinary preamp may not be enough for some ribbons. Some ribbon-specific preamps like those by AEA have more than 80 dB of gain and available high input impedance [29]. If the preamp input impedance is too low, the ribbon resonance may be over-damped, and the low end may not be fully captured. The usual rule of thumb is that preamplifier input impedance should be five to ten times the microphone impedance. For a 300-Ω microphone, the preamp input impedance should be at least 1500 Ω, and 3000 Ω is better. At a resonance of 80 Hz, an ordinary 250-Ω ribbon microphone could have its impedance peak at 1900 Ω [8]. The AEA ribbon microphones are tuned to 16.5 Hz. Their impedance can go as high as 900 Ω at resonance [8]. The input impedance of the AEA ribbon preamp can be set as high as 9 kΩ [8].

One should not enable phantom power in the microphone preamp to which the ribbon microphone will be connected, just to be safe. In principle, the presence of phantom power should not damage a ribbon in the absence of a wiring fault or short, but the microphone should never be plugged into a jack with hot phantom power. A resulting transient could blow or deform the ribbon.

Noise Sources

A sensitive ribbon microphone motor that produces 50 μV at 94 dB SPL with self-noise of 14 dB-A would create an A-weighted noise voltage 80 dB below 50 μV, or 5.0 nV. For reference, the thermal noise of a 0.2-Ω ribbon motor is 0.058 nV/√Hz. In the 13-kHz effective noise bandwidth of A-weighting, there are 114 √Hz, so the A-weighted noise voltage from the motor will be 6.6 nV. A noise level of 5 nV would be 80 dB below 50 μV, corresponding to self-noise of 14 dB-A. The 6.6 nV of self-noise is 2 dB higher, so electrical self-noise from the motor would be about 16 dB-A. After step-up of 1:37 by an ideal transformer, the 0.058 nV/√Hz at the motor becomes 2.1 nV/√Hz and the output voltage at 94 dB SPL becomes 1.85 mV. Any additional noise sources not significantly below this number, such as transformer winding resistances, will degrade the self-noise of the microphone or of the combined microphone and preamplifier further. This illustrates how challenging the noise issue is.

Consider a sensitive ribbon microphone having 14 dB-A self-noise. If the sensitivity of that microphone is −54 dBV/Pa, it will produce 2 mV at 94-dB SPL. This means that the A-weighted self-noise voltage must be 80 dB below 2 mV. That is 0.2 μV, or 200 nV, or −134 dBV. With 114 √Hz in the A-weighting noise bandwidth, this means that the noise density must be 1.75 nV/√Hz. A microphone preamplifier with the same EIN of 1.75 nV/√Hz will degrade the SNR by 3 dB, to the point where the microphone looks like it has 17 dB-A of self-noise with noiseless electronics. This poses a challenge for the microphone preamplifier. A more ordinary microphone with sensitivity of −60 dBV/Pa will make things 6 dB worse for the preamplifier.

Just as with MC cartridges combined with transformers, there are several noise sources in the passive ribbon microphone arrangement. Those include the ribbon resistance, the transformer primary resistance and the transformer secondary resistance. Consider the Lundahl LL2912 1:37 transformer [28]. Its primary resistance is 0.05 Ω and its secondary resistance is 59 Ω. If the ribbon motor resistance is a low 0.1 Ω, the sum of resistances of the ribbon and primary of the transformer is 0.15 Ω. The thermal noise of this resistance will be 0.05 nV/√Hz. When multiplied by a turns ratio of 1:37, that becomes 1.85 nV/√Hz. The secondary resistance of 59 Ω will add 1.0 nV/√Hz on a power basis, for a total of 2.1 nV/√Hz.

With A-weighting, the 2.1-nV/√Hz noise density will result in noise of 239 nV, or −132 dBV. If the microphone sensitivity is −54 dBV/Pa, this self-noise is 78 dB below the 94 dB reference, translating to 22 dB-A of self-noise. It can be difficult to do better than this and easy to do worse. The input noise of the microphone preamplifier has not yet been taken into account. With microphone output noise density of 2.1 nV/√Hz, a preamplifier with input noise of 2.1 nV/√Hz would degrade the effective self-noise of the combination by a full 3 dB. A preamplifier with input noise of only 1.0 nV/√Hz will degrade this by 0.9 dB.

Notice that in this example the transformer multiplies the 0.15-Ω primary circuit impedance by a factor of 1369 to an impedance of 205 Ω. With the addition of the secondary resistance of 59 Ω, the total output impedance becomes 264 Ω.

There is an additional way to look at the noise issue for dynamic and ribbon microphones without much detailed knowledge of the microphone. Assume a microphone with −60 dBV/Pa sensitivity and output impedance of 300 Ω. This can be resistive, whether it is the direct resistance of a dynamic microphone coil or the transformed impedance of a ribbon microphone motor. The thermal noise of the 300-Ω resistive component of the impedance is 2.25 nV/√Hz. In the A-weighted ENBW of 114 √Hz, this corresponds to 257 nV, or −132 dBV. With sensitivity of −60 dBV/Pa, self-noise is 72 dB below the 94-dB reference, or 22 dB. It is quite difficult for a microphone with −60 dBV/Pa sensitivity to have low self-noise.

There appears to be a tradeoff between self-noise from the ribbon resistance and transient response. A thinner ribbon will be able to react more quickly, but it will have greater resistance, causing greater thermal noise. This will largely apply if the ribbon resistance is the major noise source.

Finally, consider what it would take to make an active transformerless ribbon microphone. A ribbon resistance of 0.2 Ω will produce thermal noise of 0.058 nV/√Hz, or 6.6 nV A-weighted. If the ribbon motor sensitivity is 50 μV/Pa, the SNR will be 77 dB and self-noise will be 17 dB-A. If the solid-state preamplifier has input-referred noise of 0.058 nV/√Hz, the self-noise will be increased by 3 dB to 20 dB-A. An ideal transistor with no base or emitter resistance could achieve 0.058 nV/√Hz if it is biased so that *re′* is 0.4 Ω. This corresponds to a bias current of 65 mA. That is quite high, and underscores the difficulty of dispensing with the transformer in a ribbon microphone.

If we allow a current draw of 7 mA from the phantom power, the available voltage will be 24 V and the available power will be 168 mW. If one ideal transistor is operated at 65 mA, the available power would allow it to operate at 2.6 V. This is barely viable, and would require a DC-DC Buck converter to deliver the high current at the low voltage. The Buck converter would obviously have to be very quiet. A perforated emitter power transistor might get one close to small enough base and emitter resistances.

External Amplification

Even with a built-in step-up transformer, the output signal level from a passive ribbon microphone is sometimes inadequate for a typical mixing console microphone preamplifier. The preamp may not have enough gain, may have too much noise, and may have input impedance that is low enough to affect the signal level or the sound from a ribbon microphone.

For these reasons, an external phantom-powered pre-preamplifier can be used, usually in close proximity to the ribbon microphone. These are sometimes called microphone activators. They are used with both passive ribbon and dynamic microphones. An example is a product called the *Cloud Lifter* [26]. It is arguably the most popular and best-performing such device. It presents high impedance to the passive ribbon microphone and has low output impedance. It provides gain of up to 25 dB. Input and output are both balanced XLR format. As a result, a robust signal of good amplitude is delivered to the microphone cable that runs to the microphone preamplifier. These booster preamplifiers must have low input noise if they are not to degrade the self-noise of the microphone. The devices do not pass the phantom power through to the microphone. Many similar microphone activators are available, ranging from very simple to more complex and some with features like selectable input impedance.

The Cloud Lifter employs a clever JFET-based patented circuit [27]. Consider a well-known self-biased single-ended JFET cascode amplifier. The circuit comprises a JFET cascode transistor J2 on top of a similar JFET common-source amplifying transistor J1, the latter usually being degenerated. With the gate of J2 connected to the source of J1, the drain of J1 will be V_{gs} above the source of J1 and $2V_{gs}$ above the gate of J1. The Cloud Lifter circuit extends this concept to a degenerated differential-in, differential-out long-tailed pair, the cascode drains of which are loaded and powered by the phantom-powered hot and cold microphone lines from the main preamplifier that have 6.81-kΩ pull-ups to +48 V. The circuit requires little more than two pairs of JFETs and a few resistors.

In practice, each pair of JFETs should be a dual monolithic pair like the LSK389. Matching the amplifying pair limits offset and aids common-mode rejection. Matching the cascode pair provides equal V_{gs} for the amplifying pair, important when they are operating just out of the triode region. If the J1 and J2 transistors have different operating V_{gs}, it is desirable to use the pair with greater V_{gs} for the cascodes so as to maximize V_{ds} for the amplifying pair.

As with the circuit of Figure 18.15, driving the microphone lines with the high-impedance outputs of BJT collectors or JFET drains improves common-mode rejection compared with driving the lines with emitter followers. However, without a differential shunt resistor to establish a reduced output impedance, the gain of circuits like these will be directly affected by the input impedance of the preamplifier, as may be the frequency response when driving long microphone cables. A minor disadvantage of the circuit is the need for two dual monolithic matched pairs that are not inexpensive.

The limited amount of source degeneration used for the JFET differential pair to obtain the needed gain will result in some distortion, but the input signals are very small. A ribbon microphone with sensitivity of 2 mV/Pa (−54 dB) will produce about 200 mV at an SPL of 134 dB. In this worst-case scenario, distortion from the activator will be that which is created by a 200-mV differential input.

Figure 18.22 shows a different phantom-powered external amplifier. This amplifier provides gain of 20 dB. It has a balanced XLR input for the microphone and a balanced XLR output to feed the microphone cable. Its input impedance is very high at 191 kΩ. A differential load resistor should be placed across the balanced inputs to establish the desired microphone loading. In some designs this resistor may be switched to different values. The downstream microphone preamplifier provides standard phantom power to operate the circuit.

The design uses a differential version of the JFET CFP circuit. The CFP greatly increases the effective *gm* of the JFETs. The high effective transconductance of the JFETs reduces distortion that is due to the use of small-value degeneration resistors R6 and R7. The JFETs are operated at 2 mA and the CFP transistors Q3 and Q4 are operated at about 1.5 mA. With LSK389 JFETs running at 2 mA each in a differential pair, input-referred noise will be about 1.5 nV/√Hz or less. This is important, since this stage is now taking the low-noise amplification responsibility instead of the main microphone preamplifier.

Shunt resistors R10 and R11 establish a reasonably low output impedance of 294 Ω if the preamplifier load is ignored. With a preamplifier load of 1.5 kΩ, the net load impedance on the differential pair is about 250 Ω. In this case, gain will be about 8.5, or about 18.6 dB.

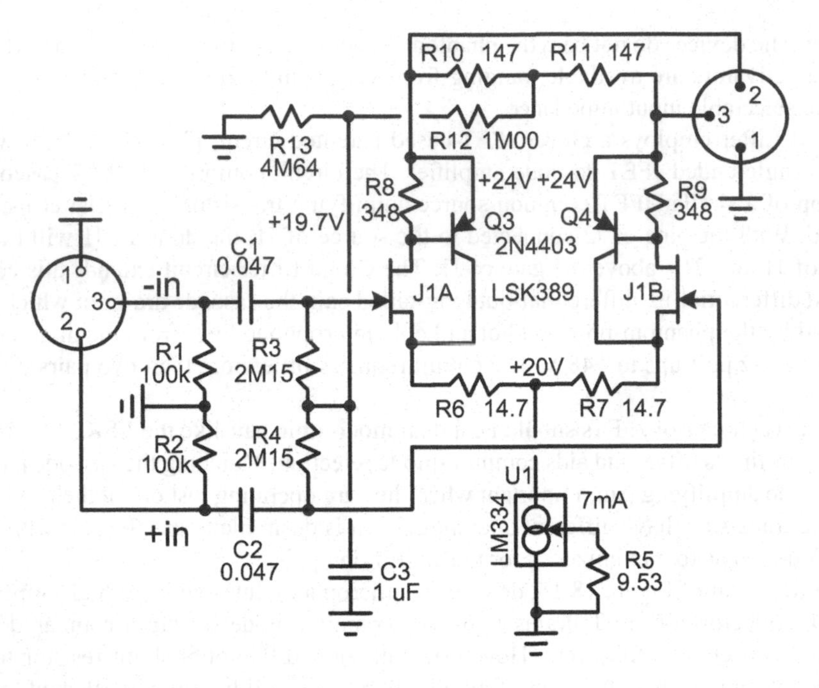

Figure 18.22 External Phantom-Powered Amplifier.

The tail current is set to 7 mA by the LM334 programmable current source, but a discrete current source circuit using a JFET or other circuit can be used. When the amplifier is drawing 7 mA from the phantom supply, the output lines will sit at about +24 V. The circuit requires only one modestly expensive dual monolithic pair. In an SMT implementation the footprint is very small.

Active Ribbon Microphone with Buffer

There are active ribbon microphones that incorporate amplifier circuitry to make the signal more suitable for transmission to a standard microphone preamp. Recall that the transformer multiplies the impedance at its primary to a value at its secondary that is larger by the square of the turns ratio. Using a transformer with a 1:74 turns ratio would increase the signal level by 6 dB compared to 1:37 transformer, but make its output impedance four times as high. This would be way too high for connection to a typical microphone preamplifier. In this case, including an active unity-gain buffer will solve that problem. The Royer Labs *R*-122 MKII is an example of such an approach [30]. The active ribbon microphone uses phantom power to operate the buffer. In some cases the buffer can be as simple as a pair of source followers or emitter followers configured to produce a balanced output.

Figure 18.23 shows an active ribbon microphone with a buffer. Here a Lundahl high-ratio transformer is used. The LL1927 can be configured for ratios of 1:55 or 1:110, with primary resistance of 0.1 and 0.025 Ω, respectively [31]. In both cases the secondary DC resistance is 364 Ω. An advantage of this approach is that the noise of the unity-gain differential buffer need not be as low because of the larger signal produced by the transformer. The circuit uses a Schoeps-like unity-gain differential emitter follower, like that described in Figure 18.14(b).

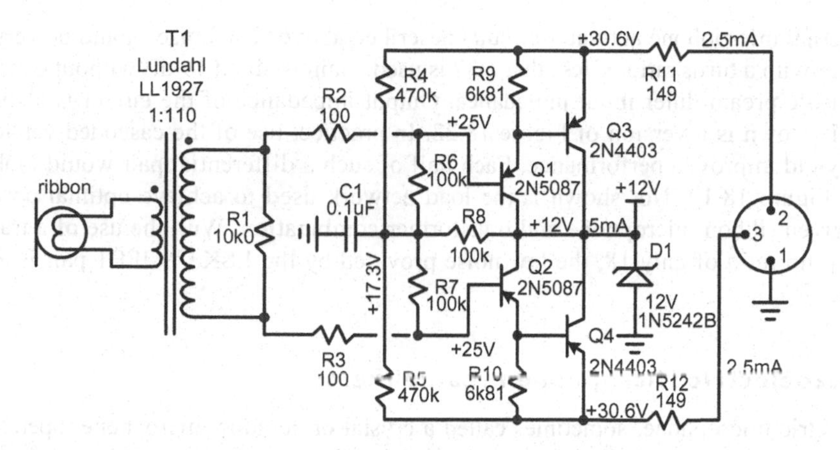

Figure 18.23 Active Ribbon Microphone with Unity-Gain Buffer.

Active Ribbon Microphone with Amplifier

There is an alternative implementation of an active ribbon microphone. In this case a conventional 1:37 transformer can be used and can be followed by an active amplifier with voltage gain of 6 dB or more and a balanced output impedance on the order of 300 Ω, somewhat like that of a dynamic microphone. Indeed, a transformer with a smaller turns ratio could be used if the amplifier has low enough noise and provides more gain. Sometimes a transformer with a smaller turns ratio can perform better.

Figure 18.24 shows an active ribbon microphone using an active circuit with 12 dB of gain and a 1:18 Cinemag CM-2670 transformer [32]. The amplifier circuit can be virtually the same

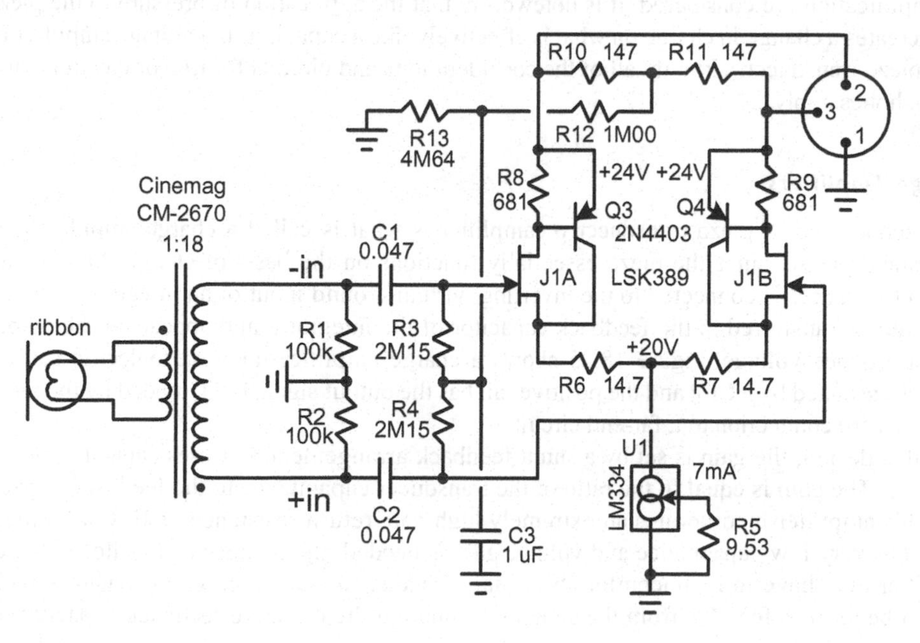

Figure 18.24 Active Ribbon Microphone with Amplifier.

as the external microphone activator circuits described above. The noise should be very low if a transformer with a turns ratio of less that 1:37 is used. Gain is about 19 dB without consideration of the console preamplifier input impedance. Output impedance of the circuit is about 300 Ω. The circuit shown is a version of Figure 18.22. In practice, use of the cascoded version would probably yield improved performance. Each half of such a differential pair would look like the design in Figure 18.12. Not shown is the load network used to achieve optimal performance with the given ribbon microphone and transformer combination. With the use of a transformer with a step-up ratio of only 18, the low noise provided by the LSK389 JFET pair is especially important.

18.8 Piezoelectric Microphone Preamplifiers

A piezoelectric microphone, sometimes called a crystal or ceramic microphone, operates on the principle of piezoelectricity, wherein some materials, like crystals or ceramics, create a voltage when subjected to sound pressure. They are often used as so-called "contact microphones" where they are in contact with a vibrating surface, sometimes in an electric guitar. They are also the basis for other types of transducers. In sensor applications they are robust and can withstand high pressure. The output impedance is very high, so they also may be associated in close proximity with a JFET amplifier with high input impedance.

Voltage Amplifiers

A piezoelectric microphone creates a voltage when pressure is applied to its crystal, in this case sound pressure. The piezo transducer is modeled as a capacitance, and its output impedance is very high, that of its capacitance. This makes it much like a condenser microphone when the needs for preamplification are considered. It is noteworthy that the application of pressure to the piezo element creates a change in charge on what is effectively like a capacitor. If a voltage amplifier is used for a piezo transducer, virtually all of the considerations and circuits for true or electret condenser microphones apply.

Charge Amplifiers

The second type of piezo transducer preamplifier is what is called a charge amplifier [33, 34]. This makes sense, since the piezo essentially functions on the basis of charge. In this case, the piezeo transducer is connected to the inverting, virtual ground input of an integrator. Any change in charge is transferred to the feedback capacitor of the integrator and appears at the output as a low-impedance voltage. Figure 18.25 shows a charge amplifier for a piezoelectric sensor. The JFET is cascoded by a CFP and the positive rail has the output signal on it, loaded by the resistance of the 2-wire connection to a far-end circuit.

In this design, the gain is set by a shunt feedback arrangement that uses capacitors instead of resistors. The gain is equal to the ratio of the transducer capacitance to the feedback capacitance C1. This amplifier incorporates an extremely high gate return resistance of 10 GΩ. If the transducer has very low capacitance and voltage gain is needed, the feedback capacitor in the charge amplifier may have to be uncomfortably small in value. In such a case, a capacitance attenuator can be used to feed C1 from the signal rail, making the effective feedback capacitance look smaller.

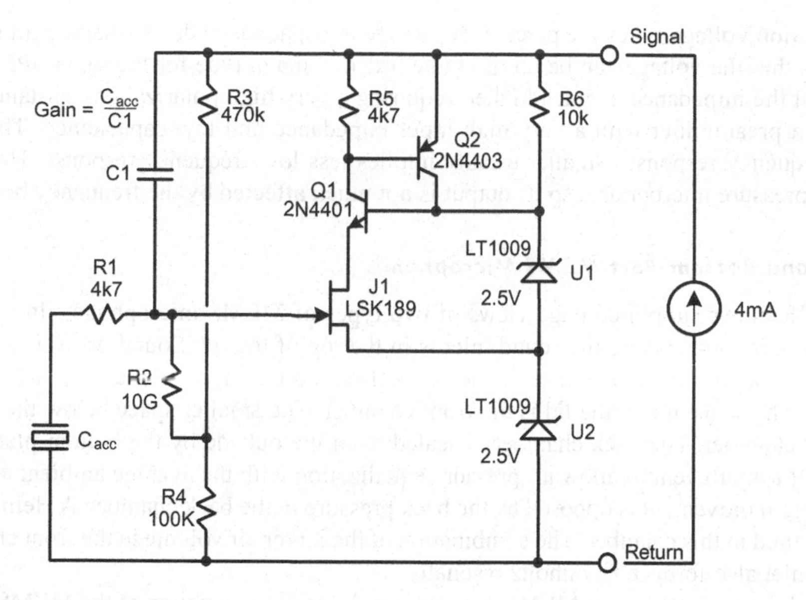

Figure 18.25 Piezoelectric Sensor Charge Amplifier.

18.9 MEMS Microphones

No contemporary discussion of microphones and microphone preamplifiers would be complete without touching on MEMS condenser microphones (MCMs) [35–41]. The term MEMS stands for Micro-Electro-Mechanical System. MEMS microphones derive from the semiconductor industry and have rapidly grown in use in recent years as a competitor to electret microphones. They are very small and can be attached to circuit boards with surface-mount technology. They are ubiquitous and have replaced electret microphones in many applications, particularly in products like cell phones. Billions of MCMs are made every year. They are used extensively in cell phones and the like. MEMS microphones are extremely small and that is perhaps their biggest attraction. Most are still not of sufficient quality to play a large-role in high-performance audio, but they are a very important technology.

Most MEMS microphones are condenser microphones that rely on the constant charge of a varying capacitance to create a signal voltage. These are essentially like the "true" condenser microphones discussed earlier, and like them, they need a polarizing voltage to make them work. The MEMS microphone is a product of silicon integrated circuit technology where silicon is etched to provide a diaphragm and a backplate. The diaphragm is extremely close to the backplate (about 1 μm) and does not move very much. However, it is close enough to create a capacitance on the order of 1 pF. The 1-pF capacitance is polarized to typically 5–20 V. The MEMS microphone can be very small, at a few millimeters on a side, and with a low profile. Their small size is perhaps their biggest attraction. They are also remarkably inexpensive. MCMs are omni-directional.

It may be non-intuitive to some that a tiny condenser microphone with a diaphragm typically less than 2 mm in diameter or on edge can have decent sensitivity or SNR, much less low-frequency response. The concept of scaling explains this. While the area of the diaphragm has been reduced by a factor of perhaps 100, the air gap, on the order of 1 μm, has decreased by a very large factor. This means that the capacitance is still about 1 pF. Keep in mind that the signal voltage is essentially the polarization voltage times the percentage change in capacitance due to diaphragm movement. This means that the voltage can be on the same order of magnitude for the same SPL. The challenge is that the impedance is much higher, requiring a very high polarization resistance, perhaps a TΩ, and a preamplifier with a very high input impedance and low capacitance. Then there is the low-frequency response. Smaller usually implies less low-frequency response. However, the MCM is a pressure microphone, so its output is not much affected by the frequency being low.

Top-Port and Bottom-Port MEMS Microphones

Figure 18.26 shows simplified edge views of two types of MEMS microphones. In (a) is shown a top-port microphone where the sound inlet is in the top of the lid. Sound pressure enters at the top, passes through the perforated stationary backplate and impinges on the moveable diaphragm membrane. The space under the lid is the front chamber. The smaller space below the diaphragm is the back chamber. The back chamber is sealed from the outside by the bottom plate, with the exception of a small vent to allow air pressure equalization with the average ambient air pressure. The diaphragm movement is opposed by the back pressure in the back chamber. A Helmholtz resonance is formed in this chamber. The combination of the larger air volume in the front chamber and the sound inlet also forms a Helmholtz resonator.

In (b) is shown a bottom-port MEMS microphone. Here the orientation of the MEMS chip is the same, but the sound pressure enters through a hole in the substrate and/or the PWB on the bottom. The sound pressure impinges directly on the diaphragm from below in the front chamber. Above the diaphragm again is the stationary perforated backplate. The sound passes from the sound inlet through the smaller front chamber to reach the diaphragm. The front chamber forms a Helmholtz resonator, but at a higher frequency than that of the larger front chamber of the top-port device. The enclosed back chamber of the bottom-port microphone occupies virtually all of the space under the lid, and its larger volume allows for more air compliance in opposing motion of the diaphragm.

Most top-port devices are inferior in performance to bottom-port devices for two reasons. The much smaller air volume and reduced air compliance in the back chamber of the top-port devices more strongly resists motion of the diaphragm, reducing microphone sensitivity. At the same time, the smaller volume of the front chamber in connection with the sound inlet creates a Helmholtz resonance at a higher frequency, disturbing frequency response.

STMicroelectronics makes a top-port device (MP34DT01) that mitigates these two issues by mounting the MEMS chip and the application-specific integrated circuit (ASIC) preamplifier upside down on the inner side of the lid, providing a back chamber of larger volume as in the bottom-port devices [42]. The device is 3 × 4 × 1 mm in dimensions.

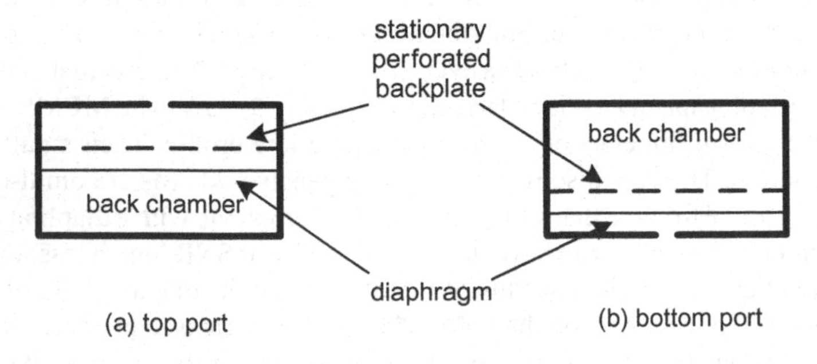

Figure 18.26 Top-Port and Bottom-Port MEMS Microphones.

Silicon Fabrication

The MEMS microphone device is a three-dimensional silicon chip structure that is fabricated on a silicon wafer using advanced directional etching processes. There are two etching processes that are of particular importance in MEMS fabrication. The anisotropic etch is a highly directional etch that takes advantage of the different etch rates for different silicon crystal orientations. It is a key process for micromachining to form microstructures. Etching occurs along crystal planes. It is a subtractive microfabrication technique involving preferential removal of material.

These wet etch techniques exploit the crystalline properties of a material to etch in directions governed by the crystallographic orientation. As an example, a wet etch can form vertical sidewalls. The selectivity of the process is the ratio of the etch rate of the target material to the etch rate of the material not to be removed. Anisotropic etching is often used to remove silicon of the (100) orientation. Etching of silicon of the (110) orientation is fairly new, and it is capable of etching ratios of greater than 600:1.

Another micromaching technology is the reactive ion etch (RIE), a highly anisotropic dry etching technique. It employs a plasma that is chemically reactive. Plasma ions with high energy attack the surface of the wafer and react with the material to be removed. An RF electric field at about 13.6 MHz ionizes the gas to create the plazma. Different gasses used to form the plazma behave differently. Sulphur hexaflouride is often used as the gas for etching silicon. An example would be the etching of a silicon dioxide layer on a silicon substrate that has been defined by a protective photoresist. RIE produces very vertical walls in the silicon dioxide because of the vertical way in which the reactive ions are delivered to and strike the surface to be etched. A wet etch will produce walls that are somewhat angled, resulting in some undercut.

Diaphragm Size and Spacing

The ratio of the diaphragm diameter to the spacing controls the capacitance. As an example, a diaphragm with a 600-μm dimension might have a 2-μm thickness and be supported by 15 polysilicon springs to achieve a 2.25-μm spacing [43].

Self-Noise, Sensitivity and SNR

Given the earlier discussions of condenser microphone capsule capacitance and GΩ gate return resistor thermal noise, it is easy to see that noise for such a small-capacitance diaphragm is a great challenge. Similarly, stray capacitance poses a big issue.

The SNR of currently available MEMS microphones generally lies between 55 and 65 dB, with some outliers on the high side. Diaphragm noise is created by the random Brownian motion of the air molecules in the enclosed chamber behind the diaphragm. Noise in the ASIC preamplifier has multiple sources, largely the same as those in a conventional condenser microphone, one of which is the kT/C noise from the diaphragm polarizing bias resistor, which for a MEMS microphone must be at least 100 GΩ or several TΩ as a result of the very small diaphragm capacitance on the order of 1 pF.

The MEMS microphone air gap is 1–10 μm and the polarizing voltage is about 6 V. Sensitivity is about 10-mV/Pa open circuit. Self-noise is about 32 dB-A. This is not as good as studio microphones, but is more than adequate for many applications. The acoustic overload point (AOP) for 10% THD is on the order of 120–130 dB SPL.

Frequency Response and Distortion

The frequency response of a MEMS microphone is affected by Helmholtz resonances and other acoustic or electrical properties. Although in theory the response of an MCM can extend to quite low frequencies, in practice the response of most MCMs falls off below about 100 Hz. High-frequency response can be strongly affected by Helmholtz resonances, and this may result in frequency response peaks in the 2–10 kHz range followed by substantial roll-off. Again, there are exceptions to the rule, and some MCMs can be relatively flat to 100 kHz.

Assuming that the MCM is being operated sufficiently below its acoustic overload point (AOP), distortion in the MCM chip itself (not the preamplifier) can result from stray capacitance shunting the diaphragm capacitance. One source of capacitance is the gate capacitance of the front-end MOSFET in the preamplifier. Stray shunt capacitance does not change much with signal, but the change of capacitance of the diaphragm with signal is fundamental to the MCM's operation. This effect was also described in the section on conventional condenser microphones. The stray capacitance forms a voltage divider with the source capacitance of the diaphragm. Because the source capacitance is changing with signal, the amount of attenuation of this voltage divider changes with signal, resulting in distortion. Put another way, parasitic capacitance makes the relationship of ΔV to Δd nonlinear.

Microphone Parallelism

MEMS microphones are small and inexpensive, so using a good number of them is economic and compact in area. This makes them ideal for operating numerous MCMs in parallel to increase SNR. This approach takes advantage of the fact that noise in the microphones is random and uncorrelated among multiple microphones, while the desired external sound is correlated among multiple microphones that are very close together. If the outputs of two MCMs are summed, the desired signal will increase by 6 dB, while the uncorrelated noise of the microphones will only increase by 3 dB. The result is a 3-dB increase in SNR. In fact, for every doubling of the number of microphones in parallel, a 3-dB improvement in SNR will be realized. Four microphones will fetch a 6-dB improvement, while a 4 × 4 array of 16 microphones will yield a 12-dB improvement.

With their very small size, a massively parallel array of microphones can be employed to achieve a very high SNR. The challenge is to sum all of those outputs in an efficient way that does not contribute a noise penalty itself. This can be achieved in the analog domain or the digital domain. In the analog domain, this can be achieved by routing the outputs of multiple MCMs to the virtual ground of an op amp, as can be done in a mixing console. For larger arrays, multiple-tier mixing can be done. Of course, the details of summing using resistors must minimize noise contributed by those resistors. In the digital domain, the outputs of addressable digital MCMs can be scanned and multiplexed onto a bus. Those outputs are then summed by DSP.

Polarization Voltage

The MEMS condenser microphone requires the application of a polarizing voltage, just like a conventional condenser microphone. The current requirement is essentially zero, and a charge pump integrated with the amplifier in the associated ASIC is usually employed to provide the polarization voltage. Typical polarization voltages range from 5 to 20 V. The polarization voltage can be applied to the backplate.

A Dickson voltage multiplier is usually employed in the CMOS ASIC to create the polarization voltage [44]. This multiplier usually consists of a string of diodes with capacitors driving the connection nodes with two square waves that are 180° out of phase on an every-other node basis. Odd-numbered capacitors are driven with clock, while even-numbered capacitors are driven with an inverted clock. In the MCM ASIC, the diodes are replaced by MOSFETs with various means of gate biasing. This arrangement is attractive because the MOSFETs do not have to withstand the higher voltages created. Only the capacitors are required to do so [43].

The Amplifier

The amplifier in the ASIC is implemented in CMOS technology, which has virtually no leakage current. The integrated preamp ASIC typically has an output impedance of a few hundred ohms. The polarizing and/or gate resistance must be extremely high −100 GΩ to several TΩ. A reverse-biased silicon diode is often used to apply bias.

Packaging

MEMS microphones are compatible with surface-mount technology. This is an advantage over electret microphones that can be damaged by exposure to the high re-flow temperatures used in surface-mount technology. MEMS microphones can withstand vibration, heat, re-flow, pick-place, etc. The MEMS microphone device must be in extremely close proximity to the ASIC. The analog MEMS microphone module has three leads from the unit: power, ground and signal.

Digital MEMS Microphones

Both analog and digital MEMS condenser microphones (MCM) are available, with a CMOS ASIC inside. The use of a CMOS ASIC provides a great opportunity for digital signal processing (DSP). The ASIC can include thousands of transistors and provide significant digital processing locally.

Digital MCMs usually employ a sigma delta A/D converter and put out a pulse density modulated signal. Alternatively, the ASIC can include an internal decimation filter that can enable it to output a PCM I^2S signal. The microphone can include a select pin to choose or multiplex microphones. They can be made addressable for use in microphone arrays.

Beam-Forming Microphone Arrays

MEMS microphones are small and inexpensive, so using a large number of them in a one- or two-dimensional array is practical, economic and compact in area. This makes them ideal for the use of beam-forming techniques to achieve controlled directionality. Arrays of 2–100 or more microphones are practical. While the outputs of a parallel array of MCMs are merely summed to achieve increased SNR, in a beam-forming array the different outputs are treated differently and then summed. The use of different delays and other DSP techniques before summing creates and controls the directivity of the array. This is not unlike phased array radar, where many radar element pixels are scanned and processed to scan the area of view in both dimensions.

Because correlation comes into play, improvements in SNR like those in parallel arrays are also achieved. Sensitivity matching among the numerous MCMs in an array is important to its proper functioning in forming the receive beam and its direction. Sensitivity matching of ±1 dB is desirable. This is achievable due to uniformity of the semiconductor manufacturing process [45].

One-dimensional microphone arrays can be either of the broadside array or the endfire array orientation. The former are perpendicular to the source of the desired sound. Sounds from non-perpendicular sources will arrive at the microphones with different delays, allowing DSP/delay processing to form a beam of directionality. The endfire microphone arrays consist of a line of microphones parallel to the direction of the desired sound.

They take advantage of differing arrival delays combined with DSP/delay to line up the signals in time from the different microphones. They capture sound only from the front entrance with sensitivity along the desired axis. They achieve a narrow pickup angle, like a shotgun microphone.

Applications

Applications include smartphones, speech recognition devices, laptop computers and hearing aids. Additional applications include:

- earbuds
- directional determination
- gunshot detection
- home digital assistants and voice-enabled devices
- noise cancellation

An estimated 10 billion MCMs were used in 2021 and the numbers appear to nearly double every four years.

MEMS vs Electret Microphones

MEMS are smaller, lower in cost and have better matching of sensitivity. Using a local ASIC, they achieve relatively low output impedance. Digital output is available and they are easily multiplexed. They are completely compatible with SMT manufacturing, being unaffected by SMT soldering temperatures. In comparison, electret microphones have potentially higher performance, a wide voltage range and can also be made directional.

Piezoelectric MEMS Microphones

The piezoelectric MEMS microphone (MPM) is a form of piezoelectric microphone that is built with IC/MEMS technology. A piezoelectric material is attached to the diaphragm. Common piezoelectric materials used include Zinc Oxide (ZnO) and Aluminum Nitride (AIN). Movement of the diaphragm creates a voltage via the piezoelectric effect. The diaphragm comprises a piezoelectric film surrounded by metal on both sides. Piezoelectric MEMS microphones do not require a polarization bias, have a wider dynamic range and are dustproof and waterproof. For these reasons, they are enjoying increased popularity for some applications. They feature low power consumption, but have a fairly high noise level and low sensitivities.

A typical MPM may have a sensitivity of −38 dBV/Pa and self-noise of 62 dB-A and an AOP of 125 dB SPL. The Vesper Technology VM2000 low-noise bottom-port MPM achieves an AOP of 135 dB SPL [46].

References

1. Glen Ballou, *Handbook for Sound Engineers*, 5th edition, Focal Press, Burlington, MA, Chapter 20, Microphones, 2015. Also, Chapter 25, Preamplifiers and Mixers.

2. Francis Rumsey and Tim McCormick, *Sound and Recording Applications and Theory*, 7th edition, Focal Press, New York, NY, 2014.
3. Ethan Winer, *The Audio Expert*, 2nd edition, Focal Press, New York, NY, 2018.
4. Shure Incorporated, *Can a Dynamic Microphone Handle Really Loud Sounds?* Shure, Inc., April 6, 2021, service.shure.com/s/article/can-a-dynamic-microphone-handle-really-loud-sounds-maximum-spl?language=en_US
5. Harvey Fletcher and Wilden Munson, Loudness, Its Definition, Measurement and Calculation, *Journal of the Acoustic Society of America*, vol. 5, 1933.
6. Phantom Power, IEC 61938. IEC 61938:2018 – *Guide to the Recommended Characteristics of Analog Interfaces to Achieve Interoperability*, Edition 3.0, 2018.
7. Neumann U87A schematic, https://www.scribd.com/document/U87ai-Schenatic.
8. *What Is the Big Ribbon Sound?*, AEA Ribbon Mics, aearibbonmics.com/tricks-of-the-trade/what-is-the-big-ribbon-mic-sound/.
9. Acoustic Measurement Sensors & Instrumentation, PCM Piezotronics Inc., PCB.com/Acoustics.
10. Jensen Twin Servo 990 Mic Preamp schematic, jensen-transformers.com/wp-content/uploads/2014/08/as083.pdf.
11. Jim Brown, *Pin 1 Revisited,* RaneNote165, ranecommercial.com.
12. Website, pin1problem.com.
13. Hypex Electronics, *Dealing with Legacy Pin 1 Problems*, Application Note, troelesgraveesen.dk/DUSION-BI-AMP/AN_Legacy_pin_1_problems.pdf.
14. Bill Whitlock, *An Overview of Audio System Grounding & Shielding*, Audio Engineering Society Tutorial T-2, bennettprescott.com/downloads/grounding_tutorial.pdf.
15. Linear Integrated Systems LSK170, LSK189, LSK489, LSK389 JFET Datasheets.
16. Bob Cordell, *Dynamic MIC Preamp*, LSK489 Application Note, p. 11, Linear Integrated Systems.
17. Bob Cordell, *Dynamic Microphone Preamp*, Cordell Audio, cordellaudio.com.
18. THAT 1512 *Low-Noise, High Performance Audio Preamplifier IC*, datasheet, THAT Corporation, thatcorp.com.
19. *Configuring Gain with the THAT1510 & THAT1512*, Design Note 138, THAT Corporation, thatcorp.com.
20. Wikipedia, *Voltage Multiplier*, wikiperia.org.
21. Multi-Pattern Condenser Microphone, recordinghacks.com/pdf/miktek/C7_om_ENG_v1_2.pdf.
22. Christian Hugonnet and Pierre Walder, translated by Patrick R.W. Roe, *Stereophonic Sound Recording – Theory and Practice*, John Wiley & Sons, 1998, pp. 118–122.
23. My New Microphone, *The Complete Guide to Electret Condenser Microphones*, https://mynewmicrophone.com/the-complete-guide-to-electret-condenser-microphones/.
24. Texas Instruments, *LMV1031-20 Amplifier for Internal 3-Wire Analog Microphones and External Preamplifier*, LMV1031 datasheet, ti.com.
25. John Conover, Using the WM61A as a Measurement Microphone, johncon.com/john/wm61a/.
26. Cloudlifter CL-1, Cloud Microphones, cloudmicrophones.com.
27. US Patent 9,668,045, *Integrated Phantom-Powered Circuit Module in Portable Electronic Device for Creating Hi-Fidelity Sound Characteristics*, May 30, 2017, Rodger Cloud.
28. Lundahl LL2912 Ribbon Microphone transformer, lundahltransformers.com/ribbon-mic-transformer-II2912-added-webshop.
29. *AEA TRP2 Preamp Overview*. aearibbonmics.com/products/trp2-2/mp.
30. Royer Labs R-122 MKII Active Ribbon Microphone, royerlabs.com.
31. Lundahl LL1927 Ribbon Microphone Transformer.
32. Cinemag CM2670 Ribbon Microphone Transformer.
33. Bob Cordell, *Piezo Accelerometer Charge Amplifier*, LSK489 Application Note, pg. 13, Linear Integrated Systems, linearsystems.com.
34. Ron Quan, *Higher-Performance Militray Sensor Signal Chains*, Linear Integrated Systems application note, pp. 7–11, linearsystems.com.
35. John Widder and Alessandro Morcelli, *Basic Principles of MEMS Microphones*, EDN, May 2014.

36. Joey Mulqueen, *Electret Condenser (ECM) vs MEMS Microphone*, DK Employee, August 2017.
37. Lanna Cooper, *Design Considerations When Choosing a MEMS or ECM Microphone*, CUI Devices, January 2019, cuidevices.com.
38. Bruce Rose, *Analog or Digital: How to Choose the Right MEMS Microphone Interface*, CUI Devices, November 2020, cuidevices.com.
39. InvenSense Inc., *Using a MEMS Microphone in a 2-Wire Microphone Circuit*, AN-1181, December 2013, invensence.com.
40. Jerad Lewis, *Analog and Digital MEMS Microphone Design Considerations*, Analog Devices Inc, MS-2472, 2013, analog.com.
41. Sparkfun, *MEMS Microphone Hookup Guide*, sparkfun.com.
42. STMicroelectronics, *MEMS Audio Sensor Omnidirectional Digital Microphone*, Datasheet, February, 2015, st.com.
43. Axel Thomsen, *How Does a MEMS Microphone Work?*, cirrus Logic video, May 2019, youtube.com/watch?v=JRtLeTJdOYA.
44. Andrea Ballow et al., A Review of Charge Pump Topologies for the Power Management of IoT Nodes, *Electronics Journal*, 2019, mdpi.com/journal/electronics.
45. InvenSense, *Microphone Array Beamforming*, Application Note AN-1140, invensense.tdk.com/wp-content/uploads/2015/02/Microphone-Array-Beamforming.pdf.
46. Vesper Technologies, *VM2000 Low-Noise Bottom Port Piezoelectric MEMS Microphone*, Datasheet, 2017.

Chapter 19

Balanced Inputs and Outputs

Balanced is beautiful – that's what a mentor of mine told me many years ago. In this chapter balanced inputs and outputs for preamplifiers and other line-level audio equipment will be discussed. Preamplifiers and mixers that have single-ended internal circuitry and balanced inputs will first be considered. These are the most common variant of balanced operation in consumer and professional audio. Many of these devices can also accommodate a single-ended input as well. Most professional equipment, and a number of hi-end audio devices, incorporate balanced inputs and outputs using XLR connectors with pin 2 hot. This provides for improved rejection of external common-mode noise [1–5].

Implementing balanced inputs for equipment with a single-ended (SE) internal signal chain is not always straightforward, especially if the highest sound quality is to be maintained and if the device must allow the user to operate it with either XLR balanced inputs or RCA single-ended inputs or single-ended ¼″ jack inputs (with or without the need for a mode switch). The purpose of balanced inputs is to achieve a good common-mode rejection ratio (CMRR), independent of the nature of the source, as long as the output impedance of the source is the same to ground for both signal polarities. The impedance to ground for both signal polarities at the destination receiver must also be the same for best CMRR. For a balanced receiver, CMRR is the gain from a differential-mode input divided by gain from a common-mode input. Often, differential-mode signals are referred to as normal-mode signals. Transformers have been used to implement balanced inputs, but solid-state circuitry is preferred for purposes of space, performance and cost if the desired amount of isolation and common-mode rejection can be achieved. Circuits that must have bullet-proof galvanic isolation still use transformers.

Balanced outputs are also needed if balanced interconnect is to be driven. Signals from single-ended internal circuitry are often converted to balanced outputs merely by passing the signal through an inverter to create the opposite phase of the signal. In this case, both signals have relatively low output impedance to ground, often in the range of 100 Ω. This is in contrast to what is achieved with a transformer, in which the balanced output from the secondary is floating. A floating balanced signal can be more immune to imperfections that degrade the CMRR of a balanced link between two pieces of equipment. As will be discussed, there are ways to electronically simulate this behavior of an output transformer, providing a floating balanced output without resort to a transformer. Such sources have finite output impedance to ground, called the common-mode output impedance.

19.1 XLR Connectors

The XLR connector is the standard connector for balanced interconnects in audio applications. Figure 19.1 shows the pin connections for an XLR connector. In most cases pin 2 is *hot*, meaning

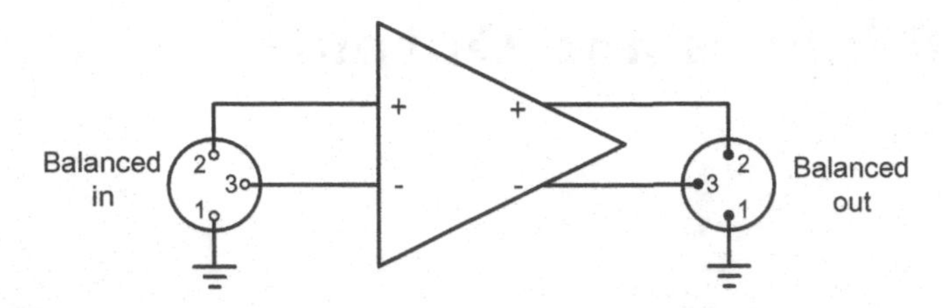

Figure 19.1 XLR Connector Pin Assignments.

positive signal polarity (e.g., the non-inverted signal polarity). Pin 3 is often referred to as *cold*, meaning negative signal polarity (e.g., inverted signal polarity). In this book, female XLR inputs are drawn with open circles for the pins. Male XLR connectors are drawn with filled circles for the pins.

Sometimes 1/4-inch tip-ring-sleeve (TRS) jacks are used instead of XLR connectors because they take up less space. The tip carries the hot signal and the ring carries the cold signal while the sleeve carries the ground (shield).

Common-Mode Rejection

The whole concept of balanced audio interconnect revolves around the rejection of common-mode (CM) signals or noise. A single-ended signal and its ground reference shield will be subject to picking up common-mode signals that will intrude on both the shield and signal conductor. With a differential-mode (DM) signal, the hot and cold signal-carrying conductors will be affected in the same way. At the receive end, the balanced receiving circuit amplifies the differential signal and ignores, to the extent possible, any common-mode signal that is identical on the hot and cold signal conductors.

Balanced interconnections consist of a twisted pair of wires comprising the hot path and the cold path from the source to the receiver. Usually in audio applications, there is a shield which constitutes a third path. In some cases, the ground path may be open, with the shield not connected at one end or the other.

The common-mode rejection ratio (CMRR) for a balanced receiver is the DM gain from a differential input divided by CM gain from a common-mode input. Precision resistors are required to achieve good CMRR in a receiver. In rough terms, 1% resistor tolerances might yield CMRR of only 40 dB. At the interconnect level, CMRR is the ratio of differential gain in the path to the common-mode gain in the path, recognizing that the path itself can be the origin of common-mode signals. If the source is not floating (like a transformer secondary) precise symmetry and impedance matching at both ends of the interconnect path are usually required to achieve high CMRR.

Why Common-Mode Rejection Is Important

Signal voltages and currents in the common mode are very undesirable because they usually include hum and noise, degrading SNR. The CMRR of a balanced receiving circuit is defined as the ratio of differential-mode gain to common-mode gain. Ideally, a differential input circuit should not be affected by, or pass on, input signals arriving in the common mode. This feature is why differential signaling is superior to single-ended signaling. This can be particularly important in noisy environments where noise can be induced in the two signal conductors in the common mode. Hum is a good example.

Common-mode signals can be converted to differential-mode signals if the loss of the hot path to ground is different than the loss of the cold path to ground in the differential signal path from source to destination. The signal loss with respect to ground being discussed is largely created by the attenuator formed by the source resistance to ground and the input's load resistance to ground. For this reason, the source resistance to ground (mainly determined by the series output resistors) should be precisely the same, requiring the use of 1% tolerance (or better) source output series resistors. For the same reason, the input resistors to ground at the receive end must be by precision resistors.

Since this is an attenuation-match issue for the hot and cold paths, the balanced input impedance should be high, since there will be less attenuation mismatch if there is less attenuation in the first place. If the source series resistors are 100 Ω and the receiver input resistance to ground is 10 kΩ, the resulting signals across the source resistors will be down by 40 dB. If the source or receiver resistances to ground are off by less than 1%, this adds another 40 dB of improvement for the issue of mismatched loss between the hot and cold paths. Of course, if the four resistances comprising the source and receiver resistances to ground are off by 1% in the worst directions, this additional improvement will be reduced by 12–28 dB.

For those output circuits that create the cold signal by using an inverter, the resistors in the inverter should be precisely the same so that the gain of the inverter is exactly unity. There are numerous other ways in which common-mode to differential-mode conversion can happen.

Effects of Source Impedance

If the impedances of a balanced source from hot and cold to ground are not the same, induced common-mode voltage in the cable will cause current to flow and create a conversion because resistances at the source are of different values. Note that if a source is floating, common-mode currents cannot flow. This also applies to the input impedances of a receiving device.

Consider an ideal differential receiver where there is some load resistance to ground for each of the hot and cold differential inputs. Assume for the moment that these input impedances to ground are identical. Assume that a common-mode signal is induced into the balanced interconnect. Assume that the balanced source also has output resistance to ground (in other words, it is not floating). If the hot and cold resistances to ground at the source are not identical, the gain in the hot and cold paths to the receiver input loads to ground will be different. This will result in common-mode to differential-mode conversion. If the hot and cold gains differ by 1%, CMRR can be reduced to about 40 dB.

A similar example is that of applying phantom power to the input hot and cold of a microphone preamp. The 48-V phantom supply is fed to the hot and cold signal wires through 6.81-kΩ tightly matched resistors. If these resistors are not matched, the resistances from hot and cold to ground will be slightly different.

Finally, the nature of the "balanced" output of the source can influence CMRR. For example, if the cold side of the source is grounded, transmission on the balanced pair is largely single-ended. Many sources provide an impedance-balanced output where the cold side goes to ground through a resistance that is the same as the source resistance of the hot side. This helps preserve CMRR, even though the balanced signal is not symmetrical (i.e., hot and cold not driven with equal and opposite signal polarities), as long as the destination has balanced inputs with good CMRR. This approach gives up 6 dB of source headroom since both hot and cold are not driven. This approach also results in the transmission of a common-mode signal of the same amplitude as the differential-mode signal. Depending on the design of the destination receiver input stage, this could create some CM-to-DM conversion and/or distortion.

A great many pieces of equipment create their balanced output by driving the cold side through an inverter. In this case, as in the others, the common-mode output impedance is low. There is nothing at the source to discourage common-mode current flow.

Ground Loops and Noise Pickup

Avoidance of ground loops and other means of noise pickup is important in a preamplifier if best sound quality and SNR are to be achieved. Best practices for grounding and interconnect should be observed. Bill Whitlock of Jensen Transformers has published numerous excellent articles on this topic [1–3]. Some of those best practices include:

- precision identical impedances to ground at both ends
- highest possible common-mode impedances to ground at both ends
- an open shield at one end
- prevention of currents from flowing in the shield

Mechanisms for Common-Mode Intrusion

There are two mechanisms of greatest interest for common-mode signals to intrude in a signal path. The first is electromagnetic induction into the interconnect cable from external sources. The simplest example of this is when a signal cable is in close proximity to a power cable, picking up hum and its harmonics and electromagnetic interference (EMI). This interference will be induced into both the hot and cold conductors of a balanced interconnect in equal amounts, creating a common-mode signal whose hum and noise voltage can be ignored by a receiver with good CMRR and equal source and receiver impedances.

The second mechanism can occur when the local grounds of the source and destination have different AC common-mode potentials. The simplest example of this is when the source and destination are powered from different branch outlets, causing the safety grounds to have different interference voltages on them. This situation is especially troublesome for single-ended interconnects, since the source and destination references for the single-ended signals, the local grounds, have a difference signal corrupted by hum and noise that will be interpreted as a differential signal. Note that in this situation, hum and noise current will flow in the shield. This issue is greatly reduced when the interconnect is balanced, although it may not be completely eliminated.

Even when the two devices are connected to the same outlet, some interfering current can flow in the shield, creating a noise voltage between the source and receiver local grounds. For example, AC line leakage at the source due to stray capacitance or Y capacitance in an EMI filter can flow through the shield to the local ground of the destination device. Coaxial cable with only a foil shield and a drain wire will not perform as well in this respect. Tightly woven copper braded shields are superior.

Star-Quad Cable

So-called *star-quad* cable consists of four wires twisted together inside a shield with diagonally opposite wires connected together at both ends. Manufacturers of star-quad cable include Belden, Mogami and Canare. Two ubiquitous examples are Mogami 2534 and Canare L4E6S. This

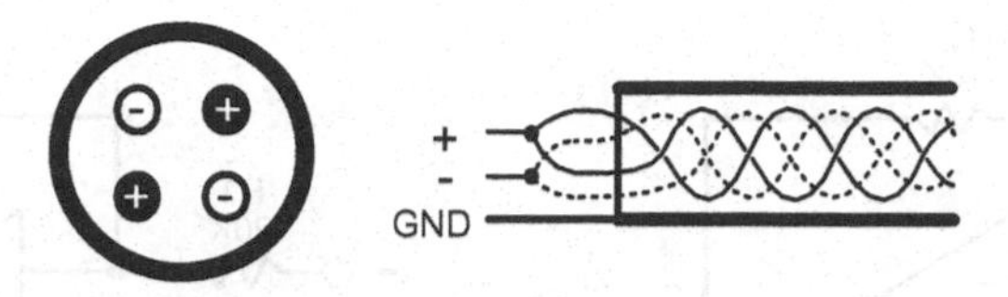

Figure 19.2 Star-Quad Cable Construction.

construction results in far superior immunity to externally induced noise from magnetic fields as compared to conventional shielded twisted pair cable [6, 7]. Magnetically induced noise is insidious and difficult to suppress. Cable shielding blocks noise from electric fields but does little for noise originating from magnetic fields. Twisted pairs, whether shielded or not, suppress differential noise induction from magnetic fields by subjecting each lead of the pair to the same magnetic induction of voltages. However, this equality of exposure is not perfect. The geometry of star-quad cable takes this balancing act to a much higher level, sometimes by as much as 20 dB [6].

Although most often associated with standard-size microphone cable, star-quad cable is available in miniature sizes as well, down to as small as 30 AWG conductors in a cable of 1/8-inch OD that is very flexible. These cables can be especially useful for short runs inside preamplifiers for improved noise immunity from noisy magnetic fields (e.g., from transformers, mains wiring, power supply wiring, etc.). Figure 19.2 illustrates how star-quad cable is constructed. In some instances, the star-quad geometry has been emulated on printed wiring boards for improved noise immunity. This is accomplished with a particular arrangement of traces on multiple layers.

19.2 Balanced Input Circuits

Most professional, and a number of hi-end audio power amplifiers and preamplifiers incorporate balanced inputs using XLR connectors with pin 2 hot. This provides for improved rejection of external common-mode noise.

Implementing balanced inputs for a single-ended (SE) preamplifier is not always straightforward, especially if the highest sound quality is to be maintained and if the preamplifier must allow the user to operate with either XLR balanced inputs or RCA single-ended inputs (with or without the need for a mode switch). The purpose of balanced inputs is to achieve good CMRR, independent of the nature of the source, as long as the output impedance of the source is the same to ground on both sides.

Extremely high common-mode input impedance to ground helps CMRR by reducing the flow of common-mode current. We cannot assume that the source has a floating output or ground-referenced output with perfectly matched hot and cold impedances to ground. Importantly, if the common-mode impedance to ground is infinite at either end, then common-mode current cannot flow to cause common-mode to differential-mode conversion (CMDMC) to occur.

Single Op Amp Differential Amplifier

An op amp can be used to implement a differential-to-single-ended converter, as shown in Figure 19.3(a) [8, 9]. Notice that this circuit presents different load impedances to single-ended sources on the positive and negative inputs. A single-ended source on the positive input sees an input impedance of 20 kΩ. A single-ended source on the negative input sees an input impedance of 10 kΩ. This is undesirable. For common-mode signals, the situation is different. A common-mode

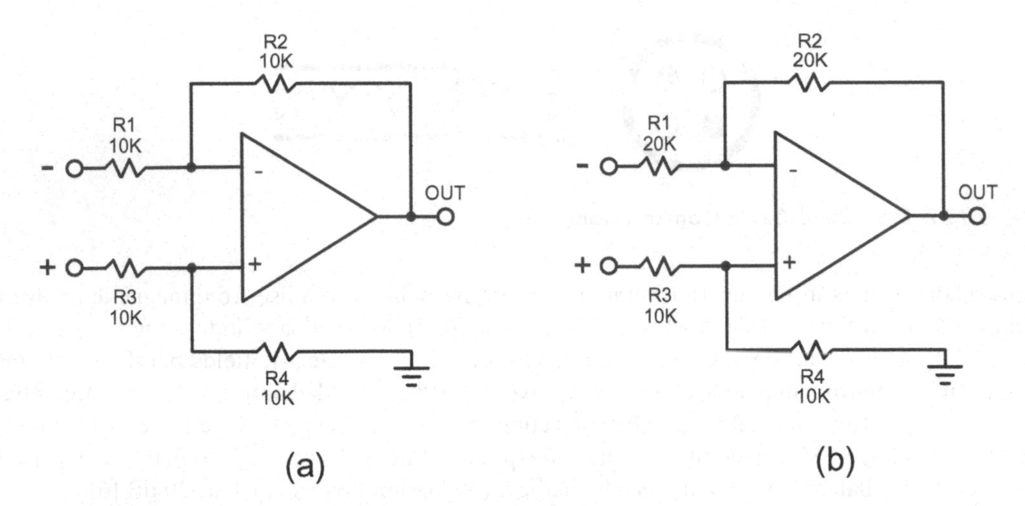

Figure 19.3 Single Op Amp Differential-to-Single-Ended Input Circuits.

source sees 20 kΩ on each side for a net of 10 kΩ. If the source impedance on both sides of the source is the same, common-mode rejection will be acceptable if resistor tolerances are tight. A useful check, in simulation or reality, is to apply a common-mode signal at the source end with source resistances that are identical and see how much CMDMC occurs.

In Figure 19.3(b) the resistor values are chosen so that the positive and negative inputs exhibit the same input impedance to a single-ended source (the resistors on the inverting side have been doubled to 20 kΩ). However, when a balanced signal is applied (with each of its signals referenced to ground at the source), the signal current flowing in the plus and minus inputs is not the same. Similarly, if a common-mode signal is applied to both inputs, the current flowing into the positive input is larger than the current flowing into the negative input. This happens because the virtual short at the op amp inputs forces 1/2 of the voltage at the positive input to appear at the negative terminal of the op amp. This seriously compromises CMRR in proportion to the output impedance of the source. If the positive and negative source impedances are not zero, CMRR will suffer.

The single op amp solution also forces a compromise between input impedance and noise. The circuit as shown has single-ended input impedances of only 20 kΩ and yet puts 20-kΩ resistors in the signal path, generating significant noise. The 20-kΩ input impedance also forces the use of large-value input-blocking capacitors if they are to be used.

Instrumentation Amplifier

The three op amp instrumentation amplifier shown in Figure 19.4 does not have these limitations [8–11]. It can be used to achieve high common-mode rejection and symmetrical input impedances under all drive conditions. Input impedance on each side is the same for both single-ended and common-mode signals. It also provides high input impedance while allowing the use of relatively low-value resistors in its implementation. Input impedances can be made especially high if JFET op amps are used. However, the circuit is more complex, and concerns about passing the signal through multiple op amps are magnified for some audio engineers.

The circuit is composed of unity-gain input buffers followed by a differential-to-SE converter with a gain of 1 [12–15]. Very high input impedances can be obtained by using low-noise JFET

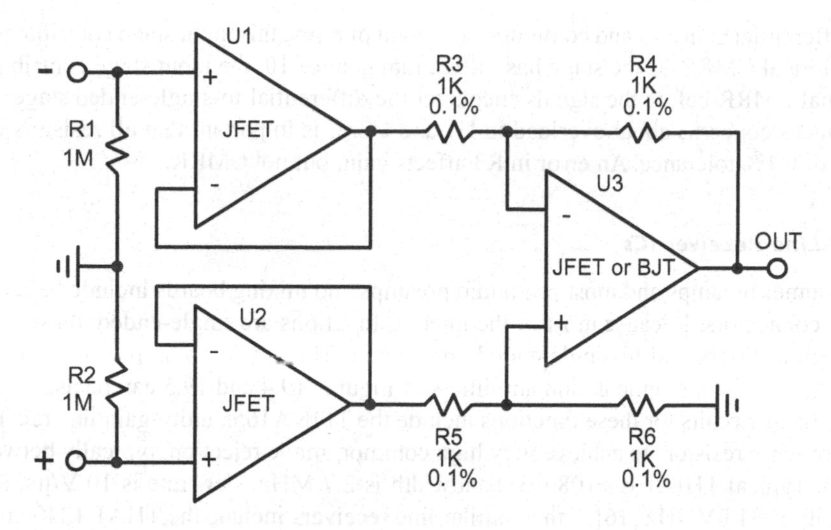

Figure 19.4 Triple Op Amp Instrumentation Amplifier.

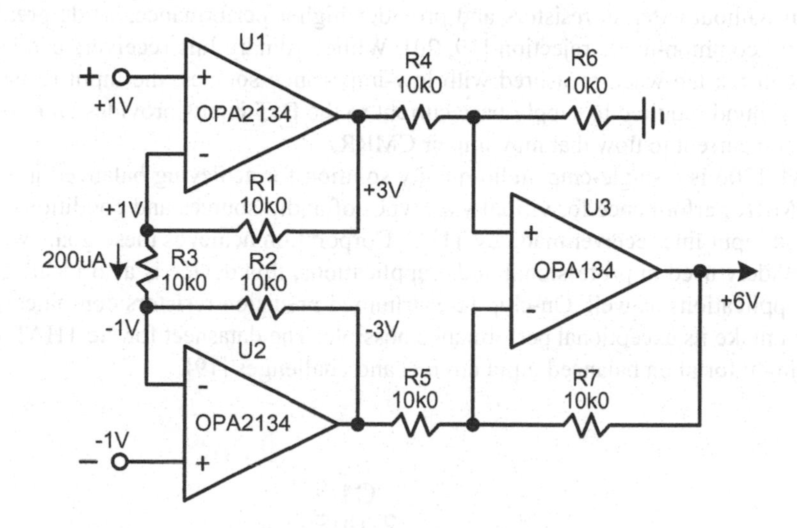

Figure 19.5 Triple Op Amp Instrumentation Amplifier with Gain.

op amps for the input buffers. BJT op amps will often require smaller base return resistors R1 and R2 due to input bias current. That in turn will require larger input coupling capacitors if they are used. The resistors surrounding U3 should have 0.1% tolerance in order to maintain high CMRR.

An important variant of the instrumentation amplifier is shown in Figure 19.5. Here the input buffers are configured to have gain of 3 for differential signals but only unity for common-mode signals. Because the feedback shunt resistor floats, the gain remains unity for common-mode signals. The indicated node voltages illustrate the operation of the amplifier when a 2-V_{p-p} differential input signal is applied.

With differential gain of 3 and common-mode gain of unity, this input stage contributes about 10 dB of additional CMRR. If the stage has differential gain of 10, the input stage contributes 20 dB of additional CMRR before the signals encounter the differential-to-single-ended stage. However, this introduces concerns about overload in U1 and U2. It is important that all resistors except R3 should be of 0.1% tolerance. An error in R3 affects gain, but not CMRR.

Balanced Line Receiver ICs

Many consumer preamps and most pro-audio preamps and mixing boards include balanced inputs with XLR connectors. Because most of the internal functions are single-ended, these inputs must pass through a differential-to-single-ended conversion. The single op amp differential amplifier of Figure 19.3 or the instrumentation amplifiers of Figures 19.4 and 19.5 can be used for this purpose. Integrated circuits for these functions include the TI INA165x unity-gain line receiver. It has precision on-chip resistors to achieve very high common-mode rejection, typically between 85 dB and 95 dB. Typical THD-1 is −108 dB, bandwidth is 2.7 MHz, slew rate is 10 V/μs, and output noise density is 31 nV/√Hz [16]. Other similar line receivers include the THAT 1240 and the ADI SSM2141 [17, 18].

The *InGenious*® 1206 chip from THAT Corp. takes this concept a step further. It implements this function without external resistors and provides higher performance, while greatly improving real-world common-mode rejection [19, 20]. While ordinary line receivers can achieve very high CMRR in the lab when measured with low-impedance sources, the input resistance of the receivers to ground required to supply base current to the BJT inputs provides an opportunity for common-mode current to flow that may impair CMRR.

The THAT 1206 is a single-chip audio-quality solution for achieving balanced inputs with exceptional CMRR performance for virtually all types of audio sources and conditions. The Ingenious balanced input line receiver made by THAT Corporation achieves these goals with very low distortion. Widely used in professional audio applications, this device is also ideal for consumer audiophile applications as well. On-chip laser-trimmed precision resistors combined with a very clever circuit make its exceptional performance possible. The datasheet for the THAT 1206 is also an outstanding tutorial on balanced input circuits and challenges [19].

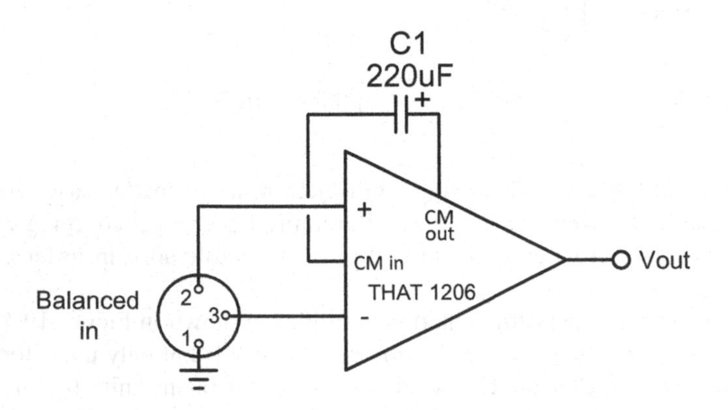

Figure 19.6 THAT 1206 Balanced Input Circuit.

It differs from an instrumentation amplifier in that it has an additional op amp that provides common-mode bootstrapping feedback to the input return resistors R1 and R2 in Figure 19.4, resulting in very high common-mode input impedance. The common-mode signal to be fed back is obtained by summing resistors at the outputs of unity-gain input amplifiers U1 and U2 in Figure 19.4.

A balanced input circuit using the 1206 IC is shown in Figure 19.6 [19]. The circuit inside the 1206 is akin to an instrumentation amplifier where the input amplifiers are operating at unity gain. However, a common-mode signal is created from the outputs of U1 and U2 of Figure 19.4. It is fed back through a fourth op amp and C1 to bootstrap bias resistors R1 and R2 (located inside the chip) with the common-mode signal. This largely takes them out of the picture in respect to their creating a path to ground that can reduce common-mode rejection due to source impedance imbalances.

The circuit of Figure 19.4 has very high input impedances to ground because the use of a JFET op amp obviates the need for input bias current to flow in R1 and R2. For low-noise BJT op amps this is not the case, and smaller resistor values must be used for R1 and R2 to carry the input bias current while not creating much voltage drop. The Ingenious® circuit was developed and patented by Bill Whitlock of Jensen Transformers [20].

Discrete Balanced Input Amplifiers without Feedback

A simple long-tailed pair differential amplifier can serve as a balanced differential input amplifier without the complication of global negative feedback. It can provide very low noise if it is not degenerated, and quite low distortion if signal levels are small. If implemented with JFETs, it can be implemented with or without source degeneration. If the JFETs are connected as complementary feedback pairs, distortion can be made quite low as a result of the greatly increased effective transconductance. Figure 19.7 illustrates these three approaches. Similar approaches can be used with BJTs, but distortion will be higher in (a) without degeneration.

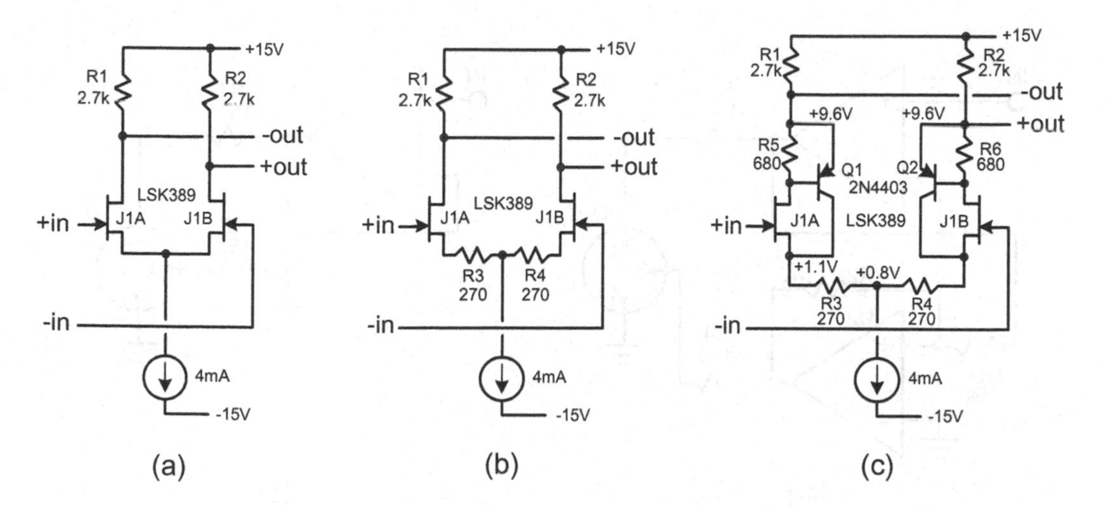

Figure 19.7 Balanced Input Stages without Negative Feedback.

19.3 Balanced Output Circuits

Balanced interconnections require the driving source to have a balanced output circuit for best performance. Because the internal circuits are usually single-ended, a single-ended-to-balanced output circuit is needed. The best in terms of common-mode rejection and reduction of ground loops is a transformer that will provide a floating balanced output. Transformers are expensive, take up space, do not have an outstanding frequency response and introduce some distortion. For this reason, most equipment employs some type of active balanced output circuit. These range from simple to sophisticated.

Balanced Output Using Inverter

The simplest approach is to take the single-ended signal and make it the hot signal and pass it through an inverter and call that the cold signal, as shown in Figure 19.8(a). This is very common, and requires just an inverter. It does, however, have low impedance to ground, so the benefits of a floating output are lost. The impedance to ground of the hot and cold lines will usually be set by series resistors R3 and R4 on the outputs, often on the order of 150 Ω. The finite impedance to ground at the outputs allows common-mode current to flow if the receiver at the far end has finite impedances to ground. Common-mode rejection can be impaired by paths to ground at the transmitter or receiver.

Impedance-Balanced Outputs

The impedance-balanced output is a cost-reduced version of the circuit above, and it is functionally not anywhere near a balanced output. As seen in Figure 19.8(b), it eliminates the inverter and merely connects the cold line to ground through a terminating resistor of the same value as the series resistor on the hot output. Only the impedances to ground for the hot and cold outputs are the same. Transmission of the signal is essentially single-ended, although there is some benefit to the fact that a receiver with CMRR is used at the other end. Notice that, for a given output op

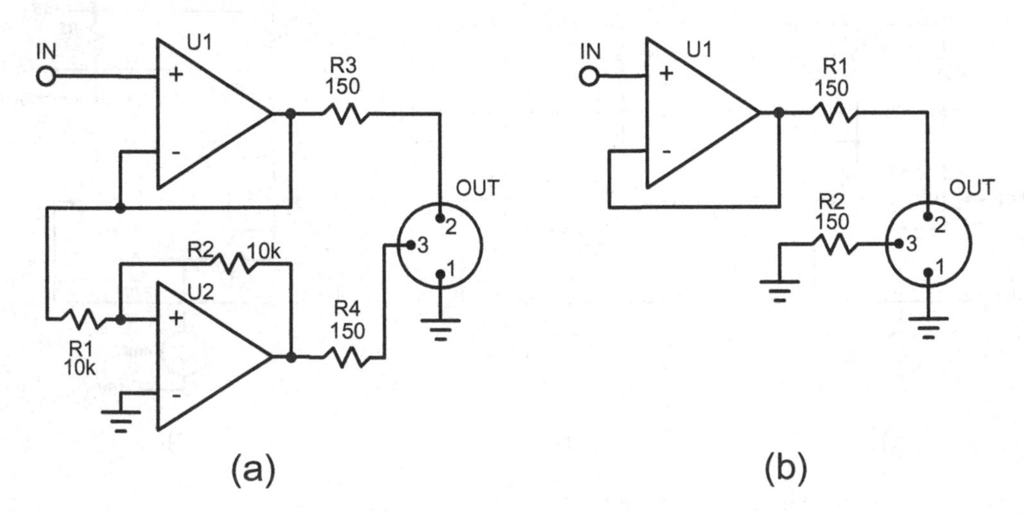

Figure 19.8 Balanced Output Circuits

amp voltage overload point, the circuit in (a) can produce 6 dB more signal. This alone means that overall achievable CMRR will be improved by 6 dB.

Balanced Output Using Differential Amplifiers

For those who prefer symmetry, a balanced output can be implemented with two differential amplifiers driven to opposite output phases by connecting the single-ended input to opposite-polarity inputs of the differential amplifiers. With the opposite inputs of each differential amplifier connected to ground, the circuit devolves into an inverter on one side and a follower on the other side.

Cross-Coupled Balanced Drivers

The circuit in Figure 19.9 is a cross-coupled output stage (CCOS) design employing differential unity-gain positive feedback to emulate the floating differential output of a transformer [21–25]. Circuit operation can be understood by recognizing the necessary behavior of a transformer and analyzing the voltages and currents in the circuit.

Recall that a transformer output has infinite impedance in the common mode (Z_{cm}). This means that if we apply equal currents to the hot and cold outputs of an active transformer emulator, those currents should see infinite impedance. Similarly, if the hot and cold outputs are connected together and a voltage is forced on them, no current should flow if there is no input signal. Furthermore, if a transformer is putting out +1 V on the hot side and −1 V on the cold side, and the cold side is then shorted to ground, +2 V will appear on the hot side. An active circuit properly emulating a

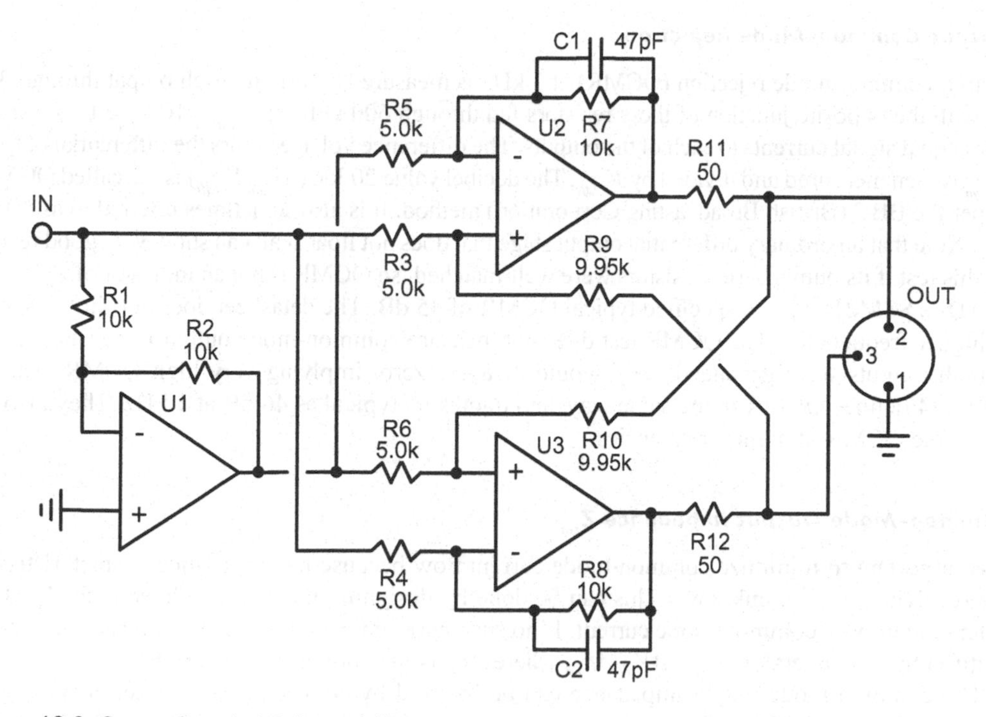

Figure 19.9 Cross-Coupled Balanced Driver

transformer will do the same – it will allow one output to be shorted to ground and have the other polarity of output increase by 6 dB.

Consider the following experiment on the cross-coupled circuit of Figure 19.9: set the input at zero and force both output polarities to +1 V. Now calculate the voltage that will appear at the output of either U2 or U3. If it turns out to be +1 V, then no common-mode current is flowing and the conditions for transformer emulation have been satisfied.

For the moment, assume that R9 and R10 are 10 kΩ rather than 9.95 kΩ. With +1 V applied to both outputs, it is clear that exactly 1/3 V will appear at the positive inputs of U2 and U3. Now notice that U2 is, with no input, configured by R5 and R7 to have non-inverting gain of exactly 3. This will result in +1 V at the output of U2. There will thus be no current flowing in R11, so no common-mode current is flowing and the desired floating output condition has been satisfied. The same holds true for the path through U3.

There is, however, a little bit of current flowing in R10 that has been overlooked. That current must be supplied by the output of U2 being ever so slightly positive. After some brief analysis, it becomes clear that the true operating condition is that the sum of R11 and R10 must be exactly 10 kΩ. With R11 being 50 Ω, R10 must be 9.95 kΩ. The signal paths through U2 and U3 each have gain of +1, corresponding to unity positive feedback. That is what makes this circuit work. If the gain around each loop is not exactly unity, the common-mode output impedance will go down by a significant amount. For this reason, precision resistors must be used in this circuit. If, for example, R9 is 0.1% high, common-mode output resistance falls to 69 kΩ.

The theory behind the operation of the circuit in Figure 19.9 is applied in the SSM2142 balanced driver from Analog Devices, Inc., and in the DRV134 balanced driver from Texas Instruments [23, 24].

Output Common-Mode Rejection

Output common-mode rejection (OCMR) at 1 kHz is measured by driving each output through 300 Ω, with the opposite junction of these resistors fed through 600 Ω from $V_{CMR} = 10\ V_{p\text{-}p}$. This test applies equal signal currents to each of the outputs. The difference voltage across the differential outputs ΔV_{out} is then measured and divided by V_{CMR}. The decibel value 20 log ($\Delta V_{out}/V_{CMR}$) is the called OCMR, as per the BBC (British Broadcasting Corporation) method. It is also sometimes referred to as $CMR\text{-}R_{out}$. Note that an ordinary differential output stage that does not float at all can show very good results on this test if its output series resistors were well-matched, so OCMR is not an indicator of Z_{cm}.

ADI's SSM2142 has a specified typical OCMR of 45 dB. The datasheet does not show OCMR at higher frequencies. The OCMR test does not measure common-mode output impedance. Even if both outputs were grounded, ΔV_{out} would measure zero, implying very high OCMR. The TI DRV134 defines OCMR in the same way, and quotes its typical as 46 dB at 1 kHz. They also do not quote OCMR at higher frequencies.

Common-Mode Output Impedance Z_{cm}

It is important to minimize common-mode current flow because the impedance characteristics at the receiving end are unknown. This can be done by designing the driver to have high Z_{cm}. This deters the flow of common-mode current. If no such current flows, there can be no common mode to differential conversion. $Z_{cm} = \Delta V_{out}/\Delta I_{cm}$, where ΔI_{cm} is the applied signal current.

The common-mode output impedance can be inferred by connecting a common-mode signal source through a 1-MΩ resistor to the junction of a pair of 300-Ω resistors in series across the output. The attenuation at this node is then measured. If attenuation is 40 dB, then Z_{cm} is about 10 kΩ. Greater attenuation indicates lower common-mode output impedance. Remember, the circuit is trying to bootstrap 50 Ω up by a factor of 1000 or more, so great precision is needed in the component values. In a simulation of the circuit, an error of 0.1% in R9 and R10 dropped the common-mode output resistance to 38 kΩ, still a good value.

Limited gain-bandwidth product in U2 and U3 in combination with the stabilizing capacitors C1 and C2 reduces common-mode output impedance at high frequencies, making it look like there is a common-mode capacitance to ground in parallel with the common-mode output resistance. Depending on design details, this effective capacitance can exceed 2000 pF. It is thus important to use op amps with high gain-bandwidth product and to minimize the values of C1 and C2 to maintain good Z_{cm} to the high end of the audio band.

The circuit in Figure 19.9 fails at one of the transformer behaviors. If the cold output is shorted to ground, significant signal current flows in R12. The hot signal output does double, however, as expected. It is unclear to what extent this shortcoming matters in the real world. This situation occurs when the balanced output is connected to a single-ended destination (like RCA) where the cold line of the balanced output is grounded. If the amplifier outputs can handle the large current through R12 at high signal levels, and R12 can handle the resulting dissipation, this might not be a serious problem. This behavior can be mitigated if series resistors of about 300 Ω are placed external to the driver chip, assuming that a balanced source impedance of 600 Ω is sufficiently low. Some manufacturers instruct the user to not make the balanced-to-single-ended connection this way, citing increased power dissipation and temperature in the output drivers and possibly increased distortion. Instead, they instruct the user to leave pin 3 floating. Manufacturers will often include a separate single-ended RCA output for single-ended use. This problem can occur in the simple balanced output circuit of Figure 19.8(a) and in many cross-coupled line driver circuits.

Output Common-Mode Rejection vs Output Common-Mode Impedance

OCMR is a measure the differential voltage across the differential outputs when a common-mode voltage is applied equally to both outputs through a relatively low resistance on the order of 750 Ω, causing a current to flow into those outputs. Common-mode output impedance Z_{cm} can be measured by applying a relatively constant common-mode signal current to the outputs through a large resistance and measuring the average common-mode voltage at the outputs.

These two specifications are related, but not the same. A higher value of Z_{cm} will increase OCMR because less current can flow into the outputs to create a differential voltage. However, a balanced output device with low and accurately equal output impedances to ground will yield a very high reading of OCMR even though its Z_{cm} is low. In other words, a non-floating output can have very high OCMR as well. Thus, a high value of OCMR says little or nothing about common-mode output impedance. OCMR thus does not tell the whole story.

A device measured for OCMR will show high OCMR if the device has high Z_{cm} because the high Z_{cm} will minimize the amount of common-mode current that can flow into the outputs to create a differential voltage. The flow of common-mode current into the outputs is the cause of conversion to a differential voltage. Thus, high Z_{cm} is more important than high OCMR.

THAT 1646 OutSmarts® Balanced Output IC

THAT Corporation provides a single-chip balanced output solution. The THAT 1646 *OutSmarts®* Balanced Line Driver provides a transformer-like floating balanced output [26–31].

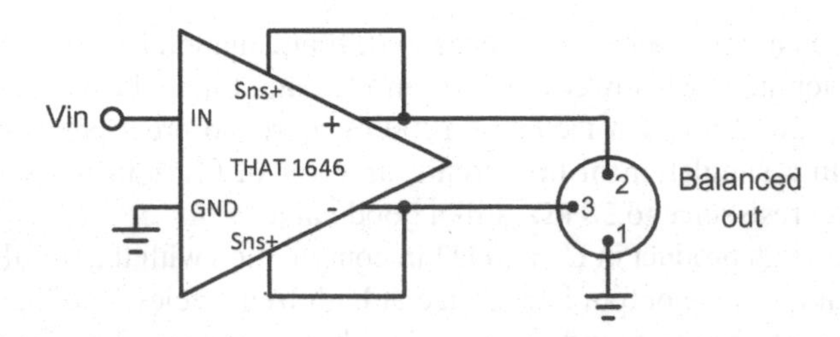

Figure 19.10 THAT 1646 Balanced Output Circuit.

Balanced outputs can be provided by transformers or active circuits like those described above. The 1646 *OutSmarts*™ chip from THAT Corp. provides the best of both worlds in a single chip that requires no external components. A balanced output circuit using this chip is shown in Figure 19.10.

The operating principle of the 1646 is different from that of the cross-coupled circuit in Figure 19.9. It uses summing resistors across the differential outputs on both ends of the output series resistors, deriving *before* and *after* common-mode signals, CM_1 and CM_2. If $CM_1 = CM_2$, then there is no common-mode output current and the common-mode output impedance is high. Common-mode feedback is used to drive the difference between CM_1 and CM_2 to zero. This is done by applying CM_1 and CM_2 to the differential inputs of an op amp, creating a common-mode error signal. The error signal drives the positive inputs of the two inverting input amplifiers to create identical common-mode outputs that drive CM_1–CM_2 to zero. The gain of the 1646 is 2 from the input to the differential output.

Unlike other cross-coupled balanced output drivers, the 1646 can drive its balanced output into a single-ended load where pin 3 is connected to ground, without the problems discussed earlier.

Output DC offset of the 1646 can be reduced substantially if optional coupling capacitors are placed in the common-mode feedback paths to the sense input pins. Without this AC coupling, common-mode output DC offset can be as high as 250 mV [26]. These capacitors should be on the order of 10 μF or more, meaning they should be non-polarized electrolytics. Because these capacitors are in the signal path, they may create distortion. Non-polarized (NP) electrolytic capacitors create less distortion when they have a larger value and a larger voltage rating [9]. The Nichicon ES series is a good choice.

An excellent article on using the THAT 1646 balanced line driver and the THAT 1206 balanced line receiver can be found at Ref. [32].

19.4 Transformers

A chapter on balanced I/O would not be complete without some mention of transformers, as they are used in some audio equipment to this day. They are large, heavy, expensive and lacking in some aspects of performance. Functionally, however, they solve a lot of problems within and between pieces of equipment. Transformer coupling is still preferred by some for balanced interconnects at equipment inputs and outputs. They can also be used for (step-up) voltage gain within some pieces of equipment, like for moving coil cartridges and ribbon microphones. In fact, some prefer the sound of transformer coupling, just as they some prefer vacuum tube electronics to solid-state electronics. Transformers can solve a lot of problems, but it is important to understand the details of their operation and construction [33–37].

Galvanic Isolation

While true galvanic isolation is not usually needed in audio applications, such as for safety, it makes life much easier if transformer shortcomings can be overlooked. Without transformers, very good true-balanced interconnections can be achieved using the THAT 1646 *OutSmarts*™ chip and the THAT the *InGenious*™ 1206 chip [19, 26]. These are rightly called true balanced (floating) interfaces, but they do not technically provide galvanic isolation. Where transformers may come in handy is when one doesn't know whether the piece of equipment at the other end of the interconnect has these kinds of very good balanced interfaces. Nevertheless, even using one of these chips just at one end of a connection can provide many of the advantages of balanced interconnect.

Avoidance of Ground Loops

One of the most noteworthy advantages of using a transformer is the general absence of ground loops. They are an insidious source of trouble in interconnecting audio hardware. It is important to understand that the troubles originate from unwanted magnetic fields that induce a voltage into the connecting wires. That voltage is usually quite small, but a current large enough to create problems can flow in a ground loop because the resistance around the loop can be quite small. Ground loops can be especially troubling if the mains safety ground is involved as part of the loop. It can be worse if the two pieces of equipment are powered through different branch circuits. This is often the case in sound reinforcement application where cable lengths may be long. One should never lift AC ground to eliminate a hum problem.

Most IEC ingress RFI filters include *Y* capacitors from line and neutral to ground. They conduct some current, even at 60 Hz. If that current is too high, it may trip a GFCI breaker. This can especially be the case if many pieces of equipment are connected to the same outlet or line. In the worst case, this can interfere with the desirability of having all interconnected equipment powered through the same mains circuit.

Preservation of Common-Mode Rejection

Common-mode rejection is very important for maintaining clean signals, especially in noisy environments. Common-mode rejection is best when there is symmetry between the hot and cold signal paths, meaning that they are true-balanced. Moreover, it is best when there is no path for common-mode current to flow in either of the signal lines. The transformer accomplishes this, even to quite high frequencies if it employs properly screened construction. It is notable that electronic balanced inputs can suffer a reduction in common-mode rejection at high frequencies, especially with respect to high-frequency EMI. Similarly, some electronic balanced drivers may have their floating behavior degraded at high frequencies. Although not always needed, a bullet-proof balanced interconnection can virtually always be provided by a transformer.

Attenuation of Conducted Emissions

Even if excellent common-mode rejection has been achieved with electronically balanced approaches, there remains the issue of conducted emissions. The standards for allowed amounts of conducted emissions at high frequencies can sometimes be difficult to meet, especially if switching power supplies, class-D amplifiers or significant digital circuitry are included in the equipment. Once again, a screened transformer of proper construction can largely stop conducted emissions.

Transformer Shortcomings

Apart from the well-understood transformer shortcomings of size, weight and cost, there are performance limitations like frequency response and distortion. It is far too easy to overlook transformer distortion due to the approach of core saturation at high signal levels, especially at low frequencies. It is also very important to avoid DC current flow in audio transformer windings – it can cause or hasten the onset of core saturation.

Not all transformers are created equal, and it is important to understand the subtleties of their design. It is also very important to understand the relationship between the transformer inductances and the impedances of the circuits in which they are used. Finally, it is important to understand the role of primary and secondary winding resistances in some applications. This is exemplified in the use of transformers for moving coil cartridges and ribbon microphones. Although it is tempting to think of a transformer as a noiseless way of obtaining voltage gain, they do in fact contribute thermal noise due to their winding resistances. All of this is a bit beyond the scope of this chapter. There are numerous excellent references where much more detail about these issues can be found [33–37].

19.5 Grounding and Shielding

Even with good balanced input and output circuits in the equipment, it is still very important to observe best practices in grounding, shielding and interconnect techniques [38–40]. Grounding considerations can also be found in Chapter 14.

Ordinary balanced patch cables have the shield connected at both ends. Patch cords can be made with the shield connected at only one end. This can be indicated with an arrow on the cord pointing in the preferred direction of signal flow, often with the shield left unconnected at the receiving end.

References

1. Bill Whitlock, *An Overview of Audio System Grounding and Signal Interfacing*, Tutorial T5, 135th Convention of the Audio Engineering Society, October 2013, New York City.
2. Bill Whitlock, *An Overview of Audio System Grounding and Shielding*, Tutorial T2, Convention of the Audio Engineering Society, Star-Quad Page 126.
3. Bruno Putzeys, The G Word, or How to Get Your Audio Off the Ground, *Linear Audio*, vol. 5, April 2013, pp. 105–126.
4. Mike Rivers, *Balanced and Unbalanced Connections*, ProSoundWeb, January 18, 2019, prosoundweb.com/2019/01/18/.
5. Rod Elliott, *Balanced Inputs & Outputs – The Things No-One Tells You*, Elliott Sound Products, November 2020, sound.au.com.
6. Belden, *How Starquad Works*, belden.com/blog/broadcastav/How-Starquad-Works.cfm.
7. Canare, *The Starquad Story*, canare.com/UploadedDocuments/Cat11_p.35.pdf.
8. Walter G. Jung, *Op Amp Applications Handbook*, Elsevier/Newnes, Burlington, MA, 2005, pp. 121–148.
9. Bob Cordell, *Designing Audio Power Amplifiers*, 2nd edition, Routledge, New York, NY, 2019.
10. Thomas Kugelstadt, *Getting the Most Out of Your Instrumentation Amplifier Design*, Texas Instruments, Inc., Analog Applications Journal, 4Q2005, ti.com/aaj.
11. Texas Instruments Inc., *INA828 Low-Power Instrumentation Amplifier*, datasheet, 2018, ti.com.
12. Wayne Kirkwood *Audio Line Receiver Impedance Balancing Using a 2nd Diff Amp*, EDN, January 30, 2013.
13. Rod Elliott, *Instrumentation Amplifiers vs Op Amps*, Elliott Sound Products, June 2017, sound-au.com.
14. Rod Elliott, *Fully Differential Amplifier*, ESP Project 176, 2018, http://sound.whsites.net/project176/htm.

15. Rod Elliott, *Balanced Transmitter and Receiver II*, Project 87, Elliott Sound Products, April 1, 2002, sound-au.com.
16. INA165x SoundPlus High Common-Mode Rejection Line Receivers, Datasheet, November 2018.
17. THAT Corp., *Balanced Line Receiver ICs*, THAT 1240 datasheet, thatcorp.com.
18. SSM2141 Balanced Line Receiver, datasheet, Analog Devices, Inc., analog.com.
19. THAT Corp., *Ingenious® Balanced Input Line Receiver ICs*, THAT 1206 datasheet, THAT Corporation, thatcorp.com.
20. Bill Whitlock and Fred Flouru, *New Balanced-Input Integrated Circuit Achieves Very High Dynamic Range in Real-World Systems*, Convention Paper 6261, presented at the 117th Convention of the Audio Engineering Society, October 2004.
21. George Pontis, Floating a Source Output, *Hewlett-Packard Journal*, vol. 31, no. 8, August 1980, pp. 12–13.
22. Leslie Tyler, THAT Corporation, *Cross-Coupled Output stages for Balanced Audio Interfaces*, EE Times, September 4, 2008.
23. DRV13x Audio-Balanced Line Driver, datasheet, Texas Instruments, ti.com.
24. SSM2142 Balanced Line Driver, datasheet, Analog Devices, Inc., analog.com.
25. Rod Elliott, *Balanced Line Driver and Receiver*, ESP Projects 51 & 87, http://sound.whsites.net/project51/htm.
26. THAT Corp., *OutSmarts® Balanced Line Driver ICs*, THAT 1646 datasheet, THAT Corporation, thatcorp.com.
27. Gary K. Hebert, *An Improved Balanced Floating Output Driver IC*, presented at the 108th AES Convention, February 2000.
28. Chris Strahm, *Balanced Output Circuit*, US Patent 4,979,218, December 18, 1990.
29. Gary Hebert, *Floating Balanced Output Circuit*, US Patent 6,316,970, November 13, 2001.
30. Rod Elliott, *Balanced Line Driver with Floating Output*, Elliott Sound Products, February 10, 2012, sound-au.com.
31. Bill Whitlock and Rod Elliott, *Design of High-Performance Balanced Audio Interfaces*, Elliott Sound Products, 2010.
32. Gary Galo, THAT's Balanced Line Drivers and Receivers, audioXpress, June 2010, also available at https://thatcorp.com/aes-2007-convention-report.
33. Bill Whitlock, Balanced Lines in Audio – Fact Fiction and Transformers, *JAES*, vol. 43, no. 6, June 1995.
34. Bill Whitlock, *Answers to Common Questions About Audio Transformers*, Jensen Application Note AN-002.
35. Lundahl Transformers, *Mixed Feedback Drive Circuits for Audio Output Transformers*, lundahltransformers.com/wp-content/uploads/datasheets/feedback.pdf.
36. Bill Whitlock, *Audio Transformers*, Jensen Transformers, originally published as Chapter 11, *Handbook for Sound Engineers*, 3rd edition, 2001 Glenn Ballou, Editor, available at jensen-transformers.com.
37. Bill Whitlock, *Audio Transformers, Handbook for Sound Engineers*, 5th edition, Glenn Ballou, Editor, Chapter 15, Focal Press, New York, NY, 2015, pp. 367–401.
38. Bill Whitlock, *Design of High Performance Audio Interfaces*, Jensen Transformers, Inc.
39. Bill Whitlock, *Interconnection of Balanced and Unbalanced Equipment*, Jensen Application Note AN-003.
40. Rane Commercial Audio Products, *Sound System Interconnection*, Rane Note 110, 2015, http://www.rane.com/note110.html.

Chapter 20

Equalizers and Tone Controls

Equalizers and tone controls are used to achieve a desired tonal balance from source material played through a given system, including its loudspeakers. Tone controls and equalizers can be used to remove anomalies or to compensate for non-flat frequency regions. They are usually adjusted to taste. They can also be used to compensate for frequency response effects due to the listening room. They are a key tool in professional audio applications, such as in mixers, recording consoles, stage and concert hall acoustic compensation [1–7].

The traditional bass and treble tone controls are routinely used to adjust the sound to one's taste in consumer equipment where they are available. Often, they represent an attempt to make up for some shortcomings in the loudspeakers or in the room acoustics. In more sophisticated equipment, bass, midrange and treble controls are available to better target different areas in the frequency spectrum. In still more sophisticated equipment, multi-band equalizers that cover five or seven or ten bands or even 1/3 octave-band equalizers (with 30 bands over 10 octaves) are used to better pinpoint areas in the spectrum for correction. These are often referred to as graphic equalizers [8–11]. These more sophisticated equalizers are used in professional applications as well. Finally, parametric equalizers focus on a single frequency band and are able to adjust boost or cut and the sharpness of the equalization. The spectral sharpness of the correction is referred to as the "Q" or quality factor. Multi-band or parametric equalizers are often used to tame feedback in sound reinforcement systems.

At the heart of every equalizer is one or more simple filter circuits like those that have been described in Chapter 11. The key parameters of a bandpass equalizer are amplitude (amount of boost or cut), center frequency and Q. The sharpness of the response is represented by Q. It was dealt with at length in Chapter 11 on filters. Low-pass and high-pass shelving equalizers are characterized by amount of boost or cut and the frequency at which the boost or cut falls by 3 dB when maximum boost or cut is selected. A tone control with bass and treble settings is a form of shelving equalizer. High-cut and low-cut filters are characterized by the frequency at which they cut by 3 dB and the steepness of their slope (e.g., 6 dB/octave, 12 dB/octave).

20.1 Baxandall Tone Control

The Baxandall tone control shown in Figure 20.1 is by far the most popular tone control [12]. Two variations of the design are shown in the figure. In (a) an inverting op amp receives a varying mixture of feedforward and feedback signals at high frequencies as dictated by treble control P1 and high-pass capacitor C1. It receives this signal at its inverting input virtual ground. When the wiper of P1 is toward the input, the amount of high-frequency feedforward exceeds the amount of high-frequency feedback and gain is increased. The converse occurs when the wiper is toward the

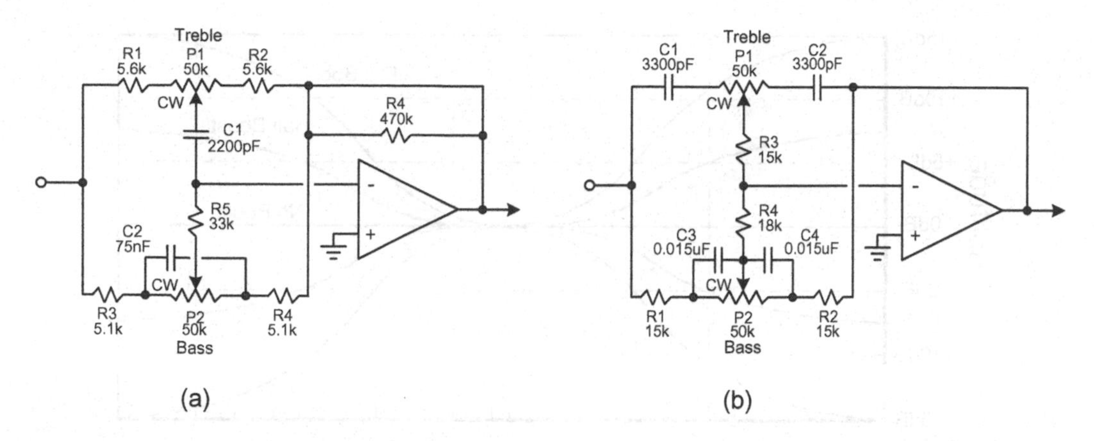

Figure 20.1 Baxandall Tone Controls.

right. At very high frequencies, the full amount of boost or cut set by P1 occurs. At very low frequencies C1 is an open circuit and the position of P1 has no effect. At very high frequencies, where C1 is a short, an extra load is placed on the virtual ground when the pot is near either end. This can reduce loop gain a bit. The typical Baxandall tone control can provide up to ±15 dB of boost or cut.

It is worth noting that when a pot's wiper feeds a frequency-selective circuit, like C1, the position of the wiper affects the source impedance to that element and may change the frequency response slightly. The effective series resistance added by the pot will be zero at either end of the track, but it will be 1/4 of the pot resistance when the pot is centered.

The bass control operates in a similar manner, except that the low-pass frequency selectivity is created by C2 shorting together the ends of bass control P2 at higher frequencies so that the wiper position of P2 only has effect at low frequencies. Since the frequency selectivity has been accomplished at P2, only resistor R5 is needed to connect to the op amp virtual ground to influence bass boost or cut. The beauty of the Baxandall tone control is that the signals from the bass and treble pots are added in the current domain at the virtual ground of the op amp, preventing some interaction between the bass and treble control settings. However, there is still some interaction because the setting of the bass control affects the net resistance of R5, which in turn can affect the behavior of the treble control.

Figure 20.1(b) shows a more complex implementation that is sometimes seen. Here the action of treble control P1 is confined to high frequencies by capacitors C1 and C2 in series with each end of the pot. A resistor R3 then connects the wiper to the op amp virtual ground. This arrangement, requiring an extra capacitor, has little effect and little benefit. The capacitors do reduce the loading by the pot from input to output at lower frequencies. The use of a resistor instead of a capacitor to connect the wiper to the virtual ground may improve stability in some cases and guarantee that the net resistance seen by the virtual ground is no lower than the value of that resistance, regardless of the remainder of the pot circuitry.

Similarly, two capacitors are employed in the bass control circuit. Instead of shorting out the ends of the pot at high frequencies with a single capacitor, C3 and C4 short each end of P2 to its wiper. The effect of Figure 20.1(b) on behavior is subtle but helpful. At high frequencies it prevents the wiper resistance from appearing in series with R4, which connects the wiper to the virtual ground. The main feedback path in the tone control is through the bass control, and pot position can affect this degree of feedback, resulting in some influence on the action of the treble control.

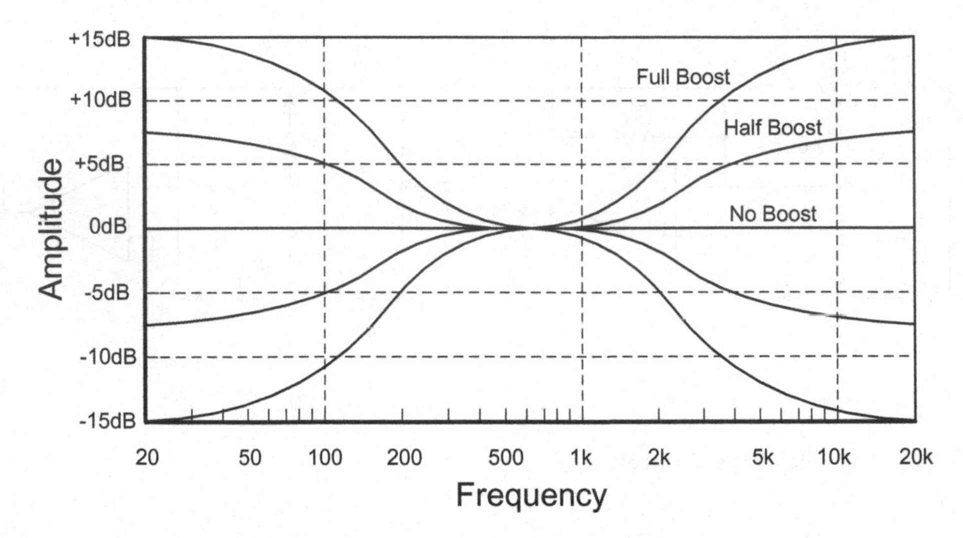

Figure 20.2 Baxandall Tone Control Frequency Characteristics.

The use of C3 and C4 prevents the amount of feedback from changing at high frequencies with adjustment of the bass control. Equal end resistors R1 and R2 default the equalizer to an inverter at frequencies not being controlled. In my experience, the circuit of (b) is easier to adjust with component values to achieve a satisfactory set of curves.

Figure 20.2 illustrates the boost and cut characteristics of an idealized Baxandall tone control. Notice that at the extreme low and high frequencies the amount of boost or cut levels off to be a relatively constant value as a function of frequency. For this reason, these are often called shelf equalizers because the frequency response characteristic transitions to a constant-height shelf. The Baxadall frequency-shaping networks are of only first order, so the equalization slopes will always be less than 6 dB/octave. This means that the bass and treble controls affect the frequency response over a broad spectrum.

20.2 Three-Band Tone Control

The three-band tone control adds a mid-frequency control to the Baxandall two-band arrangement. As shown in Figure 20.3, it is implemented with the addition of a third network with a mid-band pot (here P2) whose ends are shunted by C2, preventing its influence at high frequencies, somewhat like the bass control in Figure 20.1(a). Similarly, P2's wiper output passes through a series capacitor C3 to the virtual ground of the op amp, preventing its influence at low frequencies, somewhat like the treble control of Figure 20.1(a). In this design, the mid-band peak is at about 500 Hz, providing up to ±15 dB of mid-band boost or cut.

Figure 20.4 illustrates the frequency response characteristics of an idealized three-band tone control, boost or cut, one band at a time. The mid-frequency band has a bandpass type of characteristic, although not very sharp (low Q). Due to the shape of the mid-band equalizer, this is sometimes referred to as a bell type equalizer characteristic. As with the Baxandall tone controls, many different implementations of the three-band tone control can be found in the literature.

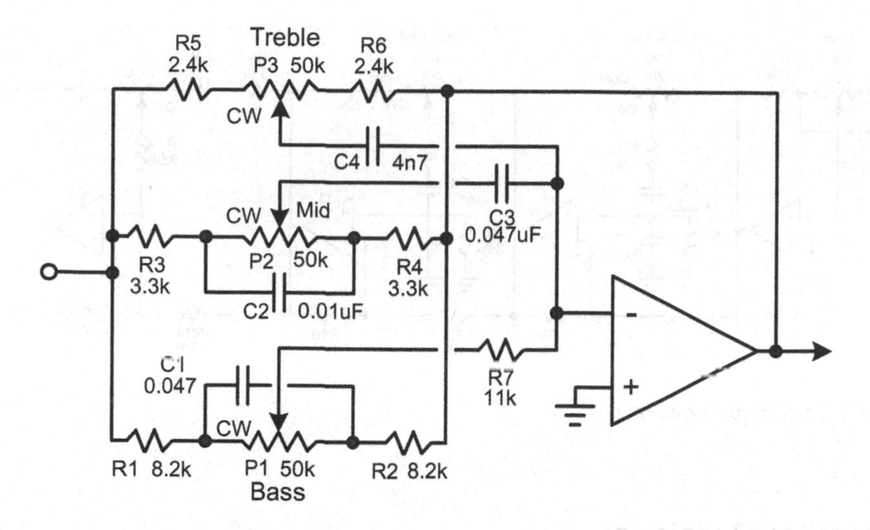

Figure 20.3 Three-Band Tone Control.

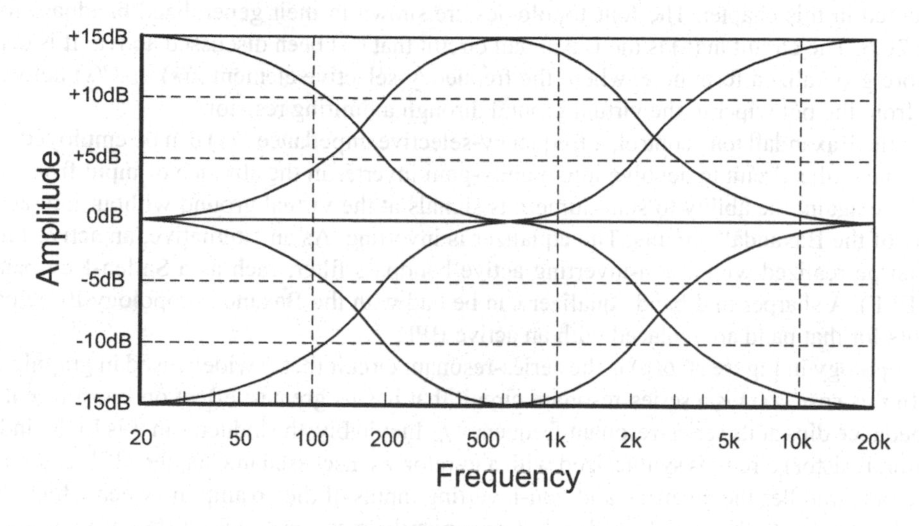

Figure 20.4 Frequency Response of a 3-Band Tone Control.

20.3 Four-Band Equalizer

Figure 20.5 shows a four-band equalizer. These are used in many compact professional mixers. Here each band is served by a separate op amp and network. The low and high bands are each like half of a Baxandall tone control, preventing any interaction between the two as a result of changing equalization settings. The low-mid and high-mid bandpass equalizers are realized with Wien bridge equalizers to be described later. They can provide ±15 dB of boost or cut at each of the two center frequencies 400 Hz and 2.2 kHz.

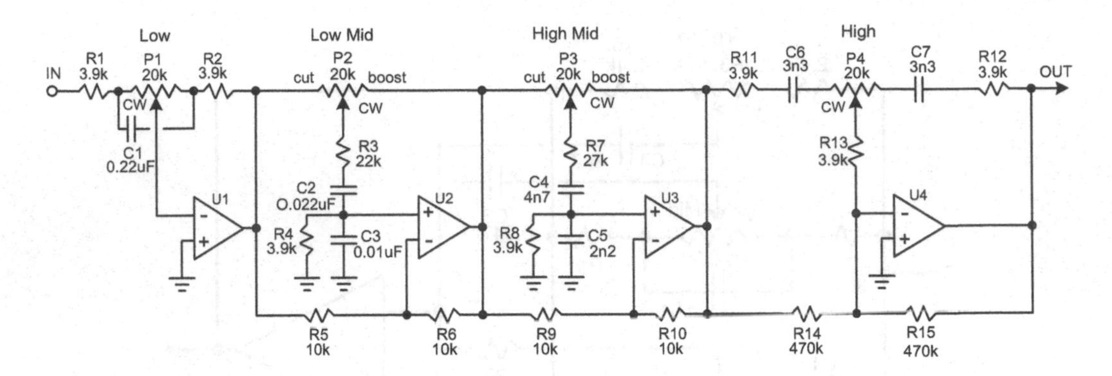

Figure 20.5 Four-Band Equalizer.

20.4 Equalizer Topologies

There are many different equalizer topologies, but there are four of greatest interest, and each is represented in this chapter. The four topologies are shown in their generalized bandpass form in Figure 20.6. The circuit in (a) is the Baxandall circuit that has been discussed above. It is depicted in a more generalized form here, where the frequency-selective element $Z(s)$ or $H(s)$ delivers the signal from the pot wiper to the virtual ground through a limiting resistor.

As in the Baxandall tone control, a frequency-selective impedance $Z(s)$ can be employed. Resistor R1 causes the circuit to devolve into a unity-gain inverter in the absence of input from $H(s)$ or $Z(s)$. Once again, the ability to sum numerous signals at the virtual ground without interaction is a virtue of the Baxandall circuits. The equalizer is inverting. As an alternative, an active function $H(s)$ can be realized with a non-inverting active bandpass filter, such as a Sallen-Key bandpass filter (BPF). A sharper mid-band equalizer can be had with the Baxandall topology if the passive elements for that band are replaced with an active BPF.

The topology in Figure 20.6(b) is the series-resonant circuit that is widely used in graphic equalizers. In this case, $Z(s)$ is a series-resonant circuit that has its greatest effect on boost or cut when its impedance dips at its series-resonant frequency f_0. Inevitably, the inductor in this LCR (inductor, capacitor, resistor) circuit is synthesized with a gyrator, as discussed in Chapter 11. It is interesting that the pot straddles the inverting and non-inverting inputs of the op amp, between which there is a virtual short due to the overall feedback. Conveniently, many pots with different LCR networks can be placed across the positive and negative inputs of the single op amp that can serve all of the frequency bands.

The maximum Q of the equalizer occurs when the pot is at one extreme. The effective series resistance of the LCR network will limit the maximum Q. That resistance need not be a physical resistance in series with the network, since the effective resistance can be controlled by the design of the gyrator, as described in Chapter 11. The effective resistance as seen at the wiper of the pot is at its maximum value when the pot is centered. At this point, that resistance is 1/4 the value of the pot, here 2.5 kΩ. This adds to the resistance of the series LCR network, reducing Q as the wiper moves toward its center. It is not practical to make the pot value arbitrarily small to minimize this effect in an equalizer with numerous bands because all of the pots have their ends in parallel. This can increase the noise gain of U1, and force other compromises.

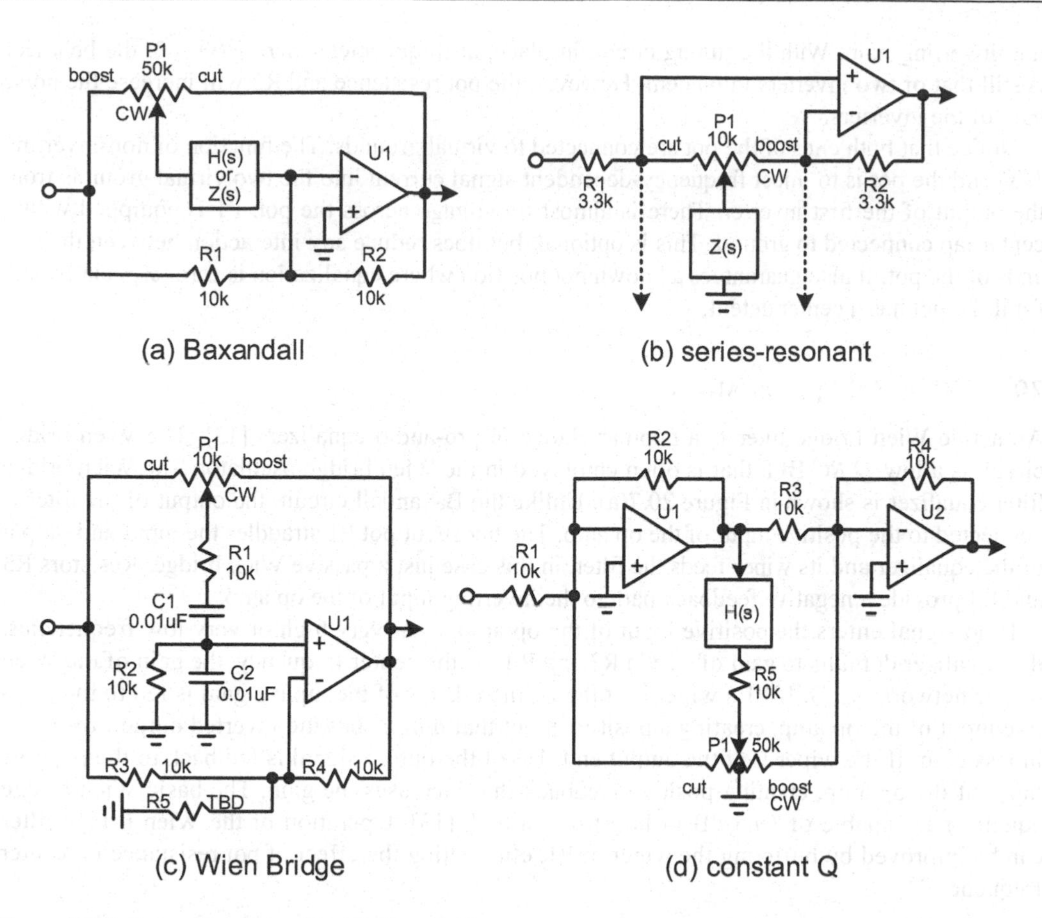

Figure 20.6 Four Equalizer Circuit Topologies.

In many of the schematics going forward, an inductor will often be shown for convenience, with the understanding that it is implemented with a gyrator. The series-resonant topology is non-inverting. In the absence of admittance from $Z(s)$ the pot is just a resistor connecting the op amp inputs together, with virtually no significance other than to increase the noise gain. The circuit devolves to a unity-gain buffer that happens to have resistors in series with its input and feedback paths.

The topology in Figure 20.6(c) is often associated with the Wien Bridge equalizer that will be discussed later, but in the general case it is not limited to having $H(s)$ be a passive Wien bridge network. How it achieves controlled boost and cut with a given $H(s)$ is what is important here. Resistor R5 is optional, and its shunting effect to ground increases the equalizing effect. As R5 becomes smaller, both the equalizing range and Q increase.

The constant-Q topology in Figure 20.6(d) is special in that it provides equalization whose Q (sharpness) is constant, independent of amount of boost or cut [8]. This circuit was invented by Dennis Bohn of Rane Corporation. This will be discussed further at a later point. Without the tuning element $H(s)$, the circuit just looks like two inverters (U1 and U2) in tandem, yielding unity non-inverting gain. With the tuning circuit in place, at frequencies where $H(s) = 0$, the behavior is still that of two inverters in tandem. However, the pot resistance and R5 will increase the noise gain of the inverters.

Notice that both ends of the pot are connected to virtual grounds. The function of non-inverting $H(s)$ and the pot is to inject frequency-dependent signal current into the two virtual grounds from the output of the first inverter. There is almost no voltage across the pot. P1 is equipped with a center tap connected to ground. This is optional, but does reduce any interaction between the two ends of the pot. It also guarantees a known pot position where equalization is zero, especially useful if the pot has a center detent.

20.5 Wien Bridge Equalizer

An active Wien bridge filter is a popular choice in pro-audio equalizers [13]. The Wien bridge circuit is a low-*Q* *RC* BPF that is often employed in the Wien bridge oscillator. The Wien bridge filter/equalizer is shown in Figure 20.7(a). Unlike the Baxandall circuit, the output of the filter is connected to the positive input of the op amp. The boost/cut pot P1 straddles the input and output of the equalizer and its wiper feeds the filter, in this case just a passive Wien bridge. Resistors R3 and R4 provide a negative feedback path to the inverting input of the op amp.

If no signal enters the positive input of the op amp, as at very high or very low frequencies, the equalizer defaults to gain of −1 via R3 and R4. At the center frequency, the gain of the Wien bridge network is 1/3. If P1's wiper is at the input end, 1/3 of the input signal is fed to the positive input of the op amp, creating a positive input that diminishes the inverted output, resulting in less gain. If the wiper is at the output end, 1/3 of the output signal is fed back to the positive input of the op amp, creating positive feedback that increases the gain. The basic Wien bridge equalizer is capable of ±9.5 dB of boost or cut at f_0 [13]. Operation of the Wien bridge filter can be improved by buffering the wiper of P1, eliminating the effect of pot resistance on center frequency.

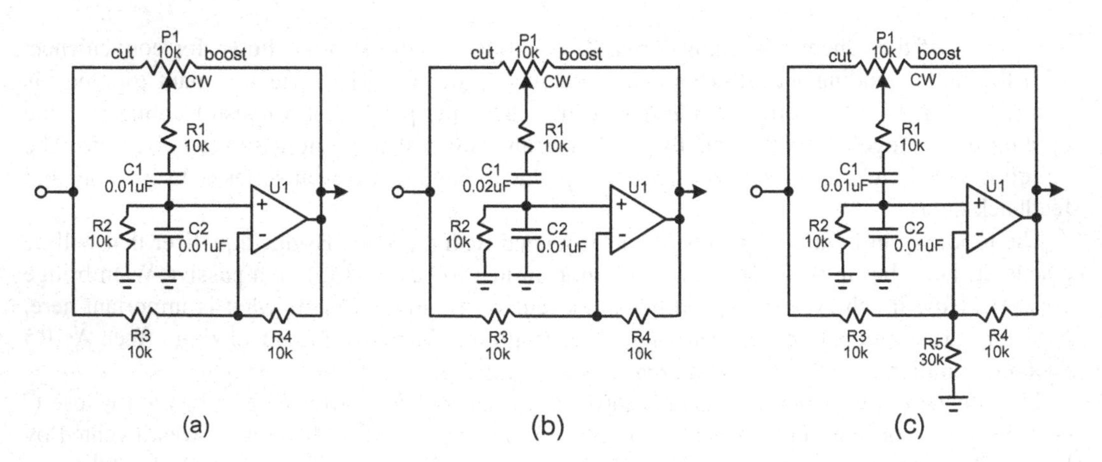

Figure 20.7 Wien Bridge Bandpass Equalizers.

Wien Bridge Equalizers with Higher Q and Greater Boost/Cut

A range of only ±9.5 dB is often too small compared to the typical ±15-dB range provided by other equalizers like the Baxandall. Sometimes it is also desirable to have a higher-*Q* characteristic than is provided by the conventional Wien bridge equalizer. The range and *Q* of the equalizer can be increased if the ratio $K = C1/C2$ in the Wien bridge network is made larger than unity, while keeping R1 = R2, as illustrated in Figure 20.7(b) [13, 14]. Cal Perkins used this technique with C2/C1 = 2.6 in the Mackie Onyx mixer series [14]. The so-called Perkins Equalizer was very well received and is well-known in pro-audio circles.

Tuning resistors R1 and R2 should remain equal, but their value may have to be decreased because the increased capacitance of C1 will reduce the center frequency otherwise. Reducing R1 and R2 maintains the same f_0 if the capacitance of C1 is increased without reducing the capacitance of C2. If C1 is doubled, f_0 will be reduced by 0.707. All resistors can remain at their nominal value. If the ratio *K* is made too large, oscillation can result. The Mackie Perkins equalizer was among the first to employ values of *K* greater than unity to achieve greater equalization range [14].

There is another way to increase equalizer *Q* and equalization range shown in Figure 20.7(c) [13]. Recognizing that positive feedback is involved in the functioning of the equalizer, increasing the gain of the amplifier can amplify the effects of the equalizer, resulting in greater range and *Q*. The introduction of shunt resistor R5 attenuates equally the input and output signals passing through R3 and R4, so the nominal inverting gain of unity remains unchanged. However, the shunting action of R5 increases the gain of the amplifier as seen by the network at the op amp's positive input. In doing so, it increases the positive feedback action. Only a small increase is needed to substantially increase the equalization range and *Q*.

Assume that R3 and R4 in (c) are 10 kΩ. If R5 is absent, range will be ±9.5 dB. If R5 = 39 kΩ, range = ±12 dB. If R5 = 27 kΩ, range = ±14 dB. If R5 =18 kΩ, range = ±17 dB [10]. In principle, R5 could be made variable, allowing *Q* to be adjusted for a range from ±9.5 to ±17 dB.

20.6 Equalizer Q

For many equalizer designs, the *Q* of their bandpass characteristic at their maximum boost will not be maintained as the pot wiper is moved toward the center to reduce the amount of equalization. In other words, the sharpness and selectivity of the response decreases as the amount of equalization is decreased. Such equalizers are sometimes called proportional-*Q* equalizers, since the *Q* changes somewhat in proportion to the amount of boost or cut [7, 8]. This is not always desirable.

In some cases, it is preferable for the *Q* to remain the same as the amount of equalization is decreased. A constant-*Q* equalizer meets this need. Figure 20.8 shows the responses of a proportional-*Q* equalizer and a constant-*Q* equalizer for settings of full boost and half boost or cut. Both equalizers were designed to have the same *Q* at their maximum settings, where their responses overlap. Response of the proportional-*Q* design at reduced equalization can be seen to be broader in frequency. The constant-*Q* design maintains its sharpness at lower levels of boost/cut.

The constant-*Q* equalizer circuit topology of Figure 20.5(d) provides bandpass equalization whose *Q* is independent of amount of boost or cut [8]. The mid-frequency bandpass Baxandall equalizer of Figure 20.6(a) has a proportional-*Q* behavior.

The series-resonant equalizer of Figure 20.6(b) in theory approaches constant *Q* if the pot's wiper resistance is very small compared to the resistance in the series-resonant circuit, but that is

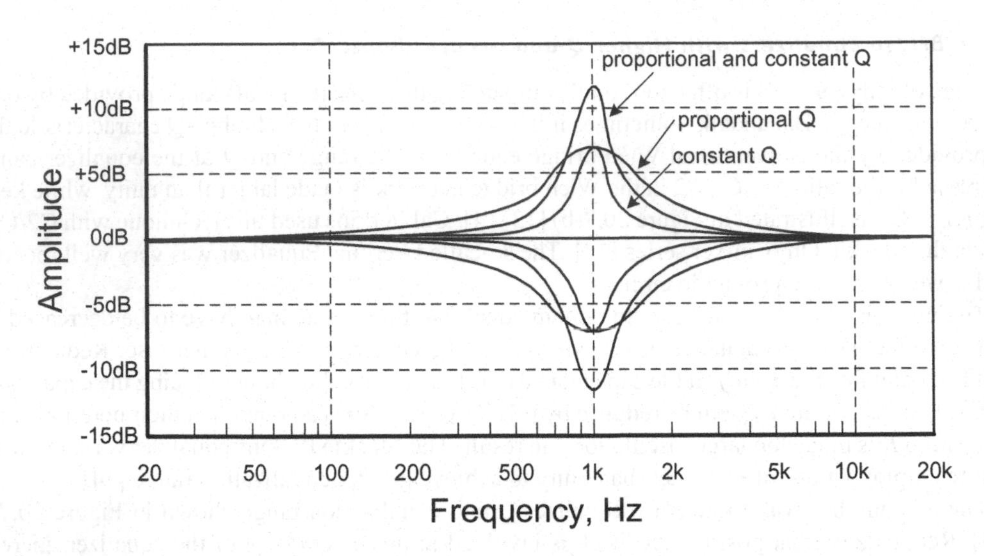

Figure 20.8 **Constant-Q and Proportional-Q Equalizer Responses.**

not a practical design. In practice, the total resistance in series with the *L* and *C* elements determines the *Q*. As amount of equalization decreases, the pot wiper moves toward the center and the wiper resistance increases. That resistance adds to the series resistance of the resonant circuit and reduces *Q*. As a result, the behavior is proportional *Q*.

Assume a 10-kΩ pot and a 5-kΩ network resistance. The maximum wiper resistance R_W for the pot is ¼ its value, here 2.5 kΩ. The resistance setting the *Q* increases from 5 to 7.5 kΩ at the pot's center. The *Q* is reduced by 33%. This is not so bad in equalizers with modest LCR *Q*, but becomes serious in equalizers that require high *Q* and thus smaller network resistance for the same *L* and *C* values.

The Wien bridge topology of Figure 20.6(c) is a proportional-*Q* design. The constant-*Q* topology of Figure 20.6(d) is the only topology of the four that has a truly constant-*Q* behavior.

Constant-Q Equalizers

The constant-*Q* equalizer of Figure 20.6(d) is revisited in Figure 20.9 [7, 8]. Recall that the circuit is essentially two inverters in tandem with the pot P1 straddling the two virtual grounds. Notice that the pot has a center detent and a center tap connected to ground, preventing any signal leakage between the two virtual grounds. It also makes the analysis easy, since when the wiper is to one side, the other side of the pot acts like a fixed resistor. The four resistors labeled R1–4 are of equal value. Resistor R5, in combination with the non-inverting gain of $H(s)$ at f_0, determines maximum boost or cut. Keep in mind that the portion of the pot resistance that is a fixed resistance merely appears as a resistance across the inputs of the associated op amp and thus has no effect on the gain (−1) of that stage.

Similarly, if the output of $H(s)$ is zero, the resistance on the active side of the pot just appears as a resistance from the associated virtual ground to ground, and also has no effect on that side's gain. Gain is affected on one side of the equalizer only when $H(s)$ is delivering signal current to that side's virtual ground through R5. Also note that if $H(s)$ has gain greater than unity at its center frequency, the effective resistance of R5 gets divided by that amount of gain.

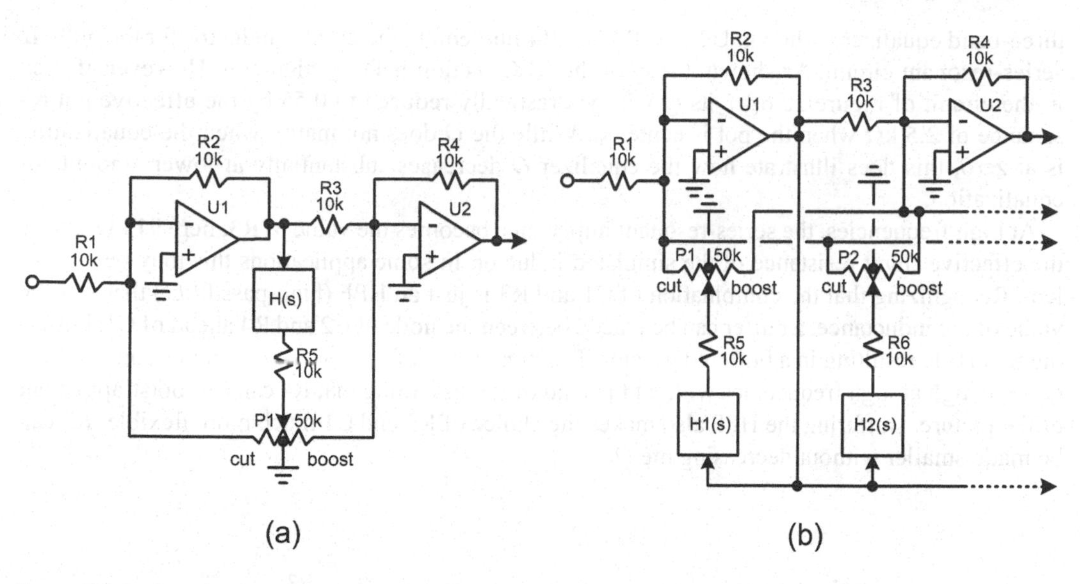

Figure 20.9 Constant-Q Equalizers.

Assume that R1–5 = 10 kΩ and that P1 = 50 kΩ. Assume that $H(s)$ has unity gain at f_0. If the pot wiper is fully to the input side, the effective feedback resistor for U1 is the parallel combination of R1 and R5, or 5 kΩ. The gain of the first stage will thus drop to −6 dB. If the pot wiper is fully to the output side, the effective feedforward resistance to U2 is the parallel combination of R3 and R5, or 5 kΩ. This makes the gain of the U2 arrangement +6 dB, fully complementary to the action of the wiper when it is at the other end of the pot. While this design example has a range of only ±6 dB, R5 can be reduced or $H(s)$ can be designed to have gain, a greater equalization range is available.

If the gain of $H(s)$ at f_0 is 3, it is easy to see that the gain of the affected inverter will be 1/4 when the wiper is to the left and 4 when the wiper is to the right, corresponding to a range of ±12 dB. If the wiper is halfway to the right, the pertinent pot resistance will be 12.5 kΩ and the effective feedforward resistance will be the parallel combination of R3 and 22.5 kΩ/3, or 10 kΩ||7.5 kΩ = 4.3 kΩ, resulting in gain of −2.3, or about +7 dB. The gain vs rotation curve is influenced by the pot resistance seen at the wiper, as this resistance is effectively in series with R5.

20.7 Graphic (Multi-band) Equalizers

Graphic equalizers (multi-band equalizers) usually comprise a number of bandpass circuits whose outputs are summed on a summing bus. This can give rise to some noise increase because the loading on the associated op amp increases its noise gain as the number of bands increases. The number of bands is often 7, 10, 20 or 30, the latter corresponding to a 1/3 octave-band equalizer.

Bandpass Tuning Circuits for Graphic Equalizers

Any of the filters in Figure 20.10 can be used as the basis of the multiple tuning circuits in a multi-band (graphic) equalizer. In Figure 20.10(a) is the series-resonant bandpass circuit using a gyrator inductor. It can be used in an equalizer configuration where a decreasing shunt impedance at f_0 can cause loss or gain. This popular application is illustrated in Figure 20.11, where a three-band equalizer is shown. U2, C1, R3 and R4 implement the gyrator inductor for the 500-Hz series-resonant circuit. As shown, the Q of the RLC section is fairly high at 6. However, if used in the circuit of Figure 20.6(b), its Q will be drastically reduced to 0.55 by the effective pot resistance of 2.5 kΩ when the pot is centered. While the Q does not matter when the equalization is at zero, this does illustrate how the equalizer Q decreases substantially at lower amounts of equalization.

At high frequencies, the series-resonant impedance becomes the value of R3, here 91 kΩ; this is the effective shunt resistance of the simulated inductor. In some applications this may be a problem. Recognizing that the combination of C1 and R3 is just an HPF (high-pass filter) that sets the value of the inductance, a buffer can be placed between the node of C2 and R4 ahead of C1, buffering this HPF, resulting in a buffered gyrator. The impedance of the series-resonant circuit can then go very high at high frequencies without the load of R3, assuming that R4 can be bootstrapped out of the picture. Buffering the HPF also makes the choice of R3 and C1 much more flexible; R3 can be made smaller without decreasing the Q.

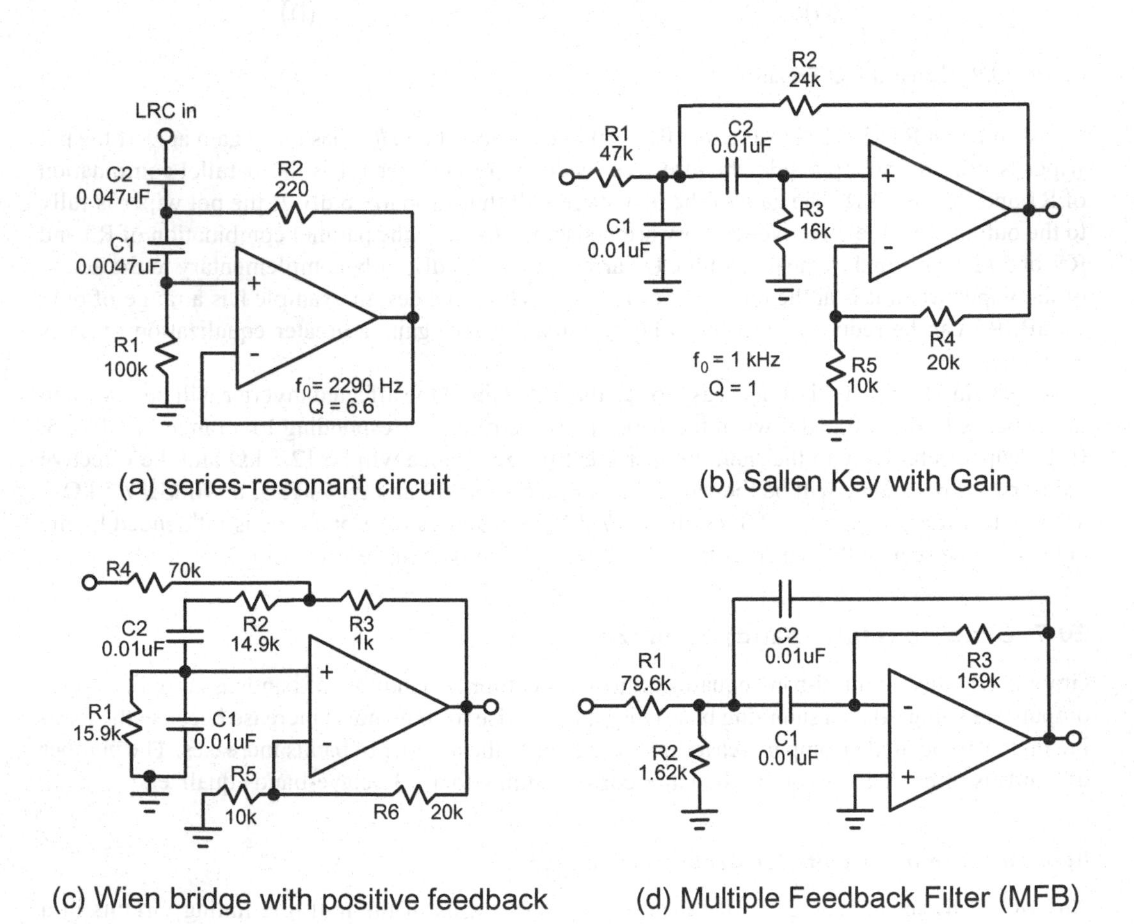

Figure 20.10 Bandpass Filters for Multi-band Equalizers.

In Figure 20.10(b) is a 1-kHz Sallen-Key BPF. It uses an active stage with gain of 3 in order to achieve a Q of unity. It can be used in an equalizer configuration where a non-inverting BPF is needed. As shown, its gain is unity, but it can be designed with higher or lower gain. See Chapter 11.2.

In Figure 20.10(c) is a 1-kHz Wien bridge equalizer that employs positive feedback controlled by R6 to increase Q well above that which a passive Wien bridge provides. With R6 = 20 kΩ, gain of the amplifier is 3, close to what would be used for a Wien bridge oscillator (but this circuit has some losses in the network). Here the input signal is injected at a tap between R2 and R3 by R4. The sum of R2 and R3 should be about equal to the value of R1, while taking into account the shunting effect of R4. The value of R4 controls the gain, here unity non-inverting at f_0. It can be used in the same way as the BPF in Figure 20.10(b). As shown, this filter's Q is 1.4 and its gain at f_0 is 0 dB non-inverting. Values of R6 ranging from 10 to 25 kΩ control Q over a range of 0.71–3.8. As with many active filters, Q is fairly sensitive to component values.

Figure 20.10(d) is an inverting multiple feedback (MFB) BPF that can be used in a multi-band equalizer arrangement where an inverting BPF can be used. An inverter can be added if it is used in a circuit that requires a non-inverting BPF. As shown, its Q is 5 and its gain at f_0 is 0 dB.

Series-Resonant Multi-band Equalizer

A graphic equalizer with three bands is shown in Figure 20.11. The heart of the circuit is a unity-gain amplifier comprising R1, R2 and U1. Its gain is unity if there is no loading to ground at the op amp's positive and/or negative inputs. Resistance that just straddles the positive and negative inputs has no effect on gain (other than to increase the noise gain). Each band has a pot that straddles the positive and negative inputs of U1. The wiper of each pot is connected to a series-resonant circuit whose impedance dips to a low value, R_0, at f_0. If the pot is centered, the resulting equal resistances creating a loading path to ground for the op amp inputs has no effect. If the wiper is at the left end, the input resistor R1 will be loaded to ground and gain will be attenuated at f_0 for that

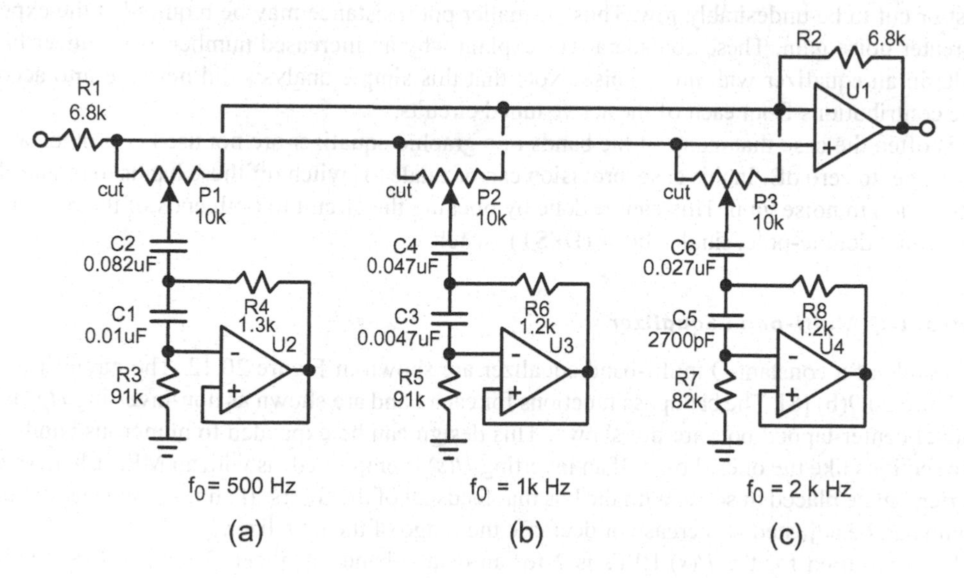

Figure 20.11 Multi-band Equalizer with Gyrators.

band. If the wiper is at the right end, negative feedback from R2 will be attenuated, increasing gain at f_0 for that band. The Q of the resulting filter function depends on the Q of the gyrator and the net resistance seen by the wiper of the pot. As shown, the maximum equalizer Q is 2.4 and the range is ±12 dB. with the linear pot at 3:00, boost is 3.4 dB.

Noise Gain of Multi-Band Equalizers

The presence of the multiple pots wired across the positive and negative inputs of U1 increases the noise gain of the circuit. Without the pots, the circuit is simply a unity-gain buffer, with noise gain of unity. Here we are ignoring the thermal noise of the resistors and just focusing on what happens to the input voltage noise of the op amp. Placing the net resistance R_{shunt} of all of the pots across the op amp inputs attenuates the negative feedback as seen by the op amp, since it shunts the feedback signal from R2. In fact, it forms an attenuator to the feedback against the sum of R1 and R2. We have noise gain $G_n = (R1 + R2 + R_{shunt})/R_{shunt}$.

If R1 and R2 are each 6.8 kΩ, and R_{shunt} from ten 10-kΩ pots in parallel is 1 kΩ, then noise gain is (6.8+6.8+1)/1 = 14.6. The lesson here is to employ a low-noise op amp and be mindful of the ratio of pot resistance to that of R1 and R2. At some point, the noise gain becomes almost proportional to the number of bands, such as with a ten-band equalizer. If two five-band equalizers are put in tandem, each with half the total number of 10 bands, the total noise only increases by 3 dB over that of a five-band equalizer. In equalizers with many bands, it may be helpful to split the bands into two equalizers in tandem, with even bands in one equalizer and odd bands in the other. This can reduce noise accumulation with number of bands and also reduce interaction between adjacent bands [7].

Maximum Q for a band occurs when the wiper is at either extreme of the pot, and is controlled by the Q of the L-C network. Minimum Q occurs when the wiper is at or near the center of the pot, where gain or loss of the band is minimal. This equalizer thus falls into the family of proportional-Q equalizers. The minimum Q is determined by the inverse of $R_0 + R_{pot}/4$. For a 10-kΩ pot, the $R_{pot}/4$ term is 2.5 kΩ. That much series resistance will cause the Q at lesser amounts of boost or cut to be undesirably low. Thus, a smaller pot resistance may be required at the expense of greater noise gain. These considerations explain why an increased number of equalizer bands results in an equalizer with more noise. Note that this simple analysis did not take into account noise contributions from each of the active tuned circuits.

It is often the case that many of the bands of a graphic equalizer are not used – in other words, they are set to zero dB. In this case, provision can be made to switch off these bands to reduce their contribution to noise gain. This can be done by opening the circuit to both ends of the pot for that band with a double-pole, single-throw (DPST) switch.

Constant-Q Multi-band Equalizer

Two bands of a constant-Q multi-band equalizer are shown in Figure 20.12. The circuit is based on Figure 20.9(b) [8]. The bandpass functions for each band are shown as non-inverting $H_n(s)$. The optional center-tapped pots are not shown. This design can be expanded to numerous bands with more sections like the ones shown. If an inverting $H(s)$ is employed, as with an MFB filter, a single inverter can be placed in series with the bus that feeds all of the filters. If an inverting stage is used, its gain can be adjusted to increase or decrease the range of the equalizer.

The Q required for the $H(s)$ BPFs is 2 for an octave-band equalizer, 3 for a 1/2 octave-band equalizer and 4 for a 1/3 octave-band equalizer [10]. As the number of bands increases, so does the loading on the virtual grounds of U1 and U2, increasing noise gain.

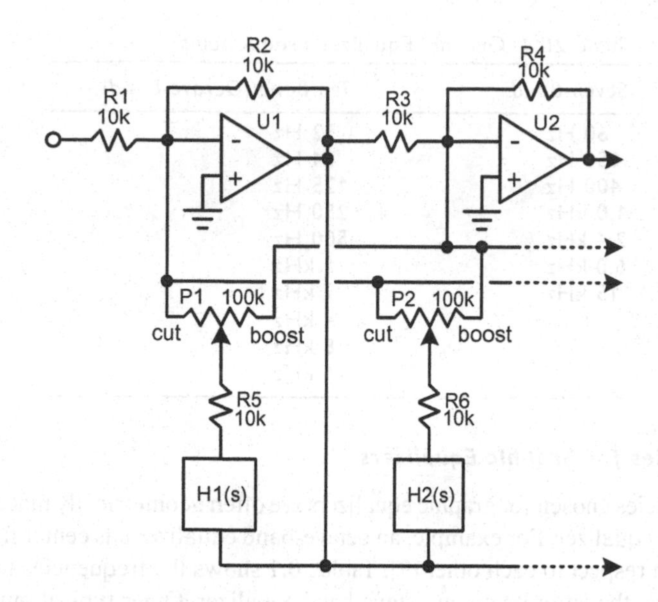

Figure 20.12 Constant-Q Multi-band Equalizer.

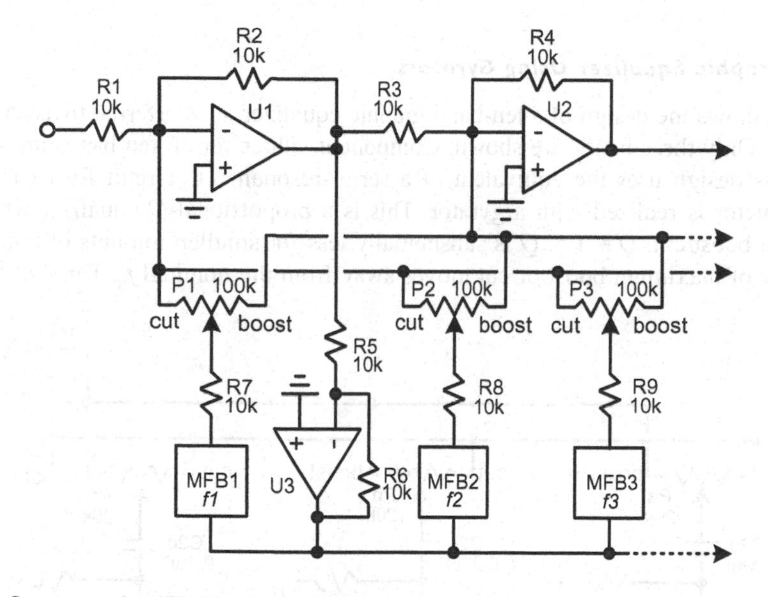

Figure 20.13 Constant-Q MFB Multi-band Equalizer.

Constant-Q MFB Multi-band Equalizer

Figure 20.13 shows a constant-Q multi-band equalizer that uses inverting MFB bandpass filters as the tuning element [8, 10]. This uses the constant-Q arrangements shown in Figure 20.9. Three bands are shown, and it can be easily expanded with more bands.

Table 20.1 Graphic Equalizer Frequencies

Seven-Band	*Ten-Band (Octave-Band)*
60 Hz	32 Hz
150 Hz	64 Hz
400 Hz	125 Hz
1.0 kHz	250 Hz
2.4 kHz	500 Hz
6.0 kHz	1 kHz
15 kHz	2 kHz
	4 kHz
	8 kHz
	16 kHz

Center Frequencies for Graphic Equalizers

The center frequencies chosen for graphic equalizers are often geometrically placed across the number of bands of the equalizer. For example, an octave-band equalizer has center frequencies that are multiples of 2 with respect to each other [9]. Table 20.1 shows the frequencies for seven-band and ten-band equalizers, the latter being an octave-band equalizer. Other typical multi-band equalizer frequencies are spaced by 1/2 octave and 1/3 octave. The higher the number of bands, the higher the Q of the filters must be.

Ten-Band Graphic Equalizer Using Gyrators

Figure 20.14 shows the design of a ten-band graphic equalizer, also referred to as an octave-band equalizer [9]. Only three bands are shown. Component values for all ten bands are shown in Table 20.2. This design uses the equivalent of a series-resonant LC circuit for each band. However, the inductor is realized with a gyrator. This is a proportional-Q equalizer wherein at ±12 dB maximum boost/cut, $Q = 1.5$. Q is substantially less for smaller amounts of boost or cut and the frequency of maximum boost or cut moves away from the nominal f_0. Three of the ten tuned

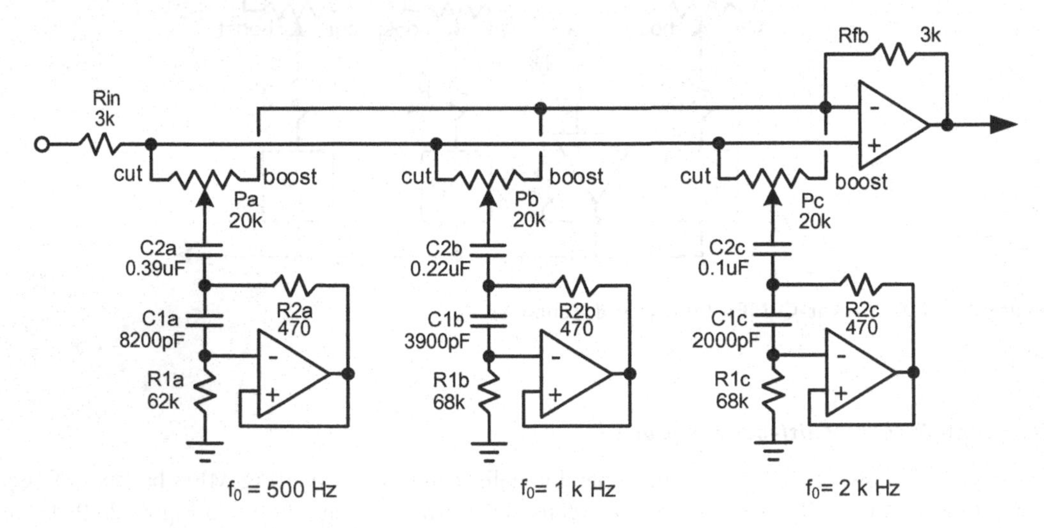

Figure 20.14 Ten-Band Graphic Equalizer Using Gyrators.

Table 20.2 Component Values for Graphic Equalizer

Frequency	*C1(n)*	*C2(n) (μF)*	*R1(n) (kΩ)*	*R2(n) (Ω)*
32	0.12 μF	4.7	75	500
64	0.056 μF	3.3	68	510
125	0.033 μF	1.5	62	510
250	0.015 μF	0.82	68	470
500	8200 pF	0.39	62	470
1 kHz	3900 pF	0.22	68	470
2 kHz	2000 pF	0.1	68	470
4 kHz	1100 pF	0.056	62	470
8 kHz	510 pF	0.022	68	510
16 kHz	330 pF	0.012	51	510

circuits are shown, with f_0 of 500 Hz, 1 kHz and 2 kHz. These LCR-based graphic equalizers are very popular, but tend to fall short in performance as a function of boost or cut. If you are looking for superior performance in a graphic equalizer, there are better choices [10].

The P1, C1, C2, R1, R2 network is repeated nine times, resulting in ten such circuits. The CCW and CW ends of all of the 20-kΩ pots are connected in parallel, as shown in Figure 20.14. This results in a combined 2-kΩ load across the inputs of the main op amp. For purposes of noise gain, the 3-kΩ input resistor R_{in} also must be considered. The net load is then about 1.2 kΩ. The feedback attenuation factor is about 4.2/1.2 = 3.5, and noise gain is 3.5. An op amp with low input voltage noise should be used.

20.8 Parametric Equalizers

A parametric equalizer focuses on a single frequency band f_0 whose frequency is adjustable over a wide range. It can create boost or cut at and around f_0. Its bandwidth, or Q, can be adjusted so that the band of frequencies affected is wider or narrower. The bandwidth affected follows the inverse of the Q. In fact, the bandwidth of a parametric equalizer (PE) is sometimes denoted in octaves, such as 0.5 octave to 3 octaves. The PE thus provides very targeted equalization [7]. It's boost/cut, f_0 and Q are all adjustable. A PE is most often based on a state variable filter (SVF).

The PE provides remarkably flexible equalization that can with precision affect a particular band of frequencies. The PE can provide a very sharp cut or notch to deal with a resonance. They typically can provide an equalization range of ±12 dB. The PE is more complex and expensive than one band of a graphic equalizer, but few of them are needed due to their great flexibility. Sometimes two or more PE's with limited and overlapping f_0 range will be employed. It is difficult to build a single PE that can be adjusted over the full audio range (at least without range selection switches).

One can implement a PE by just substituting an SVF for the BPF in one of the equalizer circuits previously discussed, such as in Figure 20.6(c) and (d). Like other bandpass equalizers, some PEs provide a proportional-Q frequency characteristic, where the sharpness (Q) of the response decreases as less *EQ* is selected. Others provide a constant-Q characteristic where the relative sharpness is unchanged as amount of equalization is changed. The SVF gain at f_0 must not change with changes in its Q setting for either design. In equalizer circuits where the boost-cut pot feeds the SVF, as in Figure 20.6(c), pot loading by the SVF should be kept light. In the constant-Q circuit of Figure 20.6(d), with two inverters in tandem, the SVF is fed by an op amp output, so loading by the input of the SVF is not an issue.

A PE is more complex than one band of a graphic equalizer, but it allows the frequency and bandwidth of each band to be independently controlled. Because of the greatly improved targeting and flexibility of the PE, one will likely require far fewer bands to obtain the correction that is needed. Sometimes as few as two parametric equalizes are needed, each with different frequency ranges that have some overlap. The equalizers will usually be put in tandem. Using fewer parametric equalizers instead of a graphic equalizer with many bands means that lower noise will likely be achieved.

Parametric Equalizer Bandpass Filter

Although there are exceptions, the BPF in a PE is usually a state variable filter (SVF). The gain at f_0 of an ordinary SVF often increases in proportion to its Q. This is common in numerous other BPFs as well. The key attributes of the SVF for use in the constant-Q circuit of Figure 20.6(d) are that it be non-inverting and that its gain does not change as Q is changed. The first attribute is easy to achieve, even if an inverter is needed to make it so. There are many SVF topologies, but the gain for many of them increases as Q is set to a higher value. There are several ways to correct this, but some require a dual-gang pot for adjusting Q, where the second gang is used as a compensating level control on the output of the SVF to make the gain independent of Q. This works because gain at f_0 is proportional to Q. Note that the BPF gain ahead of the Q pot in such an arrangement will still increase in proportion to Q, so the possibility of overload at high Q settings must be considered.

A better approach is shown in Figure 20.15. One inverting op amp stage U4 is added in the damping path, and Q is set by varying the gain of this stage. Greater gain means more damping and lower Q. The output of U4 has the desired constant-gain BPF characteristic. This version of the SVF is also conveniently non-inverting.

Gain at f_0 is 3 due to the combination of R3 and R5 loading the feedback resistor around U1. In some cases an attenuator may be placed at the input of the BPF to reduce gain at f_0 to unity. As shown, this SVF has a high-impedance input and a low-impedance output.

If an inverting SVF is needed, as in the equalizer arrangement of Figure 20.6(c), the SVF input can be applied to the inverting input of U1 though a resistor. Bear in mind that the Q of an equalizer at different boost/cut settings can be evaluated in the usual way, using the gain at f_0 for a particular

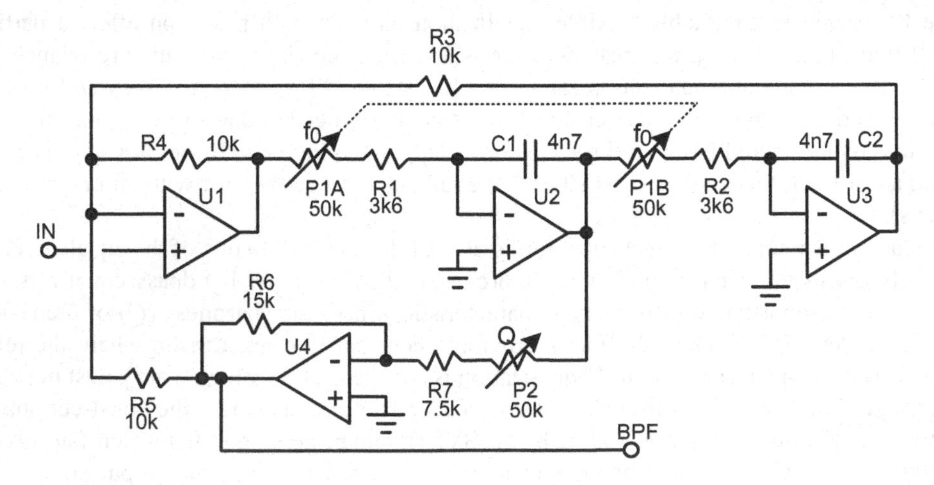

Figure 20.15 Parametric Equalizer BPF Employing an SVF.

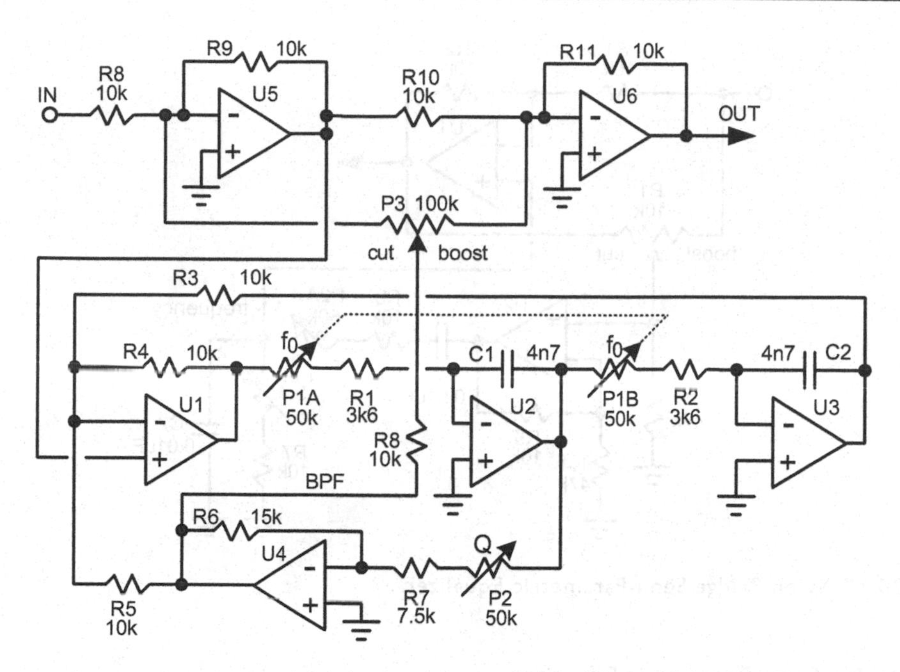

Figure 20.16 Constant-Q Parametric Equalizer.

boost/cut setting and dividing f_0 by the bandwidth span at the frequency where the amplitude is down 3 dB from the peak. At less than maximum boost/cut, the skirts of the proportional-Q equalizer will be wider than those of the constant-Q design.

A similar SVF constant-gain arrangement can be achieved if U4 is replaced with a non-inverting buffer and Q is controlled by a pot in front of the non-inverting input of U4. The change in damping feedback polarity can be accommodated by connecting the damping signal to U1's positive input. There are many possibilities.

Constant-Q Parametric Equalizer

A complete constant-Q PE is shown in Figure 20.16. It has an f_0 range of 750 Hz to 10 kHz and a Q range of 0.5–3.8. Boost/cut range is ±12 dB. It would usually be accompanied by another PE in tandem whose frequency range was 50 Hz to 1 kHz. Together the two PEs with overlapping frequency ranges would cover the most important parts of the audio band that might need a PE.

20.9 Semi-Parametric Equalizers

A less expensive compromise is a so-called semi-parametric equalizer (sometimes called a quasi-parametric equalizer) whose frequency and boost/cut can be adjusted, but whose Q is fixed. Eliminating the Q control also saves space on the front panel. Such an equalizer may have two to four switched Q values available, in which case it is often called a quasi-parametric equalizer. A PE without Q control is largely the same as a swept-band of a graphic equalizer. The term "quasi-parametric" equalizer is also often used to describe such a PE. However, a PE with a few fixed Q values that can be selected by a switch is also referred to as a quasi-parametric equalizer. If the Q is fixed at one value, $Q = 1.5$ is often used.

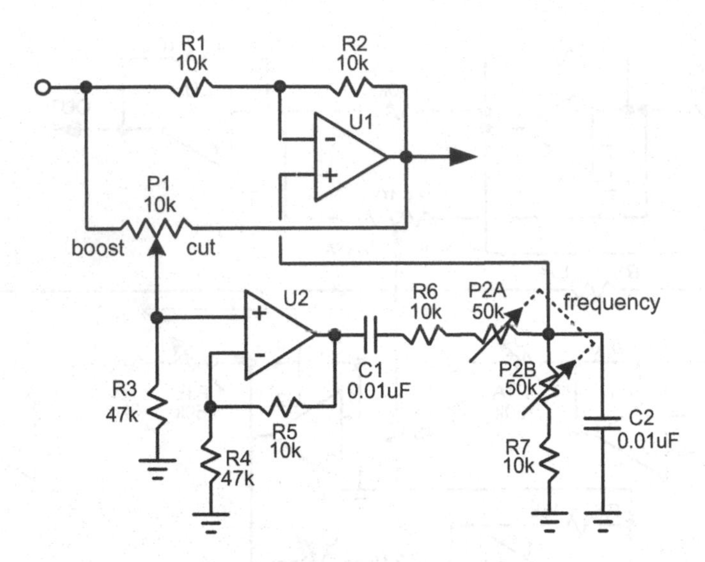

Figure 20.17 Wien Bridge Semi-Parametric Equalizer.

Wien Bridge Semi-Parametric Equalizer

Figure 20.17 shows a semi-parametric Wien bridge equalizer, where only frequency and boost/cut can be controlled. It employs the equalizer architecture of Figure 20.6(c) and only requires 2 op amps. U2 is a buffer for the Wien bridge and supplies some gain to achieve the desired equalization range.

20.10 Shelving Equalizers

Shelving equalizers operate at both ends of the audio band, providing boost or cut to a portion of the spectrum called a shelf. A low shelf increases or decreases low frequency amplitude beginning at f_2 as frequency goes down and stops increasing or decreasing amplitude at some lower frequency f_1. It is usually a first-order function with an ultimate slope of 6 dB/octave.

A low shelf has a pole followed by a zero when boosting low frequencies, with unity gain at high frequencies. Shelf equalizers can usually provide about ±15 dB of boost or cut. Apart from having adjustable turnover frequency where the shelf begins (f_2), they are not much different than a bass control whose turnover frequency is adjustable.

A high shelf increases or decreases high-frequency amplitude beginning at f_1 as frequency goes up and stops increasing or decreasing amplitude at some higher frequency f_2. It is usually a first-order function with an ultimate slope of 6 dB/octave. A high shelf has a zero followed by a pole when boosting high frequencies, with unity gain at low frequencies. When cutting high frequencies, it has a pole followed by a zero. Apart from having adjustable frequency, the high shelf is not much different than a treble control.

Simple first-order shelving equalizers are essentially like bass and treble controls, and can be implemented with Baxandall circuits whose turnover frequency can be continuously adjusted. However, that is difficult to do with a Baxandall circuit, even if the low shelf and high shelf circuits are made separate. Instead, first-order shelf equalizers can be implemented with the circuit of Figure 20.6(c), which is drawn with a Wien bridge frequency shaping circuit. The Wien bridge network is replaced with a simple *RC* low-pass filter for a low shelf or a high-pass filter for a high shelf.

The simple first-order shelving equalizers have a pole followed by a zero or a zero followed by a pole as frequency is increased. They have an asymptotic 6-dB per octave slope, but often the actual incremental slope is less than 6 dB per octave because the pole and zero are not sufficiently separated in frequency to achieve full slope. The amount of boost or cut depends on how far apart the zeros are from the poles. They can boost or cut with a shelf-like frequency response.

The boost or cut corresponds to movement of the poles and zeros, with greater equalization corresponding to the poles and zeros being further apart. Depending on the particular design, changing the amount of equalization may affect one or the other of the poles or zeros, or both. With large separation of the poles and zeros, the equalization slope gets closer to 6 dB/octave.

Second-Order Shelf Equalizers

Sometimes a steeper shelf response is needed. In this case a second-order shelf equalizer can be created by employing a second-order filter. Second-order shelves can provide an equalization slope that asymptotes to a maximum of 12 dB/octave when the poles and zeros are spaced far apart in frequency. The transfer function of such an equalizer has one pair of complex poles and one pair of complex zeros. Such a filter is called a biquad, since there are quadratic polynomials in both the numerator and denominator of the transfer function. The biquad function is usually realized with an SVF (see Sections 11.6 and 11.8). It is notable that the Linkwitz equalizer for loudspeakers is a second-order low shelf equalizer for closed-box woofers with a second-order roll-off. The Linkwitz equalizer is a biquad shelf equalizer that uses only one op amp, but it requires four capacitors and six resistors, making it difficult to implement variable tuning. Second-order shelving equalizers are more complex and less common, and will not be discussed here for lack of space.

20.11 Low-Cut and High-Cut Filters

Low-cut and high-cut filters are just high-pass and low-pass filters, respectively. They are usually of second or third order, and often implemented with Sallen-Key filters. Some have a fixed cutoff frequency, but many have a variable or switch-selected cutoff frequency, making them more useful. Most microphone preamplifiers have a fixed-frequency low-cut filter that can be switched in or out. Unlike shelving equalizers, these filters only attenuate low and high frequencies, respectively, and the attenuation continues to increase indefinitely out of the audio band.

A continuously variable second-order set of low-cut and high-cut filters using Sallen-Key filters is described in Ref. [15]. The Sallen-Key filters are swept in frequency by use of a dual-gang pot, with both gangs of the same value. The low-pass filter is easily made with variable resistors of equal value, with the shunting capacitors having differing values that determine the Q.

The high-pass Sallen-Key filter usually requires differing values for the shunt resistors R1 and R2. As discussed in Chapter 11 (Figure 11.1), the first shunt resistor R1 in these filters is often connected to the output of a VCVS (voltage-controlled voltage-source) with gain of +1. R1 is basically bootstrapped by the output. By adding some gain to the VCVS (about 4 dB) the Q of the filter can be made to be that of a Butterworth filter while using tuning resistors of equal value [15].

References

1. Steve Dove, *Designing a Professional Mixing Console – Part Seven – Equalizers 1*, Studio Sound, April 1981.
2. Steve Dove, *Designing a Professional Mixing Console – Part Eight – Equalizers 2*, Studio Sound, May 1981.

3. Steve Dove, *Designing a Professional Mixing Console – Part Nine – Equalizers 3*, Studio Sound, June 1981.
4. Ethan Winer, *The Art of Equalization*, Popular Electronics Magazine, August 1979, available at ethanwiner.com/equalizers.html.
5. Ethan Winer, *Audio Filters – Theory and Practice*, Recording/engineer/producer magazine, August 1981, available at ethanwiner.com/filters.html.
6. Craig Anderson, *The 5 Types of Mixer EQ – And How to Use Them*, Full Compass Live, January 2020.
7. Dennis A. Bohn, *Operator Adjustable Equalizers: An Overview*, Rane Corporation RaneNote 122, August 1997, ranecommercial.com/legacy/note122.html.
8. Dennis A. Bohn, Constant-Q Graphic Equalizers, Rane Corporation, *Journal of the Audio Engineering Society*, vol. 34, no. 9, pp. 611–626, September 1986. Also at ranecommercial.com/legacy/note101.html.
9. TI/National Semiconductor, *10 Band Graphic Equalizer*, LM-833-N datasheet, May 2012.
10. Rod Elliott, *Expandable Graphic Equalizer*, Project 75, Elliott Sound Products, sound-au.com/project75.htm, February 2001.
11. Ethan Winer, *Spectrum Analyzer and Equalizer Designs*, Recording-engineer/Producer Magazine, February 1982, available at ethanwiner.com/spectrum.html.
12. Peter Baxandall, Negative Feedback Tone Control – Independent Variation of Bass and Treble without Switches, *Wireless World*, vol. 58, October 1952, p. 402.
13. Rod Elliott, *Wien Bridge Based Parametric Equalizer*, Project 150, Elliott Sound Products, 2014, sound-au.com/project150.htm.
14. Mackie Onyx Mixer, Mackie Corporation, Mackie.com.
15. Rod Elliott, *Variable High-Pass and Low-Pass Filters*, Project 155, Elliott Sound Products, 2015, sound-au.com/project155.htm.

Chapter 21

Headphone Amplifiers

High-performance headphones are capable of reproducing exceptional sound quality. For this reason, the headphone amplifier (HPA) often included in preamplifiers, mixers and other equipment should be more than just an afterthought. In fact, a good HPA shares many things in common with a quality power amplifier. Although it does not need to produce as much power, it must operate with a wide range of load impedances and with a reactive load whose impedance may vary over frequency and headphone models.

Conversely, the HPA is much like a line-level amplifier, but with a higher output current capability. This suggests the use of op amps with a discrete output buffer inside the feedback loop. Because of the lower voltages involved, there is usually less to be gained by making an all-discrete HPA as is usually the case with a power amplifier [1]. The HPA should be able to produce peak sound pressure levels (SPLs) of at least 116–128 dB SPL into headphones with widely varying nominal impedance and sensitivity. The reason for such high peak power capability is the high crest factor of well-recorded music. Rated headphone impedance ranges from about 16 Ω to about 600 Ω. An HPA must be designed for extremely low noise, including noise immunity from EMI. This is especially true for low-impedance headphones that have high voltage sensitivity.

21.1 Headphone Types

Most headphones are dynamic, but there are planar magnetic and electrostatic ones as well. Dynamic headphones are reasonably priced and ubiquitous, while planar magnetic headphones tend to be more expensive, but promise lower distortion and greater accuracy. Electrostatic headphones are very expensive and are usually limited to the stratospheric hi-end of the breed. The key characteristics of headphones that pertain to design of the HPA are their nominal impedance, their impedance characteristic and their sensitivity.

Dynamic Headphones

Dynamic headphones are the most popular type and typically the lowest in cost. They employ an electromagnetic transducer with many of the same properties as loudspeakers, but of course they do not include a crossover. As with a loudspeaker, they have a diaphragm that is pushed and pulled by a voice coil attached at the center of the diaphragm. Because the driving force originates at only one place, the diaphragm can distort in shape, causing distortion and nonlinearity. Also like a loudspeaker, the dynamic headphone has a low-frequency resonance dictated by the mass of the diaphragm and voice coil in combination with the compliance of the diaphragm and the enclosed air. The resonance results in a peak in the impedance. The dynamic headphone model for the Sony

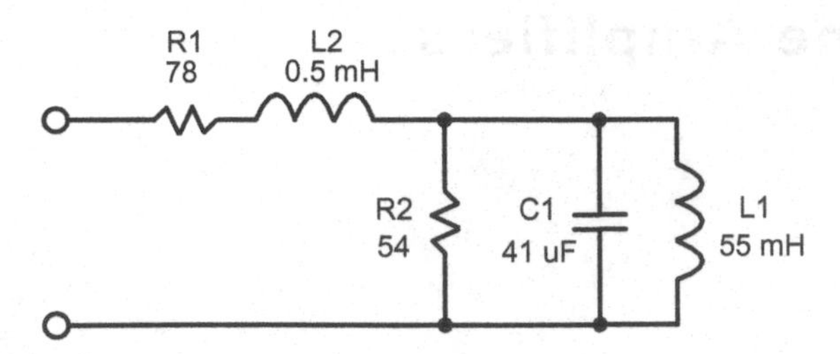

Figure 21.1 Electrical Model for Sony MDR-XD200 Dynamic Headphones.

MDR-XD200 in Figure 21.1 illustrates this. Because of the varying impedance with frequency, dynamic headphones perform most faithfully when driven by an HPA with very low output impedance. This is especially the case for low-impedance headphones.

Dynamic headphones have a very wide range of nominal impedance and sensitivity. Impedance ranges from 16 to 600 Ω, with most falling between 32 and 300 Ω. This means that an HPA able to drive all dynamic headphones to high peak SPL must be capable of both high voltage swing and high current swing.

The Sennheiser HD 650 headphone is an excellent example of a high-impedance dynamic headphone, with rated impedance of 300 Ω and sensitivity of 103 dB/mW. Its measured impedance ranges from 360 to 540 Ω from 20 Hz to 20 kHz. This open-back headphone has a long-term power rating of 200 mW, corresponding to 7.75 V_{RMS} into 300 Ω. This corresponds to 115 dB continuous SPL [2].

Planar Magnetic Headphones

Planar magnetic headphones, sometimes referred to as "isodynamic," also operate on the principle of current flowing through a wire connected to a diaphragm and fixed permanent magnets. However, the diaphragm in this case is flat (planar) and is driven across its full surface. The wires are implemented as traces on the thin diaphragm. The magnets are on one or both sides of the diaphragm, the latter making possible a push-pull arrangement with lower distortion. The diaphragm being uniformly driven across its full surface results in much less deformation and allows the use of a thinner and lighter diaphragm. These headphones are typically more expensive than dynamic headphones. They also tend to be bulkier and heavier in some cases.

Planar magnetic headphones can be more detailed and accurate, especially in the mid- and high frequencies. They are "faster" because of the more intimate relationship between the coil and the diaphragm. This results in better transient response. They are sometimes a little harder to drive than dynamic headphones and benefit from an HPA with good output current capability. This is partly due to the often-low resistance of the "coil" – copper or aluminum traces on the diaphragm. They can also sometimes be less power-efficient than dynamic headphones. On the other hand, the load they present is almost entirely resistive with largely uniform impedance across the audio band. This makes them easier to drive in order to achieve a flat frequency response with an HPA whose output impedance is not as low as might be desired.

Audeze and HIFIMAN are two of the bigger names in planar magnetic headphones. Some planar magnetic headphones can be bulky and heavy because more magnets must be placed on one or both sides of the film diaphragm. The open-back HIFIMAN Sundara is a good example of a mid-priced low-impedance planar magnetic headphone, with rated impedance of 37 Ω and sensitivity of 94 dB/mW [3]. Its measured impedance is flat with frequency at 44 Ω from 20 Hz to 20 kHz. They require about 120 mW to achieve 115 dB. At their 37-Ω impedance, this corresponds to 2.1 V_{RMS} and 57 mA_{RMS}.

The Audeze LCD-X planar magnetic headphones have rated impedance of only 22 Ω and power sensitivity of 96 dB SPL/mW. However, their voltage sensitivity is 112.5 dB/V_{RMS}. This means that to drive them to 115 dB, they require voltage of about 2.5 dBV, or about 1.33 V_{RMS}. This corresponds to 60 mA_{RMS}, or 80 mW.

Electrostatic Headphones

Electrostatic headphones are prized for their high-frequency response and "speed". These are the most expensive headphones, and are revered for the transparency of their sound. They operate on the principle of a thin 1.5-μm Mylar diaphragm with an electrostatic charge suspended between two electrodes or "stators". The electrodes are energized with signal so as to either attract or repel the diaphragm. One electrode is in front and one is in the rear. They have opposite signal voltages applied to them and they thus operate in a push-pull fashion. The extremely thin diaphragm does not have to bear the mass of a coil and so is extremely light. Operation is essentially "touch-less".

Electrostatic headphones require high signal voltages and a high-voltage bias, provided by a so-called "energizer". Signal voltages can easily be on the order of 100 V_{RMS}, sometimes nearly 400 V_{RMS}, and bias voltages can be over 200 V_{DC}, sometimes nearly 600 V_{DC}. For this reason, electrostatic headphones require the energizer. Passive versions of this device usually comprise a step-up transformer and a bias voltage supply. The bias is applied to the center tap of the secondary, typically through a 1 MΩ resistor. The high-voltage audio signals appear in both polarities at the ends of the secondary, and they drive the electrodes, sometimes called stators. The diaphragm film is charged to the bias voltage level, typically through a very high resistance, like 27 MΩ. It can actually take seconds to charge the diaphragm to its nominal operating voltage. The primary of the transformer is fed from a small audio power amplifier with rated output power of at least 5 W.

The need for a step-up transformer with a significant turns ratio opens the door to distortion, especially at low frequencies. A typical Stax electrostatic headphone (they like to call them "earspeakers") requires 100 V_{RMS} to produce SPL of 100 dB. The sensitivity of electrostatic headphones is usually quoted as dB SPL/100 V_{RMS}. The step-up ratio can be as great as 100, allowing a 1-V input to produce 100 dB SPL. A 1:100 step-up ratio will result in a 1:10,000 impedance transformation. The typical 120-pF headphone capacitance can thus be transformed to a rather significant 1.2 μF.

Active energizers are more common in modern electrostatic headphone systems. These are vacuum tube amplifiers or solid-state amplifiers using high-voltage transistors. This approach eliminates the need for a transformer. As a result, they are capable of much better performance, especially at low frequencies. These amplifiers can produce as much as 400 V_{RMS} for versions of the headphones that employ a 580-V_{DC} bias. Such a system can produce 112-dB SPL. A corresponding peak voltage of 566 V_{PK} at 20 kHz will require a slew rate of 71 V/μs. To achieve this slew rate, 8.5 mA is required to pass through the 120 pF of headphone capacitance.

The Stax SR-009S electrostatic headphones have an impedance of 110 pf and 145 kΩ. They require 580 V_{DC} bias and their sensitivity is 101 dB/100 V_{RMS}. In order to reach its maximum rated SPL of 188 dB SPL, the driving voltage must be 17 dB above 100 V_{RMS}, which comes to 707 V_{RMS} which corresponds to 1000 V_{pk} [4].

Given the realities of driving a step-up transformer with a minimum of 5 W or using a high-voltage amplifier, there appears to be no role for conventional HPAs in driving electrostatic headphones.

21.2 Sound Levels and Load Impedance

The maximum SPL target for the amplifier-headphone combination is subject to a great deal of judgment and estimates of dynamic range and crest factor of the music to be played. Most headphones are capable of producing 115–130 dB SPL, at least for brief transients. The HPA should not clip below the power required to drive the headphone to its maximum power level. It is important to emphasize that this is for reasons of operating margin against brief transient peaks in SPL due to large dynamics in the source material. It should be clear that this is not being discussed in the context of hearing damage for an extended period of time, which will occur at much lower SPL [5, 6]. The National Institute for Occupational Safety and Health (NIOSH) specifies a safe exposure time of only 500 ms for an SPL level of 130 dB.

For context, the safe distance from gunfire without hearing protection has been characterized as being between 50 and 200 m for large-caliber weapons, corresponding to a safe peak SPL of 140 dB [7]. The peak SPL of an assault rifle was estimated to be 154 dB at a distance of 4 m from the muzzle. As with power amplifiers for loudspeaker systems, there should always be a healthy amount of power margin for best sound, especially for solid-state amplifiers. It is also important to bear in mind that an amplifier should never clip. For this reason, the nominal peak SPL target for this discussion will be 120 dB SPL, with the HPA operating at least 1 dB below clipping.

Sound Pressure Levels and Sensitivity

The SPL for a loudspeaker in the context of loudspeaker sensitivity is usually referenced to a distance of 1 m, so a typical specification might be 88 dB at 1 m for 1 W (at 8 Ω). More specifically, the reference power level is often considered to be occurring at an applied voltage level of 2.83 V_{RMS}, which corresponds to 1 W into a resistive 8-Ω load. For headphones, there is no appreciable listening distance, so power sensitivity is rated as SPL for 1 mW of input power to the rated load resistance. Power sensitivity will usually range from 97 to 107 dB SPL/mW [8].

Headphone voltage sensitivity is another way of specifying sensitivity. It is specified as dB SPL at 1 V_{RMS}. It often ranges from 112 to 120 dB. Note that 1 V_{RMS} fed to a typical headphone resistance of 70 Ω will result in a current of 14.3 mA_{RMS}, corresponding to 14.3 mW. Put another way, 1 V_{RMS} results in 1 mW when applied to a load of 1000 Ω. The voltage sensitivity of a headphone is not related to the power sensitivity in a linear way.

The use of two different ways of specifying headphone sensitivity is a constant source of confusion for many. In fact, some manufacturers will specify sensitivity as something like 97 dB without indicating whether it is 97 dB/mW or 97 dB/V_{RMS}. Specifying sensitivity with respect to 1 mW is somewhat more traditional, while some manufacturers now specify sensitivity with respect to 1 V_{RMS}. Most modern HPAs are voltage amplifiers with a low-impedance output. What they produce is voltage, not power. For this reason, it makes more sense to specify headphone sensitivity with respect to voltage. With modern HPAs, this is a far better indicator of perceived loudness of different headphones at a given volume control setting. For consistency, if a headphone's sensitivity is quoted in dB SPL at 1 mW, that number should be converted to the more useful voltage sensitivity for purposes of sensitivity comparison.

Consider a headphone rated at 100-dB SPL/mW having impedance of 100 Ω. It will require 0.316 V_{RMS} to produce 100-dB SPL. If that headphone is supplied with 1 V_{RMS}, it will produce 10 dB greater SPL, having a voltage sensitivity of 110 dB/V_{RMS}. Now consider a headphone rated at 100 dB SPL/mW having impedance of 10 Ω. It will require 0.1 V_{RMS} to produce 100-dB SPL. If that headphone is supplied with 1 V_{RMS}, it will produce 120-dB SPL, for a voltage sensitivity of 120

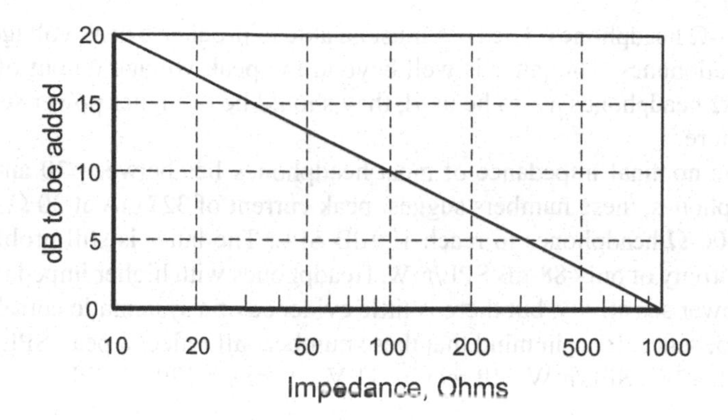

Figure 21.2 Voltage Sensitivity as a Function of Impedance for 100 dB/mW.

dB/V_{RMS}. For most values of headphone impedance, voltage sensitivity in dB/V_{RMS} will be greater than power sensitivity in dB/mW. Only at an impedance of 1000 Ω are voltage and power sensitivity numbers the same. A useful relationship here is:

$$V = \sqrt{P \times R}$$

where V is the RMS voltage required to produce power P into R ohms. A sensitivity specified in dB/mW can be converted to dB/V_{RMS} by the following formula [8]:

$$\text{dB/V}_{\text{RMS}} = \text{dB/mW} + 20 \times \log\left(\sqrt{1000 / R}\right)$$

The graph in Figure 21.2 shows how many dB to add to power sensitivity to obtain voltage sensitivity as a function of impedance.

Peak Power, Voltage and Current Requirements

A compilation of headphone sensitivity and power requirements is provided in Ref. [9]. Due to inconsistencies in how manufacturers rate headphone sensitivity, some numbers in this compilation may be questionable.

Nominal load impedance for headphones can range from 16 to 600 Ω. This means that a headphone amplifier fit to drive all such headphones must be capable of producing fairly high voltages as well as fairly high currents. Headphone power sensitivity also covers a wide range spanning a low of about 88 dB SPL/mW to a high of 110 dB SPL/mW.

Consider two headphones with average sensitivity of 100 dB SPL at 1 mW. Assume that the target peak SPL is 120 dB. Both headphones thus require 100 mW of input power to reach the target SPL. The first pair has a nominal impedance of 16 Ω. It requires only 1.26 V_{RMS}, but requires 79 mA_{RMS}, corresponding to peak current of 112 mA. The second pair has a nominal impedance of 600 Ω. It requires 7.7 V_{RMS}, corresponding to peak voltage of 10.9 V_{PK}, but only 13 mA_{RMS}.

Now consider a pair of low-sensitivity headphones with 88-dB SPL/mW sensitivity and having the same nominal impedances. The 12-dB drop in sensitivity corresponds to a factor of 16 in power. What previously required 100 mW now requires 1.6 W. Correspondingly, the required peak voltages and currents increase by a factor of 4. The HPA must be able to produce peak current of

448 mA for the 16-Ω headphones. The HPA must be able to produce a peak voltage swing of 43.6 V for the 600-Ω headphones. The latter is well beyond the peak voltage output of most solid-state HPAs, so if 600-Ω headphones are to be used, they should be of higher power sensitivity, close to 100 dB/mW or more.

Fortunately, the nominal impedance of most headphones lies between 30 and 300 Ω. For the 88-dB/mW headphones, these numbers suggest peak current of 327 mA at 30 Ω and peak voltage of 31 V for the 300-Ω headphones to reach 120 dB SPL. The latter is still problematic for headphones with sensitivity of only 88-dB SPL/mW. Headphones with higher impedance may be likely to have higher power sensitivity, but there is little evidence of a systematic correlation of sensitivity with rated impedance. Bear in mind that these numbers all reflect a peak SPL target of 120 dB. A 300-Ω pair with 97 dB SPL/mW will require 11 V_{PK} to reach 120 dB SPL.

Headphone Impedance

Because the actual impedance of dynamic headphones can vary with frequency, the output impedance of the HPA should be very low compared to the minimum impedance presented to the HPA by the headphones (usually at mid-band frequencies). Otherwise, frequency response may be uneven and the resonance of dynamic headphones may be under-damped. This requirement is analogous to having a high damping factor for a power amplifier. Because rated headphone impedance can be as low as 16 Ω, the damping factor of an HPA should be not unlike that of a power amplifier. A damping factor of 50 with respect to 16 Ω corresponds to output impedance of 0.32 Ω over the full audio frequency range. Some headphone impedances may drop below the nominal rated value, just as do most loudspeakers.

Figure 21.3 shows the measured impedance as a function of frequency for three different headphones. The top curve shows the impedance of the Sennheiser HD-650 dynamic headphone that

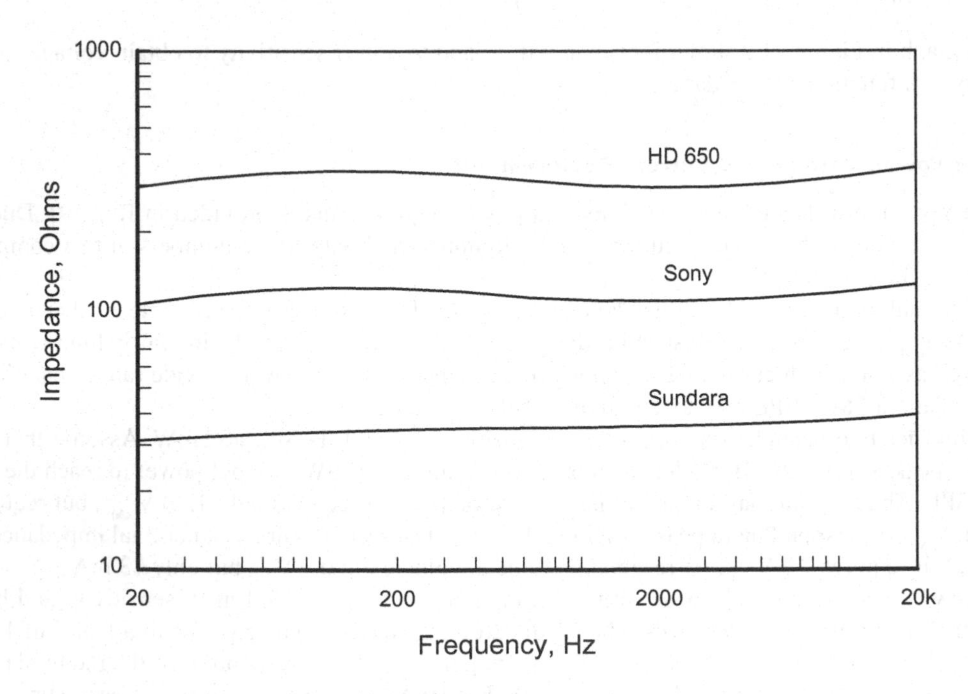

Figure 21.3 Measured Impedance of Three Headphones.

is rated at 300 Ω. The middle curve shows the impedance of a pair of older Sony dynamic headphones (MDR-XD200) whose nominal impedance measures about 70 Ω. The bottom curve depicts the impedance of the HIFIMAN Sundara planar magnetic headphones rated at 37 Ω. The HD-650 shows a gentle impedance peak at about 100 Hz, while the Sony shows a sharper peak at about 100 Hz. Both show a bit of a rise in impedance heading toward 20 kHz as a result of some coil inductance. There is little evidence of either one presenting a difficult reactive load as would often be the case with a loudspeaker. The Sundara impedance curve is remarkably flat at about 44 Ω out to well beyond 100 kHz. The planar magnetic headphones present a virtually perfect resistive load, albeit with fairly low impedance. If the Sundara is driven with a peak voltage of 15 V, about 340 mA will be required to drive it.

Headphone impedances as a function of frequency do not typically vary as much as those for loudspeakers. Headphones do not have the low-end box resonances that loudspeakers do, nor do they have crossovers. This does not mean to say that headphones are without any resonances.

21.3 Headphone Sensitivity and Drive Requirements

The wide range of both sensitivity and impedance of headphones means that the HPA must be capable of driving high impedances or low impedances to significant power levels. This can be a tall order, stressing both the voltage and current delivery capabilities of the amplifier. Given a 100-dB SPL/mW sensitivity and a more generous target peak SPL of 130 dB, the HPA must be able to drive 1 W into the impedance of the headphone being used, which can reasonably range from 16 to 300 Ω.

The former requires peak current of 354 mA from the HPA. This is definitely achievable on a transient basis with a hefty power supply or at least one that has a substantial amount of reservoir capacitance.

The latter requires peak voltage swing of 24.5 V, requiring about ±27-V rail voltages in practice. This amount of rail voltage is greater than what most op amps can handle, and is not readily available in small-signal electronics power supplies unless the unregulated supply voltage is unusually high. Unless discrete circuitry is used, the best one can usually do is to use regulated ±18-V supplies and achieve 16-V peak output swing, corresponding to about 426 mW into 300 Ω. If the headphones have sensitivity of 100 dB/mW, maximum SPL will be about 126-dB SPL.

All of this assumes that the power sensitivity of the 300-Ω headphone is not greater than that of the 16-Ω headphone. In practice, it is likely that the sensitivity in terms of power for the 300-Ω dynamic headphone is higher. There is little evidence of a systematic correlation of sensitivity with rated impedance, however. Headphones with high voltage sensitivity make low noise in the amp more important. Such headphones have become more common with the rise in mobile device use.

21.4 Headphone Amplifier Design Considerations

The design of a solid, high-performing HPA is not unlike that of a small power amplifier, and more on that can be found in Ref. [1]. However, the smaller amount of required power output and the higher impedances to be driven make available more choices in the design. These can make the HPA more compact and less expensive. Most notably, the lower voltages allow the use of operational amplifiers for most headphones of reasonable sensitivity, even if the output stage is still discrete for reasons of current delivery.

There is also a greater opportunity to operate the output stage in pure class A up to an average listening level for many headphones, while transitioning into class AB when higher output current is required. All of the designs discussed here have single-ended outputs (not balanced or bridged). Although balanced or bridged designs could enable more SPL with high-impedance headphones, they would often double the output current requirements when driving low-impedance headphones to the point of being too large. Such amplifier designs are discussed in Ref. [1].

Power

HPA power output is often specified into 32 Ω, and usually ranges from 0.3 to 2 W. Even 2 W into 32 Ω requires only 11.3 V_{PK} and 353 mA_{PK}. However, based on the discussions above, driving 2 W into significantly higher or lower impedances can become difficult. For that reason, even an HPA rated at 2 W into 32 Ω may be significantly less capable driving 16- or 64-Ω headphones, not to mention 300-Ω headphones. A fully discrete design powered from unregulated 24-V rails, as might be the case in a power amplifier, would provide more power into high-impedance headphones, but would be more costly.

The Rickey Lee Jones album *Flying Cowboys* contains a track called "Ghetto of My Mind" that has a whopping 24-dB dynamic range on the originally-mastered CD [10]. This is a factor of 250:1 in peak-to-average power. If you listen to this cut at an average level of 96-dB SPL for a short period of time (less than 30 minutes) [5], your HPA and headphones will have to deliver 120 dB SPL cleanly and without clipping. For comparison, a high-dynamic-range rendition of the 1812 Overture clocks in at 18 dB. Unfortunately, I've heard that *Flying Cowboys* was re-mastered several years ago with less dynamic range. If you have to turn up your volume control so that its average sound level is comparable to most CDs, it is likely the original one with high dynamic range. They had to make room in the dynamic range for the peaks.

A clone of the MOSFET Audio-Technica AT-HA5000 HPA produced 2.4 W into a 16-Ω load and 240 mW into a 300-Ω load. Its output impedance measured 0.2 Ω. The Benchmark Media HPA4 is rated at 6 W into a 16-Ω load and 11.5 V_{RMS} into a 300-Ω load (440 mW) [11].

Gain

Because the maximum volume line level to a power amplifier is often on the order of 1.0 V_{RMS}, and since maximum level into headphones is usually no more than 10 V_{RMS}, the HPA should probably need gain of about 15–20 dB. Note that 10 V_{RMS} will put 330 mW into a 300-Ω headphone. Sensitivity of 100 dB SPL/mW would then yield about 125 dB SPL. This argues for a maximum gain of 20 dB for a HPA.

The large variation in voltage sensitivity of headphones suggests that in some cases a gain switch capable of dropping the gain by 10 dB would be valuable for an HPA. This would enable better use of the volume control range on the preamplifier driving the HPA. As discussed below, many HPAs may be fed directly from an analog source that has not been sent through a preamplifier, such as a CD player or DAC. In this case, the HPA may be equipped with a volume control.

Hum and Noise

Because of the sensitivity of the headphone and proximity to the ear, hum and noise from the HPA must be extremely low. This is especially important because the voltage sensitivity of some low-impedance headphones can be very high. Imagine putting your ear right to the drivers of a loudspeaker fed by a power amplifier. This argues for an HPA input stage with very low noise if a volume control is put in front of it at the very input.

It is not practical to attenuate the output of an HPA to reduce noise because that would ruin the damping factor. However, attenuation can be achieved by reducing the amplifier's closed-loop gain. This is sometimes the way gain is reduced in an HPA with a gain switch. This is different than an ordinary volume control placed in front of the amplifier. A good HPA should be able to operate with different amounts of closed-loop gain, either variable or in fixed steps (such as high, medium, and low). This can have implications for the feedback compensation. Note that a conventional volume control may still be required to achieve silence if there is not one elsewhere in the signal chain.

Noise should be put in context by considering the SPL threshold of hearing (0 dB SPL) and the level of equivalent self-noise of very quiet microphones. Such microphones have self-noise between 10 and 15 dBA (A-weighted SPL), and a few claim to reach 5 dBA.

Consider a 32-Ω headphone with 110-dB SPL/mW power sensitivity. This corresponds to 179 mV into the headphones for 1 mW and 110 dB SPL. One V_{RMS} is greater by a factor of 5.6, corresponding to increased power by a factor of 31.3. This corresponds to a power increase of about 15 dB, bringing voltage sensitivity up to about 125 dB/SPL/V. If we drop down by 120 dB, to what voltage does this correspond? This will correspond to 1 μV for 5 dB SPL at the ear. If we set the audible noise target at 15 dBA, then we have about 3.2 μV, A-weighted.

Now consider an HPA with input-referred noise of 5 nV/√Hz and a gain of 10. Output will be 50 nV/√Hz. The equivalent noise bandwidth (ENBW) of A-weighting is about 13 kHz, corresponding to 114 √Hz. A-weighted noise voltage will be 5.7 μV. This misses the 15 dBA target by about 5 dB. This illustrates how hard it can be to get the noise down to an inaudible level in a reasonable worst-case scenario. A very low noise amplifier with input-referred noise of 2 nV/√Hz and fixed gain of 10 could beat this target by about 4 dB.

Many quality HPAs employ a 50-kΩ ALPS audio taper pot right at the input. The 50-kΩ input impedance does not load most preamplifiers or other audio sources too much. However, note that when the pot is at its −6 dB setting, the resistance seen at its wiper is 12.5 kΩ, which creates thermal noise of about 14 nV/√Hz. Things get much better at greater attenuation settings. At −14 dB, resistance at the wiper is 8 kΩ and noise is 11.9 nV/√Hz. Even at −20 dB, the resistance is 4 kΩ and noise is about 8.9 nV/√Hz. This all assumes that there is no noise created by input current noise of the amplifier stage being driven. Fortunately, when the volume is set to zero, the pot creates no thermal noise.

Volume Control Considerations

As described above, how the volume control on an HPA is implemented is important, especially in regard to noise. As an aide in a deeper understanding of volume controls, you are referred to Chapter 22. Not discussed here, but pertinent is the use of a Baxandall volume control as discussed in Chapter 22. It provides attenuation nearly linear in dB with pot rotation while using a linear-taper pot. The use of relay-switched volume controls is another option that can be attractive as an inter-stage control. These two options also allow the use of smaller resistances to reduce noise. As mentioned earlier, a switch that reduces the closed-loop gain of the output stage can also be effective in keeping the noise down. Since few people change headphone types often, putting such a switch on the back panel may make sense.

Volume Control in a Two-Stage Amplifier

As an alternative to placing the volume control at the very input, one can employ two stages of gain and put a volume control between them, as illustrated in Figure 21.4. The two stages together may

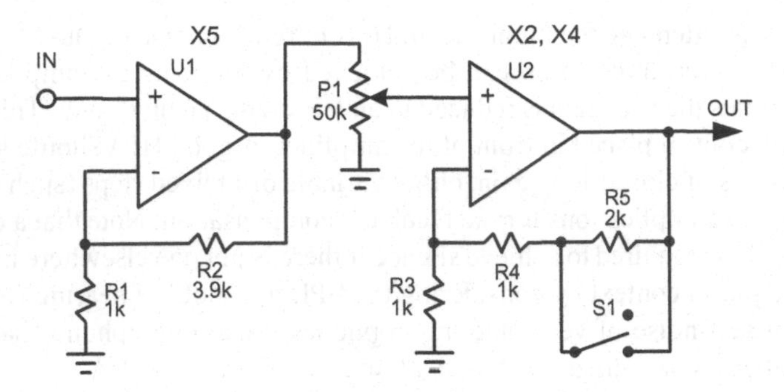

Figure 21.4 Headphone Amplifier with Two Stages.

provide the maximum HPA gain of 10 (20 dB). The second stage, the non-inverting HPA output amplifier U2, may have low gain of 2–4. This keeps the noise down when the volume control is at zero or a low setting. Even with a gain of only 2, the first-stage op amp can easily provide enough voltage swing to drive the output stage to full output with 15- or 18-V rails. Optional switch S1 can select output stage gain of 2 or 4.

The input stage is a non-inverting gain stage with high input impedance and low output impedance. It buffers the volume control, allowing it to be a low-value pot for low noise. The gain of this stage should be 5, 3 or 2.5 to obtain overall gain of 10× with output stage gain of 2, 3 or 4, respectively. This choice of input stage gain can be important in avoiding overload in the first stage. It is undoubtedly safe to make it 2, but at the other extreme, a fixed value of 4 may overload with some hot sources that don't themselves have a volume control. The two-stage approach also supports switched volume controls.

Distortion

Distortion should be well below 0.01% across the full audio band driving any headphone to at least 115 dB SPL, for nominal headphone impedances between 16 and 300 Ω. The HPA should be able to provide high current as needed by low-impedance headphones without increased distortion. Output stage operation can be class AB or class A. For a pair of 32-Ω headphones with 100 dB SPL/mW, 56 mW is required to produce 115 dB SPL. This requires 1.34 V_{RMS} and 42 mA_{RMS}, or 59 mA_{pk}. Quiescent output stage bias of only 30 mA will permit this power to be achieved while still in class A. This is not difficult. Output stage idle dissipation with ±18-V rails will only be 1.1 W. Operating in class A to satisfy this reasonable criteria for 32-Ω headphones of typical power sensitivity is thus quite practical.

Power Requirements and Rail Voltages

Performance driving high-impedance headphones to high peak SPL will be limited by output voltage swing, which will in turn be limited by rail voltages. This suggests that the HPA should operate from ±18-V rails if operating from a regulated power supply. This is the normal maximum supply voltage for operational amplifiers. If operational amplifiers are not in the signal chain, as in a discrete HPA or one using high-voltage ICs like IC power amplifiers, higher voltages can be used, and may not need to be regulated (as is done in audio power amplifiers). An example might be a ±24-V unregulated supply that is used for the HPA and which is regulated down to ±15 or ±18 V for other devices in the same enclosure.

An HPA with ±18-V rails should be able to put out peak swings of ±16 V without distortion, corresponding to 11.3 V_{RMS}. This corresponds to 37.7 mA_{RMS} into a 300-Ω headphone load, or 426 mW. For headphone with average power sensitivity of 100 dB SPL/mW, this will produce 126 dB SPL, which is plenty. For headphones with a somewhat low 90 dB SPL/mW sensitivity, 116 dB SPL will result. Even 600-Ω headphones with the same sensitivity can be driven to 110 dB SPL.

Driving low-impedance headphones to high SPL may be limited by the current available from the power supply. However, if this high SPL is only a result of musical peaks, the power supply may have no problem delivering this high current for brief intervals, especially if it includes good-sized reservoir capacitors. A conservative approach means driving both channels with continuous sine waves into test loads.

Driving a 16-Ω headphone channel to 2 W, as some hi-end HPAs can, requires peak current of 500 mA. For both channels, a total brief peak current requirement of 1 A from the power supply could result. It would be very important that a regulated power supply not fall out of regulation under these conditions. For best performance, the power supply should be capable of at least 1 A continuous current and be equipped with good-sized reservoir capacitors, probably at least 4700 μF. Headphones with 16-Ω impedance and 94 dB SPL/mW sensitivity can be driven to 127 dB SPL with 2 W. One high-end commercial HPA can drive 6 V_{RMS} into 16 Ω, corresponding to 2.4 W [11]. That same HPA can deliver 11.9 V_{RMS} into 300 Ω, corresponding to 472 mW. For 100 dB SPL/mW headphones, this would produce 127 dB SPL.

DC Coupling and DC Servos

Just as in audio power amplifiers, it is important to keep output DC offset small while avoiding the use of electrolytic capacitors in the signal path if possible. It is also desirable to minimize the use of large-value film capacitors, as they can be expensive and bulky. This suggests the adoption of DC servo circuits for control of output offset in otherwise DC-coupled circuits. The design and use of DC servos was discussed in detail in Ref. [1]. Figure 21.5 shows a simplified circuit of an HPA using a DC servo.

A conventional circuit without a DC servo requires an electrolytic capacitor in the shunt part of the feedback network to reduce the gain of the amplifier to unity at DC. This prevents input offset of the first stage from being multiplied by the gain of the amplifier. If the feedback network impedances are kept small to reduce noise, the value of that capacitor can become quite large. A 100-Ω feedback network impedance requires the capacitance to be about 150 μF to achieve a −3-dB frequency of 10 Hz. A non-polar electrolytic capacitor with a significant voltage rating is preferred for lowest distortion, but this does not fully eliminate the distortion problem [1]. It is desirable to have a lower 3-dB point of 3 Hz to achieve low droop at 20 Hz and keep the signal voltage across the electrolytic capacitor very small to achieve low electrolytic capacitor distortion. This would require about 470 μF.

The circuit of Figure 21.5 employs a DC servo to monitor output DC offset, integrate its value and feed an offset error back to the input stage. This arrangement keeps the DC offset to the small value of input-voltage offset of the op amp. Note that a 2-op amp circuit is used for the integrator. The integrator itself is inverting, so an inverter is added to make the circuit overall non-inverting. Non-inverting integrators can be implemented with a single op amp, but they require two integrator capacitors. The integrator capacitor (C1) value is usually between 0.1 and 1.0 μF and should be a quality polypropylene film type. The DC servo is effectively in the signal path, so an audio-grade JFET op amp should be used for the integrator [1].

Notice also that the circuit is able to employ a film capacitor for input coupling if needed, because the input impedance has been kept high. The easiest way to keep the input impedance high

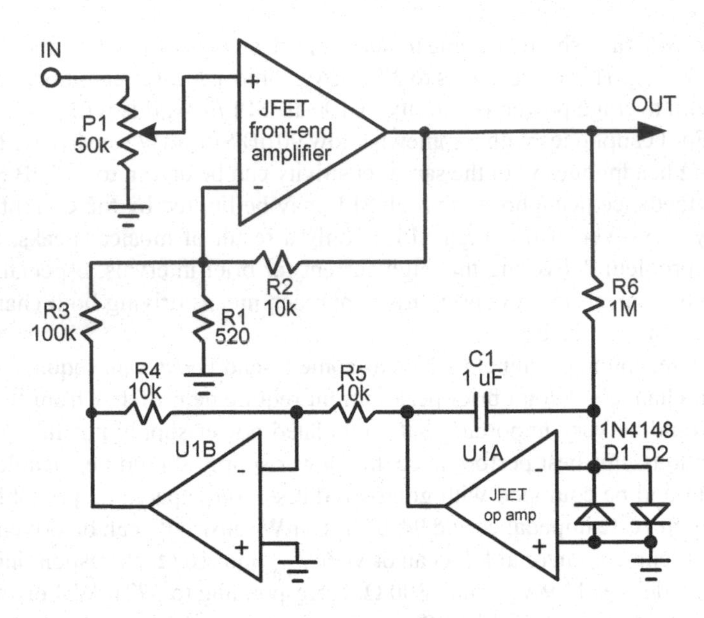

Figure 21.5 Use of a DC Servo to Minimize DC Offset.

is to use a JFET input op amp (or discrete amplifier) at the input, since it does not draw any input bias current through the input circuit, here a 50-kΩ high-quality audio taper pot. Alternatively, if a BJT input stage is used, more DC offset due to input bias current is allowed there because it will be corrected by the DC servo. The usual practice of having the DC resistance to ground be the same on both inputs to cancel input bias current offset need not be applied because of the correction afforded by the DC servo.

Muting and Protection Circuits

Today's more powerful HPAs with low output impedance can present a real danger to headphones, especially under some fault conditions. Some can put over 2 W into a 32-Ω headphone. DC on the output of the HPA, especially at the rails due to a circuit fault, can destroy the headphones. Most quality HPAs are DC-coupled at the output. Many headphones may be damaged if exposed to significant DC or turn-on/turn-off pulses that may be created by some HPAs. This is not unlike the issue with audio power amplifiers [1]. In some cases, an HPA can create significant offset voltage on power-up before the power supplies and amplifier circuits have stabilized. A muting relay at the output can be implemented that protects the headphone from DC due to an amplifier fault or large thumps that may occur during turn-on or turn-off. The need for such a feature is very dependent on the particular HPA design.

In some cases, a large-amplitude, low-frequency bass signal may falsely trigger the protection circuit. It is important for the circuit to be able to distinguish between a high-amplitude brief bass note and an undesirable low-amplitude continuous DC offset. Most DC offset protection circuits use a first-order low-pass filter (LPF) to help distinguish the two. However, some use a second- or third-order LPF filter to allow better discrimination between the two situations.

A simple protection circuit might have a 3-second time delay that starts when the power supply voltages reach 75% and there is very little filtered DC offset. The output is then disabled when the rails fall below 70% when power is turned off. This protection circuit just requires a quad comparator and some other minor circuitry that may protect against a significant DC offset.

No discussion of HPA protection circuits would be complete without mention of the popular uPC1237 integrated circuit expressly designed for the protection of power amplifiers, loudspeakers and headphones. It is equally appropriate for HPAs and protects both stereo channels. The device is a nine-pin SIP for controlling a relay that is in series with the output. It performs the following functions:

- Over-current protection
- DC protection for the headphones for two channels
- Delayed turn-on muting
- Fast turn-off muting
- Retry after about 3 seconds

Figure 21.6 shows a typical application circuit employing the uPC1237. The IC is a convenient collection of transistors that provides the above functions. Most of its action is controlled by about 15 external passive components. The output at pin 6 controls the output relay with an open-collector Darlington driver. When the relay is energized, the headphone is connected to the amplifier. Pin 6 can tolerate up to +60 V and can sink up to 130 mA through the relay coil. A miniature 12-V DPDT relay can be used in an HPA. The relay driver is controlled by a Schmitt trigger that provides hysteresis to prevent relay chatter.

Positive supply current is sourced to pin 9, where the voltage is shunt-regulated internally to +3.1 V. R4 limits the current to the shunt regulator to about 2.5 mA. R5 provides about 2.8 mA of negative bias current from a negative power supply.

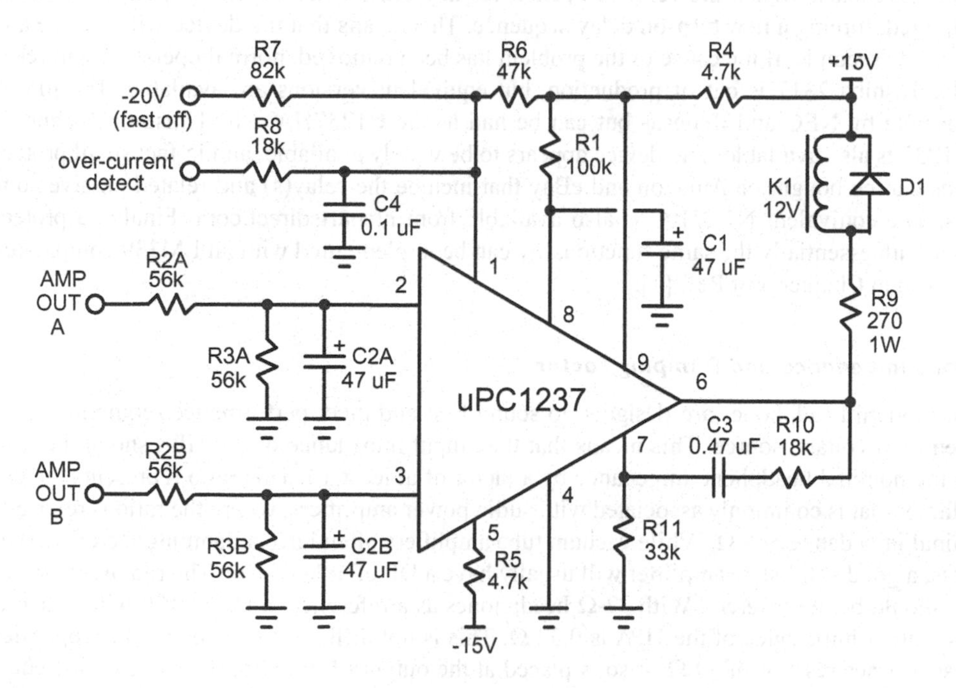

Figure 21.6 Headphone Amplifier Protection Circuit.

The voltage applied to pin 8 controls muting. When power is applied, the voltage at pin 8 rises toward the 3.1 V internal supply at a rate controlled by R1 and C1. When the voltage on pin 8 reaches about +1.3 V, the headphone relay will be closed. If C1 = 47 μF and R1 is 100 kΩ, initial mute time is 2 seconds.

Pin 1 controls fast-off muting and over-current protection. It is connected to the base of an internal NPN transistor whose emitter is connected to ground. When turned on, that transistor will discharge the pin 8 node and open the relay. If for any reason current is allowed to flow into pin 1, the relay will be opened. R6 sources a current of 50–170 μA to pin 1, depending on the voltage at pin 1. R7 sinks about 250 μA from pin 1, overcoming the current from R6.

Fast muting at turn-off is controlled by R7. It is powered from a simple half-wave −20 V supply connected to the power transformer. This supply comprises a rectifier, small filter capacitor, and bleeder resistor (typically 1 μF and 15 kΩ). Its output will fall very quickly when power is removed. On shutoff, the current through R6 will drive pin 1 positive and quickly open the relay.

Over-current shutdown is controlled by current-limiting resistor R8. It sources current into the pin 1 node to open the relay. R8 is fed from an over-current detection circuit within the amplifier (if available in a discrete design) that supplies positive rail voltage when the current limit has been exceeded. When the over-current detect voltage goes to the positive rail, pin 1 goes to +1 V_{be} and the relay is opened.

Pins 2 and 3 are for DC detection for each of two channels. If the voltage at either pin goes high or low by 1 V_{be}, the relay will be opened. External components R2 and C2 filter the signal and provide a delay for DC detection. R2 and R3 form an attenuator from the amplifier output signal to pin 2 (or 3). C2 must be large enough to prevent legitimate low-frequency signals from activating the protection circuit. If R2 = R3 = 56 kΩ and C2 = 47 μF, a DC level of 10 V will trigger shutdown in about 70 ms. A DC level of about 1 V will trigger the circuit in about 1 second.

R10, R11 and C3 act as a speedup circuit for opening of the relay by providing some initial positive feedback. When the relay is opened for any reason, C1 of the turn-on mute circuit is discharged, forcing a new turn-on delay sequence. This means that the device will execute a retry after a few seconds. If the cause of the problem has been removed, normal operation will resume.

The Toshiba 7317 is out of production, but equivalent versions are available. The μPC1237 is obsolete by NEC and Renesas but can be had as the C1237HA. The Unisonic Technologies μPC1237 is also available. The device appears to be widely available, and in fact small protection boards can be bought on Amazon and eBay that include the relay(s) and related passive components. The equivalent NTE7100 is also available from ntepartsdirect.com. Finally, a protection circuit with essentially the same functionality can be implemented with an LM339 comparator, as discussed in Chapter 4 of Ref. [1].

Output Impedance and Damping Factor

Most modern headphones are designed to sound best and meet performance requirements when driven by a voltage source. This means that the output impedance of the HPA should be smaller than the nominal headphone impedance by a factor of at least 10. This ratio represents the damping factor that is commonly associated with audio power amplifiers, where the ratio is referred to a nominal impedance of 8 Ω. While vacuum tube amplifiers often have a damping factor (DF) of 20 or less, a good solid-state amplifier will usually have a DF of at least 100. The requirements for an HPA should be no different. With 32-Ω headphones as a reference, a DF of 100 will be achieved if the output impedance of the HPA is 0.32 Ω. This is not difficult to achieve with proper design unless a series resistor of 10 Ω or so is placed at the output of the HPA to assure high-frequency stability. There are better ways to assure stability, just as used in power amplifiers [1].

Output Resistance

Not all headphones are designed or voiced to sound best when fed with an amplifier with low or extremely low source resistance in the 0–10 Ω range. Indeed, there is an older IEC industry standard that headphones of some types be driven with a source resistance of 120 Ω [12]. Adherence to this standard for quality headphone listening is essentially obsolete. The standard was largely a reflection of the fact that some consumer equipment, like *A*/*V* receivers, used a resistive divider from the power amplifier outputs to obtain the headphone output as a cost-saving measure. A high output impedance like 120 Ω can, of course, affect frequency response, which could mean that the manufacturer voiced the headphones for 120-Ω series resistance. In such a case, the bass response might be altered by a higher source resistance. Such a headphone might sound thin when driven by a very low resistance. Optionally, an HPA can have a switch or a second output jack to put a 120-Ω resistor in series with the output. One can then check to see which source resistance makes the headphones sound best.

Stability

High-frequency stability with capacitive loads is an issue with HPAs just as it is in audio power amplifiers. This is particularly the case for some HPAs that employ wideband ICs or wideband output stages that drive the load. Some HPAs ensure stability in such cases by inserting a resistor at the output of the HPA. As discussed earlier, this is not a good idea, as the best HPAs should have very low output impedance. This issue is no different than for a power amplifier, and the solutions are similar [1]. If a series resistor is needed for stability, it should not be greater than 4.7 Ω.

Where such high-frequency stability measures are needed in HPAs, a coil with a parallel resistor can be placed in series with the output, just as done in an audio power amplifier [1]. This provides the needed high-frequency isolation without increasing output impedance in the audio band. A 2-W metal oxide film resistor with about 15 turns of #24 AWG magnet wire wound on it will do the job. The resistor can have a value of 10 Ω. The use of a shunt Zobel network at the output is also advisable. It can comprise 33 Ω in series with 0.01 μF. This will ensure stability when driving headphone cables and headphones themselves, where total capacitance can be 500 pF or more. HPAs should be tested for stability with different capacitive loads. Poor bypassing of the output stage can also make an HPA prone to instability with certain loads. Just keep in mind that an HPA is much like a power amplifier with significantly lower power capability.

21.5 Power Supply Requirements

As touched on above, in order for the HPA to supply high voltage and high current to cover a broad range of headphone impedances and sensitivities, the power supply must be able to provide adequate rail voltages and current. If the HPA is integrated into a preamplifier, CD player or DAC, it may be the most demanding part of the unit in terms of voltage and current. In many HPAs, one or more op amps will be involved, so the rails should be set to ±18 V, which is the maximum for most modern op amps. This will permit peak swing of about ±16 V, corresponding to 11.3 V_{RMS}, which will deliver 426 mW into 300-Ω headphones. Some rail-to-rail op amps can provide a little more output voltage.

Driving low-impedance headphones will dominate the continuous current delivery requirement. However, this can be tricky because average safe listening levels will require significantly lower current than the peaks. The safest approach is to design the power supply to support simultaneous peak current for both channels with low-impedance headphones driven to fairly high power. This will result in over-design, but will contribute to very solid sound. Putting this in perspective, delivering 6 V_{RMS} into a 16-Ω headphone load will result in peak current of 530 mA. For two channels this would result in about 1-A peak. This is probably a worst case, but could occur under demanding HPA testing conditions, especially at low frequencies where the current demand will be near its peak for longer periods of time.

Unlike in a power amplifier, most quality HPAs will be powered from regulated rail voltages. If one specifies no more than 1-V_{p-p} ripple at the input to the regulator under a 1-A current draw, then the required reservoir capacitance for each rail will be on the order of 8000 μF. This is a lot of reservoir capacitance, but this is a worst-case scenario. However, if one was to require only 100 mV_{p-p} ripple ahead of the regulator in pursuit of extra-clean rails under less demanding current requirements of only 100 mA_{p-p}, one would still arrive a such a large amount of capacitance. This is still only about 1/5 the reservoir capacitance for a very good power amplifier. Don't let the perceived lower-power amplification in an HPA lure you into complacency about the needed size of the reservoir capacitors. Putting 1000 μF on the downstream side of the regulator is also a good idea as long as stability is not impaired. Finally, giving the other electronics their own voltage regulator in an HPA integrated with other functions is not a bad idea.

A less demanding worst case would involve driving 32-Ω headphones with 90 dB SPL/mW efficiency to 120 dB SPL. This requires 1 W, with peak current of 250 mA per side. This would cut in half the reservoir capacitance requirements cited above.

Quiescent Current and Power Dissipation

A discrete class AB output stage with 1-Ω emitter resistors will obey the Oliver condition for crossover distortion if it is biased at 26 mA [1]. This is not too bad. Top NPN and bottom PNP power transistors will both be in conduction up to a peak output current of 52 mA. This corresponds to 81 mW into 32 Ω. With 100 dB SPL/mW headphones, they will produce 116 dB SPL. Most listening will fall in the class-A range. Continuous quiescent draw for two channels will be 52 mA, and output stage power dissipation will total 1.9 W with 18-V rails.

Alternatively, consider a class-A design that can deliver 1 W into 32 Ω, requiring peak current of 250 mA. This can be achieved with quiescent current of 125 mA in each channel, for total HPA output stage quiescent current of 250 mA, corresponding to power dissipation of 9 W with ±18-V rails.

21.6 Headphone Amplifier Designs

There are a great many choices for HPAs out there, all the way from single-chip stereo HPAs to fully discrete HPAs [13–21]. With lower voltages and very good op amps available, there is a bit less incentive to implement fully discrete HPAs. Rail voltages of ±18 V are able to deliver about 11 V_{RMS}, enough to drive a 300-Ω load to over 400 mW, for SPL of 126 dB if the headphones have 100 dB SPL/mW sensitivity. Peak current in this case is just 52 mA.

Low-impedance headphones present a bit more of a challenge, and a great many of the quality headphones out there now are of low impedance, many in the range of 32 Ω. The current demands can be a challenge to meet with some of the headphone ICs. Most headphones in the range of 32 Ω have power sensitivities between 90 and 100 dB SPL/mW, with some outliers above and below. To serve a broad range of these headphones, the target SPL used here will be 120 dB SPL for 32-Ω headphones with 90 dB SPL/mW sensitivity. This comes to 1 W into 32 Ω, corresponding to peak current of 250 mA. Many of the designs discussed below will meet this target, but some will miss the mark by a bit.

High-current performance in the amplifiers below is based on rated continuous signal current for the sake of reliability and conservative design. Peak current output, usually governed by short-circuit protection, will sometimes be higher.

OPA1622 Stereo Headphone Amplifier

This is a fairly popular choice for HPAs because it has two complete HPAs in one chip [22]. It has very low noise of 2.8 nV/√Hz, very low distortion driving a 32-Ω load to 100 mW and is stable driving a significant 600-pF capacitive load.

It can be tempting to estimate the power a chip can deliver to a 32-Ω load by looking at its maximum rated output current, here 130 mA. This corresponds to 92 mA_{RMS} and 2.9 V_{RMS} into 32 Ω, corresponding to power of 271 mW. However, this can lead to overly optimistic results. The datasheet shows clipping at 2 V into 32 Ω, corresponding to only 127 mW. The maximum output current of 130 mA is not a reliable way to estimate the clipping point into a low-impedance load, at least for this device. The stated power of 100 mW will drive a 90 dB/mW headphone to 110 dB SPL, a bit shy of the minimum target of 115 dB SPL previously stated.

Figure 21.7(a) shows a typical implementation of an HPA using the OPA1622 [23]. The circuit is designed for a maximum gain of 4 (12 dB). With an input of 1 V, a maximum of 4 V can be applied to a 300-Ω headphone with sensitivity of 100 dB SPL/mW, delivering 53 mW and SPL of about 114 dB SPL. Gain can be increased by reducing R2 or increasing R3. Note that the device has a ground pin.

The device includes an enable pin which, when brought above 0.82 V, turns on the amplifier. Components R4, R5, R6, C2 and D1 control the enable pin to implement a slow turn-on and a fast turn-off (a mute function). Turn-on delay is about 2.5 seconds.

The OPA1622 has low input-voltage noise of 2.8 nV/√Hz, but noise here is dominated by the 50-kΩ volume control thermal noise in the worst case. When P1 attenuates by 6 dB and has net wiper resistance of 12.5 kΩ, it contributes 14.8 nV/√Hz of thermal noise. This contribution

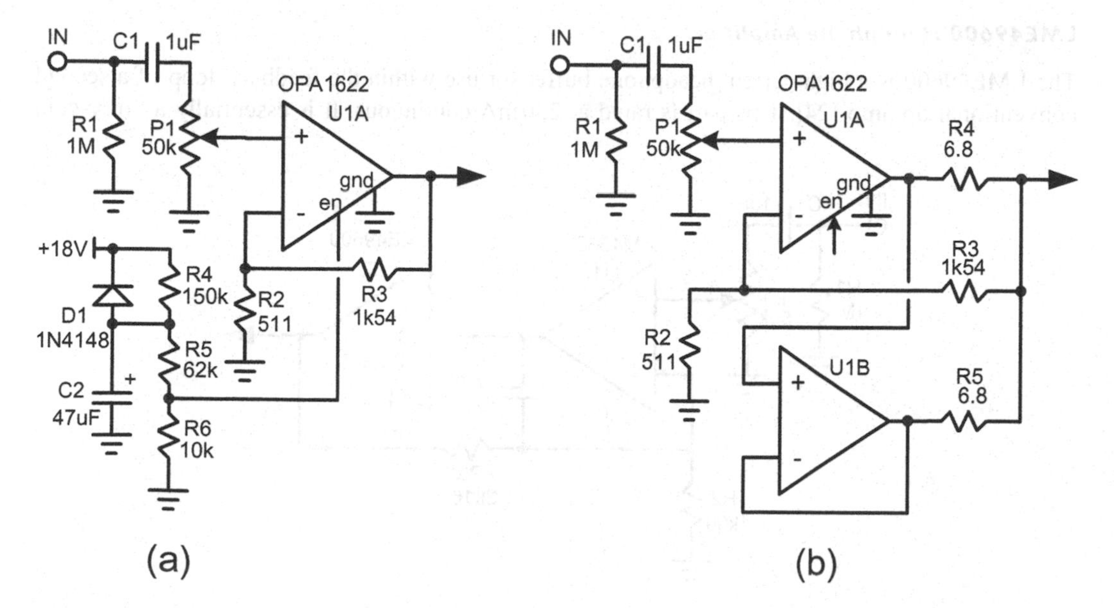

Figure 21.7 Headphone Amplifiers Using the Stereo OPA1622.

becomes smaller at larger amounts of attenuation. The choice of 50 kΩ for P1 reflects the need to keep input impedance high enough for a wide range of sources that may be AC-coupled. A 1-μF coupling capacitor loaded by 50 kΩ causes frequency response to be down 3 dB at 3.2 Hz. Noise can be reduced by going to a two-stage HPA as described in Figure 21.4, using an input stage with 6 dB of gain and reducing the value of P1 to 10 kΩ. Together, these two changes can reduce the noise contribution of the volume control by about 12 dB.

Dual OPA1622 Headphone Amplifier

The output current shortcoming of the OPA1622 can be mitigated by running the pair in a package in parallel. This is worthwhile because the OPA1622 has superior performance to many single headphone chips that have higher current output. The price paid is a couple of additional passive components as illustrated in Figure 21.7(b). The devices are connected in parallel by using resistors at the outputs to prevent the devices from fighting. U1B acts as a non-inverting slave. Feedback for U1A is taken from the summed output to preserve a low-impedance output. With twice the output current available for low-impedance loads, the combination can deliver four times the rated power of 100 mW into a 32-Ω, yielding a 6-dB increase to 400 mW, corresponding to 116 dB SPL into a 32-Ω headphone with sensitivity of 90 dB/mW.

Note that there is more than one way to parallel op amps to achieve higher current capability. The simplest way is to build two identical closed-loop amplifiers and connect the outputs in parallel with a pair of resistors to enable current sharing. This has the disadvantage of increasing the output impedance from near zero to about ½ the value of each output resistor. The second way is to implement one amplifier as the main closed-loop amplifier with gain and to augment it with a second unity gain op amp. The outputs are connected together by current-sharing resistors. The output node is used for the feedback for the main amplifier. This configuration maintains low output impedance and does not require two matched feedback networks.

LME49600 Headphone Amplifier

The LME49600 is a high-current headphone buffer for use within the feedback loop of a second conventional op amp [24]. This part is rated at 250 mA continuous. It is essentially a unity-gain

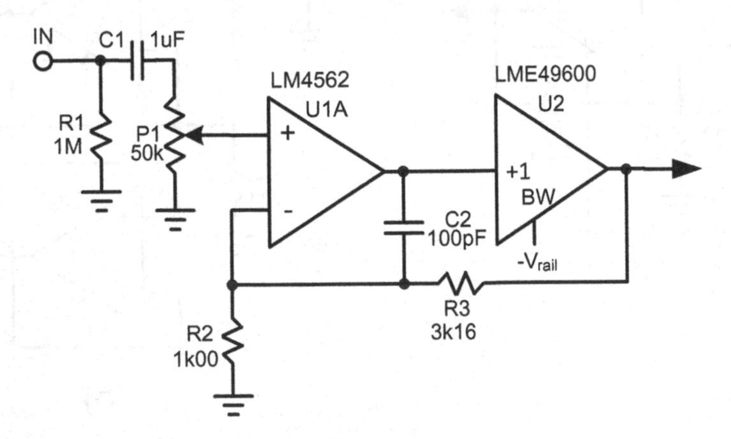

Figure 21.8 Headphone Amplifier Using the LME49600.

diamond buffer with sufficiently wide bandwidth to be used inside the feedback loop of another op amp. Figure 21.8 illustrates an HPA using this two-chip approach.

A stereo HPA can be made with only three packages, with the driving op amp being a high-performance dual OPA1642 FET op amp [25]. The LME49600 buffer can deliver 1 W into a 32-Ω load, corresponding to 120 dB SPL with 90 dB/mW headphones. The LME49600 has a bandwidth of over 100 MHz, so bypassing and layout is important.

Headphone Amplifier with a Discrete BJT Output Stage

Figure 21.9 illustrates an HPA with a discrete output stage. Here the concept is the same as the above circuit, but a discrete BJT diamond buffer output stage is employed. This requires more parts, but they are inexpensive and the amount of current that can be supplied is quite substantial. Perhaps more importantly, it allows the freedom to choose the output stage bias and the corresponding emitter resistors. You can run the stage as hot as you want, including in pure class A up to whatever output current you choose (with adequate heat sinking). As shown, the output stage is biased at 40 mA, providing class-A operation up to significant SPL. At 11 V_{RMS} output, it can deliver 3.8 W into a 32-Ω at 340 mA_{RMS}.

The TO-220 output transistors can easily handle the 720 mW of dissipation without heat sinks. A less expensive version can be made if the current sources are replaced with resistors, at some sacrifice in performance. For further improved PSRR, R12 can be replaced by a floating current source. The high-performance dual OPA1642 FET op amp is employed, but other audio-grade op amps like the OPA2134 can serve as well [25].

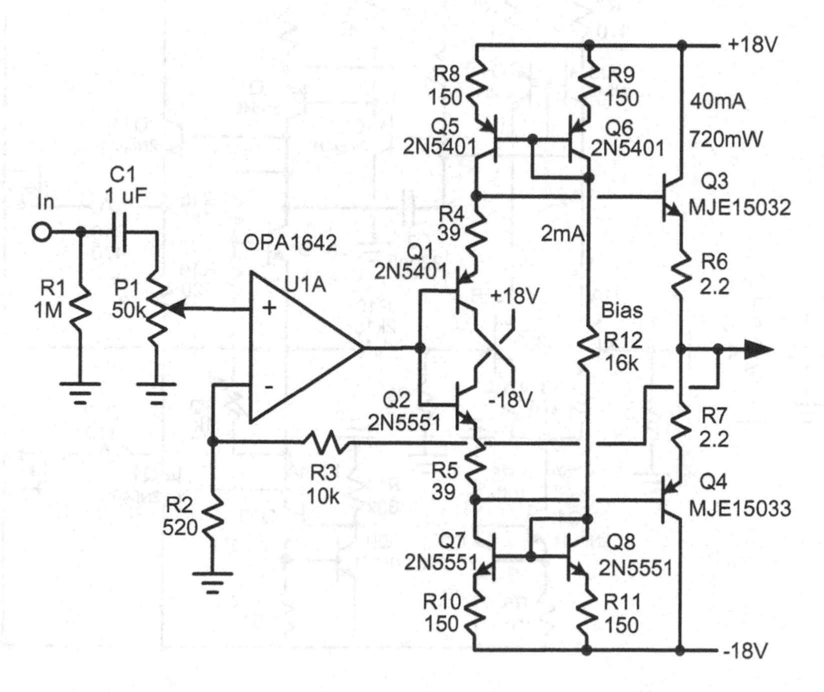

Figure 21.9 Headphone Amplifier with Discrete Output Stage.

Discrete Lateral MOSFET Headphone Amplifier

Some prefer the sound of lateral power MOSFET amplifiers, so the HPAs considered should be no different. It is also the case that an all-discrete HPA does not have the ±18-V limit on rail voltages that exists for most op amps. Somewhat higher rail voltages, like ±24 V provide extra voltage margin for 300-Ω headphones, allowing up to 22 V peak, corresponding to 15.6 V_{RMS}, corresponding to 811 mW into 300 Ω. This will produce 119 dB SPL with 90 dB/mW headphones.

The design is shown in Figure 21.10. Exicon lateral MOSFETs ECX10N20 and ECX10P20 may seem like overkill, but they make getting rid of the heat easy [26]. They come in a TO3P plastic package, and can be run with little or no heat sinking in a headphone application. With quiescent bias of 50 mA, each device dissipates 1.2 W with ±24-V rails. This provides class-A operation up to 160 mW even into 32 Ω. In the class AB region, as much as 6 W could be provided. You have complete freedom to choose the output stage bias current. Also, MOSFETs transition from class A into class AB much more smoothly than BJTs.

The design is shown in Figure 21.10. It looks much like a quality MOSFET power amplifier scaled down in voltage and power dissipation. It employs a dual monolithic JFET input stage loaded by a current mirror. The 2T BJT VAS is loaded by a 10 mA current source. The bias spreader is a simple V_{be} multiplier employing Q8. Bias current is set by pot P2. BJT emitter followers drive the lateral MOSFET gates through gate stopper resistors for stability. Maximum gain is about 10 dB.

The turn-on voltage for the enhancement-mode MOSFETs can be quite small, so the bias spreader is designed with enough range on the low side to achieve a 0-V spread at the gates. Forward bias at 100 mA drain current can range from 0.15 to 1.5 V for these devices. Exicon lateral

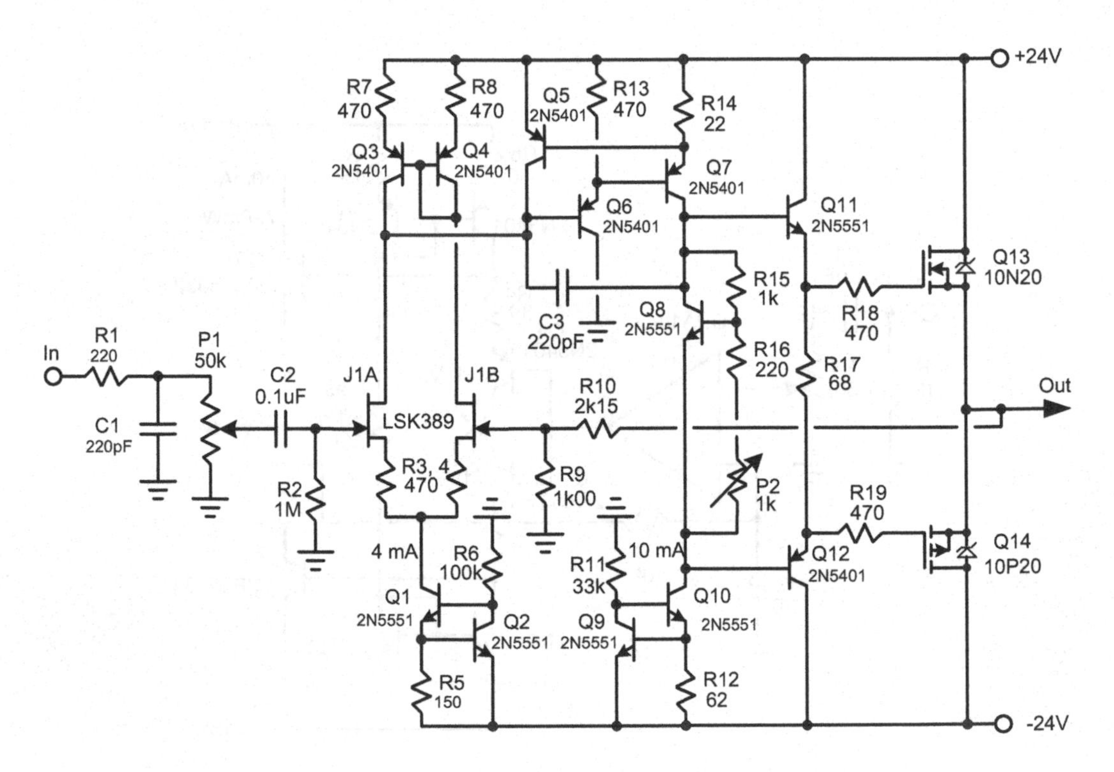

Figure 21.10 Lateral MOSFET Headphone Amplifier.

MOSFETs are available in binned values of operating current for a given V_{gs} as shown below, with drain currents for the 10N20 at $V_{gs=}$0.5 V and 10P20 at $V_{gs=}$1.0 V:

Exicon 10N20 and 10P20 Operating Current

Color	*Min*	*Max*
Red	105	125
Orange	125	140
Yellow	140	155
Green	155	170
Blue	170	185
White	185	205

Other Headphone Amplifier Design Choices

There are many other circuit combinations that can be used to build HPAs. We touch upon a few of those other possibilities here.

The LM3886 power amplifier chip can be used as an HPA. It is capable of higher voltages and high current output. Its distortion is quite low when driving an easy 32-Ω load. It requires some additional feedback compensation to operate at lower gains than normal, such as a gain of 4 or 10. It is at the heart of many amplifiers referred to as "Gain Clones". Some examples of its use can be found in Ref. [1]. With ±35-V rails it can deliver 40 W into 8 Ω. THD under this condition is less than 0.01% across the full audio band. THD will be much lower driving a headphone load. It can operate with rails between ±12 and ±40 V.

The TPA6120 current feedback stereo HPA can be used, and it can deliver up to 700 mA. However, its PSRR and distortion are not that impressive and it requires a series output resistor for stability.

The OPA552 power operational amplifier is rated at 200 mA continuous with a typical short-circuit current of 380 mA. It can be operated with rails up to ±30 V. It is stable for gains greater than 5. The OPA551 is unity gain stable. Its noise is on the high side at 14 nV/√Hz and its PSRR is only 70 dB. Its THD-20 is 0.0015% into a light load of 300 Ω. Distortion probably increases into a 32-Ω load. It is available in a DDPAK/TO-263 surface-mount package which can dissipate power into a copper plane on the board with thermal resistance of 7.7°C/W.

Finally, the highly regarded Audio-Technica AT-HA5000 HPA has been cloned. The design is a discrete one that employs Hitachi vertical MOSFETs in the output stage (2SK2955 and 2SJ554). It can deliver 2.4 W into 16 Ω, 1.2 W into 32 Ω and 130 mW into 300 Ω. Maximum gain is 15 dB and THD is rated at 0.006% at 100 mW into 32 Ω.

21.7 Crossfeed Circuits

When listening to stereo speakers, most of the sound from the right speaker goes to your right ear, but some of it goes to your left ear as well. That signal is delayed by what is called the interaural time delay (ITD). Its amplitude and frequency response is also altered in traveling past the head, corresponding to interaural level differences (ILD). ITD is one of the main cues for determining the spatial location of sound sources. ITD varies with the angle of the head with respect to the sound source. It will be zero when directly facing the sound source. It can be on the order of 600 μs when the head is at 90° with respect to the source. This angle is referred to as the azimuth. With a typical listening azimuth of 45°, ITD will be about 300 μs.

Most music is recorded assuming that such "air mixing" will happen in the stereo listening experience. Indeed, that is what is happening when the mastering engineer is tailoring the sound when listening to his monitor speakers in the studio. However, there isn't any such air mixing when you're listening with headphones. The left channel is sent only to your left ear and the right channel only to your right ear. An unnatural and fatiguing listening experience can result as the brain struggles to localize using amplitude and time cues that do not reflect the real world. Sometimes this phenomenon is referred to as "super-stereo" or an "in-head" listening experience.

A crossfeed circuit can be used to mitigate the effect by emulating the air mixing that occurs in the real world. It does that by mixing some of the signal from one channel into the sound heard in the opposite ear. It does this by delaying, attenuating and low-pass filtering the signal from one channel and crossfeeding it into the other channel. The crossfeed signal is manipulated so that it simulates what happens to the air-mixed signal as it travels past one's head to the opposite ear. In most implementations, the only delay is the group delay introduced by the LPF. Often, the LPF is of only first order. In more sophisticated implementations, additional delay may be introduced into the cross-feed paths by employing all-pass filters.

The greatest improvement contributed by crossfeed occurs when recorded material has been panned hard left or hard right. Crossfeed mitigates the "super-stereo" effect that results. It can also reduce the effect of centered sounds seeming to come from within your head. Crossfeed is very personal and not for everyone. Its effect can also depend on the type of music. In all designs there should be the ability to bypass crossfeed, and some designs might incorporate a control to vary the amount of crossfeed.

The falling frequency response of the crossfeed channel emulates the low-pass filtering that the passage of the signal across the head experiences. This is called head shadowing. In other words, the delayed signal does not have a flat frequency response. This effect is largely absent at low frequencies, but does come into play at mid-band and higher frequencies. Benjamin Bauer first experimented with crossfeed in the early 1960s [27, 28]. A conceptual diagram showing three approaches to implementing a crossfeed circuit is shown in Figure 21.11.

The approach in (a) is the most conceptual and is often implemented with active circuits like op amp summers. It also leaves a lot of room for manipulating the crossfeed signal, such as adding some "real" group delay with all-pass filters. The simple passive approach in (b) effectively allows

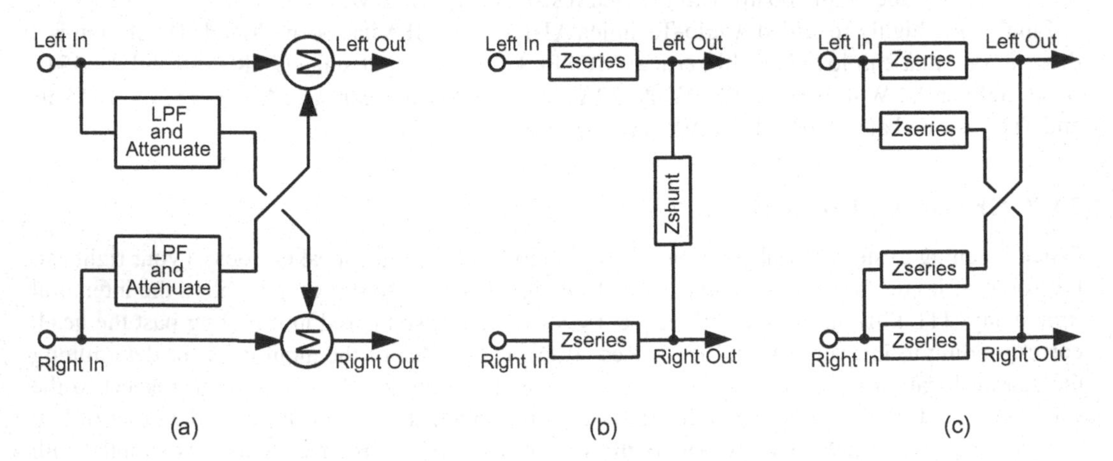

Figure 21.11 Conceptual Crossfeed Circuits.

frequency-dependent crosstalk. The passive approach in (c) achieves frequency-dependent crossfeed through the interconnection of series impedances from each input channel.

It is important to recognize the difference between phase delay (actually phase difference or phase shift) and group delay. Group delay pertains to interaural delay, but phase difference and the impact on L-R amplitude matters as well, since phase affects the soundstage. For example, phase inversion of one stereo channel kills center localization.

The delay in a crossfeed circuit emulates the interaural delay, which may be on the order of a few hundred microseconds. This simulates what is heard from a loudspeaker. A signal from the right channel speaker is heard by the left ear a few hundred microseconds later than when it is heard in the right ear. In a typical stereophonic listening arrangement the interaural time delay may be on the order of 300 μs. The difference in arrival time of a sound depends on the angle of each ear with respect to the loudspeaker from which the sound arrived.

Frequencies much above about 1 kHz need not be delayed and crossfed because stereophonic perception at higher frequencies appears to be a function mainly of sound intensity. Indeed, crossfeed at high frequencies can dull the sound and decrease the sparkle, so it is often desirable to attenuate crossfeed at high frequencies above 5–10 kHz. The crossfeed signal may thus be fed through an LPF down about 3 dB at 700 Hz compared to its level at lower frequencies. At very low frequencies delay is inconsequential and shadowing of the head is not an issue, since low-frequency SPL is largely equal-pressure around the head.

Key Parameters and Types of Crossfeed

Four characteristics help distinguish most crossfeed circuits. The first is the transition frequency that defines the low-frequency and high-frequency regions. This is often around 700 Hz, and is usually the frequency where the amplitude of the crossfeed signal is down 3 dB from its nominal low-frequency value. The second and third characteristics are left-right separation at low frequencies and separation at higher frequencies. Crossfeed circuits often have small or no stereo separation at low frequencies where the crossfeed amplitude is in full effect. At high frequencies, crossfeed circuits usually allow substantially larger separation, if not full stereo separation.

The fourth parameter is monophonic low-frequency boost. In some crossfeed circuits, the crossfeed signal, with its largest amplitude at low frequencies, augments the signal level in the channel into which it is being crossfed. This means that a sound positioned in the center of the sound stage (monophonic) has its amplitude increase at low frequencies, affecting the frequency response of a signal originating from center stage. One way to evaluate this is to feed a monophonic signal into both channels of the crossfeed circuit and measure the outgoing frequency response of the sum of the two channels. The amount of monophonic low-frequency lift depends on the amount of crossfeed, and can range between 0 dB and sometimes as high as 3–5 dB.

There are many ways to implement crossfeed, but most fall into two categories. The first category can be called additive crossfeed. It simply means that some of the signal from one channel is altered and added to the signal of the opposite channel, as in Figure 21.11(a). The alteration usually includes attenuation, low-pass filtering and delay. This is the type of crossfeed that most often causes monophonic low-frequency lift, and it is fairly easy to see why this happens. The Linkwitz crossfeed circuit and its many variants fall into this category [29]. Most approaches to crossfeed are based on the Linkwitz crossfeed approach. That approach is discussed in detail in a later subsection.

The second category is crossfeed by frequency-dependent signal merging, as in Figure 21.11(b). This is not unlike merging the left and right signals in a phono preamp to eliminate the L-R vertical cartridge output at low frequencies so as to reduce rumble and flutter in the vertical dimension of stylus movement. This merge is implemented by passing the signals of each channel through coupling capacitors and introducing a crosstalk resistor from one channel to the other. At low frequencies this causes the desired loss of separation as frequency goes down. The "crosstalk" flows in both directions in the resistor, and no monophonic low-frequency lift occurs. The original crossfeed scheme in the Bauer patent employed the merging approach, using inductors across the channels to effectively short the channels together at low frequencies [28]. The arrangement in Figure 21.11(c) is a more generalized depiction of passive crossfeed.

Implementation Issues

The crossfeed addition of low-frequency material to the opposite channel can increase the perceived loudness at frequencies below the transition frequency. This can affect the overall frequency balance in such a way as to reduce the overall perception of higher frequencies. To correct this, a high-frequency boost is sometimes added to the direct-receiving ear, so that the perceived overall frequency response is approximately returned to flat.

With identical left and right signals panned to the center, the crossfeed circuit should provide a flat response corresponding to the equal incidence of signals from both speakers on both ears. This may not always be the case with all crossfeed circuits, and can influence the timbre of some vocals.

While crossfeed makes complete sense for signals coming primarily from the left or right, its effect on signals from the center can be of some concern. Consider crossfeed that has a pure delay, corresponding to linearly increasing phase lag with frequency. When that phase lag passes through 180°, there will be cancellation and consequent attenuation (dip) of the center channel signal in both ears. This will repeat at higher frequencies where the phase passes through 180° points again. When the phase passes through 360° points, the signal heard in both ears will be augmented. This is a comb filtering effect. A pure delay of 300 μs corresponds to ½ cycle of a 1.67-kHz signal, so the first dip will occur at 1.67 kHz. This will be followed by a boost at 3.34 kHz and another dip at 5 kHz.

The amount of amplitude ripple depends on the amount of crossfeed. With small amounts of crossfeed, this might not be much of a problem. Many vocalists are panned to dead center, so this effect could alter the sound of a singer's voice. In reality, most analog crossfeed circuits are simple minimum phase designs without pure delay. This, coupled with reduced crossfeed amplitude at higher frequencies, reduces any comb filtering effect. Often, in fact, the delay provided by simple first-order crossfeed filters is less than the approximate 300 μs that one might want. A higher-order minimum phase crossfeed filter might help. With care, an all-pass filter in the crossfeed path could play a useful role. The notion that any of these crossfeed circuits can be flat under all listening arrangements is fantasy. However, that is also the case with listening to loudspeakers in a room. Crossfeed is very imperfect and very subjective.

Early passive designs drove the headphones directly and interacted with headphone impedance variations. They also required more input power. Most passive designs that drive the headphones directly suffer from circuit interactions and imperfections in achieving the desired responses.

Benjamin Bauer Crossfeed Circuit

Benjamin Bauer was basically the pioneer of headphone crossfeed. Circa 1960, he recognized the shortcomings of headphone listening and decided to do something about it. Thus was born crossfeed, and Bauer obtained a patent for it [27]. He also described it in the literature [28]. His circuit is a passive L-R-C design that feeds the headphones directly.

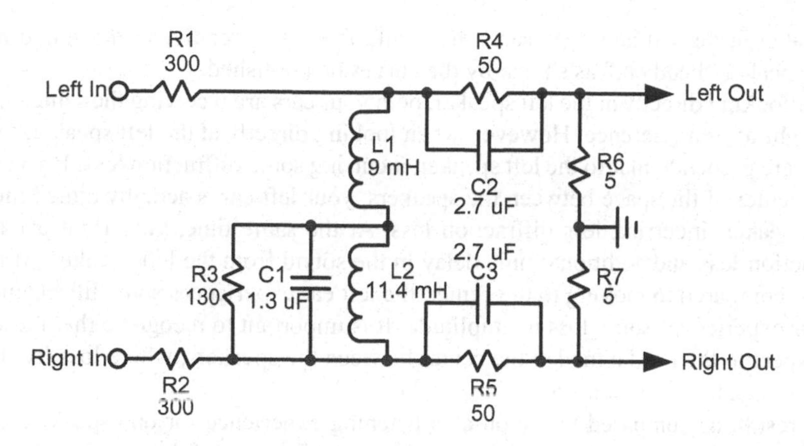

Figure 21.12 Bauer Crossfeed Circuit.

The circuit example in the Bauer patent for low-impedance headphones is shown in Figure 21.12. Each headphone transducer is fed the input signal through R1 and R2 in series with R4 and R5, the latter each in parallel with C2 and C3. With low-impedance headphones (5 Ω or less, not readily available today), this network forms an attenuator with a little bit of high-frequency lift. With higher impedance headphones, 5-Ω resistors R6 and R7 can be included to provide the necessary loading. The crossfeed path is implemented by L1 and L2 connected in series across the junctions of R1 and R2 of the left and right channels.

The network allows crossfeed signals to flow in both directions between the channels. Indeed, at very low frequencies, the left and right headphone signal paths are shorted together by the inductors, resulting in the signals becoming monophonic at low frequencies. At higher frequencies, where the impedance of the inductors has increased, left-right separation is permitted. C1 and R3 in parallel with L2 tweak the crossfeed response somewhat, but are not fundamental to the basic operation.

Because the crossfeed signal must pass through inductors, it is low-pass filtered and delayed. This provides the main part of the emulation of the interaural delay. At the same time, the load on the main signal path from the inductors is lightened as frequency increases, lifting high frequencies and contributing some phase lead. This adds to the phase difference of the signals fed to the two headphone inputs, and thus adds to the time delay difference. Notice that the inductors create a frequency-dependent attenuation of the L-R signal by gradually shorting the left and right signals together with decreasing frequency.

Of great interest in crossfeed designs is a set of three frequency response curves, as reported in a paper by F. M. Weiner [30]. The first response curve of interest shown in the Weiner paper depicts the crossfeed frequency response. This is the signal that makes its way from the right channel input to the left input of the headphone. The reduced amplitude and the roll-off of the crossfeed signal above about 700 Hz are shown. The second curve of interest is the frequency response for signals from the right channel input that are sent to the right headphone. As expected, these signals are higher in amplitude and relatively flat in frequency response. They are plotted along a 0-dB reference line. The third curve is the frequency response when a monophonic signal is sent from both channels to the left or right headphone input. This represents what happens when the crossfeed signal augments the main signal when sounds are placed in the center.

The curve amplitudes and shapes used by Bauer are based on data reported in the paper by Weiner [30]. As reported by Weiner, at a 45° angle with the speakers, sound pressure produced by the left speaker at the left and right ears varies with frequency relative to the sound pressure for facing one speaker "head on," as shown by the curves he published.

If you are looking directly at the left speaker, both your ears are receiving the same signal, so this can be thought of as a reference. However, when looking directly at the left speaker, the openings of both ears are perpendicular to the left speaker, incurring some diffraction loss. If you are looking toward the center of the space between the speakers, your left ear is actually aimed more directly at the left speaker, incurring less diffraction loss. At the same time, your right ear is incurring more diffraction loss and additional time delay in the sound from the left speaker. At mid to high frequencies, compared to looking to the center, the left ear experiences some lift in amplitude and the right ear experiences some loss in amplitude. It is important to recognize that these responses are with respect to the head aimed at the center between the speakers, where loss from diffraction is equal for both ears from their same-side speaker.

The end result, as compared to the on-axis listening experience for one speaker, is that when facing the center, the mid and high-frequency response of the ear of the same side as the speaker is experiencing a bit of mid- and high-frequency lift, while the opposite ear is experiencing significant loss at mid and high frequencies plus a delay. This explains why the data suggests that the left ear should experience some lift of the signal from the left speaker at mid and high frequencies when the head is facing forward to the center. The importance of these off-axis issues is mitigated by the fact that usually when listening to loudspeakers, both the mixing engineer and the serious audiophile are facing the center or off-axis by only a small angle.

Disadvantages of the Bauer circuit include the expense of the inductors, the significant power needed to drive it and that its performance can be influenced by the impedance of the headphones. The design was not a commercial success, but paved the way for crossfeed designs.

Sigfried Linkwitz Crossfeed Circuit

Circa 1970, Sigfried Linkwitz decided to resurrect the crossfeed concept in a more practical and economic form to mitigate the super-stereo effect [29]. His design is a passive RC circuit that drives the headphones directly. It is shown in Figure 21.13.

R1 and R2 act as an attenuator for the main signal of each channel in directly driving the headphones. C1 and R7 across R1 provide a bit of high-frequency lift whose purpose is discussed below. The left input signal is also low-pass filtered by R3 and C3. The resulting signal is then injected into the opposite channel by R5. Notice that this circuit, in contrast to the Bauer circuit, reduces separation at low frequencies but does not make the low-frequency signals fully monophonic. Low-frequency separation in this circuit is about 3 dB, rising to 7.5 dB at 700 Hz and 21 dB at 4 kHz. Low frequencies are perceived by both ears in virtually the same way, so the loss of separation at very low frequencies makes little difference in practice.

In this design, the crossfeed amount at low frequencies is down by 3 dB as compared to the main path. This contrasts with the Bauer design where at low frequencies the contributions from the left and right paths become equal. Unlike the Bauer circuit, this circuit implements crossfeed by augmenting the opposite channel, making L+R larger than L−R by making L+R larger than it otherwise would be.

Notice that the crossfeed signal augments the low frequencies of either channel, since at low frequencies each channel receives low-frequency signals by two paths that tend to add. In this design, C1 and R7 provide about 2 dB of high-frequency lift in the main signal path in an attempt to make the overall frequency response flat.

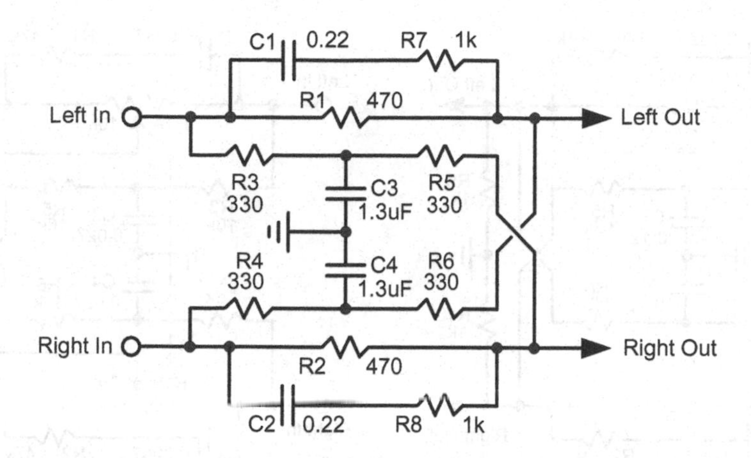

Figure 21.13 Linkwitz Crossfeed Circuit.

This passive circuit also suffers power loss in driving headphones directly. Performance is also influenced by the impedance of the headphones and any frequency dependence in it. If this circuit is to be simulated to observe its performance, loads to ground must be placed at the outputs, perhaps on the order of 100 Ω, to emulate a headphone load.

Chu Moy Crossfeed Circuit

Chu Moy realized that the Linkwitz circuit would fare much better if it was paired with an HPA, thereby enjoying the buffering provided. He modified the Linkwitz circuit to improve its sound and to optimize the circuit for use with HPAs [31]. Implementing the circuit to drive an HPA with high input impedance makes the Linkwitz-like approach independent of the headphone impedance and avoids wasted power. The circuit is largely a scaled version of the Linkwitz circuit with impedances scaled up by a factor of 10. A revised version of the circuit is shown in Figure 21.14(a).

Notice that the relative values of R7,8 in this circuit are only 0.43 times the scaled value of R1,2 in the Linkwitz circuit, meaning that the amount of attenuation in the main path is significantly less. This causes the relative amount of crossfeed to be smaller. Low-frequency separation increases from 3 dB in the Linkwitz design to 10 dB in this design as a result. Separation is about 14 dB at 700 Hz, where crossfeed is down an additional 3 dB.

Moy was concerned that the amount of crossfeed in the Linkwitz circuit was too much. He felt that the Linkwitz circuit introduced some other sonic problems. The high frequencies seemed attenuated compared to the original signal in spite of the high-frequency lift. Voices took on a thick quality. The low-frequency blending had raised the perceived low-frequency response. After Moy's change to reduce low-frequency crossfeed, he claimed that just a trace of emphasis in the lower mid-range remained. This highlights the subjective nature of crossfeed.

He also included a "perspective" switch (not shown) that allowed R7,8 to be made 1.5 kΩ so as to provide an option with a smaller amount of crossfeed, providing a wider soundstage. Note that if R7,8 are shorted, the crossfeed is bypassed. He also discusses other modifications that can be tried, including a reduction of C1,2 that may give more depth to the soundstage or changing R1,2 to increase or decrease the amount of high-frequency lift.

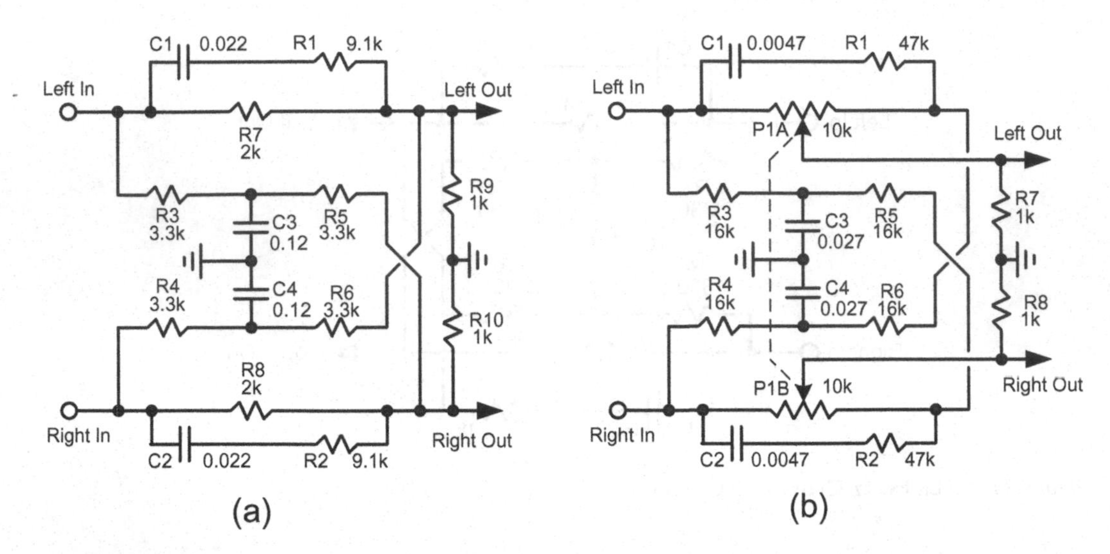

Figure 21.14 Chu Moy's Crossfeed Circuit.

Figure 21.14(b) shows a modification of the Moy design in (a) with some additional scaling and an additional feature. All impedances have been further scaled up by a factor of 5 to reduce loading on the signal source to make its response more independent of source impedance, especially at low frequencies where a preamp coupling capacitor might cause a significant rise in output impedance (the impedance of a 1-μF capacitor is 8 kΩ at 20 Hz). The net DC crossfeed path resistance of 4.3 kΩ can attenuate the L-R signal content from the source at low frequencies.

With the additional 5:1 scaling, the 2-kΩ input-output resistors R7,8 have been replaced with a two-gang 10 kΩ potentiometer P1 that picks off the output at a location between the input and output of the crossfeed network. This allows the amount of crossfeed to be controlled from zero to the full amount. This reflects the fact that some musical content may sound better with no crossfeed or less crossfeed. It also makes it easy to evaluate the difference in soundstage produced by crossfeed. The passive nature of the Moy circuit makes it attractive, but it can be improved by adding active buffers.

Adding Delay

Most of the passive circuits do not add as much group delay as is needed to approximate the intreraural time delay, often said to be about 300 μs. The Moy crossfeed circuit delivers only 113 μs at 700 Hz. However, its circuit arrangement makes it easy to add delay with active all-pass filters. One can easily add a first-order or second-order all-pass filter ahead of R3 and R4 in each crossfeed filter.

As described in Chapter 11, all-pass filters add phase lag and group delay to a signal without changing its amplitude. Their group delay is not constant, but group delay in the crossfeed circuit doesn't matter much at frequencies well below and above the usual crossfeed center frequency of 700 Hz. A simple all-pass filter includes a resistor and a capacitor whose R-C time constant determines its turnover frequency f_0, where half of its group delay has occurred. The resistor and capacitor are usually arranged to form an LPF at the positive input of an op amp, as shown in Figure 11.26(a). The turnover frequency of the all-pass filter is where the LPF is down 3 dB.

The group delay of an all-pass filter is at its maximum at very low frequencies and decreases as the frequency is increased. A first-order all-pass filter has a monotonically falling group delay with frequency. Its phase lag changes by 180° going from DC to high frequencies. Its phase lag is 90° at its f_0. Its group delay has fallen to half its DC value at f_0. At low frequencies, the all-pass filter is non-inverting.

At low frequencies, the group delay is 2RC. At f_0, its group delay has fallen to RC. An RC of 22.9 kΩ and 0.01 μF yields a 3-dB frequency of 700 Hz and a time constant of 229 μs, for group delay of 229 μs.

As described in Chapter 11, a second-order all-pass circuit can be realized with 2 op amps, the first implementing an MFB second-order bandpass filter and the second configured to subtract the bandpass filter output from the input signal, as shown in Figure 11.27(a). Its phase changes from 0° at low frequencies, through 180° to 360° at high frequencies. The second-order all-pass gives a flatter group delay response over the same frequency range, just like a second-order LPF gives a flatter amplitude response over the same frequency range as a first-order LPF.

Jan Meier Crossfeed Circuit

Jan Meier designed a very simple, yet effective passive crossfeed circuit for driving an HPA that is shown in Figure 21.15(a) [32–34]. A 1-kΩ series resistance R5,6 is placed in each channel's main line with the output channels shunted together by 2.2-kΩ of resistance implemented with R3 and R4. This arrangement reduces separation to 10 dB at low frequencies by attenuating the L-R signal. Each series resistor is shunted by a 0.47-μF capacitor (C1, C2) which largely shorts out the L-R attenuation at high frequencies by shorting out series resistors R5 and R6.

Separation goes from a low of 10 dB to a high of full separation as frequency increases. The crossfeed action occurs by means of the bidirectional merging action of R3 and R4. This reduces the L-R component at low frequencies by the differential high-pass filter (HPF) action of C1 and C2. R1 and R2 limit the HPF action of C1 and C2 at high frequencies. The center node between R3 and R4 is shunted to ground by C3. This eliminates the crossfeed at higher frequencies.

If only a left signal is applied, its main signal path through R5 is loaded by R3 at higher frequencies where the impedance of C3 becomes low and the combination of C2 and R2 in the right channel path. The presence of C1 creates some high-frequency lift in the left main path. At lower frequencies the impedance of C3 increases and allows some blending of the left and right channel signals (i.e., crossfeed). At very low frequencies, where all of the capacitors have high impedance, there is an effective resistance of 4.2 kΩ. This can create a presumably unintended crossfeed at low frequencies if the source driving the network is AC-coupled with smaller capacitors. However, when the low frequencies are monophonic and equal in both channels, this effect will not occur.

In the transition region, the crossfeed signal from the left channel is delayed by the shunting LPF action of C2 in the right channel, creating about 250 μs of phase delay in the crossfeed signal. At the same time, the main left path experiences a leading phase shift of about 70 μs by the phase lead effect of C1 being loaded by R3. This is elegant. As a result, a total phase difference between the main and crossfeed signals of 320 μs is achieved, primarily at frequencies below 1 kHz.

Most crossfeed circuits are simply additive, in that the signal from one channel is attenuated, low-pass filtered, delayed and added to the opposite channel. This is satisfactory and effective for signals panned to the left and to the right. However, for signals panned to the center, which are monophonic, each ear already hears the proper amplitude and (equal) phase without the additive crossfeed. The crossfeed is superfluous in this case. However, the additive nature of the crossfeed

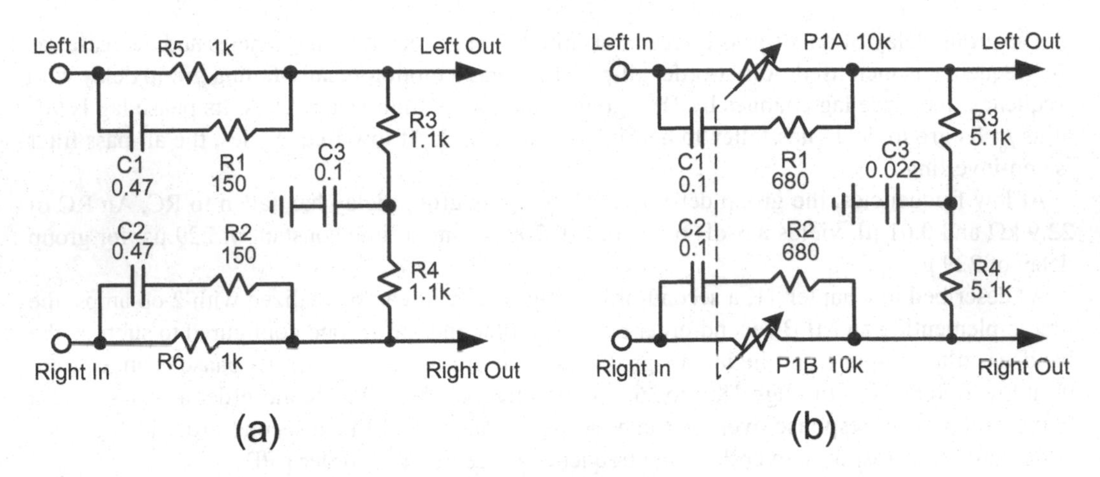

Figure 21.15 Jan Meier's Crossfeed Circuit.

process increases the total amplitude heard at low frequencies where the crossfeed is in effect. This creates a frequency response anomaly for sounds from the center.

The Meier crossfeed circuit is not additive in the same way, and does not create this problem. In essence, this is because his circuit operates in the differential domain. It is clear that if the signals in both the left and right paths are identical, there will be no alteration of the signals in this design. In other words, the circuit only attenuates the L-R signal at low frequencies by the action of R3 and R4. There is no L-R signal for a sound source panned to the center stage. There is just frequency-dependent L-R attenuation. In Figure 21.15(b), impedances are scaled up by a factor of about 5 and R5 and R6 are replaced by P1A and P1B to allow variable amounts of crossfeed.

Meier describes some other variations of his circuit in his papers, including an "enhanced bass" version and a "frequency-extended" version in which the crossfeed delay is effective to a higher frequency, closer to 2 kHz rather than just 1 kHz [34, 35]. The crossfeed amplitude roll-off is also extended to higher frequencies. This allows more of the remaining directional cues above 1 kHz to be in effect.

Active Crossfeed Circuits

The crossfeed circuits discussed so far are passive and relatively simple. Most are based on the idea of mixing a low-pass-filtered version of the signal from one channel into the signal path of the other channel, in some cases with some high-frequency boost in the main path of each channel. Active circuitry can be introduced to provide more features and greater functionality [35, 36]. The simplest use of active circuitry is to introduce a summing circuit at the output end of each channel where the main signal is summed with the crossfeed signal. This allows the crossfeed circuit to be fed by a low impedance and to be loaded by a relatively high frequency-independent impedance. The next step in going active for greater functionality is to introduce an active circuit that tailors the crossfeed signal in a more sophisticated way, such as the addition of crossfeed delay by all-pass filters. Some passive crossfeed circuits load the preamplifier too heavily, especially if they use only modest-sized output coupling capacitors.

Crossfeed can be thought of as a frequency-dependent attenuation of the L−R component of stereo signals. One can convert the L and R signals to L+R and L−R signals and then manipulate the L − R signal with one functional block. The L + R and modified L − R signals can then be added and subtracted to form the outgoing L and R signals. This approach creates an effect more like the passive Meier circuit in that the amplitude of center-stage signals is not increased or altered.

Relative Benefit of Crossfeed

The relative benefit of crossfeed depends somewhat on the nature of the music being played. Performances that include sounds that are panned hard left or hard right will benefit the most. Classical recordings where the microphones are not focused on a single sound source and strongly panned left or right will benefit less from crossfeed, partly because such recordings already possess some natural crossfeed. It will also depend on the type of crossfeed circuit used and the listener's personal preference. One can always turn it off.

21.8 Low-Frequency Equalization Circuits

Many headphones are not able to reproduce the deep bass of which many loudspeakers are capable. On some, the sound is nearly inaudible at 50 Hz. Published headphone frequency response claims at low frequencies are sometimes misleading. This brings up the matter of low-end equalization, more simply called bass boost. Many headphones have significant acoustic headroom to spare – this is why hearing damage is always a concern with headphones. This also means that some bass boost can be applied without the amount of overload risk associated with loudspeaker systems. This, of course, is a further argument for use of HPAs with adequate power. A headphone system with deep, tight bass can produce more body and realism in the sound.

The simplest form of bass boost is that of an ordinary tone control, which is basically a first-order filter. This is workable, but the equalization slope is fairly shallow, meaning that a significant amount of bass boost at, say 40 Hz, will result in excessive boost in the 100–300 Hz range that may introduce boom to the sound.

The Linkwitz Transform

Like closed-box loudspeakers, most headphones have a second-order low-end roll-off. The Linkwitz Transform [37, 38]. has often been applied to loudspeakers to effectively shift the second-order low-end roll-off to a lower frequency, perhaps as much as an octave lower (corresponding to a 12-dB boost). The Linkwitz Transform was used in the EQSS (Equalized Quasi Sealed System) equalization technique where a vented loudspeaker was tuned to approximate a closed-box second-order response that was then equalized with a Linkwitz Transform [39, 40].

A typical Linkwitz Transform circuit is illustrated in Figure 21.16. Its gain increases at an asymptotic slope of 12 dB/octave as frequency decreases, acting like a second-order biquadratic shelf equalizer. Its complex poles cancel the complex zeros that correspond to the transducer's low-end roll-off. Its complex zeros restore a normal second-order roll-off at a lower target frequency, often at a frequency lower by an octave. The equalizer circuit of U1 is inverting. Inverter U2 restores the proper phase.

The equalizer can include a first-order HPF with a capacitor in series with R7 (not shown) at about 11 Hz to bring the boost from the Linkwitz Transform back down at frequencies below 20 Hz. Net gain is returned to 0 dB at 2.6 Hz. Net boost is 2.2 dB at 100 Hz, 5.9 dB at 50 Hz and 10.3 dB at 20 Hz. The transform by itself provides total boost of 12.6 dB, which is moderated by an output HPF. Those playing vinyl with significant flutter at infrasonic frequencies may want to

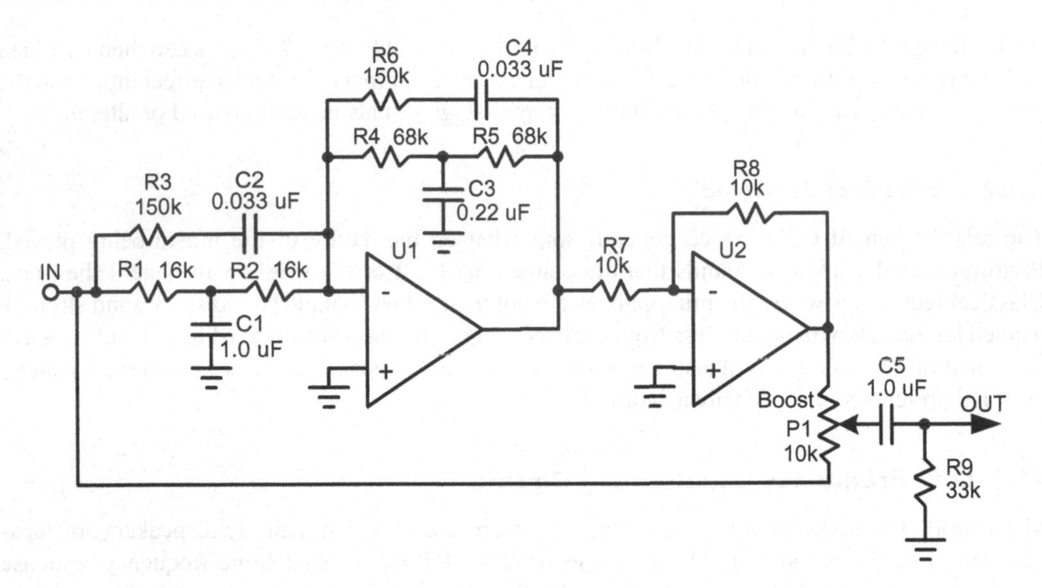

Figure 21.16 Linkwitz Transform Circuit for Headphones.

implement a third-order infrasonic filter instead of the first-order filter discussed here. Pot P1 allows the amount of transform boost to be reduced.

With about 10 dB of net boost at 20 Hz using this equalizer, ten times the power will be required to produce the same SPL as at mid-band frequencies. A headphone-amplifier combination that is capable of delivering 120 dB SPL at mid-band frequencies will then only be able to maintain 110 dB of SPL at 20 Hz. However, the equalizer produces only 7.4 dB of boost at 40 Hz, below which there is little musical energy, even in cuts with substantial bass content. The Linkwitz transform circuit is discussed in more detail in Chapter 24.

21.10 Hearing Loss Protection

Listening to headphones presents a hearing loss concern [6]. For this reason, some kind of hearing loss protection circuitry may be desirable if a suitable means of detecting the onset of unsafe hearing levels can be devised. In addition to semi-continuous music, some kind of fast muting against loud clicks, pops and unexpected loud sounds may be desirable.

A key challenge to such a circuit design is being able to discern the amount of power being delivered to the headphones and knowing what the headphone power sensitivity is. Absent detailed headphone sensitivity information, a user could manually set a threshold based on his or her perception of safety comfort. Alternatively, knowing the voltage sensitivity of the headphones makes this job easier. This implies that the headphone power sensitivity and impedance (or the headphone voltage sensitivity), must be entered by the user in some way.

An RMS voltage detector could then monitor the output of the HPA and send the result to a fairly simple microcomputer to calculate loudness-time exposure and cause a safety action to occur, such as muting or compression, with an LED indication of when such action is being taken. If data on the frequency dependence of hearing damage were available, that could be taken into account with a shaping filter ahead of the detector. In fact, A-weighting is usually employed for industrial noise measurement for OSHA compliance.

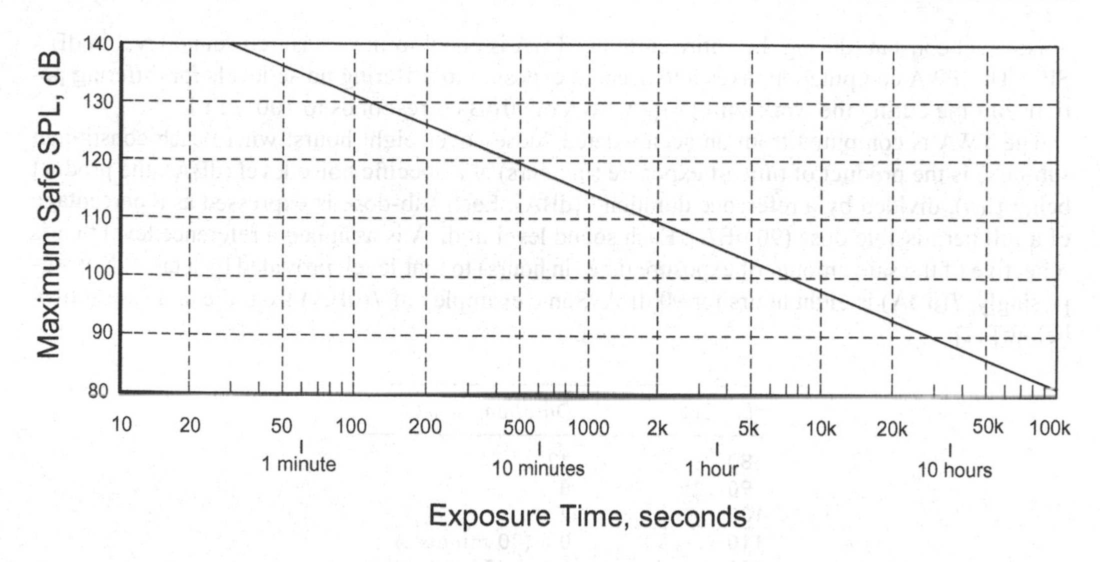

Figure 21.17 Maximum Safe SPL Exposure as a Function of Time [5].

Hearing Loss SPL Levels

Figure 21.17 is a graph that shows recommended maximum SPL exposure as a function of exposure time based on NOISH data points [5]. Allowable SPL goes down by 3 dB each time the exposure time is doubled. This relationship is referred to as a 3-dB exchange rate. On the graph, this corresponds to SPL going down at a rate of 10 dB per decade of exposure time. A gunshot creates a very high SPL level, but its duration is very short [7]. Allowing exposure to 130 dB SPL for up to 112 seconds in Figure 21.17 seems a bit too much, and using a 4-dB exchange rate is probably a safer approach.

A circuit arrangement that accumulates varying exposure and high-SPL events might act like a noise dosimeter, often worn by workers in noisy environments where the permissible noise exposure limit (PEL) is regulated [41–43]. If the HPA has a headphone protection circuit, such as one incorporating an output relay, one could incorporate some degree of hearing protection of varying degree of sophistication into the HPA to open the circuit if the accumulated SPL-time exposure limit is exceeded. It is easy to imagine this being done by a simple micro-controller (MCU). Among other tasks, the MCU might integrate the product of SPL and time in some appropriately weighted way. The hearing protection circuit should allow very high SPL levels for very brief periods of time, while at the same time taking into account that compressed music with small crest factors can be very dangerous to hearing.

Noise Exposure Computation

The Occupational Safety and Health Administration (OSHA) has very specific rules and procedures for ensuring safe noise exposure levels in industrial settings governed by 29 CFR 1910.95 [42–44]. OSHA defines a permissible exposure limit (PEL) for noise in industrial settings that is based on an 8-hour total weighted average (TWA) that corresponds to 90 dBA SPL for an 8-hour work shift where the noise level is constant. The OSHA standard implicitly assumes that frequency-dependent hearing damage is dependant on the A-weighted SPL (dBA). For an 8-hour work shift with the noise level constant during the entire shift, the TWA is equal to the measured sound level in dBA SPL. The TWA computation takes into account exposure to differing noise levels for differing periods of time during the work shift [44]. A TWA of 90 dB corresponds to 100% PEL.

The TWA is computed from an accumulated "dose" over eight hours, where each constituent sub-dose is the product of time of exposure (in hours) at a specific noise level (dBA), the product being $C(n)$, divided by a reference duration T(dBA). Each sub-dose is expressed as a percentage of a full permissible dose (90 dBA). Each sound level in dBA is assigned a reference level that is reflective of the safe amount of exposure time (in hours) to that level, provided by a table. Not surprisingly, T(dBA) is eight hours for 90 dBA. Some examples of T(dBA) from the table are shown below [42]:

dBA, SPL	*Duration, hours*
80	32
90	8
100	2
110	0.5 (30 minutes)
120	0.125 (7.5 minutes)
130	0.031 (112 seconds)

The table is based on a 5-dB "exchange rate"; when the noise level is increased by 5 dB, the amount of time a person can be exposed to that level to receive the same dose is cut in half. For example, if 90 dBA for eight hours is a full dose, 95 dB for four hours is also a full dose. Note that NIOSH uses a 3-dB exchange rate and an 85-dB 8-hour TWA exposure limit (PEL) [5]. Any duration of exposure to 140 dBA SPL or more is not permitted.

The TWA, expressed in dBA, is:

$$\text{TWA} = 16.61 \times \log(10)(D / 100) + 90$$

where D is the sum of the individual sub-doses. A 100% total dose corresponds to TWA = 90 dBA. A 200% dose corresponds to TWA = 95.4 dBA. The OSHA rational is that employees exposed to TWA greater than 90 dBA should wear hearing protectors. Employees wear commercial noise dosimeters where noise exposure is regulated. In the headphone hearing safety discussions here, it is assumed that music should be treated in the same way as noise. Anyone attending a rock concert might want to bring with them a noise dosimeter.

Accounting for Headphone Sensitivity

As touched on earlier, if the hearing protection circuit only has access to the output voltage of the HPA, it is very difficult to calculate the exposure level accurately in the absence of knowledge of the impedance and sensitivity of the headphones. If the voltage sensitivity of the headphones is known and can be dialed into the HPA, then there is a possibility that one could achieve some accuracy and reliability.

It is easy to imagine an HPA with hearing protection circuitry that includes a single switch to enter headphone voltage sensitivity in 10-dB increments, based on the owner's headphone manual and recognizing that converting from headphone power sensitivity and impedance to voltage sensitivity is an easy calculation (see Section 21.2). Most headphones fall between 90 and 125 dB SPL/V. Nominal 10-dB voltage sensitivity selections could be 95, 105, 115 and 125 dB/V. Each selection would allow for a ±5 dB range about the nominal selection, which is probably adequate granularity for this purpose. Such approximations are certainly better than having no protection at all.

Hearing Loss Protection Circuits

Because having a less than perfect hearing protection circuit is better than having none at all, there are many tradeoffs and compromises in design that can be acceptable. Central to the implementation is effectively a crude SPL dosimeter. As shown in Figure 21.18, the basic model suggested here includes four elements: (1) A-weighting of the HPA output signal; (2) SPL estimation that takes into account headphone sensitivity; (3) moving-average integrator for calculating exposure over the listening interval; and (4) means of attenuating the sound level and indicating that action has been taken.

Implementing an A-Weighting filter for the signal is quite easy with a few passive components, as described in Chapter 11. SPL estimation is best done with a fairly fast RMS-to-DC IC preceded by a selectable attenuation that takes into account headphone voltage sensitivity. Without much compromise, an average detector can be used instead. Either approach could be augmented by a sample-hold to capture transient peaks. An additional cost-saving measure with minimal impact on accuracy is to measure a single monophonic signal created by adding both HPA output channels together.

The calculating means should almost certainly be carried out by a small MCU whose ADC input is fed the DC voltage from the detector. As described in Chapter 28, these can be remarkably inexpensive, as little as $4 for a Raspberry Pi Pico complete MCU board that measures 0.8 × 2.0″. The MCU just needs to calculate the SPL corresponding to the received voltage, multiply it by the time interval being used by the MCU for each sub-dose and accumulate the result for the duration of the listening session and compare the accumulated value to the 90-dB TWA threshold. The MCU can obviously make the time interval for calculating each sub-dose adaptive if that would help.

Indeed, like most modern consumer products, some HPAs already include an MCU, if for no other reason than to operate a fancy LCD display and allow shaft encoder control means for volume. If the display is touch-sensitive, or if navigation buttons exist, it would be easy to incorporate means whereby the user could enter manufacturer-supplied impedance and power sensitivity for their headphones and have the MCU convert it to voltage sensitivity for measurement purposes. One could even have the MCU store sensitivity values for multiple headphones.

Once the MCU detects a dangerous sound exposure level, it can take action to protect the user and make the user aware. This may mean something more intrusive than just lighting a red LED. For example, if the HPA includes a mute/protect relay, that same relay can be opened and remain open until a reset button is pressed. Alternatively, a 10- or 20-dB volume attenuator can be activated until reset. Actions like these force the user to pay attention. If they intentionally want to take

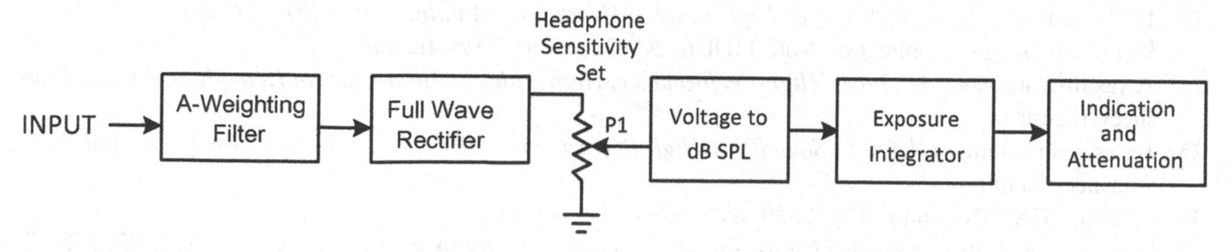

Figure 21.18 A Hearing Protection System for Headphone Listening.

more risk of hearing damage, they can fool the circuit by entering a lower value for headphone sensitivity. In our gadget-hungry consumer environment, the intelligent hearing protection "audio dosimeter" might pay for itself in increased sales for all but the least expensive HPAs.

References

1. Bob Cordell, *Designing Audio Power Amplifiers*, 2nd edition, Routledge, New York, NY, 2019.
2. Sennheiser HD 650 Dynamic Headphones, sennheiser.com.
3. HIFIMAN Sundara Planar Magnetic Headphones, hifiman.com.
4. Stax SR-009S Electrostatic Headphone Reference?
5. National Institute for Occupational Safety and Health (NIOSH), *Recommended Exposure Limit (REL), Criteria for a Recommended Standard – Occupational Noise Exposure*, 1998, U.S. Department of Health and Human Services.
6. Headwize, *Preventing Hearing Damage When Listening with Headphones*, headwize.com.
7. M. Ylikoski, J.O. Pekkarinen, J.P. Starck, R.J. Pääkkönen and J.S. Ylikoski, Physical Characteristics of Gunfire Impulse Noise and Its Attenuation by Hearing Protectors, *Scand Audiol*, vol. 24, no. 1, 1995, pp. 3–11. doi: 10.3109/01050399509042203. PMID: 7761796.
8. John Siau, *Headphone Impedance and Sensitivity*, Benchmark Media Systems, May 2014, benchmark-media.com/blogs/application_notes/14017381-headphone-impedance-and-sensitivity.
9. Dennis Bohn, *Understanding Headphone Power Requirements*, Rane Corporation, https://headwizememorial.wordpress.com/2018/03/14/understanding-headphone-power-requirements/.
10. Adrian Try, Are the Volume Wars Killing Music? Which Side are You On?, *tuts+*, November 18, 2010, musictutsplus.com/are-the-volume-wars-killing-music-which-side-are-you-on--audio-8381a.
11. Benchmark Media HPA4 Headphone Amplifier, benchmarkmedia.com.
12. International Electrotechnical Commission, *Multimedia Systems – Guide to the Recommended Characteristics of Analogue Interfaces to Achieve Interoperability*, IEC 61938, Section 12, 1996.
13. Richard Crowley and Rod Elloitt, *Hi-Fi Headphone Amplifier*, Project 24, Elliott Sound Products, sound.whsites.net.
14. Rod Elliott, *Headphone Amplifier*, Project 113, Elliott Sound Products, sound.whsites.net.
15. Pete Millett, *The Butte DIY Solid-state Headphone Amp*, pmillett.com/butte.htm.
16. Pavel Ruzicka, *TPA6120 HiFi Headphone Amplifier*, pavouk.org.
17. Texas Instruments, *A High-Power High-Fidelity Headphone Amplifier for Current Output Audio DACs Reference Design*, Application Note TIDU672C, December 2014, ti.com.
18. John Caldwell, *Headphone Amplifier for Voltage-Output Audio DACs Reference Design*, TI Designs, TIDUAW1, ti.com.
19. Dimitri Danyuk, *Linear Integrated Systems Headphone Amplifier Evaluation Board*, Application Note, linearsystems.com.
20. John Conover, *Direct Coupled Stereo Headphone Amplifier*, johncon.com.
21. Headphones.com, *OPA1622 Integrated Headamp Project*, lost33 1.2020, https://forum.headphones.com/opa1622-integrated-headamp-project/4794.
22. Texas Instruments, *OPA1622 Sound Plus High Fidelity, Bipolar Input, Audio Operational Amplifier*, datasheet, ti.com.
23. Texas Instruments, *A High-Power High-Fidelity Headphone Amplifier for Current Output Audio DACs Reference Design*, Application Note TIDU672C, December 2014, ti.com.
24. Texas Instruments, *LME49600 High-Performance, High-Fidelity, High-Current Headphone Buffer*, datasheet, ti.com.
25. Texas Instruments, *OPA1642 SoundPlus High Performance, JFET Input Audio Operational Amplifier*, datasheet ti.com.
26. Exicon, ECX10N20 and ECX10P20, datasheets, exicon.info.
27. Benjamin B. Bauer, *Stereophonic to Binaural Conversion Apparatus*, U. S. Patent 3,088,997, May 7, 1963.

28. Benjamin B. Bauer, Stereophonic Earphones and Binaural Loudspeakers, *JAES*, vol. 9, no. 2, April 1961, pp. 148–151.
29. Sigfried Linkwitz, *Improved Headphone Listening; Build a Stereo Crossfeed Circuit*, Audio Magazine, December 1971 and linkwitzlab.com.
30. F.M. Weiner, On the Diffraction of a Progressive Sound Wave by the Human Head, *JASA*, vol. 19, January 1947, pp. 143–146.
31. Chu Moy, *An Acoustic Simulator for Headphone Amplifiers*, 1999, https://astersart.net/amp/headphones/acoussim/acoustic_simulator/for_headphone.htm.
32. Jan Meier, *A DIY Headphone Amplifier with Natural Crossfeed*, https://headwizememorial.wordpress.com/2018/03/09/a-diy-headphone-amplifier-with-natural-crossfeed/.
33. Jan Meier, *Enhanced Bass Natural Crossfeed*, https://blackwidowaudio.tripod.com/articles/enhnatural-crossfeed.htm.
34. Meier Audio, "Crossfeed," https://www.meier-audio.com/crossfeed.html.
35. Dimitri Danyuk, *Adjustable Crossfeed Circuit for Headphones*, Electronics World, August 2005.
36. John Conover, *Spacial Distortion Reduction Headphone Amplifier*, January 4, 2021, http://www.johncon.com/john/SSheadphoneheadphoneAmp/.
37. Sigfried Linkwitz, *12 dB/oct Highpass Equalization*, linkwitzlab.com/filters.htm.
38. Rod Elliott, *Subwoofer Equalizer*, Elliott Sound Products (ESP) Project 71, http://sound.whsites.net/project71.htm.
39. Bob Cordell, *EQSS™ Equalized Quasi-Sealed System – Get Extended Low-Frequency Response and Higher SPL with EQSS™*, audioXpress magazine, September 2014.
40. US Patent 7,894,614, *System and Method for Achieving Extended Low-frequency Response in a Loudspeaker System*, February 22, 2011.
41. Occupational Safety and Health Administration (OSHA) Technical Manual, Section III, Chapter 5. August 2013.
42. OSHA Noise Standard 29 CFR 1910.95.
43. Occupational Noise Exposure Criteria, NOISH Publication No. 98–126, Chapter 4: *Instrumentation for Noise Measurement*, 33–35.
44. OSHA 1910.95 Appendix A – Noise Exposure Computation.

Chapter 22

Volume, Balance, Fader and Panning Controls

Volume controls are central to the operation of a preamplifier or mixer (where they are often called faders), and in fact, some preamplifiers comprise little more than a volume control augmented by input switches and buffers. The traditional audio taper potentiometer volume control is still a workhorse, but its disadvantages include imperfect matching among channels, expense in multi-channel systems, inability to be automated unless it is motorized and they wear out. Potentiometers can also create wiper noise and wiper distortion. Channel balance in audio taper stereo pots can often be off by 1 dB or more, and dependent on the pot's setting.

Balance controls are absent from many modern preamplifiers, but still add useful functionality for better imaging and placement in the face of channel mismatches among phono cartridges, volume controls and loudspeakers. Balance pots usually have a linear taper and often have a detent at the center position. Pan pots, which place an instrument or vocal input either left or right in the sound stage, are essential to mixers and studio consoles.

If a linear pot is used for control of volume, its linear law can be changed to something more logarithmic by connecting a resistor from the wiper to ground. However, the ±20% pot resistance tolerance then becomes a problem against the precision taper-altering resistance. This will introduce channel imbalance if the two sections of the potentiometer have different resistances.

Resistive loading on a pot can sometimes introduce some wiper distortion. Resistive loading on a pot wiper to alter the taper is not as much an issue for a fader if tracking with another fader for a different channel is not needed.

22.1 Potentiometer Volume Controls and Faders

Stereo volume controls and multi-channel controls in home theater and other professional applications must have good channel attenuation matching. Channel balance for audio taper pots is usually not great. Linear pots have better tracking, but the linear taper cannot adequately handle an attenuation range of many dB as a function of pot rotation.

Faders

Faders control the signal level for each channel and for the main outputs in a mixer. They must cover a wide attenuation range, from *off* to full output, with a user-friendly and convenient taper. Faders are usually implemented with slider potentiometers, some motorized, or voltage-controlled amplifiers (VCA). Most medium and large analog mixing consoles employ VCAs that are controlled by a DC voltage from a linear pot (which may be motorized) or from a DAC or some combination.

VCA-based fader arrangements are especially attractive in consoles that support automation. Of course, this function is performed with DSP (digital signal processor) in all-digital consoles.

Resistance Elements

Potentiometers are usually implemented with a rotating wiper in contact with a circular track of a conducting material, like carbon or metalized conductive plastic. Linear faders are essentially the same, just constructed with straight strips of the conducting element. Higher-quality pots, like the ALPS RK271 Blue series, use a plastic track [1].

Tapers

A volume control or fader must be conveniently variable in its attenuation from 0 to at least −60 dB, at which point thereafter attenuation should be infinite. A linear potentiometer cannot do this properly. In a conventional arrangement with a linear pot, the first 180° of rotation is devoted to a span of only 6 dB. Ideally, the number of dB of attenuation should be linear with pot rotation. With a 240° electrical rotation and 80 dB of maximum controlled attenuation, this would work out to 3° of rotation per dB of attenuation.

The conventional audio taper pot is only an approximation to this ideal. It is constructed of a conductive strip that has two or three differing resistivities at different portions of the rotation. At mid-rotation, an audio taper pot usually has between 17 and 20 dB of attenuation. Pots with better conformance to a log characteristic are available and often used in mixers. These are referred to as having a log taper, and are more expensive. Pots with a reverse-log taper are also available. Of course, even log-taper pots must depart from a logarithmic approximation when set near infinite attenuation.

Figure 22.1(a) shows the taper characteristics of a 50-kΩ audio taper pot, with resistance from the CCW terminal to the wiper vs percent of electrical rotation [2]. Mechanical rotation is usually a greater number of degrees than electrical rotation, about 270° vs 240°. At 50% rotation, the resistance of the shunt portion of an audio taper pot is usually between 10% and 15% of the rated pot resistance.

Figure 22.1(b) shows the same information, but with the y axis in dB of attenuation instead of resistance. At 50% rotation, the attenuation lies between 17 and 20 dB.

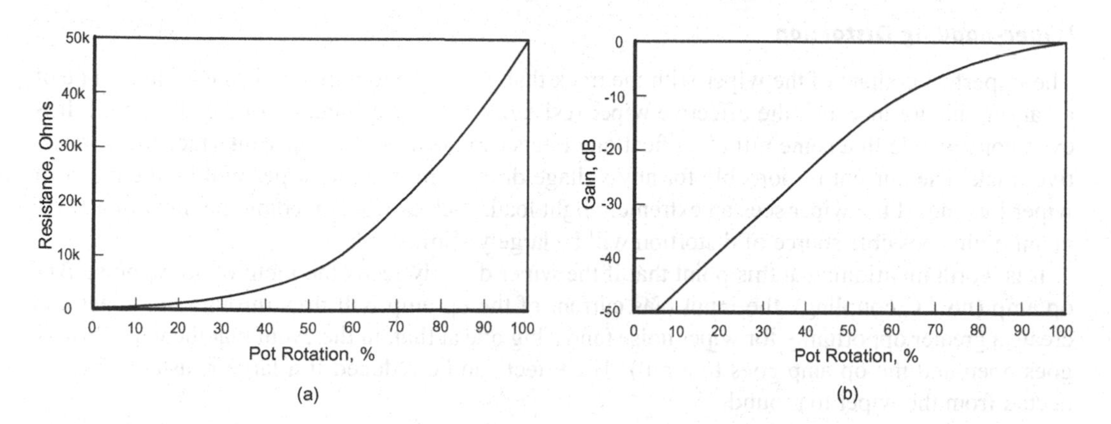

Figure 22.1 Audio Taper Pot Resistance and Attenuation vs Rotation.

Wiper Noise

Wiper noise occurs because the mechanical/electrical contact with the resistive element is not perfect, especially if the pot is dirty. The wiper can occasionally lose contact with all or part of the strip. If there is no DC across the pot, this will result in noise as a result of very brief interruptions of the signal. If there is DC across the pot, this can result in much more audible noise as the DC voltage at the wiper may change briefly by a large amount. Even a few mV of DC can make wiper noise more audible. As a rule, DC across a pot must be avoided or made very small. Wiper noise can be measured by applying DC to one end of the pot and grounding the wiper, while measuring noise on the other end of the pot as it is turned.

Wiper Resistance

The contact of the wiper with the conductive track is not perfect, so there is some resistance between the wiper and the effective electrical point on the conductive strip. This is often not a problem. However, if the wiper does not fully cover the track, there can be a path for some conductance of the signal around the wiper. This can degrade performance in some applications, like with a balance control implemented with a single pot whose wiper is connected to ground.

Depending on pot construction, there can be effective end resistance at either extreme of the rotation, where the resistance from an end terminal of the pot and the wiper is not zero. This is usually quite small.

Wiper-Loading Effect on Attenuation

If you consider a linear pot as an example, attenuation at center rotation should be 6 dB. If a 10-kΩ pot is fed from a voltage source, the impedance at the wiper will be a maximum of 2.5 kΩ at center rotation. If the wiper is loaded by the circuit it feeds by 10 kΩ, the attenuation as a function of rotation will no longer be linear, and the attenuation at mid-rotation will be greater than 6 dB by about 2 dB. The load that the pot puts on its driving circuit will also change with rotation in this case. In some cases, such loading can be deliberate in an attempt to make a linear pot act more like one with an audio taper. This can bring with it other problems, however.

Wiper-Loading Distortion

The imperfect contact of the wiper with the track that gives rise to wiper resistance can also be of a varying nature in which the effective wiper resistance may be a function of signal current. It is even conceivable that some mild rectification effects can occur at the wiper interface to the resistive track. The current responsible for any voltage drop from track to wiper will be the result of wiper loading. If the wiper sees an extremely light load, such as if it is feeding the input of a JFET op amp, this possible source of distortion will be largely eliminated.

It is worth mentioning at this point that if the wiper directly feeds the input of a low-noise BJT op amp (no AC coupling), the input bias current of the op amp will flow through the wiper and create a greater opportunity for wiper noise (and a big one at that, in the event that the wiper briefly goes open and the op amp goes to a rail). The effect can be reduced if a large resistance is connected from the wiper to ground.

In order to test for wiper-loading distortion, connect the pot as a volume control and measure it with a distortion analyzer with the pot at half position. In this part of the test, little or no signal current is flowing through the wiper, so any nonlinearity in the wiper contact resistance will cause little or no distortion because there is little or no signal voltage drop between the track and the wiper.

Next the wiper is loaded to ground with a resistor whose value is ¼ the total pot resistance. The value ¼ is chosen because this is the output impedance of the pot at its central position. This creates an additional loss of 6 dB and causes signal current to flow from the track to the wiper, creating a small signal voltage drop that may be nonlinear. The distortion in the resulting signal is then measured.

One can also run the signal into the pot at the CCW end with the wiper centered and connected to ground. The leakage signal seen at the CW end can be measured for amplitude and distortion.

Output Impedance and Thermal Noise vs Rotation

The output impedance of the pot will generally be at its greatest when the track resistances on both sides of the wiper are equal. This will occur at mid-rotation for a linear pot, but at a different rotation for an audio taper pot. The point of maximum impedance will generally be 1/4 the value of the pot for a linear pot. Since the output impedance of a pot is resistive, that resistance will create thermal noise. A 50-kΩ pot will have a maximum resistance at the wiper of 12.5 kΩ, and will generate 14 nV/√Hz of thermal noise. This is a good reason to keep pot resistances low where possible. Fortunately, when a volume control pot is set for maximum attenuation, its impedance is virtually zero.

Tracking, Matching and Trimming

In a stereo environment it is important that the left-right balance not shift with changes in the volume setting. This means that the two gangs in a stereo volume control must have good tracking, both mechanical and electrical. This can be a bit more difficult to achieve with pots that have an audio taper. The worst-case difference in attenuation between the two gangs of a stereo pot can often range from 1 to 2 dB.

However, there is a reasonably accurate solution if one is willing to customize the circuitry associated with one or both gangs of the pot. This involves measuring the attenuation of each channel when the pot is set to its mid-point (which need not be exact). One channel will have lower gain than the other channel to the degree that there is mismatch at the center of rotation. Connect a resistor of suitable value from the wiper of the channel with less attenuation to ground to bring its signal level down to equal that of the other channel. Doing this mid-rotation correction will result in quite good tracking between channels above and below the mid-point.

To get an idea of the value of resistor to use, short both ends of the pot section with less attenuation together and measure the resistance to the wiper. This provides the effective source impedance of the pot at mid-rotation. Think of this as the series resistance of an attenuator. If the gain of that section at mid-position is 1 dB higher than that of the other channel, choose a shunt resistor to create 1 dB of attenuation against the effective source impedance of the pot. Figure 22.2(a) illustrates this technique as used with a fixed resistor that will be different for different pots of the same model. In (b), a trimmer is used in conjunction with a fixed resistor to allow trimming for tracking *in situ*.

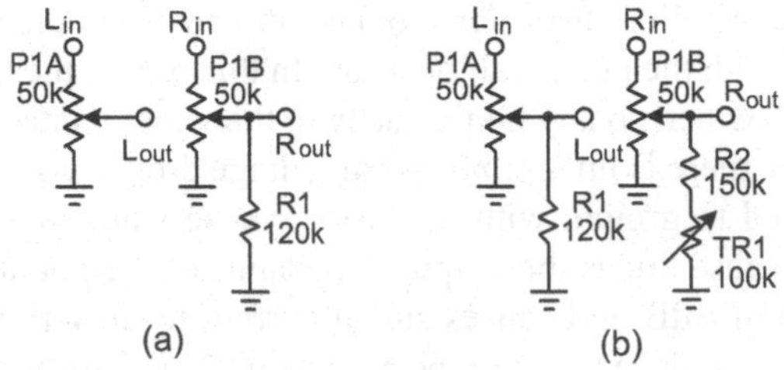

Figure 22.2 Trimming to Improve Tracking between Gangs.

Changing the Taper Law

With a similar wiper-loading technique, one can make a linear pot behave with an approximation to an audio taper [3]. This can be an attractive choice, given that linear pots exhibit better gang tracking than audio taper pots. This approach is viable if the pot exhibits low wiper distortion when the wiper is loaded. The required loading resistor will result in 17–20 dB of attenuation at 50% rotation. Without the loading, the attenuation will be 6 dB and the source resistance will be 1/4 the resistance of the pot. The loading resistor must create an additional 12 dB if the attenuation at 50% is chosen to be 18 dB. For a 50 kΩ pot the required resistance is 4 kΩ. In the general case, the loading resistance should be about 8% of the pot resistance.

Alas, there is still the ±20% tolerance on pot resistance between gangs. With wiper loading, this will create gang mismatch because the source resistance seen by the fixed loading resistance will have the same tolerance. Therefore, a trimmer should be used with one of the loading resistors as shown in Figure 22.2(b) but with smaller resistances than shown.

Motorized Potentiometers

With the advent of remote controls and multi-channel receivers came the introduction of motorized pots. These were convenient, but added some cost. In a great many cases they have been replaced by other means of electronically controlling volume, such a by voltage-controlled amplifiers, digitally stepped CMOS attenuators and arithmetic in a DSP signal path. Motorized faders are still popular in analog consoles where automation is employed, since they provide a visible indication of their setting.

In some cases a motorized fader may have a second gang that reports back a voltage to the controller that indicates its physical position. In other cases, the motorized fader is just a single-gang linear pot that controls a VCA with a DC voltage and also reports that voltage to the control circuit. Motorized pots usually operate with a motor voltage of 5 V.

22.2 Baxandall Active Volume Control

The Baxandall volume control illustrated in Figure 22.3(a) is a clever arrangement using a linear potentiometer with an op amp to achieve a quasi-logarithmic taper behavior [4–6]. It uses a combination of feedforward and feedback to accomplish this, and can cover a very wide range of attenuation and/or gain. In its simplest form, the gain would be close to unity inverting when the pot is at its mid-point if R1 was 0 Ω. R1 sets the maximum gain and influences the attenuation at the pot's mid-point.

At the normal location of a volume control in a preamp, little or no gain is required, so some measures need to be taken to obtain the desired gain scaling. Often a gain of no more than 3 is needed. Issues of headroom and SNR can come into play.

Without any scaling, it can be seen that at 10% electrical travel the gain will be approximately −20 dB. At 0% rotation, the attenuation is essentially infinite, but will depend on the open-loop gain of the op amp at any given frequency. At 90% rotation, the gain will be about +20 dB depending on selection of the gain-limiting resistor R1. If R1 is chosen to be 1/3 the value of the pot (here 16.7 kΩ), gain at full clockwise rotation will be 3. This choice will affect the nature of the taper. Figure 22.4 shows gain and attenuation for this arrangement as a function of pot rotation in comparison to an ideal log taper for R1 = 16 kΩ. By calculation, at 50% electrical rotation, and with R1 = 16 kΩ, relative attenuation will be 14.3 dB.

If instead R1 = 10 kΩ, maximum gain will be 14 dB and relative attenuation at 50% rotation will be 17 dB, closer to what was measured for an audio taper pot. If the maximum gain of the

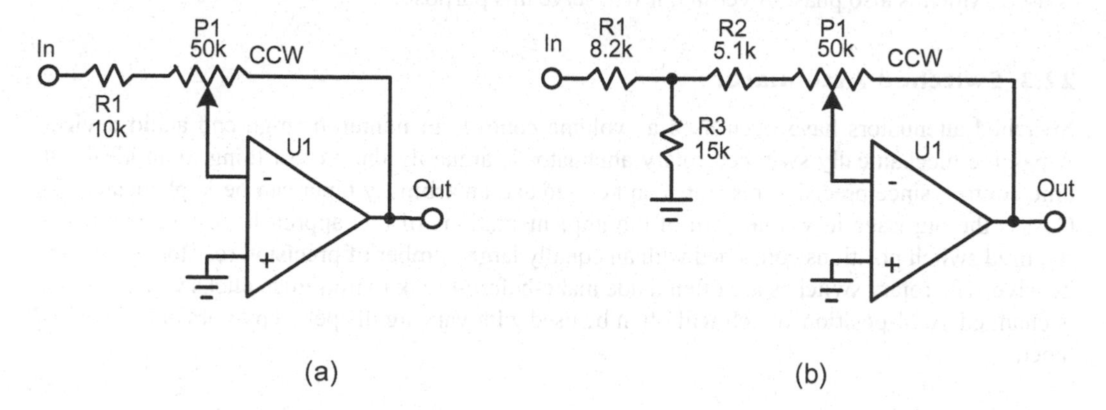

Figure 22.3 Baxandall Volume Control.

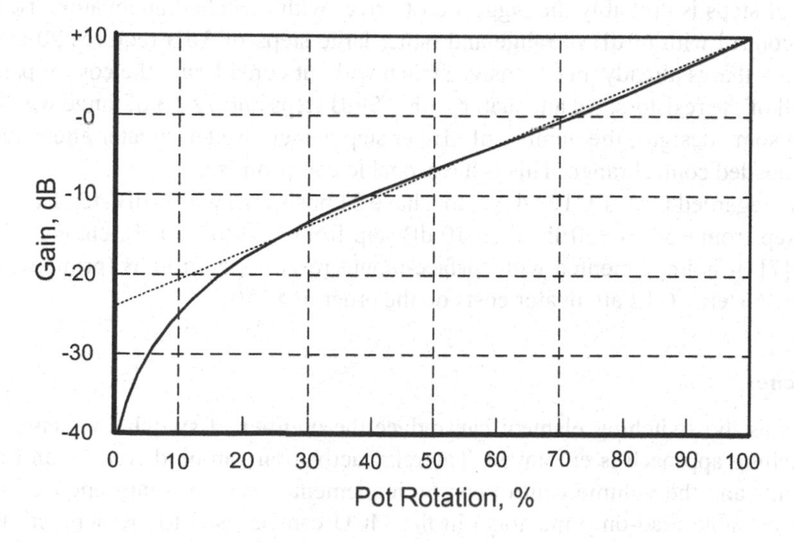

Figure 22.4 Baxandall Volume Control Gain vs Pot Rotation.

Baxandall circuit of 14 dB is too high, fixed attenuation can be put in the path by connecting a shunt resistor R3 from R1 to ground, as shown in Figure 22.3(b). Here R1 and R2 are made 8.2 kW and 5.1 kW, while R3 is made 15 kW, so a loss of 5.2 dB is inserted while keeping the source resistance to 10 kΩ and keeping relative loss at 50% rotation at 17 dB. The maximum gain is reduced to 8.8 dB, which is a more satisfactory maximum gain.

Depending on the op amp chosen, there can be a source of distortion under conditions of high attenuation. This is caused by the input signal current flowing through the pot into the output of the op amp. If the op amp has any crossover distortion, the attenuated output will have an increased percentage of distortion as the attenuation increases.

Given a possible ±20% tolerance between the gang resistances, there is once again the matter of a non-exact resistance in series with the fixed R1 resistance. This can cause gang mismatch. This can be addressed with a trimmer in series with R1 for one of the channels. Even better, pots with well-matched gang resistances can be selected. The Baxandall volume control is phase-inverting, so somewhere else in the signal chain there must be another inversion. If a Baxandall tone control is used, which is also phase-inverting, it will serve this purpose.

22.3 Switched Attenuators

Switched attenuators have been used as volume controls in numerous high-end audio devices. A passive mechanically switched rotary attenuator is arguably the closest thing to an ideal volume control, since precision resistors can be used and an arbitrary taper can be implemented [7]. Cost is the big issue in virtually all of the implementations of this approach. A large number of required switch positions combined with an equally large number of precision resistors can be expensive. The rotary switches are often made make-before-break to minimize clicks when a setting is changed. A 24-position switch will often be used with varying dB per step to obtain the desired taper.

Granularity

Granularity of steps is probably the biggest cost driver with switched attenuators used as volume controls. A control with 60 dB of range and rather large steps of 3 dB requires 20 positions on a rotary switch – that is already an expensive switch without considering the cost of parts and labor for adding all of the resistors. An attenuator with 1.5-dB steps and 72 dB of range would require 48 positions. In some designs, the number of dB per step is increased at greater attenuation levels to achieve the needed control range. This is a reasonable compromise.

The highly regarded DACT CT2 attenuators have 24 positions, with 2 dB/step from 0 to −34 dB, then 4 dB/step from −34 to −50 dB, then 10 dB/step from −50 to −60 dB, ending with complete attenuation [7]. It is implemented with surface-mount resistors on a PCB (printed circuit board) switch wafer. A stereo CT2 attenuator costs on the order of $150.

Relay Switches

Using relays as the switching element can reduce the number of switch contacts if a binary or similar switching approach is employed. The relay activation can be driven by an MCU (microcontroller unit) and the volume control can be implemented with a rotary encoder. An EEROM (electrically erasable read-only memory) in the MCU can be used to “remember” the previous setting. MCUs are covered in Chapter 28.

In a binary approach, six bits, corresponding to six DPDT relays, can provide 64 positions, yielding a range of 64 dB in 1-dB steps. Eight relays would yield a range of 128 dB in 0.5-dB steps. Other approaches can employ Gray coding so that only a single bit changes at each step.

A problem with relay-switched attenuators is that the operating time is fairly long, being in the ms range. Indeed, there are three relay operating times of concern in a double-throw relay. The obvious ones are closure and opening times, which are often different and may depend on the driver circuit (coil current and flyback control means). The third is the time when the swinger is in transition from one contact to the other, and is not connected to anything. These delays can create pops and clicks as the attenuation is changed. Care in timing of relay actuations in combination with other measures can minimize these effects [8].

Ladder Attenuators

The R2R ladder DAC is frequently used in very high-speed DACs because it is simple and the number of available steps is equal to the binary number that represents the number of switches. For example, a seven-bit number has 128 values, so in principle seven relays can select one of 128 numbers or positions. Unfortunately, the binary R2R DAC produces linearly spaced increments that cannot cover the minimal 60-dB rage in increments that all have the same dB value.

With a similar topology, where the resistors are not in a 1:2 relationship, the arrangement can be made to be logarithmic, where each step corresponds to the same dB value. Figure 22.5 is a simple example of such an attenuator using six relays and covering a range of 64 dB in 1-dB steps [9]. Each section of the attenuator provides a different attenuation in dB in a binary order. Each section consists of a two-resistor attenuator with a 25-kΩ input impedance and is designed to feed a subsequent load of 25 kΩ. Each section of attenuation can be enabled or bypassed by its relay. Each relay either bypasses the series resistor or adds the shunt resistor. The use of latching relays allows it to remain in its previous attenuation setting after power-down. The attenuation of each section is binary weighted from 1 to 32 dB. Use of an MCU with non-volatile memory can allow the use of ordinary relays.

As shown, the ladder attenuator should be fed from a low-impedance source and loaded by a very high impedance. The input impedance is a constant 25 kΩ, independent of attenuation setting. Its output impedance changes with attenuation setting, being much lower at high values of attenuation. Some measures may need to be taken to avoid clicks when the relay positions are changed. Such an attenuator will usually be controlled by an MCU and the physical volume control may just be a shaft encoder.

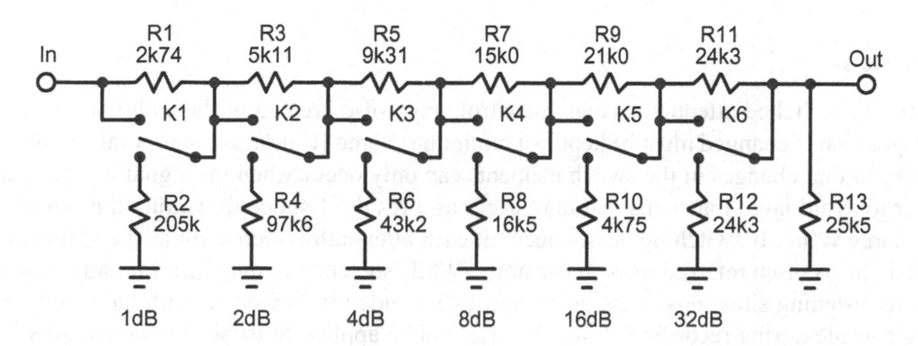

Figure 22.5 Logarithmic Ladder Attenuator.

The AMB Laboratories $\delta 1$ relay-based stereo ladder attenuator is a good example of such an attenuator [9]. It is driven by an I^2C interface from an MCU.

Active Switched Attenuators

The attenuators discussed thus far in this section have all been passive. However, the switched approach, be it with relays or other means, can be applied to active amplifier stages with some significant advantages. There are many ways to do this, but the Baxandall approach, using feed-forward and feedback, is particularly attractive because of its inherent log-like behavior. Any combination of relays or other switches that can reasonably emulate a linear potentiometer can be used. Of course, the constant end-to-end resistance of a real potentiometer need not be preserved. If constant end-to-end resistance were required, the number of available steps might be reduced. If a relatively small variation in "pot" resistance were permitted, the reduction in available steps would be mitigated.

The ratiometric behavior is what is important. As an example, one group of attenuators can implement the forward path and a second group can implement the feedback path. With four relays in each path, 256 steps in attenuation/gain can be realized. In principle, a range of over 100 dB could be realized in 0.5-dB steps. Choices would have to be made in the arrangement and weighting would have to be made to avoid the gain of the stage exceeding about 10 dB. Importantly, the feedback path must never be allowed to be opened during a relay transition.

22.4 Integrated Circuit CMOS Volume Controls

A number of companies make integrated circuit switched attenuators that can be controlled by a digital input, as from an MCU. These devices usually have a SPI or similar serial interface to the controller. Some of these devices are controlled with rotary shaft encoders. Some others can be controlled manually with up/down buttons. Most devices are fabricated in a CMOS process and use CMOS transmission gates to control the amount of attenuation. Quite low values of distortion are achievable, and the repeatability of the IC process helps achieve smooth and accurate switching. Many are best suited for MCU control, and independent MCU control of each channel permits balance and panning functions. Some use a tapped series string, others use a ladder topology, and still others use a ladder topology with taps along series segments. Many have poor tolerance on input impedance. Four stereo examples of IC volume controls are discussed below.

Zipper Noise

As with any switched attenuator volume control, any noise created in the audio signal when the switch position is changed must be kept to a minimum. Some IC designs incorporate zero-crossing detectors so that changes in the switch elements can only occur when the signal is at or near zero in order to avoid large transients that may occur as a result of suddenly cutting the signal to zero from a large value. If switching noises occur at each attenuation increment as the volume control is varied, this is often referred to as *zipper* noise. While such noises may be minimally annoying to consumer listening situations, these noises must be avoided if they occur with fader controls on a mixing console during recording. Also, DC must not be applied to these volume controls in order to minimize zipper noise.

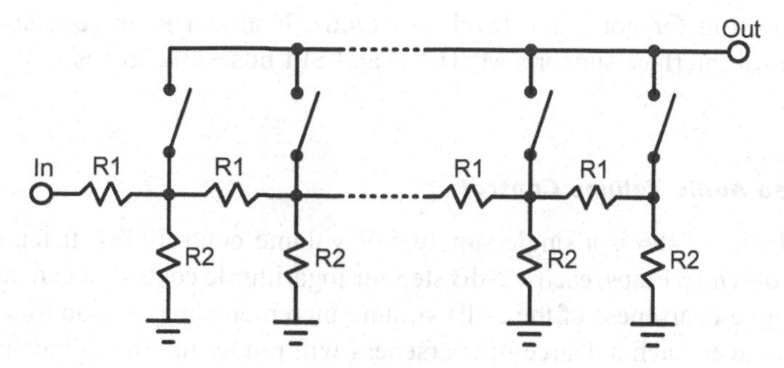

Figure 22.6 LM1972 Attenuator Architecture.

LM1972 μPot Two-Channel 78-dB Audio Attenuator

The LM1972 from TI is an economical passive digital attenuator with very good performance. The audio path is essentially passive with the exception of the switching multiplexers. The device provides attenuation from 0 to 78 dB in 0.5-dB steps at low/medium attenuation with 1-dB steps for high amounts of attenuation, a good compromise to make the most effective use of the available taps [10]. Below −78 dB, the device enters a mute condition.

The attenuator uses a hybrid R2R-like architecture where the tap switches select taps along each of the series resistors in the network, as shown in Figure 22.6. Distortion is only 0.003% maximum at a signal level of 0.5 V_{pk} at 0 dB attenuation. Distortion rises to 0.007% at 2 V_{RMS}. The input impedance is nominally 40 kΩ, but this number can range from a minimum of 20 kΩ to a maximum of 60 kΩ. This is not uncommon for ICs that do not use thin-film resistors. The device is controlled by a 3-wire serial interface, such as SPI, from an MCU and can be daisy-chained. It boasts pop- and click-free attenuation changes. Its output impedance varies with attenuation and can be as high as about 15 kΩ; for this reason its output should feed a JFET op amp buffer. It operates from ±6-V supplies and costs in the vicinity of $4 in small quantities.

PGA2311 Stereo Audio Volume Control

The TI PGA2311 volume control chip includes an output op amp and can cover a very wide gain range of +31 to −95 dB in 0.5-dB steps with gain matching of better than ±0.1 dB [11]. Distortion is very low, at less than 0.001% at 2 V_{RMS}. It is intended for professional and high-end applications; cost is in the vicinity of $11 in small quantities. Its attenuator architecture is depicted as a multiplexer choosing a tap on a series attenuator string followed by variable feedback gain control of the output amplifier. Its input resistance is nominally 10 kΩ. It operates from ±5-V analog supplies plus a +5-V digital supply. Maximum output voltage is ±3.75 V. It features zero-crossing detection for noise-free level transitions. It is controlled by an MCU SPI interface and can be daisy-chained. The PGA4311 is the four-channel version.

PGA2320 Stereo Audio Volume Control

The high-performance TI PGA2320 is very similar to the PGA 2311, but is fabricated in a more expensive mixed-signal BiCMOS process to allow operation with ±15-V analog power rails, providing maximum input and output levels of ±14 V [12]. The PGA2320 includes high-performance operational amplifiers whose gain is controlled by both feedforward and feedback. It features zero-crossing detection for noise-free level transitions. Nominal input resistance is 12 kΩ. Its 3-wire serial control interface supports MCU I²C and SPI buses. Its cost is in the vicinity of $20 in small quantities.

MAX5486 Stereo Audio Volume Control

The Maxim/ADI MAX5486 is a single-supply 5-V volume control [13]. It features de-bounced pushbutton control with 31 taps, each a 2-dB step for logarithmic control. It can also control stereo balance. The relative coarseness of the 2-dB volume increments may be too much for audiophile applications. Moreover, such a degree of coarseness will render the stereo balance adjustment to be quantized by 2 dB.

22.5 Voltage-Controlled Amplifiers

The term voltage-controlled amplifier (VCA) is usually associated with integrated circuits based on what are called translinear techniques, most often implemented with bipolar transistors in integrated circuits like the THAT 2180 series [14]. These are also referred to as log-antilog circuits. Gain is usually dependent on a control voltage in a fashion that is linear in dB, so the control function for gain is described in dB/volt or volts/dB. This type of control characteristic is sometimes referred to as a *decilinear* characteristic. The 2180 has control gain of 6 mV/dB. As shown in Figure 22.7, the control input impedance must be very small, here less than 22 Ω. The resistive divider consisting of R2 and R3 scales the control gain to 100 mV/dB. The control voltage also must have very low noise, so a large capacitor is usually placed across R3 or elsewhere in the control circuit.

VCAs can control gain over a very wide range, often 80 dB or more, so they are well-suited to volume control applications. The 2180 boasts a 130-dB control range. The logarithmic control characteristic is an almost ideal form of audio taper. The input and output of the 2180 are audio currents, not voltages. As shown in Figure 22.7, an audio input signal is converted to a current by R1, which feeds a low input impedance presented by the 2180. The output signal current from the 2180 is converted to a signal voltage by U2, which is configured as a transresistance amplifier.

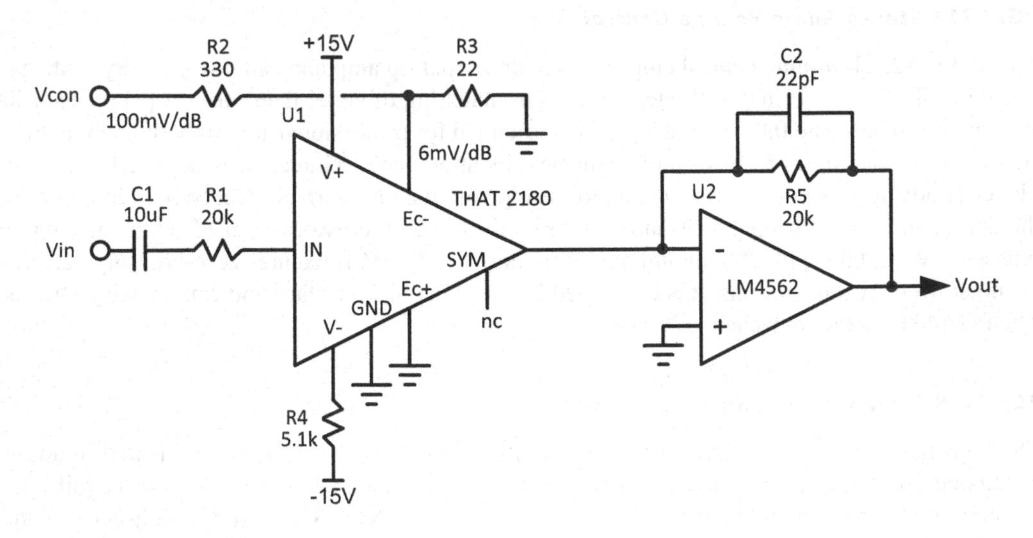

Figure 22.7 THAT 2180 Voltage-Controlled Amplifier.

VCAs like the THAT 2180 are used widely in professional audio recording consoles, but are also valuable in replacing multi-gang potentiometers in consumer stereo and home theater applications because the matching between gain control elements can be very good [14, 15]. VCA volume controls can also be arranged to function as a balance control by feeding different control signals to the left and right VCAs. Because VCAs are a fairly specialized technology, they are discussed in detail later in Chapter 25.

22.6 Balance Controls

Accurate balance of left and right signal levels is important to good imaging, yet many phono cartridges specify level balance to only within ±1 dB or so (sometimes worse). Similarly, many loudspeaker pairs have sensitivity matched to ±1 dB or worse. Placement in a room or room acoustics can also introduce differences in perceived loudness that can be mitigated by some adjustment of a balance control. On top of this, many volume control pots have attenuation errors between channels of 1-3 dB. For this reason, a preamplifier should include a balance control. The balance control need only have a range of ±3 dB for most purposes, but many allow the level of one channel to go to zero at the CW or CCW pot extreme. Once a common control on preamplifiers, balance controls have all but disappeared. The simple BC-2 preamplifier described in Chapter 2 includes a balance control with ±3 dB of range. Full left-right control is not needed for correction of minor channel imbalance in the signal chain.

Balance controls are not very complicated. They either have a small range or a full range. Many are implemented with a two-gang linear-taper potentiometer controlling the signal levels going to the volume control pot, as shown in Figure 22.8(a). Optional resistors R1 and R2 limit the range. If the balance control is instead fed by the volume control, that control's taper will be altered by the balance control load.

Other balance controls use a single pot to control the level in each channel by forming the shunt element to ground in a resistive divider, as shown in Figure 22.8(b), where optional resistors R3 and R4 limit the range. The single pot with its wiper grounded can also be incorporated into the shunt leg of the feedback network in a non-inverting amplifier stage. Range in this approach is limited to maximum gain reduction approaching unity. An advantage of this approach is that the stage can serve as a high-impedance buffer for the preamplifier inputs. This approach is used in the simple BC-2 preamplifier discussed in Chapter 2. A single-pot balance control is also used in the VinylTrak phono preamp [16]. There are numerous ways to implement a balance control [6].

Balance Control Behavior

When a balance or pan control setting is changed, it is desirable that the perceived loudness not change. This is made a bit of a challenge because what affects perceived loudness depends on whether the sound as perceived is being added on a sound power basis or a sound pressure basis. This is analogous to whether monophonic signals mixed together add on a power basis or a voltage basis. The nature of the addition depends on the degree to which the signals heard from the loudspeakers are correlated. Low frequencies common to both channels are more likely to be perceived as correlated and add on a voltage basis. High frequencies common to both channels are more likely to be perceived on a power basis, since minor phase differences can reduce correlation. Of course, left and right stereo signals not panned somewhat toward the center will obey different rules.

If a compromise 4-dB increase of perceived level when both channels are equal and monophonic, over that when only one channel is active, then the level of a channel may be set to be

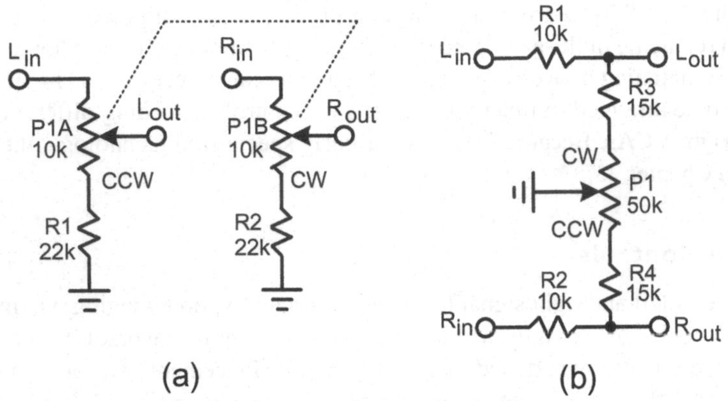

Figure 22.8 Balance Controls.

4 dB below that which it sees when the control is to an extreme where the other channel is highly attenuated. If the balance control range is limited to ±3 dB, these issues are rather insignificant.

Wiper Crosstalk

Wiper crosstalk occurs because the finite resistance of the wiper allows a small amount of the signal from one channel to escape to the other side of the wiper. This can affect the performance of single-pot balance controls if the effective wiper resistance is significant. Wiper resistance on typical pots of decent to outstanding quality is often in the range of 0.5%–2% of the total pot resistance. Bear in mind that even if the wiper makes "perfect" contact with the surface of the resistance element, the track is three-dimensional and has depth, allowing some signal to bypass the wiper and create some crosstalk.

The wiper crosstalk of a given pot can also be evaluated by connecting the wiper to ground, centering the pot, and applying a 1-V signal to one end. The open-circuit signal voltage that appears at the other end is measured. An example 100-kΩ pot whose effective wiper resistance was about 1.5% had about 23 mV of signal at the other end when a 1-V signal was applied to one end of the pot with the wiper grounded. The presence of any wiper distortion can also be evaluated by measuring the THD of the above-measured wiper crosstalk signal. Finally, the latter approach to measurement shows remarkable sensitivity to rotational pot noise, since pot noise originates from changes in wiper resistance. This can be a good measure of pot quality.

As mentioned earlier, if a VCA is used for the volume control, the balance control function can be easily added just by adding a DC control voltage for balance to the VCA control input. A stereo volume control using relays can also incorporate the balance control function if an MCU controls the level of each channel independently. Unfortunately, this cannot be done if DPDT relays are used to control both channels together.

22.7 Pan Pots

Pan pots are used in mixers to place a channel's signal between left and right in the stereo space. Unlike a balance control, a pan pot has one input and two outputs. It directs the signal to the left and right channels in different proportions. However, they share many of the same technical issues. The panning function thus has quite a bit in common with the balance control.

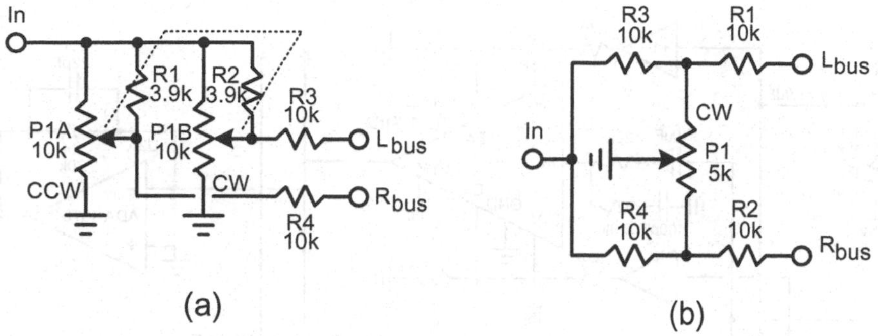

Figure 22.9 Two Pan Pot Implementations.

The Pan Law

When a signal in a mixer channel is panned from left to center to right and all locations in between, it is desirable that the perceived loudness not change. This is made a bit of a challenge because what affects perceived loudness depends on whether the sound as perceived is being added on a sound power basis or a sound pressure (SPL) basis. This is analogous to whether voltage signals mixed together add on a power basis or a voltage basis. The nature of the addition depends on the degree to which the signals are correlated. If the signals are being added on an SPL basis, then the panning pot should create 6 dB of loss when the signal is centered in the soundstage. On the other hand, if the signals are uncorrelated, the panning pot should create just 3 dB of loss when centered. As a compromise, panning circuits are often designed to be down 4.5 dB in each channel when centered.

As with balance controls, the panning function can be implemented with a single-gang pot or a dual gang pot. In the single-pot case, the issue of wiper crosstalk comes into play, as discussed above for balance controls. In the case of panning, a single signal is sent to two channels in varying amounts from zero to the full amount.

Figure 22.9 shows two approaches to panning, one using a single pot and the other using a dual pot. Notice in both cases the output of the pan circuit goes through a resistor into the virtual ground of a summing bus, so in both cases there is some wiper loading.

Voltage-Controlled Pan Pot

A pan pot can also be implemented with a pair of VCAs, as illustrated in Figure 22.10. This is especially convenient for automated applications. The VCAs also provide the fader function [17].

Designing pan pot circuits with a pair of potentiometers that implement the desired signal distribution law can be a challenge involving numerous compromises. Using a pair of VCAs, with their convenient linear decibel control law, can make the design easier and better-performing. The THAT 2162 dual VCA is a good choice for this application. The two VCAs are driven with opposing control voltages arranged so that the DC voltage provided by the pan pot when set for center causes the output to each channel to be down by 3.0 dB for a constant-power law. The DC control voltage to each VCA can be created with nonlinear circuits to implement different pan pot control laws. Once again, the current output of the VCA makes it convenient to drive the virtual ground right and left summing buses directly without summing resistors that increase mixer noise gain.

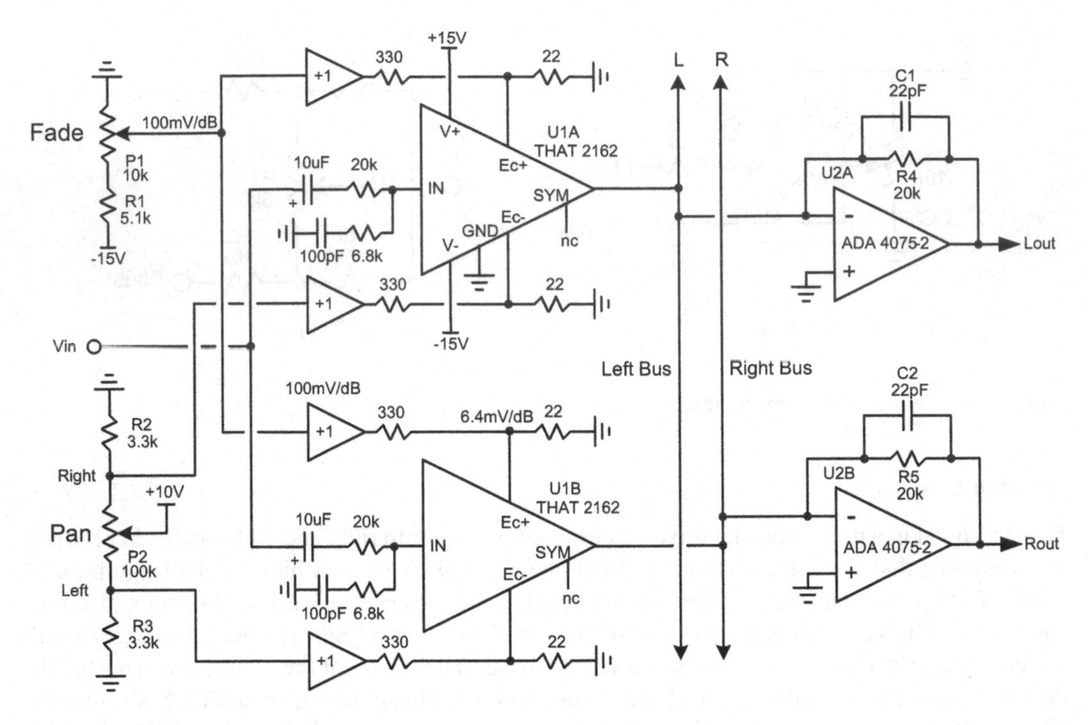

Figure 22.10 A Pan Pot Using a VCA.

The VCA has two control inputs, one of each control polarity. They can be driven differentially or independently. The negative control port can be driven with the pan pot control signal while the positive control port can be driven by the fader signal, implementing the fader and pan pot functions together with a single dual VCA. The ability to do this while keeping the pan position the same with different fader levels is made possible by the linear decibel control law of the VCA.

A simplified diagram of such a circuit is shown in Figure 22.10. The +1 unity gain control buffers must be able to deliver up to 27 mA at −10 V to drive the low-impedance VCA control port resistive divider. Once again, better approaches to buffering the control port signals can be found. The resistive dividers of the panning potentiometer and its associated resistors create nonlinear control voltages as a function of rotation that are remarkably close to what is needed to implement the constant-power panning law. Power is constant within a ±0.2 dB window over the full rotation of the pan pot, with maximum attenuation of the low side approaching 100 dB. The panning law can be changed to a 4.5-dB compromise law merely by changing R2 and R3 to 5200 Ω.

22.8 Loudness Controls

Fletcher and Munson discovered long ago that the frequency response of the ear is different for tones of different loudness levels [18]. Put another way, the perceived loudness of a tone always has a particular frequency response with respect to 1 kHz at a given SPL. However, that frequency response curve is not the same at different tone amplitudes. Tones at low frequencies and high frequencies have lower relative perceived loudness when the signal level is lower. This partly

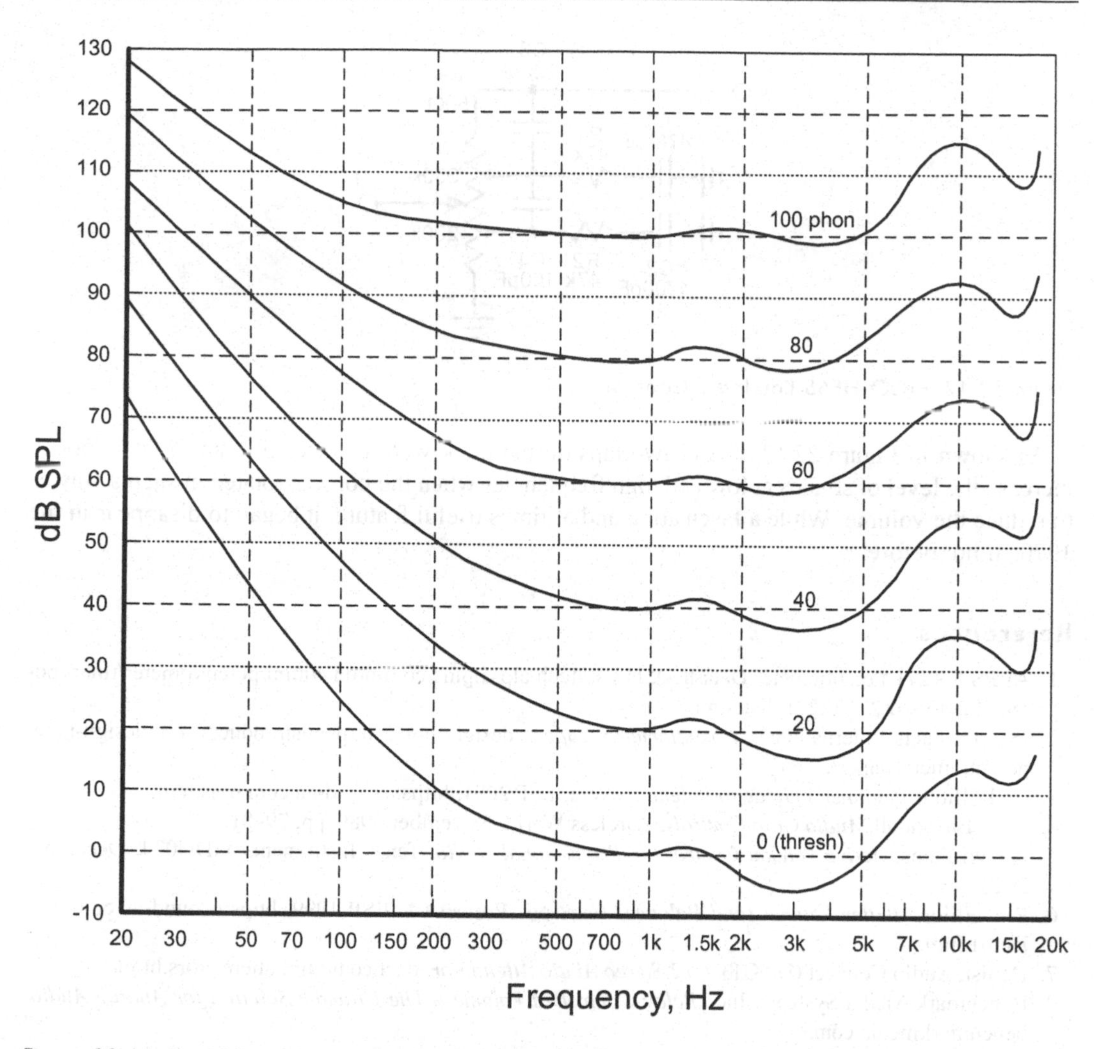

Figure 22.11 Equal-Loudness Contours (Simplified).

explains why music often sounds better when played louder. Figure 22.11 is a very simplified illustration of the Fletcher-Munson curves.

The curves are referred to as the equal-loudness contours because the height of the curve represents how much higher in level a tone at a given frequency must be to have the same perceived loudness as a 1-kHz version of that tone. As a curve moves upward at a lower frequency, that is essentially indicative of a fall-off in the ear's frequency response. The contour curve is somewhat like the inverse of the ear's frequency response.

In the past, some preamplifiers were equipped with what were referred to as *loudness* controls that acted as the volume control [19]. Since the Fletcher-Munson curves depend on level, a level control sometimes accompanied the loudness control to normalize the action of the loudness control [20]. While the level control was like an ordinary volume control, the loudness control had a frequency response that was a function of pot rotation.

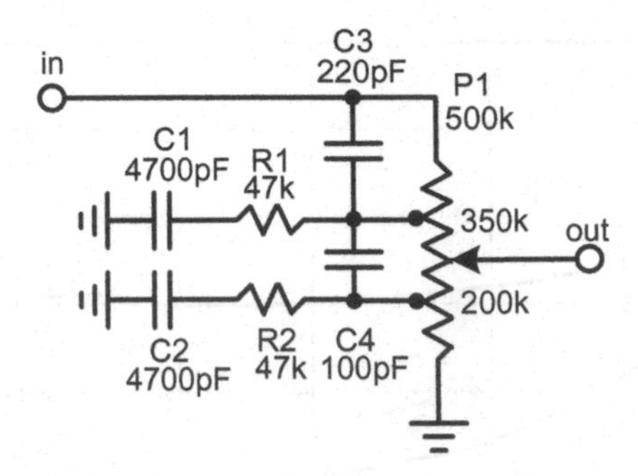

Figure 22.12 EICO HF65 Loudness Control.

As shown in Figure 22.12, one or two taps on the track were connected to an *RC* network to increase the level of sounds at low and high frequencies when the pot was rotated in such a way as to reduce the volume. While a fascinating and at times useful feature, it began to disappear in the 1970s if not before.

References

1. ALPS RK271 Potentiometer Datasheet, https://tech.alpsalpine.com/prod/e/html/potentiometer/rotarypotentiometers/rk271/rk271_list.html.
2. ISL Products International, *Potentiometer Taper*, design note, https://islproducts.com/design-note/potentiometer-taper/.
3. Rod Elliott, *Beginner's Guide to Potentiometers*, ESP 2001, https://sound-au.com/pots.htm.
4. Peter Baxandall, *Audio Gain Controls*, Wireless World, November 1980, pp. 79–81.
5. Ian Williams, *Active Volume Control for Professional Audio*, Texas Instruments TIDU034, December 2013.
6. Rod Elliott, *Better Volume (and Balance) Controls*, Project 01, ESP 1999, https://sound-au.com/project01.htm.
7. Danish Audio Connect (DACT), *CT2 Stereo Audio Attenuator*, dact.com/html/attenuators.html.
8. Benchmark Media Systems, Inc., *Relay-Controlled Volume – The Ultimate Solution for Analog Audio*, benchmarkmedia.com.
9. AMB Laboratories, The δ1 Relay-based R2R Stereo Attenuator, amb.org/audio/delta1/.
10. LM1972 μPot 2-Channel 78 dB Audio Attenuator, Datasheet, Texas Instruments.
11. PGA2311 Stereo Audio Volume Control Datasheet, Texas Instruments.
12. PGA2320 Stereo Audio Volume Control Datasheet, Texas Instruments.
13. MAX5486 Stereo Volume Control with Pushbutton Interface, Analog Devices, Inc./Maxim Integrated, maximintegrated.com.
14. THAT 2180 Datasheet, THAT Corporation, thatcorp.com.
15. Rod Elliott, *VCA Preamplifier*, Project 141, ESP 2012, https://sound-au.com/project141.htm.
16. Bob Cordell, VinylTrak – A Full-Featured MM/MC Phono Preamp, *Linear Audio*, vol. 4, pp. 131–148, September 2012.
17. Bob Cordell, *VCA-Based Faders and Panning Pots*, cordellaudio.com.
18. Harvey Fletcher and Wilden Munson, Loudness, Its Definition, Measurement and Calculation, *Journal of the Acoustic Society of America*, vol. 5, pp. 82–108, 1933.
19. Tomlinson Holman and Frank Kampmann, Loudness Compensation: Use and Abuse, *JAES*, vol. 26, issue 7/8, July/August 1978.
20. EICO HF 65 High Fidelity Preamplifier, schematic.

Chapter 23

Digital-to-Analog Converters

With more and more program sources being digital, it only makes sense that the digital-to-analog converter (DAC) takes its place among preamplifier input interfaces like the phono preamp. The ability to download or stream high-resolution audio further underlines the DAC as a mainstay of audiophile reproduction. The increasing use of music servers has done the same. For quite some time, some audio systems have incorporated DACs that can accept an external digital input. Many preamplifiers and headphone amplifiers now incorporate built-in DACs. The role of stand-alone DACs may be migrating to the high-end audiophile space, but they may be replacing/including preamps in predominantly digital setups, some even including a headphone amplifier [1].

At the heart of most DACs is the DAC chip. Numerous high-performance DAC chips are available from several manufacturers, most being able to operate at up to 24 bits at 192-kHz sample rates and many up to 768 kHz. However, there is much more to producing a good DAC than just selecting a good DAC chip and wiring it up in accordance with the manufacturer's reference design. The external analog output circuit, noise sources, grounding and digital signal jitter can all conspire to keep the DAC chip from performing to its full capabilities. Details of implementation can make a big difference.

This chapter contains two basic parts. The early sections treat the DAC chip largely as a black box and focus on the DAC system, which includes input interfaces, the DAC chip and the output amplifiers. This section is of greatest interest to those who might design a DAC product. The later sections go into some details about how DACs work and the different types of DAC architecture, like R2R ladders and the Sigma-Delta modulators. Over-sampling and noise-shaping concepts are covered as part of this overview [2–5].

23.1 DAC Design

Here we are talking about DAC design in terms of the box, not the chip. It is difficult to achieve better sound or performance than the reference design provided by the chip manufacturer, but is easy to do much worse. The best advice is to follow the manufacturer's reference design, right down to the printed circuit board layout. At the same time, things that can obviously be improved within the context of the reference design can be considered, such as the use of higher-quality components and pristine power supply and reference sources.

The output signal from many DAC chips is a differential current that must be converted to a single-ended voltage signal. This is accomplished with an I/V converter, usually referred to as a transimpedance amplifier (TIA). Some other DACs produce a differential output voltage. In both cases, the output must be low-pass filtered by what is called a reconstruction filter. These filters are often of first, second or third order.

The external I/V converter circuit and op amp(s) offer more opportunity for designers to customize and optimize. However, when in doubt, go with the manufacturer's design. If a reference board is available, by all means obtain it and compare results with your design by both listening tests and measurement. Indeed, measurement of DACs can be difficult and require instrumentation that may not be available to everyone.

Grounding noise can seriously degrade the performance of a DAC chip. It is important to bear in mind that the DAC chip comprises both digital and analog circuitry, often with separate supplies. Do not let current flow through the grounding interconnect local to the DAC chip. Consider slowing down applied digital signals to a bandwidth of ten times the highest digital rate to be transmitted. Consider using active, low-noise power rail filtering that re-references the rail to the ground interconnect local to the DAC chip. This generally means using a capacitance multiplier filter, but beware of the consequent voltage drop of at least 1 V_{be} from input to output of such a circuit. A higher voltage rail that is dropped down to the needed voltage of the DAC may work better. Be certain that the TIA is stable. Be certain that the clocks are of exceptionally high stability with very low jitter. Operate the clock oscillator from an extremely quiet power supply.

How Audio DACs Work

There once was a time when DACs were simple. Beginning in the late 1960s they began to come into use in industrial automation and telecommunications. Their first major use in audio was in telephone systems where 24 voice channels were carried digitally on the 1.5 Mb/s T1 digital transmission system. Eight-bit converters clocked at 8 kHz were used at both ends of the transmission system that carried 24 voice channels on a twisted pair. Those DACs were very simple by today's standards. Modern audio DACs were ushered in by digital tape recording and CDs, traditionally at 16 bits with sample rates of 44.1 or 48 kHz.

The DAC

The signal-processing part of a DAC mainly consists of three sections, as shown in Figure 23.1. The incoming signal from a CD player or other source arrives in the S/PDIF format (Sony/Philips Digital Interface), which is essentially the same as what has been called AES/EBU or AES3 [6–9]. It is a single digital bit stream in which the clock timing is embedded by using *Manchester* coding. In consumer equipment, the bit stream arrives via coaxial cable or optical Toslink [10, 11]. This signal enters the first block, called the Digital Audio Interface (DAI).

The DAI is a chip that recovers clock and outputs clock(s) and data in the I^2S format (Inter IC Sound) [12–21]. The I^2S format separates the data and clock signals. This format includes data for both stereo channels in a single stream, along with several different clocks to implement a synchronous interface. The quality of the clock recovery function in the DAI largely determines the amount of jitter in the signals fed to the DAC. The DAI clock recovery circuit must tolerate and filter out jitter in the incoming steam while not contributing any jitter of its own. It can have a large effect on the sound quality.

The I^2S bus feeds the DAC chip, which outputs the stereo signal as two differential pairs, either in voltage mode or current mode. These quasi-digital signals are in a quantized form, either as a stair-step signal or as a pulse density modulation (PDM) signal.

The third block is the analog interface. It is responsible for converting the differential quasi-digital signals to a robust analog single-ended audio signal suitable for interfacing to the line-level inputs of a preamplifier or similar device. The analog block is also responsible for reconstruction

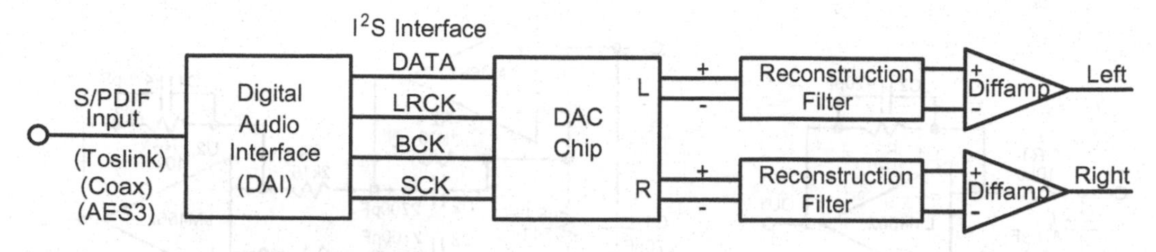

Figure 23.1 Block Diagram of a DAC.

filtering with a low-pass filter (LPF) that is usually of first, second or third order. Reconstruction is essentially a signal smoothing process that eliminates content at frequencies above the audio band. It converts discrete-time signals from the DAC to continuous-time signals suitable for audio output. The analog interface block is the one over which the designer has the most control in implementation. It can have a large influence on the sound quality of the DAC.

This chapter will begin with the analog interface, treating the DAC chip largely as a black box that outputs differential current or voltage as pulses or stair steps whose average value represents the analog audio signal. Later sections will include how DAC chips work, especially the widely used over-sampled sigma-delta (ΣΔ) designs, and the issues of clock recovery, jitter and the Digital Audio Interface (DAI) chip [14, 22].

23.2 The Analog Interface

The analog part of the DAC system is perhaps the most important and the one a DAC designer can do the most about. One has to pick the most suitable DAC chip and then largely treat it as a black box. A given DAC chip, no matter how good, may sound quite different in different DAC designs depending on the analog output circuitry, the jitter management and the power supply and grounding. The analog parts of a DAC can be the most important to the sound.

Output Amplifier

The output of most DAC chips is differential, but it may be in voltage mode or current mode. Many will work either way, meaning that if they see a high-impedance load they will produce a known voltage with known source impedance. These make for simple DAC output circuits, needing little more than a single-op amp output differential amplifier, as shown in Figure 23.2(a). However, the voltage output configuration does not always provide the best performance. For example, if there is any nonlinearity in the DAC output impedance, some distortion can result. Some benefit may result if an instrumentation amplifier arrangement with 3 op amps is used to perform the differential-to-single-ended output conversion.

Voltage Output DACs

Consider the TI/Burr-Brown PCM1793 voltage output DAC. Full-scale differential voltage output is 3.2 $V_{p\text{-}p}$ when it drives a single-op amp differential-to-single-ended converter (DSEC) when lightly loaded by a few kΩ, as shown in Figure 23.2(a).

The single-op amp DSEC presents an asymmetrical load to the DAC, which may cause distortion in some cases. Ideally, the two differential outputs of the DAC should be loaded identically.

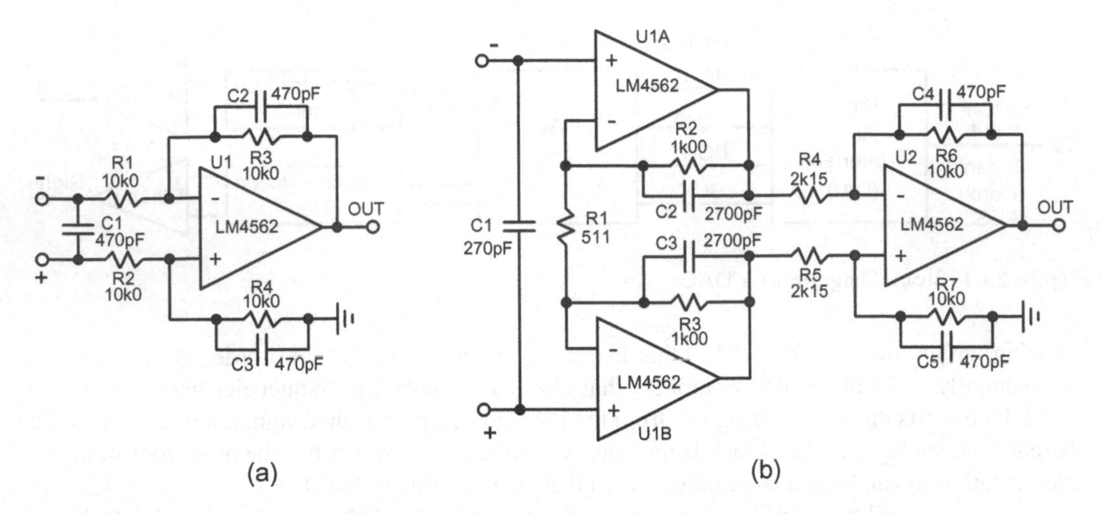

Figure 23.2 Voltage Output DAC Circuits.

Unequal DSEC input impedances to ground degrade performance. The circuit shown provides a second-order LPF at about 77 kHz for reconstruction [23]. Gain is 1.83 to the amplifier output, yielding 5.9 $V_{p\text{-}p}$. The single-op amp DSEC is usually implemented with fairly high resistances in the range of 5–20 kΩ that can introduce thermal noise. In Figure 23.2(b), an instrumentation amplifier comprising 3 op amps implements the DSEC function. It presents identical high-impedance loads to both DAC chip output polarities, along with good common-mode rejection.

Current-Output DACs

Internally, most DAC chips produce current, and work best when their outputs are operated in current mode into the virtual ground of a TIA. In these cases, the voltage at the output pins of the DAC chip do not change with signal. The common approach is to feed each differential output to a TIA that converts the signal into a voltage of usually more than 1 V peak, as shown in Figure 23.3. If the maximum signal current output swing from the DAC chip is 3 $mA_{p\text{-}p}$ and the TIA feedback resistor is 1 kΩ, 3 V_{pk} will be produced at the output of each of the TIAs. Each TIA usually has a feedback capacitor that low-pass filters the signal. The feedback capacitor also helps to stabilize the op amp in the face of DAC chip output capacitance. The output signals from the DAC can have fast transitions with significant current, sometimes on the order of 3 mA. For this reason, an op amp with good gain bandwidth and low output impedance should be used.

The reference voltage fed to the non-inverting input of each TIA forces the output pins of the DAC chip to be at that voltage, due to the virtual short at the input of the TIA. This voltage at the output pins of the DAC is called the compliance voltage. Sometimes, the lowest distortion produced by the DAC is dependent on the compliance voltage value. The outputs of the TIAs are then sent through a single-op amp DSEC to the output. Consider a current output DAC that sources pulses of current into the TIAs in average amounts. This DAC cannot sink current and its output impedances are high. If the reference for the TIAs is connected to ground, the outputs of both TIAs will be driven negative, one more than the other with signal not at digital zero. A digital zero, representing the idle state, will have each output sourcing about half the maximum current, on average. The TIA outputs will both be negative by equal amounts, and will thus have a negative common-mode offset. Alternatively, the reference connected to the TIA non-inverting inputs can be made positive by an amout that will bring the TIA outputs at digital zero up to 0 V. This will also force the TIA outputs to be a positive voltage if this is a good compliance voltage for the DAC chip.

An output circuit for the current-output TI/Burr-Brown PCM1794 DAC is shown in Figure 23.3 [24]. The DAC outputs feed the virtual grounds of two TIAs, which in turn feed a single-op amp DSEC to provide a single-ended output. Capacitors C1 and C2 implement the first-order portion of the reconstruction filter. A second-order Sallen-Key filter employing U2B completes the reconstruction filtering for the DAC. The positive inputs of the TIA op amps are at ground potential in this circuit, forcing the DAC outputs to the ground potential. The DAC sources current into the TIAs, causing their outputs to be at a negative potential of about −5.1 V. The 1794's bipolar zero output current is stated as 6.2 mA. Full-scale swing is 2.3 mA for one polarity and 10.1 mA for the opposite polarity. Each output from the DAC thus swings 7.8 $mA_{p\text{-}p}$. This is a significant amount of signal current. The subsequent DSEC does not pass the negative common-mode voltage from the TIAs. The output circuit delivers up to 4.5 V_{RMS}.

In other designs, the reference voltage applied to the TIAs may be at a positive potential so that the common-mode level at the outputs of the TIAs is near ground, allowing their outputs to swing the maximum amounts positive and negative. The quiescent current flowing out of the DAC chip output pins will play a big role in determining the best common-mode level at the TIA outputs.

The op amp outputs will rest at the reference voltage. For some DACs, there is an optimal reference voltage at which their current output feeds into the TIA. For maximum op amp output headroom in both output polarities, their digital zero quiescent output voltage should be zero volts. This may conflict with the optimal DAC output reference voltage.

The virtual ground provided by the TIA keeps the voltage at the output of the DAC from changing. However, the impedance of the virtual ground is not zero, and in fact it increases with frequency as op amp open-loop gain falls. This means it can look inductive. The low impedance of the virtual ground depends on the feedback from the output of the op amp being very fast. An incoming current spike from the DAC chip with a very fast edge can drive the input stage to some

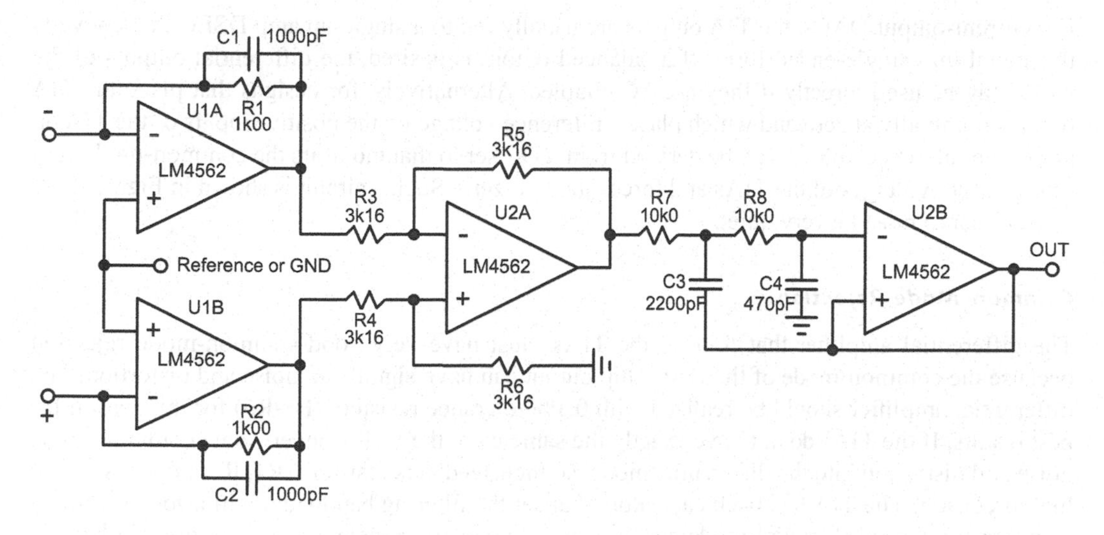

Figure 23.3 Current-Output DAC Circuit with Filter.

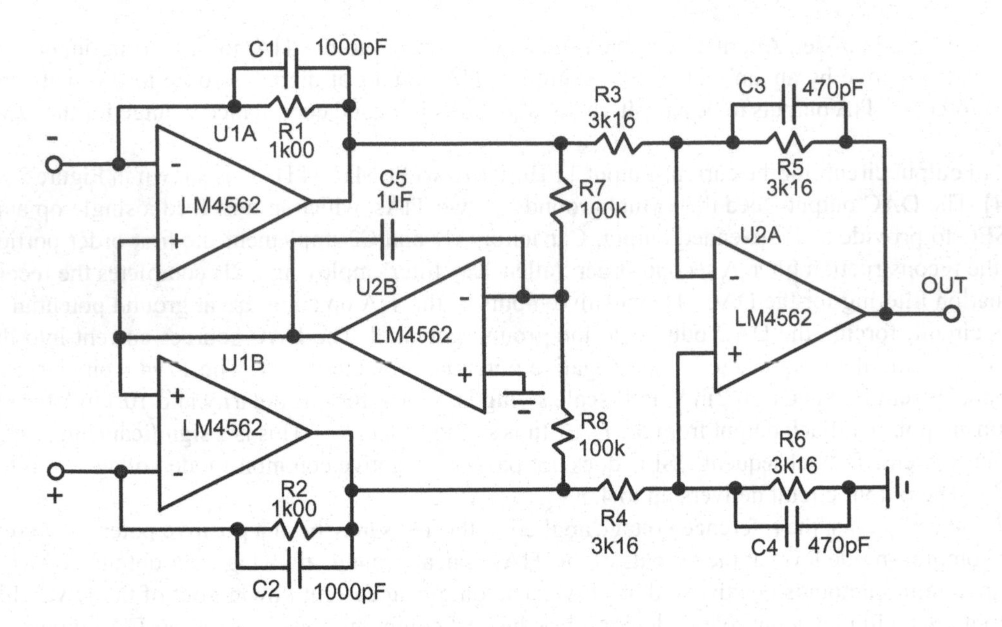

Figure 23.4 DC Servo for Control of Output Common-Mode Voltage.

number of mV before the feedback voltage can change to restore the voltage to the reference value. This can cause some distortion. However, the feedback filter capacitors C1 and C2 mitigate this by shunting the fast edge current right into the output stage of the op amp before the op amp has a chance to react anyway. This current from the DAC chip will be seeing the open-loop output impedance of the op amp, which should be relatively small. The reference voltage applied to the TIAs must be very stable and quiet.

DC Servo for Common-Mode Offset Control

For current-output DACs, the TIA outputs are usually fed to a single-op amp DSEC that converts the signal into single-ended form. If a balanced output is desired, the differential outputs of the TIAs may be used directly if they are AC coupled. Alternatively, for designs that place the TIA outputs nominally at zero and which place a reference voltage on the positive inputs of the TIA op amps, the reference voltage can be derived from a DC servo that monitors the common-mode sum of the output voltages of the TIAs and forces them to zero. Such a circuit is shown in Figure 23.4. The DC servo must be very quiet.

Common-Mode Rejection

The differential amplifier that follows the TIAs must have very good common-mode rejection because the common-mode of the DAC chip output can have significant noise and distortion. The differential amplifier should be realized with 0.1% tolerance resistors (R3-R6) for this reason for best results. If the TIAs do not have exactly the same gain, they will convert some common-mode noise and distortion into the differential mode, so their feedback resistors (R1, R2) must also be of high precision. The TIA feedback capacitors that set the filtering bandwidth will also cause some common-mode to differential-mode conversion at higher frequencies if they are not of identical values as well. They should have a tolerance of 1% if possible.

Other DAC Chips

There are many different DAC chips available, including the TI/Burr-Brown PCM1796 [25], the AKM AK4396 [26] and the ESS Technology chips [27, 28]. The Analog Devices AD1955 and the Cirrus Logic CS4398 are also high-performance contenders. New DAC chips come out almost every year, so the DAC chips mentioned here may be superceded with newer parts.

TIA Noise and Speed

The TIAs must have very low noise – they will often set the noise floor. Of course, the signal current from the DAC is fairly healthy, so this helps SNR. This is especially so because this means that the TIA feedback resistors can be of lower resistance, creating less thermal noise. As mentioned earlier, it is also very important that the TIA op amps have wide bandwidth, good RFI (radio frequency interference) immunity and good slew rate. This is because they will be called upon to process fast edges from the differential current outputs of the DAC chip. The feedback capacitor helps this, but be aware that the signal output current from the DAC at high edge rates gets dumped right into the output of the TIA op amp. Although the excellent LM4562 is shown in many of the circuit examples here, the LME49860 is probably the best op amp available for these applications. It can operate at rail voltages as high as ±22 V and can output up to 37 mA.

TIA Distortion

The typically low TIA feedback resistance that makes possible good noise performance may also place a relatively heavy load on the TIA op amp output. In circuits where the output of the TIA can swing in both polarities (unlike Figure 23.3) the op amp output stage will likely be forced into its class B region. For these reasons, op amps that have low distortion when driving heavier loads are desirable. Notice that in the design of Figure 23.3 the output is always at a negative potential and that the DAC is always sourcing some current into the TIA. This means that the op amp output stage will tend to operate in class A, with only the pull-down half of the output stage active.

Output Filtering

The TIA feedback capacitors are the first point of filtering in a current-output design, and sometimes the only point. They may be set for a pole frequency on the order of 200 kHz. A first-order roll-off occurs as a result. These capacitors should be closely matched to avoid common-mode to differential-mode conversion at high frequencies. There are also several different ways to incorporate additional first-order or second-order low-pass filtering into the output DSEC [29, 30]. If two complex poles can be implemented in the DSEC, then an overall third-order Butterworth or Bessel filter can be realized. However, filtering beyond the first pole is best done in a separate stage after the DSEC rather than within the DSEC circuit, so that mismatch in the filtering capacitors will not compromise CMRR at high frequencies. This approach is taken in Figure 23.3, where a Sallen-Key second-order LPF section follows the DSEC.

DAC Output Capacitance

The differential output pins of the DAC may have substantial capacitance to ground, due in part to many current source transistors connected to each output. Even apart from its LPF function, the TIA feedback capacitor helps stabilize the TIA feedback loop in the presence of this input capacitance that it sees. Without the feedback capacitor, there would be an LPF at the inverting input pin formed by the feedback resistance and the output capacitance of the DAC, and instability would result.

The minimum recommended value of this capacitance is that which forms a capacitance voltage divider with the output capacitance of the DAC that sets the closed-loop gain at high frequencies (where the feedback capacitor dominates the feedback resistor) to a reasonable value. Setting the capacitor too large probably does not cause a problem, as it just puts the op amp into a unity-gain closed-loop mode at high frequencies, as long as it does not cut the amplifier/DAC bandwidth by too much.

23.3 Power Supplies, References and Grounding

The power supplies and DAC references must be very stable and clean, while dumping very little noise into the local grounds. In some cases, an op amp can be used to supply these voltages. In other cases a well-designed capacitance multiplier can work well. The latter re-references the voltage produced to the local quiet ground without dumping much current into the ground, as would a large local bypass capacitor.

Op Amp Power supply

The ±15-V power supply to the op amps must have extremely low noise, but the amount of regulation or needed precision is a little less important. A related question here is the voltage supply to the DAC, which is often +5 V and +3.3 V. These voltages can be supplied by a low-dropout regulator (LDO) or even by an op amp that creates the filtered voltage, but there will then be some voltage drop. The lower voltages that feed the DAC will often be supplied by a local low-noise regulator that drops the voltage from the ±15-V supply. The required power supply accuracy and absence of noise will depend on the particular DAC chip. The regulator's power supply rejection is quite important.

The use of a discrete capacitance multiplier should not be ruled out. It is simple and can provide extremely low noise by filtering of the pass transistor's base voltage, maybe even to second order. Just as importantly, it re-references the supply to the local analog ground without dumping filtering current into that quiet ground. No bypass capacitors are required from the main supply rail to the sensitive analog ground. There is no significant reason why its regulation and accuracy cannot be the same as the rail voltage supplied to it. However, if a BJT emitter follower is used for the capacitance multiplier, one V_{be} will be dropped from the main supply, requiring the main supply to be higher in voltage or requiring that the op amps operate from the slightly reduced 14.4-V supply.

The advantage of the capacitance multiplier is that the supply voltage is re-filtered after being supplied a quiet supply from the main regulator. If the main regulators are set for ±16 V, achieving ±15-V supplies for the op amps with a capacitance multiplier is easy. The capacitance multiplier may not provide as much absolute voltage accuracy as a high-performance regulator chip.

DAC Chip Analog Power Supply

Just as with the op amp power supply, the DAC chip analog power supply should be extremely clean. This is a different, lower supply voltage than that for the op amps, often +5 or +3.3 V.

Overall performance can benefit from creating the analog supply with a capacitance multiplier or a low-noise op amp run from the ±15-V op amp supply. Unless the output of the op amp is further filtered, this power supply can be no quieter than the op amp, which will typically be between 2 and 5 nV/√Hz. This supply must have very low output impedance across the full audio band.

Reference Supply

The reference supply to the TIAs in Figure 23.3 sets the quiescent voltage at the output pins of a current-output DAC chip by way of the virtual grounds of the TIAs. It is also a common-mode input signal to the circuit. It must be extremely quiet. It can be produced by an audio op amp, just as if it were an important audio signal. It should be highly filtered to the analog ground. It may be derived from the DAC chip +3.3-V analog power supply.

Grounding

Grounding is critical and ground noise must be kept very small. The analog ground should have very little if any current flowing through it. Even bypass capacitors that can dump noise into the ground from a less-quiet ground or a supply must be considered. If any op amps are operating in class AB, that must also be treated carefully so that their class AB currents do not get into the ground via their bypass capacitors.

23.4 Clock Recovery and Jitter

In most digital signal transport systems, the clock must be recovered from the data so that the data can be cleaned up and sent to whatever digital processing that follows. In almost all cases, the clock signal is inferred from the time of the transitions in the data signal. This function is usually performed by a phase locked loop (PLL). Of course, with random data, there will not always be a transition for each data bit. The randomness of the transitions can affect the clock recovery circuit, resulting in jitter in the recovered clock. Jitter is the enemy. When there is jitter, the right data value is stored at the wrong time, or the wrong data value is stored at the right time. Jitter also modulates the DAC conversion process, producing noise sidebands.

Phase Locked Loops

The clock recovery for a DAC is usually done in a digital audio interface (DAI) chip. That chip includes PLL clock recovery. The DAI takes in the serial SPDIF (Sony/Philips digital interface), recovers the clock and puts out the I^2S data bit stream. It also outputs the bit clock, right-left word clock and system clock to the DAC chip.

A very simple PLL is shown in Figure 23.5. It is a feedback loop that consists of a phase detector (PD) a low-pass filter (LPF) and a voltage-controlled oscillator (VCO). The feedback around the loop adjusts the VCO so that its frequency is identical to that of the input signal, and that the phase of the output is locked to a known relationship to that of the input signal (often 90°). The PD compares the phase of the input signal to that of the output signal and creates an error signal to correct any error. The PD can be as simple as an exclusive OR (XOR) circuit. In this case the DC value of the error signal will be zero when the input and output signals are 90° apart. The loop LPF attenuates the ripple coming from the output of the phase detector. When the loop is locked,

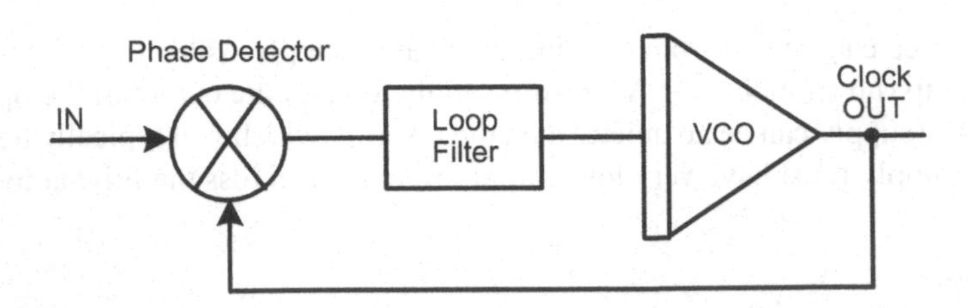

Figure 23.5 Phase Locked Loop.

the ripple will be at twice the clock frequency if an XOR PD is used. The LPF often includes an integrator to drive the phase error to zero (i.e., 90° with an XOR PD).

The PLL is essentially an LPF in the phase domain. Input signal phase changes at low frequencies will pass through the PLL with little attenuation. Input signal phase changes at higher frequencies will be attenuated as the frequency of the phase change increases (e.g., jitter). Because the VCO acts as an integrator in the phase domain, it introduces a pole in the feedback loop. The LPF introduces another pole in the loop. The simple PLL is thus a second-order system. This means that it will have a second-order LPF filter response that may result in peaking, where output jitter can exceed input jitter over a band of frequencies near the filter's cutoff frequency. This is called jitter peaking.

The PLL is also characterized by its capture range and its lock-in range. If the input and output frequencies differ, the PLL is out of lock and the PD will output a beat frequency which largely does not have a DC component to correct the VCO frequency. However, the amount of the beat frequency that does get through modulates the instantaneous VCO frequency in a nonlinear way. It causes the output sinusoid of the VCO to be asymmetrical, resulting in a small DC component that slowly builds up in the LPF. The polarity of that small DC voltage drives the VCO toward the input frequency. As the VCO frequency gets closer to the input signal frequency, the magnitude of the DC voltage becomes larger, causing a regenerative lock-in process. If the input and output frequencies are too far apart, this DC component of the error signal will be too small to initiate the regenerative capture process. The lock range of the PLL describes the range over which the locked PLL can maintain lock, and is related to the maximum DC output from the PD and the maximum frequency deviation that can be had from the VCO.

Some PLLs incorporate a phase and frequency detector that will output an unambiguous DC signal that will explicitly indicate the direction of the frequency error in a data signal [31–35]. The capture range and speed of capture for the PLL is thus greatly increased. PLLs whose input is a continuous square wave can be implemented with a much simpler phase-frequency detector.

The VCO in a PLL is very important to the center frequency accuracy and the amount of phase noise at the output of the PLL. In systems where the clock rate is known with some precision, good accuracy of the (free-running) center frequency of the VCO can make the capture of the input signal fast and reliable. Different types of VCOs have significantly different amounts of phase noise that is not attenuated by the feedback process in the PLL at higher phase noise frequencies that lie above the closed-loop bandwidth of the PLL. Simple IC PLLs that do not employ some kind of resonator like an LC tuned circuit, ceramic resonator or crystal can create significant phase noise and have limited power supply rejection.

A common example of an IC VCO is a ring oscillator consisting of an odd number of inverters in a closed loop. The frequency of oscillation is controlled by changing the bias in the inverter stages to vary their switching speed. For critical applications where the required frequency is well known, crystal-controlled VCOs (VCXO) are often used. A varactor diode is used to pull the crystal frequency up or down, often by as much as 100 ppm.

In many applications, the input and output frequencies of the PLL are not the same. A higher output frequency can be obtained by placing a frequency divider in the feedback path between the VCO and phase detector. Similarly, a frequency divider can be placed in the input path to achieve a lower output frequency. Often both dividers are employed to obtain an output frequency ratio of *m*/*n*, where m and n are the integer divider factors. When *m* and *n* are made controllable, the PLL can be made to adapt to different input and output frequencies [36]. PLLs play a central role in frequency synthesizers.

Clock Recovery

The input signal to the PLL is not always a square wave or sinusoid. This is usually the case when the function of the PLL is to recover clock from an incoming data signal. The PLL can lock to such a signal if there is an adequate amount of center frequency spectrum in the input signal. The polarity transitions in the input signal convey the frequency and phase of the data signal. A signal with more transitions on average has a higher transition density. A square wave has 100% transition density.

Data signals are coded in different ways to maintain adequate transition density to lock the PLL. A simple one-zero binary data sequence with a bit rate of half the clock rate has zero average transition density, since the transitions can occur in either direction at any given time. Such an un-coded signal is called a non-return-to-zero (NRZ) signal. It can be differentiated and rectified to create a signal with an adequate spectral component at the clock frequency. Note, however, that a long string of 1 or 0 bits will not produce any useable clock information during that period without data transitions. The transition density can become very low during such a string.

A very simple coded signal is one that returns to zero during half the duration of each data bit. This is called a return-to-zero (RZ) coded signal. It has a fairly strong spectral component at the clock frequency. The RZ signal has clock information embedded in it. RZ coding requires greater channel bandwidth for transmission, so it trades transmission efficiency for being able to transport clock information.

Jitter

Jitter will have the largest effect on the audio waveform when the signal rate of change is greatest. For a sine wave, this is at the zero crossing. At 20 kHz, a sine wave changes at a rate of 0.125 V/μs/V_{pk}. If the edge timing is off by 1 ns, and the signal is changing at a rate of 0.125 V/μs for a 1-V_{pk} audio signal, distortion can be characterized by how many μV the output is wrong when the sample is taken. If the sampling time is off by 1 ns, then the voltage error at the zero crossing is 125 μV. If this error is all considered distortion, the resulting distortion is 125 ppm of the 1-V_{pk} signal. That corresponds to 0.0125% (this is not THD, but is a ballpark number).

It is desirable that the jitter will not cause more error than one LSB. Consider a 16-bit system wherein full-scale is ±1 V. One LSB = 2/65,536 = 30.5 μV. At 0.125 μV/ps, the signal will change by 30.5 μV in 244 ps. In a 24-bit system, the maximum peak jitter must be 256 times smaller, or 0.95 ps.

Transition Density Jitter

While a repeating 1, 0, 1, 0 data stream has a constant transition density of 100%, random data has pattern-dependent transition density with density less than 100%, and often averaging 50%, depending on how the data has been encoded. For example, the differential Manchester coding used in S/PDIF guarantees a minimum transition density [37]. Varying transition density is thus a form of random noise that can result in jitter when clock is recovered from such a data stream. Proper clock recovery explicitly depends on data transitions. The more transitions, the more information the PLL has to help it stay in the right phase relationship with the data.

Indeed, varying transition density causes the loop gain of the PLL to be changed or modulated with transition density noise. Dynamically changing PLL loop gain can result in jitter as a result of static offset in the free-running frequency of the VCO, since the loop gain is what takes the phase error and makes a correction to the VCO frequency and phase. With lower transition density, it takes more phase error to make the correction for the VCO. Similarly, if the VCO has a higher free-running frequency offset, the phase error needed to correct that offset is greater. This kind of clock recovery jitter can occur even if the incoming data stream itself has zero jitter. A perfectly tuned VCO will result in clock recovery with much less of this type of jitter.

VCO Phase Noise

Some VCOs have considerable phase noise and limited power supply rejection for jitter. This can especially be the case for simple on-chip CMOS VCOs. A better solution is to employ an L-C oscillator/resonator whose frequency is controlled over a small range by a varactor diode. Going a step further, crystal oscillators can have extremely low jitter. In critical PLL applications, a voltage-controlled crystal oscillator, called a VCXO, will be employed. The frequency of the VCXO can only be controlled over a very small frequency range, called the absolute pull range (APR). This means that VCXO PLLs are mainly suitable for applications where the bit rate is fixed with close tolerance, as in digital audio applications.

As an example, the Vectron VVC4 CMOS VCXO can be ordered with center frequencies from less than 4 MHz to greater than 77 MHz [38]. Devices can be ordered with APR up to ±100 ppm. Jitter is less than 1 ps_{RMS} in a jitter frequency spectrum from 12 kHz to 20 MHz. Of course, using a VCXO adds cost and some functional constraints. For digital audio, the center frequency must be an integer ratio of 44.1 and 48 kHz (e.g., 882:960 => 42.336 MHz). Otherwise, two VCXOs will be required.

Re-Clocking I²S Signals

It is usually impractical to incorporate a VCXO into an existing DAI's PLL. Moreover, a PLL that is tasked with both clock recovery and strong attenuation of jitter will not excel in either task. Instead, the I²S from a conventional DAI can be re-clocked with a VCXO PLL, as shown in Figure 23.6.

For an audio DAC, the PLL needs only to operate at 44.1 kHz and multiples of 48 kHz. The control logic sets the dividers to provide an integer ratio divide of the 42.336 MHz VCXO frequency to achieve output frequencies of 44.1 kHz and multiples of 48 kHz. A divide ratio of 960 provides 44.1 kHz, while a ratio of 882 provides 48 kHz.

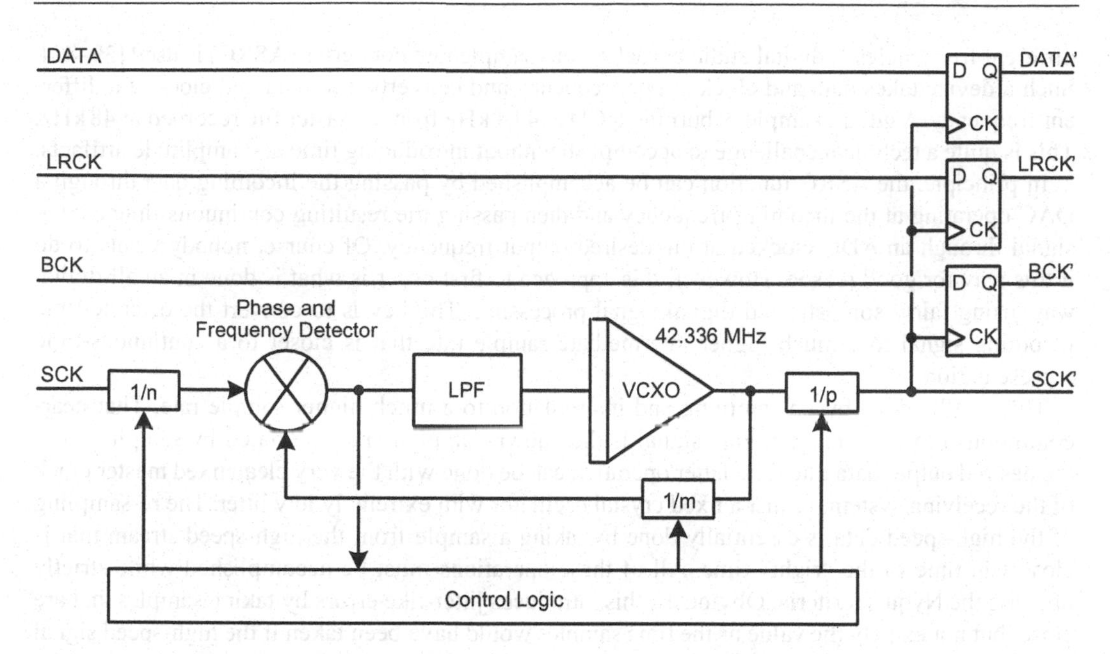

Figure 23.6 VCXO PLL I²S Re-Clocking.

Clocking the DAC Chip with a Fixed Clock

If you clock the DAC with a precision fixed clock, as from a fixed-frequency crystal oscillator, there will not be a detectable jitter problem during some time interval. Eventually the data and clock will drift out of sync because the transport clock received and the independent precision DAC clock are not exactly at the same frequency. The terms wander and creep are sometimes used to describe the changing phase relationship between the bits received and when those bits are clocked into the device. If the incoming and local clock frequencies differ by 50 ppm, then one clock might be 48.0000 kHz and the other 48.0024 kHz, a difference of 2.4 Hz. This means that the wander will be one clock cycle every 417 ms. A bit will be dropped or repeated every 417 ms. This observation is useful in DAC chip designs that can use a crystal oscillator for their clock while mitigating the phase creep problem from slightly mismatched clock frequencies, thus avoiding the effects of jitter in the incoming clock recovered by the DAI.

ESS Technology uses an internal asynchronous sample rate converter-like technique to provide data to the DAC, which is clocked with a precision, low-jitter fixed clock [27]. The clock will not wander out of sync with the data because of the small corrections made to the average sample rate of the data. ESS Technology claims that they alter approximately every 600th bit to make this work [28]. The details of how they implement the approach are not public. However, it is not hard to imagine that the average sample rate of the data can be tweaked by occasional bit-stuffing or de-stuffing at the multiplied sample rate as part of the up-sampling and interpolation.

Asynchronous Sample Rate Converters

There is an alternative to depending completely on PLLs for reduction of the jitter from a conventional DAI. If one needs to interface one digital audio system to another, where their clocks are at different frequencies, a digital audio asynchronous sample rate converter (ASRC) is used [39–42]. Such a device takes data and clock at one frequency and converts it to data and clock at a different frequency. A good example is burning a CD at 44.1 kHz from a master file recorded at 48 kHz. This is quite a technical challenge to accomplish without introducing time and amplitude artifacts.

In principle, the ASRC function can be accomplished by passing the incoming data through a DAC operating at the incoming frequency and then passing the resulting continuous-time analog signal through an ADC clocked at the desired output frequency. Of course, nobody wants to do it this way for good reason. However, this approach to first order is what is done in an all-digital way, using fairly sophisticated digital signal processing. The key is to convert the discrete-time incoming signal to a much higher intermediate sample rate that is closer to a continuous-time representation.

This can be done by up-sampling and interpolation to a much higher sample rate. That near-continuous-time intermediate-rate signal is then down-sampled and decimated by sampling it at the desired output data rate. The latter operation can be done with the very clean fixed master clock of the receiving system, or just a fixed crystal oscillator with extremely low jitter. The re-sampling of the high-speed data is essentially done by taking a sample from the high-speed stream that is closest in time to the "right" time. All of these operations must be accomplished while strictly obeying the Nyquist criteria. Obviously, this introduces jitter-like errors by taking samples that are close, but not exactly the value as the time samples would have been taken if the high-speed signal was in continuous time (analog).

Minimizing this error means minimizing the time between the high-speed intermediate samples, meaning that the high-speed samples are at a much higher rate with very fine time granularity. Things get ugly when you do the math, resulting in GHz intermediate sample rates and digital filtering with an enormous number of taps. Many digital signal-processing simplifications are therefore employed to make the ASRC implementation practical while achieving low distortion and jitter artifacts. These techniques are way beyond the scope of this book, but they do work well [43, 44].

In an audio DAC application, the *nominal* input and output sampling rates can be the same, but they are not exactly the same. They still represent two different frequency domains, even if the rates differ by only a few Hz. The key advantage of using an ASRC in a DAC is that the output sample rate can be fixed by using a low-jitter crystal oscillator clock source. A PLL is still needed in the DAI to recover the incoming clock, but its output jitter is substantially reduced by the ASRC. Of course, adding the ASRC to the signal chain between the DAI and the DAC chip adds cost.

ASRCs are not perfect and completely absent of jitter-related phenomena. This is because the up-sampled stream is only a very good approximation to a continuous-time stream. Not all ASRCs are alike in this respect. Different time granularity in the intermediate stream and different algorithms and digital filtering can make a significant difference. In general, an ASRC with more transistors and more sophisticated digital signal processing (DSP) will more closely approximate perfection.

S/PDIF Manchester Coding

The Sony/Philips Digital Interface (S/PDIF) signal is widely used for transporting stereo digital audio signals. It is a serial data stream that has the clock information embedded in it. It uses differential Manchester coding (DMC), which provides high and reliable average transition density [37]. High transition density helps the PLL lock onto the clock embedded in the data and reduces jitter. This is in contrast to a simple NRZ data stream where a long string of zeros or ones has no

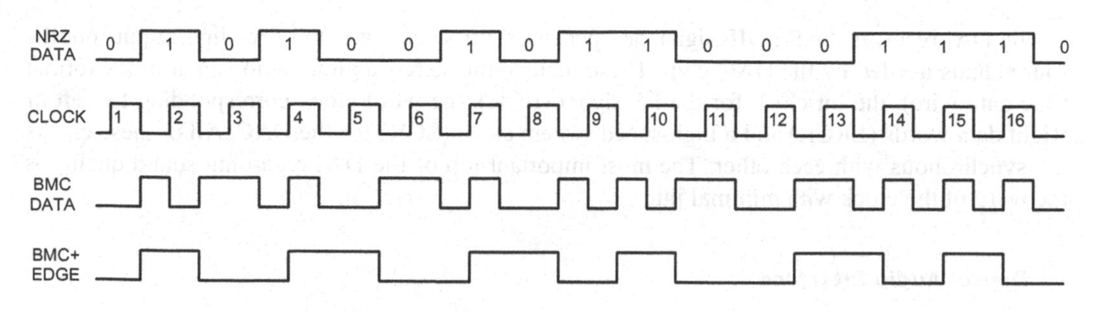

Figure 23.7 NRZ, RZ and Manchester Data Coding.

transitions. The differential Manchester code, also called the biphase mark code (BMC), is unaffected by AC coupling or signal inversion. This is the audio signal that travels over your coax or Toslink.

Figure 23.7 illustrates data, clock and BMC coding. At every other clock tick, there is a potential transition conditional on the data. At other clock ticks, the polarity changes unconditionally. One version of the code makes a transition for zero and no transition for 1. In S/PDIF, a transition is made for a 1 and no transition for zero. In other words, a logical one is represented by a transition that would not be there for a logical zero.

Some transitions are thus separated by only 1/2 bit interval, so the equivalent clock rate is twice that of NRZ, requiring a wider bandwidth channel. The additional bandwidth of the signal is not a problem for short-distance transmission by twisted pair (AES3), coax or Toslink optical fiber. In the DMC format, bandwidth is traded for ease of reliable low-jitter clock recovery.

With DMC there is a transition at the end of each bit. Furthermore, if the bit is a one, there is an additional transition at the center of the bit period. For a 48-ks/sec signal that is all binary zeros, this would amount to a 24 kHz square wave. For a 48-ks/sec signal that is all binary ones, a 48-kHz square wave will result. Phase transitions thus define the clock signal and its timing. This eliminates the need for DC coupling of the signal. The DMC encoding has the following advantages:

- A transition is guaranteed at least once every bit.
- Detecting transitions is less subject to noise than detecting voltage.
- Only the presence of a transition is important, not the polarity.
- The average voltage around each unconditional transition is zero.

23.5 Digital Audio Interfaces

Here we discuss how the S/PDIF digital audio input signal via twisted pair, coax or Toslink gets to the DAC chip, and the related signal formats involved [6–22]. In other words, how the signal from a CD player or other digital audio source gets transformed from the outside world single-wire S/PDIF (coax or TOSLINK) to the I^2S signal that drives the DAC with data, bit clock, L-R word clock and usually the DAC system clock. AES3 (AES/EBU) and S/PDIF use the same data format, just on different physical paths [6–9]. AES3 employs shielded 110-Ω twisted pair with XLR connectors while S/PDIF uses an unbalanced signal on TOSLINK or 75-Ω coax. This path from the incoming S/PDIF to the I^2S signal that drives the DAC is implemented by a Digital Audio Interface (DAI) chip in most DACs [14, 15].

The DAI receives the S/PDIF signal and performs clock recovery and decoding. It puts out the four signals needed by the DAC chip. These include the stereo digital audio signal in I^2S format (one data wire), the bit clock for the I^2S signal (BCK), the word clock corresponding to Left or Right data words (LRCK) and a high-speed system clock (SCK) for the DAC. All of these clocks are synchronous with each other. The most important job of the DAI regarding sound quality is recovery of the clock with minimal jitter.

I^2S Digital Audio Interface

I^2S is a synchronous standard 3-wire audio data format [12, 13]. It stands for Inter-IC Sound. It is an electrical serial bus interface standard for communication of audio signals among ICs on a circuit board. It is used in CD players, DACs and other digital audio applications. It includes DATA, BCK, LRCK and usually SCK. BCK is the bit clock for the synchronous transmission of the data. LRCK is a word clock that distinguishes information for the left and right channels of the stereo signal. SCK is a high-speed system clock that is used as the clock source for the receiving digital circuitry. These signals are recovered by a DAI chip that receives S/PDIF or AES3.

Digital Audio Interface Receivers

The DAI receiver chip takes in a serial asynchronous S/PDIF signal and outputs the signal in the I^2S synchronous format [16–20]. More specifically, it puts out an I^2S signal, comprising DATA, BCK, LRCK, plus a recovered system clock for the DAC. The DAI uses an on-board PLL to recover the clocks from the differential Manchester-coded S/PDIF signal, including a system clock for the DAC. The PLL often requires only an external passive R-C filter. An external clock or resonator is not usually required.

The PLL clock recovery circuit in the DAI is the most critical element. It is responsible for recovering a jitter-free clock from incoming S/PDIF signals that may have jitter or have bit patterns that can cause jitter in the clock recovery process. In most DACs the I^2S output of the DAI is sent directly to the DAC, so any kind of jitter from a conventional DAI will degrade DAC chip performance. In some arrangements a VCXO may be used as the heart of the PLL. In other arrangements the DAC will be fed with a fixed crystal oscillator signal and sophisticated techniques will be used to make up for bit creep that results from the SPDIF clock being different by the small frequency tolerance allowed at the two ends of the system. Just throwing out a number, a worst-case 100 ppm difference in the incoming S/PDIF clock frequency and the crystal oscillator frequency would imply a bit-creep rate of 4.41 Hz, corresponding to adding or deleting a bit every 227 ms.

The Cirrus Logic CS8416 is a popular DAI chip that does well in keeping jitter small [15, 16]. It can handle 24 bits up to 192 kHz, and claims typical clock jitter of 200 ps_{RMS} [15]. The TI DIR9001 is another popular DAI [21]. It can handle 24 bits up to 96 kHz, and claims typical clock jitter of 50 ps_{RMS}. The AKM4115 is yet another DAI chip with good performance [18].

USB Digital Audio Inputs

The Universal Serial Bus (USB) is another transmission format capable of transporting digital audio. Chips for receiving digital audio over USB bridge the USB inputs to I^2S outputs that can drive a DAC chip. The Microchip USB249xx and the Silicon Labs CP2114 are two examples [45, 46].

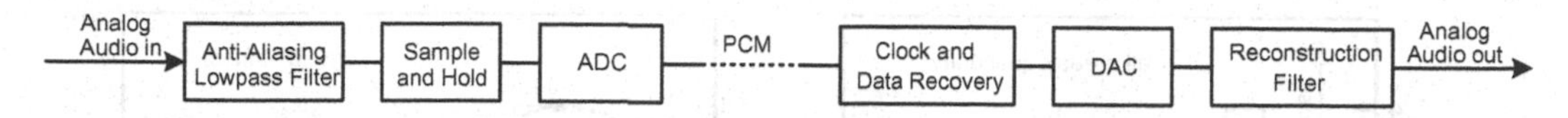

Figure 23.8 Digital Audio Transmission.

23.6 Digital Audio Sampling and Transmission

Before getting into the nuts and bolts of DACs, it is helpful to look at digital audio sampling and transmission in a more abstract and theoretical way. This does not mean a lot of math. It means understanding the Nyquist criteria and the sampling theorem, and related matters like aliasing, imaging, down-sampling (decimation) and up-sampling (interpolation). The common term for up-sampling is over-sampling.

Figure 23.8 illustrates a simple view of the path that the signal goes through from analog source to the received analog signal, all via digital transmission as PCM (pulse code modulation) through the band-limited channel.

Sampling and Reconstruction

The Nyquist-Shannon sampling theorem conclusions are based on sampling with an impulse that has amplitude of infinity, width of zero and area of unity. This pulse is called a Dirac delta function or unit impulse [2]. Such an impulse cannot occur in the real world. However, it is useful to know that sampling with a narrow pulse, perhaps of duration one tenth the unit interval (UI), can provide a fair approximation to that which would result from impulse sampling. The real-world departure from an impulse sample to versions of a stream of rectangular pulses does not void the fact that all of the information is conveyed if the Nyquist criteria is met. However, it can affect the frequency response at the DAC end. Pulse widths of one UI can result in a high-frequency roll-off of about 4 dB at 20 kHz after reconstruction.

Figure 23.9(a) illustrates how the samples can be conveyed, from a Dirac delta to the usual one-UI flat-top pulse. All of these have unit area. The 0.5- and 0.25-UI pulses are obviously a closer approximation to the unit impulse, and mitigate any frequency response effects from the departure from an impulse. These represent the kinds of pulses that the DAC chip can deliver to the analog reconstruction filter. The width of these pulses is referred to as the sampling aperture. A higher sampling frequency corresponds to a smaller aperture time. A 2X over-sampled DAC would deliver a stair step consisting of 0.5-UI pulses. A 4× over-sampled DAC would deliver 0.25-UI pulses. Even a 2X over-sampled DAC is down only 0.5 dB at 20 kHz for $f_{s=}$ 48 kHz.

Figure 23.9(b) illustrates the stair-step illustration of the raw data coming out of a simple DAC, without presuming anything about the duration of each step compared to a UI. The important thing is that these are rectangular pulses that begin at a sample time and persist until the next sample time, at whatever sample rate is coming out of the DAC. This representation of the data as a series of rectangular pulses is called a zero-order hold (ZOH) function [47]. The ZOH modulates the information by the Fourier transform of the sampling function.

The Fourier transform of an impulse is flat at 0 dB. The Fourier transform of a flat-topped pulse is a so-called *sinc* function, sinc(x), more formally called a *sine cardinal function* [48]. It is depicted in Figure 23.10(a). It has a frequency response that goes through zero at multiples of the UI. The frequency response looks like (sin f)/f = sinc(f). With $f_{s=}$ 44.1 kHz, the frequency response of the ZOH is down 4 dB at 20 kHz.

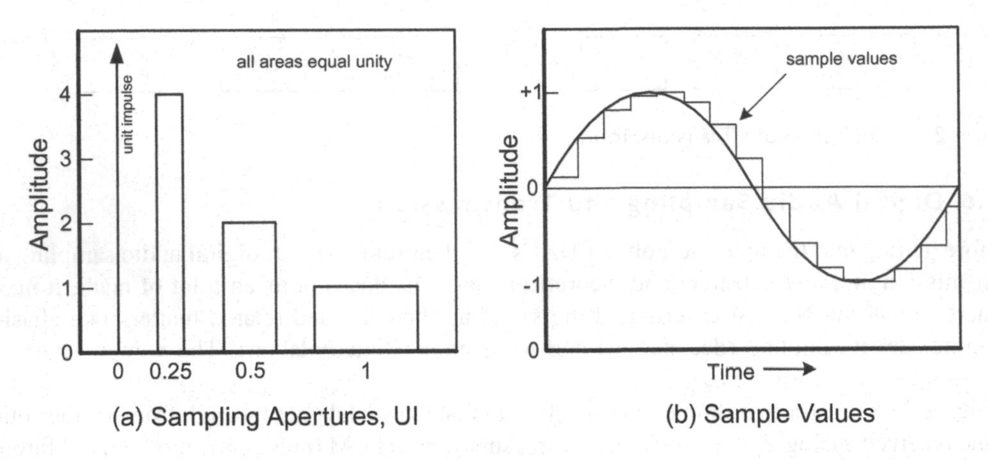

Figure 23.9 Sampling and Holding.

Practical ADCs are preceded by a sample-and-hold (S/H) that actually samples the incoming signal at an instant and holds the value for the entirety of the UI. This gives the ADC time to create its digital PCM output. It is easy to see that the S/H process produces a delay of the signal of 1/2 UI.

The *sinc(x)*=sin(x)/x function is also called the *sampling function*. In reconstructing a signal with a DAC, we normally assume that the DAC puts out a value for the duration of time until the next sample arrives. However, a true sampling process at the ADC end would take an instantaneous sample, as with an impulse. In practice, the ADC includes a sample-and-hold function. The signal is being multiplied by the ZOH function, which itself has a frequency response in the shape of a *sinc* function. Similarly, the DAC and reconstruction filter decode a stair-step function rather than a series of impulses.

The original analog frequency spectrum has been multiplied by a sinc function, which has a modest roll-off as frequency approaches $f_{Nyquist}$. The frequency response of the *sinc* function goes to zero at integer multiples of f_s. An analog filter can be designed to compensate for the *sinc* function effect on the frequency response.

The sampling and holding for a UI creates a signal-processing function called the ZOH. This does not distort the baseband signal, but it does delay it by 1/2 UI and creates some high-frequency attenuation.

The frequency response of the ZOH is unity at DC and falls by 0.15 dB at $f_s/10$, which is 4.8 kHz for $f_{s=}$ 48 kHz. At the Nyquist frequency $f_{Nyquist}$ (24 kHz) it is down by 3.92 dB. It is often quoted as being down 4 dB at 20 kHz, but this is based on a 44.1-kHz sampling rate. For completeness, the ZOH frequency response goes to zero at f_s.

The sinc Function and Zero-Order Hold

A conventional DAC does not output a stream of Dirac impulses (e.g., unit impulse), but rather a staircase that consists of a stream of rectangular pulses of unit-interval duration. For this reason, the zero-order hold (ZOH) comes into play in determining the frequency response of the output [47]. The ZOH response is like an LPF function at frequencies up to f_s. The response is down 3.9224 dB at $f_{Nyquist}$, corresponding numerically to $\text{sinc}(x)=\text{sinc}(0.5)=2/\pi$. The ZOH frequency response for $f_{s=}$ 48 kHz is shown below in Figure 23.10(b). It is down 0.63 dB at 10 kHz, 2.64 dB at 20 kHz and

3.92 dB at $f_{Nyquist=}$24 kHz. The below numbers are for $f_{sig=}$20 kHz. The *x* columb represents the value of *x* in the expression sine(*x*)/*x*, here the signal frequency divided by the sampling frequency. OSR (over-sampling ratio) is the ratio of the sampling frequency f_s to the nominal sampling frequency that satisfies the Nyquist criteria.

f_s	*x*	*OSR*	*Amplitude*	*dB*
44.1	0.454	1	0.694	−3.17
48	0.417	1	0.738	−2.64
96	0.208	2	0.930	−0.63
192	0.104	4	0.982	−0.16
480	0.0417	10	9.997	−0.025

The Fourier transform of an impulse is flat at 0 dB. The Fourier transform of a flat-topped pulse is a function of frequency that looks like (sin *f*)/*f*=sinc(*f*), as shown in Figure 23.10(a). With $f_{s=}$44.1 kHz, the frequency response of the ZOH is down 3.17 dB at 20 kHz, as shown in Figure 23.10(b).

The *sinc* function is depicted in Figure 23.10(a). The term *sinc* is an abbreviated version of the *sine cardinal*. It is also referred to as the *sampling function* [48]. Let T=UI:

$$\operatorname{sinc}(x) = \sin(\pi x)/(\pi x)$$

$$\operatorname{sinc}(0.5) = 2/\pi \Rightarrow -3.9224 \text{ dB at } fNyquist$$

$$H(s) = (1 - e^{-\omega T})/(\omega T) = (e^{-\pi f T}) \times \operatorname{sinc}(fT)$$

$$H_{ZOH}(f) = (e^{-\pi f T}) \times \operatorname{sinc}(fT)$$

The ZOH frequency response can be corrected before the DAC with a digital filter or after the DAC with an analog filter. This is called aperture correction.

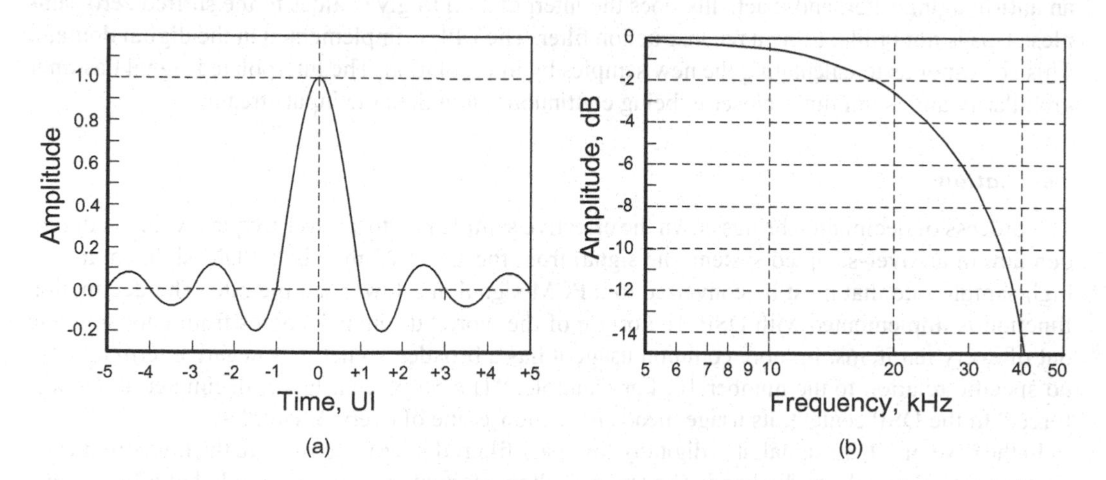

Figure 23.10 The *sinc* Function and ZOH Frequency Response.

The effect of the ZOH is reduced significantly if f_s is increased beyond twice $f_{Nyquist}$. This increase is referred to as the over-sampling ratio (OSR)$=f_s/2\,f_{Nyquist}$. Over-sampling, even by a factor of 2, largely eliminates the *sinc* function frequency response, since doubling f_s doubles the available $f_{Nyquist}$ to $f_s/2$ for the process, and the *sinc* function is quite flat up to half the available $f_{Nyquist}$. If OSR$=2$, droop due to the ZOH is reduced to about 0.9 dB at $f_s/2$. If OSR$=10$, droop at $f_s/2$ is only 0.036 dB.

The baseband audio information is not distorted or lost by the ZOH function, but its frequency response has a high-frequency roll-off which is down about 3.9 dB at 24 kHz if $f_s=48$ kHz. The frequency response of the ZOH falls to zero at f_s. The ZOH results in a PCM stream that, when converted to analog by a DAC, is a stair-step function that is an approximation to the sampled signal. The stair-step approximation must be low-pass filtered to smooth it out. This is called reconstruction. It is very important to recognize that the ZOH function and its frequency response is a function of the signal path in the DAC, not in the ADC. The DAC outputs 1-UI rectangular pulses that are subject to the ZOH function upon reconstruction. The second-generation Philips CD players used 4× over-sampling with the Philips TDA1541 current-weighted DAC.

23.7 Interpolation and Decimation

The functions of interpolation and decimation are fundamental to the DSP functionality in ΣΔ ADCs, DACs and many other functions. For that reason they will be explained a bit further here. Up-sampling and down-sampling depend on these DSP functions and will also be discussed.

Interpolation

Interpolation is used to take information clocked at a given sample rate and convert it into the identical information at a higher sample rate. A digital PCM signal sampled at $f_{s=}$48 kHz can be converted to a sample rate of 480 kHz with an OSR of 10. There are $N=10$ samples created for each of the original samples. In the process, only one of the original samples is taken and the other $(N-1)=9$ samples are replaced with zeros. This is a form of bit-stuffing. The signal content is still there, but it needs to be low-pass filtered to smooth the stuffed samples with the original samples. This process also creates spurious image signals at higher frequencies. The low-pass filter is called an anti-imaging filter, and it actually does the interpolation to give values to the stuffed zero samples. This is not unlike using a reconstruction filter. The LPF is implemented in the digital domain. This corresponds to calculating the new samples by interpolation. The interpolated signal has finer granularity and is ten times closer to being continuous-time than the input stream.

Decimation

The process of decimation brings down the effective sample rate to a lower frequency, in the digital domain. In an over-sampled system, the signal from the ΣΔ DAC may be a PDM signal at a very high sample rate that must be converted to a PCM signal at a lower sample rate. The decimation function is implemented with DSP. The origin of the word "decimate" comes from choosing one out of every ten items. In more common usage it has a broader, sometimes negative term and has no specific relation to the number 10. For example, "The 50 percent layoff decimated the workforce." In the DSP context, its usage means is to choose one of every N samples.

In the case of a 1-bit signal, it is digitally low-pass filtered in order to remove the high-frequency components above the audio band. The signal is then decimated – i.e., re-sampled at a lower rate.

Here decimation means keeping one out of N samples, where the OSR may equal the fraction $1/N$. As with any sampling, the lower-frequency re-sampling must obey the Nyquist criteria, so the preceding digital LPF is actually an anti-aliasing filter. In other words, it is OK to resample the digital signal at a lower rate as long as that digital signal has been passed through an appropriate anti-alias filter. This is all done in the digital domain.

Consider a stream of samples at 10 Ms/s (and containing signal frequencies of less than $f_{Nyquist=}$5 MHz). We want to get it down to 1 Ms/s. We resample that signal at the new 1 Ms/s rate, taking every tenth sample. If we just do this, we have aliasing, since the signal being sampled has information extending to 5 MHz. If we pass the 10-Ms/s data through a 500-kHz low-pass anti-aliasing filter before re-sampling we now have a 10-Ms/s digital stream that has no frequencies above $f_{Nyquist=}$500 kHz. This means that we can take every tenth sample at a regular rate as long as that new sampling rate satisfies the Nyquist criteria.

Just as with over-sampling, the anti-alias filter slope requirement can be relaxed if the new f_s is considerably higher than twice the highest frequency in the signal. This gives more room in the frequency domain to achieve the required amount of anti-alias filter roll-off. For example, we may do anti-alias filtering at 500 kHz in the above digital filter in preparation for re-sampling at a lower frequency, but we may do the re-sampling at 2 MHz instead of 1 MHz, more than satisfying the Nyquist criteria by a factor of 2. This means that the roll-off of the anti-alias filter need not be as steep, resulting in a simpler filter that does less damage to phase linearity. One can over-sample and create a wastefully wide-bandwidth signal as long as one is not trying to transmit it as a finished product in a strictly band-limited channel.

Up-Sampling and Down-Sampling

Up-sampling (interpolation) requires a low-pass reconstruction filter after the up-sampling. Down-sampling (decimation) requires a low-pass anti-aliasing filter before the re-sampling. The key is that any re-sampling process, be it up or down, must satisfy the Nyquist criteria, meaning that reconstruction filters and anti-alias filters must be used. This means that image frequencies must be removed in the process of reconstruction and that alias frequencies must be eliminated in the process of sampling.

Nyquist proved that sampling a signal at twice its highest bandwidth would retain all the information and permit full recovery of the signal. He did not say that the sampling process would not create additional unwanted information. In fact, sampling is a process of modulation of a signal by an impulse. This means that sidebands of the signal will be created on either side of integer multiples of the sampling frequency. These are image frequencies. A 10-kHz sine wave sampled at 48 kHz will create image frequencies at 38 and 58 kHz. Images will also be created at 96 ±10 kHz and so on. If none of these images has been filtered out, no reconstruction has occurred.

Full reconstruction, though not entirely required, means that all of the image frequency energy is filtered out. In particular, the image frequencies on the low side of the sampling frequency must not overlap the highest frequency of the baseband signal. Consider f_s=48 kHz where $f_{Nyquist=}$24 kHz. If a 24-kHz baseband signal is sampled, the image on the low side of the sampling frequency will be at 24 kHz, just barely intersecting the upper end of the baseband signal. There will be no overlap of signals of meaningful amplitude if perfect brick-wall anti-alias and reconstruction filters are employed. However, no filters have a brick-wall characteristic, so overlap of small amplitude remnants of the signals can occur unless an anti-alias filter begins roll-off at a slightly lower frequency and reaches essentially zero transmission by 24 kHz. The anti-alias filter is important. If

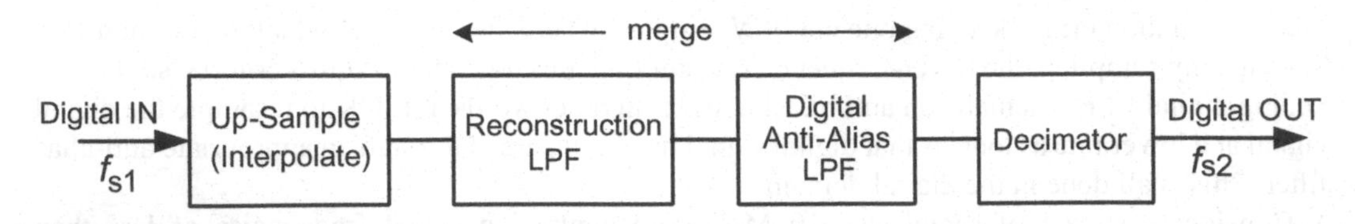

Figure 23.11 Interpolation and Decimation.

there is no signal in the baseband at and above $f_s/2$, there will be no image frequency to go that low in frequency.

Figure 23.11 is a block diagram of an up-sampling and down-sampling sequence, showing that in this special case the reconstruction LPF following the up-sampling and the anti-alias LPF preceding the down-sampling are adjacent and can be consolidated.

23.8 R2R DACs

A simple DAC can be made by turning on or off a series of binary-weighted current sources whose outputs are summed. In practice, this is difficult to do with accuracy, partly because the MSB is so much larger than the LSB in a DAC with a significant number of bits [49]. In particular, when the signal value changes by one LSB to a number that changes the MSB, there will be an error that is significant if the MSB is not the proper exact multiple of the LSB. Resistor matching of so many different values is difficult. The problem could be mitigated on an integrated circuit by using an enormous number of resistors of equal value, each contributing one LSB of current to the output. An 8-bit DAC would require 128 resistors to build the MSB, so about 256 resistors would be needed for such a DAC. Imagine how many resistors would be needed for a 16-bit DAC.

R2R DACs are simple and popular, in part because they use only two values of resistor. The problem of requiring many resistors and switches is mitigated by implementing the R2R DAC, which is composed of resistors of value R and $2R$, as shown in Figure 23.12. A reference voltage is applied to the shunt resistors in the ladder that are selected by ones in the binary word. It is easily seen how the effect of the reference as applied to a resistor in the ladder is attenuated as it passes each node in reaching the output. Its effect is attenuated by a factor of 2 as it progresses each step in the ladder. The output is a current that flows into a TIA to create the output voltage. The $2R$ resistors can be formed from two resistors of value R. This means that the entire DAC can be made from resistors of identical values, easing the matching problem. However, the resistors still must be extremely well-matched. It is practical to get to 16 bits with this technique. The R2R DAC was the mainstay of DACs for many years in the digital audio world.

These are sometimes operated as non-over-sampled (NOS) DACs in which f_s is twice the highest frequency of interest. However, R2R DACs are fast, and can be operated in over-sampled systems as well. Some audiophiles prefer R2R DACs, including ones that are NOS [50]. Most non-ΣΔ DACs are referred to as R2R DACs, but they may also employ other techniques for creating some of the bit values.

The Philips TDA1540 DAC chip was the first audio DAC for use in CD players and was widely used in first-generation CD players [51]. The TDA1541A was the most-used DAC in second-generation CD players [52]. Long out of production, it is still prized by some audiophiles.

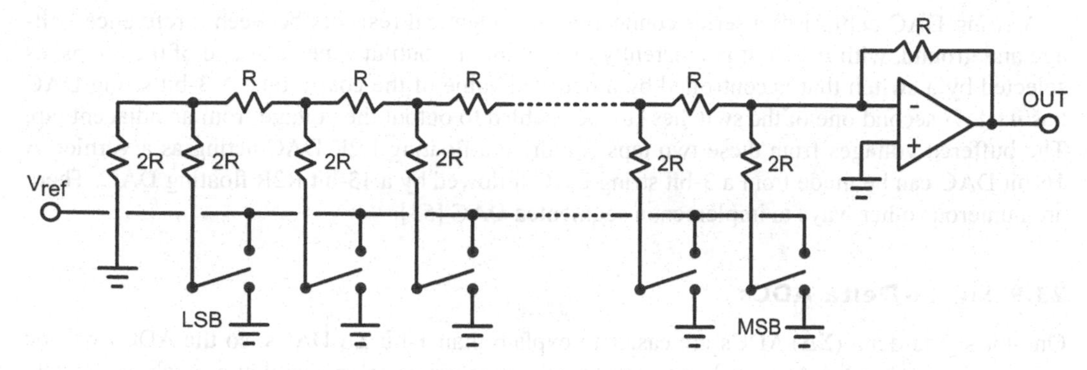

Figure 23.12 R2R DAC.

Quantization Noise

With a limited number of bits, the sample is not an exact representation of the sampled analog value. The sampled value is quantized to within ±½ LSB. This error is quantization noise. It is noise whose power is uniformly distributed with frequency, and the noise spectral density is flat with frequency.

The SNR when considering only the contribution of quantization noise for an *N*-bit ADC or DAC is:

$$\text{SNR} = 1.761 + N \times 6.02\text{ dB}$$

A 16-bit PCM stream thus has an SNR of 98.08 dB. Notably, a 1-bit stream starts out with an SNR of only 7.8 dB, so it has a long way to go before reaching audio-grade SNR.

Over-Sampling to Increase Effective Bits

R2R DACs can be fast, and can be operated in an over-sampling mode to obtain a noise reduction of 3 dB for each doubling of f_s. This is so because the noise power is spread over twice the bandwidth if f_s is doubled. The noise spreading thus halves the noise power density in the audio spectrum. If 4× over-sampling is used, corresponding to OSR of 4, a 6-dB improvement in SNR results, corresponding to the addition of one effective bit. The over-sampling ratio $\text{OSR} = 2f_s/f_{Nyquist}$.

Over-sampling also makes reconstruction filtering easier. As shown earlier in Figure 23.10(b), sampling at the Nyquist rate corresponds to relative unity aperture time, which results in zero-order-hold roll-off of about 4 dB at 20 kHz. It can be seen in the figure that over-sampling by OSR = 2 results in the attenuation shown for 10 kHz.

Over-sampling in a PCM DAC is accomplished by up-sampling the incoming PCM with an interpolator followed by a digital LPF. The higher-speed PCM stream is passed through the DAC. The output of the DAC can be reconstructed by an analog external LPF.

Segmented DACs

Increased DAC resolution can be obtained with a cascade of two DACs, one implementing the coarse bits and one implementing the fine bits with a vernier function between the values of two adjacent coarse bits. The bits of the binary input are thus segmented into coarse and fine bit groups.

A string DAC comprises a series connection of n identical resistors between a reference voltage and ground, with n taps. It is inherently monotonic. Its output appears at one of the n taps, as selected by a switch that is controlled by a decoded value of the coarse bits. A 3-bit string DAC has $n=8$. A second one of the switches can be enabled to output the voltage from an adjacent tap. The buffered voltages from these two taps can drive a floating R2R DAC acting as a vernier. A 16-bit DAC can be made from a 3-bit string DAC followed by a 13-bit R2R floating DAC. There are numerous other ways to implement a segmented DAC [53].

23.9 Sigma-Delta ADCs

One-bit sigma-delta (ΣΔ) ADCs are easier to explain than 1-bit ΣΔ DACs, so the ADCs will be described first. The ΣΔ ADCs rely on sampling the incoming analog signal at a much higher rate than the Nyquist rate. This rate is higher than $2\,f_{Nyquist}$ by the OSR. The output of the ADC is a high-speed 1-bit digital stream at the OSR rate that contains all of the information of the original analog signal. The fact that the sampling rate is way over $2f_{Nyquist}$ means that the analog anti-alias filter for the input signal can have a much shallower roll-off rate.

Sampling Theorem, Interpolation and Decimation

The Nyquist-Shannon sampling theorem holds that if a signal is sampled at twice its highest frequency, no information will be lost. That highest allowable signal frequency is $f_{Nyquist}$. The Nyquist rate is twice the Nyquist frequency. The Nyquist frequency for f_s=48-kHz is 24 kHz. The basic theorem assumes that the signal is sampled with a Dirac delta function whose duration is zero, whose amplitude is infinite and whose area is unity [2]. In the real world, the signal is sampled with a sample-and-hold circuit, which achieves a very small effective sampling aperture, as illustrated in Figure 23.9(a).

If one has a series of samples of a signal and wants to have many more samples in the same time interval, one can create new samples between the original samples by interpolation. As described in Section 23.6 this is what happens when you enter several Excel data point y values as a function of time on the x axis and have the program create a smooth curve. This smooth curve represents a continuous-time version of the data points that originally consisted of samples at discrete times.

If you have a great many samples of a signal, many more than necessary to meet the Nyquist criteria, and wish to have fewer samples at a lower sampling rate, you sample the samples at the desired lower rate. You might take one out of every ten samples to create a digital signal at a lower sample rate. This is no different than sampling a continuous-time analog waveform at a rate that satisfies the Nyquist criteria. Taking occasional samples of a digital stream of many more samples is referred to as decimation, as described in Section 23.6.

Delta Modulation

A delta modulator is shown in Figure 23.13(a). It operates based on the use of a very high sampling rate, much higher than the Nyquist rate for the signal being sampled. Its output represents differences between samples at very closely spaced sample times, thus the term delta [2]. The output is a 1-bit binary code that indicates whether the current analog sample of the data is larger or smaller than the data value that has previously been transmitted. Notice that the 1-bit logic level is converted to an analog value with a 1-bit DAC in the feedback path. The 1-bit output stream is a form of pulse density modulation. It is easy to see that the 1-bit DAC and integrator in the feedback path have effectively demodulated the digital stream and that the feedback has acted to make the resulting

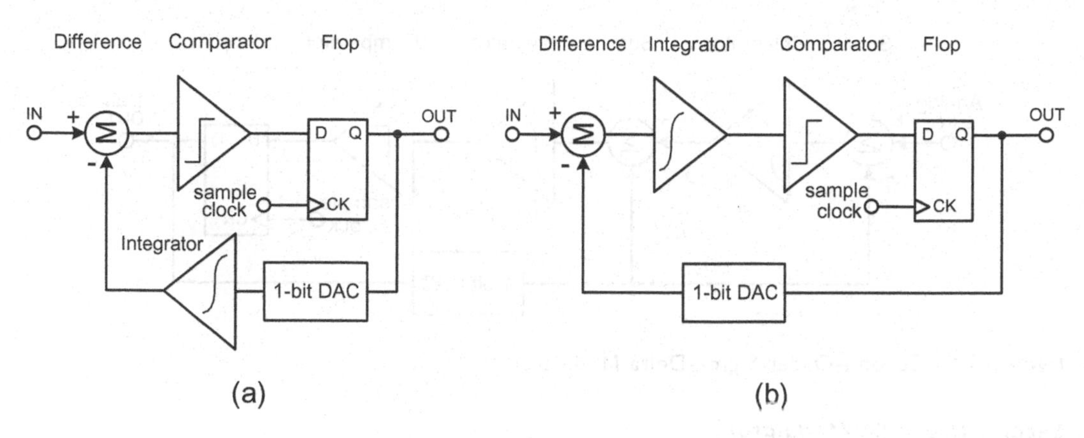

Figure 23.13 Delta and Sigma-Delta Modulators.

demodulated signal almost equal to the analog input signal. This means that the recovery of the analog signal at the far end need only duplicate the 1-bit DAC and integrator in the transmitter.

The very coarse 1-bit sampling results in a large amount of quantization noise in the signal, with SNR of only about 7.8 dB. However, as the sampling frequency is increased, the quantization noise at the in-band frequencies is reduced by noise spreading. Doubling the sample rate spreads the quantization noise power over twice the bandwidth, diluting it and reducing the noise in a given baseband spectrum by 3 dB.

Sigma-Delta Modulator and Noise Shaping

A ΣΔ modulator is shown in Figure 23.13(b) [2–4]. It largely represents a simple re-arrangement of the elements of a delta modulator. It also outputs a 1-bit code that represents the input signal. With the integrator now placed after the input subtractor, it will act to drive the difference in signals from the input and output to zero at in-band frequencies. The output of the subtractor is now seen as an error signal. Notice that now it is the output of the 1-bit DAC that is made to be the same average value as the analog input signal, meaning that the demodulation at the far end need not include the integrator.

The operation of the ΣΔ modulator is not unlike that of an analog feedback amplifier with an input differential pair followed by an integrator in the forward path. Like the analog feedback amplifier, the integrator in the forward path drives the error down by 20 dB/decade as frequency decreases. The "error" in the ΣΔ modulator is the large quantization noise that results from 1-bit coding. The reduction of quantization noise density by 20 dB/decade is called noise shaping [4]. The total of the quantization noise power remains the same, but most of it is pushed out to higher frequencies above the audio band. Its noise spectrum is thus shaped.

The greater the OSR, the further out into the out-of-band high-frequency region the noise is pushed, by a rate of 20 dB/decade of OSR. Recall that the noise spreading caused by over-sampling decreases the noise density by 10 dB/decade of increased sample rate. The combination of noise spreading and noise shaping in the ΣΔ modulator thus reduces in-band noise by a total of 30 dB/decade of OSR. An OSR of 100 with f_s 4.8 MHz can result in an SNR on the order of 7.8 + 60 = 67.8 dB in a 24-kHz bandwidth.

Although the ΣΔ modulator here has been described as a hybrid analog-digital system, it is important to recognize that all of the functions in the modulator can be implemented digitally, wherein the input is a PCM digital signal instead of an analog signal.

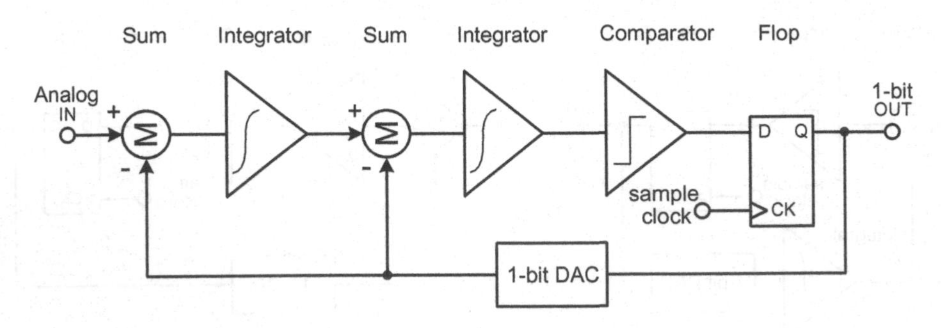

Figure 23.14 Second-Order Sigma-Delta Modulator.

Second-Order ΣΔ Modulator

A second-order ΣΔ modulator is shown in Figure 23.14. It also outputs a 1-bit representation of the input signal. However, a second integrator is put in the forward path and a portion of the output is subtracted from the input of the second integrator. This becomes a second-order feedback system. As a result of the double integration in the forward path, the quantization error is driven down by 40 dB/decade as frequency decreases. This results in the quantizing noise being pushed further out to high frequencies at a rate of 40 dB/decade of OSR, leaving still less noise in the audio band.

The combination of noise spreading and noise shaping in the second-order ΣΔ modulator thus reduces in-band noise by a total of 50 dB/decade of OSR. An OSR of 100 with $f_s =$ 4.8 MHz can result in an SNR on the order of 7.8 + 100 = 107.8 dB in a 24-kHz bandwidth.

Higher-Order ΣΔ ADCs

The noise-shaping properties of ΣΔ modulators can be more aggressively exploited by employing high-order loops with more integrator sections in the forward path. These modulators often fall in the range of third to fifth order. The order of the loop is generally the same as the number of integrators in the loop. Higher-order loops provide much more in-band SNR for a given clock frequency. One price paid for using modulators of third order and above is stability, especially under overload conditions. This makes the design approach more complex.

Multibit ΣΔ ADCs

While greater OSR and higher-order modulators can make significant reductions in quantization noise, the design can only go so far in terms of clock frequency and modulator order. The next tool that can be applied is taking the 1-bit ΣΔ modulator to multibit operation. The concept is illustrated in Figure 23.15. The comparator in a 1-bit ΣΔ design is actually a 1-bit ADC. In the example here it is replaced by a 3-bit flash ADC. This means that there is a 3-bit code circulating in the digital portion of the modulator feedback loop, yielding eight codes instead of just two codes. Correspondingly, the feedback DAC function is replaced with a very linear 3-bit thermometer DAC that is inherently monotonic [54].

As is well known in PCM transmission, quantization noise is reduced by 6 dB each time an additional bit is added to the PCM code. Going from a 1-bit representation of the signal in the loop to a 3-bit representation thus corresponds to the addition of 2 bits of resolution, for a quantization noise reduction of 12 dB. While the multibit ΣΔ ADC seems like an intuitive next step, its high-performance implementation in light of component tolerances requires a sophisticated implementation well beyond the scope of this book [55–59].

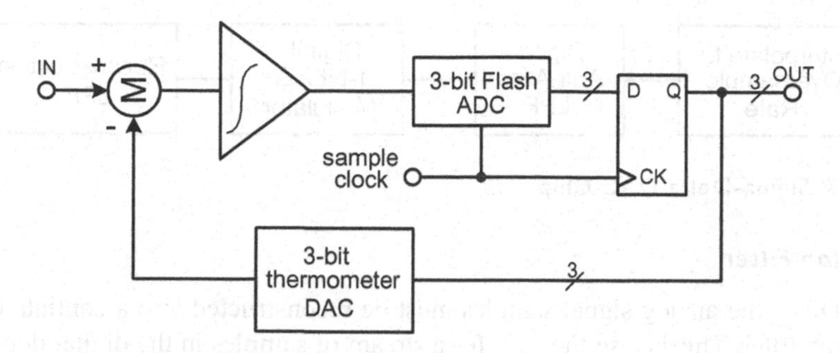

Figure 23.15 **Multilevel ΣΔ ADCs.**

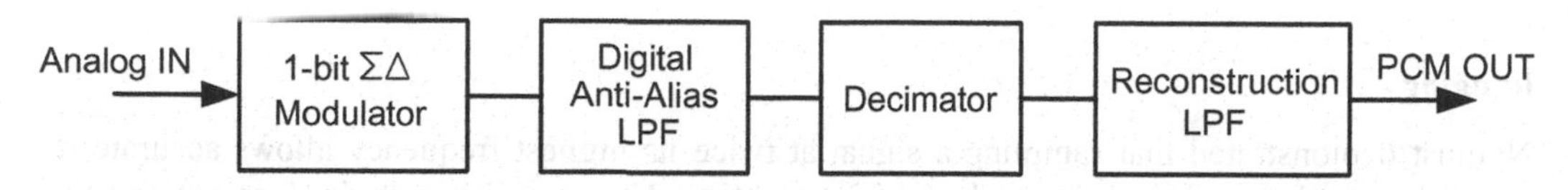

Figure 23.16 **Conversion of One-Bit PDM to PCM.**

Conversion to PCM

The over-sampled 1-bit code from a conventional ΣΔ modulator must be converted to PCM at the channel rate using DSP techniques. As shown in Figure 23.16, the output of the modulator must be passed through a digital anti-aliasing filter before it can be down-sampled to the channel rate in order to satisfy the Nyquist criteria. The signal is then decimated to reduce the rate to the channel rate. Finally, the process of decimation requires a reconstruction filter to provide a PCM output suitable for the channel.

23.10 Sigma-Delta DACs

Operation of the 1-bit sigma-delta DAC is somewhat the inverse of the ADC. The incoming 16-bit PCM stream is interpolated and filtered by a large OSR to create a high-rate PCM stream. That signal is then passed through a digital ΣΔ modulator that converts the PCM to a noise-shaped 1-bit stream. That stream then passes through a digital LPF to eliminate image frequencies. The high-OSR 1-bit stream is then applied to a 1-bit DAC that produces a PDM output stream that is then filtered by an external analog low-pass reconstruction filter.

The primary role of the ΣΔ modulator in the DAC is to provide noise shaping that acts like a high-pass filter for quantizing noise and an LPF for the signal. Recall that the sigma-delta modulator in the ADC did virtually the same thing. The difference here is that the input to the first summer in the modulator is in the form of a PCM number rather than a continuous-time analog signal value (Figure 23.17).

In the summer of the modulator, the value of the 1-bit PDM signal is subtracted from the value of the PCM stream using a digital subtractor. The rest of the sigma-delta modulator in the DAC is essentially the same as that in the ADC. The 1-bit PDM stream is its noise-shaped output. That output is digitally filtered (reconstructed) to remove images and the resulting 1-bit stream is converted into high-rate analog high-low levels that can be filtered with an external analog reconstruction filter.

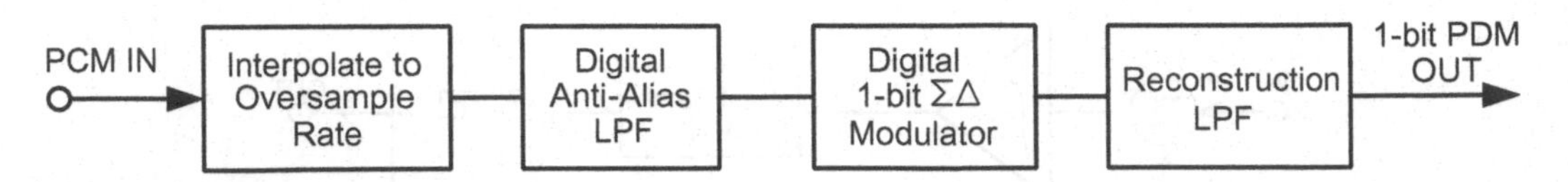

Figure 23.17 A Sigma-Delta DAC Chip.

Reconstruction Filter

Following a DAC, the analog signal samples must be reconstructed into a continuous stream by a reconstruction filter. This is also the case for a stream of samples in the digital domain that will be up-sampled to a higher rate. In this case, the reconstruction filter is performing interpolation. In both cases the reconstruction filter is performing a low-pass filtering function akin to averaging.

Imaging

Nyquist demonstrated that sampling a signal at twice its highest frequency allows accurate reconstruction of the signal without loss of information. However, his criteria does not say that new frequencies are not created by the process. These new frequencies are called images, and the reconstruction filter is tasked with eliminating them. For this reason, the reconstruction filter is sometimes called an anti-imaging filter. The out-of-band images created by the sampling process lie above and below the sampling frequency and its integer multiples.

The presence of such images can interfere with the accuracy of reconstruction, so they must be eliminated, especially if the result is to be sampled again. If the sidebands on the low side of the sampling frequency overlap the baseband signal, tones may be created. This is more likely if the signal has not been over-sampled. If the signal to be reconstructed was over-sampled, this is much less likely to happen. In this case, the reconstruction filter need not have as sharp a cutoff. It can be gentler and create less phase distortion and group delay variation in the reconstructed signal.

Some DACs have little or no analog reconstruction filtering for creating the analog output signal. In this case, the image frequencies may not be heard, but they may disturb some downstream circuits. Distortion in those circuits can re-modulate the high-frequency images to create audible frequency components. For this reason, it is important that any remaining image frequencies be of low amplitude if they are present. It is an interesting exercise to examine the output of a CD or SACD player with a spectrum analyzer with large enough bandwidth to display the spectrum that includes the image frequencies.

Analog reconstruction filters with a sharp cutoff can be hard to build and may create extreme phase distortion and group delay problems. It is much better to do any sharp filtering with digital filters, then use gentle roll-off in the analog reconstruction filter to preserve uniform delay up to frequencies well above the Nyquist frequency.

Intra-Sample Signal Peaks

When an audio signal is sampled at the Nyquist rate, the highest-amplitude sample may not reflect the highest amplitude peak of the signal, since the highest analog amplitude may have occurred at some time between 44.1 or 48 kHz samples. If the highest recorded PCM samples are at digital full scale (DFS), then subsequent interpolation up-sampling and digital reconstruction filtering can fill in the intra-sample signal peak values, which may lie above DFS (digital full-scale) of the DSP hardware and cause digital overload [44]. This may require extra digital headroom if the reconstruction is done in the digital domain with DSP. Alternatively, digital attenuation of the incoming signal can be implemented to achieve adequate amplitude margin in the digital processing [1]. Of course, this may cost a little bit of SNR. The worst-case example occurs with a full-scale sine wave at 1/4 the sampling frequency, which is 11.025 kHz for $f_{s=}$ 44.1 kHz. The maximum intra-sample overage in this worst case is +3 dB [44].

23.11 Control Interfaces

Like many modern ICs, the DAI and DAC can be, or sometimes must be, controlled by a microcontroller or microcomputer. There are two very popular control interfaces in use, the I²C bus and the SPI bus. They are discussed in greater detail in Chapter 28. The I²C bus is more popular in DAC applications [60–62].

I²C Bus

The I²C (Inter Integrated Circuit) bus was conceived by Philips (now NXP), and is a synchronous peer-to-peer inter-chip signaling protocol [60, 61]. It is a 2-wire protocol, carrying data (SDA) and clock (SCK). It is often used for enabling a microcontroller to communicate with one or more slave peripherals. It supports half-duplex communications. It is good for communicating with a large number of peripherals at modest data rates. A large number of peripherals can be attached to the single 2-wire bus. Peripherals are pre-assigned fixed addresses by their manufacturers.

SPI Bus

The most common alternative to I²C, the SPI bus (Serial Peripheral Interface), was conceived by Motorola [62]. It is a synchronous master-slave protocol that includes master output, slave input (MOSI) and master input, slave output (MISO) data lines and can support full duplex synchronous communications with the host processor. The SPI interface also includes clock (SCK) and a chip select (CS) for each peripheral slave. SPI is faster than I²C, but requires more wires. It is good for communicating with a smaller number of peripherals at much higher data rates. It is especially useful for communicating with external FLASH memory.

References

1. Benchmark Media Systems, DAC3 HGC.
2. Ken C. Pohlmann, *Principles of Digital Audio*, 6th edition, McGraw Hill, New York, NY, ISBN 978-0-07-166346-5, 2011.
3. Richard Schreier and Gabor Temes, *Understanding Delta-Sigma Data Converters,* Wiley-IEEE Press, 2004.
4. Walt Kester, *The Data Conversion Handbook*, Elsevier/Nunes, Boston, MA, 2005.
5. Dennis Bohn, *Digital Dharma of Audio A/D Converters*, https://headwizememorial.wordpress.com/2018/03/14/digital-dharma-of-audio-a-d-converters/.
6. STMicroelectronics, *Receiving S/PDIF Audio Stream with the STM32F4 Series*, Application Note AN5073.
7. Wikipedia, *S/PDIF*, wikipedia.org.
8. Audio Engineering Society, AES3–2009: *AES Standard for Digital Audio Engineering – Serial Transmission Format for Two-Channel Linearly Represented Digital Audio Data*, aes.org/publications/standards [AES3].

9. WikiAudio, *AES EBU,* April 5, 2018, wikiaudio.org/aes-ebu.
10. TOSLINK, techopdia, November 19, 2012, techopedia.com.
11. TOSLINK (Toshiba Link), Wikipedia, wikipedia.org.
12. Robert Keim, *Introduction to the I2S Interface*, All About Circuits, March 4, 2020, allaboutcircuits.com/technical-articles.
13. Jenny List, *All You Need to Know About I2S*, Hackaday, April 18, 2019, hackaday.com/2019/04/18/all-you-need-to-know-about-i2s/.
14. Walt Kester, *DAC Interface Fundamentals*, Analog Devices MT-019 Tutorial.
15. Cirrus Logic, *CS8416 192 kHz Digital Audio Interface Receiver*, Datasheet.
16. Jonathan Schwartz, *CS8416 Delivers Performance Gains Over CS8413/14*, Cirrus Logic AN339, 2009, cirrus.com.
17. Randy McAnally, *S/PDIF Digital to Analogue Converter*, Elliott Sound Products, Project 85, http://sound.whsites.net/project85.htm.
18. AKM, *AK4115 High Feature 192 kHz 24-Bit Digital Audio Interface Transceiver*, Datasheet.
19. Texas Instruments, *DIR9001 96-kHz, 24-Bit Digital Audio Interface Receiver*, Datasheet.
20. Electronics DIY, *24-bit 192 kHz PCM1793 DAC with DIR9001 Receiver and OPA2134 Op Amp*, http://electronics-diy.com/electronic_schematic.php?id=806.
21. Electronics DIY, *DIR9001 SPDIF Decoder*, electronics-diy.com/electronic_schematic.php?id=830.
22. Cirrus Logic, *CS8416 192 kHz Digital Audio Interface Receiver*, Datasheet.
23. Burr-Brown/TI, *PCM1793 24-Bit, 192 kHz Sampling, Advanced Segment, Audio Stereo DAC*, Datasheet, January 2004.
24. Burr-Brown/TI, *PCM1794A 24-Bit, 192-kHz Sampling, Advanced Segment, Audio Stereo Digital-to-Analog Converter*, Datasheet, December 2015.
25. Burr-Brown/TI, *PCM1796 24-Bit, 192-kHz Sampling, Advanced Segment, Audio Stereo Digital-to-Analog Converter*, Datasheet, November 2006.
26. Asahi Kasei (AKM Semiconductor), *AK4396 Advanced Multi-Bit 192 kHz 24-Bit ΣΔ DAC*, Datasheet.
27. Martin Mallinson and Dustin Forman, *Technical Details of the Sabre Audio DAC*, ESS Technology document, December 2007.
28. ESS Application Note, *Maximizing DAC Performance for Every Budget*, ESS Technology, esstech.com.
29. Michael Steffes, *Design for a Wideband, Differential Transimpedance DAC Output*, Texas Instruments Application Report SBAA150A, October 2016.
30. Wayne Liu, *HiFi Audio Circuit Design*, Texas Instruments Application Report SBOA237, August 2017.
31. R. Cordell, J. Forney, W. Dunn and W. Garrett, *A 50-MHz Phase and Frequency-Locked Loop*, 1979 IEEE International Solid-State Circuits Conference. Digest of Technical Papers, 1979, pp. 234–235, doi: 10.1109/ISSCC.1979.1155935.
32. Jules A. Bellisio, *New Phase-Locked Timing Recovery Method for Digital Regenerators*, International Conference on Communications Record, Philadelphia, June 1976, pp. 10–17.
33. Jules A. Bellisio, *Timing Recovery Circuit for Digital Data*, U.S. Patent 4,015,083, March 29, 1977.
34. Robert R. Cordell, *Phase and Frequency Detector Circuits*, U.S. Patent 4,773,085, September 20, 1988.
35. David G. Messerschmitt, *Frequency Detectors for PLL Acquisition in Timing and Carrier Recovery*, IEEE Transactions on Communications, vol. COM-27, no. 9, September 1979.
36. Burr-Brown/TI, *PLL1707 3.3-V Dual PLL Multiclock Generator*, Datasheet.
37. Wikipedia, *Differential Manchester Coding*. wikipedia.org.
38. Vectron VVC4 Voltage Controlled Crystal Oscillator, Datasheet, vectron.com/products/vcxo/VVC4.pdf.
39. Analog Devices AD1896, *192 kHz Stereo Asynchronous Sample Rate Converter*, Datasheet, analog.com.
40. Robert Adams and Tom Kwan, A Stereo Asynchronous Digital Sample-Rate Converter for Digital Audio, *IEEE Journal of Solid State Circuits*, vol. 29, no. 4, April 1994, pp. 481–488.
41. Cirrus Logic, *CS8421 32-Bit, 192 kHz Asynchronous Stereo Sample-Rate Converter*, Datasheet.
42. Texas Instruments, *SRC4392 Two-Channel Asynchronous Sample Rate Converter with Integrated Digital Audio Interface Receiver and Transmitter*, Datasheet.

43. John Siau, *Asynchronous Upsampling to 110 kHz*, Benchmark Media, Application Notes, July 1, 2010, benchmarkmedia.com.
44. John Siau, *Inside the DAC2 – Part 2 – Digital Processing*, Benchmark Media, June 2016, benchmarkmedia.com.
45. Jeffrey Hunt, *AN3387 USB-to-I2S Bridging with Microchip USB249xx Hubs*, Application Note, Microchip Technology Inc.
46. Silicon Labs, *CP2114 USB Audio to I2S Digital Audio Bridge*, Datasheet.
47. Stefano Di Gennaro, *Zero-Order Hold – Fundamentals of Linear Systems*, Ics-VC-marcy.syr.edu.
48. Wikipedia, *sinc function*, wikipedia.org.
49. Walt Kester, *Basic DAC Architectures 1: String DACs and Thermometer (Fully Decoded) DACs*, MT-014, Analog Devices, 2009.
50. Audiophile Inventory, Non-Oversampling DAC – Advantages, Disadvantages, samplerateconverter.com/educational/nos-dac.
51. Philips TDA1540 datasheet.
52. Philips TDA1541 datasheet.
53. Walt Kester, *Basic DAC Architectures III: Segmented DACs*, MT-016, Analog Devices, 2009.
54. Walter Kester, *ADC Architectures III: Sigma-Delta ADC Basics*, Analog Devices MT-022 Tutorial.
55. Walt Kester, *ADC Architectures IV: Sigma-Delta ADC Advanced Concepts and Applications*, Tutorial MT-023, Analog Devices.
56. Robert W. Adams, Design and Implementation of an Audio 18-Bit Analog-to-Digital Converter Using Oversampling Techniques, *Journal of the Audio Engineering Society*, vol. 34, March 1986, pp. 153–166.
57. Walt Kester and James Bryant, *DACs for DSP, Part 2: Interpolating and Sigma-Delta DACs*, EE Times, January 17, 2008.
58. Analog Devices, *AD1955 High Performance Multibit Σ-Δ DAC with SACD Playback*, analog.com.
59. Analog Devices, *Sigma-Delta ADCs and DACs*, Application Note AN-283.
60. ExostivLabs, *Introduction to I²C and SPI Protocols*, exostivlabs.com/introduction-to-i2c-and-spi-protocols/.
61. *I²C Bus*, i2c-bus.org.
62. Piyu Dhaker, *Introduction to SPI Interface*, Analog Devices, Analog Dialog 52–09, September 2018, analogdialog.com.

Chapter 24

Active Crossovers and Loudspeaker Equalization

The great majority of consumer loudspeaker systems employ passive crossovers to direct signals in different frequency bands to the appropriate drivers, like woofers (LF) and tweeters (HF) and often mid-range (MF) drivers as well. Passive crossovers have limitations, including large passive components like capacitors and inductors, all potential contributors to distortion and frequency response anomalies. At lower frequencies, larger inductance values are needed in combination with low resistance, often leading to the need for ferrite or iron cores. These can become nonlinear if the current through them pushes them toward saturation.

Similarly, at low frequencies large-value capacitors are needed, leading often to the use of non-polar electrolytic capacitors. Even in line-level circuits their use is often frowned upon. It is worse in crossovers, since they are performing a frequency-shaping function, implying audio signal voltage across them. Driver-to-driver differences in impedance characteristics and sensitivity can lead to frequency response errors in the crossovers as well. Good passive crossovers are difficult to design and can be fairly expensive if high-quality components are used.

Active crossovers solve many of the problems faced by passive crossovers. All of the frequency shaping is done at line level by circuits that employ small components that can be very linear, and rarely depend on inductors at all. Moreover, there is a separate power amplifier for each driver frequency band. This means that the speaker is driven by a very low impedance-voltage source, and its response is more predictable and under better control. There is no interaction between the driver and the crossover. There is a large amount of instructive material out there that goes into greater depth on many of the topics discussed here [1–4].

Active crossovers are not without their challenges, and sources of added expense. First, and most obviously, more power amplifiers are needed. Moreover, the system is no longer "plug and play." Even though the crossover must intimately match the frequency response needs of the drivers, it is a separate component separated from the drivers by amplifiers. In other words, the active crossover must essentially be a package with the speaker, and likely the amplifiers as well. This is a completely different paradigm than the normal one where one can go out and buy a different pair of speakers and use them with one's existing power amplifier with little worry.

Even though higher performance can be achieved with an active approach, active crossover systems are not common in the traditional consumer environment. However, active speakers have taken over a large portion of the professional audio market. The key is that the active crossover, amplifiers and speaker drivers are together as a unit in a single box. This allows the necessary optimization of the whole system.

There is a major exception in the consumer environment. Most people own an active loudspeaker with an active crossover and amplifier built into it. It is their subwoofer, be it in their home theater system or their audiophile system. There is also increasing use of active loudspeaker systems in additional consumer applications, such as sound bars and surround speakers, some of the latter of which are conveniently wireless. In any consumer application where the device includes both the power amplification and the loudspeaker drivers, there is a likelihood that active crossovers will be used. This is especially the case since modern implementations of such devices employ compact and economical class D power amplifiers. This chapter is not about loudspeakers, and yet it is impossible to discuss loudspeaker crossovers without discussing some aspects of loudspeaker design.

24.1 Subwoofer Crossovers

The subwoofer crossover has long been the most common active crossover in the consumer environment because most subwoofers include the power amplifier (the so-called *plate* amplifier) as part of the subwoofer. These crossovers usually include a low-pass filter (LPF), a crossover frequency control, a level control and a phase switch or control. They are inevitably integrated as part of the plate amp. In some cases, these subwoofer crossovers may also include some equalization for the subwoofer. Surround-sound receivers also often include some crossover means associated with the subwoofer output low-frequency effect (LFE), sometimes called bass management.

Crossover Frequency and Slope

In many stereo systems to which a subwoofer has been added, the acoustic output of the subwoofer is just added to that of the main speakers. In other words, there is no corresponding high-pass filter (HPF) added in the signal path to the main speakers. The left and right signals are summed, filtered and amplified in the plate amplifier. The subwoofer just augments the low-frequency output of the main speakers. The subwoofer crossover frequency will thus often be set in the vicinity of the frequency where the main speakers are losing their LF effectiveness. If the subwoofer cutoff frequency is too high, the sound may be too boomy. The crossover frequency can usually be set somewhere between about 40 and 120 Hz [5–8]. The crossover slope is often fixed at 12 or 18 dB/octave. A simple subwoofer crossover is shown in Figure 24.1.

This subwoofer operates in the augmentation mode where the left and right speaker signals are passed through to the left and right speakers and the summed signals are picked off, attenuated and added to form a monophonic signal that is then low-pass-filtered. The low-pass filter (LPF) 3-dB frequency can be adjusted and the filter is usually of order 2, 3 or 4. Less expensive subwoofers are often of order 2, since only a two-gang frequency-adjust pot is needed. Alternatively, line-level L/R signals are summed at the virtual ground. The summer U1 inverts, and the necessary inversion in the path to compensate is not shown. The monophonic bass signal can also be taken from the LFE output of a receiver. Variable and/or fixed 180° phase controls follow, with variable phase usually implemented with a simple first-order all-pass filter whose turnover frequency is varied with a pot. An optional 20-Hz infrasonic filter may be included to mitigate cone flutter.

Phase Control

The phase control is needed because the subwoofer is often several feet away from the main speakers, so phase delay may cause cancellation and actually weaken the bass in some frequency regions. Room effects can also influence how the low-frequency outputs of the main speakers and subwoofers add. Often, the phase control is just a switch that inverts the signal. In other cases, it is a variable phase control that sets the cutoff frequency of an all-pass filter. It is useful to bear in mind that one-fourth wavelength at 100 Hz corresponds to 2.5 feet.

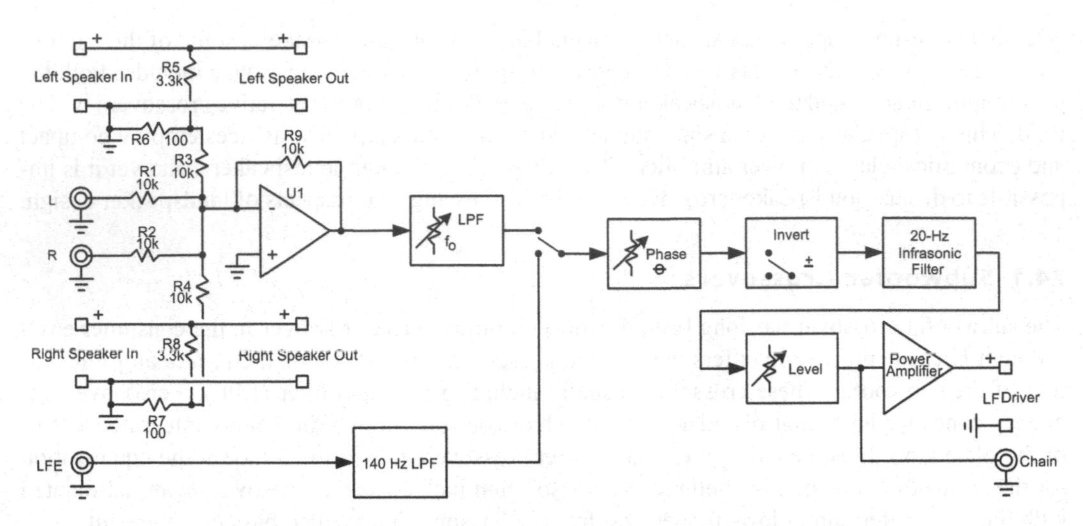

Figure 24.1 Typical Subwoofer Crossover.

24.2 Advantages of Active Crossovers

In addition to the advantages of active crossovers described in the introduction in broad-brush terms, there are additional advantages as well. It is important to distinguish between active crossover systems assembled from separate components and what we call active loudspeakers. The latter generally combine all of these functions into one unit, and are optimized together as a system. This is the most practical way to incorporate active crossovers into a system.

Most active crossovers are associated with two-way or three-way loudspeakers, and the use of two or three amplifiers is commonly referred to as a bi-amp or tri-amp system. Having separate amplifiers for each of two or three frequency bands has the immediate advantage of reduced opportunity for intermodulation distortion. The more fragile high frequencies are not being amplified in the presence of low frequencies of much larger amplitude that are constantly changing the operating points within the amplifier. Low-frequency drivers often have lower efficiency and require more power, so it is also the case that amplifiers of different power capability may be employed for different frequency bands to better match the needs of the drivers in that band.

Active Crossover Example

The advantages of active crossovers are best discussed in the context of an example active crossover, as shown in Figure 24.2, where a generic three-way crossover is shown. It consists of two 2-way crossovers, the first splitting the bass and higher frequencies at 750 Hz. The second crossover splits the higher frequencies into mid-range and tweeter signals at 3000 Hz. Each crossover shown is a second-order crossover with 12 dB/octave slopes. Each two-way crossover consists of an LPF and a HPF implemented with second-order Sallen-Key filters with $Q=0.5$. Notice that below 750 Hz the tweeter actually sees a 24 dB/octave high-pass slope, further reducing its exposure to low-frequency signal excitation.

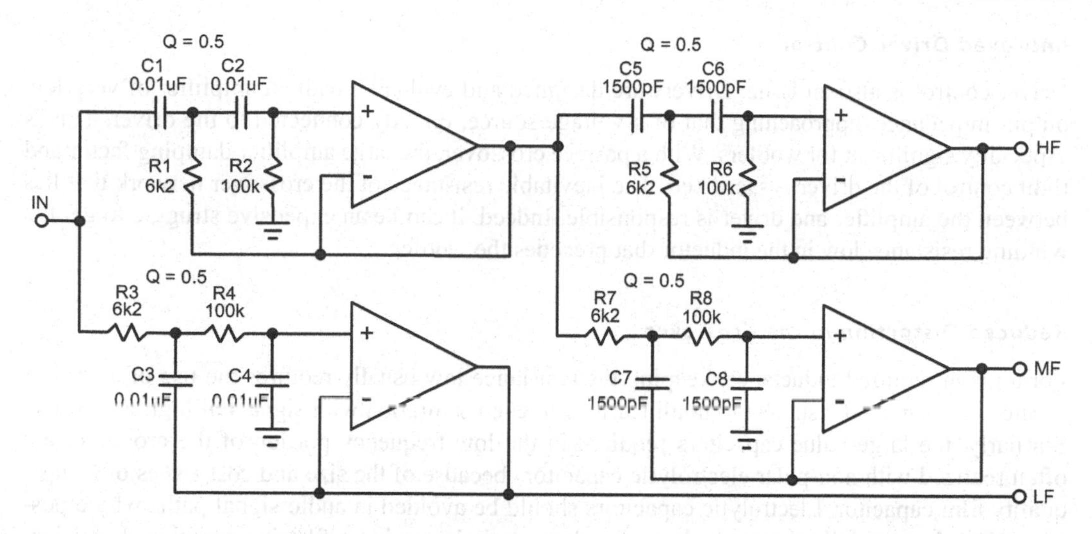

Figure 24.2 Three-Way Active Crossover Example.

Two-way systems are simple and widely used, employing just one of the two-way crossovers shown above, often with a crossover frequency in the vicinity of 2kHz. Two-way systems fall short of high performance in several ways, but are the basis for designing and understanding most loudspeaker crossovers. For this reason, most of the crossover discussions will be in the context of a two-way crossover.

In pro audio, many of the best near-field studio monitors are two-way powered loudspeaker systems with active crossovers and built-in amplifiers [9].

Reduced Dependence on Amplifier Performance

In a conventional passive crossover system, all of the audio frequencies must pass through a single amplifier, making the overall sound more influenced by intermodulation distortion or clipping in the amplifier. In a three-way system, the mid and high frequencies are riding on top of the strong low frequencies, for example. In the worst case, if the amplifier clips on a strong bass note, the harsh edges of the clipped waveform will be reproduced by the mid-range and tweeter drivers. Such clipping can even damage delicate tweeters.

Amplifiers Tailored to the Drivers

The type of amplifier for each frequency band might also be different. The simplest example here would be the use of a powerful class D amplifier for the woofer and a class AB (or class A, or even vacuum tube) amplifier for the mid or upper frequencies. In an active crossover system, each amplifier is connected to a known load. The single-amplifier conventional passive system must cope with the large back EMF (electromotive force) from the woofer, load phase angle, crossover impedance variations and safe operating area issues posed by the loudspeaker. The mid- and high-frequency amplifiers in a system with active crossovers will not need to cope nearly as much with such difficult loads, and may in fact need less invasive protection circuits.

Improved Driver Control

Driver control is also an issue. Drivers are designed and evaluated with an amplifier of very low output impedance, approaching that of a voltage source, directly connected to the driver. This is especially significant for woofers. With a passive crossover, the large amplifier damping factor and tight control of the driver is sacrificed. The inevitable resistance of the crossover network that lies between the amplifier and driver is responsible. Indeed, it can be an expensive struggle to get the winding resistance low in the inductor that precedes the woofer.

Reduced Distortion in the Crossover

For a given required inductance, keeping the resistance low usually requires the use of a steel or ferrite core that can be subject to nonlinearity and even saturation with signals of high amplitude. Similarly, the large-value capacitors required in the low-frequency portion of the crossover are often realized with non-polar electrolytic capacitors because of the size and cost issues of using a quality film capacitor. Electrolytic capacitors should be avoided in audio signal paths where possible. This is especially the case where the electrolytic is serving a filtering function. Precision, low-distortion passive crossovers are more expensive than their active equivalents.

Reduced Impedance Interactions

The frequency-dependent impedance of the drivers interacts with the non-zero frequency-dependent output impedance of the crossover, making crossover design more difficult and complex. For example, sometimes a Zobel network is connected across a tweeter to compensate for its rising impedance with frequency that results from its voice coil inductance. For amplitude matching, resistive pads often must be placed in front of tweeters and mid-range drivers because of their higher efficiency. These obviously introduce series resistance that may allow additional impedance interactions. With pads, precious amplifier power is wasted. The bottom line here is that there is a big advantage in an active crossover system as a result of directly connecting each driver to an amplifier. The active crossover will introduce virtually no distortion and there will be no unwanted interaction between crossover and driver.

Crossover Flexibility, Precision and Channel Matching

The active crossover affords more flexibility in choosing various crossover characteristics and adapting them to the driver being driven. Indeed, when implemented actively, it is much less expensive to implement a crossover of higher order or one that includes some driver equalization. Importantly, it becomes very easy to equalize driver acoustic output in spite of widely varying driver efficiencies.

Active Circuit Phase Correction

The phase relationships among drivers must be correct for correct acoustic addition to occur. The most obvious case is driver offset due to different driver effective radiation depths. Getting this right at and near the crossover frequency is crucial to obtaining the performance achievable with the crossover. The best way to introduce driver phase correction is with active all-pass filter delays, and this is enabled by the use of active crossovers [10].

Improved Imaging

Active crossovers can be implemented with much closer frequency and amplitude tolerances. Identical frequency and phase response in the left and right channels of a system can be important for good and stable imaging. It is especially important that the left and right system responses be maintained the same in the critical crossover regions, where amplitude and phase can be changing quickly with frequency. The active crossover affords the opportunity to make the left and right loudspeaker responses almost identical with far less need to hand-match drivers.

System Voicing and Driver Equalization

More precise voicing of the loudspeaker system is also made easier by the use of active crossovers, where driver levels, crossover frequencies and other parameters are easily controlled. Driver frequency response anomalies, such as resonance are more easily dealt with, given the flexibility afforded by active crossovers. Active filters that perform such corrections are economically implemented. Similarly, phase correction can be implemented if needed. This is especially important for non-coincident drivers. With sound traveling at about 1 foot per millisecond, 3 inches of non-coincidence between a woofer and a mid-range will correspond to about 250 μs. This represents 90° of phase shift at 1 kHz. This can seriously alter the way acoustic addition of the outputs of the two drivers will take place, causing a suck-out or a bump if the crossover was designed assuming the sound from each of the drivers would radiate in the same vertical plane.

Isolation of Clipping Events

Amplifiers clip more than we realize when well-recorded music with high crest factors is being reproduced. With separate amplifiers for each of the two or three frequency bands, clipping events are isolated to the driver whose amplifier is clipping. Tweeters will not be exposed to the sharp clipping edges that can sometimes damage them, especially in systems where amplifiers are under-powered. Tweeters will also not reproduce those awful clipping sounds in a system with active crossovers.

Clipping will also likely occur less often. Consider the loud transient thwack of a snare drum riding on top of a strong bass note. The woofer and the tweeter may each require 100 W for the same brief interval (especially in a conventional system where the tweeter efficiency has been padded down to match that of the woofer). Amplifiers are often voltage-limited. Each of the woofer and tweeter waveforms will require 40 V_{pk}, adding to 80 V_{pk} if the phase and time of the peaks are the same. This corresponds to 400 W into an 8-Ω load, *not* just 200 W. The snare drum thwack will be mercilessly clipped should it occur at the wrong time.

Active Baffle Step Compensation

Diffraction effects come into play at low frequencies when the width of the loudspeaker baffle is on the order of the wavelength of the sound [11–14].

If the width of the baffle is large compared to the wavelength, the loudspeaker tends to emit sound in a hemispherical way (half-space). This is the case of over most of the frequency range for most baffle sizes. However, if the baffle width is smaller compared to the wavelength, the speaker emits sound in a more omnidirectional way (full-space). This means that at low frequencies there is relatively less SPL (sound pressure level) radiated in a way that is audible at the listening position.

At frequencies between the two extremes, the effective amplitude increases gradually from one step level to a higher step level as frequency increases. This is called the *baffle step*.

The difference in dB between the two steps can theoretically be up to 6 dB, but is more often in the 1–4 dB range. This variation is in part due to the wall behind the loudspeaker redirecting some of the diffracted energy forward. The baffle step can be easily compensated in an active crossover with a 6-dB/octave low-frequency step. The mid-point frequency of the step will usually be in the vicinity of 200–400 Hz, depending on the width of the baffle. Active baffle step compensation permits the size of the compensation step to be adjusted, and sometimes the frequency. There are some loudspeakers that incorporate an approximation to baffle step compensation in their passive crossover. Others add an additional woofer that cuts off at a lower frequency than the main woofer, thus augmenting the low-frequency response.

Bass Response Extension

The bass response of a closed-box loudspeaker can often be extended (equalized) to a lower frequency. Such an equalizer puts in a theoretical 12 dB/octave step in the frequency range from approximately the minus 3-dB point of the loudspeaker down to the desired new low-frequency 3-dB point. This equalizer is known as the Linkwitz transform and will be discussed in detail in a later section of this chapter [15–16]. It is easily added to the low-frequency section of an active crossover.

Absence of Speaker Cables

Most active crossovers live inside the speaker cabinet along with the power amplifiers. This is called an active loudspeaker, often called a self-powered loudspeaker. Those who are into the nuances of different-sounding speaker cables will surely be disappointed with an active loudspeaker system, since there is no speaker cable. The amplifier is usually connected directly to the speaker driver. The best speaker cable is no speaker cable.

24.3 Disadvantages of Active Crossovers

Conventional loudspeakers with passive crossovers are designed as a system, with an intimate relationship between the passive crossover and the drivers. The same degree of system integration and customization is generally not possible with general-purpose active crossovers that are not part of the physical loudspeaker. Although external active crossovers exist, it is difficult to properly fit their crossover characteristics to a loudspeaker system with individual drivers. The loudspeaker and active crossover must be designed as a system, with the same amount of care, measurement and listening to achieve the desired performance as a whole. This involves a more multi-disciplinary approach that involves loudspeaker design, amplifier design and active filter design.

This encourages one to integrate the amplifiers and active crossovers into the same enclosure as the loudspeakers. Unless class D amplifiers are used, getting the heat out can be a challenge. While class D amplifiers are usually suitable for driving the woofers, many might still argue for the use of traditional class AB amplifiers for the mid-range and tweeters.

Obviously, an additional disadvantage of active crossovers is the need for more hardware, like the active crossover and the greater number of power amplifiers. These are difficult obstacles to the use of active crossovers in the audiophile consumer market. It is natural for an audiophile to want to be able to put together a system by picking loudspeakers and amplifiers separately.

24.4 Self-Powered Loudspeakers

Self-powered loudspeakers include the active crossover, power amplifiers and loudspeaker drivers all in the loudspeaker enclosure. This allows for tight system integration and high performance [9, 14]. Subwoofers are the most prolific examples of self-powered loudspeakers. Many self-powered loudspeakers automatically turn themselves on when a signal is detected.

Self-powered loudspeakers are widely used in professional audio applications. The powered loudspeaker is fed a line-level signal, often from an XLR cable. One of the first professional powered loudspeakers was the Meyer Sound HD-1 near-field monitor, introduced in 1989. It is a two-way system with two class AB power amplifiers [9]. Today, virtually all near-field monitors are self-powered, and a great many other pro-audio loudspeakers are as well, from stage wedges to large theater speaker systems.

One of the challenges in early self-powered loudspeakers was getting the heat out from the class AB amplifiers that were typically used [3]. With advances in class D amplification, many employ that technology for the power amplifiers and the heat issue is greatly diminished. Hybrid designs using class AB amplifiers for mid-range and tweeters can also be implemented. The typically higher efficiency of mid-range and tweeter drivers may mitigate the resulting heat issue somewhat by allowing the use of somewhat smaller class AB amplifiers.

24.5 Types of Crossovers, Roll-Off Slopes and Phase Relationships

The primary job of the crossover is to ensure that each driver handles only the part of the frequency spectrum that it is capable of reproducing faithfully, without frequency response anomalies, distortion and cone breakup. Here we cover the meat of active crossover design. Much of it mirrors the design issues faced in passive crossover design, but instead are approached with active circuits usually based on active filters. Important issues include crossover slopes and phase characteristics that affect how the acoustic output of two or more drivers add.

In fact, the role of phase delay and acoustic time delay differences between drivers is a prevailing issue in crossover design [2]. Most crossover design issues can be covered thoroughly in the context of two-way systems. Much has to do with the crossover slope and phase relationships between the two drivers at the crossover frequency f_o. How the acoustic addition of the two drivers results in a flat response is of paramount importance. Improper acoustic addition can result in a bump or a dip at f_o. The nature of the acoustic addition also influences the directivity of the summed response at f_o as a result of differences in time delay from the tweeter and woofer to the listener at and near f_o.

One reality is that a woofer and a tweeter mounted in the same vertical plane on a baffle have differing acoustic time delay, largely due in part to the fact that the voice coil of the woofer is recessed behind the plane of the baffle, while such recess for a tweeter is minimal. The points of acoustic radiation are not coincident, so this arrangement is referred to as one of non-coincident drivers.

First-Order Crossover

The first-order crossover is the simplest. It comprises a first-order LPF in the woofer path and a first-order HPF in the tweeter path. Each filter is down 3 dB at the crossover frequency f_c and have 45° of lagging and leading phase shift at f_c, respectively. The driver signals are 90° apart (in quadrature) and, absent other sources of phase shift, add to 0 dB in both voltage and power, which is nice. The result is a minimum-phase crossover, with no phase or group delay distortion. If the woofer and tweeter outputs of the crossover are added, the result is a unity transfer function with no phase shift.

However, the 6-dB/octave roll-off slopes are usually too shallow for the drivers, especially in a two-way system. Such shallow slopes force them to perform beyond their capable frequency range. This is especially a problem for the tweeter. An octave away from f_c the drive signal is only down 5 dB. A three-way system with small woofers and higher crossover frequencies mitigates this a bit. The above arrangement works only for coincident drivers with no other sources of phase shift. Additional sources of phase shift will disturb the quadrature phase relationship and cause the signal at f_c to be magnified or attenuated.

Time Alignment, Non-Coincident Drivers and Lobing Error

Loudspeaker crossover analysis and design almost always *starts* with the well-intentioned but over-simplified assumption that both drivers emit as a point source sharing exactly the same location, which is impossible. Two drivers are inevitably non-coincident. If for any reason there is a difference in distance to the listener between one driver and the other, there will be a time delay difference, resulting in an alteration of phase relationships at the crossover frequency. This changes their acoustical addition. This is easy to see and understand even with something as simple as a first-order crossover.

As shown in Figure 24.3, two drivers mounted on the same flat baffle will have different points of acoustic radiation if the drivers have a different recess from the baffle to the voice coil. This is usually the case with a woofer and a tweeter. The woofer acoustic center is recessed. The difference in time delay corresponds to a phase difference at f_c that interferes with proper acoustic addition. Acoustically, the two drivers are thus non-coincident and the acoustic centers of the drivers are not time-aligned. This issue must be taken into account in crossover design. Sometimes this is addressed by using a sloped baffle. With the tweeter on top, a sloped-back baffle can correct some of the delay difference. In other cases, some of the phase difference can be made up at f_c by tweaking the crossover frequencies or slopes. In other cases, electronic delay can be added to the tweeter to correct the phase difference.

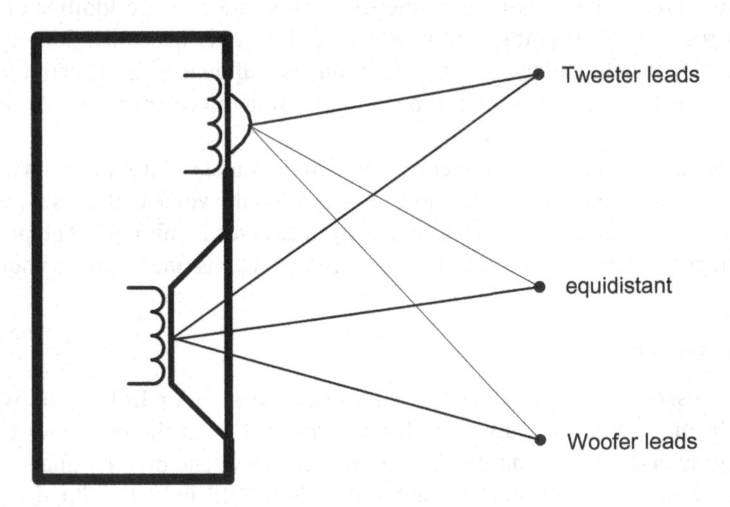

Figure 24.3 Woofer and Tweeter(s) Mounted on a Vertical Baffle.

It takes sound about 76 μs to travel 1 inch. If the woofer's point of radiation is recessed by 2 inches, its signal will be delayed by 152 μs. The wavelength of a 2 kHz acoustic signal is about 6.6 inches. The 2-inch woofer recess will correspond to 109° at 2 kHz. This is a huge phase error and is made so large because a two-way loudspeaker must typically have $f_c = 2$ kHz.

Vertical Lobing Error

There is another form of non-coincidence that influences the acoustic addition of the drivers at locations above and below the horizontal axis. This is due to the vertical distance between the drivers. It is easy to see that above the loudspeaker the listening distance to the woofer becomes greater than that to the tweeter, ultimately by the distance between the radiation centers of the ideal drivers. Similarly, below the horizontal axis the distance from the tweeter to the listener becomes greater, causing phase shift in the opposite direction. Consider an odd-order crossover where the time-aligned drivers are driven 90° out of phase at f_c when listening on the horizontal axis.

For a first-order crossover, the woofer has a 45° lag and the tweeter has a 45° lead at f_c. The tweeter signal leads the woofer signal by 90°. At some point above the horizontal axis, the woofer will incur additional delay of 90° compared to the tweeter signal due to the added geometric distance to the woofer as compared to the tweeter. The drivers will then be 180° out of phase, and there will be complete cancellation (destructive interference). At some point below the horizontal axis, the tweeter will incur an additional 90° of phase lag relative to the woofer, causing the drivers to be in-phase at f_c, resulting in a 3-dB bump from constructive interference. The acoustic radiation thus has a lobe where its maximum amplitude is tilted downward. Mounting the tweeter below the woofer will reverse the direction of the lobe, causing its maximum to be above the center axis.

Consider an 8″ woofer mounted as close as possible to a small 1″ dome tweeter, for a distance of about 5.5 inches between the acoustic centers. Assume for the moment that there is no woofer recess. At the extreme position vertically on the baffle, the time delay will correspond to 5.5″, or about 417 μs, almost equal to the period of a 2 kHz signal, corresponding to about 300° of phase shift. With the crossover creating a 90° electrical difference between the drivers, it only takes an additional 90° of error to create a null or a 6-dB bump. Of course this is an extreme case, but it points out how being off axis in the vertical direction can create significant lobing error.

MTM Driver Arrangement

A popular driver arrangement has two mid-range drivers situated above and below the tweeter, thus the term MTM. This arrangement, also referred to as the D'Appolito design, reduces lobing error through the symmetry of the driver arrangement [17, 18]. It is advantageous to have the drivers as close together as possible. However, there will still be lobes situated symmetrically about the center of the tweeter. With the MTM arrangement, a single woofer will usually be mounted below the MTM arrangement. In a two-way system, there will be two woofers situated above and below the tweeter, and the larger size of the woofers will force the acoustic centers further apart, exacerbating the lobing problem.

24.6 Conventional Crossovers

Drivers misbehave in many ways when they are called upon to deliver signals beyond their capability. This is one of the main reasons to have a crossover. If the crossover overlap region where both drivers must perform is larger, the drivers will be called upon to function at frequencies further from the crossover frequency. Steeper crossover slopes are thus desirable.

Second-Order Butterworth Crossovers

The second-order Butterworth high-pass and low-pass filters each provide a 12-dB/octave roll-off. Each filter is down 3 dB at f_c and leading or lagging the input signal by 90° at f_c. The acoustic addition of the signals that are out of phase by 180° will result in a notch. If the signal to the tweeter is inverted, the acoustic signals will be in-phase at f_c and their sum will be up 3 dB, resulting in a bump. Douglas Self has shown that separating the low-pass and high-pass filter frequencies by a factor of 1.3 can significantly flatten the bump [4]. Because the acoustic signals are in-phase at f_c, the lobe will be horizontal and there will be less lobing error. At frequencies well above f_c, the tweeter signal will be 180° out of phase with the input signal because it was inverted. This results in a non-minimum-phase crossover that has phase distortion. But for the 3-dB bump, it would constitute an all-pass filter. Waveform preservation is not achieved in such crossovers. The group delay of the signal changes in the vicinity of the crossover frequency. The audibility of such phase distortion remains a controversy.

Third-Order Butterworth Crossovers

The popular third-order Butterworth filter has an 18-dB/octave roll-off that is down 3 dB at f_c and has a phase shift of 135°. This means that the HF and LF crossover filters have outputs that are 270° out of phase. This quadrature relationship means that the woofer and tweeter outputs will add on a power basis, leading to a flat response on the horizontal axis when acoustically summed. However, the signal to the tweeter should be inverted. This ultimately means that the combined response will be that of a first-order 180° all-pass function. Like the odd-order first-order crossover with quadrature addition, it will exhibit lobing error, with a 3-dB peaking axis and a cancellation axis. Without inversion of the tweeter signal, the summed response is still flat, but the summed response becomes that of a 360° second-order all-pass filter [4].

Fourth-Order Butterworth Crossovers

The fourth-order low-pass Butterworth filter has a 24-dB/octave roll-off that is down 3 dB at f_c and has 180° of phase lag at f_c. The HPF has 180° of phase lead at f_c. The LF and HF signals are thus 360° apart and in phase. The two signals, each with 0.707 amplitude, add in voltage to produce amplitude of 1.414, resulting in a +3 dB bump at f_c. Because the drivers are driven in-phase, and if they are time-aligned, there will be minimal lobing error. Douglas Self has shown that separating the LPF and HPF frequencies by a factor of 1.128 replaces the 3-dB bump with a maximally flat response with ripple of ±0.47 dB [4].

24.7 Linkwitz-Riley Crossovers

As we have seen, it is difficult enough to implement a crossover that provides a smooth, flat response on the horizontal axis. Yet that is only part of the story. Add to that the acoustic addition anomalies in the vertical direction that create vertical "lobing error" [2]. The Linkwitz-Riley (LR) crossovers mitigate both of those problems [19–22].

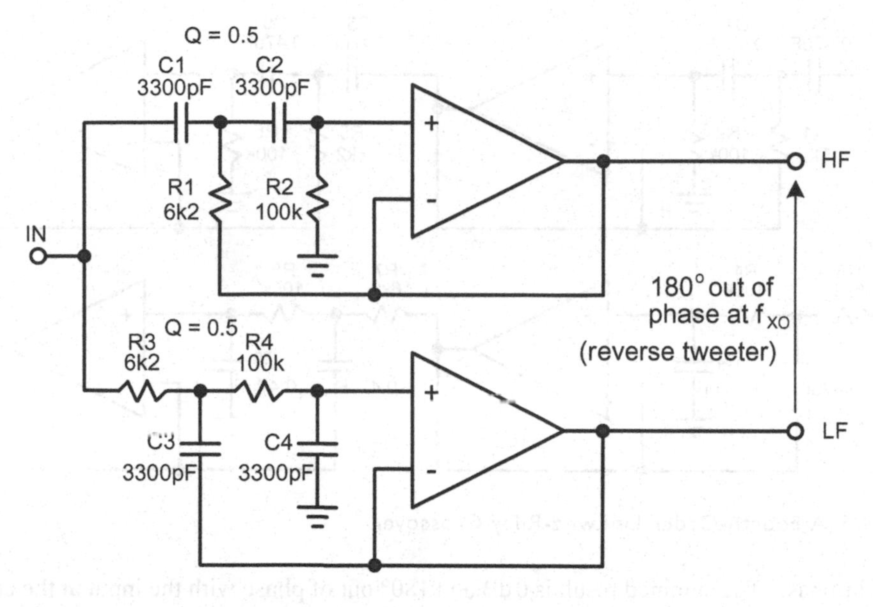

Figure 24.4 A Second-Order Linkwitz-Riley Crossover.

Second-Order Linkwitz-Riley Crossover

As shown in Figure 24.4, the 12 dB/octave high-pass and low-pass sections of the second-order Linkwitz-Riley (LR-2) crossover are simply over-damped second-order Sallen-Key filters, each with Q=0.5. The woofer is fed with a 90° lag and the tweeter is fed with a 90° lead, so they are 180° out of phase at f_c. Their outputs are down 6 dB at f_c. This allows the acoustic addition to take place correctly, adding to unity, if the polarity of the tweeter is reversed so that the drivers are in phase at f_c. If instead it is not desired to physically reverse the tweeter polarity, an inverter can be put in its signal path. The overall response is thus that of a 180° all-pass filter where the tweeter is 180° out of phase at high frequencies. The LR crossover depends on the drivers being time-aligned for proper operation.

A key advantage insofar as lobing error is concerned is that the LR crossovers have the drivers operating in phase at all frequencies (in theory) rather than in quadrature. This reduces lobing error because distance errors between the drivers must be significantly larger to create enough phase shift to introduce a null or a 6-dB bump.

The LR-2 response being that of a 180° all-pass filter means that it does not preserve waveform and introduces phase distortion. This has been studied, and there is not universal agreement on its audibility [23]. However, it is fair to say that it is not noticeable to most people and that its advantages well outweigh the matter of its phase distortion. Most other crossovers create phase distortion anyway. There appears to be no perfect crossover.

Fourth-Order Linkwitz-Riley Crossover

As shown in Figure 24.5, a fourth-order LR crossover is formed if two second-order Butterworth filters are put in series for the high-pass and low-pass sections of the crossover. Each section is down 6 dB at f_0. With four poles, each section has 180° of phase shift, and the acoustic summing

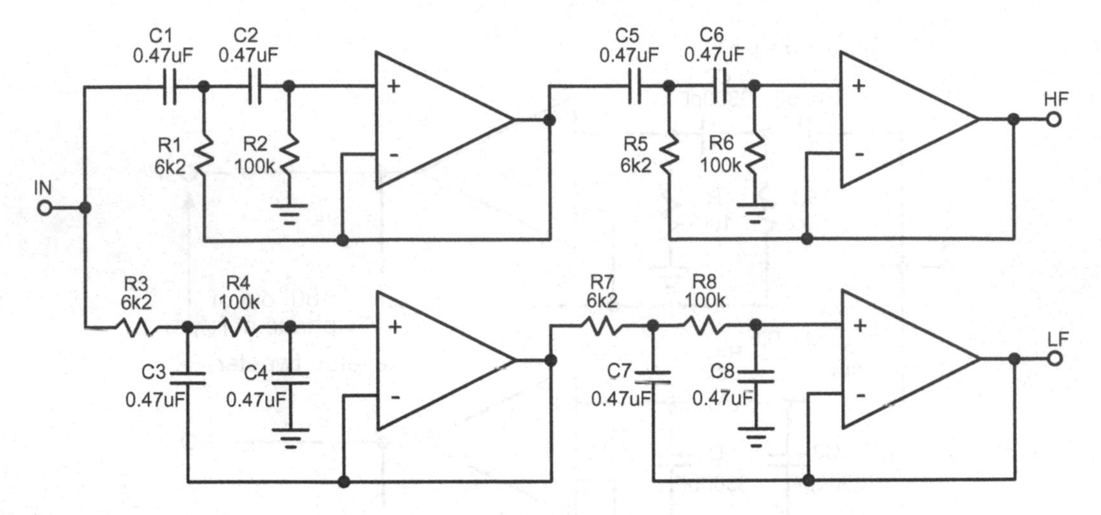

Figure 24.5 A Fourth-Order Linkwitz-Riley Crossover.

result is in-phase. The summed result is 0 dB and 180° out of phase with the input at the crossover frequency. The phase relationship continues to twist as frequency is increased, with the phase of the summed output approaching 360° at high frequencies. This a second-order all-pass filter. The fourth-order Linkwitz-Riley (LR-4) is probably the most popular crossover filter. It forms a 360° all-pass function and provides a healthy 24-dB/octave crossover slope.

24.8 Three-Way Crossover Architectures

There are two popular crossover architectures for three-way loudspeaker systems. One arrangement was shown in Figure 24.2. It consists of two 2-way crossovers in tandem. As shown in Figure 24.6(a), the source feeds the 500-Hz crossover whose LF output feeds the woofer. The HF output of the 500-Hz crossover feeds the input of the 4000-Hz crossover. Its LF output feeds the mid-range driver and its HF output feeds the tweeter. The mid-range driver covers the frequency range from 500 to 4000 Hz, a range covering 3 octaves. This is where much of the sound lies. Note that in this arrangement, the signal passes through two HPFs to get to the tweeter, one at 500 Hz and the other at 4000 Hz. The tweeter roll-off is thus fourth order below 500 Hz if both two-way crossovers are of second order. This can further help keep large low-frequency signals out of the tweeter.

The second approach has each of the LF, MF and HF drivers fed directly from the source after each passing through a filter, as shown in Figure 24.6(b). This can be called the parallel approach. The woofer is fed through a 500-Hz second-order LPF. The mid-range is fed through a bandpass filter comprised of a second-order HPF at 500 Hz followed by a 4000-Hz second-order LPF. The tweeter is fed through a second-order HPF. The electrical and acoustic summation for this arrangement is not flat [24]. Some tweaking of the filter frequencies can make it closer to being flat.

Both of these arrangements are readily implemented as active three-way crossovers. However, there are some differences in how their outputs add acoustically from perfectly time-aligned drivers. This results from phase differences at the crossover frequencies that affect how the addition takes place at and near the crossover frequencies.

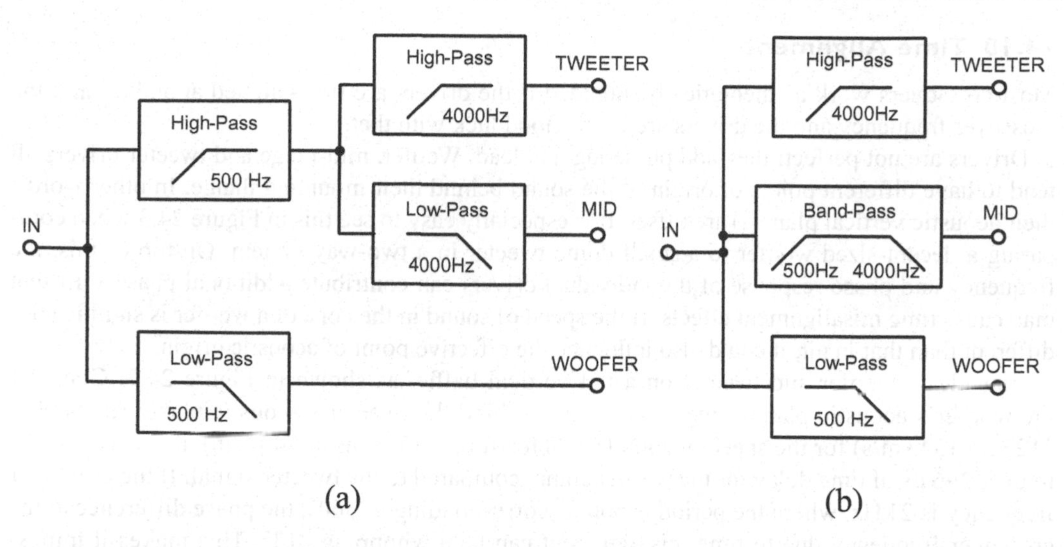

Figure 24.6 Three-Way Crossover Architectures.

24.9 3.5-Way Loudspeakers

The term "3.5-Way" describes a three-way system with a particular form of bass management that usually delivers deeper bass and higher bass SPL levels. The system includes two woofers, often identical. While a four-way system would likely have two LF drivers, they are usually different, with one being of substantial size to deliver low-bass. The 3.5-way system has some advantages. These include lower cost and smaller drivers that permit narrower enclosures while maintaining good bass response. They take advantage of the fact that SPL increases by 6 dB when a pair of drivers are delivering the same frequency in-phase. The 6-dB increase in SPL corresponds to four times the acoustic power.

The main woofer closest to the mid-range operates full-range up to its crossover frequency to the mid-range driver. The purpose of the other (low) woofer is to augment the bass at lower frequencies where it is harder to get higher SPL. Sometimes the lower frequency woofer includes a 6 dB/octave HPF at between 100 Hz and 200 Hz to provide baffle step compensation by adding a 6 dB/octave low-frequency shelf to the overall response. A typical "half" crossover might be in the vicinity of 100–200 Hz. At the low frequencies where both woofers are working together, the 6-dB advantage in SPL comes into play. If not using the arrangement for baffle step compensation, in order to keep the frequency response from increasing by 6 dB in the low-frequency region, half the voltage driving the low-frequency woofer can be subtracted from the voltage driving the main woofer, implementing a simple constant-voltage crossover. For a given low-frequency SPL, each speaker now operates at 1/4 of the power at low frequencies had there been only one conventional woofer of the same size.

The 3.5-way arrangement can be implemented with a passive crossover. With an active crossover system it is easily accomplished with an active circuit, but the additional driver(s) will require an additional power amplifier, for a total of four power amplifiers. The author built a narrow 3.5-way tower speaker in 2005. The bass was handled by two main woofers and two low woofers, each of the four being only 5-1/2-inch drivers. Four 125-W power amplifiers were included in the cabinet of each loudspeaker [14, 25].

24.10 Time Alignment

Most crossovers work as theoretically intended if the drivers are time-aligned at and around the crossover frequency and the drivers are ideal. Good luck with that.

Drivers are not perfect; they add phase lag and lead. Woofer, mid-range and tweeter drivers all tend to have different points of origin of the sound behind their mounting flange. In other words, their acoustic vertical plane(s) are offset. It is especially easy to see this in Figure 24.3 when comparing a decent-sized woofer to a small dome tweeter in a two-way system. On top of this, the frequency and phase response of the individual drivers can contribute additional phase shift that may cause time misalignment effects. If the speed of sound in the cone of a woofer is significantly different than that in air, it could also influence the effective point of acoustic origin.

Consider a woofer and tweeter on a flat vertical baffle, as shown in Figure 24.3. Consider the woofer's acoustic plane being 4″ (102 mm) behind the tweeter's acoustic plane. At roughly 1125 ft/s (343 m/s) for the speed of sound (0.88 ms/ft or 1.13 ft/ms or 74 μs/in), this corresponds to over 296 μs of time delay for the woofer signal compared to the tweeter signal. If the crossover frequency is 2 kHz, where the period is 500 μs corresponding to 360°, the phase difference at the crossover frequency due to time misalignment can be a whopping 213°. This makes it impossible to make a crossover perform its intended function without taking the time difference into account. Admittedly, this example is a bit of a worst case. The coincidence error will be smaller if the loudspeaker is a three-way design and the difference in radiating depth between the drivers is smaller.

Keep in mind that the difference in acoustic plane may not be the same as the difference in physical position of the voice coils, since sound may travel faster in the solid material of the cone.

A small two-way system with a smaller woofer may have a physical difference of only 1″ in the driver acoustic planes, resulting in a phase difference of only 53° at 2 kHz if perfect drivers are used. However, phase lag of the woofer driver operating at its high-frequency end and phase lead of the tweeter driver operating at its low-frequency end will act together to create additional net phase difference at the 2-kHz crossover frequency.

A three-way system improves matters by reducing the differences in the acoustic plane recess from the baffle among drivers adjacent in frequency. It also allows the drivers to operate in a region where they may not create as much excess phase lag or lead by operating so close to the edge of their frequency response limits where response is down 3 dB. Consider a system where the woofer acoustic plane is recessed 2″ (51 mm) behind that of a cone mid-range driver and crossed over at 400 Hz. The time difference will be about 148 μs and the phase difference at 400 Hz will be about 21°. If the mid-range is crossed to the tweeter at 4 kHz and the acoustic plane offset between the two is 1″, the time difference will be about 74 μs and the phase difference will be about 107°.

There are at least three possible approaches to dealing with unwanted phase differences at the crossover frequency that are caused by driver non-coincidence and driver phase shift. The first is to slope the baffle back at an appropriate angle so that the smaller drivers lying above the larger drivers are physically further back. This can add expense and may cause other acoustic anomalies.

The second approach is to allow the phase difference to just be part of the crossover design and tweak the crossover slopes and frequencies to get a reasonably smooth crossover that does not conform to theory. For example, the second-order crossover that in theory created a null without adding an inversion to the tweeter might become flatter with the introduction of 90° of phase shift resulting from time misalignment. Alternatively, making the woofer crossover a little bit higher in frequency than that of the mid-range or tweeter might reduce the phase difference caused by time

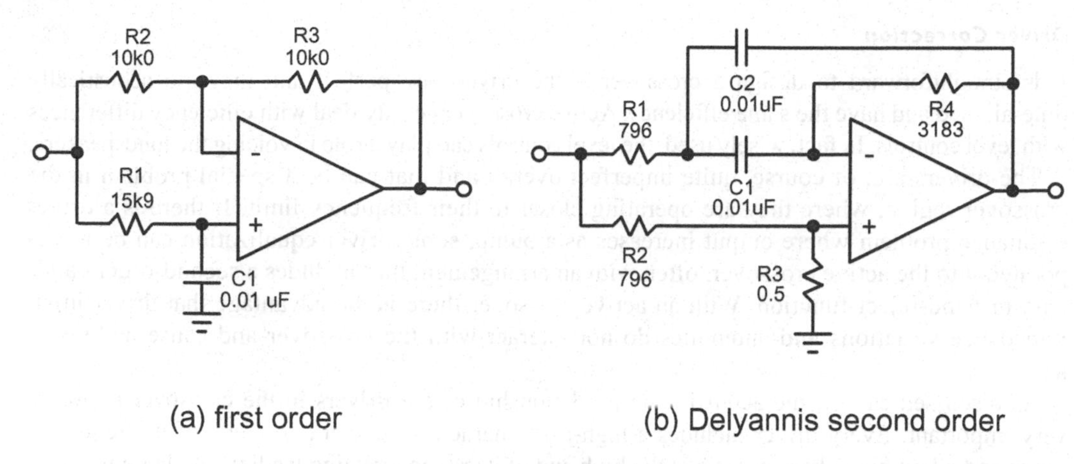

Figure 24.7 **First- and Second-Order All-Pass Delay Networks.**

misalignment. Thus, there is the art and magic of crossover design. Neither of these approaches produces a flat response in accordance with theory, but loudspeakers are very imperfect anyway.

The third approach is to add electrical time delay that is an approximation to pure delay, at least in the vicinity of the crossover frequency. This can be done in the passive regime, but is much more easily done in the active regime. This is an area in which active crossovers have an advantage. Adding delay is especially straightforward in the digital domain using a digital signal processor (DSP).

Active Delay Networks for Time Alignment

All-pass networks (APF) can be used to create delay, at least in the crossover region. The all-pass networks create lagging phase shift that is a function of frequency, and this corresponds to group delay. However, this delay is not constant with frequency, but listening has shown that this phase distortion is largely inaudible to most listeners with real music.

First- and second-order all-pass filters are illustrated in Figure 24.7. The design of APFs and delay lines is covered in Chapter 11. Delay will usually be put in the HF signal path [10].

The group delay is a maximum at DC and very low frequencies, and falls to half that at the −3-dB frequency of the LPF in Figure 24.7(a). At that frequency, the group delay equals the R-C time constant. At higher frequencies, group delay falls to zero. A second-order Delyannis APF is shown in Figure 24.7(b). It requires only one op amp.

Bear in mind that the phase relationship between the drivers is of great importance at f_0, but has little effect on frequency response and lobing at frequencies well away from f_0. This means that even though the group delay of the APF is changing with frequency, it is effective at creating a local group delay that can be helpful to time alignment in the vicinity of f_0. Indeed, by changing the frequency of the APF, some in-situ trimming of the time alignment between the drivers is possible.

24.11 Crossover Filter Design

The individual crossover filters can all be designed based on the material previously presented in Chapter 11. This includes any all-pass sections that are added to make phase/delay corrections (typically adding delay to the tweeter path).

Driver Correction

It is straightforward to design a crossover if the drivers are perfect and they are acoustically time-aligned and have the same efficiency. Active crossovers easily deal with efficiency differences with level controls. In fact, wisely used, the level controls can play a role in voicing the loudspeakers.

The drivers are, of course, quite imperfect overall and that can be a special problem in the crossover region, where they are operating closer to their frequency limit. If there is a driver resonance problem where output increases as a bump, some driver equalization can be incorporated into the active crossover, often with an arrangement that includes a second-order bandpass or band-reject function. With an active crossover there is the advantage that driver input impedance variations and anomalies do not interact with the crossover and cause additional headaches.

As discussed earlier, the acoustic phase relationship of the drivers in the crossover region is very important. Every driver includes a high-pass characteristic at the low end of its frequency range and a low-pass characteristic at the high end of its range, creating leading and lagging phase shift, respectively. Their phase contributions at the crossover frequency can be important. If at f_0 the woofer has a net leading phase due to its low-frequency roll-off and the tweeter has a net lagging phase due to its high-frequency roll-off, an additional phase difference between the two can alter the crossover behavior. On the other hand, if the tweeter's net phase shift is leading due to its low-frequency roll-off, then the phase difference between the drivers at the crossover frequency will be reduced. A leading phase shift in a driver can be canceled or reduced by an all-pass filter. Keep in mind that the group delay of a conventional all-pass filter is greatest at the lowest frequencies and falls to 1/2 its maximum value at its turnover frequency (in a first-order APF, the turnover frequency is the 3-dB frequency of its R-C section).

Crossover Effect on Imaging

Lateral stereo imaging depends on the phase relationships of the two stereo channels being as intended. We all know that reversing the phase of one channel destroys the center imaging. Most crossovers create frequency-dependent phase shift in the summed signal in the crossover region. If the crossover frequencies of the left and right channels are not exactly the same, there will be an image shift and image instability in that part of the frequency spectrum.

Achieving this consistency between left and right crossover frequencies is a big advantage of active crossovers. With passive crossovers, the expense of using components of sufficient precision can be prohibitive. Think of the tolerance of a non-polar electrolytic used in portions of a lower-frequency crossover. With a steep crossover, even a 10% difference in crossover frequency can create a significant phase difference between the left and right channels at f_0. This kind of problem can be exacerbated by high-order crossovers where both amplitude and phase are changing rapidly with frequency at the crossover frequency.

24.12 Active Baffle Step Compensation

As described earlier, diffraction effects come into play at low frequencies when the width of the loudspeaker baffle is on the order of the wavelength of the sound [1]. If the width of the baffle is large compared to the wavelength, the loudspeaker tends to emit sound in a hemispherical way (half-space). This is the case over most of the frequency range for most baffle sizes. However, if the baffle width is smaller compared to the wavelength, the speaker radiates sound in a more omnidirectional way (full-space). This means that at low frequencies there is relatively less SPL

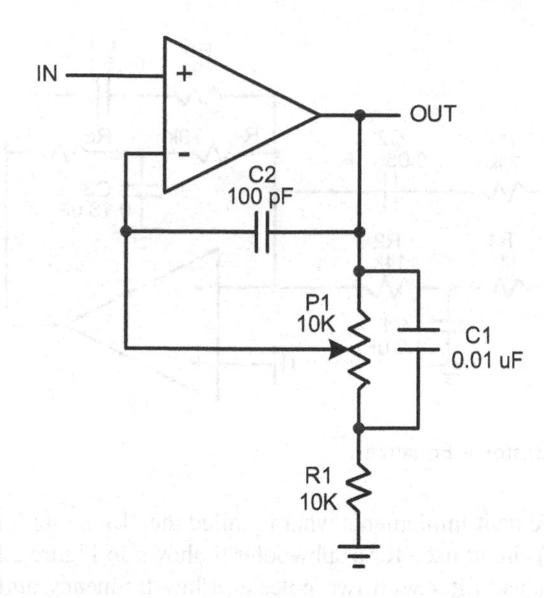

Figure 24.8 Baffle Step Compensation Circuit.

emitted in a way that is audible at the listening position. At frequencies between the two extremes, the effective amplitude transitions gradually from one step level to another, higher, step level as frequency increases.

The difference in dB between the two steps can theoretically be up to 6 dB, but is more often in the 1–4 dB range. This variation is in part due to the wall behind the loudspeaker redirecting some of the diffracted energy forward. The baffle step can be easily compensated in an active crossover with a 6-dB/octave frequency response step. The mid-point frequency of the step will usually be in the vicinity of 200–400 Hz, depending on the width of the baffle [2–4]. Active baffle step compensation (BSC) permits the size of the compensation step to be adjusted, and sometimes the frequency. There are some loudspeakers that incorporate an approximation to BSC passively by adding an additional woofer that cuts off at a lower frequency than the main woofer, thus augmenting the low-frequency response. These are so-called 3.5-way loudspeakers.

Figure 24.8 shows the simple active BSC circuit used in the Athena active loudspeaker [3]. It allows both the amount and frequency range of the baffle step to be selected. It should be noted that the Athena has a narrow baffle, only 8 inches wide. It is a 3-1/2-way system that includes four 5.25-inch woofers. Each pair of woofers is driven by a 125-W MOSFET power amplifier. The mid-range and tweeter are each directly driven by a dedicated 125-W amplifier. The baffle is sloped back to improve driver coincidence.

The design includes a six-position switch to select the values comprising P1 above and below the "wiper." P1 is not a pot, but is depected as one for convenience. Amounts of baffle step from 0 to 6 dB are available. Similarly, the value of C1 is selectable with six different values.

24.13 Woofer Equalization – Linkwitz Transform

Properly designed closed-box woofers have a relatively predictable 12 dB/octave roll-off at their low end. The low-frequency 3-dB point can be moved to a lower frequency by appropriate equalization as long as there is adequate amplifier power and cone excursion available to achieve the required SPL levels at the lowest frequency. Sigfried Linkwitz created an active filter equalizer

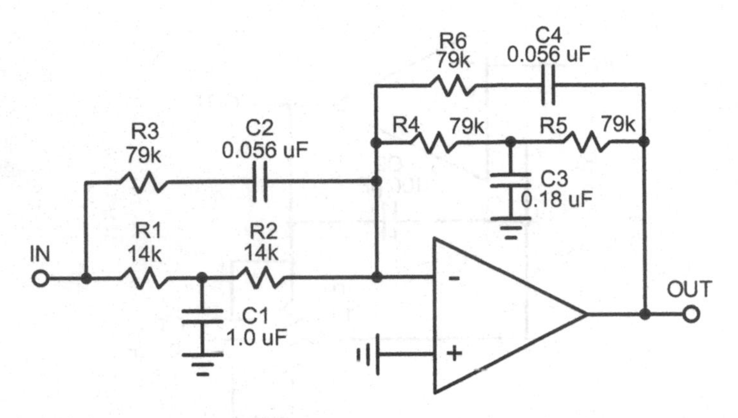

Figure 24.9 Linkwitz Transform Equalizer.

to accomplish this. The circuit implements what is called the "Linkwitz transform" [5, 15, 16]. A Linkwitz transform (LT) circuit used for a subwoofer is shown in Figure 24.9.

The LT circuit is a biquad filter with two poles at a low frequency and two zeros at a higher frequency. This forms a low-frequency shelf function with an asymptotic slope of 12 dB/octave. A closed-box woofer has two zeros at its low-end cutoff frequency, after which it falls with a slope of 12 dB/octave at lower frequencies. If one sets the frequency and Q of the high-frequency poles of the LT biquad to be the same as those zeros of the closed-box woofer, the pole and zero pairs will cancel. What remains is the pair of zeros in the LT at a lower frequency. The net result is that the LT has shifted the low-frequency cutoff to a lower frequency.

Indeed, this technique can be applied to closed-box frequency responses with different values of Q, both less than and greater than the Q of a Butterworth filter. This ability to alter the LF frequency response of a closed-box system allows one to use driver Q_{ts} and or box size that normally would not yield a satisfactory response. Thus greater flexibility in choosing driver characteristics and box size results. For example, one could use a woofer with low Q_{ts} in a box where the response would be over-damped. The stronger woofer magnet creating the lower Q_{ts} would tend to increase efficiency. The LT would be used to create a response with higher Q, perhaps with a Butterworth shape, while extending the LF response.

Using the LT, one can buy up to an octave of deeper bass in a mathematically sound way. Bear in mind that this does not change the reality that much greater power must be fed to the woofer at those lower frequencies. The LT is just a second-order shelving equalizer. Calculators are available online for the design of the LT circuit [21, 22]. The LT in Figure 24.9 implements a 15-dB low-frequency shelf centered at 33 Hz. It is down 1 dB from its maximum at 11 Hz. It is up 1 dB from its nominal gain of 0 dB at 106 Hz.

24.14 Equalized Quasi-Sealed System

There is another use for the Linkwitz transform. It can be used to equalize the frequency response of a vented system that uses a specialized alignment that causes its frequency response to look like that of a sealed system, but which retains some of the important benefits of the vented system. A vented system with an alignment that makes it behave much like a sealed system, with a mostly 12-dB/octave roll-off, is called a quasi-sealed system (QSS). Coupled with an LT equalizer, it is called an equalized quasi-sealed system (EQSS™) [14, 25–27].

A big challenge in loudspeaker system design is in achieving extended low-frequency performance without large drivers and cabinets. We'd like to have a system that is flat (−3 dB) to or below the 30–50 Hz range. A more difficult challenge is to achieve high SPL at these low frequencies, owing to the need to move large amounts of air. This is made more difficult when it must be achieved in a small enclosure, or with small loudspeaker drivers, or both.

The maximum cone excursion of the driver, in combination with the driver's effective cone area, determines the amount of air that can be moved. This in turn sets a limit on SPL that can be achieved at low frequencies. The low-frequency SPL limitation is referred to as excursion-limited SPL, or ELSPL. The ELSPL of a driver is a function of frequency, and decreases at lower frequencies because a correspondingly larger amount of air must be moved to achieve a given SPL.

Sealed systems tend to have higher 3-dB cutoff frequencies (f_3) and lower ELSPL as compared to vented systems. The cutoff frequency of a sealed system can be reduced, but at the expense of a larger box or a woofer with a heavier cone and lower efficiency. In either case, however, low-frequency ELSPL is not increased for a driver with a given cone area and maximum excursion.

Vented systems include a port in the box, forming a Helmholtz resonator. When properly designed, the box resonance f_b helps produce a lower f_3 and a higher ELSPL at low frequencies. These systems have a fourth-order high-pass frequency response. At frequencies below f_3, the frequency response begins to fall off rapidly, at a rate approaching 24 dB/octave. Although ported systems provide increased ELSPL at frequencies above f_3, the ELSPL of ported systems falls off severely at frequencies below f_b, the box tuning frequency, providing virtually no useful output at such frequencies. Ported systems actually produce *less* ELSPL than that of a comparable sealed system at frequencies below f_b.

As described earlier, sealed systems can employ LT equalization in order to achieve an extended response with a reduced f_3. This approach does not suffer from the need for larger cabinets or heavier woofer cones, but requires high amplifier power and does not provide higher ELSPL. Ported systems can be equalized, but they virtually never are. This is because the f_3 of a conventional ported system usually lies near the box tuning frequency, and the ELSPL drops off severely at frequencies below f_b.

A Vented System Acting Like a Sealed System

A ported system can be made to act much like a sealed system in both frequency response shape and roll-off slope over an extended band of low frequencies when the box tuning frequency is chosen to be unusually low. By this we mean substantially lower than the f_b commonly used with a given driver-box combination. The low-frequency SPL capability of such a differently-ported system is greatly improved compared to that of a similar sealed system. This is true even at frequencies well below the 3-dB frequency response point of this "low-tuned" ported system. We refer to such a ported system that has a frequency response similar to that of a sealed system as a quasi-sealed system (QSS).

In a nutshell, EQSS™ includes a low-frequency driver in a ported box with an unconventionally low box tuning frequency f_b, and an LT equalizer that makes the frequency response flat to the target 3-dB frequency. The box is tuned so that the response of the driver-box combination at f_3 is substantially below the reference response level (e.g., −6 to −12 dB). The resulting response then approximates a second-order response down to frequencies at least one-half octave below f_b. This arrangement then comprises the QSS [27].

Equalizing the Quasi-Sealed System

It is also useful to define a virtual sealed system (VSS) as a sealed system design whose box volume and driver parameters have been manipulated so that its frequency response accurately models that of a QSS over the frequency range of interest. This comes into play when an LT calculator is used to design the equalizer.

The equalizer parameters are calculated in accordance with proper equalization of the VSS whose frequency response accurately models that of the QSS. This approach is therefore referred to as an EQSS™.

The QSS, although ported, acts like a sealed system, but with increased efficiency at low frequencies. The frequency response of the QSS may be equalized in the same way as the VSS, using the same LT equalizer. This is so because their frequency responses are essentially the same in the frequency band of interest. If the response of the QSS is equalized to become the more desirable response, the EQSS™ system will be the result.

Achieving Higher SPL at Low Frequencies

The maximum undistorted SPL that can be reproduced by a non-ported loudspeaker at low frequencies is limited by the distance that the loudspeaker's cone can move (excursion). In order to reproduce sound at low frequencies at high levels, a loudspeaker must move a large amount of air (displacement). A loudspeaker's displacement is proportional to the product of its cone area and its excursion.

A ported system provides substantially larger ELSPL over the frequency range of interest than the sealed system. This is due to the action of the port, which loads the loudspeaker driver and produces substantial SPL output at frequencies in the vicinity of the box tuning frequency f_3. The ported system described in Ref. [14] is capable of at least 102 dB SPL down to 60 Hz. This is fully 15 dB more ELSPL than the sealed system at 60 Hz.

In the system of Ref. [14], the QSS is over-damped and capable of 92 dB ELSPL down to 60 Hz. This is 5 dB better than the sealed system over the same frequency range. It is less than the 102-dB ELSPL capability of the ported system at 60 Hz, but the larger ELSPL of the ported system is not in a frequency range where it is really needed, given the objectives here. Now we begin to see the engineering tradeoff made possible by the EQSS™ technique.

The QSS here is capable of 91-dB ELSPL over this entire frequency range extending to 35 Hz. This is 13 dB better than the 78 dB ELSPL capability of the sealed system at 35 Hz. It is a remarkable 21 dB better than the 70 dB ELSPL of the ported system at 35 Hz.

The response of the over-damped QSS falls with decreasing frequency, being down 3 dB at approximately 90 Hz. The response of the equalizer, on the other hand, rises with decreasing frequency, being up approximately 3 dB at 90 Hz. Similarly, at 35 Hz the response of the QSS is down 14 dB while the response of the equalizer is up 11 dB. The combination of the equalizer and the QSS provides a system response with extended low-frequency response and a −3 dB frequency of approximately 35 Hz.

The Equalizer

The Linkwitz transform is used to equalize the QSS to be flat over the system's operating frequency range, down to 35 Hz, as shown in Figure 24.10. The response of the QSS (labeled speaker) falls with decreasing frequency, being down 3 dB at approximately 90 Hz. The response of the

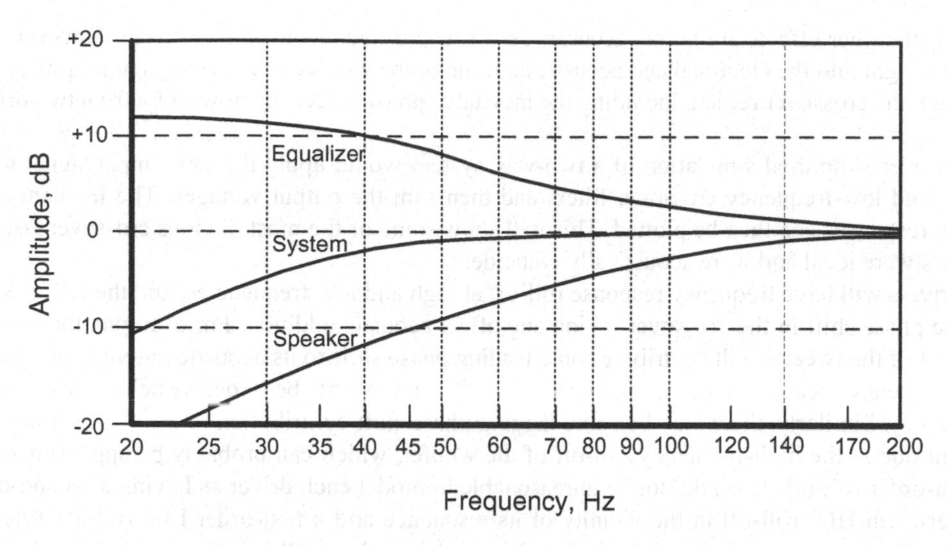

Figure 24.10 **Calculated Frequency Responses of an EQSS Loudspeaker.**

equalizer, on the other hand, rises with decreasing frequency, being up approximately 3 dB at 90 Hz. Similarly, at 35 Hz the response of the QSS is down 14 dB while the response of the equalizer is up 11 dB. The combination of the equalizer and the QSS provides a system response with extended low-frequency response and a −3 dB frequency of approximately 35 Hz.

A big advantage of EQSS™ is that it allows great freedom in picking a driver and its Thiele-Small parameters. Because equalization is fundamental to EQSS™, the issue of frequency response is divorced from matters of box size, tuning frequency and achievable ELSPL. For example, if a bigger magnet is used to achieve higher efficiency in the driver, there is no problem created by the reduced Q_{ts}. Similarly, if a larger driver is used in a given cabinet, there will be no problem with the higher V_{as}. Most drivers can be tuned to a suitable QSS alignment that can be equalized with the LT. The EQSS™ technique is a patented technology (US 7,894,614) [27]. EQSS has also been used to implement a powerful subwoofer with two 10″ woofers. More on the EQSS™ technology, its applications and design procedure can be found at cordellaudio.com.

24.15 DSP Crossovers

Digital signal processing can obviously be used to implement active crossovers, even if an ADC conversion must be made ahead of it and a DAC must be used after it. Of course, in an increasing number of applications the input signal to the active crossover is already PCM.

In the digital domain it is straightforward to implement pure delay and linear phase filters, often with an FIR (finite impulse response) filter structure. Delay correction to achieve time alignment of drivers is thus very straightforward. Of course, it is also possible to implement the traditional Linkwitz-Riley crossover in the digital domain by using IIR (infinite impulse response) digital filters. Finally, the programmability of the DSP makes tweaking the crossover easy.

24.16 SPICE Simulation

Applying SPICE simulation to gain insight regarding the performance of a crossover-driver-baffle arrangement can be helpful as long as expectations are not overly optimistic; models for the drivers and some other effects are not as accurate as we would like. The simulation can, however, give good insight into the electrical and acoustic addition of the woofer and tweeter signals as they pass through the crossover region, including the inevitable phase effects on how SPLs from two drivers add.

An over-simplified simulation of a two-way system would apply the same input signal to the high- and low-frequency crossover filters and then sum the output voltages. The frequency and phase responses can then be plotted. This will show some of the most obvious crossover issues if drivers were ideal and were acoustically coincident.

Drivers will have frequency response roll-off at high and low frequencies, and the roll-offs will create phase shift in the crossover region and affect acoustic addition. For example, the low-end roll-off of the tweeter will contribute some leading phase shift to its acoustic output at the crossover frequency because it has a second-order roll-off that may only be an octave below the crossover frequency. Similarly, there may be some lagging phase shift contribution to the woofer's acoustic output due to the high-frequency roll-off of the woofer, which can probably be approximated as a first-order roll-off. It would not be unreasonable to model each driver as having a second-order Butterworth HPF roll-off in the vicinity of its resonance and a first-order LPF roll-off a decade above. These are very rough estimates, but they are better than nothing.

References

1. Vance Dickason, *Loudspeaker Design Cookbook*, 7th edition, KCK Media Corporation, Chase City, Virginia, 2006.
2. Dennis Bohn, *Linkwitz-Riley Crossovers: A Primer*, RaneNote 160, Rane Corporation, 2005.
3. Dennis Bohn, *Signal Processing Fundamentals*, RaneNote 134, Rane Corporation, 1997, ranecommercial.com/legacy/pdf/ranenotes/Signal_Processing_Fundamentals.pdf.
4. Douglas Self, *The Design of Active Crossovers*, Focal Press, New York, NY, 2011.
5. Rod Elliott, *Subwoofer Equalizer*, Elliott Sound Products (ESP) Project 71, https://sound-au.com/project71.htm.
6. Rod Elliott, *The Subwoofer Conundrum*, Elliott Sound Products (ESP), 2004, https://sound-au.com/subcom.htm.
7. Rod Elliott, *Eight Band Sub-Woofer Graphic Equalizer*, Elliott Sound Products (ESP), Project 84, 2002, https://sound-au.com/project84.htm.
8. Rod Elliott, *Subwoofer Phase Controller*, Elliott Sound Products (ESP), Project 103, 2004, https://sound-au.com/project103.htm.
9. Joao Martins, 25 Years after the HD-1, Meyer Sound Unveils Amie Precision Studio Monitor, *audioXpress*, October 2015.
10. Charlie Hughes, *Using All-Pass Filters*, Voice Coil, February 2010.
11. John Vanderkooy, A Simple Theory of Cabinet Edge Diffraction, *JAES*, vol. 39, no. 12, December 1991, pp. 923–933.
12. John L. Murphy, *Loudspeaker Diffraction Loss and Compensation*, True Audio, trueaudio.com/st_diff1.htm.
13. Rod Eliott, *Baffle Step Compensation*, Elliott Sound Products (ESP), http://sound-au.com/bafflestep.htm.
14. Bob Cordell, *The Athena Active Loudspeaker*, PowerPoint presentation, cordellaudio.com.
15. Sigfried Linkwitz, *12 dB/oct Highpass Equalization*, linkwitzlab.com/filters.htm.
16. Bob Cordell, EQSS™ Equalized Quasi-Sealed System – Get Extended Low-Frequency Response and Higher SPL with EQSS™, *audioXpress magazine*, September, 2014.
17. Joseph D'Appolito, "A Geometric Approach to Eliminating Lobing Error in Multiway Loudspeakers," Audio Engineering Society 74th Convention, New York City, October 9, 1983.

18. Joseph D'Appolito, Active Realization of Multiway All-Pass Crossover Systems, *Journal of the Audio Engineering Society*, vol. 35, no. 4, April 1987, pp. 239–245.
19. S.H. Linkwitz, Active Crossover Networks for Non-coincident Drivers, *Journal of the Audio Engineering Society*, vol. 24, January/February 1976, pp. 2–8.
20. Rod Elliott, *24 dB/Octave 2/3-Way Linkwitz-Riley Electronic Crossover*, Elliott Sound Products (ESP), Project 09, 2018, https://sound-au/project09.htm.
21. HiFi Audio Design, *Linkwitz Transformation*, mh-audio.nl/Calculators/LinkwitzTransformation.html.
22. True Audio's *Linkwitz Transform Circuit Design Spreadsheet*, https://trueaudio.com/downloads/linkx-frm.xls.
23. Stanley Lipshitz, Mark Pocock and John Vanderkooy, On the Audibility of Midrange Phase Distortion in Audio Systems, JAES, vol. 30, no. 9, September 1982, pp. 580–595.
24. Sigfried Linkwitz, "Crossover Topology Issues," www.linkwitzlab.com/frontiers_5.htm#V.
25. Bob Cordell, *Athena Active Loudspeaker System*, cordellaudio.com. See also AX article.
26. Bob Cordell, *EQSS™ Equalized Quasi-Sealed System*, White Paper, cordellaudio.com.
27. Bob Cordell, *System and Method for Achieving Extended Low-frequency Response in a Loudspeaker System*, US Patent 7,894,614, February 22, 2011.

Chapter 25

Voltage-Controlled Amplifiers

The term voltage-controlled amplifier (VCA) is usually associated with integrated circuits whose variable-gain functionality is based on what are called translinear techniques. These designs are most often implemented with bipolar transistors in integrated circuits. These functions are also referred to as log-antilog circuits. The gain of a VCA is usually dependent on a control voltage in a fashion that is linear in dB, so the control function for gain is described in dB/volt. This type of control characteristic is often referred to as a *decilinear* characteristic [1–3]. These circuits can control gain over a very wide range, often 80 dB or more, so they are well-suited to volume control applications. The decilinear control characteristic is an almost ideal type of audio taper.

VCAs are used widely in professional audio-recording consoles, but are also valuable in replacing multi-gang potentiometers in consumer stereo and home theater applications because the matching between VCA gain control elements is very good and they are well-suited to remote control [4–7].

VCAs are a staple of audio consoles. They permit automation, replace motor-controlled pots and enable voltage-controlled functions like compressors. They permit fairly precise gain control and can be controlled by a microcomputer if a DAC is used to create the control voltage.

25.1 Translinear Circuits

The late Barrie Gilbert of Gilbert Cell fame coined the term "translinear" to describe the functionality of circuits that he implemented that took advantage of the fact that V_{be} of a bipolar transistor follows the log of collector current [8–13]. So-called *translinear* circuits take advantage of the fact that V_{be} of a transistor changes as the log of the collector current and the collector current changes as the antilog (exponential) of base-emitter voltage. Functions like squaring and multiplication can be done in the log domain by addition. Input signals in the linear domain are logged to put them in the log domain. Those signals are then manipulated in the log domain, sometimes by just adding a DC control voltage. Those processed signals are then put through an antilog circuit that restores them to the linear domain.

The V_{be} of a transistor increases by about 60 mV each time the collector current is increased by a factor of 10, meaning that V_{be} is a log function of collector current. The degree to which this relationship is accurate over several decades of current value is referred to as the transistor's log conformance. Base resistance is a major factor in degrading a transistor's log conformance because it adds a somewhat linear term to V_{be}, namely base current times the base resistance. The ratio of beta to $r_{bb'}$ is a reasonable figure of merit for log conformity.

We all know that the addition of log values corresponds to multiplication, and in fact is the basis for analog multiplier chips. Although an analog multiplier can be used as a form of VCA, it has fairly limited control range in dB because both the X and Y inputs have a linear relationship with the output. Similar circuitry can inject the gain control voltage into the circuit in a different way, causing the gain to depend on the exponential of the control voltage, meaning that the gain becomes linear in dB with changes in the control voltage. There are numerous variations on this technique used in different VCAs, but the basic principle is the same in most cases.

Put another way, the transconductance (*gm*) of a BJT is proportional to collector current. If we increase the current, the small-signal gain (transconductance) increases by the factor of the current increase. But the amplification is nonlinear due to the exponential relationship between base voltage and collector current. We can pre-distort the applied signal's base voltage by making it logarithmic by applying signal current to a base-emitter junction. The output then becomes linear and is an undistorted multiple of the input signal. The following simplified transistor equations are relevant:

$$I_c = I_0 e^{(qVbe/kT)} = I_0 e^{(Vbe/V_T)}, \quad \text{where } V_T = KT/q$$

$$V_{be} = V_T \ln(I_c / I_s)$$

The un-degenerated current mirror is the simplest translinear circuit. The voltage V_{be} at the bases of Q1 and Q2 in Figure 25.1(a) follows the log of the input current. The current at the output of Q2 follows the antilog of its base-emitter voltage. If a value is added to a log value and then the antilog is taken, the effect is to multiply the result. We know that adding 60 mV to a base-emitter voltage multiplies the current by a factor of 10. If a 60-mV voltage source V_{ctrl} is placed between the base of Q1 and the base of Q2, the current in Q2 will be ten times as large. This follows for both DC current and AC current. A signal current applied to Q1 will thus be multiplied by ten in this case, instead of unity in the case of no inter-base voltage applied. We thus have a voltage-controlled amplifier whose gain is controlled by the voltage V_{ctrl} put between the bases. The control voltage relationship is 20 dB for a 60 mV change (3 mV/dB).

A more practical version of such a circuit is shown in Figure 25.1(b). Input resistor R1 converts the input signal voltage to a current by driving the virtual ground of U1. Then U1 forces a linear version of the input current to flow in Q1. This results in a voltage at the emitter of Q1 that represents a logged version of the input voltage. With this same logged version of the input applied to the emitter of Q2, the collector current of Q2 becomes a linear current representation of the input current. U2 then converts this current into an output voltage via negative feedback and R4. Notice that if $V_{ctrl}=0$ and the base of Q2 is essentially at ground, the collector signal currents of Q1 and Q2 will be the same and the circuit will act as a non-inverting unity gain circuit. If the base of Q2

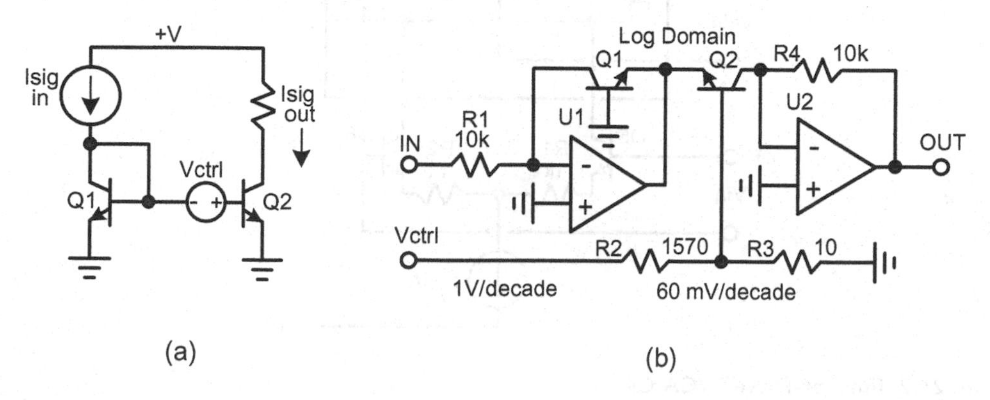

Figure 25.1 Simple Translinear Circuits.

is non-zero with a control voltage, this is equivalent to the additions of a voltage between the bases, as was done in (a), resulting in an addition to the logged voltage from the emitter of Q1. This circuit can only operate for positive inputs. A negative input will cause the output of U1 to go positive and reverse bias the emitters of Q1 and Q2. A feedback diode should be added to prevent the output of U1 from going more positive than a diode drop.

Emitter-Driven VCA

In Figure 25.2, differential pairs are used to allow the input and control signals to be differential and to allow both polarities of the input signal to be used. The output is also differential. The differential input pair Q5–Q6 is degenerated by 1 kΩ resistors R1 and R2. It converts the differential input signal voltage to a differential current applied to the tails of the upper differential pairs Q1–Q2 and Q3–Q4. The upper differential pairs are thus emitter-driven by the signal current. The logged version of the input signal appears at the emitters of those upper differential pairs. The gain control voltage V_{ctrl} is applied to the bases of the upper differential pairs and causes the signal current at their emitters to be directed more (or less) to Q1 and Q4 than to Q2 and Q3. The signal collector current of Q2 and Q3 is just dumped into the supply rails, while that from Q1 and Q4 creates a differential output voltage via 2-kΩ load resistors R3 and R4. With $V_{ctrl} = 0$, half the signal current flows in Q1 and Q4, so the gain is unity. Maximum gain is 2. Even without looking at log and antilog behavior in this circuit, one can see that the signal current from Q5 and Q6 just follows the DC current in the same proportion. Note that this circuit has significant common-mode control voltage feedthrough.

A control voltage of +60 mV will cause Q1 and Q4 to conduct ten times as much DC and signal current as Q2 and Q3, resulting in near-maximum gain. A control voltage of −60 mV will cause Q2 and Q3 to conduct ten times as much as Q1 and Q4, resulting in attenuation of about 15 dB. Gain

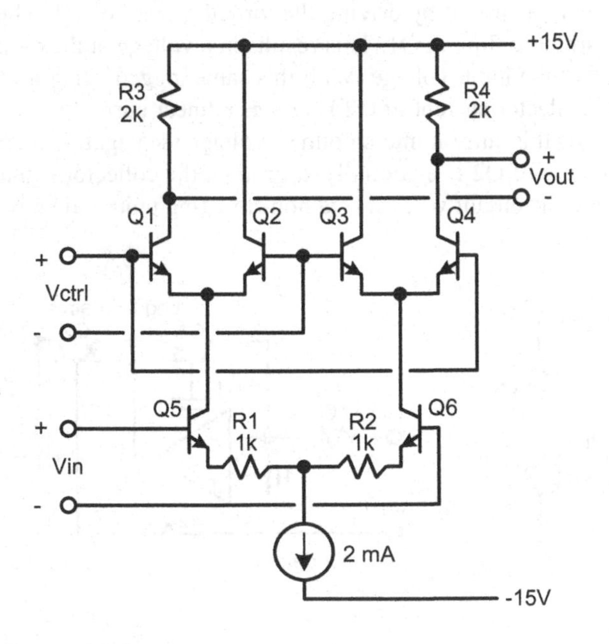

Figure 25.2 Emitter-Driven VCA Circuit.

with V_{ctrl} = −120 mV is about −34 dB. Maximum attenuation is a very large number of dB, so in this region, the change in gain is largely linear, at about 3 mV/dB. Maximum gain as shown is about 2, as determined by the ratio of (R3 + R4)/(R1 + R2).

While the gain is approximately linear in dB at high attenuation, the behavior is not decilinear for higher gains approaching 2, since no matter how high a positive control voltage is applied, the gain cannot go above 2. We will see that some of the other VCA circuits do not have this departure from decilinear behavior of gain in dB as a function of control voltage.

At V_{ctrl} = 0 V, output noise is quite low, being 6.2 μV in a 20 kHz bandwidth, corresponding to SNR of 105 dB. THD is −57 dB (0.17%) at 1 V_{RMS} output. Distortion is dominated by the third harmonic and is mainly due to the V-I converter comprising R1, R2, Q5 and Q6.

Base-Driven VCA

The VCA shown in Figure 25.3 is base-driven, where a pre-distorted input signal is applied to the bases of an un-degenerated differential pair that performs the antilog function. It is the design that was described and analyzed in the seminal VCA paper by Sansen and Meyer [4]. It shares much of the concept of the four-quadrant multiplier.

The differential input voltage is converted to a differential current by the degenerated input differential pair Q5 and Q6 (the V-I converter). The differential current is pre-distorted by a $\tanh^{-1}$ logging circuit at the emitters of Q1 and Q2 where it is converted to a differential voltage. The pre-distorted differential voltage at the emitters of Q1 and Q2 is applied to the differential pair Q3–Q4, which performs the antilog function. The collector signal currents of Q3 and Q4 are thus linear, and are converted to a differential output voltage by 2-kΩ load resistors R3 and R4.

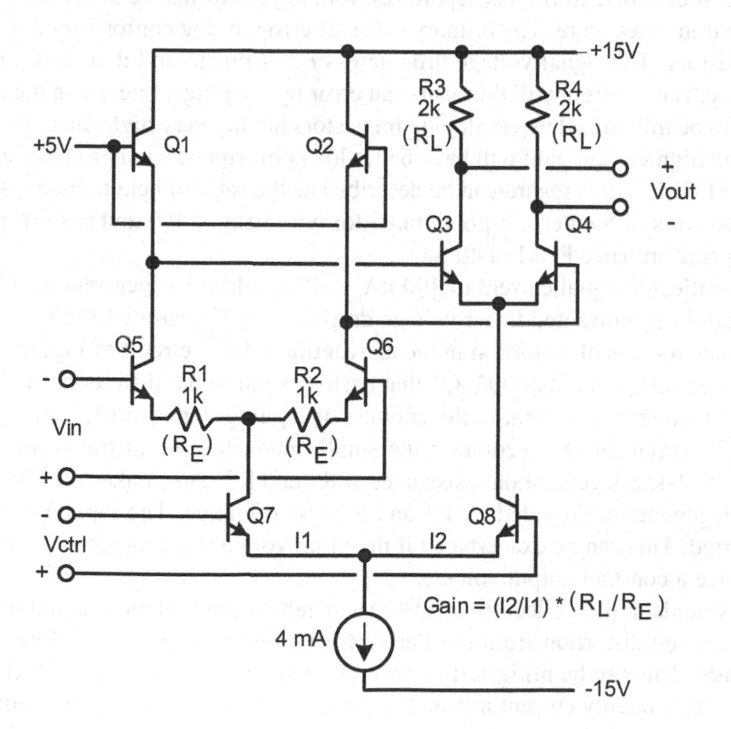

Figure 25.3 Base-Driven VCA Circuit.

Notice that the bottom differential pair Q7 and Q8 routes the 4-mA tail current to the differential pairs Q5–Q6 and Q3–Q4 in accordance with the differential gain control voltage applied to the bases of Q7 and Q8. Current I1 is the tail current for differential pair Q5–Q6 while current I2 is the tail current for differential pair Q3–Q4. V_{ctrl} controls the ratio of I1 to I2 in an antilog fashion.

Gain of the circuit is controlled by the ratio of I1/I2. If I1 is greater than I2, the current through logging transistors Q1–Q2 will be greater and the current through antilog transistors Q3–Q4 will be less. The current gain from the emitters of Q1–Q2 to the collectors of Q3–Q4 will be reduced by the ratio I1/I2 in this case because the transconductance of Q1–Q2 will be greater than the transconductance of Q3–Q4. The resulting linear signal current at the collectors of Q3–Q4 will then be converted by R3 and R4 to the differential output voltage. At $V_{ctrl}=0$ V, gain is 2. Output noise is 151 nV//√Hz, corresponding to 21 μV in a 20 kHz bandwidth, resulting in SNR of 94 dB. Distortion simulates at 0.001% at 1 V_{RMS} output with $V_{ctrl}=0$ V. Note that this circuit will have common-mode control signal feedthrough as a result of the tail current (I2) changing with control voltage. At low gain, I2 will be small and the voltage drops across R3 and R4 will be small.

Distortion

Many different VCA arrangements are possible, as we shall see, all depending on the log-antilog operation wherein a number is added to the log value before the antilog operation. If the log and antilog behavior of the transistors is imperfect, distortion will result. This source of distortion, resulting from imperfect log conformance of the transistors Q1–Q4, is common to all translinear VCAs.

The common source of distortion is departures from log conformance of the transistors Q1 to Q4 comprising the translinear core. The primary cause of errors in log conformance is base resistance $r_{bb'}$ in the transistors. The signal voltage drop across $r_{bb'}$ is linear, and it creates a departure from log conformance. Base current will thus cause an error by creating a linear voltage drop across $r_{bb'}$. Base current can be minimized by employing transistors having very high current gain. Transistors with low $r_{bb'}$ and high current gain will have better log conformance and produce less distortion. A figure of merit (FOM) in this regard can be described as the ratio of beta to base resistance ($\beta/r_{bb'}$), which will have units of Siemens. A good transistor with beta of 400 and base resistance of 10 Ω will have a log conformance FOM of 40 S.

If $\beta=400$, a collector signal current of 100 μA_{pk} will result in base current of 0.25 μA_{pk}, which will in turn result in an unwanted linear voltage drop of 2.5 μV_{pk} across 10-Ω $r_{bb'}$.

There are other sources of distortion in the conventional VCA circuit of Figure 25.3 [4]. One is in the degenerated differential pair Q5–Q6 that performs the initial differential voltage-to-current conversion that feeds signal current to the emitters of logging transistors Q1 and Q2. At low gain settings, the tail current of Q7 becomes quite small, increasing *re′* of transistors Q5 and Q6 in comparison to the 1-kΩ degeneration resistances of R1 and R2. This makes the distortion-reducing action of the degeneration provided by R1 and R2 less effective. The input V-I conversion thus becomes distorted. This can be exacerbated if the input voltages are larger when lower gain is required to produce a constant output voltage.

The output signals at the collectors of Q3–Q4, if left in the voltage domain using load resistors, can create some distortion from the Early effect since β of Q3–Q4 will be a mild function of output voltage. This can be mitigated by cascoding the collectors of Q3–Q4. If R3 and R4 are replaced with a high-quality current mirror, the common-mode control signal feedthrough will be suppressed and a single-ended output signal current will be produced.

Noise

The other common performance issue with VCAs like the one in Figure 25.3 is noise. The differential pair Q3,4 is an un-degenerated gain stage with significant gain if its tail current is high. With $V_{ctrl}=0$, all four transistors Q1–Q4 will operate at 1 mA. Voltage gain of Q3–Q4 will be about (R3+R4)/2*re'* or 77. This means that the sum of the noise of all of the noise sources in the circuit (except thermal noise of R3 and R4) will be multiplied by 77. When simulated with LM394 models, with $V_{ctrl}=0$, SNR is 94 dB with respect to 1 V_{RMS} output.

Control Linearity

The design of Figure 25.3 provides a reasonably linear relationship of dB gain as a function of control voltage over a decent gain/attenuation range. This is mainly because the circuit can provide high gain when I1 is small and I2 is large, with the transconductance of the Q1–Q2 transistors being small and the transconductance of Q3–Q4 being large. The control coefficient is 3.15 mV/dB within 5% over an 80-dB range up to +20 dB.

Control Feedthrough

A key issue in most of these VCA circuits is control voltage feedthrough. Changing the control voltage makes significant changes is the operating currents of the transistors comprising the VCA, especially in the antilog output stage. Ideally, in most of these designs, the effects of the control voltage will be confined to the common mode if the output is differential. However, if resistors and transistors are not well-matched this cancellation will not be perfect and a DC differential-mode voltage will appear at the output that changes as the control voltage changes.

In some AC-coupled applications, such as in a manual volume control that does not change very fast, control feedthrough may not be a problem. However, in applications where the control voltage can change quickly, as in a compressor, this control feedthrough can interfere with the signal and cause audible problems. While the particular architecture of the VCA influences susceptibility to control feedthrough, it can be execution of the circuit design that matters most. Good execution basically means very good matching of transistors and resistors in the circuit. This can often be achieved when the VCA is implemented on an integrated circuit.

25.2 VCA Circuits

Voltage-controlled amplifiers are employed in a large number of applications, and there are many circuit approaches to implementing them. While it is true that gain can be controlled by voltage in some way with circuits that are not based on translinear techniques, like using a JFET's voltage-controlled resistance property, the circuits discussed here are all based on translinear techniques implemented with bipolar junction transistors. The usual use of the term VCA refers to this type of implementation.

The introductory circuits discussed thus far operate in class A; all transistors remain in an *on* state during nominal operation of the VCA. There are some VCAs that operate in class AB, like a power amplifier's output stage, where, depending on signal polarity, some transistors are off and other transistors are on and carrying the signal. A VCA using the class AB approach will be discussed last. Interestingly, that one is the most widely used VCA design for audio applications [14–21].

Baskind-Rubin VCA

The VCA developed by Baskind and Rubin is much like the base-driven VCA illustrated in Figure 25.3, but is improved in several ways [22–24]. Most significantly, large-area transistors, each composed of a multiplicity of identical transistors effectively connected in parallel, are used for the two transistor pairs that comprise the VCA core. This greatly reduces $r_{bb'}$ base resistance while allowing operation at low current densities in each device. As a result, log conformity is improved and noise is reduced.

Additional improvements include a highly linear V-I converter that drives input signal current into the emitters of the pre-distorting transistor pair. The differential outputs of the core are also buffered by highly linear emitter followers and feed an op amp configured as a differential-to-single-ended converter.

Feedback VCA

The VCA circuit shown in Figure 25.4 is a somewhat different arrangement. It is a very early example of a VCA that employs negative feedback. Here it is called a feedback VCA. It was implemented in the early 1970s by this author to create a four-channel voltage-controlled volume control for a quadraphonic preamplifier (remember that?). It was later used as part of the autoscaling circuit in a total harmonic distortion (THD) analyzer design published in 1981 [26].

This VCA uses feedforward and feedback to control the gain, much like in the Baxandall volume control design from a later point in time. In essence, the core of this feedback VCA is a voltage-controlled pot, where different amounts of signal from either end of the pot are selected by the electronic wiper. Depending on the control voltage, the gain in the circuit as shown can range from zero to a very large number in a decilinear way. It was originally built using the

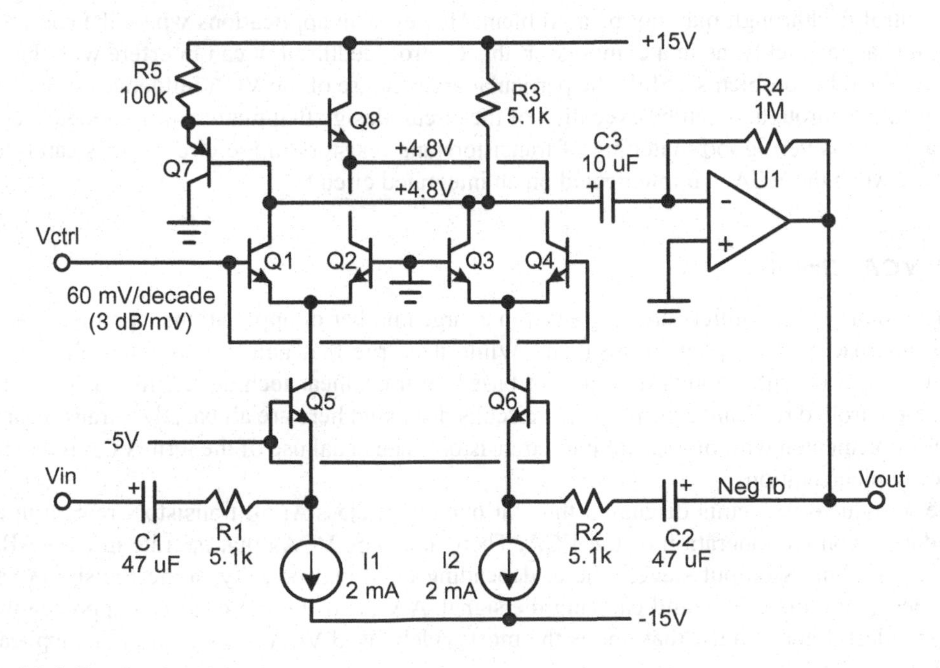

Figure 25.4 Feedback VCA.

LM1496 double-balanced modulator IC. That IC essentially implements Q1–Q6 and the two current sources. Performance is limited by the mediocre log conformity of the transistors in this IC. If implemented with LM394-class transistors, with a high log conformance FOM, noise and distortion performance can be quite good.

The circuit employs negative feedback from op amp U1 as part of its operation [25]. As shown, gain is unity when V_{ctrl} is zero. The ratio R2/R1 controls nominal gain at $V_{ctrl}=0$. A drawback in this circuit is its single-ended output, which results in some control feedthrough, although it is small if Q1–Q4 are well-matched. C3 blocks this largely DC control feedthrough from the op amp.

Two differential pairs comprise the VCA core. They are fed their tail currents from cascode transistors Q5 and Q6, each with a pull-down current source. The voltage input signal is fed to the emitter of Q5 through R1 where it is converted into a current signal that is applied to the differential pair Q1–Q2. Similarly, the output voltage of the op amp is fed to the emitter of Q6 through R2, where it is converted to a signal current that is applied as negative feedback to the tail of differential pair Q3–Q4.

The collectors of Q1 and Q3 are connected together and are loaded by R3 to the positive rail. The output of these collectors is AC-coupled to the virtual ground of op amp U1. R4 provides bias stabilization at low frequencies. There is thus virtually no signal voltage at the collectors of Q1 and Q3 – only a tiny error signal, since it is a virtual ground. This preserves headroom and eliminates the Early effect in Q1 and Q3. When V_{ctrl} is positive, most of the input signal current flows through Q1 to the input of U1, while less of the feedback signal current flows into U1. Gain will thus be greater than unity.

If the DC currents from Q5 and Q6 are the same, the sum of the currents through R3 from Q1 and Q3 will be approximately constant at about 2 mA, independent of the control voltage. The collectors of Q1 and Q3 will thus reside at about +4.8 V regardless of the control voltage. Input and feedback signal currents flow in proportion to the DC currents of the corresponding transistors Q1 and Q3.

The voltages at the emitters of Q5 and Q6 are logged versions of the input and feedback signal currents. These nonlinear signals subtract from the input and feedback voltages across R1 and R2, causing some distortion in the V/I conversions. Suppose the current in Q5 swings from 1.5 to 2.5 mA with input signal voltage of 5.1 $V_{p\text{-}p}$. On a small-signal basis *re'* of Q5 is 13 Ω, implying a logged signal amplitude at the emitter of very roughly 13 $mV_{p\text{-}p}$. The nonlinear log signal is about 0.06% of the signal swing. If the log signal were all distortion, THD would be about 0.06%, but it is not, so things are not quite that bad. A reduction in the THD of the V/I conversion could be had by making Q5 and Q6 CFPs so that the dynamic impedance at the emitters is very small.

Complementary emitter followers Q7 and Q8 place the same voltage on the collectors of Q2 and Q4 as on Q1 and Q3. This keeps V_{cb} of the transistors the same and helps log conformance. Power supply rejection for this circuit is poor, since noise on the power supply has a path through R3 to the op amp virtual ground. The power supply must be very quiet, perhaps by use of a capacitance multiplier.

As V_{ctrl} increases, the DC current in Q1 increases and that in Q3 decreases, providing a degree of control feedthrough cancellation that depends on matching, particularly between I1 and I2.

When $V_{ctrl=}0$, all transistors are conducting the same amount, 1.0 mA. Half of the input signal current goes through Q1 to the op amp and half of the feedback signal goes through Q3 to the op amp. Gain is then unity.

With V_{ctrl} strongly positive, Q1 and Q4 are on and Q2 and Q3 are off. The full input signal from Q1 goes to the op amp and no feedback signal from Q3 goes to the op amp, and gain is extremely high. With V_{ctrl} strongly negative, Q2 and Q3 are on and Q1 and Q4 are off. Full feedback signal through Q3 goes to the op amp and no input signal from Q1 goes to the op amp. Gain is essentially zero.

This circuit easily provides a gain range of ±40 dB as shown, but only a portion of the positive gain range is useful in most applications, such as a volume control. The available range can be shifted by changing the ratio of R2 to R1. If R1 = 10R2, gain would be shifted down by 20 dB, allowing 20 dB more of the positive gain range to be used. Range might then be −60 to +20 dB. In practice, numerical gain can go to zero with sufficient negative V_{ctrl}. Gain would then be −20 dB with V_{ctrl} = 0 but this will impact the available SNR under certain gain conditions.

The transistors in the LM1496 do not have a great log conformance FOM (β/r_b), so distortion of this circuit, as implemented, is not as good as it can be. Similarly, the base resistance of these transistors is not particularly low, so noise performance is also not as good as it can be. This circuit has decilinear gain vs control voltage behavior over a very wide gain range and is capable of providing a very wide gain range, easily from −40 to +40 dB. The single-ended nature of the core of this circuit allows noise from current sources I1 and I2 to contribute to the noise of this VCA.

Noise originates mainly from the undegenerated Q1−Q2 and Q3−Q4 differential pairs, which have fairly high transconductuce with each transistor operating at 1 mA and unity VCA gain. Under these conditions, the effective noise gain is controlled by feedback resistor R2; a larger value of R2 results in higher noise gain multiplying the noise from the two differential pairs. At V_{ctrl} = 0 V, simulated output noise is about 383 nV/√Hz, corresponding to SNR = 86 dB in a 20-kHz bandwidth relative to 1 V_{RMS} output when using transistors with 40-Ω base resistance. Distortion is about −75 dB (0.02%) at 1 V_{RMS} output, and results mainly from distortion in the V-I conversion where R1 and R2 connect to the emitters of the differential pairs. This distortion is almost entirely 2nd harmonic and is greatly reduced if CFP cascodes are used instead of Q5 and Q6. Simulations show that the introduction of the CFP cascodes reduces distortion to about 0.001%. A more sophisticated version of the VCA would include degenerated PNP differential pairs loaded with current mirrors providing the input and feedback currents to the emitters of Q5 and Q6.

Frey VCA

Recall that in the emitter-driven VCA of Figure 25.2, the differential output was taken from the outer pair of core transistors, Q1 and Q4. They conduct more current than the inner pair of transistors if V_{ctrl} is positive in the case of higher gain. The signal from the inner pair was thrown away, with the collectors connected to the rail. The signal from the outer pair can be sent to a differential output amplifier. This amounts to feedforward gain control. If the output of the inner pair is instead equipped with load resistors and applied to a differential amplifier, a signal is created that can be used as negative feedback. The core acts as a router for the signal currents applied as feedback or feedforward (not unlike the feedback VCA just discussed).

Figure 25.5 shows a simplified version of the Frey VCA. It can be seen that the outer pair Q1–Q4 feeds a portion of the emitter-driven signal to the output differential amplifier U2. At the same time, the inner pair Q2–Q3 feeds the remainder of the emitter-driven signal through differential amplifier U1 to create a negative feedback signal that is routed back to the input.

The NFB signal is subtracted from the input signal by U3, forming a negative feedback loop that will tend to reduce gain. At higher gain settings, the feedback signal is smaller and there will be less gain reduction from the feedback, thus enhancing the increased gain from the feedforward path. There are now two mechanisms that work together to change the gain of the VCA. This is the basic idea behind the Frey VCA [26–31].

The control voltage determines the ratio G of the output signal to the feedback signal as an exponential. The ratio changes by one decade for each 60 mV change in the control voltage. If V_{ctrl} is zero, the ratio G is unity and the feedforward and feedback signal amplitudes are equal. VCA gain

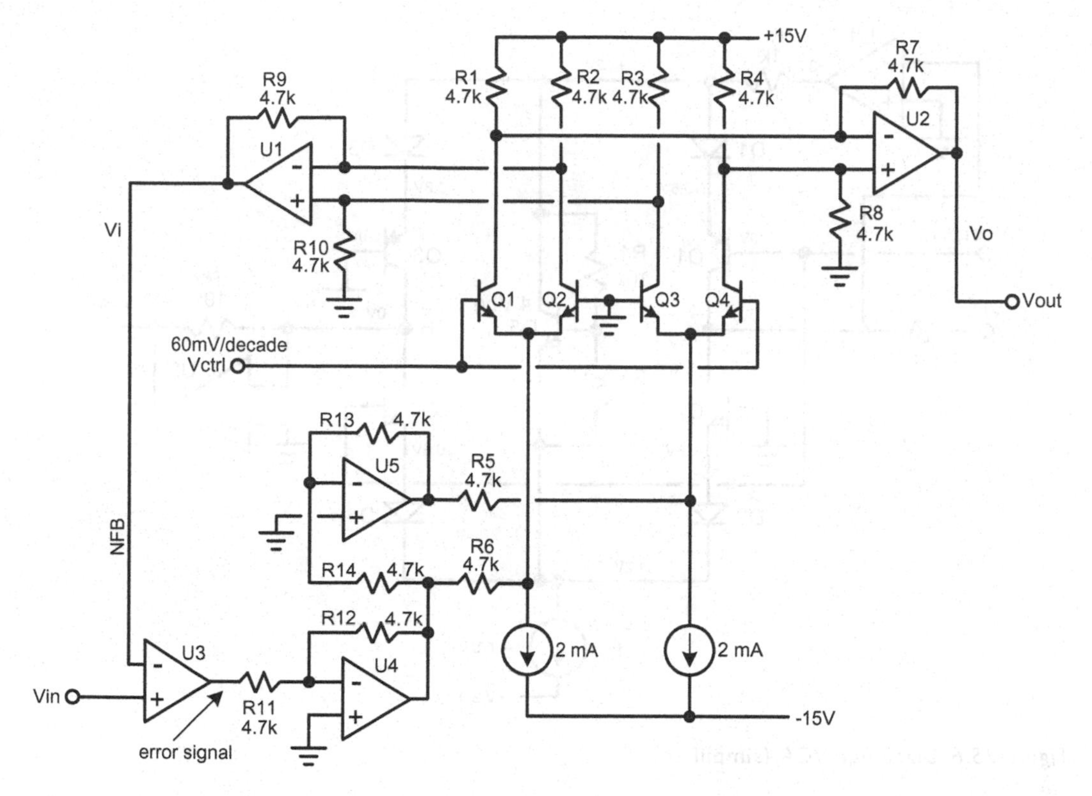

Figure 25.5 Frey VCA.

is then unity. If the control signal is positive by +60 mV, the feedforward signal will be ten times as large as the feedback signal, and *G* will be 10. If the control signal is −60 mV, the feedback signal will be ten times larger than the feedforward signal, and the gain will be 0.1.

The input signal and the feedback signal are subtracted at U3 to form an error signal that is applied differentially to the LTP emitters by U4 and U5. Note that U3 has no local feedback so that its gain is very high. This means that the feedback signal is made to equal the input signal by the high gain of the feedback loop. The output signal must then be *G* times the input signal.

Versions of the Frey VCA circuit can be found in several commercial VCA chips, but which are in limited manufacture [32–36].

Blackmer VCA

The Blackmer VCA is arguably the most popular VCA used in audio systems. It is made by THAT Corporation, and has very good performance and is well-supported [2, 3, 7, 15–21]. All of the VCAs discussed thus far have operated in class A, meaning that none of the transistors ever completely shuts off and that both polarities of the signal excursion follow the same path through the VCA. In the Blackmer VCA there are two parallel signal paths, each with a log-antilog functionality. One path, using PNP transistors, handles positive signal swings. The other path, using NPN transistors, handles negative signal swings. During most of a signal swing, only one of these paths is passing signal. For this reason this is called a class AB VCA.

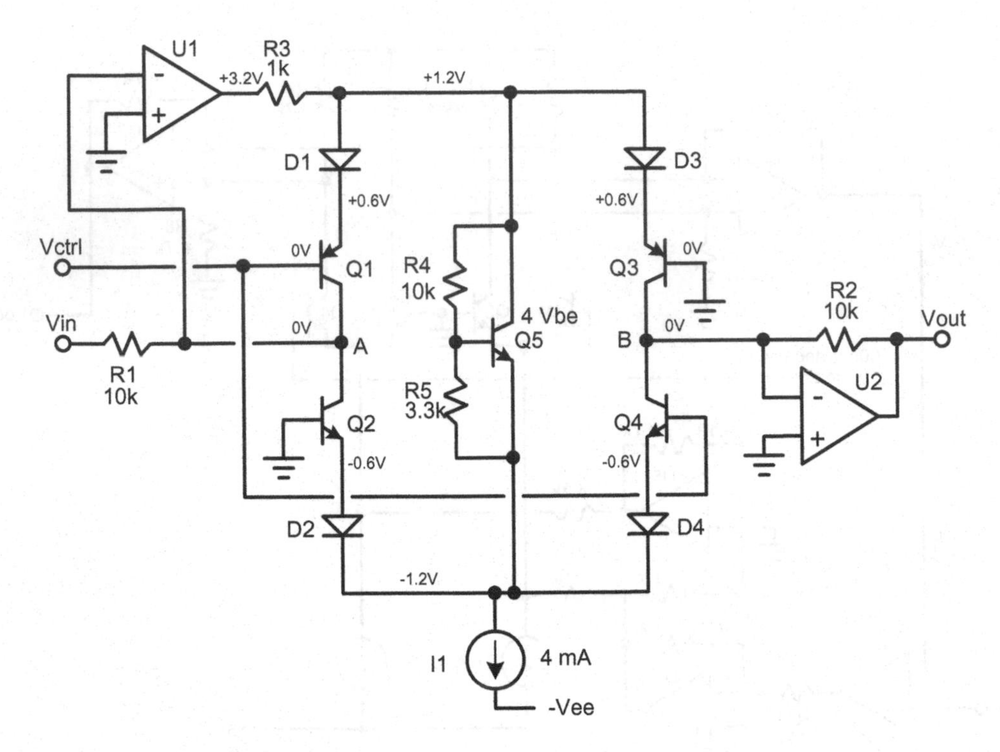

Figure 25.6 Blackmer VCA (simplified).

This is not unlike the situation in a class AB power amplifier, where there are two power stage paths, one NPN and the other PNP. Figure 25.6 is a highly simplified schematic that illustrates its operation.

As shown in Figure 25.6, the input signal is applied as a current to Q1 and Q2, the PNP and NPN input logging transistors. Op amp U1 forces the sum of the PNP and NPN collector currents to equal the input signal current. The V_{be} multiplier bias spreader Q5 forces some quiescent current to flow through Q1 and Q2 when there is no signal or a very small input signal.

On positive input signal swings, most of the signal current flows through Q2, the NPN logging transistor. The bias spreader, whose impedance is low, forces the same logged input signal swing to be the same on the upper and lower signal paths. As a result of the current in Q2, the upper and lower signal paths from the emitters of Q1 and Q2 tend to move negative in accordance with the log value of the signal, with the lower signal rail pulling down on the emitter of Q2 through diode D2. Assume that all of the diodes are actually diode-connected transistors.

This causes the antilog cascode transistor Q3 to conduct a linear signal current into the transimpedance amplifier U2, resulting in a negative output swing. Assume that $V_{ctrl}=0$, so that the bases of Q3 and Q4 are at ground potential. The logged voltage swing on the bottom signal rail will pull down the emitter of Q4, causing it to perform the antilog function and conduct a linear current. This current causes the output of transimpedance amplifier U2 to swing positive. Note that with the signal rails swinging negative in log fashion under these conditions, the current in D3 and Q3 is greatly diminished.

The control voltage applied to the bases of Q1 and Q4 causes differences in the operating currents between Q1 and Q2 and between Q3 and Q4. Note that the collectors of Q1 and Q2 are forced to be at 0 V by the feedback action of U1. If the control voltage is zero, then all of the bases are at 0 V. This means that under these conditions the transistors are operating at 0 V collector-to-base.

Operating the transistors at low current in the quiescent state reduces VCA noise. Similarly, under signal swing, with one half of the transistors off, noise is reduced because they are not wastefully conducting current under those conditions.

The Blackmer VCA is widely used in the professional audio industry. First introduced by dbx, various high-performance versions of this circuit have been manufactured by THAT Corporation over the years with a high-performance complementary PNP-NPN process employing laser trimming.

25.3 VCA Applications

Volume Control

The gain of a signal channel is controlled by a DC voltage supplied from a pot or a DAC. The number of dB change in gain is linear with the voltage delivered to the VCA if a decilinear VCA is used, and linear with pot rotation if a pot is used to create the control voltage. This allows a pot to control gain over an 80+ dB range conveniently with the same rotational gain sensitivity regardless of the gain amount. Obviously, multiple identical VCAs can be used to form the equivalent of a multi-gang potentiometer where matching among "gangs" is very good. This application was covered in Chapter 22. VCA applications were discussed in Refs. [37, 38].

Balance Control

If a stereo volume control is implemented by 2 VCAs, it is trivial to incorporate a balance control by merely inserting a differential DC control voltage between the control inputs of the 2 VCAs. This application was covered in Chapter 22 [39].

Panpot

A panning potentiometer (pan pot) is used in a mixer to direct a monophic signal from a mixer channel to left and right output circuits in different proportions to place the signal from left to right in the stereo soundstage. In similar fashion to a volume and balance control arrangement, a pan pot can deliver opposite control voltages to 2 VCAs in the signal path of a stereo signal. This application was covered in Chapter 22 [39].

Compressors

The ability to control gain with an electronic control signal is fundamental to the operation of a compressor. In the simplest compressor, the signal output amplitude from the VCA is used to increase VCA attenuation as the signal level increases. The sharpness of the compression behavior depends on the loop gain of the DC control loop. With high gain in the control loop, the compressor becomes a limiter that does not alter the wave shape of the signal. A side-chain compressor controls the VCA gain with a DC control signal that represents the amplitude of the input signal (feedforward) or of another signal (cross-compression). In a compressor, it is all about how the control for the VCA is implemented. This application is covered in Chapter 26 [40].

Automatic Gain Control Circuit

Automatic gain control circuits (AGC) are usually employed to keep the signal level at the output of a circuit to a fixed value or a small range of values. The AGC detects the amplitude at the output of the circuit and feeds back a DC control voltage to alter the gain of the circuit electronically, using negative feedback to achieve the desired output amplitude. If the rectified output of the VCA is fed back to the control input with substantial control loop gain, and the rectified voltage is compared to a reference level, then the output voltage will be at the reference level regardless of its amplitude [25].

Autoscaling

In a THD analyzer, the percentage readout should not change as a result of signal fundamental amplitude change. In essence, the distortion residual should be divided by the fundamental amplitude and thus referenced to the fundamental amplitude. This autoscaling can be achieved by using a VCA to AGC the fundamental amplitude to a reference level, like 1 V_{RMS}. The control voltage for that VCA that is required to adjust the gain of the VCA to put the fundamental amplitude at the reference level is then used to control a second identical replica VCA in the path of the distortion residual. Such an arrangement, using two feedback VCAs as described in Figure 25.4, was used in the author's THD analyzer design published in 1981 [25].

AC Voltmeter with Linear dB Readout

If a decilinear VCA is connected as an AGC, so that its rectified output is always at a reference level, then its control voltage is an indication of the signal level in dB with respect to the reference level. Such an arrangement is shown in Figure 25.7, where a version of the feedback VCA in

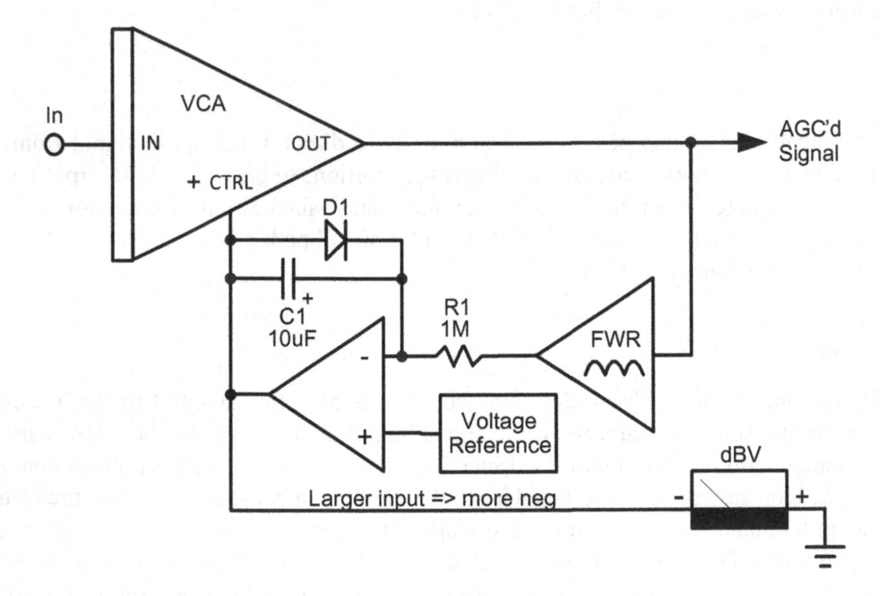

Figure 25.7 AC Voltmeter with Readout Linear in dB.

Figure 25.4 is employed. The VCA is configured as an AGC circuit producing a reference level and the control voltage fed to the VCA is read out on a linear scale.

Tunable State Variable Filters and Oscillators

A state variable filter (SVF) comprises two integrators in tandem with an inverting summer stage completing the loop. The Q of the filter is determined by feeding back some of the output of the first integrator to the inverting summer. The center frequency can be controlled by placing variable gain elements in both of the integrator paths. This was discussed in Chapter 11. Three VCAs can be employed to accomplish this task. One is placed in front of each integrator, to set the integrator gain and thus the center frequency. The center frequency will then be adjusted by some fraction of a decade per mV of control voltage. Similarly, the Q of the filter can be controlled by a third VCA in the path of the feedback from the first integrator to the summer. A voltage-controlled parametric equalizer can thus be formed. Parametric equalizers were discussed in Chapter 20.

In like fashion, a state variable oscillator can have its frequency controlled over a large frequency range with logarithmic behavior. In the case of the oscillator, an amplitude control circuit with limited range is used in the path from the output of the first integrator to the summer in such a way that slightly positive or slightly negative feedback can stabilize the oscillator output amplitude.

Automation

Automation of the gain or filtering of different signal paths can be accomplished with a microcontroller by sending a digital representation of a control signal to a DAC that subsequently creates an analog DC control voltage for the specified VCA.

References

1. THAT Corporation, *A Brief History of VCAs*, www.thatcorp.com/History_of_VCAs.shtml.
2. David Blackmer, *Multiplier Circuits*, U.S. Patent 3,714,462. January 30, 1973.
3. Wikipedia, *Blackmer Gain Cell*, wikipedia.org.
4. Willy M.C. Sansen and Robert G. Meyer, Distortion in Bipolar Transistor Variable Gain Amplifiers, *IEEE Journal of Solid State Circuits*, vol. SC-8, no. 4, August 1973, pp. 275–282.
5. Ben Duncan, *VCAs Investigated*, Parts 1–3, Studio Sound and Broadcast Engineering, June, pp. 82–88, July, pp. 58–62 and August, 1989, pp. 59–66, worldaudiohistory.com/Archive-All-Audio/Archive-Studio-Sound/80s/Studio-Sound-1989-06.pdf and -07.pdf and -08.pdf.
6. Glen Ballou, *Handbook for Sound Engineers*, Chapter 16.3.4, Focal Press, New York, NY, 2015.
7. THAT Corporation, *THAT 2181 Voltage Controlled Amplifier*, Datasheet, 2016.
8. Barrie Gilbert, A New Wide-Band Amplifier Technique, *IEEE Journal of Solid State Circuits*, vol. SC-3, no. 4, December 1968, pp. 363–365.
9. Barrie Gilbert, A Precise Four-Quadrant Multiplier with Subnanosecond Response, *IEEE Journal of Solid State Circuits*, vol. SC-3, December 1968, pp. 365–373.
10. Barrie Gilbert, *Translinear Circuits: A Proposed Classification*, Electronics Letters, January 9, 1975.
11. Evert Seevinck, *Analysis and Synthesis of Translinear Integrated Circuits*, Elsevier, Amsterdam, 1988.
12. Evert Seevinck and Remco J. Wiegerink, Generalized Translinear Principle, *IEEE Journal of Solid-Sate Circuits*, vol. 26, no. 8, August 1991, pp. 1098–1102.
13. Paul R. Gray, Paul J. Hurst, Stephen H. Lewis and Robert G. Meyer, *Analysis and Design of Analog Integrated Circuits*, 4th edition, John Wiley & Sons, 2002, pp. 708–720.

14 THAT Corporation, *THAT 2180 Blackmer Pre-Trimmed IC Voltage Controlled Amplifiers*, Datasheet, 2008.
15. Robert W. Adams, *Gain Control Systems*, U.S. Patent 4,331,931, dbx, Inc. November 1, 1979.
16. Daniel B. Talbot, *Multiplier Circuit*, U.S. Patent 4,316,107, dbx Inc., February 16, 1982.
17. David E. Blackmer, *Gain Control Systems*, U.S. Patent 4,403,199, dbx, Inc. September 6, 1983.
18. David R. Welland, *Multiplier Circuit*, U.S. Patent 4,454,433, dbx, June 12, 1984.
19. David R. Welland, *All NPN Variably Controlled Amplifier*, U.S. Patent 4,471,324, dbx, September 11, 1984.
20. Gary K. Hebert, *An Improved Monolithic Voltage-Controlled Amplifier*, 99th Convention of the Audio Engineering Society, New York, October 6–9, 1995.
21. THAT Corporation, *VCA Symmetry Auto-Trim Circuit*, 2010, https://thatcorp.com/datashts/dn121.pdf
22. David Baskind and Harry Rubens, *Techniques for the Realization and Applications of Voltage-Controlled Amplifiers and Attenuators*, Presented at the 60th Convention of the Audio Engineering Society, *Journal of the Audio Engineering Society* (Abstracts), vol. 26, July/August 1978, p. 572, preprint 1378.
23. David Baskind and Harry Rubens, The Design and Integration of a High-Performance Voltage-Controlled Attenuator, Presented at the 64th Convention of the Audio Engineering Society, *Journal of the Audio Engineering Society* (Abstracts), vol. 27, December 1979, pp. 1018, 1020, preprint 1555.
24. Harry Rubens et al., *Voltage Controlled Attenuator*, U.S. Patent 4,155,047, May 15, 1979.
25. Robert R. Cordell, Build a High Performance THD Analyzer, *Audio*, vol. 65, July, August, September 1981, available at www.cordellaudio.com.
26. Douglas R. Frey, An Integrated Voltage-Controlled Building Block, Presented at the 81st Convention of the Audio Engineering Society, *Journal of the Audio Engineering Society* (Abstracts), vol. 34, December 1986, p. 1031, preprint 2403.
27. Douglas R. Frey, *The Design of High-Performance Voltage-Controlled Equalizers*, presented at the 83rd Convention of the Audio Engineering Society, *Journal of the Audio Engineering Society* (Abstracts), vol. 35, December 1987, p. 1059, preprint 2527.
28. Douglas Frey, The Operational Voltage-Controlled Element: Generalizing the VCA, *Journal of the Audio Engineering Society*, vol. 39, no. 10, October 1991, pp. 775–784.
29. Douglas R. Frey, *Voltage Controlled Element*, U.S. Patent 4,471,320, September 11, 1984.
30. Douglas R. Frey, *Monolithic Voltage Controlled Element*, U.S. Patent 4,560,947, December 24, 1985.
31. Douglas R. Frey, *Voltage Controlled Amplifier*, U.S. Patent 5,528,197, June 18, 1996.
32. Precision Monolithics, *SSM-2014 Operational Voltage Controlled Element*, Datasheet, 1987.
33. Analog Devices, *SSM-2018 Trimless Voltage Controlled Amplifier*, Datasheet, 2013.
34. Derek F. Bowers, *Voltage Controlled Amplifier with a Negative Resistance Circuit for Reducing Nonlinearity Distortion*, U.S. Patent 5,587,689, December 24, 1996. ADI.
35. Robert G. Meyer, *Wide Dynamic Range Variable Gain Amplifier*, U.S. Patent 6,049,251, April 11, 2000, assigned to Maxim.
36. Analog Devices, *Precision Variable Gain Amplifiers* (*VGAs*), MT-072, 2009.
37. Rod Elliott, *VCA Techniques Investigated*, December 29, 2012, https://sound-au.com/articles/vca-techniques.html.
38. Rod Elliott, *VCA Preamplifier*, Project 141, December 28, 2012, https://sound-au.com/project141.htm.
39. Cordell, Bob, *VCA-Based Faders and Panning Pots*, cordellaudio.com.
40. Leslie B. Tyler, *Signal Amplitude Compression System*, U.S. Patent 4,182,993, January 8, 1980.

Chapter 26

Compressors and Other Dynamic Processors

Compressors can be used to prevent overload or to increase the average perceived loudness level. They reduce dynamic range in chosen regions of the signal amplitude range. Limiters provide a "hard" form of compression that has a steep onset and will prevent the output signal from exceeding a predetermined level no matter how high the input. Limiters generally do not distort the signal waveform, as clippers do.

Compressors are also used to make artistic changes to the sound, especially in recording studios. Compressors will sometimes be used to reduce the dynamic range of a given band of frequencies. Compressors are also valuable tools in managing gain levels and audibility among multiple channels. Sometimes compressors are linked together in some way, or the control circuitry of a compressor will respond to the audio level in a different channel while compressing its channel. The use of a compressor can also improve intelligibility. Indeed, multi-band compressors play a central role in modern hearing aids.

26.1 Compressors

A simple compressor decreases the gain of a channel as the input signal level increases, so as to reduce the dynamic range of the signal at its output. A simple compressor monitors its output and feeds back to a voltage-controlled gain element to reduce gain as the output level increases. In the limit, this action is not unlike that of an AGC (automatic gain control) circuit. Other compressors are arranged so that the voltage-controlled gain element is responsive to the input amplitude in a feedforward fashion. Compressors can be used in a great many ways and have numerous architectures [1–5].

Versatility

It is important to recognize that compressors are not used just to reduce dynamic range and thereby increase the perceived loudness of the material, as is often done in radio stations and in the *loudness wars* [6, 7]. They are used to alter the sound in an artistic way in recording studios, for example. This is often accomplished by changing attack and release times. If the attack time is made longer, overshoot in the signal may occur, but this effect may be desired in order to bring out the transient of an instrument, like the thwack of a drum.

Dynamically Changing the Gain

At the heart of compressors, limiters and other dynamic processing functions is an electronically controlled gain element, often implemented with a voltage-controlled amplifier (VCA) integrated circuit [8–10]. Other compressors are implemented with a JFET voltage-controlled resistor (VCR) or light-dependent resistor (LDR). In the past, vacuum tubes with a remote cutoff characteristic were used to control gain with a variable bias voltage. Even today some prefer the sound of variable mu® compressors, such as those from Manley, often based on the 5670 dual triode. The legendary Fairchild 660/670 from the 1950s is still one of the most revered compressors based on the variable mu concept. It was based on the 6386 vacuum tube for the variable gain element.

The remainder of the compressor largely involves the circuitry that controls the gain in a way that is responsive to signal amplitudes. The details of how the control works define the differences among compressors and their settings. Although the compressor is just one form of dynamic processing, much of this discussion will concern compressors because most dynamic processing has its roots in how compressors work.

In the most basic sense, a compressor reduces the dynamic range of a signal, making soft sounds louder and making loud sounds softer. More commonly gain is reduced when a signal exceeds a certain level, called the *threshold*. In the simplest case, the level is measured at the output of the compressor and a control signal is fed back to the variable gain element to reduce the signal level. As the output signal rises further above the threshold, the control signal for reducing gain is increased and gain is further reduced. For signal levels below the threshold there are no changes in gain, and the compressor is transparent. For a compressor with nominal gain of unity, the output level will always be equal to, or less than, the input level. This functionality has a lot in common with AGC. There are also feedforward compressors in which the gain element is controlled by the amplitude of the input signal.

In most cases, the output level will continue to increase as the input level increases when the signal level is above threshold, but the output amplitude will not increase as fast as the input amplitude increases. This behavior describes the so-called *compression ratio*. Above threshold, if the input signal increases by 2 dB and the output signal increases by only 1 dB, then the compression ratio at that point is said to be 2:1. Below threshold, the compression ratio is obviously 1:1.

A very high compression ratio, like 10:1, corresponds to what is called a limiter. The limiter will preserve the signal shape (unlike a clipper), but will rather strictly limit the maximum output amplitude. A limiter is thus merely a compressor with a very high compression ratio. Figure 26.1 illustrates the input-output characteristic of compressors with different compression ratios. A simple 2:1 compressor with a threshold at 0 dBV is shown in (a), while a limiter with a compression ratio of 10:1 and a threshold of +5 dBV is shown in (b). Above threshold, its output increases by only 1 dB when the input increases by 10 dB. The threshold and compression ratio are two of the most fundamental controls on a compressor.

In the simple feedback compressor example curve labeled (a), the gain of the compressor will be unity as long as the signal level at the output of the compressor is below 0 dBV. No control signal or change in DC control voltage will be fed back to the variable gain element under these conditions. For output signals above 0 dBV, a control signal will be fed back to the gain element to decrease its gain as the output signal level increases further. For example, if the output signal is at +1 dBV, the gain element will be controlled to have 1 dB of loss. This means that a 2-dB increase in input signal level will have been required to reach an output level of +1 dBV. The change in incremental gain of the compressor here happens quickly as the input signal level goes above 0 dB. The change in incremental gain at 0 dBV is called the *knee* of the compression characteristic. A compressor that suddenly transitions from a 1:1 compression ratio below threshold to a 10:1 compression ratio above threshold, as in (b), is said to have a very sharp knee. Limiters often operate with a sharper knee.

Assume that a decilinear VCA is used with control gain of 10 dB/Volt in a feedback compressor. Attenuation is governed by output voltage when the input voltage is above the threshold. The

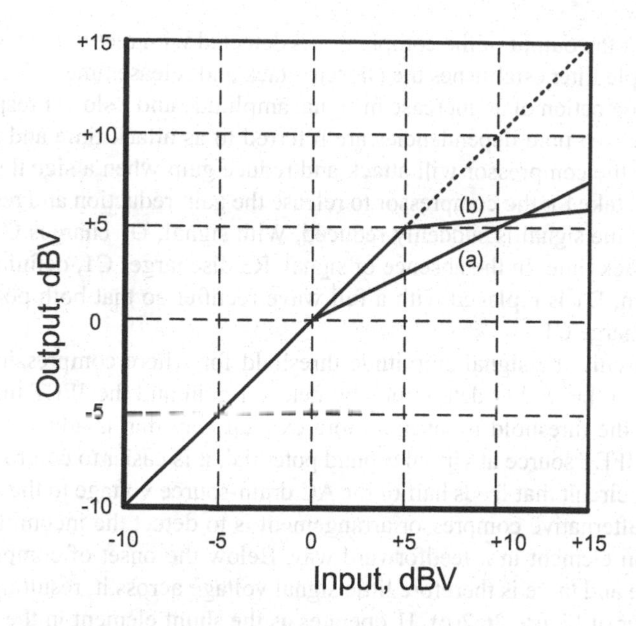

Figure 26.1 **Compressor Output Amplitude vs Input Signal Level.**

compression ratio is then determined by the gain from the output to the control input of the VCA. If the output level is measured in Volts/dB, the compression behavior is easier to understand, since a feedback loop is formed whose loop gain is constant. An input-referred analysis can be used to obtain the compression curve. We assume an output voltage in dBV and determine what input voltage in dBV would be necessary to achieve that amplitude at the output. This is useful in analyzing a feedback compressor. The input-referred approach basically breaks the feedback loop for analysis. In a feedforward compressor, the VCA is controlled by the input amplitude above threshold. If the input goes up 10 dB above threshold, attenuation goes up by K*10 where K equals the VCA control gain. If the signal amplitude is detected as a linear voltage, rather than in V/dB, as is often the case, analysis is less straightforward.

26.2 A Simple Compressor

Figure 26.2 illustrates three very simple feedback compressor designs. The circuit in (a) employs an N-channel JFET as the voltage-controlled gain element. Its variable resistance determines the gain. The JFET J1 is connected as a shunt in the input path, forming an attenuator with R1. The function labeled "bias" provides a negative voltage at the gate of J1 sufficient to pinch it off in the absence of any positive voltage at the output of the full-wave rectifier. A threshold for turning on J1 is established by the bias voltage and the threshold voltage of J1. As the output signal exceeds a threshold, it reduces the reverse bias provided by the function labeled "bias." This moves the JFET into the on state and causes attenuation of the input signal based on the amplitude of the output signal.

The simple compressor in Figure 26.2(b) configures J1 as a variable resistance in series with the virtual ground input of the op amp. As the output signal amplitude exceeds a certain voltage, rectifier D1 creates enough negative voltage to begin and continue to reverse bias J1, increasing its resistance and decreasing the gain. In the absence of signal at the output, the gain is equal to the ratio of R2 to the sum of R1 and the R_{DS_ON} resistance of J1. R1 limits the maximum gain to less than 10.

The signal level at the output of the compressor is detected with rectifier D1, creating a negative gate voltage. Its ripple filter establishes the chosen attack and release times. The filter is usually arranged to have fast detection of an increase in signal amplitude and a slower response to a decrease in amplitude. These two time dependencies are referred to as attack time and release time. They define how quickly the compressor will attack and reduce gain when a signal suddenly increases and how long it will take for the compressor to release the gain reduction and restore the gain to its nominal value after the signal is suddenly reduced. With signal, D1 charges C1 negative through R2, defining the attack time. In the absence of signal, R3 discharges C1, defining the release time. In a practical design, D1 is replaced with a full-wave rectifier so that both positive and negative signal excursions charge C1.

In this simple circuit, the signal amplitude threshold for where compression action begins is not explicitly defined, but rather determined by detector gain and the JFET threshold voltage. In most compressors, the threshold is set in a more explicit way that is adjustable with reasonable accuracy. With the JFET source at virtual ground potential, it is easier to control the gate and add a distortion-reducing circuit that feeds half of the AC drain-source voltage to the gate. Later, we will see that a popular alternative compressor arrangement is to detect the incoming signal amplitude and control the gain element in a feedforward way. Below the onset of compression, J1 is in its low-resistance state and there is therefore little signal voltage across it, resulting in low distortion.

In the compressor of Figure 26.2(c), J1 operates as the shunt element in the feedback loop of a non-inverting stage. When the output signal is zero, J1 is in its R_{DS_ON} state and the gain is equal to the ratio of R2 to the sum of R1 and the R_{DS_ON} resistance of J1, resulting in gain less than 11, depending on R_{DS_ON} of J1. When J1 is reverse-biased into the off state by a large output signal, the gain ultimately falls to unity. A disadvantage of this circuit is that the gain of the compressor can never go below unity.

These three simple compressor circuits are not very practical in the simple implementations shown, but they illustrate the principles of JFET-based compressors. Practical implementations require the use of a trimmer to account for variations in JFET threshold voltage. Selection of JFETs for threshold voltage and R_{DS_ON} may be necessary. Some additional circuitry is also needed to properly establish the compressor threshold.

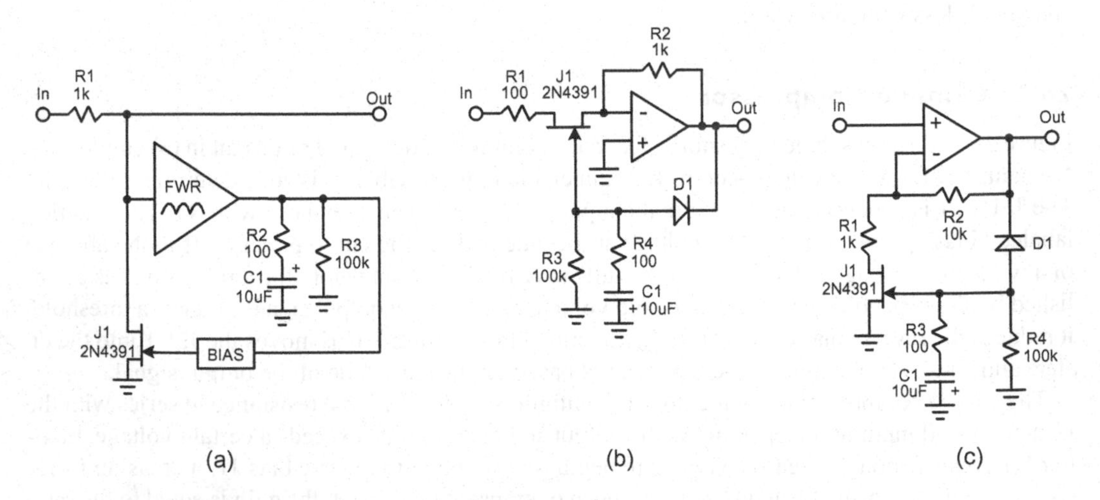

Figure 26.2 Simple Feedback Compressor Designs.

Makeup Gain

Although most people think of the compressor as being there to make the apparent sound louder, as in the *loudness wars*, the compressor itself is technically viewed as being a device that reduces sound level, if its no-signal reference gain can never be greater than unity; it attenuates loud sounds. Alone, it will thus act to reduce the average loudness if the signal amplitudes are above threshold. The apparent sound level at the output of the compressor will decrease, albeit perhaps only infrequently, depending on where the threshold is set. Because the average gain of the compressor is less than unity, in some applications *makeup gain* will be added after the compressor to restore the perceived loudness to its uncompressed level. The net effect here is to increase the sound level of signals below threshold by the amount of the selected makeup gain. This means that softer sounds will be more audible. Of course, taken to an extreme with low thresholds, high compression ratios and significant makeup gain, the program material becomes victim to the all-too-familiar program loudness wars.

Usage Context

Keep in mind that compressor behavior and functionality must be viewed in the context of the purpose for which the compressor or limiter is being used. For example, it is obvious that one important use is to avoid signal levels that would cause clipping or over-modulation downstream. In a similar context, the compressor, in conjunction with gain controls, may be used to enforce maximum permissible perceived loudness levels as dictated by government regulations. In these cases, the compressor will be acting on the full content of the program material in the outgoing stereo channels.

In sharp contrast, a compressor may be used on individual tracks in the multi-track recording stage of a studio recording. This is a more artful application of compression to achieve a better overall sound. In this case, a compressor might be used to tame an instrument or group of instruments that have high dynamic range and can be dominatingly loud. On the other hand, it might be used to effectively boost the sound of a singer's voice when it might otherwise fade too low to be heard in the presence of the background music. At a later stage in production, a compressor might be used to act on the complete contents of a stereo track to improve the overall presentation.

26.3 Compressor Attributes

Four attributes largely define the behavior of a conventional compressor, as defined below. Most compressors allow adjustment of all of these features.

Compression Ratio

When the compressor is compressing, the output will increase more slowly than the input increases. If a 1-dB increase in input signal causes a 0.5-dB increase in output signal the compression ratio is said to be 2:1. Compression ratio is simply $\Delta dB_{in}/\Delta dB_{out}$. It is important to recognize that the compression ratio is not necessarily constant as a function of signal amplitude.

Compressor Threshold

As the input signal increases from zero to the compressor threshold, no compression is taking place and the compression ratio is unity. At signal amplitudes above the threshold, the compressor attenuates the signal in accordance with the set compression ratio. It is important to realize that the gain transition usually does not take place immediately, as often depicted in simple compression curves like those in Figure 26.3, for both practical reasons and because there may be deliberate softening of the knee.

Attack and Release Time

As described earlier, the attack time is a measure of how quickly the compressor can react to an instantaneous increase in signal amplitude. Attack time cannot be zero and is usually set to be in the millisecond (ms) range. Release time describes the gradual increase in gain back to the uncompressed value after the input signal becomes small. It is always longer than the attack time and is usually set to be in the tens to hundreds of millisecond range. The choice of attack and release time can have a big effect on dynamics and on the proclivity for pumping and breathing artifacts.

Compression Knee

The compression knee determines the onset behavior of compression. With a hard knee, signals above threshold immediately experience the full compression ratio as the input signal amplitude rises through the threshold. With a soft knee, the compression ratio gradually increases to its target value as the signal amplitude approaches and goes beyond the threshold. The threshold is thus not well defined as a particular signal voltage. Here we will designate the threshold as the point at which any reduction in gain begins.

Often, a soft knee is preferred. A soft knee can be implemented so that there is a compression transition region where the compression ratio changes from 1:1 to the full compression ratio over some range of signal level below and above the designated threshold. This behavior is illustrated in Figure 26.3 by the curve labeled "soft." Notice that the threshold as defined above lies at a

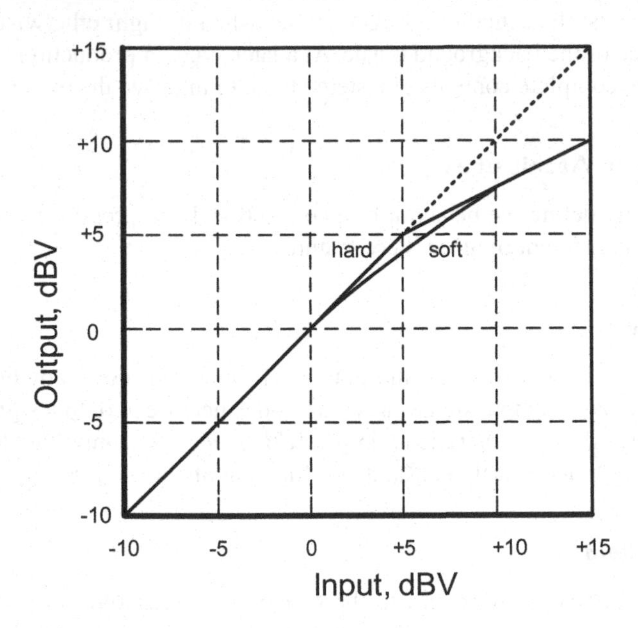

Figure 26.3 Soft-Knee and Hard-Knee Compressor Behavior.

poorly-defined level of about +2 dBV, while the threshold for the hard-knee curve lies at an obvious level of +5 dBV. Some compressors allow adjustment of the width of that range, so that the knee can be softer (wide range) or harder (narrow range).

Breathing and Pumping

A common problem with any dynamic gain-controlled circuit is audible breathing and pumping. When the gain of a system having noise is changed as a function of the signal dynamics, what is called breathing can occur and be heard as a result of the varying noise level. This happens because the perceived noise level is going up and down. It is most noticeable during the release time, when the gain begins to come back up after a compression gain reduction interval.

Pumping is somewhat like breathing, but it is associated with the changing volume of legitimate audio whose amplitude has been modulated by other material in the signal, perhaps like a bass drum or bass guitar. The audibility of such pumping depends a lot on the chosen attack and decay times.

Threshold Circuit and Attack/Release Timing

In a simple feedback compressor, a detector circuit provides a DC level that is indicative of the output signal amplitude. The threshold circuit ignores the output of the detector circuit until its amplitude reaches the threshold level. Above that, the threshold circuit can have a nominal gain of unity or some larger value. An attenuator after the threshold amplifier sets the control path gain and thus the compression ratio.

In a simple feedback compressor arrangement, the attack and release times may be set by the filtering time constants in the detector, as in Figure 26.2. Such an arrangement is shown in Figure 26.4(a), where for clarity the attack/release control signal conditioning is shown as a separate block. In this case the control signal from the attack/release conditioning circuit, with its likely exponential rise and fall characteristics, must pass through the nonlinearity of the threshold circuit before it reaches the VCA. Here the VCA is assumed to reduce its gain with a positive input control signal. When the input signal amplitude suddenly rises as with a tone-burst, the output of the attack/release circuit will begin to rise at a rate in accordance with its timing capacitor and related resistance. Until this voltage reaches the threshold value, no compression will occur. This earlier portion of the rise time just acts like a time delay before compression begins.

As the control signal goes above the threshold, and is still rising as a function of the attack charging time, compression begins and the rate of change of the attack/release circuit output will now affect the rate of change of the compression ratio until the target compression ratio is reached (if the signal transient lasts long enough for the detector signal to go through this whole process). Compression timing is thus a delay plus a time constant. Release is complete once the absence of

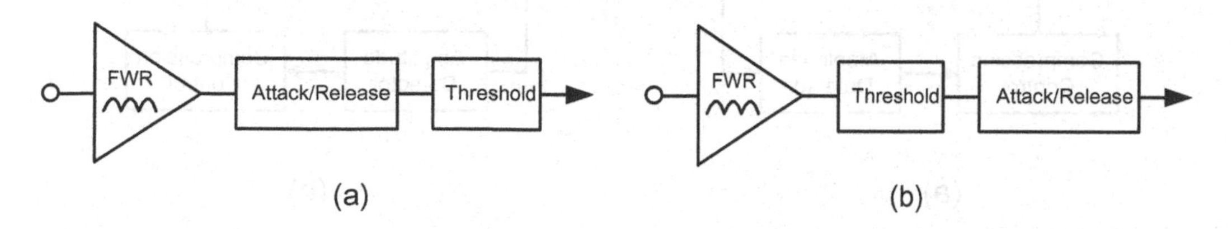

Figure 26.4 Different Placements of the Attack/Release Control Circuit.

signal causes the output of the attack/release circuit to fall below the threshold. Further falling of that voltage to its final value has no effect on compression (which is not occurring at that point).

In other compressor designs, the main action of attack and release time occurs in an attack/release conditioning circuit that comes after the threshold circuit [11]. This arrangement is illustrated in Figure 26.4(b). In such cases, the detector's ripple filter will usually be set to have a short time constant that is equal to or less than the desired fastest attack time. The delay to reaching threshold will thus be short and much less consequential. The charging of the timing capacitor in the attack/release circuitry located after the threshold circuit will have a more immediate effect on the rate at which compression ratio increases to its target value. On release, the falling value of the control signal will continue to affect gain all the way down to its final value.

Whether the bulk of the attack-release time conditioning comes before or after the threshold circuit matters. It can make a difference in how the compressor sounds when transitioning into and out of the compression regime.

26.4 Compressor Architecture

There are numerous ways to build a compressor and many types of functionalities to be had. For example, compression may be frequency-dependent or it may involve interaction between two or more channels. The family of compressors more broadly includes devices with signal-dependent gain, and thus expandors and compandors will also be discussed.

Feedback and Feedforward Compressors

The DC level fed to the gain control circuit can come from detection of the signal amplitude at either the output or input of the compressor, as shown in Figure 26.5. In the *feedback* compressor of (a), the output amplitude is detected and fed back to the control circuit, where threshold, compression ratio, attack time and release time are determined. The output of the control circuit is then fed to the VCA or other type of gain control element. Feedback compressors have been discussed so far to keep things simple in an introductory way. Similar in some ways to an AGC circuit, they were the earliest compressor topologies to be used, and are still used today in some applications, such as compandors and where their sound is preferred.

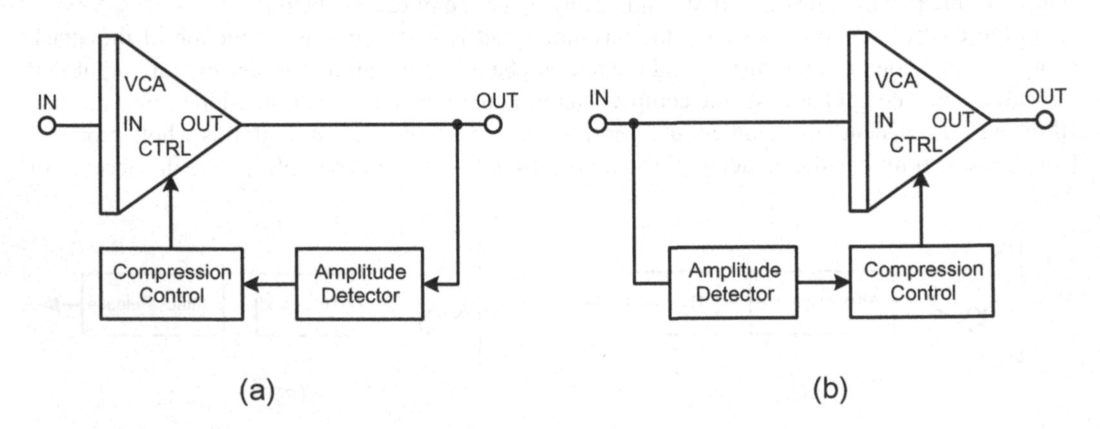

Figure 26.5 Feedback and Feedforward Compressors.

In (b), the amplitude of the input signal is detected and fed to the control circuit, thus forming a *feedforward* compressor. The behaviors will tend to be different for these two arrangements, as one is a feedback circuit while the other is a feedforward circuit.

Feedforward Compressors and Side-Chains

The feedforward compressor is more dominant in modern usage, partly because it is much more versatile. In this case, the amplitude of the input signal is detected, and if it increases above a threshold a control signal is sent to the gain element to reduce gain. Going forward, we will refer to the voltage-controlled gain element as a VCA, even though other types of voltage-controlled gain elements may be used.

As shown in Figure 26.5(b), the input signal is fed to both the VCA and to the compression control circuit in parallel, with the latter being referred to as the *side-chain*. Operation of the control circuit thus is not influenced by the output of the compressor. This can result in different compressor behavior and a different sound. Attack and release behavior is different because there is no feedback at work. Feedforward compressors are made more practical and consistent by having more precise signal detection and controlled gain elements than were available many decades ago. A key difference is that, in the absence of feedback, the control circuit is not aware of the effect it has had on the signal. They are best implemented with VCAs, since the VCA is a more precise controlled gain element.

The feedforward compressor, with its side-chain, is more flexible than the feedback compressor. Indeed, the input to the side-chain need not be the same as the input to the gain control element. It can be processed with a filter or it can come from a different channel. For example, there are compressor configurations whose action is dependent on the amplitude of a different signal, or on the amount of compression being applied to another signal.

The Side-Chain

The chain of circuits that create the control signal for the VCA is usually called the side-chain, since it is not directly in the signal path. It usually consists of the amplitude detector, the threshold circuitry and the nonlinear filter that establishes the attack and release times. This circuitry also controls the compression ratio. This is where the behavior and personality of the compressor is largely determined. The semantics here are a bit muddy, with some reserving the term *side-chain* for feedforward compressors. Others reserve the term *side-chain* for arrangements where the control circuit is fed from signals from a different channel elsewhere in the system, as opposed to the signal being processed. For simplicity, here we will use the term in its broader context to describe the chain of control circuits that feed the VCA in a feedforward compressor, even if the signal comes from an external input. Side-chain compressors are best implemented with VCAs for consistency and precision.

Decilinear Compressors and Log Domain Processing

The term *decilinear* was coined by the company dbx to describe compressors and related devices which use VCAs that respond to a control voltage by causing a decibel gain change that is linearly related to a change in control voltage. Similarly, decilinear amplitude detectors produce a DC voltage that is linearly related to dB changes in the input signal amplitude. A decilinear VCA can thus be described as having a gain control characteristic that is constant when described in dB/mV, such as 0.17 dB/mV. Similarly, a decilinear (log-responding) amplitude detector has a detection characteristic that is constant when described in mV/dB. A typical number for both of these operations is about 6.2 mV/dB. This is the case in THAT's Analog Engine® which contains both a decilinear VCA and a log-responding RMS (root mean squared) detector [12–14].

The analysis of such compressors is greatly simplified if you think in dB rather than in linear quantities [15]. This is especially so when determining compression ratio for either a feedforward compressor or a feedback compressor. It can be shown that for a feedforward compressor the compression ratio C_r above threshold is simply:

$$C_r = dB_{in} / dB_{out} = 1/(1-G),$$

where G equals the gain of the side-chain [15].

The gain of the side-chain is the product of G1, G2 and G3, where G1 is the gain of the log detector in mV/dB, G2 is the gain of the control circuit when the signal is above threshold and G3 is the control gain of the VCA in dB/mV. As expected, G is dimensionless. It is easy to see that if $G=0.5$, then the compression ratio is 2:1. Notice that if $G=1$ in a feedforward compressor the compression ratio is infinite and we have a hard limiter.

For a feedback compressor, it can be shown that the compression ratio for signals above threshold is:

$$C_r = dB_{in} / dB_{out} = 1 + G,$$

where it can be seen that if $G=1$, the compression ratio is 2:1. Notice that even for large G the compression ratio can only approach a very large number. For $G=19$, the compression ratio will be only 20:1, which is adequate for most limiter applications. Keep in mind that the feedback compressor involves a feedback loop; the loop gain is equal to G. For both types of compressors, if G is constant (above threshold) then the compression knee is hard. If a nonlinear circuit is inserted into the chain to make G an increasing function of signal level, then a soft-knee compressor will result.

For many compressors in the real world, especially earlier ones that may not use a decilinear VCA and log-responding detector, the compression ratio is not dictated cleanly by the above math, and behavior in some or all of the side-chain is not decilinear, meaning that G, and thus the compression ratio, is not constant as a function of signal level above threshold. In a simple example, a compressor uses a decilinear VCA, but uses a linear detector. It is easy to see that the detector gain is not constant in the log domain. A 1-dB change in an input voltage of 1 V might cause an output voltage change of (1−0.89)=110 mV. A 1-dB change in an input voltage of 0.1 V might cause an output voltage change of (0.1−0.089)=11 mV. Thus, the log domain gain of the detector is proportional to signal amplitude (rather than constant). Since side-chain gain here is proportional to signal level, we have a natural soft-knee compression characteristic. For a feedback compressor, the compression ratio then becomes something like:

$$C_r = dB_{in} / dB_{out} = 1 + \left(V_{in} / V_{ref}\right) G2G3$$

where V_{ref} is the input signal amplitude at the point where the incremental log domain gain of the detector is unity. For a circuit where a threshold circuit follows a linear detector, the equation might be more like:

$$C_r = dB_{in} / dB_{out} = 1 + \left(\left(V_{in} / V_{ref}\right) - V_{th}\right) G2G3$$

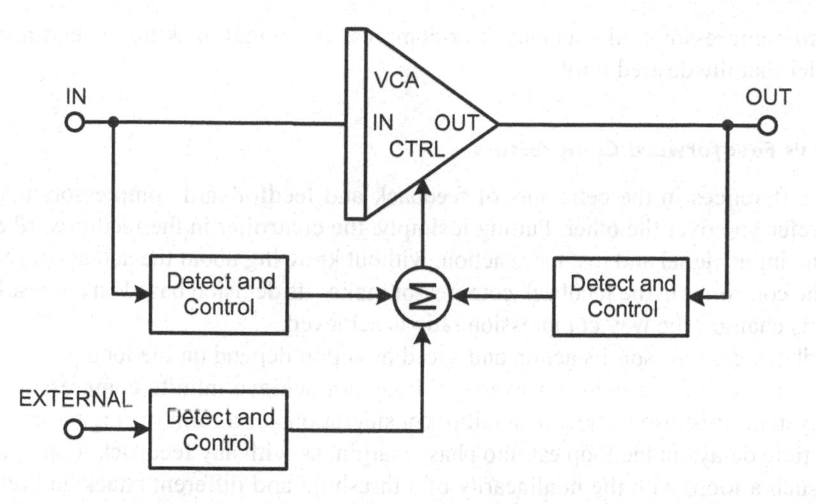

Figure 26.6 A Generalized Compressor/Limiter with Multiple VCA Inputs.

where V_{th} is the amplitude threshold. Similar relationships will hold for a feedforward compressor with a linear detector. For many compressors, V_{th} is set at a healthy signal level that is easily dealt with by a linear detector, so a soft knee and a small cost reduction may result. This is especially the case for feedback compressors, where the dynamic range of the signal at the output has already been compressed.

Generalized Compressors

If one thinks of the controlled gain element as having three control voltage inputs that are summed, it can act as a feedback compressor, a feedforward compressor or an external side-chain compressor, and even operate in these different modes at the same time. This opens up possibilities too numerous to count. A simple example is shown in Figure 26.6 for a compressor/limiter. The compressor function is implemented with feedforward and the limiter function is implemented with a feedback path. In this way a single VCA can accomplish more functions.

Limiters

Although a limiter is really just a compressor with a very high compression ratio, a sharp knee and a fast attack time, it functions as the ultimate gate-keeper of signal amplitude, preventing clipping and damage in downstream equipment. It should rarely come into play. Often, a conventional compressor will be followed by a limiter, so there are actually two compressor functions in tandem. Alternatively, a feedback compressor can perform both functions with separate amplitude detectors whose combined control voltage is fed back to the gain control element. The limiter might have a detector with a higher threshold and faster attack time than the conventional detection path. It might also detect amplitude with a peak detector instead of an RMS detector.

A limiter is simply a compressor with a very high compression ratio approaching infinity. The output signal should not rise more than a small amount above the threshold. Note that such a limiter does not destroy the waveform, as does a clipper.

Limiters are often implemented with a feedback compressor, since the end result of the compression action is what is most important. It is easy to see that with high compression ratios a feedforward compressor could actually over-compress the signal, making the end result numerically smaller that the desired limit.

Feedback vs Feedforward Compressors

There are differences in the behaviors of feedback and feedforward compressors that may lead some to prefer one over the other. Putting it simply, the controller in the feedforward compressor analyzes the input signal and just takes action, without knowing about the actual consequences. In contrast, the controller in the feedback compressor makes its decision based on the result that it has caused. This changes the way compression ratio is achieved.

In a feedback compressor, its action and speed of action depend on the loop gain of the control feedback loop. With finite gain, for example, it cannot achieve infinite compression ratio. As a feedback system, it is also subject to stability considerations or a form of ringing in the compression if the time delays in the loop eat into phase margin, as with any feedback loop. The nonlinear nature of such a loop, with the nonlinearity of a threshold and different attack and release times, can make behavior more complex. The effective loop gain in the log domain may also be different depending on what kind of a level detector is used, depending on whether the detector is linear-responding or log-responding. These issues are usually of much less concern if the compression ratio is not extreme.

Another difference between feedback and feedforward is the way that the compressors react in time to a sudden transient that goes above threshold. Think of a tone-burst input. Neither type of compressor can react instantly to such an event due to attack delay, and so in either case there will be an overshoot in the amplitude of the output signal. The difference in behavior lies in the fact that the feedback compressor sees and responds to the overshoot, tending to speed up its attack because of the feedback loop gain. In contrast, the feedforward compressor's side-chain control is completely unaware of the overshoot. Its attack behavior is unaffected by the overshoot at the output.

External Side-Chain Inputs

Using an external input to the side-chain makes it possible for the output of one track to control the action of a compressor on a completely different track. For example, such an arrangement can be used to reduce the level of a music track when there is dialog to make room for the dialog within the available dynamic range. If the dialog level is detected to go up in one channel, the gain in the music channel is made to go down. It is a form of feedforward compression, where the gain of the compressing element is not made to depend on the amplitude of the signal going into the compressor, but rather an external signal.

Linked Compressors

If a stereo signal is being compressed, it is desirable for both channels to experience the same gain change so that there is no shifting of the stereo image. The left and right gain elements should be matched and receive the same control signal. This is referred to as *linking* of compressors. That control signal might be derived from the side-chain of one compressor and be fed as a control signal to the gain element of the other compressor. Alternatively, an independent amplitude detector that measures the sum of the two channels and feeds forward the same control signal to the left and right compressors' external side-chain inputs could be used. The input to the single side-chain might just be the monophonic sum of the left and right input signals. Such an approach can be used

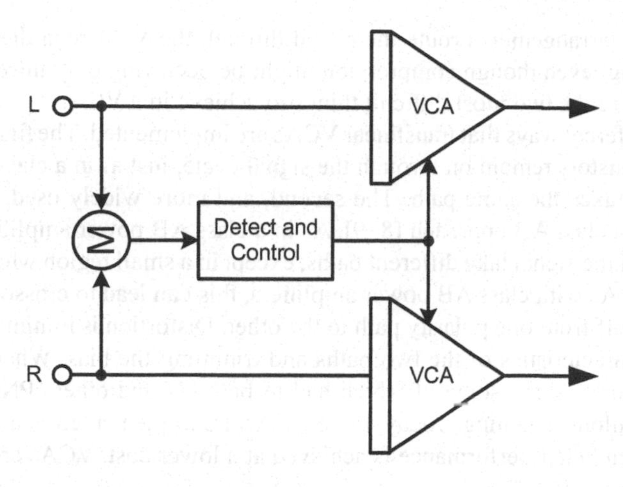

Figure 26.7 Linked Compressors for Stereo.

to preserve the relative balance of the two signals when compression is at work. Such an arrangement is illustrated in Figure 26.7.

26.5 Voltage-Controlled Gain

At the heart of every compressor is the voltage-controlled gain element, often a VCA. It is directly in the signal path, and so affects audio quality. Its control law (the way in which the control voltage changes gain) affects the functionality of the compressor.

Required Gain Range of Control Element

While a voltage-controlled volume control or fader requires a very large gain range, the required gain control range in an above-threshold compressor operating range is often only modest. A volume control or fader requires at least 60–80 dB of gain range, and must be able to be controlled semi-logarithmically to be practical. This means that the control characteristic is reasonably constant in dB/V, perhaps 10–20 dB per volt. Most of the action in a compressor application will be in going from 0 dB attenuation to perhaps 20 dB of attenuation with the possible exception of going to greater attenuation in a limiter.

Voltage-Controlled Amplifiers

VCAs are probably the most widely used gain control element in the pro-audio industry. Their wide gain control range exceeding 80 dB and their linear-in-dB control characteristic makes them unbeatable for volume controls and faders. As described in Chapter 25, they rely on predictable bipolar transistor characteristics in what are called translinear circuits. Matching between devices is thus naturally quite good.

However, for some compressor applications it may be a bit of overkill. When not in the compression region, the compressor should be absolutely transparent, contributing no noise and no distortion. A compressor based on a VCA always has the VCA in the signal path, able to contribute noise and distortion. Even the best VCAs are not perfect in terms of noise and distortion. Most modern compressor arrangements route the signal through the VCA regardless of whether compression is occurring, even though compression might be occurring only infrequently. High SNR and low distortion are the two most difficult things to achieve in a VCA.

There are two different ways that translinear VCAs are implemented. The first is class A, wherein all of the VCA transistors remain on through the signal cycle, just as in a class A power amplifier. The signal always takes the same path. The second, and more widely used, is of the Blackmer topology, which is a class AB approach [8, 9]. As in a class AB power amplifier, the positive and negative portions of the signal take different paths, except in a small region where the signal passes through both paths. As with class AB power amplifiers, this can lead to crossover distortion in the region of the hand-off from one polarity path to the other. Distortion is minimized through careful matching of the characteristics of the two paths and trimming the bias. When these VCAs were built with discrete transistors, some of which had to be NPN and others PNP, this was difficult and involved meticulous matching. Today, these VCAs are implemented in a full-complementary IC process and much better performance is achieved at a lower cost. VCAs are discussed in more detail in Chapter 25.

Using other elements, like photoresistors (LDR) or JFETs can result in virtually no noise or distortion increase when compression is not being applied, which may be most of the time [16, 17]. Such gain control elements are suitable for use in compressors if their limited attenuation range (about 40 dB) and non-uniform dB/volt control characteristics are acceptable. In some compressor applications, matching of two or more compressors may be necessary. This may be less important if only a single channel is being compressed. In a stereo application, matched dual monolithic JFETs can be used to implement a pair of compressors [17].

JFET Gain Control Elements

JFETs are quite popular for gain control because they are widely available, low in cost and easy to control [16–19]. In this application they act as voltage-controlled resistors (VCR). However, they will contribute some distortion. Also, part-to-part uniformity, such as threshold voltage and R_{DS_ON} is not as great as one would like. This can lead to the need for a trim pot or to selection of devices. The LS26VNS has typical R_{DS_ON} of 26 Ω and typical threshold voltage of –4 V. The dual monolithic VCR11 is of particular interest where matched JFETs are required [17]. Its typical R_{DS_ON} is 150 Ω and its typical pinchoff voltage is −10 V. The use of JFETs for variable gain elements is described in Chapter 25.

Greater transparency can be had when there is no compression occurring if a JFET gain control element is used and it is essentially out of the circuit when compression is not occurring. For example, in the simplified arrangement of Figure 26.2(c), the JFET can be used as the feedback shunt element in a non-inverting amplifier. It will be at its R_{DS_ON} value when compression is not happening. The slight distortion introduced by the JFET during intervals of compression will go largely un-noticed. This is also true of any noise contribution its action might make, since it will be masked by the larger signals present in a time of compression. In a different circuit, the JFET can be arranged to be in pinch off when there is no compression.

The downside of using a JFET is that its variable gain behavior is not accurate, not having the reliable constant dB/mV characteristic of a translinear VCA. In addition, the variance of threshold voltage of different JFETs of the same type may require device selection or the addition of a trimmer potentiometer.

Although not a JFET, the MOSFET 3N163 DMOS P-channel enhancement-mode FET is also useful as a VCR gain control element [20]. Its typical R_{DS_ON} is 180 Ω and its typical threshold voltage is –2.5 V. The voltage on the insulated gate can swing in either polarity, and this can be an advantage in some circuits. As the gate voltage swings from –4 V to –12 V, its typical drain-source resistance goes from 10 kΩ to 200 Ω.

Optical Gain Control Elements

A light-dependent resistor (LDR) can be packaged with an LED to form an electrically controlled resistor. The LED creates light in proportion to its drive current. That light impinges on the LDR and reduces its resistance as light intensity increases. Devices of this type are called Vactrols, after the once-trademarked Vactrol device [21–23]. They have adequate gain range and introduce very little distortion. The resistance ranges somewhat logarithmically, often from about 30 kΩ at 0.1 mA down to about 150 Ω at 10 mA. The change in resistance can be fairly slow, with an attack time constant of about 6 ms and a decay time constant of about 300 ms for a decay to 100 times the on resistance.

A nice feature is that the LDR provides galvanic isolation between the controlled side (LDR) and the controlling side (LED). There are, however, some differences in control sensitivity and a lack of uniformity from unit to unit. Vactrols were largely in use before high-quality VCAs were available at a reasonable price and with good noise and distortion performance. Vactrols can be expensive and hard to find. Many are no longer in manufacture.

The NSL-32 made by Luna is also an LED-LDR optocoupler variable resistor [24]. It has an on resistance of about 40 Ω at 16 mA. Its attack risetime is about 50 ms, while decay to 100 kΩ in about 80 ms.

Using a more modern optical approach, there are photo-FET optocouplers where an LED is packaged with a MOSFET, taking advantage of the well-known behavior wherein the resistance of a MOSFET exposed to light goes from essentially infinity to a controlled resistance that decreases with increased light intensity [25]. These devices are optimized for small-signal variable resistance control. The H11F1M made by ON Semiconductor can be a bit power-hungry, requiring about 16 mA through the LED to drop the resistance to its minimum rated 200 Ω. Resistance at 2 mA is about 2.5 kΩ.

Log resistance vs log current is fairly linear up to about 5 mA, where resistance is about 1.8 times its full-current value. Resistance changes by a factor of about 15 for a one-decade change in current. There is virtually no delay in resistance change. As with other FET-based resistors, resistance increases with peak voltage of either polarity across the device, caused by self-pinch-off of the channel. This can cause odd-order distortion. Peak signal voltage should be kept well below 50 mV.

Vacuum Tube Gain Control Elements

Vacuum tubes were used in some of the earliest compressors, like Fairchild's legendary 670 variable-mu compressor, where the bias on one of the grids of a 6386 medium-mu remote cutoff twin triode vacuum tube was used to control gain (mu meaning amplification factor). [26–29]. The 5670 medium-mu twin triode has also been used for variable gain control in the Manley Laboratorys Stereo Variable Mu® Limiter Compressor.

26.6 Amplitude Detection

How the signal amplitude is detected for action by a compressor is also a distinguishing feature. For example, the default level detector is often an RMS detector. However, a lower cost, or preferable, detection means may be a common average detector. In other situations, such as with limiters, the preferred means of detection may be a peak detector. Sometimes a combination detector might be used, where the detector with the greatest reason for acting determines the control signal. An RMS detector might be used for the ordinary softer compression action, but a peak detector might also be combined with it to enter a limiter regime if there is danger of overload.

Log-Responding RMS Detector

A very useful type of detection is a log-responding RMS detector. This comprises a translinear circuit that performs RMS detection in the log domain and whose output remains in the log domain. As a result, it has an output that changes by a fixed voltage per dB of detected input voltage change. Its sensitivity can be described in mV/dB. As such, the detector can function over a large dynamic range. As a bonus, it can have the numerical inverse of the VCA's sensitivity that is measured in dB/mV. THAT Corp makes such a detector, available in their Analog Engine® that also includes a VCA [12, 13].

VCA-Based Amplitude Detection

VCAs are often used as the controlled gain element in compressors since their action is predictable, precise and relatively easily matched to the behavior of other compressors in a system. VCA compressors can be controlled with RMS detectors or peak detectors or other linear detectors (where the output is linearly related to the signal amplitude). However, a log-responding RMS detector can bring out the full advantages of using translinear VCAs in compressors. There are advantages if the controlling detector has a log response that is complementary to the exponential response of the VCA. The detector responds with mV/dB of change in the input signal while the VCA exerts gain control with a dB/mV way. Such a log detector is available in the THAT Analog Engine® IC [12]. A stand-alone log-responding RMS detector is no longer available.

Alternatively, an identical VCA can be used as a replica VCA in a detection circuit, so a form of tracking in behavior can be achieved between the detection process and the gain control process. A simple example is shown in Figure 26.8. A replica VCA is used as the gain element in a feedback

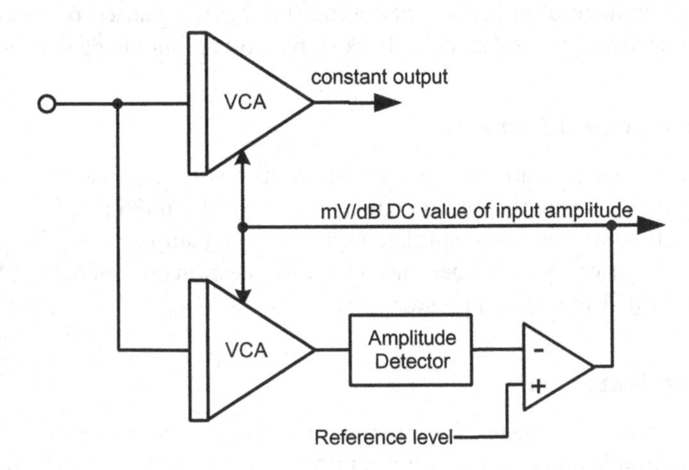

Figure 26.8 A Log-Responding Detector Using a Replica VCA.

loop where a suitable detector, like an RMS one, is used to feed a control signal back to the replica VCA to keep the measured signal level constant, as in an AGC arrangement. This results in a control signal that represents the log value, or dBV value of the input signal. This makes it more practical to achieve known compression ratios in a feedforward or side-chain compressor where a log-responding RMS detector is not available. This approach is especially convenient when implemented with a dual VCA like the THAT 2162 [30].

26.7 Frequency-Dependent Compression

In some cases certain frequencies in a signal from an instrument or a voice will be perceived as harsh or just have unusually large amplitude. In this case, a filter can be placed at the input to the side-chain that emphasizes those frequencies to increase compression of those signals by effectively lowering the threshold [1]. If the filter increases the gain of the signal as applied to the side-chain by 3 dB, the compression threshold will be reduced by 3 dB for those frequencies. In this case, a broadband decrease in the gain of the channel will occur when those frequencies are present at significant amplitude. The side-chain filter can be simple or complex.

26.8 Multi-Band Compression

Compressors can be used on individual frequency bands, just as a graphic equalizer can be used selectively for certain frequency bands. Such compressors are referred to as *multi-band compressors*. The objective here is often to avoid harshness in a mix where certain frequencies in a source may be irritable or distracting. This applies to both voice and instruments. A softer, more pleasing sound may result. The Fletcher-Munson curves of perceived loudness as a function of level and frequency may play a psycho-acoustic role in applying such compression. Compression may also be applied in low-frequency bands where drums and bass guitar might otherwise dominate over other program material or be unintentionally loud due to some resonance.

As illustrated in Figure 26.9, the most obvious way to implement a multi-band compressor is to split the signal into two to four or more bands, apply a separate compressor to each band's signal and then recombine the signals [1]. There are many multi-band equalizers that do it this way, but the challenge is in putting the compressed signals back together without audible artifacts. This is not unlike crossover arrangements in loudspeaker systems, and indeed in this approach there are crossovers. It is especially important that a flat frequency response be maintained when compression is not occurring. This is more easily accomplished if first-order filters are used, assuming that

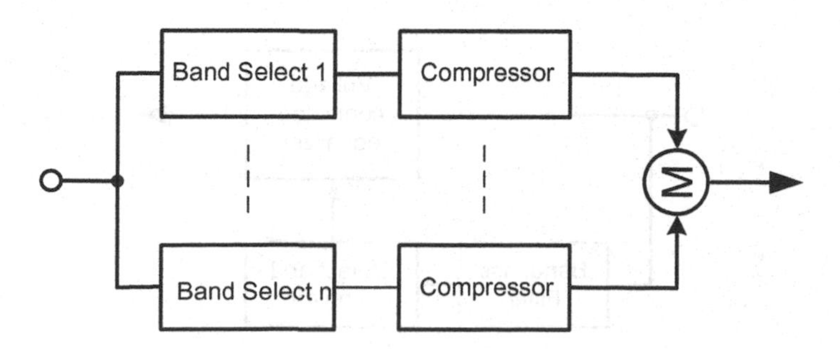

Figure 26.9 A Multi-Band Compressor.

they are sufficiently sharp to define the band for compression. In some cases, phase-compensated crossovers may be necessary if steeper slopes are needed.

The action of a multi-band compressor can be targeted to the optimum compression needs of a particular frequency band. In particular, different attack and release times can be used for different frequency bands, with shorter times often being more appropriate and practical for higher frequency bands.

26.9 Dynamic Equalization

An alternative to multi-band compression that performs a similar function is dynamic equalization, which can be even more flexible and surgical [1]. This is somewhat like configuring the signal-processing chain as a multi-band equalizer whose controls are changed dynamically in response to the frequency-dependent commands of the side-chain. In essence, the voltage-controlled gain element is replaced with a voltage-controlled filter, often a parametric equalizer. A bandpass filter with the same frequency and Q is placed in the input to the side-chain. The control signal is responsive to the same frequencies as the equalizer in the signal path. Dynamic equalization allows one to equalize the loud parts of a track in the selected frequency band differently than the softer parts. Such an arrangement is illustrated in Figure 26.10.

There are occasions where two or more instruments have spectral overlap and may mask each other or fight for space in that busy part of the spectrum. The clarity of one of the instruments may be seriously compromised as a result. One of the most common examples cited is conflict between a bass instrument and a kick drum. In this case, during the brief interval of the kick, it may be desirable to carve out a space in the spectrum for the kick drum by attenuating a certain frequency band of the bass instrument during only the drum kick. This can be accomplished with a dynamic equalizer in the signal path of the bass instrument. The dynamic equalizer's external side-chain input is fed the kick drum signal. The presence of that signal triggers the dynamic equalizer to attenuate a band of frequencies in the path of the bass instrument.

In a typical dynamic EQ application, boost of certain frequency bands is often not necessary. This can lead to some circuit simplification in some dynamic EQ architectures if only attenuation of the selected band is needed (a cut-only voltage-controlled parametric equalizer). As described in Chapter 20 there are types of equalizers that operate by feeding forward or feeding back a bandpass filtered version of the signal for boost or cut of the main signal. If one uses only feedback to reduce the amplitude, a VCA in the filter path can control the amount of feedback and thus the amount of compression. When no compression is in play, the gain of that filter path is made to be zero so the frequency response in that part of the spectrum remains flat. Often, the filter element can be a state variable filter that can be set to a chosen center frequency with a desired Q.

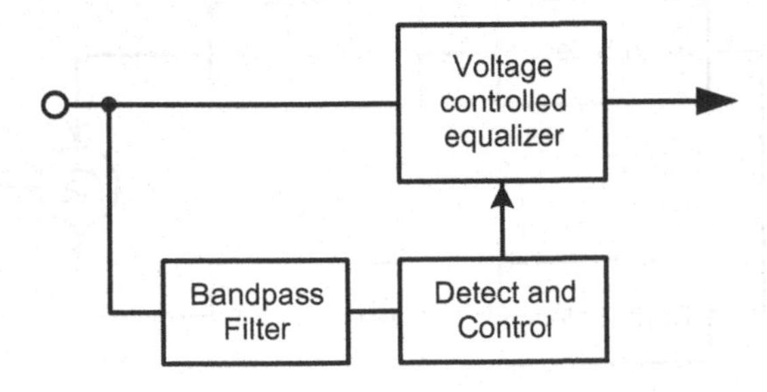

Figure 26.10 A Dynamic Equalizer.

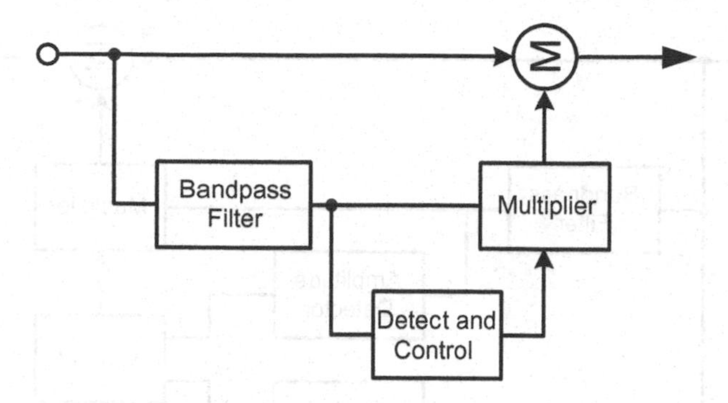

Figure 26.11 A Dynamic Equalizer with a Single Filter.

Consider an arrangement involving channels A and B. As in other dynamic processing arrangements, the input to the side-chain for channel A need not come from the input to channel A. It can come from channel B so that certain content in channel B can trigger equalization action in channel A. In such a case, if an instrument in channel B should not compete with a signal from channel A in the same frequency band, the signal from channel B may be input to the side-chain of channel A so as to attenuate the gain in channel A when the signal in channel B is present in the same frequency band. For example, during the kick interval of a kick drum in channel B, one might want to briefly attenuate the signal of a bass guitar in channel A. A similar fix might be used if a lead vocal is competing with the instruments for the same frequency band in a mix.

Common-Filter Dynamic Equalizer

In some arrangements, it may be possible to use the same filter for detection as for gain control. The input signal is fed through a state variable bandpass filter (SVF BPF) to control the side-chain. The output of that filter is also sent to the main signal path through an analog multiplier (or other voltage-controlled device) whose output is added to the main signal path. Depending on polarity of the added filtered signal, boost or cut can be introduced into the signal path at the selected frequencies. Numerous circuit configurations using different kinds of voltage-controlled elements can be used to achieve this. In such an approach the same equalizer frequency and shape can be used to operate the control circuit as is used to implement the boost or cut. Figure 26.11 illustrates such an arrangement.

26.10 De-Essers

The human voice can on occasion create a burst of high-frequency noise that corresponds to the sound of the letter *s* (as in "ess") and other like consonants. This is annoying and distracting to the listener and can create clipping or other forms of overload. The general approach to a de-esser is to detect signal energy in the high-frequency band, often 5–10 kHz, and use this detected amplitude as a compressor control signal to reduce gain. In a sense, this is a compressor that is responsive to only signals in that frequency range. A de-esser would normally only be used on a microphone channel.

Some de-essers may use that control signal to reduce gain across the frequency spectrum, as in an ordinary compressor. Other de-essers, however, are more surgical in their approach. In fact, those would fall into the multi-band compression or dynamic equalization category, where the

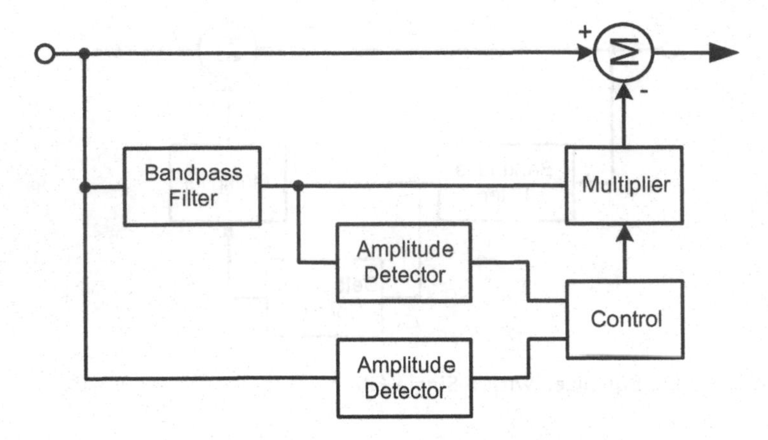

Figure 26.12 De-Esser with Relative Signal Amplitude Control.

de-esser will reduce the amplitude of the signal only in the band where the unwanted energy exists. Multi-band compression or dynamic equalization can also be used to suppress plosives that are often associated with a letter like *p* when a vocalist is close to the microphone. Some de-essers affect only the gain of the *essing* frequencies, and only when their amplitudes exceed a fixed threshold, as with a broadband compressor.

A better de-esser will operate on all signal levels of the *essing* band, since *essing* can occur at any voice level. Rather than have a threshold at which de-essing engages, the control circuit can operate on how the amplitude of the *essing* spectrum compares with the amplitude of the overall spectrum. If the amplitude of the *essing* frequency is abnormally higher than the broadband level, *essing* is assumed and the frequencies in that band are attenuated. This is somewhat the equivalent of a signal-dependent threshold for compression or dynamic equalization in the frequency band associated with *essing*. Figure 26.12 illustrates such a de-esser [1].

26.11 Downward Expansion and Upward Compression

Compression is often thought of as operating on the signal when it is of higher amplitude and above threshold, reducing signal gain under those conditions. However, it is sometimes desirable to bring up soft sounds without changing normal sound levels. In other words, to reduce the dynamic range at the low-amplitude end of the spectrum. At a still-lower threshold, gain may be returned to its nominal value or even zero to prevent amplification of noise (like a noise gate). Upward compression operates when the signal amplitude is below threshold, where it acts to increase gain in proportion to how far below threshold the amplitude is. This action is illustrated in Figure 26.13(a).

An expandor is the inverse of a compressor. It is used to increase the dynamic range of a signal. Often the expandor increases the downward amplitude change of a signal that is below a threshold. In essence, it adds attenuation by causing the signal path to decrease its gain. It can operate in an analogous way to a compressor, having threshold, expansion ratio, attack time and release time. If, below the threshold, the amplitude of a signal decreases by 1 dB and the expansion ratio is set to 2:1, then the output of the expander will decrease by 2 dB. This behavior is illustrated in Figure 26.13(b). Of course, the expander need not have a threshold and can expand the full dynamic range of the signal. In either case, the expandor action will tend to reduce noise in the input signal if the noise is not masked.

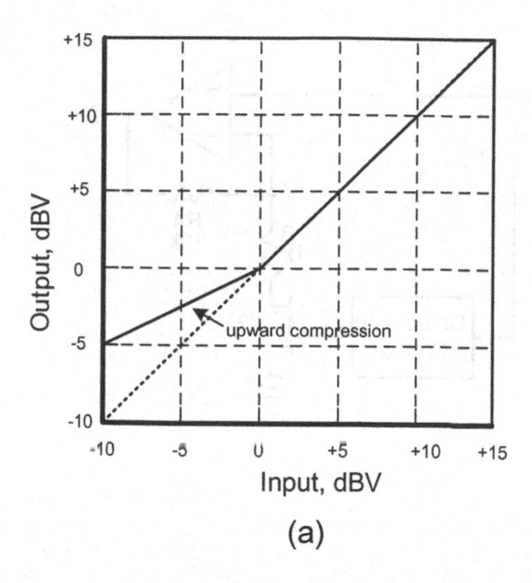

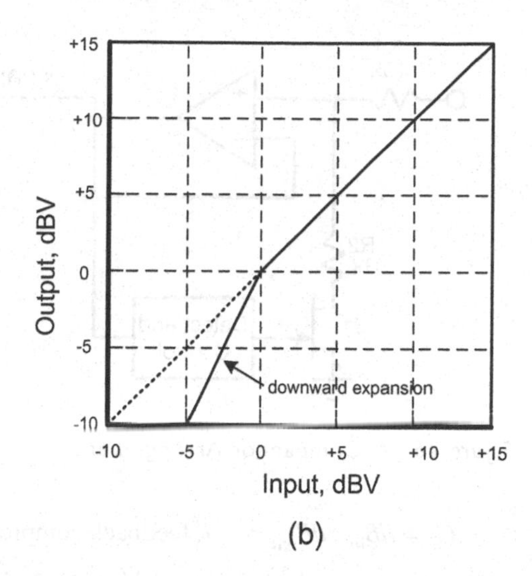

Figure 26.13 Upward Compression (a) and Downward Expansion (b).

Upward Compression

Most of the compressors discussed so far cause some form of signal attenuation when the signal amplitude exceeds a threshold, implementing a gain reduction or *downward compression*. However, there may be cases where the low-end dynamic range should be reduced so that signals that are fainter can be better heard. Thus, there is a threshold below which a signal is amplified rather than attenuated. Such a function is referred to as *upward compression*. There usually must be a limit in place with regard to how much gain can be employed when the signal level falls to a very low value, since unwanted signals in the form of noise should not be enhanced. In fact, if the signal falls below a second lower threshold, noise gating might be used. A graph of output vs input for such a compressor is shown in Figure 26.13(a).

26.12 Compandors

A compandor is shown in Figure 26.14. It comprises a compressor followed by an expander, the former at the transmission end of a noisy signal channel and the latter at the receiving end of the channel. It is usually used as a noise reduction technique. The compressor reduces the dynamic range of the signal to be passed through the signal channel, often boosting the lower-level signals. The expandor restores the dynamic range, often reducing the gain of smaller signals and the noise at the same time. The companding action may take place over the full dynamic range of the signal, or may occur only when the signal is above or below a threshold, as in a compressor. Thus, some portions of the dynamic range may be transported without being companded. In a popular architecture, the compressor is in a feedback arrangement and the expandor is in a feedforward arrangement. This means that the control circuits at both ends of the channel see the same signal to exercise control over the variable gain element.

Recall the equations for compression ratio for feedback and feedforward compressors discussed in Section 26.4, where C_r is the compression ratio, E_r is the expansion ratio and G is the control path gain (which is the product of the log detector gain, the control circuit gain and the VCA control gain):

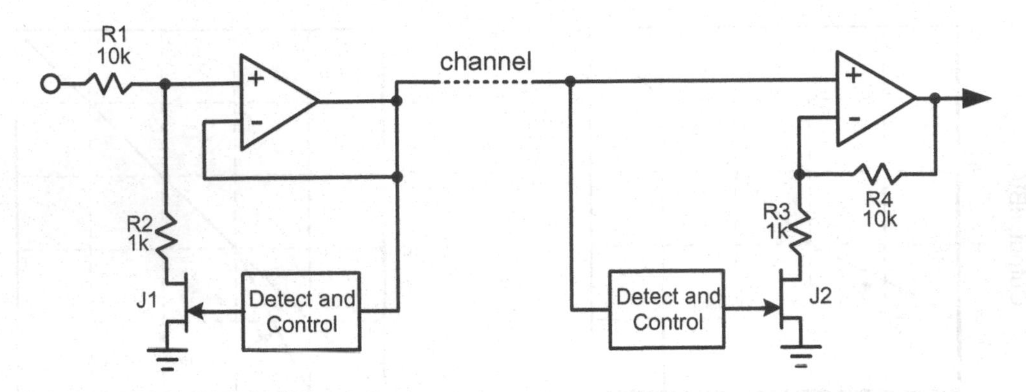

Figure 26.14 Compandor Arrangement.

$$C_r = dB_{in} / dB_{out} = 1\ G(\text{feedback compressor})$$

$$C_r = dB_{in} / dB_{out} = 1/(1-G)(\text{feedforward compressor})$$

$$E_r = dB_{out} / dB_{in} = (1+G)(\text{feedforward expandor})$$

We can see that if the feedback compressor and the feedforward expandor are fed the same signal, the compression and expansion ratios are the same and cancel each other out. More significantly, this happens whether or not the compressor and expandor are decilinear in nature. Their characteristics just need to be complementary. If G changes as a function of level due to non-decilinear behavior, things are fine as long as G, determined by the signal in the channel, is the same for both the compressor and expandor.

Figure 26.14 illustrates a simple compandor that uses JFETs as the control element. The compressor uses the JFET as an attenuator controlled by the output signal amplitude at the transmit end, while the expandor uses the JFET in a circuit where it creates attenuation in an amplifier feedback path at the receive end. The JFET is controlled by the input signal, which is essentially the same as the signal controlling the JFET at the transmit end if the transmission path gain is unity. This results in complementary compression and expansion characteristics if the JFETs are the same and are fed by identical side-chains. The key to cancellation of the compressor effect is that the gain control elements at both ends see essentially the same signal.

The NE570/SA571 was introduced in the late 1970s by signetics as a versatile compandor IC. Due to its popularity, the NE570 is still in production by OnSemi. Applications for compandors can be found in Philips/Signetics NE57-/571/SA571 AN174 and OnSemi AND8159, AN8227.

Frequency-Dependent Compandors and Noise Reduction

Compandors can be frequency-dependent. We all know that static pre-emphasis followed by de-emphasis is a noise-reducing technique. It increases the high-frequency parts of the signal on transmission and decreases them on reception, restoring the frequency response and decreasing the noise from the channel. However, its use is limited because too much increase of the high frequencies can lead to overload. Recognizing that it is only the low-level signals that need noise reduction, frequency-dependent companding can be used, where high-frequency signals of a lower amplitude are increased at the transmit end and decreased at the receive end. This is a form of

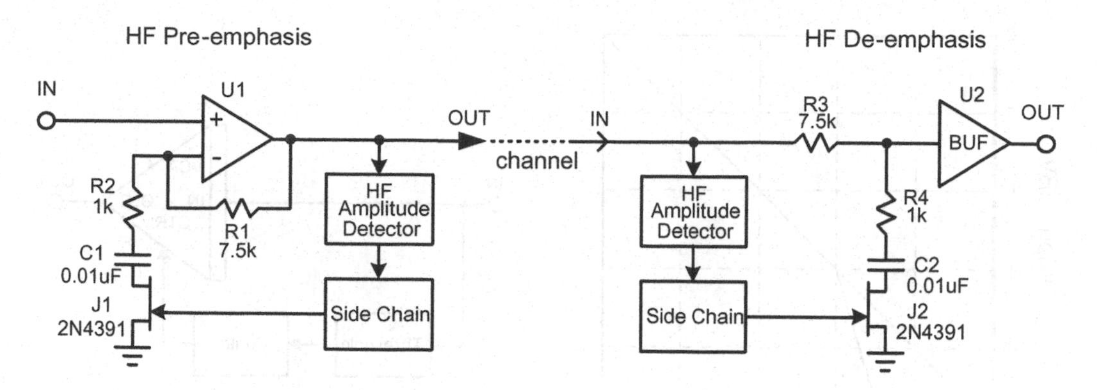

Figure 26.15 Simplified Noise Reduction Circuit.

dynamic pre-emphasis. The most well-known example of this is the Dolby B noise reduction system used in cassette recorders and many other applications requiring noise reduction [31, 32]. A highly simplified version of this kind of arrangement is shown in Figure 26.15.

The variable-gain action of the JFETs is made to affect only the high frequencies by placing a high-pass filtering capacitor C1 in series with the JFET. Lower frequencies see unity gain at both ends of the channel. The amplitude detector only responds to the higher frequencies to control the JFETs at both ends in the same way. In the absence of significant high-frequency content, the JFETs have zero gate voltage and are in their on state, allowing normal pre-emphasis by increasing the gain at higher frequencies through R2 and C1 in the feedback path of U1. The value of C1 and resistance of R1 determine the frequency where pre-amphasist boost begins. C1 and R2 determine the higher frequency where the boost levels off. R2 and R1 together set the size of the maximum amount of pre-emphasis boost. The R_{DS_ON} of the JFET is assumed to be fairly small compared to the resistance of R2. When there is significant high-frequency content, the HF amplitude detector and side chain create a negative JFET gate voltage so that its resistance increases and pre-emphasis is reduced. The de-emphasis at the receive end works in an inverse was so that overall transmission frequency response is flat.

26.13 Other Types of Dynamic Gain Control

Several other dynamic gain control circuits are worth mentioning here. They include AGC, noise gating and ducking.

AGC and AGC Hold

The goal of an automatic gain control (AGC) circuit is to maintain constant output amplitude, as measured by whatever kind of detector is used. This is not much different than a stiff limiter with a low threshold and a high compression ratio, but will instead usually operate to add gain when the detected signal level is below the target amplitude.

If the signal being AGC'd goes away or below a certain threshold, the voltage controlling the AGC gain element can be held at its last value. Pumping and breathing can be reduced in some situations if the AGC gain is prevented from increasing due to silent intervals. Think of a follow-and-hold circuit. As long as the signal is above a certain threshold, the control signal follows the command of the detector and control circuit. However, if the amplitude of the signal source falls below a threshold, the circuit holds the last value of the control voltage until signal reappears.

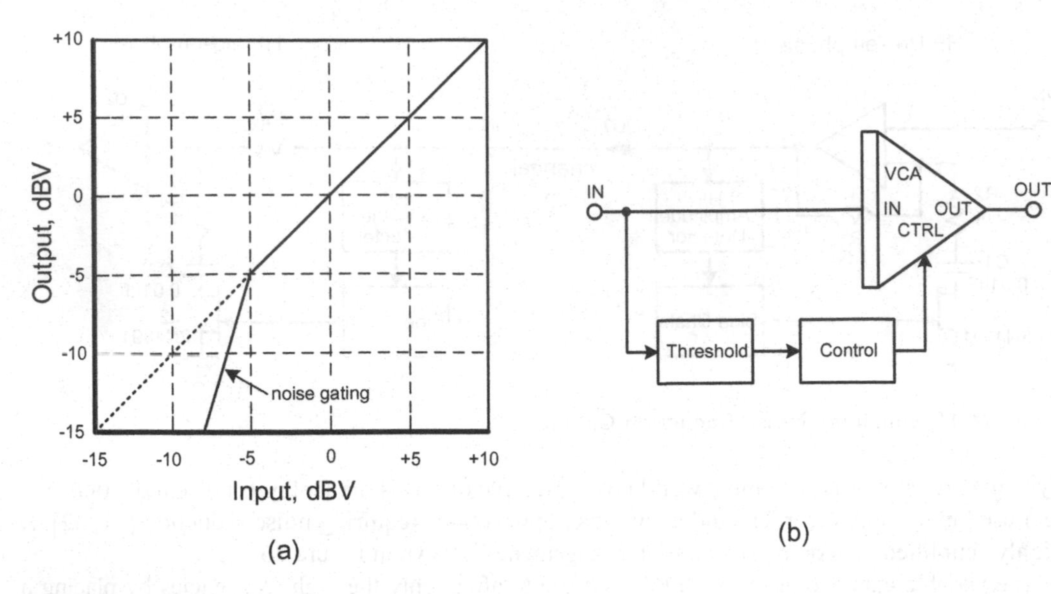

Figure 26.16 Noise Gating Circuit.

Noise Gating

Another form of dynamic processing is called *noise gating* [33]. In typical usage, the gain of a channel is reduced or cut to zero when that channel is determined to not be actively producing desired material, or when a different channel is determined to be producing desired material that should not be interfered with or clouded. Figure 26.16(a) illustrates the input-output behavior of a noise gating circuit. At input levels above –5 dBV, the signal is passed without alteration of its gain. At lower signals, there is a sharp reduction of gain as the input signal level falls. In a live interview, for example, when either party is not talking, it is desirable that background noise be reduced, or at least not be allowed to rise through the action of a compressor or AGC circuit. In this case, the gating reduces the gain for low-level signals, thus acting like a dynamic range expandor with a low-amplitude threshold. A simple noise gate circuit is illustrated in Figure 26.16(b).

Ducking

There are situations where the presence of a signal in a particular channel A should cause a reduction in the amplitude of other signals for improved clarity. This is a form of what is called *ducking* [34, 35]. A frequent use of this action is when a singer's voice or PA announcement is present and it is desirable to attenuate the background music or noise. This is another example of routing an external signal into the side-chain of a compressor channel to control (reduce) the other channel's gain. Keep in mind that the use of a different signal to control the gain of a given signal may not be considered to be compression in the strictest sense.

26.14 Clippers Hard and Soft

A hard clipper places an absolute limit on the peak amplitude of a signal, and does not preserve waveform. It is generally harsh and creates a spray of harmonics with high order. Symmetrical clipping of a sine wave generally creates odd-order harmonics, which tend to sound worse. However, a symmetrical clipping characteristic does not necessarily create odd harmonic distortion with asymmetric signals like audio. We do not live in a world of symmetrical sine waves. Symmetrical clippers can create even-order distortion terms if the signal being clipped is asymmetrical – like music. One polarity of the signal waveform may be clipped while the other side is not; this is essentially the same as asymmetrical clipping that causes second harmonic distortion on a sine wave. In other words, because music signals are asymmetrical, a symmetrical clipper will not clip both polarities at the same time.

Diodes are frequently used as clippers, and their clipping characteristic is of medium sharpness when distortion is plotted as a function of signal amplitude. Their clipping characteristic can be made sharper by increasing the signal level at which clipping occurs by adding threshold voltages that reverse bias the diodes. Negative feedback can also be used to alter the sharpness of the clipping. The clipping level is easily controlled in a temperature-independent way from a control circuit that employs a similar diode.

A clipper often uses a pair of silicon diodes, one for each polarity of the signal, to shunt the signal to AC ground when the voltage exceeds the forward drop of the diode, usually on the order of 0.6 V. Due to the exponential nature of the diode current as a function of forward bias, the dynamic resistance of the diode decreases as the forward bias voltage increases. It decreases by a factor of 10 for each 60 mV of increase in forward voltage. This means that the clipping is not extremely hard, in the sense that the incremental gain does not suddenly go to zero when the signal voltage reaches the turn-on voltage of the diode. In some cases, a diode-connected transistor will be employed, but signal levels must be kept small enough to avoid the relatively low (about 4 V) reverse base-emitter breakdown region.

Consider a diode whose turn-on voltage is defined as when it conducts 100 μA, at which point its dynamic resistance is 260 Ω. Assume that the forward voltage at 100 μA peak signal current is 500 mV. If the source resistor is 2300 Ω, incremental gain will be reduced by about 20 dB at this point. At a smaller input signal level, where peak diode current is only 10 μA, the dynamic resistance of the diode will be about 2300 Ω at the peak, and incremental gain will be reduced by only 6 dB. At a still smaller signal level whose peak causes 1 μA to flow, the dynamic resistance of the diode will be about 26,000 Ω, and incremental gain will be reduced by only about 1 dB. Over the 2-decade range of the peak diode current in this example, the signal amplitude across the diode has changed by a full 120 mV, making it easy to see that clipping behavior is fairly soft.

As shown in Figure 26.17(a), each diode can be supplied with a reverse bias voltage that creates a threshold at which clipping will occur. If this reverse bias is set to 10 V, it will take a large signal voltage to reach the point where the peaks cause the diode to conduct 100 μA, where incremental gain is down 20 dB. This effectively makes the clipping behavior sharper, keeping in mind that the output voltage at the onset of clipping is much larger than the forward voltage of the diode required to cause it to conduct 100 μA. A smaller reverse voltage applied to the diodes will result in softer clipping at a lower signal amplitude. The circuit driving this clipper must be able to supply enough signal amplitude to obtain the desired amount of signal attenuation during clipping.

Clipping can also be made sharper by putting the limiting diodes in the forward path of a feedback circuit that has modest loop gain. This is why clipping of a feedback amplifier is quite sharp. It is thus possible to control both the threshold and sharpness of a clipper. Such a circuit is shown in Figure 26.17(b).

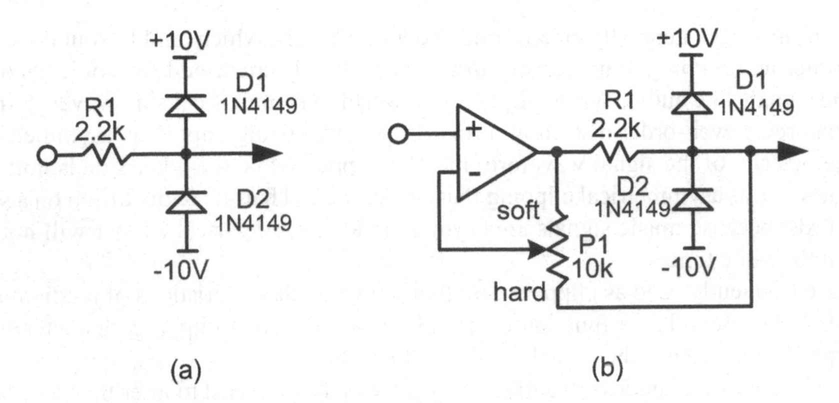

Figure 26.17 Clipper Circuits.

26.15 The Klever Klipper

The best power amplifier is one that never clips. If clipping must occur, it is best accomplished with a passive circuit ahead of the amplifier input. Such a circuit can clip the signal quickly and cleanly, without creating sharp edges. Ordinary silicon diodes like the 1N4149 can fulfill this task. Of course, with soft clipping, there will be some gradual rise in THD prior to clipping. As a result, such amplifiers might not measure as well as amplifiers of the same rated power that are not preceded by a soft-clip circuit.

Soft Clipping with Dynamic Threshold

While diodes clipping to fixed thresholds can provide the desired soft-clipping behavior, they will not allow for the dynamic headroom normally available for brief signal bursts in normal power amplifiers. Thus there is the need for dynamic clipping thresholds that track the rail voltages of the amplifier in real time. This approximation accounts for rail voltage sag under high-power conditions. Figure 26.18 shows what I call the Klever Klipper circuit [36].

Diodes D1 and D2 provide the soft-clip function in combination with R1. The soft-clip threshold voltages are created at op amps U1A and U1B. These voltages track the short-term average power supply rail voltage. Adaptive soft clipping occurs just shy of output amplifier hard clipping at any given power supply rail voltage condition. As a result, little or no dynamic headroom is sacrificed.

The Klever Klipper control circuit takes the negative rail supply voltage and scales it down to approximately 1/20 of its value, to roughly match the gain of the power amplifier. This moving reference voltage is filtered by C1 and C2. The U1A buffer then adds one diode drop (D3) to this voltage to account for the approximate conducting diode drop of the soft-clip diodes when they begin to clip. The output of U1A is the negative clipping voltage that is applied to D1. U1B creates a positive version of this voltage for application to D2. This simple version of the circuit assumes that the positive and negative amplifier rail voltages are of the same magnitude.

Clip adjustment potentiometer P1 is set with the amplifier operating at full power just at clipping into an 8-Ω load. Assuming that the pot starts at its maximum CW end, meaning least likelihood of causing soft clipping, the pot is then adjusted in a CCW direction until it just begins to soften the clipping to the point where it is controlling the clipping rather than clipping of the power amplifier itself. Figure 26.19 shows THD-1 as a function of power for a 50 W/8Ω amplifier with and without the Klever Klipper engaged [36].

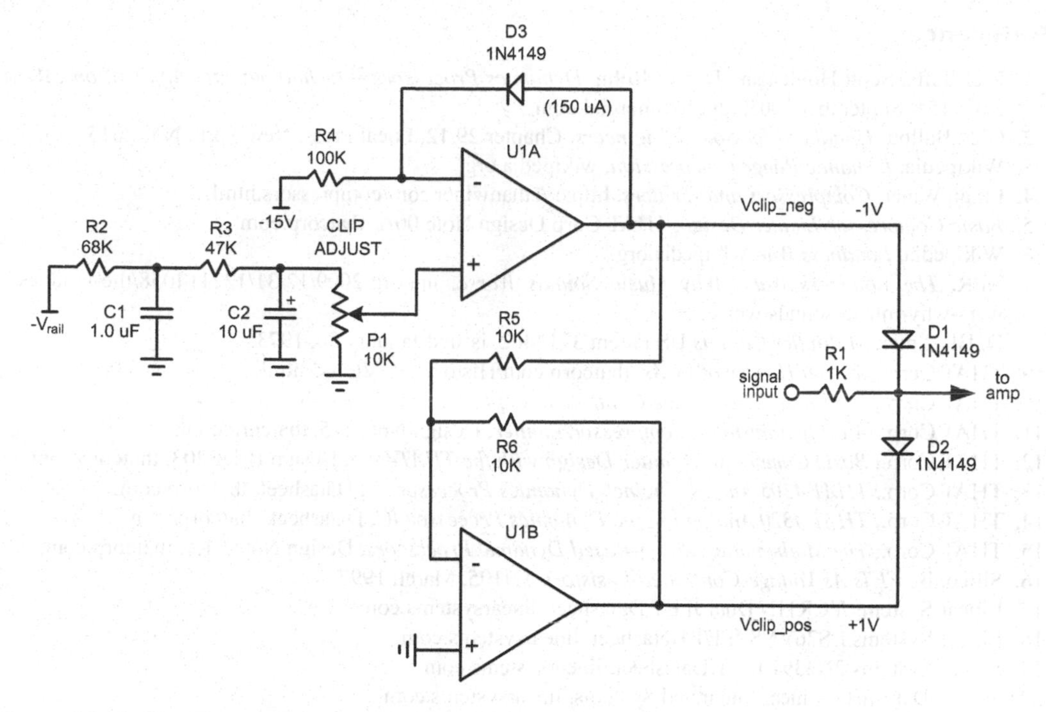

Figure 26.18 The Klever Klipper Soft-Clip Circuit.

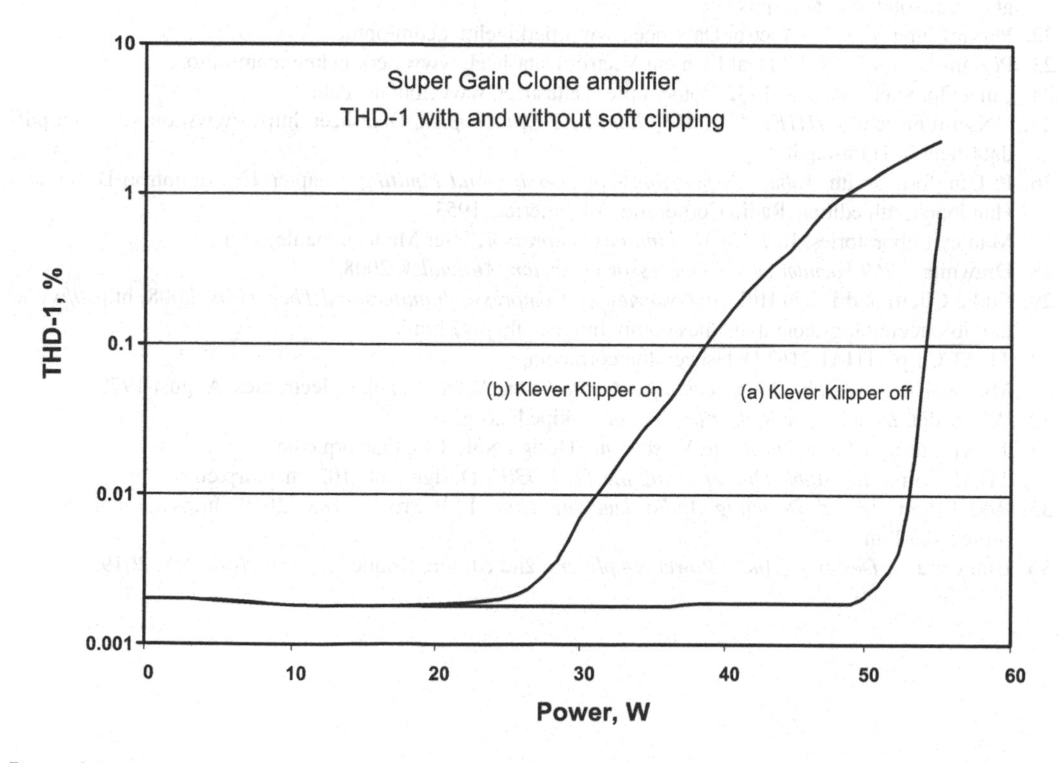

Figure 26.19 1-kHz THD+N vs Power Using the Klever Klipper.

References

1. Rick Jeffs, Scott Holden and Dennis Bohn, *Dynamics Processors – Technology and Applications*, Rane Note 155, September 2005, ranecommercial.com.
2. Glen Ballou, *Handbook for Sound Engineers*, Chapter 29.12, Focal Press, New York, NY, 2015.
3. Wikipedia, *Dynamic Range Compression*, wikipedia.org.
4. Ethan Winer, *Compressors and Limiters*, https://ethanwiner.com/compressors.html.
5. *Basic Compressor/Limiter Design*, THAT Corp Design Note 00A, thatcorp.com.
6. Wikipedia, *Loudness War*, wikipedia.org.
7. NPR, *The Loudness Wars: Why Music Sounds Worse*, npr.org/2009/12/31/122114058/the-loudness-wars-why-music-sounds-worse.
8. D. Blackmer, *Multiplier Circuits* US Patent 3714462, issued January 30, 1973.
9. THAT Corp., *A Brief History of VCAs*, thatcorp.com/History_of_VCAs.shtml.
10. THAT Corp., *2181 and 2162 Voltage Controlled Amplifier*, Datasheets, thatcorp.com.
11. THAT Corp., *A Fully Adjustable Compressor/Limiter*, Design Note 115, thatcorp.com.
12. THAT Corp., *Basic Compressor/Limiter Design with the THAT4305*, Design Brief 203, thatcorp.com.
13. THAT Corp., *THAT 4305 Analog Engine® Dynamics Processor IC*, Datasheet, thatcorp.com.
14. THAT Corp., *THAT 4320 Analog Engine Dynamics Processor IC*, Datasheet, thatcorp.com.
15. THAT Corp., *The Mathematics of Log-Based Dynamic Processors*, Design Note 01A, thatcorp.com.
16. Siliconix, *FETs As Voltage-Controlled Resistors* AN105, March 1997.
17. Linear Systems VCR11N Dual JFET Datasheet, linearsystems.com.
18. Linear Systems LS26VNS JFET Datasheet, linearsystems.com.
19. Linear Systems 2N4391 JFET Datasheet, linearsystems.com.
20. 3N163 Datasheet, Linear Integrated Systems, linearsystems.com.
21. Perkin Elmer, *Analog Optical Isolators*, perkinelmer.com/CMSResources/Images/44–3429APP_AnalogOpticalIsolatorsAudioApps.pdf.
22. PerkinElmer VTL5C3 Vactrol Datasheet, www.perkinelmer.com/opto.
23. PerkinElmer VTL5C4/2 Dual Element Vactrol Datasheet, www.perkinelmer.com/opto.
24. Luna Optoelectronics NSL-32 Optocoupler Datasheet, www.lunainc.com.
25. ONsemi/Fairchild, *H11F1M/2M/3M Photo FET Optocoupler*, Datasheet, https://www.onsemi.com/pdf/datasheet/h11f3m-d.pdf
26. F. Langford Smith, *Volume Expansion, Compression and Limiting*, Chapter 16, Radiotron Designer's Handbook, 4th edition, Radio Corporation of America, 1953.
27. Manley Laboratories, *Variable Mu Limiter Compressor*, User Manual, manley.com.
28. Drawmer, *1969 Vacuum Tube Compressor Operators Manual*, ©2008.
29. Eddie Ciletti and David Hill, *An Overview of Compressor/Limiters and Their Guts*, 2008, https://www.tangible-technology.com/dynamics/comp_lim_ec_dh_pw2.html.
30. THAT Corp., THAT 2162 Datasheet, thatcorp.com.
31. Mannie Horowitz, *The Dolby Technique for Reducing Noise*, Popular Electronics, August 1972.
32. Wikipedia, *Dolby Noise Reduction System*, wikipedia.org.
33. THAT Corp., *A Fully Adjustable Noise Gate*, Design Note 100, thatcorp.com.
34. THAT Corp., *Adjustable Ducker Using the THAT4301*, Design Note 102, thatcorp.com.
35. Rod Elliott, *Signal Detecting Audio Ducking Unit*, ESP Project 183, 2019, https://sound-au.com/project183.htm.
36. Bob Cordell, *Designing Audio Power Amplifiers*, 2nd edition, Routledge, New York, NY, 2019.

Chapter 27

Level Displays and Metering

Level displays are crucial to the broadcasting, recording and sound reinforcement industries. They help prevent clipping and help ensure reasonable listening levels. They are especially important in setting signal levels in broadcasting to prevent over-modulation and signal levels exceeding legal limits. Similarly, level displays are crucial in setting recording levels so as not to create overload, as with analog tape recorders and DACs. The most common display is the VU (volume units) meter, often using a traditional mechanical meter movement or a bar graph display. VU meters do not respond instantly to signal level; they respond to a full-wave average rectified level with a defined time constant and are calibrated in VU from −20 VU to +3 VU. For this reason, various other forms of peak, average and RMS displays are used in some cases as well. The most well-known of these is the peak program meter (PPM).

Metering technology has been heavily impacted by the needs of digital audio, and new peak-measuring approaches have evolved. Finally, the importance of having a more accurate measure of perceived loudness has led to new metering standards and the LU (loudness units), a measure of loudness intended as a tool to make the perceived loudness of different programs more adherent to a common standard, thwarting the "loudness wars."

27.1 The VU Meter

The venerable VU meter has served us well for over 80 years. Originally known as the standard volume indicator (SVI) meter, it was born at a time when many different metering devices were in use for broadcast, film and recording. The VU meter is the most common level meter found on audio equipment. The term *VU* stands for volume units, in accordance with its goal of displaying the perceived volume of a signal in an approximate way.

The VU meter was defined in 1939 by telephone and radio broadcast experts so that there would be some commonality in measuring signal levels across audio applications and equipment. It was jointly developed by engineers representing CBS, NBC and Bell Laboratories. The concept, rationale and specifications for the VU meter were publicized in a 1940 IRE paper, "A New Standard Volume indicator and Reference Level" [1]. It was based on a remarkable amount of objective and subjective research, including human factors like eyestrain. In 1942 it was made a standard, ANSI C16.5-1942. The definition of the VU and how it should be displayed was strongly influenced by the technology available at the time, including the use of copper-oxide rectifiers.

The intent of the VU meter was to have a dynamic characteristic that responded in a way similar to how the ear perceives sound level, while also giving a very rough idea of how high the signal levels went. It is a roughly linear indication of average voltage level on a nonlinear dB scale. The dynamic responsiveness of the combination of the signal rectifier and mechanical time constant of the meter movement is referred to as the VU meter's ballistics.

Zero VU Reference Level

VU meters display VU in dB with respect to a reference level. The scale runs from marked values of −20 to +3 VU. The 0-VU point is set to be +4 dBu, where 0 dBu is the voltage corresponding to power of 1 mW (0 dBm) into an impedance of 600 Ω. This corresponds to 0.775 V_{RMS}. It is important not to confuse dB_{VU}, dB_u and dB_V. The level 0 dB_{VU} corresponds to 1.228 V_{RMS}. This is the 0-VU reference level [2–4].

Meters marked as VU don't always conform to the +4 dBu reference level of 1.228 V_{RMS}. Sometimes just an arbitrary reference level suitable for the application or the particular location in the system is used. Of course, VU meters do not respond to power, but to an average rectified signal voltage. Meters on a power amplifier are often designated as VU, but don't represent VU, but rather something that has a loose relationship to the maximum power output of the amplifier into an 8-Ω load. They may or may not have VU ballistics in the meter movement. In other cases, VU meters have a relationship to the amplitude of a signal with respect to the safe maximum amplitude before clipping or overload.

The nonlinear VU meter dB scale covers a wide range, typically from −20 to +3 dB VU. Although the meter responds rather linearly to average voltage at the higher levels like 0 VU, it responds in a somewhat nonlinear way at low signal levels due to the nonlinear voltage drop of the copper-oxide (or germanium) rectifying diodes. This causes the response to act more like a log characteristic at low signal levels and allows the scale to be reasonably readable to a lower signal level. For most signal levels, it is a log-calibrated linear-responding meter. The most important point of meter accuracy is at the 0-VU level.

Meter Ballistics and Response Time

The intent of the VU meter standard was to provide a rough indication of perceived loudness. Since human perception of loudness depends somewhat on exposure time, the ballistics of the VU indicator are important; a very brief burst of sound is perceived to be of less loudness than a longer or continuous burst of sound. This, of course, comes at the expense of its ability to accurately indicate peak levels.

VU meters employ a 200-μA D'Arsonval meter movement. Meter needles cannot move instantaneously and that would not be the objective of the VU meter application anyway. The meter movement responds as a damped second-order system. The meter ballistics were defined so that, in response to the sudden application of a tone, the meter would reach 99% of its final deflection in 300 ms, with overshoot of not less than 1% and not more than 1.5% [ANSI C16.5-1961] [5]. This defines a slightly under-damped second-order response that was found to be the optimum for readability and minimal eyestrain given the other objectives. The VU meter reaches a level 2 dB down from the final value in about 210 ms. The falling time constant is the same as the rising time constant [1–3].

Response to Peaks and Maximum Signal Level

The VU meter ballistics represent a compromise between correlation with perceived loudness and the display of peak response. Because the VU measurement is average-responding with an approximate 300-ms rise time to 99%, peak levels are usually assumed to be 10 dB higher. Moreover, 6 dB of additional headroom margin against clipping or allowed peak level is usually assumed. This means that the system full-scale level should be at least 16 dB above 0 VU. With 0 VU corresponding to the +4 dBu reference level, this comes to +20 dBu, corresponding to a voltage level of 7.75 V_{RMS}, or 10.96 V_{PK}.

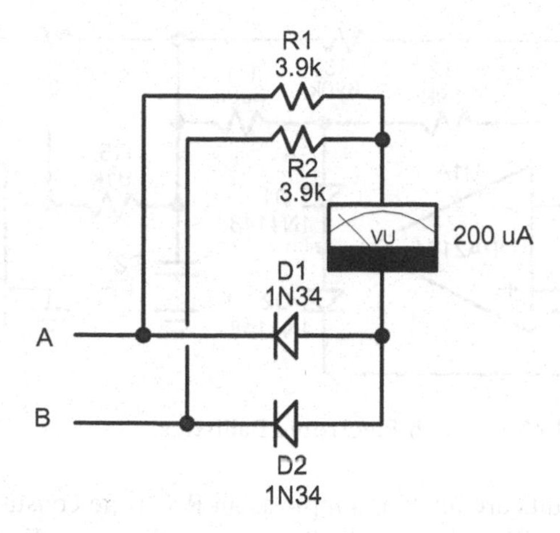

Figure 27.1 Typical VU Meter Circuit.

Typical Passive VU Meter Circuit

Figure 27.1 illustrates a typical traditional VU meter circuit. It consists of little more than a 2-diode copper-oxide full-wave bridge rectifier driven by inputs A and B, and followed by a 200-µA D'Arsonval meter movement whose resistance is 3900 Ω. The remaining two legs of the bridge are 3.9-kΩ resistors that establish the correct sensitivity and play a role in meter damping. The total VU meter input resistance is 7800 Ω at the 0-VU signal level. The scale is marked from −20 VU up +3 VU to display the logarithmic value of the relatively linear meter deflection. The 50% deflection point of the meter is at about −2.5 VU.

The use of the 2-diode bridge keeps voltage drop to 1 diode drop and supplies the necessary series resistance external to the meter movement. The copper-oxide diodes begin to turn on at about 0.15 V. As with many rectifiers, the copper-oxide rectifier imparts a bit of a log-like characteristic at small voltage levels. For example, if its stated turn-on voltage is 150 mV (at some value of current), it does not instantly stop conducting if the applied voltage is less than 150 mV; rather, the rectified voltage decreases in a nonlinear way, not unlike a logarithmic way. Keep in mind, only an average of 20 µA needs to flow in the rectifier to deflect the needle by 10% of full scale. This property actually makes the markings at the low end of the scale of a traditional VU meter a bit more linear in dB than one might expect. Similarly, this property allows signals smaller than expected to cause some perceptible deflection. Later VU meter designs often employed germanium diodes, which were close enough in their characteristics.

The primary intent of the VU meter is to indicate when a particular signal's volume is at the reference level and to serve as a guide to the voltage present at that point. The VU meter should be accurate at that point when fed with a sine wave. VU meters tend to be less accurate at the lower ends of the scale, and this is acceptable, since the intent is an indication of relative volume. The VU meter should not be confused with a measuring instrument.

Electronic Ballistics and Bar Graph Displays

A great many VU meters since the 1970s use some form of electronic display, like an LED bar graph or a graphic display on an LCD screen. These displays respond almost instantaneously and have no ballistic characteristics. In these cases, electronic circuits are used to create the ballistic

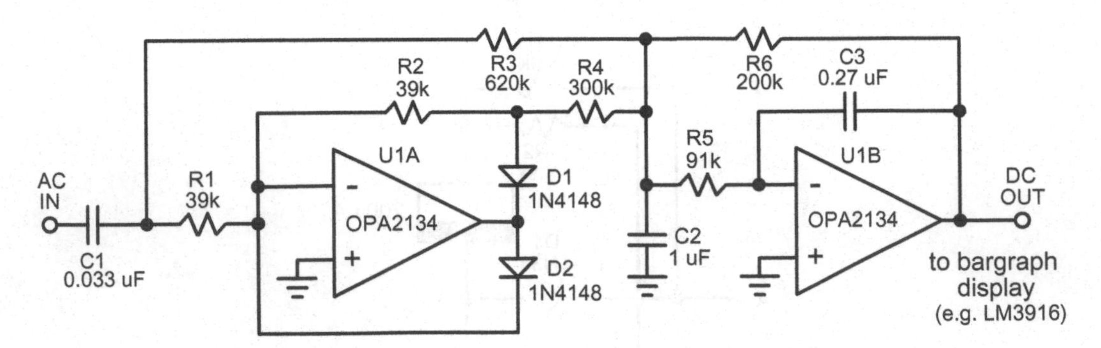

Figure 27.2 Bar Graph VU Meter with Electronic Ballistics.

characteristics. These circuits are often as simple as an R-C time constant placed after an average detector. The displays are still usually marked in the traditional nonlinear dB scales. However, a more accurate approach is available for the ballistics. The VU meter movement is a second-order system with particular cutoff frequency (2.1 Hz) and Q (0.62) that define the ballistics. This can be duplicated by a slightly under-damped low-pass filter following the rectifier [6].

Figure 27.2 shows a simple dot/bar graph LED display VU meter circuit. It employs the LM3916 and incorporates circuitry to approximate the proper VU meter ballistics [3]. The signal is first full-wave rectified by U1A, a precision full-wave rectifier with high input impedance. U1B, which completes the full-wave rectification, also incorporates a multiple feedback second-order filter with the required cutoff frequency and Q. The LM3916 is no longer in manufacture, but an equivalent is available as the NTE1549. Many other displays are in use, including LCD devices operated by a microcontroller.

Microcontroller-Based VU Meters

Today's inexpensive microcontrollers (MCU) can readily implement a VU meter with minimal hardware and a simple program. The fact that virtually all of these MCUs include several A/D converters makes it very easy, and the large number of available logic I/Os can drive LEDs or other display devices. A very capable surface-mountable MCU board that can be had for as little as $4 is the Raspberry Pi Pico [7]. The tiny surface-mountable board has 26 multi-function GPIOs, three of which can be assigned to be 12-bit ADCs (multiplexed across the three channels). For the same reasons, many of the other meters described here can be readily implemented at low cost, whether they exist in an analog environment or a digital environment.

27.2 Peak Program Meter

The fairly slow rise time of the VU meter does not enable it to display peaks in a way that is adequate for avoiding circuit clipping or signal peaks that are beyond allowed limits. This was less of a problem in the 1940s and 1950s when signal chains implemented with vacuum tubes were a bit more forgiving of clipping. A big part of the VU meter's value was that it could capture the essence of what was going on with the signal in a single, rather simple, instrument.

As time passed it became clear that a much faster meter was also needed to display a more accurate indication of peak levels. In fact, a peak-reading meter of sorts had been in use in Germany

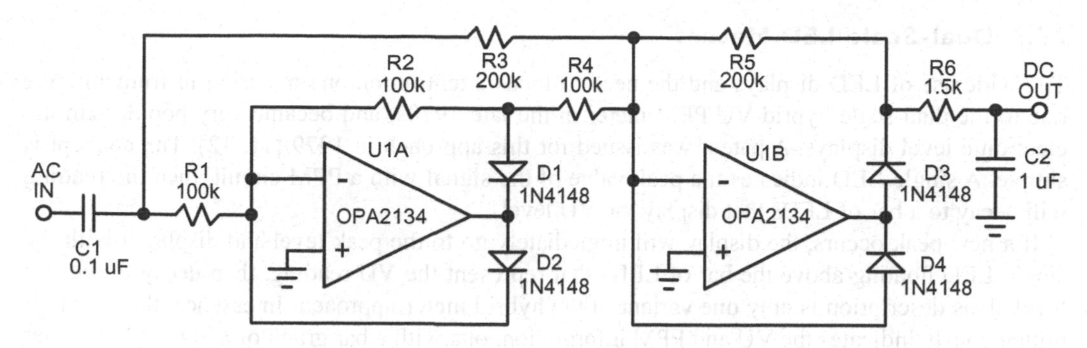

Figure 27.3 Peak Program Meter (PPM).

prior to the development of the VU meter. The popular peak program meter (PPM) was born. It is aimed more at the electronic realities of clipping, overload and distortion than the perceived volume of the audio. For this reason, it is more responsive to peak signal values and has a much faster rise time. It also incorporates a rather slow fall time so that the peaks can be more easily read. The characteristics of PPMs are specified in DIN45406 [2]. Figure 27.3 shows a PPM circuit.

Attack and Release Time

The PPM responds quickly to the peak voltage of a full-wave-rectified input signal. Attack ballistics are specified as the PPM being able to display a reading only 1 dB down from its final value in the space of 10 ms. This is about 30 times faster than a VU meter. The PPM should be able to display a value 4 dB down from the final value in 3 ms. This attack behavior corresponds to an R-C time constant of 1.7 ms. The return rate is much slower, requiring 1.5 seconds for the displayed value to drop by 20 dB. This corresponds to an R-C time constant of about 650 ms. Even with the faster 10-ms rise time, the PPM is not capable of reading the actual peak value of a fast transient, so it is more appropriately called a Quasi PPM (QPPM) [8–10].

Type I and Type IIb PPMs

There are a few different types of PPM defined by the IEC. These differ primarily in the way the scale is marked and calibrated. The Type I PPM scale runs from −25 to +5 dB, with 0 dB corresponding to the permitted maximum level (PML), corresponding to full scale (0 dBFS). Even with the faster response time, use of the QPPM requires some margin against fast transients. In a digital audio system, 9 dB of headroom against clipping or digital full scale is recommended. The reference signal level then becomes −9 dBFS. This is also called the alignment level (AL), which is the target for the maximum displayed program level [10].

The Type IIb PPM, commonly used in the United States, has a scale running from −12 to +12 dB, with 0 dB being the alignment level and +9 dB being the PML. It can be seen that PML is 9 dB greater than AL in both cases, reflecting the needed amount of headroom.

The simple PPM in Figure 27.3 employs a precision full-wave rectifier that provides a positive unidirectional output to a 1-μF filter capacitor through a 1.5-kΩ resistor. This combination establishes the attack time. The 200-kΩ resistor from the rectifier output back to the op amp's virtual ground establishes the time constant for the return [6].

27.3 Dual-Scale LED Meters

The wide use of LED displays and the need to focus attention on one metering instrument gave rise to the dual-scale hybrid VU/PPM meter in the late 1970s, and became very popular among electronic level displays. A patent was issued for this approach in 1979 [11, 12]. The concept is simple. A single LED indicates the peak value of the signal with a PPM circuit, then the reading will decay to a bar of LEDs that display the VU level.

If a new peak occurs, the display will immediately go to the peak level and display it with the single LED floating above the bar of LEDs that represent the VU reading, then decay to the VU level. This description is only one variant of the hybrid meter approach. In essence, the meter simultaneously indicates the VU and PPM information, one with a bar graph of LEDs and the other with a single LED. These displays are especially valuable because incorporation of the fast peak display gives a better indication of the possibility of clipping without losing the VU function.

Figure 27.4 shows such a display circuit. It is much like the bar graph VU circuit of Figure 27.2, but with the addition of a peak-detecting parallel path driving the LEDs in dot mode. Both metering paths use the same rectification circuit for the incoming signal. The two paths differ in their attack and release time constants.

27.4 The Dorrough Meter

Even together on the same scale, VU meters and PPMs fall short in displaying accurately the most important aspects of the audio signal. The PPM was still not fast enough to catch brief peaks and the VU meter, being a compromise instrument, was not as good at correlating with perceived loudness as desired. The dual-scale Dorrough meter was developed and patented to address these shortcomings [13–15]. The meter is a bit like the VU/PPM hybrid, but both functions are refined, with different time constants and a readout that is linear in dB.

Freed from the compromise of having to give some indication of highest signal levels, the VU-like functionality of the meter for persistent signal was free to be optimized for better correlation with perceived loudness. The average persistent signal level is measured by a precision full-wave rectifier with a filter time constant of 270 ms. This results in a rise time of about 600 ms, about twice that of the VU meter. This rise time was found to provide the best correlation with perceived loudness. The first-order filtering of the average signal contrasts with the second-order response of the VU meter.

Both the peak and average measurements are passed through logging circuits so that the display is linear in dB. This makes the meter easily readable over a wide 39-dB range, with one LED for each dB of level display. It is worth noting that the logarithmic display affects the visually perceived

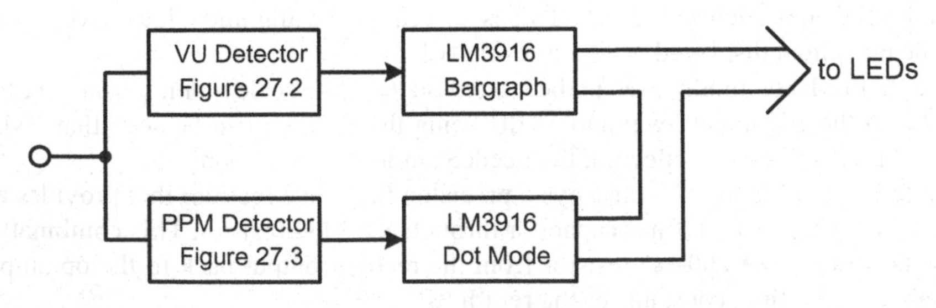

Figure 27.4 Hybrid VU/PPM Metering Circuit.

rise and fall times of the peak and average displays in a beneficial way. The Dorrough meter displays the sum of both channels in order to indicate the dynamics of both channels simultaneously.

27.5 Digital Peak Meter

With the advent of digital audio, a new, and much faster peak-indicating meter was implemented for use in signal path locations that contain digital PCM (pulse code modulation). This meter monitors every digital sample with nearly instant response and has a decay time of 1.5 seconds for a drop of 20 dB in indicated level [2, 16]. It is sometimes referred to as a peak sample meter. With digital clipping much less forgiving, this meter indicates how big the peak voltage is with respect to digital full scale in dBFS. In the numeric world of samples, this meter can be very accurate in displaying the peak value of every sample. This meter is used in a great many digital consoles and Digital Audio Workstations (DAW).

27.6 True Peak Meters

Even the speed and precision of the peak sample meter can fall short of truly adequate peak monitoring. This is because the continuous time inter-sample peak magnitude of a signal can be larger than the value of the signal at the sampling instant. A transient signal can have a slew rate fast enough to change its value significantly during the 20.8-µs interval between 48-kHz samples. The playback reconstruction filter correctly creates the larger signal value that lies between samples in the continuous-time domain [17–22].

Consider a 12-kHz sinusoid whose zero-crossing times are aligned with every other sample time. The signal will be exactly at a peak at the sampling instant and the peak will be read correctly. At the next sample time it will read zero, but that is of no consequence for a peak meter. At the next sample time after that, it will properly read its most negative peak.

Now delay the sine wave by 45°. The peak will now occur between sample times and the sine wave amplitude will be 3 dB down at the sample times on either side of the peak. The 12-kHz signal will be under-read by a full 3 dB. This error becomes less serious as the frequency of the sinusoid goes down or as the sampling frequency goes up. At a 4× sampling frequency of 192 kHz, the error will be less than about 0.6 dB for the same 12-kHz sine wave. This inter-sample under-read (or *digital over*) can have dire consequences for downstream digital processing.

To avoid this problem, the so-called *True Peak* meter was defined [17–22]. It achieves a much closer approximation to the true peak by using a peak sample meter at a minimum sample rate of 192 kHz. It will under-read a 12-kHz sinusoid by no more than 0.6 dB. The term *dBTP* was coined to describe a peak reading relative to digital full scale corresponding to the improved approximation of the continuous time peak of the signal. For some margin, the standard practice required that peak readings not exceed −1 dBTP.

It should be pointed out that conventional peak sample metering at digital audio locations in the recording signal path that are already at a 4× or greater sampling rate naturally conform to the True Peak definition.

27.7 Loudness Metering – the LU, the LUFS and LKFS

In response to "loudness war" complaints, the more stringent demands of all-digital audio chains and government legislation, means of measuring loudness as perceived by the human ear became crucial. Moreover, advances in technology made such measurement much more practical. After extensive work on measuring perceived loudness, in 2006 the ITU published its ITU BS.1770 recommendation for measuring loudness in an objective and consistent way [17].

In the United States, the *Commercial Advertisement Loudness Mitigation Act* (CALM) was passed in 2010. It requires the FCC to regulate permitted loudness levels of TV commercials so that they do not exceed the loudness of the program material [23]. This follows the earlier 2009 recommendation by the Advanced Television Systems Committee (ATSC) in their Recommended Practice A/85 [24]. That in turn was based on the ITU BS.1770 recommendation published in 2006. The FCC adopted the CALM Act rules and those rules took effect in 2012. Similarly, the European Broadcast Union (EBU) adopted their R.128 standard based on BS1770 [20]. The ITU-R BS.1770-4 standard is the most widely used and accepted standard [17].

Loudness meters display a measure of the perceived loudness of program material that is more accurate than previous metering approaches. This is especially important in the broadcast world, where there are strict standards for perceived loudness of program material. We are all too familiar with the commercials on TV sounding louder than the program. This involves many factors in psychoacoustics that can be manipulated to make material sound louder even when it registers the same on a VU meter or PPM. One factor is the degree of compression of the audio. Highly compressed material delivers far more time-averaged audio power than normal material (which is also often compressed to a much lesser degree). Even for normal program material, FM radio stations are famous for using sophisticated compressors to sound louder than their competitors even while staying within the FCC-mandated modulation limits. These competitions for greatest loudness are sometimes referred to as the "loudness wars."

Terminology

The loudness unit, *LU*, was defined to provide a numerical basis for loudness, not unlike the Volume Unit, *VU*, defined long ago. Both units are in dB, but the LU measures a much more accurate version of perceived loudness as defined in BS.1770. One feature of the LU measurement is that it is frequency-weighted with what is called K-weighting. The signal amplitude is described in LU with respect to digital full-scale, as with dBFS. The amplitude designation is *LUFS*, which stands for LU units with respect to full scale. So, for example, a standard broadcast level is stated as −23 LUFS. An alternative term, technically the same as LUFS, is *LKFS*, which stands for LU, K-weighted, with respect to digital full scale. The LU measurement is always K weighted, so the latter term is a bit redundant.

K-Weighting

The perception of loudness is frequency-dependent. This is fairly obvious, given the widespread use of A-weighting for signal-to-noise ratio specifications. However, prior to BS.1770, most audio metering did not include frequency weighting. Such weighting is a significant feature of BS.1770, where the weighting is designated as *K-weighting*.

The K-weighting filter consists of a pre-filter and a high-pass filter known as the revised low-frequency B curve, commonly referred to as RLB weighting [25]. The pre-filter comprises a +4 dB second-order shelf centered near 1500 Hz. It takes account of head effects and the increased sensitivity of loudness perception in that frequency range. The pre-filter curve is shown in BS.1770-4 in Figure 2 [17]. It is +0.6 dB at 1 kHz, +2.0 dB @ 1.5 kHz and +3.2 dB @ 2 kHz. This pre-filter

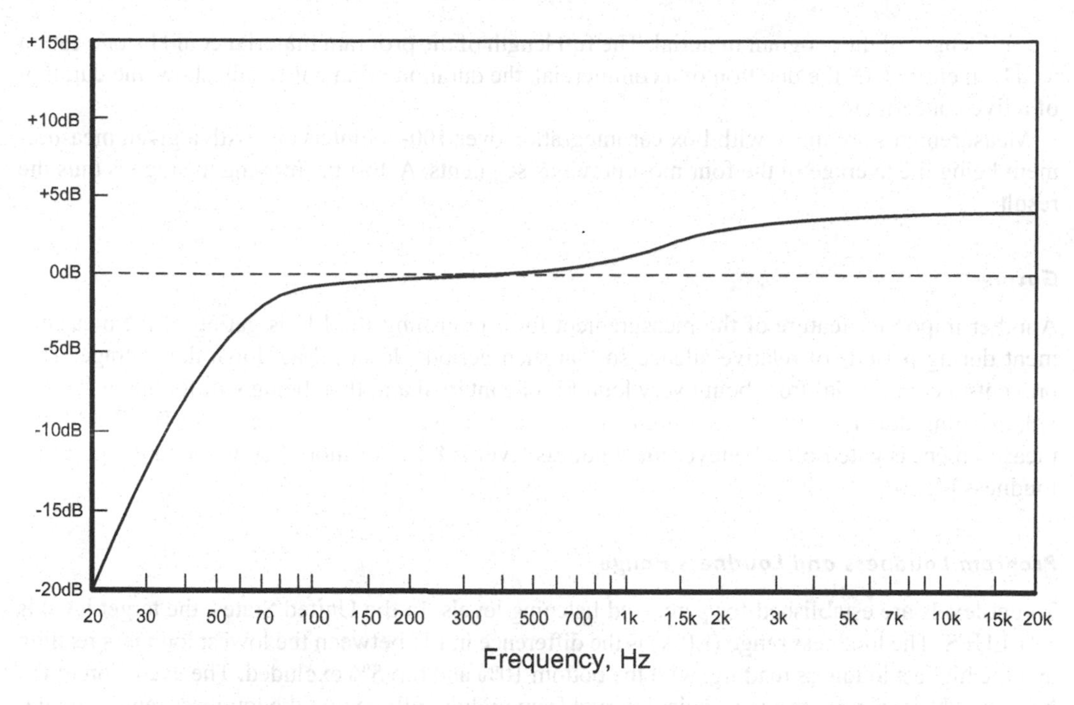

Figure 27.5 K-Weighting Curve.

response is shown in Figure 27.5 for frequencies above 400 Hz. The high-pass filter response (RLB) is seen in the portion of that figure below 400 Hz.

The RLB portion of the K-weighting curve at low frequencies is a modified version of the old B-weighting curve. The revision eliminates the high-frequency roll-off portion of the B curve so the curve is flat at high frequencies. It also moves the 3-dB HPF point down to 60 Hz. The result is a second-order Butterworth HPF down 3 dB at 60 Hz. The RLB curve is shown in BS.1770-4 [17].

The K-weighting curve is shown in Figure 27.5. It can be implemented with an analog filter comprising a biquadratic second-order shelf followed by a Sallen-Key second-order Butterworth high-pass filter.

The LUFS Measurement

The measurement of LUFS is essentially the power sum of all of the channels' K-weighted signals. In the case of material with more than two channels, the additional channels' contributions may be weighted. More specifically, the mean-square value (the squared signal voltage) of each of the signals is summed to form a single number. This value is then passed through a logging function to arrive at a dB value in LUFS.

Measurement Duration and Timing

Three temporal loudness measurements are specified in EBU-R128 [20]. These correspond to Momentary (M), Short (S) and Integrated (I). The momentary measurement is for 400-ms samples, the short measurement is for 3-second samples, and the integrated measurement is carried out over the full length of the program material. The full length of the program material could be one cut on a CD, an entire CD, the duration of a commercial, the duration of an entire talk show, the duration of a live concert, etc.

Measurements are made with box car integration over 100-ms intervals, with a given measurement being the average of the four most previous segments. A 400-ms moving average is thus the result.

Gating

Another important feature of the measurement for establishing the LU is gating of the measurement during periods of relative silence so that such periods do not drag down the average. This prevents a commercial from being very loud for one interval and then being soft for another interval, bringing down the LUFS measurement into compliance. In accordance with EBU R 128, the measurement is gated off whenever the loudness level is 8 LU or more below the target program loudness [4, 20].

Program Loudness and Loudness Range

Target levels are established for perceived listening levels. In the United States, the target level is −21 LUFS. The loudness range (LRA) is the difference in LU between the lowest loudness reading and the highest loudness reading, with the bottom 10% and top 5% excluded. The exclusion of the bottom 10% is to prevent a very quiet interval from unduly influencing the loudness range calculation. The top 5% is excluded to prevent a brief, but very loud, event from unduly influencing the loudness range calculation. The gating and computation for the loudness levels for LRA are a bit different than those used for LUFS and are defined in EBU Technical Document 3342 [4, 26, 27].

27.8 Broadcast vs Cinema Loudness Models

With CCIR 468 weighting, now BS.468-1, the level is measured with a quasi-peak meter. The high-frequency portion of the weighting curve is 0 dB at 1 kHz, +12 dB at 6 kHz and −21 dB at 20 kHz. The low-frequency weighting is −20 dB at 100 Hz and −34 dB at 20 Hz, with a 20 dB/decade, first-order slope.

With CCIR-ARM, the measurement is made with an average-responding meter instead of a quasi-peak meter. There is a 5.6-dB offset from the CCIR-468 measurement. The weighting curve crosses 0 dB at 2 kHz and −40 dB at 20 Hz. It is usually associated with noise measurement [28].

27.9 Other Audio Level Displays

Other level displays that are found less often include displays based on true RMS detection, sometimes combined with peak detection. For example, RMS detection is a better indication of actual average power from a heating point of view. This might be appropriate for a level meter on a power amplifier that is intended to better reflect the amount of power presented to a loudspeaker that will reflect its voice coil temperature. Loudspeakers can tolerate very large brief peak power levels, but must not be subjected to average power levels over a period of time that would exceed their thermal handling ability. This can be especially important if the program material is highly compressed. At the same time, the peak-detecting portion of the display will indicate how close the amplifier is to clipping.

Peak-Average Meter

A similar kind of display with separate peak and average power displays was constructed to illustrate the likely incidence of clipping in a power amplifier under conditions of realistic (but not ear-bleed) conditions in a consumer environment [29]. Its average meter uses true RMS detection to display medium-term average power calibrated for an 8-Ω load. This is intended to be a rough approximation of the perceived loudness (i.e., the average power in watts that it took to create that average level of perceived loudness).

The peak meter displays the momentary average-calibrated power that occurs on a musical peak. It is the power that a conventionally-rated power amplifier would have to briefly deliver into an 8-Ω load in order to produce that musical peak. It will detect and display a musical peak that lasts as little as 20 μs. The peak reading is held for 1 second before any decay begins. A continuous sine wave input voltage will yield the same reading on the average and peak meters.

In several listening presentations it was illustrated that power amplifiers may clip more than we think when listening to well-recorded music in the home at realistic listening levels, even in a relatively small room. This, of course, is driven by the crest factor of the program material.

Power Amplifier Metering

Several of the sections above pertain to the meters found on many power amplifiers. One thing of great importance to keep in mind is that these displays are indicating a measure of the voltage swing at the output of the amplifier, not the power actually being delivered. The reason for this is that these meters are measuring voltage, and they can be calibrated for only one load impedance, usually a resistive 8-Ω load. In other words, 40 V_{RMS} output into an 8-Ω resistor is indeed 100 W delivered into that resistor. However, if the load is 4 Ω, the actual power delivered into the resistor will be twice that, or 200 W.

To get more technical, the true power delivered into the load depends on the load impedance, not just the load resistance. The magnitude and phase angle of the impedance of a loudspeaker typically goes all over the place, so the detection of output voltage as a metric for delivered power is very questionable. Real power delivered into a purely reactive load, like a capacitor or inductor, is theoretically zero. Such meters are really only a good indicator of how close the amplifier is to voltage clipping. It is also the case that the 0-dB VU point on a power amplifier meter may not be the rated output of the amplifier.

Bar Graph Spectrum Displays

It is often desirable to see what the program level is in different portions of the frequency spectrum. These displays often incorporate 5–12 separate bar graph displays, each driven by a bandpass filter whose center frequency corresponds to the assigned frequency of each display. These displays can have either a single-average display or a peak-average display as described above. They are often found on multi-band equalizers.

References

1. H. A. Chinn, D. K. Gannett and R. M. Morris Chinn, et al., A New Standard Volume Indicator and Reference Level, *Proceedings of the I.R.E*, vol. 28, January 1940, pp. 1–17, https://www.aes.org/aeshc/pdf/chinn_a-new-svi.pdf.

2. Glen Ballou, Chapter 30, Audio Output Meters and Devices, *Handbook for Sound Engineers*, edited by Glen M. Ballou, 5th edition, Focal Press, Oxford, 2015.
3. John G. (Jay) McKnight, *Some Questions and Answers on the Standard Volume Indicator ("VU meter")*, 2006, aes.org/aeshc/pdf/mcknight_qa-on-the-svi-6.pdf.
4. B. Eddy Bogh Brixton, *Audio Metering, Measurements, Standards and Practices*, 2nd edition, Focal by Taylor & Francis, 2014.
5. ANSI C16.5-1961, *Volume Measurements of Electrical Speech and Program Waves*, Audio Engineering Society. Also IEEE 152-1953, *IEEE Recommended Practice for Volume Measurements of Electric Speech and Program Waves*.
6. *National Semiconductor LM3916 Dot/Bar Display Driver*, Datasheet, January 2000.
7. Raspberry Pi Pico, Technical Specification, raspberrypi.com/documentation/microcontrollers/raspberry-pi-pico.html.
8. Sound System Equipment – Part 10: Programme Level Meters, IEC 60268-10:1991.
9. Wikipedia, *Peak Programme Meter*, wikipedia.org.
10. John Emmett, *Audio Levels in The New World of Digital Systems*, EBU Technical Review, January 2003, tech.ebu.ch/docs/rechreview/trev_293-emmett.pdf.
11. John H. Roberts, *Simultaneous Display of Multiple Characteristics of Complex Waveforms*, US Patent 4,166,245, August 28, 1979.
12. John H. Roberts, *Linear Array Level Meter Displaying Multiple Audio Characteristics*, US Patent 5,119,426, June 2, 1992.
13. Michael L. Dorrough, *Dual Loudness Meter and Method*, US Patent 4,528,501, July 9, 1985.
14. Mike Dorrough, *Reading a Dorrough Meter*, dorrough.com/readmeters.html.
15. Dorrough, *Dorrough Loudness Monitor Model 40-A2*, Technical Instruction Manual, dorrough.com/Usermanuals.html.
16. Sound System Equipment – Part 18: Programme Level Meters, IEC 60268-18:1995.
17. ITU-R BS.1770-4, *Algorithms to Measure Audio Programme Loudness and True-peak Audio Level*, International Telecommunication Union, October 2015.
18. ITU-R BS.1771, *Requirements for Loudness and True-Peak Indication Meters*, International Telecommunication Union, 2006.
19. E. M. Grimm et al., Towards a Recommendation for a European Standard of Peak and LKFS Loudness Levels, *SMPTE Motion Imaging Journal*, vol. 119, no. 3, April 2010, pp. 28–34.
20. EBU R 128, *Loudness Normalization and Permitted Maximum Level of Audio Signals*, Geneva, June 2014.
21. Owen Green, *Loudness and Metering (Part 1)*, February 2013, https://designingsound.org/2013/02/06/loudness-and-metering-part-1/.
22. Owen Green, *Loudness and Metering (Part 2)*, February 2013, https://designingsound.org/2013/02/14/loudness-and-metering-part-2/.
23. Commercial Advertisement Loudness Mitigation Act (CALM), H.R. 1048/S. 2847, 111th Congress, congress.gov/111/bills/s2847/BILLS-111s2847enr.pdf.
24. ATSC Recommended Practice: Techniques for Establishing and Maintaining Audio Loudness for Digital Television, A/85:2013, Advanced Television Systems Committee, March 12, 2013, atsc.org.
25. RLB-weighting (revised low-frequency B curve), AES Pro Audio Reference (W), AES.org/par/w/. See also ITU-R BS.1770-1, p. 4.
26. EBU Technical Document 3342, *Loudness Range: A Measure to Supplement EBU R 128 Loudness Normalization*, Version 3.0, 2016.
27. TC Electronic, *Loudness Explained*, https://tcelectronic.com/loudness-explained.html.
28. R. Dolby, D. Robinson and K. Gundry, A Practical Noise Measurement Method, *JAES*, vol. 27, no. 3, 1979, pp. 149–157.
29. Bob Cordell, *The Peak-Average Meter*, cordellaudio.com.

Chapter 28

Microcontrollers and Microcomputers

Microcontrollers (MCUs) are in everything, from kid's toys to home theater receivers to automobiles. Many modern automobiles have dozens of them spread throughout the vehicle. The MCU is just another low-cost IC. Most consumer items that you purchase today have some kind of MCU in them. If nothing else, they implement outstanding human interfaces at much less cost than traditional approaches. A decent MCU can cost less than $8, and sometimes just $1 [1]. They can be as small as an 8-pin SOIC. Non-volatile flash memory built into many modern MCUs makes them especially useful and flexible. Such flash memory can often be programmed *in situ*. The majority of MCUs are 8-bit.

The scope of this chapter will be primarily focused on MCUs, as microcomputers (μC) are largely an extension of the capabilities of MCUs, but with fewer on-chip peripherals to interface the outside world. μCs often run an operating system [2]. MCUs are more hardware-centric and μCs are more computing-centric.

28.1 Embedded Systems

MCUs are critical to the human interface in even simple applications. They respond to pushbuttons and rotary encoders and they display settings, information and messages via LCD displays and LEDs. They are also used to detect signal levels and sensor values. Even a digital volume control often requires an interface to an MCU or a microcomputer. Indeed, MCUs are referred to as physical computing devices, as their primary job is to usually take in information from physical devices and control physical devices. These are referred to as *embedded systems* because the processor is embedded in application-specific hardware [3, 4]. In more sophisticated applications, the MCU may communicate with other intelligent devices via USB, Ethernet, WIFI or Bluetooth interfaces. They play a big role in the Internet of Things (IoT) [5, 6]. My refrigerator has its own MAC address for its WIFI connectivity. MCUs also provide efficient and convenient interfaces to other peripheral chips. These interfaces are often bus oriented and require few wires. Finally, MCUs are sometimes used to implement self-test routines in their application hardware.

28.2 System-on-Chip

As the scale of integration increased with Moore's Law from the 1970s to the present, many chips began to take on the properties of what we now refer to as System-on-Chip (SoC), usually containing a microprocessor core [7]. Other chips used the added transistor count to build larger and more powerful microprocessors. Early microprocessors like the Intel 4004, 8008 and 8080 required quite a bit of external hardware to make them do anything useful. What is largely considered to be the first MCU (and SoC) was the Intel 8048-8051 series of devices. These 8-bit devices brought the necessary RAM, non-volatile EPROM and other interface circuitry onto the chip, allowing the chip to perform very useful functions with very little external hardware overhead. The keyboards for the IBM PC used one of these chips.

28.3 Microcomputers

Microcomputers (μC) are used for more complex tasks and more computing power. While MCUs are often limited to running one thread, microcomputers usually have an operating system that allows them to perform many more complex tasks [2]. Some, for example, can run the Linux operating system or one derived from it [8]. Microcomputers often have more resources, like RAM and flash memory. Some or all of their RAM is usually off-chip and their program memory is almost always off-chip. In fact, a microcomputer is often a microprocessor (μP) plus some peripheral devices with external program and data memory. Only a small portion of this chapter will be devoted to microcomputers, mainly to distinguish them from MCUs.

28.4 Board-Level MCUs and μCs

Many MCUs and microcomputers are available on PCBs with the necessary support circuitry and are pretty much ready to go. This is not only helpful for prototyping, but also for DIY and low-run products. Two well-known vendors of these are Arduino and Raspberry Pi Foundation (Rπ) [9–13]. Hardware includes small, inexpensive boards like the Nano, Pico and Zero [14–16].

These organizations provide a tremendous amount of support for these products, including software and peripherals. Most manufacturers of MCU chips also make evaluation boards for their chips as well [17]. A μC almost always requires a board and is sometimes referred to as a single-board computer (SBC), even though the board may be as small as 1×2 inches [16].

28.5 CPU Architectures, Instructions and Execution

This chapter just scratches the hardware surface of MCUs and μCs, and is certainly not aimed at making you a programmer or computer scientist. However, understanding some key terminology can provide useful context.

An important aspect of CPU architecture is the type of instruction set. Most traditional machines, like the X86 PC, employ what is called a complex instruction set (CISC) with a great many instructions that can accomplish many functions in one fell swoop. Intuitively, this would normally lead to requiring fewer instructions to accomplish a task and result in higher speed. Alternatively, many machines use what is called a reduced instruction set (RISC) where there are far fewer instructions that are simpler. On the average, more than one instruction may be needed to accomplish a given function. However, at the hardware level the RISC machines are much simpler and faster as a result. Virtually all MCUs are of the RISC variety and most μCs employ a RISC architecture. The ARM processor is probably the most well-known RISC architecture [7].

Another architectural term defining a computing machine concerns the way that program and data memory are accessed. In the more traditional Von Neumann designs, the program and data are accessed over the same data/address bus. The fetching of an instruction and the writing of data to memory cannot occur at the same time. In the Harvard architecture, the program and data memories have their own address/data buses, allowing simultaneous program and data memory accesses. This alone can result in higher effective processing speed. The Harvard architecture is especially important for digital signal processing (DSP) chips. This architecture is also well-suited to MCUs, and most MCUs are based on the Harvard architecture.

A third concept to know about concerns how instructions in the written program become machine language that is actually stored in the program memory. Traditionally, the human language instructions are sent through a compiler that creates machine language. This process occurs in non-real-time and is done once. The machine language is what is stored in the program memory. Often, the program is written in a so-called high-level language. Compiled languages include Fortran, COBOL and C. There is compiled BASIC such as BASCOM-AVR for the Atmel AVR series of MCUs.

In many MCU applications a simple low-level language is adequate and preferred. In this case, the written program instructions are stored in the program memory and pass through an "interpreter" instead of a compiler to run the machine at its physical level. The interpreter converts the instructions into machine commands on the fly, in real time. Every time the computer is run, the code passes through the interpreter software on the MCU. Because of the need to interpret on-chip, interpreted code usually runs more slowly. Examples of interpreted languages include Lisp, BASIC and Python. The C language can also be interpreted.

Traditional computers execute out of RAM that has been loaded from a disk drive or other non-volatile mass memory. This allows for high-speed execution, since RAM can be very fast. In contrast, most MCUs execute directly out of flash memory that is on-chip. This incurs a speed penalty, but speed of execution is not as important in most MCU applications.

28.6 Clocks

The allowed clock frequencies for MCUs span the range of about 4 MHz to over 100 MHz. Most MCUs allow clocking them in many different ways, some of which are quite economical. Two pins are usually allocated to clocking, and the most familiar arrangement is to clock the device with a crystal across these pins, creating a clock signal with a precise frequency. Of course, an external clock can also be connected to one of these pins. However, cost is a major issue in many applications, especially those where the MCU itself is a couple of dollars or less. In many cases a ceramic resonator can be substituted for the crystal at a reduced cost and with only a modest reduction in frequency accuracy.

Still more economically, many MCUs allow the use of a clock circuit created by just a single resistor and capacitor connected to the external pins where fairly low frequency accuracy is acceptable. Finally, some MCUs can clock themselves with an on-chip RC oscillator. In this case, frequency accuracy may span a range of 2:1. Notably, many MCU applications can be served with a very inaccurate clock frequency at modest frequencies. Many children's toys fall into this category.

Some MCUs have on-chip PLLs (phase-locked loops) to allow internal clocks to be at a higher frequency than that of the external clock. On the other hand, some MCUs can function with a clock that is of a very low frequency, like 30 kHz in order to consume very little power (the power consumption of most digital CMOS circuits increases with the clock speed).

28.7 General Purpose I/O

Pins are precious. Often the cost of an MCU rises with the number of pins, as does the occupied board space. At the same time, transistors are almost free. The answer is to spend more transistors to make the most of the available I/O pins. The number of package pins among MCUs typically range from 8 to 100, with corresponding chip costs from less than $1 to close to $20, even in low quantity. Apart from support pins like power supply and clock, these pins are usually referred to as general purpose I/0 (GPIO). While one might think that these pins are able to act as ordinary digital outputs or inputs, tri-state outputs or open-drain outputs, they are typically capable of much more than that. This approach provides great flexibility in applications. It also allows efficient communication with inexpensive external chips that can expand the MCUs I/O capability. Not all GPIO pins on a device are able to serve all of the available functions on a device.

Two of the most important members of the category are the serial peripheral interface (SPI) and I²C serial bus interfaces that will be discussed later. Finally, many MCUs include one or more ADC (analog-to-digital converter) inputs, and often the pin or pins associated with the corresponding analog input(s) can be programmed to accept these analog inputs. As a result, some of the circuit modulcs associated with GPIO pins involve quite a bit of circuitry in supporting so many different pin functionalities. Even devoting a thousand transistors to support each pin is no big deal.

Analog-to-Digital Conversion

Many MCUs contain an on-chip analog-to-digital converter (ADC) with one or more selectable input channels. These converters are usually of the successive approximation register type (SAR) and are not very fast, requiring more than one clock cycle per bit. They are often used to monitor slowly moving input voltages, as in a DVM (digital voltmeter) measurement or a sensor input. Surprisingly, some of these converters have 10–12 bits of resolution. For many applications, the speed is quite adequate, when one realizes that the clock speed is often in the range of 4–20 MHz. Some use the supply voltage as their reference, while others have an on-chip fixed reference of greater accuracy. Depending on their circuit details, the accuracy of some of these ADCs depends on the source impedance being low. These include those that employ switched-capacitor SARs.

Digital-to-Analog Conversion

Very few MCUs have traditional on-chip digital-to-analog converters (DAC). However, they often have pulse-width modulation (PWM) circuits that accomplish the DAC function, at least at lower frequencies and assuming that the PWM output signal is externally low-pass filtered so as to extract the average duty-cycle value of the square wave. Many DACs use the supply rail as their reference, with the PWM output swinging rail to rail. However, some incorporate an internal reference for the high value of the signal.

28.8 Microcontroller Examples

One of the earliest, and still one of the most popular MCUs is the PIC family from Microchip Technology, founded in 1987. Many MCUs require very little external support to carry out their task. Most are aimed at low-cost applications, where dedicating a computing device to a simple task is like what long ago was dedicating an integrated circuit to a simple task. They are so convenient and inexpensive that many consumer devices employ numerous MCUs. There are many MCUs in most modern automobiles. Among the very large number of MCU providers, another important one is Atmel, founded in 1984 and acquired by Microchip Technology in 2016. Atmel's AVR series of MCUs are quite popular.

Microchip Technology PIC

The acronym PIC had its origins as a *peripheral interface controller*. It was one of the first IC MCUs, and is very popular. Two of the family members include the PIC12F615 and the PIC16F1788. The 64-pin PIC18F66J60 is an example of a more powerful device in the PIC family, including a 64 kB flash program memory, 3808 bytes of SRAM and Ethernet. It includes 39 I/O pins. It costs about $6 in quantities of 25. The larger 100-pin 18F97J60 has 128 kB of flash and 70 I/O pins. It costs about $8 in quantities of 25. This 18F series includes SPI and I^{2}C interfaces.

The very small 8-pin PIC 12F615 includes 1.75 kB of flash program memory and 64 Bytes of SRAM with a 5-MIPS RISC CPU. Its 5-pin I/O functionality includes a 4-channel 10-bit ADC, a 10-bit PWM DAC, three timers and one comparator. Many of its I/O pins are multi-functional to provide a wide variety of functions within a limited pin count. It is about $1 in quantities of 25.

The 28-pin PIC 16F1788 includes 28 kB of flash program memory and 2 kB of SRAM, plus 256 bytes of EEPROM. It includes an internal 32-MHz clock oscillator for the 8-MIPS CPU. Its I/O capabilities include SPI, I^{2}C and UART ports, an 11-channel 12-bit ADC, four DACs, four comparators and three timers. Many of its I/O pins are multifunctional to provide a wide variety of functions within a limited pin count. It is less than $3 in quantities of 25.

Arduino UNO

The Arduino UNO MCU is based on the Atmel AVR series and is quite popular. It includes the Atmega328 8-bit MCU on which the Arduino is based. The UNO is popular and well-known. It includes 32 kB of flash program memory, 2 kB SRAM and 1 kB EEPROM. The EEPROM makes it especially suited to applications that must be able to write small amounts of data into a non-volatile memory to remember settings, for example.

The UNO includes 22 digital I/O pins whose function can be programmed in order to make best use of available GPIO pins for the needs of a given application. It supports USB, SPI, I^{2}C and serial UART interfaces. Six analog I/O pins provide six channels of 10-bit analog-to-digital conversion (ADC). Six digital pins can be used for 8-bit PWM DAC outputs. Two interrupt pins are also provided. A 16-MHz crystal oscillator is included. The UNO can be powered from the 5-V USB interface or an external 7–12 V supply. An onboard regulator converts the external supply voltage to the 5-V operating voltage.

The Arduino integrated development environment (IDE) is simple and easy to use. It resides on a PC connected to the UNO via USB. It is based on the C/C++ language. The Arduino benefits from a large user community and software library.

Arduino Nano

The Arduino Nano is basically a smaller version of the UNO with the same capabilities. It is based on the 8-bit Atmega328p operating at 16 MHz. The 30-pin Nano MCU board has a very small footprint of 18×45 mm (0.7 × 1.8 in), yet has 32 kB of flash, 1 kB of EEPROM, 2 kB of SRAM and 22 digital I/O pins. Each pin is assigned with multiple functions as determined by the program. It includes USB, SPI, I^{2}C and UART interfaces. A nano clone can be had for about $6, but the cost is about $20 for a genuine nano.

Raspberry Pi Pico

The Raspberry Pi Pico is a remarkable little MCU with a lot of bang for the buck. It is a surface-mountable board that can withstand reflow temperatures and costs only about $4–6. It is about the same size as a 40-pin DIP, measuring 7/8" × 2".

It is available with a pre-soldered 40-pin header for about $12. The Pico WH with a pre-soldered header includes WIFI using an onboard Infineon CYW43439 wireless chip. This makes it especially useful for IoT applications. It costs about $15.

It is based on the RP2040 MCU chip designed by Raspberry Pi. It includes a 32-bit dual-core ARM Cortex-M0+ processor operating with a 133 MHz clock. The Pico has 2 MB of flash, 264 kB RAM and 26 multi-function GPIO pins. Its digital I/O interfaces include 2 SPI, 2 I^2C and 2 UART ports and a USB connector. It also supports up to 16 controllable PWM DAC channels. Its analog I/O pins support three 12-bit ADC channels. Its small 21×51 mm (0.83×2.0 inches) footprint makes it attractive as a turnkey microcomputer module on a system board. Compared to the small Arduino Nano, it has a faster processor, more flash and more GPIO pins, but no EEPROM.

The Pico comes with complete software support for the MicroPython and C/C++ languages and includes the Raspberry Pi SDK (system development kit) software for both languages.

In situ *Programming*

The non-volatile on-chip flash memory included on most MCUs contains the program code and fixed data. That information must be programmed before the device is useful. This is almost always accomplished with the board fully assembled, including the MCU itself. The board is then powered up into a special mode where a programmer can be connected to the board to accomplish *in situ* programming.

28.9 Microcomputers

Although distinctions get blurred as the silicon technology advances, with both MCUs and microcomputers really being systems on a chip (SoC) in modern times, the μC usually relies more on external peripherals for support, like flash memory for non-volatile storage of program and fixed data, and external RAM. The distinguishing factor that will be used here is that microcomputers are able to run a full-fledged operating system and are capable of desktop computer functionality. More often than not, their RAM and program memory are off-chip. Their focus is on computing rather than interfacing to external hardware to be monitored or controlled. Most have SPI or I^2C ports that enable significant interface functionality with peripheral chips [16, 18–22]. Many MCUs also feature similar I/O interfaces. The line of Raspberry Pi single-board computers is a good example of what is available at low cost.

Raspberry Pi 3 Model B

The Raspberry Pi 3 Model B is a single-board computer that runs the Linux operating system. It employs the Broadcom BCM2837 1.2 GHz quad-core 64-bit CPU and has 1 GB of RAM. It provides 100Base-T Ethernet and includes the BCM4348 to support wireless LAN and Bluetooth communications. It has four USB 2.0 ports and an HDMI port. It includes a 40-pin GPIO header. It has an 85 × 56 mm footprint. The GPIO pins are programmed to act as 3.3-V inputs or outputs. SPI, I^2C and serial interfaces are supported. Four hardware PWM outputs are also supported. Unlike most MCUs, the Rπ does not include an ADC.

Raspberry Pi Zero

The Raspberry Pi Zero is a small and affordable single-board computer that includes a 1 GHz single-core CPU and 512 kB of RAM. It has micro USB and mini HDMI ports.

Operating Systems

The Raspberry P1 employs an ARM processor and uses an operating system called Raspberry Pi OS. It has its roots in the Debian Linux OS. As a result it is pretty much like Unix. It is optimized for use with the ARM processor running on the line of Raspberry Pi MCUs. There is a Lite version of the operating system that is adequate for many applications. The MicroC/OS operating system can run on MCUs and its name stands for Microcontroller Operating System. It is also referred to as µC/OS. It is a compact real-time operating system (RTOS). It can run a number of instructions from the C programming language.

There are numerous bare-bones operating systems suitable for the Arduino processors. They include FreeRTOS, Zephyr, NuttX, Riot, HeliOS and Simba. One of the most popular is FreeRTOS.

28.10 FPGAs and Embedded Processors

Yet another programmable approach is to have hard or soft processors embedded in FPGAs (field-programmable logic arrays) [23]. The processor can be more like an 8-bit MCU or can be more like a fairly powerful µC that can run an operating system. A soft processor on an FPGA is just another group of programmable gates implementing the design of a processor. A hard processor is a portion of the FPGA whose logic implements a processor with fixed, custom logic. Being able to program both the processor and 20,000 gates surrounding it leads to a phenomenal capability. In many cases, the processor implemented on an FPGA is more than powerful enough to run an operating system, thus falling into the category of microcomputer. Reasonable choices include the Atmel AVR910 and the Xilinx Spartan 6 series.

Configuration

An FPGA has to be configured (programmed) to perform its function. The interconnections of the logic elements (LEs) are controlled by a memory. There are three types of FPGA in terms of how they are configured. This first is somewhat like the way a computer boots up. The program is stored in an external non-volatile device, like a hard drive or a flash memory, and the FPGA retrieves the configuration information from that device and stores it on-chip in RAM. The RAM being volatile, the FPGA has to boot up every time after power has been removed.

A convenient approach is to have the FPGA fetch the configuration information from a SPI flash memory. This requires few pins, is adequately fast in most cases and employs inexpensive memory for storage of the configuration. When the FPGA is first powered up, it starts in configuration mode and acts as the master for the SPI flash, fetching all of the configuration information. Alternatively, many FPGAs can be configured as a peripheral by another device, such as an MCU, to be loaded at power-up.

The second approach has a similar configuration model, but the flash is located on-chip [24]. At boot-up (configuration) the FPGA transfers the information from the flash into the RAM. The configuration model for the third type stores the configuration in the on-chip flash and the on-chip flash controls the logic interconnections directly. In every case, wherever it is located, the flash memory must initially be loaded with the configuration data. Prominent FPGA vendors include Xilinx (now part of AMD), Altera (now part of Intel), Lattice Semiconductor and Microchip Technology.

Soft and Hard IP

As mentioned above, there can be a soft or hard processor embedded in an FPGA. In fact, a great many other pre-designed functions, soft or hard, can be dropped into an FPGA to form a system. These designs are thus referred to as soft and hard intellectual property (IP). The hard IP content is a fixed feature of the given FPGA. This can include processors, memories, PLLs, 1 Gb/sec high-speed I/O transceivers and many other functions targeted to the market for that FPGA. Obviously, hard IP will usually outperform soft IP in speed and or power consumption. The variety of soft IP functions available is very large, and need not come only from the manufacturer of the FPGA.

Xilinx Spartan 6

But one example line of FPGAs is the Xilinx Spartan-6 family. It is a good example of what can be found in FPGAs. Members of the family provide between 3840 and 147,443 logic cells in a 45-nm process technology. The devices include several hard IP functions including SDRAM controllers, high-speed LVDS (low-voltage differential signaling) serial transceivers, 18 kB RAM blocks, clock PLLs, a fast 18×18 multiplier and DSP slices. Many soft-IP reference designs are also available. Configuration data is stored in SRAM-like latches.

The smallest member of the family, the XC6SLX4, includes 3840 logic cells, 4800 flip flops, 12 18-kB RAM blocks, 8 DSP slices and 132 user I/O pins. Each DSP slice includes an 18×18 multiplier, an adder and an accumulator. The FPGA can store 3 Mbit of configuration data. The configuration data is loaded upon power-up. The configuration data can be loaded serially from an external non-volatile memory device like a small flash memory. The PLL clock VCO can operate from 400 to 1080 MHz. In most applications it will be followed by one or more programmable frequency dividers. This powerful device costs about $30.

Verilog

The digital circuitry of FPGAs is created in a hardware description language (HDL) that is user-friendly and shares some similarities with the program language C. One such language is Verilog [25, 26]. Others include VHDL and SystemC.

28.11 Microcontroller Resources

Here we discuss some of the on-chip resources that are often found on MCUs. In many cases, these resources are implemented as part of a multifunctional GPIO's pin circuitry. The most common resource is a GPIO that can act as a logic output or input. As an output, these can usually be configured as tri-state or open-drain as well. Some of the other functions for which certain GPIOs can be configured are described below.

RAM Memory

Static RAM is an important resource on every MCU chip, often in the range of 2–8 kB or more. It is volatile, but it is much faster than non-volatile flash memory. It is addressable at the byte level and is preferred for storing information that will change quickly and many times during a session.

Flash Memory

The availability of on-chip non-volatile flash memory for program and fixed data storage is a feature that distinguishes most MCU chips. Flash memory depends on floating-gate transistors to store each bit. The charge on the floating gate determines the binary state of the cell and that state remains indefinitely due to the extremely high insulation of the floating gate. Programming the memory usually depends on first erasing the cells by bringing them all to the same state. Actually programming the erased cells depends on selective charge injection to the cells that will be taken to the non-erased state. The erase and programming actions depend on phenomena that can overcome the insulation of the gate to influence the charge on the gate. These phenomena usually include hot-electron injection (HEI) and Nordheim-Fowlert tunneling (NFT).

The program and erase processes can degrade the insulating oxide ever so slightly each time they are carried out, so the flash memory can only remain reliable for a finite number of write cycles, usually between 10,000 and 1 million. Erasure and programming operations are also relatively slow, while read operations are quite fast. This makes flash memory best suited to applications involving infrequent write operations, like program storage. In order to achieve high storage density, flash memory is erased and programmed in large or small blocks of bytes [27, 28]. Typical MCU flash memory size ranges from 1 to 256 kB.

EEPROM Memory

EEPROM uses similar program-erase techniques, but data can usually be written on a byte-by-byte basis. The tradeoff here is much smaller density for EEPROM functionality. Some MCUs have a small amount of EEPROM, often on the order of 256 bytes to 4 kB. It is especially useful for settings that may need to be changed occasionally, such as setup from a remote control on a smart TV.

Analog-to-Digital Converter

Embedded systems are not all digital, so a 10- or 12-bit ADC is usually available on an MCU. This will often take the form of a single SAR-type ADC that has an analog multiplexer in front of it to enable multi-channel ADC capability. For the ADC inputs, certain GPIOs are configured to accept an analog input that can range from ground to +Vcc.

PWM Outputs

It is rare to find MCUs with DACs, but the equivalent functionality is achieved with GPIOs configured with a PWM output capability, where the average value of the PWM waveform represents the analog output. It is crude, but it works. It is not unlike the way class-D audio power amplifiers work. Most MCUs have multiple GPIOs that are capable of PWM.

Depending on the nature of the load they may be driving, the PWM outputs may require external analog low-pass filtering to extract the average value. However, if they are just controlling the brightness of an LED or display, they may need little or no low-pass filtering. This will often be the case when driving motors to variable speeds. PWM outputs can also be used to create tones and other sounds, and in some cases synthesized speech. The PWM waveform will transition from ground potential to the positive rail potential or in some cases to a reference potential.

Timers

Most MCUs are equipped with 1–8 timers. These can be valuable in creating control sequences, polling of inputs, creation of output waveforms and measuring the duration of digital input signals. Their accuracy is tied directly to that of the system clock. The timers are usually 8- or 16-bit timers, with the former usually being more plentiful. The timers often have numerous modes for which they can be configured. Many also have a watchdog timer to enable recovery after the program stops or goes off into the weeds.

Serial Ports

Most MCUs provide one or more asynchronous serial ports with UARTs, enabling RS-232 and similar types of communication. These are usually implemented with configurable GPIOs.

SPI and I²C Bus Interfaces

Synchronous serial interfaces are crucial to efficient communications with external peripheral chips. Virtually every MCU has at least one group of GPIOs that can be configured to support the SPI and/or I²C communications described below. Each of these types of buses can support numerous peripherals with a great variety of functionality, as described later.

Other Resources

Some MCU chips have on-chip Ethernet or USB interfaces. The PIC 18F66J60 is a good example of an economical MCU equipped with Ethernet that can be had for about $5. The PIC 18F2455 is an example of an MCU equipped with a USB interface. It can be had for about $7.

28.12 SPI Bus

The serial peripheral interface (SPI) bus is one of the most popular buses for communications between an MCU controller peripherals on a PCB [18–22]. It is fast, synchronous, full duplex and relatively simple. It employs simple protocols with no overhead for sending bytes. Designed by Motorola in the early 1980s, it is now a de facto standard; there is no formal standard for the SPI protocol. In fact, the SPI peripheral defines the protocol. The SPI interface is a 4-wire bus with the following signals:

SCK (clock generated by the controller)
MOSI (controller output to peripheral input)
MISO (controller input from peripheral output)
CS (chip select, active low)

There can only be one controller, but there can be numerous peripherals. Each peripheral must have its own chip select (CS) to address it and enable it onto the bus. All peripherals that are not selected must put their MISO output pins into a tri-state mode. If you have N peripherals, then you need N+3 wires for the arrangement. A pull-up resistor is usually placed on the MISO line so as to define the logic state if no peripheral is selected. Pull-ups are sometimes also placed on the CS lines. Technically, SPI might better be described as a 3-wire bus, since the chip select is not really

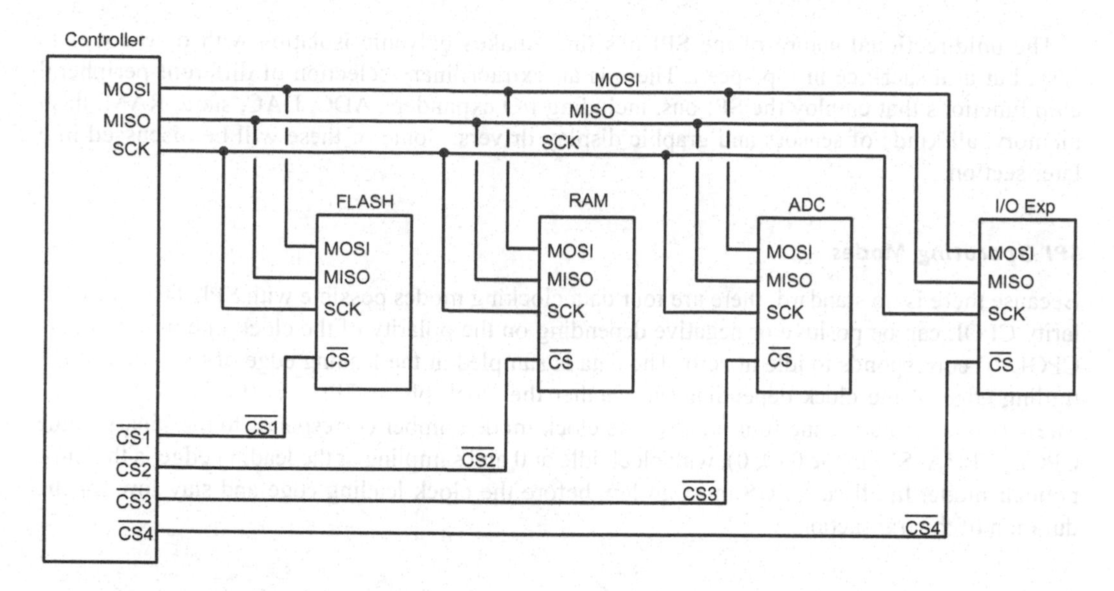

Figure 28.1 SPI Bus Example.

a bus connection. Some MCUs have dedicated pins for SCK, MOSI and MISO, and use GPIOs for chip selects. Others have certain GPIOs that can be configured to operate as SPI bus pins.

The full-duplex SPI bus can support simultaneous communication in both directions. Depending on the controller and peripherals, the number of bits sent in a transaction is quite flexible and can be as high as 32. However, most transactions are 8 bits or 16 bits. Figure 28.1 shows a typical SPI bus arrangement where an MCU communicates with four SPI devices: a flash memory, a RAM, an I/O expander and an ADC.

The SPI bus is fast because the lines can be driven in push-pull (no open collector lines needing pull-ups), there is no physical layer protocol overhead, the signals are unidirectional and the operation is synchronous. There are no set speeds for the SPI bus, but SCK speeds up to 50 MHz are possible. A SPI bus clocked at 50 MHz will have a transfer speed of 50 Mb/s during a transaction. In general, the SPI clock is sent in bursts, but is often continuous for the duration of a transaction with a given peripheral. Because pull-up resistors are not required, there is a minor power advantage of SPI over some other protocols. The price paid for SPI bus performance is three wires for the bus plus numerous chip selects. Thus there is quite a bit more controller pin and wiring usage compared to the much slower 2-wire I^{2}C bus described below. The I^{2}C bus does not require chip selects.

At the heart of every SPI peripheral is a shift register whose length is often 8 bits. At the SPI controller there is also a similar shift register. They are both clocked by SCK. The input to the peripheral's shift register receives data from the MOSI line. The output of the shift register is connected to the MISO line. The MISO line is connected to the input of the controller's shift register. Thus the controller and peripheral shift registers form a ring. With every group of eight clock pulses, the two shift registers exchange their contents.

For input to the peripheral, an 8-bit latch reads the parallel content of the peripheral's shift register at the right time. For output, a an 8-bit latch applies its signal to parallel-load the shift register at the right time. When the SPI chip is reading, the data on MOSI line passes right through the shift register to the MISO output unmodified and delayed by eight clock cycles.

The unidirectional nature of the SPI bus lines makes galvanic isolation with opto-couplers easy, but at a sacrifice in top speed. There is an extraordinary selection of different peripheral chip functions that employ the SPI bus, including I/O expanders, ADC, DAC, static RAM, flash memory, all kinds of sensors and graphic display drivers. Some of these will be discussed in a later section.

SPI Operating Modes

Because there is no standard, there are four data clocking modes possible with SPI. The clock polarity CPOL can be positive or negative depending on the polarity of the clock line in idle mode. CPOL=0 corresponds to idle at zero. The data is sampled at the leading edge of the clock or the trailing edge of the clock depending on whether the clock phase CPHA is 0 or 1, respectively. These two attributes define four modes. The clock mode number corresponds to the binary value CPOL, CPHA. SPI mode 0 (0, 0), with clock idle at 0 and sampling at the leading edge is the most popular mode. In all cases, CS must go low before the clock leading edge and stay low for the duration of the transaction.

SPI Payload Throughput

The raw information bit rate for SPI is generally equal to the clock speed; i.e., SPI operating at a clock speed of 30 MHz will transfer bits at 30 Mb/s during a burst of clock cycles. Bytes are the most commonly transferred chunk of information and it takes only eight clock cycles to transmit a byte. However, there is overhead in most data transactions because more than one byte may need to be transferred per byte of payload data sent or received from a peripheral. Often an op code, address and data byte need to be sent, reducing payload throughput in this case to 10 Mb/s if those three chunks of information are each only one byte. Writing a single byte to a 256 kB RAM might require an op code, 3 bytes for the 18-bit address and one byte for the byte to be written, totaling 5 bytes. The SPI peripheral defines this protocol.

Throughput can vary tremendously depending on the details of the peripherals and the nature of the transaction. Finally, there is nothing sacred about sending data in byte chunks; a peripheral could define its protocol as chunks of 12 bits, for example. So-called Dual SPI and Quad SPI arrangements can be used to achieve higher data throughput in some applications. For example, two or four MISO lines might be provided for reading a flash memory.

28.13 I²C Bus

The I²C bus, developed by Philips Semiconductor (now NXP) circa 1982, is a synchronous 2-wire half-duplex interface between a controller and multiple peripherals [21, 22]. Its I²C acronym stands for Inter-Integrated Circuit protocol. Its most important feature is that it requires only two wires, clock and data, called SCL and SDA. Both wires are bi-directional in nature and both are open-drain, so that pull-up resistors are required. The open-drain arrangement prevents two devices from driving the bus in such a way that there would be a hard fight that might damage the chips. Any device on the bus can pull down either SCL or SDA, although a peripheral pulling down SCL is somewhat of a special case (multi-master). A key feature of I²C is that any device on the bus can participate as a master. The bus is a peer-to-peer communication system. A typical I²C arrangement with an MCU communicating with three peripherals is illustrated in Figure 28.2.

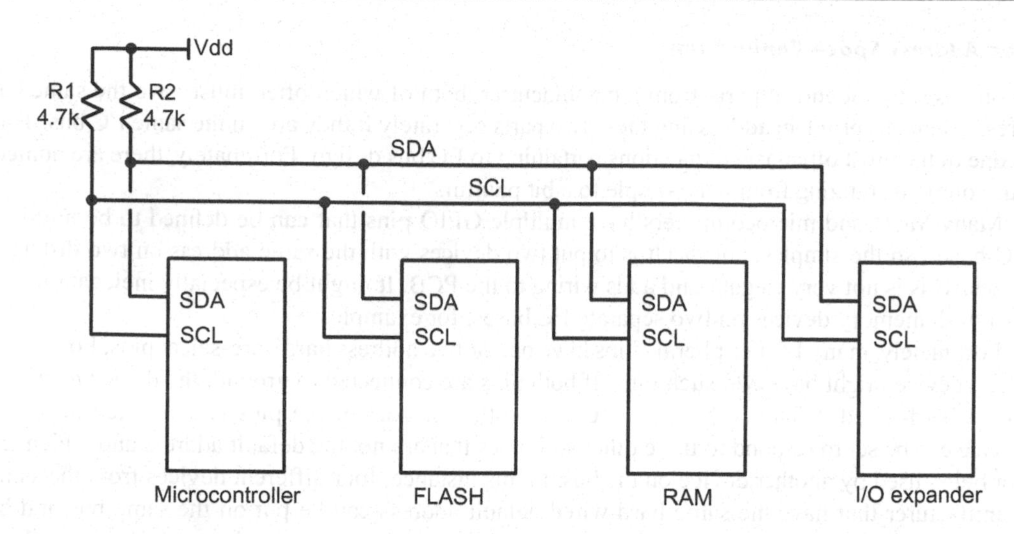

Figure 28.2 I²C Bus Example.

The protocol is the key to addressing peripherals and controlling the direction of data flow. An important thing to keep in mind is that the SCL clock is not generated by the master on a continuous basis, and that the clock is not always a uniform square wave when it is active. When the clock polarity changes, it does so at regular timing intervals. The unique protocol, relying on the relationship of the data to the clock polarity and edges, makes the exchange of information on only two wires possible. Details of the I²C protocol and timing requirements are widely available in publications and datasheets for I²C chips [21, 22].

The I²C bus exchanges speed for this simplicity. Standard I²C operates at 100 kHz, while fast I²C operates at 400 kHz. There are faster versions, but they are less common. Controller-peripheral communication that requires high speed is more often done with a SPI interface. Two things limit I²C speed. The first is the use of pull-up resistors for transitions to the one state for both SCL and SDA. This, combined with an allowed bus capacitance of as much as 400 pF, and pull-up resistances that can only be so small, limit the speed. A typical value used for pull-up resistors is 4.7 kΩ. The worst-case time constant might be an RC on the order of 4.7 kΩ and 400 pF, or about 2 μs. Pull-up resistors of smaller size can be used to the extent that all of the devices can pull the bus down and that the increased power dissipation is acceptable.

The second limitation is that of protocol overhead. The protocol necessary to make I²C work takes up additional bits in the bit stream. Depending on particulars, the actual communication bit rate can be as low as half the clock rate. As with SPI, additional efficiency is consumed if it is necessary to send op codes and addresses to the peripheral for a complete transaction.

The I²C bus can be operated over a range of voltages from 5 V down to 1.5 V in some cases. However, it is not always possible to put devices designed for different voltages on the same bus. Very commonly, devices that operate at 5 and 3.3 V are compatible. The pull-up will usually be connected to the smaller voltage, but the inputs of many devices are 5-V tolerant.

The original standard I²C address space had 7 bits. An extended 10-bit address protocol was later added, but the 7-bit protocol is most often used. The 7-bit protocol allows for 128 unique addresses, of which a few are reserved. A given manufacturer licenses one or more unique addresses for their products, and those default 7-bit addresses are hard-coded into the integrated circuits by the manufacturer.

The Address Space Conundrum

If one uses two identical parts from a manufacturer, both of which often must have the same address, there is a problem addressing these two parts separately if they are on the same I²C bus. This is one of the most often-asked questions pertaining to I²C bus design. Fortunately, there are numerous solutions, ranging from very simple to a bit painful.

Many MCU and microcomputers have multiple GPIO pins that can be defined to be multiple I²C buses, so the simplest approach is to put two devices with the same address on two different buses. This is not very elegant and adds wiring to the PCB. It might be especially inelegant to put two flash memory devices on two separate I²C buses, for example.

Fortunately, many I²C peripheral chips have one or two address hardware-select pins. For example, a device might have two such pins. If both pins are connected to ground, the device responds to the hard-wired default address. If one or the other or both of the pins is connected high, the device can be set to respond to three other addresses that are not the default address and which are not being used by another device on the bus. In this instance, four different devices from the same manufacturer that have the same hard-wired default address can be put on the same bus and be uniquely addressed. In a slightly different approach, one of these pins can be used as a chip select; when that pin is made high, the device is configured for an address that is never used on that bus.

Yet another way to expand the useable I²C address space is to use parts made by different manufacturers for different functions. For example, one might use I/O expanders from manufacturer A and A/D converters from manufacturer B. Most functions can be found in equivalent incarnations made by more than one manufacturer.

There also exist I²C bus transceivers that can split an I²C bus into two buses, depending on which buffer has an active chip select. I²C bus buffering usually requires special ICs due to the bi-directional nature of both the SCL and SDA wires.

28.14 SPI Bus Peripherals

The following is a small sampling of the enormous variety of peripheral chips that operate with the SPI bus. Always check the datasheet and in particular check the allowed operating voltage range. Where cited, costs of chips are for single quantities. In quantities of 25, these prices are often less by 15% or so.

I/O Expander

The Microchip MCP23S08 is an 8-bit I/O expander available in an 18-pin SOIC. It costs less than $1.50. For more ports, the Microchip MCP23S17 is a 16-bit I/O expander available in a 28-pin SOIC for about the same cost. It operates from 1.8 to 5.5 V.

A/D Converters

Most on-chip ADCs use the successive approximation register (SAR) technique for conversion. Multi-channel ADCs usually consist of a single SAR ADC with an analog-input multiplexer. At any given time, an analog multiplexer will select one of the eight inputs for conversion. The SAR ADCs are not very fast, with conversion times of 5 μs or more, depending on the clock frequency.

The Maxim (now Analog Devices) MAX192 is a 10-bit, 8-channel ADC. It is available in a 20-pin SOIC for about $10. The MAX1245 is a 12-bit, 8-channel ADC. The Maxim MAX11213 is a 1-channel 16-bit delta-sigma ADC. It operates from 1.7 to 3.6 V and costs about $5. The

Microchip MCP3201 12-bit ADC can operate from 2.7 to 5.5 V and is available in an 8-pin SOIC. It costs less than $3.

D/A Converters

Most MCUs provide an onboard DAC function by using PWM, and so their output must be filtered. However, external DAC peripherals can provide a more traditional DAC function.

The MAX525 is an 11-bit, 4-channel SPI DAC. The Microchip MCP4812 is a dual 10-bit DAC that can operate from 2.7 to 5.5 V. It costs less than $3. The Microchip MCP48CXBX4 is a quad DAC that can operate from 2.7 to 5.5 V. It is available in a 20-pin TSSOP package. The Microchip MCP4921 is a single 12-bit DAC that can operate from 2.7 to 5.5 V. It is available in an 8-pin SOIC and costs less than $3.

Flash

Some MCUs do not have flash memory, while some applications require far more flash memory than is available on an MCU chip. Flash memory peripherals usually operate on a SPI bus. Their access is serial, so their speed capability is not as great as a parallel-access memory. However, the high speed at which the SPI bus can operate mitigates any disadvantages caused by serial access. Writing to flash memory requires erasure before writing. This is usually done in large blocks. In other words, the memory cannot be written on a byte-by-byte basis. Flash memory uses hot electron injection (HEI) plus Nordheim Fowler Tunneling (NFT).

Flash memory has a wear-out mechanism that limits the number of lifetime writes to a given block. This is often on the order of 10,000 cycles. Of course, if it is used to store the program and fixed data, this is not a problem.

The Renesas/Dialog AT45DB161 contains 2 MB of storage, but it is good for only 1000 writes. The Microchip SST25PF040C is a 4-Mb, 3.3 V SPI NOR flash, available in an 8-pin SOIC for less than $1. The Winbond W25Q128JV is a 128-Mb, 3-V flash memory available in an 8-pin SOIC. It costs about $2.

EEPROM

Electrically erasable PROM is similar to flash, but it can be written and read on a byte basis. The price paid for this convenience is significantly smaller memory capacity. EEPROM density is typically no better than SRAM. It is also slow, taking milliseconds to erase or write. They are often found on an I^2C bus.

The EEPROM does not have a wear-out mechanism, so it is often used to store settings and data that may change frequently, such as in a remote control application. It uses NFT.

The AT25640 is an example SPI EEPROM chip. It has only 8 kB of memory. The Microchip 93LC56A 2K SPI EEPROM is available in an 8-pin SOIC package and operates from 2.5 to 5.5 V. It costs less than $0.50.

SD Card

The ubiquitous SD card is nothing more than a packaged flash memory, and it is available in large memory sizes. The SD card interface that can be used is basically the SD card socket that includes a SPI interface chip.

RAM

Microcontrollers and microcomputers often do not have sufficient RAM to meet the needs of some applications. This allows them to keep the cost down and economically serve the millions of applications that do not need much RAM. The Microchip 23LC512 is a 512k SPI serial RAM, available in a small 8-pin SOIC. It can operate from 2.5 to 5.5 V and costs less than $2.

28.15 I²C Bus Peripherals

Peripherals with many of the same functions as those for SPI are available for the I²C bus. In fact, there are some peripherals that will work with either SPI or I²C. The I²C bus is best suited to lower-speed functions. The convenience of I²C and the lesser amount of wiring is a big plus for applications where its slower speed is adequate.

I/O Expander

The Microchip MCP23008 is an I²C 8-bit I/O expander available in an 18-pin SOIC. For more I/O pins, the Microchip MCP23017 is a 16-bit I/O expander available in a 28-pin SOIC, and operates from 1.8 to 5.5 V. They both cost less than $1.50. I/O expanders are available with either I²C or SPI interfaces.

A/D and D/A Converters

The Microchip MCP3221 is an I²C 12-bit ADC available in a small 5-pin SOT-23 package. It can operate from 2.7 to 5.5 V. It costs less than $2. I²C DAC chips tend to be a bit more expensive. The Microchip MCP47CMB04 is a 4-channel (Quad) I²C 8-bit DAC that costs about $6. The Microchip MCP47CVB24 is a 4-channel 12-bit DAC that costs less than $8.

FLASH

I²C flash memory is not very popular, and SPI or parallel interfaces are strongly recommended for flash memory.

EEPROM

As mentioned earlier, EEPROM non-volatile memories tend to be small and slow, but they are accessible by byte. They are well-suited to applications where setup data does not have to be accessed fast or frequently. The Microchip 93AA56A is a 256×8 memory that costs less than $0.50. It can reliably withstand 1 million erase/write cycles and retain data for 200 years. While reading data can take up to 400 ns, an erase/write cycle can take up to 6 ms. The 8-pin Microchip 24LC64 is an 8k×8 memory that costs less than $1. If powered from 2.5 to 5.5 V, read time is less than 1 μs and write time is 5 ms.

RAM

The Microchip 47L64 is a 64-kbit I²C static RAM that can be had for about $1. It is available in an 8-pin SOIC package.

References

1. PIC12F16, Microchip Technology Inc., Device Datasheet.
2. Fredrick M. Cady, *Principles of Microcomputers and Microcontroller Engineering*, 2nd edition, Oxford University Press, Oxford.
3. Tim Wilmshurst, *Designing Embedded Systems with PIC Microcontrollers: Principles and Applications*, 2nd edition, Newnes, New York, NY, 2009.
4. John Catsoulis, *Designing Embedded Hardware: Create New Computers and Devices*, 2nd edition, O'Reilly Media, 2005.
5. Scott Shackelford, *The Internet of Things: What Everyone Needs to Know*, Oxford University Press, Oxford, 2020, ASIN: B087JYFMNQ.
6. Neil Wilkins, *Internet of Things: What You Need to Know About IoT*, self-published, 2019, ASIN: B07PG317XS.
7. Joseph Yiu, *System-on-Chip Design with ARM® Cortex®-M Processors: Reference Book*, ARM Education Media, 2019.
8. Aaron Newcomb, *Linux for Makers: Understanding the Operating System That Runs Raspberry Pi and Other Maker SBCs*, Make Community LLC, 2017.
9. C. Valens, *Mastering Microcontrollers Helped by Arduino*, 3rd edition, Elektor Press, 2016.
10. *Getting Started with Arduino - An Introduction to Hardware, Software Tools, and the Arduino API*, docs.arduino.cc/learn/starting-guide/getting-started-ardwino.
11. Simon Monk, *Programming Arduino, Getting Started with Sketches*, 2nd edition, McGraw-Hill Education, New York, 2016.
12. Raspberry Pi Foundation, *The Official Raspberry Pi Handbook 2022*, Raspberry Pi Press, 2021, magpi.cc.
13. Gareth Halacree and Ben Everard, *Get Started with MicroPython on Raspberry Pi Pico*, Raspberry Pi Press, 2021.
14. Arduino Nano Microcontroller Datasheet, store.arduino.cc/products/arduino-nano/.
15. Raspberry Pi Pico Microcontroller Datasheet, raspberrypi.com/products/raspberry-pi-pico/.
16. Raspberry Pi Zero Microcomputer Datasheet, raspberrypi.com/products/raspberry-pi-zero/.
17. Microchip SAM D21G17D Curiosity Nano Evaluation Kit.
18. Piyu Dhaker, *Introduction to SPI Interface*, Analog Devices, Analog Dialog 52-09, September 2018, analogdialog.com.
19. Freescale/NXP, *Using the Serial Peripheral Interface to Communicate Between Multiple Microcomputers*, Application Note AN991/D, 2004.
20. Motorola/Freescale/NXP, *SPI Block Guide V03.06*, 2003.
21. ExostivLabs, *Introduction to I²C and SPI Protocols*, exostivlabs.com/introduction-to-i2c-and-spi-protocols/.
22. *I²C Bus*, i2c-bus.org.
23. Andrew Moore, *FPGAs for Dummies*, 2nd Intel Special Edition, John Wiley & Sons, Hoboken, NJ, 2017.
24. Embedded Blog, *Flash FPGAs Give Designers More Flexibility*, January 25, 2015, embedded.com/flash-fpgas-give-designers-more-flexibility/.
25. Blaine Readler, *Verilog by Example: A Concise Introduction for FPGA Design*, Full Arc Press, 2011.
26. Simon Monk, *Programming FPGAs: Getting Started with Verilog*, McGraw-Hill Education TAB, 2016.
27. Micron Technology, Inc., *Small-Block vs. Large-Block NAND Flash Devices*, Technical Note TN-29-07, 2005.
28. Embedded Blog, *Flash 101; NAND Flash vs NOR Flash*, July 23, 2018, embedded.com/flash-101-nand-flash-vs-nor-flash/.

Chapter 29

Mixers and Recording Consoles

At the heart of the recording studio or the sound reinforcement system is the mixing console, be it anywhere from a simple mixer to a complex console with a large number of channels and functions. In its simplest form, a mixer processes each input channel and mixes the output of that channel into one or more output channels (buses), at different signal levels. Mixers encompass an enormous range of functionality and complexity, from the simple to the very complex. One can think of four categories of mixers for convenience. These are the live sound mixer, the home studio mixer, the recoding studio console mixer and the broadcast mixer. These all may have differences in number of channels, architecture, features and complexity, not to mention cost, but they all have similar basic functionality.

Although some mention of differences will be made here, the main focus will be on the circuits of the different functions that mixers of most types have in common. Here the primary focus will be on a stereo mixer that has left and right main outputs, with many other features like monitor outputs, auxiliary sends, inserts, etc. that will be discussed. The signal from each input channel must be processed and ultimately fed to one or both of the outputs at varying levels. Most of the processing functionality occurs in the channel strip, one of which is dedicated to each input channel.

The basic functionalities in a channel strip include a microphone preamplifier, an equalizer, a fader and a pan pot. The latter is used to place the instrument in the stereo space by applying different amounts of the processed signal to the right or left output buses, where those and similar signals are mixed to create the combined left and right output signals. The channel strip often includes a monitor output that permits that channel (or a combination of channels) to be listened to apart from the main mix at the mixer output. The channel strip also includes auxiliary outputs (sends) that can be mixed on auxiliary buses to create other mixes different from the main stereo mix. Some of these may be sent to effects processors and others might be used to drive stage monitor speakers that allow performers to hear themselves. The channel strip may also include means to insert an external device into its signal path (inserts), often either before or after the fader. Such external devices might include echo processors, compressors, limiters, additional equalizers and so on.

The back-end master module of the mixer is where the stereo mix and auxiliary mixes will be processed globally. The key element here is the main fader, but other functions include metering and headphone outputs and perhaps equalization and inserts in the mix path. Line-level inputs that need not pass through a channel strip are also available here.

As a central part of this chapter, we describe a fairly simple generic 16-channel mixer. This has the most-used types of features, but small mixers deviate greatly in the particulars. Some, for example, do not include all of the features on all of the 16 channel strips. Not all of the channels may have microphone preamps. In other cases, mixers have additional features that push them in the direction of studio console capabilities. In the generic mixer described here, all 16 channel strips are identical, with typical mid-range features. Similarly, the master output section includes the usual back-end features found in mid-range mixers.

Channel strip features include balanced microphone/line inputs, channel input level controls, auxiliary/monitor sends with level controls, low-mid-high equalization, a fader and a pan pot. Also included are pre-fade listen (PFL), solo mode select and post fader inserts.

Master section features include stereo and mono main outputs, a mono monitor output, a stereo headphone/monitor output, stereo auxiliary returns, level and balance controls, aux return solo capability, headphone output with level control, left and right main output faders and left and right level meters. This chapter also includes some discussion of mixer features, architectures and different types of mixers and consoles

29.1 A Mixer in its Simplest Form

Here we begin by describing a mixer in its simplest form, ignoring features like auxiliary sends, monitoring, inserts and so on. The simple mixer comprises a plurality of channel strips whose outputs are mixed together onto left and right output buses that share a main fader after the signals are mixed. The mix buses are virtual grounds of the inverting op amps in the master section. Each channel strip includes a microphone preamp with a wide-range gain adjustment to accommodate a wide range of microphone types and sensitivities. It accepts input microphone-level signals on an XLR connector and can provide phantom power. It can also receive a balanced line-level input on a standard 1/4" tip-ring-sleeve (TRS) jack.

A block diagram of the channel strip is shown in Figure 29.1. The microphone preamp includes a fixed-frequency 80-Hz low-cut filter (HPF) that can be bypassed. That is followed by a low-mid-hi equalizer, not much more than would be found on an audio preamplifier with three tone controls. The equalizer feeds a channel fader. Finally, a stereo output is created by a pan pot. It delivers the signal to the left and right mix buses in desired left-right proportions. A DPDT switch allows the channel output to be applied to the main stereo output bus or to the alternate bus.

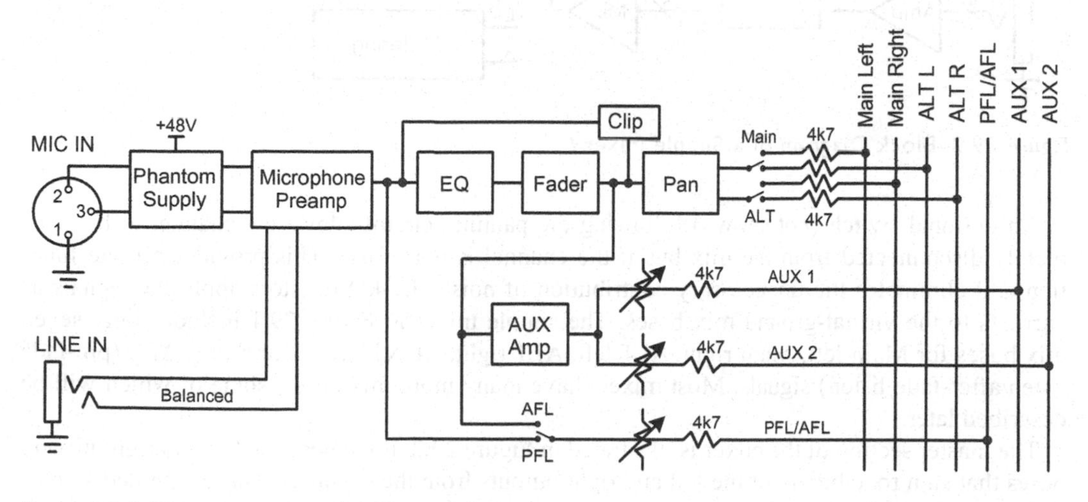

Figure 29.1 Block Diagram of a Channel Strip.

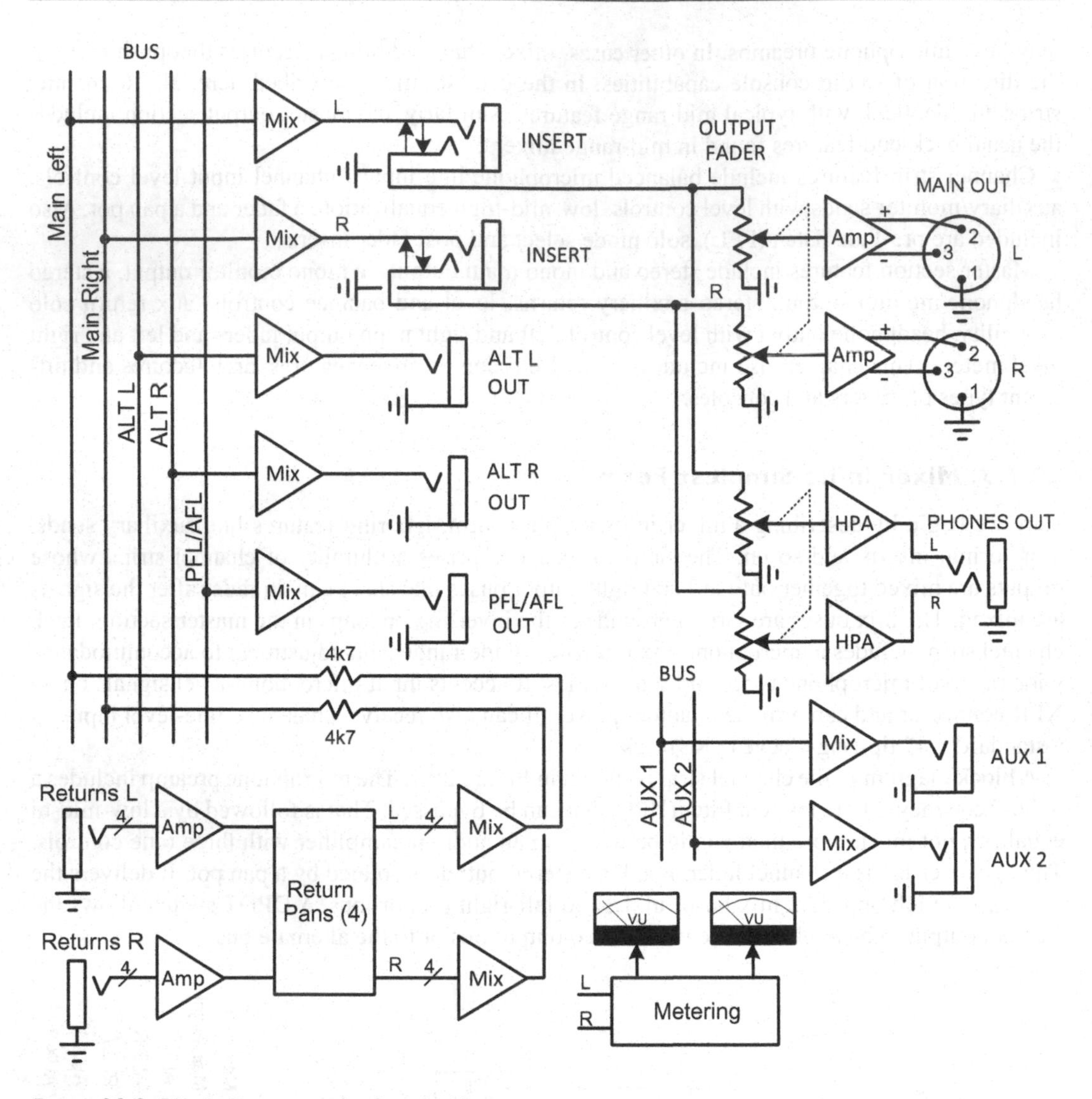

Figure 29.2 Block Diagram of a Simple Mixer.

An optional switch (not shown) following the panning circuit allows the channel to be completely disconnected from the mix bus if the channel is not in use. This provides a mute function and eliminates the unnecessary contribution of noise. 4.7-kΩ resistors apply the signals as currents to the virtual-ground mix buses. The simple mixer in Figure 29.1 includes only seven mix buses for Main left, Main right, ALT left, ALT right, AUX1, AUX2 and PFL/AFL (pre-fade listen/after-fade listen) signals. Most mixers have many more mix buses, some of which will be described later.

The master section of the mixer is illustrated in Figure 29.2. It includes the left and right mixing buses that sum together all of the left and right outputs from the channels. Those summed signals then go through a stereo fader to the stereo outputs. The stereo right and left output levels are also metered. A great many of the technical issues in a mixer are captured in this simple signal chain. Numerous other functions shown will be discussed. Most mixers include many more buses than the seven shown. The mix amplifiers are op amps connected as inverters, creating a virtual ground into which the mix bus currents are fed.

Microphone Preamp

Microphone preamplifiers were discussed at length in Chapter 18. In a mixer, the microphone preamplifier most importantly exhibits low noise and good headroom. It must accept XLR inputs from different types of microphones like dynamic, condenser, ribbon and electret. These microphones may create widely different signal levels, depending on their sensitivity and the instrument with which they are associated. The preamp includes a level control, and sometimes attenuators (pads) that can be switched in to accommodate higher microphone signal levels or line-level signals.

Figure 29.3 illustrates a microphone preamp arrangement based on a simple low-noise instrumentation amplifier. The input stage is a low-nose instrumentation amplifier with nominal gain of 20 dB. For larger microphone input signals, the input stage gain can be reduced to 0 dB by using a switch (not shown) to open R9. The level control can span continuous gain range of no signal (gain=0) to 40 dB. Including the input stage, maximum gain is 60 dB. In this simple design, microphone signals from –60 dBu to as much as 0 dBu can be accommodated (0 dBu corresponds to 0.774 V_{RMS}). The gain control is realized with a Baxandall volume control circuit, as discussed in Chapter 22. It provides a decent approximation to gain in dB being proportional to clockwise rotation of the linear taper pot. The mic preamp must not only be capable of very high gain, but also must have very high common-mode rejection (CMRR). On top of all of this, it must be able to accept a line-level balanced input from a 1/4" tip-ring-sleeve (TRS) input jack. The latter feature is usually accomplished with a 20 dB pad in the line input path.

Input-referred noise is about 3.3 nV/√Hz if the LME49860 is used for U1. A-weighted E_{in} with a 150-Ω source resistance is a respectable −125 dBu. Differential input impedance is quite high, at about 12 kΩ. It can optionally be reduced by adding a shunt resistor across the balanced input. Common-mode input impedance is about 24 kΩ when phantom power is disabled.

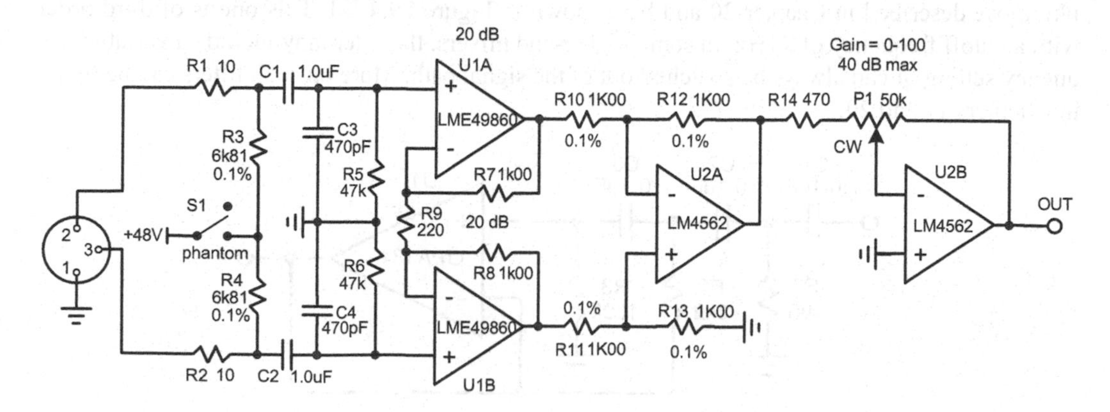

Figure 29.3 A Simple Microphone Preamp.

Phantom Power

Condenser microphones include an active microphone preamp that requires power to operate. Other devices like active ribbon microphones and some mic splitters and direct inject (DI) boxes may need phantom power as well. So-called "phantom" power is delivered by the microphone cable. Phantom power is applied to the balanced microphone cable pair at the mixer preamplifier in the common mode by R3 and R4 to the microphone inputs, as shown in Figure 29.3 [1]. A quiet +48-V supply is applied to the hot and cold XLR microphone inputs (pins 2 and 3, respectively) through a pair of precision 6.81-kΩ resistors. These should be of 0.1% tolerance so as not to degrade the common-mode rejection of the interconnection between the microphone and the microphone preamp.

Phantom power can be switched on or off, and generally should not be applied to microphones that do not require it. Phantom power enabling here is on a per-channel basis by switch *S*1 in series with the +48-V phantom source. Some mixers only have one switch to control phantom power to all microphone channels. This will likely create a noise penalty and introduce a conflict with microphones that do not need phantom power or should not be phantom powered, such as passive ribbon microphones.

Phantom current available to a microphone is generally limited to 10 mA, based on 5 mA flowing in each of the 6.81-kΩ resistors and a voltage drop of 34 V, leaving 14 V available for operation of the microphone. This corresponds to 140 mW. Microphones requiring less than 10 mA will have a bit more voltage available for their circuits. Notably, a microphone that draws 7.1 mA will be power-matched to the 3.4-kΩ common-mode phantom source impedance. In this case, the microphone will have 24 V available to it, corresponding to 170 mW. Some devices that use phantom power employ switch-mode converters to bring the voltage down and increase the available current.

Low-Cut Filter

It is usually desirable to cut out low frequencies from a channel that is not associated with instruments that create low frequencies. This reduces the amount of low-frequency (LF) grunge that gets into the mix. Such a filter is often a second-order or third-order high-pass filter set to a fixed frequency like 80 Hz. Often, the filter is a straightforward Sallen-Key Butterworth high-pass filter like those described in Chapter 20 and here shown in Figure 29.4 [2]. This one is of third order with a cutoff frequency of 80 Hz. In some higher-end mixers, the filter may include a variable frequency setting. It can always be switched out of the signal path. More on such filters can be found in Chapters 11 and 20.

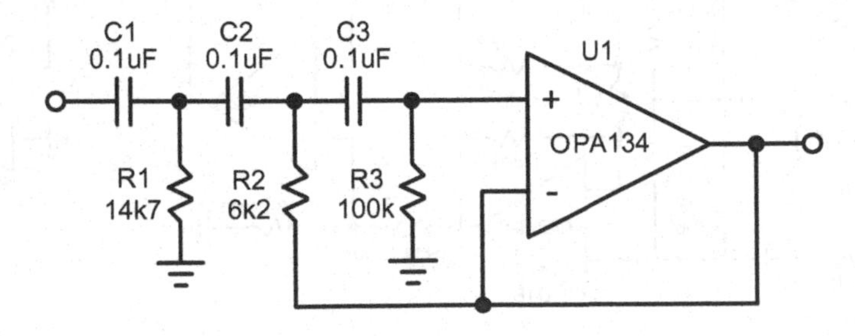

Figure 29.4 80-Hz Sallen-Key Low-Cut Filter.

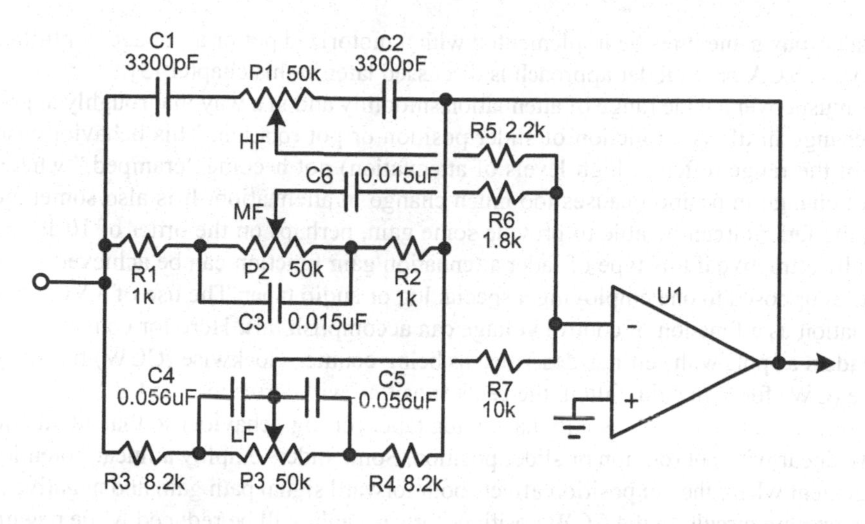

Figure 29.5 Three-Band Equalizer.

Equalizer

The equalizer (EQ) in a basic mixer is often little more than a three-band tone control that might be found in an audio preamp, as described in Chapter 20. The equalizer is shown in Figure 29.5. Here the conventional Baxandall bass control is referred to as a low-frequency shelving EQ. Beginning at some low frequency and ending at some lower frequency, it typically can provide boost or cut up to ±10 or 15 dB at the lowest frequency of interest. Here the LF shelf EQ reaches ±12 dB of boost or cut at its specified filter turnover frequency of 80 Hz (3 dB less than the maximum amount of equalization). The filter is a simple first-order 6-dB/octave design.

Similarly, the treble control is referred to as the high-frequency shelf EQ. It can provide its maximum boost or cut of ±15 dB at the highest frequency of interest. It provides 3 dB less than that (i.e., ±12 dB at its specified turnover frequency of 12 kHz). It is also a first-order filter here. The shelving equalizers in some mixers can be of second-order so as to provide steeper slopes and have less effect on mid-band frequencies.

The mid-band control can provide up to ±12 dB of boost or cut at its center frequency of 2.5 kHz. As mentioned above, these simple equalizers are based on the Baxandall tone control circuit and do not provide a lot of slope.

In more costly equalizers, the EQ controls may be more sophisticated and plentiful. For example, two mid-band control(s) may cover two different frequency ranges whose center frequency can be adjusted. There may also be a parametric equalizer, where the center frequency and sharpness (Q) can be controlled. A so-called *quasi-parametric* equalizer allows its center frequency to be swept over some range, but does not allow its Q to be varied. More on equalizers can be found in Chapter 20.

Fader

The channel fader is little more than a very nice slider-type volume control for the main output of the channel that will be mixed on the bus with the other channels. As discussed in Chapters 22 and

25, the fader may sometimes be implemented with a motorized pot or a voltage-controlled amplifier (VCA). A VCA-based fader approach is discussed later in this chapter [3].

Faders must cover a wide range of attenuation smoothly and in a way that roughly approximates a linear change in dB as a function of slider position or pot rotation. This behavior ensures that portions of the range (often at high levels of attenuation) not become "cramped," where a small amount of change in position causes too much change in attenuation. It is also sometimes desirable that the fader circuit be able to provide some gain, perhaps on the order of 10 dB. Finally, it is especially attractive if this type of fader attenuation/gain function can be achieved with a linear taper pot, as opposed to one employing a special log or audio taper. The use of a VCA with linear dB attenuation as a function of control voltage can accomplish this. Here for convenience we will refer to faders as pots with settings described as being counter-clockwise (CCW) for low gain and clockwise (CW) for high gain. Often, the center position will correspond to 0 dB gain or less.

The conventional fader pots usually have a log taper (or log behavior) so that attenuation in dB is roughly linear with pot rotation or slider position. Some faders employ a linear potentiometer in an arrangement where the pot position affects both forward signal path gain and negative feedback gain in an op amp circuit. In the CCW position, forward gain will be reduced while negative feedback around the associated op amp will be increased, both leading to reduced gain and ultimately to no signal transmission. In the CW position, forward path gain will be increased and the amount of negative feedback around the amplifier will be decreased, both actions leading to increased gain. In most mixer channel strips the main fader is just a passive slide pot with a log taper. Here we mention some active fader circuits.

In the fader circuit of Figure 29.6(a), the input signal passes through R1 (15 kΩ) to the CCW end of the 50-kΩ linear fader pot P1 and to the positive input of the op amp. With the wiper of P1 grounded, full attenuation occurs when P1 is in its full CCW position. R3, R2 and the CW side of P1 provide negative feedback that configures U1 as a non-inverting gain stage. When P1 is at its full clockwise position, forward gain of the stage is simply R3/R2+1, here 4.4. Attenuation at the input is RP1/(RP1+R1), here 0.71, so overall gain is 3.12, or about +10 dB. At mid position, gain is about -3 dB.

A disadvantage of this arrangement is its reliance on wiper resistance and the possibility that if the wiper goes open due to dirt on the track, the gain can change to about −3 dB and noise can be created. A further limitation is that wiper resistance limits the maximum attenuation attainable in the full CCW position. If the wiper resistance is 1 Ω, attenuation will be about 80 dB.

A second fader circuit is shown in Figure 29.6(b). The input signal is passed through R1 (15 kΩ) to the wiper of 50-kΩ linear fader pot P1. The CCW end of the pot is connected to ground, shorting out the forward path when the pot is full CCW. The CW end of the pot is connected to the virtual ground input of an op amp with a 56-kΩ feedback resistor (R2). When the pot is full CW, the gain

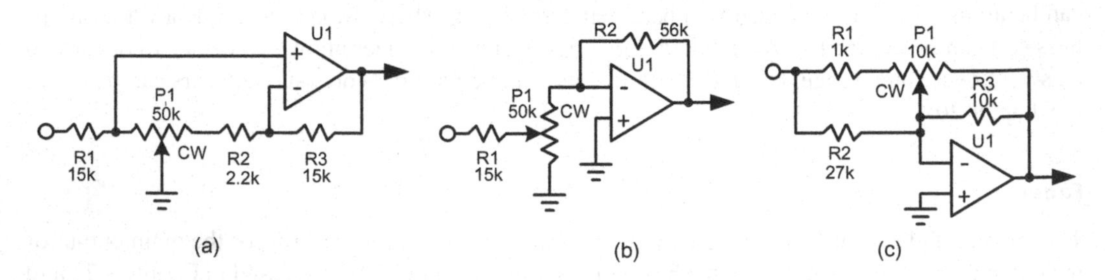

Figure 29.6 Fader Circuits.

of the circuit is the ratio of the 56-kΩ feedback resistance to the 15-kΩ input resistance. This corresponds to 3.7 or about +11 dB. Wiper resistance again limits maximum achievable attenuation. In the case of a 1-Ω end-stop resistance, attenuation will be about 84 dB. Note that in this condition, the gain of the remaining inverting amplifier will be about unity, since its input resistance is 50 kΩ from fader P1 and its feedback resistance is 56 kΩ from R2.

The Baxandall volume control can also implement a fader with the desired characteristics using a linear pot [4, 5]. It also depends on the action of the pot affecting both the forward path gain and the gain attributable to negative feedback. In this circuit, shown in Figure 29.6(c), the input signal is fed through R1 (1 kΩ) to the CW end of 10-kΩ linear fader pot P1. The wiper of P1 is connected to the inverting input of op amp U1. It is configured as an inverting amplifier with a virtual ground and whose feedback resistance is the track resistance of P1 to the CCW side of the wiper. The gain of the stage is the ratio of the feedback resistance to the forward path resistance. In the CCW position, the forward path resistance is high and the feedback resistance is ideally zero, resulting in extremely high attenuation. Ignoring R2 and R3 for the moment, in the CW position, the forward path resistance is only 1 kΩ and the feedback resistance is 10 kΩ, resulting in 20 dB of gain. In the center position, the forward path resistance is 6 kΩ and the feedback resistance is 5 kΩ, resulting in –2 dB of gain. A 27-kΩ resistor (R2) can be connected as shown from the input to the wiper of P1 to bring down the forward path resistance at the center position to 5 kΩ, resulting in 0 dB of gain at the center position, if desired.

Two advantages of the fader in Figure 29.6(c) are that there is little current flowing in the pot wiper and that the positive input of the op amp is at ground, eliminating common-mode distortion in the op amp. A disadvantage of this circuit is that if the wiper momentarily goes open, the op amp will go open-loop, causing a large-amplitude noise pop at minimum. This behavior is mitigated if a 10-kΩ resistor (R3) is connected from the output of the op amp to the wiper of P1. In this case, an open wiper will result in zero gain. This assumes that the above-mentioned 27-kΩ forward path shunt resistor R2 is not included; if included, the default open-wiper gain will be about -10 dB. A final slight concern is that the end-stop resistance of P1 may not be perfectly zero. This means that the maximum CCW attenuation of the fader is not infinite, but rather is the ratio of the end-stop resistance to about 11 kΩ. A 1-Ω end-stop resistance would result in gain of about –80 dB. This is probably acceptable. The high maximum attenuation assumes that the closed-loop output impedance of the op amp is very low over the full audio band. This means that an op amp with a high gain-bandwidth product, like the LM4562 with its typical value of 55 MHz, should be used.

Pan Pot

The pan pot is responsible for placing the output of the channel at its desired place in the stereo image, from left to right. It thus distributes the channel signal to left and right output lines from the channel strip for mixing into the left and right mix buses. A desirable feature of the pan pot is that it distributes the same total amount of power to the left and right channels as it is adjusted. This means that if the gain to one channel is unity at a pot extreme, the gain to both channels when the pot is at its center is 0.707 (i.e., down 3 dB). In this way, the perceived listening level of that channel changes little with the position of the pan pot. A ganged pot with log and reverse-log tapers, each configured as a volume control, is sometimes used to achieve an approximation to this behavior.

It is more desirable if the pan pot function can be accomplished with linear taper potentiometers. Some approaches use a single potentiometer while others use a ganged pair of pots. The common circuit in Figure 29.7(a) implements a pan pot with a single linear potentiometer. R1 and

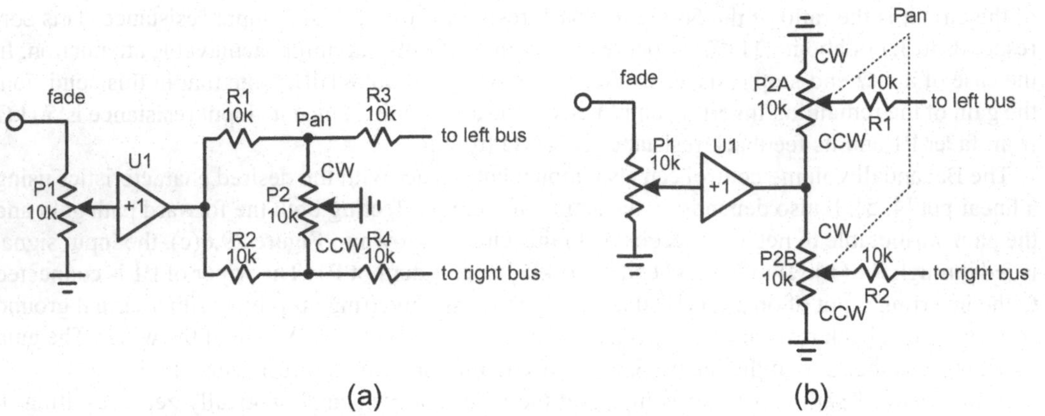

Figure 29.7 Pan Pot Circuits.

R2, each 10 kΩ, feed the mono channel strip signal to the clockwise (CW) and counter-clockwise (CCW) ends of 10-kΩ pan pot P2, respectively. The wiper of the pan pot is connected to ground, so that the pan pot acts as an attenuating shunt from R1 and R2 to ground. At the CW position, when panned full right, the wiper grounds R1 and there is no signal sent to the left channel. At the CCW position, the wiper grounds R2 and no signal is fed to the right channel. At the center position, equal signals are sent to the left and right channels through the CW and CCW ends of the pot, respectively.

The left channel signal at the CW end of the pot is connected through R3 (10 kΩ) to the virtual ground of the mix bus. R3 thus loads the CW end of the pot to virtual ground. R4 provides the right channel output to the mix bus in the same way. The panning circuit could load the fader pot wiper, possibly altering the attenuation of the fader as the signal is panned. For this reason, buffer U1 is inserted in the path. If some gain is needed as a result of the passive loss in the pan circuit, U1 can easily be configured to have some gain with a pair of resistors in the feedback path.

The values of the resistors are chosen so that when the pot is at either extreme, the gain to one of the outputs is unity and the gain of the other output is zero. Moreover, the values are chosen so that when the pot is at its central position, the gain to both left and right outputs is –3 dB. This means that the power sum of the two outputs is the same at the pot extremes and at the center position. The power sum at other positions is a close approximation to the same constant value. Resistors R3 and R4 drive the left and right mix buses. A DPST switch can be placed in series with R3 and R4 to take the channel off the bus when that channel is not in use. This reduces the noise gain of the mix amplifier.

A dual-gang pan pot is shown in Figure 29.7(b). As in (a), the panning circuit is preceded by a buffer to avoid interaction between P2A and P2B. Here the pots P2A and P2B are configured as complementary volume controls. Loading of the wipers by R1 and R2 to the virtual ground mix bus drops the gain at the mid position by 1 dB, for a total of 7 dB; with a linear pot, this is more than the desired 4 to 6-dB range. There are numerous other ways to create a pan pot circuit, including some active ones employing a VCA. Figure 29.8 shows idealized left and right channel pan pot gain when gain in each channel is chosen to be -6 dB, assuming acoustic summation on a pressure basis.

In the design above, in (b), there is unity gain to one channel when the pot is at one extreme. Gain to each channel is down about 7 dB when a linear pot is centered. When this condition is met, the power sum at other positions is a close approximation to the same constant value.

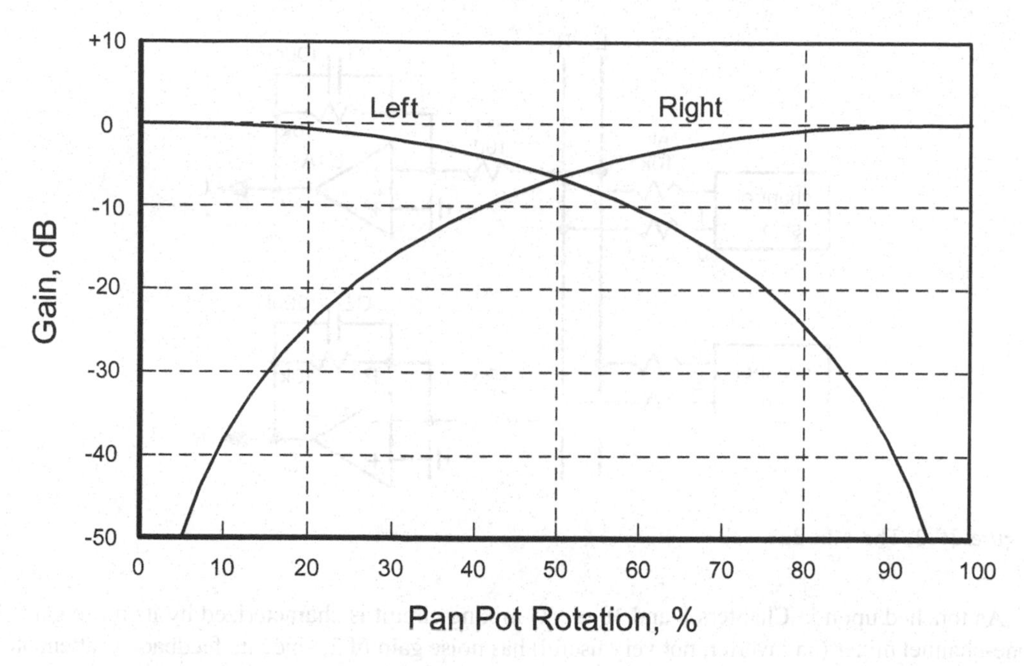

Figure 29.8 **Right and Left Gain as a Function of Pan Pot Rotation.**

The Mixing Bus and the Master Section

The output side of the mixer comprises all of the functions that are global to the channel strips, most importantly including the mix amplifiers, the left and right mix buses and the mixer output controls like the main fader.

Stereo Mix Bus

The remainder of the mixer is the master control section that is common to all of the channels. In its simplest form, the main stereo mix bus is implemented with two op amps, one for the left channel and one for the right channel, as shown in Figure 29.9. These op amps are configured as inverting virtual ground (VG) summers with 10-kΩ feedback resistors. Each of the left and right outputs of all of the channels is added to the virtual ground bus through a 10-kΩ series resistor for each channel. The gain of any channel to the mixed output is thus almost unity. The output of a channel can be deleted (muted) at the channel strip from addition to the bus if it is not to be included in the mix. This reduces noise accumulation on the mix bus.

Capacitors C1 and C2 stabilize the mix amplifiers in light of the significant capacitance of the bus that shunts the inverting inputs. The 100-Ω resistors further help stabilize the amplifier by providing some isolation from the bus. These resistors also provide a form of transmission-line termination for the bus.

The mix bus and mixing amplifier are a key design challenge in keeping the noise of the mixer low. This is particularly and progressively important for mixers with larger numbers of channels. The so-called “noise gain” of a virtual ground mixing circuit was discussed in Chapter 10. These challenges and related circuit approaches are described in a later section in this chapter. Notice that muting unused channels by opening their connections to the mix bus reduces the noise gain.

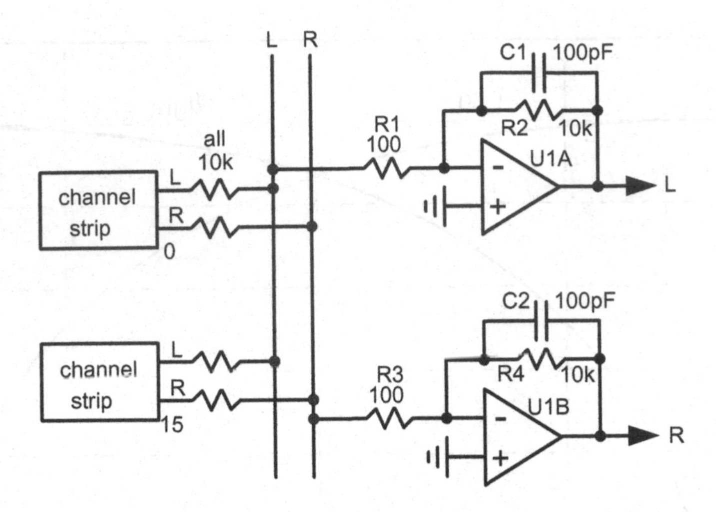

Figure 29.9 The Mix Bus.

As touched upon in Chapters 8 and 10, a VG mixing circuit is characterized by its noise gain. A one-channel mixer (an inverter, not very useful) has noise gain of 2, since its feedback is attenuated by a factor of 2 in getting back to the inverting input of the op amp. Put another way, a signal injected at the positive input will see a gain of 2. Similarly, a two-channel mixer will have noise gain of 3. In the general case, a VG mixing circuit will have noise gain equal to the number of inputs plus one.

The noise gain of a 16-channel mixer of the arrangement in Figure 29.9 will be 17, or about 24 dB if all channels are engaged. This is quite large. Imagine the noise gain of a 64-channel mixing circuit, at 65 or about 36 dB. A mix bus with 64 10-kΩ inputs will have a very low bus impedance of 156 Ω. With a source impedance of 156 Ω working against a 10-kΩ feedback resistor, it is easy to see why the noise gain is high under these conditions. Even without considering resistor thermal noise, the input voltage noise of the summing operational amplifier, perhaps on the order of 2 nV/√Hz, will become 130 nV/√Hz at the output of a conventionally-implemented 64-channel VG mixing amplifier. This becomes about 16 μV in a 20-kHz bandwidth, corresponding to SNR of 96 dB with a 1-V_{RMS} reference. Use of a low-noise op amp for the mix amplifiers is crucial to good noise performance of the mixer as a whole.

There are a couple of more details shown in Figure 29.9. Capacitors C1 and C2 are connected in parallel with feedback resistors R1 and R2 to attenuate high frequencies, compensate for mix bus capacitance and improve mix amplifier stability. Finally, note that the closed-loop bandwidth and feedback loop gain will be reduced by the noise gain factor of 65 in a 64-channel mixer. At 20 kHz, the open-loop gain of an op amp with a 10-MHz gain-bandwidth product is only about 500. If you now reduce this by a noise-gain factor of 65, you are left with loop gain of only about 7.7, which is on the order of 18 dB. This makes it important to use an op amp with a high gain bandwidth product of about 50 MHz, which would yield loop gain of about 32 dB at 20 kHz for a 64-channel mixer with all channels engaged.

Master Fader and Line Output Amplifier

The master fader is basically a stereo volume control to control the signal level of the main stereo outputs of the mixer. It should have a log taper. The line output amplifier will usually provide left and right outputs that can drive a 600-Ω load to the full output level of which the mixer is capable. These outputs are usually balanced, with positive and negative polarity voltage outputs referred to ground. The balanced outputs are made available at an XLR or TRS jack.

The line output circuit is shown in Figure 29.10. The hot side of the output is buffered by U1A and passes through R3 to pin 2 of the XLR connector. The cold side of the balanced output is simply created by an inverter, U1B. In more expensive mixers, these balanced outputs may be floating balanced outputs implemented by a transformer or electronic means, such as the THAT 1646 OutSmarts® balanced line driver [6]. In less expensive mixers a so-called *impedance-balanced* output is provided by eliminating U1B and grounding the left end of R4. This results in a 6-dB loss of headroom for balanced outputs. Balanced outputs are discussed at length in Chapter 19.

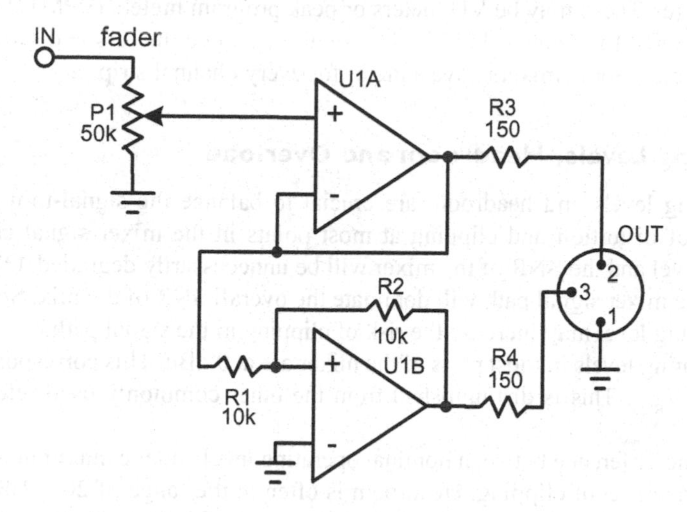

Figure 29.10 Line Output Amplifier.

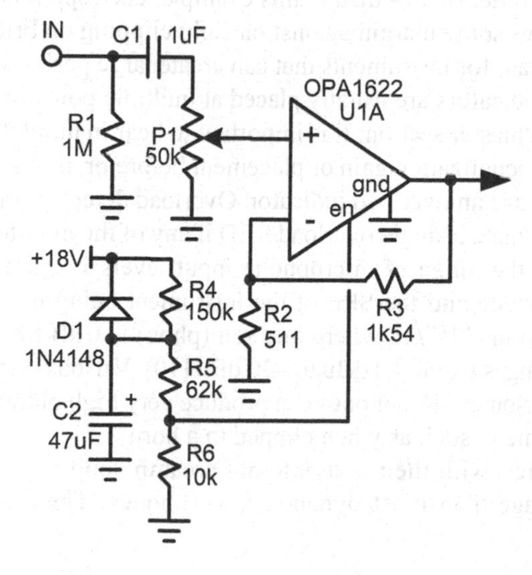

Figure 29.11 Headphone Amplifier.

Headphone Output

Virtually every mixer has a headphone output for monitoring the stereo mix or some other combination of signals in the mixer. This can be implemented with a stereo level control and a headphone amplifier (see Chapter 21). This output can alternatively be used to drive self-powered monitor loudspeakers. Figure 29.11 shows an implementation of the headphone amplifier. It is based on an OPA1622 stereo headphone amplifier chip [7]. One channel is shown.

Level Meters

At minimum, a pair of level meters or displays must be included to show the levels on the stereo outputs of the mixer. These may be VU meters or peak program meters (PPM). They may also be hybrid dual-function LED meters, like the Dorrough [8]. Level meters and indicators were described in Chapter 27. Some mixers have a meter for every channel strip.

29.2 Operating Levels, Headroom and Overload

Nominal operating levels and headroom are crucial to balance the signal-to-noise ratio (SNR) against the risk of distortion and clipping at most points in the mixer signal chain. Too low a nominal signal level and the SNR of the mixer will be unnecessarily degraded. Often, the point of lowest SNR in the mixer signal path will dominate the overall SNR of the mix. Similarly, too high a nominal operating level may increase the risk of clipping in the signal path.

Nominal operating levels in most parts of the mixer are at 0 dBu. This corresponds to 1 mW into 600 Ω, or 0.774 V_{RMS}. This is distinguished from the other commonly used reference of 0 dBv, which represents 1 V_{RMS}.

Headroom is the difference between nominal operating level and the maximum level that can be handled before the onset of clipping. Headroom is often in the range of 20–22 dB. A circuit with 20 dB of headroom with respect to a nominal operating level of 0 dBu will clip at 7.74 V_{RMS} or 10.9 V_{pk}. Overload is often specified as when a signal level reaches a point 6 dB below the maximum level. This will be on the order of +14 dBu in this example, corresponding to 3.9 V_{RMS}, or 5.5 V_{pk}. The 6-dB number provides some margin against outright clipping on brief peaks. Adequate headroom is especially important for instruments that can create large peak levels, such as snare drums. Overload detectors and indicators are usually placed at multiple points in the signal chain, both in the channel strips and the master section. It is important to bear in mind that amplitudes may differ in different places due to equalization gain or placement before or after an attenuator. At minimum the channel strip should have an overload indicator. Overload detectors monitoring multiple points in a signal path may illuminate a single overload LED if any of the monitored points is in overload.

As mentioned earlier, the range of microphone input levels is quite large, depending on the sensitivity of the microphone and the SPL of the instrument being mic'd [9]. The sensitivity of a microphone is specified in *dBV/Pa*, where one son (phon) is 0 dB SPL. Sensitivity of dynamic microphones typically ranges from –60 dBu to –30 dBu [10]. Various microphone signal levels are discussed in Chapter 18. Some microphones can produce very high output levels if they are placed close to a musical instrument, such as when clipped to a horn.

Condenser microphones, with their own internal preamp, tend to have higher sensitivity and thus higher output voltage than most dynamic microphones. Their sensitivity is often in the range of −35 dBV/Pa [11]. Ribbon microphones are often of fairly low sensitivity. The voltage from the ribbon itself is very low, and is boosted by a transformer in the microphone. The turns ratio of the transformer determines the output voltage and thus the sensitivity. However, the impedance transformation of the transformer goes up with the square of the turns ratio, meaning that there is a practical limit to the turns ratio and thus the sensitivity of the microphone, lest the output impedance become too high for driving the microphone cable. Sometimes the output of a ribbon microphone (or low-output dynamic microphone) will be boosted by an external active amplifier like the Cloud Lifter [12], which is located close to the microphone and is powered by phantom power. Active ribbon microphones include an internal amplifier to boost their sensitivity.

The microphone preamplifier will usually be called upon to boost the microphone input level to about 0 dBu on average. Typical line-level inputs are generally in the range of 0 dBu to +4 dBu. The former is considered to be the consumer level, while the latter is considered to be the professional level. Microphone preamp gain will often be set to achieve an average level of 0 dBu, as seen by 0 dB on a VU meter. The VU meter produces an average reading that is easy to view because of its slower ballistics, but it does not respond to brief peaks. For this reason, headroom comes into play, allowing for occasional peaks that exceed the 0 dB point on the VU meter without overload or clipping.

29.3 A Generic 16-Channel Mixer – The Channel Strip

Here we describe a more complete 16-channel generic mixer that includes the basic functions and circuits described above plus the features and functions that are usually included in virtually every real-world mixer. There are a great many mid-range mixers in the marketplace and some will have more features than included here while others will lack some of these features, or may not include all of the features on every channel. For example, some mixers may not have a microphone preamp in every channel strip, and may just have unbalanced line inputs on the rest of the channels.

The Channel Strip

In this mixer example, all channel strips are identical. Figure 29.12 is a block diagram of the generic mixer channel strip that includes these additional features. Many of the additional features in a mixer are implemented by numerous other mix buses in addition to the main stereo bus discussed so far. Some of the additional buses comprise stereo pairs, while other buses are monophonic. This example 16-channel mixer has a total of 16 buses, three stereo and ten mono. Among the buses that can be found in a mixer are the following:

- Stereo (L&R)
- ALT (L&R)
- SOLO (L&R)
- AUX (1–8)
- PFL
- MON

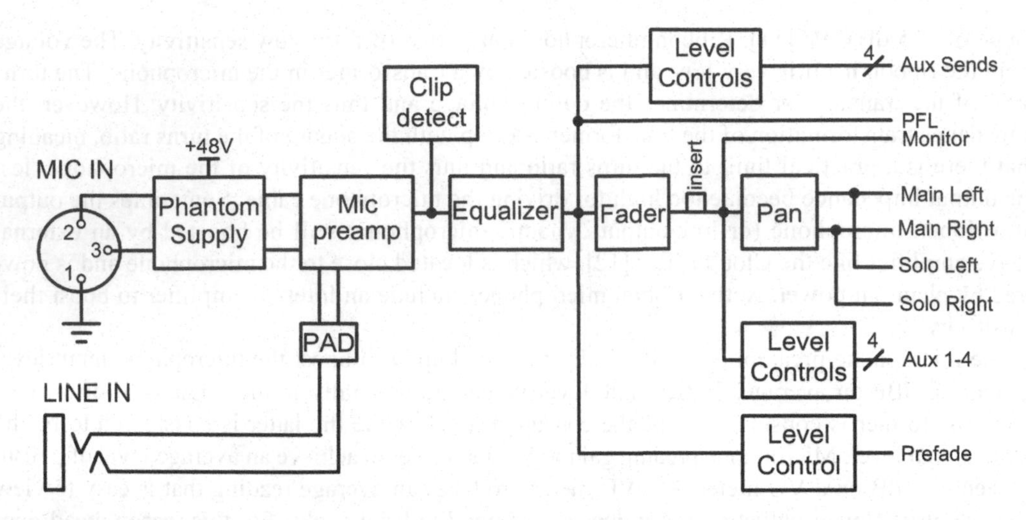

Figure 29.12 Block Diagram of the Generic Mixer Channel Strip.

Alternate Left and Right Buses

In addition to the main mix stereo buses, another pair of *ALT* stereo buses is provided. The alternate left and right buses allow the panned stereo outputs of a channel strip to be directed to the alternate (ALT) bus so they can be used in a different way.

Solo Bus

A solo button on the channel strip allows only that channel to be monitored and mutes the rest of the mixer's outputs. In some ways it is like the inverse of the mute condition for the chosen channel. The channel can be listened to in stereo through the complete path of the mixer, right to the front-of-house (FOH) speakers. It is panned identically to the main bus. The solo button should never be activated during a performance for obvious reasons.

Auxiliary Sends

Each channel strip usually has four or more additional monophonic outputs called auxiliary sends (AUX). These sends each have a level control, and can originate from a point in the signal path before the fader (pre-fade) or after the fader (post-fade). Each AUX is sent to a separate auxiliary mix bus. This means that four or more separate monophonic mixes can be created on the auxiliary mix buses. The auxiliary mix buses operate in the same way as the stereo main mix buses, using virtual ground mix amplifiers. The figure shows four post-fade AUX sends and four prefade AUX sends.

The resulting AUX bus mixes are often sent to outboard processors or effects devices. One may be used for headphone monitoring. Others may be used to drive stage monitors that help performers hear themselves. Post-fade sends are often used for outboard processors so that the signal through the processor tracks changes in the channel fader position. Pre-fade sends are often used for stage monitors so that the listening level for the performer does not change with changes in the mix fader. This generic mixer example has eight AUX buses, four pre-fade and four post-fade. Depending on the manufacturer, the pre-fade sends may come ahead of the equalizer or after the

prefade or postfade signal — P1 10k — S1 — R1 10k — to AUX bus

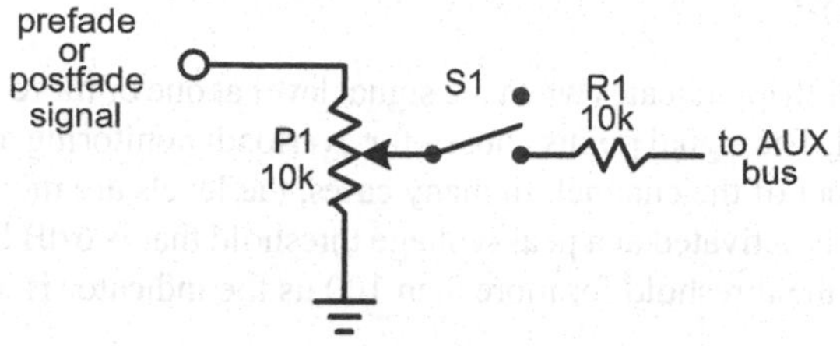

Figure 29.13 Auxiliary Send Circuit.

equalizer. In some mixers, each AUX send can be set to pre-fade or post-fade mode. The auxiliary send circuit is shown in Figure 29.13.

Pre-Fade Listen

It is often desirable to listen to the output of one channel and perhaps check its level as well. This is accomplished with a pre-fade listen (PFL) bus. If the PFL button is pressed on a channel, the signal prior to the channel fader is sent to the PFL bus. Mixing of multiple channels is not done on the PFL bus; it is just a convenient way of giving access to individual channels for their PFL signals. Importantly, activating a PFL does not interfere with the mix in any way. The PFL is often used in setting the microphone preamp gain control (not the fader) on the channel so that a reasonable signal level is presented to the channel fader.

Monitor Bus

Any channel can be sent to the monitor bus without affecting the mix. Those channels selected will be mixed together. The monitor signal for the bus is usually taken after the fader. A useful example is where a drum kit is assigned several microphones, and those channels are sent to the monitor bus to adjust those channels to establish the sound of the drums kit as a whole.

Channel Inserts

Each channel has an insert jack after the fader (post-fade insert) that allows an external device to be inserted into the signal path at that point. An example of such an external device would be a graphic equalizer or parametric equalizer for the channel. The jack is a 1/4" TRS jack. The outbound signal is connected to the tip and the returning inbound signal is connected to the ring. The sleeve is ground. When no plug is inserted, the tip and ring are connected. Some mixers with more features have pre-fader inserts as well. In the design shown, all of the post-fade outputs are affected by the insert.

Mute

A mute button mutes all outputs of the channel. Unused channels should usually be muted. Some channels that are connected to unused microphones in a certain part of a performance should also be muted. It is best if the mute button disconnects the outputs of the channel strip to the buses to maintain the best mix-bus SNR when not all channels are being used. If not all buses are disconnected during mute, the main left and right outputs should be disconnected in order to preserve SNR on these most important buses.

Overload Indicators

An LED on each channel strip indicates when the signal level at one or more points in the signal chain exceeds some peak level. The signal points chosen for overload monitoring are often the output of the mic preamp and the output of the channel. In many cases, the levels are monitored in both locations. The overload indication is activated at a peak-voltage threshold that is 6 dB below the actual clipping point. If a peak exceeds the threshold for more than 100 μs the indicator is activated for 1 second.

29.4 A Generic 16-Channel Mixer – The Master Section

The master section shown in Figure 29.14 includes the virtual ground mix amplifiers that create the outputs from the buses and the main stereo mix fader. Not all of the buses shown in the channel strip are included in the figure. The master section also includes the headphone amplifier, metering, etc. Most of the master section pertains to what is done with the mix signal from each bus, how metering and monitoring is accomplished and any provision for additional line-level inputs (such as from a CD player or DAC). The features and routing in the master section below are examples only, as there are greater complexities in routing and greatly varying architectures in different mixers.

Mixing Amplifiers

Each mix bus in the mixer has its inputs applied via resistors, and the bus is the virtual ground node of the bus mix amplifier. A mix amplifier and its interface to a bus was shown in Figure 29.9. Each bus also has provision for additional line-level inputs from TRS jacks. As mentioned earlier, noise accumulation can occur in the mix amplifier when a large number of channels is connected to its bus, increasing its noise gain. For this reason, it is important to use an op amp with very low input noise, less than 3 nV/√ Hz, for example. A mix bus also has a considerable amount of stray capacitance that can cause instability in the mix amplifier. For this reason, a stabilizing capacitor must be connected from the op amp's output back to its input, as shown in Figure 29.9.

Main Stereo Outputs

The stereo signals from the *L* and *R* mix amplifiers pass through the main output fader and balanced output amplifiers. The pre-fader stereo signals also pass to a stereo level control and the stereo headphone amplifier.

Master Insert

The master section also includes left and right TRS insert jacks ahead of the main stereo mix fader. These can be used to insert devices that will influence the aggregate of all of the channels in the main stereo mix. A good example here might be the insertion of a graphic equalizer to tailor the room frequency response.

Auxiliary Outputs

The master portion of the mixer includes two of the auxiliary outputs from the auxiliary bus mixes, AUX1 and AUX2. Two master auxiliary level controls (not shown) set the signal levels of these outputs, as shown in Figure 29.14. As mentioned earlier, these signals often are routed to effects devices or to stage monitors.

Return Inputs

The master portion of the generic mixer includes eight auxiliary return jacks. This is where external effects devices can return the signals that they received from the auxiliary sends. In contrast to inserts, these signals go back to the master section of the mixer where they can enter the mix. These returns can also be used to input signals that come from entirely separate line-level sources.

The four stereo pairs of line-level return inputs can be faded and panned for application to the stereo mix buses. The inputs need not be in stereo pairs. These signals are often the return paths from external devices whose inputs originated from the AUX1 and AUX2 outputs. Alternatively these jacks can receive stereo inputs from a source like a CD player or DAC.

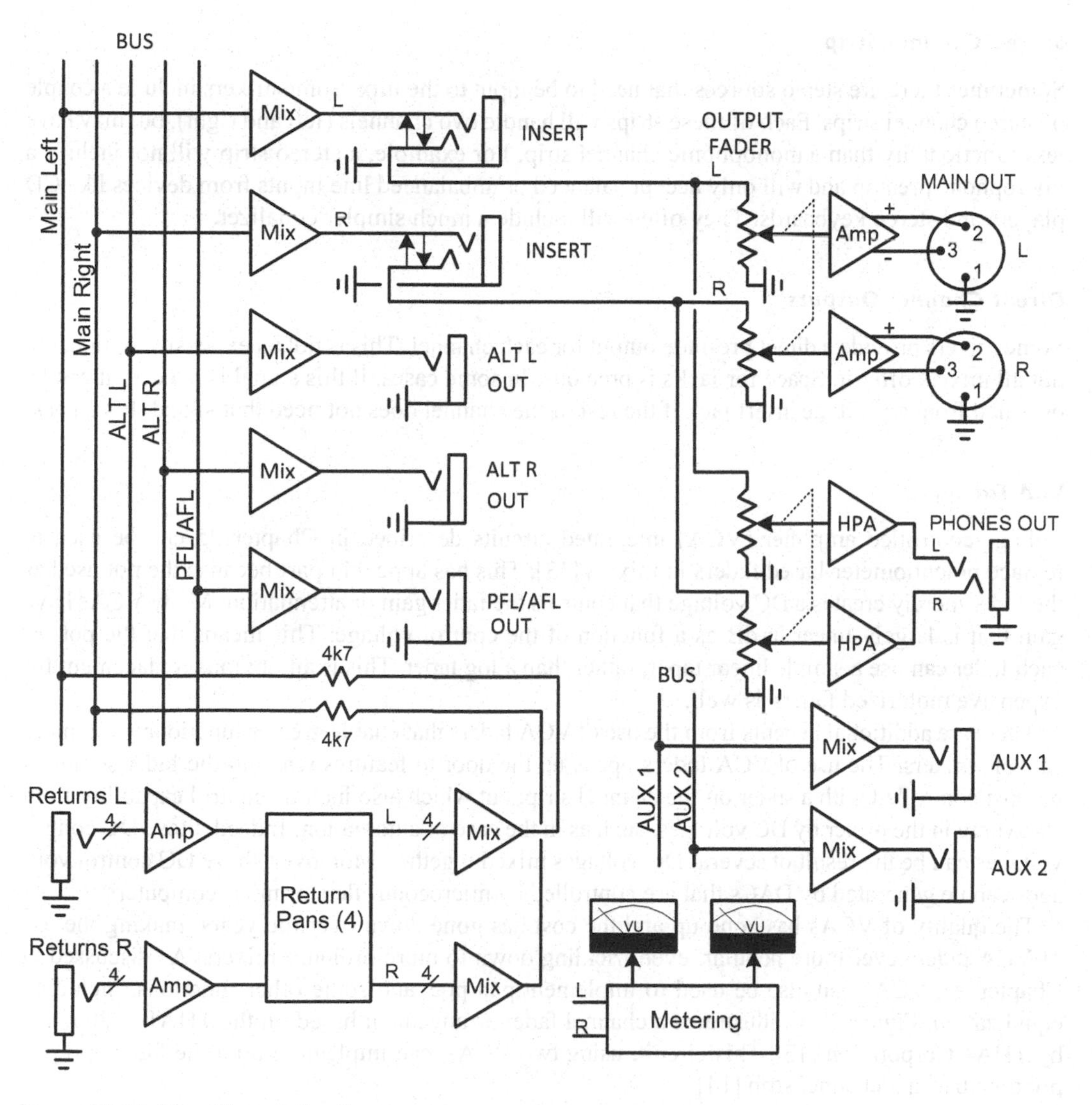

Figure 29.14 The Master Section.

Foldback

Foldback is a different name for the on-stage monitors that have their own mixes. As mentioned earlier, these monitors allow performers to hear themselves better, and the corresponding monitors are fed mixes that are customized for that performer. These mixes of auxiliary sends from AUX send buses usually do not include all channels in the FOH mix. These monitor mixes are usually taken from pre-fade auxiliary sends.

29.5 More Capable Mixers

As requirements for flexibility and functionality increase, many more capable mixers are available, especially for live-sound venues.

Stereo Channel Strip

Sometimes there are stereo sources that need to be input to the mix. Some mixers include a couple of stereo channel strips. Each of these strips will handle two channels (left and right), but may have less functionality than a monophonic channel strip. For example, a stereo strip will not include a microphone preamp and will only accept balanced or unbalanced line inputs from devices like CD players and stereo keyboards. They often will include a much simpler equalizer.

Direct Channel Outputs

Some mixers provide a direct pre-fade output for each channel. This is not an expensive option, but not all mixers offer it. Space for jacks is precious. In some cases, if this signal is needed, it can be obtained from a pre-fade insert jack if the rest of the channel does not need that signal. It's a hack.

VCA Faders

Voltage-controlled amplifier (VCA) integrated circuits described in Chapter 25 can be used to replace potentiometer-based faders in mixers [13]. This has appeal in part because the pot used as the fader merely creates a DC voltage that controls the fader gain or attenuation. Many VCAs have gain that is largely linear in dB as a function of the control voltage. This means that the pot for each fader can use a simple linear taper, rather than a log taper. This is an obvious replacement for expensive motorized faders as well.

There are additional benefits from the use of VCA faders that enable greater functionality in more capable mixers. The use of VCA faders opens up the door to features wherein the fader setting is not just controlled with a slider on the channel strip, but which also include control capability from elsewhere in the mixer by DC voltages, such as in the case of automation. Indeed, those DC control voltages can be the result of several DC voltages mixed together. Moreover, those DC control voltages can be generated by DACs that are controlled by microcontrollers or microcomputers.

The quality of VCAs has gone up and the cost has gone down over the years, making the use of VCA faders ever more popular, even trickling down to more ordinary mixers. As discussed in Chapter 25, VCAs can also be used to implement pan pots and some other functions, including equalization. Figure 29.15 illustrates a channel fader arrangement based on the THAT 2180 made by THAT Corporation [13]. This circuit, using two VCAs, can implement both the fader and pan pot together in a channel strip [14].

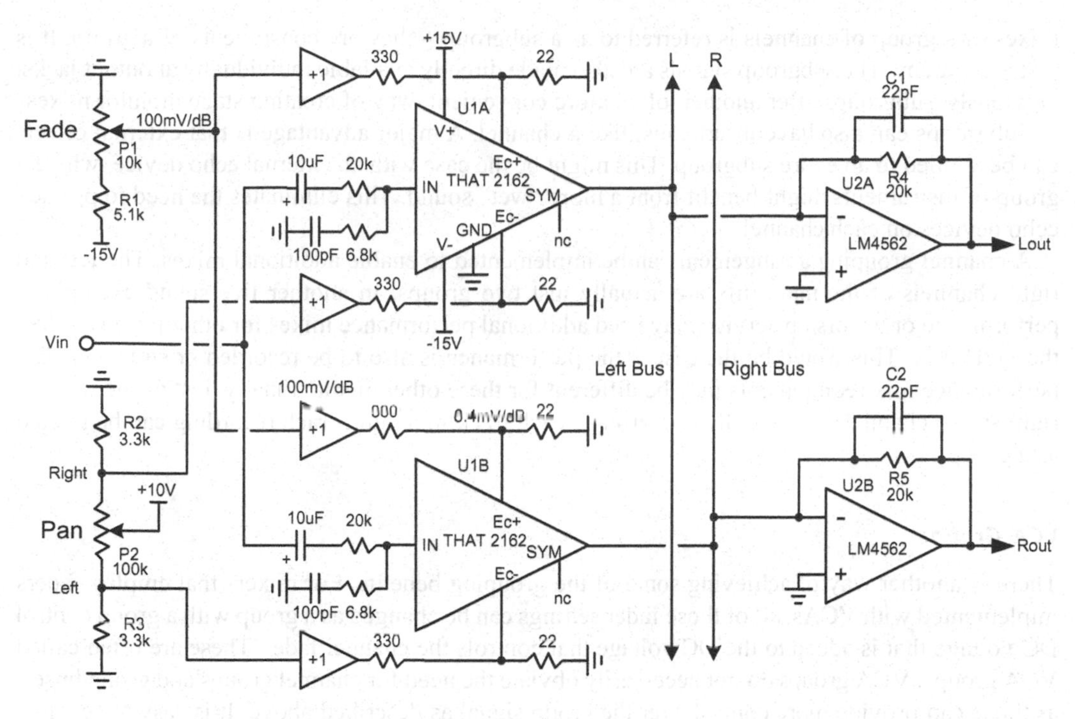

Figure 29.15 Channel Fader and Panner Implemented with a VCA.

Number of Auxiliary Sends and Flexibility

A fairly obvious extension to mixer capability is to have an increased number of auxiliary sends and auxiliary buses. Note that there can be more send buses than the number of sends available in a channel, as long as the available sends can be routed to the desired auxiliary bus. Flexibility in making the auxiliaries pre-fade or post-fade also increases their usefulness.

Channel Grouping and Subgroups

Sometimes it is desirable to treat several channels together as a group. For example, if there are four microphones on a drum kit, it might be desirable for the associated channels to be treated as a subgroup so that a single fader and pan pot can more conveniently control them together as a group. It might then be desirable to control the group with a single fader or treat those channels in other ways as a group, perhaps with a single pan pot. This can be a major convenience for the mixing engineer.

Four individual channels will tailor the signals with EQ and fading from each microphone or line input. The outputs from those channels will then be applied to a subgroup mixing bus. This is not completely unlike the treatment of auxiliary sends. However, the group signals originate after the fader in each channel. The constituents of the resulting subgroup have thus been premixed on a group bus. It is not unusual to have four subgroup buses in a more capable mixer. The mixed outputs of the subgroup bus will then be further manipulated in a master group section in the mixer where they can be faded and panned as a subgroup.

Think of the main stereo mix as a group, whose inputs can be from subgroups, or even other groups, or individual channels. Think of the alternate bus as a group as well. This view is why the mixes of a group of channels is referred to as a subgroup – they are constituents of a group. It is just a hierarchy. The subgroup signals are also made directly available individually at output jacks. Obviously, subgroups offer another, often more convenient, way of creating stage monitor mixes.

Subgroups can also have inserts, just like a channel. A major advantage is that external effects can be applied to an entire subgroup. This might be the case with an external echo device, where a group of instruments might benefit from a more "wet" sound. This eliminates the need to provide echo devices on each channel.

A channel grouping arrangement can be implemented to enable additional mixes. The left and right channels of the main mix are actually just two groups. In another live sound example, a performance or a worship service may need additional performance mixes for other purposes than the FOH mix. This would be the case if the performance is also to be recorded or streamed. The performance mix requirements may be different for these other media. Finally, just as the left and right stereo channels are actually two groups, each track in a multi-track recording can be treated as a group.

VCA Groups

There is another way of achieving some of the grouping benefits. For mixers that employ faders implemented with VCAs, all of those fader settings can be changed as a group with a group control DC voltage that is added to the DC voltage that controls the channel fader. These are often called VCA groups. VCA groups do not necessarily obviate the need for channel groups and group buses, as those can provide more control over the group signal as described above. It is easy to see how several VCA faders can be configured to implement a VCA group by connecting the included additional DC control input to a DC VCA group bus that is controlled by a VCA group DC fader.

Simple Matrix Mixer

A matrix mixer takes a multiplicity of inputs and routes them to a multiplicity of outputs. An $n \times m$ matrix can take in n inputs and route any of them to any of m outputs. A given input can be routed simultaneously to any number of available outputs. In a simple matrix mixer, the mixing takes place on the m output lines, each of which is a virtual ground mix bus.

In a more complex matrix, there is a level control at every crosspoint in the mixer, allowing each of the n inputs to be mixed onto each of the m outputs with different levels. This becomes the equivalent of m different N-channel mixers. Such a matrix mixer would have n times m level controls. Some would say the ultimate in flexibility. It also has the potential to create a very large number of different output mixes. For example, a 16-channel mixer with a symmetrical 16×16 mixer matrix would be able to create 16 different output mixes – but that matrix function would consume 256 level controls. At the other end of the scale, 4×4 matrix mixers can often be found in the master section of a mixer or in a stand-alone desktop format. So-called matrix patch mixers without level controls and just an illuminated pushbutton at each intersection can sometimes replace a patch panel, allowing quick and convenient reconfiguration of connections between pieces of equipment.

A mix matrix need not be symmetrical. For example, a mixer might contain a very modest 6×2 matrix mixer. Indeed, it might not even be called a matrix at this scale, since combinations of AUX buses and SUB buses could accomplish largely the same functionality. The true use of a

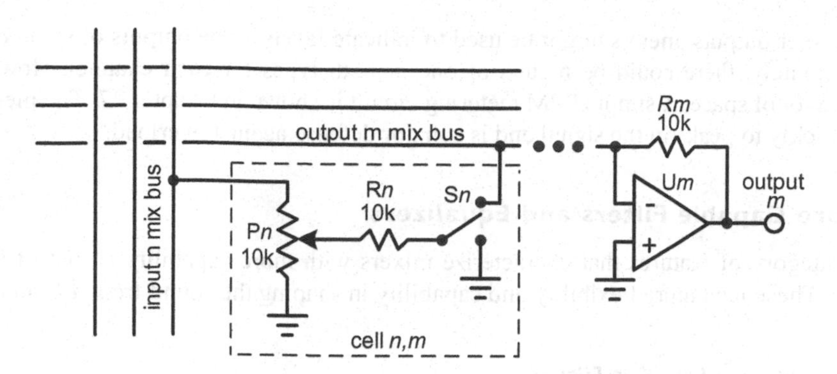

Figure 29.16 A Matrix Mixing Circuit.

matrix really comes into play when a large number of output mixes is needed, such as when there are a large number of stage monitors and in-ear monitors for performers, and in addition there may be a large number of additional fill speakers in the venue. Sixteen such outputs is not an unthinkable number under these conditions. Achieving that number of output mixes with ordinary AUX and SUB buses would be difficult if not inconvenient.

In most matrix mixers, only the outputs of buses in the mixer are able to be inputs to the matrix. Few or no individual channel signals are able to be put into the matrix. The matrix is thus mainly dealing with signal entities that have already been mixed, like subgroups or groups. For example, the main FOH stereo mix might be input to the matrix where some augmenting signal is added for some reason. It is important to not lose sight of the fact that a matrix is largely an extension of the concepts of AUX and SUB buses. They add to convenience, functionality, flexibility and scale.

A routing matrix allows several different signals to be routed to a matrix-based mix bus that may be one of several matrix buses. For example, a 6×2 matrix will allow any of six inputs to be routed to either of two mix buses. An input to a mix bus is either enabled or disabled by a switch circuit in the matrix.

Figure 29.16 shows the circuit for a switching and routing crosspoint cell of a matrix mixer. The *n* input buses have low-impedance voltage signals on them. The *m* output buses are virtual ground mix buses terminated in a virtual ground mix amplifier, U*m*. Op amp U*m* and its associated feedback resistor R*m* are not part of the matrix cell, and are shown only to illustrate the virtual ground mix amplifier that terminates the bus. The drawing implies that the cell depicted is that for the *n*th input bus.

The contributing input channel from input bus *n* enters the matrix cell through level control pot P*n* and 10-kΩ resistor R*n*, and passes through switch S*n* to be applied to the virtual ground output bus *m* when the crosspoint is enabled. When the crosspoint is disabled, S*n* shorts R*n* to ground to minimize crosstalk.

Metering

Metering was discussed in Chapter 27. In a more capable mixer it might include mechanical VU meters and LED bar graph peak program meters (PPM) for indicating levels at various points in the mixer signal path. VU and PPM meters each serve a different purpose. In addition to metering the main mixer outputs, meters might be used to indicate levels at the outputs of some of the mix buses. Ultimately, there could be meters of one or both types for each channel. However, this consumes a lot of space. A simple PPM metering circuit is shown in Chapter 27. The meter should respond quickly to peaks in the signal and is a major defense against overload.

29.6 More Capable Filters and Equalizers

Another category of features that characterize mixers with more capability is that of filters and equalizers. These lend more flexibility and capability in shaping the sound from a channel.

Variable Frequency Low-Cut Filter

A useful enhancement is to have a low-cut filter in the microphone preamp whose turnover frequency is variable. One mixer has such a filter that can be varied from 20 to 400 Hz [15]. The typical fixed frequency of 80 or 100 Hz is not always optimal. Such a variable low-cut filter can be implemented with a version of a Sallen-Key second-order HPF as shown in Figure 29.17. Here the filter is arranged to have the two tuning resistors be of equal value so that they can be varied with a two-gang pot with identical sections. It is not common to implement such a filter with equal-value resistors, so some modification is needed. In this case, the gain of the amplifier section, which is usually unity, is set to be higher to obtain the necessary Q. As shown, the filter can be tuned over a decade, from 30 to 300 Hz.

Steeper Shelving Equalizers

First-order shelving equalizers have shallow slopes that can never exceed 6 dB/octave. They may create too much effect in the mid-range, interfering with mid-range equalization. A first-order shelving equalizer is little more than a Baxandall tone control [4]. Shelving equalizers with greater slope are thus often desirable.

A bandpass shelving equalizer operates by creating a temporary shelf in frequency by using a second-order bandpass filter. The shelf is created by one side of the bell of the bandpass characteristic. The other side is outside the frequency range of interest for shelf equalization, going

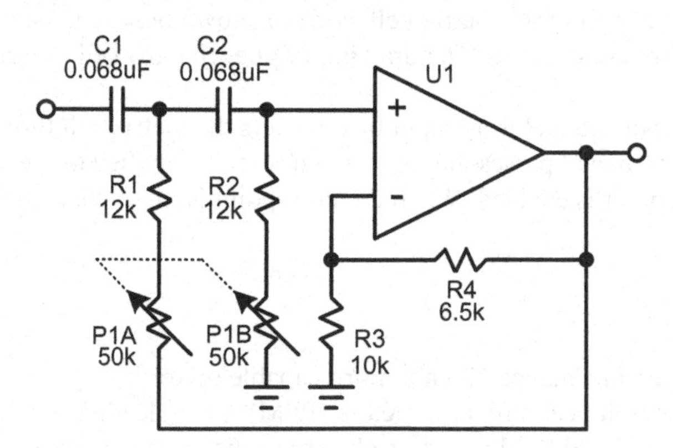

Figure 29.17 Variable Frequency Low-Cut Filter.

beyond the high-frequency limit for an HF shelf and below the LF limit for an LF shelf. The center frequency of the LF shelf in this case might be 25 Hz, while the center frequency of the HF shelf might be 15–20 kHz. The steepness of the shelf is established by the *Q* of its bandpass filter. Its equalization thus returns to zero dB outside the extreme of the audio spectrum. The steepness of the shelf is achieved by the *Q* of the second-order bandpass filter; it can exceed the theoretical 6 dB per octave if the *Q* is higher. This also has the advantage of not extending the boost in frequency ranges well outside the audio spectrum. It can be realized by a series-resonant filter using a gyrator for the inductor, as is often done in graphic equalizers.

Series-Resonant Mid-Band Equalizers

The Baxandall mid-band equalizer has very low *Q*. Its bell-shape is very broad and affects a wide range of frequencies. The simple 3-band 6 dB/octave tone control EQ discussed earlier cannot be very surgical, to say the least. A higher-*Q* mid-band EQ is often desirable, especially if more than one mid-band EQ is implemented. Often a four-band EQ will include two mid-band equalizers, one called a low-mid and one called a high-mid. These fixed-frequency equalizers are often series-resonant equalizers using a gyrator to synthesize the required inductor. These equalizers are much like the equalizers employed in graphic equalizers.

Wien Bridge Equalizers

The Wien bridge, most well-known for its use in audio oscillators, can be used to implement a mid-band bell-type equalizer. A conventional Wien bridge equalizer is shown in Figure 29.18(a) [16]. Sometimes it is desirable to have greater *Q* or boost/cut range in a Wien bridge equalizer. As described in Chapter 20, this can be accomplished by the circuit modifications in (b) and (c). The circuit in (b) is similar to the Mackie Perkins equalizer, where capacitor C1 is made larger than C2 by a ratio lying between 2 and 3. In (c), the *Q* is increased by adding feedback shunt resistor R5. It increases the gain of the amplifier surrounding the Wien bridge network. This arrangement shares a similar arrangement and operating theory to that of a Wien bridge oscillator.

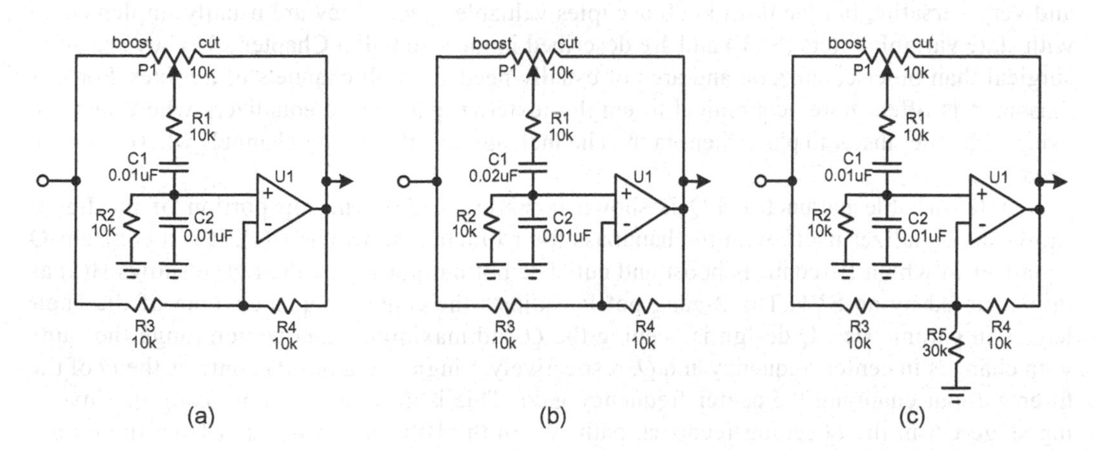

Figure 29.18 Wien Bridge Mid-Band Equalizer.

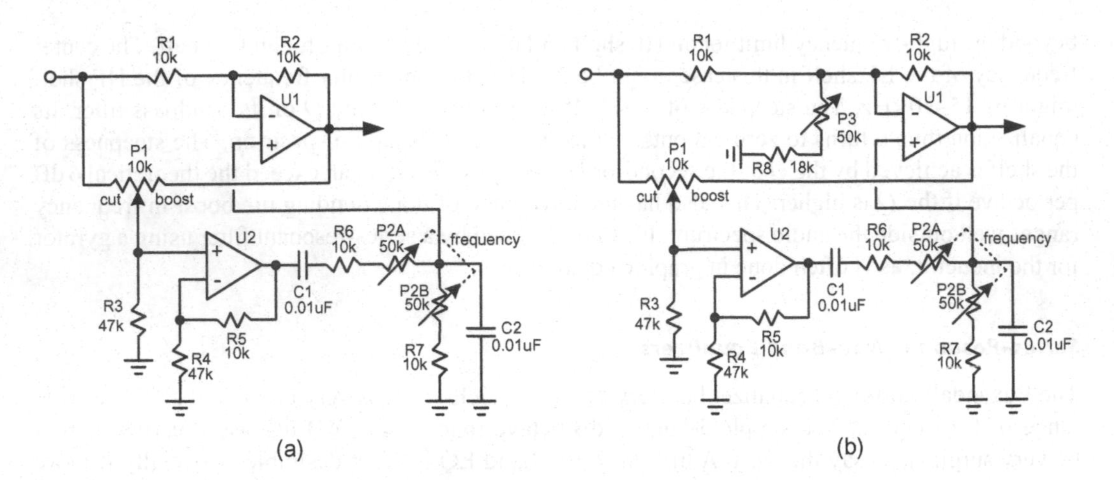

Figure 29.19 Quasi-Parametric Equalizer.

Quasi-Parametric Equalizers

A mid-band EQ whose frequency can be swept, and thus has both boost/cut and frequency controls is called a quasi- or semi-parametric equalizer. It is like a parametric equalizer, but it has fixed *Q*. *A quasi-parametric* Wien bridge equalizer circuit is shown in Figure 29.19(a). The tuning resistors are simply made variable by the use of a dual-gang pot. The fact that the two resistors in a Wien bridge equalizer network are equal makes this straightforward. In Figure 29.19(b), some modest amount of *Q* control can be had by adding some variable positive feedback to the equalizer circuit U1 via P3 and R8.

Parametric Equalizers

Parametric equalizers are bandpass equalizers where the boost/cut amplitude, the center frequency and the *Q* can all be controlled. They thus have three controls. They are more complex and very versatile, but the third knob occupies valuable space. They are usually implemented with state variable filters (SVF) and are described in more detail in Chapter 20. They are more surgical than other equalizers, and are not usually needed on all channels of a mixer. For this reason, it is often more economical to employ external parametric equalizers where needed, using a channel insert to place them in the channel signal path only on channels where they are needed.

A state variable parametric EQ is shown in Figure 29.20. The top portion of the figure shows the equalizer itself, with the bandpass filter function shown as H(s). It is a constant-Q equalizer in which P1 controls boost and cut. The bottom portion of the figure shows H(s) as implemented by an SVF. The 2-gang pot P2 adjusts the center frequency. One of the challenges in parametric EQ design is keeping the *Q* and maximum equalization range the same with changes in center frequency and *Q*, respectively. Single-gang pot P3 controls the *Q* of the filter without changing the center-frequency gain. This is accomplished by using the inverting stage U6 in the *Q*-setting feedback path. When the BPF output is taken from the output of U6 (instead of the usual output of U4), the gain of the BPF does not change when the *Q* is changed.

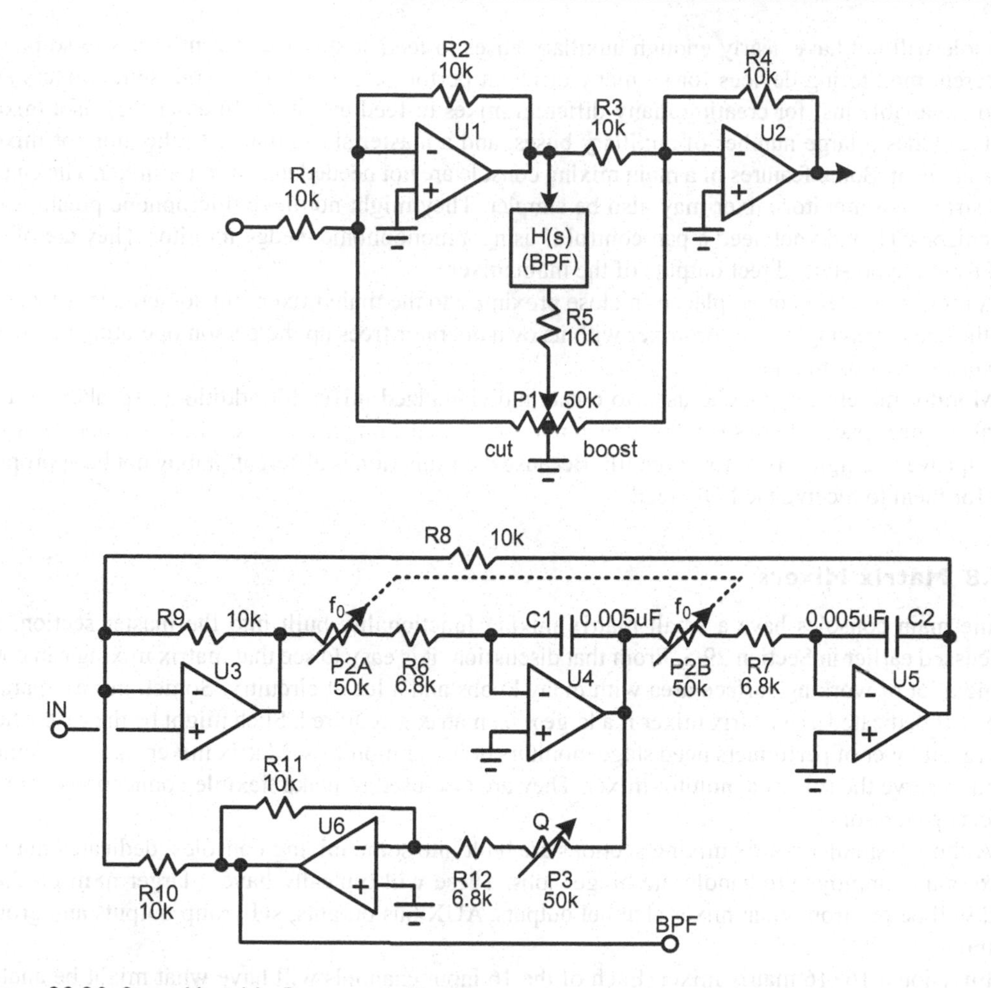

Figure 29.20 State Variable Parametric Equalizer.

Constant-Q Equalizers

The parametric equalizer of Figure 29.20 is a constant-Q equalizer. As explained in Chapter 20, constant-Q equalizers maintain the same value of Q as the amount of equalization is increased or decreased. Conversely, the Q of proportional-Q equalizers increases as the amount of equalization increases [17]. The series-resonant equalizer often used in graphic equalizers is a proportional-Q equalizer. The conventional Wien bridge equalizer of Figure 29.18 is also a proportional-Q equalizer. It is often the case that it is less desirable to have an equalizer whose selectivity decreases as amount of equalization decreases.

29.7 Monitor Mixers

It is easy to imagine a situation where there are many performers on the stage that need to be able to listen to themselves or a select group of feeds. They may be listening to on-stage monitor speakers (so-called *wedges*) or may have a wireless in-ear monitoring device. In some cases the mixing console will not have nearly enough auxiliary buses to feed so many different mixes to so many different monitoring devices for so many different performers. For this reason, some mixers are made available just for creating many different mixes to feed monitors. In essence, it is a mixer that includes a large number of auxiliary buses, and a master side suitable to the monitor mixer arrangement. Some features of a main mixing console are not needed in a monitor mixer. The channel strips in a monitor mixer may also be simpler. They might not have microphone preamps or equalizers. They do not need a pan control if using a monophonic wedge monitor. They are often fed from the pre-fade direct outputs of the main mixer.

A monitor mixer is often placed in close proximity to the main mixer, but sometimes is located on the stage. Having a monitor mixer with its own operator frees up the person operating the main mixer to focus on that job.

Monitor mixers can also be used to create individualized mixes for additional speakers in the performance space. These speakers, which may be located along the sides or the back, may be used to improve intelligibility or for room fill. Because their function is different, it may not be appropriate for them to receive the FOH feed.

29.8 Matrix Mixers

Some main consoles have a small matrix mixing functionality built into the master section, as discussed earlier in Section 29.5. From that discussion, it is easy to see that matrix mixing can consume a lot of working surface area with many knobs and a lot of circuitry. Sometimes a separate mixer is dedicated as a matrix mixer if a large $n \times m$ matrix is required. Such might be the case when a large number of performers need stage monitors or in-ear monitors. Matrix mixers are sometimes used to serve the role of a monitor mixer. They are also used to make flexible connections among effects processors.

Although smaller matrix mixing sections are built into some mixing consoles, dedicated matrix mixers are employed to handle the bigger jobs. These will typically have a larger n×m product and will be fed from main mixer channel outputs, AUX bus outputs, subgroup outputs and group outputs.

Envision a 16×16 matrix mixer. Each of the 16 input channels will have what might be analogous to a channel strip. Unlike a conventional mixer channel strip, there will not generally be microphone inputs, equalizers etc. Rather, each matrix strip will primarily contain a column of 16 level control pots. Each of these pots controls the level of the input channel's signal that is sent to the output mix bus for that strip's assigned output bus. This is not unlike the column of four or more pots on a conventional mixer channel strip that direct the output of that channel to four or more aux send buses. The level pot at each matrix intersection (crosspoint) may be accompanied by a switch that enables the connection at that junction. If an input will not be routed to an output bus, disabling that input may improve the SNR for that bus. A 16×16 matrix mixer of this arrangement will have 16 matrix strips and a total of 256 crosspoints and 256 gain pots.

Figure 29.21 shows a simple matrix mixer crosspoint circuit. It is fed input signals from an n-wide set of input buses. Figure 29.21 depicts the crosspoint with S*n* shown as an SPDT switch. Every vertical array of crosspoint cells in a matrix mixer receives every matrix mixer input. The cell routes its output to its associated output bus. Its level control pot, often fairly simple, attenuates the signal and feeds it to the output mixing bus through R*n*, and S*n*. For completeness, U*m* and its feedback resistor R*m* represent the output mix bus amplifier where all of the signals on a given output bus are summed. U*m* is not part of the crosspoint.

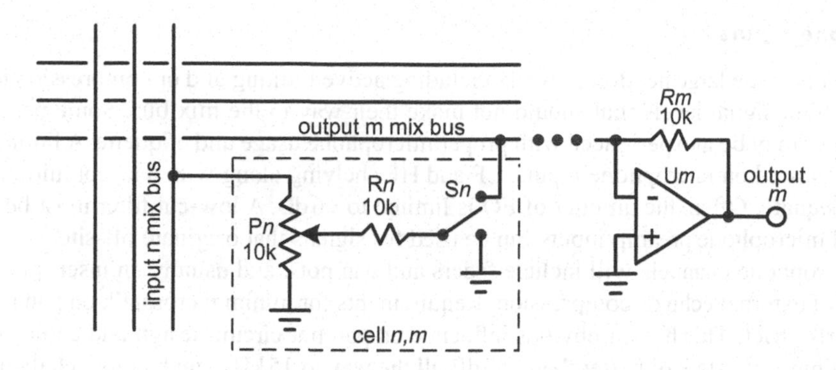

Figure 29.21 A Matrix Mixer Crosspoint.

As with most mix buses, the bus ultimately feeds a virtual ground mix amplifier, here depicted as U*m*. Between R*n* and the bus is an SPDT pushbutton switch with R*n* connected to its swinger. When the crosspoint is enabled, the switch connects R*n* to the bus. When disabled, the switch connects R*n* to ground. This minimizes crosstalk from that channel. When the crosspoint is disabled, the absence of R*n* being connected to the bus reduces the noise gain of the bus mixer and improves SNR.

Each output channel of the matrix mixer may include a fader. Additional functionality of a more sophisticated matrix mixer might be of the sort associated with the master side of a conventional mixer, such as monitoring, PFL and solo controls. Such additional functionality will require additional mix buses. A 16×16 matrix mixer might thus have as many as 20 mix buses or more. Additional functionality in the output section might include inserts, some EQ, headphone monitoring and metering.

One can see that a matrix mixer has many knobs and many mix buses. However, referring again to the channel strip analogy, a 16×16 matrix mixer need not be larger than a conventional 16-channel mixer, with 16 channel strips. It will have at least 16 pots and perhaps 16 pushbutton switches on each matrix strip. This is not vastly different than the number of controls on a conventional mixer's channel strip. The 16×16 matrix mixer will have at least 16 buses. Even a conventional 16-channel mixer will have a large number of mix buses when all of the main, alternate, aux send, monitor, group and PFL buses are considered. The same mix bus technical challenges exist with a matrix mixer as with a conventional mixer. The ability to disable any given crosspoint can optimize SNR performance.

29.9 Broadcast Consoles

For a broadcast or live sound mixer, there may be no recording machine, since all is live. By comparison with live sound mixers and studio consoles, the signal path in a broadcast mixer is usually quite straightforward. There are no stage monitors, no extra house speakers, little or no multi-track recording, little EQ and few effects processors. This leads to relatively few mix buses and aux buses. Extra features in a broadcast mixer may include telephone call-in capability, processing, metering and monitoring that is essential to staying within strict broadcast modulation limits. Ergonomic ease of use is important, since results may often be in real time with no re-do. Compressors, de-essers, and other dynaimc speech control functions may also be included.

Microphone Inputs

These must provide large headroom while including active limiting and or compression to manage unusually high signal levels that should not make their way to the mix bus. Some persons using microphones may be inexperienced with proper microphone usage and etiquette. A limited amount of EQ is required on microphone inputs. LF and HF shelving along with a swept mid-band EQ is usually adequate. Often, the amount of EQ is limited to ±6 dB. A low-cut filter must be included. Line-level microphone preamp inputs can be used for signals that originate off-site.

The microphone channels will include faders and pan pots, and usually an insert point for perhaps a bit of external echo or compression. Requirements for minimal crosstalk on pan pot circuits can be fairly strict. This has an obvious influence on pan pot circuit design and choice of quality pots. Obtaining crosstalk of better than –50 dB all the way to 15 kHz can be more challenging than one might think.

Line-Level Inputs

Line-level stereo inputs, as often from a CD player, DAC, tape machine or cartridge player, are usually tamer and more predictable than a microphone input. Their levels are often fairly normalized and their dynamic range more limited. Much of it is often MP3-derived, so it is quite tame. The need for EQ and effects processors for these sources is minimal.

Phone-in line-level inputs will usually be mono and will often arrive from an external phone interface that provides the hybrid function. The signal sent back to the caller will often be the main mix less the caller's contribution to that mix. The phone interface may be included in the broadcast mixer.

Buses

In addition to the main stereo mix buses there will be at least one secondary bus (like the ALT bus), which may be used to record the program if for some reason the main bus is not desired for that function. Other buses may include a stereo PFL bus and a pair of mono auxiliary buses. There may also be four or more so-called "clean feed" monophonic buses.

On-Location Live Broadcasts

On-location broadcasts with a crowd will often require a fairly simple PA system and mix, and of course may include microphones for pickup of audience sounds and interviewees.

Loudness Metering

Metering of the main feed to the transmitter is especially important for compliance with governmental limits on modulation. This metering is usually governed by specific broadcast standards. In particular, loudness metering in loudness units (LU) and amplitudes of LUFS or LKFS, as described in Chapter 27, may be mandatory.

29.10 Recording Consoles

The most complex mixers are generally multi-track recording studio consoles. These consoles are used to record numerous tracks onto a recording machine, or to mix down tracks from the recording machine, or to perform over-dubbing.

Multi-track Recording

A recording console may often be used to lay down tracks on a multi-track machine in groups. There is usually no need at this point to mix groups together, process them and assign them to stereo channels.

Mix-Down

The next step in making a recording may be the mix-down, where the multi-track recording is played back and the groups on the different tracks are fed into the same console, but with the console configured slightly differently so that groups can be mixed, using the channel strips, into final stereo (or multi-channel) outputs.

29.11 Mix Bus Technical Challenges

The conventional virtual ground mix bus encounters some technical challenges when a large number of channel sources are connected to it. These challenges include noise, distortion, mix amplifier stability and susceptibility to noise and EMI (electromagnetic interference).

Noise Gain

Suppose we have a 32-channel mixer and we want to place all of the 32 channels somewhere in the stereo space. All 32 channels will have an active pan pot output that must be mixed with all of the other 31 outputs on each of the left and right output buses. This presents some technical challenges in regard to noise, distortion and headroom.

Recall from Chapter 8 that the noise gain of a virtual ground (VG) mixing amplifier stage is equal to the number of channels being mixed plus one if each input channel can achieve unity gain through the stage. For example, the channel input resistors and the VG amplifier feedback resistors might both be 10 kΩ. The noise gain is the inverse of the attenuation that the negative feedback sees at the virtual ground node. With 32 channels being mixed onto one bus, where each channel has a 10-kΩ input resistor and the mixing op amp has a 10 kΩ feedback resistor, the feedback attenuation is 10 kΩ against a net shunt of 32 10-kΩ resistors in parallel, or 313 Ω.

If the op amp has input noise of 5 nV/√Hz, the output noise will be 165 nV/√Hz, or 23.3 μV in a 20-kHz noise bandwidth. This does not include the thermal noise contributions of the resistors, but keeps the concept simple. The noise is directly proportional to the number of channels (plus 1). In a kinder world, the noise would go up on a power basis, which is the way that the individual noises from the 32 channels will add. If this were the case, the noise would go up in approximate proportion to the square root of the number of channels. Mix bus amplifier noise goes up directly with the number of channels.

The LM4562 is an excellent choice for the mix amplifier, with typical input noise of 2.7 nV/√Hz, gain-bandwidth product (GBP) of 55 MHz and maximum supply voltage of ±18 V. The similar LME49860 is a better choice, with typical input noise of 2.7 nV/√Hz, GBP of 55 MHz and maximum supply voltage of ±22 V. For some, this is the best-sounding op amp available. The extra headroom provided by the higher allowed operating voltage of ±22 V is especially helpful for use in mix amplifiers where there is a greater possibility of multiple channels adding together to create an unusually large peak-signal swing. Typical maximum output swing into a 600- Ω load is about ±20 V with a ±22-V supply, corresponding to a clipping point of 14 V_{RMS}. With supply voltage of only ±18 V, the clipping point would only be about 11.8 V_{RMS}, lower by 1.5 dB. Looking at it differently, the higher operating voltage can improve SNR by 1.5 dB.

Other Problems Mixing Many Channels

Noise is not the only problem. Distortion contributed by the op amp also tends to go up in proportion to its noise gain. This is simply a result of the fact that the negative feedback loop gain goes down as the noise gain goes up. Suppose we have a very good op amp whose distortion at high output levels at 20 kHz is only 0.001%. With noise gain of 33, that distortion number is now degraded to 0.032%, no longer that impressive.

Without care in the design, summing amplifier stability can also be a concern. The bus capacitance can be quite high with many contributing resistors hanging on the virtual ground bus over some significant physical distance. It is possible to imagine 5 pF per contributor, for a total of 165 pF for 32 channels. Such a capacitive load on a virtual ground will create a pole in the feedback path that may eat into phase margin. If all channels are connected to the bus and the bus resistance is 313 Ω, the bus pole frequency will be at about 3.1 MHz. This is not too bad, since the noise gain is 33. If the op amp is an LM4562 or LME49860 with 55-MHz GBP, the feedback gain crossover frequency will be at about 1.7 MHz with noise gain of 32. Still, some phase margin will be stolen by the bus pole that lies only about an octave above the VG amplifier unity gain frequency.

To make matters worse, suppose that most of the channels are disabled from the bus by being open-circuited. Suppose only two channels are enabled, but that the total bus capacitance remains unchanged, at 165 pF. Now the bus resistance on the VG node is 5 kΩ and the bus pole is down at about 190 kHz (actually the pole of interest is more like 290 kHz when we take the resistance of the 10-kΩ feedback resistor into account). Moreover, the noise gain is now only 3, meaning that the feedback gain crossover frequency could be as high as 18 MHz. We have very strong recipe for oscillation.

Often, this will be addressed by adding a capacitor across the 10-kΩ feedback resistor of the summing amplifier, forming a capacitance voltage divider that will keep the feedback attenuation flatter with frequency when the noise gain is 33. In this case a value of about 5 pF might suffice to form an adequate capacitance voltage divider against the 165-pF bus capacitance. The 5-pF capacitor across the 10-kΩ feedback resistor will result in a closed-loop bandwidth on the order of 3.2 MHz, which is more than enough. Indeed, a lower closed-loop bandwidth is desirable, leading to the use of a larger feedback capacitor, perhaps 100 pF. This also helps the bus to be more immune to EMI.

Finally, headroom could be a problem if 32 channels are routed to a VG mixer and they all contribute significant signal. Just because each contributing channel does not clip does not mean that the output of the bus mix amplifier will not clip. This might call for a clip indicator at the output of the bus amplifier.

References

1. Glenn Ballou, editor, *Handbook for Sound Engineers*, 5th edition, Focal Press, New York, NY, 2015.
2. Arthur B. Williams and Fred J. Taylor, *Electronic Filter Design Handbook*, McGraw-Hill, New York, NY, 1995.
3. THAT Corp., *1642 Voltage Controlled Amplifier*, Datasheet, thatcorp.com.
4. Peter Baxandall, *Audio Gain Controls*, Wireless World, November 1980, pp. 79–81.
5. Ian Williams, *Active Volume Control for Professional Audio*, Texas Instruments TIDU034, December 2013.
6. THAT Corp., *1646 Outsmarts Balanced Line Driver*, Datasheet, thatcorp.com.
7. Texas Instruments, *A High-Power High-Fidelity Headphone Amplifier for Current Output Audio DACs Reference Design*, Application Note TIDU672C, December 2014, ti.com.

8. Mike Dorrough, *Reading a Dorrough Meter*, dorrough.com/readmeters.html.
9. Shure Incorporated, *Can a Dynamic Microphone Handle Really Loud Sounds?* https://www.shure.com/en-US/support
10. Francis Rumsey and Tim McCormick, *Sound and Recording Applications and Theory*, 7th Edition, Focal Press, New York, NY, 2014.
11. Neumann U87A condenser microphone specification.
12. Cloudlifter CL-1, Cloud Microphones, cloudmicrophones.com.
13. THAT Corp., *THAT 2181 Voltage Controlled Amplifier*, Datasheet, 2016, thatcorp.com.
14. Bob Cordell, *VCA-Based Faders and Panning Pots*, cordellaudio.com.
15. Soundcraft, *Spirit Live 42 Users Guide*, http://www.soundcraft.com.
16. Dennis A. Bohn, *Operator Adjustable Equalizers: An Overview*, Rane Corporation RaneNote 122, August 1997.
17. Dennis A Bohn, Constant-Q Graphic Equalizers, Rane Corporation, *Journal of the Audio Engineering Society*, vol. 34, no. 9, September 1986, pp. 611–626.

Chapter 30

DI Boxes and Microphone Splitters

The need to drive the input of a monitor mixer brings up the issues of signal splitting, impedance transformation, buffering and ground isolation. A microphone splitter is often required to drive both a front-of-house (FOH) mixer and a monitor mixer. A guitar or other instrument may require a type of splitter to enable it to drive both its own amplifier and the FOH mixer. Although not always directly applicable to splitting off a signal for a monitor mixer, understanding the well-known DI box for this latter kind of splitting is a good starting point.

30.1 Passive DI Boxes

A DI box is a "direct inject box," sometimes called a direct input box. It takes an unbalanced high-impedance signal from a line-level 1/4" TR (tip and ring) plug and converts it to a balanced low-impedance signal with an XLR connector [1]. DI boxes usually attenuate the signal by about 20 dB. This makes it possible to send guitar pickup signals and the like directly into mixer microphone inputs and/or over long cable runs without signal degradation. Passive DI boxes use transformers to make the impedance and unbalanced-to-balanced conversions. The galvanic isolation provided by the transformers also eliminates ground loops. Many DI boxes also connect the input signal directly to a 1/4" TR output jack as a so-called *through* signal, permitting its normal use as well. In the case of a guitar, this would mean feeding the signal to the guitar amplifier.

Some DI boxes provide a ground-lift switch that interrupts the XLR pin 1 screen path to break a ground loop. Some provide an attenuator that can be switched in to reduce high signal levels. Some also provide a polarity reverse switch.

For electric guitars, direct inject is an alternative to putting a microphone in front of the guitar amp speaker to feed the main mixer. The DI box should provide galvanic ground isolation or the equivalent to prevent ground loops between the mixer and guitar amplifier (or other active electronics). The DI box must also not degrade CMRR of the balanced interface to the mixer. Some guitar amplifiers have an XLR output to obviate the need for a DI box. It should be transformer coupled or have a floating active-balanced output to minimize the possibility of a ground loop.

The issue of transparency and non-intrusiveness of the DI box is crucial, given that the unbalanced source may be a high-impedance source that can be easily affected by loading. This can be especially the case with some types of guitar pickups. The high-impedance output of a guitar pickup expects to see load impedance as high as 1 MΩ [2]. Indeed, a simple model of a guitar pickup would comprise an inductance of 1.9 H in series with 6 kΩ and shunted by 220 pF [2]. The guitar pickup's signal would also thus be degraded by the capacitance of a long cable run to a mixer. Microphone cable capacitance is about 25 pF per foot; a 300-foot run would total 7500 pF. Even ignoring the pickup's series inductance, this amount of capacitance shunting a source

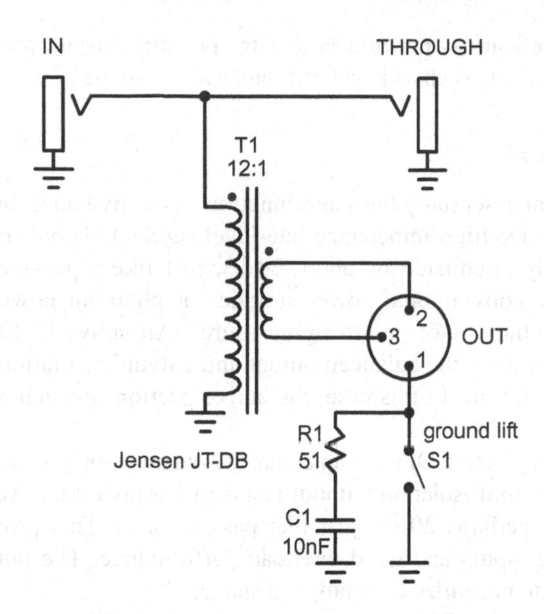

Figure 30.1 A Passive Direct Inject Box.

resistance of 6 kΩ would cause a frequency response roll-off beginning at 3.5 kHz. Star quad microphone cable will have about twice as much capacitance [3].

Other unbalanced line-level signal sources that are active, like keyboards, will not fare as badly because their output impedance is lower. However, they still need the conversion to balanced transmission and the galvanic isolation or equivalent to perform well over a long cable run. Figure 30.1 shows a transformer-based passive DI Box [4]. Here the 1/4" input jack is wired directly to a 1/4" output jack to provide a through path to a guitar amplifier. The high-impedance primary of the DI transformer is bridged across the input and its secondary is connected to the XLR jack. A ground-lift switch allows the ground from the XLR jack to float in order to prevent a ground loop from being formed between the instrument electronics (e.g., guitar amplifier or keyboard) and the console microphone preamp. The grounds can be optionally connected if appropriate.

The impedance transformation is achieved by using a transformer with a primary-to-secondary turns ratio of about 11:1 [5]. A transformer converts the impedance in accordance with the square of the turns ratio. For an 11:1 turns ratio, the impedance is transformed by a factor of 121. In the case of a fairly low microphone preamplifier input impedance of 1 kΩ, the signal source would see an impedance of 121 kΩ at the primary.

The quality of the DI transformer is crucial to the sound quality. A good DI box transformer can cost $75. The primary inductance is typically 100 to 800 H for a quality DI transformer. A transformer primary inductance of 100 H has impedance of only 13 kΩ at 20 Hz. With the standard source test input resistance of 6.81 kΩ, primary inductance needs to be 250 H for 0.2 dB loss at 20 Hz. Insufficient primary inductance can especially be a problem for bass guitars. Some quality DI transformers include the Cinemag CM-DBX, Lundahl 1935, Jensen JT-DB and the Sowter 4243 [6–10].

A less expensive transformer will have lower inductance and will degrade the low end. It may also have a smaller core that will more easily saturate, and much less shielding and Faraday screening between windings. The bottom line is that a passive DI box has relatively low input impedance compared to what some sources perform best with. For this reason, passive DI boxes are best suited to active stage sources, such as keyboards and active guitars.

30.2 Active DI Boxes

Active DI boxes perform essentially the same function as passive ones, but with active circuitry. They take in a single-ended high-impedance line-level signal and convert it to a low-impedance balanced output, usually attenuated by about 20 dB, just like a passive DI box. They can be powered with batteries, conventional power supplies or phantom power. A key benefit is the extremely light loading they place on the signal source. An active DI box can employ a transformer at its output to achieve the balanced output and galvanic isolation, but this adds expense and possible signal coloration. In this case, the active portion just acts as a buffer to drive the transformer.

More commonly, a fully active DI box can create the balanced output electronically and achieve a very high degree of ground isolation without resort to a transformer. Active DI boxes will still attenuate the signal by perhaps 20 dB, just like passive ones. This provides for compatibility with mixer microphone inputs and good overload performance. The output level is preferably microphone-level so as to minimize crosstalk in a snake.

The most convenient active DI boxes are phantom powered from the microphone preamp in the mixer. Phantom power is provided in the common mode by two precision 6.81-kΩ resistors in the mixer preamp from a +48-V phantom supply to each of the hot and cold microphone lines. The net phantom supply resistance to its load is thus 3.4 kΩ. This means that a phantom-powered active device will see a single-supply voltage of about 24 V if it draws 7 mA. Most phantom-powered circuits should and do draw less than 10 mA. Virtually all signal sources that feed a DI box do not need phantom power, so the phantom power is available for use by the active DI box electronics.

Some active DI boxes without transformers actually provide unity gain, but do not provide galvanic isolation and typically do not have a through output for connection to something like a guitar amplifier. If the mixer is the only (non-battery) electronic device connected to the DI box, galvanic isolation is not needed because there is not a ground loop issue. Such DIs are small and convenient in applications where they fill the need. An example is the phantom-powered Avenson Audio Small DI [11].

An active DI box should have good overload capability. It should be able to handle a maximum input level from the single-ended instrument source. This level might be comparable to what a mixer microphone preamp can take. This is because single-ended devices, such as keyboards, can produce up to 0 dBu and more. The ability to take in +10 dBu with low distortion (2.4 V_{RMS}) is laudable, but may require a pad.

Some active DI's use a transformer to achieve galvanic isolation, while others are all-active and achieve near-galvanic isolation that is more than adequate. There are two ways to implement a transformer-based active DI. Both have the advantage of presenting very high input impedance and improving the performance of a given transformer. Both provide true galvanic isolation.

The first approach is to employ an active unity-gain buffer followed by a DI transformer. The transformer provides the impedance transformation, the unbalanced-to-balanced conversion and the galvanic isolation. The active front-end eliminates transformer loading that results from the reflected load of the mixer's microphone preamp and from the finite inductance of the transformer. It greatly reduces the need for very high inductance in the transformer. Transformers also perform better and create less distortion when driven from lower impedance, as provided by the active

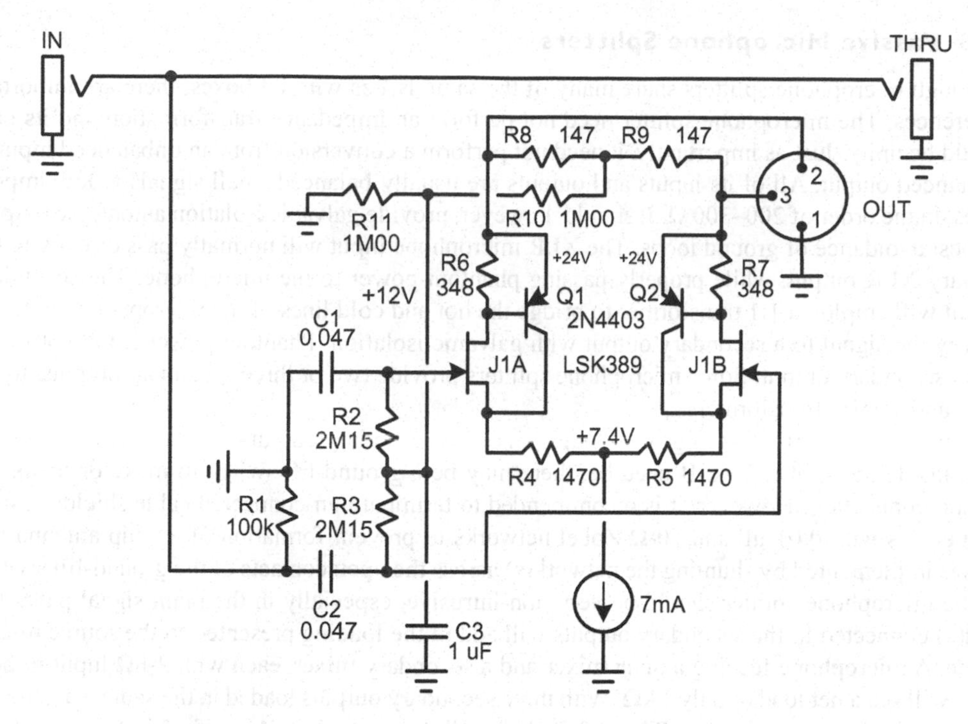

Figure 30.2 An Active Direct Inject Box.

front-end buffer. High sound quality can usually be achieved with a less expensive transformer in this case.

The second approach is to precede a 1:1 microphone splitter transformer with an active buffer that may provide some attenuation, on the order of the usual 20 dB. The impedance transformation is provided by the active buffer, while the unbalanced-to-balanced conversion and galvanic isolation is provided by the transformer. In either case, the active DI box operates from phantom power supplied by the mixer preamp to which it is connected.

Many popular low-cost active DI boxes use a dual JFET op amp. The first op amp forms a unity-gain buffer and the second op amp inverts the output of the first to provide the cold rail of the balanced output. Some include a 150-Ω resistor in series with each output to create a differential output impedance like that of a microphone. These outputs do not have high common-mode output impedance, which is a desirable feature. Some also include selectable input attenuation and a ground-lift switch. The op amp is usually powered between ground and +24 V that is derived from the phantom power and a Zener regulator. A voltage divider provides the needed center-supply 12-V bias voltage for the op amp inputs.

Figure 30.2 shows a transformerless active DI Box that provides a true-balanced output with high common-mode output impedance. It is implemented with a degenerated JFET differential pair wherein each JFET is configured as a complementary feedback pair using a PNP BJT. Each JFET is biased at 2 mA and the differential gain is 0.1 (−20 dB). The differential load resistance is about 300 Ω, making it look like a typical dynamic microphone. This design can handle up to +10 dBu at its input with THD less than 0.02%. At 0 dBu input, distortion is less than 0.002%. Output noise is only 2.5 nV/√Hz. The DI box does not pass phantom power through, and protects connected devices.

30.3 Passive Microphone Splitters

Although microphone splitters share many of the same issues with DI boxes, there are important differences. The microphone splitter need not perform an impedance transformation and its gain should be unity. Just as importantly, it need not perform a conversion from an unbalanced input to a balanced output. All of its inputs and outputs are usually balanced small signals at low impedances on the order of 200–300 Ω. It should, however, provide galvanic isolation among the outputs for best avoidance of ground loops. The XLR microphone input will normally pass directly to the primary XLR output, while properly passing phantom power to the microphone. The secondary output will employ a 1:1 transformer to bridge the hot and cold lines of the microphone input and convey the signal to a secondary output with galvanic isolation. Phantom power is not conveyed to the secondary output. Some microphone splitters provide two or three secondary outputs, using 1:1:1 and 1:1:1:1 transformers.

To prevent the formation of ground loops, all of the secondary outputs should have their shields open inside the splitter box. If need be, there may be a ground-lift switch to make or break the ground connection. However, it is recommended to terminate un-connected cable shields at high frequencies with 0.01 μF and 50-Ω Zobel networks to prevent formation of a whip antenna [4]. This is implemented by shunting the network(s) across the open contacts of the ground-lift switch.

The microphone splitter should be very non-intrusive, especially in the main signal path. The load(s) connected to the secondary outputs will add to the loading presented to the source microphone. A microphone feeding a main mixer and a secondary mixer, each with 2-kΩ input impedance, will see a net load of only 1 kΩ. With three secondary outputs loaded in the same way, the net load seen by the microphone will be 500 Ω. This will attenuate the output of a 300-Ω microphone and likely color its sound.

The quality of the transformer is an issue for frequency response, coloration and distortion, particularly at low frequencies and high signal levels. Transformers can contribute distortion and can overload at low frequencies. Just because they are passive does not mean that they have very large signal-handling ability. The cores will saturate at surprisingly low signal levels at 20 Hz, for example. This is not just an issue of passing the signal to the secondary output with wide frequency response and low distortion. Transformer saturation will cause distortion in the main through signal as well. This also pertains to frequency response of the through path, since the inductance of the transformer will increase loading of the microphone signal at low frequencies.

Transformer nonlinearity can be especially troublesome with a high-output microphone, like a condenser microphone. As an example, one quality splitter transformer is specified to have a maximum input at 20 Hz of +2.0 dBu (less than 1 V_{RMS}) for 1% THD. The same transformer is rated at 0.05% at –20 dBu, or only 77 mV. These ratings assume 150-Ω microphone source impedance. Microphones with a more typical source impedance of 300 Ω will not fare as well. Just as with passive DI box transformers, high-quality microphone splitter transformers can be quite expensive, in the $60–$90 range. Less expensive transformers can do much more damage to the sound. An alternative is to add a 20-dB input attenuator that can be inserted with a switch when a high-output source is used.

Passive ribbon microphones may especially be affected, since their nominal output impedance is in the range of 300 Ω, but actually can increase at lower frequencies as a result of ribbon resonance. The transformer loading of a passive splitter is likely to have a significant effect on the frequency response and sound of a passive ribbon microphone. Active ribbon microphones have a more uniform low output impedance and are less prone to such degradation. Passive ribbon microphones are best used with an external activator/buffer if they are to feed a passive microphone splitter.

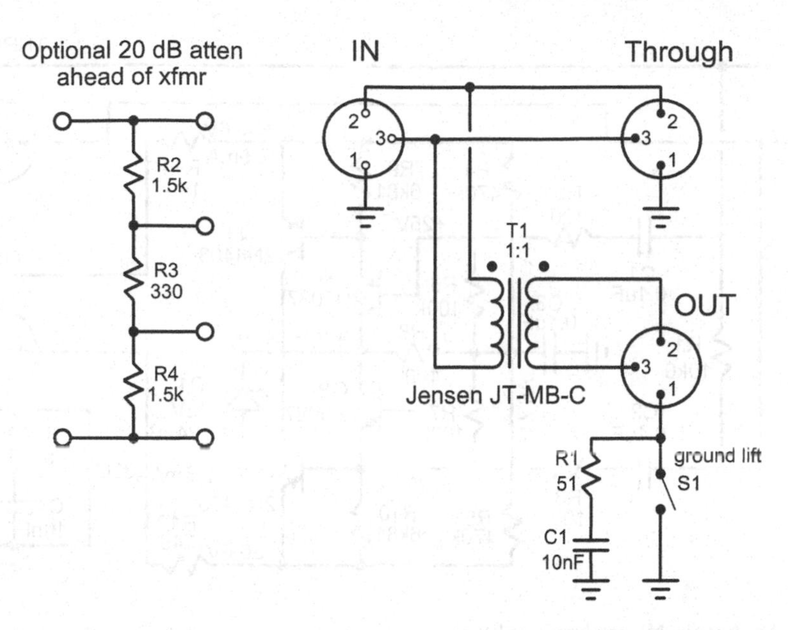

Figure 30.3 A Passive Microphone Splitter.

Figure 30.3 shows a passive microphone splitter [12, 13]. The primary path is just XLR-to-XLR and carries phantom power. The secondary path bridges the primary path with a 1:1 microphone transformer that provides the balanced secondary output and galvanic isolation. Connection of the cable shields of the secondary output(s) and the inclusion of a ground-lift switch is largely the same as that described for the DI box.

30.4 Active Microphone Splitters

Just as with a DI box, microphone splitters can be actively implemented and phantom powered. Of course, some are battery-powered or line-powered. Hybrid splitters employ an active unity-gain buffer that feeds one or more 1:1 transformers to achieve galvanic ground isolation [14]. Active microphone splitters often cannot be phantom powered if the microphone requires phantom power. Others are fully active, employing unity-gain buffers for each secondary output that achieve high common-mode output impedance. In this case, the microphone can be wired straight through to a primary output while each of the secondary outputs is implemented as a unity-gain buffer using phantom power from its destination microphone preamplifier, which may be that of a monitor mixer.

An active microphone splitter must have low noise in its amplifier(s) that drive the secondary output(s), since it is dealing with small microphone-level signals. The noise and overload requirements are similar to those of a microphone preamplifier in a mixer. Fortunately, the microphone splitter needs little more gain than unity.

Figure 30.4 shows an active microphone splitter with the through output and one phantom-powered secondary output. The secondary output is implemented with a unity-gain phantom-powered differential follower using Darlington connected emitter followers for each signal polarity. This

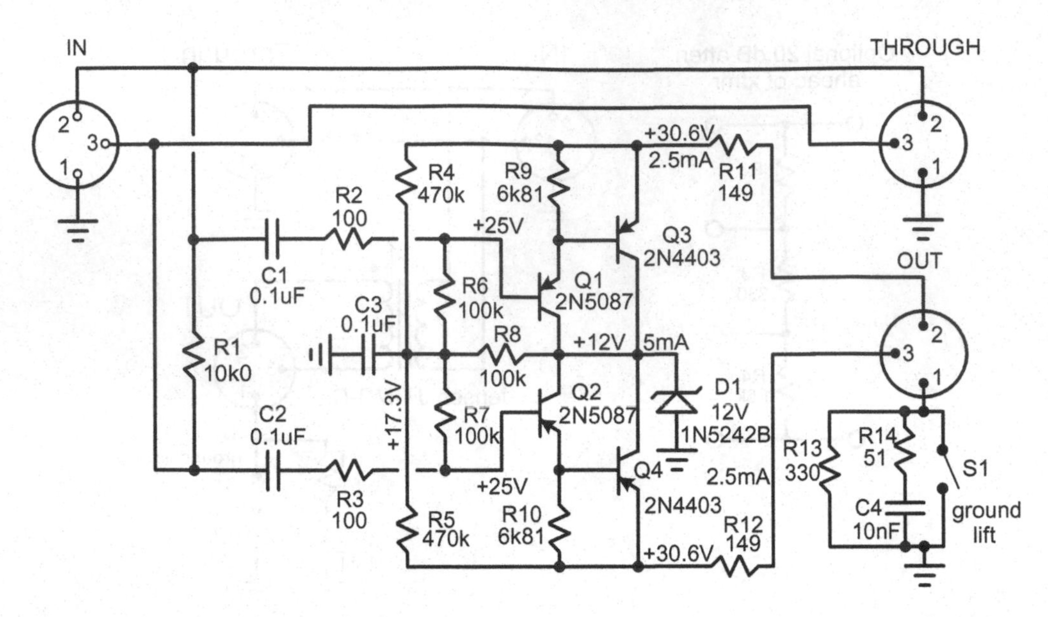

Figure 30.4 An Active Microphone Splitter.

circuit is much like the modified Schoeps condenser microphone preamplifier output stage described in Chapter 18 in Figre 18.14(b).

References

1. Francis Rumsey and Tim McCormick, *Sound and Recording Applications and Theory*, 7th edition, Focal Press, New York, NY, 2014.
2. Guitar pickup model.
3. Bill Whitlock, *An Overview of Audio System Grounding and Signal Interfacing*, Tutorial T5, 135th Convention of the Audio Engineering Society, October 2013, New York City.
4. Bill Whitlock, *An Overview of Audio System Grounding and Shielding*, Tutorial T2, Convention of the Audio Engineering Society, May 2009.
5. Bill Whitlock, *Design of High Performance Audio Interfaces*, Jensen Transformers, Inc.
6. Bill Whitlock, Balanced Lines in Audio – Fact, Fiction, and Transformers, *Journal of the AES*, vol. 43, no. 6, June 1995, pp. 454–464.
7. Cinemag model CM-DBX Direct Box Transformer, Cinemag Inc, cinemag.biz/direct_box/PDF/CM-DBX.pdf.
8. Lundahl Model LL1935 DI Transformer, lundahltransformers.com/wp-content/uploads/datasheets/1935.pdf.
9. Jensen Transformers, JT-DB direct box transformer, jensen-transformers.com/wp-content/uploads/2014/08/jt-db-e.pdf.
10. Sowter Transformers, Type 4243 DI Box Transformer specification sheet, souter.co.uk/specs/4243.php.
11. Avenson Audio Small DI, avensonaudio.com/small-di/.
12. Bill Whitlock, *Theory and Construction of Mic Splitters*, Jensen AN-005.
13. Jensen Transformers, *JT-MB-C Microphone Bridging Transformer*, jensen-transformers.com/wp-content/uploads/2014/08/jt-mb-c.pdf.
14. Elliott Sound Products (ESP), *Microphone Splitters for Live Sound & Recording*, 2016, https://sound-au.com/articles/mic-splitting.htm.

Index

Printed in the United States
by Baker & Taylor Publisher Services